LAMBDA II

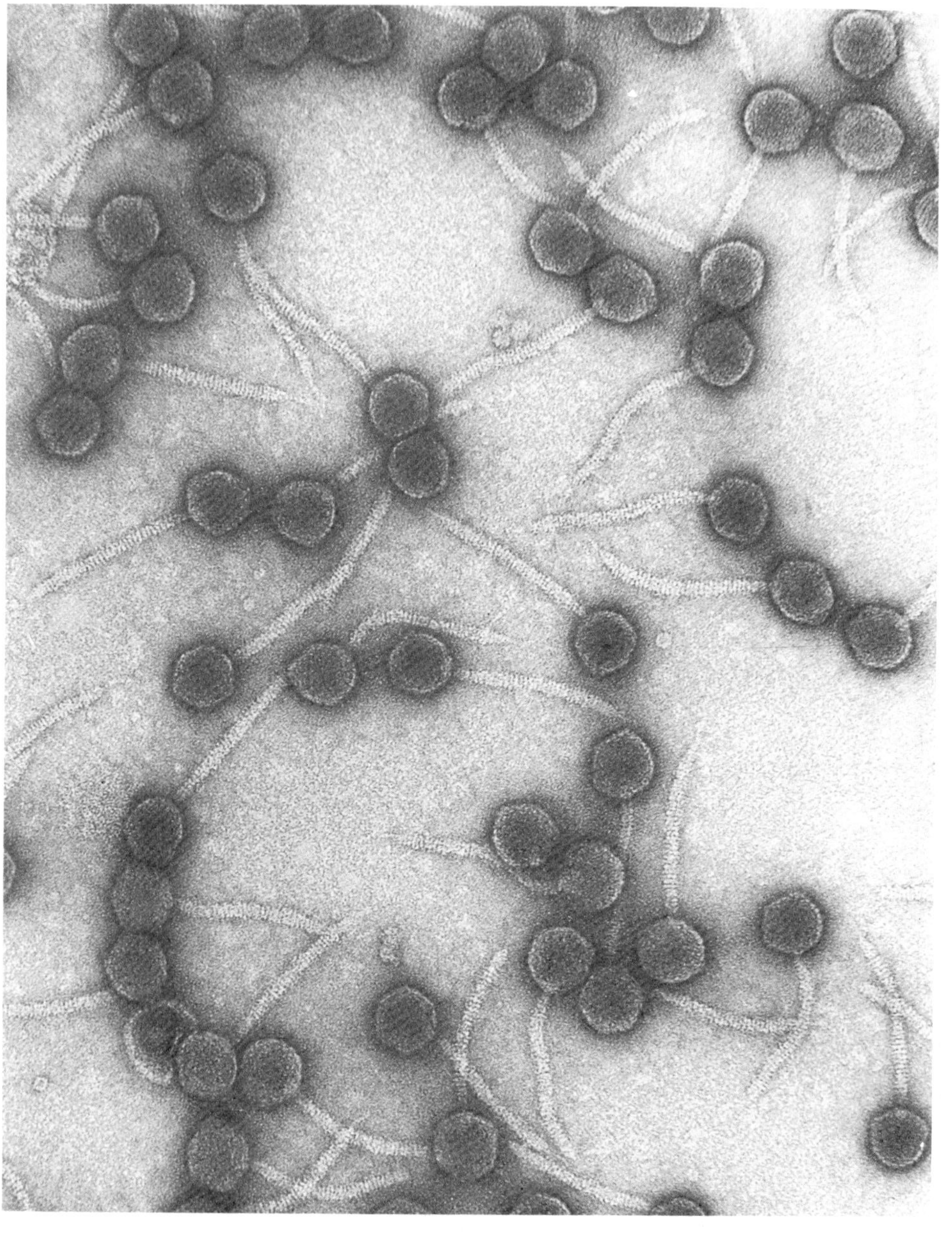

Phage virions with uranyl acetate negative staining. Magnification, ~150,000×. (Photo courtesy of Roger Hendrix.)

LAMBDA II

Edited by

Roger W. Hendrix
University of Pittsburgh

Jeffrey W. Roberts
Cornell University

Franklin W. Stahl
University of Oregon

Robert A. Weisberg
National Institutes of Health

Cold Spring Harbor Laboratory
1983

COLD SPRING HARBOR
MONOGRAPH SERIES

The Lactose Operon
The Bacteriophage Lambda
The Molecular Biology of Tumour Viruses
Ribosomes
RNA Phages
RNA Polymerase
The Operon
The Single-Stranded DNA Phages
Transfer RNA:
  Structure, Properties, and Recognition
  Biological Aspects
Molecular Biology of Tumor Viruses, Second Edition:
  DNA Tumor Viruses
  RNA Tumor VIruses
The Molecular Biology of the Yeast Saccharomyces:
  Life Cycle and Inheritance
  Metabolism and Gene Expression
Mitochondrial Genes
Nucleases
Lambda II

LAMBDA II
MONOGRAPH 13

Printed in the United States of America

Book design by Emily Harste

Cover: Beyond the world of λ by Sydney Roark.

Library of Congress Cataloging in Publication Data
Main entry under title:

Lambda II.

(Cold Spring Harbor monograph series)
Includes index.
1. Bacteriophage lambda. I. Hendrix, Roger W.
II. Title: Lambda 2. III. Title: Lambda two.
IV. Series. [DNLM: 1. Phage Lambda. W1 CO133c v. 13/
QW 161.5.C6 L219]
QR342.L35 1983   576′.6482   81-70528
ISBN 0-87969-150-6
ISBN 0-87696-177-8 (pbk.)

All Cold Spring Harbor Laboratory publications are available through booksellers or may be ordered directly from Cold Spring Harbor Laboratory, Box 100, Cold Spring Harbor, New York 11724.

SAN 203-6185

Contents

Preface, vii

Introduction to Lambda, 3
A. D. Hershey and W. Dove

Progress since 1970, 13
R. W. Hendrix

Lytic Mode of Lambda Development, 21
D. Friedman and M. Gottesman

Establishment of Repressor Synthesis, 53
D. Wulff and M. Rosenberg

Control of Integration and Excision, 75
H. Echols and G. Guarneros

Repressor and Cro Protein: Structure, Function, and Role in Lysogenization, 93
G. Gussin, A. Johnson, C. Pabo, and R. Sauer

Lysogenic Induction, 123
J. Roberts and R. Devoret

Lambda DNA Replication, 145
M. Furth and S. Wickner

General Recombination, 175
G. Smith

Site-specific Recombination in Phage Lambda, 211
R. Weisberg and A. Landy

Phage Lambda's Accessory Genes, 251
D. Court and A. Oppenheim

Lambdoid Phage Head Assembly, 279
C. Georgopoulos, K. Tilly, and S. Casjens

DNA Packaging and Cutting, 305
M. Feiss and A. Becker

Tail Assembly and Injection, 331
I. Katsura

Bacteriophage P22 Antirepressor and Its Control, 347
M. Susskind and P. Youderian

Evolution of the Lambdoid Phages, 365
A. Campbell and D. Botstein

A Beginner's Guide to Lambda Biology, 381
W. Arber

Phage Lambda and Molecular Cloning, 395
N. Murray

Experimental Methods for Use with Lambda, 433
W. Arber, L. Enquist, B. Hohn, N. Murray, and K. Murray

Appendices

I. A Molecular Map of Coliphage Lambda, 467
D. Daniels, J. Schroeder, W. Szybalski, F. Sanger, and F. Blattner

II. Complete Annotated Lambda Sequence, 519
D. Daniels, J. Schroeder, W. Szybalski, F. Sanger, A. Coulson, G. Hong, D. Hill, G. Petersen, and F. Blattner

III. Lambda Vectors, 677
N. Murray

Subject Index, 685

Preface

Lambda II, like its predecessor, *The Bacteriophage Lambda,* attempts to put all the available information about λ into the form of coherent and easily accessible review articles. Research on λ has illuminated and continues to illuminate many aspects of gene regulation and function. We therefore hope that *Lambda II* will be a useful source of information for the student of molecular biology as well as for the hard-core lambdologist and his ilk. In particular, the papers by Hershey and Dove, by Hendrix, and by Arber should make the more specialized reviews accessible to the novice. Another major current interest in λ concerns its role as a vector for molecular cloning. Murray's paper on the use of λ in cloning, together with those by Arber and by Arber et al. address this interest. These papers provide most of the information required for the effective use of λ in cloning experiments.

We are deeply indebted to Al Hershey, editor of *The Bacteriophage Lambda.* His book gave us a model to emulate and made recruiting authors for this volume an easy task. We thank Jim Watson for encouragement and for his continuing support of the kind of science represented here. The efforts of Nancy Ford, Nadine Dumser, Michaela Cooney, Mary Cozza, and Gail Anderson of the Cold Spring Harbor Publications Department made this book a practical undertaking and are greatly appreciated.

R. Hendrix
J. Roberts
F. Stahl
R. Weisberg

A Gallery of Early λ Workers

These photographs, taken from the Cold Spring Harbor Archives and from a scrapbook compiled by M. Lieb, include many (though certainly not all) of the major figures involved in λ research during its first 10–15 years.

(Above left) J. and E. Lederberg (1951), (above right) W. Szybalski (1951) (photos courtesy of Cold Spring Harbor Laboratory Research Library Archives); (right) F. Jacob (1953) (photo courtesy of M. Lieb); (below, from left to right) J. Weigle, O. Maaløe, E. Wollman, G. Stent, M. Delbrück, and G. Soli (1949) (photo from Ross Madden/Black Star).

(Top) D. Kaiser and J.-I. Tomizawa (1968) (photo courtesy of Cold Spring Harbor Laboratory Research Library Archives); (above right) A. Lwoff (1953) (photo courtesy of M. Lieb); (left) A.D. Hershey (1951), (below left) G. Bertani (1949), (below center) L. Siminovitch (1967), (below right) M. Lieb (1948) (photos courtesy of Cold Spring Harbor Laboratory Research Library Archives).

LAMBDA II

Unlike those that follow it, the paper by Hershey and Dove was not written for Lambda II; *it is reprinted from* The Bacteriophage Lambda, *published in 1971. This has been done partly for historical continuity, but mainly because, after 12 years, it is still an excellent introduction to λ. The novice should appreciate the clarity with which Hershey and Dove present the fundamentals of λ biology. The expert may wish to savor the prose and imagine what subtle changes of emphasis might appear if this paper were rewritten now. For historical interest, the paper has been reprinted directly from the original plates. Thus, the references to other chapters are for* The Bacteriophage Lambda *and not for this volume.*

Introduction to Lambda

A. D. HERSHEY[1] AND WILLIAM DOVE[2]
[1]*Carnegie Institution of Washington,*
Genetics Research Unit, Cold Spring Harbor, New York
and
[2]*McArdle Laboratory, University of Wisconsin, Madison, Wisconsin*

THE bacteriophages form a diverse collection of viruses that multiply in bacterial cells. This book describes the variety called λ and some of its relatives, which multiply in *Escherichia coli*. Chapter 1 presents general characteristics of the λ group, and explains how λ came to be the subject of intensive study.

The classical phages destroy their host cells by lysis. Lambda causes lysis too, but can also propagate in a form that permits joint multiplication of phage and host. Phages possessing this added potentiality are called temperate, in distinction from the broad class of intemperate or virulent species, which cannot reproduce without destroying their host cells.

Lambda is also one of a few well known genetic elements, called episomes, that are able to multiply in the cell either autonomously or as part of the bacterial chromosome. In fact λ and the fertility agent F of *E. coli* are historical prototypes of the class. Campbell (1969) presents the comparative biology of episomes.

For newcomers to phage research, Stent's *Molecular Biology of Bacterial Viruses* (1963) can be recommended as a source book.

THE LIFE CYCLE

Lambda phage particles are about half protein and half DNA. Each contains one double-stranded DNA molecule encapsulated in an icosahedral head, 0.05 μ in diameter, from which projects a tubular tail, 0.15 μ long (Chapter 14). The DNA molecule is made from the usual four bases and weighs about 31 million daltons. Phage particles can be prepared in milligram quantities for analysis of various sorts.

Lambda is an obligatory parasite; the phage particle itself is metabolically inert. Growth begins when a particle adsorbs by the tip of its tail to a host cell and injects its DNA molecule. Electron micrographs of λ and its host, *E. coli*, are shown at the same magnification on the facing page, upper section. Phage particles are shown at a higher magnification on the same page, lower section.

Lambda multiplies in *E. coli* in either of two ways. In productive growth, the injected DNA molecule directs the synthesis of numerous gene products (Chapter 13). These promote autonomous replication of the phage DNA, its packaging into phage particles, and lysis of the cell to release about one hundred progeny particles—all within 50 minutes at 37°C. This sequence of events is called the productive or lytic cycle of phage growth.

Lambda DNA can also replicate as part of the bacterial chromosome. To pursue this option, the injected DNA must first direct the synthesis of gene products that promote insertion of the phage chromosome into the DNA of the host, and then express other genes that act promptly to repress autonomous DNA replication and most phage functions. When these processes mesh properly, the phage DNA is inserted into the bacterial chromosome at a characteristic site. The inserted phage DNA is called prophage. Its replication contributes to viral growth indirectly, because every cell carrying prophage is a potential center for production of phage particles. A bacterial strain carrying a prophage is called a lysogen. The process beginning with infection and leading to the establishment of prophage is called lysogenization.

Not all prophages replicate as part of the bacterial chromosome. For this reason some of the words used above have a more general meaning than a description of λ can convey. Lysogeny (originally lysogenicity) signifies the hereditary power to produce phage particles. The word derives from the frequent connection between phage production and lysis. A lysogenic bacterium, or lysogen, is one that regularly transmits to its progeny the power to produce phage particles (Jacob et al., 1953). The word prophage signifies the hereditary structure present in a lysogen and absent in the corresponding nonlysogenic bacterium. In lysogens carrying λ, the prophage is part of the bacterial chromosome. In lysogens carrying phage P1, the prophage location is unknown. P1 does not occupy a characteristic site, if any, in the bacterial chromosome.

The frequency with which an infecting λ phage particle initiates prophage replication rather than productive phage growth depends on the temperature, the state of the bacteria, and the genotypes of the phage and the host. For wild-type λ, the frequency of lysogenization can vary from nearly zero to nearly unity, depending only on the conditions of infection.

Lambda typically forms turbid plaques because it does not kill all the bacteria it infects. The ability to form turbid plaques depends on three genes called *c*I, *c*II, and *c*III, which were the first genes to be identified in λ (Kaiser, 1957). Gene *c*I alone suffices for maintenance of prophage; all three are necessary for efficient lysogenization. The genes were called *c* because they were identified in mutants forming clear plaques.

The prophage multiplies in the same passive way as other elements in the bacterial chromosome. The gene *c*I ensures production of a repressor that prevents expression of other prophage genes and thus serves as the mainstay of lysogeny. The *c*I protein also represses genes of superinfecting λ phage particles. For this reason the lysogenic cell is specifically immune to reinfection by λ.

If the repressor ceases to act for any reason, the prophage enters productive growth. This process requires excision of the prophage from the host chromosome, autonomous DNA replication, phage production, and lysis. In a lysogenic culture, spontaneous phage production occurs in about one cell per thousand cell generations. As a result, the culture typically contains some phage particles to which the cells are immune.

Phage production can also be induced by experimental means. For example, exposure of a lysogenic culture to ultraviolet radiation can initiate the lytic cycle of phage growth in nearly every cell. When the prophage carries a mutation in gene *c*I that makes the repressor thermolabile, the lysogenic cells can grow normally at low temperature but lyse as a consequence of phage growth after the temperature is raised. Induction by ultraviolet light also involves inactivation of repressor. Here the inactivation is indirect since it depends on a bacterial gene (*recA*) as well as on the structure of the repressor. Mutations ($recA^-$, $c\mathrm{I}ind^-$) that interfere with induction by irradiation also lower the frequency of spontaneous phage production in lysogenic cultures. Neither mutation interferes with thermal induction of a prophage whose repressor is thermolabile.

Integration and repression can be separated experimentally. When a phage is deficient in a genetic element (*int* or *att*) necessary for insertion of prophage into the bacterial chromosome, repression can nevertheless be established and the cell can survive. Since repressed DNA cannot multiply on its own, it is transmitted only to one daughter at each cell division. This situation is called abortive lysogeny and may correspond to a normal prelysogenic phase (Chapter 6).

Induction of phage development can also be abortive. The repressor of some *c*I mutants can be inactivated reversibly by brief thermal treatment. Then if the timing is right the prophage is excised, repression is established again, and the repressed phage DNA fails to replicate as the cells multiply. The nonlysogenic bacterial progeny are said to be cured.

Under special conditions λ DNA can be carried in cells as an unstable plasmid. This is true of a deletion mutant, called λdv, that retains only 15% of the λ genome in terms of DNA length (Matsubara and Kaiser, 1968). It is true also of *N* mutants of λ (Signer, 1969; Lieb, 1970). In λ the plasmid condition depends on escape from repression (because repressed DNA cannot replicate) and on quiescence of "late" genes (genes responsible for phage-particle formation and cell lysis, which are lethal when expressed). In λdv these conditions are met by mutations (in the antecedent phage) making DNA function indifferent to repressor, and by deletion of late genes. In *N* mutants, they are met because the *N* protein of λ indirectly regulates both *c*I and late genes (Chapters 12 and 13). Cultures carrying λdv or λN^- DNA contain ten or more copies per bacterium (Lieb, 1970).

In summary: Lambda DNA can replicate in cells in either of two mutually exclusive modes, as an autonomous element or as part of the bacterial chromosome. It can persist in either of two inactive states, in cells when it is made quiescent by its own repressor, or in specially designed DNA vectors called phage particles.

DISCOVERY OF LAMBDA

Lambda is carried as a prophage in *E. coli* strain K12, with which Gray and Tatum (1944) and J. Lederberg and Tatum (1946) began their work in bacterial genetics—work leading, among other things, to the discovery in K12 of the fertility agent F (Cavalli et al., 1953; Hayes, 1953; Wollman et al., 1956). Lambda went unnoticed until Esther Lederberg (1951) found a derivative of K12 that had lost its prophage as an incidental result of mutagenic treatment with ultraviolet light, and had thus become sensitive to infection with λ.

Lysogeny was discovered in 1921, but the subject remained controversial for many years (see Stent, 1963, for a historical account). In 1950, Lwoff and Gutmann reported microscopic observations of single-cell cultures of lysogenic *Bacillus megaterium*, which showed clearly that the lysogenic condition could be perpetuated without intervention of phage particles. They followed a cell progeny through nineteen cell divisions, removing one daughter cell from the microculture to an agar medium after each division. All of the transplanted cells produced lysogenic colonies. At no time could phage particles be recovered from the fluid portion of the microdrop. (In other microdrops, phage particles were recovered, but only in bursts and only after a cell had been seen to lyse.) Shortly afterward, Lwoff, Siminovitch, and Kjeldgaard (1950) reported that phage production could be induced by ultraviolet light in nearly every cell of a lysogenic culture, thus showing in a different way that every lysogenic cell contains prophage. The work of Lwoff and his colleagues presented lysogeny as three problems: one concerned with the inheritance of prophage, another with the induction of phage growth, and a third with the phenomenon of immunity, which was clearly connected with the other two (Lwoff, 1953).

In the meantime, Bertani (1951) had begun his work with phage P2, directed toward an understanding of prophage inheritance. In a review (Bertani, 1953) he pointed out that all hypotheses concerning inheritance of prophage could be reduced to two: those postulating many or a variable number of identical prophages per cell or cell nucleus, and those postulating one or a fixed number. Bertani concluded that the first hypothesis was inconsistent with the results of his work showing substitution of one prophage for another on reinfection of lysogens. He also showed that recombination could occur between prophages in double lysogens, and that it never gave rise to lysogens carrying three prophage genotypes, contrary to expectation if the prophage were carried in numerous copies.

Owing to the happy conjunction of the work of the Lederbergs, Lwoff, and Bertani, the discovery of λ in a fertile bacterial strain was seen at once to present some novel opportunities. First, Lederberg and Lederberg (1953) found that λ prophage segregates with the bacterial gene *gal* in crosses between lysogenic and nonlysogenic bacteria. Their finding gave the first direct clue to a mechanism of hereditary transmission of a virus. Second, pursuing similar crosses, Wollman and Jacob (1954) discovered that phage growth is induced when prophage enters the cytoplasm of a nonlysogenic recipient during bacterial mating, a phenomenon called zygotic induction that in due course explained immunity in terms of repressor action. Third, Morse et al. (1956) discovered preferential transduction of *gal* genes by λ, which provided another clue to the mode of inheritance of the prophage.

Ten years of intensive work followed before bacterial crosses in general and λ-*gal* linkage in particular were understood. Eventually it was shown that λ prophage is inserted into the bacterial chromosome between the *gal* and *bio* genes by a locus-specific and phage-specific mechanism of genetic recombination (Chapter 6). The prophage location, together with another novel kind of genetic recombination (Chapter 8), were also to account for the propensity of λ to form transducing phage particles that disseminate *gal* and *bio* genes. Finally, immunity was seen not merely as a device serving the needs of the prophage, but as a typical example of gene regulation (Jacob and Monod, 1961).

It is worth noting that lambda's way of ensuring perpetuation of lysogeny is a special solution to a more general problem. The fertility agent F in its autonomous phase, and probably also P1 prophage, can be maintained at the rate of one copy per bacterial chromosome without insertion into it. Coordinated replication and segregation of these elements and of the bacterial DNA cannot be understood even in principle (Campbell, 1969). The same may be said, of course, of the coordinated replication and segregation of chromosomes in general.

The plasmid state in λ furnishes an illuminating contrast to coordinated replication of chromosomes. When an *N* mutant phage particle injects its DNA into a cell, the DNA quickly replicates to produce ten or twenty copies. To keep pace with bacterial growth, further replication must be very slow. Replication and segregation must also be poorly regulated, as shown by the rapid accumulation of uninfected daughter cells.

Historical matters are dealt with by Stent (1963), Campbell (1969), and various chapters of the present book. In the bibliography of Chapter 1 we cite separately the first publications concerning λ. The list of titles may convey some impression of how this book got its start.

TAXONOMY

Lambda is one of a number of phages that have three properties in common: they recombine when intercrossed, their DNA molecules possess identical pairs of cohesive ends (Chapter 5), and their prophages are inducible by ultraviolet irradiation. The best known members of the group, often called lambdoid phages, are listed in Table 1. As shown in the table, they differ from one another in many respects.

Another group, whose best known member is P2, contains temperate phages that are not inducible by ultraviolet irradiation, fail to recombine

TABLE 1. PROPERTIES OF LAMBDA AND SOME RELATIVES

Phage	Immunity	Host range	Tail antigen	Prophage location[a]	*N* gene product	DNA ends
λ	unique[b]	—	—	*gal–bio*[c]	—	—
21	unique[b]	like λ[b]	like λ[b]	near *trp*[b]	unlike λ[d]	like λ[e]
ϕ80	unique[f]	like T1[f]	unlike λ[f]	*tdk–trp*[c,g]	unlike λ[h]	like λ[i]
ϕ81	unique[j]	like T1[j]	?	*gal–bio*[j]	?	like λ[i]
82	unique[b]	like 434[b]	like λ[b]	*gal–bio*[c]	?	?
424	unique[b]	like 434[k]	like λ[b]	near *his*[b]	?	like λ[e]
434	unique[b]	unlike λ[b]	like λ[b]	*gal–bio*[c]	like λ[d]	like λ[e]

[a] Where two bacterial loci are mentioned, the prophage is inserted between them. For more detail, see Taylor (1970) and Chapter 2.
[b] Jacob and Wollman (1961)
[c] Taylor (1970).
[d] Thomas (1970).
[e] Baldwin et al. (1966).
[f] Matsushiro (1963).
[g] Igarashi et al. (1967).
[h] Szpirer and Brachet (1970).
[i] Yamagishi et al. (1965).
[j] Takeda et al. (1970).
[k] J. S. Parkinson (personal communication).

with λ, and possess cohesive DNA ends of a second class (Chapter 5). The two groups probably represent major families of temperate coliphages. Among 42 nonidentical strains isolated by the Bertanis in Los Angeles some years ago, 12 proved to be serologically related to λ and 16 to P2 (Bertani and Bertani, 1971).

Phages resembling λ are ubiquitous. Jacob and Wollman (1956) isolated several in France. They found that 32 of 500 strains of *E. coli* carried temperate phages active on a nonlysogenic derivative of K12. Among them three were superficially identical with λ, and at least six others were inducible and could recombine with λ. Phages ϕ80 and ϕ81 were isolated in Japan by Matsushiro (1963). *E. coli* K12, presumably already carrying λ, was isolated in the United States in 1922 (J. Lederberg, 1951).

The phages listed in Table 1 were selected for study because they differ with respect to specificity of immunity. Each can infect lysogens carrying any of the other phages, but cannot successfully reinfect its own lysogen. Such phages are said to be heteroimmune with respect to each other. J. S. Parkinson (personal communication) collected data concerning the prevalence of homoimmune phages. He studied seventeen UV-inducible phages active on *E. coli* K12. Of these, seven resembled λ and ten resembled 434 with respect to host specificity. Among the first group, six possessed the immunity of phage 82 and one possessed an unidentified immunity (different from that of λ, 21, 82, 424, and 434). Among the second group, two possessed the immunity of λ, two that of 434, one that of 82, and five possessed an unknown immunity. These data show that phages with similar immunity regions occur frequently, and confirm the impression gained from the study of heteroimmune phages that there is little correlation between type of immunity and other characteristics among relatives of λ.

Members of the λ group vary in the readiness with which they recombine with λ and in their ability to supply missing gene products in cells coinfected with λ mutants (Dove, 1968; Thomas, 1970; Szpirer and Brachet, 1970). Heteroduplex DNA molecules constructed from one strand of λ DNA and one strand of the DNA of a related phage (21, 80, 82, or 434) have been mapped by electron microscopy, which permits recognition of regions of complementary nucleotide sequence (Chapter 3). The data support three notable conclusions. First, DNA molecules of the several phages contain segments that are either identical with segments of λ DNA or completely different; only a few partially matching segments are found. Second, the identical sequences in a given pair of DNAs add up to 35% to 60% of the total molecular length, depending on the pair. Third, identical segments occupy positions characteristic of the pair and lie approximately equidistant from the left molecular end in both members of the pair (Simon et al. and Fiandt et al., this volume). These facts, together with the common structure of the molecular ends and the ability to form intervarietal hybrids, suggest the existence of a common genome that is nearly but not quite independent of specific nucleotide sequences.

USEFUL PROPERTIES

The following list aims to connect the natural history of λ with its technical advantages as an experimental organism. Lambda, and the contents of this book, are best appreciated in terms of that connection.

1. Lambda's host, *E. coli* K12, is genetically well known (Taylor, 1970). It is fertile and potentially diploid, and its suppressors make amber mutations accessible to study (Chapter 2).

2. Owing to its passive replication as part of the bacterial chromosome, λ prophage can survive almost any mutation or deletion, including those that prevent autonomous replication of the phage DNA. This circumstance permits genetic analysis that is unlimited in principle (Jacob et al., 1957).

3. Lambda prophage is inducible by various experimental treatments. The inducible character is crucial to exploitation of defective prophages (item 6 below).

4. The DNA of phage λ is easily extracted as uniform, double-stranded molecules representing 30 to 40 genes (Chapter 3). Each of the complementary strands, as well as a few molecular segments that differ from each other in composition, can be isolated in pure form.

5. Recombination frequency is high enough to be measured and low enough to permit isotopic density analysis of primary recombinants (Chapter 7).

6. The transducing phages and other chromosomal aberrations provide well-defined DNA segments that are useful for deletion mapping (Chapters 2 and 3) and analysis of transcription (Chapter 13).

7. The linkage map and the DNA molecule are linear and congruent (Chapter 3).

8. The main regulator of gene action, the λ repressor, has been isolated and studied in vitro (Chapter 11).

9. The lambdoid phages carry homologues of genetic determinants present in λ, and can form hybrids with λ. One hybrid has guided the study of immunity (Chapters 10, 11, and 12).

REFERENCES

BALDWIN, R. L., P. BARRAND, A. FRITSCH, D. A. GOLDTHWAIT, and F. JACOB. 1966. Cohesive sites on the deoxyribonucleic acids from several temperate coliphages. J. Mol. Biol. *17:* 343.

BERTANI, G. 1951. Studies on lysogenesis. I. The mode of phage liberation by lysogenic *Escherichia coli*. J. Bacteriol. *62:* 293.

BERTANI, G. 1953. Lysogenic versus lytic cycle of phage multiplication. Cold Spring Harbor Symp. Quant. Biol. *18:* 65.

BERTANI, L. E., and G. BERTANI. 1971. Genetics of P2 and related phages. Advan. Genet. (Demerec Memorial Volume) (in press).

CAMPBELL, A. M. 1969. Episomes. Harper and Row, New York. 193 p.

CAVALLI, L. L., J. LEDERBERG, and E. M. LEDERBERG. 1953. An infective factor controlling sex compatibility in *Bacterium coli*. J. Gen. Microbiol. *8:* 89.

DOVE, W. F. 1968. The genetics of the lambdoid phages. Ann. Rev. Genet. *2:* 305.

GRAY, C. H., and E. L. TATUM. 1944. X-ray induced growth factor requirements in bacteria. Proc. Nat. Acad. Sci. *30:* 404.

HAYES, W., 1953. Observations on a transmissible agent determining sexual differentiation in *Bacterium coli*. J. Gen. Microbiol. *8:* 72.

IGARASHI, K., S. HIRAGA, and T. YURA. 1967. A deoxythymidine kinase deficient mutant of *Escherichia coli*. II. Mapping and transduction studies with phage ϕ80. Genetics *57:* 643.

JACOB, F., C. R. FUERST, and E. L. WOLLMAN. 1957. Recherches sur les bactéries lysogènes défectives. II. Les types physiologiques liés aux mutations du prophage. Ann. Inst. Pasteur *93:* 724.

JACOB, F., A. LWOFF, L. SIMINOVITCH, and E. WOLLMAN. 1953. Définition de quelques termes relatifs à la lysogénie. Ann. Inst. Pasteur *84:* 222.

JACOB, F., and J. MONOD. 1961. Genetic regulatory mechanisms in the synthesis of proteins. J. Mol. Biol. *3:* 318.

JACOB, F., and E. L. WOLLMAN. 1956. Sur les processus de conjugaison et de recombinaison chez *Escherichia coli*. I. L'induction par conjugaison ou induction zygotique. Ann. Inst. Pasteur *91:* 486.

JACOB, F., and E. L. WOLLMAN. 1961. Sexuality and the genetics of bacteria. Academic Press, New York. 374 p.

KAISER, A. D. 1957. Mutations in a temperate bacteriophage affecting its ability to lysogenize *Escherichia coli*. Virology *3:* 42.

LEDERBERG, E. M. 1951. See "first papers" below.

LEDERBERG, E. M., and J. LEDERBERG. 1953. See "first papers" below.

LEDERBERG, J. 1951. Genetic studies with bacteria, p. 263–289. *In* L. C. Dunn [ed.] Genetics in the 20th century. Macmillan, New York.

LEDERBERG, J., and E. L. TATUM. 1946. Novel genotypes in mixed cultures of biochemical mutants of bacteria. Cold Spring Harbor Symp. Quant. Biol. *11:* 113.

LIEB, M. 1970. λ mutants which persist as plasmids. J. Virol. *6:* 218.

LWOFF, A. 1953. Lysogeny. Bacteriol. Rev. *17:* 269.

LWOFF, A., and A. GUTMANN. 1950. Recherches sur un *Bacillus megatherium* lysogène. Ann. Inst. Pasteur *78:* 711. [English translation in Stent (1960).]

LWOFF, A., L. SIMINOVITCH, and N. KJELDGAARD. 1950. Induction de la lyse bactériophagique de la totalité d'une population microbienne lysogène. Compt. Rend. Acad. Sci. 231: 190. [English translation in Stent (1960).]

MATSUBARA, K., and A. D. KAISER. 1968. λ dv: an autonomously replicating DNA fragment. Cold Spring Harbor Symp. Quant. Biol. *33:* 769.

MATSUSHIRO, A. 1963. Specialized transduction of tryptophan markers in *Escherichia coli* K12 by bacteriophage φ80. Virology *19:* 475.

MORSE, M. L., E. M. LEDERBERG, and J. LEDERBERG. 1956. Transduction in *Escherichia coli* K12. Genetics *41:* 142.

SIGNER, E. R. 1969. Plasmid formation: a new mode of lysogeny by phage λ. Nature *223:* 158.

STENT, G. S. 1960. Papers on bacterial viruses. Little, Brown and Co., Boston. 365 p.

STENT, G. S. 1963. Molecular biology of bacterial viruses. W. H. Freeman and Company, San Francisco. 474 p.

SZPIRER, J., and P. BRACHET. 1970. Relations physiologiques entre les phages tempérés λ et φ80. Mol. Gen. Genet. *108:* 78.

TAKEDA, Y., T. MORISHITA, and T. YURA. 1970. Specialized transduction of the galactose genes of *Escherichia coli* by phage φ81. Virology *41:* 348.

TAYLOR, A. L. 1970. Current linkage map of *Escherichia coli*. Bacteriol. Rev. *34:* 155.

THOMAS, R. 1970. Control of development in temperate bacteriophages. III. Which prophage genes are and which are not *trans*-activable in the presence of immunity? J. Mol. Biol. *49:* 393.

WOLLMAN, E. L., and F. JACOB. 1954. Lysogénie et recombinaison génétique chez *Escherichia coli* K12. Compt. Rend. Acad. Sci. *239:* 455.

WOLLMAN, E. L., F. JACOB, and W. HAYES. 1956. Conjugation and genetic recombination in *Escherichia coli* K-12. Cold Spring Harbor Symp. Quant. Biol. *21:* 141.

YAMAGISHI, H., K. NAKAMURA, and H. OZEKI. 1965. Cohesion occurring between DNA molecules of temperate phages φ80 and lambda or φ81. Biochem. Biophys. Res. Commun. *20:* 727.

THE FIRST PAPERS ABOUT LAMBDA

APPLEYARD, R. K. 1953. Segregation of lambda lysogenicity during bacterial recombination in *E. coli* K-12. Cold Spring Harbor Symp. Quant. Biol. *18:* 95.

BERTANI, G., and J. J. WEIGLE. 1953. Host controlled variation in bacterial viruses. J. Bacteriol. *65:* 113.

BOREK, E. 1952. Factors controlling aptitude and phage development in a lysogenic *Escherichia coli* K-12. Biochim. Biophys. Acta *8:* 211.

JACOB, F., and E. L. WOLLMAN. 1953. Induction of phage development in lysogenic bacteria. Cold Spring Harbor Symp. Quant. Biol. *18:* 101.

JACOB, F., E. WOLLMAN, and L. SIMINOVITCH. 1953. Propriétés inductrices des mutants virulents d'un phage tempéré. Compt. Rend. Acad. Sci. *236:* 544.

LEDERBERG, E. M. 1951. Lysogenicity in *E. coli* K-12. Genetics *36:* 560.

LEDERBERG, E. M. and J. LEDERBERG. 1953. Genetic studies of lysogenicity in *Escherichia coli*. Genetics *38:* 51.

LIEB, M. 1953. The establishment of lysogenicity in *Escherichia coli*. J. Bacteriol. *65:* 642.

LIEB, M. 1953. Studies on lysogenization in *Escherichia coli*. Cold Spring Harbor Symp. Quant. Biol. *18:* 71.

SIMINOVITCH, L., and F. JACOB. 1952. Biosynthèse induite d'un enzyme pendant le développement des bactériophages chez *Escherichia coli* K12. Ann. Inst. Pasteur *83:* 745.

WEIGLE, J. J. 1953. Induction of mutations in a bacterial virus. Proc. Nat. Acad. Sci. *39:* 628.

WEIGLE, J. J., and M. DELBRÜCK. 1951. Mutual exclusion between an infecting phage and a carried phage. J. Bacteriol. *62:* 301.

WOLLMAN, E. L. 1953. Sur le déterminisme génétique de la lysogénie. Ann. Inst. Pasteur *84:* 281.

Progress Since 1970

Roger W. Hendrix
Department of Biological Sciences
University of Pittsburgh
Pittsburgh, Pennsylvania 15260

This paper reviews progress in λ biology since the predecessor to this book, *The Bacteriophage Lambda*, was written in 1970. References are given to other papers in the volume, and these should be consulted for detailed discussions and access to the primary literature.

LYSIS/LYSOGENY DECISION

In the past decade gratifying progress has been made in our understanding of the regulatory circuits of the phage, particularly those dealing with the "decision" between lytic growth and lysogeny (Friedman and Gottesman; Wulff and Rosenberg; Echols and Guarneros; Gussin et al.). Stripped to its essentials, the decision can be understood as a competition between the products of genes *c*I and *cro* for occupancy of the operator o_R. These two proteins, both of which are transcriptional repressors, compete for the operator and repress each other's synthesis. The most important modulator of this balance is the *c*II protein, which is itself regulated by both *c*I and *cro*, as well as by *c*III and the host gene *hflA*.

Studies on the structures of operators have elucidated how the *c*I and *cro* proteins act. o_R (and o_L) contains three adjacent binding domains, and the rules for how *c*I and *cro* proteins bind to them (order of binding, cooperativity, affinity) are different for the two repressor proteins in ways that can be understood as contributing to the observed behavior of the regulatory circuits. The recent determinations by X-ray diffraction of the structures of *cro* protein and the DNA-binding portion of *c*I protein reveal detailed structural information about the protein-DNA interactions.

The demonstration that there are two separate promoters for the *c*I gene has also been central to our understanding of this system. One of these promoters, p_{RM} (also called p_M), maintains repressor levels in a lysogen; it is activated by *c*I protein and repressed by *cro* protein. The second, p_{RE} (also called p_E), is active in establishing repression and is inactive in a lysogen; it is activated by gp*c*II.

Our current understanding of how the various elements in this scheme interact to produce the observed behavior is described by Friedman and Gottesman, Wulff and Rosenberg, Echols and Guarneros, Gussin et al., and the fold-out poster (Herskowitz). The system assays the environment in the infected cell, in particular, the physiological condition of the cell and the multiplicity of infection, and feeds this information into the binary decision between the lytic and lysogenic pathways. As the decision is being made, *int* protein, the protein that catalyzes integration of λ DNA into the host chromosome, is made in sufficient quantity to insure integration in the event that the decision is made for lysogeny. Although the switch is sensitive to minor changes in the input information, once it has been thrown toward one of the alternative pathways, feedback loops insure efficient operation by locking the switch in that position. The correct functioning of this system depends not only on interactions between proteins and sites on the DNA molecule but also on the detailed ordering and orientation of genes and sites along the DNA, on protein-protein interactions, and on differential protein stability.

INTEGRATION AND EXCISION

Mechanism

By 1970 it had been established that integration requires the function of the *int* gene, whereas excision requires both *int* and *xis*, and the efficiencies of recombination between different pairs of attachment sites had been determined. Since then, gp*int* and gp*xis* have been purified, and both integration and excision can be carried out in vitro. Both reactions require the host protein IHF (integration host factor). Enough is known about the biochemical details of the reactions that models for their mechanisms can profitably be entertained (Weisberg and Landy). DNA sequencing studies have revealed that the phage and bacterial attachment sites share a 15-bp identity and that breakage and rejoining of the DNAs takes place within this region, at precise positions staggered 7 bp apart on the two DNA strands of each identity region. The phage attachment site (*att*P) extends considerably beyond the 15-bp identity; protection by gp*int*, gp*xis*, and IHF binding defines a phage attachment site 240 bp long. In contrast, the bacterial attachment site (*att*B) extends only a few bases beyond the identity.

Regulation

In 1970 it was known that the *int* and *xis* genes are expressed from the early promoter p_L. Shortly thereafter, it was shown that *int* can also be expressed from a second promoter, p_I, lying just upstream from the *int*-coding sequence (Echols and Guarneros). Transcription from p_I is stimulated by the same factors that stimulate *c*I synthesis from p_{RE} during the establishment of repression. Unlike p_L, this promoter does not allow expression of the adjacent *xis* gene; thus, expression of the p_I transcript favors integration, and expression of the p_L transcript favors excision. Recently, a posttranscriptional feature of *int* regulation has emerged. The probability that an mRNA containing the *int* sequence will be translated depends on whether the transcript was initiated at p_L or at p_I.

This difference in expression requires the presence of λ DNA sequences 300 bp downstream from the *int* gene, on the opposite side of the attachment site. The basis of this retroregulation, as it has been called, is now understood in terms of RNA polymerase termination specificity, mRNA secondary structure, and mRNA processing by RNase III. Since retroregulation makes the relative rates of expression of *int* and *xis* dependent on which DNA is present on the other side of the attachment site, it provides a means for the phage to modulate expression of *int* and *xis* (and therefore to control whether integration or excision predominates) according to whether the phage is free or is integrated into the host chromosome.

INDUCTION

It has been known from the earliest days of λ research that the λ prophage is induced by ultraviolet irradiation; in fact, the discovery of λ depended on this property (see Arber). Nevertheless, it is only in the past few years that we have come close to an understanding of how ultraviolet light inactivates λ repressor to cause induction (Roberts and Devoret). Progress in understanding λ induction has gone hand in hand with progress in understanding the complex regulation of the *Escherichia coli* "SOS functions," with which λ induction is intimately entangled. The picture that has emerged is that the damage to cellular DNA caused by ultraviolet light activates both λ and the SOS genes. The host protein that mediates this cellular response is the *recA* protein, a multifunctional protein that, under the conditions in an irradiated cell, cleaves both λ repressor and the cellular repressor of SOS genes. Cleavage destroys the repressor's ability to oligomerize and, as a consequence, its ability to bind cooperatively to the operators, and repression of the prophage is lost.

LYTIC GROWTH

The regulation of λ lytic growth was well outlined by 1970, but the intervening time has seen the elucidation of interesting details, particularly with respect to how RNA polymerase works and how its action can be modified by λ. The main theme of λ lytic growth is antitermination of transcription, which is now known to be carried out by two phage proteins: gp*N*, the positive regulator of early functions, and gp*Q*, the positive regulator of late genes. Antitermination by gp*N* has been shown to require sites called *nut* (*N ut*ilization), which lie downstream from the promoter and upstream from the site of antitermination and at which gp*N* is thought to alter the properties of the transcription complex. To date there is no direct evidence for such a site in antitermination by gp*Q*, but sequence similarity with the *nut* sites suggests that a "*qut*" site probably lies within the first 20 bp downstream from the late promoter $p_R{}'$. Genetic studies have revealed a number of host functions that are required for gp*N* function, including an RNA polymerase factor and ribosomal proteins. At least one of these host factors, the *nusA* protein, is also required for gp*Q* function. Although the timing of λ gene expression and much of its quantitative regulation occurs at a transcriptional level, there are many examples of differential translation. This is particularly apparent with the head and tail genes, which are all transcribed equally but produce proteins that vary in number over a several-hundredfold range.

ASSEMBLY

A major advance in our understanding of λ head assembly that came shortly after publication of *The Bacteriophage Lambda* was the demonstration that the DNA molecule is packaged into a preformed protein shell. Previously, it had been thought that the DNA condensed and was then surrounded by protein. In the process of establishing the mode of DNA packaging, it became clear that the small empty protein shells called *petit* λ, which had been observed in lysates much earlier and were assumed to be dead-end assembly by-products, are the precursor shells into which DNA is inserted. This realization, together with the development of in vitro assembly systems, made it possible to study the steps of head assembly that occur prior to DNA packaging. Georgopoulos et al. review our current view of head assembly, which includes the participation of host proteins, a scaffolding protein that acts transiently and is then degraded, protein cleavage and fusion, and a concerted conformational change in the major head subunit during DNA packaging.

Most steps of λ virion assembly can now be carried out in vitro, in most cases in the presence of a crude extract. DNA packaging is one of the best studied of these reactions (Feiss and Becker). During packaging, DNA is cut from a concatemer. The responsible phage-coded enzyme, terminase, has been purified, and genetic and biochemical experiments have begun to reveal how the terminase acts, what the important features of its DNA substrate are, and what the nature of the linkage between DNA cutting and DNA packaging is. Besides being of considerable scientific interest, the DNA packaging reaction occupies an important place in molecular cloning technology.

The entire tail assembly sequence (Katsura) can now be carried out in vitro in crude extracts, and this has made it possible to determine the order in which the tail genes act. As with other long-tailed phages, a structure that will become the tail tip assembles first, and this serves as an initiator for assembly of the tail shaft. Some progress has been made in understanding the biochemical details of individual steps; e.g., one minor protein that is assembled into the nascent structure early in the assembly pathway is cleaved to a slightly smaller form several steps later. The long-standing question of how tail length is determined is still under debate but shows signs of yielding to experimental attack.

In parallel with advances in our understanding of assembly, knowledge of the structure of the mature virion has become considerably more detailed over the past decade. (Georgopoulos et al.; Feiss and Becker; Katsura).

DNA REPLICATION

λ DNA replication has two phases. In the early phase, monomeric circular molecules replicate bidirectionally from a defined origin to give θ form intermediates and monomeric circular daughters. The late phase, which starts 10 minutes after infection, produces concatemeric molecules that serve as the substrate for packaging. The mode of late replication, which was not known in 1970, now appears nearly certainly to be the rolling-circle mechanism. Late replication can go in either direction and starts from the same origin as early replication, at least a great part of the time. The origin of replication, *ori*, has been identified as a region within the coding sequence of gene *O*, one of the two phage genes required for replication. Exactly how *ori* functions in replication is not yet clear, but gp*O* has been found to bind to it and to protect a region containing four nearly exact 19-bp repeats in the sequence. Genetic evidence argues that gp*P*, the other λ-coded replication protein, interacts with gp*O*, as well as with several host DNA replication proteins that are required for λ replication. In vitro systems for λ DNA replication have been established recently and will soon reveal biochemical details of the process.

RECOMBINATION

The three main recombination systems that act on λ—Red (λ-coded, homology-dependent), Int (λ-coded, site-specific), and RecBC (host-coded, homology-dependent)—were known in 1970. Studies over the past several years show that Int- and RecBC-mediated recombinations go by a break-join mechanism in which DNA replication plays no detectable role (Smith). The Red system, however, requires extensive prior or concomitant DNA replication. This and other features of Red-mediated recombination have spawned the proposal that Red is a break-copy system. Considerable progress has been made in understanding the enzymatic properties of several phage- and host-coded recombination proteins, and detailed mechanisms have been proposed for all three recombination systems. It seems certain that when the actual mechanisms are established, they will be different from one another.

An unexpected entry into the cast of recombination characters was the Chi site, an 8-base-long DNA sequence that stimulates recombination by the host RecBC pathway. Chi in λ causes increased recombination over a region of several kilobases, located asymmetrically around the Chi sequence. Curiously, this enhancement of recombination depends on the correct orientation of the asymmetric Chi sequence with respect to the *cos* sites in the DNA. Although Chi was first found in a λ mutant, there are no Chi sites in wild-type λ. The *E. coli* genome, on the other hand, has a Chi site about once every 5000 bp. There is reason to think that Chi plays a major role in *E. coli* recombination and that other organisms may have Chi-like sites as well.

ACCESSORY GENES

A large set of λ genes is traditionally grouped together because the genes share the property that they are not required for either the lytic or the lysogenic cycle, at least under laboratory conditions. "Nonessential genes," the name usually applied to these, certainly underestimates their interest, and Court and Oppenheim have renamed them "accessory genes." Most of the accessory genes lie in the third of the genome between genes *J* and *N*, with a few scattered to the right of *N*. For the most part, the functions of the accessory genes are either totally unknown or known only in general terms; nevertheless, a large body of tantalizing information has accumulated. For example, the several proteins encoded in the *b* region of the genome have no known functions, but they include three distinct DNase activities, and another *b*-region protein is found in the outer membrane. Some of the accessory genes between *att* and *N* influence the turn-on and turn-off of host genes following infection, the uptake of certain macromolecular precursors, the levels of cAMP and ppGpp, and the functioning of host restriction systems. Two other accessory genes are responsible for functions that exclude superinfecting phages.

COMPARATIVE STUDIES

From its earliest days, λ research has benefited from comparisons with studies on related phages. This has continued to be true over the past several years, particularly with respect to the *Salmonella* phage P22. The genetic similarity between λ and P22 has turned out to be considerably greater than had been appreciated. P22 has an immunity region that is strikingly similar and almost certainly homologous to λ's, and this has led to informative comparative studies of the regulatory circuits of the two phages (Susskind and Youderian), as well as of the constituent regulatory proteins, especially the repressors (Gussin et al.). In addition to its λ-like immunity region, P22 has a second immunity region that encodes, among others, a protein that antagonizes both the P22 repressor and, if given the opportunity, the λ repressor. Another similarity between λ and P22 has been found in their virion assembly pathways (Georgopoulos et al.), and a comparison of the two pathways has led to a better understanding of which features are fundamental and which are embellishments on the basic theme. The level of comparative detail about λ, P22, and other related phages is now sufficient to allow tasteful speculation about evolutionary relationships among lambdoid phages and their hosts (Campbell and Botstein).

MOLECULAR CLONING

λ has played a central role in the molecular cloning methodology that has revolutionized biology over the past decade. This is largely because λ is so well studied that it became an obvious choice for development as a cloning vector; indeed, it is doubtful that in the absence of 25 years worth of accumulated λ biology, cloning technology could have developed nearly as soon or as rapidly as it has. The advanced knowledge of λ biology continues to facilitate modification and manipulation of λ vectors. Strategies for the use of λ as a vector are discussed by Murray.

DNA SEQUENCE

DNA sequences have been available for some parts of the λ genome for several years, and these have substantially advanced our knowledge of the corresponding parts of λ biology. Familiar examples include the sequences of the attachment site, the replication origin, and the promoters and operators of the right arm. These and other examples of the value of DNA sequencing in deciphering biological function appear throughout the book. Recently, the sequence of the entire λ genome has been completed, and it is given here (Appendix II). This (by a small margin over phage T7) is the largest genome for which the DNA sequence is completely known. Much of this sequence is only beginning to be assimilated, and its analysis can be expected to produce many new insights.

HOST FACTORS

An increasingly important theme of λ research over the past decade has been the role of host proteins in the phage life cycle. Examples of this occur throughout the book. Although some of these, such as RNA polymerase, ρ factor, *dnaB* protein, and *lamB* protein (the phage receptor), have been known for many years, the number of known host factors has increased substantially since *The Bacteriophage Lambda* was written. The size of the list of host factors depends somewhat on the definition accepted for a host factor, but a current list might include, in addition to those already mentioned, *recA* protein (recombination, induction); *recBC* protein (recombination); *nusA-nusE* proteins (transcription); *groEL* and *groES* proteins (head assembly); *hflA*, *hflB*, and *lon* proteins (lytic/lysogenic decision); IHF (integration, lytic/lysogenic decision); gyrase, ligase, *ssb*, *polA* (recombination, DNA replication); *dnaE, G, J, K, X, Y,* and *Z* proteins (DNA replication); *ptsM* protein (injection); and RNase III (mRNA processing). Investigation of these connections between λ and *E. coli*, which occur at all levels of the life cycle, has enriched our knowledge both of the phage and of its host.

Lytic Mode of Lambda Development

David I. Friedman
Department of Microbiology and Immunology
University of Michigan School of Medicine
Ann Arbor, Michigan 48109

Max Gottesman
Laboratory of Molecular Biology
National Cancer Institute
National Institutes of Health
Bethesda, Maryland 20205

Lytic growth of bacteriophage requires the concerted expression of a number of functions, products of both viral and host genes. These functions promote transcription, replication, DNA processing, and packaging. Additionally, temperate phage such as λ, which can grow either lytically or lysogenically, must insure that once the decision to adopt one of the alternative modes of existence has been made, functions that can interfere with that mode are not expressed. In the lytic mode, the phage functions must be sequentially expressed, i.e., replication must precede packaging which, in turn, must precede lysis. Thus, although phage propagation occurs in a relatively short time—some 45 minutes for λ—it follows a strictly coordinated developmental plan.

In this paper we review current knowledge of λ lytic development. The lysogenic program is discussed by Wulff and Rosenberg and by Echols and Guarneros (both this volume). Our discussion focuses on (1) gene organization, (2) patterns of mRNA synthesis, and (3) regulatory controls of gene expression. The role of the λ*N* function in regulating the lytic cycle is stressed. This is, in part, because *N* function is among the best characterized regulatory proteins and, also, its action typifies a mechanism of regulation based upon transcription termination-antitermination. For further background material and discussions from different perspectives, see Echols (1971), Weisberg et al. (1977), and Herskowitz and Hagen (1980).

GENETIC ORGANIZATION

The λ genome expresses approximately 50 proteins (Szybalski and Szybalski 1979). In this section we discuss the arrangement of the λ genes with respect to the dynamics of phage development.

The isolation of a variety of λ mutations led to the construction of a fine-structure genetic and physical map (Szybalski and Szybalski 1979) and promoted the detailed analysis of the regulation of gene expression. The various sets of λ genes could be designated according to their roles in either lysis or lysogeny: (1) solely lytic, (2) solely lysogenic, (3) those participating in both, and (4) genes whose role is unclear. Genes in sets 2 and 4 can be considered nonessential, since λ can propagate exclusively by the lytic mode. Genes in sets 1 and 3, which will be our primary concern, were identified by the isolation of conditionally lethal mutations: either temperature-sensitive or, more importantly, amber mutations (Campbell 1961). Analysis of lysogens with defective prophage (Eisen et al. 1966; Gottesman and Weisberg 1971) led to the identification of λ promoters, as well as sites of action of the λ regulatory gene products.

The genetic map of λ is based on recombination frequencies between various mutations, as well as marker rescue experiments using deletion and deletion-substitution phages. More recently, direct sequencing of mutant phage has become possible. The total recombination distance for distal markers is 15% (Kaiser 1955). The smallest recombination distance measured is 0.05% (McDermit et al. 1976); sequence analysis shows that these mutations are separated by 4 bp (Rosenberg et al. 1978).

A number of temperate phage species can recombine with λ to form viable hybrids (Kaiser and Jacob 1957; Liedke-Kulke and Kaiser 1967; Szpirer et al. 1969; Gemski et al. 1972; Botstein and Herskowitz 1974). Construction of such hybrids is possible because these "lambdoid" phages have some sequence homology and share the same genomic arrangement (Davidson and Szybalski 1971). With a single exception, all lambdoid phages exhibit unique immunity specificities. Thus, phages ϕ80, 434, 21, and λ all express and recognize unique repressors. The one exception is the case of coliphage 21 and *Salmonella* phage P22 (Botstein and Herskowitz 1974), which carry the same repressor gene. The hybrid heteroimmune phage have proved extremely useful in studying gene control, since they differ in only a few regulatory genes, e.g., *c*I, *cro*, and *N*, in an otherwise identical genetic background.

Figure 1 shows the arrangement of genes and sites pertinent to this paper (for a more detailed map, see Appendix I). A striking feature of the λ map is the arrangement of genes in functional units. Moreover, the placement of these units reflects their temporal order of expression. The units can be classified as follows.

Immunity region: This region includes important regulatory functions, encoding both the *c*I and *cro* repressors and their cognate operator sequences o_R and o_L (Ptashne et al. 1976; Johnson et al. 1981). The operators are located on either side of *c*I. In λ, gene *rex* intervenes between o_L and *c*I. Interpenetrating the operator regions are the early leftward (p_L) and early rightward (p_R) promoters. The *c*I repressor maintenance promoter, p_{RM}, is also contained within the o_R sequence. (For a detailed discussion, see Gussin et al. [this volume].)

Although the major phenotypic characteristic of the immunity region involves repressor activity, this region is defined structurally as the total nonhomology between λ and other lambdoid phages in the region of the *c*I gene. Thus, the extent of the immunity region depends on the particular phage being compared with λ (see Fig. 1). The immunity region defined by comparing λ and phage 434 genomes is relatively small, including the *c*I, *cro*, and *rex* genes, as well as the

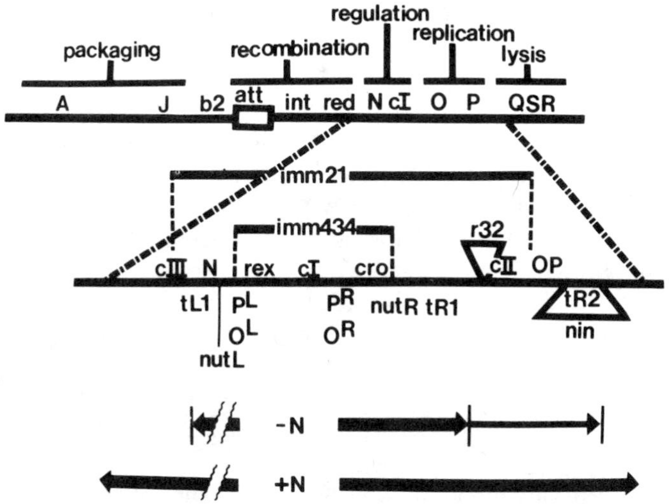

Figure 1 Map of λ (not drawn to scale). (*Top line*) The complete genome showing the organization by functional units and the location of a few key genes. (*Middle three lines*) An expansion showing the organization of the λ immunity region and the extents of the nonhomologies with phages 21 and 434. (*Bottom two lines*) Transcription patterns in the presence and absence of p*N*.

operators and promoters discussed above (Kaiser and Jacob 1957). A larger immunity region is defined when λ is compared with phage 21; here the nonhomology in the *c*I region includes *c*I, *cro*, and *rex*, as well as the *N* and *c*II genes (Liedke-Kulke and Kaiser 1967).

Early and delayed early genes: After infection or induction, transcription initiates within the immunity region and proceeds leftward from p_L or rightward from p_R, defining two operons (see Fig. 1). The only essential gene in the p_L operon is *N*; the others are involved in generalized recombination, site-specific recombination, and a variety of other, less well-understood phenomena (see Court and Oppenheim, this volume).

The first cistron transcribed in the p_R operon is the essential gene *cro* (Neubauer and Calef 1970; Echols et al. 1973; Eisen et al. 1975; Folkmanis et al. 1977). Next transcribed is *c*II, encoding a dispensable function whose primary role is to establish repression in the lysogenic response (see Wulff and Rosenberg, this volume). Downstream of *c*II are the genes required for autonomous λ DNA replication, *O* and *P* (see Furth and Wickner, this volume).

The *cro*- and *N*-gene products are synthesized early after infection, whereas the other products of the p_L and p_R operons are produced after the delay required for synthesis and action of *N* protein (Luzzati 1970; Kourilsky et al. 1971a; Lozeron et al. 1976; Rosenberg et al. 1978; Court et al. 1980).

Regulation of late genes: The gene promoter-distal to *P* is *Q*, whose product is required for late gene expression (Thomas 1966; Herskowitz and Signer 1970; Murialdo and Siminovitch 1972; Ray and Pearson 1974). The *Q* protein acts on transcription initiating at p_R' to suppress termination (Blattner and Dahlberg 1972; Roberts 1975; Sklar et al. 1975). All of the late genes are transcribed from p_R' (Oda et al. 1969; Chowdhury and Guha 1973; Gariglio and Green 1973).

Late genes, lysis: Two genes, *R* and *S*, encode products required for cell lysis (Harris et al. 1967; Goldberg and Howe 1969). The *R*-gene product has been identified as a 17K protein with murein transglycosylase activity (Bienkowska-Szewczyk and Taylor 1980). It has been proposed that the product of the *S* gene is a 15K protein that is modified to an active 11K form; the active form of S protein interacts with the cell membrane, causing pore formation (Wilson 1982). Phage carrying a transposon downstream from the *R* gene are lysis-defective. This has led to the suggestion that λ has a third lysis gene (Young et al. 1979). However, no protein associated with such a gene has been identified (R. Young, pers. comm.).

Mutations in the *S* gene are useful in isolating large amounts of phage, since lysis does not occur and phage production can continue long past the normal lysis time. Phage release is accomplished by addition of chloroform to the culture (Goldberg and Howe 1969).

Late genes, morphogenesis: The products of genes *Nu1* through *J* are required for head and tail synthesis and assembly (see Georgopoulos et al.; Feiss and Becker; Katsura; all this volume).

b region: The *b* region encodes unessential genes, including a DNase, a membrane protein, and a site regulating Int expression (see Court and Oppenheim this volume).

Transcription

The lytic program begins with the initiation of transcription at p_L and p_R. The early p_L transcript terminates at t_{L1} to generate an RNA (12S RNA) about 1000 nucleotides long (Roberts 1969; Lozeron et al. 1976), whereas the early p_R transcript ends distal to the *cro* gene at t_{R1}, to generate a 9S transcript (Roberts 1969). The early p_L transcript encodes the *N* gene (Franklin and Bennett 1979), whose product suppresses transcription termination (Roberts 1969). The early p_R transcript encodes the *cro* gene; *cro* negatively regulates repressor synthesis (Heinemann and Spiegelman 1971; Kourilsky et al. 1971a,b; Kumar et al. 1971), as well as the p_R and p_L transcripts (see Gussin et al., this volume).

In the case of p_L-promoted transcription, the presence of p*N* overcomes the t_{L1} and subsequent terminators and permits transcription to proceed into the *b* region, at least several thousand nucleotides beyond *att* (Lozeron et al. 1976). In the *b* region, a p*N*-resistant terminator prevents the p_L-promoted transcript from entering the antisense strand of the λ left arm (Gottesman et al. 1980; Honigman 1981). Although termination at t_{L1} depends on the *Escherichia coli* transcription termination factor, ρ (Roberts 1969), at least one other terminator in the p_L operon, t_{L2}, near *kil*, is ρ-independent. Thus, p*N* suppresses termination at both ρ-dependent and ρ-independent terminators (Salstrom and Szybalski 1978b; Gottesman et al. 1980).

The p_L transcript is subject to extensive processing (Kourilsky et al. 1971b; Lozeron et al. 1976, 1977). The p_L proximal portion is cleaved at three sites by RNase III: (1) about 70 nucleotides from p_L, to generate l_1^0, which contains the 5′-ppp terminus. This cleavage appears to cross the two strands of a stem-loop structure, with loss of nucleotides 71–89; (2) about 120 nucleotides to the left of site 1, generating l_1 (nucleotides 89 to ~190); (3) at or near t_{L1}, yielding l_2. Sequence analysis of the *N*-gene region shows that the *N* protein could begin at two alternative initiation codons, at nucleotide positions 145 or 223; RNase site 2 lies between them. Estimations of the size of the *N* protein, and the absence of an internal methionine suggest that the second ATG is likely to be the *N* initiation codon (Franklin and Bennett 1979; Greenblatt and Li 1982). The l_2 RNA segment, which probably encodes p*N*, is highly unstable ($t_{1/2}$=20 sec); the nuclease responsible for its degradation is as yet unidentified. The l_1 segment is, in contrast, very resistant to decay.

The p_R message, unlike the p_L message, is not processed (H.A. Lozeron, pers. comm.). Termination at t_{R1} is ρ-dependent and is suppressed by *N* function (Rosenberg et al. 1978; Rosenberg and Court 1979). The t_{R1} terminator differs, however, in lacking the five or six characteristic T residues found in other terminators (Rosenberg et al. 1978). Although t_{R1} has a possible stem-loop structure, the region of dyad symmetry is small, and the structure, if formed, would be rather unstable. These features may account for the relative inefficiency of t_{R1} as a terminator. Even in the absence of p*N*, about 50% of the p_R transcripts will, after a lag period, continue through t_{R1} and into the delayed early *c*II-*O*-*P* region (Court et al. 1980; Dambly-Chaudière et al. 1983). For this reason, λ replication is not fully *N*-dependent.

Although the function of t_{R1} is not known, mutations that affect its structure do have physiological consequences. The *cin* mutation increases the efficiency of termination at t_{R1} (Wulff 1976). Pseudorevertants of λ*cin* carrying a second mutation, *cnc*, which reduces termination at t_{R1}, have been isolated (Wulff 1976). The two mutations have not been separated. Nucleotide sequence analysis shows that both the *cin* and *cnc* mutations influence the possible base-pairing in the t_{R1} stem-loop structure. The *cin* mutations increase, whereas the *cnc* mutations reduce pairing. These experiments suggest that the stem-loop structure is important in termination (Rosenberg et al. 1978). However, *cin* creates a new leftward promoter whose role in termination has not been evaluated (Court et al. 1980).

In addition to *c*II, *O*, and *P*, the set of p_R operon delayed early genes also includes *Q*. Between genes *P* and *Q* is a second terminator, t_{R2}. The t_{R2} terminator is of greater efficiency than t_{R1} and is also suppressed by p*N*; in the absence of p*N*, no *Q* function is expressed (Couturier and Dambly 1970). Deletion of t_{R2} (*nin*) creates a phage that can, to some extent, propagate lytically even in the absence of p*N* (Court and Sato 1969).

A third terminator, t_{6S}, separates *Q* from the late gene region. An efficient ρ-independent terminator, t_{6S} is also suppressed by p*N* (Sklar et al. 1975; Gottesman et al. 1980).

Several minutes after λ infection, the synthesis of early and delayed early gene products diminishes. This reflects an inhibition of transcription from p_L and p_R due to the binding of the phage *cro*-gene product to o_L and o_R. *cro*-mediated inhibition is essential for the λ lytic program. Excessive p_L activity with overproduction of the λ*ea*10-gene product results in several defects in host and phage metabolism (Georgiou et al. 1979). Unmodulated p_R activity is also lethal to λ, for reasons as yet unknown.

The next stage in λ lytic development is the expression of the viral late genes. Late gene expression is analogous to early gene regulation. A positive regulatory function, the product of the delayed early λ gene *Q*, acts to antiterminate transcription initiated at $p_R{}'$, located between *Q* and *S* (Roberts 1975). Unlike the cases of p_R or p_L, transcription from $p_R{}'$ is constitutive. The 6S transcript corresponding to the region $p_R{}'$ to t_{6S} can be detected in a repressed lysogen (A. Oppenheim, pers. comm.).

Under the influence of p*Q*, the $p_R{}'$ transcript extends throughout the late gene region and terminates somewhere in *b* (Bouvre and Szybalski 1969; Burt and Brammer 1982). The absence of polar mutations in the late region and the observation that, in a *b*2 prophage, $p_R{}'$-promoted transcrition enters the *bio* operon, are consistent with an antitermination action of p*Q* (Krell et al. 1972). Suppression by p*Q* of bacterial amber-ochre or insertion-sequence (IS) mutations supports this model (Forbes and Herskowitz 1982). Hybridization studies indicate that the late region is uniformly transcribed (Ray and Pearson 1974). In contrast, transcription from p_L under the influence of p*N* diminishes rapidly with distance from the promoter (Adhya et al. 1976). Perhaps this difference explains why a second positive regulatory system, in addition to p*N*, is needed for λ lytic growth.

LYSIS VERSUS LYSOGENY

The ability of λ to choose between a lytic and a lysogenic mode of propagation has probably evolved in response to the fluctuations in habitats of its host, *E. coli* (M. Savageau, pers. comm.). In the colon, *E. coli* grows in a relatively rich environment (Rosebury 1962), whereas in sediment and soil, bacteria are comparatively deprived (Van Donsel and Geldreich 1971). λ lytic replication is most productive when the host is growing well in abundant nutrient (Kourilsky 1973); however, the lysogenic mode allows the prophage to maintain itself indefinitely in poorer growth conditions.

How is the growth rate of the host signaled to the infecting λ? Some evidence suggests that the level of 3′-5′-cAMP influences the lysis-lysogeny decision (Hong et al. 1971; Grodzicker et al. 1972). When cAMP levels are low, as is the case in rich medium, λ favors the lytic pathway. In host *cya* mutations lacking cAMP, the lytic pathway is chosen over the lysogenic, independently of the growth medium. Although cAMP is known to stimulate or repress bacterial promoters, no activity of cAMP has been reported for any phage promoter. It is likely, therefore, that the effect of cAMP on λ regulation is indirect, mediated through some bacterial gene product. A candidate for such a gene is *hfl*, whose product appears to reduce the activity of *c*II function (Belfort and Wulff 1974). Other possibilities include *himA-hip*, whose products are required both for λ integration and for the synthesis of *c*II protein (Miller and Friedman 1980; Miller 1981; Oppenheim et al. 1982); *lon*, which decreases the stability of *c*II protein (Gottesman et al. 1981); and RNase III. λ is directed exclusively into the lytic pathway in RNase-III-deficient mutants (A. Oppenheim, pers. comm.).

The frequency of lysogenization is also influenced by the multiplicity of infection. At low multiplicities, the phage is channeled toward lysis (Boyd 1951; Lieb 1953). Again, the levels of *c*II function could be the critical factor in this response. A low initial phage copy number might yield less active *c*II product (Kourilsky 1973). Teleologically, a low ratio of phage to bacteria signifies a large supply of host for lytic growth, whereas a high ratio indicates an imminent loss of new hosts on which to propagate lytically.

REGULATION OF THE LYTIC PROGRAM

N-gene Product

The product of the *N* gene is a 12,200-dalton protein (Franklin and Bennett 1979; Greenblatt and Li 1982) that activates gene expression by antagonizing transcription termination (Roberts 1969; Luzzati 1970; Nijkamp et al. 1970, 1971; Heinemann and Spiegelman 1971; Kourilsky et al. 1971b; Kumar et al. 1971; Lozeron et al. 1976). Our current concept of p*N* action originated with Roberts (1969), who identified transcription termination sites on the λ chromosome and postulated that these sites were suppressed by λ*N*-gene product. By this means, p*N* permits the expression of the λ early genes that lie promoter-distal to these sites.

A remarkable feature of regulation by p*N* is its specificity. Only transcription originating at the λ p_L or p_R promoters is rendered resistant to termination (Thomas 1966; Dambly et al. 1968; Courturier and Dambly 1970; Herskowitz and Signer 1970; Luzzati 1970). Transcription initiating at bacterial promoters[1] or at the promoters of other lambdoid phages (except 434, which has the λ*N* gene), does not become termination-resistant in the presence of λ*N* protein

[1] A possible exception is the *rpoD* gene, whose expression is stimulated by p*N* (Nakamura and Yura 1976).

(Friedman et al. 1973b; Adhya et al. 1974; Franklin 1974; Segawa and Imamoto 1976). Lambdoid phages 21, ϕ80 and P22, in fact, have been shown to encode their own specific *N*-like functions (Friedman et al. 1973b; Franklin 1974; Hilliker and Botstein 1976).

The specificity elements for p*N* regulation are not the terminators, since viable λ-21 hybrids can be constructed (Liedke-Kulke and Kaiser 1967). These phages have the immunity region of 21, including N^{21} but carry the t_{R2} terminator of λ (Friedman et al. 1973b). Furthermore, provided transcription initiates at p_L or p_R, terminators in bacterial DNA can also be suppressed by p*N* (Adhya et al. 1974; Franklin 1974; Segawa and Imamoto 1976). These considerations led to the idea that the sites of p*N* action, the terminators, and of p*N* recognition, where p*N* modifies transcription, were not the same (Friedman et al. 1973b; Adhya et al. 1974; Franklin 1974; Szybalski 1974). The fact that the p_L and p_R promoters of λ and 434 lie in a region of nonhomology yet the phages share the same *N* gene suggested that the promoters were not the sites of p*N* recognition. The p*N* recognition sites must lie between the promoters and the first terminators, t_{L1} and t_{R1}.

Salstrom and Szybalski (1978a) identified a site (*nut*) in the p_L operon between the promoter and t_{L1} that fulfilled the requirements for a p*N* recognition site. Mutations in this site prevented the modification of the p_L transcript by p*N*, so that transcription terminated at t_{L1} despite the synthesis of active *N* product. DNA sequence analysis (Fig. 2) reveals that the *nut* mutations lie in a region of hyphenated dyad symmetry, consistent with the formation of a DNA or RNA stem-loop structure (Rosenberg et al. 1978). These studies also identified a similar (16-bp/17-bp) sequence that lies between *cro* and t_{R1} (see Fig. 2). Presumably this sequence lies at the site at which p*N* modifies rightward transcription.

That this latter sequence, in fact, identifies the *nut*R site is shown by several different types of experiments. First, a 395-bp fragment extending from within *cro* to within *c*II, which included this site as well as t_{R1}, was cloned downstream from the promoter for the *gal* operon. In this construct, transcription initiating at the *gal* promoter was modified by p*N* and failed to terminate at t_{R1} (de Crombrugghe et al. 1979). This finding suggests that no p*N*-specificity elements reside in the promoter regions. Second, Reyes et al. (1979) fused *gal* to the p_R operon downstream of *nut*R and t_{R1}. The expression of *gal* was stimulated by *N* function. Analysis of deletions within this fusion that eliminated the response to p*N* placed the *nut*R site between *cro* and t_{R1} (Dambly-Chaudière et al. 1983). Third, the mapping of λ*nut*R$^-$ mutations is also consistent with this assignment (Olson et al. 1982).

Although the experiments of de Crombrugghe et al. (1979) give an upper limit for the size of *nut*, they do not define the minimum number of nucleotides that constitute a p*N*-recognition region. We point out, in Figure 2, homologies in addition to the 17-bp sequence near the *nut*L and the putative *nut*R. Arguing for the significance of these additional sequences is the observation that the 17-bp

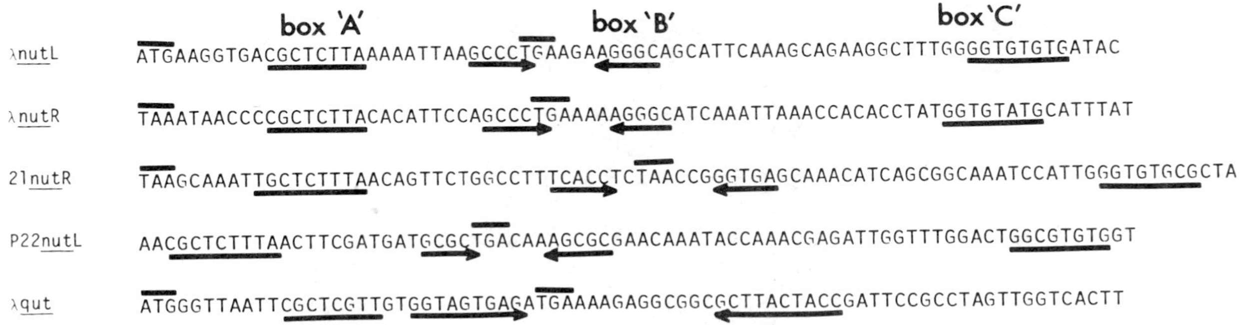

Figure 2 Compilation of *nut* regions or, in the case of box B, the region of dyad symmetry likely to be the site of action of the various *N* products. The putative site for p*Q* action is also shown and labeled as *qut*. Significant sequences are underlined. (For details, see text.)

*nut*L fragment, when cloned into a plasmid constructed similarly to that of de Crombrugghe et al., showed little response to p*N* compared to larger fragments (Drahos et al. 1982).

N independence

λ*N*⁻ mutations fail to propagate lytically because essential portions of the phage chromosome are not transcribed (Skalka et al. 1967; Heinemann and Spiegelman 1971; Kourilsky et al. 1971b; Kumar et al. 1971). Deletion of the t_{R2} region, *nin* (Court and Sato 1969), permits some limited growth of λ*N*⁻, possibly because only about 50% of the transcripts initiating at p_R terminate at t_{R1} (Court et al. 1980). Oddly, even when termination at t_{R1} is presumably eliminated by the incorporation of a *cnc-cin* double mutation into a λ*N*⁻*nin* phage, there is no increase in the efficiency of plating (Court et al. 1980). Perhaps leftward transcription initiating at *cin* reduces opposite-strand transcription from p_R. Another possible solution to this paradoxical finding is offered by the recent in vitro studies of Lau et al. (1982), which show that the t_{R1} region has three distinct termination signals. Only one of these signals is eliminated by the *cnc* mutation. If these remaining terminators are active in vivo, then partial integrity of t_{R1} would be maintained in the mutant phage. p*N* independence can also be achieved by a combination of a constitutive promoter in *cy*, distal to t_{R1} (*c*17), and a second mutation, distal to t_{R2} (*byp*) (Butler and Echols 1970; Hopkins 1970; Sternberg and Enquist 1979). Expression of the p_L operon in the absence of *N* function (or with a *nut* mutant phage) likewise can be established by terminator deletions (Salstrom and Szybalski 1978b). Additionally, some, but not all, *cro* mutations permit the expression of the p_L operon in λ*N*⁻ infection (Oppenheim et al. 1977; Salstrom et al. 1979). Possibly, the elevated rate of p_L transcription in *cro* mutants may exceed the functional capacity of the p_L operon terminators.

ρ Protein

Although p*N* suppresses both ρ-dependent and ρ-independent terminators, it is the action of p*N* at the former that plays an essential role in the λ lytic cycle. Thus, *N*⁻ mutations propagate, albeit inefficiently, on ρ-defective host mutations (Brunel and Davison 1975; Inoko and Imai 1975; Das et al. 1976; Richardson et al. 1977). How p*N* might suppress ρ-dependent terminators is suggested by an analysis of the mechanism of action of ρ.

The *E. coli* ρ protein is a hexamer composed of identical 50,000-dalton subunits (Roberts 1969). The protein catalyzes the termination of transcription in a purified in vitro system at unique sites on the DNA template (Roberts 1969). The termination reaction requires a ribonucleoside triphosphate, which is hydrolyzed in an RNA-dependent reaction whose products are ribonucleoside diphosphate and inorganic phosphate (Howard and de Crombrugghe 1975; Galluppi et al. 1976). During a transcription reaction, the molecules of triphosphate cleaved far exceed the number of RNA chains terminated, and, in fact, the hydrolysis reaction continues after termination is completed (Lowery-Goldhammer and Richardson 1974). How hydrolysis is used is not clear. Perhaps ρ moves along the RNA chain to seek out an RNA polymerase molecule paused at a terminator (Adhya and Gottesman 1978). Alternatively, the energy may be expended in pulling the nascent RNA chain from the template (Richardson and Conway 1980).

Evidence for a protein-protein interaction between ρ and RNA polymerase comes from a study of ρ-deficient mutations, conditionally lethal for growth (Das et al. 1976). Suppressors that restore transcription termination and viability in these mutations reside in *rpoB* (Das et al. 1978; Guarente and Beckwith 1978). Allele specificity between the *rpoB* and ρ mutations is consistent with the idea of a direct interaction between ρ and the β subunit of RNA polymerase.

E. coli ρ mutations were first isolated as suppressors of amber or ochre polarity (Beckwith 1963; Richardson et al. 1975; Korn and Yanofsky 1976; Ratner 1976). Polarity occurs when a polypeptide-chain-terminating codon in a promoter-proximal cistron prevents the expression of the promoter-distal cistrons of the operon (Franklin and Luria 1969). The effect is due to premature transcription termination within the operon at ρ-dependent terminators (Adhya et al. 1974). Translation termination is thought to activate these transcription terminators by creating untranslated RNA to which ρ can affix. Thus, ribosomes, by blocking the access of ρ to RNA, suppress transcription termination. Since not all amber or ochre mutations are polar (Newton et al. 1965), other factors influence the extent to which transcription terminates distal to the nonsense codon. These may be (1) the distance between the nonsense codon and the next cistron, (2) the existence of translation restart codons (Adhya and Gottesman 1978), (3) the location of terminators in the operon (Adhya and Shapiro 1969; de Crombrugghe et al. 1973), and (4) the presence or absence of specific RNA binding sites for ρ protein. This last consideration is based on the finding of a site in gene *cro* RNA that is protected from RNase by ρ protein (Bektesh and Richardson 1980). As expected for a ρ antagonist, p*N* suppresses amber and ochre polarity (Adhya et al. 1974; Franklin 1974).

Host Factors Affecting p*N* action

A number of host factors have been demonstrated to be necessary for effective p*N* activity. These factors, termed *nus* (for p*N* *u*tilization *s*ubstance, also known as *groN*) were initially identified genetically. *E. coli* mutations were selected that failed to support λ growth in general (Georgopoulos 1971) or p*N* action specifically (Friedman 1971). These mutant hosts all allow growth of p*N*-independent variants such as λ*nin*. The *nus* and *nus*-like mutations are located in five loci: *nusA* (min 69), *nusB* (min 11). *nusC* (min 88), *nusD* (min 84), and *nusE* (min 72).

The various *nus* mutations, although widely scattered on the *E. coli* genome (see Fig. 3) have very similar phenotypes (Friedman et al. 1981). They are recessive, indicating the loss of a function. They appear to block the action of p*N*, rather than its synthesis. And finally, the *nusA*1, the *nusE*71, and many of the *nusB* mutations are temperature-sensitive; i.e., they are more restrictive for λ growth at 42°C than at 32°C. The widespread appearance of temperature-sensitive *nus* mutations raises the possibility that the mutant proteins are not themselves temperature-sensitive but that some aspect of host physiology is different at the elevated temperatures. For example, it has been reported that ρ-dependent transcription termination may be more efficient at 42°C (Das et al. 1976), and therefore more p*N* may be required for antitermination. In fact, the growth of λ variants thought to require more p*N* activity than wild type, e.g., a λ bearing an IS2 terminator in the p_R operon, is restricted in *nus* mutations even at 32°C (Tomich and Friedman 1977). However, the protein identified as the product of the mutant *nusA*1 gene is temperature-sensitive in vitro (see below), and it remains possible that the selection procedure could account for the repeated isolation of temperature-sensitive *nus* mutations.

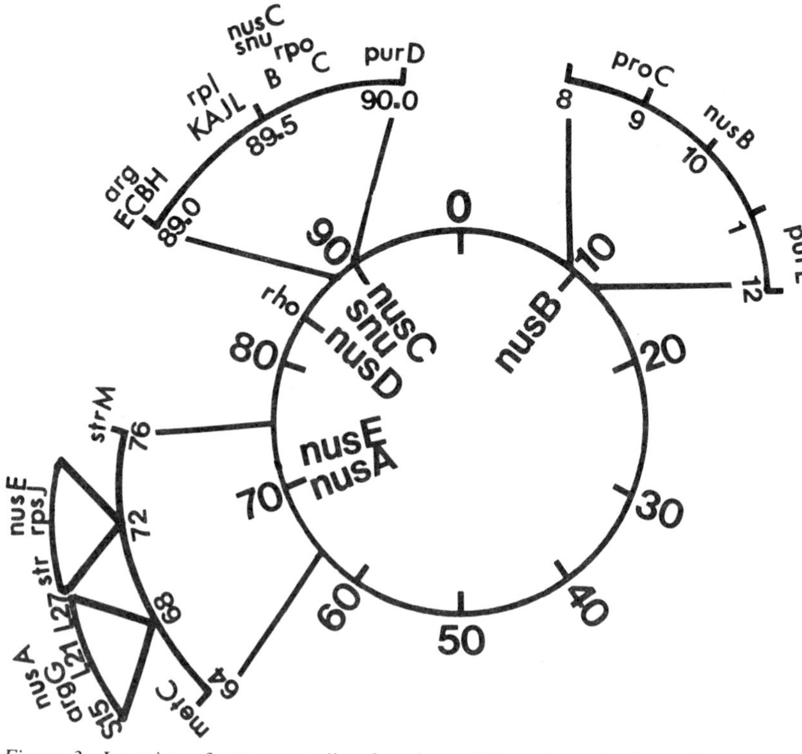

Figure 3 Location of genes encoding functions of *E. coli* involved in p*N* action.

NusA

Only one Nus$^-$ mutation in the *nusA* gene has been extensively characterized (*nusA*1; Friedman and Baron 1974). Recently, conditionally lethal *nusA* amber mutations (Y. Nakamura; R. Haber and S. Adhya; both pers. comm.), as well as conditionally lethal *nusA* missense mutations (Y. Nakamura, pers. comm.; D.I. Friedman, unpubl.), have been isolated. In addition, a hybrid bacterium containing primarily *E. coli* genetic material, but the *nusA* gene from *Salmonella typhimurium*, exhibits a Nus$^-$ phenotype; it fails to support growth of p*N*-dependent λ (Baron et al. 1970; Friedman and Baron 1974).

The protein that appears to be the product of the *nusA* gene has been identified by affinity chromatography of *E. coli* proteins on an *N*-protein-agarose column (Greenblatt and Li 1981). Two proteins specifically bind to p*N*, a 69-kD protein and a 25-kD protein. The 69-kD protein is thought to be NusA, since the wild-type and the *nusA*1-encoded proteins display a charge difference by isolectric focusing. In addition, the mutant protein is more thermolabile than wild type.

The NusA protein has been shown to be identical to L factor (Greenblatt et al. 1980), a protein active in stimulating in vitro synthesis of β-galactosidase in a coupled transcription-translation system (Kung et al. 1975). The role of L factor in this system is not clear, although it appears to increase the synthesis of *lac* mRNA. On the other hand, the expression of the *lac* operon in *nusA*1 mutants in vivo is not impaired (Friedman et al. 1973a).

The NusA protein acts as a transcription termination factor, similar to ρ but with different site-specificity (Greenblatt et al. 1981; Kingston and Chamberlain 1981). In the p_R operon, it appears to act at t_{R2}, distal to the ρ-dependent t_{R1} site (J. Greenblatt, pers. comm.). At some sites, NusA appears to act synergistically with ρ in stimulating termination (Farnham et al. 1982).

This in vitro termination activity of NusA is correlated with a partial relief, in vivo, of *trp* attenuation and amber polarity by the *nusA*1 mutation (Ward and Gottesman 1981).

The isolation of conditionally lethal *nusA* mutations implies that the *nusA* function is essential for bacterial viability. Study of the metabolic regulation of *E. coli* protein synthesis shows that the NusA protein is controlled in the same manner as several proteins involved in transcription and translation (F.C. Neidhardt and D.I. Friedman, unpubl.). Consistent with this regulation is the evidence (presented below) of an interaction between NusA and RNA polymerase.

NusB

In selecting for *nus* mutations, by far the most common class recovered is in the *nusB* gene (min 11) (Keppel et al. 1974; Friedman et al. 1976). On the basis of studies employing *nusB*-transducing phages, the *nusB* product has been identified as a 16-kD basic polypeptide (Strauch and Friedman 1981; Swindle et al. 1981). Some *nusB* mutations are cold-sensitive for growth (Georgopoulos et al. 1980). Revertants that grow at 30°C also lose their Nus$^-$ character, which suggests that a single mutation is responsible for both phenotypes. The metabolic defect responsible for the cold-sensitive character is not known. The *nusB5* mutation partially relieves amber polarity (Ward and Gottesman 1981), suggesting that the *nusB* product may also be involved in transcription termination. NusB may also act as an antitermination factor (J. Greenblatt; F. Imamoto; both pers. comm.).

NusD

A unique class of *rho* mutations, *hdf*, isolated as resistant to phage T4, display a Nus⁻ phenotype (Simon et al. 1979). The *hdf* mutations (such as *hdf*026), like other *nus* mutations, support the lytic growth of wild-type λ at 32°C but fail to grow λ variants that seem to require elevated levels of p*N* activity. Wild-type λ grows on *hdf* mutations at 32°C or 42°C. The effect of the *hdf* mutation on p*N* activity can also be demonstrated in lysogens with p_L-*gal* fusions. In cases where p*N* is required for activity, the *hdf*026 mutation reduces *gal* expression by about 70% (A. Das, pers. comm.).

Like the *nusA*1 and *nusB*5 mutations *hdf* mutations show an incomplete relief of amber polarity. The polarity suppression is far less than that observed with a truly deficient *rho* mutation, such as *rho*15, and is restricted to partially polar mutations.

Efforts to explain the Nus⁻ phenotype of *hdf* mutations are still incomplete. *hdf*026, like other *rho* mutations, is hyperdegradative for abnormal proteins such as puromycil peptides or canavanine-containing proteins. Since *hdf*026 is suppressed by *lon* mutations, which stabilize abnormal proteins, it was suspected that the Nus⁻ and hyperdegradative phenotypes might be related. *N* protein, like abnormal *E. coli* proteins, is highly unstable in wild-type strains ($t_{1/2} = 1$–2 min) and considerably more stable in *lon* mutations ($t_{1/2} = 10$ min) (Gottesman et al. 1981). However, the stability of p*N* in wild-type strains and *hdf*026 is identical (S. Gottesman, pers. comm.). The suppression of *hdf*026 by *lon* mutations could reflect an increased p*N* stability or, alternatively, some Nus protein proteolyzed in *hdf*026 might be stabilized by the *lon* mutation. More recently, experiments by A. Das (pers. comm.) suggest the possibility that the *hdf*026 mutation creates a ρ protein that is at least partially resistant to p*N*. Das finds that although *hdf*026 inhibits suppression by p*N* of termination at ρ-dependent terminators, it has no effect on suppression at the ρ-independent terminator, t_{L2}.

NusE

The *nusE*71 mutation is an allele of *rpsJ* (min 72), which encodes ribosomal protein S10 (Friedman et al. 1981). The mutant S10 protein has an altered isoelectric point.

The isolation of a mutation in a ribosomal protein codon that affects p*N* activity is consistent with the interference by translation of transcription termination. However, it may be that the entire ribosome does not play a role in p*N* action. The ribosomal protein S1, e.g., acts independently of the ribosome as part of the *Qβ* replicase (Kamen 1975). Other ribosomal proteins act as repressors of translation (Yates and Nomura 1981). Also, the capacity of a λ*rpsJ*$^{+}$-transducing phage to form plaques on *nusE*71 mutations is suggestive of a ribosome-independent role for S10 (Friedman et al. 1981). It is unlikely that enough *rpsJ*$^{+}$ ribosomes would form de novo after infection with λ*rpsJ*$^{+}$ to permit phage growth. The possibility of an exchange between ribosomes containing mutant S10 and the wild-type S10, synthesized by the transducing phage, is not excluded.

rpoB

Several mutations in *rpoB* (encoding the *β* subunit of RNA polymerase) that affect p*N* action have been isolated. The *rpoB* mutation *ron* fails to support the growth of λ derivatives carrying a *mar* mutation in *N*; wild-type λ grows in *ron* strains (Ghysen and Pironio 1972). N.C. Franklin (pers. comm.) suggests that the *mar* mutations cause a reduction in p*N* activity, and *ron* creates an RNA polymerase that requires higher levels of p*N* for antitermination. Sternberg (1976) has isolated *rpoB* mutations in which λ production is delayed. The defect appears to be in the p*N*-mediated suppression of t_{R2}. The *nusC*60 mutation (D.I. Friedman et al., in prep.) carries mutation(s) located in or close to *rpoB* and fails to permit p*N* activity at 32°C or 42°C; it is thus the most restrictive Nus^{-} mutation yet isolated. An *rpoB* mutation, *lycA*, has been isolated that confers the opposite phenotype; the growth of λ*N*$^{-}$ mutants is permitted under conditions of low cell density (Lecocq and Dambly 1976).

Interactions between Host Functions That Affect p*N* Activity

An interaction between RNA polymerase core enzyme and the NusA protein has been demonstrated (Greenblatt and Li 1981). When *E. coli* extracts are passed through an affinity column to which the NusA protein is fixed, the core enzyme, but not the holoenzyme, is bound. Moreover, the *σ* subunit can displace the NusA protein from this complex. One interpretation of these data is that NusA protein may be a normal constituent of RNA polymerase, replacing the *σ* polypeptide during elongation.

The biochemical evidence for RNA polymerase interaction with NusA protein is consistent with the isolation of *rpoB* mutations, *snu*, that potentiate the Nus^{-} phenotype of the *nusA*1 mutation. λ fails to grow on the double mutant even at

32°C (Baumann and Friedman 1976). However, the *snu* mutation also increases the restrictiveness of the *nusB*5 and *nusE*71 mutations (Baumann and Friedman 1976; E. Mashni and D.I. Friedman, unpubl.).

Other interactions between host functions that influence p*N* action are suggested by the behavior of various multiple mutant strains. The *nusA*1 mutations are more restrictive, i.e., fail to plate λ at 32°C, when they also carry the *nusB*5 mutation (Friedman et al. 1976) or the *nusE*71 mutation (A.T. Schauer and D.I. Friedman, unpubl.). In contrast, the *nusB*101 allele suppresses the p*N*-inhibitory effect of *nusA*1. (D.F. Ward and M. Gottesman, unpubl.). *nusB*101 does not display, on its own, a characteristic phenotype. Interestingly, although *nusB*101 suppresses *nusA*1 for the growth of λ, it does not do so for phage 21 (D.F. Ward and S. Gottesman, unpubl.). Other *nusB* mutations that specifically permit the growth of λ*imm*^{21} in *nusA*1 hosts have also been isolated (D.F. Ward and M. Gottesman, unpubl.). It should be stressed that this type of genetic evidence for interactions between genes does not necessarily mean that the corresponding gene products interact.

N Mutations

Mutations in *N* that allow λ growth on Nus$^-$ mutations have also been isolated. Two types of *N* mutations have been obtained. The first, *punA*1, selected on a *nusA*1 host, is located between two *N* mutations, *Nam*7 and *Nam*53 (Friedman and Ponce-Campos 1975), and introduces a substitution of a single amino acid in the *N* protein (N. Franklin, pers. comm.). λ*punA*1 also grows on strains carrying *nusE*71, but not on *nusB*5, *nusC*60, or the *E. coli-Salmonella* hybrids referred to above. The fact that *punA*1 lies in the *N* structural gene suggests that the synthesis of p*N* is not affected by the mutation, although the possibility that it increases p*N* stability has not been tested. The idea that the affinity of p*N* for the *nusA*1- and *nusE*71- gene products is improved by the *punA*1 mutation (Friedman et al. 1981) has also not been tested experimentally.

A second class of *N* mutations, which suppress the effects of both *nusA*1 and *nusB* mutations on λ growth, have recently been reported (C.P. Georgopoulos, pers. comm.).

Model of *N* Action

Any model of p*N* action must explain the roles of the various components of the regulation system discussed above: (1) RNA polymerase; (2) the *N* protein; (3) host factors other than RNA polymerase, e.g., *ρ*, NusA, NusB, and ribosomal proteins; and (4) the *nut* sites. Since no purified in vitro system responding to p*N* has yet been established, we cannot assume that this list is exclusive. In speculating about how p*N* might work, based on the evidence at hand, our point of departure is the notion that p*N* acts at the *nut* sites to modify, in some manner,

RNA polymerase so that it becomes resistant to termination. The nature of this reaction can be approached by examining individually each element of the p*N* control system.

RNA Polymerase

The possibility that RNA polymerase is modified at *nut* is supported by the isolation of mutations located in *rpoB* that reduce or eliminate the action of p*N*. Two types of modification could occur at *nut*: (1) A reaction with p*N* could catalytically alter one or more subunits of RNA polymerase. This alteration is likely to be transient, however, since the antitermination effect of p*N* decreases with distance from the promoter (Gottesman et al. 1980). (2) An additional component(s) could be added to the polymerase, forming a termination-resistant complex. This complex, also of limited stability, might include one or more of the host factors listed above.

N *Protein*

Comparisons of the various lambdoid phages have demonstrated that different p*N*s are specific for their homologous *nut* sites. This implies that the *N* protein plays an essential role in *nut* recognition. The isolation of mutations in *N* that permit λ growth on hosts with *rpoB*, *nusA*1, and *nusE*71 mutations is consistent with an interaction between p*N* and the products of those genes. Biochemical evidence shows that a complex is formed between p*N* and NusA protein, although *nut* is not required for this interaction.

Nus Proteins

A number of host factors, in addition to RNA polymerase, influence p*N* action. For some of these, a physiological role is known, suggesting a mechanism for their role in the p*N* reaction. In addition to binding p*N*, the NusA protein promotes transcription termination on a variety of templates and competes with σ subunit for attachment to RNA polymerase core. The NusB protein may also be involved in transcription termination. The *nusD* mutation, *hdf*026, is located in *rho* and partially relieves mutational polarity. The *nusE* mutation alters the *rpsJ* gene, encoding ribosomal protein S10, whose role in translation has not been determined.

nut

Analyses of the *nut* sites have been facilitated by the cloning and sequencing of the *nut* region. These studies show that a functional *nut* site must be downstream from a promoter but that the distance between the promoter and the *nut* site can vary. Thus, although *nut*L lies less than 60 bp from p_L (Franklin and Bennett 1979), *nut*R is some 250 bp from p_R (Rosenberg et al. 1978). The nature of the promoter does not appear to play a vital role in *nut* action, since *nut*R can be cloned away from p_R and placed, in an active form, downstream to the *gal* promoter (de Crombrugghe et al. 1979).

The size of the *nut* sequences could indicate the nature of the reaction taking place at these sites. Recently, a 25-bp fragment of the *nut*L region, consisting of the 17-bp *nut*L symmetrical sequence and extending 7 bp upstream and 1 bp downstream of this sequence, has been cloned next to a plasmid promoter. In the presence of p*N*, the cloned fragment suppresses termination at t_{L1} (Drahos and Szybalski 1981). A small size for *nut* is consistent with the notion that only a single protein, presumably p*N*, recognizes the *nut* sequence. However, our examination of the sequences reported for λ*nut*L (Franklin and Bennett 1979) and λ*nut*R (Rosenberg et al. 1978) reveals extensive homologies outside the 17-bp core sequence. These homologies can be found in the putative *nut*R site of phage 21 (Schwartz 1980) and the putative *nut*L site of *Salmonella* phage P22[2] (A. Poteete, pers. comm.). Studies showing that a deletion of 1 bp in the 3′ end of *cro* results in a NutR⁻ phenotype led to the identification of the first of these homologies (Olson et al. 1982). These sequences (designated A in Fig. 2) lie 8–14 bp promoter-proximal to the core region of dyad symmetry (designated B). In λ, the sequence CGCTCTTA is found; in P22, the sequence is CGCTCTTTA, and in 21, it is TGCTCTTTA. A second homologous sequence (designated C) is found 21–31 bp promoter-distal to the dyad symmetry. The sequences for the four C regions are as follows: λ*nut*L, GGTGTGTG; λ*nut*R, GGTGTATG; 21*nut*R, GGTGTGCG; P22*nut*L, GGCGTGTG. All sequences contain the translation start codons GTG or AUG.

With respect to the region of dyad symmetry that constitutes the *nut* core sequences (B, Fig. 2), there is similarity between P22*nut*L and λ*nut*L and λ*nut*R sites. The potential stem-loop structure is the same size (15 bp) with considerable sequence homology. The promoter-proximal portion of the stem in λ is GCCCT; in P22, it is GCGCT. The loop sequence for λ*nut*L, λ*nut*R, and P22 *nut*L is GAAGA, GAAAA, and GACAA, respectively. The initial G residue is mutant in λ*nut*L44 (A) and in λ*nut*L63 (T). In contrast, the B region of 21*nut*R bears little resemblance to λ or P22. The 5-bp stem is TCACC, and the 7-bp loop is TCTAACCG. All the loop regions thus far sequenced contain a translation stop signal. In the cases of λ*nut*R, λ*nut*L, and P22*nut*L there is a UGA, whereas in 21*nut*R there are both UAA and UGA codons.

The presence of two discrete regions of near homology in the *nut* regions flanking the dyad symmetry suggests that a host protein involved in p*N* action may interact at these sequences, either at the RNA or at the DNA level. The NusA protein is a good candidate for such a host factor: First, the p*N* activity of all three lambdoid phages is reduced in a *nusA*1 mutation (Hilliker et al. 1978). Second, λ*imm*^{21} and λ*imm*^{P22}, which have runs of three Ts in their box-A sequences, grow well in the hybrid bacterium with the *nusA* gene of *Salmonella*,

[2] It should be noted that the sequences of 21*nut*L and P22*nut*R are not available.

whereas λ with two Ts in box A fails to grow in this hybrid. A mutant of λ that grows in the hybrid has been isolated. Analysis of the *nut*R region of this phage shows that it has a T substituted for an A in box A, resulting in an identically placed set of three Ts (D. Friedman and E. Olson, unpubl.). This alteration is essential for λ to grow with the *Salmonella* NusA.

Analysis of the sequences at the beginning of the 6S mRNA, the leader message extended by action of the *Q* protein, also suggests that box A is important for NusA protein activity. As shown in Figure 3, near the 5′ end of the 6S message is the sequence CGCTCGTT, which differs only by one added G from the initial 7 bases of box A at λ*nut*. Studies by Grayhack and Roberts (1982) show that the NusA protein is required for p*Q* action.

Since the NusA protein promotes transcription termination (Greenblatt et al. 1981), we examined regions identified as termination sites for sequences resembling that of box A. We have been able to identify box-A-like sequences both in the "leader" regions of biosynthetic operons, where regulation occurs by attenuation, and at the end of operons. The similarities and differences in these sequences may be seen in Table 1.

Modification Reaction

A relatively simple model for the p*N*-catalyzed modification reaction can be postulated, in which p*N* attaches to RNA polymerase at *nut*. The p*N*-RNA-

Table 1 Box-A-like sequences identified in regions of transcription termination

	Consensus box-A sequence C/TGCTCTT(T)A	
Terminators		
crp	CGCGTTA	(Aiba et al. 1982)
trp	CGCAGTTA	(Wu and Platt 1978)
Attenuators		
his	ATGACACGCGTTC	(Barnes 1978; Di Nocera et al. 1978)
trp	GTGGCGCACTTC	(Yanofsky 1981)
phe	CGCATTC	(Zurawski et al. 1978)
thr	ATGAAACGCATTA	(Gardner 1979)
ampc	ATGGCGGGCCGTTT ↓ A (change reduces attenuation)	(Jaurin et al. 1981)
ilv	GTGGCGGGCTGCACTT	(Lawther and Hatfield 1980)
rrnB	TGCTCTTTA	(Brosius et al. 1981)

Compilation of box-A-like sequences found in regions of transcription termination. Note that the sequences observed vary from the paradigm box-A sequence but always retain the GC (1 or 2 bp) TT arrangement. Moreover, in many of the attenuators there are nearby translation initiation codons located promoter-proximal to the sequence. The initiation codons and the box-A-like sequences are underlined. In the case of the *ampC* attenuator, a change in the consensus sequence that reduces attenuation is shown. The references for the sequences are located in parenthesis to the right of each sequence.

polymerase complex is termination resistant because it no longer pauses at terminators. Alternatively, p*N* might block the access of host termination proteins, such as NusA, NusB, and ρ. According to this model, the various *nus* mutations would encode Nus proteins altered so as to be active even with a p*N*-modified RNA polymerase. In another version of this model, the NusA, NusB, and ρ proteins might bind to RNA polymerase, but with *N* present in the complex, termination is inhibited. The role of the ribosomal proteins in this scheme would be indirect. The *nusE* mutation, e.g., might reduce the amount of NusA protein; some ribosomal proteins are known to regulate the translation of specific transcripts (Jinks-Robertson and Nomura 1982). This possibility is consistent with the observation that the λ*punA* mutations grow on both the *nusA*1 and *nusE*71 strains.

A different molecular mechanism by which p*N* converts RNA polymerase into a nonterminating form has been presented by Friedman et al. (1981). This model likewise assumes that p*N* interacts with host factors and RNA polymerase at the *nut* site, creating a termination-resistant complex. They propose that these factors include ribosomal proteins or the ribosome itself. The proposal rests on the following observations: (1) There are numerous instances in prokaryotic operons where translation and transcription are coupled (Ames and Hartman 1964; Chakrabarti and Gorini 1975). In attenuation, ribosomes are thought to affect the secondary structure of a leader transcript, preventing the formation of a terminator structure (Johnston et al. 1980; Yanofsky 1981). In mutational polarity, the release of ribosomes at a nonsense codon permits the access of ρ protein to the untranslated mRNA and the subsequent release of a prematurely terminated transcript. (2) A mutation in *rpsJ* confers a Nus$^-$ phenotype. (3) The presence of translation initiation or termination codons in the vicinity of all known *nut* sites suggests an active involvement of the ribosome in RNA polymerase modification. (4) It has been suggested that the placement of ribosomes terminating translation of *cro* might influence the p*N* reaction (Olson et al. 1982). These investigators have shown that a deletion in the 3′ terminus of *cro* that places the terminating ribosome 4 bp downstream of the normal termination site results in a NutR$^-$ phenotype.

A direct role for ribosomes in the p*N* reaction has been challenged by in vitro experiments using a p*N*-responsive S100 system (Ishii et al. 1980). Furthermore, the translation of *cro* may not be required for the interaction of p*N* at *nut*R. Amber mutations in *cro* or mutations reducing the efficiency of the ribosome attachment site of *cro* do not eliminate the antitermination activity of p*N* on t_{R1} (de Crombrugghe et al. 1979; C. Debouck and M. Rosenberg; E. Olson; both pers. comm.). Since only one ribosome would be needed for each modified polymerase, residual translation of *cro* might, nevertheless, suffice for this activity. Contaminating ribosomes in the S100 extract might account for the p*N* activity of that system.

Therefore, a definitive assessment of the role that the ribosomal proteins play in p*N* activity awaits the development of a reconstituted cell-free system responsive to p*N* action.

*c*II- and *c*III-gene Products

The *c*II-gene product belongs to the class of λ gene products that play roles in both the lytic and lysogenic cycles. The action of *c*II in lysogeny is reasonably well understood, whereas the action of *c*II in lytic growth is still the subject of lively debate (Herskowitz and Hagen 1980).

The *c*II gene, which is located in the p_R operon (see Fig. 1) (Pereira de Silva and Jacob 1967), encodes a 10.5-kD protein (Epp and Pearson 1976) that is active as an oligomer (M. Rosenberg, pers. comm.). The *c*II protein is highly unstable (Belfort and Wulff 1974; Reichardt 1975; Jones and Herskowitz 1978), especially in *lon*⁻ strains (Gottesman et al. 1981), but its half-life, as suggested by Belfort and Wulff (1974), is extended by the *c*III-gene product (A. Rattray and A.B. Oppenheim; H. Echols; both pers. comm.). The transcription of two genes in the lysogenic pathway is stimulated by *c*II protein, *c*I from the p_E promoter and *int* from the p_I promoter (see Echols and Guarneros; Gussin et al; both this volume). Stimulation of p_E occurs when *c*II protein binds to the -35 region of the promoter (Wulff et al. 1980; Shimatake and Rosenberg 1981). No promoters other than p_E and p_I are known to be stimulated by *c*II protein (Oppenheim et al. 1982).

*c*II protein appears to act in two ways to retard lytic growth: (1) by inhibiting DNA synthesis (Kourilsky and Gros 1976) and (2) by reducing expression of late genes (McMacken et al. 1970; Court et al. 1975). The inhibition of late gene expression by *c*II protein is consistent with a reduction in *Q*-gene activity. Low p*Q* levels would result in reduced expression of all late genes (see next section).

The *c*II-mediated inhibition of DNA synthesis and *Q*-gene expression can most simply be explained by an interference of rightward transcription from p_R by leftward *c*II-stimulated transcription (Court et al. 1975). Inhibition by convergent transcription has previously been observed in the case of the p_L and the *trp* promoters (Ward and Murray 1979). In fact, a twofold reduction in in vivo p_R-initiated transcription due to *c*II product has been noted by Schmeissner et al. (1980). The involvement of p_E in this inhibitory effect of *c*II is indicated by the behavior of defective p_E promoter mutations. These are of two types: *cy*L, which represent mutations in the Pribnow box, and *cy*R, which are mutations in the -35 sequence, the site of *c*II binding (Wulff et al. 1980). Either of these mutations results in a relative insensitivity to *c*II regulation of late gene expression (Court et al. 1975). The residual inhibitory activity of *c*II protein in *cy* mutations is reduced in a *c*III⁻ and increased in a *cro*⁻ infection (Oppenheim et al. 1977); these conditions decrease or increase, respectively, *c*II protein levels. The ability of *c*II protein to act on the lytic cycle in *cy* mutations probably reflects the fact that the *cy* mutations tested, *cy*42 and *cy*2002, are "leaky" and will respond to elevated *c*II-protein levels. Alternative mechanisms have not been excluded. (1) There may be direct inhibition of p_R and perhaps $p_R{}'$ activity by *c*II protein. Consistent with this idea is the observation that when *c*II is present on a multicopy plasmid, the resulting *c*II-gene product is lethal to *E. coli* (Shimatake and Rosenberg 1981). At high *c*II levels, no proteins other than *c*II-gene product are synthesized (A.B. Oppenheim et al., in prep.). The nature of this inhibition is not known, nor has it been established whether it is related to *c*II regulation of the λ lytic cycle. (2) *c*II protein may activate a promoter in the p_R-*Q* region other than p_E. A sequence resembling p_E has in fact been identified near base 44,200 in gene *Q*. Leftward transcription from this promoter might also block p_R (D. Court, pers. comm.).

Gene *Q*

Late gene expression of λ requires the action of the p*Q* regulatory protein. Included in the group of late functions positively regulated by p*Q* are those involved in lysis, head and tail formation, and in the process of DNA packaging and morphogenesis. The requirement for *Q* protein is not absolute; a Q^- mutation shows about 10% of the normal level of expression of late functions (Dambly and Couturier 1971). Certain *cro* mutations permit partial *Q* independence (Herskowitz and Signer 1970). The *cro* mutations result in an increased level of transcription from p_R. Presumably, some of the p_R transcripts continue into late genes. In vivo studies demonstrate that p*Q* acts to stimulate gene expression at a site just to the right of gene *Q* (Herskowitz and Signer 1970; Roberts 1975).

The *Q*-gene product, like that of p*N*, regulates gene expression by interfering with transcription termination. The constitutively synthesized 6S message, which initiates at the $p_R{}'$ promoter becomes the leader for the late gene transcript in the presence of p*Q* (Daniels and Blattner 1982). Studies with $p_R{}'$-*gal* promoter genetic fusions show that p*Q*-stimulated transcription, like that of p*N*, can overcome polarity caused by termination barriers (Forbes and Herskowitz 1982).

The different lambdoid phages have p*Q*-like functions. Those of λ and P22 are interchangeable (Roberts et al. 1976), as are the p*Q*-like functions of ϕ80 and phage 82 (Schlief 1972). The former pair do not substitute for the latter. Although the leader RNA of each phage is analogous to the 6S λ transcript, these are quite distinct in size and sequence in all four phage, and they show no cross-hybridization (Roberts 1975; Roberts et al. 1976; Schechtman et al. 1980). λ-phage 82 hybrids have been constructed that carry the *QSR* region of phage 82; thus p*Q* and its site of action are restricted to a relatively small region that contains the leader RNAs (Schechtman et al. 1980). Evidently, the λ left arm contains no specificity elements for p*Q*. If there exists a p*Q*-utilization site, *qut*, analogous to the p*N*-recognition site, *nut*, it has not yet been identified by mutation. Nor have host mutations that block p*Q* activity been isolated. Since λ*nin* plates on all known *nus* mutations (Friedman et al. 1981) and is still p*Q*-

dependent for growth, it would seem that mutations that block p*N* function do not affect p*Q*. However, we note that the procedure employed to identify *nus* mutations would eliminate any that failed to support p*Q* activity. As previously discussed, recent in vitro studies have demonstrated that p*Q* acts to extend the 6S transcript. This reaction requires the *nusA*-gene product (Grayhack and Roberts 1982).

Translation Control

Although the λ late gene region is uniformly transcribed, the rate of synthesis of the late gene products varies over a wide range (Ray and Pearson 1974). The basis of this posttranscriptional control is unknown. It is conceivable that differential processing of the p_R'-*b* transcript determines the level of expression of a particular gene. Alternatively, the ribosome attachment sites of the various cistrons might be of different efficiencies. Not excluded is the possibility that one or more phage gene products may directly act at the translational level to modulate gene expression.

CONCLUDING REMARKS

In considering this picture of the lytic cycle of λ, we find especially striking the fact that expression of most of the λ genes is controlled at the level of transcription termination. λ, in fact, carries two transcription-antitermination functions, p*N* and p*Q*. The product of the immediate early *N* gene is required for expression of both delayed early and late genes. *N* synthesis is reduced soon in the lytic cycle, and the principal antitermination activity is then assumed by the delayed early *Q*-gene product. Other than these temporal differences, the *N* and *Q* products also differ in their recognition sites, *nut* and *qut*, on the λ genome.

Although the regulation of transcription termination is one means to accomplish temporal control of gene expression, it is not an exclusive one. Many phages, e.g., use different promoters that are stimulated or inactivated at various times in the life cycle (Rabussay and Geiduschek 1977). We can imagine, however, two advantages of a termination control system over a system that regulates transcription initiation.

First, "retroregulation" of *int*-gene expression (see Wulff and Rosenberg; Echols and Guarneros; both this volume) requires two modes of transcription through the same gene, one subject to termination and another not.

Second, the possession of antitermination functions might aid in the acquisition of new genes. The evolution of the lambdoid phages may have depended upon their ability to acquire blocks of genes with related functional capacities (Hershey 1971; Botstein and Herskowitz 1974). A complete operon could thus be obtained by one recombinational event. However, such exchanges would be likely to include termination signals that prevented the expression of the newly acquired structural genes. Here, an antitermination function such as p*N* would not only allow the transcription of terminator-distal genes but could exploit the terminators to regulate expression of these genes.

ACKNOWLEDGMENTS

We thank Emma Williams for her excellent help in preparing this manuscript and Lisa Mashni for designing the figures. Naomi Franklin, Jack Greenblatt, and Waclaw Szybalski are thanked for their helpful comments. Work in the laboratory of D.I.F. was supported by grants from the National Institutes of Health.

REFERENCES

Adhya, S. and M. Gottesman. 1978. Control of transcription termination. *Annu. Rev. Biochem.* **47:** 967.

Adhya, S. and J. Shapiro. 1969. The galactose operon of *E. coli* K-12. I. Structural and pleiotropic mutations of the operon. *Genetics* **62:** 231.

Adhya, S., M. Gottesman, and B. de Crombrugghe. 1974. Release of polarity in *Escherichia coli* by gene N of phage λ: Termination and antitermination of transcription. *Proc. Natl. Acad. Sci.* **71:** 2534.

Adhya, S., M. Gottesman, B. de Crombrugghe, and D. Court. 1976. Transcription termination regulates gene expression. In *RNA polymerase* (ed. R. Losick and M. Chamberlain), p. 719. Cold Spring Harbor Laboratory, Cold Spring Harbor, New York.

Aiba, H., S. Fujimoto, and N. Ozaki. 1982. Molecular cloning and nucleotide sequencing of the gene for *E. coli* cAMP receptor protein. *Nucleic Acids Res.* **10:** 1345.

Ames, B.N. and P.E. Hartman. 1964. The histidine operon. *Cold Spring Harbor Symp. Quant. Biol.* **28:** 349.

Barnes, W.M. 1978. DNA sequence from the histidine operon control region: Seven histidine codons in a row. *Proc. Natl. Acad. Sci.* **75:** 4281.

Baron, L.S., E. Penido, I.R. Ryman, and S. Falkow. 1970. Behavior of coliphage lambda in hybrids between *Escherichia coli* and *Salmonella*. *J. Bacteriol.* **102:** 221.

Baron, L.S., I.R. Ryman, E.M. Johnson, and P. Gemski, Jr. 1972. Lytic replication of coliphage lambda in *Salmonella typhosa* hybrids. *J. Bacteriol.* **110:** 1022.

Baumann, M.F. and D.I. Friedman. 1976. Cooperative effects of bacterial mutations affecting λ*N* gene expression. *Virology* **73:** 128.

Beckwith, J.R. 1963. Restoration of operon activity by suppressors. *Biochim. Biophys. Acta* **76:** 162.

Bektesh, S.L. and J.P. Richardson. 1980. A ρ-recognition site on phage λ *cro*-gene mRNA. *Nature* **283:** 102.

Belfort, M. and D. Wulff. 1974. The roles of the lambda *c*III gene and the *Escherichia coli* catabolite gene activation system in the establishment of lysogeny by bacteriophage lambda. *Proc. Natl. Acad. Sci.* **71:** 779.

Bienkowska-Szewczyk, K. and A. Taylor. 1980. Murein transglycosylase from phage lysate purification and properties. *Biochem. Biophys. Acta* **615:** 489.

Blattner, F. and J. Dahlberg. 1972. RNA synthesis start points in bacteriophage lambda: Are the promoters and operators transcribed? *Nat. New Biol.* **237:** 232.

Botstein, D. and I. Herskowitz. 1974. Properties of hybrids between *Salmonella* phage P22 and coliphage λ. *Nature* **251:** 584.

Bouvre K. and W. Szybalski. 1969. Patterns of convergent and overlapping transcription within the b_2 region of coliphage λ. *Virology* **38:** 614.

Boyd, J. 1951. "Excessive" dose phenomenon in virus infection. *Nature* **167:** 1061.

Brosius, J., T.J. Dulls, D.D. Sleeter, and H. Noller. 1981. Gene organization and primary structure of a ribosomal operon from *Escherichia coli. J. Mol. Biol.* **148:** 107.

Brunel, F. and J. Davison. 1975. Bacterial mutants able to partly suppress the effect of *N* mutations in bacteriophage λ. *Mol. Gen. Genet.* **136:** 167.

Burt, D.W. and W.J. Brammer. 1982. Transcription termination sites in the b2 region of bacteriophage lambda that are unresponsive to antitermination. *Mol. Gen. Genet.* **185:** 462.

Butler, B. and H. Echols. 1970. Regulation of bacteriophage λ development by gene *N*: Properties of a mutation that bypasses *N* control of late protein synthesis. *Virology* **40:** 212.

Campbell, A. 1961. Sensitive mutants of bacteriophage λ. *Virology* **14:** 22.

Chakrabarti, S.L. and L. Gorini. 1975. A link between streptomycin and rifampicin mutation. *Proc. Natl. Acad. Sci.* **72:** 2084.

Chowdhury, D.M. and A. Guha. 1973. Characterization of the mRNA transcribed from the head and tail genes of the phage lambda chromosome. *Nat. New Biol.* **241:** 196.

Court, D. and K. Sato. 1969. Studies of novel transducing variants of lambda: Dispensability of genes *N* and *Q*. *Virology* **39:** 348.

Court, D., L. Green, and H. Echols. 1975. Positive and negative regulation by the *c*II and *c*III gene products of bacteriophage λ. *Virology* **63:** 483.

Court, D., C. Brady, M. Rosenberg, D. Wulff, M. Behr, M. Mahoney, and S. Izumi. 1980. Control of transcription termination: A Rho-dependent termination site in bacteriophage lambda. *J. Mol. Biol.* **138:** 231.

Couturier, M. and C. Dambly. 1970. Activation sequentielle des fonctions tardives chez les bacteriophages temperes. *C.R. Acad. Sci. Paris* **270:** 428.

Dambly, C. and M. Couturier. 1971. A minor Q-independent pathway for the expression of late genes in bacteriophage λ. *Mol. Gen. Genet.* **113:** 244.

Dambly, C., M. Couturier, and R. Thomas. 1968. Control of development of temperate bacteriophages. II. Control of lysosyme synthesis. *J. Mol. Biol.* **32:** 67.

Dambly-Chaudière, C., M. Gottesman, C. Debouck, and S. Adhya. 1983. Regulation of the *pR* operon of bacteriophage lambda. *J. Mol. Appl. Genet.* **2:** 45.

Daniels, D.L. and F.R. Blattner. 1982. The nucleotide sequence of the *Q* gene and the *Q* to *S* intergenic region of bacteriophage lambda. *Virology* **117:** 81.

Das, A., D. Court, and S. Adhya. 1976. Isolation and characterization of conditional lethal mutants of *Escherichia coli* defective in transcription termination factor *rho*. *Proc. Natl. Acad. Sci.* **73:** 1959.

Das, A., C. Merril, and S. Adhya. 1978. Interaction of RNA polymerase and *rho* in transcription termination: Coupled ATPase. *Proc. Natl. Acad. Sci.* **75:** 4828.

Davidson, N. and W. Szybalski. 1971. Physical and chemical characteristics of lambda DNA. In *The bacteriophage lambda* (ed. A.D. Hershey), p. 45. Cold Spring Harbor Laboratory, Cold Spring Harbor, New York.

de Crombrugghe, B., S. Adhya, M. Gottesman, and I. Pastan. 1973. Effect of rho on transcription of bacterial operons. *Nat. New Biol.* **241:** 260.

de Crombrugghe, B., M. Mudryj, R. Di Lauro, and M. Gottesman. 1979. Specificity of the bacteriophage lambda *N* gene product (N): *Nut* sequences are necessary and sufficient for the antitermination by N. *Cell* **18:** 1145.

Di Nocera, P.P., F. Blasi, R. Di Lauro, R. Frunzio, and C.B. Bruni. 1978. Nucleotide sequence of the attenuator region of the histidine operon of *Escherichia coli* K-12. *Proc. Natl. Acad. Sci.* **75:** 4276.

Drahos, D. and W. Szybalski. 1981. Antitermination and termination functions of the cloned *nut*L, *N* and t_{L1} modules of coliphage lambda. *Gene* **16:** 261.

Drahos, P., G.R. Galluppi, M. Caruthers, and W. Szybalski. 1982. Synthesis of the *nut*L DNA segments and analysis of antitermination and termination functions of coliphage lambda. *Gene* **18:** 343.

Echols, H. 1971. Regulation of lytic development. In *The bacteriophage lambda* (ed. A.D. Hershey), p. 247. Cold Spring Harbor Laboratory, Cold Spring Harbor, New York.

Echols, H., L. Green, A.B. Oppenheim, A. Oppenheim, and A. Honigman. 1973. Role of the *cro* gene in bacteriophage λ development. *J. Mol. Biol.* **80:** 203.

Eisen, H., M. Georgiou, C.P. Georgopolous, G. Selzer, G. Gussin, and I. Herskowitz. 1975. The role of gene *cro* in phage λ development. *Virology* **68:** 266.

Eisen, H.A., C.R. Fuerst, L. Siminovitch, R. Thomas, L. Lambert, L. Pereira da Silva, and F. Jacob. 1966. Genetics and physiology of defective lysogeny in K12(λ): Studies of early mutants. *Virology* **30:** 224.

Epp, C. and M.L. Pearson. 1976. Association of bacteriophage lambda *N* gene protein with *E. coli* RNA polymerase. In *RNA polymerases* (ed. M. Chamberlin and R. Losick), p. 667. Cold Spring Harbor Laboratory, Cold Spring Harbor, New York.

Farnham, P.J., J. Greenblatt, and T. Platt. 1982. Effects of NusA protein on transcription termination in the tryptophan operon of *Escherichia coli. Cell* **29:** 945.

Folkmanis, A., W. Maltzman, P. Mellon, A. Skalka, and H. Echols. 1977. The essential role of the *cro* gene in lytic development by bacteriophage λ. *Virology* **81:** 352.

Forbes, D. and I. Herskowitz. 1982. Polarity suppression by the *Q* gene product of phage lambda. *J. Mol. Biol.* **160:** 549.

Franklin, N.C. 1974. Altered reading of genetic signals fused to the N operon of bacteriophage λ: Genetic evidence for the modification of polymerase by the protein product of the *N* gene. *J. Mol. Biol.* **89:** 33.

Franklin, N.C. and G.N. Bennett. 1979. The *N* protein of bacteriophage lambda, defined by its DNA sequence, is highly basic. *Gene* **8:** 107.

Franklin, N.C. and S.E. Luria. 1961. Transduction by bacteriophage P1 and the properties of the lac genetic region in *E. coli* and *S. dysenteriae. Virology* **15:** 299.

Friedman, D.I. 1971. A bacterial mutant affecting λ development. In *The bacteriophage lambda* (ed. A.D. Hershey), p. 733. Cold Spring Harbor Laboratory, Cold Spring Harbor, New York.

Friedman, D.I. and L.S. Baron. 1974. Genetic characterization of a bacterial locus involved in the activity of the *N* function of phage λ. *Virology* **58:** 141.

Friedman, D.I. and R. Ponce-Campos. 1975. Differential effect of phage regulator functions on transcription from various promoters: Evidence that the P22 gene and the λ gene *N* products distinguish three types of promoters. *J. Mol. Biol.* **98:** 537.

Friedman, D.I., M. Baumann, and L.S. Baron. 1976. Cooperative effects of bacterial mutations affecting λ *N* gene expression. I. Isolation and characterization of a *nus*B mutant. *Virology* **73:** 119.

Friedman, D.I., C.T. Jolly, and R.J. Mural. 1973a. Interference with the expression of the *N* gene product of phage λ in a mutant of *Escherichia coli. Virology* **51:** 216.

Friedman, D.I., G.S. Wilgus, and R.J. Mural. 1973b. Gene N regulator function of phage λ*imm*21: Evidence that a site of N action differs from a site of N recognition. *J. Mol. Biol.* **81:** 505.

Friedman, D.I., A.T. Schauer, M.R. Baumann, L.S. Baron, and S.L. Adhya. 1981. Evidence that ribosomal protein S10 participates in the control of transcription termination. *Proc. Natl. Acad. Sci.* **78:** 1115.

Galluppi, G., C. Lowery, and J.P. Richardson. 1976. Nucleoside triphosphate requirement for termination of RNA synthesis by Rho factor. In *RNA polymerase* (ed. R. Losick and M. Chamberlain), p. 657. Cold Spring Harbor Laboratory, Cold Spring Harbor, New York.

Gardner, J. 1979. Regulation of the threonine operon: Tandem threonine and isoleucine codons in the control region and translation control of transcription termination. *Proc. Natl. Acad. Sci.* **76:** 1706.

Gariglio, P. and M.H. Green. 1973. Characterization of polycistronic late lambda messenger RNA. *Virology* **53:** 392.

Gemski, P., Jr., L.S. Baron, and N. Yamamoto. 1972. Formation of hybrids between coliphage λ and *Salmonella* phage P22 with a *Salmonella typhimurium* hybrid sensitive to these phages. *Proc. Natl. Acad. Sci.* **69:** 3110.

Georgiou, M., C.P. Georgopoulos, and H. Eisen. 1979. An analysis of the Tro phenotype of bacteriophage λ. *Virology* **94:** 38.

Georgopoulos, C.P. 1971. A bacterial mutation affecting *N* function. In *The bacteriophage lambda* (ed. A.D. Hershey), p. 639. Cold Spring Harbor Laboratory, Cold Spring Harbor, New York.

Georgopoulos, C., J. Swindle, F. Keppel, M. Balliver, R. Bisig, and H. Eisen. 1980. Studies on the *E. coli groNB* (nusB) gene which affects bacteriophage λ *N* gene function. *Mol. Gen. Genet.* **179:** 55.

Ghysen, A. and M. Pironio. 1972. Relationship between the *N* function of bacteriophage λ and host RNA polymerase. *J. Mol. Biol.* **65:** 259.

Goldberg, A.R. and M. Howe. 1969. New mutations in the S cistron of bacteriophage lambda affecting host cell lysis. *Virology* **38:** 200.

Gottesman, M.E. and R.A. Weisberg. 1971. Prophage insertion and excision. In *The bacteriophage lambda* (ed. A.D. Hershey), p. 113. Cold Spring Harbor Laboratory, Cold Spring Harbor, New York.

Gottesman, M.E., S. Adhya, and A. Das. 1980. Transcription antitermination by bacteriophage lambda *N* gene product. *J. Mol. Biol.* **140:** 57.

Gottesman, S., M. Gottesman, J.E. Shaw, and M.L. Pearson. 1981. Protein degradation in *E. coli*: The *lon* mutation and bacteriophage lambda *N* and *c*II protein stability. *Cell* **24:** 225.

Grayhack, E.J. and J.W. Roberts. 1982. The phage λ *Q* gene product: Activity of a transcription antiterminator *in vitro*. *Cell* **30:** 637.

Greenblatt, J. and J. Li. 1981. Interaction of the sigma factor and the *nusA* gene protein of *E. coli* with RNA polymerase in the initiation-termination cycle of transcription. *Cell* **24:** 421.

——. 1982. Properties of the *N* gene transcription antitermination protein of bacteriophage λ. *J. Biol. Chem.* **257:** 362.

Greenblatt, J., M. McLimont, and S. Hanly. 1981. Termination of transcription by *nus*A gene protein of *Escherichia coli. Nature* **292:** 215.

Greenblatt, J., J. Li, S. Adhya, D.I. Friedman, L.S. Baron, B. Redfield, H. Kung, and H. Weissbach. 1980. L factor that is required for β-galactosidase synthesis is the *nusA* gene product involved in transcription termination. *Proc. Natl. Acad. Sci.* **77:** 1991.

Grodzicker, T., R.R. Arditti, and H. Eisen. 1972. Establishment of repression by lambdoid phage in catabolite activator protein and adenylate cyclase mutants of *Escherichia coli. Proc. Natl. Acad. Sci.* **69:** 366.

Guarente, L. and J. Beckwith. 1978. Mutant RNA polymerase of *Escherichia coli* terminates transcription in strains making defective *rho* factor. *Proc. Natl. Acad. Sci.* **75:** 294.

Harris, A.W., D.W.A. Mount, C.R. Fuerst, and L. Siminovitch. 1967. Mutations in bacteriophage lambda affecting host cell lysis. *Virology* **32:** 553.

Heinemann, S. and W. Spiegelman. 1971. Role of the gene *N* product in phage lambda. *Cold Spring Harbor Symp. Quant. Biol.* **35:** 315.

Hershey, A.D. 1971. Comparative molecular structure among related phage DNAs. *Carnegie Inst. Wash. Year Book* **1970:** 3.

Herskowitz, I. and D. Hagen. 1980. The lysis-lysogeny decision of phage λ: Explicit programming and responsiveness. *Annu. Rev. Genet.* **14:** 399.

Herskowitz, I. and E.R. Signer. 1970. A site essential for expression of all late genes in bacteriophage λ. *J. Mol. Biol.* **47:** 545.

Hilliker, S. and D. Botstein. 1976. Specificity of genetic elements controlling regulation of early functions in temperature bacteriophages. *J. Mol. Biol.* **106:** 537.

Hilliker, S., M. Gottesman, and S. Adhya. 1978. The activity of *Salmonella* phage P22 gene 24 product in *Escherichia coli. Virology* **86:** 37.

Hong, J., G.R. Smith, and B.N. Ames. 1971. Adenosine 3′5′-cyclic monophosphate concentration in the bacterial host regulates the viral decision between lysogeny and lysis. *Proc. Natl. Acad. Sci.* **68:** 2258.

Honigman, A. 1981. Cloning and characterization of a transcription termination signal in bacteriophage λ unresponsive to the *N* gene product. *Gene* **13:** 299.

Hopkins, N. 1970. Bypassing a positive regulator: Isolation of a λ mutant that does not require *N* product to grow. *Virology* **40:** 223.

Howard, B. and B. de Crombrugghe. 1975. ATPase activity required for termination of transcription by the *E. coli* protein factor *rho*. *J. Biol. Chem.* **251:** 2520.

Howard, G.A., J. Gordon, K. Farnung, and D. Richter. 1976. Stringent factor binds to *Escherichia coli* ribosomes only in the presence of protein L11. *FEBS Lett.* **68:** 211.

Inoko, H. and M. Imai. 1976. Isolation and genetic characterization of the *nit*A mutants of *E. coli* affecting the termination factor Rho. *Mol. Gen. Genet.* **143:** 211.

Ishii, S., J. Salstrom, Y. Sugino, W. Szybalski, and F. Imamoto. 1980. A biochemical assay for the transcription-antitermination function of the coliphage λ*N* gene product. *Gene* **10:** 17.

Jaurin, B., T. Grundstrom, T. Edlund, and S. Normark. 1981. The *E. coli* β-lactamase attenuator mediates growth rate-dependent regulation. *Nature* **290:** 222.

Jinks-Robertson, S. and M. Nomura. 1982. Ribosomal protein S4 acts *in trans* as a translational repressor to regulate expression of the α operon in *Escherichia coli*. *J. Bacteriol.* **151:** 193.

Johnson, A.D., A.R. Poteete, G. Lauer, R.T. Sauer, G.K. Ackers, and M. Ptashne. 1981. λ repressor and Cro⁻ components of an efficient molecular switch. *Nature* **294:** 217.

Johnston, H.M., W.M. Barnes, F. Chumley, L. Bossi, and J.R. Roth. 1980. Model for regulation of the histidine operon of *Salmonella*. *Proc. Natl. Acad. Sci.* **77:** 508.

Jones, M.O. and I. Herskowitz. 1978. Mutants of bacteriophage λ which do not require the *c*III gene for efficient lysogenization. *Virology* **88:** 199.

Kaiser, A.D. 1955. A genetic study of temperature phage λ. *Virology* **1:** 424.

Kaiser, A.D. and F. Jacob. 1957. Recombination between related temperate bacteriophages and the genetic control of immunity and prophage localization. *Virology* **4:** 509.

Kamen, R.I. 1975. Structure and function of the Qβ replicase. In *RNA phages* (ed. N. Zinder), p. 203. Cold Spring Harbor Laboratory, Cold Spring Harbor, New York.

Keppel, F., C. Georgopoulbs, and H. Eisen. 1974. Host interference with expression of the lambda *N* gene product. *Biochimie* **56:** 1503.

Kingston, R.E. and M.J. Chamberlain. 1981. Pausing and attenuation of *in vitro* transcription in the *rrnB* operon of *E. coli*. *Cell* **27:** 523.

Korn, L.J. and C. Yanofsky. 1976. Polarity suppressors defective in transcription termination at the attenuator of the tryptophan operon of *Escherichia coli* have altered rho factor. *J. Mol. Biol.* **106:** 231.

Kourilsky, P. 1973. Lysogenization by bacteriophage lambda. I. Multiple infections and the lysogenization response. *Mol. Gen. Genet.* **122:** 183.

Kourilsky, P. and D. Gros. 1976. Lysogenization by bacteriophage lambda. IV. Inhibition of phage DNA synthesis by the products of genes *c*II and *c*III. *Biochimie* **58:** 1321.

Kourilsky, P., M.F. Bourguignon, and F. Gros. 1971a. Kinetics of viral transcription after induction of prophage. In *The bacteriophage lambda* (ed. A.D. Hershey), p. 647. Cold Spring Harbor Laboratory, Cold Spring Harbor, New York.

Kourilsky, P., M. Bourguignon, M. Bouquet, and F. Gros. 1971b. Early transcription control after induction of prophage λ. *Cold Spring Harbor Symp. Quant. Biol.* **35:** 305.

Krell, K., M.E. Gottesman, J.S. Parks, and M.A. Eisenberg. 1972. Escape synthesis of the biotin operon in induced λ*b*2 lysogens. *J. Mol. Biol.* **68:** 69.

Kumar, S., E. Calef, and W. Szybalski. 1971. Regulation of the transcription of *Escherichia coli* phage λ by its early genes *N* and *tof*. *Cold Spring Harbor Symp. Quant. Biol.* **35:** 331.

Kung, H., C. Spears, and H. Weissbach. 1975. Purification and properties of a soluble factor required for the deoxyribonucleic acid-directed *in vitro* synthesis of β-galactosidase. *J. Biol. Chem.* **250:** 1556.

Lau, L.F., J. W. Roberts, and R. Wu. 1982. Transcription terminates at λt_{R1} in three clusters. *Proc. Natl. Acad. Sci.* **79:** 6171.

Lawther, R.P. and G.W. Hatfield. 1980. Multivalent translation control of transcription termination at attenuator of *ilvGEDA* operon of *Escherichia coli* K-12. *Proc. Natl. Acad. Sci.* **77:** 1862.

Lecocq, J.P. and C. Dambly. 1976. A bacterial RNA polymerase mutant that renders growth independent of *N* and *cro* function at 42°C. *Mol. Gen. Genet.* **145:** 53.

Lieb, M. 1953. The establishment of lysogenicity in *Escherichia coli*. *J. Bacteriol.* **65:** 642.

Liedke-Kulke, M. and A.D. Kaiser. 1967. Genetic control of prophage insertion specificity in bacteriophage λ and 21. *Virology* **32:** 465.

Lowery-Goldhammer, C. and J.P. Richardson. 1974. An RNA-dependent nucleoside triphosphate phosphohydrolase (ATPase) associated with Rho termination factor. *Proc. Natl. Acad. Sci.* **71:** 2003.

Lozeron, H.A., P.J. Anevski, and D. Apirion. 1977. Antitermination and absence of processing of the leftward transcript of coliphage lambda in the RNase III-deficient host. *J. Mol. Biol.* **109:** 359.

Lozeron, H.A., J.E. Dahlberg, and W. Szybalski. 1976. Processing of the major leftward mRNA of coliphage lambda. *Virology* **71:** 262.

Luzzati, D. 1970. Regulation of λ exonuclease synthesis: Role of the *N* gene product and repressor. *J. Mol. Biol.* **49:** 515.

McDermit, M., M. Pierce, D. Staley, M. Shimaji, R. Shaw, and D. Wulff. 1976. Mutations masking the lambda *cin*-1 mutation. *Genetics* **82:** 417.

McMacken, R., N. Mantei, B. Butler, A. Joyner, and H. Echols. 1970. Effect of mutations in the *c*II and *c*III genes of bacteriophage λ on macromolecular synthesis in infected cells. *J. Mol. Biol.* **49:** 639.

Miller, H.I. 1981. Multilevel regulation of bacteriophage λ lysogeny by the *E. coli himA* gene. *Cell* **25:** 269.

Miller, H.I. and D.I. Friedman. 1980. An *Escherichia coli* gene product required for lambda site specific recombination. *Cell* **20:** 711.

Murialdo, H. and L. Siminovitch. 1972. The morphogenesis of bacteriophage lambda. IV. Identification of the gene products and control of the expression of morphogenetic information. *Virology* **48:** 785.

Nakamura, Y. and T. Yura. 1976. Induction of sigma factor synthesis in *Escherichia coli* by the *N* gene product of bacteriophage lambda. *Proc. Natl. Acad. Sci.* **73:** 4405.

Neubauer, Z. and E. Calef. 1970. Immunity phase-shift in defective lysogens; nonmutational hereditary change of early regulation of λ prophage. *J. Mol. Biol.* **51:** 1.

Newton, W.A., J.R. Beckwith, D. Zipser, and S. Brenner. 1965. Nonsense mutants and polarity in the *lac* operon of *Escherichia coli. J. Mol. Biol.* **14:** 290.

Nijkamp, H.J.J., K. Bøvre, and W. Szybalski. 1970. Control of rightward transcription in coliphage λ. *J. Mol. Biol.* **54:** 599.

Nijkamp, H.J.J., W. Szybalski, M. Ohashi, and W.F. Dove. 1971. Gene expression by constitutive mutants of coliphage λ. *Mol. Gen. Genet.* **114:** 80.

Oda, K., Y. Sakakibara, and J. Tomizawa. 1969. Regulation of transcription of the lambda bacteriophage genome. *Virology* **39:** 901.

O'Farrell, P.H. 1978. The suppression of defective translation by ppGpp and its role in the stringent response. *Cell* **14:** 545.

Olson, E.R., E.L. Flamm, and D.I. Friedman. 1982. Analysis of *nut*R: A region of phage lambda required for antitermination of transcription. *Cell* **31:** 61.

Oppenheim, A., S. Gottesman, and M. Gottesman. 1982. Regulation of bacteriophage λ*int* gene expression. *J. Mol. Biol.* **158:** 327.

Oppenheim, A.B., N. Katzir, and A. Oppenheim. 1977. Regulation of protein synthesis in bacteriophage λ: Restoration of gene expression in λN^- strains by mutations in the *cro* gene. *Virology* **79:** 405.

Pereira Da Silva, L.H. and F. Jacob. 1967. Induction of *c*II and *O* functions in early defective lambda prophages. *Virology* **33:** 618.

Ptashne, M., K. Backman, Z. Humayan, A. Jeffrey, R. Maurer, B. Meyer, and R.T. Sauer. 1976. Autoregulation and function of a repressor in bacteriophage lambda. *Science* **194:** 156.

Rabussay, D. and E.P. Geiduschek. 1977. Regulation of gene action in the development of lytic bacteriophages. *Comp. Virol.* **8:** 1.

Ratner, D. 1976. Evidence that mutations in the *su*A polarity suppressing gene directly affect termination factor *rho*. *Nature* **259:** 151.

Ray, P.N. and M.L. Pearson. 1974. Evidence for post-transcriptional control of the morphogenetic genes of bacteriophage lambda. *J. Mol. Biol.* **85:** 163.

Reichardt, L.F. 1975. Control of bacteriophage lambda repressor synthesis after phage infection: The role of the *N*, *c*II, *c*III, and *cro* products. *J. Mol. Biol.* **93:** 267.

Reyes, O., M. Gottesman, and S. Adhya. 1979. Formation of lambda lysogens by IS2 recombination: *gal* operon-lambda p_R promoter fusions. *Virology* **94:** 400.

Richardson, J.P. and R. Conway. 1980. Ribonucleic acid release activity of transcription termination protein Rho is dependent on the hydrolysis of nucleotide triphosphates. *Biochemistry* **19:** 4293.

Richardson, J.P., C. Grimley, and C. Lowery. 1975. Transcription termination factor *rho* activity is altered in *E. coli* with *su*A gene mutations. *Proc. Natl. Acad. Sci.* **72:** 1725.

Richardson, J.P., P. Fink, K. Blanchard, and M. Macy. 1977. Bacteria with defective Rho factors suppress the effects of *N* mutations in bacteriophage λ. *Mol. Gen. Genet.* **153:** 81.

Roberts, J.W. 1969. Termination factor for RNA synthesis. *Nature* **224:** 1168.

——. 1975. Transcription termination and late control in phage lambda. *Proc. Natl. Acad. Sci.* **72:** 3300.

Roberts, J.W., C.W. Roberts, S. Hilliker, and D. Botstein. 1976. Transcription termination and regulation in bacteriophages P22 and lambda. In *RNA polymerase* (ed. R. Losick and M. Chamberlain), p. 707. Cold Spring Harbor Laboratory, Cold Spring Harbor, New York.

Rosebury, T. 1962. *Microorganisms indigenous to man*. McGraw-Hill, New York.

Rosenberg, M. and D. Court. 1979. Regulatory sequences involved in the promotion and termination of RNA transcription. *Annu. Rev. Genet.* **13:** 319.

Rosenberg, M., D. Court, H. Shimatake, C. Brady, and D.L. Wulff. 1978. The relationship between function and DNA sequence in an intercistronic regulatory region of phage λ. *Nature* **272:** 414.

Salstrom, J.S. and W. Szybalski. 1978a. Coliphage λ*nut*L: A unique class of mutants defective in the site of gene *N* product utilization for antitermination of leftward transcription. *J. Mol. Biol.* **124:** 195.

——. 1978b. Transcription termination sites in the major leftward operon of coliphage lambda. *Virology* **88:** 252.

Salstrom, J., M. Fiandt, and W. Szybalski. 1979. *N*-independent leftward transcription in coliphage lambda: Deletions, insertions and new promoters bypassing termination functions. *Mol. Gen. Genet.* **168:** 211.

Schechtman, M.G., J.D. Snedeker, and J.W. Roberts. 1980. Genetics and structure of the late regulatory region of phage 82. *Virology* **105:** 393.

Schleif, R. 1972. The specificity of lambdoid phage late gene induction (lambdoid phage late gene specificity). *Virology* **50:** 610.

Schmeissner, U., D. Court, S. Shimatake, and M. Rosenberg. 1980. The promoter for the establishment of repressor synthesis in bacteriophage lambda. *Proc. Natl. Acad. Sci.* **77:** 3191.

Schwartz, E. 1980. "Sequence analyses des DNA lambdoide bacteriophages." Ph.D. thesis, University of Freiburg, Federal Republic of Germany.

Segawa, T. and F. Imamoto. 1976. Evidence of read-through at the termination signal for transcription of the *trp* operon. *Virology* **70:** 181.

Shimatake, H. and M. Rosenberg. 1981. Purified λ regulatory protein *c*II positively activates promoters for lysogenic development. *Nature* **292:** 128.

Simon, L.D., M. Gottesman, K. Tomczak, and S. Gottesman. 1979. Hyperdegradation of proteins in *Escherichia coli rho* mutants. *Proc. Natl. Acad. Sci.* **76:** 1623.

Skalka, A., B. Butler, and H. Echols. 1967. Genetic control of transcription during development of phage λ. *Proc. Natl. Acad. Sci.* **58:** 576.

Sklar, J.L., P. Yot, and S.M. Weissman. 1975. Determination of genes, restriction sites and DNA sequences surrounding the 6S RNA template of bacteriophage lambda. *Proc. Natl. Acad. Sci.* **72:** 1817.

Sternberg, N. 1976. A class of rif^R polymerase mutations that interfere with the activity of coliphage λ*N* gene product. *Virology* **73:** 139.

Sternberg, N. and L. Enquist. 1979. Analysis of coliphage lambda mutations that affect *Q* gene activity: *puq*, *byp* and *nin*5. *J. Virol.* **30:** 1.

Strauch, M. and D.I. Friedman. 1981. Identification of the *nus*B gene product of *Escherichia coli*. *Mol. Gen. Genet.* **182:** 498.

Swindle, J., J. Ajioka, and C. Georgopoulos. 1981. Identification of the *E. coli groNB* (*nus*B) gene product. *Mol. Gen. Genet.* **182:** 409.

Szpirer, J., R. Thomas, and C.M. Radding. 1969. Hybrids of bacteriophage λ and φ80: A study of nonvegetative functions. *Virology* **37:** 585.

Szybalski, W. 1974. Initiation and regulation of transcription in coliphage lambda. In *Proceedings of the Calcutta Symposium on the Control of Transcription*, India, February 12-15, 1973, (ed. B.B. Biswas et al.), p. 201. Plenum Press, New York.

Szybalski, E. and W. Szybalski. 1979. A comprehensive molecular map of bacteriophage lambda. *Gene* **7:** 217.

Thomas, R. 1966. Control of development in temperate bacteriophages. I. Induction of prophage genes following heteroimmune super-infection. *J. Mol. Biol.* **22:** 79.

Tomich, P.K. and D.I. Friedman. 1977. Isolation of mutations in insertion sequences that relieve IS induced polarity. In *DNA insertion elements, plasmids, and episomes* (ed. A. Bukhari et al.), p. 99. Cold Spring Harbor Laboratory, Cold Spring Harbor, New York.

Van Donsel, D.L. and E.E. Geldreich. 1971. Relationships of *Salmonella* to fecal coliforms in bottom sediments. *Water Res.* **5:** 1079.

Ward, D.F. and M.E. Gottesman. 1981. The *nus* mutations affect transcription termination in *Escherichia coli*. *Nature* **292:** 212.

Ward, D.F. and N.E. Murray. 1979. Convergent transcription in bacteriophage λ—Interference with gene expression. *J. Mol. Biol.* **133:** 249.

Weisberg, R.A., S. Gottesman, and M.E. Gottesman. 1977. Bacteriophage λ: The lysogenic pathway. *Comp. Virol.* **8:** 197.

Wilson, D. 1982. Effect of the lambda *S* gene product on properties of the *Escherichia coli* inner membrane. *J. Bacteriol.* **151:** 1403.

Wu, A.M. and T. Platt. 1978. Transcription termination: Nucleotide sequence at 3′ end of tryptophan operon in *Escherichia coli*. *Proc. Natl. Acad. Sci.* **75:** 5442.

Wulff, D.L. 1976. Lambda *cin-1*, a new mutation which enhances lysogenization by bacteriophage lambda and the genetic structure of the lambda *cy* region. *Genetics* **82:** 401.

Wulff, D.L., M. Beher, S. Izumi, J. Beck, M. Mahoney, H. Shimatake, C. Brady, D. Court, and M. Rosenberg. 1980. Structure and function of the *cy* control region of bacteriophage lambda. *J. Mol. Biol.* **138:** 209.

Yanofsky, C. 1981. Attenuation in the control of expression of bacterial operons. *Nature* **289:** 751.

Yates, J.L. and M. Nomura. 1981. Feedback regulation of ribosomal protein synthesis in *E. coli*: Localization of the mRNA target sites for repressor action of ribosomal protein L1. *Cell* **24:** 243.

Young, R., J. Way, S. Way, J. Yin, and M. Syvanen. 1979. Transposition mutagenesis of bacteriophage lambda. *J. Mol. Biol.* **132:** 307.

Zurawski, G., K. Brown, D. Killingly, and C. Yanofsky. 1978. Nucleotide sequence of the leader region of the phenylalanine operon of *Escherichia coli*. *Proc. Natl. Acad. Sci.* **75:** 4271.

Establishment of Repressor Synthesis

Daniel L. Wulff
Department of Biological Sciences
State University of New York at Albany
Albany, New York 12222

Martin Rosenberg*
Laboratory of Biochemistry, National Cancer Institute
National Institutes of Health
Bethesda, Maryland 20205

INTRODUCTION

mRNA encoding the *c*I repressor of bacteriophage λ can be initiated at either of two promoters, which function at different stages in the λ life cycle. In the lysogenic state, repressor transcription initiates from the p_{RM} promoter, which lies immediately adjacent to the *c*I gene (Fig. 1). This promoter is positively regulated by the *c*I-gene product and both structurally and functionally overlaps with the promoter-operator signals ($p_R o_R$) controlling the major rightward operon of λ. In contrast, when λ infects a sensitive cell, repressor transcription originates from the p_{RE} promoter, which lies several hundred base pairs upstream of the *c*I gene in the λ*y* region (Fig. 1). This promoter structurally overlaps the region encoding the aminoterminal amino acid sequence of the *c*II gene. Transcription from p_{RE} is also positively regulated and, in this case, the activator is the *c*II protein. The *c*II protein also activates transcription from p_I,

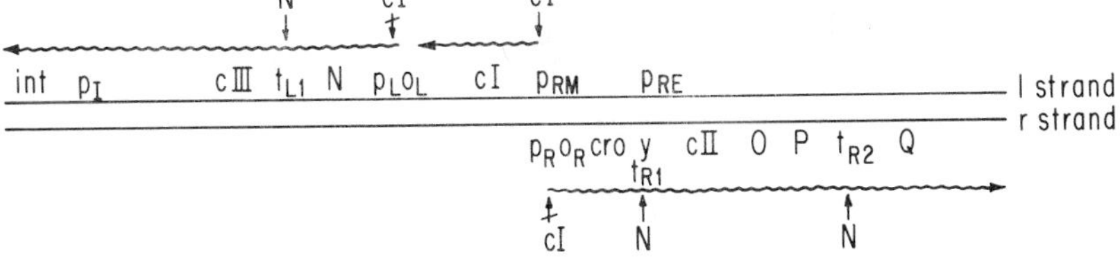

Figure 1 Partial genetic map of λ. Arrow-headed wavy lines indicate the origin, direction, and extent of various transcripts. Vertical arrows indicate sites of action of the *N*-, *c*I-, *c*II-, and *c*III-gene products. Arrows with slashes indicate that the gene product represses transcription. Arrows without slashes at the start of a wavy line indicate that the gene product activates a promoter. Arrows in the middle of a wavy line indicate that the gene product prevents transcription from terminating at a termination site.

*Present address: Smith Klein and French Laboratories, Philadelphia, Pennsylvania 19101.

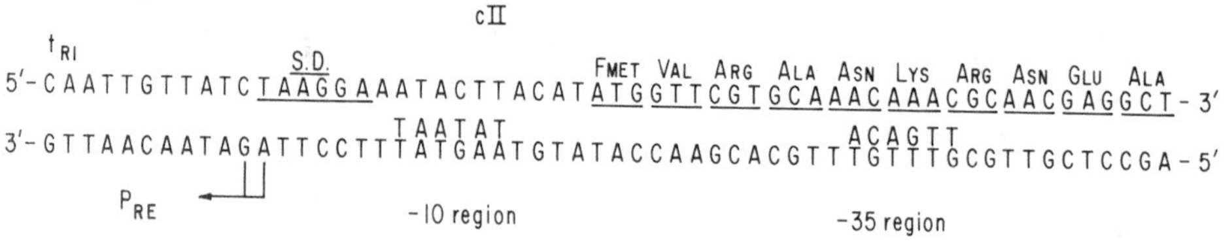

Figure 2 Nucleotide sequence of the p_{RE} promoter and aminoterminal region of the *c*II gene. The 6-base consensus sequences for the −10 and −35 regions of prokaryotic promoters are depicted in the space between strands. The p_{RE} message is initiated at either of 2 nucleotides, as indicated. The t_{R1} termination site for rightward transcription from p_R is at the extreme left end, as indicated. The region of hyphenated dyad symmetry preceding t_{R1} is to the left of the sequence. (S.D.) The Shine and Delgarno homology for the *c*II gene (i.e., the region where p_R mRNA is homologous with the 3′ end of 16S rRNA [see Shine and Delgarno 1974].)

the promoter for the λ integrase gene. Thus, the *c*II protein coordinately regulates the two transcription units required for the lysogenic response and thereby acts as a major determinant in controlling the balance between lytic and lysogenic development. The *c*II activity is itself regulated at multiple levels (transcription, translation, processing, stability, and so on), and a variety of phage and host-encoded functions take part in this regulation.

In this paper we examine the p_{RE} transcription unit, with particular emphasis on the p_{RE} promoter signal and its mode of activation by *c*II protein. We examine the *c*II protein both genetically and biochemically and explore the various factors that affect the course of phage development by regulating the intracellular levels of this protein.

p_{RE} TRANSCRIPTION UNIT

The concept of a p_{RE} promoter in the *y* region was first explicitly proposed by Reichardt and Kaiser (1971) and Echols and Green (1971). A key experiment toward proving the p_{RE} postulate was reported by Jones et al. (1979). Using defective prophages with extensive deletions, they showed that deletions that eliminated the *y* region were incapable of *c*II-dependent expression of the *c*I gene, whereas deletions contained entirely to the right of *y* did not eliminate the *c*II-dependent expression of the *c*I gene. The definitive demonstration of a p_{RE} promoter in *y* came when Schmeissner et al. (1980) isolated and characterized the *c*II-dependent p_{RE} transcript from infected cells. They showed that repressor transcription initiates specifically in the *y* region, 403 nucleotides preceding the *c*I gene. The precise positioning of the 5′ end of this mRNA defined the p_{RE} promoter region, which was seen to overlap with the sequence encoding the ribosome-binding signal and aminoterminal end of the *c*II gene (Fig. 2).

These same studies demonstrated that, during the time of establishment of repression, no leftward transcription through *y* occurs from promoters located upstream of p_{RE} (i.e., *c*I transcription derives exclusively from p_{RE}). Moreover, it was shown that activation of *c*I transcription from p_{RE} caused a significant reduction in the levels of converging rightward transcription from p_R. Previous work had indicated that *c*II activation of repressor expression resulted in a concomitant reduction in (i.e., delay of) late gene expression (McMacken et al. 1970; Court et al. 1975). This reduction is explained, at least in part, by the effects of this convergent transcription and suggests that convergent transcription from p_{RE} and p_R plays an important regulatory role in phage development.

p_{RE} PROMOTER

The DNA structure of the p_{RE} promoter region exhibits little homology with other promoters that are recognized by *Escherichia coli* RNA polymerase (Rosenberg and Court 1979; Fig. 2). These promoters generally exhibit two regions of strong sequence homology: one centered at approximately 10 bp before the transcription start site (the −10 region), and the other centered at approximately 35 bp before the start site (the −35 region). The −10 region of p_{RE} is identical in only three positions to the hexamer consensus sequence for prokaryotic promoters. However, two of these positions, the second and sixth, coincide with the most strongly conserved bases in the −10 regions of prokaryotic promoters. In addition to the rather poor −10 region similarity, the −35 region of p_{RE} shows essentially no homology to a 6-base consensus sequence found in this region of other promoters. Perhaps the lack of structural homology between p_{RE} and other promoters is not surprising, since p_{RE} function is totally dependent upon an activator protein. The p_{RE} promoter does not function with RNA polymerase alone in a purified in vitro transcription system (Shimatake and Rosenberg 1981). Moreover, even glycerol, which stimulates transcription in vitro from other positively regulated promoter sites in the absence of their activators, has no effect on p_{RE} (M. Rosenberg, unpubl.).

Although p_{RE} lacks sequence homology with other promoters, it shares a striking structural similarity to the p_I promoter signal, the other phage promoter regulated by *c*II protein (Abraham et al. 1980; Davies 1980; Hoess et al. 1980). By aligning the two signals with respect to their in vivo start sites (Schmeissner et al. 1980, 1981), one observes a homology in 11 out of 14 base pairs, positioned identically in the −35 region of both sites (Fig. 3). In contrast, both the −10 region sequences and the start site sequences of these two *c*II-dependent promoters show little similarity.

It has been proposed that the extensive −35 region homology exhibited by p_{RE} and p_I defines a site involved in the positive regulation of these two promoters (Abraham et al. 1980; Hoess et al. 1980; Schmeissner et al. 1980, 1981; Wulff et al. 1980). The nature and positions of a set of −35 region mutations that have been characterized for p_{RE} support this contention (Rosenberg et al. 1978a; Wulff et al. 1980, and in prep.). As we describe below, experiments by

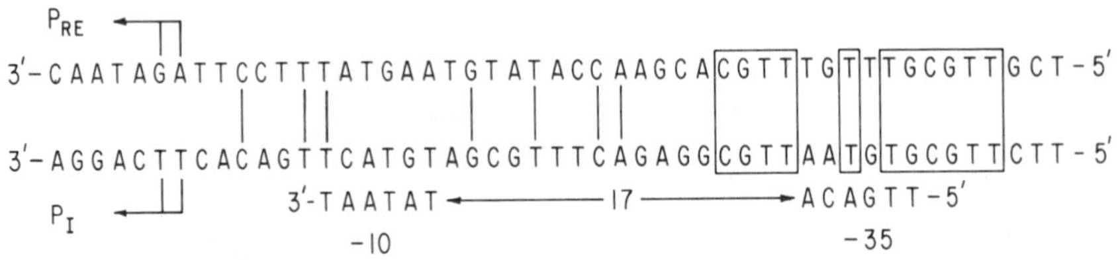

Figure 3 Nucleotide sequences of the p_{RE} and p_I promoters. The regions of sequence identity are indicated. The 6-base consensus sequences for the −10 and −35 regions of prokaryotic promoters are shown underneath.

Shimatake and Rosenberg (1981) and Ho and Rosenberg (1982) have now demonstrated directly that *c*II protein is the positive regulator and acts by specifically binding to the −35 region homologies of the p_{RE} and p_I signals.

MUTATIONS AFFECTING p_{RE} FUNCTION

Two types of clear plaque mutations may be isolated in the *y*-*c*II region of λ: *cy* mutations, which are p_{RE} promoter-down mutations (Brachet and Thomas 1969; Wulff et al. 1980), and *c*II mutations (Kaiser 1957). λ*cy* and λ*c*II mutations may be distinguished from other clear-plaque mutations by genetic mapping, as well as by complementation tests. (λ*cy* mutations complement like λ*c*I⁻ mutations, even though they are several hundred base pairs from the *c*I gene.) Analogous to most other promoter mutations, the *cy*s are clustered in either the −10 (*cy*L) or the −35 region (*cy*R) of the p_{RE} sequence (Fig. 4). Certain *c*II mutations lie within the region that structurally overlaps the p_{RE} promoter site. These mutations are different from *cy* mutations (or from mutations isolated in almost any other promoter region) in that although they occur within the promoter region, they have no effect on promoter function. Most of these mutations lie within the 17-bp spacer region positioned between the important −10 and −35 regions of p_{RE}. A few, however, occur within the −10 and −35 regions of p_{RE} and presumably indicate specific base changes in these important regions that are acceptable to promoter function.

Mutations that inactivate promoter function usually alter the highly conserved bases in the −10 and −35 region hexamer sequences (Rosenberg and Court 1979). The nature of the −10 region p_{RE} mutations is consistent with this idea. Four of the five *cy*L mutations affect the highly conserved second position A-T pair and the sixth position T-A pair found in p_{RE}. The fifth *cy*L mutation (*cy*3019) alters a G-C pair to an A-T pair 2 bases upstream from the conserved hexamer. Although G-C, C-G, and T-A pairs are found at this position in other promoters, rarely does an A-T pair occur at this site (Rosenberg and Court 1979). Apparently, an A-T pair at this position has a negative influence on promoter function. It is of interest that, unlike the other *cy* mutations, this *cy* muta-

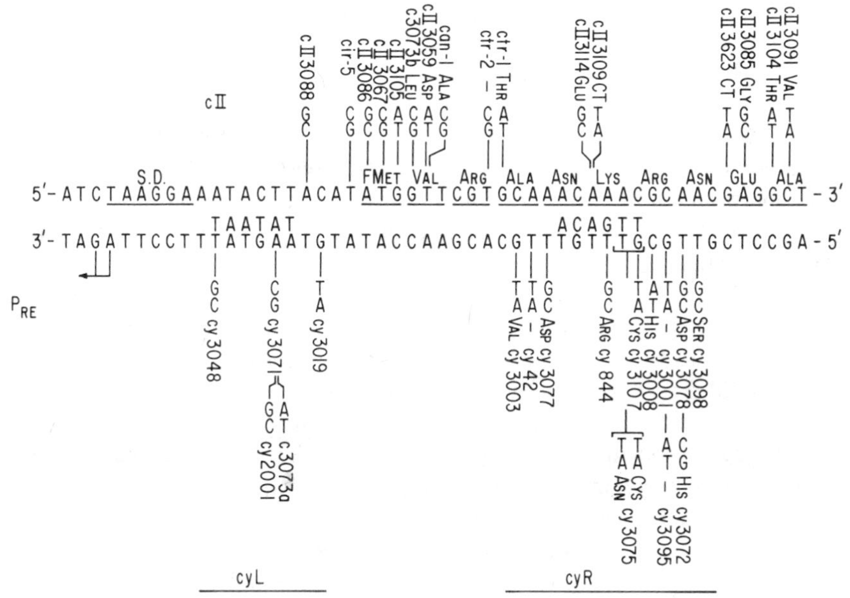

Figure 4 Nucleotide sequence changes in the p_{RE} promoter and aminoterminal region of the *c*II gene. Changes for p_{RE} mutations in the −10 (*cy*L) and −35 (*cy*R) regions of the promoter are shown below the line. Changes for *c*II⁻ mutations are shown above the line. The *cir*5, *can*1, *ctr*1, and *ctr*2 mutations are all *c*II⁺. The *cir*5 mutation restores partial *c*II function to λ*c*II3086. The *can*1 mutation results in a more stable *c*II protein. The *ctr*1 and *ctr*2 mutations appear to be up-translation mutations. These are described more fully in the text. Amino acid substitutions are indicated for mutations that lie to the right of codon 1. If no amino acid change is shown, then the altered codon does not result in an amino acid change. Other conventions are as in Fig. 2.

tion does not exhibit the typical clear-plaque phenotype but, rather, forms lightly turbid plaques. This phenotype is indicative of a partial defect in function, consistent with its location in a more weakly conserved region of the promoter.

In contrast to the *cy*L mutations, the mutations that occur in the −35 region of p_{RE} (*cy*R) do not affect bases that are usually conserved in the promoter. Moreover, these mutations span some 13 bp, a region far more extensive than the corresponding RNA polymerase contact region in other promoters as defined by either homology or mutational analysis. The region encompassed by these p_{RE} mutations corresponds exactly to the homologous region in the other *c*II-dependent promoter, p_I (Fig. 3). In the following sections we show that the *cy*R mutations can be subdivided into two classes: those that destroy promoter function by inactivating a site of action for the *c*II protein, and those that directly eliminate RNA polymerase binding to the promoter without affecting the interaction with *c*II protein.

Promoter-up mutations in the p_{RE} promoter have not been reported. Attempts to isolate such mutations (as *c*II-independent promoter mutations) have resulted in the *cin*1 mutation (Wulff 1976), which is located some 40 nucleotides to the left of the p_{RE} start site (Rosenberg et al. 1978a). The *cin*1 mutation creates a new leftward transcription start site 7 bp to its left. Transcription from this new promoter (called p_{CIN}) does not require *c*II protein (H. Shimatake and M. Rosenberg, unpubl.). Transcription from the p_{CIN} promoter in repressed prophages results in about 40–50% as much repressor as transcription from p_{RM} (Gussin et al. 1980).

The −10 region of p_{CIN} has the hexamer sequence TACACT, which coincides with the ideal consensus sequence of TATAAT at four positions, including the highly conserved first, second, and sixth positions (Rosenberg and Court 1979). Wild-type λ has C in place of T at the sixth position of this sequence, which explains the inactivity of this promoter in λ^+. Mutations that inactivate p_{CIN}, called *cnc* mutations (McDermit et al. 1976), have additional base changes in the first and second positions of the hexamer sequence (Rosenberg et al. 1978a; Rosenberg and Court 1979). The *cin*1 and *cnc* mutations lie in the t_{R1} termination region for p_R-directed rightward transcription and alter the structure and function of the t_{R1} signal, as discussed elsewhere in this volume.

PURIFIED *c*II PROTEIN ACTIVATES TRANSCRIPTION FROM p_{RE} AND p_I

A major hurdle in studying the *c*II protein and defining its precise role in the activation of the p_{RE} and p_I transcription units was overcome when *c*II protein was obtained in pure form (Shimatake and Rosenberg 1981). This made it possible to reconstitute the *c*II-dependent activation system in vitro. Shimatake and Rosenberg cloned the *c*II gene into a plasmid vector in such a way that its expression was placed under the control of the repressible, strong λ promoter p_L. Using this construction, they were able to produce *c*II protein in a bacterial lysogen at levels approaching 5% of cellular protein. A simple purification procedure allowed isolation of pure *c*II protein at a yield of 1.0 mg/g of wet weight of cell culture (Ho et al. 1982).

The ability to obtain large quantities of pure *c*II protein allowed extensive physical characterization of the protein. To date, the protein has been characterized as to its amino acid composition, aminoterminal amino acid sequence (see below, *c*II Protein Is Processed), molar extinction coefficient, molecular weight, oligomeric structure, isoelectric point, α-helical content, and antigenic capability (Table 1; Ho et al. 1982). In addition, analytical ultracentrifugation analysis indicates that in solution, at concentrations above 4×10^{-7} M (monomer m.w. = 11,000), *c*II protein exists predominantly as a tetramer, with an equilibrium constant $K_{1,4} = 3.2 \times 10^{18}$ (Ho et al. 1982). The quantities of pure protein now available have allowed crystallization studies to be initiated.

Filter-binding studies carried out with the pure *c*II protein indicate that *c*II pro-

Table 1 Physical and chemical properties of *c*II protein

Molecular weight	
glycerol gradient ultracentrifugation	43,600
sedimentation equilibrium	44,000
SDS-polyacrylamide gel electrophoresis	11,300
$S^0_{20,w}$	3.4S
$\epsilon 280$	7.2×10^4
Subunit structure	tetramer
Aminoterminal sequence	(fMet-Val)-Arg-Ala-Asn
Isoelectric point	8.8
α-Helix content	50–60%

tein allows RNA polymerase to bind selectively to both the p_{RE} and p_I sites (Shimatake and Rosenberg 1981). Moreover, in vitro transcription experiments prove that this binding leads to active transcription from both promoters. Activation of p_{RE} and p_I requires only template DNA, RNA polymerase, and *c*II protein. Therefore, *c*II protein is both necessary and sufficient to activate transcription from both promoters (Shimatake and Rosenberg 1981). In these same studies it was found that fully activated transcription from the p_{RE} promoter was at least four times more efficient that that obtained from p_I. This difference was shown later by Ho and Rosenberg (1982) to be due to differences in the intrinsic ability of RNA polymerase to interact at the two signals. λ*c*II protein was found to have identical affinities for the p_{RE} and p_I sites and thus did not appear to contribute to the differences seen in promoter strengths. It was concluded that the differences in the −10 region sequences of p_{RE} and p_I and the radically different starting nucleotides for the two promoters (ATP for p_{RE} and UTP for p_I) were the major factors responsible for the different promoter efficiencies.

INTERACTION OF *c*II PROTEIN WITH p_{RE} AND p_I

Using both chemical and enzymatic DNA probe techniques, Ho and Rosenberg (1982) have characterized in detail the interaction of *c*II protein and RNA polymerase with the p_{RE} and p_I sites. DNase protection studies (i.e., footprinting; Galas and Schmitz 1978) demonstrated that *c*II protein interacted specifically in the −35 regions of both p_{RE} and p_I, protecting a region of some 20–25 bp (Fig. 5). RNA polymerase alone showed no interaction with either promoter. However, in the presence of *c*II protein, polymerase bound tightly and protected the entire promoter region.

Ho and Rosenberg (1982) carried out similar binding analyses using 17 different p_{RE} promoter mutations that had been previously characterized (Rosenberg et al. 1978a; Wulff et al. 1980, and in prep.). Their results indicate that mutations in the −10 region have no effect on binding by *c*II protein. In-

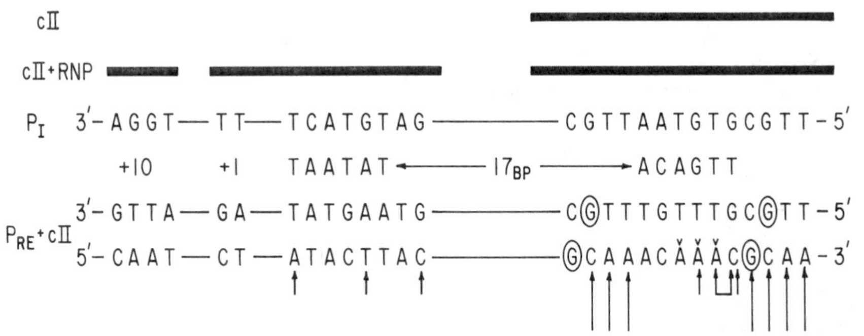

Figure 5 A comparison of the DNA sequences of the p_I and p_{RE} promoters aligned with respect to their *c*II-dependent transcription start sites (position +1). The consensus −10 and −35 region sequences recognized by RNA polymerase are also shown for comparison. The regions of the promoter protected from nuclease digestion are overscored by thick solid bars. Long-stemmed arrows indicate those bases within the p_{RE} promoter which, when mutated, affect *c*II binding. Short-stemmed arrows indicate those positions which, when mutated, have no effect on *c*II binding. Residues protected by *c*II protein from methylation (Ⓖ) or whose methylation was enhanced Ǎ) are also indicated. (Adapted from Ho and Rosenberg 1982.)

terestingly, *cy*R mutations in the −35 region divided into two groups according to their effect on the binding of *c*II protein. Those *cy*R mutations that eliminated *c*II binding fell into two clusters positioned 6 bp apart (Figs. 4 and 5). Three other *cy*R mutations positioned between the two clusters (i.e., within the 6-bp separation) had no effect on binding by *c*II protein. The two *c*II⁻ mutations in this same area also do not affect binding since they are phenotypically p_{RE}^{+}. Most importantly, this 6-bp region is positioned exactly 17 bp upstream from the −10 region sequence and is thus exactly analogous in both size and position to the conserved −35 region hexamer sequence found in other promoter signals. Ho and Rosenberg concluded that this sequence in p_{RE} likewise functions as the −35 region hexamer involved in RNA polymerase interaction and that the p_{RE} mutations found here interfere with RNA polymerase contacts, rather than *c*II protein contacts.

Most strikingly, the two clusters of p_{RE} mutations that eliminate binding by *c*II protein fall in the repeated nucleotide sequences $TTGCN_6TTGC$ (Fig. 5). Each corresponding position of this repeat is positioned 10 bp apart, exactly one turn of the DNA helix. Thus, when the *c*II protein recognizes the two TTGC sequences, recognition occurs on one face of the DNA helix. Moreover, the 6 nucleotides in between (N6) comprise the −35 region contact sites for RNA polymerase discussed above.

Ho and Rosenberg (1982) obtained additional support for these conclusions by studying the close apposition of *c*II protein and DNA by chemical protection methods. End-labeled DNA was methylated with dimethylsulfate in the presence or absence of *c*II protein. The DNA was then cleaved at the methylated purine bases and analyzed on sequencing gels. Those residues protected by *c*II protein from methylation and cleavage were monitored. Their results showed that all four G residues occurring in the two TTGC repeat sequences are completely protected from methylation by *c*II protein (Fig. 5). λ*c*II protein did not protect any of the other purine residues between the TTGC repeats. Binding of both RNA polymerase and *c*II protein to the p_{RE} DNA resulted in strong protection of other purine residues positioned both outside and between the TTGC repeats. These results are completely consistent with the conclusions drawn from the footprinting data. They further indicate that interaction with *c*II protein occurs mainly, if not exclusively, within the major groove of the DNA, since G methylation occurs at the N7 position in the major groove. Thus, it seems likely that two identical dimer subunits of the *c*II tetramer recognize and bind to the repeating DNA structure in the major groove on one face of the DNA helix and that this interaction allows RNA polymerase to recognize the −35 region sequence positioned between the repeat sequences (Ho and Rosenberg 1982).

GENETICS OF THE *c*II GENE

The DNA sequence of the *c*II gene shows that it codes for 97 amino acids (Schwarz et al. 1978). D.L. Wulff et al. (in prep.) have isolated more than 500 *c*II mutations, and identified and determined DNA sequence changes for *c*II$^-$ mutations at the aminoterminal end of the gene (Fig. 4). The mutations occurring at the extreme aminoterminal end (i.e., up to the third codon) are discussed below in the sections on translational control and processing. In this section we discuss the remaining mutations.

Only two *c*II$^-$ alterations have been identified in codons 3–8. One of these is an ochre (chain termination) mutation, and the second changes a basic amino acid to an acidic amino acid (Lys → Glu at codon 6) (Fig. 4). Most strikingly, nine of the *cy*R mutations result in amino acid substitutions in this region of the protein (Fig. 4). Complementation studies indicate that all of these altered *c*II proteins are functionally active, although *cy*3075, a Lys-Arg → Asn-Cys double change at codons 6 and 7, has only partial activity. We conclude that *c*II function is not particularly sensitive to amino acid sequence in this region. (Even the Arg → Cys change in *cy*3107 has no apparent effect on *c*II activity.) Apparently, evolution has placed a premium on an optimal promoter sequence for this region of DNA, with the amino acid sequence determined by this region of the *c*II gene being a much less strongly selected characteristic.

Although most of the *c*II gene has not been genetically characterized as well as the aminoterminal end, we know that *c*II$^-$ mutations exist in many different locations along the *c*II gene (D.L. Wulff and M. Mahoney, unpubl.). For example, genetic and DNA sequencing studies indicate the existence of ten different *c*II$^-$ base changes in codons 9–15. In contrast, very few *c*II$^-$ mutations with classical

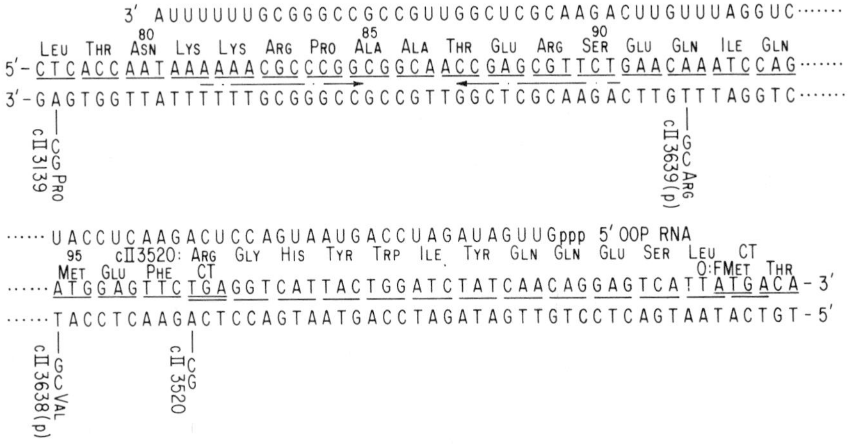

Figure 6 Sequence of the carboxyterminal end of the *c*II gene, the intercistronic region between the *c*II and *O* genes, and the first 2 codons of the *O* gene. The *OOP* RNA transcript is also shown, and the region of hyphenated dyad symmetry preceding the *OOP* RNA termination site is indicated by arrows. (Adapted from Schwarz et al. 1978.) Amino acid sequence changes are indicated for the four *c*II⁻ mutations that have been located in this region. Two of these mutations retain partial *c*II activity (p).

clear-plaque morphologies are found in the carboxyterminal region of the *c*II gene. With one exception, the collection of more than 500 *c*II⁻ mutations isolated by D.L. Wulff et al. (in prep.) contains no clear-plaque *c*II⁻ mutations to the right of *c*II3139, a CTC → CCC (Leu → Pro) mutation at codon 78 (Fig. 6) (D.L. Wulff and M. Mahoney, unpubl.). The lone exception, *c*II3250, is a UGA → CGA mutation in the termination codon that results in extension of the *c*II protein by 12 amino acids. The elongated *c*II gene in λ*c*II3250 overlaps the *O* gene by 4 bp (Fig. 6).

It is possible to isolate *c*II⁻ mutations that form lightly turbid plaques, indicative of partial *c*II activity. Five such mutations have been mapped, and two lie to the right of *c*II3139, the clear-plaque *c*II⁻ mutation at codon 78. One of these lightly turbid *c*II⁻ mutations is a CAA → CGA (Gln → Arg) change at codon 92, and the second is a ATG → GTG (Met → Val) change at codon 95 (Fig. 6) (D.L. Wulff and M. Mahoney, unpubl.). No *c*II⁻ mutations have been identified between codons 78 and 92.

The carboxyterminal end of the *c*II gene overlaps the DNA that codes for the 77-base *OOP* RNA sequence (Fig. 6; Blattner and Dahlberg 1972; Schwarz et al. 1978). *OOP* RNA is transcribed leftward from a start site located in the intercistronic region between the *c*II and *O* genes. The function of the *OOP* RNA in phage λ development is unknown. The 3′ end of *OOP* RNA contains the sequence U_6A, and the DNA sequence preceding the *OOP* RNA termination site contains a region of hyphenated dyad symmetry, features that are typically found at termination sites (Rosenberg and Court 1979). The *OOP* RNA termination region (including the region of hyphenated dyad symmetry) overlaps codons 80–90 of the *c*II gene (Fig. 6), and a *c*II⁻ mutation in this region could affect termination of the *OOP* transcript. Failure to terminate transcription of *OOP* RNA might be lethal to the phage, and this may explain the absence of *c*II⁻ mutations in this region. Alternatively, amino acid alterations in codons 80–90 may result in partially active *c*II proteins (as do the mutations at codons 92 and 95) and therefore may not have been included in our collection of *c*II⁻ mutations.

EFFICIENCY OF *c*II mRNA TRANSLATION

As is discussed in the following sections, the concentration of active *c*II protein in a cell is determined by the level of *c*II mRNA, the rate of translation of *c*II mRNA, and the rates of processing, tetramer formation, and breakdown of the *c*II protein. These processes are subject (or potentially subject) to various controlling elements. The p_R promoter, which is subject to negative regulation by the *c*II- and *cro*-gene products, and the t_{R1} termination site, which is subject to regulation by the *N*-gene product, are two sites at which transcription control occurs. These control points are discussed elsewhere in this volume. An additional transcriptional control, the lowering of transcription from p_R as a result of converging leftward transcription from p_{RE}, is discussed above, in p_{RE} Transcription Unit. No other transcriptional controls on *c*II expression have been found. In this section we discuss translation controls on *c*II activity and, in the following two sections, the processing and stability of the *c*II protein.

The host HimA protein appears to affect the rate of translation of *c*II mRNA into protein. Lysogenization is poor, and very little repressor is made when *E. coli himA* strains are infected with λ^+ (Miller 1981; Miller et al. 1981). Efficient translation of *c*II mRNA in vivo appears to require the HimA protein (Hoyt et al. 1982), suggesting a possible direct interaction of HimA protein with *c*II mRNA. No interaction between HimA protein and mRNA has been established. However, N. Craig and H. Nash (pers. comm.) have shown that the *E. coli* HimA and Hip proteins bind specifically to λ*y* region DNA, in a region between the t_{R1} termination site and the p_{RE} transcription start site. In vitro transcription experiments suggest that this binding represses transcription from p_{RE} in the presence of limiting amounts of *c*II protein (M. Rosenberg, unpubl.). Thus, the HimA protein may play some role in both the translation of the *c*II gene and transcription from p_{RE}. We point out that little is understood about the mechanism of HimA action on repressor synthesis.

A number of mutations at the aminoterminal end of the *c*II gene affect translation efficiencies. λ*c*II3088 was shown by C. Debouck and M. Rosenberg (unpubl.) to reduce dramatically the rate of translation of *c*II mRNA in an in vitro RNA-dependent S30 system. λ*c*II3088 is an A→G change that lies outside of the *c*II-coding region, 4 bp to the left of the initial AUG *N*-formylmethionine codon and 7 bp to the right of the region of complementarity with the 3′ end of

16S rRNA (the Shine and Delgarno sequence) (Fig. 4). By comparing mRNA sequences near the initiation regions of many genes, Stormo et al. (1982) found that G (and the dinucleotide UG) is not often present in the region between the Shine and Delgarno sequence and the initial AUG codon. They note that AUG, GUG, and UUG codons are recognized by initiator tRNA (Steege 1977; Napoli et al. 1981) and propose that the presence of these codons in the wrong places could interfere with initiation at the proper initiation codons. This would explain the *c*II$^-$ phenotype of the *c*II3088 mutation, for it creates a UUG codon between the Shine and Delgarno sequence and the proper AUG initiation codon. The five *cy*L mutations also lie between the Shine and Delgarno sequence and the initial ATG codon. These mutations are all *c*II$^+$ by complementation, and C. Queen and M. Rosenerg (unpubl.) have shown normal translation efficiencies with these mutations in vitro using an RNA-dependent S30 system.

Transition mutations have been found in each of the three positions of the initial AUG codon of the *c*II gene. These generate GUG, ACG, and AUA codons. GUG is reported to be an active initiation codon in other prokaryotic systems (see Stormo et al. 1982), and it is surprising that the mutation to GUG (*c*II3086) is *c*II$^-$. C. Debouck and M. Rosenberg, using an in vitro RNA-dependent S30 system, found no *c*II expression from this GUG codon (unpubl.). Interestingly, this mutation also creates an AUG codon 2 nucleotides upstream from the normal start site, and this out-of-phase AUG codon may interfere with normal *c*II translation. The translation defect in λ*c*II3086 is partially restored by the *cir*5 mutation, a T→C change immediately preceding the GUG initiation codon (Fig. 4) (L. Strano et al., unpubl.). Thus, the out-of-phase AUG codon in λ*c*II3086 becomes an ACG codon in λ*cir*5*c*II3086, which supports the proposed explanation for *c*II3086 inactivity.

The two *c*II$^-$ mutations in the second codon might exert their effect by altering the functional activity of the protein or by changing the translation efficiency. Y. Ho and M. Rosenberg (unpubl.), using pulse labeling in vivo, have shown that the second codon mutation *c*II3059 is translated at markedly reduced rates. Similar results were obtained in vitro using an S30 translation system (C. Debouck, unpubl.). Pseudorevertants of λ*c*II3059 have been isolated, which may reverse the translation deficiency of *c*II3059 mRNA. One of these, *ctr*2, is a T→C change in the third position of the third *c*II codon, and the other, *ctr*1, is a G→A change in the first position of the fourth codon (Fig. 4) (J. Place et al., unpubl.). λ*c*II3059*ctr*2, which by plaque morphology is indistinguishable from λ$^+$, has the same amino acid composition as λ*c*II3059, since the *ctr*2 mutation does not result in an amino acid change. This suggests that the *c*II3059*c*II protein is a *c*II$^+$ protein, since it can be fully reverted without alteration. λ*c*II3059*ctr*1, which by plaque morphology is a partial revertant, has an additional Ala→Thr change at codon 4. Moreover, this mutation affects a base in one of the two TTGC sequences thought to define the binding site for *c*II protein (see above). The partial revertant phenotype of this strain may result from the change in this TTGC sequence. Both mutations destabilize a stem and loop structure that can

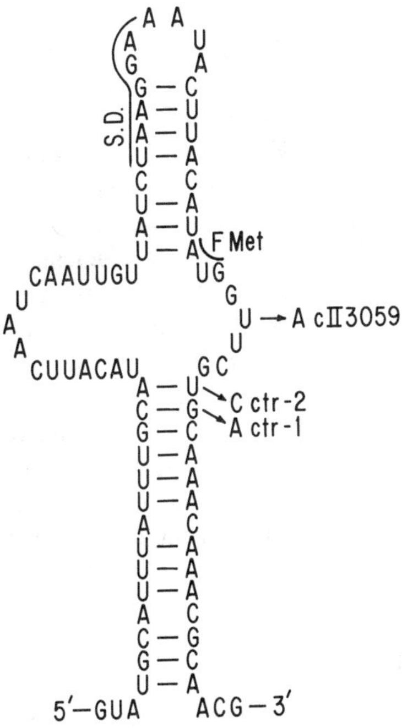

Figure 7 Stem and loop structure that can be formed by p_R mRNA in the aminoterminal region of the *c*II gene. The Shine and Delgarno sequence and the initial formylmethionine codon are indicated. The sequence changes for *c*II3059, *ctr*1, and *ctr*2 are indicated. Strains with the *ctr*1 and *ctr*2 mutations were isolated as *c*II pseudorevertants of λ*c*II3059. These strains retain the original *c*II3059 mutation.

be formed by *c*II mRNA in the initiation region (Fig. 7) (Rosenberg et al. 1978b; Wulff et al. 1980), and this destabilization might result in an increased translation efficiency. This would imply that, with λ^+ mRNA, the stem structure reduces the rate of translation initiation, possibly by restricting the access of ribosomes to the initiation region. In contrast to *c*II3059, the other second codon mutation, *c*3073b, appears to synthesize *c*II protein at nearly normal levels. Hence, this mutation must alter the function of the protein.

*c*II PROTEIN IS PROCESSED

The first two amino acid residues of the nascent *c*II protein are removed in vivo to form the mature, active protein (Ho and Rosenberg 1982). Both the *N*-formylmethionine and second amino acid, valine, predicted from the DNA sequence of the *c*II gene, are not present on the protein as it is isolated. Instead the aminoterminal sequence of the *c*II protein begins with the third amino acid: Arg-Ala-Asn. . . . The removal of the two amino acids from the nascent *c*II protein may be an important event in maturation of the *c*II protein and may strongly influence the level of active intracellular *c*II protein. This is suggested by the *can*1 mutation, a Val → Ala change in the second amino acid codon that results in a more stable *c*II protein (Epp 1978; Jones and Herskowitz 1978; Herskowitz and Hagen 1980; Wulff et al. 1980; M.A. Hoyt and H. Echols, unpubl.; Y. Ho and M. Rosenberg, unpubl.). The *can*1 mutation could not have this effect if processing were normal in *can*1 strains, and aminoterminal amino acid analyses of the *can*1*c*II protein indicate that only the terminal *N*-formylmethionine is removed from the nascent *can*1 protein (Y. Ho and M. Rosenberg, unpubl.).

The observation that the *can*1*c*II protein is processed abnormally suggests that processing may also be affected in other second codon mutations. We have not yet examined the processing step with either of the two *c*II⁻ mutations in the second codon (*c*II3059 and *c*3073b). λ*c*II3059 is translated at lower efficiencies (see above), and one need not invoke abnormal processing to explain the *c*II⁻ phenotype of this mutation. However, *c*3073b is translated at nearly normal levels, and we conclude that the *c*3073b mutation must interfere with the processing event. It is as yet unclear whether the resulting *c*II defect is caused by the interference with processing, and/or the Leu substitution. It will be necessary to determine the effect of each of the second codon mutations on *c*II synthesis, processing, oligomerization, DNA binding, transcriptional activation, and turnover to elucidate the role of the second amino acid and the processing event in the biological activity of the *c*II protein.

CONTROL OF *c*II ACTIVITY: THE HOST *hflA* GENE

The level of *c*II protein in an infected cell is determined both by its synthesis (i.e., transcription, translation, processing, and assembly of tetramers) and by its turnover. The λ*c*II protein is known to be metabolically unstable and must be continuously synthesized for transcription from p_{RE} to occur (Reichardt 1975). The rate of its degradation is a key factor in determining the level of *c*II activity present in infected cells. The host HflA protein is partially responsible for degradation of the *c*II protein, and the λ*c*III protein protects the *c*II protein from degradation (Hoyt et al. 1982). This section summarizes the experiments that have led to this understanding.

The λ*c*III protein greatly enhances lysogenization of wild-type hosts. However, *c*III function is not required under all conditions. For example, host strains that carry an *hflA* mutation are lysogenized at high frequencies by both λ⁺ and λ*c*III mutants (Belfort and Wulff 1973b). Host *hflA* mutations are recessive to the *hflA*⁺ allele, suggesting that the *hflA* gene codes for a polypeptide (Gautsch and Wulff 1974). The phage *c*III function is also not required when the phage carries a *can* mutation (Jones and Herskowitz 1979). Most *can* mutations are *cro*⁻ mutations that cause overproduction of the *c*II protein, whereas the *can*1 mutation lies in the *c*II gene itself and increases the stability of the *c*II protein (see the preceding section). These observations suggest that the host HflA protein is an antagonist of *c*II function and that *c*III protein protects the *c*II protein from the HflA protein.

The host cAMP-activated catabolite repression system is also required for normal lysogenization by phage λ. Both *E. coli cya*⁻ and *crp*⁻ strains are lysogenized at lower levels by λ (Hong et al. 1971; Grodzicker et al. 1972; Belfort and Wulff

1974). Yet *cya*$^-$ and *crp*$^-$ strains are lysogenized efficiently when they are HflA$^-$. This observation suggests that the catabolite repression system exerts its influence on phage lysogenic development by controlling HflA activity (e.g., by directly repressing synthesis of the HflA protein). Hence, both the phage *c*III gene and the host *cya* and *crp* genes influence lysogeny indirectly by affecting HflA activity. Although λ^+ lysogenizes *hflA*$^+$*cya*$^-$*crp*$^-$ hosts to an appreciable extent (ca. one-half the normal frequency), and although λ*c*III mutants lysogenize *hflA*$^+$*cya*$^+$*crp*$^+$ hosts to a small extent (ca. one-tenth the λ^+ frequency), λ*c*III mutants do not lysozenize *hflA*$^+$*cya*$^-$*crp*$^-$ hosts at all (Belfort and Wulff 1974). This indicates that the host HflA protein completely blocks lysogenization unless either the *c*III protein or the cAMP-activated catabolite repression system intervenes. Maximum lysogenization in *hflA*$^+$ hosts occurs when both are active.

The *hflA* locus lies at 93 minutes on the *E. coli* genetic map and is tightly linked to the *purA* locus (Belfort and Wulff 1973a). The *hfl* mutations that are near *purA* comprise a single complementation group, although some intracistronic complementation is seen (Gautsch and Wulff 1974). The *hflA* locus therefore has been presumed to code for a single polypeptide. Strains carrying presumptive null mutations of *hflA* (insertions of Mu or *lacZ*) have been isolated (Hoyt et al. 1982; F. Banuett and I. Herskowitz, pers. comm.), indicating that the *hflA*$^+$ gene is not essential for *E. coli* growth. Cloning of the *hflA* locus into a high copy number plasmid and analysis of its polypeptide indicates that *hflA* may in fact contain two distinct genes, tentatively denoted *hflK* and *hflC* (F. Banuett and I. Herskowitz, pers. comm.). Insertions of transposon Tn*1000* into this plasmid were used to define the region responsible for complementing the *hflA*1 mutation. Maxicell analysis of polypeptides produced by *hfl*$^+$ and *hfl*::Tn*1000* plasmids indicate that the *hflA* region codes for two polypeptide chains of 46,000 daltons and 37,000 daltons. This analysis provides the first biochemical information on Hfl proteins. The genes coding for these polypeptide chains probably comprise an operon.

The discussion up to now indicates that the HflA protein reduces *c*II protein activity but gives no real evidence on how this happens. Experiments of Hoyt et al. (1982) indicate that this reduction of *c*II activity occurs by means of proteolytic action. They examined the decay of *c*II protein made from a *c*II gene on a high copy number plasmid. In an *hflA*$^+$ host (and with no source of *c*III protein) the *c*II protein decayed at a rate twofold greater than in an isogenic *hflA* host. Hoyt et al. used a second plasmid containing the *c*III gene as a source of *c*III protein to show that *c*III protein protects the *c*II protein from HflA-mediated degradation. A second set of experiments indicated that these same general relationships occur under conditions of actual λ infection.

Detailed mechanisms remain to be elucidated. The HflA protein might be a protease, it might control the synthesis or activity of a protease, or it might modify the *c*II protein to make it more susceptible to proteolytic attack. The *c*III protein might serve as an inhibitor or an alternate substrate of this proteolytic activity, or it might bind to *c*II protein directly (Hoyt et al. 1982).

OTHER HOST FACTORS CONTROLLING *c*II ACTIVITY

The HflA protein cannot be the only host factor controlling *c*II activity, since *c*II protein is still degraded in an *hflA*$^-$ host, albeit at a slower rate (Hoyt et al. 1982). F. Banuett and I. Herskowitz (pers. comm.) have shown that the *hfl*29 mutation of Gautsch and Wulff (1974) lies at a single site on the *E. coli* genetic map, which they call *hflB*. The *hflB*-gene product, like the HflA protein, at least partially bypasses the requirement for the λ*c*III protein (Gautsch and Wulff 1974) and the host catabolite gene activation system (F. Banuett and I. Herskowitz, pers. comm.). Knoll (1980) finds that infection of an *hflA*$^-$ host by λ$^+$ results in twice as much repressor as infection of the same *hflA*$^-$ host by λ*c*III$^-$, and Hoyt et al. (1982) find that *c*II protein is degraded less rapidly in a λ$^+$ infection of an *hflA* strain than in a λ*c*III infection. Both observations indicate that the *c*III protein has other protective functions besides protecting against the HflA protein and are consistent with the idea that *c*III protects against the *hflB* function as well.

M.A. Hoyt and H. Echols (pers. comm.) have shown that *c*II protein is degraded less rapidly in λ$^+$ infection of an *hflB* strain than of an *hflB*$^+$ strain and that *c*II protein is more resistant to degradation in a λ$^+$ infection of an *hflA hflB* double mutant strain than infections of either *hflA* or *hflB* strains. However, *c*II protein is degraded at a finite rate even in the *hflA hflB* host.

In their studies of the degradation of *c*II protein made from a high copy number plasmid, Hoyt et al. (1982) found that the protease inhibitors *N*-ethylmaleimide (NEM) and phenylmethylsulfonyl fluoride (PMSF) both protect *c*II protein from degradation in *hflA*$^+$ cells (with the greatest degree of protection occurring in the presence of both inhibitors), whereas only NEM protects *c*II protein in an *hflA* host. Since NEM and PMSF inhibit cysteine and serine proteases, respectively, Hoyt et al. propose that the *hflA* gene might control a serine protease and that the *hflA*-independent degradative pathway could involve a cysteine protease. The whole phenomenon of protein turnover in *E. coli* is not well understood. Sreedhara Swamy and Goldberg (1981) have obtained evidence for eight different proteolytic activities in extracts of *E. coli*, but nothing is known about the roles of these different enzymes in *c*II protein turnover. Since the *c*II protein is interconverted between a monomeric and tetrameric form, it is possible that only the monomer is degraded (analogous to the specific turnover of *c*I monomers; Mount 1980; Phizicky and Roberts 1980). This would yield yet another level of complexity to the control of *c*II degradation in the cell and would predict that mutant *c*II proteins that cannot form tetramers might be degraded more rapidly.

KINETICS OF p_{RE}-DIRECTED REPRESSOR SYNTHESIS

The last two sections examine the role of the p_{RE} promoter in the overall course of λ development. Repressor is synthesized after a 5–15 minute lag in λ wild-type infections and is shut off after 20–30 minutes (Reichardt 1975). The initial lag reflects the time needed to accumulate sufficient quantities of *c*II and *c*III protein to activate p_{RE}. The p_{RM} promoter, which is activated by the repressor protein itself, is not active early in infection (Yen and Gussin 1980). p_{RE} directs the synthesis of repressor at much higher rates than p_{RM} (Reichardt and Kaiser 1971), and newly infected cells acquire much higher levels of repressor than those present in established lysogens.

Repressor synthesis is not shut off at the normal time in infections by λ*cro*⁻ phages, and this leads to considerable overproduction of repressor at late times (Reichardt and Kaiser 1971; Echols et al. 1973; Reichardt 1975; Yen and Gussin 1980). λ*cro*⁻ mutations have striking effects on plaque formation. λ*c*I*ts*857*cro*27 forms plaques at 37°C but does not form plaques at 32°C or 42°C. λ*c*I*ts*857*cro*27*c*II⁻ and λ*c*I*ts*857*cro*27*c*III⁻ strains form plaques at 32°C and 37°C but not at 42°C (Eisen et al. 1975; Folkmanis et al. 1977). Overproduction of the *c*II and *c*III proteins in *c*III⁺*c*I⁺*cro*⁻*c*II⁺ infections may result in the channeling of infective centers exclusively into the lysogenic pathway (Jones and Herskowitz 1978; Winston and Botstein 1981), but this is not always the case. The percentage of infective centers entering the lysogenic pathway in an infection by λ*c*I*ts*857*cro*27 at 32°C is no greater than with λ*c*I*ts*857 (Folkmanis et al. 1977). But infective centers that do not complete the lysogenic pathway in these infections do not complete the lytic pathway either, and it is not known what goes "wrong" in these abortively infected cells.

λ*cro*⁻ strains cannot complete the lytic pathway under stringent *c*I⁻ conditions, regardless of the *c*II or *c*III genotype (e.g., lytic growth does not occur in infections by λ*c*I*ts*857*cro*27 at 42°C, or by λ*c*I*am*14*cro*27*c*II2002 in a *su*⁻ host at any temperature [Folkmanis et al. 1977]). The primary action of the *cro* gene is thus to facilitate lytic growth, presumably by binding to p_L and p_R to turn off early functions late in the growth cycle (see Georgiou et al. 1979). It is appropriate to consider the action of *cro* in detail elsewhere (Gussin et al., this volume).

LYSIS-LYSOGENY DECISION

When a population of sensitive bacteria is infected by λ⁺, some infective centers develop by the lysogenic pathway, and some, by the lytic pathway. This implies that in each infective center a "decision" is made during development to enter either the lysogenic or the lytic pathway. The lytic and lysogenic pathways are interconnected. Transcription from p_L results in *N* protein, which is necessary for both the lytic and lysogenic pathways. Transcription from p_L beyond the *N* gene yields both the *c*III protein, which facilitates the lysogenic pathway, and the Red and Gam proteins, which facilitate lytic growth. Transcription from p_R yields the *c*II protein, which is required for the lysogenic pathway, the *O* and *P* proteins, which are needed both for lytic growth and for efficient lysogenization, and the *Q* protein, which is required for expression of the late genes in the lytic pathway. Upon infection of a normal host strain with λ⁺, it is impossible to proceed down one pathway without going at least part way down the other as well.

When growing cells are infected, the proportion of infected cells entering the lysogenic pathway is about the same, regardless of the growth medium or the cell division time (Echols et al. 1975). Conditions that slow entrance into the lytic pathway would seem to slow entrance into the lysogenic pathway by about the same amount, apparently because key proteins in each pathway are under common transcriptional control. (However, starvation of sensitive cells prior to infection increases the likelihood of the lysogenic response [Kourilsky 1973].)

The multiplicity of infection greatly influences the ratio of the lysogenic to the lytic response, with high multiplicities of infection favoring the lysogenic response and low multiplicities favoring the lytic response (Kourilsky 1973; Knoll 1979a; Yen and Gussin 1980). This may be a result of lower levels of *c*II protein and, especially, of *c*III protein in cells infected at low multiplicities of infection (see Kourilsky 1974; Reichardt 1975). Mutations that appear to increase the activity of either the *c*II protein (e.g., *can* mutations, see above, Control of *c*II Activity: The Host *hflA* Gene) or the *c*III protein increase the extent of lysogenization at low multiplicities. The *c*IIIs mutation, an alteration in the *c*III structural gene that results in a higher level of *c*III protein, increases lysogenization frequencies and rates of repressor synthesis in singly infected cells (Knoll 1979a,b).

What determines whether a particular infective center enters the lysogenic or the lytic pathway? The level of *c*II protein is clearly a crucial determinant, and Herskowitz and Hagen (1980) argue that it is *the* crucial determinant. We have shown here that there are many levels at which *c*II activity is regulated, including transcription, mRNA stability, translation, processing of the nascent *c*II protein, tetramer formation, and breakdown. The level of *c*II protein in an infected cell is determined by the interplay of the various forces controlling these parameters. Clearly, the interactions are intricate and often difficult to disentangle. In addition to the *c*II protein, a variety of other factors, such as the DNA synthesis and the activity of *Q* protein, may also influence the frequency of phage lysogenization. A precise evaluation of the relative contribution of each factor that influences the lysis-lysogeny decision must await future experiments. These should continue to be productive in uncovering new gene regulatory mechanisms.

ACKNOWLEDGMENTS

Work from D.L.W.'s laboratory was supported by U.S. Public Health Service grants GM-25438 and GM-28370. We thank I. Herskowitz, G. Gussin, W. McClure, and H. Echols for critically reading the manuscript and Mrs. K. Schuff for her help in preparing it.

REFERENCES

Abraham, J., D. Mascarenhas, R. Fischer, M. Benedik, A. Campbell, and H. Echols. 1980. DNA sequence of the regulatory region for the integration gene of bacteriophage λ. *Proc. Natl. Acad. Sci.* **77:** 2477.

Belfort, M. and D.L. Wulff. 1973a. Genetic and biochemical investigation of the *Escherichia coli* mutant *hfl*-1 which is lysogenized at high frequency by bacteriophage lambda. *J. Bacteriol.* **115:** 299.

——. 1973b. An analysis of the processes of infection and induction of *E. coli hfl*-1 by bacteriophage lambda. *Virology* **55:** 183.

——. 1974. The roles of the lambda *c*III gene and the *Escherichia coli* catabolite gene activation system in the establishment of lysogeny by bacteriophage lambda. *Proc. Natl. Acad. Sci.* **71:** 779.

Blattner, F.R. and J.E. Dahlberg. 1972. RNA synthesis startpoints in bacteriophage λ: Are the promoter and operator transcribed? *Nat. New Biol.* **237:** 227.

Brachet, P. and R. Thomas. 1969. Mapping and functional analysis of *y* and *c*II mutants. *Mutat. Res.* **7:** 257.

Court, D., L. Green, and H. Echols. 1975. Positive and negative regulation by the *c*II and *c*III gene products of bacteriophage λ. *Virology* **63:** 484.

Echols, H. and L. Green. 1971. Establishment and maintenance of repression by bacteriophage lambda: The role of the *c*I, *c*II, and *c*III proteins. *Proc. Natl. Acad. Sci.* **68:** 2190.

Echols, H., L. Green, R. Kurdra, and G. Edlin. 1975. Regulation of phage λ development with the growth rate of host cells: A homeostatic mechanism. *Virology* **66:** 344.

Echols, H., L. Green, A.B. Oppenheim, A. Oppenheim, and A. Honigman. 1973. Role of the *Cro* gene in bacteriophage λ development. *J. Mol. Biol.* **80:** 203.

Eisen H., M. Georgiou, C.P. Georgopoulos, G. Selzer, G. Gussin, and I. Herskowitz. 1975. The role of gene *cro* in phage development. *Virology* **68:** 266.

Epp, C. 1978. "Early protein synthesis and its control in bacteriophage lambda." Ph.D. thesis, University of Toronto, Ontario.

Folkmanis, A., W. Maltzman, P. Mellon, A. Skalka, and H. Echols. 1977. The essential role of the *cro* gene in lytic development by bacteriophage λ. *Virology* **81:** 352.

Galas, D. and A. Schmitz. 1978. DNase footprinting: A simple method for the detection of protein DNA binding. *Nucleic Acids Res.* **5:** 3157.

Gautsch, J.W. and D.L. Wulff. 1974. Fine structure mapping, complementation, and physiology of *Escherichia coli hfl* mutants. *Genetics* **77:** 435.

Georgiou, M., C.P. Georgopoulos, and H. Eisen. 1979. An analysis of the Tro phenotype of bacteriophage λ. *Virology* **93:** 38.

Grodzicker, T., R.R. Arditti, and H. Eisen. 1972. Establishment of repression by lambdoid phage in catabolite activator protein and adenylate cyclase mutants of *Escherichia coli. Proc. Natl. Acad. Sci.* **69:** 366.

Gussin, G.N., K. Matz, and D. Wulff. 1980. Suppression of λp_{RM} mutations by *cin*-1, a mutation creating a new promoter for leftward transcription of the *c*I gene. *Virology* **103:** 465.

Herskowitz, I. and D. Hagen. 1980. The lysis-lysogeny decision of phage λ: Explicit programming and responsiveness: *Annu. Rev. Genet.* **14:** 399.

Ho, Y. and M. Rosenberg. 1982. Characterization of the phage λ regulatory protein cII. *Ann. Microbiol. (Paris)* **133A:** 215.

Ho, Y. and M. Rosenberg. 1982. Characterization of the phage λ regulatory protein cII. *Ann. Microbiol.* (Paris) **133A:** 215.

Hoess, R.H., C. Foeller, K. Bidwell, and A. Landy. 1980. Site-specific recombination functions of bacteriophage λ: DNA sequence of the regulatory regions and overlapping structural genes for Int and Xis. *Proc. Natl. Acad. Sci.* **77:** 2482.

Hong, J., G.R. Smith, and B.N. Ames. 1971. Adenosine 3′:5′-cyclic monophosphate concentration in the bacterial host regulates the viral decision between lysogeny and lysis. *Proc. Natl. Acad. Sci.* **8:** 2258.

Hoyt, M.A., D.M. Knight, A. Das, H.I. Miller, and H. Echols. 1982. Control of phage λ development by stability and synthesis of *c*II protein: Role of the viral *c*III and host *hfl*A, *him*A, and *him*D genes. *Cell* **31:** 565.

Jones, M.O. and I. Herskowitz. 1978. Mutants of bacteriophage λ which do not require the *c*III gene for efficient lysogenization. *Virology* **88:** 199.

Jones, M.O., R. Fischer, I. Herskowitz, and H. Echols. 1979. Location of the regulatory site for establishment of repression of bacteriophage λ. *Proc. Natl. Acad. Sci.* **76:** 150.

Kaiser, A.D. 1957. Mutations in a temperate bacteriophage affecting its ability to lysogenize *Escherichia coli. Virology* **3:** 42.

Knoll, B.J. 1979a. Isolation and characterization of mutations in the *c*III gene of bacteriophage λ which increase the efficiency of lysogenization of *Escherichia coli* K-12. *Virology* **92:** 518.

——. 1979b. An analysis of repressor overproduction by the λ *c*IIIs mutant. *J. Mol. Biol.* **132:** 551.

——. 1980. Interactions of the λ*c*IIIs1 and the *E. coli hfl*-1 mutations. *Virology* **105:** 270.

Kourilsky, P. 1973. Lysogenization by bacteriophage lambda. I. Multiple infection and the lysogenic response. *Mol. Gen. Genet.* **122:** 183.

——. 1974. Lysogenization by bacteriophage lambda. II. Identification of genes involved in the multiplicity dependent processes. *Biochimie* **56:** 11.

McDermit, M., M. Pierce, D. Staley, M. Shimaji, R. Shaw, and D.L. Wulff. 1976. Mutations masking the lambda *cin*-1 mutation. *Genetics* **82:** 417.

McMacken, R., N. Mantei, B. Butler, A. Joyner, and H. Echols. 1970. Effect of mutations in the *c*II and *c*III genes of bacteriophage λ on macromolecular synthesis in infected cells. *J. Mol. Biol.* **49:** 639.

Miller, H.I. 1981. Multilevel regulation of bacteriophage λ lysogeny by the *E. coli himA* gene. *Cell* **25:** 269.

Miller, H.I., J. Abraham, M. Benedik, A. Campbell, D. Court, H. Echols, R. Fischer, J.M. Galindo, G. Guarneros, T. Hernandez, D. Mascarenhas, C. Montanez, D. Schindler, U. Schmeissner, and L. Sosa. 1981. Regulation of the intergration-excision reaction by bacteriophage λ. *Cold Spring Harbor Symp. Quant. Biol.* **45:** 439.

Mount, D.W. 1980. The genetics of protein degradation in bacteria. *Annu. Rev. Genet.* **14:** 279.

Napoli, C., L. Gold, and B.S. Singer. 1981. Translation reinitiation in the rIIB cistron of bacteriophage T4. *J. Mol. Biol.* **149:** 133.

Phizicky, E.M. and J.W. Roberts. 1980. Kinetics of *recA* protein-directed inactivation of repressors of phage λ and phage P22. *J. Mol. Biol.* **139:** 319.

Reichardt, L. 1975. Control of bacteriophage lambda repressor synthesis after phage infection: The role of the *N*, *c*II, *c*III and *cro* products. *J. Mol. Biol.* **93:** 267.

Reichardt, L. and A.D. Kaiser. 1971. Control of λ repressor synthesis. *Proc. Natl. Acad. Sci.* **68:** 2185.

Rosenberg, M. and D. Court. 1979. Regulatory sequences involved in the promotion and termination of RNA transcription. *Annu. Rev. Genet.* **13:** 319.

Rosenberg, M., D. Court, H. Shimatake, C. Brady, and D.L. Wulff. 1978a. The relationship between function and DNA sequence in an intercistronic regulatory region in phage λ. *Nature* **272:** 414.

——. 1978b. Structure and function of an intercistronic regulatory region in bacteriophage lambda. In *The operon* (ed. J.H. Miller and W.S. Reznikoff), p. 345. Cold Spring Harbor Laboratory, Cold Spring Harbor, New York.

Schmeissner, U., D. Court, K. McKenney, and M. Rosenberg. 1981. Positively activated transcription of λ integrase gene initiates with UTP *in vivo. Nature* **292:** 173.

Schmeissner, U., D. Court, H. Shimatake, and M. Rosenberg. 1980. Promoter for the establishment of repressor synthesis in bacteriophage λ. *Proc. Natl. Acad. Sci.* **77:** 3191.

Schwarz, E., G. Scherer, G. Hobom, and H. Kössel. 1978. Nucleotide sequence of *cro*, *c*II, and part of the *O* gene in phage λ DNA. *Nature* **272:** 410.

Shimatake, H. and M. Rosenberg. 1981. Purified λ regulatory protein *c*II positively activates promoters for lysogenic development. *Nature* **292:** 128.

Shine, J. and L. Delgarno. 1974. The 3′-terminal sequence of *Escherichia coli* 16S ribosomal RNA: Complementarity to nonsense triplets and ribosome binding sites. *Proc. Natl. Acad. Sci.* **71:** 1342.

Sreedhara Swamy, K.H., and A.L. Goldberg. 1981. *Escherichia coli* contains eight soluble proteolytic activities, one being ATP-dependent. *Nature* **292:** 652.

Steege, D.A. 1977. 5′-Terminal nucleotide sequence of *Escherichia coli* lactose repressor mRNA: Features of translation initiation and reinitiation sites. *Proc. Natl. Acad. Sci.* **74:** 4163.

Stormo, G., T. Schneider, and L. Gold. 1982. Characterization of translational initiation sites in *E. coli. Nucleic Acids Res.* **10:** 2971.

Winston, F. and D. Botstein. 1981. Control of lysogenization by phage P22. I. The P22 *cro* gene. *J. Mol. Biol.* **152:** 209.

Wulff, D.L. 1976. Lambda *cin-1*, a new mutation which enhances lysogenization by bacteriophage lambda, and the genetic structure of the lambda *cy* region. *Genetics* **82:** 401.

Wulff, D.L., M. Beher, S. Izumi, J. Beck, M. Mahoney, H. Shimatake, C. Brady, D. Court, and M. Rosenberg. 1980. Structure and function of the *cy* control region of bacteriophage lambda. *J. Mol. Biol.* **138:** 209.

Yen, K.M. and G.N. Gussin. 1980. Kinetics of bacteriophage λ repressor synthesis directed by the p_{RE} promoter: Influence of temperature, multiplicity of infection, and mutation of p_{RM} or the *cro* gene. *Mol. Gen. Genet.* **179:** 409.

Control of Integration and Excision

Harrison Echols
Department of Molecular Biology
University of California
Berkeley, California 94720

Gabriel Guarneros
Departmento de Genética y Biología Molecular
ĈINVESTAV-IPN, A.P. 14-740
México 14, D.F., Mexico

INTRODUCTION

Phage λ controls integration and excision as a crucial aspect of regulated development along the lysogenic or lytic pathways (Echols 1980; Herskowitz and Hagen 1980). The lysogenic response to infection requires two coordinated events: establishment of repression and integrative recombination. The lytic response to infection utilizes an alternative switch to a virus-producing late stage of productive growth, and integration and *c*I-mediated repression do not occur. The lytic pathway after prophage induction requires, in addition, the excision of the viral DNA. In terms of these three aspects of λ development, we can define three states of the integration-excision switch: integration, excision, and no response.

Control of the integration-excision reaction depends on the existence of two pathways for catalysis of site-specific recombination: (1) the forward (insertion) reaction, which requires only Int among phage-specified proteins (Gingery and Echols 1967; Zissler 1967; Gottesman and Yarmolinsky 1968), and (2) the reverse (excision) reaction, which requires the phage-coded Int and Xis proteins (Fig. 1) (Guarneros and Echols 1970; Kaiser and Masuda 1970). Thus, differential synthesis of Int with respect to Xis should favor insertion. In this paper we summarize information about mechanisms for control of Int and Xis production that explain quite well the three stages of the integration-excision switch noted above. We also consider some additional regulatory features of the reaction. In closing, we comment on properties of this biological system that may be pertinent to other creatures.

GENERAL FEATURES OF DIRECTIONAL CONTROL

Phage λ uses differential expression of the tightly linked, partially overlapping *int* and *xis* genes to control the direction of the integration-excision reaction (Fig. 2). In this section, we point out the major features of this gene regulation;

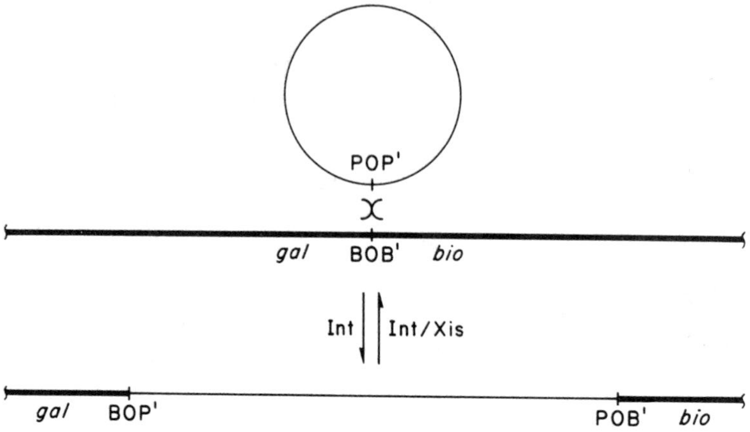

Figure 1 The integration-excision reaction by phage λ. Integrative recombination between the attachment site on the circular λ DNA (POP′) and the attachment site on the host genome (BOB′) provides for prophage insertion. Excisive recombination between the prophage attachment sites (BOP′ and POB′) detaches the λ DNA if the prophage is induced to lytic development. Integrative recombination requires only Int among phage-coded proteins; excisive recombination depends on Int and Xis. Both reactions require, in addition, the integrative host factor (IHF), specified by two bacterial genes.

the experimental basis for these concepts and ideas about molecular mechanisms are considered in more detail in the following sections.

For the lysogenic response to infection, more Int than Xis should be produced. This is accomplished by activation of the p_I promoter by the *c*II protein (Fig. 2). The p_I transcript includes the complete *int*-coding sequence but only a portion of *xis*; thus, *c*II stimulates production of Int but not Xis. The *c*II protein has two additional functions: positive regulation of the *c*I gene, which encodes the repressor that maintains lysogeny, and negative regulation of lytic functions. The multiple activities of *c*II provide for coordinate regulation of the lysogenic response; thus, *c*II functions as the primary partition function for the determination of lysogenic or lytic development. The level of *c*II activity depends on the viral *c*III protein and at least three host proteins, HflA, HimA, and Hip (also called HimD) (for reviews, see Echols 1980; Herskowitz and Hagen 1980; Wulff and Rosenberg, this volume).

For the lytic response, production of Int protein should be minimal. This regulatory response is achieved in part by the failure of *c*II to act effectively, resulting in very limited expression of the *int* gene from the p_I promoter. The *int* gene is also transcribed from the p_L promoter under positive regulation by the λ*N*[1] protein, and this transcription unit will be active during the lytic response. However, a *cis*-acting negative regulatory element, *sib*, functions at a post-transcriptional level to turn off expression of the *int* gene from the p_L RNA

[1]In this volume, the product of a single-letter gene is italicized, and the product of a three-letter gene is romanized.

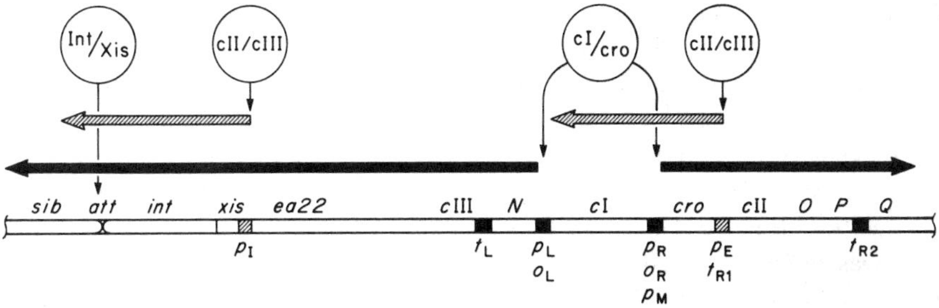

Figure 2 Regulation of transcription during the early stage of λ development. Immediately after infection, transcription from p_L and p_R is terminated mainly at t_L and t_{R1}, with some RNA chains continuing to t_{R2}. The *N* protein eliminates these termination events and provides for transcription of the remainder of the early gene region (solid arrows). The *c*II and *c*III proteins partition λ development toward the lysogenic response because *c*II stimulates RNA synthesis from p_E and p_I (hatched arrows) and delays the expression of late lytic functions; *c*III stabilizes *c*II. The *c*I protein binds to the operators o_L and o_R, shutting off the early λ transcripts from p_L and p_R; *c*I also maintains its own further synthesis by regulating the p_M promoter for the *c*I gene. The alternative late stage of productive development utilizes the Cro and *Q* proteins; *Q* turns on the genes for head, tail, and lysis proteins, and Cro turns off transcription of the *c*I gene from p_M and reduces the synthesis of the p_L and p_R transcripts for early proteins.

(Fig. 2). For the lytic response during prophage induction, the Int and Xis proteins must be produced in appropriate balance to turn the recombinational switch to excision. The p_L transcript can supply this need because the negative regulator *sib* is separated from its target *int* gene by the integrative recombination that inserts the viral DNA (Fig. 2). Regulation by *sib* is novel with respect to classical prokaryotic mechanisms because *sib* is distal with respect to the *int* promoter; this mode of control has been termed retroregulation.

Other mechanisms may exist for directional control of integration-excision, although they are less defined so far than those just described. The partial overlap of the *xis* and *int* genes might provide an additional limitation of Int production from the p_L transcript because efficient translation of *int* might be impeded by ribosomes in transit synthesizing Xis protein (Davies 1980; Hoess et al. 1980; Miller et al. 1981b). There may also be control at the level of protein activity, involving an inhibition of integrative recombination by Xis (Nash 1975a,b; G. Guarneros and J.M. Galindo, unpubl.). Another regulatory feature may involve the functional instability of Xis (Weisberg and Gottesman 1971).

POSITIVE CONTROL BY *c*II AT p_I

Genetics and Physiology

The role of *c*II in the establishment of repression was inferred long before its action on the *int* gene was suspected. We summarize this work briefly here because comparison of *c*I- and *int*-gene regulation is important. For a detailed account, see the reviews by Echols (1980), Herskowitz and Hagen (1980), and Wulff and Rosenberg (this volume). Kaiser (1957) showed that the *c*II- and *c*III-gene products promote the establishment of lysogeny. Subsequent assays of protein and RNA synthesis defined the role of *c*II and *c*III as positive regulators of λ's *c*I and negative regulators of late lytic functions (McMacken et al. 1970; Echols and Green 1971; Reichardt and Kaiser 1971). The site of these regulatory activities (p_E) was defined by analysis of *cis*-dominant point mutations and deletions (Echols and Green 1971; Reichardt and Kaiser 1971; Court et al. 1975; Jones et al. 1979) (Fig. 2). A variety of experiments indicated a more central role for *c*II than *c*III: less severe defect of *c*III⁻ mutations than *c*II⁻ mutations (Kaiser 1957; Reichardt and Kaiser 1971); dispensability of *c*III activity in the *hflA*⁻ host mutant (Belfort and Wulff 1973).

The fact that the *int* and *xis* genes are transcribed from the p_L promoter initially obscured the existence of p_I as a *c*II-activated promoter for *int* (Signer 1970; Gottesman and Weisberg 1971). The p_I promoter was discovered by studies of *int*-gene expression from a prophage (measured by read-through into adjacent host genes) (Shimada and Campbell 1974a,b). This low-level, constitutive expression of the *int* gene from a prophage is sufficient for a small amount of integrative recombination (Echols 1975); however, a normal prophage is stable because Xis is not produced from the p_I transcript.

The capacity of *c*II to activate the p_L promoter became evident from measurements of the *c*II dependence of Int protein synthesis or integrative recombination (Katzir et al. 1976; Chung and Echols 1977; Court et al. 1977; Kotewicz et al. 1977). Since excisive recombination is not stimulated by *c*II, it was possible to infer that positive regulation by *c*II provides for differential regulation of *int* with respect to *xis* (Chung and Echols 1977; Enquist et al. 1979b). This idea was supported by properties of the *int-c* mutations, which were isolated by higher *int*-gene expression in a lysogen and which also confer a Xis⁻ phenotype; the *int-c* mutations could be interpreted as an alteration of the p_I promoter located within the *xis* gene (Shimada and Campbell 1974b) (Fig. 2). However, another finding suggested that at least part of p_I is outside of the *xis* gene: the existence of a deletion that eliminates *c*II activation but not *xis* expression (Heffernan et al. 1979).

In summary, genetic and physiological experiments pointed to regulation of the *int* gene by *c*II protein, acting at a p_I promoter close to and perhaps overlapping with the *xis* gene. The biochemical studies described in the next section have confirmed this view and defined the mechanism.

Biochemistry

The relationship between the *int* and *xis* genes and the p_I promoter has been clarified by three complementary biochemical approaches: (1) determination of the DNA sequence of the p_I region and the base changes of regulatory mutations

(Abraham et al. 1980; Hoess et al. 1980), (2) analysis of the p_I transcript in vivo and in vitro (Abraham and Echols 1981; Schmeissner et al. 1981; Shimatake and Rosenberg 1981), and (3) study of the interaction of *c*II protein with p_I in vitro (Ho and Rosenberg 1982). The results of this work are summarized in Figure 3.

The DNA sequence analysis identified a region with notable homology (11 out of 15 bases) with a "consensus" (computer-generated) interaction site (−10 region) for RNA polymerase (Scherer et al. 1978) (left boxed sequence of Fig. 3); the particularly critical "Pribnow box" subsequence (Pribnow 1975) is indicated by heavy lines. The *int-c* mutations generate a *c*II-independent promoter for transcription of the *int* gene (Fischer et al. 1979; Abraham and Echols 1981); the base change of three of these mutations is an identical transition from a disfavored G to an A, which is one of the most conserved bases of promoter sites (Abraham et al. 1980; Hoess et al. 1980). Thus, from sequence data, the left boxed region is likely to be the interaction sequence for RNA polymerase subject to activation by *c*II.

Because *c*II acts at p_E as well as p_I, a common sequence essential for *c*II activity should exist in both promoter regions. The right boxed sequence of Figure 3 exhibits a striking homology (11 out of 14 bases) with a sequence in the p_E region determined by Rosenberg et al. (1978) and Schwarz et al. (1978). Moreover, a deletion in the p_I region that removes most of the right boxed sequence is *c*II insensitive, whereas a deletion that leaves a few more λ bases responds to *c*II activation (Abraham et al. 1980) (Fig. 3). The homologous sequence in p_E is the site of a number of *c*II-insensitive point mutations (Wulff et al. 1980; Wulff and Rosenberg, this volume). Thus, the right boxed region is likely to be a site necessary for *c*II activity, perhaps a *c*II-binding site. The presumptive *c*II site is in the −35 or recognition region for RNA polymerase, indicating that positive regulation by *c*II occurs within a DNA region that is occupied by the enzyme during its binding interaction. The inferences about the *c*II site from sequence data have been verified directly by DNase protection experiments, using purified *c*II protein; the right boxed region is protected from DNase by *c*II (Ho and Rosenberg 1982).

If the left boxed sequence of Figure 3 is the −10 interaction sequence for RNA polymerase, the p_I transcript should begin about 5 bases downstream (Fig. 3). This prediction has been confirmed by an analysis of transcripts from three sources: (1) constitutive RNA produced in vitro from p_I with an *int-c* mutation (Abraham and Echols 1981), (2) *c*II-activated RNA in vivo (Schmeissner et al. 1981), and (3) *c*II-dependent RNA in vitro (Shimatake and Rosenberg 1981). In each case, the major RNA start is the U nucleotide noted in Figure 3; there is also a less prevalent chain initiation with the adjacent U. These experiments also confirm that *int-c* is a mutation of p_I.

The final piece of information necessary for a regulatory interpretation of the p_I region is the location of the coding sequence for Xis protein. The sequence change of a nonsense mutation in the *xis* gene has determined the reading frame and located the probable initiation codon for Xis as the ATG within the interac-

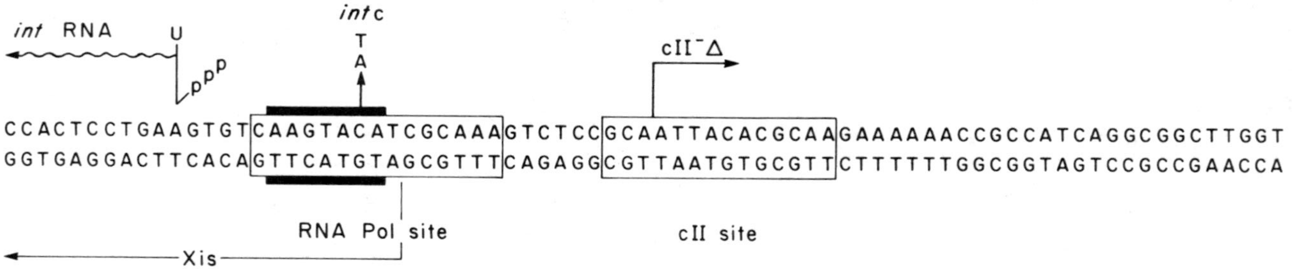

Figure 3 The p_I promoter region and the mechanism of regulation by *c*II. The *c*II protein recognizes the right boxed sequence (cII site) and promotes interaction of RNA polymerase with the left boxed sequence (RNA Pol site). The *c*II-activated RNA chain begins with UTP at adjacent bases, with the major start at the point noted above the sequence. Positive regulation of Int but not Xis ensues because the coding sequence for Xis begins with the ATG indicated, and the *c*II-stimulated RNA thus lacks the translation start and other aminoterminal codons for Xis. The base change of an *int-c* mutation to a constitutive promoter is shown above the sequence, as is the beginning point of a deletion that eliminates *c*II activity (see text).

tion sequence of p_I (Abraham et al. 1980; Hoess et al. 1980) (Fig. 3). The *int-c* base change alters the initiation codon for Xis, an observation that explains the Xis^- phenotype of this mutation. The *c*II-insensitive, Xis^+ deletion removes most of the *c*II activity site but leaves an intact *xis* gene (Fig. 3). Thus, the genetic and biochemical evidence together indicate strongly that the p_I promoter partially overlaps the *xis* gene, with the *c*II site outside and the RNA polymerase interaction sequence mostly within *xis*.

The regulatory consequence of the p_I-*xis* overlap is that *c*II will provide for differential synthesis of Int with respect to Xis because the *c*II-activated RNA has a complete coding sequence for Int but not Xis. Thus, we can explain how *c*II regulates the *int-xis* region to favor the lysogenic response. As a consequence of the arrangement of *att-int-xis*, we can consider this region as a "recombination module" that might be subject to different regulation for alternative choices of external sequence (Campbell et al. 1979; Abraham et al. 1980; Miller et al. 1981b).

NEGATIVE RETROREGULATION BY *sib*

sib Region

The *b* region of the λ genome specifies a site, *sib* (for *s*itio *i*nhibidor en *b*), that inhibits *int*-gene expression. The existence of the *sib* regulatory locus was first inferred from in vivo experiments. Phage carrying a large *b* deletion, λ*b*2, promoted substantially more integrative recombination than λ^+ (Lehman 1974; Roehrdanz and Dove 1977). These experiments were carried out before the action of the *c*II-gene product on the p_I promoter was known. A reinvestigation of this phenomenon, under more controlled conditions for *c*II function, led to the conclusion that *sib*-mediated inhibition of *int* expression is observed only in the absence of a functional *c*II gene (Guarneros and Galindo 1979). This regulatory capacity of *sib* for *int* has been confirmed biochemically: The presence of the *b* region severely reduces the synthesis of Int protein, as shown by gel electrophoresis of radioactive proteins from UV-irradiated cells infected by phage λ (Belfort 1980; Epp et al. 1981; Schindler and Echols 1981; J.M. Galindo and G. Guarneros, unpubl.).

The retroregulator *sib* has been located close to the right end of the *b* region, separated from the *int* gene by the λ attachment site (Fig. 2). Deletion mapping of the *sib* locus has shown that a 200-bp segment adjacent to the attachment site is sufficient to provide full retroregulation of *int* (Mascarenhas et al. 1981; Guarneros et al. 1982; C. Montanez and G. Guarneros, unpubl.; D. Court, pers. comm.). The use of point mutations has defined more precisely the region of the λ genome within this 200-bp segment that is essential for *sib* activity. The DNA sequence of the *sib* region has been determined for four mutants defective in retroregulation; the mutations involve single base-pair substitutions within a DNA stretch of 50 bp with dyad symmetry. In each of the four cases, the base change reduces the stability of a possible stem-and-loop structure in RNA produced by transcription through the symmetrical region (Guarneros et al. 1982) (Fig. 4). A shorter RNA sequence from the *sib* region might form a less extensive stem and loop followed by a row of uridines at the 3′ end, a common feature of transcription terminators (Rosenberg and Court 1979) (Fig. 4). The possible significance of these RNA structures for *sib* regulation is discussed below.

Properties of *sib* Retroregulation

The retroregulator *sib* acts in *cis*; i.e., *int*-gene expression is inhibited only when *sib* and a functional *int* gene are on the same DNA molecule. The sequence

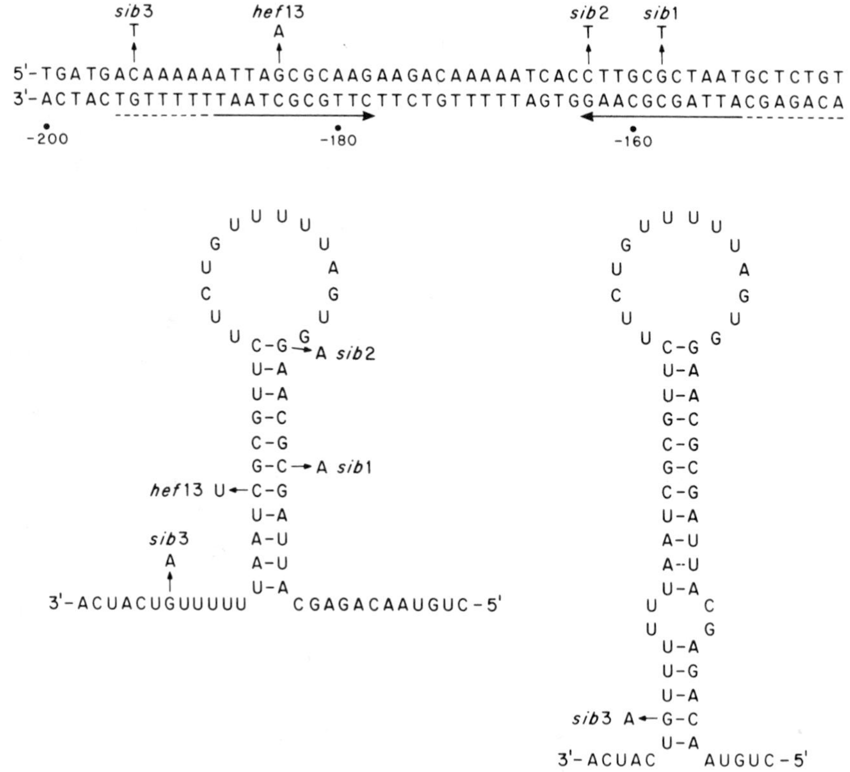

Figure 4 The *sib* region. The locations of four *sib*⁻ mutations are shown in the DNA sequence of the *sib* region (*top*). Possible secondary structures of the RNA from this region are shown below. The RNA structure on the right can form from p_L RNA but not p_I RNA because the p_I RNA terminates at the last of the six consecutive U bases (position −193) (see text).

in the 50-bp *sib* region shows no open reading frame for a long polypeptide; therefore, the hypothesis that a *cis*-acting protein mediates retroregulation is unlikely. The presence of the *sib* region provides for a severe inhibition of Int synthesis; however, there is little or no retroregulation on the synthesis of Xis protein from the adjacent, partially overlapping *xis* gene or the neighboring *ea*22 gene (Fig. 2) (Schindler and Echols 1981). Proximity between *sib* and the target gene seems necessary for retroregulation because inhibition of Ea22 protein synthesis does occur if the *b* region is moved closer by deletion of the *int* and *xis* genes (Schindler and Echols 1981) (Fig. 2).

Evidence suggests that the regulation of *int* expression by *sib* is posttranscriptional. The synthesis of the complete Int protein is severely reduced by *sib*, but production of aminoterminal fragments of Int protein generated by nonsense mutations escape efficient *sib* regulation (Schindler and Echols 1981). Since the *N*-activated p_L transcript proceeds past *int* for a substantial distance into the *b* region, these results indicate posttranscriptional regulation. This inference has been confirmed by experiments measuring stability of *int*-gene RNA; the regulatory effect of *sib* is associated with a decreased overall chemical stability of the *int*-gene transcript from p_L (Guarneros et al. 1982).

The effect of *sib* on stability of *int*-gene RNA suggests the involvement of RNase activities. Experiments in vivo have supported this view: Bacterial mutants defective in RNase III do not show *sib* regulation, as judged by measurement of Int protein (Belfort 1980; Schindler and Echols 1981) or of integrative recombination (Guarneros et al. 1982). In addition, biochemical experiments in vivo and in vitro indicate that the p_L transcript is cleaved by RNase III in the *sib* region (Fig. 4) (Rosenberg and Schmeissner 1982; U. Schmeissner et al., pers. comm.).

As noted above, *sib* inhibits *int*-gene expession from the p_L transcript but not from the p_I transcript. What is the basis for this selectivity? Gene expression from p_L is regulated by the λ*N* protein, which renders the transcription complex resistant to most termination signals. Indeed, the *c*II-activated p_I transcript stops at the transcription terminator in the *sib* region noted in Figure 4, but the p_L transcript does not (U. Schmeissner et al., pers. comm.). Thus, the p_L RNA (but not p_I) might form the longer stem and loop structure with a bubble of mispaired bases shown in Figure 4. Since gene expression from p_L and not from p_I is regulated by *sib*, the longer stem-and-loop structure might be the preferential substrate for RNase III, allowing for retroregulation (Guarneros et al. 1982).

The following model explains retroregulation of *int*-gene expression consistent with the characteristics discussed above. The p_L transcript is cleaved at the long stem-and-loop structure in the *sib* region and is then either inactivated for Int synthesis by a limited processive degradation in the 3′ to 5′ direction or associates in a secondary structure inhibitory for the full translation of the *int* transcript (Guarneros and Galindo 1979; Belfort 1980; Schindler and Echols 1981; Guarneros et al. 1982). The processive degradation mechanism seems most likely to be correct because a host mutant defective in a 3′ exonuclease

(RNase II) does not carry out normal *sib* retroregulation of *int* (D. Schindler and H. Echols, unpubl.).

Although distal regulatory properties identical to *sib* have not yet been described in other systems, it would be surprising if this capacity for an added level of control were unique to λ. There is a strong indication for a distal regulatory function for the *ant* gene of phage P22 (Susskind and Youderian, this volume) and for gene *1.2* of phage T7 (Saito and Richardson 1981).

CATALYTIC REGULATION

Integration and Excision Pathways: Directional Control

As noted in the Introduction, the basis for directional control of integration and excision is the different catalytic requirements for the forward and reverse reactions. Integration requires only Int; excision requires Int and Xis. Both reactions require integration host factor (IHF) (Nash 1977; Weisberg et al. 1977; Echols 1980; Miller et al. 1981b). The overall reaction may thus be written as follows:

$$\underset{\text{POP}'}{att\text{P}} + \underset{\text{BOB}'}{att\text{B}} \underset{\text{Int/Xis/IHF}}{\overset{\text{Int/IHF}}{\rightleftharpoons}} \underset{\text{BOP}'}{att\text{L}} + \underset{\text{POB}'}{att\text{R}}$$

P, P′, B, and B′ refer to structurally distinct segments, and *O* refers to the crossover region common to both.

The direction of the reaction is determined by the quantity of functional recombination proteins and by the chemical nature of the substrate (attachment) sites. The breaking and joining event appears to involve a concerted phosphodiester exchange without a free intermediate; however, the detailed mechanism is not understood (Nash 1981; Weisberg and Landy, this volume). Some properties of the integration and excision reactions that may play a role in directional control are reviewed in this section.

The requirement of Int protein in the integration and excision reactions appears to be absolute. Most mutants defective in the *int* gene are unable to mediate integration or excision (Gingery and Echols 1967; Zissler 1967; Gottesman and Yarmolinsky 1968; Enquist and Weisberg 1977a). The *xis* gene is required only for excisive recombination (Guarneros and Echols 1970; Kaiser and Masuda 1970). However, excisive recombination proceeds at a very low level with xis^- mutations in vivo (and to a greater extent in vitro under appropriate conditions) (Gottesman and Abremski 1980; Abremski and Gottesman 1981). These observations indicate that Int is sufficient to carry out the breaking and joining reactions. A biochemical paradox thus ensues as to why the phosphodiester exchange reaction is so far toward integration under xis^- conditions in vivo (see below). Under other conditions, the integration-excision reaction swings far toward excision. Very little integrative recombination takes place at elevated temperature (42°C), whereas excisive recombination proceeds efficiently (Guarneros and Echols 1973). Integrative recombination in vitro is also thermolabile (Nash 1975a), but excisive recombination is not (Gottesman and Gottesman 1975).

The very pronounced directional effects on integration-excision indicate that the overall reaction may be best described in terms of two distinct individual reactions: integrative recombination and excisive recombination.

$$\text{Integrative:}\quad \text{POP}' + \text{BOB}' \overset{\text{Int}}{\rightleftarrows} \text{BOP}' + \text{POB}' \text{ (thermolabile)}$$

$$\text{Excisive:}\quad \text{BOP}' + \text{POB}' \overset{\text{Int/Xis}}{\rightleftarrows} \text{POP}' + \text{BOB}' \text{ (thermostable)}$$

The notion of distinct reaction pathways is also consistent with additional observations on classes of *int*⁻ mutations. A large fraction of *int*⁻ mutations selected for inability to integrate excise with nearly normal efficiency (Guarneros and Echols 1970). Although the very large *int*⁻ collection of Enquist and Weisberg (1977a) was selected for inability to excise, some temperature-sensitive mutants are also excision proficient at intermediate temperatures (Enquist and Weisberg 1977b). One mutation in the *int* gene is defective in excision but not integration (Enquist and Weisberg 1977a). The existence of the three *int*⁻ mutation types—defective in integration and excision or defective in integration or excision—is consistent with the idea that Int protein participates in a different way in the two pathways (Guarneros and Echols 1970; Enquist and Weisberg 1977a,b).

The two-reaction description for integration-excision provides at the least a formal way to describe the shift to integration or excision. An ''Int state'' (e.g., Int excess) provides for preferential integration; an ''Int/Xis state'' leads to preferential excision. The Int/Xis state might be a protein complex (Echols 1970) or a more subtle functional interaction (Enquist et al. 1979a). Any selective effects on Int or Xis protein levels will shift the reaction accordingly. We have described above mechanisms for differential synthesis of Int in a lysogenic response and Xis in a lytic response. An additional feature is likely to be the stability of Xis; Xis activity decays in vivo after synthesis is stopped (Weisberg and Gottesman 1971). Thus, a cell in which repression is established will eventually accomplish stable integration even if Int is not initially present in excess (Echols and Court 1971; Weisberg and Gottesman 1971).

Biochemical Aspects of Directional Control

An understanding of the biochemical mechanism for directional control of integration versus excision must await a molecular description of the reaction mechanism and the role of the distinct proteins and substrates. However, some inferences are possible from the genetic and physiological studies described above and the limited biochemical data available.

The phosphodiester exchange occurs in the sequence common to the phage and host attachment sites (the O of POP′ and BOB′) and most likely is catalyzed by Int protein (Nash 1981; Weisberg and Landy, this volume). Since the expected equilibrium for this exchange reaction should be about 50%, we expect directional control to come from interactions of Int or Int/Xis at the flanking sites, P, P′, B, or B′. Studies of integrative recombination in vitro indicate that extensive sequences in the P and P′ sites are important but not in B or B′ (Hsu et al. 1980; Mizuuchi and Mizuuchi 1980). Experiments on binding of Int protein to the *att*P (POP′) region reveal multiple binding sites (Davies et al. 1979; Ross et al. 1979); these involve two in the P region, the common core, and the P′ segment, involving altogether about 230 bp (Ross et al. 1979; Hsu et al. 1980, also see Weisberg and Landy, this volume). Studies with the electron microscope show that the binding of Int to *att*P DNA results in the condensation of the 230-bp POP′ region into a tight complex (Better et al. 1982). Thus, the interaction of Int in the P′ and P sequences is likely to be important in overall directional control of integration-excision.

A special role of the P′ interaction for integrative recombination is indicated by three sets of data. First, the P′ site appears to be the highest affinity binding site for Int, as judged by heparin resistance of binding and by direct binding studies (Davies et al. 1979; Ross et al. 1979). Second, a mutation that blocks integrative but not excisive recombination is within the P′-binding region (A. Winoto et al., unpubl.). Third, only substrates with the P′ region form condensed complexes with Int protein (Better et al. 1982). Possibly, tight binding by Int at the P′ site initiates a cooperative interaction with Int at O and P that generates a reactive, condensed structure able to pair with *att*B (Better et al. 1982). In terms of this formulation, Int/Xis might favor excisive recombination by a cooperative interaction initiated at P that generates a reactive structure for *att*R (POB′) able to pair efficiently with *att*L (BOP′). Indeed, the addition of Int and Xis to *att*R provides for a stable interaction visible by electron microscopy and allows the formation of stable paired complexes of *att*L and *att*R (Better et al. 1983).

Although the biochemical discussion is necessarily highly speculative, the fact that sufficient data are now available for such proposals indicates that a solution to the problem of directional control is not likely to be far off.

IHF as a Dual Regulator of Gene Expression and Protein Activity

The IHF is essential for integrative recombination (Nash 1981; Weisberg and Landy, this volume). IHF is a protein of two subunits, α and β. The IHFα subunit is specified by the *himA* gene (Miller and Nash 1981); the IHFβ subunit is probably the product of an unlinked gene designated as *hip* (Miller et al. 1979) or *himD* (Miller et al. 1981a). (Recent work indicates that *hip*$^-$ and *himD*$^-$ mutations affect the same gene; see Weisberg and Landy, this volume.)

Study of the *himA* gene has revealed a remarkable diversity of function. As judged by the properties of *himA*$^-$ mutations, the HimA protein is not only a catalytic component of site-specific recombination by λ but is also required for effective growth of bacteriophage Mu, for precise excision of certain transposable elements, for normal bacterial motility, and for efficient production of the *c*I and Int proteins in the lysogenic response to infection (Miller and Friedman 1980; Miller 1981; H.I. Miller and M. Simon, pers. comm.). For bacteriophage λ, the action of HimA is notable because HimA participates in lysogenization at two levels: catalysis of the integration reaction and production of the proteins required for the lysogenic response, *c*I and Int. The regulatory activity of HimA is probably at the level of synthesis of *c*II protein (Hoyt et al. 1982); for *c*II control, HimA most likely acts with Hip (as IHF) because *himA*$^-$ and *himD*$^-$ (*hip*$^-$) mutations have similar effects on *c*II levels.

The HimA protein is also subject to multiple regulatory pathways. The *himA* gene is negatively regulated by HimA protein itself and is induced by the SOS system activated by DNA damage (Miller et al. 1981a). The response to DNA damage may indicate the participation of HimA in an enhanced cellular capacity for genetic variation that is part of a population escape mechanism for a hostile environment (Echols 1981).

GENERAL COMMENTS ON CONTROL OF INTEGRATION-EXCISION

Multifaceted Regulation: Transcription, Translation, Protein Stability, and Catalysis

The control of integration-excision is notable for general considerations of regulatory strategies because of the diverse mechanisms used to provide effective multiple inputs into this single reaction. Positive regulation by *c*II of the initiation stage of transcription provides for differential synthesis of Int over Xis in the lysogenic response to infection. Positive regulation by *N* at the termination stage of transcription gives properly timed production of both Int and Xis in prophage induction. Negative regulation by *sib* at a posttranscriptional level prevents unwanted Int synthesis during the lytic response to infection; in turn, recombinational control of *sib* allows Int production when the prophage is induced to lytic development.

Additional regulatory inputs from the phage and host are provided by the viral *c*III protein and the host HflA, HimA, and Hip proteins. These proteins control the amount of active *c*II in the infected cell (see Herskowitz and Hagen 1980; Wulff and Rosenberg, this volume). HflA and *c*III control the stability of *c*II; proteolytic degradation of *c*II is enhanced by HflA and diminished by *c*III (Hoyt et al. 1982; C. Epp and M.L. Pearson, pers. comm.). As noted in the previous sections, HimA and Hip probably act as IHF to control *c*II synthesis, as well as to participate in catalysis of the recombination event itself.

The use of a single protein, *c*II, as the critical positive regulator or partition function provides for an efficiently coordinated regulatory response (Echols 1980; Herskowitz and Hagen 1980). The *c*I and Int proteins are produced coordinately so that repression does not occur without integration or the converse.

The cellular control proteins HimA, Hip, and HflA act at a single regulatory step, allowing efficient transmission of environmental signals (e.g., the cAMP system for HflA). Once the initial partition between the lysogenic and lytic pathways occurs, the chosen pathway is stabilized by subsequent regulatory interactions between *c*I and Cro and between Int and Xis (Echols 1980). The control of *c*II (and *N*) stability is an effective way to remove the activity of a regulatory protein for which further action is undesirable for a given choice of developmental pathway.

The *att-int-xis* Region as an Insertion Module

The *att-int-xis* region of λ can be thought of as an insertion module capable in principle of regulation in a variety of ways (Campbell et al. 1979; Miller et al. 1981b; Campbell and Botstein, this volume). For λ, the p_{I} promoter within the module is subject to positive regulation by *c*II, and p_{L}-mediated transcription of the *int* gene is under negative regulation by *sib*. For another biological system, other regulatory interactions might control the same module in a different way.

For the transposon Tn*3*, the resolution stage of transcription also involves a site-specific recombination event. The single recombination protein, resolvase, is subject to negative autoregulation (Gill et al. 1978; Reed et al. 1981; Sherrat et al. 1981). The relative simplicity of this system compared with λ is not surprising because regulation of the λ module has evolved in response to the need for the multiple controls described in the preceding section. Remarkably, the enormous diversity of the λ system is provided by relatively small additions to a Tn*3*-like base of a promoter and the gene for the breaking and joining enzyme. The externally coded regulatory proteins of λ allow for diverse inputs, and the existence of the Xis protein provides for differential control of the direction of the reaction.

Since the publication of *The Bacteriophage Lambda*, the study of the λ integration-excision reaction and its regulation has continued to provide new insights into mechanisms of viral and cellular regulation and the mobility of temperate phage and transposable elements. Many additional insights no doubt remain for the next 10 years.

ACKNOWLEDGMENTS

We thank the multitude of "lambdologists" who have contributed published and unpublished data and ideas to the topics covered in this paper. We also thank Terri DeLuca for editorial assistance. The preparation of this paper has been aided by grants from the National Institute of General Medical Sciences (grant GM-17078 to H.E.) and by the Pan American Health Organization and the Consejo Nacional de Ciencia y Tecnología, Mexico (to G.G.).

REFERENCES

Abraham, J. and H. Echols. 1981. Regulation of *int* gene transcription by bacteriophage λ: Location of the RNA start generated by an *int*-constitutive mutation. *J. Mol. Biol.* **146:** 157.

Abraham, J., D. Mascarenhas, R. Fischer, M. Benedik, A. Campbell, and H. Echols. 1980. DNA sequence of regulatory region for integration gene of bacteriophage λ. *Proc. Natl. Acad. Sci.* **77:** 2477.

Abremski, K. and S. Gottesman. 1982. Site-specific recombination: *xis*-independent excisive recombination of bacteriophage lambda. *J. Mol. Biol.* **153:** 67.

Belfort, M. 1980. The cII-independent expression of the phage λ *int* gene in RNase III-defective *E. coli. Gene* **11:** 149.

Belfort, M. and D.L. Wulff. 1973. An analyis of the processes of infection and induction of *E. coli hfl*-1 by bacteriophage lambda. *Virology* **55:** 183.

Better, M., C. Lu, R.C. Williams, and H. Echols. 1982. Site-specific DNA condensation and pairing mediated by the Int protein of bacteriophage λ. *Proc. Natl. Acad. Sci.* **79:** 5837.

Better, M., S. Wickner, J. Auerbach, and H. Echols. 1983. Role of the Xis protein of bacteriophage λ in a specific reactive complex at the *att*R prophage attachment site. *Cell* **32:** 161.

Campbell, A., M. Benedik, and L. Heffernan. 1979. Viruses and inserting elements in chromosomal evolution. In *Concepts of the structure and function of DNA, chromatin and chromosomes* (ed. A.S. Dion), p. 51. Symposia Specialists, New York.

Chung, S. and H. Echols. 1977. Positive regulation of integrative recombination by the *c*II and *c*III genes of bacteriophage λ. *Virology* **79:** 312.

Court, D., L. Green, and H. Echols. 1975. Positive and negative regulation by the *c*II and *c*III gene products of bacteriophage λ. *Virology* **63:** 484.

Court, D., S. Adhya, H. Nash, and L. Enquist. 1977. The phage λ integration protein (Int) is subject to control by the *c*II and *c*III gene products. In *DNA insertion elements, plasmids, and episomes* (ed. A.I. Bukhari et al.), p. 389. Cold Spring Harbor Laboratory, Cold Spring Harbor, New York.

Davies, R.W. 1980. DNA sequence of the *int-xis-*p_I region of the bacteriophage lambda; overlap of the *int* and *xis* genes. *Nucleic Acids Res.* **8:** 1765.

Davies, R.W., P.H. Schreier, M.L. Kotewicz, and H. Echols. 1979. Studies on the binding of lambda Int protein to attachment site DNA; identification of a tight binding site in the P′ region. *Nucleic Acids Res.* **7:** 2255.

Echols, H. 1970. Integrative and excisive recombination by bacteriophage λ: Evidence for an excision-specific recombination protein. *J. Mol. Biol.* **47:** 575.

——. 1975. Constitutive integrative recombination by bacteriophage λ. *Virology* **64:** 557.

——. 1980. Bacteriophage λ development. In *Molecular genetics of development* (ed. T. Leighton and W.F. Loomis), p. 1. Academic Press, New York.

——. 1981. SOS functions, cancer, and inducible evolution. *Cell* **25:** 1.

Echols, H. and D. Court. 1971. The role of helper phage in *gal* transduction. In *The Bacteriophage lambda* (ed. A.D. Hershey), p. 701. Cold Spring Harbor Laboratory, Cold Spring Harbor, New York.

Echols, H. and L. Green. 1971. Establishment and maintenance of repression by bacteriophage lambda: The role of the *c*I, *c*II, and *c*III proteins. *Proc. Natl. Acad. Sci.* **68:** 2190.

Enquist, L.W. and R.A. Weisberg. 1977a. A genetic analysis of the *att-int-xis* region of coliphage lambda. *J. Mol. Biol.* **111:** 97.

——. 1977b. Flexibility in attachment site recognition by λ integrase. In *DNA insertion, elements, plasmids*, and *episomes* (ed. A. Bukhari et al.), p. 343. Cold Spring Harbor Laboratory, Cold Spring Harbor, New York.

Enquist, L.W., A. Kikuchi, and R.A. Weisberg. 1979a. The role of λ integrase in integration and excision. *Cold Spring Harbor Symp. Quant. Biol.* **43:** 1115.

Enquist, L.W., A. Honigman, S.-L. Hu, and W. Szybalski. 1979b. Expression of λ *int* gene function in ColEl hybrid plasmids carrying the C fragment of bacteriophage λ. *Virology* **92:** 557.

Epp, C., M. Pearson, and L.W. Enquist. 1981. Downstream regulation of *int* gene expression by the *b2* region in phage lambda. *Gene* **13:** 327.

Fischer, R., Y. Takeda, and H. Echols. 1979. Transcription of the *int* gene of bacteriophage λ. New RNA polymerase binding site and RNA start generated by *int*-constitutive mutations. *J. Mol. Biol.* **129:** 509.

Gill, R., F. Heffron, G. Dougan, and S. Falkow. 1978. Analysis of sequences transposed by complementation of two classes of transposition-deficient mutants of Tn3. *J. Bacteriol.* **136:** 742.

Gingery, R. and H. Echols. 1967. Mutants of bacteriophage λ unable to integrate into the host chromosome. *Proc. Natl. Acad. Sci.* **58:** 1507.

Gottesman, M.E. and R.A. Weisberg. 1971. Prophage insertion and excision. In *The bacteriophage lambda* (ed. A.D. Hershey), p. 113. Cold Spring Harbor Laboratory, Cold Spring Harbor, New York.

Gottesman, M.E. and M.B. Yarmolinski. 1968. Integration-negative mutants of bacteriophage lambda. *J. Mol. Biol.* **32:** 487.

Gottesman, S. and K. Abremski. 1980. The role of HimA and Xis in lambda site-specific recombination. *J. Mol. Biol.* **183:** 503.

Gottesman, S. and M. Gottesman. 1975. Excision of prophage λ in a cell-free system. *Proc. Natl. Acad. Sci.* **72:** 2188.

Guarneros, G. and H. Echols. 1970. New mutants of bacteriophage λ with a specific defect in excision from the host chromosome. *J. Mol. Biol.* **47:** 565.

——. 1973. Thermal asymmetry of site-specific recombination by bacteriophage λ. *Virology* **52:** 30.

Guarneros, G. and J.M. Galindo. 1979. The regulation of integrative recombination by the *b2* region and the *c*II gene of bacteriophage λ. *Virology* **95:** 119.

Guarneros, G., C. Montanez, T. Hernandez, and D. Court. 1982. Posttranscriptional control of bacteriophage λ *int* gene expression from a site distal to the gene. *Proc. Natl. Acad. Sci.* **79:** 238.

Heffernan, L., M. Benedik, and A. Campbell. 1979. Regulatory structure of the insertion region of bacteriophage λ. *Cold Spring Harbor Symp. Quant. Biol.* **43:** 1127.

Herskowitz, I. and D. Hagen. 1980. The lysis-lysogeny decision of phage λ: Explicit programming and responsiveness. *Annu. Rev. Genet.* **14:** 399.

Ho, Y. and M. Rosenberg. 1982. Characterization of the phage λ regulatory protein cII. *Ann. Microbiol.* **133A:** 215.

Hoess, R.H., C. Foeller, K. Bidwell, and A. Landy. 1980. Site-specific recombination functions of bacteriophage λ: DNA sequence of regulatory regions and overlapping structural genes for Int and Xis. *Proc. Natl. Acad. Sci.* **77:** 2482.

Hoyt, M.A., D.M. Knight, A. Das, H.I. Miller, and H. Echols. 1982. Control of phage λ development by stability and synthesis of *c*II protein: Role of the viral *c*III and host *hfl*A, *him*A, and *him*D genes. *Cell* **31:** 565.

Hsu, P.L., W. Ross, and A. Landy. 1980. The λ phage *att* site: Functional limits and interaction with Int protein. *Nature* **285:** 85.

Jones, M., R. Fischer, I. Herskowitz, and H. Echols. 1979. Location of the regulatory site for establishment of repression by bacteriophage λ. *Proc. Natl. Acad. Sci.* **76:** 150.

Kaiser, A.D. 1957. Mutations in a temperate bacteriophage affecting its ability to lysogenize *Escherichia coli*. *Virology* **3:** 42.

Kaiser, A.D. and T. Masuda. 1970. Evidence for a prophage excision gene in λ. *J. Mol. BIol.* **47:** 557.

Katzir, N., A. Oppenheim, M. Belfort, and A.B. Oppenheim. 1976. Activation of the lambda *int* gene by the *c*II and *c*III gene products. *Virology* **74:** 324.

Kotewicz, M., S. Chung, Y. Takeda, and H. Echols. 1977. Characterization of the integration protein of bacteriophage λ as a site-specific DNA-binding protein. *Proc. Natl. Acad. Sci.* **74:** 1151.

Lehman, J.F. 1974. λ site-specific recombination: Local transcription and an inhibitor specified by the *b2* region. *Mol. Gen. Genet.* **130:** 333.

Mascarenhas, D., R. Kelley, and A. Campbell. 1981. DNA sequence of the *att* region of coliphage 434. *Gene* **15:** 151.

McMacken, R., N. Mantei, B. Butler, A. Joyner, and H. Echols. 1970. Effect of mutations in the *c*II and *c*III genes of bacteriophage λ on macromolecular synthesis in infected cells. *J. Mol. Biol.* **49:** 639.

Miller, H.I. 1981. Multilevel regulation of bacteriophge λ lysogeny by the *E. coli him*A gene. *Cell* **25:** 269.

Miller, H.I. and D.I. Friedman. 1980. An *Escherichia coli* gene product required for site-specific recombination. *Cell* **20:** 711.

Miller, H.I. and H.A. Nash. 1981. Direct role of the *him*A gene product in phage λ integration. *Nature* **290:** 523.

Miller, H.I., M. Kirk, and H. Echols. 1981a. SOS induction and autoregulation of the *him*A gene for site-specific recombination in *Escherichia coli. Proc. Natl. Acad. Sci.* **78:** 6754.

Miller, H.I., A. Kikuchi, H.A. Nash, R.A. Weisberg, and D.I. Friedman. 1979. Site-specific recombination of bacteriophage lambda: The role of host gene products. *Cold Spring Harbor Symp. Quant. Biol.* **43:** 1121.

Miller, H.I., J. Abraham, M. Benedik, A. Campbell, D. Court, H. Echols, R. Fischer, J.M. Galindo, G. Guarneros, T. Hernandez, D. Mascarenhas, C. Montanez, D. Schindler, U. Schmeissner, and L. Sosa. 1981b. Regulation of the integration-excision reaction by bacteriophage λ. *Cold Spring Harbor Symp. Quant. Biol.* **45:** 439.

Mizuuchi, M. and K. Mizuuchi. 1980. Integrative recombination of bacteriophage λ: Extent of the DNA sequence involved in attachment site function. *Proc. Natl. Acad. Sci.* **77:** 3220.

Nash, H.A. 1975a. Integrative recombination of bacteriophage lambda DNA *in vitro. Proc. Natl. Acad. Sci.* **72:** 1072.

——. 1975b. Integrative recombination in bacteriophage lambda: Analysis of recombinant DNA. *J. Mol. Biol.* **91:** 501.

——. 1977. Integration and excision of bacteriophage λ. *Curr. Top. Microbiol. Immunol.* **78:** 171.

——. 1981. Integration and excision of bacteriophage λ. The mechanism of conservative site specific recombination. *Annu. Rev. Genet.* **15:** 143.

Pribnow, D. 1975. Nucleotide sequence of an RNA polymerase binding site at an early T7 promoter. *Proc. Natl. Acad. Sci.* **72:** 784.

Reed, R.R. 1981. Resolution of cointegrates between transposons γδ and Tn3 defines the recombination site. *Proc. Natl. Acad. Sci.* **78:** 3428.

Reichardt, L. and A.D. Kaiser. 1971. Control of λ repressor synthesis. *Proc. Natl. Acad. Sci.* **68:** 2185.

Roehrdanz, R.L. and W.F. Dove. 1977. A factor in the *b*2 region affecting site-specific recombinations in lambda. *Virology* **79:** 40.

Rosenberg, M. and D. Court. 1979. Regulatory sequences involved in the promotion and termination of RNA transcription. *Annu. Rev. Genet.* **13:** 319.

Rosenberg, M. and U. Schmeissner. 1982. Regulation of gene expression by transcription termination and RNA processing. In *Translational/transcriptional regulation of gene expression* (ed. B. Safer and M. Grunberg-Manago), p. 1. Elsevier North Holland, New York.

Rosenberg, M., D. Court, H. Shimatake, C. Brady, and D.L. Wulff. 1978. The relationship between function and DNA sequence in an intercistronic regulatory region in phage λ. *Nature* **272:** 414.

Ross, W., A. Landy, Y. Kikuchi, and H. Nash. 1979. Interaction of Int protein with specific sites on λ att DNA. *Cell* **18:** 297.

Saito, H. and C.C. Richardson. 1981. Processing of mRNA by ribonuclease III regulates expression of gene 1.2 of bacteriophage T7. *Cell* **27:** 533.

Scherer, G.E.F., M.D. Walkinshaw, and S. Arnott. 1978. A computer aided oligonucleotide analysis provides a model sequence for RNA polymerase-promoter recognition in *E. coli. Nucleic Acids Res.* **5:** 3759.

Schindler, D. and H. Echols. 1981. Retroregulation of the *int* gene of bacteriophage λ: Control of translation completion. *Proc. Natl. Acad. Sci.* **78:** 4475.

Schmeissner, U., D. Court, K. McKenney, and M. Rosenberg. 1981. Positively activated transcription of λ integrase gene initiates with UTP *in vivo. Nature* **292:** 173.

Schwarz, E., G. Scherer, G. Hobom, and H. Kössel. 1978. Nucleotide sequence of *cro*, *c*II and part of the *O* gene in phage λ DNA. *Nature* **272:** 410.

Sheratt, D.J., A. Arthur, and M. Burker. 1981. Transposon-specified, site-specific recombination systems. *Cold Spring Harbor Symp. Quant. Biol.* **45:** 275.

Shimada, K. and A. Campbell. 1974a. Int-constitutive mutants of bacteriophage lambda. *Proc. Natl. Acad. Sci.* **71:** 237.

——. 1974b. Lysogenization and curing by Int-constitutive mutants of phage λ. *Virology* **60:** 157.

Shimatake, H. and M. Rosenberg. 1981. Purified λ regulatory protein *c*II positively activates promoters for lysogenic development. *Nature* **292:** 128.

Signer, E.R. 1970. On the control of lysogeny in phage λ. *Virology* **40:** 624.

Weisberg, R.A. and M.E. Gottesman. 1971. The stability of Int and Xis functions. In *The bacteriophage lambda* (ed. A.D. Hershey), p. 489. Cold Spring Harbor Laboratory, Cold Spring Harbor, New York.

Weisberg, R.A., S. Gottesman, and M.E. Gottesman. 1977. Bacteriophage λ: The lysogenic pathway. *Comp. Virol.* **8:** 197.

Wulff, D.L., M. Beher, S. Izumi, J. Beck, M. Mahoney, H. Shimatake, C. Brady, D. Court, and M. Rosenberg. 1980. Structure and function of the cy control region of bacteriophage lambda. *J. Mol. Biol.* **138:** 209.

Zissler, J. 1967. Integration-negative (*int*) mutants of phage λ. *Virology* **31:** 189.

Repressor and Cro Protein: Structure, Function, and Role in Lysogenization

Gary N. Gussin
Department of Zoology
University of Iowa
Iowa City, Iowa 52242

Alexander D. Johnson
Department of Biochemistry and Biophysics
University of California Medical Center
San Francisco, California 94143

Carl O. Pabo
Department of Biophysics
Johns Hopkins Medical School
Baltimore, Maryland 21205

Robert T. Sauer
Department of Biology
Massachusetts Institute of Technology
Cambridge, Massachusetts 02139

λ's choice between lysogeny and lytic multiplication is influenced directly by the relative levels of two phage-encoded regulatory proteins: repressor (*c*I-gene product) and *cro* protein. Although the two proteins bind to the same sites on λ DNA, their effects on phage development are mutually antagonistic.

Repressor is essential for lysogenization. By binding to two phage operators, o_R and o_L, it blocks transcription initiation from two promoters, p_R and p_L. In so doing, it prevents expression of phage early genes (including the *cro* gene) whose products are necessary for lytic multiplication. In addition, repressor bound at o_R stimulates p_{RM}, the promoter that directs transcription of the *c*I gene of an integrated prophage. Thus, repressor promotes lysogenization in two ways: It blocks expression of lytic genes, and it positively regulates its own synthesis.

In contrast to repressor, *cro* protein promotes lytic multiplication, primarily by inhibiting repressor synthesis both during infection of a sensitive cell and following induction of a lysogen. These effects of *cro* protein are also mediated by binding to o_R and o_L. In subsequent sections one will see that the ability of repressor and *cro* protein to bind to the same operators and yet serve distinct physiological functions is due to (1) the existence within each operator of three repressor (and *cro*)-binding sites, (2) the differing relative affinities of the two proteins for each of the sites, and (3) the spatial relationship between the operator sites and the promoters p_L, p_R, and p_{RM}.

In this review we consider the basic properties of repressor and *cro* protein, their distinct roles in the λ life cycle, and the molecular mechanisms by which they regulate transcription initiation, particularly from p_R and p_{RM}. Transcriptional control is discussed in terms of a small number of specific protein-DNA and protein-protein interactions. These in turn are related to the physiology of the maintenance of lysogeny and the induction of lytic development. (For previous reviews, see Eisen and Ptashne 1971; Ptashne 1971; Ptashne et al. 1980; Johnson et al. 1981.)

BIOLOGICAL ROLES OF REPRESSOR AND *cro* PROTEIN

Synthesis and Action of Repressor

Repressor bound at o_L and o_R prevents transcription of phage early genes *N*, *cro*, *O*, *P*, and *Q*, whose products are necessary for lytic development (Friedman and Gottesman, this volume). Repression of both the p_L and p_R operons is required for the establishment and maintenance of a lysogen (Eisen et al. 1968; Ptashne and Hopkins 1968; Hopkins and Ptashne 1971). Purified repressor blocks transcription initiation at both promoters in vitro (Steinberg and Ptashne 1971; Wu et al. 1972), and inactivation of repressor leads rapidly to expression of both operons in vivo (Kourilsky et al. 1968; Szybalski et al. 1971). Thus, to a first approximation, the level of repressor synthesized after infection of a sensitive cell determines whether or not the cell is lysogenized. Similarly, the stability of a lysogen depends on the level of repressor produced by the integrated prophage. Repressor synthesis is regulated differently in these two situations.

During the *establishment* of a lysogen following infection of a sensitive cell (Wulff and Rosenberg, this volume), repressor synthesis is directed by the promoter, p_{RE}, under positive regulation by the *c*II and *c*III proteins (Echols and Green 1971; Reichardt and Kaiser 1971; Spiegelman et al. 1972; Reichardt 1975a). However, since repressor prevents transcription from p_R and p_L, it blocks expression of the *c*II and *c*III genes, respectively (Fig. 1). Thus, these genes are inactive in lysogens (Bode and Kaiser 1965). Furthermore, since the *c*II and *c*III proteins are unstable (Reichardt 1975a), the action of repressor must ultimately lead to a turnoff of transcription from p_{RE} in a cell destined to become a lysogen. Therefore, transcription from a second promoter, p_{RM}, is necessary for the *maintenance* of a repressed prophage.

The distinction between p_{RM} and p_{RE} is based on several observations. (1) Prophage repressor synthesis is unaffected by *c*II$^-$, *c*III$^-$, or *cy*$^-$ (p_{RE}^-) mutations, all of which cause a reduction in repressor synthesis upon infection of a sensitive cell (Echols and Green 1971; Reichardt and Kaiser 1971; Reichardt 1975b). (2) The rate of repressor synthesis is much greater on infection of a sensitive cell than in a lysogen (Reichardt and Kaiser 1971). (3) Prophage repressor synthesis is regulated positively by repressor and negatively by *cro* protein (see

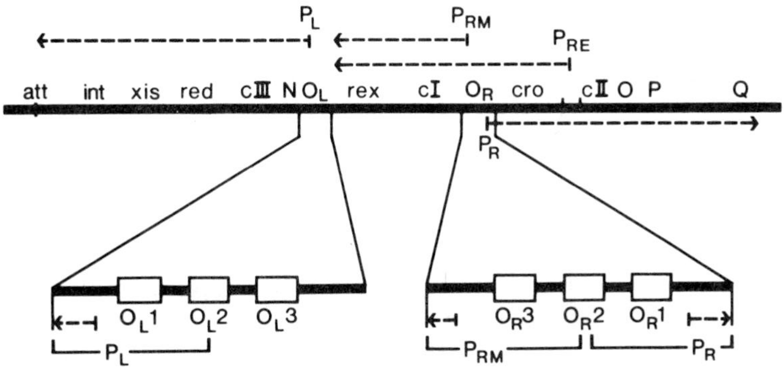

Figure 1 Regulation of early λ transcription. Transcription of early λ genes is initiated at p_L and p_R and is subject to repression by repressor acting at o_L and o_R, respectively. Transcription of the *c*I (repressor) and *rex* genes can be initiated either at p_{RM} (the maintenance promoter) or p_{RE} (the establishment promoter) (Reichardt and Kaiser 1971). Transcripts are indicated by dashed lines. The expanded diagrams illustrate spatial relationships between p_L and three repressor binding sites in o_L, and among p_{RM}, p_R, and three repressor-binding sites in o_R (Ptashne et al. 1976). *cro* protein also binds at o_R and o_L to reduce transcription from p_R, p_{RM}, and p_L (Reichardt 1975a,b; Takeda 1977; Ptashne et al. 1980).

below) even when p_{RE} is deleted (Castellazzi et al. 1972; Reichardt 1975b). (4) Mutations in p_{RM} do not reduce the rate of repressor synthesis under conditions in which p_{RE} is active (Yen and Gussin 1973; Gussin et al. 1975).

The signal for the transition from the establishment to the maintenance mode of repressor synthesis is the occupancy of o_R to an extent sufficient to fully repress transcription from p_R. The orientation of p_R and p_{RM} with respect to o_R (Fig. 1) permits bound repressor to block transcription from p_R and activate p_{RM} simultaneously (Reichardt and Kaiser 1971; Meyer and Ptashne 1980). Thus, once a lysogen is established, its maintenance depends on the continued presence of active repressor. Thermal inactivation of the *c*I857 (temperature-sensitive) repressor leads immediately to reduced rates of p_{RM}-directed *c*I mRNA and repressor synthesis (Eisen et al. 1968; Heinemann and Spiegelman 1970; Reichardt 1975b). In vitro, transcription initiation at p_{RM} is stimuated between 10-fold and 20-fold by purified repressor (Meyer et al. 1975; Walz et al. 1976).

Action of *cro* Protein

Transcription initiation from p_{RM} is also regulated by *cro* protein. A lysogen containing a prophage that is defective in early lytic functions and encodes a thermolabile repressor can be ''induced'' at high temperature without killing the host cell. Prolonged incubation of such a lysogen at high temperature leads to complete cessation of repressor synthesis, which is reflected in loss of immunity to superinfection. The ability to synthesize repressor is not regained even upon transfer back to low temperature (Eisen et al. 1968; Neubauer and Calef 1970; Spiegelman 1971). This is due to the repression of p_{RM} by *cro* protein. If the prophage is *cro*$^-$, immunity is recovered gradually, due to low-level constitutive synthesis of repressor under the direction of p_{RM} (Eisen et al. 1970; Reichardt 1975b). Thus, in normal lysogens, induction leads to a permanent shutoff of p_{RM}, both because of the inactivation of repressor and because of the synthesis of active *cro* protein.

In addition to its direct inhibition of repressor synthesis directed by p_{RM}, *cro* protein indirectly inhibits p_{RE}-directed repressor synthesis. This effect of *cro* protein is mediated by its binding at o_L and o_R to turn down expression of the *c*III and *c*II genes, which in turn leads to a shutoff of p_{RE} (Echols et al. 1973; Reichardt 1975a). Repression of transcription initiation at p_R by *cro* protein has been demonstrated in vivo and in vitro (Champoux 1971; Kumar et al. 1971; Takeda et al. 1977).

Sites of Action of Repressor and *cro* Protein

The multiple effects of repressor and *cro* protein on transcription initiated at p_{RM} and p_R can be understood most easily in terms of the relative locations of p_R, o_R, and p_{RM} (Fig. 1). As mentioned previously, o_R (as well as o_L) contains three repressor-binding sites (Fig. 1); these are similar in sequence and contain axes of twofold hyphenated symmetry (Maniatis et al. 1975; Humayun et al. 1977a) (Fig. 2). The three sites in o_R are arranged nearly symmetrically with respect to p_R and p_{RM}; o_{R1} and half of o_{R2} overlap p_R, whereas o_{R3} and half of o_{R2} overlap p_{RM} (Fig. 1).

The role of the three sites in o_R in the regulation of p_R and p_{RM} has been worked out through the construction of p_{RM}-*lacZ* and p_R-*lacZ* fusion phages, which make it possible by assaying the synthesis of β-galactosidase to monitor the activities of the two promoters in vivo (see Ptashne et al. 1980). These studies showed that at low concentrations repressor binds to o_{R1} and o_{R2} to block transcription from p_R (Meyer et al. 1980). Mutants with reduced affinity of repressor for o_R (Horiuchi et al. 1969; Hopkins and Ptashne 1971; Flashman 1978) were repressed efficiently only at much higher repressor concentrations in vivo. Based on mutational analyses, it was also shown that repressor binding *either* at o_{R1} or o_{R2} is sufficient to prevent p_R-directed β-galactosidase synthesis (Meyer et al. 1980).

The idea that repressor also activates p_{RM} by binding to o_R was based on the inability of repressor to stimulate p_{RM}-directed repressor synthesis when o_{R1} and o_{R2} were mutated (Reichardt and Kaiser 1971). Subsequently, a mutant defective both in o_{R1} and o_{R3} was used to demonstrate that the site of repressor binding required for activation of p_{RM} is in fact o_{R2} (Meyer et al. 1980). In addition, as is discussed in the next section (Repressor Structure and Function), repressor oligomers bound at o_{R1} and o_{R2} interact cooperatively to increase the overall

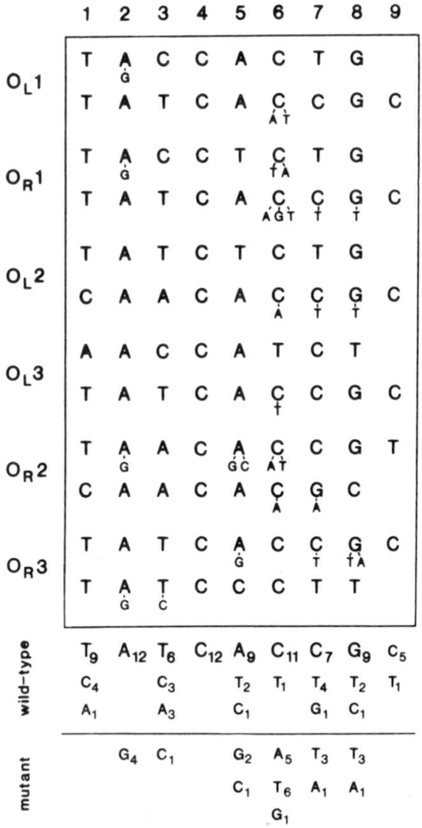
1 2 3 4 5 6 7 8 9
o_L1
T A C C A C T G
T A T C A C C G C
o_R1
T A C C T C T G
T A T C A C C G C
o_L2
T A T C T C T G
C A A C A C C G C
o_L3
A A C C A T C T
T A T C A C C G C
o_R2
T A A C A C C G T
C A A C A C G C
o_R3
T A T C A C C G C
T A T C C C T T
wild-type
T_9 A_{12} T_6 C_{12} A_9 C_{11} C_7 G_9 C_5
C_4 C_3 T_2 T_1 T_4 T_2 T_1
A_1 A_3 C_1 G_1 C_1
mutant
G_4 C_1 G_2 A_5 T_3 T_3
C_1 T_6 A_1 A_1
G_1

Figure 2 λ half-operator sites and operator-constitutive mutations. For each half-operator site, the sequence is written 5′ to 3′, with the complementary sequence omitted. In each case, the half-site is written first for which the 5′ to 3′ orientation is the same as the direction of synthesis of the corresponding mRNA (from p_L or p_R). For example, the complete sequence of o_{L1} is 5′-TACCACTGGCGGTGATA-3′ / 3′-ATGGTGACCGCCACTAT-5′. The underscored sequences are listed, with the sequence in the upper strand first. The sites are listed in descending order of affinity for λ repressor (o_{L1} is the strongest site; o_{R3} is the weakest). The partial two-fold symmetry axis for each operator site passes through the ninth base pair. Only one of the bases at position nine is shown; in each case, this base extends the half site most closely resembling the consensus sequence (TATCACCGX), though for o_{L2} the choice is arbitrary. Characterized operator-constitutive mutations are listed below each half-site sequence. (*Bottom*) The frequencies of bases found in positions 1–9 of wild-type and mutant half-sites are summarized. The mutations are o_{L1} (upper): $A_2 \rightarrow G$ (v_{101}); (lower): $C_6 \rightarrow A$ (v_2), $C_6 \rightarrow T$ (v_{003}); o_{R1} (upper): $A_2 \rightarrow G$ (v_R18), $C_6 \rightarrow A$ (v_s326), $C_6 \rightarrow T$ (BC1); (lower): $C_6 \rightarrow A$ (v_C1), $C_6 \rightarrow G$ (v_s387), $C_6 \rightarrow T$ (NR5), $C_7 \rightarrow T$ (U93, M36), $G_8 \rightarrow T$ (v_3, UV8); o_{L2} (lower): $C_6 \rightarrow A$ (v_L2688); $C_7 \rightarrow T$ (v_{169}), $G_8 \rightarrow T$ (v_{305}); o_{L3} (lower): $C_6 \rightarrow T$ (v_L668); o_{R2} (upper): $A_2 \rightarrow G$ (W1-26), $A_5 \rightarrow G$ (v_3C), $A_5 \rightarrow C$ (E104), $C_6 \rightarrow T$ (E93); $C_6 \rightarrow A$ (v_1); (lower): $C_6 \rightarrow A$ (v_N), $G_7 \rightarrow A$ (BR1); o_{R3} (upper): $A_5 \rightarrow G$ (v_C3), $C_7 \rightarrow T$ (r1), $G_8 \rightarrow T$ (r2), $G_8 \rightarrow A$ (r3); (lower): $A_2 \rightarrow G$ (c10), $T_3 \rightarrow C$ (c12). References for identification of mutants: Ordal and Kaiser (1973); Bailone and Devoret (1978); Flashman (1978); Ptashne (1978); Meyer et al. (1980); Rosen et al. (1980). NR5, BC1, W1-26, and BR1 were isolated and characterized by J. Eliason (pers. comm.).

affinity of repressor for o_{R2}; thus, mutations that reduce binding of repressor to either site can prevent activation of p_{RM} (Johnson et al. 1979; Meyer et al. 1980; Rosen et al. 1980).

At higher concentrations, repressor binds to o_{R3} (Maniatis et al. 1975; Johnson et al. 1979). This can lead to an inhibition of repressor synthesis both in vitro and in vivo, though at concentrations found in lysogens, repressor does not bind to o_{R3} at levels sufficient to inhibit p_{RM} significantly (Meyer et al. 1975; Walz et al. 1976; Maurer et al. 1980; Ackers et al. 1982).

Transcription initiation at p_R and p_{RM} is also prevented by *cro* protein (Takeda et al. 1977; Johnson et al. 1978; Takeda 1979), which recognizes the same three sites as repressor (Johnson et al. 1978). Binding to o_{R3} inhibits p_{RM}, whereas binding to o_{R2} or o_{R1} prevents transcription from p_R (Maurer et al. 1980; Meyer et al. 1980). In contrast to repressor, which binds most tightly to o_{R1} (Maniatis et al. 1975; Johnson et al. 1979), *cro* protein binds most strongly to o_{R3} (Johnson et al. 1978). The differing relative affinities of *cro* protein and repressor for the three operator sites help explain why the biological effects of the two proteins are different even though they recognize the same DNA sequences.

The overlap between sites in o_R and p_R or p_{RM} suggests steric mechanisms for repression of the two promoters by *cro* protein and repressor. However, as is discussed in a subsequent section (Interactions of RNA Polymerase with p_{RM} and p_R), the mechanism of activation of p_{RM} by repressor is not so obvious.

REPRESSOR STRUCTURE AND FUNCTION

Interdependence of Repressor Dimerization and Operator Binding

Repressor is monomeric at very low concentrations ($<10^{-9}$ M), but as its concentration is raised it forms dimers and higher-order oligomers (Pirotta et al. 1970; Chadwick et al. 1971; Brack and Pirrotta 1975; Sauer 1979). Analysis of the concentration dependence of oligomerization suggests that repressor molecules associate in two distinct ways: (1) by dimerization of repressor monomers ($K_d = 2 \times 10^{-8}$ M) and (2) by polymerization of repressor dimers ($K_d \cong 10^{-6}$ M for addition of one dimer to another or to a higher-order oligomer). Since the repressor concentration in a lysogen is about 4×10^{-7} M (Reichardt 1975b), dimers should be the major form of repressor in the cell.

Repressor oligomerization and DNA binding are linked functions (Pirotta et al. 1970; Chadwick et al. 1971; Sauer 1979). Dimers bind tightly to single operator sites, whereas monomers bind only weakly. Since monomers and dimers are in equilibrium, the binding of repressor to a single operator site is a sigmoidal function of repressor concentration (Chadwick et al. 1971).

The two-step reaction is represented as follows:

$$(1) \qquad 2R \overset{K_1}{\rightleftharpoons} R_2 \qquad \text{(dimerization)}$$

$$(2) \qquad R_2 + O \overset{K_2}{\rightleftharpoons} R_2O \qquad \text{(operator binding)}$$

with dissociation constants $K_1 = (R)^2/(R_2)$ and $K_2 = (R_2)(O)/(R_2O)$. The overall reaction $2R + O \rightleftharpoons R_2O$ is characterized by the complex dissociation constant $K = K_1K_2 = (R)^2(O)/(R_2O)$.

Thus, determination of K_2 must take into account the actual concentration of repressor dimers; this can be calculated from the equilibrium constant for the dimerization reaction which, as indicated above, is about 2×10^{-8} M. On the basis of these considerations, K_2 is estimated to be about 3×10^{-9} M for binding to o_{R1} in 0.2 M KCl at 37°C (Johnson et al. 1979).

Interaction of Repressor with Single Operator Sites

All evidence to date indicates that repressor recognizes the operator in a form that is similar to B-form DNA; e.g., cruciform structures are unlikely since repressor does not unwind the operator significantly upon binding (Maniatis and Ptashne 1973). Therefore, specific binding of repressor to operator sites must depend on the recognition of specific nucleotide sequences.

Each 17-bp operator site contains an axis of approximate twofold symmetry; thus, each can be considered to be composed of two half-sites. The nucleotide sequences of the 12 half-sites in o_R and o_L are displayed in Figure 2 (Maniatis et al. 1975; Humayun et al. 1977a). Although the half-sites are not identical, no sequence differs from the consensus sequence (TATCACCGX) by more than 3 bases. The most highly conserved base pairs are the AT pair at position 2 and the CG pairs at positions 4 and 6. The importance of the AT pair at position 2 and the CG pair at position 6 is confirmed by the isolation of numerous operator constitutive mutations at these positions (Fig. 2). No operator mutations at position 4 have been isolated; however, the CG pair at this position is undoubtedly important since methylation of the equivalent G in all six sites reduces repressor binding (Johnson et al. 1980; Johnson 1982).

The existence of operator constitutive mutations at positions 3, 5, 7, and 8 shows that these positions are also important for repressor binding (Maniatis et al. 1975; Flashman 1978), but the wild-type base pairs at these positions in many of the operator half-sites differ from the consensus sequence. Furthermore, some of these operator mutations change the wild-type base pair to a pair found at the same position in another site. Clearly, the natural sequence variation among operator sites plays an important role in determining the affinity of each site for repressor.

Of about 30 operator mutations analyzed, all but one change a wild-type base pair that originally matched the consensus sequence. This suggests strongly that repressor should bind most tightly to the consensus sequence itself.

The interaction of repressor with o_{R1} has been studied in detail using a number of chemical and enzymatic methods (Humayun et al. 1977b; Johnson et al. 1980; Johnson 1983). The results of these experiments are indicated in Figures 3 and 4. Three important points emerge from this work.

1. Bound repressor makes close contacts at or near functional groups in the major groove of operator DNA. Methylation of seven of the nine guanines in o_{R1} interferes with repressor binding; furthermore, bound repressor protects the same seven guanines from methylation by dimethylsulfate. Since it is the N7 of guanine that is methylated, and this nitrogen is exposed in the major

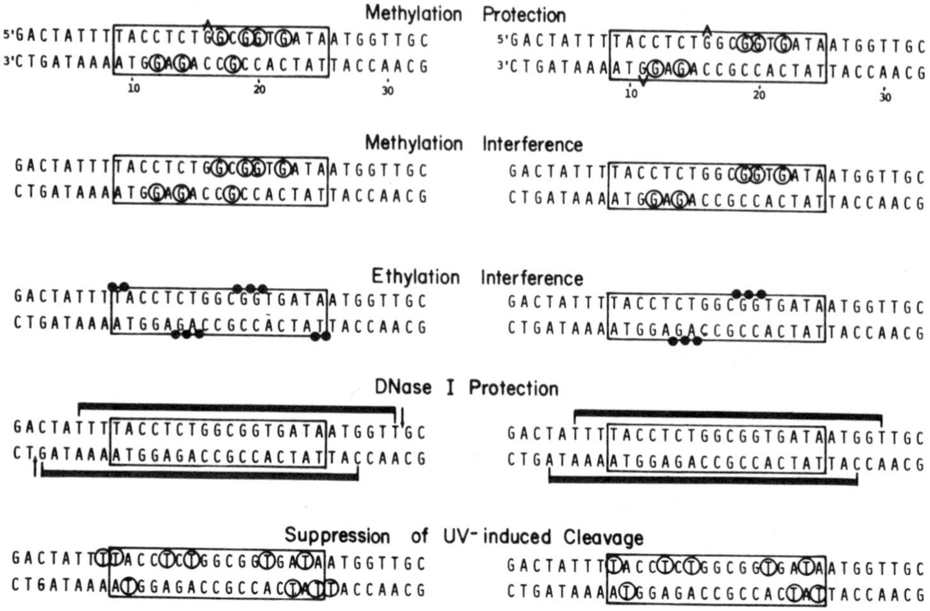

Figure 3 Contacts between repressor and *cro* protein and o_{R1}. The results of chemical and nuclease probe analyses of the interaction between repressor or *cro* protein and a DNA fragment containing o_{R1} (Johnson 1983) are outlined. (*Left*) Results obtained with repressor; (*right*) those obtained with *cro* protein. Methylation protection: Purines protected from methylation by dimethylsulfate are circled; those whose methylation is enhanced are marked by a caret. Methylation interference: Purines whose methylation reduces binding of repressor or *cro* protein are circled. Ethylation interference: Phosphates whose ethylation by ethylnitrosourea reduces the binding of repressor or *cro* protein are indicated by dots. DNase protection: Regions of protected DNA are indicated by heavy lines; enhancement of cleavage by bound repressor is indicated by arrows. UV-induced strand breakage: Each BrdU (replacing a T), whose UV-induced strand scission is suppressed by bound repressor or *cro* protein, is circled. At the position marked with an asterisk, no UV-induced breakage was observed either in the presence or absence of protein. The rationale of many of these probe experiments is outlined by Siebenlist et al. (1980).

groove, most of the contacts between the repressor and operator DNA must lie in the major groove. In no case does methylation of the N3 of adenine, which is exposed in the minor groove, affect repressor binding. Suppression of UV-induced cleavage of BrdU-substituted operator DNA by bound repressor is also consistent with the idea that repressor contacts primarily the major groove (Siebenlist et al. 1980; Johnson 1983).

2. Contacts between repressor and the sugar-phosphate backbone are symmetrically disposed about the partial twofold symmetry axis of the operator site. These contacts are identified as phosphates at which ethylation by

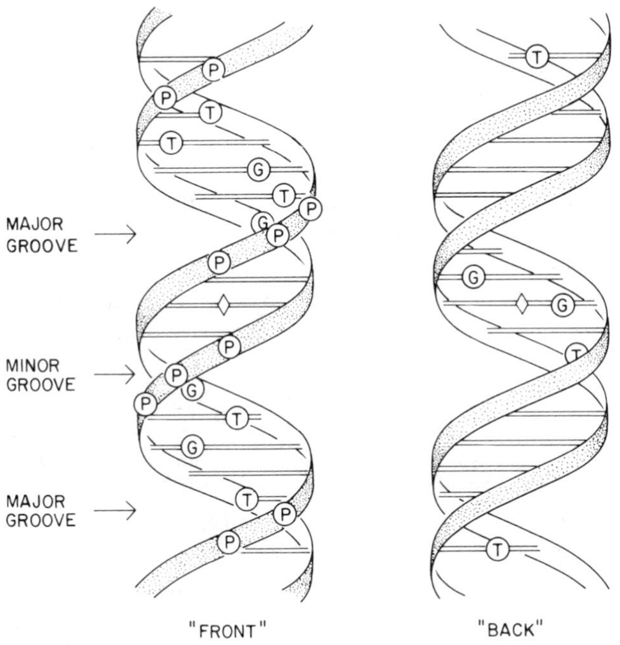

Figure 4 Contacts made by repressor with B-form DNA. Shown is o_{R1} as B-form DNA in which the locations of guanines (G), phosphates (P), and thymines (T) identified as sites of repressor contact are indicated either on the front or back face of the DNA helix. The sites of close contact are those identified in the experiments summarized in the legend to Fig. 3.

ethylnitrosourea interferes with repressor binding and as phosphodiester bonds that are protected by bound repressor from DNase I cleavage.

3. Most sites at which repressor makes close contacts with operator DNA can be contacted from one face of the DNA helix (Fig. 4, ''front''), but a few potential contact sites are accessible only from the opposite face (Fig. 4, ''back'').

Repressor Structure

The repressor monomer is a single-chain polypeptide (m.w. = 26,228) containing 236 amino acids. Both the amino acid sequence of repressor and the nucleotide sequence of the *c*I gene are known (Sauer 1978; Sauer and Anderegg 1978). Repressor can be dissected with proteases to yield protein fragments that retain folded structures, as well as specific functions of intact repressor (Pabo et al. 1979; Sauer et al. 1979). Papain cleaves repressor at three major sites, yielding fragments containing repressor residues 1–92, 93–236, 122–236, and

132–236. Studies of the isolated fragments, calorimetric analysis of the denaturation of intact repressor, and the pattern of intragenic complementation between *c*I$^-$ mutants indicate that repressor is a two-domain protein (Lieb 1976; Pabo et al. 1979; Pabo 1980).

The carboxyterminal fragments of repressor contain the major sites for repressor oligomerization but have no detectable affinity for operator DNA. A fragment containing residues 132–236 aggregates to form dimers, whereas a fragment containing residues 93–236 aggregates to form tetramers and higher-order oligomers (Pabo et al. 1979). These results suggest that the repressor dimer contacts are located between residue 132 and the carboxyl terminus of the protein, whereas the linear polymerization contacts are located at least in part between residues 93 and 131. In addition, the carboxyterminal domain of repressor (residues 93–236) is involved in two interactions important for repressor inactivation: (1) An Ala-Gly sequence at residues 111 and 112 is cleaved by activated *recA* protein during induction (Roberts and Roberts 1975; Sauer et al. 1982a; see Roberts and Devoret, this volume). (2) The carboxyterminal domain is the site of recognition by the P22 antirepressor protein (J. DeAnda and R.T. Sauer, unpubl.; see Susskind and Youderian, this volume).

The isolated aminoterminal domain (residues 1–92) retains the ability to bind operator DNA (Sauer et al. 1979). The binding energy is only about half the energy of binding of intact repressor to the operator, but in methylation or nuclease protection experiments binding of the aminoterminal fragment cannot be qualitatively distinguished from binding of intact repressor (Sauer et al. 1979; Johnson 1980). The three aminoterminal residues can be removed from the aminoterminal domain by tryptic cleavage. The resulting fragment (residues 4–92) still binds to operator DNA, but in methylation experiments, fails to protect the guanines that lie near the center and on the back face of the operator site (see Fig. 4). This experiment implies that the first three residues of repressor normally contact the back of the DNA helix (Pabo et al. 1982). The accessibility of the aminoterminal residues to tryptic cleavage suggests that this region of repressor may be less tightly folded than the rest of the aminoterminal domain, and the structure of the aminoterminal fragment in the crystal (see below) further suggests that these residues are part of an extended arm (residues 1–7) that is flexible in solution. When repressor first binds to the operator, this flexible arm may wrap around the DNA to allow contacts on the back face (Fig. 5). Dissociation of the repressor from the operator would require reversal of this process.

The papain-generated aminoterminal domain of repressor has been crystallized, and its three-dimensional structure has been determined from X-ray diffraction data (Pabo and Lewis 1982). In the crystal the fragment is dimeric, but it is probable that the observed dimer contacts are weak since dimers are not detected in equilibrium ultracentrifugation of the fragment at a starting concentration of 5×10^{-5} M (Pabo et al. 1979; Pabo 1980). The aminoterminal domain is rather simple, being composed of an extended aminoterminal "arm" and five α-helices connected by nonhelical regions. The first four helices form a globular

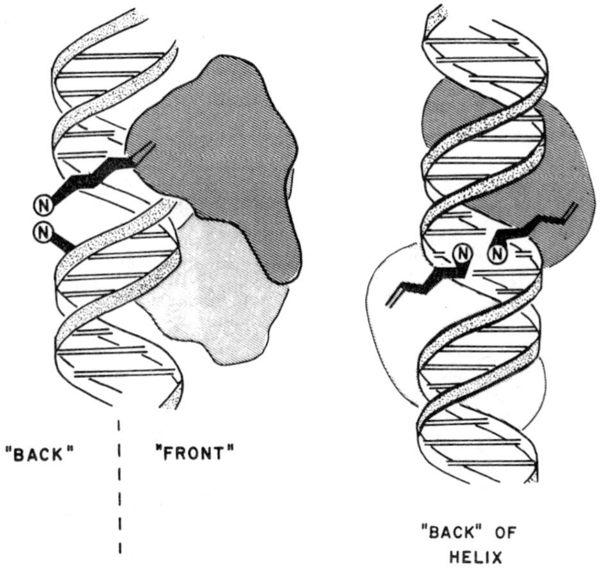

Figure 5 Proposed complexes between the aminoterminal domain dimer of repressor and a single 17-bp operator site. The two subunits of the dimer are shaded differently; only the outline of the protein structure is indicated (Pabo and Lewis 1982). These views show how the flexible aminoterminal arms of a repressor dimer might wrap around the operator site to contact bases on the opposite (back) face of the DNA helix. The position and shape of the globular domain are the same as in Fig. 6A.

structure with a well-formed hydrophobic core, whereas interactions between side chains of helix 5 (residues 79–92) appear to stabilize the dimeric form of the aminoterminal domain.

In model-building experiments, a dimer of the aminoterminal domain fits the operator well if the twofold axis of the dimer is superimposed on the axis of partial twofold symmetry of a single operator site (Pabo and Lewis 1982). Such an arrangement allows each aminoterminal domain in the dimer to interact with a symmetric half-operator sequence. This is consistent with the sequence symmetry of each 17-bp operator site and with the symmetry of the contacts between intact repressor dimers and phosphates in the sugar-phosphate backbone (see above). Oriented in this way, the central 5 bases of the 17-bp operator site are contacted by the aminoterminal arm of each aminoterminal domain (Fig. 5). Some of these contacts appear to be on the back of the DNA helix, and some may be on the side. A second set of contacts are made between the outer bases of the operator site and side chains protruding from the symmetrically related aminoterminal regions of repressor helices 2 and 3 (Fig. 6). The aminoterminal region of helix 3 projects directly into the major groove, and several hydrogen bonds between glutamine and serine residues in helix 3 and bases in the major

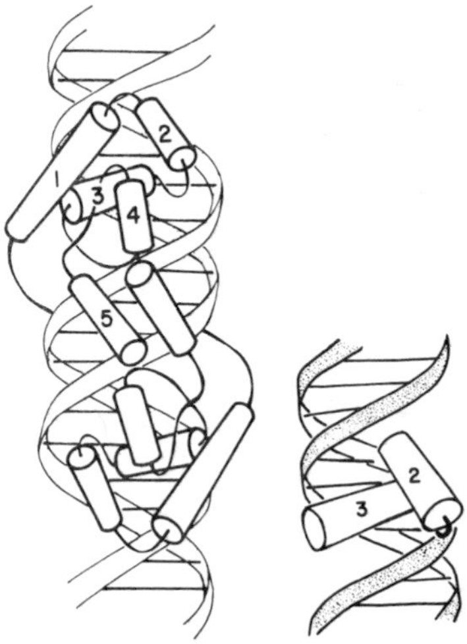

Figure 6 Binding of aminoterminal domain dimers to a single operator site. (*Left*) A rough view of how two aminoterminal fragments of λ repressor can fit against an operator site. Cylinders are used to represent α-helices, and the helices of one monomer are numbered. Helix 3 fits into the major groove, and the extended aminoterminal arm wraps around the operator. The twofold axis of the fragment dimer is perpendicular to the plane of the page. (*Right*) Enlarged view of the helix-2-turn-helix-3 region of complexes depicted at *left*. Helix 2 is above the major groove, but its aminoterminal end (top) is close to the sugar-phosphate backbone of the DNA. Helix 3 is closer to the DNA, and the aminoterminal part of this helix fits directly into the major groove.

groove are possible. Residues near the amino terminus of helix 2 may hydrogen bond to the sugar-phosphate backbone.

The model for the repressor-operator interaction depicted in Figure 6 accounts in general for the chemical protection and modification data, since repressor bound in the proposed complex is close to each DNA site implicated as a region of repressor contact by the probe experiments. The model is also strongly supported by studies of mutant repressors that retain their ability to fold into a stable three-dimensional structure but have severely reduced affinity for operator DNA. Each mutant of this type is changed in a side chain that is predicted by the model building either to contact the operator DNA or to be very close to it when repressor-operator complexes are formed (Hecht et al. 1983; Nelson et al. 1983).

This model provides a molecular explanation for the coupling of repressor dimerization and binding of repressor to operator DNA. Repressor monomers bind only to operator half-sites. If we ignore the effects of rotational and translational entropy, the energy of binding should be roughly half the energy of binding of a repressor dimer to an intact (17-bp) operator site. Dimerization, which requires interactions between repressor carboxyterminal domains, permits the binding energies of two aminoterminal domains to be coupled; thus, the dimer binds several orders of magnitude more tightly to the operator than does the monomer. During induction, *recA* protein cleaves repressor, physically separating the aminoterminal and carboxyterminal domains. Thus, functional inactivation of repressor by *recA* protein is the consequence of uncoupling strong repressor dimerization from operator binding. Even though the aminoterminal fragment created by *recA* cleavage potentially can bind to the operator, its concentration in the cell is too low to result in significant occupancy of the operator sites, and repression is relieved.

Cooperative Binding of Repressor to Adjacent Operator Sites

The three operator sites in o_R are separated by 6 bp or 7 bp (Fig. 8). A repressor dimer bound at one site affects the binding of additional dimers to adjacent sites (Johnson et al. 1979). For example, when o_{R1} is occupied, repressor binds more avidly to o_{R2} than when o_{R1} is unoccupied. This cooperation is observed only in the binding of intact repressor and not in the binding of the isolated aminoterminal domain to o_R (Johnson 1980). Thus, the cooperative interactions appear to be mediated by carboxyterminal regions of repressor dimers bound at adjacent sites. When o_{R1} is inactivated by mutation, repressor dimers bound at o_{R2} and o_{R3} interact cooperatively. These observations suggest that a repressor dimer bound at o_{R2} can interact either with a dimer at o_{R1} or a dimer at o_{R3} but not with both. Similar phenomena have been observed in repressor binding at o_L. The biological implications of these interactions are discussed below (see Opposing Roles of Repressor and *cro* Protein).

cro PROTEIN

The nucleotide sequence of the *cro* gene and the amino acid sequence of the *cro* protein have been determined (Hsiang et al. 1977; Roberts et al. 1977). *cro*-gene product is a small, one-domain polypeptide, 66 amino acids in length (m.w. = 7351). The protein has been crystallized, and its three-dimensional structure has been determined from X-ray diffraction data (Anderson et al. 1979, 1981). *cro* monomers consist of three α-helices and three strands of antiparallel β-sheet with connecting loop regions. The *cro* dimer appears to be stabilized by interactions between carboxyterminal β strands.

Interactions between *cro* Protein and the Operator

The *cro* protein recognizes the same six operator sites as does repressor (Johnson et al. 1978). Enzymatic and chemical probe experiments (see Fig. 3) suggest that the protein interacts with many of the same bases and functional groups in operator DNA that have been identified as repressor contacts (Johnson et al. 1978; Johnson 1980). Furthermore, operator mutations selected on the basis of reduced affinity of the operator for either protein reduce its affinity for the other (Ordal and Kaiser 1973; Berg 1974; Reichardt 1975b; Folkmanis et al. 1976; Flashman 1978; Johnson et al. 1979; Maurer et al. 1980; Meyer et al. 1980).

There are three important differences between the interactions of *cro* protein and repressor with operator DNA: (1) *cro* protein appears to contact only one face of the operator DNA (Johnson 1980), unlike repressor, which contacts both faces. (2) Repressor and *cro* protein have different relative affinities for the three binding sites in o_R. Repressor binds more tightly to o_{R1} and less tightly to o_{R2} and o_{R3}, whereas *cro* protein binds more tightly to o_{R3} and less tightly to o_{R2} and o_{R1} (Johnson et al. 1978). This difference contributes significantly to the opposing biological roles of *cro* protein and repressor (see below, Opposing Roles of Repressor and *cro* Protein). (3) Unlike repressor, *cro* protein binds independently to each of the three sites in o_R and o_L; there are no cooperative interactions between *cro* protein dimers bound at adjacent sites (Johnson et al. 1979).

Although the crystal structures of the *cro* protein and the repressor aminoterminal domain differ significantly, there seem to be some similarities in their interaction with the operator (Anderson et al. 1981; Pabo and Lewis 1982). Both appear to bind single operator sites as dimers, utilizing the twofold symmetry of the dimer to make symmetric contacts with operator half-sites. Moreover, model-building experiments suggest that *cro* protein, like repressor, utilizes helices 2 and 3 to make operator contacts.

REPRESSOR AND *cro* PROTEINS OF RELATED PHAGES

Repressor proteins of phages P22 and 434 have been purified and partially characterized (Ballivet et al. 1977; Sauer et al. 1981; DeAnda et al. 1983; R. Yocum and J. Anderson, unpubl.). Like λ repressor, these proteins appear to contain two domains and to bind to operator DNA as dimers. There is substantial amino acid sequence homology among these repressors, the bacterial *lexA* repressor, and λ repressor (Sauer et al. 1981, 1982b), and all are cleaved at Ala-Gly sequences by activated *recA* protein (Sauer et al. 1982a; R. Yocum, unpubl.).

The *cro* protein of phage 434 has been partially purified (Aono and Horiuchi 1977), and the amino sequences of the *cro* proteins of phages 434 and P22 have been deduced from corresponding DNA sequences (Schwarz et al. 1978; Grosschedl and Schwarz 1979; R.T. Sauer and A.R. Poteete, unpubl.). Like λ*cro* protein, 434 and P22 *cro*-gene products are small polypeptides, presumably containing a single domain. All three proteins are substantially homologous in amino acid sequence to one another, to the aminoterminal domains of the phage repressor proteins, and to the *c*II proteins of phages λ and 434 and the *c*II analog of P22 (Matthews et al. 1982; Sauer et al. 1982b). The sequence homologies in combination with the known three-dimensional structures of λ *cro* protein and repressor suggest that in all of these proteins, contacts with operator DNA are made by side chains in a conserved region composed of an α-helix, a turn, and a second α-helix. In λ repressor and λ*cro* protein, this structure includes helices 2 and 3. The region of amino acid sequence homology spanning these two helices is shown in Figure 7. Within this conserved region, the phage proteins also show strong homology with the bacterial *lac* and *gal* repressors (Beyreuther et al.

33	34	35	36	37	38	39	40	41	42	43	44	45	46	47	48	49	50	51	52		
GLN ***	GLU	SER	VAL	ALA ***	ASP	LYS	MET	GLY ***	MET	GLY	GLN ---	SER ---	GLY	VAL ***	GLY	ALA	LEU	PHE	ASN	λ	R
GLN ***	THR	LYS	THR	ALA ***	LYS	ASP	LEU	GLY ***	VAL	TYR	GLN ---	SER ---	ALA	ILE ***	ASN	LYS	ALA	ILE	HIS	λ	CRO
GLN ***	ALA	ALA	LEU	GLY	LYS	MET	VAL	GLY ***	VAL	SER ---	ASN	VAL	ALA	ILE ***	SER ---	GLN ---	TRP	GLN	ARG	P22	R
GLN ***	ARG	ALA	VAL	ALA ***	LYS	ALA	LEU	GLY ***	ILE	SER ---	ASP	ALA	ALA	VAL ***	SER ---	GLN ---	TRP	LYS	GLU	P22	CRO
GLN ***	ALA	GLU	LEU	ALA ***	GLN	LYS	VAL	GLY ***	THR	THR	GLN ---	GLN ---	SER ---	ILE ***	GLU	GLN	LEU	GLU	ASN	434	R
GLN ***	THR	GLU	LEU	ALA ***	THR	LYS	ALA	GLY ***	VAL	LYS	GLN ---	GLN ---	SER ---	ILE ***	GLN	LEU	ILE	GLU	ALA	434	CRO
THR	GLU	LYS	THR	ALA ***	GLU	ALA	VAL	GLY ***	VAL	ASP	LYS	SER ---	GLN ---	ILE ***	SER ---	ARG ---	TRP	LYS	ARG	λ	cII
GLN ***	ARG	LYS	VAL	ALA ***	ASP	ALA	LEU	GLY ***	ILE	ASN	GLU	SER ---	GLN ---	ILE ***	SER ---	ARG ---	TRP	LYS	GLY	P22	c1
THR	GLU	LYS	THR	ALA ***	GLU	ALA	VAL	GLY ***	VAL	ASP	LYS	SER ---	GLN ---	ILE ***	SER ---	ARG ---	TRP	LYS	ARG	434	cII
ILE	LYS	ASP	VAL	ALA ***	ARG	LEU	ALA	GLY ***	VAL	SER	VAL	ALA	THR	VAL ***	SER	ARG	VAL	ILE	ASN	GAL	R
LEU	TYR	ASP	VAL	ALA ***	GLU	TYR	ALA	GLY ***	VAL	SER	TYR	GLN	THR	VAL ***	SER	ARG	VAL	VAL	ASN	LAC	R

HELIX 2 (33–39) | TURN (40–43) | HELIX 3 (44–52)

Figure 7 Alignment of DNA-binding protein sequences in regions corresponding to α-helices 2 and 3 of λ repressor and λ*cro* protein. (***) Highly conserved residues. (---) Residues predicted to contact the DNA in proteins (like λ repressor and λ*cro* protein) that recognize the same or similar binding sites (Sauer et al. 1982b). Note that residues 36, 40, 42, 47, and 50 in all of the aligned proteins tend to be hydrophobic. Positions 38, 43, 44, 48, and 49 tend to be hydrophilic. All of the sequences shown could readily form helix-turn-helix structures very similar to those of λ repressor and λ*cro* protein (see Fig. 6B).

1975; von Wilcken-Bergmann and Müller-Hill 1982), suggesting that these proteins also utilize a helix-turn-helix structure in binding to operator DNA.

INTERACTIONS OF RNA POLYMERASE WITH p_{RM} AND p_R

Promoter DNA Sequences

Surveys of the DNA sequences of more than 50 promoters recognized by *E. coli* RNA polymerase reveal preferred (consensus) sequences in two regions (Rosenberg and Court 1979; Siebenlist et al. 1980). One region is centered near −35 (35 bases preceding the transcription startpoint), whereas the other, the so-called Pribnow box (Pribnow 1975), is centered near −10. The consensus sequences and the corresponding sequences in p_R and p_{RM} are indicated in Figure 8. As data in Table 1 indicate (see below), p_R is a moderately strong promoter, whereas p_{RM} (in the absence of repressor) is very weak. The −10 region of p_R differs from the consensus sequence by only a single base, whereas the same region of p_{RM} differs from the consensus sequence at three positions. In the −35 region, p_R disagrees with the consensus sequence at only one site, but the −35 sequence of p_{RM} can be made to correspond with the consensus sequence only if the spacing of both sequences is interrupted by single-base gaps (Fig. 8).

Identification of nucleotides essential for p_{RM} function has been facilitated by analysis of *prm*⁻ mutations, which reduce p_{RM} activity, and an up-promoter mutation, *prmup*1 (Yen and Gussin 1973; Rosen and Gussin 1979; Meyer et al. 1980). In general, there is a strong correlation between the phenotypic effects of the mutations and their effects on the two conserved regions of the promoter: (1) *prmup*1 is an AT-to-GC change at −31. The mutant promoter is in better agreement with the conserved sequence than wild-type p_{RM} (see Fig. 8). Consistent with this change in sequence, levels of transcription from *prmup*1 in vivo are seven to eight times as great as wild-type levels in the absence of repressor and three to four times as great as wild-type levels in the presence of repressor (Meyer and Ptashne 1980; Meyer et al. 1980). (2) *prm*116, *prm*U31, and *prm*M104 are independently isolated mutations that change the same highly conserved CG pair at −33 to a TA pair; *prm*E37 is a change of a weakly conserved CG pair at −14 to a TA pair. All four mutations cause about a 20-fold decrease in p_{RM}-directed β-galactosidase synthesis in vivo and are defective in transcription initiation from p_{RM} in vitro (Meyer et al. 1975; Rosen and Gussin 1979; Rosen et al. 1980; Shih and Gussin 1982). (3) Both *prm*E104 and *prm*E93, which are located in o_{R2}, confer weak virulence on test phages. Studies of transcription in vitro (see below) indicate that *prm*E104 significantly affects the interaction of RNA polymerase with p_{RM} in the absence of repressor, but *prm*E93 appears to owe its (weak) Prm⁻ phenotype primarily to a defect in repressor activation of p_{RM}. Interestingly, *prm*E104, at −38, affects the first nucleotide of the −35 consensus sequence, whereas *prm*E93 is outside the consensus sequence altogether.

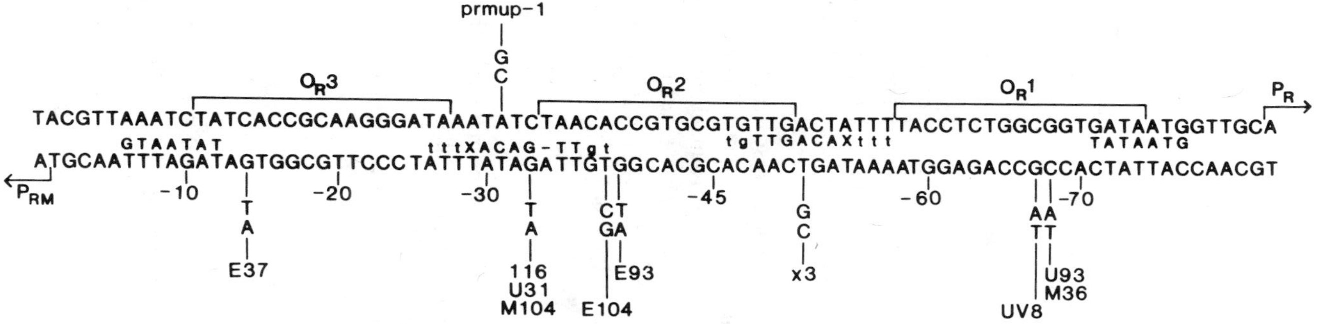

Figure 8 Nucleotide sequence of the p_{RM}-o_R-p_R region and changes caused by mutations affecting p_{RM} or p_R. The sequence shown is numbered with respect to the startpoint of the p_{RM} transcript at +1 (Ptashne et al. 1976; Walz et al. 1976) and extends to the startpoint for the p_R transcript (Blattner and Dahlberg 1972; Walz and Pirrotta 1975). The extent of the three repressor-binding sites in o_R is indicated by horizontal brackets above the sequence (Maniatis et al. 1975; Pirrotta 1975; Humayun et al. 1977a). All nucleotide sequence changes are for mutations affecting p_{RM} (Meyer et al. 1975, 1980; Rosen et al. 1980), except *x3* which is a p_R^- mutation at −32 with respect to the p_R transcription startpoint (Hawley 1982). Phenotypes of the mutations are described in the text. The consensus sequences for the Pribnow box (−10 region) and the −35 "recognition" region are indicated at the location of the corresponding sequences in p_{RM} and p_R. The hyphen in the −35 consensus sequence at p_{RM} indicates a 1-base gap that is necessary to align the consensus sequence with the actual sequence. Capital letters indicate bases that occur in promoters recognized by *E. coli* RNA polymerase with a frequency greater than 46% (three standard deviations from the mean; see Siebenlist et al. 1980). Lowercase letters indicate most frequent bases that occur with a frequency of 33–46%. X can be any base.

Table 1 Calculated kinetic parameters for transcription initiation at p_{RM}

Promoter	K_B (M^{-1})	k_2 (sec^{-1})
A. p_{RM} (no repressor)	9.9×10^6	7.0×10^{-4}
p_R	6.7×10^8	1.0×10^{-2}
T7 A1	1×10^9	1×10^{-1}
T7 A2	2×10^8	4×10^{-2}
B. p_{RM} (+ repressor)	9.2×10^6	7.8×10^{-3}
C. *prmup*1 (no repressor)	1.2×10^8	3.7×10^{-3}
*prmup*1 (+ repressor)	4.0×10^7	2.3×10^{-2}

Data pertaining to p_{RM} wild type, *prmup*1, and p_R were reported by Hawley and McClure (1982); data pertaining to T7 A1 and T7 A2 were taken from Hawley et al. (1982).

Three mutations, UV8, U93, and M36, which were isolated as Prm$^-$ mutations, are located in o_{R1} (Fig. 8). They confer partial virulence on test phages deleted for o_L and apparently owe their Prm$^-$ phenotypes to failure of repressor to activate p_{RM} (Rosen and Gussin 1979).

Several mutations in p_R have also been isolated (Eisen et al. 1970). One p_R^+ mutation, *x*3, has been studied in detail; it causes about a 20-fold reduction in transcription initiation at p_R in vivo and in vitro (Heinemann and Spiegelman 1970; Maurer et al. 1974; Hawley and McClure 1980). Like the *prm*$^-$ mutations, *x*3 is a single-base change (at -32), which decreases the agreement between p_R and the consensus sequence (Fig. 8) (Hawley 1982).

Biochemical Analysis of p_{RM}

According to current models there are two major steps in transcription initiation: binding (or recognition) to form a *closed* complex, followed by the essentially irreversible isomerization of the closed complex to form an *open* complex (see Chamberlin 1974, 1976). The two-step reaction is represented by the following scheme:

$$(1) \qquad E + D \rightleftharpoons E\text{-}D_c$$

$$(2) \qquad E\text{-}D_c \rightleftharpoons E\text{-}D_o$$

where enzyme (E) is RNA polymerase holoenzyme, DNA (D) is a specific promoter, and E-D_c and E-D_o denote closed and open complexes, respectively. Step 2, which appears to involve DNA strand separation (Chamberlin 1976; Hsieh and Wang 1977; Siebenlist 1979), is followed by very rapid initiation of RNA chain synthesis in the presence of ribonucleoside triphosphates (Mangel and Chamberlin 1974; McClure 1980).

A method recently developed by McClure (1980) permits the determination of K_B, the equilibrium constant for reaction 1, and k_2, the forward rate constant for reaction 2, in vitro. In studies of p_{RM} in the absence of repressor, values for K_B and k_2 (Table 1A) were found to be as much as 10–100 times lower than values found for other promoters, including p_R (Hawley et al. 1982; Hawley and McClure 1982). Thus, in the absence of repressor, p_{RM} is weak both because the initial binding of RNA polymerase is weak and because the isomerization of closed to open complexes is slow. These results are consistent with the observation that p_{RM} produces detectable levels of *c*I mRNA in vitro in the absence of repressor only when very high concentrations of RNA polymerase are used (Meyer et al. 1975; Meyer and Ptashne 1980). Under these conditions, saturation of the promoter with polymerase increases the overall level of RNA synthesis even though the slow isomerization step may still be rate-limiting.

This kind of analysis can also be used to determine which step in initiation is affected by repressor binding (Table 1). In the presence of repressor, the rate of isomerization increases 11-fold, whereas the initial binding of RNA polymerase to the promoter is unchanged (Hawley and McClure 1982). Thus, repressor activates p_{RM} by accelerating the transition from closed to open complexes. The magnitude of the effect is similar to the degree of stimulation of p_{RM} activity by repressor in vitro and in vivo (Meyer et al. 1975, 1980; Meyer and Ptashne 1980).

A similar analysis of the effects of *prm*$^-$ mutations on p_{RM} function is summarized in Table 2 (Shih and Gussin 1982, 1983). In the absence of repressor, *prm*U31 and *prm*E104, which are changes in the −35 region, decrease K_B substantially, without affecting k_2. On the other hand, *prm*E37, which is located near, but not in, the Pribnow box, decreases k_2 without affecting K_B (The mutation *prm*E93 affects K_B very slightly in the absence of repressor and has no ef-

Table 2 Calculated kinetic parameter for transcription initiation at wild-type and mutant p_{RM}

Mutant	Repressor	K_B (M^{-1})	k_2 (sec^{-1})
A. wild-type	0	3.0×10^7	1.7×10^{-3}
*prm*E37	0	2.7×10^7	*4.5×10^{-4}*
*prm*U31	0	*3.1×10^6*	2.3×10^{-3}
*prm*E104	0	*4.8×10^6*	2.1×10^{-3}
*prm*E93	0	2.1×10^7	1.7×10^{-3}
B. wild-type	+	2.5×10^7	1.2×10^{-2}
*prm*E37	+	2.5×10^7	*1.3×10^{-3}*
*prm*U31	+	*1.8×10^6*	1.3×10^{-2}
*prm*E104	+	*4.7×10^6*	1.2×10^{-2}
*prm*E93	+	1.2×10^7	*4.9×10^{-3}*

Parameters were determined by the method of McClure (1980). Data were taken from Shih and Gussin (1983). Significant differences from wild-type values are indicated by italics.

fect on k_2 [Table 2A]; thus, information essential for the interaction of RNA polymerase with p_{RM} may extend only as far as the *prm*E104 site [−38].)

The effects of the *prm*⁻ mutations on K_B and k_2 are virtually the same in the presence as in the absence of repressor (Table 2B). That is, *prm*U31 and *prm*E104 affect only K_B, whereas *prm*E37 affects only k_2. Comparison of the values of k_2 obtained in the presence of repressor (Table 2B) with the corresponding values obtained in its absence (Table 2A) indicates that *prm*U31 and *prm*E104 are stimulated normally by repressor, but *prm*E37, which is defective in the isomerization step, is not. For *prm*E93, k_2 does not respond very well to repressor, presumably because the mutation affects repressor binding at o_{R2}.

The results in Table 2 are consistent with the idea that the information required for transcription initiation from *E. coli* promoters is partitioned between the −35 and −10 regions (Gilbert 1976; Siebenlist et al. 1980); i.e., DNA sequences in the −35 region may be required specifically for initial recognition by RNA polymerase (the binding step), whereas sequences in the −10 region may be essential for formation of *open* complexes (the isomerization step).

However, the results of a similar analysis of *prmup*1 are not so straightforward. This mutation was originally selected as one that permitted p_{RM}-directed β-galactosidase synthesis in the absence of repressor (Meyer et al. 1980). Since repressor activation of wild-type p_{RM} involves enhancement of the isomerization rate, it might have been expected that *prmup*1 would map near the −10 region of the promoter and affect only the isomerization step in transcription initiation. Surprisingly, *prmup*1 is located at −33 and apparently affects both binding (K_B) and isomerization (k_2) (Table 1C). This behavior is not explained by the simple model discussed above but can be accommodated in a slightly revised model in which the −35 region of the promoter mediates two sequential steps that comprise the binding of RNA polymerase to the promoter (Hawley and McClure 1982). (In this model, the −10 region still provides information only for DNA strand separation [isomerization].) With this modification, the effect of *prump*1 can be ascribed entirely to the recognition reaction, affecting only K_B and not k_2. Since repressor stimulates the *prmup*1 promoter nearly as well as wild type, the activation of p_{RM} by repressor is apparently independent of the stimulatory effect of *prmup*1 (Meyer and Ptashne 1980; Meyer et al. 1980; Hawley and McClure 1982). Thus, since p_{RM} is a weak promoter as a consequence of defects in both isomerization and RNA polymerase binding, p_{RM} can be activated either by a repressor-mediated increase in the isomerization rate or by a mutation that increases the affinity of RNA polymerase binding to the promoter. In principle, it should also be possible to isolate p_{RM} mutations that create repressor independence primarily as a consequence of an increase in the intrinsic isomerization rate.

Molecular Basis of Repressor Activation of p_{RM}

How does repressor bound at o_R activate p_{RM}? One possibility is that repressor binding induces a conformational change in promoter-operator DNA which, in turn, affects the interaction of RNA polymerase with the promoter. Although this possibility cannot be completely ruled out, the available data suggest an alternative model in which specific protein-protein contacts between bound repressor and RNA polymerase play the major (if not exclusive) role in turning on p_{RM}.

Two sets of experiments support this model: First, repressor bound to o_{R2} and RNA polymerase bound to p_{RM} (in *open* complexes) appear to be mutually stabilizing (Johnson 1980; Hawley and McClure 1982; A.D. Johnson et al., unpubl.). Mutual stabilization of bound repressor and polymerase can be explained most simply by a direct interaction between the two proteins. Second, repressor missense mutations defective in positive control (*pc* phenotype) have been isolated and characterized (Guarente et al. 1982; Hawley and McClure 1983; Hochschild et al. 1983). These mutations prevent repressor activation of p_{RM} without significantly affecting the ability of repressor to bind to o_{R2}. This indicates that by itself, occupancy of o_{R2} by repressor is not sufficient to activate p_{RM} and suggests that specific protein-protein interactions must be required. The *pc* phenotype is also inconsistent with models in which repressor activates p_{RM} indirectly by preventing polymerase binding to p_R (see Johnson et al. 1981).[1]

Therefore, two functions of repressor are distinguishable by mutation: recognition of o_{R2} and activation of p_{RM}. Several lines of evidence indicate that the aminoterminal domain of repressor mediates both functions. (1) The isolated aminoterminal domain, at sufficiently high concentrations, activates p_{RM} both in vitro and in vivo (Sauer et al. 1979). (2) Mutual stabilization of the binding of repressor and RNA polymerase is observed when the isolated aminoterminal domain is substituted for intact repressor (A.D. Johnson, unpubl.). (3) The *pc* mutations change amino acids 34, 38, and 43 in the aminoterminal domain (Guarente et al. 1982; Hochschild et al. 1983). The positions of these changes are consistent with the idea that activation of p_{RM} is mediated by a direct interaction between bound repressor and RNA polymerase (Johnson et al. 1981; Hochschild et al. 1983).

Thus, it appears that binding of repressor dimers to o_{R2} positions repressor so that direct contacts can be made between an aminoterminal domain and RNA polymerase bound at p_{RM}. These contacts facilitate the transition between closed and open complexes. A repressor-dilution experiment indicates that once the transition has taken place, repressor is no longer needed (Hawley and McClure 1982). This observation is consistent with the fact that the transition (isomerization) reaction is essentially irreversible.

[1]If RNA polymerase binding at p_R interfered with transcription initiation from p_{RM}, repressor might activate p_{RM} merely by preventing polymerase binding at p_R. There is ample evidence that rules out this explanation as the sole basis for repressor activation of p_{RM}: (1) p_{RM} activity is repressor-dependent even when p_R contains the *x*3 mutation (Fig. 3), which reduces polymerase binding about 20-fold (Heinemann and Spiegelman 1970; Hawley and McClure 1980); (2) transcription from p_{RM} on templates deleted for most of p_R (but not o_{R2}) is stimulated by repressor to about the same extent as transcription from wild-type templates (Meyer and Ptashne 1980; Meyer et al. 1980); (3) the *pc*1 mutation prevents activation of p_{RM} by repressor, even though the ability of the repressor to bind to the operator is unimpaired (Guarente et al. 1982).

Why does repressor bound to o_{R2} activate p_{RM} but repress p_R? One possibility is that bound repressor just touches (and activates) RNA polymerase bound at p_{RM} but sterically interferes with polymerase binding at p_R. In fact, measured from each transcription start point, p_R is one nucleotide closer to o_{R2} than is p_{RM}; however, there is no direct evidence that the differences in distance are mechanistically significant (see Ptashne et al. 1976, 1980). Chemical probes have not been used to study the interaction of RNA polymerase with p_{RM} or p_R, but some information can be derived from studies of the Lac and T7 A3 promoters (Siebenlist et al. 1980). On the basis of sequence homologies with these promoters, it has been inferred that repressor bound to o_{R2} may contact only one phosphate (between -38 and -39) that is also contacted by polymerase bound to p_{RM} but as many as three phosphates that are contacted by polymerase bound to p_R (Johnson 1980; Johnson et al. 1981). These conclusions are speculative but nevertheless support the idea that the spatial relationships between p_R or p_{RM} and o_{R2} are different. Chemical and enzymatic protection experiments also suggest that the region of repressor binding to o_{R2} overlaps the region of polymerase binding to both promoters. However, the two proteins could bind to opposite faces of the DNA helix without actually touching or interfering with each other.

OPPOSING ROLES OF REPRESSOR AND *cro* PROTEIN

Although repressor and *cro* protein bind to the same three sites in o_R, they play very different roles in λ physiology. Their individual effects on transcription from p_R and p_{RM} can be understood in terms of their affinities for o_{R1}, o_{R2}, and o_{R3}, combined with the knowledge of how occupancy of each site affects either promoter. However, at first glance the organization of the entire system seems needlessly complex. Since many operons are controlled by a single repressor binding to a single operator, why has λ chosen to use both a repressor and a *cro* protein that bind to three operator sites? In the following discussion, we argue that much of this apparent complexity plays an important role in λ's ability to exist as a tightly repressed and stably maintained lysogen, which can nevertheless be rapidly and "quantitatively" induced by modest decreases in intracellular repressor levels (see Johnson et al. 1981).

Repressor exists, as far as one knows, only to maintain the lysogenic state; i.e., a *c*I$^-$ phage cannot lysogenize, but it exhibits normal or even improved lytic growth (Kaiser 1957). Most of the time in a lysogen, repressor dimers are bound to o_{R1} and o_{R2} (as well as o_{L1} and o_{L2}). However, since a repressor dimer bound at either o_{R1} or o_{R2} effectively represses p_R, why are two sites used to mediate repression? Theoretical calculations of the activity of p_R as a function of repressor monomer concentration (R) provide a plausible answer to this question (Ackers et al. 1982). A plot of the percentage of repression of p_R as a function of R is sigmoidal, with an unusually steep rise over a narrow range of repressor concentrations. Three factors contribute to this unusual steepness. (1) The concentration of active repressor dimers rises as a function of the square of the monomer concentration. (2) Occupancy of either o_{R1} or o_{R2} suffices to repress p_R. (3) Cooperative interactions between repressor dimers bound at o_{R1} and o_{R2} increase the probability that the two sites are simultaneously filled. At the repressor concentrations found in lysogens, these three effects combine to produce greater than 99% repression of p_R. Thus, repression is very tight. However, since the "repression curve" is steep, a modest decrease in repressor concentration should result in a significant increase in promoter activity. The calculations suggest that 50% maximal p_R activity can be induced by a tenfold reduction in the repressor concentration. In a system controlled by a single operator, much greater reduction in repressor levels would be required to achieve 50% induction. The degree of p_R activity required for induction of a significant proportion of a population of lysogens is not known, but experimental measurements show that a tenfold reduction in repressor levels results in virtually complete induction (Bailone et al. 1979); thus, a twofold or threefold reduction in repressor levels may be significant. Whatever the actual value, the system seems to be designed to allow very tight repression but to be easily inducible.

In addition to repression at p_R, repressor has a second role in maintaining the lysogenic state; it stimulates transcription of its own gene. As discussed above, this effect is mediated directly by a repressor dimer bound at o_{R2} and, indirectly, through cooperative stabilization, by a repressor dimer bound at o_{R1}. Why is repressor required for its own synthesis? Again, it seems that this design facilitates phage induction. When the intracellular repressor concentration falls below the threshold required for o_{R1} and o_{R2} occupancy, not only is p_R derepressed but repressor synthesis from p_{RM} also decreases. Thus, repression cannot be reestablished easily. These properties can be contrasted with those of the *lac* operon in which repressor continues to be synthesized constitutively when the operon is induced. Thus, expression of the *lac* operon requires the continuous presence of inducer, and repression can be immediately reestablished when the inducer is removed.

Since a modest reduction in repressor levels potentially can lead to induction, the rate of spontaneous λ induction might be expected to be high. In fact, it is *low* (~ 1 in 10^5), suggesting that repressor levels rarely drop below the threshold concentration required for induction. How are intracellular repressor levels maintained at a low level (~ 200 monomers/cell) without random fluctuations to inducing levels? The answer seems to be that fully stimulated p_{RM} directs the synthesis of a reasonable level of *c*I mRNA but that the resulting transcript is inefficiently translated (Ptashne et al. 1976). Repressor mRNA is translated only about 5% as efficiently as *lacZ* mRNA. This results in a low but steady rate of synthesis of repressor that presumably just balances the rate of repressor turnover. Again, this can be contrasted with the *lac* operon, in which the repressor gene is inefficiently transcribed but efficiently and repeatedly translated. Although this yields an *average* low level of *lac* repressor, repressor levels in individual cells probably fluctuate substantially, since each transcription event is followed by a burst of repressor synthesis.

The major functions of repressor in maintaining the lysogenic state are effected by its binding to o_{R1} and o_{R2}. In lysogens, o_{R3} is occupied by repressor only about 20% of the time (Maurer et al. 1980). In principle, if an excess of repressor accumulated in the cell, o_{R3} would become significantly occupied, and p_{RM} would be shut off until repressor levels dropped. Thus, a mechanism exists for imposing a ceiling on repressor levels. However, it is not clear whether this mechanism is physiologically significant. For example, a fivefold increase in intracellular repressor concentration only turns down p_{RM} by about 20%.

When repressor levels in a lysogen begin to fall, e.g., as a consequence of *recA* cleavage, p_R is the first phage promoter that is derepressed. (This is due to the fact that repressor dimers bind more tightly to the operator sites in o_L than to the operator sites in o_R.) As p_R is derepressed, *cro* protein is synthesized, and when sufficient concentrations accumulate, it first binds to o_{R3}, its strongest binding site. This prevents any further p_{RM}-directed *c*I transcription and probably commits the phage to lytic development. Two experiments support the idea that the action of *cro* protein at o_{R3} is required for efficient induction (Johnson 1980). First, an o_{R3}^- prophage is induced less efficiently by ultraviolet light than is a wild-type prophage. Second, a wild-type prophage can be induced at high efficiency by *cro* protein supplied in *trans*, but a prophage mutant that is defective in o_{R3} does not undergo this *cro*-mediated induction. The mechanism of the *cro*-mediated induction of the wild-type prophage is probably turnoff of p_{RM}. Then, as repressor levels fall, either as a consequence of turnover or dilution caused by cell division, p_R becomes derepressed and the lytic program begins.

At some point in the induction process, *cro* protein must be bound at o_{R3}, but o_{R1} and o_{R2} must remain substantially unoccupied. This state is normally a transient intermediate in lytic development, but a prophage can be frozen in this state if it carries mutations that prevent expression of lytic genes other than *cro*. In this state *cro* protein continues to be synthesized, but repressor synthesis is turned off. This new state, called the "antiimmune" state, can be stably inherited by subsequent generations (Eisen et al. 1968, 1970; Neubauer and Calef 1970; Spiegelman 1971). The levels of *cro* protein in such strains are not sufficient to render the cell immune to λ infection but, rather, the infecting phage all multiply lytically.

cro protein has a second, and perhaps equally crucial, role in λ physiology. It is required for turnoff of early genes when the continued synthesis of the products of these genes is no longer necessary. *cro* protein bound at o_{R1} and o_{R2} represses early rightward synthesis from p_R, whereas *cro* protein bound at o_{L1} and o_{L2} represses early leftward synthesis from p_L. In λ*cro*⁻ phage, overexpression of an early leftward gene product (*tro*) poisons lytic multiplication, whereas unchecked p_R transcription may also be lethal, perhaps because it interferes with DNA replication (Eisen et al. 1975).

In summary, the rightward operator-promoter region of phage λ has been designed to allow both the stable existence of the lysogenic state, as long as sufficient repressor concentrations are maintained, and an efficient and effectively irreversible switch to lytic development when repressor levels fall. The state of the system depends on the concentrations of both repressor and *cro* protein and on the interactions of these proteins with the three sites in o_R. Repressor bound at o_{R1} and o_{R2}—the lysogenic configuration—turns off p_R-directed synthesis of *cro* and other early proteins and stimulates transcription of its own gene, *c*I, from p_{RM}. A modest decrease in repressor levels causes a decrease in p_{RM}-directed repressor synthesis and an increase in p_R-directed *cro* gene expression. The newly synthesized *cro* protein binds to o_{R3}, where it prevents further repressor synthesis and effectively commits the phage to lytic development. This allows a λ prophage to be induced efficiently in response to a transient inducing signal.

REFERENCES

Ackers, G.K., M.A. Shea, and A.D. Johnson. 1982. Quantitative model for gene regulation by λ phage repressor. *Proc. Natl. Acad. Sci.* **79:** 1129.

Anderson, W.F., D.H. Ohlendorf, Y. Takeda, and B.W. Matthews. 1981. Structure of the cro repressor from bacteriophage λ and its interaction with DNA. *Nature* **290:** 754.

Anderson, W.F., Y. Takeda, H. Echols, and B.W. Matthews. 1979. The structure of a repressor: Crystallographic data for the cro regulatory protein of bacteriophage λ. *J. Mol. Biol.* **130:** 507.

Aono, J. and T. Horiuchi. 1977. Purification and some properties of presumptive *tof* gene product of *Coli* phage 434. *Mol. Gen. Genet.* **156:** 195.

Bailone, A. and R. Devoret. 1978. Isolation of ultravirulent mutants of phage λ. *Virology* **84:** 547.

Bailone, A., A. Levine, and R. Devoret. 1979. Inactivation of prophage λ repressor *in vivo. J. Mol. Biol.* **131:** 553.

Ballivet, M., L.F. Reichardt, and H. Eisen. 1977. Purification and properties of coliphage 21 repressor. *Eur. J. Biochem.* **73:** 601.

Berg, D.E. 1974. Genes of phage λ essential for λdv plasmids. *Virology* **62:** 224.

Beyreuther, K., K. Adler, E. Fanning, C. Murray, A. Klemm, and N. Geisler. 1975. Amino acid sequence of the *lac* repressor from *Escherichia coli. Eur. J. Biochem.* **59:** 491.

Blattner, F.R. and J.E. Dahlberg. 1972. RNA synthesis startpoints in bacteriophage λ: Are the operator and promoter transcribed? *Nat. New Biol.* **237:** 227.

Bode, V.C. and A.D. Kaiser. 1965. Repression of the *c*II and *c*III cistrons of phage lambda in a lysogenic bacterium. *Virology* **25:** 111.

Brack, C. and V. Pirrotta. 1975. Electron microscopic study of the repressor of bacteriophage λ and its interaction with operator DNA. *J. Mol. Biol.* **96:** 139.

Castellazzi, M., P. Brachet, and H. Eisen. 1972. Isolation and characterization of deletions in bacteriophage λ. *Mol. Gen. Genet.* **117:** 211.

Chadwick, P., V. Pirrotta, R. Steinberg, N. Hopkins, and M. Ptashne. 1971. The λ and 434 phage repressors. *Cold Spring Harbor Symp. Quant. Biol.* **35:** 283.

Chamberlin, M. 1974. The selectivity of transcription. *Annu. Rev. Biochem.* **43:** 721.

——. 1976. Interaction of RNA polymerase with the DNA template. In *RNA polymerase* (ed. R. Losick and M. Chamberlin), p. 159. Cold Spring Harbor Laboratory, Cold Spring Harbor, New York.

Champoux, J.J. 1971. The sequence and orientation of transcription in bacteriophage λ. *Cold Spring Harbor Symp. Quant. Biol.* **35:** 319.

DeAnda, J., A.R. Poteete, and R.T. Sauer. 1983. P22 *c*2 repressor: Domain structure and function. *J. Biol. Chem.* (in press).

Echols, H. and L. Green. 1971. Establishment and maintenance of repression by bacteriophage lambda: The role of the *c*I, *c*II, and *c*III proteins. *Proc. Natl. Acad. Sci.* **68:** 2190.

Echols, H., L. Green, A.B. Oppenheim, A. Oppenheim, and A. Honigman. 1973. Role of the *cro* gene in bacteriophage λ development. *J. Mol. Biol.* **80:** 203.

Eisen, H. and M. Ptashne. 1971. Regulation of repressor synthesis. In *The bacteriophage lambda* (ed. A.D. Hershey), p. 239. Cold Spring Harbor Laboratory, Cold Spring Harbor, New York.

Eisen, H., L. Pereira da Silva, and F. Jacob. 1968. Sur la règulation prècoce du bacteriophage λ. *C. R. Acad. Sci. Paris* **266:** 1176.

Eisen, H., P. Brachet, L. Pereira da Silva, and F. Jacob. 1970. Regulation of repressor expression in λ. *Proc. Natl. Acad. Sci.* **66:** 855.

Eisen, H., M. Georgiou, C.P. Georgopoulos, G. Selzer, G. Gussin, and I. Herskowitz. 1975. The role of gene *cro* in phage development. *Virology* **68:** 266.

Flashman, S.M. 1978. Mutational analysis of the operators of bacteriophage lambda. *Mol. Gen. Genet.* **166:** 61.

Folkmanis, A., Y. Takeda, J. Simuth, G. Gussin, and H. Echols. 1976. Purification and properties of a DNA-binding protein with characteristics expected for the Cro protein of bacteriophage λ, a repressor essential for lytic growth. *Proc. Natl. Acad. Sci.* **73:** 2249.

Gilbert, W. 1976. Starting and stopping sequences for the RNA polymerase. In *RNA polymerase* (ed. R. Losick and M. Chamberlin), p. 193. Cold Spring Harbor Laboratory, Cold Spring Harbor, New York.

Grosschedl, R. and E. Schwarz. 1979. Nucleotide sequence of the cro-cII-oop region of bacteriophage 434 DNA. *Nucleic Acids Res.* **6:** 867.

Guarente, L., J.S. Nye, A. Hochschild, and M. Ptashne. 1982. Mutant λ repressor with a specific defect in its positive control function. *Proc. Natl. Acad. Sci.* **79:** 2236.

Gussin, G.N., K.-M. Yen, and L. Reichardt. 1975. Repressor synthesis *in vivo* after infection of *E. coli* by a p_{rm}^- mutant of bacteriophage lambda. *Virology* **63:** 273.

Hawley, D. 1982. "Control of transcription initiation frequency from the p_R and p_{RM} promoters of bacteriophage λ." Ph.D. thesis, Harvard University, Cambridge, Massachusetts.

Hawley, D.K. and W.R. McClure. 1980. *In vitro* comparison of initiation properties of bacteriophage λ wild-type P_R and *x*3 mutant promoters. *Proc. Natl. Acad. Sci.* **77:** 6381.

———. 1982. Mechanism of activation of transcription initiation from the λ P_{RM} promoter. *J. Mol. Biol.* **157:** 493.

———. 1983. The effect of a lambda repressor mutation on the activation of transcription initiation from the lambda P_{RM} promoter. *Cell* **32:** 327.

Hawley, D.K., T.P. Malan, M.E. Mulligan, and W.R. McClure. 1982. Intermediates on the pathway to open complex formation. In *Promoters: Structure and function* (ed. R. Rodriguez and M. Chamberlin), p. 54. Praeger Press, New York.

Hecht, M.H., H.C.M. Nelson, and R.T. Sauer. 1983. Mutations in lambda repressor's amino-terminal domain: Implications for protein stability and DNA binding. *Proc. Natl. Acad. Sci.* **80:** (in press).

Heinemann, S. and W. Spiegelman. 1970. Control of transcription of the repressor gene in bacteriophage lambda. *Proc. Natl. Acad. Sci.* **67:** 1122.

Hochschild, A., N. Irwin, and M. Ptashne. 1983. Repressor structure and the mechanism of positive control. *Cell* **32:** 319.

Hopkins, N. and M. Ptashne. 1971. Genetics of virulence. In *The bacteriophage lambda* (ed. A.D. Hershey), p. 571. Cold Spring Harbor Laboratory, Cold Spring Harbor, New York.

Horiuchi, T., H. Koga, H. Inokuchi, and J. Tomizawa. 1969. Lambda phage mutants insensitive to temperature-sensitive repressor. I. Isolation and genetic analysis of weak-virulent mutants. *Mol. Gen. Genet.* **104:** 51.

Hsiang, M.W., Y. Takeda, R.D. Cole, and H. Echols. 1978. Amino acid sequence of the Cro regulatory protein of bacteriophage λ. *Nature* **270:** 275.

Hsieh, T. and J.C. Wang. 1977. Physiochemical studies on interactions between DNA and RNA polymerase. Ultraviolet absorption measurements. *Nucleic Acids Res.* **5:** 3337.

Humayun, Z., A. Jeffrey, and M. Ptashne. 1977a. Completed DNA sequences and organization of repressor-binding sites in the operators of phage lambda. *J. Mol. Biol.* **112:** 265.

Humayun, Z., D. Kleid, and M. Ptashne. 1977b. Sites of contact between λ operators and λ repressor. *Nucleic Acids Res.* **4:** 1595.

Johnson, A.D. 1980. "Mechanism of action of the lambda cro protein." Ph.D. thesis, Harvard University, Cambridge, Massachusetts.

——. 1983. Interactions of the λ repressor and the λ *cro* protein with operator DNA. *J. Mol. Biol.* (in press).

Johnson, A.D., B.J. Meyer, and M. Ptashne. 1978. Mechanism of action of the *cro* protein of bacteriophage λ. *Proc. Natl. Acad. Sci.* **75:** 1783.

——. 1979. Interactions between DNA-bound repressors govern regulation by the λ phage repressor. *Proc. Natl. Acad. Sci.* **76:** 5061.

Johnson, A.D., C.O. Pabo, and R.T. Sauer. 1980. Bacteriophage λ repressor and cro protein: Interactions with operator DNA. *Methods Enzymol.* **65:** 839.

Johnson, A.D., A.R. Poteete, G. Lauer, R.T. Sauer, G.K. Ackers, and M. Ptashne. 1981. λ repressor and cro—Components of an efficient molecular switch. *Nature* **294:** 217.

Kaiser, A.D. 1957. Mutations in a temperate bacteriophage affecting its ability to lysogenize. *Virology* **3:** 42.

Kourilsky, P., L. Marcaud, P. Sheldrick, D. Luzzati, and F. Gros. 1968. Studies on the messenger RNA of bacteriophage λ. I. Various species synthesized early after induction of the prophage. *Proc. Natl. Acad. Sci.* **61:** 1013.

Kumar, S., E. Calef, and W. Szybalski. 1971. Regulation of the transcription of *Escherichia coli* phage λ by its early genes N and *tof. Cold Spring Harbor Symp. Quant. Biol.* **35:** 331.

Lieb, M. 1976. λ *c*I mutants: Intragenic complementation and complementation with a *c*I promoter mutant. *Mol. Gen. Genet.* **146:** 291.

Mangel, W. and M. Chamberlin. 1974. Studies of ribonucleic acid chain initiation by *Escherichia coli* ribonucleic acid polymerase bound to T7 deoxyribonucleic acid. I. An assay for the rate and extent of RNA chain initiation. *J. Biol. Chem.* **249:** 2995.

Maniatis, T. and M. Ptashne. 1973. Structure of the λ operators. *Nature* **246:** 133.

Maniatis, T., M. Ptashne, K. Backman, D. Kleid, S. Flashman, A. Jeffrey, and R. Mauer. 1975. Recognition sequences of repressor and polymerase in the operators of bacteriophage lambda. *Cell* **5:** 109.

Matthews, B.W., D.H. Ohlendorf, W.F. Anderson, and Y. Takeda. 1982. Structure of the DNA-binding region of *lac* repressor inferred from its homology with *cro* repressor. *Proc. Natl. Acad. Sci.* **79:** 1428.

Maurer, R., T. Maniatis, and M. Ptashne. 1974. Promoters are in the operators of phage lambda. *Nature* **249:** 221.

Maurer, R., B.J. Meyer, and M. Ptashne. 1980. Gene regulation at the right operator (o_R) of phage λ. I. o_R3 and autogenous negative control by repressor. *J. Mol. Biol.* **139:** 147.

McClure, W.R. 1980. Rate-limiting steps in RNA chain initiation. *Proc. Natl. Acad. Sci.* **77:** 5634.

Meyer, B.J. and M. Ptashne. 1980. Gene regulation at the right operator (o_R) of phage λ. III. λ repressor directly activates gene transcription. *J. Mol. Biol.* **139:** 195.

Meyer, B.J., D.G. Kleid, and M. Ptashne. 1975. λ Repressor turns off transcription of its own gene. *Proc. Natl. Acad. Sci.* **72:** 4785.

Meyer, B.J., R. Maurer, and M Ptashne. 1980. Gene regulation at the right operator (o_R) of bacteriophage λ. II. o_R1, o_R2, and o_R3: Their roles in mediating the effects of repressor and cro. *J. Mol. Biol.* **139:** 163.

Nelson, H.C.M., M.H. Hecht, and R.T. Sauer. 1983. Mutations defining the operator binding sites of bacteriophage λ repressor. *Cold Spring Harbor Symp. Quant. Biol.* **47:** 441.

Neubauer, Z. and E. Calef. 1970. Immunity phase-shift in defective lysogens: Non-mutational hereditary change of early regulation of λ prophage. *J. Mol. Biol.* **51:** 1.

Ordal, G.W. and A.D. Kaiser. 1973. Mutations in the right operator of bacteriophage lambda: Evidence for operator-promoter interpenetration. *J. Mol. Biol.* **79:** 709.

Pabo, C.O. 1980. "Molecular structure of the λ repressor." Ph.D. thesis, Harvard University, Cambridge, Massachusetts.

Pabo, C.O. and M. Lewis. 1982. The operator-binding domain of λ repressor-structure and DNA recognition. *Nature* **298:** 443.

Pabo, C.O., W. Krovatin, A. Jeffrey, and R.T. Sauer. 1982. N-terminal arms of λ repressor wrap around the operator DNA. *Nature* **298:** 441.

Pabo, C.O., R.T. Sauer, J.M. Sturtevant, and M. Ptashne. 1979. The λ repressor contains two domains. *Proc. Natl. Acad. Sci.* **76:** 1608.

Pirrotta, V. 1975. Sequence of the O_R operator of phage λ. *Nature* **254:** 114.

Pirrotta, V., P. Chadwick, and M. Ptashne. 1970. Active form of two coliphage repressors. *Nature* **227:** 41.

Pribnow, D. 1975. Bacteriophage T7 early promoters: Nucleotide sequences of two RNA polymerase binding sites. *J. Mol. Biol.* **99:** 419.

Ptashne, M. 1971. Repressor and its action. In *The bacteriophage lambda* (ed. A.D. Hershey), p. 221. Cold Spring Harbor Laboratory, Cold Spring Harbor, New York.

——. 1978. λ Repressor function and structure. In *The operon* (ed. J.H. Miller and W.S. Reznikoff), p. 325. Cold Spring Harbor Laboratory, Cold Spring Harbor, New York.

Ptashne, M. and N. Hopkins. 1968. The operators controlled by the λ phage repressor. *Proc. Natl. Acad. Sci.* **60:** 1282.

Ptashne, M., K. Backman, M.Z. Humayun, A. Jeffrey, R. Maurer, B. Meyer, and R.T. Sauer. 1976. Autoregulation and function of a repressor in bacteriophage lambda. *Science* **194:** 156.

Ptashne, M., A. Jeffrey, A.D. Johnson, R. Maurer, B.J. Meyer, C.O. Pabo, T.M. Roberts, and R.T. Sauer. 1980. How the λ repressor and cro work. *Cell* **19:** 1.

Reichardt, L.F. 1975a. Control of bacteriophage lambda repressor synthesis after phage infection: The role of the *N*, *c*II, *c*III, and *cro* products. *J. Mol. Biol.* **93:** 267.

——. 1975b. Control of bacteriophage λ repressor synthesis: Regulation of the maintenance pathway by the *cro* and *c*I products. *J. Mol. Biol.* **93:** 289.

Reichardt, L. and A.D. Kaiser. 1971. Control of λ repressor synthesis. *Proc. Natl. Acad. Sci.* **68:** 2185.

Roberts, J.W. and C.W. Roberts. 1975. Proteolytic cleavage of bacteriophage lambda repressor in induction. *Proc. Natl. Acad. Sci.* **72:** 147.

Roberts, T.M., H. Shimatake, C. Brady, and M. Rosenberg. 1977. The nucleotide sequence of the *cro* gene of bacteriophage lambda. *Nature* **270:** 274.

Rosen, E.D. and G.N. Gussin. 1979. Clustering of Prm⁻ mutations of bacteriophage λ in the region between 33 and 40 nucleotides from the *c*I transcription start point. *Virology* **98:** 393.

Rosen, E.D., J.L. Hartley, K. Matz, B.P. Nichols, K.M. Young, J.E. Donelson, and G.N. Gussin. 1980. DNA sequence analysis of *prm*⁻ mutations of coliphage lambda. *Gene* **11:** 197.

Rosenberg, M. and D. Court. 1979. Regulatory sequences involved in the promotion and termination of RNA transcription. *Annu. Rev. Genet.* **13:** 319.

Sauer, R.T. 1978. DNA sequence of the bacteriophage λ *c*I gene. *Nature* **276:** 301.

——. 1979. "Molecular characterization of the λ repressor and its gene *c*I." Ph.D. thesis, Harvard University, Cambridge, Massachusetts.

Sauer, R.T. and R. Anderegg. 1978. Primary structure of the λ repressor. *Biochemistry* **17:** 1092.

Sauer, R.T., M.J. Ross, and M. Ptashne. 1982a. Cleavage of the λ and P22 repressors by RecA protein. *J. Biol. Chem.* **257:** 4458.

Sauer, R.T., C.O. Pabo, B.J. Meyer, M. Ptashne, and K.C. Backman. 1979. Regulatory functions of the λ repressor reside in the amino terminal domain. *Nature* **279:** 396.

Sauer, R.T., R. Yocum, R. Doolittle, M. Lewis, and C. Pabo. 1982b. Homology among DNA-binding proteins suggests use of a conserved super-secondary structure. *Nature* **298:** 447.

Sauer, R.T., J. Pan, P. Hopper, K. Hehir, J. Brown, and A.R. Poteete. 1981. Primary structure of the phage P22 repressor and its gene *c*2. *Biochemistry* **20:** 3591.

Schwarz, E., G. Scherer, G. Hobom, and K. Kössel. 1978. Nucleotide sequence of *cro*, *c*II and part of the *O* gene in phage λ DNA. *Nature* **272:** 410.

Shih, M.-C. and G.N. Gussin. 1982. Mutations that affect two different steps in transcription initiation at the p_{RM} promoter of bacteriophage lambda. In *Promoters: Structure and function* (ed. R. Rodriguez and M. Chamberlin), p. 96. Praeger Press, New York.

——. 1983. Mutations affecting two different steps in transcription initiation at the phage P_{RM} promoter. *Proc. Natl. Acad. Sci.* **80:** 496.

Siebenlist, U. 1979. RNA polymerase unwinds an 11 base pair segment of a T7 promoter. *Nature* **279:** 651.

Siebenlist, U., R.B. Simpson, and W. Gilbert. 1980. *E. coli* RNA polymerase interacts homologously with two different promoters. *Cell* **20:** 269.

Spiegelman, W.G. 1971. Two states of expression of genes *c*I, *rex*, and *N* in lambda. *Virology* **43:** 16.

Spiegelman, W.G., L.F. Reichardt, M. Yaniv, S.F. Heinemann, A.D. Kaiser, and H. Eisen. 1972. Bidirectional transcription and the regulation of phage λ repressor synthesis. *Proc. Natl. Acad. Sci.* **69:** 3156.

Steinberg, R.A. and M. Ptashne. 1971 *In vitro* repression of RNA synthesis by purified λ phage repressor. *Nat. New Biol.* **230:** 76.

Szybalski, W., K. Bøvre, M. Fiandt, S. Hayes, Z. Hradecna, S. Kumar, H.A. Lozeron, H.J.J. Nijkamp, and W.F. Stevens. 1971. Transcriptional units and their controls in *Escherichia coli* phage λ: Operons and scriptons. *Cold Spring Harbor Symp. Quant. Biol.* **35:** 341.

Takeda, Y. 1979. Specific repression of *in vitro* transcription by the cro repressor of bacteriophage lambda. *J. Mol. Biol.* **127:** 177.

Takeda, Y., A. Folkmanis, and H. Echols. 1977. Cro regulatory protein specified by bacteriophage λ. *J. Biol. Chem.* **252:** 6177.

von Wilcken-Bergmann, B. and B. Müller-Hill. 1982. Sequence of *galR* gene indicates a common evolutionary origin of *lac* and *gal* repressor in *Escherichia coli. Proc. Natl. Acad. Sci.* **79:** 2427.

Walz, A. and V. Pirrotta. 1975. Sequence of the P_R promoter of phage λ. *Nature* **254:** 118.

Walz, A., V. Pirrotta, and K. Ineichen. 1976. λ repressor regulates the switch between P_R and P_{RM} promoters. *Nature* **262:** 665.

Wu, A.M., S. Ghosh, H. Echols, and W.G. Spiegelman. 1972. Repression by the *c*I protein of phage λ: *In vitro* inhibition of RNA synthesis. *J. Mol. Biol.* **67:** 407.

Yen, K.-M. and G.N. Gussin. 1973. Genetic characterization of a *prm*⁻ mutant of bacteriophage lambda. *Virology* **56:** 300.

Lysogenic Induction

Jeffrey W. Roberts
Section of Biochemistry, Molecular and Cell Biology
Cornell University
Ithaca, New York 14853

Raymond Devoret
Laboratoire d'Enzymologie
Centre Nacional de la Recherche Scientifique
91190 Gif-sur-Yvette, France

Thus, the problem of the mechanism of induction is now close to its solution. (Lwoff 1966)

INDUCIBLE PROPHAGES

Most bacteria found in nature are lysogenic (Lwoff 1953). They carry the genome of a virus in a dormant state, which is then called a prophage. The term lysogenic derives from the ability of these bacteria to generate the lysis of related species. The fact that the supernatant of a culture of lysogenic bacteria (or lysogens) causes lysis of other bacteria implies two properties of the lysogenic state. (1) In a growing culture of a lysogen, some cells spontaneously release phage particles. (2) Lysogenic bacteria are constantly exposed to the released phage but are not themselves lysed; they are said to be immune. Released phage can be revealed by infection of other sensitive bacteria that serve as indicator.

In trying to understand spontaneous production of phage by lysogens, Lwoff et al. (1950) sought ways to increase phage production artifically. Their first success followed exposure of lysogenic *Bacillus megaterium* to UV light. The culture lysed en masse, liberating hundreds of phage per cell. Since vegetative growth of a virus had been generated from within the bacterium by some as yet unknown action of UV light, this process was called lysogenic induction.

Soon after the discovery of lysogenic induction in *Bacillus, Escherichia coli* K12 was found to carry prophage λ, a prophage also inducible by UV light (Weigle and Delbrück 1951). Since *E. coli* K12 lent itself to genetic analysis, phage λ immediately became a model temperate bacteriophage for the study of lysogenic induction. About half of the temperate phages of distinct immunity types that lysogenize enteric bacteria are inducible by UV light (Table 1; also see Hershey and Dove, this volume).

PROPHAGE INDUCTION IS CAUSED BY REPRESSOR INACTIVATION

The λcI Gene Is A Repressor That Maintains the Prophage State and Determines Immunity

Whether or not they are inducible by UV light, prophages are all dormant viruses. Genetic and physiological studies first suggested that this state is maintained by a repressor.

Table 1 Inducibility of some prophages in three Enterobacteriaceae

Enterobacteriaceae	Inducible prophages	Noninducible prophages
Escherichia coli	λ, 434, 21, 80, 82, 170, 186[a], 424[b]	18, Mu, 62, 299
Salmonella typhimurium	P22, L	
Shigella dysenteriae	P1[c]	P2, P4

[a]Fully inducible by UV light, prophage 186 does not display zygotic induction, because the conjugating female is temporarily refractory to phage infection. The delay enables repression to be established before the cell recovers its sensitivity to phage 186 infection (Woods and Egan 1981).

[b]Prophage 424 has properties similar to phage 186 (P. Morand and R. Devoret, unpubl.).

[c]Data from Melechen and Skaar (1962).

When prophage λ is introduced by conjugal transfer into a nonlysogenic recipient, it is immediately induced to produce progeny phage, a process called zygotic induction. In contrast, it is not induced when the recipient bacterium is a λ lysogen. The prophage is not induced by the reciprocal cross, in which DNA of a nonlysogenic Hfr is transferred into a λ lysogenic recipient. The explanation for this lack of symmetry is that a cytoplasmic substance represses gene expression in *cis* to ensure the dormant state of the prophage and in *trans* to confer immunity to the host by preventing the development of a superinfecting phage homologous to the prophage (Jacob and Wollman 1961). The phenomenon of zygotic induction provided strong genetic evidence in favor of the concept of negative regulation (Jacob and Monod 1961).

The behavior of various λ*c*I mutants has suggested that this gene encodes a repressor protein that confers immunity. The most direct evidence was the existence of temperature-sensitive *c*I mutations such as *c*I857, which allow prophage carrying them to be maintained stably only at low temperature and cause development of phage to begin immediately when the temperature is raised (Sussman and Jacob 1962). Such mutations exist in many temperate phages, whether or not they are inducible by UV light, and provide a means to prepare phage easily. (Although this process is called "induction," in this paper we mean by lysogenic induction the cellular process initiated by agents such as UV light.)

Ptashne (1967) isolated the λ*c*I-gene product and showed that it is a repressor. The λ*c*I repressor binds to two operators flanking genes *c*I and *rex* (Maniatis et al. 1974), preventing transcription of other prophage genes and thus blocking phage development (Friedman and Gottesman, this volume). Purification of the *c*I repressor of phage 434, a temperate phage related to λ, confirmed the common regulatory properties of repressors of the so-called lambdoid phages (Pirrotta et al. 1970). (For the sake of brevity, we will call the *c*I repressor of prophage λ the λ repressor.)

The Repressor Is the Target of the Cellular Inducer

Given that the prophage state is maintained by a repressor coded by the prophage, lysogenic induction must result from the action of a cellular inducer that either destroys repressor or interferes with its function. The existence of mutations in the λ repressor gene that affect induction but not repression suggested that the repressor is the target of the cellular inducer. First, *cIind*$^-$ mutations, e.g., *cIind*1 (Jacob and Campbell 1959), render phage λ noninducible by UV light (Eshima et al. 1972; Lieb 1981). Second, the mutation λ*cIind*^{s} (Horiuchi and Inokuchi 1967) has the contrasting effect of allowing induction at much lower doses of UV light than are required for lysogens of wild-type prophage.

Kinetics of λ Repressor Inactivation Promoted by UV Light

The first biochemical demonstration of repressor inactivation in lysogenic induction was by Shinagawa and Itoh (1973), who used the repressor-operator filter binding assay to show that repressor activity disappears in cells after an inducing treatment. Repressor activity disappears by 30 minutes (Tomizawa and Ogawa 1967; Bailone et al. 1979). This is consistent with the fact that lysis after UV irradiation is delayed 30 minutes relative to lysis after abrupt inactivation of repressor by temperature shift of a λ*c*I857 lysogen.

In a lysogen there are about 200 λ repressor monomers (Lévine et al. 1979). Lysogenic induction occurs when less than 10% of these remain in the cell (Bailone et al. 1979) to block the prophage operator sites (Johnson et al. 1981). Since inactivation of repressor occurs over nearly a cell generation, induction requires that a few hundred molecules be inactivated per cell in 30 minutes.

The Inducer Is Generated by a Cellular Process

Development of the inducer occurs by a cellular mechanism, independent of the prophage. This was indicated early by an ingenious experiment (Ogawa and Tomizawa 1967) using an abortive lysogen (λ*b*2) that allows the prophage to segregate away, leaving nonlysogenic cells containing repressor (see Weisberg and Landy, this volume). Wild-type repressor in these cells, detected by immunity to infection, is destroyed after UV irradiation in a process requiring protein synthesis and metabolic energy. In contrast, inactivation of *c*I857 repressor by heat does not require protein synthesis. The phenotypes of *ind*^{s} and *ind*$^-$ lysogens are maintained in cells that have segregated away the prophages, showing clearly that both mutations affect induction by modifying the repressor protein (Tomizawa and Ogawa 1967).

Agents or Treatments That Induce Lambdoid Prophages

It was found soon after the discovery of lysogenic induction that inducing agents are numerous and varied (Lwoff 1953). In addition to physical carcinogens (e.g., radiation) and chemical carcinogens (e.g., metabolized aflatoxin B_1 or benzopyrene) (Goze et al. 1975; Moreau et al. 1976, 1980b), other treatments of lysogenic bacteria like thymine deprivation of thymine-requiring bacteria (Korn and Weissbach 1962; Sicard and Devoret 1962) promote lysogenic induction. This observation pointed to the common property of inducing agents or treatments: They all affect DNA replication either by the production of DNA lesions or by preventing the replication fork from operating properly (see Tables 2 and 3). The target of an inducing agent or treatment is DNA, either the chromosome of the lysogen or a replicon that is transmitted to the lysogen (see below, Cellular Activation of RecA Protein), and is not the λ repressor itself or the prophage DNA.

Induction of prophage λ provides a sensitive test for carcinogens (Moreau et al. 1976; Moreau and Devoret 1977) and for antitumor drugs that act by damaging DNA (Heinemann 1971; Bradner 1978; Anderson et al. 1980) (see references in the footnote to Table 2).

The *recA* Gene Controls Induction of Prophage λ

It was found unexpectedly that *recA* bacterial mutants, isolated to be deficient in genetic recombination, are also unable to support lysogenic induction (Brooks and Clark 1967; Hertman and Luria 1967), showing neither spontaneous production of phage nor DNA damage-provoked induction. The mutants have other striking defects, including high sensitivity to killing by radiation and inability to be mutagenized by radiation (Howard-Flanders and Boyce 1966; Miura and Tomizawa 1968; Witkin 1969). These phenotypes suggested immediately that

Table 2 Examples of agents or treatments that induce prophage λ

Agents that damage DNA
physical carcinogens: UV light, X-rays, α-rays
activated chemical carcinogens: benzapyrene, 7,12-dimethylbenzanthracene, aflatoxins
antitumor drugs: bleomycin, daunorubicin, mitomycin C, neocarzinostatin
Agents or treatments that disrupt DNA replication
antifolates: aminopterin, trimethoprim
antigyrase drugs: nalidixic acid, oxolinic acid
inhibitors of thymidylate synthetase synthesis: 5-fluorouracil
thymine deprivation of *thy*⁻ mutants
growth of mutants in nonpermissive conditions (see Table 3)
Introduction of replicons into the lysogen
UV-damaged plasmids and phages: F sex factor, colicinogenic factor I, R factors, phageP1
intact plasmids: *lynR* mutants of mini F

For general references, see Heinemann (1971), Borek and Ryan (1973), Moreau and Devoret (1977), Bradner (1978), and Anderson et al. (1980).

the *recA*-gene product might directly relate lysogenic induction to other cellular functions induced by DNA damage. The existence of another RecA phenotype, produced by mutation *tif*1 located in the *recA* gene (McEntee 1977), suggested that *recA* regulates induction. *tif*1 mutants induce prophage λ in the absence of any DNA damage, during growth at high temperature in a medium containing adenine (Kirby et al. 1967; Castellazzi et al. 1972a, b). Bacterial mutations other than *recA* also affect lysogenic induction (see Table 3).

RecA Protein Is the Inducer

It was found that inactivation of λ repressor during lysogenic induction is accompanied by its cleavage into two fragments (Roberts and Roberts 1975; Roberts et al. 1977, 1978a). Although proteolysis in vivo might follow any inactivation mechanism, it is in fact the primary mechanism of inactivation. Furthermore, the protein encoded by the *recA* gene is the protease and is therefore the cellular inducer (Roberts et al. 1978; Craig and Roberts 1981).

RecA protein is also the central enzyme in general genetic recombination, acting to catalyze annealing of single-stranded DNA to homologous sequences in duplex DNA (McEntee et al. 1979; Shibata et al. 1979; West et al. 1980). Besides promoting recombination, RecA protein also plays an important role in regulating DNA repair by the so-called SOS genes, which specify functions normally repressed by LexA protein (Little and Mount 1982). Following DNA damage, RecA protein inactivates the LexA repressor by cleavage (Little et al. 1980) and thus increases expression of the SOS genes. The SOS genes include *recA* itself, so that induction leads to amplification of RecA protein (Gudas and Pardee 1975; Gudas and Mount 1977; Little and Mount 1982), necessary for efficient DNA repair processes (Quillardet et al. 1982).

FUNCTION AND REGULATION OF THE PROTEASE ACTIVITY OF RecA PROTEIN

Repressor Cleavage by RecA Protein in Lysogenic Induction

When it is activated by binding single-stranded DNA and a nucleoside triphosphate, RecA protein binds λ repressor and inactivates it by a single enzymatic cut (Roberts and Roberts 1975; Roberts et al. 1978; Craig and Roberts 1980; Phizicky and Roberts 1981; Sauer et al. 1982). The activity of highly purified RecA protein in this reaction indicates that no other proteins are required, although DNA and small molecules are (see below). The cutting reaction with RecA protein purified from mutants reflects their phenotypes, confirming that this is the inactivation reaction. Repressor altered by the cI*ind*1 mutation is not cut in vitro (Roberts et al. 1977), whereas λ repressor from the hyperinducible mutant cI*ind*^{s} is cut five to ten times as fast as wild-type λ repressor (Cohen et al. 1981; D. Burbee and J. Roberts, unpubl.). A λ repressor fragment of the same electrophoretic mobility as one of the fragments produced in vitro is found

Table 3 Bacterial mutations that affect lysogenic induction

Name of gene and typical allele	Phenotype	Biochemical change	Map site	Reference
Mutations affecting the activity of RecA protease				
recA			58	
*recA*36	deficient in recombination, inducible repair, and prophage λ induction	defect in all activities of RecA protein	id	Van de Putte et al. (1966); Morand et al. (1977b)
*recA*430 (*lexB*30)	deficient in induction of prophage λ and inducible repair	λ repressor recognition by RecA protease, unknown defect in activation	id	Blanco et al. (1975); Devoret et al. (1982)
*recA*142	deficient in recombination and inducible repair, delayed induction of prophage λ	defective binding of RecA protein to DNA, of strand-pairing activity	id	Clark (1973); Roberts and Roberts (1981)
*recA*441 (*tif*1)	prophage λ induced after temperature shift and addition of adenine	more effective activation of protease by DNA and nucleoside triphosphate	id	Castellazzi et al. (1972a); Phizicky and Roberts (1981)
Mutations affecting the amount of RecA protein				
recA				
*recA*543 (*zab*53)	deficient in inducible repair, suppressor of RecA441 phenotype	down-promoter mutation	id	Castellazzi et al. (1972b); Jo Rebollo et al. (in prep.); McEntee (1977)
*recA*o98	suppressor of *lexA*3, λ inducible	*recA*-operator-constitutive	id	Clark (1982); Quillardet et al. (1982)
lexA			90.5	
*lexA*3	deficient in induction of RecA protein	LexA repressor noncleavable by RecA protein: LexA(Ind$^-$)	id	Mount et al. (1972)
*spr*51	derepression of *recA* and other SOS genes	deficient in LexA repressor: LexA(def)	id	Mount (1977)
*tsl*1	derepression of *recA* and other SOS genes by temperature shift	thermoinactivable repressor: LexA(ts)	id	Mount et al. (1975)

Genes coding for products involved in DNA replication				
*polA*12	deficient in DNA replication, prophage λ induced at nonpermissive temperature	deficient 5′→3′ exonuclease function of polymerase I	86	Monk and Kinross (1972); Blanco and Pomes (1977); Uyemura and Lehman (1976)
*dnaB*252	deficient in DNA replication, prophage λ induced at nonpermissive temperature	defect of portable promoter protein	90.5	D'Ari et al. (1979)
*dnaC*2	deficient in DNA replication, prophage λ induced at nonpermissive temperature	defect of protein that complexes with DnaB protein	99	Schuster et al. (1973); Wickner and Hurwitz (1975)
*dnaE*486	deficient in DNA replication, prophage λ induced at nonpermissive temperature	defect of component of polymerase III holoenzyme	4	Schuster et al. (1973); Kornberg (1980)
*dnaG*308	deficient in DNA replication, prophage λ induced at nonpermissive temperature	defect of primase	66	Schuster et al. (1973); Rowen and Kornberg (1978)
*ligA*7	deficient in DNA replication, prophage λ induced at nonpermissive temperature	defect of polynucleotide ligase	52	Gottesman et al. (1973)
*ssb*1	deficient in prophage λ induction at nonpermissive temperature	defect of single-stranded binding protein	91	Meyer et al. (1979); Meyer et al. (1982); Baluch et al. (1980a)
Uncharacterized mutations				
*recF*143	prophage induction delayed	unknown	82	Horii and Clark (1973); Armengod and Blanco (1978)
*infA*3	deficient in prophage λ induction and filamentation	unknown	60–73	Bailone et al. (1975)
*infB*1	deficient in prophage λ induction and filamentation	unknown	66–83	Huisman et al. (1980)

For map references, see Bachmann and Low (1980).

in cells after inducing treatments (Roberts et al. 1977), providing further evidence that the enzymatic activity of purified RecA protein is responsible for repressor inactivation in vivo.

Requirements for Repressor Cleavage and DNA Strand Exchange in Vitro

RecA protein must bind single-stranded DNA to cleave repressor (Craig and Roberts 1980), just as it binds single-stranded DNA in initiating DNA strand exchange. Both reactions occur in vitro (and presumably in vivo) when RecA protein complexes stoichiometrically with single-stranded DNA in the presence of ATP (or dATP) and Mg^{++}. dATP binds RecA protein more tightly than does ATP, suggesting that the cellular dATP level may be important to *recA* function in vivo (Phizicky and Roberts 1981). In DNA strand exchange, the enzyme activated by a bound single strand melts duplex DNA to find a sequence complementary to the bound strand; it then anneals these, displacing a strand from the duplex (Cunningham et al. 1979; McEntee et al. 1979). Although RecA protein can catalyze reciprocal strand exchange between two homologous duplexes, such exchange must be initiated from a single-stranded region in one of the duplexes (DasGupta et al. 1981). Thus, recognition of single-stranded DNA is fundamental to the function of RecA protein in recombination. The DNA requirement of about 5 nucleotides per RecA monomer to activate cleavage is also found for strand exchange (Shibata et al. 1979; Craig and Roberts 1981). Any single-stranded DNA, even an oligonucleotide as small as dA(5), both promotes repressor cleavage and causes RecA protein to unwind duplex DNA. The occurrence of single-stranded DNA as a consequence of damage thus may lead both to recombinational DNA repair and repressor inactivation.

Presumably a conformational change in RecA protein as it enters a ternary complex with nucleoside triphosphate and DNA activates it for both DNA strand exchange and repressor cleavage. In both reactions, binding but not hydrolysis of ATP is required for association of RecA protein with DNA in the active complex. This is shown clearly by the activity of ATP-γ-S, an ATP analog that is hydrolyzed at a negligible rate by the RecA ATPase activity (Craig and Roberts 1981). DNA, RecA protein, and ATP-γ-S form a tight complex detectable by nitrocellulose filter binding. RecA protein in this ternary complex promotes homologous pairing, although hydrolysis of ATP is required for the branch migration that produces extensive strand exchange (Cox and Lehmann 1981).

The *E. coli* single-stranded DNA binding (SSB) protein may be involved in activation of RecA protein by DNA in vivo. SSB protein is required for efficient DNA strand exchange in vitro (Cox and Lehmann 1981). Mutations in the *ssb* gene inhibit lysogenic induction (Baluch et al. 1980a; Vales et al. 1980) (Table 3); this may be due to a requirement for SSB protein to participate along with RecA protein in binding DNA or, alternatively, due to the SSB defect inhibiting DNA replication, preventing development of the activating signal (see below).

SSB protein stimulates repressor cleavage in vitro in some conditions and can prevent inhibition of cleavage by higher than stoichiometric concentrations of DNA (Resnick and Sussman 1982), arguing that it may mediate the activation of RecA protease activity by DNA.

RecA Polypeptide

RecA protein sediments as an oligomer (perhaps a tetramer) of the 37-kD polypeptide (Roberts et al. 1978a; Ogawa et al. 1979) and forms filaments upon binding DNA (McEntee et al. 1981). However, small oligonucleotides support both cleavage and DNA unwinding, so that polymerization on a long polynucleotide is not essential to either reaction. Genetic complementation studies indicate that the active species is a multimer (J.E. Rebollo et al., in prep.).

Little is known about the catalytic mechanism of RecA protease action. No specific inhibitor of the protease has been found. Both results of genetic mapping and complementation (Castellazzi et al. 1977; Glickman et al. 1977; Morand et al. 1977a,b) and perusal of the primary structure of the RecA polypeptide (Sancar et al. 1980) suggest that is has distinct domains, one of which might have the protease activity.

Cleavage by Mutationally Altered RecA Protein

The activity of RecA protein encoded by several *recA* mutants in cleaving repressor confirms the central role of cleavage in repressor inactivation and suggests that the ability of RecA protein to interact with DNA and nucleoside triphosphate determines its activity in cleavage.

recA*430*

This mutant was selected for its inability to induce prophage λ by thymine starvation (Devoret and Blanco 1970), one of several inducing treatments to which a *recA*430 (λ) lysogen does not respond. It is proficient in genetic recombination and displays radiation resistance intermediate between wild-type and "classical" *recA*$^-$ mutants (Clark 1973), suggesting some capacity for recombinational DNA repair (Morand et al. 1977a). Purified RecA430 protein does not cleave λ repressor detectably in vitro (Roberts and Roberts 1981). However, it has nearly normal ATPase activity and catalyzes strand exchange in vitro (C.W. Roberts and J.W. Roberts, unpubl.). Thus, its biochemical activities match its phenotypes for λ induction and genetic recombination. Although RecA430 protein does not cleave λ repressor, it has not lost the protease activity; a ϕ80 lysogen of *recA*430 is induced and RecA protein is amplified slightly (Devoret et al. 1983), corresponding to the fact that LexA protein is cleaved by RecA430 protein in vitro (D. Burbee and J.W. Roberts, unpubl.). Presumably, RecA430 protein does not cleave LexA repressor efficiently enough in vivo to induce most SOS genes.

recA*142*

This other atypical *recA* mutant is recombination-defective but is partially proficient in lysogenic induction (Clark 1973). Unlike classical *recA* mutants, a *recA*142 (λ) lysogen shows almost normal spontaneous phage production and elaborates the λ repressor fragment R_1 associated with spontaneous phage production in wild-type lysogens (Roberts and Roberts 1975), although it is induced inefficiently and after a delay in response to UV irradiation (M. Dutreix et al., in prep.). Purified RecA142 protein cleaves λ repressor in vitro, but it interacts inefficiently with DNA and nucleoside triphosphate (Roberts and Roberts 1981). In particular, it cleaves repressor only with the nonhydrolyzed analog ATP-γ-S or dATP (but not ATP) and has a salt-sensitive DNA-dependent ATPase activity. This suggests a defect in efficiently interacting with DNA to catalyze strand exchange, despite having protease activity that is expressed in some conditions.

tif *Mutants*

In the *tif*1 mutant (now renamed *recA*441), repressors are inactivated constitutively, i.e., in the absence of inducing treatments (Kirby et al. 1967; Castellazzi et al. 1972a). Purified RecA441 protein is more active than wild-type RecA protein in cleaving λ repressor in vitro, particularly in conditions of limiting DNA and nucleoside triphosphate (Roberts et al. 1978a; Phizicky and Roberts 1981). This advantage of the RecA441 protein is due to its more efficient formation of the ternary complex with DNA and nucleoside triphosphate. Two other *tif* mutants, *recA*447 and *recA*448 (Mount 1979), share this property (Phizicky and Roberts 1981). The constitutivity of *tif* mutants may result from the ability of mutant RecA protein to be activated by intact cellular DNA, such as single-stranded DNA that occurs at a replication fork or is exposed by negative supercoiling, rather than requiring more extensive damage-dependent single-stranded regions.

Repressor Substrate

It appears that a single proteolytic cut is made in the four known substrates of RecA protease: λ, phage 434, P22, and LexA repressors (Horii and Ogawa 1981; Sauer et al. 1982). In each case, the cut occurs between alanine and glycine, in comparable positions in the four polypeptides; the λ repressor monomer of 236 amino acids is cut between positions 111 and 112. The cleavage site in λ and P22 repressors is located in a hinge region between the two domains—one binding to operator, and the other, to repressor monomers (Sauer et al. 1979; see Gussin et al., this volume).

What part of the repressor molecule is recognized by RecA protein in the cleavage reaction? Experiments with repressor fragments suggest that RecA protein does not recognize only the local amino acid sequence at the cleavage site but, instead, a larger structure that includes the carboxyterminal domain (Sauer et al. 1982). The carboxyl terminus is the most conserved region in the amino acid sequences of repressors cleaved by RecA protein (Sauer et al. 1982), as might be expected of a common recognition site.

It has been suggested that repressor might bind DNA along with RecA protein in the cleavage reaction (Baluch and Sussman 1978), a possibility that has not been tested. The fact that λ repressor lacking the amino terminus, and thus its DNA-binding activity, is cleaved tends to argue against an obligatory binding of repressor to DNA in the reaction.

To explain the kinetics of repressor cleavage, Phizicky and Roberts (1980) proposed that the monomer of λ repressor is the primary substrate recognized by RecA protein, whereas the dimer is much less susceptible to cleavage. The fraction of wild-type λ repressor cleaved per unit time in vitro decreases continously as repressor concentration is increased in the range of 0.1 μM to about 10 μM. Although this effect superficially resembles a classical approach to V_{max} for an enzymatic reaction, it is not, because increasing λ repressor concentration does not inhibit cleavage of a second repressor (of phage P22) present in the same reaction mixture (Phizicky and Roberts 1980). Thus, it is not that the active site of RecA protein is saturated with increasing λ repressor concentration, but, instead, that repressor at higher concentrations is intrinsically less sensitive to cleavage. In this concentration range, which is well above the dissociation constant of the λ repressor dimer, 0.02 μM (R. Sauer, pers. comm.), free monomer increases as the square root of total repressor concentration. The observed rate of cleavage also increases approximately as the square root of repressor concentration, suggesting that only free monomer is a substrate. This could be true if the recognition site of repressor for RecA protein were covered or distorted by dimerization, or if the dimer bound elsewhere (e.g., nonspecifically to DNA) and were removed from the reaction. Preferential cleavage of repressor monomers also explains an exactly corresponding effect of high repressor concentration on inactivation in vivo. Only a negligible fraction of λ repressor made in high concentration by an overproducing plasmid is inactivated after inducing treatments, yet a second repressor present at its normal low concentration in the same cell is inactivated at the normal rate (Bailone et al. 1979).

λ*c*I Mutations That Change the Rate of Repressor Cleavage

Two sorts of λ repressor mutations have been described that change the rate of repressor inactivation. (1) *c*I*ind*$^-$ mutations (Eshima et al. 1972) prevent *recA*-dependent repressor inactivation in vivo. At least one *c*I*ind*$^-$ mutation, *ind*1 (Jacob and Campbell 1959), which is located in the middle of the *c*I gene (Lieb 1966), is not cleaved by RecA protein in vitro (Roberts et al. 1977). Others have not been tested. (2) Repressor modified by *ind*^{s} mutations is inactivated in vivo by inducing doses lower than that required by wild-type repressor (Horiuchi and Inokuchi 1967). In vitro, repressor modified by the *ind*^{s} mutation is cleaved five to ten times as fast as wild-type repressor (Cohen et al. 1981; D. Burbee and J.W. Roberts, unpubl.). These results further confirm that cleavage is the cause of lysogenic induction.

How could *ind*⁻ and *ind*ˢ mutations modify repressor to change the cleavage rate? First, they might change the interaction between repressor and RecA protein, which depends upon at least two factors: the local structure of the cleavage site and the structure of distant regions of repressor involved in recognition. These could be altered by *ind*⁻ mutations that map in distinct regions: in the carboxyl terminus, which may carry the recognition specificity for RecA protein, and near the cleavage site (Sauer 1978; Lieb 1981).

If repressor monomers are the substrate, there is a second way by which repressor mutations could change the rate. A different equilibrium constant for dimerization would change the fraction of repressor in cleavable form (Phizicky and Roberts 1980). There is evidence that Indˢ repressor does have less tendency to dimerize than the wild-type, although its dissociation constant has not been measured accurately (Cohen et al. 1981). It is consistent with these results that the *ind*ˢ mutation modifies the carboxyl terminus of repressor, which contains the contacts by which repressor dimerizes (R. Sauer, pers. comm.). Some *ind*⁻ mutations, such as *ind*543 mapping in the carboxyl terminus (Eshima et al. 1972), could have the opposite effect of decreasing the dissociation constant and lowering the concentration of monomer (M. Dutreix et al., in prep.).

Rate of Repressor Cleavage

Is the rate of repressor inactivation found in vitro fast enough to explain induction? The rate of cleavage in vitro measured at the repressor concentration in a lysogen, about 0.3 μM, is one cleavage event per RecA monomer per several hours (Craig and Roberts 1981; Phizicky and Roberts 1981) (although the maximum rate measured in vitro, with Indˢ repressor, is about one per 5 minutes). A lysogen makes about 200 monomers of repressor per generation, and because the promoter that maintains this level, *prm*, is normally fully expressed (Gussin et al., this volume), autoregulation should not increase the rate of repressor synthesis during induction. Cleavage leading to induction occurs over a generation time (30 min), so that about 200 cleavage events are required in 30 minutes. Since the basal concentration of RecA protein is several thousand molecules per cell (and this increases during induction) (Paoletti et al. 1982), the rate measured in vitro is clearly sufficient, assuming that most RecA molecules initially present in a lysogen are activated by the inducing signal.

CELLULAR ACTIVATOR OF RecA PROTEIN

Lysogenic Induction and SOS Functions

SOS functions are a set of cellular responses induced by agents or treatments that damage DNA (Witkin 1967, 1976; Radman 1975; Kenyon and Walker 1980). They include amplification of RecA protein (Gudas and Pardee 1975), cell filamentation (George et al. 1975), error-prone repair (George et al. 1974; Witkin 1974; Defais et al. 1976), stable DNA replication (Kogoma et al. 1979), and excision repair (Kenyon and Walker 1981). SOS functions do in fact aid both cell survival (Quillardet et al. 1982) and the survival of infecting phages (Weigle 1953).

About a dozen genes that govern SOS functions, called SOS genes, have been identified over the past 15 years (for review, see Little and Mount 1982). Under normal growth conditions, the expression of SOS genes is repressed by LexA protein, the *lexA*-gene product (Little and Harper 1979; Brent and Ptashne 1980, 1981; Little et al. 1980). Induction of SOS genes (Bagg et al. 1981) results from the inactivation of LexA protein through its cleavage by activated RecA protein (Little et al. 1980). This is the molecular event that relates induction of SOS functions to lysogenic induction.

The concept of SOS functions implies that a damaged cell survives and resumes its normal development after their operation. By enhancing DNA repair, SOS functions can reverse fully the cellular change that activates RecA protein. LexA repressor is no longer cleaved, and SOS genes become repressed again (Little and Mount 1982).

In contrast to induction of *lexA*-controlled functions, lysogenic induction is an irreversible phenomenon that necessarily causes cell death and thus is a phage, not a cellular SOS function. Prophage λ is still induced if derepression of SOS genes, in particular amplification of RecA protein (Baluch et al. 1980b; Moreau et al. 1980a), is prevented by the presence of a *lexA* (*ind*$^-$) mutation (Crowl et al. 1981; Sedgwick et al. 1981; Quillardet et al. 1982). The only consequence to the prophage, if cellular SOS genes are not induced after DNA damage, is its failure to be fully repaired and therefore to develop efficiently. All bacteria appear to possess SOS genes, which are required for them to cope with a hazardous environment; in contrast, only some prophages are inducible.

Although both LexA and λ repressor are cleaved by activated RecA protein, LexA repressor is cleaved at least ten times as fast as λ repressor. These different rates lead to full induction of some SOS genes (e.g., *recA* itself) at low UV doses but to insufficient cleavage of λ repressor to induce prophage λ at all (subinduction) (Bailone et al. 1979; Moreau et al. 1982). Two factors affect the induction of the separate SOS genes. First, the various SOS operators have different affinities for LexA repressor (Little and Mount 1982). Second, the concentration of LexA repressor depends upon the rate at which it is destroyed by activated RecA protein. The interplay of these factors provides substantial cell survival at low UV dose but yields prophage induction at much higher UV doses. Thus, induction of wild-type prophage occurs when the probability of cell survival is slight.

Figure 1 shows schematically the pathway by which cellular processes dependent upon RecA protein are invoked by aborted DNA replication.

Activator of RecA Protein in Vivo

What is the primary cellular event that activates RecA protein to a protease? In vitro evidence shows that RecA protein is activated to cleave repressor when it

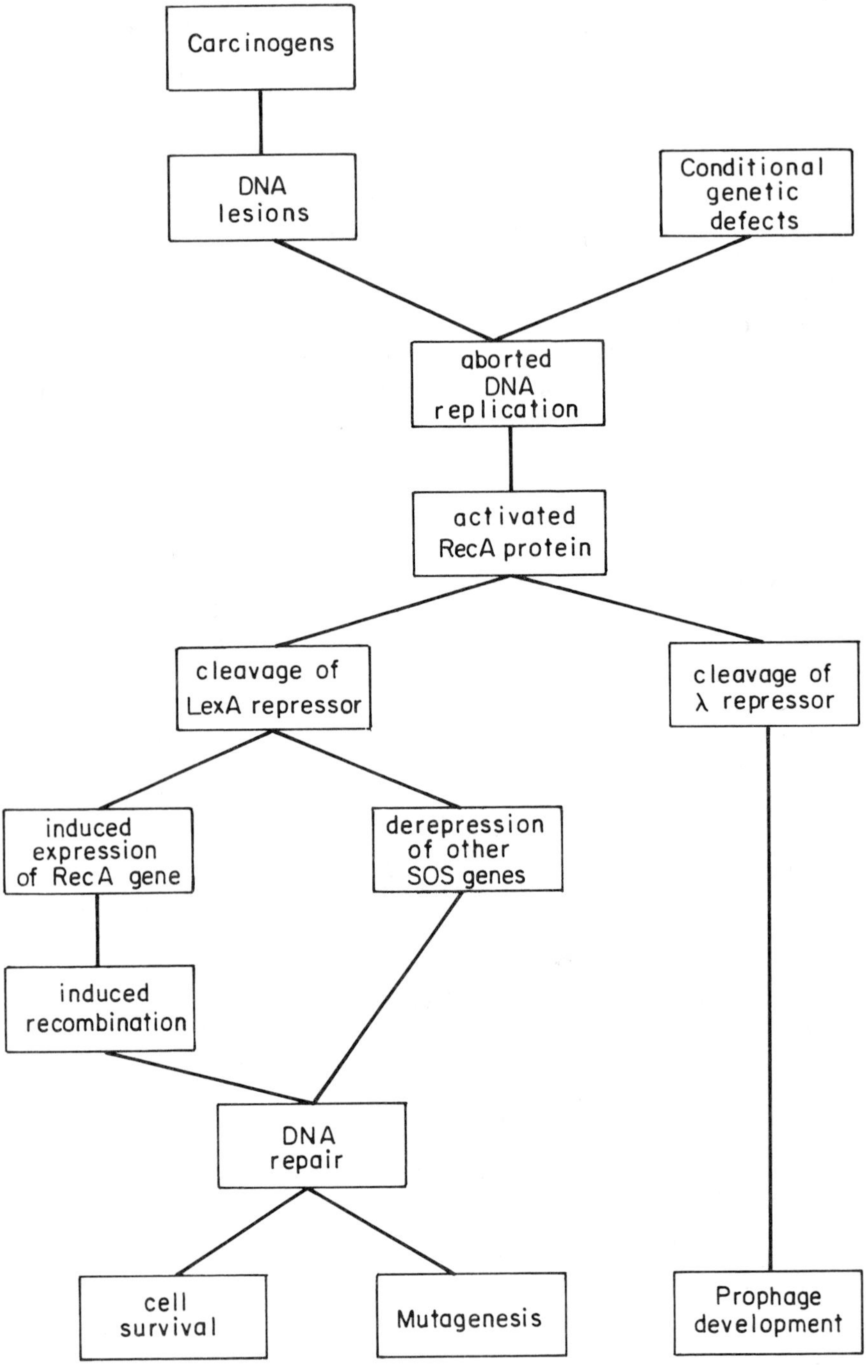

Figure 1 Central role of activated RecA protein.

forms a ternary complex with single-stranded DNA and a nucleoside triphosphate. We thus presume that the result of all inducing treatments is to make single-stranded DNA accessible to RecA protein. How does this occur? All inducing treatments abort replication, although in different ways: (1) by introducing a noncoding lesion such as a pyrimidine dimer in the template strand; (2) by inactivation of regulatory or structural proteins of replication, like *dnaB*, and (3) by removal of a precursor substrate for replication, such as thymine.

The simplest event to consider is a replication fork encountering a noncoding lesion that has not been removed by excision repair. It is thought that single-stranded DNA is exposed in the damaged strand as the helix is unwound in advance of the lesion without concurrent DNA synthesis; a gap is then formed as DNA synthesis reinitiates beyond the lesion. By hypothesis, RecA protein binds the emerging gap and is activated thereby to cleave repressors. This binding also initiates recombinational repair of the lesion, one important DNA repair function of RecA protein (West et al. 1981).

We speculate that RecA protein is activated to cleave repressor as it assembles into the gap, but the protease activity is quenched when this bound RecA protein engages duplex DNA in search of homology. This sequence of events is suggested by the fact that excess single-stranded DNA inhibits cleavage in vitro, a phenomenon we take to simulate engagement of the duplex DNA-binding activities of RecA protein that are required for strand exchange. We imagine that as the growing fork progresses, it repeatedly encounters lesions. The concentration of activated RecA protein and the overall rate of repressor cleavage would depend upon the linear density of lesions in DNA which, in turn, is a function of the dose of inducing treatment. We thus view activation of RecA protein as occurring transiently at each lesion. A high density of lesions assures that the signal is sufficiently intense and enduring so that the protease formed will cleave all of the λ repressor in a lysogen in 30 minutes.

How do treatments that abort replication but do not introduce lesions induce? We suggest that the mechanism is essentially the same, except that the inducing structure is static, generated by stalled and immovable replication forks from which unwinding gives single-stranded DNA.

A more extreme but related case occurs in cells carrying the *tif*1 mutation of *recA*. An active replication fork containing single-stranded DNA in lagging strand gaps or in a region unwound in advance of the fork might suffice to activate the RecA (Tif) protein, which has a higher affinity for DNA (Phizicky and Roberts 1981). In effect, the altered RecA (Tif) protein would compete with replication proteins such as SSB for naturally occurring single-stranded DNA, whereas the wild-type RecA protein has an affinity balanced such that it is activated only in aberrant conditions that provide more extensive single-stranded regions.

An Experimental Model for Aborted Replication

It is difficult to study directly the structure of the bacterial chromosome and the gene products involved in aborted replication and, therefore, in the activation of

RecA protease in vivo. A good substitute is the F plasmid, for several reasons. (1) The F plasmid and host chromosome share a common regulation, both replicons being maintained at the same low copy number in the cell and segregating together at cell division (Jacob et al. 1964). (2) A UV-irradiated F sex factor, when introduced into an intact lysogen, induces the resident prophage (indirect induction), a process that is about half as efficient as inducing directly with UV light (Borek and Ryan 1958, 1973; Devoret and George 1967; George and Devoret 1971). (3) Upon indirect induction, RecA protein is activated and amplified as it is in direct induction (Moreau et al. 1982). (4) The cloning of the replication region of F (mini-F), whose size is 10% of the whole F plasmid, has facilitated manipulation of the replicon and its transfer to recipient cells (Couturier et al. 1979). (5) Mutations, including lethal mutations, can be easily studied in F since it is dispensable for the cell. Phage P1, whose replication and segregation properties are similar to plasmid F, also promotes indirect induction when UV-damaged, although with a much lower efficiency than the F plasmid (Rosner et al. 1968).

Two genetic loci in the F plasmid, *lynR* and *lynA*, involved in indirect lysogenic induction have been identified recently by deletion mapping (Brandenburger et al. 1983; A. Bailone et al., in prep.). The *lynA* locus, situated between the two F origins of replication, *oriP* and *oriS*, appears to encode a replication function required for indirect prophage induction. The *lynA* product may stabilize a DNA structure that leads to activation of RecA protein. The *lynR* locus controls the establishment and the cosegregation of the F plasmid with the host chromosome. Impairment of its function causes prophage to be induced even in the absence of UV damage to mini F. No indirect prophage induction occurs when initiation of DNA replication of mini F is blocked, showing that replication functions are involved. Presumably the change in expression of *lynA* and *lynR* loci gives rise to aborted replication structures.

ACKNOWLEDGMENTS

We thank our colleagues and associates for their advice and criticism. J.W.R. acknowledges support by grants from the National Institutes of Health, and R.D. acknowledges support from the Centre Nacional de le Recherche Scientifique.

This paper was written while one of us (R.D.) was spending a year at the National Cancer Institute-Frederick Cancer Research Facility as a visiting scientist in the Genetic Engineering and Molecular Genetics Sections of the Cancer Biology Program. R.D. is grateful to the Association pour le Dévelopment de la Recherche sur le Cancer and to the Eleanor Roosevelt Cancer Fund (IUCC) for providing a fellowship and a travel grant, respectively.

The Ligue Française contre le Cancer and the Fondation pour la Recherche Médicale are gratefully acknowledged for providing funds for computer-assisted word processing.

REFERENCES

Anderson, W.A., P.L. Moreau, R. Devoret, and R. Maral. 1980. Induction of prophage λ by daunorubicin and derivatives. Correlation with antineoplastic activity. *Mutat. Res.* **77:** 197.

Armengod, M.E. and M. Blanco. 1978. Influence of the *recF143* mutation of *E. coli* K12 on prophage λ induction. *Mol. Gen. Genet.* **52:** 37.

Bachmann, B.J. and K.B. Low. 1980. Linkage map of *Escherichia coli* K12, edition 6. *Microbiol. Rev.* **44:** 1.

Bagg, A., C.J. Kenyon, and G.C. Walker. 1981. Inducibility of a gene product required for UV and chemical mutagenesis in *Escherichia coli. Proc. Natl. Acad. Sci.* **78:** 5749.

Bailone, A., M. Blanco, and R. Devoret. 1975. *E. coli* K12 *infA*: A mutant deficient in prophage λ induction and cell filamentation, *Mol. Gen. Genet.* **136:** 291.

Bailone, A., A. Lévine, and R. Devoret. 1979. Inactivation of prophage λ repressor in vivo. *J. Mol. Biol.* **131:** 553.

Baluch, J. and R. Sussman. 1978. Correlation between UV dose requirement for lambda bacteriophage induction and lambda repressor concentration. *J. Virol.* **26:** 595.

Baluch, J., J.W. Chase, and R. Sussman. 1980a. Synthesis of *recA* protein and induction of bacteriophage lambda in single-strand deoxyribonucleic acid-binding protein mutants of *Escherichia coli. J. Bacteriol.* **144:** 489.

Baluch, J., R. Sussman, and J. Resnick. 1980b. Induction of prophage λ without amplification of RecA preotein. *Mol. Gen. Genet.* **178:** 317.

Blanco, M. and L. Pomes. 1977. Prophage induction in *E. coli* K12 cells deficient in DNA polymerase. *Mol. Gen. Genet.* **154:** 287.

Blanco, M., A. Lévine, and R. Devoret. 1975. *lexB*: A new gene governing radiation sensitivity and lysogenic induction of *E. coli* K12. In *Molecular mechanisms for repair of DNA* (ed. P.C. Hanawalt and R.B. Setlow), p. 379. Plenum Press, New York.

Borek, E. and A. Ryan. 1958. The transfer of irradiation-elicited induction in a lysogenic organism. *Proc. Natl. Acad. Sci.* **44:** 374.

——. 1973. Lysogenic induction. *Prog. Nucleic Acid Res. Mol. Biol.* **13:** 249.

Bradner, W.T. 1978. New prescreens for antitumor antibiotics. *Antibiot. Chemother.* **23:** 4.

Brandenburger, A., A. Bailone, A. Lévine, and R. Devoret. 1983. Transinduction: Prophage induction by undamaged DNA of λminiF. *J. Mol. Biol.* (in press).

Brent, R. and M. Ptashne. 1980. The *lexA* gene product represses its own promoter. *Proc. Natl. Acad. Sci.* **77:** 1932.

——. 1981. Mechanism of action of the *lexA* gene product. *Proc. Natl. Acad. Sci.* **78:** 4204.

Brooks, K. and A. Clark. 1967. Behavior of λ bacteriophage in a recombination deficient strain of *E. coli. Virology* **1:** 283.

Castellazzi, M., J. George, and G. Buttin. 1972a. Prophage induction and cell division in *E. coli*. I. Further characterization of the thermosensitive mutation *tif-1* whose expression mimics the effect of UV irradiation. *Mol. Gen. Genet.* **119:** 139.

——. 1972b. Prophage induction and cell division in *E. coli*. II. Linked (*recA, zab*) and unlinked (*lex*) suppressors of *tif-1* mediated induction and filamentation. *Mol. Gen. Genet.* **119:** 153.

Castellazzi, M., P. Morand, J. George, and G. Buttin. 1977. Prophage induction and cell division in *E. coli*. V. Dominance and complementation analysis in partial diploids with pleiotropic mutations (*tif, recA, zab* and *lexB*) at the *recA* locus. *Mol. Gen. Genet.* **153:** 297.

Clark, A.J. 1973. Recombination deficient mutants of *Escherichia coli* and other bacteria. *Annu. Rev. Genet.* **7:** 67.

——. 1982. *recA* operator mutations and their usefulness. *Biochimie* **64:** 669.

Cohen, S., B.J. Knoll, J.W. Little, and D.W. Mount. 1981. Preferential cleavage of phage lambda repressor monomers by recA protease. *Nature* **294:** 182.

Couturier, M., J. Janssens, F. Bex, A. Desmyter, and I. Bonnevalle. 1979. Construction *in vivo* of phage-plasmid chimerae: A new tool to analyse the mechanism of plasmid maintenance. *Mol. Gen. Genet.* **169:** 113.

Cox, M.M. and I.R. Lehman. 1981. Directionality and polarity in recA protein-promoted branch migration. *Proc. Natl. Acad. Sci.* **78:** 6018.

Craig, N.L. and J.W. Roberts. 1980. *E. coli recA* protein-directed cleavage of phage λ repressor requires polynucleotide. *Nature* **283:** 26.

——. 1981. Function of nucleoside triphosphate and polynucleotide in *Escherichia coli recA* protein-directed cleavage of phage λ repressor. *J. Biol. Chem.* **256:** 8039.

Crowl, R.M., R.P. Boyce, and H. Echols. 1981. Repressor cleavage as a prophage induction mechanism: Hypersensitivity of a mutant λcI protein to RecA-mediated proteolysis. *J. Mol. Biol.* **152:** 815.

Cunningham, R.P., T. Shibata, C. DasGupta, and C.M. Radding. 1979. Single strands induce RecA protein to unwind duplex DNA for homologous pairing. *Nature* **281:** 191.

D'Ari, R., J. George, and O. Huisman. 1979. Suppression of *tif*-mediated induction of SOS functions in *Escherichia coli* by an altered DnaB protein. *J. Bacteriol.* **140:** 381.

DasGupta, C., A.M. Wu, R. Kahn, R.P. Cunningham, and C.M. Radding. 1981. Concerted strand exchange and formation of Holliday structures by *E. coli* RecA protein. *Cell* **25:** 507.

Defais, M., P. Caillet-Fauquet, M.S. Fox, and M. Radman. 1976. Induction kinetics of mutagenic DNA repair activity in *E. coli* following ultraviolet irradiation. *Mol. Gen. Genet.* **148:** 125.

Devoret, R. and M. Blanco. 1970. Mutants of *Escherichia coli* K12 $(\lambda)^+$ noninducible by thymine deprivation. I. Method of isolation and classes of mutants obtained. *Mol. Gen. Genet.* **107:** 272.

Devoret, R. and J. George. 1967. Induction indirecte du prophage λ par le rayonnement ultraviolet. *Mutat. Res.* **4:** 713.

Devoret, R., M. Pierre, and P.L. Moreau. 1983. Prophage φ80 is induced in *Escherichia coli* K12 *recA430*. *Mol. Gen. Genet.* **189:** 199.

Eshima, N., S. Fujii, and T. Horiuchi. 1972. Isolation of λ*ind*⁻ mutants. *J. Genet.* **47:** 125.

George, J. and R. Devoret. 1971. Conjugal transfer of UV-damaged F-prime sex factors and indirect induction of prophage λ. *Mol. Gen. Genet.* **111:** 103.

George, J., M. Castellazzi, and G. Buttin. 1975. Prophage induction and cell division in *E. coli*. III. Mutations *sfiA* and *sfiB* restore division in *tif* and *lon* strains and permit the expression of mutator properties of *tif*. *Mol. Gen. Genet.* **140:** 309.

George J., R. Devoret, and M. Radman. 1974. Indirect ultraviolet-reactivation of phage λ. *Proc. Natl. Acad. Sci.* **71:** 144.

Glickman, B.W., N. Guijt, and P. Morand. 1977. The genetic characterization of *lexB32, lexB33* and *lexB35* mutations of *Escherichia coli*: Location and complementation pattern for UV resistance. *Mol. Gen. Genet.* **157:** 83.

Gottesman, M.M., M.L. Hicks, and M. Gellert. 1973. Genetics and function of DNA ligase in *Escherichia coli*. *J. Mol. Biol.* **77:** 531.

Goze, A., A. Sarasin, Y. Moulé, and R. Devoret. 1975. Induction and mutagenesis of prophage λ in *E. coli* K12 by metabolites of aflatoxin B1. *Mutat. Res.* **28:** 1.

Gudas, L.J. and D.W. Mount. 1977. Identification of the *recA* (*tif*) gene product of *Escherichia coli*. *Proc. Natl. Acad. Sci.* **74:** 5280.

Gudas, L.J., and A.B. Pardee. 1975. Model for regulation of *Escherichia coli* DNA repair functions. *Proc. Natl. Acad. Sci.* **72:** 2330.

Heinemann, B. 1971. Prophage induction in lysogenic bacteria as a method of detecting potential mutagenic, carcinogenic, carcinostatic, and teratogenic agents. In *Chemical mutagens, principles and methods for their detection.* (ed. A. Hollander), vol. 1, p. 235. Plenum Press, New York.

Hertman, I. and S. Luria. 1967. Transduction studies on the role of a *rec*⁺ gene in the ultraviolet of prophage λ. *J. Mol. Biol.* **23:** 117.

Horii, Z.I. and A. J. Clark. 1973. Genetic analysis of the *recF* pathway to genetic recombination in *Escherichia coli*: Isolation and characterization of mutants. *J. Mol. Biol.* **80:** 327.

Horii, T. and H. Ogawa. 1981. Nucleotide sequence of the *lexA* gene of *E. coli*. *Cell* **23:** 689.

Horiuchi, T. and H. Inokuchi. 1967. Temperature-sensitive regulation system of prophage lambda induction. *J. Mol. Biol.* **23:** 217.

Howard-Flanders, P. and R.P. Boyce. 1966. DNA repair and recombination: Studies on mutants of *Escherichia coli* defective in these processes. *Radiat. Res.* (Suppl.) **6:** 156.

Huisman, O., R. D'Ari, and J. George. 1980b. Dissociation of *tsl tif*-induced filamentation and RecA protein synthesis in *Escherichia coli* K12. *J. Bacteriol.* **142:** 819.

Jacob, F. and A. Campbell. 1959. Sur le système de répression assurant l'immunité chez les bactéries lysogènes. *C. R. Acad. Sci.* **248:** 3219.

Jacob, F. and J. Monod. 1961. Genetic regulatory mechanisms in the synthesis of proteins. *J. Mol. Biol.* **3:** 318.

Jacob, F. and E.L. Wollman. 1961. *Sexuality and the genetics of bacteria.* Academic Press, New York.

Jacob, F., S. Brenner, and F. Cuzin. 1964. On the regulation of DNA replication in bacteria. *Cold Spring Harbor Symp. Quant. Biol.* **28:** 329.

Johnson, A.D., A.R. Poteete, G. Lauer, R.T. Sauer, G.K. Ackers, and M. Ptashne. 1981. λ repressor and cro—Components of an efficient molecular switch. *Nature* **294:** 217.

Kenyon, C.J. and G.C. Walker. 1980. DNA-damaging agents stimulate gene expression at specific loci in *Escherichia coli. Proc. Natl. Acad. Sci.* **77:** 2819.

——. 1981. Expression of the *E. coli uvrA* gene is inducible. *Nature* **289:** 808.

Kirby, E.P., F. Jacob, and D.A. Goldthwait. 1967. Prophage induction and filament formation in a mutant strain of *Escherichia coli. Proc. Natl. Acad. Sci.* **58:** 1903.

Kogoma, T., T.A. Torrey, and M.J. Connaughton. 1979. Induction of UV-resistant DNA replication in *Escherichia coli*: Induced stable DNA replication as an SOS function. *Mol. Gen. Genet.* **176:** 1.

Korn, D. and A. Weissbach. 1962. Thymineless induction in *Escherichia coli* K12(*l*). *Biochim. Biophys. Acta.* **61:** 775.

Kornberg, A. 1980. DNA replication. W.H. Freeman, San Francisco.

Lévine, A., A. Bailone, and R. Devoret. 1979. Cellular levels of the prophage λ and 434 repressors. *J. Mol. Biol.* **131:** 655.

Lieb, M. 1966. Studies of heat inducible lambda bacteriophage. I. Order of genetic sites and properties of mutant prophages. *J. Mol. Biol.* **16:** 149.

——. 1981. A fine structure map of spontaneous and induced mutations in the λ repressor gene, including insertions of IS elements. *Mol. Gen. Genet.* **184:** 364.

Little, J.W. and J.E. Harper. 1979. Identification of the *lexA* gene product of *Escherichia coli* K12. *Proc. Natl. Acad. Sci.* **76:** 6147.

Little, J.W., and D.W. Mount. 1982. The SOS regulatory system of *Escherichia coli. Cell* **29:** 11.

Little, J.W., S.H. Edmiston, L.Z. Pacelli, and D.W. Mount. 1980. Cleavage of the *Escherichia coli* LexA protein by the RecA protease. *Proc. Natl. Acad. Sci.* **77:** 3225.

Lwoff, A. 1953. Lysogeny. *Bacteriol. Rev.* **17:** 269.

——. 1966. The prophage and I. In *Phage and the origins of molecular biology* (ed. J. Cairns et al.), p. 88. Cold Spring Harbor Laboratory, Cold Spring Harbor, New York.

Lwoff, A., L. Siminovitch, and N. Kjeldgaard. 1950. Induction de la production de bactériophages chez une bactérie lysogène. *Ann. Inst. Pasteur.* **79:** 815.

Maniatis, T., M. Ptashne, and R. Maurer. 1974. Control elements in the DNA of bacteriophage λ. *Cold Spring Harbor Symp. Quant. Biol.* **38:** 857.

McEntee, K. 1977. Protein X is the product of the *recA* gene of *Escherichia coli. Proc. Natl. Acad. Sci.* **74:** 5275.

McEntee, K., G.M. Weinstock, and I.R. Lehman. 1979. Initiation of general recombination catalyzed *in vitro* by the RecA protein of *Escherichia coli. Proc. Natl. Acad. Sci.* **76:** 2615.

——. 1981. Binding of the recA protein of *Escherichia coli* to single- and double-stranded DNA. *J. Biol. Chem.* **256:** 8835.

Melechen, N.E. and P.D. Skaar. 1962. The provocation of an early step of induction by thymine deprivation. *Virology* **16:** 21.

Meyer, R.R., J. Glassberg, and A. Kornberg. 1979. An *Escherichia coli* mutant defective in single-strand binding protein is defective in DNA replication. *Proc. Natl. Acad. Sci.* **76:** 1702.

Meyer, R.R., D.C. Rein, and J. Glassberg. 1982. The product of the *lexC* gene of *Escherichia coli* is a single-stranded DNA-binding protein. *J. Bacteriol.* **150:** 433.

Miura, A. and J.I. Tomizawa. 1968. Studies on radiation-sensitive mutants of *E. coli.* III. Participation of the Rec system in induction of mutation by ultraviolet irradiation. *Mol. Gen. Genet.* **103:** 1.

Monk, M. and J. Kinross. 1972. Conditional lethality of *recA* and *recB* derivatives of a strain of *Escherichia coli* K12 with a temperature-sensitive deoxyribonucleic acid polymerase I. *J. Bacteriol.* **109:** 971.

Morand, P., M. Blanco, and R. Devoret. 1977a. Characterization of *lexB* mutations in *Escherichia coli* K12. *J. Bacteriol.* **131:** 572.

Morand, P., A. Goze, and R. Devoret. 1977b. Complementation pattern of *lexB* and *recA* mutations in *Escherichia coli* K12; mapping of *tif-1, lexB* and *recA* mutations. *Mol. Gen. Genet.* **157:** 69.

Moreau, P. and R. Devoret. 1977. Potential carcinogens tested by induction and mutagenesis of prophage λ in *Escherichia coli* K12. In "Origins of Human Cancer" (ed. H.H. Hiatt et al.). *Cold Spring Harbor Conf. Cell Proliferation* **4:** 1451.

Moreau, P., A. Bailone, and R. Devoret. 1976. Prophage λ induction in *Escherichia coli* K12 *envA uvrB*: A highly sensitive test for potential carcinogens. *Proc. Natl. Acad. Sci.* **73:** 3700.

Moreau, P.L., M. Fanica, and R. Devoret. 1980a. Induction of prophage λ does not require full induction of RecA protein synthesis. *Biochimie* **62:** 687.

——. 1980b. Cleavage of λ repressor and induction of RecA protein synthesis elicited by aflatoxin B1 metabolites in *Escherichia coli. Carcinogenesis* **1:** 837.

Moreau, P.L., V. Pélico, and R. Devoret. 1982. Cleavage of λ repressor and synthesis of RecA protein induced by transferred UV-damaged F sex factor. *Mol. Gen. Genet.* **186:** 170.

Mount, D.W. 1977. A mutant of *E. coli* showing constitutive expression of the lysogenic and error-prone DNA repair pathways. *Proc. Natl. Acad. Sci.* **74:** 300.

——. 1979. Isolation and characterization of mutants of λ*recA* which synthesize a hyperactive RecA protein. *Virology* **98:** 484.

Mount, D.W., K.B. Low, and S.J. Edmiston. 1972. Dominant mutations (*lex*) in *Escherichia coli* K12 which affect radiation sensitivity and frequency of ultraviolet light-induced mutations. *J. Bacteriol.* **112:** 886.

Mount, D.W., A.C. Walker, and C. Kosel. 1975. Effect of *tsl* mutations in decreasing radiation sensitivity of a *recA*⁻ strain of *Escherichia coli* K12. *J. Bacteriol.* **121:** 1203.

Ogawa, T. and J.-I. Tomizawa. 1967. Abortive lysogenization of bacteriophage lambda *b2* and residual immunity of non-lysogenic segregants. *J. Mol. Biol.* **23:** 225.

Ogawa, T., H. Wabiko, T. Tsusimoto, T. Horii, H. Masukata, and H. Ogawa. 1979. Characteristics of purified RecA protein and the regulation of its synthesis *in vivo. Cold Spring Harbor Symp. Quant. Biol.* **43:** 909.

Paoletti, C., B. Salles, and P. Giacomoni. 1982. An immunoradiometric quantitative assay of *Escherichia coli* RecA protein. *Biochimie* **64:** 239.

Phizicky, E.M. and J.W. Roberts. 1980. Kinetics of RecA protein-directed inactivation of repressors of phage λ and phage P22. *J. Mol. Biol.* **139:** 319.

——. 1981. Induction of SOS functions: Regulation of proteolytic activity of *E. coli* RecA protein by interaction with DNA and nucleoside triphosphate. *Cell* **25:** 259.

Pirrotta, V., P. Chadwick, and M. Ptashne. 1970. Active form of two coliphage repressors. *Nature* **227:** 5253.

Ptashne, M. 1967a. Isolation of the λ phage repressor. *Proc. Natl. Acad. Sci.* **57:** 306.

Quillardet, P., P.L. Moreau, H. Ginsburg, D.W. Mount, and R. Devoret. 1982. Cell survival, UV-reactivation and induction of prophage λ in *Escherichia coli* K12 overproducing RecA protein. *Mol. Gen. Genet.* **188:** 37.

Radman, M. 1975. SOS repair hypothesis: Phenomenology of an inducible DNA repair which is accompanied by mutagenesis. In *Molecular mechanisms for repair of DNA* (ed. P.C. Hanawalt and R.B. Setlow), p. 335. Plenum Press, New York.

Resnick, J. and R. Sussman. 1982. *Escherichia coli* single strand DNA binding protein from wild type and *lexC113* mutant affect *in vitro* proteolytic cleavage of phage λ repressor. *Proc. Natl. Acad. Sci.* **79:** 2832.

Roberts, J.W. and C.W. Roberts. 1975. Proteolytic cleavage of bacteriophage lambda repressor in induction. *Proc. Natl. Acad. Sci.* **72:** 147.

——. 1981. Two mutations that alter the regulatory activity of *E. coli* recA protein. *Nature* **290:** 422.

Roberts, J.W., C.W. Roberts, and N.L. Craig. 1978. *Escherichia coli recA* gene product inactivates phage λ repressor. *Proc. Natl. Acad. Sci.* **75:** 4714.

Roberts, J.W., C.W. Roberts, and D.W. Mount. 1977. Inactivation and proteolytic cleavage of phage λ repressor *in vitro* in an ATP-dependent reaction. *Proc. Natl. Acad. Sci.* **74:** 2283.

Rosner, J.L., L.R. Kass, and M.B. Yarmolinsky. 1969. Parallel behavior of F and P1 in causing indirect induction of lysogenic bacteria. *Cold Spring Harbor Symp. Quant. Biol.* **33:** 785.

Rowen, L. and A. Kornberg. 1978. Primase, the *dnaG* protein of *Escherichia coli:* An enzyme which starts DNA chains. *J. Biol. Chem.* **253:** 758.

Sancar, A., C. Stachelek, W. Konigsberg, and W.D. Rupp. 1980. Sequences of the *recA* gene and protein. *Proc. Natl. Acad. Sci.* **77:** 2611.

Sauer, R.T. 1978. DNA sequence of the bacteriophage λcI gene. *Nature* **276:** 301.

Sauer, R.T., M.J. Ross, and M. Ptashne. 1982. Cleavage of the phage λ and P22 repressors by RecA protein. *J. Biol. Chem.* **257:** 4458.

Sauer, R.T., C.O. Pabo, B.J. Meyer, M. Ptashne, and K.C. Backman. 1979. Regulatory functions of the λ repressor reside in the amino terminal domain. *Nature* **279:** 396.

Schuster, H., D. Beyersmann, M. Mikolajczyk, and M. Schlicht. 1973. Prophage induction by high temperature in thermosensitive *dna* mutants lysogenic for bacteriophage lambda. *J. Virol.* **11:** 879.

Sedgwick, S.G., G.T. Yarranton, and R.W. Heath. 1981. Lysogenic induction of lambdoid phages in *lexA* mutants of *Escherichia coli. Mol. Gen. Genet.* **184:** 457.

Shibata, T., C. DasGupta, R.P. Cunningham, and C.M. Radding. 1979. Purified *Escherichia coli* RecA protein catalyzes homologous pairing of superhelical DNA and single-stranded fragments. *Proc. Natl. Acad. Sci.* **76:** 1638.

Shinagawa, H. and T. Itoh. 1973. Inactivation of DNA-binding activity of repressor in extracts of λ-lysogen treated with mitomycin C. *Mol. Gen. Genet.* **126:** 103.

Sicard, N. and R. Devoret. 1962. Effets de la carence en thymine sur des souches d'*Escherichia coli* lysogènes K12T$^-$ et colicinogènes 15T$^-$. *C.R. Acad. Sci.* **255:** 1417.

Sussman, R. and F. Jacob. 1962. Sur un système de répression thermosensible chez le bactériophage d'*Escherichia coli. C. R. Acad. Sci.* **254:** 1517.

Tomizawa, J.I. and T. Ogawa. 1967. Effect of ultraviolet irradiation on bacteriophage lambda immunity. *J. Mol. Biol.* **23:** 247.

Uyemura, D. and I.R. Lehman. 1976. Biochemical characterization of mutant forms of DNA polymerase I from *Escherichia coli*. I. The *polA12* mutation. *J. Biol. Chem.* **251:** 4078.

Vales, L.D., J.W. Chase, and J.B. Murphy. 1980. Effect of *ssbA1* and *lexC113* mutations on lambda prophage induction, bacteriophage growth, and cell survival. *J. Bacteriol.* **143:** 887.

Van de Putte, P., H. Zwenk, and A. Rorsch. 1966. Properties of four mutants of *Escherichia coli* defective in genetic recombination. *Mutat. Res.* **3:** 381.

Weigle, J.J. 1953. Induction of mutations in a bacterial virus. *Proc. Natl. Acad. Sci.* **39:** 628.

Weigle, J.J. and M. Delbrück. 1951. Mutual exclusion between an infecting phage and a carried phage. *J. Bacteriol.* **62:** 301.

West, S.C., E. Cassuto, and P. Howard-Flanders. 1981. Mechanism of *E. coli* RecA protein directed strand exchanges in post-replication repair of DNA. *Nature* **294:** 659.

West, S.C., E. Cassuto, J. Mursalim, and P. Howard-Flanders. 1980. Recognition of duplex DNA containing single-stranded regions by RecA protein. *Proc. Natl. Acad. Sci.* **77:** 2569.

Wickner, S. and J. Hurwitz. 1975. Interaction of *Escherichia coli dnaB* and *dnaC* gene products *in vitro. Proc. Natl. Acad. Sci.* **72:** 921.

Witkin, E.M. 1967. The radiation sensitivity of *Escherichia coli* B: A hypothesis relating filament formation and prophage induction. *Proc. Natl. Acad. Sci.* **57:** 1275.

——. 1969. The mutability toward ultraviolet light of recombination-deficient strains of *Escherichia coli. Mutat. Res.* **8:** 9.

——. 1974. Thermal enhancement of ultraviolet mutability in a *tif-1 uvrA* derivative of *Escherichia coli* B/r: Evidence that ultraviolet mutagenesis depends upon an inducible function. *Proc. Natl. Acad. Sci.* **71:** 1930.

——. 1976. Ultraviolet mutagenesis and inducible DNA repair in *Escherichia coli. Bacteriol. Rev.* **40:** 869.

Woods, W.H. and J.B. Egan. 1981. The transient inability of the conjugating female cell to host 186 infection explains the absence of zygotic induction for 186. *J. Virol.* **40:** 335.

Lambda DNA Replication

Mark E. Furth
Laboratory of Molecular Oncogenesis
Memorial Sloan-Kettering Cancer Center
New York, New York 10021

Sue H. Wickner
Laboratory of Molecular Biology
National Cancer Institute
Bethesda, Maryland 20205

INTRODUCTION

The replication of bacteriophage λ DNA depends on interactions with the bacterial host. As a repressed prophage, λ replicates passively, integrated into the host chromosome. As an active virus or as a plasmid, λ channels host enzymes to replicate its chromosome.

A single cycle of productive growth generates approximately 100 copies of the viral genome. During the latent period, distinct forms of λ DNA appear in an orderly progression (Fig. 1). The replication program can be divided into two phases: early and late. In the early phase, DNA synthesis initiates at a unique origin region on covalently closed circular templates and generates more circles. In the late phase, replication by a rolling-circle mechanism generates multiple-length λ DNA molecules, which serve as substrates for encapsidation. In contrast to lytic growth, some derivatives of λ can replicate indefinitely as circular plasmid DNA molecules.

The phage and host gene products essential for λ DNA replication are listed in Table 1. Most of the host requirements have been identified by challenging λ to replicate at the nonpermissive temperature in *Escherichia coli* mutants temperature-sensitive for replication of the bacterial chromosome. With the important exception of some functions thought to participate specifically in initiation at the *E. coli* replication origin, λ utilizes most of the replication machinery of its host. These proteins presumably carry out similar functions in the synthesis of λ DNA as they do in bacterial replication (for present understanding of biochemistry and control of DNA replication, see Kornberg 1980, 1982). Only two phage proteins, encoded by genes *O* and *P*, appear to participate directly in the initiation and/or propagation of replication forks. Mutants defective in these genes fail to replicate detectably (Ogawa and Tomizawa 1968). A small fragment of the λ genome, carrying only *O*, *P*, the replication origin region, the promoter p_R, and the *cro* (or *tof*) gene, replicates autonomously as a plasmid called λdv (Matsubara and Kaiser 1969; Matsubara 1981).

Two phage regulatory genes are required for normal replication. However, in each case the effect appears to be indirect, through control of transcription of

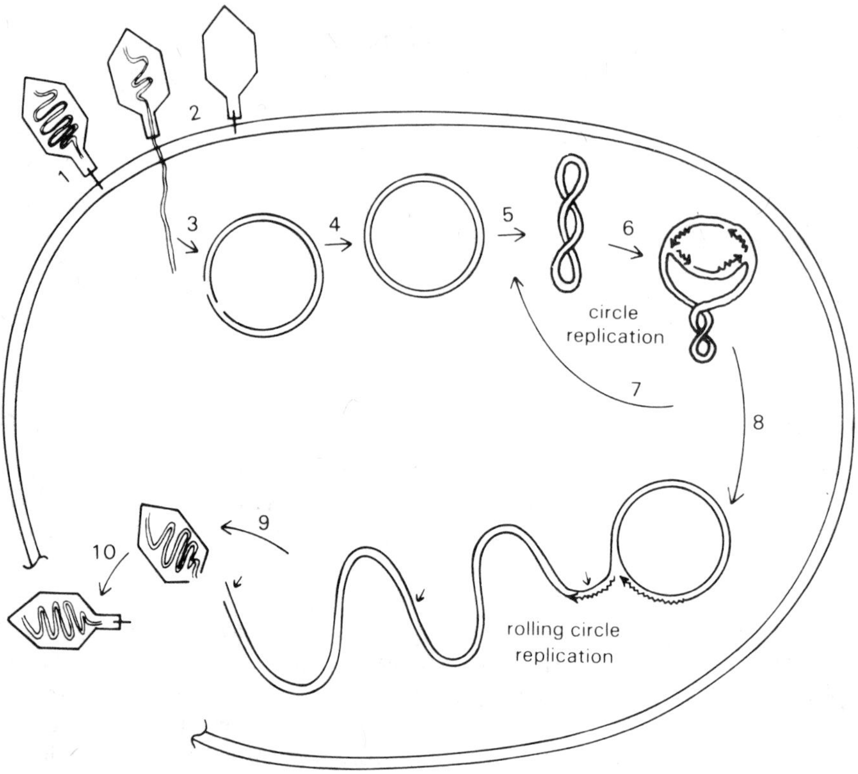

Figure 1 DNA replication cycle of λ. A λ phage particle attaches to the host cell (*1*) and injects its linear, duplex DNA molecule (*2*). The molecule circularizes by the base-pairing of complementary single-stranded ends (*3*). The resulting nicked circle is closed by DNA ligase (*4*), and supercoils are introduced by DNA gyrase (*5*). DNA replication initiates on the supercoiled circle, and at early times replication of most molecules proceeds bidirectionally to generate θ structures (*6*). The termination of a round of θ replication yields two daughter circles (*7*). At later times most molecules replicate as σ forms (rolling circles) (*8*). The concatemeric DNA produced by rolling-circle replication can be cut at the *cos* sites (arrows) and packaged into phage heads (*9*). The addition of tails (*10*) completes the maturation of phage particles capable of initiating a new cycle of infection. Wavy lines indicate the most recently synthesized DNA at active replication forks, and arrowheads on such lines indicate the 3′ ends of growing DNA strands.

other genes. λN^- mutants replicate slowly because they express limited amounts of *O* and *P* (Dove et al. 1971); they can persist as plasmids for many generations (Signer 1969). The *cro*-gene product is necessary for optimal synthesis of phage DNA, especially in the late phase after infection (Folkmanis et al. 1977). In addition, *cro* plays a critical role in the replication of λ-derived plasmids, both λN^- and λdv (Berg 1974; Matsubara and Takeda 1975; Kleckner and Signer 1977). The replication defects in the absence of *cro* protein may result from over-

Table 1 Proteins required for λ DNA replication

Gene	Essential for replication of *E. coli*	Essential for replication of λ	Biochemical function
			E. coli *initiation proteins*
dnaA	yes	no	unknown
dnaC	yes	no	interacts physically and functionally with *dnaB* protein; involved in prepriming reaction
dnaB	yes	yes[a]	ribonucleoside triphosphatase activity stimulated by single-stranded DNA; involved in prepriming reaction
rpoB	yes	yes[a]	RNA polymerase β-subunit; transcriptional activation of replication
			E. coli *elongation proteins*
dnaC	yes	no	see above
dnaB	yes	yes[a]	see above
dnaG	yes	yes[a]	primase, synthesizes ribonucleotides, deoxyribonucleotides, and mixed oligonucleotides (primers for DNA synthesis)
ssb	yes	n.t.[a,b]	binds to single-stranded DNA
dnaE	yes	yes	α-component of DNA polymerase III
dnaZ	yes	yes	γ-component of DNA polymerase III elongation complex
dnaN	yes	n.t.[b]	β-component of DNA polymerase III elongation complex
gyrA, gyrB	yes	n.t.[a]	subunits of DNA gyrase, catalyzes ATP-dependen negative supercoiling of relaxed closed circular DNA
lig	yes	yes	DNA ligase, covalently closes nicks in double-stranded DNA
*polA*ex	yes	n.t.	5′-3′ exonuclease of DNA polymerase I
dnaJ	yes	yes	unknown
dnaK	yes	yes	DNA-independent ATPase
dnaQ	yes	n.t.	ε-component of DNA polymerase III elongation complex
dnaX	yes	yes	δ-component of DNA polymerase III holoenzyme
dnaY	yes	yes	unknown
			E. coli *dispensable proteins*
grpD	no	yes	unknown
grpE	no	yes	unknown
			λ *replication proteins*
O	no	yes[a]	origin-specific DNA-binding protein
P	no	yes[a]	interacts with *dnaB* protein

[a]Shown to be required in vitro.
[b]n.t. indicates not tested.

production of the *O* and *P* proteins and, possibly, other phage gene products. The function of *cro* in regulating the replication of λdv has been reviewed in detail by Matsubara (1981).

Several dispensable λ genes influence DNA synthesis during the late phase. In some hosts the proteins encoded by the λ general recombination genes *exo* and *bet* (or *red*, to indicate either or both; see Smith, this volume) and the *gam* gene are required for the efficient production of packageable replication intermediates (Zissler et al. 1971; Enquist and Skalka 1973).

The λ replication program and features of its genetic control have been studied for many years (for review, see Kaiser 1971; Skalka 1977). Recent work has focused on the transitions between replicative forms, on the molecular interactions that determine initiation of DNA synthesis at the origin of replication, on the purification and functional analysis of essential proteins, and on the development of in vitro systems for the replication of λ DNA.

EARLY REPLICATION

After linear λ DNA molecules are injected from phage particles into bacteria, they are converted into covalently closed circular molecules; the complementary single-stranded ends base pair, the remaining nicks are ligated, and negative superhelical twists are introduced. These supercoiled circles are substrates for the initiation of DNA replication (Tomizawa and Ogawa 1969).

Molecules participating in the first round of replication have been visualized by electron microscopy. They were isolated by density gradient centrifugation after incorporation of heavy isotopes. The majority of replicating molecules are in the shape of the Greek letter theta (θ). They have two branches of equal length, corresponding to the portion of the molecule that has just undergone duplication, and a third segment that has not yet been replicated (Fig. 2a,c; Ogawa et al. 1968; Tomizawa and Ogawa 1969; Schnös and Inman 1970). A minority of the replicating molecules are circles with a single branch point and a tail, in the shape of the Greek letter sigma (σ) (Schnös and Inman 1970; Better and Freifelder 1983).

Replication begins at a unique origin region and proceeds bidirectionally (Schnös and Inman 1970; Stevens et al. 1971). By partially denaturing the molecules before examination in the electron microscope, Schnös and Inman (1970) were able to locate the branch points of θ structures with respect to the characteristic denaturation map of λ DNA (Fig. 2c). This analysis showed that replication forks move away from a single region and progress around the circle in opposite directions at roughly equal rates. Schnös et al. (1982) found a slight bias favoring progression to the right. This bias was increased dramatically after infection of a thymine-requiring host in medium with low concentrations of thymidine (Valenzuela and Inman 1981).

Reinitiation at the origin normally does not occur until a full round of replication has been completed. This rationing does not result simply from limitations

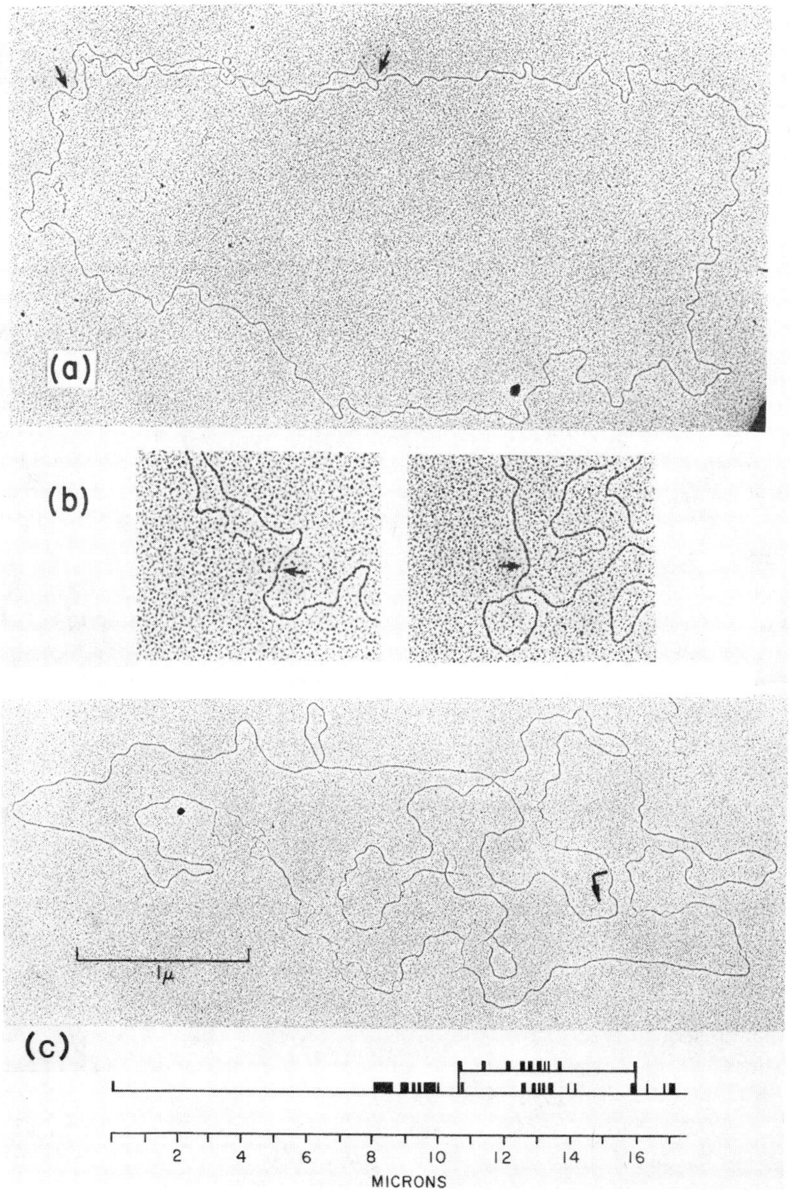

Figure 2 Electron micrographs of replicative intermediates in the first round of replication of λ*c*IIλ*c*III (Schnös and Inman 1970). (*a*) Native θ structure. (→) Branch points. (*b*) Single-stranded connections at branch points of θ structures (high magnification). (→) Branch points. Thin regions are single-stranded DNA; thicker regions are double-stranded DNA. (*c*) Partially denatured θ structure. The estimated position of *cos* is marked; (→) points from left to right with respect to physical map. The diagram below shows alignment of this molecule on the linear map, with the position of the left end as marked in the electron micrograph above. Bars indicate denatured segments. The region ~10.7 μm to 16 μm from the left end has replicated. The origin is at ~13.7 μm, implying that replication of this molecule was bidirectional. (Courtesy of Maria Schnös and Ross B. Inman, University of Wisconsin, Madison.)

in amounts of critical proteins, because in phages containing two copies of the λ origin region, initiation can occur simultaneously at both (Schnös et al. 1982). However, reinitiation has been observed at relatively high frequency (up to 20%) under some unusual conditions, most notably infection in the presence of 2 mM caffeine (Schnös and Inman 1982). The new replication forks appear to start at the normal origin. The basis for this effect of caffeine is not known, but it may result from a structural change in the DNA template induced by the drug, which is known to lower the helical stability of DNA.

Although both parental DNA strands are copied, high-magnification electron micrographs of θ structures revealed short single-stranded regions at many of the branch points (Fig. 2b; Inman and Schnös 1971). These single-stranded connections invariably were observed on the replicated (daughter) side of the branch point. When a θ structure contained a single-stranded region on each daughter segment, these regions were located at opposite branch points. These observations can be explained by an asymmetry inherent in the progression of replication forks. Known DNA polymerases carry out synthesis only by extension from the 3′-hydroxyl end of a polynucleotide chain (5′ to 3′ direction). Because the strands of the DNA double helix are antiparallel, one strand (referred to as the lagging strand) must be replicated discontinuously by the synthesis and joining of short chains to achieve overall 3′ to 5′ growth (Okazaki et al. 1969; Kornberg 1980). The single-stranded connections seen in λ replicative forms presumably correspond to regions in which a segment of the lagging strand has not yet been made.

Rounds of early bidirectional replication terminate roughly halfway around the circle, independently of specific DNA sequences. Many λ derivatives have DNA substitutions or deletions in the region normally located opposite the replication origin (the antipode). In other derivatives the normal antipode is located at a different position on the λ chromosome. All of these phages replicate normally, and in several cases termination of replication has been shown to occur halfway around the circle (Valenzuela et al. 1976). Similarly, no specific termination sequence seems to be required for the replication of λdv plasmids (Matsubara 1981). Termination may simply result from the convergence of two replication forks (Kaiser 1971; Skalka 1977).

The products of early replication are circular molecules, appearing first as nicked circles and then as supercoiled closed circles (Young and Sinsheimer 1968; Sogo et al. 1976). At the end of a round of replication, the two daughter circles may be catenated (Sogo et al. 1976). The interlocked rings could be separated by a DNA topoisomerase such as DNA gyrase (Gellert 1981).

Early replication proceeds exponentially, so that at 42°C the number of intracellular molecules doubles in 2–3 minutes (Stevens et al. 1971; Better and Freifelder 1983). About 50 circular λ DNA molecules, roughly half of which are supercoiled, accumulate within 15 minutes in an infected cell (Young and Sinsheimer 1968; Carter and Smith 1970; Better and Freifelder 1983). The time required to replicate a λ chromosome at 42°C is approximately 30 seconds

(Greenstein and Skalka 1975). At any moment, fewer than 20 percent of the intracellular λ DNA molecules, perhaps two or three per cell, are undergoing replication; the majority of intracellular molecules are monomeric circles (Better and Freifelder 1983).

LATE REPLICATION

During lytic growth, the production of covalently closed circles ceases 15 minutes after infection (Joyner et al. 1966; Young and Sinsheimer 1968; Carter et al. 1969; Carter and Smith 1970; Bastia et al. 1975). The most striking feature of replication at later times is the synthesis of concatemers, multiple units of λ covalently joined head to tail. Unlike the monomeric circular DNA that accumulates early in the latent period, this concatemeric DNA is an immediate precursor to the DNA of functional phage particles (see Szpirer and Brachet 1970; Stahl et al. 1972b; Feiss and Margulies 1973; Freifelder et al. 1974; Syvanen 1974; Hohn 1975; Dawson et al. 1976; Feiss and Becker, this volume).

Synthesis of Concatemeric DNA

The first evidence for synthesis of concatemers came from sedimentation of radiochemically labeled intracellular DNA. Molecules from two to eight times the length of a λ monomer were observed from about 15 minutes after infection until the end of the latent period and behaved as precursors to the mature DNA found in phage particles (Smith and Skalka 1966; Salzman and Weissbach 1967; Young and Sinsheimer 1967; Carter et al. 1969; Kiger and Sinsheimer 1969; Tomizawa and Ogawa 1969; Carter and Smith 1970; Enquist and Skalka 1973; McClure and Gold 1973). Denaturation of these molecules released linear single strands of greater than monomer length. Observation of partially denatured molecules by electron microscopy confirmed the presence of concatemeric λ DNA, especially in cells infected with mutants defective in head assembly (Skalka et al. 1972; Wake et al. 1972; McClure et al. 1973).

In principle, concatemeric molecules might arise either by replication or by the joining of monomers. An obligatory role for genetic recombination was ruled out by the demonstration that concatemers are synthesized in recombination-deficient (*recA*$^-$) host cells infected by λ mutants defective in both general (*red*$^-$) and site-specific (*int*$^-$) recombination (Segawa and Tomizawa 1971; Skalka 1971). Because the reduction of concatemers is coupled to packaging, it is extremely unlikely that they arise by the joining of free intracellular linear monomers. Thus, there must be some replicative mechanism for the production of concatemers.

The rolling-circle model of replication provides a possible mechanism for the synthesis of concatemeric DNA (Eisen et al. 1969; Gilbert and Dressler 1969; for reviews, see Tomizawa and Selzer 1979; Kornberg 1980). The model proposes that in the replication of a circular DNA molecule, one parental strand remains covalently closed and serves as a template, whereas the complementary strand is nicked. DNA synthesis initiates at the 3′-hydroxyl end of the nicked strand. As synthesis proceeds, this strand is displaced, starting from its 5′ end, and is spun off from the circle as a single-stranded tail. The tail can then be replicated to yield duplex DNA. If the replication fork travels around the circle repeatedly, the tail continues to grow. Thus, this mode of replication can generate long concatemers (Fig. 1). Most studies of λ replication have not demonstrated covalent joining of parental DNA from phage particles to newly synthesized DNA (Carter et al. 1969; Tomizawa and Ogawa 1969; Sogo et al. 1976; for a puzzling exception, see Ihler and Kawai 1971). However, this does not rule out a rolling-circle mechanism.

Possible rolling-circle intermediates in λ DNA replication have been observed by electron microscopy (Takahashi 1974; Bastia et al. 1975; Better and Freifelder 1983). At late times after infection, the majority of replicative forms were single-branched circles (σ forms). Molecules that had replicated to the greatest extent, as judged by incorporation of isotopically dense precursors, had the longest tails, often several times greater than monomer length (Bastia et al. 1975). Partial denaturation mapping confirmed the expected concatemeric structure. These studies also revealed that rolling-circle replication, like early θ replication, is bidirectional; on some circles the fork moves clockwise, whereas on others it moves counterclockwise (Takahashi 1974, 1975c; Bastia et al. 1975; Better and Freifelder 1983). Some σ forms with tails of greater than unit length were found even when genetic recombination was eliminated by phage and host mutations (Takahashi 1977; Better and Freifelder 1983).

The partial denaturation mapping of σ forms with tails of greater than unit length provided evidence that rolling-circle replication can initiate within the same origin region as does θ replication. The free ends of approximately 40 percent of such molecules lie near to the position of the normal origin of early replication (Bastia et al. 1975; Takahashi 1975c; Better and Freifelder 1983). Apparently, in such molecules a parental DNA strand is nicked at or near the origin to generate the free end. Possibly, all rolling circles initiate at the early origin, but many tails are broken at random sites during preparation for electron microscopy. Alternatively, some σ forms may arise directly from θ forms by inactivation of one fork during replication. In this case, the position of the free end would depend on the precise location of the fork at the time of breakage. Another possibility is that some σ forms initiate by a mechanism that does not utilize the normal origin. For example, new replication forks may be generated by the action of a recombination system (see below; Boon and Zinder 1969; Stahl et al. 1973; Skalka 1974).

Control of Late Replication

The mechanism of the transition between the early and late phases of λ DNA replication is not known. Searches for a hypothetical essential gene required to mediate a switch between the θ and σ modes have been fruitless. The idea of a switch may be somewhat misleading. An alternative view is that both θ and σ forms initiate early after infection but that θ replication shuts off, whereas σ replication persists at later times. Among molecules participating in the first round of replication, approximately one-third were σ forms (Schnös and Inman 1970). The density selection used to isolate these molecules biased against any with tails of greater than unit length, so it is not clear whether these were true rolling circles. Another method permitted the isolation of the entire population of intracellular λ DNA molecules, relatively free of host DNA, for examination by electron microscopy. Bacteria that had incorporated bromodeoxyuridine (a dense thymidine analog) into their DNA were infected with λ, the complexes were incubated in medium containing thymidine, and light phage DNA was separated from heavy bacterial DNA by equilibrium density centrifugation (Better and Freifelder 1983). A kinetic study of replication of a λ*A*⁻ mutant (defective in packaging DNA) showed that both θ forms and some σ forms with long tails accumulate early but that θ forms disappear, whereas σ forms can be seen throughout the latent period. The best estimate from small numbers of replicating molecules obtained in these studies is that approximately 3 rolling circles per cell are present from about 15 minutes after infection until the end of the latent period (Better and Freifelder 1983; see also Murialdo 1974).

One important factor in regulating the transition from the early to the late phase of replication may be the availability of the *O* and *P* proteins. Temperature-shift experiments with *tsO* and *tsP* mutants show that both proteins are essential for late, as well as for early, replication (Takahashi 1975a; Klinkert and Klein 1978). At late times the requirement for *O* appears more stringent than that for *P*, perhaps implying that *O* participates in both the initiation and progression of replication forks, whereas *P* participates only in initiation (Klinkert and Klein 1978). The expression of *O* and *P* is limited at late times by the *cro* protein. λ*cro*⁻ mutants synthesize aberrant replicative structures (Folkmanis et al. 1977), possibly because extra initiation events occur on replicating molecules (Better and Freifelder 1983). When DNA synthesis is delayed by limiting the amount of active *O* and *P* proteins made after infection, the first replicating molecules observed are almost all σ forms (Bastia et al. 1975; Takahashi 1975b). However, the level of expression of the *O* and *P* proteins is not sufficient to account for the shutoff of θ replication and the synthesis of σ forms, because some λ mutants replicating in selected *E. coli* hosts remain locked in the θ mode even when *O* and *P* should be regulated normally (see below).

The *gam*-gene product clearly plays a major role in the synthesis of concatemeric DNA, and the *red* proteins may also be important. λ*gam*⁻*red*⁻ mutants fail to form plaques on recombination-deficient *recA*⁻ hosts (Zissler et al. 1971). Under these conditions, substantial DNA synthesis occurs, but very little concatemeric DNA accumulates (Enquist and Skalka 1973). Replication continues for extended periods in the θ mode, and relaxed and closed circular DNA molecules are synthesized throughout the latent period (Enquist and Skalka

1973; Reuben et al. 1974; Greenstein and Skalka 1975; Sogo et al. 1976; Reuben and Skalka 1977). Some σ forms can be detected, but they have very short tails (Better and Freifelder 1983). The defect in rolling-circle replication does not result simply from the absence of recombination functions and *gam* protein. The λ*red*⁻*gam*⁻ phages grow well in *E. coli* double mutants defective in both *recA* and in a second recombination gene, *recB* or *recC* (Zissler et al. 1971). Under these conditions, concatemers are synthesized efficiently (Skalka 1971; Enquist and Skalka 1973), apparently by rolling-circle replication (Takahashi 1977).

The critical function of the *gam* protein is to inhibit the activity of exonuclease V, encoded by the *recB* and *recC* genes (Unger and Clark 1972; Unger et al. 1972; Enquist and Skalka 1973; Sakaki et al. 1973; Karu et al. 1975). When this nuclease can act freely on replicating λ DNA, it attacks the tails of σ forms and perhaps attacks some (unidentified) transient structure required for the initiation of rolling-circle replication (Greenstein and Skalka 1975). Thus, the *gam* protein protects the rolling-circle λ replicative forms from degradation.

The function of the *red* proteins and host recombination functions in λ replication is less clear. One possible role is to provide multimeric molecules by promoting recombination between circular monomers. Because multimers carry two *cos* sites, they would be efficient substrates for encapsidation (see Feiss and Becker, this volume). In fact, if normal replication is blocked, recombination becomes necessary for the maturation of λ chromosomes (Stahl et al. 1972b). Another possible role for recombination functions is to create new replication forks (Stahl et al. 1973; Skalka 1974). Some DNA synthesis is required for the recovery of recombinants from events mediated by the λ*red* proteins (Stahl et al. 1972a, 1982; also see Smith, this volume). Stahl and his colleagues have proposed models to account for the DNA synthesis associated with recombination. They suggest a mechanism by which recombination enzymes and a DNA polymerase might generate from two monomeric circles a structure with the topology of a rolling circle (Stahl et al. 1973, 1978; Stahl and Stahl 1974; Stahl 1979). If such a structure acquired a full complement of replication proteins, it might then replicate extensively.

Consistent with the idea that the λ recombination proteins help to create rolling-circle replication intermediates, Enquist and Skalka (1973) reported that both the rate of synthesis and the net synthesis of λ DNA are decreased in *red*⁻ mutants. However, Better and Freifelder (1983) observed no substantial defect in the accumulation of λ DNA after infection by a *red*⁻ mutant. Furthermore, neither the number nor the structure of σ forms was altered. In particular, the *red*⁻ mutation did not dramatically change the distribution of the map positions of the free ends of the tails of σ forms, as might have been expected. Thus, the study of Better and Freifelder gave no support to the hypothesis that the *red* proteins help to initiate new replication forks for rolling-circle replication. The only obvious difference in the population of intracellular molecules associated with the *red*⁻ mutation was a decrease in the number of circular oligomers present at late times. It is not known whether these oligomers are the products of *red*-mediated recombination and, if so, whether their formation involves DNA synthesis.

INITIATION OF λ DNA REPLICATION

Although the λ replication program is not completely understood, it is clear that all θ forms and at least 40 percent of σ forms initiate in a unique region of the chromosome. This region contains a regulatory site, *ori*, for the control of replication. Interactions involving *ori*, the *O* and *P* proteins, and host replication proteins determine the specific initiation of DNA synthesis.

Origin of Replication

The position of the replication origin region was determined by mapping the branch points of many θ structures and extrapolating to zero replication. It lies 81.7 ± 2.9% from the left end of the genome, in the vicinity of the replication genes *O* and *P* (Schnös and Inman 1970). It is not known whether this region contains a single startpoint or many possible startpoints for new DNA chains.

Genetic studies identified a control site for initiation, called *ori*. This site lies within the *O* gene, at approximately 80.5–80.6% from the left end of the genome. A map of the λ replication control region is shown in Figure 3. The *ori*

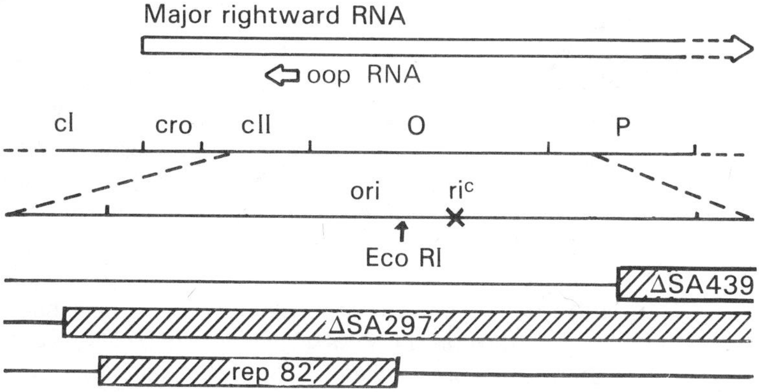

Figure 3 Replication control region of λ. Genes *cro*, cII, *O*, and *P* are shown roughly to scale. RNA transcripts synthesized in this region are indicated by open boxes; arrowheads denote the 3′ ends. The lower half of the figure is an expanded map of gene *O*. The origin control site, *ori*, maps within *O* to the left of a site for cleavage by restriction endonuclease *Eco*RI (→). The mutations *ri*ᶜ5b and *ri*ᶜD (×) create new promoters for rightward transcription. The λ DNA sequences removed by the deletions SA439 and SA297, which were used to define the *ori* region (see text) and substituted in the hybrid phage λ*rep*⁸², are indicated by the hatched boxes.

site was delimited by testing the ability of deletion mutants to replicate. Prophage strains carrying deletions entering the replication region from the right were induced and superinfected with a helper phage to provide the *O* and *P* proteins. Replication of the prophage DNA was measured by hybridization with labeled RNA probes complementary to various segments of the genome. A prophage with a deletion ending in *c*II failed to replicate (SA297, Fig. 3). However, a prophage with a deletion ending in the carboxyterminal half of *O* replicated well under identical conditions (SA439, Fig. 3). Thus, some essential *cis*-acting element must lie between the ends of these two deletions (Stevens et al. 1971).

The mapping of *ori* was refined by using cloned DNA fragments. Segments of the λ replication region were inserted into the nonessential region of a related lambdoid phage (Moore et al. 1977) or of a ColE1-type plasmid (Lusky and Hobom 1979a,b). The chimeric molecules were challenged to replicate in the presence of the λ*O* and λ*P* proteins when initiation from the origin of the vector was blocked by repression or appropriate mutations. In the phage test system, replication was measured by the virus yield in a single cycle of growth, and the *O* and *P* proteins were provided by a helper phage (Furth et al. 1977). In the plasmid system, replication was measured by the maintenance of a test plasmid carrying an antibiotic resistance gene, as judged by colony formation on selective media, and the *O* and *P* proteins were provided by a second plasmid (Lusky and Hobom 1979a,b).

A cloned DNA fragment covering the region from p_R to an *Eco*RI restriction endonuclease site within gene *O* permitted efficient λ-type origin function in the phage test system (Furth et al. 1977; Moore et al. 1979). Partial denaturation mapping of replicating molecules confirmed that bidirectional replication can initiate from the transposed *ori* region (Schnös et al. 1982). A λ segment with the same *Eco*RI endpoint gave some replication in the plasmid system, but a segment with an endpoint located an additional 34 bp to the right of the *Eco*RI site appeared to enhance replication by 20- to 30-fold (Lusky and Hobom 1979b). Tests with deletions and subfragments of the origin region localized the left-hand boundary of *ori* to less than 200 bp from the *Eco*RI site (Lusky and Hobom 1979b; Moore et al. 1979).

Colony formation promoted by *ori* in the plasmid system was observed only if the test plasmid carried a second λ segment from within gene *c*II, containing a site termed *ice* (for inceptor sequence) (Lusky and Hobom 1979b). This site lies approximately 600 bp to the left of *ori*. Remarkably, the *ice* segment without *ori* permitted the replication of a ColE1-type plasmid under conditions nonpermissive for the vector alone (Lusky and Hobom 1979a). The maintenance of the plasmid carrying *ice* depended on the λ*P* protein but was independent of the *O* protein. A similar *ice* fragment from lambdoid phage 434 promoted plasmid replication independent of any λ-gene products. In contrast, when *ori* and *ice* were contained in the same plasmid, both the *O* and *P* proteins were obligatory for its maintenance (Lusky and Hobom 1979b).

It was speculated that *ice* is a site at which an RNA primer for DNA synthesis terminates and DNA elongation begins (Hobom et al. 1979; Lusky and Hobom 1979a,b). However, it is not clear whether *ice* plays any role in λ phage replication, and it is conceivable that its observed activity reflects some peculiarity of plasmid replication, stability, or compatibility (for discussion, see Hobom et al. 1979, 1981; Moore et al. 1979, 1981). There exist viable λ transducing phages in which the *ice* site has been deleted (λ*spi vir* phages in which a substitution extends through *c*II; Belfort et al. 1975; Smith 1975) and a viable insertion mutant in which *ice* has been separated from *ori* by 4000 bp (λpk35; Young et al. 1980; Moore et al. 1981). The ability of these derivatives to replicate efficiently suggests that *ice* is not essential for phage growth or that it can be replaced by other sequences. Thus far, *ori* is the only site shown to be indispensable for normal λ DNA replication.

Essential components of the *ori* sequence have been defined by localized mutations that are *cis*-dominant to wild type and map within the origin region. The ori^- mutations were selected from an inducible λ*c*I*ts*N^- prophage that cannot excise from the host chromosome but can undergo some replication. Since replication of the integrated prophage is lethal to the host, lysogenic cells that survive induction carry phage or host mutations that block autonomous λ replication (Eisen et al. 1969). Selection from a partial diploid carrying one copy of the λ*c*I*ts*N^- prophage in the host chromosome and a second copy in an F′ episome preferentially elicited *cis*-dominant replication defects, and some mutants could express active *O* and *P* proteins (Dove et al. 1971). Similar ori^- mutations were isolated by screening among survivors of induction of a λ*c*I*ts*N^- prophage for those that could provide active *O* and *P* proteins to a λ$imm^{21}O^-P^-$ test phage (Rambach 1973).

Despite their ability to supply active *O* protein, the ori^- mutations map within gene *O* (Denniston-Thompson et al. 1977; Furth et al. 1977). Nucleotide sequence analysis of the origin regions of wild-type λ and of four ori^- mutations showed that the mutations lie in a small region just to the left of the *Eco*RI restriction site in the middle of *O* (Fig. 4A). Three of the ori^- mutations (*r*93, *r*99, *r*96) are small deletions, of 24 bp, 12 bp, and 15 bp, respectively, whereas one (*ti*12) is a point transversion (Denniston-Thompson et al. 1977; Scherer 1978). Because each of the deletions removed a multiple of 3 bp, none caused a shift in the translational reading frame of *O*. The amino acids lost from the *O* protein by these mutations must not be essential for its function (Denniston-Thompson et al. 1977; Furth and Yates 1978; see also Moore et al. 1981). The sequence analysis of the ori^- mutations defines a minimum segment of 63 bp that is important for replication origin function.

One striking structural feature of ori^λ is the presence of four nearly perfect repeats of a 19-bp sequence, which itself contains an inverted repeat (Fig. 4A; Grosschedl and Hobom 1979; Hobom et al. 1979; Moore et al. 1979). The units are separated by a few base pairs. A region with four repeated units, analogous to that in ori^λ, is present in the *ori* region of ϕ80 (ori^{80}). However, the unit in ori^{80} differs significantly in sequence from that in ori^λ (Fig. 4A; Grosschedl and

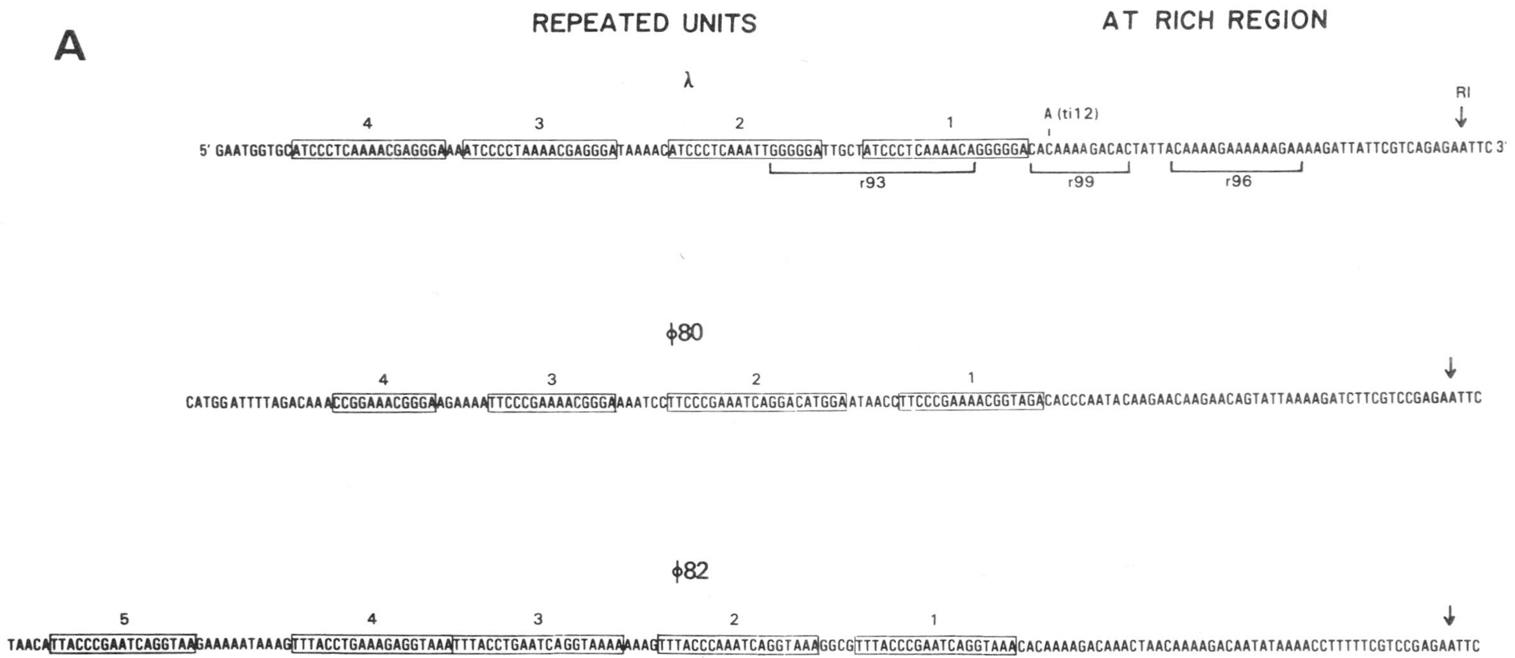

Figure 4 (*A*) DNA sequence (*l* strand) of *ori* regions of phages λ, ϕ80, and 82 (Grosschedl and Hobom 1979; Hobom et al. 1979; Moore et al. 1979, 1981). (RI) Site of cleavage within gene *O* by restriction endonuclease *Eco*RI. Repeated sequences within each *ori* region are boxed, and the segment rich in A•T base pairs is indicated. *O* protein binds to repeated units in *ori*$^{\lambda}$ (Tsurimoto and Matsubara 1981b). Brackets indicate deletion mutations *ori*$^{-}$*r*93, *ori*$^{-}$*r*99, and *ori*$^{-}$*r*96 (Denniston-Thompson et al. 1977). Transversion in *ori*$^{-}$*ti*12 is also shown (Scherer 1978). (*B*) (see facing page) Potential cloverleaf structure of *ori*$^{\lambda}$. (RI) *Eco*RI cleavage site within gene *O*. (Data from Grosschedl and Hobom 1979.)

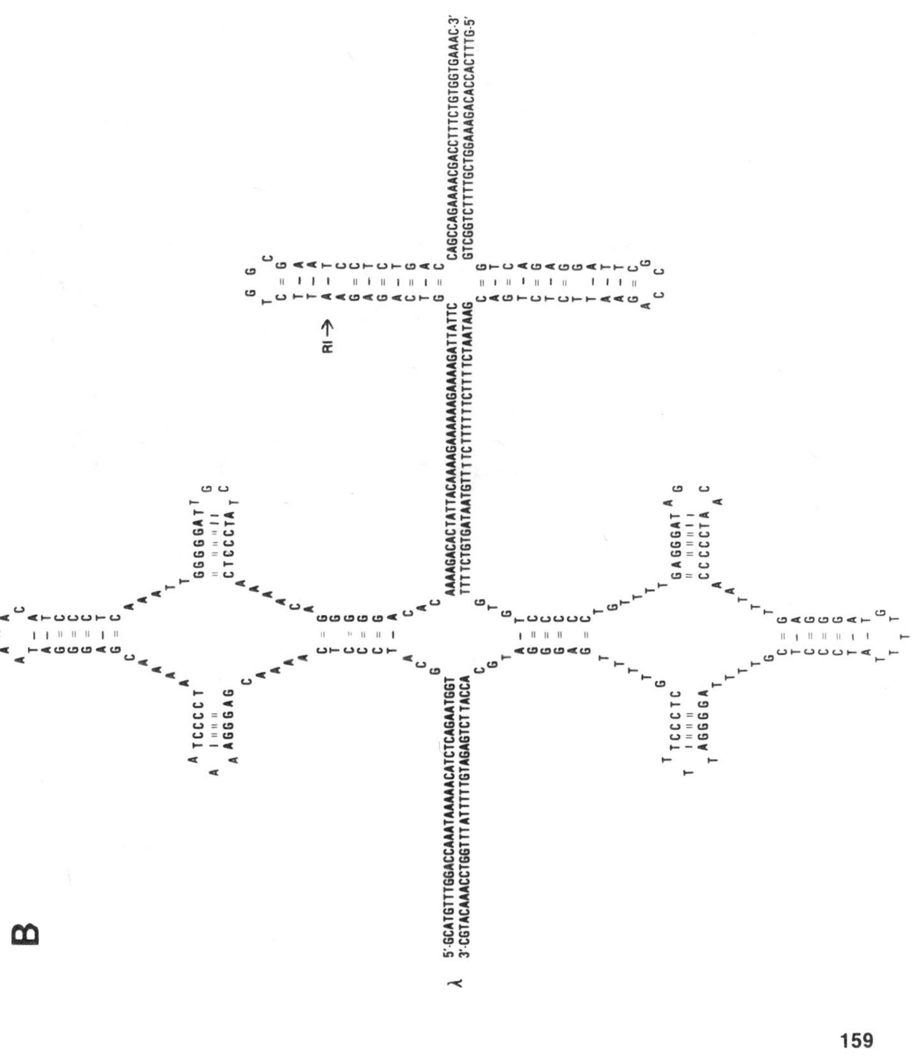
B
λ
RI

Hobom 1979; Moore et al. 1979, 1981). Phage 82 DNA has five repeats of a sequence that differs both from that of λ and that of ϕ80 (Fig. 4A; Moore et al. 1979, 1981). Thus, the presence of repeated units has been conserved, but the particular sequence has diverged among λ, ϕ80, and 82.

The portion of the *ori*$^{\lambda}$ and *ori*^{80} sequences containing the repeat units might fold into an elaborate cloverleaf structure (Fig. 4B). It has been suggested that this represents an activated state of the origin and that the structure might be stabilized by a bound initiator protein (Grosschedl and Hobom 1979; Hobom et al. 1979). Although no available experimental evidence directly tests this model, some data mitigate against it. The existence of functional lambdoid phage origins with odd numbers of repeated segments (wild-type *ori*^{82} and *ori*$^{\lambda}$ in a second-site revertant of λ*ori*$^-$*r*93) is difficult to reconcile with the hypothesized role of a highly symmetrical cloverleaf (Moore et al. 1979, 1981). Furthermore, the pattern of nuclease digestion of *ori* sequences protected by bound *O* protein (discussed below) differs from the simplest predictions of the cloverleaf model (Tsurimoto and Matsubara 1981b).

The *ori* region has other unusual structural features. The segment of 40 bp immediately to the right of the repeated units is rich in A•T pairs and displays an asymmetric distribution of purines and pyrimidines (more purines on the *l* strand, including a run of 18 in *ori*$^{\lambda}$). These properties might facilitate local denaturation of the double helix (Denniston-Thompson et al. 1977). The deletions *ori*$^-$*r*99 and *ori*$^-$*r*96 and the point mutation *ori*$^-$*ti*12 lie in this region (Fig. 4A; Denniston-Thompson et al. 1977; Scherer 1978). At the right-hand end of this A•T-rich segment, there is a short stretch in which pyrimidines predominate on the *l* strand. The next segment to the right contains an inverted repeat and can be drawn as a stem and loop (Grosschedl and Hobom 1979) (Fig. 4B). Both *ori*^{80} and *ori*^{82} strongly resemble *ori*$^{\lambda}$ in these segments, although the sequence of each is unique (Grosschedl and Hobom 1979; Hobom et al. 1979; Moore et al. 1979, 1981). More limited sequence homology has also been noted with other replication origin regions, including that of *E. coli* (Denniston-Thompson et al. 1977; Fiddes et al. 1978; Sims and Dressler 1978; Hobom et al. 1979, 1981; Hirota et al. 1979; Messer et al. 1979; Moore et al. 1981). Until the mechanism of initiation of replication is determined, the functional significance of the structure of the lambdoid phage *ori* regions and of homologies with other origins remains a matter of speculation.

Interaction of *O* Protein with *ori*

A specific interaction between the *O* protein and a segment of *ori* has been identified by genetic and biochemical analysis. The specificity was revealed initially by testing whether the replication proteins of other lambdoid phages could substitute for the *O* and *P* proteins in promoting replication of λ. For example, a hybrid phage carrying the replication genes of *Salmonella* phage P22 (λ*rep*^{22}) failed to help a λ*O*$^-$ or λ*P*$^-$ mutant to replicate in mixed infection. Conversely, λ failed to help a rep^{22} hybrid defective in a P22 replication gene (Hilliker and Botstein 1976). Similarly, coliphages 82 and ϕ80 failed to help λO^- mutants to replicate, although both of these phages did help λP^- mutants (Dove 1968; Szpirer and Brachet 1970; Dove et al. 1971).

The basis for type-specific gene-*O* complementation was analyzed using hybrids between λ and ϕ80 or phage 82. Some hybrids were derived from recombination near the middle of gene *O*, just to the right of *ori* (Furth et al. 1978). These recombinants encode hybrid *O* proteins, with the aminoterminal portion characteristic of one parent and the carboxyterminal portion characteristic of the other (Furth and Yates 1978; Furth et al. 1978, 1979). The gene *O* hybrids were used to map the specificity determinants for lambdoid phage replication. When an O^+ phage was challenged to help the replication of an O^- tester, it promoted replication only if the aminoterminal portion of its *O* protein matched the *ori* segment of the recipient. The carboxyterminal segment of the *O* protein, in contrast, could derive from another phage type. In one hybrid, approximately 430 bp from phage 82 was substituted for 400 bp of λ (rep^{82}, Fig. 3) and conferred both the *O* donor and receptor specificity of phage 82 (Furth et al. 1978). Genetic and DNA heteroduplex mapping (Furth et al. 1978), synthesis of the hybrid protein in vitro (Furth and Yates 1978), and DNA sequence analysis (Moore et al. 1981) showed that the rep^{82} substitution lies entirely within the aminoterminal segment of *O* and includes *ori*. This suggested strongly that the aminoterminal portion of the *O* protein interacts directly with *ori*.

Biochemical studies have proved that the *O* protein binds to specific DNA sequences in *ori* (Tsurimoto and Matsubara 1981a,b,c). The protein was purified from *E. coli* strains designed to produce it at high levels (Tsurimoto and Matsubara 1981a; Tsurimoto et al. 1982). On the basis of genetic evidence for a specific interaction between *O* and *ori*, protein fractions were assayed for the ability to bind λdv DNA but not ϕ80dv DNA to nitrocellulose filters. Fractions containing the *O* protein displayed the predicted λ-specific DNA-binding activity, and this assay guided the purification of the protein to apparent homogeneity (Tsurimoto and Matsubara 1981a; Tsurimoto et al. 1982). The monomeric molecular weight of the purified *O* protein, 34,000 daltons, agrees well with the identification of the *O*-gene product by radiochemical labeling in vitro and in vivo (Raab et al. 1977; Yates et al. 1977; Epp 1978; Lipinska et al. 1980; Tsurimoto and Matsubara 1981a) and with DNA sequence analysis (Scherer 1978). The nucleotide sequence reveals two possible polypeptides, depending on a choice between two initiation codons in the same translational reading frame. The amino acid sequence of the amino terminus of the purified *O* protein was determined and was consistent with initiation at the first of these start codons, giving a polypeptide of 298 amino acids after removal of the terminal formylated methionine residue (Tsurimoto et al. 1982).

Predictions of secondary structure from the deduced amino acid sequence suggest that the *O* polypeptide folds into two functional domains, with the central region corresponding to *ori* serving as a flexible bridge (Scherer 1978; Moore et al. 1981). Presumably, the aminoterminal domain would contain determinants essential for recognition of specific sequences in *ori* (Denniston-Thompson et al. 1977; Furth et al. 1978, 1979). The importance of the aminoterminal portion of the *O* protein was confirmed by the demonstration that a mutant *O* protein, lacking an internal segment of 20 amino acids near the amino terminus, fails to bind *ori* DNA (Tsurimoto and Matsubara 1981a). Furthermore, an aminoterminal *O*-protein fragment of 16,000 daltons, synthesized from a plasmid containing the λ *Eco*RI DNA fragment terminating in gene *O*, does bind *ori* DNA (K. Zahn, unpubl.).

Where in the origin region does the *O* protein bind? When fragments of λdv DNA generated by digestion with selected restriction endonucleases were tested in the binding assay with purified *O* protein, a single fragment of 164 bp was retained (Tsurimoto and Matsubara 1981a). This fragment lies immediately to the left of the *Eco*RI site in gene *O* and covers the positions of all of the well-characterized *ori*⁻ mutations. The sequences bound by the *O* protein were localized more precisely by nuclease protection experiments. Saturating levels of *O* protein protected a single DNA segment of about 95 bp from attack by either an endonuclease or an exonuclease (Tsurimoto and Matsubara 1981b,c). This segment corresponds exactly to the portion of the *ori* region containing the four 19-bp repeat units (Fig. 4A). Low levels of *O* protein protected a DNA segment of about 45 bp, containing the two inner copies of the repeat unit (units 2 and 3, Fig. 4A). Protection was symmetrical for *l*-strand and *r*-strand sequences. This symmetry and the presence of inverted repeats within each 19-bp unit suggest that the *O* protein binds as a dimer (Tsurimoto and Matsubara 1981b).

The binding of *O* protein to the four repeated units was demonstrated in another way by using dimethyl sulfate as a probe for local DNA structure. This chemical methylates purine residues exposed in the grooves of the DNA double helix. Incubation of DNA containing ori^{λ} with *O* protein enhanced the methylation of specific guanine residues in each of the four units (K. Zahn, F. Blattner, and S. Wickner, unpubl.).

The biochemical data are supported by genetic evidence that the 19-bp repeated sequences are important for origin function. The deletion $ori^{-}r93$ eliminates one copy of the repeated unit, possibly as a consequence of recombination between units 1 and 2 (Denniston-Thompson et al. 1977; Fig. 4A). The number of copies of the repeated unit may be less critical than their spacing or precise nucleotide sequence, because in one second-site revertant of $\lambda ori^{-}r93$, two compensating single-base-pair frameshift mutations restore phage replication (Moore et al. 1981). The observation that the nucleotide sequence of the repeated units differs among λ, ϕ80, and 82 (Fig. 4A) probably explains the type specificity seen in complementation tests for *O*-protein function among these phages. It is likely that the repeated units are binding sites for the specific replication initiator protein of each phage.

The interaction of the *O* protein with *ori* is one in a chain of interactions that controls λ DNA replication (Furth et al. 1978). The *P* protein and a number of host proteins are also required for the initiation of DNA synthesis. Genetic evidence suggests that the *O* and *P* proteins interact and that the *P* protein and several host proteins interact.

Interaction of *P* Protein with *O* Protein and Host Proteins

Two observations suggested that the *O* and *P* proteins act together. First, a temperature-sensitive mutation in *O* could be suppressed by mutations mapping in *P* (Tomizawa 1971). Second, substitutions in the carboxyterminal portion of *O* determined whether a λ*P* or a φ80-type *P* protein could be utilized for replication. A hybrid *O* protein carrying the aminoterminal segment of λ and the carboxyterminal segment of φ80 did not function effectively with the *P* protein of λ and required the analogous gene product of φ80 (Furth et al. 1978, 1979). To date there is no biochemical evidence for a direct interaction of the *O* and *P* polypeptides.

The *P* protein, in turn, interacts with a number of host replication proteins. These interactions were suggested from studies of *E. coli* mutants selected for their inability to replicate λ DNA under conditions permissive for replication of the bacterial chromosome. The host mutations map at five distinct loci. Phage mutants able to overcome these blocks were obtained, and in every case the compensating λ mutation mapped in gene *P* (Georgopoulos and Herskowitz 1971; Saito and Uchida 1977). Some of the *E. coli* mutants fail to replicate their own DNA at high temperature. Among these are mutants defective in the well-characterized *dnaB* protein (Georgopoulos and Herskowitz 1971; Table 1). Mutations in *dnaJ* and *dnaK* impair host replication, but the functions of the corresponding gene products are not yet fully characterized (Georgopoulos and Herskowitz 1971; Georgopoulos 1977; Saito and Uchida 1977, 1978; Sunshine et al. 1977; Yochem et al. 1978; Georgopoulos et al. 1979, 1980). Both proteins are essential in some *O* and *P* protein-dependent in vitro replication systems (C. Georgopoulos, pers. comm.). The *dnaK* protein is one of a class of "heat shock" proteins induced by incubation of *E. coli* at high temperature (Georgopoulos et al. 1982). The purified *dnaK* protein displays DNA-independent ATPase activity and undergoes autophosphorylation when incubated with ATP (C. Georgopoulos, pers. comm.). Mutants mapping in two additional bacterial genes, *grpD* and *grpE*, define functions essential for λ replication but have not yet been shown to influence host replication (Saito and Uchida 1977; Saito et al. 1978).

A direct interaction between the *P* protein and the *dnaB* protein has been demonstrated biochemically (Wickner 1979; Klein et al. 1980; Tsurimoto et al. 1982). The *P* protein was partially purified using an in vitro complementation assay. Receptor lysates were prepared from *E. coli* infected with a λP^- mutant, and λ DNA synthesis was stimulated by addition of *P* protein (Klein et al. 1978).

Active *P* protein could be copurified with *dnaB* protein as a complex containing a molar ratio of roughly two *dnaB* to one *P* polypeptide (Klein et al. 1980). The association with *P* protein affects the function of *dnaB* protein. The induction of high levels of *P* protein was found to inhibit *E. coli* DNA replication (Klinkert and Klein 1979). Extracts of λ-infected *E. coli* failed to promote replication of single-stranded φX174 phage DNA, because *P* protein masked the activity of the *dnaB* protein (Wickner 1979). This inhibitory activity of *P* protein afforded an assay for its purification to near homogeneity. The polypeptide molecular weight of the purified protein, 26,000, agrees well with estimates from radiochemical labeling experiments in which the *P*-gene product was eliminated by substitutions or nonsense mutations (Oppenheim et al. 1977; Raab et al. 1977; Lipinska et al. 1980). Both the size and the aminoterminal amino acid sequence (Tsurimoto et al. 1982) are consistent with predictions from DNA sequence analysis (Schwarz et al. 1980). The purified *P* protein specifically inhibited the DNA-stimulated ATPase activity of *dnaB* protein (Wickner 1979; Tsurimoto et al. 1982) and also protected *dnaB* protein from heat inactivation (Wickner 1979). The association of the *P* protein and *dnaB* protein and the modification of *dnaB* protein activity may allow the *dnaB* protein to enter a complex capable of replicating λ DNA.

The nature of the interactions between other host proteins and the *O* and *P* proteins and *ori* remains unknown. The mechanism by which some *P* mutations overcome the blocks to replication in mutant hosts is similarly obscure, but the level of functional *P* protein (Georgopoulos and Herskowitz 1971), perhaps determined in part by its ability to form oligomers (Klinkert and Klein 1979), may be critical. In general, once primary initiation has occurred at the replication origin, the establishment and propagation of replication forks probably proceed by essentially the same biochemical mechanisms and are catalyzed by the same proteins as in replication of the bacterial chromosome.

Another possible interaction between replicating λ DNA and host components involves the cell membrane. In *E. coli* minicells infected with λ, both the *P* protein and covalently closed circular phage DNA associate preferentially with the membrane, as judged by fractionation of gently lysed cells by centrifugation to equilibrium through metrizamide density gradients (Zylicz and Taylor 1981). A number of studies by velocity centrifugation have shown that replicating λ DNA sediments in a complex with host components. These observations are usually taken to indicate a specific association of λ DNA with the cell membrane (Salivar and Gardinier 1970; Sakakibara and Tomizawa 1971; Hallick and Echols 1973; Siegel and Schaechter 1973; Valenzuela 1975). However, a recent examination suggested that λ DNA actually associates with the *E. coli* nucleoid and that this association with the bacterial chromosome persists even after removal of membranes by treatment with detergents (Better and Freifelder 1982). The chemical nature of the interaction with the bacterial chromosome is unknown. To date, there is no compelling evidence that λ replication requires interaction with the bacterial cell membrane.

Transcriptional Activation of *ori*

In addition to the replication proteins and an intact *ori* site, the initiation of λ DNA replication requires transcriptional activation by RNA synthesis near the replication origin (Dove et al. 1969, 1971; Klein and Powling 1972; Inokuchi et al. 1973). When immune lysogens are infected with λ and a heteroimmune helper phage, to provide *O* and *P* proteins, the repressed λ genomes fail to replicate (Thomas and Bertani 1964; Ptashne 1965; Green et al. 1967). Mutations that prevent transcription rightward from p_R to *ori* also impose a *cis*-dominant block to λ replication (Dove et al. 1969, 1971; Brachet and Green 1970; Furth et al. 1979).

No specific sequence has been identified that must be transcribed to activate the origin. Transcription in the vicinity of *ori* appears to suffice. Mutants of λ have been isolated that can overcome repression to replicate, when provided with the *O* and *P* proteins. They are called *ri*^{c}, after the *r*eplication *i*nhibition *c*onstitutive phenotype (Dove et al. 1969, 1971). These *ri*^{c} mutants have acquired new promoter sites that allow constitutive rightward transcription in the replication control region (Nijkamp et al. 1971). Two of the *ri*^{c} mutations have been mapped within gene *O*, perhaps at the same site, to the right of all of the essential components of *ori*. Analysis of transcripts synthesized by one of these mutants (*ri*^{c}5b) in vitro showed that initiation occurs at least 95 bp to the right of any sequence that has been implicated in origin function and proceeds rightward away from *ori* (Fig. 3; Furth et al. 1982). The determination of the nucleotide sequence changes in *ri*^{c}5b confirmed this result and showed that the new promoter arose by at most two point mutations within gene *O* (Moore and Blattner 1982).

Genetic analysis showed that the *ri*^{c} mutations do not bypass normal components of *ori*. A demonstration that *ori* remains essential when the *ri*^{c} character is active took advantage of the type specificity of the *ori-O* protein interaction. The *ri*^{c}5b character was introduced into a λ*rep*^{82} phage, carrying the *ori* specificity determinant of phage 82, and the hybrid was challenged to replicate in an immune lysogen coinfected with helpers of different replication type. The repressed *ori*^{82}*ri*^{c} phage replicated only if the helper phage provided an *O* protein of the 82 type. Thus, the normal *O*-protein-binding site in *ori* was still required (Furth et al. 1982).

The mechanism by which local transcription activates *ori* is not known. One possibility is that RNA polymerase separates the strands of the DNA double helix and so promotes the folding of each strand into structures required for initiation (Denniston-Thompson et al. 1977; Grosschedl and Hobom 1979). If this is the case, it must be possible for activation to occur even if the startpoint for transcription lies up to 95 bp outside of *ori*. Another possibility is that transcription provides an RNA primer for the initiation of DNA synthesis. If the priming hypothesis is correct, then the start site for replication can be physically distinct from the control site for initiation.

In addition to the major rightward transcript initiated at p_R, a small leftward transcript termed *oop* is synthesized near the *ori* region (Fig. 3). It was proposed that this transcript might serve as a primer for DNA replication (Blattner and Dahlberg 1972; Hayes and Szybalski 1975). However, deletion analysis with cloned DNA fragments has shown that the *oop* sequence is dispensable for normal λ origin function (Moore et al. 1979). It is not clear what role, if any, the *oop* transcript plays in λ DNA replication or gene regulation.

The control of *ori* by transcription serves an important function in lysogenization of *E. coli* by λ, because repression by the *c*I protein can shut off autonomous replication very rapidly. Even though the λ*ri*^{c} mutants appear to establish repressor synthesis normally, they kill most infected cells. This "lethal lysogenization" presumably results because *ori* remains competent to initiate replication after integration of the phage genome into the host chromosome (Ohashi and Dove 1976). Such replication would be lethal to the host (Eisen et al. 1969). Apparently, even the rapid decay of *O* protein, which has a half-life of less than 2 minutes (Wyatt and Inokuchi 1974; Epp 1978; Kuypers et al. 1980; Lipinska et al. 1980; Gottesman et al. 1981), is not sufficient to protect the presumptive lysogen.

λ DNA REPLICATION IN VITRO

A more precise understanding of λ DNA replication awaits the characterization of in vitro replicating systems. Such systems have proved to be of tremendous value in dissecting the molecular mechanisms of replication of a number of genomes, e.g., those of the small icosahedral DNA phages, of phages T4 and T7, and of the plasmid ColE1 (for review, see Kornberg 1980; Tomizawa and Selzer 1979).

Klein and colleagues found that mitomycin-treated cells infected with λ could replicate the phage DNA when gently lysed on cellophane discs (Klein and Powling 1972). This system facilitated the purification of *P* protein (Klein et al. 1978) but was less useful for other applications. More recently, the development of a soluble enzyme system that initiates replication at the *E. coli* chromosomal origin (Fuller et al. 1981; Kaguni et al. 1982) provided the basis for similar studies with λ. Several groups have found that extracts of *E. coli*, when supplemented with the *O* and *P* proteins, either pure or in crude extracts from overproducer cells, catalyze the replication of exogenously added circular λdv DNA molecules (Anderl and Klein 1982; Tsurimoto and Matsubara 1982; Wold et al. 1982). Similar plasmid molecules carrying *ori*^{80} or *ori*^{82} instead of *ori*$^{\lambda}$ do not respond to the λ*O* and λ*P* proteins in these systems.

As would be expected for faithful initiation at the λ replication origin, the replication of λdv DNA in the soluble in vitro systems is semiconservative and requires the *O* and *P* proteins (Anderl and Klein 1982; Tsurimoto and Matsubara 1982; Wold et al. 1982). Replication requires the four ribonucleoside triphosphates and is inhibited by rifampicin, an antibiotic that blocks transcription by

RNA polymerase. The participation of RNA polymerase in the in vitro replication of λdv may well be analogous to transcriptional activation of *ori* in vivo. DNA gyrase is essential, as inferred from inhibition of replication by the specific antibiotic inhibitors nalidixic acid and coumermycin (Tsurimoto and Matsubara 1982; Wold et al. 1982). This suggests that the substrate for initiation in these systems is supercoiled DNA. DNA binding protein, *dnaB* protein, and *dnaG* protein are required, since antibodies to these proteins inhibit λ DNA synthesis in the crude extracts (Wold et al. 1982). By this criterion, *dnaC* protein is not required. These observations on host protein requirements are consistent with the results of studies of replication of λ in *E. coli* temperature-sensitive mutants (Table 1).

Tsurimoto and Matsubara (1982) demonstrated that replication of λdv in vitro initiates near the *ori* region and proceeds bidirectionally. They added a dideoxynucleoside triphosphate to block chain elongation after varying amounts of replication and then mapped the replicated regions with respect to known sites of cleavage by DNA restriction endonucleases. Initiation was also shown to occur within a cloned ori^{λ} DNA fragment contained in a ColE1-derived plasmid. The origin function was abolished when the A·T-rich segment of *ori* was deleted from the cloned fragment (Tsurimoto and Matsubara 1981c, 1982). In contrast, deletion of DNA segments covering the *ice* site and the *oop* region had no effect. Thus, in this system neither *ice* nor *oop* appears to be required for initiation of replication at the λ origin. However, the effect of the active ColE1 origins present in these in vitro systems remains to be determined.

The soluble in vitro systems should be suitable for detailed biochemical analysis of the initiation of λ DNA replication. Among many tasks of interest will be to determine the specific sites at which DNA synthesis starts, the mechanism of priming, and the precise roles of the *O* and *P* proteins and host proteins in λ replication.

ACKNOWLEDGMENTS

We thank Marc Better, Costa Georgopoulos, Kenichi Matsubara, Roger McMacken, and Ken Zahn for access to data prior to publication; Ross Inman and Maria Schnös for electron micrographs; and Gerald Selzer, Frank Stahl, and Ken Zahn for comments on the manuscript. M.F. thanks William Dove for the chance to share in explorations of this corner of the biological universe.

REFERENCES

Anderl, A. and A. Klein. 1982. Replication of λ dv DNA in vitro. *Nucleic Acids Res.* **10:** 1733.

Bastia, D., N. Sueoka, and E. Cox. 1975. Studies on the late replication of phage lambda: Rolling-circle replication of the wild type and a partially suppressed strain, *O*am29 *P*am80. *J. Mol. Biol.* **98:** 305.

Belfort, M., D. Noff, and A. Oppenheim. 1975. Isolation, characterization and deletion mapping of amber mutations in the *c*II gene of phage λ. *Virology* **63:** 147.

Berg, D. 1974. Genes of phage lambda essential for λ dv plasmids. *Virology* **62:** 224.

Better, M. and D. Freifelder. 1982. Studies on the association of *E. coli* phage λ DNA and the host chromosome: Lack of a role of membranes. *Virology* **119:** 159.

———. 1983. Studies on the replication of *Escherichia coli* phage λ DNA. I. The kinetics of DNA replication and requirements for the generation of rolling circles. *Virology* **126:** 168.

Blattner, F. and J. Dahlberg. 1972. RNA synthesis startpoints in bacteriophage λ: Are the promoter and operator transcribed? *Nat. New Biol.* **237:** 227.

Boon, T. and N. Zinder. 1969. A mechanism for genetic recombination generating one parent and one recombinant. *Proc. Natl. Acad. Sci.* **64:** 573.

Brachet, P. and B. Green. 1970. Functional analysis of early defective mutants of coliphage λ. *Virology* **40:** 792.

Carter, B. and M. Smith. 1970. The intracellular pools of bacteriophage λ DNA. *J. Mol. Biol.* **50:** 713.

Carter, B., B. Shaw, and M. Smith. 1969. Two stages in the replication of bacteriophage λ DNA. *Biochim. Biophys. Acta* **195:** 494.

Dawson, P., B. Hohn, T. Hohn, and A. Skalka. 1976. Functional empty capsid precursors produced by a lambda mutant defective for late λ DNA replication. *J. Virol.* **17:** 576.

Denniston-Thompson, K., D. Moore, K. Kruger, M. Furth, and F. Blattner. 1977. Physical structure of the replication origin of bacteriophage lambda. *Science* **198:** 1051.

Dove, W. 1968. The genetics of the lambdoid phages. *Annu. Rev. Genet.* **2:** 305.

Dove, W., H. Inokuchi, and W. Stevens. 1971. Replication control in phage lambda. In *The bacteriophage lambda* (ed. A.D. Hershey), p. 747. Cold Spring Harbor Laboratory, Cold Spring Harbor, New York.

Dove, W., E. Hargrove, M. Ohashi, F. Haugli, and A. Guha. 1969. Replicator activation in lambda. (Suppl.) *Jpn. J. Genet.* **44:** 11.

Eisen, H., L. Pereira da Silva, and F. Jacob. 1969. The regulation and mechanism of DNA synthesis in bacteriophage lambda. *Cold Spring Harbor Symp. Quant. Biol.* **33:** 755.

Enquist, L. and A. Skalka. 1973. Replication of bacteriophage λ DNA dependent on the function of host and viral genes. I. Interaction of *red*, *gam*, and *rec*. *J. Mol. Biol.* **75:** 185.

Epp, C. 1978. "Early protein synthesis and its control in bacteriophage lambda." Ph.D. thesis, University of Toronto, Canada.

Feiss, M. and T. Margulies. 1973. On maturation of the bacteriophage lambda chromosome. *Mol. Gen. Genet.* **127:** 285.

Fiddes, J., B. Barrell, and G. Godson. 1978. Nucleotide sequences of the separate origins of synthesis of bacteriophage G4 viral and complementary DNA strands. *Proc. Natl. Acad. Sci.* **75:** 1081.

Folkmanis, A., W. Maltzman, P. Mellon, A. Skalka, and H. Echols. 1977. The essential role of the *cro* gene in lytic development by bacteriophage λ. *Virology* **81:** 352.

Freifelder, D., L. Chud, and E. Levine. 1974. Requirement for maturation of *Escherichia coli* bacteriophage lambda. *J. Mol. Biol.* **83:** 503.

Fuller, R., J. Kaguni, and A. Kornberg. 1981. Enzymatic replication of the origin of the *Escherichia coli* chromosome. *Proc. Natl. Acad. Sci.* **78:** 7370.

Furth, M. and J. Yates. 1978. Specificity determinants for bacteriophage lambda DNA replication. II. Structure of O proteins of λ-φ80 and λ-82 hybrid phages and of a λ mutant defective in the origin of replication. *J. Mol. Biol.* **126:** 227.

Furth, M., W. Dove, and B. Meyer. 1982. Specificity determinants for bacteriophage λ DNA replication. III. Activation of replication in λ*ri*^{c} mutants by transcription outside of *ori*. *J. Mol. Biol.* **154:** 65.

Furth, M., C. McLeester, and W. Dove. 1978. Specificity determinants for bacteriophage lambda DNA replication. I. A chain of interactions that controls the initiation of replication. *J. Mol. Biol.* **126:** 195.

Furth, M., J. Yates, and W. Dove. 1979. Positive and negative control of bacteriophage lambda DNA replication. *Cold Spring Harbor Symp. Quant. Biol.* **43:** 147.

Furth, M., F. Blattner, C. McLeester, and W. Dove. 1977. Genetic structure of the replication origin of bacteriophage lambda. *Science* **198:** 1046.

Gellert, M. 1981. DNA topoisomerases. *Annu. Rev. Biochem.* **50:** 879.

Georgopoulos, C. 1977. A new bacterial gene (groPC) which affects lambda DNA replication. *Mol. Gen. Genet.* **152:** 201.

Georgopoulos, C. and I. Herskowitz. 1971. *Escherichia coli* mutants blocked in lambda DNA synthesis. In *The bacteriophage lambda* (ed. A.D. Hershey), p. 553. Cold Spring Harbor Laboratory, Cold Spring Harbor, New York.

Georgopoulos, C., A. Lundquist-Heil, J. Yochem, and M. Feiss. 1980. Identification of the *E. coli dnaJ* gene product. *Mol. Gen. Genet.* **178:** 583.

Georgopoulos, C., K. Tilly, D. Drahos, and R. Hendrix. 1982. B66.0 protein of *Escherichia coli* is the product of the *dnaK*+ gene. *J. Bacteriol.* **149:** 1175.

Georgopoulos, C., B. Lam, A. Lundquist-Heil, C. Rudolph, J. Yochem, and M. Feiss. 1979. Identification of the *E. coli dnaK* (*groPC*756) gene product. *Mol. Gen. Genet.* **172:** 143.

Gilbert, W. and D. Dressler. 1969. DNA replication: The rolling circle model. *Cold Spring Harbor Symp. Quant. Biol.* **33:** 473.

Gottesman, S., M. Gottesman, J. Shaw, and M. Pearson. 1981. Protein degradation in *E. coli*. The *lon* mutation and bacteriophage lambda N and cII protein stability. *Cell* **24:** 225.

Green, M., B. Gotchel, J. Hendershott, and S. Kennell. 1967. Regulation of bacteriophage lambda DNA replication. *Proc. Natl. Acad. Sci.* **58:** 2343.

Greenstein, M. and A. Skalka. 1975. Replication of bacteriophage lambda DNA: *In vivo* studies of the interaction between the viral *gamma* protein and the host *recBC* DNase. *J. Mol. Biol.* **97:** 543.

Grosschedl, R. and G. Hobom. 1979. DNA sequences and structural homologies of the replication origins of lambdoid bacteriophages. *Nature* **277:** 621.

Hallick, L. and H. Echols. 1973. Genetic and biochemical properties of an association complex between host components and lambda DNA. *Virology* **52:** 105.

Hayes, S. and W. Szybalski. 1975. Role of *oop* RNA primer in initiation of coliphage lambda DNA replication. In *DNA synthesis and its regulation* (ed. M. Goulian et al.), p. 486. W.A. Benjamin, Menlo Park, California.

Hilliker, S. and D. Botstein. 1976. Specificity of genetic elements controlling regulation of early functions in temperate bacteriophages. *J. Mol. Biol.* **106:** 537.

Hirota, Y., S. Yasuda, M. Yamada, A. Nishimura, K. Sugimoto, H. Sugisaki, A. Oka, and M. Takanami. 1979. Structural and functional properties of the *Escherichia coli* origin of DNA replication. *Cold Spring Harbor Symp. Quant. Biol.* **43:** 129.

Hobom, G., M. Kröger, B. Rak, and M. Lusky. 1981. Primary and secondary replication signals in bacteriophage λ and IS5 insertion element initiation systems. In *The initiation of DNA replication* (ed. D. Ray), p. 245. Academic Press, New York.

Hobom, G., R. Grosschedl, M. Lusky, G. Scherer, E. Schwarz, and H. Kössel. 1979. Functional analysis of the replicator structure of lambdoid bacteriophage DNAs. *Cold Spring Harbor Symp. Quant. Biol.* **43:** 165.

Hohn, B. 1975. DNA as substrate for packaging into bacteriophage lambda, *in vitro*. *J. Mol. Biol.* **98:** 93.

Ihler, G. and Y. Kawai. 1971. Alternate fates of the complementary strands of lambda DNA after infection of *Escherichia coli*. *J. Mol. Biol.* **61:** 311.

Inman, R. and M. Schnös. 1971. Structure of branch points in replicating DNA: Presence of single-stranded connections in λ DNA branch points. *J. Mol. Biol.* **56:** 319.

Inokuchi, H., W. Dove, and D. Freifelder. 1973. Physical studies of RNA involvement in bacteriophage λ DNA replication and prophage excision. *J. Mol. Biol.* **74:** 721.

Joyner, A., L. Isaacs, H. Echols, and W. Sly. 1966. DNA replication and messenger RNA production after induction of wild-type λ bacteriophage and λ mutants. *J. Mol. Biol.* **19:** 174.

Kaguni, J., R. Fuller, and A. Kornberg. 1982. Enzymatic replication of the *E. coli* chromosomal origin is bidirectional. *Nature* **296:** 623.

Kaiser, A. 1971. Lambda DNA replication. In *The bacteriophage lambda* (ed. A.D. Hershey), p. 195. Cold Spring Harbor Laboratory, Cold Spring Harbor, New York.

Karu, A., Y. Sakaki, H. Echols, and S. Linn. 1975. The γ protein specified by bacteriophage λ. Structure and inhibitory activity for the *recBC* enzyme of *Escherichia coli*. *J. Biol. Chem.* **250:** 7377.

Kiger, J. and R. Sinsheimer. 1969. Vegetative lambda DNA. V. Evidence concerning single-strand elongation. *J. Mol. Biol.* **43:** 567.

Kleckner, N. and E. Signer. 1977. Genetic characterization of plasmid formation of N^- mutants of bacteriophage λ. *Virology* **79:** 160.

Klein, A. and A. Powling. 1972. Initiation of λ DNA synthesis in vitro. *Nat. New Biol.* **239:** 71.

Klein, A., E. Lanka, and E. Schuster. 1980. Isolation of a complex between the *P* protein of phage λ and the *dnaB* protein of *Escherichia coli*. *Eur. J. Biochem.* **105:** 1.

Klein, A., B. Bremer, H. Kluding, and P. Symmons. 1978. Initiation of lambda DNA replication *in vitro* promoted by isolated P-gene product. *Eur. J. Biochem.* **83:** 59.

Klinkert, J. and A. Klein. 1978. Roles of bacteriophage lambda gene products *O* and *P* during early and late phases of infection cycle. *J. Virol.* **25:** 730.

——. 1979. Cloning of the replication gene *P* of bacteriophage lambda: Effects of increased P-protein synthesis on cellular and phage DNA replication. *Mol. Gen. Genet.* **171:** 219.

Kornberg, A. 1980. *DNA replication*. W.H. Freeman, San Francisco.

——. 1982. *Supplement to DNA replication*. W.H. Freeman, San Francisco.

Kuypers, B., W. Reiser, and A. Klein. 1980. Cloning of the replication gene *O* of *E. coli* bacteriophage lambda and its expression under the control of the *lac* promoter. *Gene* **10:** 195.

Lipinska, B., A. Podhajska, and K. Taylor. 1980. Synthesis and decay of λ DNA replication proteins in minicells. *Biochem. Biophys. Res. Commun.* **92:** 120.

Lusky, M. and G. Hobom. 1979a. Inceptor and origin of DNA replication in lambdoid coliphages. I. The λ DNA minimal replication system. *Gene* **6:** 137.

——. 1979b. Inceptor and origin of DNA replication in lambdoid coliphages. II. The λ DNA maximal replication system. *Gene* **6:** 173.

Matsubara, K. 1981. Replication control system in lambda dv. *Plasmid* **5:** 32.

Matsubara, K. and A. Kaiser. 1969. λ dv: An autonomously replicating DNA fragment. *Cold Spring Harbor Symp. Quant. Biol.* **33:** 769.

Matsubara, K. and Y. Takeda. 1975. Role of the *tof* gene in the production and perpetuation of the λ dv plasmid. *Mol. Gen. Genet.* **142:** 225.

McClure, S. and M. Gold. 1973. Intermediates in the maturation of bacteriophge λ DNA. *Virology* **54:** 19.

McClure, S., L. MacHattie, and M. Gold. 1973. A sedimentation analysis of DNA found in *Escherichia coli* infected with phage λ mutants. *Virology* **54:** 1.

Messer, W., M. Meijer, H. Bergmans, F. Hansen, K. von Meyenburg, E. Beck, and H. Schaller. 1979. Origin of replication, *oriC*, of the *Escherichia coli* K12 chromosome: Nucleotide sequence. *Cold Spring Harbor Symp. Quant. Biol.* **43:** 139.

Moore, D. and F. Blattner. 1982. Sequence of λri^c5b. *J. Mol. Biol.* **154:** 82.

Moore, D., K. Denniston, and F. Blattner. 1981. Sequence organization of the origins of DNA replication in lambdoid coliphages. *Gene* **14:** 91.

Moore, D., K. Denniston-Thompson, M. Furth, B. Williams, and F. Blattner. 1977. Construction of chimeric phages and plasmids containing the origin of replication of bacteriophage lambda. *Science* **198:** 1041.

Moore, D., K. Denniston-Thompson, K. Kruger, M. Furth, B. Williams, D. Daniels, and F. Blattner. 1979. Dissection and comparative anatomy of the origins of replication of lambdoid phages. *Cold Spring Harbor Symp. Quant. Biol.* **43:** 155.

Murialdo, H. 1974. Restriction in the number of infecting lambda phage genomes that can participate in intracellular growth. *Virology* **60:** 128.

Nijkamp, H., W. Szybalski, M. Ohashi, and W. Dove. 1971. Gene expression by constitutive mutants of coliphage lambda. *Mol. Gen. Genet.* **114:** 80.

Ogawa, T. and J. Tomizawa. 1968. Replication of bacteriophage DNA. I. Replication of DNA of lambdoid phage defective in early functions. *J. Mol. Biol.* **38:** 217.

Ogawa, T., J. Tomizawa, and M. Fuke. 1968. Replication of bacteriophage DNA. II. Structure of replicating DNA of phage lambda. *Proc. Natl. Acad. Sci.* **60:** 861.

Ohashi, M. and W. Dove. 1976. Lethal lysogenization by coliphage lambda. *Virology* **72:** 299.

Okazaki, R., T. Okazaki, K. Sakabe, K. Sugimoto, R. Kainuma, A. Sugino, and N. Iwatsuki. 1969. In vivo mechanism of DNA chain growth. *Cold Spring Harbor Symp. Quant. Biol.* **33:** 129.

Oppenheim, A., N. Katzir, and A.B. Oppenheim. 1977. The product of gene *P* of coliphage λ. *Virology* **79:** 2137.

Ptashne, M. 1965. The detachment and maturation of conserved lambda prophage DNA. *J. Mol. Biol.* **11:** 90.

Raab, C., A. Klein, H. Kluding, P. Hirth, and E. Fuchs. 1977. Cell free synthesis of bacteriophage lambda replication protein. *FEBS Lett.* **80:** 275.

Rambach, A. 1973. Replicator mutants of bacteriophage λ: Characterization of two subclasses. *Virology* **54:** 270.

Reuben, R. and A. Skalka. 1977. Identification of the site of interruption in relaxed circles produced during bacteriophage lambda DNA circle replication. *J. Virol.* **21:** 673.

Reuben, R., M. Gefter, L. Enquist, and A. Skalka. 1974. New method for large scale preparation of covalently closed λ DNA molecules. *J. Virol.* **14:** 1104.

Saito, H. and H. Uchida. 1977. Initiation of the DNA replication of bacteriophage lambda in *Escherichia coli* K12. *J. Mol. Biol.* **113:** 1.

——. 1978. Organization and expression of the *dnaJ* and *dnaK* genes of *Escherichia coli* K12. *Mol. Gen. Genet.* **164:** 1.

Saito, H., Y. Nakamura, and H. Uchida. 1978. A transducing lambda phage carrying *grpE*, a bacterial gene necessary for lambda DNA replication, and two ribosomal protein genes, *rpsP* (S16) and *rpl8* (L19). *Mol. Gen. Genet.* **165:** 247.

Sakaki, Y., A. Karu, S. Linn, and H. Echols. 1973. Purification and properties of the γ-protein specified by bacteriophage λ: An inhibitor of the host recBC recombination enzyme. *Proc. Natl. Acad. Sci.* **70:** 2215.

Sakakibara, Y. and J. Tomizawa. 1971. Gene *N* and membrane association of lambda DNA. In *The bacteriophage lambda* (ed. A.D. Hershey), p. 691. Cold Spring Harbor Laboratory, Cold Spring Harbor, New York.

Salivar, W. and J. Gardinier. 1970. Replication of bacteriophage λ DNA associated with the host cell membrane. *Virology* **41:** 38.

Salzman, L. and A. Weissbach. 1967. The formation of intermediates in the replication of phage lambda DNA. *J. Mol. Biol.* **28:** 53.

Scherer, G. 1978. Nucleotide sequence of the *O* gene and of the origin of replication in bacteriophage lambda DNA. *Nucleic Acids Res.* **5:** 3141.

Schnös, M. and R. Inman. 1970. Position of branch points in replicating lambda DNA. *J. Mol. Biol.* **51:** 61.

——. 1982. Caffeine-induced reinitiation of phage λ DNA replication. *J. Mol. Biol.* **159:** 457.

Schnös, M., K. Denniston, F. Blattner, and R. Inman. 1982. Replication of bacteriophage λ DNA: Examination of variants containing double origins and observation of a bias in directionality. *J. Mol. Biol.* **159:** 441.

Schwarz, E., G. Scherer, G. Hobom, and H. Kossel. 1980. The primary structure of phage λ *P* gene completes the nucleotide sequence of the plasmid λ dvh93. *Biochem. Int.* **1:** 386.

Segawa, T. and J. Tomizawa. 1971. Formation of concatemers of lambda phage DNA in a recombination-deficient system. *Mol. Gen. Genet.* **111:** 197.

Siegel, P. and M. Schaechter. 1973. The role of the host cell membrane in the replication and morphogenesis of bacteriophages. *Annu. Rev. Microbiol.* **27:** 261.

Signer, E. 1969. Plasmid formation: A new mode of lysogeny in phage λ. *Nature* **223:** 158.

Sims, J. and D. Dressler. 1978. The site specific initiation of a DNA fragment: DNA sequence of the initiator region. *Proc. Natl. Acad. Sci.* **75:** 3094.

Skalka, A. 1971. Origin of DNA concatemers during growth. In *The bacteriophage lambda* (ed. A.D. Hershey), p. 535. Cold Spring Harbor Laboratory, Cold Spring Harbor, New York.

——. 1974. A replicator's view of recombination (and repair). In *Mechanisms in recombination* (ed. R. Grell), p. 421. Plenum Press, New York.

——. 1977. DNA replication—Bacteriophage lambda. *Curr. Top. Microbiol. Immunol.* **78:** 201.

Skalka, A., M. Poonian, and P. Bartl. 1972. Concatemers in DNA replication: Electron microscopic studies of partially denatured intracellular lambda DNA. *J. Mol. Biol.* **64:** 541.

Smith, G., 1975. Deletion mutations of the immunity region of coliphage λ. *Virology* **64:** 544.

Smith, M. and A. Skalka. 1966. Some properties of DNA from phage-infected bacteria. *J. Gen. Physiol.* (Suppl.) **49:** 127.

Sogo, J., M. Greenstein, and A. Skalka. 1976. The circle mode of replication of bacteriophage lambda: The role of covalently closed templates and the formation of mixed catenated dimers. *J. Mol. Biol.* **103:** 537.

Stahl, F. 1979. *Genetic recombination, thinking about it in phage and fungi.* W.H. Freeman, San Francisco.

Stahl, F. and M. Stahl. 1974. Red-mediated recombination in bacteriophage lambda. In *Mechanisms in recombination* (ed. R. Grell), p. 407. Plenum Press, New York.

Stahl, F., I. Kobayashi, and M. Stahl. 1982. Distance from cohesive endsite *cos* determines the replication requirement for recombination in phage λ. *Proc. Natl. Acad. Sci.* **79:** 6318.

Stahl, F., M. Stahl, and R. Malone. 1978. Red-mediated recombination of phage lambda in a $recA^-$ $recB^-$ host. *Mol. Gen. Genet.* **159:** 207.

Stahl, F., K. McMilin, M. Stahl, and Y. Nozu. 1972a. An enhancing role for DNA synthesis in formation of bacteriophage lambda recombinants. *Proc. Natl. Acad. Sci.* **69:** 3598.

Stahl, F., K. McMilin, M. Stahl, R. Malone, Y. Nozu, and V. Russo. 1972b. A role for recombination in the production of "free-loader" lambda bacteriophage particles. *J. Mol. Biol.* **68:** 57.

Stahl, F., S. Chung, J. Crasemann, D. Faulds, J. Haemer, S. Lam, R. Malone, K. McMilin, Y. Nozu, J. Siegel, J. Strathern, and M. Stahl. 1973. Recombination, replication, and maturation in phage lambda. In *Virus research* (ed. C. Fox and W. Robinson), p. 487. Academic Press, New York.

Stevens, W., S. Adhya, and. W. Szybalski. 1971. Origin and bidirectional replication in coliphage lambda. In *The bacteriophage lambda* (ed. A.D. Hershey), p. 515. Cold Spring Harbor Laboratory, Cold Spring Harbor, New York.

Sunshine, M., M. Feiss, J. Stuart, and J. Yochem. 1977. A new host gene (groPC) necessary for lambda DNA replication. *Mol. Gen. Genet.* **151:** 27.

Syvanen, M. 1974. *In vitro* genetic recombination of bacteriophage λ. *Proc. Natl. Acad. Sci.* **71:** 2496.

Szpirer, J. and P. Brachet. 1970. Relations physiologiques entre les phages tempérés λ et φ80. *Mol. Gen. Genet.* **108:** 78.

Takahashi, S. 1974. The *rolling-circle* replicative structure of a bacteriophage λ DNA. *Biochem. Biophys. Res. Commun.* **61:** 607.

——. 1975a. Role of genes *O* and *P* in the replication of bacteriophage λ DNA. *J. Mol. Biol.* **94:** 385.

——. 1975b. Physiological transition of a coliphage λ DNA replication. *Biochim. Biophys. Acta* **395:** 306.

——. 1975c. The starting point and direction of rolling-circle replication intermediates of coliphage λ DNA. *Mol. Gen. Genet.* **142:** 137.

——. 1977. Rolling circle replication: Replicative structure of bacteriophage lambda DNA in a recombination deficient system. *Mol. Gen. Genet.* **151:** 27.

Thomas, R. and L. Bertani. 1964. On the control of the replication of temperate bacteriophages superinfecting immune hosts. *Virology* **24:** 241.

Tomizawa, J. 1971. Functional cooperation of genes *O* and *P*. In *The bacteriophage lambda* (ed. A.D. Hershey), p. 549. Cold Spring Harbor Laboratory, Cold Spring Harbor, New York.

Tomizawa, J. and T. Ogawa. 1969. Replication of phage lambda DNA. *Cold Spring Harbor Symp. Quant. Biol.* **33:** 501.

Tomizawa, J. and G. Selzer. 1979. Initiation of DNA synthesis in *Escherichia coli. Annu. Rev. Biochem.* **48:** 999.

Tsurimoto, T. and K. Matsubara. 1981a. Purification of bacteriophage λ*O* protein that specifically binds to the origin of replication. *Mol. Gen. Genet.* **181:** 325.

——. 1981b. Purified bacteriophage λ*O* protein binds to four repeating sequences at the λ replication origin. *Nucleic Acids Res.* **9:** 1789.

——. 1981c. Interaction of bacteriophage lambda *O* protein with the lambda origin sequence. In *The initiation of DNA replication* (ed. D. Ray), p. 263. Academic Press, New York.

——. 1982. Replication of λ*dv* plasmid *in vitro* promoted by purified λ O and P proteins. *Proc. Natl. Acad. Sci.* **79:** 7639.

Tsurimoto, T., T. Hase, H. Matsubara, and K. Matsubara. 1982. Bacteriophage lambda initiators: Preparation from a strain that overproduces the O and P proteins. *Mol. Gen. Genet.* **187:** 79.

Unger, R. and A. Clark. 1972. Interaction of the recombination pathways of bacteriophage λ and its host *Escherichia coli* K12: Effects on exonuclease V activity. *J. Mol. Biol.* **70:** 539.

Unger, R., H. Echols, and A.J. Clark. 1972. Interaction of the recombination pathways of bacteriophage λ and host *Escherichia coli*: Effects on λ recombination. *J. Mol. Biol.* **70:** 531.

Valenzuela, M. 1975. Intermediates of the first round of λ DNA replication are preferentially found in a rapidly sedimenting complex. *Biochem. Biophys. Res. Commun.* **65:** 1221.

Valenzuela, M. and R. Inman. 1981. Direction of bacteriophage λ DNA replication in a thymine requiring *Escherichia coli* K-12 strain. Effect of thymidine concentration. *Nucleic Acids Res.* **9:** 6975.

Valenzuela, M., D. Freifelder, and R. Inman. 1976. Lack of a unique termination site for the first round of bacteriophage lambda DNA replication *J. Mol. Biol.* **102:** 569.

Wake, R., A. Kaiser, and R. Inman. 1972. Isolation and structure of phage λ head-mutant DNA. *J. Mol. Biol.* **64:** 519.

Wickner, S. 1979. DNA replication proteins of *Escherichia coli* and phage λ. *Cold Spring Harbor Symp. Quant. Biol.* **43:** 303.

Wold, M., J. Mallory, J. Roberts, J. LeBowitz, and R. McMacken. 1982. Initiation of bacteriophage λ DNA replication *in vitro* with purified λ replication proteins. *Proc. Natl. Acad. Sci.* **79:** 6176.

Wyatt, W. and H. Inokuchi. 1974. Stability of lambda *O* and *P* replication functions. *Virology* **58:** 313.

Yates, J., W. Gette, M. Furth, and M. Nomura. 1977. Effects of ribosomal mutations on the read-through of a chain termination signal: Studies on the synthesis of bacteriophage λ *O* gene protein *in vitro. Proc. Natl. Acad. Sci.* **74:** 689.

Yochem, J., H. Uchida, M. Sunshine, H. Saito, C. Georgopoulos, and M. Feiss. 1978. Genetic analysis of two genes, *dnaJ* and *dnaK*, necessary for *Escherichia coli* and bacteriophage lambda DNA replication. *Mol. Gen. Genet.* **164:** 9.

Young, E. and R. Sinsheimer. 1967. Vegetative bacteriophage λ DNA. II. Physical characterization and replication. *J. Mol. Biol.* **30:** 165.

——. 1968. Vegetative λ DNA. III. Pulse-labeled components. *J. Mol. Biol.* **33:** 49.

Young, R., D. Grillo, R. Isberg, J. Way, and M. Syvanen. 1980. Transposition of the kanamycin resistance transposon, Tn*903*. *Mol. Gen. Genet.* **178:** 681.

Zissler, J., E. Signer, and F. Schaefer. 1971. The role of recombination in growth of bacteriophage lambda. I. The gamma gene. In *The bacteriophage lambda* (ed. A.D. Hershey), p. 455. Cold Spring Harbor Laboratory, Cold Spring Harbor, New York.

Zylicz, M. and K. Taylor. 1981. Interactions between phage λ replication proteins, phage λ DNA and minicell membrane. *Eur. J. Biochem.* **113:** 303.

General Recombination

Gerald R. Smith*
Institute of Molecular Biology
University of Oregon
Eugene, Oregon 97403

Recombination has been studied in phage λ for two principal reasons: (1) Recombination plays a central role in the life cycle of λ, and (2) the experimental tractability of λ has provided a fruitful means for studying the molecular mechanism of this process shared by nearly all organisms. There are close connections between recombination and replication in λ, as well as numerous interactions between the viral and host functions mediating these processes. Sorting out these complex relations has led to our understanding these processes at least as thoroughly in λ and *Escherichia coli* as in any other organism.

This paper reviews several recombination pathways available to λ, the genes and enzymes associated with these pathways, and the roles of recombination and replication in the growth of λ. Study of these processes led to the discovery of special sites in λ that enhance general recombination. After consideration of studies bearing on the molecular mechanisms of recombination, some current models of recombination are discussed in terms of enzymes and sites promoting recombination.

Signer (1971) has reviewed many of the early investigations of recombination in λ. Several subsequent reviews have dealt with various aspects of recombination, many of which involve λ and *E. coli* (Clark 1973; Radding 1973, 1978, 1982; Broker and Doermann 1975; Miller 1975; Fox 1978; Low and Porter 1978; Stahl 1979a,b; Dressler and Potter 1982; Whitehouse 1982).

PATHWAYS OF RECOMBINATION

Several pathways of recombination are available to λ. Two of these, the Int and Red pathways, require phage-encoded functions, whereas the others appear to require only host-encoded functions. Under the appropriate circumstances, each pathway can promote either phage or host recombination. These pathways were originally defined by mutations blocking one of the pathways. Subsequent identification of enzymes coded by the corresponding genes has helped support the notion of separate pathways.

Genes

Clark and Margulies (1965) isolated *recA* mutants of *E. coli*, in which recombination following conjugation or P1-mediated transduction was reduced to

*Present address: Fred Hutchinson Cancer Research Center, Seattle, Washington 98104.

about 10^{-4} of the level found in wild type. The finding by Brooks and Clark (1967) that λ recombines nearly as frequently in these mutants as in wild-type bacteria suggested that λ codes for its own recombination functions.

Subsequently, several groups identified phage mutants that recombined at reduced frequency in *recA*-mutant hosts (Franklin 1967; Echols and Gingery 1968; Signer and Weil 1968). The mutations were in two closely linked genes; the first is variously referred to as *exo, redX, redα*, or *redA*, and the second, as *bet, β, redβ*, or *redB* (Fig. 1). (*exo* and *bet* are used here to indicate the specific genes, and *red*, to indicate either or both.) At about the same time, phage mutants deficient in site-specific recombination were identified. The mutations were found to be in two closely linked genes, *int* and *xis*. Studies with various combinations of these host and phage mutations led to the view that the *red*-gene products participate in a general recombination pathway (Red) distinct from the host pathway (Rec), which is dependent on *recA*. The *int*- and *xis*-gene products are involved in recombination only at the prophage attachment site *att* (Signer et al. 1969). The Int pathway is discussed by Weisberg and Landy (this volume).

Soon after identification of *recA* mutants, additional *E. coli* mutants were isolated with lesions in either of two genes, *recB* or *recC*, which are closely linked to each other but distant from *recA* (Fig. 2) (Emmerson and Howard-Flanders 1967; Low 1968; Willetts and Mount 1969; Willetts et al. 1969). These mutants have a residuum of about 10^{-2} of the wild-type level of recombination in Hfr-mediated conjugation or P1-mediated transduction. The many alleles known define two complementation groups (Willetts and Mount 1969) and have similar phenotypes. The greater residual recombination activity in *recB* or *recC* mutants, compared to that in *recA* mutants, is due at least in part to the activity of two pathways whose activities can be increased by *E. coli* mutations.

Clark and his colleagues isolated recombination-proficient strains as pseudorevertants of *recB recC* double mutants. The mutations restoring recombination proficiency fell into two classes, *sbcA* and *sbcB*, located far from the *recB recC* locus and far from each other (Barbour et al. 1970; Kushner et al. 1971). (*sbc* stands for suppressor of *recB recC*.) The *sbcA* mutations lead to high-level expression of the *recE* gene, which appears to be part of a cryptic

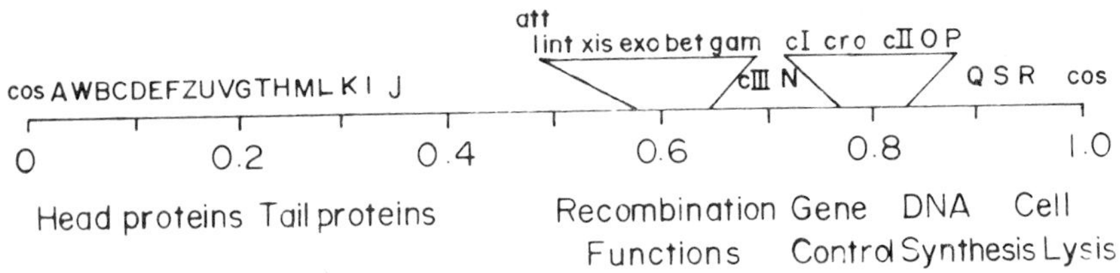

Figure 1 Genetic and physical map of phage λ. The genetic map, de ved from recombinant frequencies in vegetative crosses, is nearly congruent ʷ ⁱᵗh the physical map, derived from electron microscopy of DNA heteroduplexes employing deletions and substitutions (see Davidson and Szybalski 1971).

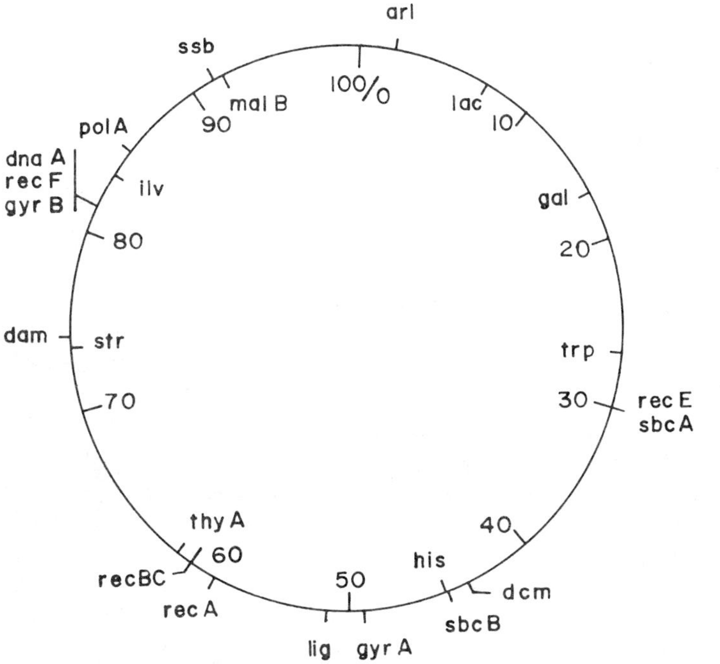

Figure 2 Genetic map of *Escherichia coli*, showing the locations of recombination-related genes discussed in this paper (see Bachman and Low 1980).

lambdoid prophage and encodes an enzyme similar to the λ*exo* product (Kushner et al. 1974; Gillen et al. 1977). The *recE* gene can be derepressed either by mutations (*sbcA*) inactivating the presumptive prophage repressor gene (Kaiser and Murray 1979) or by zygotic induction during Hfr-mediated conjugation into a female cell fortuitously free of the cryptic prophage (Low 1973). The latter type of experiment identified a locus *rac* (for *r*ecombination *ac*tivation) that transiently converted a *recBC* mutant female to recombination proficiency. The *rac* locus, which is present in some lines of *E. coli* K12 but not in others (Kaiser and Murray 1979), was subsequently identified with the *recE-sbcA* locus (Lloyd and Barbour 1974). The recombinational pathway operating in *sbcA recE*$^+$ strains is called the RecE pathway.

The second class of recombination-proficient pseudorevertants of *recB recC* strains has mutations inactivating the gene called *sbcB*, or *xonA*, coding for exonuclease I (Kushner et al. 1971). To explain this result, Clark (1973) has hypothesized that this nuclease destroys a recombination intermediate. The recombinational pathway operating in *sbcB* strains is called the RecF pathway, since mutations in the *recF* gene lead to recombination deficiency in *recB recC sbcB* strains (Horii and Clark 1973).

On the basis of early observations, Clark (1973) hypothesized the existence of the two pathways RecE and RecF as alternatives to the RecBC pathway. The distinction between the RecE and RecF pathways has become less clear with the observation that a *recF* mutation reduces recombination proficiency about 10–30-fold in a *recB recC sbcA* strain, suggesting that the RecF product contributes to the RecE pathway (Gillen et al. 1981). (A *recF* mutation has a 1000-fold reductive effect in a *recB recC sbcB* strain.) Clark (1980) has commented further on the relations of the hypothesized RecE and RecF pathways.

The three pathways described so far (RecBC, RecE, and RecF) all require the *recA*-gene product for *E. coli* chromosomal recombination. However, RecA is not needed by the RecE pathway acting on vegetative λ (Gillen 1974) or plasmids (Fishel et al. 1981). This observation is consistent with the view that RecE is very similar to Red, which does not need RecA, at least when acting on freely replicating λ, as noted earlier (but see below, DNA Replication and Recombination). It was mentioned above that *recBC* mutants have higher residual levels of recombination than do *recA* mutants. This residual activity appears to be due to low-level activity of the RecF pathway, since *recBC sbcB recF* mutants have recombination levels about as low as that of *recA* mutants (Horii and Clark 1973; Gillen 1974).

A fourth *E. coli* pathway of recombination has been detected under special circumstances. This pathway, called Rpo, requires RNA polymerase but not RecA or RecBC (Ikeda and Kobayashi 1977; Ikeda and Matsumoto 1979). These investigators extracted λ DNA from chloramphenicol-treated infected cells and packaged the DNA in a cell-free system. They obtained recombinants between Int^- Red^- phages infecting $RecA^-$ or $RecA^-$ $RecB^-$ cells. This recombination was rifampicin-sensitive, except in cells containing a rifampicin-resistant RNA polymerase, and was greatest in the region most actively transcribed (from *N* to *cro*; see Fig. 1; Friedman and Gottesman, this volume). Subsequent work has shown that extraction of λ from separately infected cells also leads to recombinants, suggesting that a major part of the reaction occurs in the cell-free packaging system (Matsumoto and Ikeda 1983). In fact, this reaction may occur only outside the cell, since in a standard λ infection the Rpo pathway is undetectable (Ikeda and Kobayashi 1977). Perhaps removal of RNA polymerase from the transcription complex, caused by the absence of translation or by DNA extraction, or both, produces recombinogenic R loops. R loops have been found at high frequency in nonreplicating intracellular λ molecules, and their formation is blocked by rifampicin but not by mutations (*recA, recBC, recF, int, red*) abolishing recombination under normal conditions (Chattoraj and Stahl 1980).

Enzymes

The *exo*-gene product, λ exonuclease, has a marked preference for double-stranded DNA over single-stranded DNA and progressively degrades each DNA strand from the 5′ end, releasing 5′ mononucleotides (Little 1967). At the end of the first stage of degradation, single strands of DNA half the length of the initial substrate are left. Thereafter, slow degradation of the single-stranded DNA continues. The enzyme will bind to a nick but will not degrade DNA from a nick (Cassuto and Radding 1971). However, it can convert a branched structure to an unbranched, nicked duplex, presumably by degradation of the 5′-ended annealed strand and subsequent annealing of the 3′-ended branch (Fig. 3) (Cassuto and Radding 1971). This reaction, called strand assimilation, is discussed below, in Models of Recombination. The native enzyme is a dimer of identical subunits with a molecular weight of 24,000 (Little 1967; Carter and Radding 1971).

The *bet*-gene product, β protein, promotes the renaturation of complementary single-stranded DNA and binds to λ exonuclease (Radding et al. 1971; Kmiec and Holloman 1981). Presumably the latter interaction, as well as interactions between the two exonuclease subunits, accounts for the complex complementation pattern seen with some *exo* and *bet* mutations (Shulman et al. 1970). The subunit molecular weight of β protein is 28,000 (Radding et al. 1971).

The *E. coli recA*-gene product has two disparate enzymatic activities. A protease activity, discussed by Roberts and Devoret (this volume), is involved in regulation of gene expression via cleavage of certain repressors. It appears that this protease activity is not needed for recombination, because certain *recA* mutants, originally called *lexB*, have reduced protease activity but nearly normal recombination activity (Morand et al. 1977; Roberts and Roberts 1981). Evidence that *recA* protein participates directly in recombination, and not via its regulation of other gene activities, comes from the observation (Kobayashi and Ikeda 1977, 1978) that *recA*-thermosensitive mutants quickly lose their recombination proficiency upon shift to high temperature, even in the presence of inhibitors of RNA and protein synthesis.

The recombination activity of *recA* protein may result from one or more of the demonstrated DNA-DNA interactions promoted by purified *recA* in the presence of ATP and Mg^{++} ions. Basic requirements for these reactions are (1) a single-stranded end in duplex DNA, either at the end of a molecule or at a nick or gap (the requirement for an end can be bypassed by a topoisomerase) and (2) a region of single-stranded DNA on one of the substrates sharing nucleotide sequence homology with a region on the other substrate. (For reviews, see McEntee and Weinstock 1981; Radding 1982.) Examples of reactions promoted by *recA* protein are diagramed in Figure 4. Notable reactions include the annealing of complementary single strands (Weinstock et al. 1979) and their formation

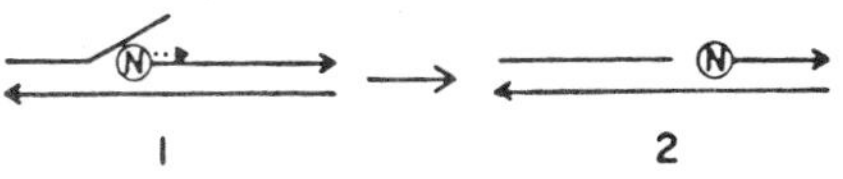

Figure 3 (N) Strand assimilation by λ exonuclease. (⟶) The 3′ end of DNA strands. (Reprinted, with permission, from Signer 1971.)

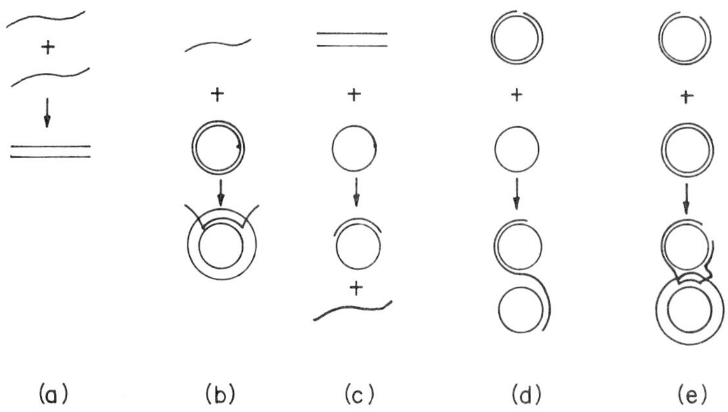

Figure 4 Representative reactions catalyzed by *recA* protein. Each line represents a single DNA strand, either linear or circular. The upper substrate in reaction *d* is a nicked circle, whereas that in reaction *e* is a gapped circle. The product of reaction *b* is a D loop.

of a D loop from single-stranded DNA complementary to double-stranded DNA (Shibata et al. 1979b). All of the reactions shown in Figure 4, except reaction a, require rather large amounts of *recA* protein. Generally, one *recA* monomer, with a molecular weight of 37,800 (Horii et al. 1980; Sancar et al. 1980), for every 5–10 nucleotides of single-stranded DNA produces maximal rates of reaction (Shibata et al. 1979a). Presumably, the single-stranded DNA must be completely coated with RecA protein for reaction to proceed. Other single-stranded binding proteins, such as the *E. coli ssb*-gene product or the phage T4 gene *32* product, both of which participate in recombination, can reduce the required quantity of *recA* protein (McEntee et al. 1980; Shibata et al. 1980; Cox and Lehman 1981a).

recA protein may play both a stoichiometric role, by coating single-stranded DNA, as well as a catalytic role, by hydrolyzing ATP and promoting DNA dissociation and association. Coating of single-stranded DNA by *recA* protein may be related to the ATP-stimulated formation of filamentous aggregates of *recA* protein both in the absence and in the presence of DNA (Ogawa et al. 1979; West et al. 1980; McEntee et al. 1981; Flory and Radding 1982). The DNA dissociation and association reactions have been separated into two kinetically distinct phases: an initial reaction forming D-loop-like structures and a subsequent reaction similar to branch migration (Cox and Lehman 1981a). The second reaction proceeds at about 5 bp/sec and is polarized. Single-strand branch migration proceeds extensively from the 3′ end of linear duplex DNA reacting with single-stranded circular DNA (Cox and Lehman 1981b; Kahn et al. 1981; West et al. 1981b). *recA* protein elongates double-stranded DNA by partially untwisting it (Di Capua et al. 1982; Dunn et al. 1982). Although *recA* protein alone does not extensively unwind double-stranded DNA (Cunningham et al. 1979; West et al. 1981a), it can partially unwind nicked circular duplex DNA in the presence of single-stranded DNA, even if the single-stranded DNA is heterologous to the duplex DNA (Cunningham et al. 1979). This observation has led Radding and his colleagues (Cunningham et al. 1979) to propose that *recA* protein can transiently unwind duplex DNA in a search for a region homologous to the single-stranded DNA. The ability of *recA* protein to promote invasion of single-stranded DNA into duplex DNA will be discussed further in Models of Recombination (below).

The *recB* and *recC* genes control the synthesis of an ATP-dependent DNase called exonuclease V, ExoV, or the *recBC* enzyme (for review, see Telander-Muskavitch and Linn 1981). When the enzyme is purified by either of two different procedures, two polypeptides with approximate molecular weights of 130,000 and 140,000 are obtained (Goldmark and Linn 1972; Eichler and Lehman 1977); these polypeptides are coded by *recC* and *recB*, respectively (Hickson and Emmerson 1981; Sasaki et al. 1982). Lieberman and Oishi (1974) observed that enzyme prepared by a third procedure could be reversibly dissociated by a high salt concentration into two subunits with approximate molecular weights of 60,000 and 170,000. The smaller subunit contained no detectable nuclease activity and was present in extracts of *recB* or *recC* mutants. The larger subunit contained a low level of nuclease activity and was absent in *recB* and *recC* mutants. Mixing the subunits restored the original activity. Possibly the smaller subunit, apparently controlled by a third as yet unidentified gene, is partially lost in the first two purification schemes (see Telander-Muskavitch 1981).

recBC enzyme has numerous activities, which can be summarized as unwinding of double-stranded DNA and hydrolysis of DNA to oligonucleotides, the limit digest being composed principally of trimers to hexamers (Goldmark and Linn 1972). Unwinding and degradation of duplex DNA appear to require either a duplex end or a gap of about 10 or more nucleotides on an otherwise circular molecule. (Nicked, closed, or supercoiled duplex circles are neither unwound nor degraded [Karu et al. 1973; Taylor and Smith 1980].) Degradation of single-stranded DNA occurs even if the molecule is circular (Goldmark and Linn 1970). This last reaction is stimulated about fivefold by ATP hydrolysis, whereas all of the other reactions have an absolute requirement for ATP hydrolysis. The ATPase activity of the enzyme is dependent on DNA but can be uncoupled from degradation, and presumably unwinding as well, by the use of duplex DNA cross-linked with psoralen (Karu and Linn 1972).

The preceding activities suggest that the enzyme first unwinds DNA, using ATP hydrolysis for energy, and then hydrolyzes the resultant single-stranded DNA by endonucleolytic cleavage to produce oligonucleotides. MacKay and Linn (1974) studied unwinding by electron microscopy and observed duplex DNA with single-stranded ends several thousand nucleotides long. Wilcox and Smith (1976), studying ExoV from *Haemophilus influenzae*, found single-stranded DNA up to 15,000 nucleotides long released from duplex DNA in the early stages of reaction. Rosamond et al. (1979) discovered that Ca^{++} ion blocks

DNA hydrolysis, but not unwinding, and observed under this condition the accumulation of single-stranded DNA, free and extending from duplex DNA. They also observed single-stranded bubbles in duplex DNA and concluded that the enzyme can transiently unwind DNA, or "track" through it, making occasional nicks and releasing long single-stranded pieces. Telander-Muskavitch and Linn (1980, 1982) observed more complex "loop-tail(s)," the single-stranded structures diagramed in Figure 5, at the end of duplex DNA formed by brief incubation in the presence of enzyme and Ca^{++} ion. Taylor and Smith (1980) observed these structures and internal single-stranded loops, or "twin loops," under similar reaction conditions. These unwinding structures proceed along the DNA at about 300 bp/sec. From the additional observation that loop-tail structures predominate over twin loops in the presence of single-stranded binding protein, Taylor and Smith (1980) proposed the scheme of DNA unwinding and rewinding, shown in Figure 5. The section on Models of Recombination (below) discusses a possible role of RecBC-produced single-stranded DNA in recombination.

The hydrolytic activity of *recBC* enzyme could be used in recombination to remove excess ends, such as the one shown in Figure 3. Alternatively, it may play no role in recombination but may be used by the cell to destroy damaged or foreign DNA. For example, *recBC* is the principal nuclease involved in the degradation of DNA following UV irradiation ("reckless" degradation) (Emmerson 1968), in the degradation of λ DNA restricted by the *E. coli* K12 *hsd* system (Simmon and Lederberg 1972), and in the degradation of T4 gene *2* mutant DNA (Darmalingham and Goldberg 1980). A later section discusses the *recBC* enzyme's blocking of the transition from θ (Cairns) replication to σ (rolling-circle) replication, presumably by degradation of a sensitive intermediate (Enquist and Skalka 1973).

The *recE*-gene product exonuclease VIII (ExoVIII) has activities similar to those of λ exonuclease. This result is in accord with the view noted earlier that the *recE* gene is part of a cryptic prophage related to λ. The two enzymes degrade DNA from the 5′ end, releasing 5′ mononucleotides, and prefer double-stranded to single-stranded DNA (Kushner et al. 1974; Gillen et al. 1977). On the other hand, there are differences between the enzymes. The en-

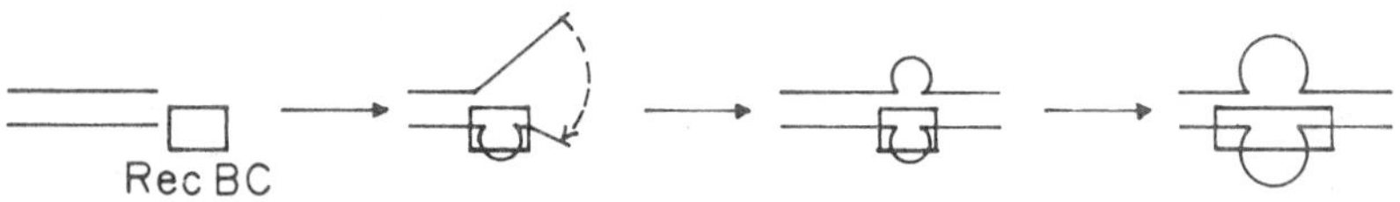

Figure 5 Model of DNA unwinding and rewinding by *recBC* enzyme, represented by the rectangular box. The first intermediate is a loop-tail structure, whose tails anneal (---→) to produce a twin-loop structure. Twin loops increase in size as *recBC* enzyme advances along the chromosome, since *recBC* enzyme unwinds DNA in front of itself at about 300 bp/sec and rewinds DNA behind itself at about 200 bp/sec (see Taylor and Smith 1980).

zymes are antigenically distinct (Gottesman et al. 1974b; Gillen et al. 1977), and λ exonuclease is a dimer of two subunits with a molecular weight of 24,000, whereas ExoVIII is a monomer with a molecular weight of about 140,000 (Kushner et al. 1974; Gillen et al. 1977).

OTHER GENES AND ENZYMES ASSOCIATED WITH RECOMBINATION

The λ*gam* gene, located just to the right of *exo* and *bet* (Fig. 1), codes for a polypeptide with a molecular weight of 16,500, which appears to exist in solution as a dimer (Karu et al. 1975). This protein binds to the *recBC* enzyme and inhibits all of the activities tested, although DNA unwinding was not examined directly. As a consequence, the activity of the RecBC pathway of recombination is reduced shortly after λ infection (Unger and Clark 1972; Unger et al. 1972). In addition, the inhibition by *recBC* enzyme of λ rolling-circle replication is counteracted by *gam* protein (Enquist and Skalka 1973) (see next section). Inhibition of the *recBC* enzyme appears to be the sole function of *gam* protein, since λgam^{+} and λ*gam* mutants behave identically in *recBC* mutant hosts (see, e.g., Stahl and Stahl 1974a).

The products of several *E. coli* genes, in addition to the *rec* genes already discussed, are implicated in recombination, but proof of their role in recombination is difficult to produce. Since many of these gene products are required for essential cellular processes such as DNA replication, mutants lacking these enzymatic activities are inviable and it is difficult to establish a specific effect on recombination in them. Nevertheless, it is likely, or at least reasonable, that the enzymes discussed below are involved in recombination.

The *E. coli ssb* gene codes for a single-stranded DNA binding protein. The *ssb*1 mutation renders growth temperature-sensitive and at intermediate temperature reduces P1-mediated transduction about fivefold (Glassberg et al. 1979). In addition, linkage between cotransduced markers is increased. As noted earlier, SSB protein facilitates D-loop formation by RecA protein (Shibata et al. 1979a; McEntee et al. 1980).

The *polA* gene codes for DNA polymerase I, initially described by Kornberg (1980). Possible roles of this enzyme in recombination are (1) the displacement, during DNA synthesis, of one DNA strand that could be used by *recA* protein in D-loop formation (see Fig. 8) and (2) the filling in of gaps remaining after strand exchange (see Fig. 9). The *lig* gene codes for DNA ligase, which seals nicks in DNA (Gellert 1967; Gottesman et al. 1974a). It seems likely that this enzymatic activity is essential for a terminal step in recombination. *E. coli* mutants having either *polA* or *lig* deficiencies recombine at increased frequency (Konrad and Lehman 1975; Konrad 1977). This initially surprising result may stem from the increased frequency of gaps in the mutants' DNA (see below).

Two genes, *gyrA* and *gyrB*, formerly designated *nal* and *cou*, respectively, code for the two subunits of DNA gyrase, which introduces negative supercoils into DNA (Gellert et al. 1976; Gellert 1981). Int-promoted, site-specific recombination proceeds best with negatively supercoiled DNA (Mizuuchi et al. 1978; Weisberg and Landy, this volume). An involvement of DNA gyrase in general λ recombination is inferred from the observation that coumermycin, an inhibitor of DNA gyrase, inhibits recombination (loss of one copy of a tandem duplication) of UV-irradiated λ (Hays and Boehmer 1978). The inhibition is due to inhibiton of DNA gyrase, since recombination was not affected in a *gyrB* mutant with a coumermycin-resistant DNA gyrase.

Mutants of *E. coli*, in which recombination occurs at a rate higher than that in wild-type bacteria, were obtained by Konrad (1977). From mutagenized cells containing two *lacZ* genes with nonoverlapping deletions, he screened for colonies containing a higher than average number of Lac^+ papillae. Among the mutants isolated were those deficient in DNA polymerase I, DNA ligase, or *dam* methylase. The first two mutants may contain increased levels of gaps in their DNA, which are thought to be recombinogenic (Rupp et al. 1971; Lin et al. 1977). The *dam* mutants may have increased capacity for correction of mismatches, which could increase recombination between markers included in a single heteroduplex. However, it is not clear that increased correction can account for the increased recombination between deletions, the basis on which the mutants were isolated (see below, Heteroduplexes and Mismatch Correction).

One class of hyperrecombinogenic mutants has a remarkable property: With increasing numbers of infection cycles in these bacteria, λ becomes increasingly recombinogenic (Hays et al. 1980; Korba and Hays 1980), and its DNA becomes increasingly sensitive to endonuclease S1, which has a preference for single-stranded DNA (Hays and Korba 1979). These properties are slowly lost upon growth in wild-type bacteria. The Arl (*a*ccumulates *r*ecombinogenic *le*sions) property is associated with partial methylation of cytosine residues in DNA, since (1) *arl dcm* double mutants are Arl^+, (2) the Arl^- mutant property is exhibited by heteroduplexes of λ constructed from one fully methylated DNA strand and one lacking methylation at CC* A/T GG sequences by the *dcm* methylase, and (3) λ and plasmid DNA from *arl* mutants contain one-third to one-half the normal level of 5-methylcytosine (Korba and Hays 1982). It should be noted that the hyperrecombinogenic property has been seen only with duplications, either the *lac* duplications with which the *arl* bacteria were isolated, or the λ tandem duplications with which the λ recombinogenicity has been measured. The frequency of recombination between λ point mutations is increased only slightly after growth in *arl* bacteria (Korba and Hays 1980).

DNA REPLICATION CYCLE AND PACKAGING IN λ INFECTION

Recombination in λ is closely associated with DNA replication and packaging. These latter processes are outlined here (Fig. 6) to allow discussion of the role of recombination in them. More detailed presentations are given by Furth and Wickner and by Feiss and Becker (both this volume).

Mature λ particles contain linear DNA molecules with unique, single-stranded

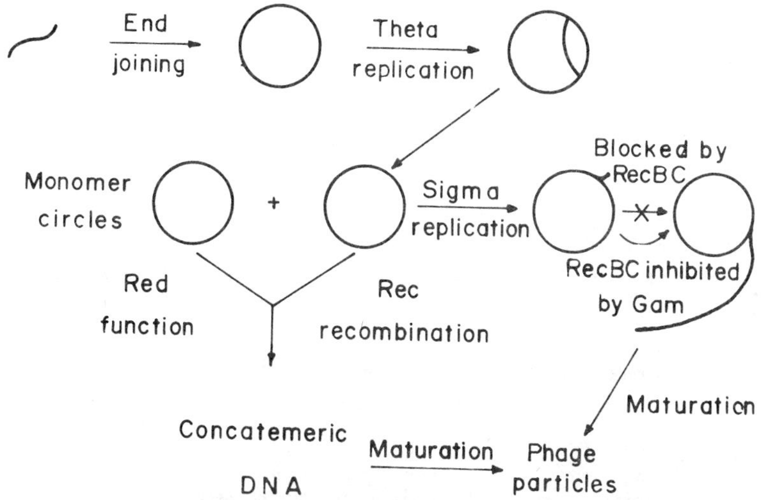

Figure 6 Replication and recombination during λ infection. Double-stranded λ DNA is represented by a single line. Linear DNA in the mature particle is injected and covalently joined by *E. coli* DNA ligase to produce a monomeric circle. These circles proceed by the indicated routes of replication and recombination to concatemeric DNA, from which linear monomeric DNA is packaged into mature phage particles. Recombination by the λ Red pathway or the *E. coli* Rec pathway produces maturable dimeric DNA, which may be either linear or circular (see below, Reciprocality; Models of Recombination; see also Enquist and Skalka 1973).

ends of complementary bases. After injection, these molecules rapidly circularize, their ends are sealed by DNA ligase, and they become supercoiled by DNA gyrase. These molecules undergo a few rounds of θ (Cairns) replication, after which replication switches to the σ (rolling-circle) mode to produce concatemeric λ DNA. Concatemeric DNA, not monomeric DNA, is the obligate immediate precursor to mature progeny. The switch from θ to σ replication is blocked by *recBC* enzyme, which in turn is inhibited by *gam* protein, as noted earlier. Thus, in Gam^+ or $RecBC^-$ infections, replication proceeds by σ replication. In Gam^- $RecBC^+$ infections, replication is limited to the θ mode, with the consequent production of monomeric DNA. This DNA can be converted to packageable DNA, such as dimers and larger forms, by recombination between monomers. In principle, any of the recombination pathways discussed earlier can accomplish this. However, the activity of the Int pathway is so limited that only a few mature phage per cell can be produced. The activity of the RecBC pathway is also limited in the absence of recombination-stimulating Chi sites (see below, Chi Sites and Their Effect on Recombination). The Red pathway aids production of packageable DNA in a Gam^- $RecBC^+$ infection in two ways: the production of recombinant dimers and the production of replication forks (Enquist and Skalka 1973; Stahl et al. 1978), which may be one and the same activity. As a consequence, more total DNA is produced, and, of that, more is packageable.

PHYSIOLOGY OF λ MUTANTS ALTERED IN RECOMBINATION

With the foregoing description, we can understand the growth of certain λ mutants on various *E. coli* strains (Table 1). λ*red gam* double mutants grow so poorly that they make no plaques on *recA E. coli* (Zissler et al. 1971a). This failure to grow, called the Fec⁻ (*fec*kless) phenotype, is not seen on *recBC* or *recA recBC* mutants. Whenever *recBC* enzyme is active (*recBC*⁺ host, *gam*⁻ phage), either the Rec or the Red recombination pathway must be active for the phage to grow (via recombinant dimer formation). Elimination of *recBC* enzyme activity, either by mutation of *recBC* or by Gam⁺-mediated inhibition, allows growth via rolling-circle replication in the absence of recombination (Enquist and Skalka 1973).

Phenotypic revertants of Fec⁻ (i.e., Red⁻ Gam⁻) phage have been obtained by selection in RecA⁻ *E. coli*. From λ deleted for *red* and *gam*, very rare pseudorevertants were obtained. These phage, called λ "reverse" or λ*rev*, proved to have incorporated the *recE* gene from the cryptic prophage discussed earlier (Zissler et al. 1971a; Gottesman et al. 1974b; Gillen et al. 1977). Their pathway of recombination, originally called Der but subsequently equated with RecE, allows growth by recombinant dimer formation in the presence of active *recBC* enzyme. λ*rev* codes for functional analogs of the *exo*- and *gam*-gene products (Gottesman et al. 1974a; Gillen et al. 1977; Reyes et al. 1982). It is not clear why λ*rev*, unlike λ⁺, can grow on P2 lysogens (see below).

λ mutants defective in either *red* or *gam* fail to grow on certain *E. coli* mutants designated Feb⁻ (*f*eckless *b*acteria) (Zissler et al. 1971a). At least two classes of

Table 1 Growth of λ mutants in *E. coli* mutants

Phage genotype		Growth on various *E. coli* hosts[a]					
red[b]	gam	wild type	RecA⁻	RecBC⁻	RecBC⁻ RecA⁻	PolA⁻ or Lig⁻	(P2)
+	+	+	+	+	+	+	−[c]
−	+	+	+	+	+	−[d]	−[c]
+	−	+	+	+	+	−[d]	−[c]
−	−	±	−[e]	+	+	−[d]	±
λ reverse (*rev*)		+	+	+	+	+	+

[a]Growth, as inferred from plaque size, is qualitatively indicated as good/fair (+), poor (±), or negligible (−).
[b]*exo*, *bet*, or double mutants give similar results.
[c]The Spi⁺ phenotype
[d]The Feb⁻ phenotype.
[e]The Fec⁻ phenotype.

Feb$^-$ mutants have been identified: *lig* and *polA*. Skalka (1977) has suggested that proper maintenance of replication forks may require both DNA polymerase I and DNA ligase and that this requirement is critical when *recBC* enzyme is destroying rolling circles (in Gam$^-$ infections) or when an alternate route to replication forks (in Red$^-$ infections) is not available.

Wild-type λ fails to grow on cells lysogenic for phage P2, a phenotype called Spi$^+$ (*s*ensitive to *P*2 *i*nterference) (Lindahl et al. 1970). λ Red$^-$ Gam$^-$, however, can grow on P2 lysogens (Spi$^-$ phenotype) (Lindahl et al. 1970; Zissler et al. 1971b). The Spi phenotype is only partially understood. P2 *old*$^+$ function, which is expressed by P2 prophage and vegetative phage, kills RecBC$^-$ *E. coli* (Sironi 1969). Presumably, the RecBC$^-$ phenocopy resulting from the action of the λ*gam* product is likewise sensitive to Old$^+$ function. However, no similarly satisfactory explanation for the role of Red in the Spi phenotype has been offered. Other mutations, *nin*5 and Chi, coupled with *gam*, allow minute plaque formation on P2 lysogens. The basis of stimulation by the *nin*5 mutation, a deletion of about 2.5 kb between genes *P* and *Q*, is not known (but see Toothman and Herskowitz 1980). On the other hand, understanding the basis of stimulation by Chi mutations has considerably enhanced our understanding of recombination in λ and *E. coli*, as discussed next.

Spi$^-$ (=Red$^-$ Gam$^-$) phages grow poorly, whether on a P2 lysogen or on wild-type *E. coli*. Mutations creating sites, called Chi, were identified in λ as mutations enhancing the growth (e.g., plaque size) of λ Spi$^-$ (Henderson and Weil 1975). Stahl and his colleagues found that one of these Chi mutations displayed recombinational hotspot activity (Lam et al. 1974) similar to that of a *bio* substitution (McMilin et al. 1974). Studies of Chi (see next section) led Stahl and his colleagues to conclude that the RecBC pathway acts poorly on λ unless the λ contains a Chi site, which may arise by mutation in λ or by insertion of foreign DNA, such as *bio*, already containing Chi.

(Many derivatives of λ currently used as cloning vectors are Spi$^-$ or become so after removal of certain restriction fragments. If the ''arms'' of the vector do not contain a Chi site, a strong selection may be made for DNA clones containing Chi. Murray [this volume] discusses the problems associated with using such vectors and solutions to these problems.)

CHI SITES AND THEIR EFFECT ON RECOMBINATION

Chi is an example of special sites that affect generalized recombination. Such sites have been inferred from genetic studies of eukaryotes, especially studies of gene conversion in fungi (see, e.g., Catcheside 1977). This section summarizes the properties of Chi; for a more extensive review of Chi, see Stahl (1979a).

Mutations creating Chi in λ have been found at four widely spaced loci: χ^+A, in or between genes *I* and *J*; χ^+B, between *xis* and *exo*; χ^+C, within the *c*II gene; and χ^+D, between *Q* and *S* (see Fig. 1) (Stahl et al. 1975; Sprague et al. 1978; Smith et al. 1980, 1981b; A. Taylor and G. Smith, unpubl.; D. Daniels and F.

Blattner, pers. comm.). At each of the three λ Chi sites analyzed, a single base-pair change created the active form, as discussed later.

The finding that the genetic region of greatest stimulation of recombination is at or near the location of the Chi mutation demonstrated that Chi is a site of action, rather than the alteration or creation of a diffusible product (Stahl et al. 1975). The enhancement of overall recombination in λ, as measured by increased burst size, is about fivefold by a single Chi site (Malone and Chattoraj 1975). This result suggests that wild-type λ has no active Chi site since, at most, a twofold enhancement would be expected if wild-type λ had one Chi.

Although wild-type λ is free of Chi, wild-type *E. coli* has an average of one Chi site per about 5000 bp, or about 10^3 Chi sites in its total genome. This point was shown by introducing various segments of the *E. coli* genome into λ with DNA cloning techniques, as well as with intracellular formation of λ transducing phages, such as the λ*bio* phages mentioned previously (McMilin et al. 1974; Malone et al. 1978). The presence of the Chi in *bio* led to some confusion in early studies of the Spi⁻ phenotype. Zissler et al. (1971b) noted that λ*bio* Red⁻ Gam⁻ phage did not have a "full" Spi⁻ phenotype (i.e., large plaque size) unless they were altered in another element called δ. Malone and Chattoraj (1975) demonstrated that the δ (i.e., large-plaque-size-determining) element was actually *chi*⁺ in the λ*bio* phages used by Zissler et al. (1971b).

Of the several pathways of recombination discussed earlier that can promote λ recombination, only the RecBC pathway is stimulated by Chi (Gillen and Clark 1974; Stahl and Stahl 1977). Chi activity is undetectable if the host is mutant in *recA, recB,* or *recC.* (The apparent Chi activity is reduced if the λ is *red*⁺, due to the high activity of the Red pathway, or if it is *gam*⁺, due to inhibition of *recBC* enzyme by *gam* protein.)

The presence of Chi sites in *E. coli* DNA and their stimulation of the *E. coli* RecBC pathway acting on λ suggested that Chi might function in *E. coli* recombination as well. Dower and Stahl (1981) demonstrated this function by carrying out P1-mediated transduction and Hfr-mediated conjugation of λ lysogens with and without Chi in the λ prophage. As in vegetative crosses, Chi altered the distribution of exchanges, in this case in the various intervals between λ markers and flanking *gal* and *bio* markers. Many of the properties of Chi in vegetative crosses were manifested in these crosses. It seems likely that Chi acts in the complete absence of λ and is an important element in bacterial recombination.

Chi stimulates recombination not just at itself but over a considerable distance. Stimulation is detectable as far away as 10,000 bp (Lam et al. 1974; McMilin et al. 1974). In this respect, Chi sites are distinctly different from the λ*att* site, which promotes recombination almost exclusively at itself (see Weisberg and Landy, this volume).

When Chi is present in only one of the two parents, stimulation is comparable to that when Chi is in both (Lam et al. 1974). In this respect, an active Chi site is dominant to inactive Chi sites. A Chi site is active even when the other (Chi⁻) parent bears a large heterology opposite the Chi (Stahl and Stahl 1975). In this case, stimulation is seen outside the region of heterology. (Although a systematic study of nonhomologous recombination has not been carried out, Chi has been observed to stimulate only homologous recombination.)

Chi has been found to manifest three sorts of asymmetry. The first asymmetry is called nonreciprocality. When Chi is present in only one parent and DNA replication is blocked, reciprocal recombinants are not obtained in equal yield; the recombinant without the active Chi site is obtained as often as ten times more frequently than the reciprocal (Stahl et al. 1980a). This bias against Chi requires Chi action since, in the same crosses, crossovers not stimulated by Chi are recovered in equal numbers. Recent studies (I. Kobayashi et al., pers. comm.) suggest that the recombinant containing the inactive Chi site is preferentially packaged because of an enhancement of Chi action by packaging from the *cos* initially located in *cis* to the parental active Chi; since Chi acts to its left (in ordinary λ), packaging proceeding from the left end of λ leads to an excess of the recombinant that lacks the active Chi (see below and further discussion in Models of Recombination). A second asymmetry, called directionality, is manifested by Chi's greater enhancement of recombination to its left (relative to the conventional λ map) than to its right (Stahl et al. 1980a). A third asymmetry of Chi, called orientation-dependence, was found with the inversion within λ of a DNA segment bearing Chi (Faulds et al. 1979). Inversion led not to rightward stimulation but to nearly total loss of Chi activity. Reinversion led to recovery of activity.

Orientation-dependence studies of Chi activity have demonstrated that Chi is an asymmetric sequence and that Chi interacts with some other site in λ. Using a transposon with a single Chi in it, Yagil et al. (1980) demonstrated that the active orientation of Chi was the same throughout a large region of λ, from 0.4 units to 0.95 units on the physical map (see Fig. 1). This result suggested that neither the direction of replication nor the direction of transcription determines the activity of Chi (see also Stahl et al. 1980b). Kobayashi et al. (1982) found that the relative orientations of Chi and *cos* (the ends of mature λ; see Feiss and Becker, this volume) determine the activity of Chi in λ. For these studies they used a derivative of λ with the normal *cos* mutated and with a secondary *cos* inserted via cloning techniques in one orientation or the other. An active Chi becomes much less active when *cos* is inverted, whereas a Chi that is weakly active because of orientation becomes fully active when *cos* is inverted. Both polarized injection and packaging of λ DNA from *cos* have been considered as the basis for this Chi-*cos* interaction, and both have been eliminated (Stahl et al. 1982b; Kobayashi et al. 1983). Instead, *cos* appears to activate Chi by virtue of being cut in a step that can, but need not, lead on to packaging. It seems probable that the asymmetric binding of the cutting enzyme, terminase (Feiss et al. 1983), is responsible for the effect of *cos* orientation on Chi activity. This point is discussed further in the section on Models of Recombination (below).

The properties of Chi just discussed are most readily understood with the supposition that Chi is a special nucleotide sequence recognized by one of the enzymes in the RecBC pathway. Support for this view has come from nucleotide sequence analyses around Chi sites, which revealed an 8-bp sequence (5′ G-C-T-G-G-T-G-G 3′) common to the six examined Chi sites arising by mutation in λ or plasmid pBR322 inserted into λ (Smith et al. 1981c). Mutations creating or inactivating these Chi sites are single base-pair changes within the Chi octamer (Schultz et al. 1981). One *E. coli* Chi site, in the *lacZ* gene, has the same sequence (Triman et al. 1982). As predicted from the orientation-dependence studies discussed earlier, the Chi octamer is asymmetric and oriented in the same direction, with 5′ G-C-T-G-G-T-G-G 3′ on the *l* strand, at all of the sites analyzed. Whether Chi is this sequence, its complement, or the duplex structure is not yet known.

Since Chi is inactive in *recB*, *recC*, and *recBC* double mutants (Stahl and Stahl 1977; D. Schultz et al., in prep.), it may be that the *recBC* enzyme recognizes Chi. Alternatively, Chi may be recognized by another protein in the RecBC pathway; in the absence of *recBC* enzyme there would be no recombination via this pathway and hence no Chi activity. Certain mutations located within *recB* or *recC* eliminate or reduce the effect of Chi, although they leave nearly normal levels of *recBC* nuclease activity and recombination proficiency in P1-mediated transductions (D. Schultz et al.; V. Lindblad et al.; both in prep.). This result suggests that the altered *recBC* enzymes in these strains no longer recognize Chi. It is reasonable to speculate that the endonuclease activity of *recBC* enzyme is activated by Chi (see further discussion in Models of Recombination).

DNA REPLICATION AND RECOMBINATION

A fundamental and long-standing question about the mechanism of recombination is whether recombination requires DNA replication. A second, related question is whether recombination is aided by, or aids, replication. An obvious difficulty in attempts to answer these questions with phage is that the two processes occur concurrently. Early studies with λ showed that some recombinants were formed by the breaking and rejoining of parental chromosomes to produce recombinant chromosomes with little or no DNA replication (Kellenberger et al. 1961; Meselson and Weigle 1961). However, in these studies replication was uninhibited, and so the bulk of the recombinants were composed of newly synthesized DNA. The question remained whether some, or even the vast majority, of the recombinants were produced by a mechanism other than by breaking and rejoining.

With the recognition that recombination in λ could proceed by different pathways (Red, Rec, Int), Kellenberger and Weisberg (1971) reinvestigated the question. They found that each of the three pathways transferred DNA from the parental chromosomes to the recombinant chromosomes and concluded that at least a portion of the recombination by each pathway involved breakage and transfer of parental DNA. The technique used (detection of radioactive label from one parent in the recombinants at higher specific activity than in the other parental type) did not indicate how much of the recombinant DNA consisted of parental DNA and, hence, how much DNA synthesis might have been associated with the recombination event.

Clearly, to determine whether DNA replication is necessary or not for recombination, DNA replication must be blocked. Stahl and his associates have carried out crosses in which DNA replication was blocked by mutational inactivation either of λ*O* (or λ*P*) function or of the host *dnaB* function, or both (Furth and Wickner, this volume). When replication was very severely limited, all of the (very few) recombinants were produced by breakage and reunion (Stahl et al. 1972). DNA synthesis must have been limited to less than 10% of the chromosomal length, judging from the density of the recombinants produced in these crosses between parents labeled with heavy isotopes. This result, obtained from Red$^+$ RecA$^-$ crosses and from Red$^+$ Rec$^+$ crosses, showed that recombination can occur with no or very limited DNA synthesis.

Examination of recombinants produced under conditions where a limited amount of DNA synthesis was permitted showed that Red-promoted recombination was most severely depressed in the middle of the λ chromosome, less severely depressed at the left end, and least severely depressed at the right end (Stahl et al. 1972). Further examination revealed that recombination in the middle of the chromosome required the most DNA synthesis, that at the left end, an intermediate amount, and that at the right end, the least. Thus, the Red pathway can produce a few recombinants with very little DNA synthesis, but replication augments recombination, particularly in the middle of the chromsome. We return to this point later (see Models of Recombination).

Examination of the Rec pathway (principally the RecBC pathway) led Stahl et al. (1974) to conclude that that pathway also produces recombinants by breakage and reunion. In contrast to the Red pathway, the Rec pathway (in the absence of λ Gam protein) shows little dependence on DNA synthesis, irrespective of the map interval examined. These results suggest that the Rec pathway may be exclusively one of breakage-reunion, perhaps with very limited amounts of DNA synthesis near the exchange.

HETERODUPLEXES AND MISMATCH CORRECTION

Heteroduplexes in λ are inferred from phage particles that give rise to two types of progeny. This inference rests on knowledge that the λ phage particles are haploid (except for duplications in certain mutants). Most easily detected are c^+/c^- heteroduplexes, since plaques resulting from such phage are mottled in appearance from the mixture of clear and turbid areas in them. Kellenberger et al. (1961) found that such heterozygous particles (hets) were more often recombinant for flanking markers than was the total phage population and concluded that hets were recombinational intermediates. Meselson (1967) found that rare c^+/c^- hets found in uninhibited-replication conditions had exchanges near the *c* gene. Stahl et al. (1972, 1974) further substantiated this conclusion by showing that when DNA replication is severely limited, c^+/c^- hets are especially prevalent (~10%) among chromosomes exchanged near the *c* locus. However, the position of the exchange in these c^+/c^- hets was not sharply defined; from crosses between parents labeled with heavy and light isotopes, c^+/c^- hets were found with a broad distribution of densities, suggesting that they contained heterozygous overlaps of variable and often considerable length.

The existence of extensive heteroduplex regions allows for multiple exchanges to occur from a single recombination event by mismatch correction within the heteroduplex region. Wildenberg and Meselson (1975), using artificially constructed heteroduplexes of λ, showed that the frequency of mismatch correction depended on the particular mismatch (presumably the base-mispairing and its surrounding sequence) and that the frequency of correction of two fairly remote markers to produce a wild-type recombinant depended not upon the distance between the markers but upon their intrinsic correction frequencies. Wagner and Meselson (1976) concluded that correction occurs with the removal of a stretch of nucleotides on one strand and that this stretch was typically about 3000 nucleotides long. A similar conclusion had been previously reached from the study of associated close genetic exchanges (localized negative interference) in λ (Amati and Meselson 1965).

Fox and his students have demonstrated that the heteroduplex region in λ recombinants can occupy nearly the entire chromosome under conditions of limited DNA synthesis. White and Fox (1974) found that either parent could contribute nearly any amount (from 10% to 90%) of the material to phages bearing (or able to give rise to) close double exchanges located about 20% of the length of λ from one end (in the *O* and *P* genes). Further analysis (Fox et al. 1979; Sodegren and Fox 1979) showed that some of the recombinants arose from particles with nearly all of one DNA strand from one parent and with the other strand from the other parent. The frequency of recombinants between two nonhomologies (large deletions or insertions) was much lower than that between two point mutations separated by a comparable distance. This result suggests that heteroduplex formation is often initiated far from a region and spreads to it unless nonhomologies intervene. Analysis of the progeny of individual incipient recombinants suggested that these initial recombinants were often heterozygous for all of the three close markers and that these recombinants were processed, presumably by mismatch correction, in the indicator bacteria in which the recombinants were detected (White and Fox 1974, 1975a). White and Fox (1975b) compared the segregation patterns of these incipient recombinants with the patterns from artificially produced heteroduplexes of known structure. They inferred that the DNA strand from the parent making the lesser material contribution had its 5′ end at the right end of the phage. The preceding results have been obtained with Red$^+$ phage in RecA$^+$ bacteria, but Stahl and Stahl (1974b) found similar results in RecA$^-$ bacteria. These features of Red-promoted recombinants have been incorporated into a model of Red-promoted recombination

(Stahl and Stahl 1974b), discussed below in the section on Models of Recombination.

Results from standard crosses (i.e., when DNA replication proceeds unhindered) have led to different conclusions regarding the effects of extensive nonhomologies on recombinant formation. In these conditions, extensive nonhomologies (deletions and substitutions) only modestly reduce the frequency of recombinants in an interval between the nonhomologies; in addition, high negative interference is observed between intervals separated by a nonhomology (Lichten and Fox 1983). (In contrast, Makin et al. [1982] report that nonhomologies in another region of λ abolish negative interference between intervals flanking the nonhomology.) M. Lichten and M.S. Fox (pers. comm.) have detected heteroduplexes that include extensive nonhomologies under conditions of limited DNA synthesis and suggest the possibility that such heteroduplexes, and recombinants issuing from them, are not packaged unless the heteroduplex is removed, e.g., by replication.

Further results from standard crosses (i.e., when phage replication proceeds unhindered) suggest that mismatch correction may not be a major factor in the production of recombinants between close markers among progeny selected to be recombinant for flanking markers. Gussin et al. (1980) observed that the frequency of recombinants between two markers m_1 and m_2, among phages selected to be recombinant for flanking markers A^+ and B^+, was a linear function of the physical distances between m_1 and m_2. These distances, ranging from 3 bp to 165 bp, were determined by nucleotide sequence analysis. The frequency of $m_1^+m_2^+$ recombinants was about tenfold less when a triple exchange was required to generate the $A^+m_1^+m_2^+B^+$ recombinant than when a single exchange was required. If mismatch correction were a major factor, one would expect, as found by Wildenberg and Meselson (1975), that the frequency of recombinants would be influenced by the particular mismatches of the markers involved and therefore would not be a simple linear function of their distances. In addition, the frequencies of recombinants in reciprocal crosses would not be expected to be always greater for single exchanges than for triple exchanges.

The preceding, seemingly inconsistent results may be reconciled with the following notions. A heteroduplex is formed at the site of an exchange. The probability that the end of the heteroduplex occurs between two markers, and hence may recombine them, is proportional to the distance between the markers. In the absence of DNA replication, the length of the heteroduplex can be increased to nearly the entire length of the λ chromosome. In these long-lived heteroduplexes, mismatch correction can occur, generating recombinants with multiple exchanges at frequencies dependent upon the nature of the markers as well as upon the distances between them. In the presence of DNA replication, however, the heteroduplex initially formed is often not subjected to mismatch correction before replication abolishes the mismatch, so that recombinant frequencies are determined primarily by the distance between the markers.

FIGURE 8 STRUCTURES

Several models of recombination (see below) postulate intermediates of recombination in which one DNA chain from each parent is broken, swapped with the other, and rejoined (see Fig. 7). Such branched structures (Holliday junctions) (Holliday 1964) take the form of a figure 8 when the parental DNAs are circular. Figure 8 molecules formed from a variety of circular DNAs in *E. coli* have been identified by electron microscopy. That they are recombinational intermediates has been inferred from one or more of the following facts: (1) Their frequency is reduced in *recA* mutant cells, (2) the junction is at a point of homology between the two circles, and (3) the loops are biparental, as indicated by their unequal sizes when the parental circles are of unequal sizes. Figure 8 molecules are most readily seen with the double-stranded form of the small, single-stranded DNA phages (Benbow et al. 1975; Thompson et al. 1975) and with plasmids (Potter and Dressler 1976). Valenzuela and Inman (1975) and Ikeda and Kobayashi (1979) observed figure 8 molecules from λ-infected cells. Ikeda and Kobayashi showed that figure 8s obtained from cells in which replication and packaging were blocked could be resolved and packaged in a cell-free system to produce genetic recombinants. Thus, figure 8s bear all the marks expected of a major recombinational intermediate.

RECIPROCALITY

Recombination is said to be reciprocal if, in a single recombination event, both complementary recombinants are produced (e.g., *a*+ × +*b*→+ + and *ab*). In certain organisms, notably fungi, it is possible to isolate all of the products of meiotic recombination and so determine if both complements are present. In fungi, meiotic recombination between distant markers is almost invariably reciprocal, although that between close markers (within a gene) is frequently

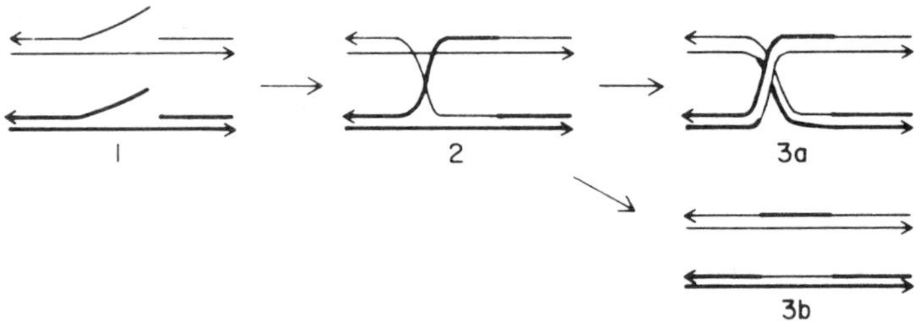

Figure 7 Model for recombination. Thick and thin lines represent single strands of DNA from the two parents. Cutting of the uncrossed strands of the Holliday junction (*2*) leads to the structures in 3a, which are recombinant for markers flanking the exchange region. Cutting of the crossed strands leads to the structures in 3b, which are parental for flanking markers (see Holliday 1964).

nonreciprocal. In phage, the classic test for reciprocality examines the progeny from individual cells that have been infected with both parental phages (single bursts). Reciprocality may be obscured, however, if one recombinant is not fairly represented in the progeny because it was not replicated or packaged, e.g., or was simply lost. Thus, a statistical correlation between complementary recombinants must be sought. Such correlation may not be due to genuine reciprocality, however, since certain cells may be more proficient than others in producing recombinants. Two methods have been used to circumvent these difficulties.

Sarthy and Meselson (1976) analyzed λ progeny from single bursts for recombinants between parents differing by three factors located more than 10,000 bp apart (i.e., greater than the estimated length of mismatch excision tracts). Of the six recombinant types, complementary types were found with one another more often than with the other types, when recombination was by the "Rec" pathway. (The phage were Int^- Red^- Gam^+, so recombination was via some combination of the Gam-inhibited RecBC pathway and the low-level RecE or RecF pathways.) No such correlation was found for recombination by the Red pathway (in a *recA* mutant host). These results indicate that "Rec" is at least often reciprocal, whereas Red may be rarely, if ever, so. (A firm conclusion that Red is not reciprocal is precluded by the alternatives mentioned in the preceding paragraph.)

A second approach to determining reciprocality takes advantage of the view that only reciprocal recombination can insert one circle into another to produce a larger circle. (It should be noted that this "view" is not the same as the definition of reciprocality. There may be ways other than reciprocal recombination by which the larger circles arise.) Three types of circle-into-circle recombination have been studied with λ: superinfecting, circularized λ into a prophage in the host's circular chromosome (Gottesman and Yarmolinsky 1968), the λdv plasmid into λ (Kellenberger-Gujer 1971), and λ into λ to form a dimeric circle (Ikeda and Kobayashi 1979; Chattoraj 1980). In each case, Rec promoted circle-into-circle recombination. In the first case, Red did not, at detectable frequency ($<10\%$ of the Rec-promoted frequency), whereas in the second case, Red was as efficient as Rec. (Only Rec was tested in the third case.) With this one exception, the available evidence *suggests* that Rec is reciprocal and that Red is not. (The exceptional case can be explained by two Red-promoted nonreciprocal exchanges between a *dimeric* λdv and λ, to insert a single copy of λdv into λ [Kellenberger-Gujer 1971].)

MODELS OF RECOMBINATION

Signer (1971) has discussed numerous molecular models of recombination. One should bear in mind that recombination by the different pathways available to λ in *E. coli* may proceed by diverse mechanisms, and one should thus keep a variety of models in mind. The following text discusses one early model and others proposed since Signer's review.

Break-join Models

Many models of break-join recombination involve three central steps: (1) the formation of single-stranded DNA regions, (2) the formation of hybrid (heteroduplex) DNA with the formation of a crossed-strand structure such as the Holliday junction, and (3) the resolution of this crossed-strand structure by cutting and ligation of exchanged ends. Before turning to specific models, let us consider mechanisms by which these steps may occur.

Single-stranded regions might be formed by any of a variety of mechanisms. DNA unwinding by the *recBC* enzyme is a plausible mechanism for the RecBC pathway. Displacement of a strand by DNA polymerase during synthesis of its equal can also produce a single-stranded tail. Digestion of one strand by λ exonuclease or by ExoVIII might produce single-stranded ends in the Red and RecE pathways. A gap (i.e., a single-stranded region flanked by double-stranded regions) is produced at high frequency during repair of UV-induced DNA damage, such as a thymine dimer, and such gaps lead to recombinants at high frequency (Rupp et al. 1971). Perhaps the *recF* enzyme is involved in production of these gaps, since RecF is involved both in repair of UV-induced damage and in recombination by the RecF pathway (Rothman and Clark 1977). Replication forks are another source of single-stranded DNA; utilization of these regions in recombination is a potential basis for the close interaction between replication and recombination.

Heteroduplex DNA can be formed from single-stranded DNA spontaneously or by RecA catalysis. Holloman et al. (1975) found that short single-stranded DNA associates with supercoiled DNA to form a D loop (Fig. 4b). *recA* protein accelerates this reaction, as well as many other associations between single-stranded DNA and double- or single-stranded DNA. At present RecA catalysis is the best candidate for the initiation of heteroduplex formation, and the extension of these heteroduplexes may proceed by *recA*-protein or by heat-driven branch migration (Warner et al. 1979; Cox and Lehman 1981b).

Resolution of crossed-strand structures could occur by default when branch migration of the structure encounters preexisting nicks. Alternatively, some endonuclease might place nicks ahead of the crossed-strand structure or at the junction itself. *recBC* enzyme has single-strand endonuclease activity but has not been shown to cut Holliday junctions, which may have no single-stranded region (Sigal and Alberts 1972). Endonuclease VII of phage T4 cleaves Holliday junctions by introducing a pair of nicks at the junction, and DNA ligase can seal these nicks to produce linear duplex DNA (Mizuuchi et al. 1982). A corresponding enzyme from uninfected or λ-infected *E. coli* has not yet been reported.

The model in Figure 7 proposed by Holliday (1964) for fungal recombination incorporated features utilized in many subsequent models and has thus become the prototype of break-join models. To account for polarity of gene conversion in fungi, the model proposes that nicks are made in both parental DNA molecules at special sites (initiators). Strands of the same polarity are exchanged to form the Holliday junction which, after migrating for some distance (on the order of a gene length), is resolved to produce molecules with either recombinant (structure 3a) or parental (structure 3b) configuration of markers flanking the heteroduplex DNA. The breaking and exchanging of one pair of strands at a time allows homologous regions to find one another, thus assuring the faithfulness of recombination and avoiding the dangers that might accompany double-strand breakage of the chromosomes. Recombination by this mechanism (in the absence of mismatch repair) would be exclusively reciprocal and could manifest close double exchanges. In the presence of mismatch repair, more numerous and nonreciprocal exchanges can occur. As noted earlier, such exchanges could account for the high localized negative interference seen in λ (Amati and Meselson 1965).

A variation of this model was proposed by Meselson and Radding (1975) (Fig. 8) to accommodate fungal data suggesting that heteroduplex DNA may sometimes be formed on only one duplex, not on both as in the Holliday model. In addition, Meselson and Radding proposed an enzymatic basis for strand transfer. DNA polymerase initiates synthesis at a nick on one duplex, with the displacement (not degradation) of one strand. This strand pairs with the other duplex (Fig. 8C), one strand of which is degraded by DNA polymerase. Following this asymmetric phase of strand transfer, symmetric (i.e., reciprocal) transfer is initiated by the joining of the 3′ end of the newly synthesized strand to the 5′ end of the strand being degraded on the other duplex. The resulting crossed-strand structure (Fig. 8D), a Holliday junction, can thereafter move along the chromosome by branch migration. Resolution occurs as in Holliday's model.

Ignoring the possibility of mismatch correction, recombination by this mechanism is reciprocal except for markers in the interval of asymmetric strand transfer. By varying the extent of such transfer more or less, reciprocal recombination can be accommodated. The model explicitly states a source of energy,

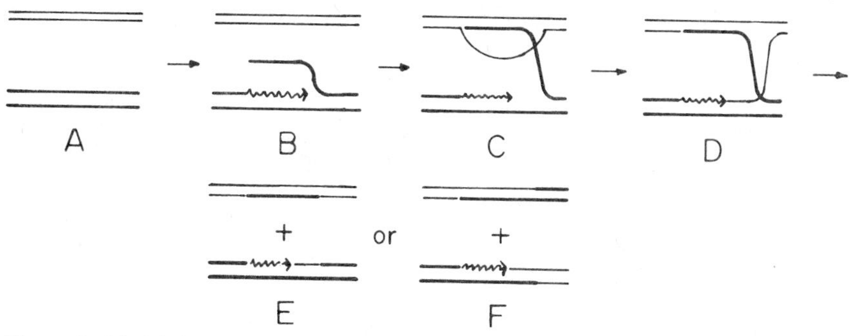

Figure 8 Model for recombination (symbols as in Fig. 7). DNA synthesis (⟿) initiated at a nick displaces one parental strand, which invades the other parental duplex to form a D loop (*C*); the D loop is later degraded. Subsequent steps proceed as in Fig. 7 (see Meselson and Radding 1975).

hydrolysis of nucleoside triphosphate during DNA synthesis, for initiation of strand transfer. Such DNA synthesis may be the small amount of DNA synthesis detected in break-join recombination by the RecBC pathway in replication-blocked λ crosses (Siegel 1974). Double exchanges and multiple exchanges by mismatch correction are proposed to occur as in the Holliday model.

A further variation of the Holliday model is designed to account for Chi-stimulated recombination catalyzed by *recA* protein and *recBC* enzyme (Smith et al. 1981a). In the model in Figure 9, *recBC* enzyme forms twin loops (Fig. 9C) on linear DNA (as described in the legend to Fig. 5). When *recBC* enzyme encounters the Chi sequence on the strand it is holding, it nicks that strand (Fig. 9D). Unwinding and rewinding continue until all of the DNA to one side of Chi has been rewound (Fig. 9E). Further rewinding is not possible since the free end of the nicked strand may now be thousands of nucleotides away from the enzyme. Continued advance of *recBC* enzyme results in collapse of the remaining loop to produce a gap (Fig. 9F). Annealing of the tail to the gap may be prevented by single-stranded binding protein. *recA* protein is postulated to form a D loop (Fig. 9G) with the extended tail and the other parental duplex. Nicking of the displaced strand of the D loop would allow that strand to anneal with the gap to form a Holliday junction (Fig. 9H). At this point, or even before nicking of the second strand, migration of the Holliday junction might be catalyzed by the combined action of *recA* protein and *recBC* enzyme. Finally, cutting and ligation resolves the Holliday junction as in other models.

The preceding model utilizes known properties of *recA* protein and *recBC* en-

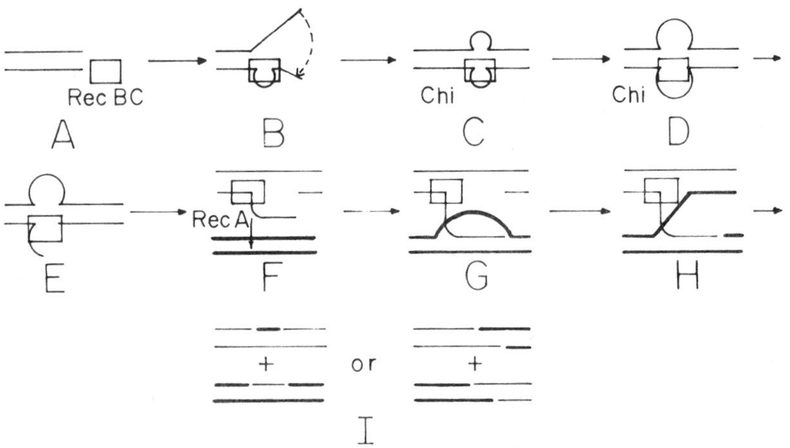

Figure 9 Model for recombination. *recBC* enzyme (rectangular box) produces twin loops (*C*) as in Fig. 5 and nicks one strand when a Chi site is encountered (*D*). The displaced strand forms a D loop (*G*), aided by *recA* protein. Subsequent steps proceed as in Fig. 8, except that branch migration (*H*) is postulated to be driven by *recBC* enzyme and *recA* protein (see Smith et al. 1981a).

zyme, except that sequence-specific endonucleolytic cleavage by *recBC* has not been demonstrated. The model accounts for properties of Chi as follows: The *recBC* dependence is implicit in the postulated nicking by *recBC* enzyme at Chi. The dominance of Chi, its action opposite a heterology, and its action over several thousand base pairs derive from the production of the very long, invasive single-stranded tail extending from Chi. Action primarily to one side (to the left in Fig. 9) results from the asymmetry of the Chi sequence and the assumption that *recBC* enzyme recognizes that sequence only when it approaches Chi from the correct direction (from the right in Fig. 9). The latter point follows from the supposition that *recBC* enzyme holds onto the two DNA strands differently (Taylor and Smith 1980). Maximal action of Chi when it is correctly oriented with respect to *cos* (Kobayashi et al. 1982) could be accounted for with the assumption that the terminase proteins, after cutting at *cos* (see Feiss and Becker, this volume), remain bound to the left end of λ and thereby allow *recBC* enzyme to attach to the right end more frequently than to the left end, as suggested by F. Stahl and I. Kobayashi (pers. comm.). Initiation of unwinding by *recBC* enzyme at the right end would, according to the model in Figure 9, result in leftward action of Chi, as observed (Stahl et al. 1980a).

Faulds et al. (1979) have proposed a very different mechanism for Chi action and its orientation dependence. In their models, Chi stimulates the resolution of figure 8s (Holliday junctions) rather than their formation as in the preceding model. *recBC* enzyme is postulated to enter the Holliday junction and to travel along a duplex until it meets Chi. If the Chi is correctly oriented, *recBC* reverses direction, travels back to the junction, degrading one strand along the way, and resolves the Holliday junction. In one of their models (Fig. 10), polarized packaging from the left end up to the Holliday junction envelops and inactivates any Chi to the left of the junction. Thus, only Chis to the right of the junction, and only those oriented correctly to reverse *recBC* enzyme's travel, are active (black arrow). Chi thereby acts leftward and is dependent on the orientation of *cos,* as observed (Stahl et al. 1980a; Kobayashi et al. 1982). Recent studies with phages containing combinations of active and inactive *cos* and Chi sites (Kobayashi et al. 1983) have shown, however, that the Chi-*cos* interaction is separable from packaging and lead to the view that if Chi activates the *recBC* enzyme to resolve Holliday junctions, it does so after *recBC* enzyme has entered the Chi-containing duplex at *cos*.

Break-copy Models

Recombination by break-copy results from the breakage of the two duplexes, the joining of one strand from one parental duplex to the other, and the synthesis of the complement of that strand to complete the recombinant duplex. (In early versions of the break-copy mechanism [Meselson 1964] *both* strands to one side of the break are newly synthesized. Break-copy as discussed here is actually a hybrid version of the early break-copy and break-join mechanisms.)

Stahl and Stahl (1974b) and Stahl et al. (1978) proposed a break-copy model

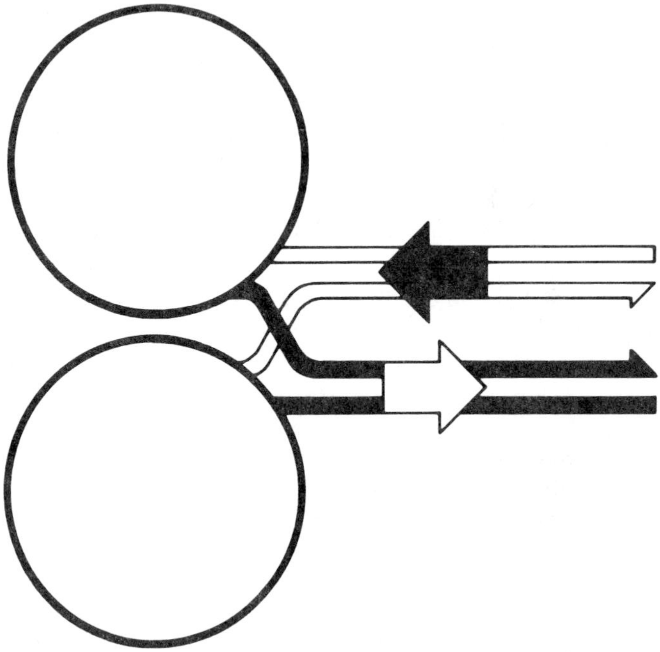

Figure 10 Model for orientation-dependence of Chi in recombination. Packaging (large circles) proceeds from the left toward a Holliday junction. Any Chi site within the package is inactive. A correctly oriented Chi (black arrow) to the right of the Holliday junction is active, whereas an incorrectly oriented one (white arrow) is inactive on that Holliday junction. (Reprinted, with permission, from Faulds et al. 1979.)

for Red-promoted recombination of λ (Fig. 11). A figure 8 is formed by an unspecified mechanism (but may be promoted by *recA* when it is available), and a nick at the Holliday junction acts as a primer for DNA synthesis on one strand. Since two *cos* sites are required to produce a maturable duplex, this DNA synthesis must extend from the Holliday junction to a *cos* site from which packaging of the recombinant commences or concludes.

This model was based on two central observations noted previously (DNA Replication and Recombination). (1) When DNA replication is severely inhibited, most recombination occurs near *cos* (Stahl et al. 1972b, 1974). This is true whether *cos* is at its normal location or in the central (*b*2) region of λ (Stahl et al. 1982a). (2) When limited DNA replication is allowed, recombinants with exchanges anywhere in λ are formed, but those with exchanges far from *cos* have more new DNA than those with exchanges near *cos* (Stahl et al. 1972b, 1974). Recent studies, however, show that a *cos* from which packaging does not occur nevertheless does stimulate Red-mediated recombination (I. Kobayashi et al., pers. comm.). These results suggest that terminase-promoted cutting at *cos* may allow λ exonuclease to initiate degradation from *cos* with the production of

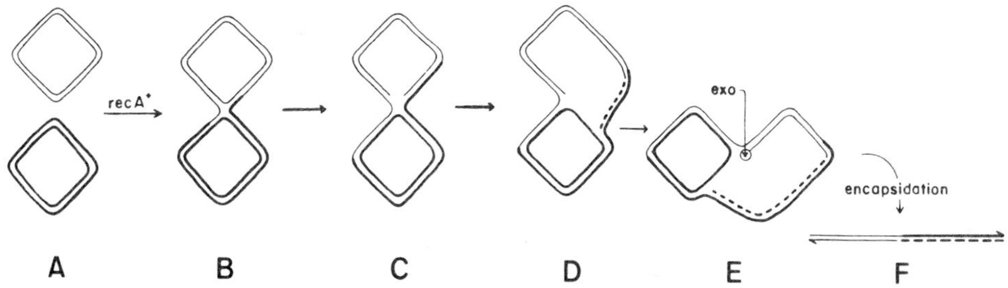

Figure 11 Model for recombination. *recA* protein catalyzes figure 8 formation (*B*) by an unspecified mechanism. A nick at the Holliday junction (*C*) allows DNA synthesis (----) to produce a maturable chromosome extending from *cos* to *cos*. Exposure of the template strand, by λ exonuclease-catalyzed degradation of a parental strand, aids synthesis and, hence, production of maturable recombinants. (Reprinted, with permission, from Stahl et al. 1978.)

a single-stranded invasive tail, which would lead to exchanges primarily near *cos*; the role of DNA replication may be to provide additional suitable substrates, such as the tail of a rolling circle, for λ exonuclease (I. Kobayashi, pers. comm.).

PERSPECTIVE

A cornerstone paper by Visconti and Delbrück (1953), now 30 years old, is titled "The Mechanism of Genetic Recombination in Phage." We are still attempting to determine that mechanism. Investigations with phage in the 1950s and early 1960s concentrated on mathematical analyses of recombinant frequencies, which demonstrated that the genetic maps of some phages are linear, whereas those of others are circular and that exchanges often occur in clusters (localized negative interference). Additional studies showed that recombination could occur by the breaking of parental DNA molecules and their rejoining with the formation of hybrid (heteroduplex) DNA. The next decade saw the elucidation of the several pathways of recombination described here, principally via analysis of mutants and enzymes. These studies showed that λ can recombine either by its own functions, Red, or by the host functions, Rec; Red is at least partially dependent on DNA replication and appears to be nonreciprocal, whereas Rec appears to be independent of DNA replication, is strongly influenced by special sites (Chi), and appears to be reciprocal.

Although we now know considerable details about the action of some recombination-promoting enzymes, integration of them into a complete pathway has not yet been achieved. Current efforts toward this goal include attempts to produce recombination in vitro. Investigations with phage λ and *E. coli* are likely to continue to make significant contributions toward that goal.

ACKNOWLEDGMENTS

I thank Frank Stahl, John Clark, Maury Fox, Hideo Ikeda, Ichizo Kobayashi, David Leach, Charles Radding, and Bob Weisberg for communicating their results before publication and for their thoughtful comments and help in preparing this paper. I am grateful to the many people in my lab for their camaraderie, their scientific endeavors cited above, and their helpful criticisms of this paper. I thank Julie Dunn for skillfully typing it. Research in my lab is supported by a grant (GM-31693) and a Career Development Award (AI-00547) from the National Institutes of Health.

REFERENCES

Amati, P. and M. Meselson. 1965. Localized negative interference in bacteriophage λ. *Genetics* **51:** 369.

Bachman, B.J. and K.B. Low. 1980. Linkage map of *Escherichia coli* K-12, Edition 6. *Microbiol. Rev.* **44:** 1.

Barbour, S.D., H. Nagaishi, A. Templin, and A.J. Clark. 1970. Biochemical and genetic studies of recombination proficiency in *Escherichia coli*. II. Rec^+ revertants caused by indirect supression of Rec^- mutations. *Proc. Natl. Acad. Sci.* **67:** 128.

Benbow, R.M., A.J. Zuccarelli, and R.L. Sinsheimer. 1975. Recombinant DNA molecules of bacteriophage φX174. *Proc. Natl. Acad. Sci.* **72:** 235.

Broker, T.R. and A.H. Doermann. 1975. Molecular and genetic recombination of bacteriophage T4. *Annu. Rev. Genet.* **9:** 213.

Brooks, K. and A.J. Clark. 1967. Behavior of λ bacteriophage in a recombination-deficient strain of *Escherichia coli*. *J. Virol.* **1:** 283.

Carter, D.M. and C.M. Radding. 1971. The role of exonuclease and β protein of phage λ in genetic recombination. II. Substrate specificity and the mode of action of λ exonuclease. *J. Biol. Chem.* **246:** 2502.

Catcheside, D.G. 1977. *The genetics of recombination.* University Park Press, Baltimore, Maryland.

Cassuto, E. and C.M. Radding. 1971. A mechanism for the action of λ exonuclease in genetic recombination. *Nat. New Biol.* **229:** 13.

Chattoraj, D.K. 1980. Dimeric intermediates of recombination in phage λ. Cell **19:** 143.

Chattoraj, D.K. and F.W. Stahl. 1980. Evidence of RNA in D loops of intracellular λ DNA. *Proc. Natl. Acad. Sci.* **77:** 2153.

Clark, A.J. 1973. Recombination deficient mutants of *E. coli* and other bacteria. *Annu. Rev. Genet.* **7:** 67.

——. 1980. A view of the RecBC and RecF pathways of *E. coli* recombination. *ICN-UCLA Symp. Mol. Cell. Biol.* **19:** 891.

Clark, A.J. and A. Margulies. 1965. Isolation and characterization of recombination-deficient mutants of *Escherichia coli* K12. *Proc. Natl. Acad. Sci.* **53:** 451.

Cox, M.M. and I.R. Lehman. 1981a. *recA* protein of *Escherichia coli* promotes branch migration, a kinetically distinct phase of DNA strand exchange. *Proc. Natl. Acad. Sci.* **78:** 3433.

——. 1981b. Directionality and polarity in recA protein-promoted branch migration. *Proc. Natl. Acad. Sci.* **78:** 6018.

Cunningham, R.P., T. Shibata, C. DasGupta, and C.M. Radding. 1979. Single strands induce recA protein to unwind duplex DNA for homologous pairing. *Nature* **281:** 191.

Darmalingham, K. and E.B. Goldberg. 1980. Restriction in vivo. V. Induction of SOS functions in

Escherichia coli by restricted T4 phage DNA, and alleviation of restriction by SOS functions. *Mol. Gen. Genet.* **178:** 51.

Davidson, N. and W. Szybalski. 1971. Physical and chemical characteristics of lambda DNA. In *The bacteriophage lambda* (ed. A.D. Hershey), p. 45. Cold Spring Harbor Laboratory, Cold Spring Harbor, New York.

DiCapua, E., A. Engel, A. Stasiak, and T. Koller. 1982. Characterization of complexes between recA protein and duplex DNA by electron microscopy. *J. Mol. Biol.* **157:** 87.

Dower, N. and F.W. Stahl. 1981. Chi activity during transduction-associated recombination. *Proc. Natl. Acad. Sci.* **78:** 7033.

Dressler, D. and H. Potter. 1982. Molecular mechanisms of genetic recombination. *Annu. Rev. Biochem.* **51:** 727.

Dunn, K., S. Chrysogelos, and J. Griffith. 1982. Electron microscopic visualization of RecA-DNA filaments: Evidence for a cyclic extension of duplex DNA. *Cell* **28:** 757.

Echols, H. and R. Gingery. 1968. Mutants of bacteriophage λ defective in vegetative genetic recombination. *J. Mol. Biol.* **34:** 239.

Eichler, D.C. and I.R. Lehman. 1977. On the role of ATP in phosphodiester bond hydrolysis catalyzed by the *recBC* deoxyribonuclease of *Escherichia coli. J. Biol. Chem.* **252:** 499.

Emmerson, P.T. 1968. Recombination deficient mutants of *Escherichia coli* K12 that map between *thyA* and *argA*. *Genetics* **60:** 19.

Emmerson, P.T. and P. Howard-Flanders. 1967. Cotransduction with *thy* of a gene required for genetic recombination in *Escherichia coli*. *J. Bacteriol.* **93:** 1729.

Enquist, L. and A. Skalka. 1973. Replication of bacteriophage λ DNA dependent on the function of host and viral genes. I. Interaction of *red, gam*, and *rec*. *J. Mol. Biol.* **75:** 185.

Faulds, D., N. Dower, M. Stahl, and F. Stahl. 1979. Orientation-dependent recombination hotspot activity in phage λ. *J. Mol. Biol.* **131:** 681.

Feiss, M., I. Kobayashi, and W. Widmer. 1983. Separate sites for binding and nicking of bacteriophage λ DNA by terminase. *Proc. Natl. Acad. Sci.* **80:** 955.

Fishel, R.A., A.A. James, and R. Kolodner. 1981. *RecA*-independent general genetic recombination of plasmids. *Nature* **294:** 184.

Flory, J. and C.M. Radding. 1982. Visualization of RecA protein and its association with DNA. A priming effect of single strand binding protein. *Cell* **28:** 747.

Fox, M.S. 1978. Some features of genetic recombination in procaryotes. *Annu. Rev. Genet.* **12:** 47.

Fox, M.S., C.S. Dudney, and E.J. Sodergren. 1979. Heteroduplex regions in unduplicated bacteriophage λ recombinants. *Cold Spring Harbor Symp. Quant. Biol.* **43:** 999.

Franklin, N. 1967. Deletions and functions of the center of the ϕ80-λ phage genome. Evidence for a phage function promoting genetic recombination. *Genetics* **57:** 301.

Gellert, M. 1967. Formation of covalent circles of lambda DNA by *E. coli* extracts. *Proc. Natl. Acad. Sci.* **57:** 148.

——. 1981. DNA topoisomerases. *Annu. Rev. Biochem.* **50:** 879.

Gellert, M., K. Mizuuchi, M.H. O'Dea, and H.A. Nash. 1976. DNA gyrase: An enzyme that introduces superhelical turns into DNA. *Proc. Natl. Acad. Sci.* **73:** 3872.

Gillen, J.R. 1974. "The RecE pathway of genetic recombination in *Escherichia coli*." Ph.D. thesis. University of California, Berkeley.

Gillen, J.R. and A.J. Clark. 1974. The RecE pathway of bacterial recombination. In *Mechanisms in recombination* (ed. R.F. Grell), p. 123. Plenum Press, New York.

Gillen, J.R., D.K. Willis, and A.J. Clark. 1981. Genetic analysis of the RecE pathway of genetic recombination in *Escherichia coli* K-12. *J. Bacteriol.* **145:** 521.

Gillen, J.R., A.E. Karu, H. Nagaishi, and A.J. Clark. 1977. Characterization of the deoxyribonuclease determined by lambda reverse as exonuclease VIII of *Escherichia coli*. *J. Mol. Biol.* **113:** 27.

Glassberg, J., R.R. Meyer, and A. Kornberg. 1979. Mutant single-strand binding protein of *Escherichia coli*: Genetic and physiological characterization. *J. Bacteriol.* **140:** 14.

Goldmark, P.J. and S. Linn. 1970. An endonuclease activity from *Escherichia coli* absent from certain *rec*⁻ strains. *Proc. Natl. Acad. Sci.* **67:** 434.

——. 1972. Purification and properties of the *recBC* DNase of *Escherichia coli* K12. *J. Biol. Chem.* **247:** 1849.

Gottesman, M. and M. Yarmolinsky. 1968. Integration-negative mutants of bacteriophage lambda. *J. Mol. Biol.* **31:** 487.

Gottesman, M., M.L. Hicks, and M. Gellert. 1974a. Genetics and function of DNA ligase in *Escherichia coli*. *J. Mol. Biol.* **77:** 531.

Gottesman, M.M., M.E. Gottesman, S. Gottesman, and M. Gellert. 1974b. Characterization of bacteriophage λ reverse as an *Escherichia coli* phage carrying a unique set of host derived recombination functions. *J. Mol. Biol.* **88:** 471.

Gussin, G., E.D. Rosen, and D.L. Wulff. 1980. Mapability of very close markers of bacteriophage λ. *Genetics* **96:** 1.

Hays, J.B. and S. Boehmer. 1978. Antagonists of DNA gyrase inhibit repair and recombination of UV-irradiated phage λ. *Proc. Natl. Acad. Sci.* **75:** 4125.

Hays, J.B. and B.E. Korba. 1979. DNA from recombinogenic λ bacteriophages generated by *arl* mutants of *Escherichia coli* is cleaved by single strand-specific endonuclease S1. *Proc. Natl. Acad. Sci.* **76:** 6066.

Hays, J.B., B.E. Korba, and E.B. Konrad. 1980. Novel mutations of *Escherichia coli* that produce recombinogenic lesions in DNA. I. Identification and mapping of *arl* mutations. *J. Mol. Biol.* **139:** 455.

Henderson, D. and J. Weil. 1975. Recombination-deficient deletions in bacteriophage λ and their interaction with *chi* mutations. *Genetics* **79:** 143.

Hickson, I.D. and P.T. Emmerson. 1981. Identification of the *Escherichia coli recB* and *recC* gene products. *Nature* **294:** 578.

Holliday, R. 1964. A mechanism for gene conversion in fungi. *Genet. Res.* **5:** 129.

Holloman, W.K., R. Wiegand, C. Hoessli, and C.M. Radding. 1975. Uptake of homologous single-stranded fragments by superhelical DNA: A possible mechanism for initiation of genetic recombination. *Proc. Natl. Acad. Sci.* **72:** 2394.

Horii, Z.-I. and A.J. Clark. 1973. Genetic analysis of the RecF pathway to genetic recombination in *Escherichia coli* K-12: Isolation and characterization of mutants. *J. Mol. Biol.* **80:** 327.

Horii, T., T. Ogawa, and H. Ogawa. 1980. Organization of the RecA gene of *Escherichia coli*. *Proc. Natl. Acad. Sci.* **77:** 313.

Ikeda, H. and I. Kobayashi. 1977. Involvement of DNA-dependent RNA polymerase in a *recA*-independent pathway of genetic recombination in *Escherichia coli*. *Proc. Natl. Acad. Sci.* **74:** 3932.

——. 1979. *recA*-mediated recombination of bacteriophage λ: Structure of recombinant and intermediate DNA molecules and their packaging in vitro. *Cold Spring Harbor Symp. Quant. Biol.* **43:** 1009.

Ikeda, H. and T. Matsumoto. 1979. Transcription promotes *recA* independent recombination mediated by DNA-dependent RNA polymerase in *Escherichia coli*. *Proc. Natl. Acad. Sci.* **76:** 4571.

Kahn, R., R.P. Cunningham, C. DasGupta, and C.M. Radding. 1981. Polarity of heteroduplex formation promoted by *Escherichia coli* recA protein. *Proc. Natl. Acad. Sci.* **78:** 4786.

Kaiser, K. and N.E. Murray. 1979. Physical characterization of the "Rac prophage" in *E. coli* K-12. *Mol. Gen. Genet.* **175:** 159.

Karu, A.E. and S. Linn. 1972. Uncoupling of the *recBC* ATPase from DNase by DNA crosslinked with psoralen. *Proc. Natl. Acad. Sci.* **69:** 2855.

Karu, A.E., V. MacKay, P.S. Goldmark, and S. Linn. 1973. The *recBC* deoxyribonuclease of *Escherichia coli* K12. Substrate specificity and reaction intermediates. *J. Biol. Chem.* **248:** 4874.

Karu, A.E., Y. Sakaki, H. Echols, and S. Linn. 1975. The γ protein specified by bacteriophage λ.

Structure and inhibitory activity for the recBC enzyme of *Escherichia coli*. *J. Biol. Chem.* **250:** 7377.

Kellenberger, G., M.L. Zichichi, and J. Weigle. 1961. Exchange of DNA in the recombination of bacteriophage λ. *Proc. Natl. Acad. Sci.* **57:** 869.

Kellenberger-Gujer, G. 1971. Recombination in bacteriophage lambda. II. The mechanism of general recombination. In *The bacteriophage lambda* (ed. A.D. Hershey), p. 417. Cold Spring Harbor Laboratory, Cold Spring Harbor, New York.

Kellenberger-Gujer, G. and R. Weisberg. 1971. Recombination in bacteriophage lambda. I. Exchange of DNA promoted by phage and bacterial recombination mechanisms. In *The bacteriophage lambda* (ed. A.D. Hershey), p. 407. Cold Spring Harbor Laboratory, Cold Spring Harbor, New York.

Kmiec, E. and W.K. Holloman. 1981. β protein of bacteriophage λ promotes renaturation of DNA. *J. Biol. Chem.* **256:** 12636.

Kobayashi, I. and H. Ikeda. 1977. Formation of recombinant DNA of bacteriophage lambda by *recA* function of *Escherichia coli* without duplication, transcription, translation, and maturation. *Mol. Gen. Genet.* **153:** 237.

——. 1978. On the role of *recA* gene product in genetic recombination: An analysis by *in vitro* packaging of recombinant DNA molecules formed in the absence of protein synthesis. *Mol. Gen. Genet.* **166:** 25.

Kobayashi, I., M.M. Stahl, D. Leach, and F.W. Stahl. 1983. The interaction of *cos* with Chi is separable from DNA packaging in *recA recBC* mediated recombination of bacteriophage lambda. *Genetics* (in press).

Kobayashi, I., H. Murialdo, J.M. Crasemann, M.M. Stahl, and F.W. Stahl. 1982. Orientation of cohesive end site *cos* determines the active orientation of sequence in stimulating recA.recBC-mediated recombination in phage λ lytic infections. *Proc. Natl. Acad. Sci.* **79:** 5981.

Konrad, E.B. 1977. Method for the isolation of *Escherichia coli* mutants with enhanced recombination between chromosomal duplications. *J. Bacteriol.* **130:** 167.

Konrad, E.B. and I.R. Lehman. 1974. A conditional lethal mutant of *Escherichia coli* K12 defective in the 5′→3′ exonuclease associated with DNA polymerase I. *Proc. Natl. Acad. Sci.* **71:** 2048.

Korba, B.E. and J.B. Hays. 1980. Novel mutations of *Escherichia coli* that produce recombinogenic lesions in DNA. II. Properties of recombinogenic λ phages grown on bacteria carrying *arl* mutations. *J. Mol. Biol.* **139:** 473.

——. 1982. Partially deficient methylation of cytosine in DNA at CC(A/T)GG sites stimulates genetic recombination of bacteriophage lambda. *Cell* **28:** 531.

Kornberg, A. 1980. DNA replication. W.H. Freeman, San Francisco.

Kushner, S.R., H. Nagaishi, and A.J. Clark. 1974. Isolation of exonuclease VIII: The enzyme associated with the *sbcA* indirect suppressor. *Proc. Natl. Acad. Sci.* **71:** 3593.

Kushner, S.R., H. Nagaishi, A. Templin, and A.J. Clark. 1971. Genetic recombination in *Escherichia coli*: The role of exonuclease I. *Proc. Natl. Acad. Sci.* **68:** 824.

Lam, S.T., M.M. Stahl, K.D. McMilin, and F.W. Stahl. 1974. Rec-mediated recombinational hot spot activity in bacteriophage lambda. II. A mutation which causes hot spot activity. *Genetics* **77:** 425.

Lichten, M. and M.S. Fox. 1983. Effects of nonhomology on bacteriophage lambda recombination. *Genetics* **103:** 5.

Lieberman, R.P. and M. Oishi. 1974. The *recBC* deoxyribonuclease of *Escherichia coli:* Isolation and characterization of the subunit proteins and reconstitution of the enzyme. *Proc. Natl. Acad. Sci.* **71:** 4816.

Lin, P.-F., E. Bardwell, and P. Howard-Flanders. 1977. Initiation of genetic exchanges in λ phage-prophage crosses. *Proc. Natl. Acad. Sci.* **74:** 291.

Lindahl, G., G. Sironi, H. Bialy, and R. Calendar. 1970. Bacteriophage lambda: Abortive infection of bacteria lysogenic for phage P2. *Proc. Natl. Acad. Sci.* **66:** 587.

Little, J.W. 1967. An exonuclease induced by bacteriophage λ. II. Nature of the enzymic reaction. *J. Biol. Chem.* **242:** 679.

Lloyd, R.G. and S.D. Barbour. 1974. The genetic location of the *sbcA* gene of *Escherichia coli*. *Mol. Gen. Genet.* **134:** 157.

Low, K.B. 1968. Formation of merodiploids in matings with a class of *rec*⁻ recipient strains of *Escherichia coli* K12. *Proc. Natl. Acad. Sci.* **60:** 160.

——. 1973. Restoration by the *rac* locus of recombinant forming ability in *recB*⁻ and *recC*⁻ merozygotes of *Escherichia coli* K12. *Mol. Gen. Genet.* **122:** 119.

Low, K.B. and R.D. Porter. 1978. Modes of gene transfer and recombination in bacteria. *Annu. Rev. Genet.* **12:** 249.

MacKay, V. and S. Linn. 1974. The mechanism of degradation of duplex deoxyribonucleic acid by the *recBC* enzyme of *Escherichia coli* K12. *J. Biol. Chem.* **249:** 4286.

Makin, G.S.V., W. Szybalski, and F.R. Blattner. 1982. Asymmetric effects of deletions and substitutions on high negative interference in coliphage lambda. *Genetics* **102:** 299.

Malone, R.E., and D.K. Chattoraj. 1975. The role of *chi* mutations in the Spi⁻ phenotype of phage λ: Lack of evidence for a gene delta. *Mol. Gen. Genet.* **143:** 35.

Malone, R.E., D.K. Chattoraj, D.H. Faulds, M.M. Stahl, and F.W. Stahl. 1978. Hotspots for generalized recombination in the *E. coli* chromosome. *J. Mol. Biol.* **121:** 483.

Matsumoto, T. and H. Ikeda. 1983. Role of R-loops in *recA*-independent homologous recombination of bacteriophage λ. *J. Virol.* **45:** 971.

McEntee, K. and G.M. Weinstock. 1981. The *recA* enzyme of *Escherichia coli* and recombination assays. In *The enzymes* (ed. P. Boyer), vol. XIV, p. 445. Academic Press, New York.

McEntee, K., G.M. Weinstock, and I.R. Lehman. 1980. *recA* protein-catalyzed strand assimilation: Stimulation by *Escherichia coli* single-stranded DNA-binding protein. *Proc. Natl. Acad. Sci.* **77:** 857.

——. 1981. Binding of the recA protein of *Escherichia coli* to single- and double-stranded DNA. *J. Biol. Chem.* **256:** 8835.

McMilin, K.D., M.M. Stahl, and F.W. Stahl. 1974. Rec-mediated recombinational hotspot activity in bacteriophage lambda. I. Hotspot activity associated with Spi⁻ deletions and *bio* substitutions. *Genetics* **77:** 409.

Meselson, M. 1964. On the mechanism of recombination between DNA molecules. *J. Mol. Biol.* **9:** 734.

——. 1967. The molecular basis of genetic recombination. In *Heritage from Mendel* (ed. R.A. Brink), p. 81. University of Wisconsin Press, Madison.

Meselson, M. and C.M. Radding. 1975. A general model for genetic recombination. *Proc. Natl. Acad. Sci.* **72:** 358.

Meselson, M. and J. Weigle. 1961. Chromosome breakage accompanying genetic recombination in bacteriophage. *Proc. Natl. Acad. Sci.* **47:** 857.

Miller, R.C. 1975. Replication and molecular recombination of T-phage. *Annu. Rev. Microbiol.* **29:** 355.

Mizuuchi, K., M. Gellert, and H.A. Nash. 1978. Involvement of supertwisted DNA in integrative recombination of bacteriophage lambda. *J. Mol. Biol.* **121:** 375.

Mizuuchi, K., B. Kemper, J. Hays, and R. Weisberg. 1982. T4 endonuclease VII cleaves Holliday structures. *Cell* **29:** 357.

Morand, P., M. Blanco, and R. Devoret. 1977. Characterization of *lexB* mutations in *Escherichia coli* K-12. *J. Bacteriol.* **131:** 572.

Ogawa, T., H. Wakibo, T. Tsurimoto, T. Horii, H. Masukata, and H. Ogawa. 1979. Characteristics of purified *recA* protein and the regulation of its synthesis in vivo. *Cold Spring Harbor Symp. Quant. Biol.* **43:** 909.

Potter, H. and D. Dressler. 1976. On the mechanism of genetic recombination: Electron microscopic observation of recombination intermediates. *Proc. Natl. Acad. Sci.* **73:** 3000.

Radding, C.M. 1973. Molecular mechanisms in recombination. *Annu. Rev. Genet.* **7:** 87.

———. 1978. Genetic recombination: Strand transfer and mismatch repair. *Annu. Rev. Biochem.* **47:** 847.

———. 1982. Homologous pairing and strand exchange in genetic recombination. *Annu. Rev. Genet.* **16:** 405.

Radding, C.M., J. Rosensweig, F. Richards, and E. Cassuto. 1971. Separation and characterization of exonuclease, β protein and a complex of both. *J. Biol. Chem.* **246:** 2510.

Reyes, O., E. Jedlicki, and J.-P. Kusnierz. 1982. Growth of λ*rev* in phage P2 lysogenic hosts. *Virology* **112:** 651.

Roberts, J.W. and C.W. Roberts. 1981. Two mutations that alter the regulatory activity of *E. coli* recA protein. *Nature* **290:** 422.

Rosamond, J., K.M. Telander, and S. Linn. 1979. Modulation of the *recBC* enzyme of *Escherichia coli* K12 by Ca^{++}. *J. Biol. Chem.* **254:** 8646.

Rothman, R.H. and A.H. Clark. 1977. The dependence of postreplication repair on *uvrB* in a *recF* mutant of *Escherichia coli* K-12. *Mol. Gen. Genet.* **155:** 279.

Rupp, W.D., C.E. Wilde, D.L. Reno, and P. Howard-Flanders. 1971. Exchanges between DNA strands in ultraviolet-irradiated *Escherichia coli. J. Mol. Biol.* **61:** 25.

Sancar, A., C. Stachelek, W. Konigsberg, and W.D. Rupp. 1980. Sequences of the *recA* gene and protein. *Proc. Natl. Acad. Sci.* **77:** 1638.

Sarthy, P.V., and M.M. Meselson. 1976. Single burst study of rec- and red-mediated recombination in bacteriophage lambda. *Proc. Natl. Acad. Sci.* **73:** 4613.

Sasaki, M., T. Fujiyoshi, K. Shimada, and Y. Takagi. 1982. Fine structure of the *recB* and *recC* gene region of *Escherichia coli. Biochem. Biophys. Res. Commun.* **109:** 414.

Schultz, D.W., J. Swindle, and G.R. Smith. 1981. Clustering of mutations inactivating a Chi recombinational hotspot. *J. Mol. Biol.* **146:** 275.

Shibata, T., R.P. Cunningham, C. DasGupta, and C.M. Radding. 1979a. Homologous pairing in genetic recombination: Complexes of *recA* protein and DNA. *Proc. Natl. Acad. Sci.* **76:** 5100.

Shibata, T., C. DasGupta, R.P. Cunningham, and C.M. Radding. 1979b. Purified *Escherichia coli recA* protein catalyzes homologous pairing of superhelical DNA and single-stranded fragments. *Proc. Natl. Acad. Sci.* **76:** 1638.

———. 1980. Homologous pairing in genetic recombination: Formation of D-loops by combined action of RecA protein and a helix destabilizing protein. *Proc. Natl. Acad. Sci.* **77:** 2606.

Shulman, M.J., L.M. Hallick, H. Echols, and E.R. Signer. 1970. Properties of recombination-deficient mutants of bacteriophage lambda. *J. Mol. Biol.* **52:** 501.

Siegel, J. 1974. Extent and location of DNA synthesis associated with a class of Rec-mediated recombinants of bacteriophage lambda. *J. Mol. Biol.* **88:** 619.

Sigal, N. and B. Alberts. 1972. Genetic recombination: The nature of a crossed strand-exchange between two homologous DNA-molecules. *J. Mol. Biol.* **71:** 789.

Signer, E.R. 1971. General recombination. In *The bacteriophage lambda* (ed. A.D. Hershey), p. 139. Cold Spring Harbor Laboratory, Cold Spring Harbor, New York.

Signer, E.R. and J. Weil. 1968. Recombination in bacteriophage λ. I. Mutants deficient in general recombination. *J. Mol. Biol.* **34:** 261.

Signer, E.R., H. Echols, J. Weil, C.M. Radding, M. Schulman, L. Moore, and K. Manly. 1969. The general recombination system of bacteriophage λ. *Cold Spring Harbor Symp. Quant. Biol.* **33:** 711.

Simmon, V.F. and S. Lederberg. 1972. Degradation of bacteriophage lambda deoxyribonucleic acid after restriction by *Escherichia coli* K-12. *J. Bacteriol.* **112:** 161.

Sironi, G. 1969. Mutants of *E. coli* unable to be lysogenized by the temperate bacteriophage P2. *Virology* **37:** 163.

Skalka, A. 1977. DNA replication—Bacteriophage lambda. *Curr. Top. Microbiol. Immunol.* **78:** 201.

Smith, G.R., D.W. Schultz, and J.M. Crasemann. 1980. Generalized recombination: Nucleotide sequence homology between Chi recombinational hotspots. *Cell* **19:** 785.

Smith, G.R., D.W. Schultz, A. Taylor, and K. Triman. 1981a. Chi sites, RecBC enzyme, and generalized recombination. *Stadler Genet. Symp.* **13:** 25.

Smith, G.R., M. Comb, D.W. Schultz, D.L. Daniels and F.R. Blattner. 1981b. Nucleotide sequence of the Chi recombinational hotspot $\chi^{+}D$ in phage λ. *J. Virol.* **31:** 336.

Smith, G.R., S.M. Kunes, D.W. Schultz, A. Taylor, and K.L. Triman. 1981c. Structure of Chi hotspots of generalized recombination. *Cell* **24:** 429.

Sodergren, E.J. and M.S. Fox. 1979. Effects of DNA sequence non-homology on formation of bacteriophage lambda recombinants. *J. Mol. Biol.* **130:** 357.

Sprague, K.U., D.H. Faulds, and G.R. Smith. 1978. A single base-pair change creates a Chi recombinational hotspot in bacteriophage λ. *Proc. Natl. Acad. Sci.* **75:** 6182.

Stahl, F.W. 1979a. Special sites in generalized recombination. *Annu. Rev. Genet.* **13:** 7.

——. 1979b. *Genetic recombination: Thinking about it in phage and fungi.* W.H. Freeman, San Francisco.

Stahl, F.W. and M.M. Stahl. 1974a. A role for *recBC* nuclease in the distribution of crossovers along unreplicated chromosomes of phage λ. *Mol. Gen. Genet.* **131:** 27.

——. 1974b. Red-mediated recombination in bacteriophage lambda. In *Mechanisms in recombination* (ed. R.F. Grell), p. 407. Plenum Press, New York

——. 1975. Rec-mediated recombinational hotspots in bacteriophage λ. IV. Effect of heterology on Chi-stimulated crossing-over. *Mol. Gen. Genet.* **140:** 29.

——. 1977. Recombination pathway specificity of Chi. *Genetics* **86:** 715.

Stahl, F.W., J.M. Crasemann, and M.M. Stahl. 1975. Rec-mediated hotspot activity in bacteriophage λ. III. Chi mutations are site-mutations stimulating Rec-mediated recombination. *J. Mol. Biol.* **94:** 203.

Stahl, F.W., I. Kobayashi, and M.M. Stahl. 1982a. Distance from cohesive end site *cos* determines the replication requirement for recombination in phage λ. *Proc. Natl. Acad. Sci.* **79:** 6318.

Stahl, F.W., M.M. Stahl, and R.E. Malone. 1978. Red-mediated recombination of phage lambda in a $recA^{-}$ $recB^{-}$ host. *Mol. Gen. Genet.* **159:** 207.

Stahl, F.W., K.D. McMilin, M.M. Stahl, and Y. Nozu. 1972. An enhancing role for DNA synthesis in formation of bacteriophage lambda recombination. *Proc. Natl. Acad. Sci.* **69:** 3598.

Stahl, F.W., M.M. Stahl, R.E. Malone, and J.M. Crasemann. 1980a. Directionality and nonreciprocality of Chi-stimulated recombination in phage λ. *Genetics* **94:** 235.

Stahl, F.W., M.M. Stahl, L. Young, and I. Kobayashi. 1982b. Injection is not essential for high level of Chi activity during recombination between λ vegetative phages. In *Proceedings of the John Innes Symposium* (ed. D.A. Hopwood). Croom Helm, London. (In press.)

Stahl, F.W., K.D. McMilin, M.M. Stahl, J.M. Crasemann, and S. Lam. 1974. The distribution of crossovers along unreplicated lambda bacteriophage chromosomes. *Genetics* **77:** 395.

Stahl, F.W., D. Chattoraj, J.M. Crasemann, N.A. Dower, M.M. Stahl, and E. Yagil. 1980b. What accounts for the orientation dependence and directionality of Chi? *ICN-UCLA Symp. Mol. Cell. Biol.* **19:** 919.

Taylor, A. and G.R. Smith. 1980. Unwinding and rewinding of DNA by the RecBC enzyme. *Cell* **22:** 447.

Telander-Muskavitch, K.M. 1981. "Exonuclease V, the *recBC* enzyme of *Escherichia coli.*" Ph.D. thesis, University of California, Berkeley.

Telander-Muskavitch, K.M. and S. Linn. 1980. Electron microscopy of *E. coli recBC* enzyme reaction intermediates. *ICN-UCLA Symp. Mol. Cell. Biol.* **19:** 901.

——. 1981. RecBC-like enzymes: Exonuclease V deoxyribonucleases. In *The enzymes* (ed. P.D. Boyer), vol. 14, p. 233. Academic Press, New York.

——. 1982. A unified mechanism for the nuclease and unwinding activities of the *recBC* enzyme of *Escherichia coli. J. Biol. Chem.* **257:** 2641.

Thompson, B.J., C. Escarmis, B. Parker, W.C. Slater, J. Doniger, I. Tessman, and R.C. Warner. 1975. Figure-8 configuration of dimers of S13 and φX174 replicative form DNA. *J. Mol. Biol.* **91:** 409.

Toothman, P. and I. Herskowitz. 1980. Rex-dependent exclusion of lambdoid phages. II. Determinants of sensitivity to exclusion. *Virology* **102:** 147.

Triman, K.L., D.K. Chattoraj, and G.R. Smith. 1982. Identity of a Chi site of *Escherichia coli* and Chi recombinational hotspots of bacteriophage λ. *J. Mol. Biol.* **154:** 393.

Unger, R.C. and A.J. Clark. 1972. Interaction of the recombination pathways of bacteriophage λ and its host *Escherichia coli* K12: Effects on exonuclease V activity. *J. Mol. Biol.* **70:** 539.

Unger, R.C., H. Echols, and A.J. Clark. 1972. Interaction of the recombination pathways of bacteriophage λ and host *Escherichia coli*: Effects on λ recombination. *J. Mol. Biol.* **70:** 531.

Valenzuela, M.S. and R.B. Inman. 1975. Visualization of a novel junction in bacteriophage λ DNA. *Proc. Natl. Acad. Sci.* **72:** 3024.

Visconti, N. and M. Delbrück. 1953. The mechanism of genetic recombination in phage. *Genetics* **38:** 5.

Wagner, R., Jr. and M. Meselson. 1976. Repair tracts in mismatched DNA heteroduplexes. *Proc. Natl. Acad. Sci.* **73:** 4135.

Warner, R.C., R.A. Fishel, and F.C. Wheeler. 1979. Branch migration in recombination. *Cold Spring Harbor Symp. Quant. Biol.* **43:** 957.

Weinstock, G.M., K. McEntee, and I.R. Lehman. 1979. ATP-dependent renaturation of DNA catalyzed by the recA protein of *Escherichia coli. Proc. Natl. Acad. Sci.* **76:** 126.

West, S.C., E. Cassuto, and P. Howard-Flanders. 1981a. Homologous pairing can occur before DNA strand separation in general genetic recombination. *Nature* **290:** 29.

——. 1981b. Heteroduplex formation by recA protein: Polarity of strand exchanges. *Proc. Natl. Acad. Sci.* **78:** 6149.

West, S.C., E. Cassuto, J. Mursalim and P. Howard-Flanders. 1980. Recognition of duplex DNA containing single-stranded regions by recA protein. *Proc. Natl. Acad. Sci.* **77:** 2569.

White, R.L. and M.S. Fox. 1974. On the molecular basis of high negative interference. *Proc. Natl. Acad. Sci.* **71:** 1544.

——. 1975a. Genetic heterozygosity in unreplicated bacteriophage λ recombinants. *Genetics* **81:** 33.

——. 1975b. Genetic consequences of transfection with heteroduplex bacteriophage λ DNA. *Mol. Gen. Genet.* **141:** 163.

Whitehouse, H.L.K. 1982. Genetic recombination: Understanding the mechanisms. John Wiley and Sons, New York.

Wilcox, K.W. and H.O. Smith. 1976. Mechanism of DNA degradation by the ATP-dependent DNase from *Haemophilus influenzae* Rd. *J. Biol. Chem.* **251:** 6127.

Wildenberg, J. and M. Meselson. 1975. Mismatch repair in heteroduplex DNA. *Proc. Natl. Acad. Sci.* **72:** 2202.

Willetts, N.S. and D.W. Mount. 1969. Genetic analysis of recombination deficient mutants of *Escherichia coli* K-12 carrying *rec* mutations cotransducible with *thyA*. *J. Bacteriol.* **100:** 923.

Willetts, N.S., A.J. Clark, and K.B. Low. 1969. Genetic location of certain mutations conferring recombination deficiency in *Escherichia coli. J. Bacteriol.* **97:** 244.

Yagil, E., N.A. Dower, D. Chattoraj, M. Stahl, C. Pierson, and F.W. Stahl. 1980. Chi mutation in a transposon and the orientation-dependence of Chi phenotype. *Genetics* **96:** 43.

Zissler, J., E.R. Signer, and F. Schaefer. 1971a. The role of recombination in growth of bacteriophage lambda. I. The gamma gene. In *The bacteriophage lambda* (ed. A.D. Hershey), p. 455. Cold Spring Harbor Laboratory, Cold Spring Harbor, New York.

——. 1971b. The role of recombination in growth of bacteriophage lambda. II. Inhibition of growth by prophage P2. In *The bacteriophage lambda* (ed. A.D. Hershey), p. 469. Cold Spring Harbor Laboratory, Cold Spring Harbor, New York.

Site-specific Recombination in Phage Lambda

Robert A. Weisberg
Section on Microbial Genetics
Laboratory of Molecular Genetics
National Institute of Child Health and Human Development
National Institutes of Health
Bethesda, Maryland 20205

Arthur Landy
Section of Microbiology and Molecular Biology
Division of Biology and Medicine
Brown University
Providence, Rhode Island 02912

INTRODUCTION

Cells lysogenic for λ carry a quiescent form of the viral chromosome called prophage. The prophage differs from the DNA of viral particles in two important ways: (1) The ends of the prophage and of the viral particle DNA are at different points in the nucleotide sequence, and (2) the prophage ends are covalently joined to host DNA. Campbell (1962) proposed that viral particle DNA is converted to prophage by the joining of the ends followed by insertion of the resulting circular molecule into the host DNA. Insertion occurs by reciprocal recombination at specific sites (attachment or *att* sites) in each chromosome (Fig. 1). This proposal has received extensive experimental support and, indeed, was generally accepted when the first edition of this book was written (see Gottesman and Weisberg 1971).

A stable lysogen is formed when an infecting viral particle succeeds both in repressing lytic functions and in inserting its DNA. The inserted prophage is then replicated as part of the bacterial chromosome. In the rare event that repression is not followed by insertion (abortive lysogeny), the viral chromosome cannot replicate and so is lost by dilution as the host cell divides. Excision of the prophage from the chromosome is rare during normal bacterial growth but occurs readily following repressor inactivation. If the repressor is inactivated only briefly (abortive or transient derepression), prophage excision can occur without lytic growth. The excised DNA is then frequently lost as the cell divides (prophage curing).

Intricate controls ensure that insertion occurs only when repression of lytic functions is also likely to occur and that excision does not compete with insertion (Echols and Guarneros, this volume).

In recent years prophage insertion and excision have been studied as examples of site-specific recombination. Site-specific recombination resembles homologous or general recombination (Smith, this volume) in its result: the movement

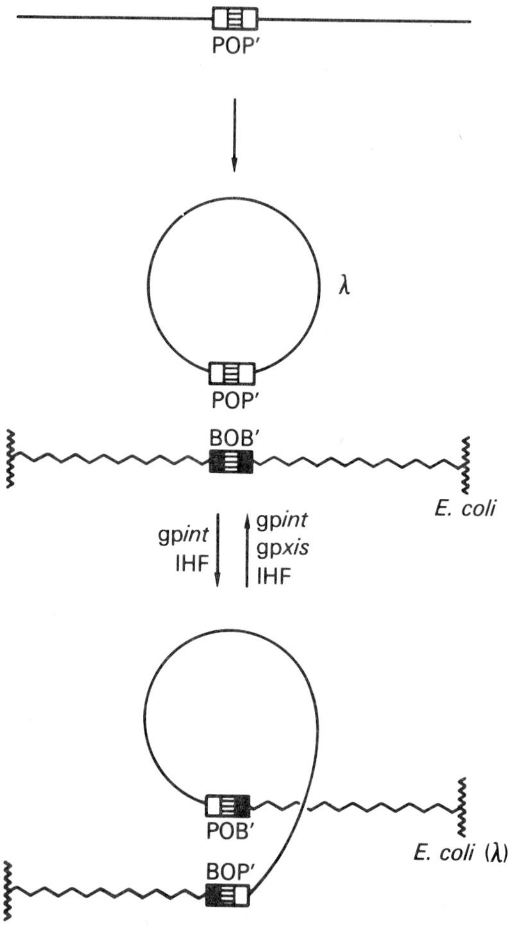

Figure 1 The Campbell (1962) model of prophage insertion and excision. Conversion of linear to circular λ DNA (straight lines) occurs by base-pairing and ligation of short complementary single-stranded segments at each end. Circular λ DNA inserts into the *E. coli* chromosome by crossing over at the *att* sites (rectangles). The hatched regions within the *att* sites represent the crossover region (O), and the solid and open regions represent the unique sequences or arms of bacterial (B and B′) or phage (P and P′) origin, respectively (see below, Attachment Sites). Insertion and excision are promoted by the proteins indicated alongside the arrows (see Genes and Proteins).

of genetic information between or within chromosomes. The two processes differ in that general recombination depends on homology between the interacting chromosomes or chromosome segments, whereas site-specific recombination requires the presence of specific nucleotide sequences (called *att* sites in the case of λ). There are four primary *att* sites: *att*P (or POP′) in the phage, *att*B (or

BOB′) in the host, and *att*L (BOP′) and *att*R (POB′) at the left and right ends, respectively, of an inserted prophage (see Fig. 1). Each of these sites has unique properties, which will be described below. In addition to these four, a heterogeneous collection of sites that resemble *att*B but have a much lower efficiency of recombination with *att*P are found in phage and bacterial DNAs. These are collectively called secondary *att* sites.

λ site-specific recombination is promoted by one host and two phage proteins. The two phage proteins are the products of the *int* and *xis* genes. The host protein (integration host factor or IHF) contains two polypeptide chains. The *himA* gene encodes one, and the *hip* gene probably encodes the other. gp*int*, gp*xis*, and IHF have been purified in active form. These proteins promote the recombination in vitro of DNA molecules that carry *att* sites. In the sections that follow, we shall describe the proteins and sites involved in λ site-specific recombination. For other recent reviews, see Mizuuchi et al. (1981), Nash (1981b), and Nash et al. (1981).

GENES AND PROTEINS

Int Protein

The *int* gene was initially defined by recessive λ mutants, including nonsense and temperature-sensitive mutants, that formed abortive but not stable lysogens (Gingery and Echols 1967; Zissler 1967; Gottesman and Yarmolinsky 1968). Subsequently, a larger collection was isolated in a technically simpler way: screening for phage variants that were unable to promote excision of a prophage inserted in a host gene (the red plaque test; Enquist and Weisberg 1976; 1977a). This screen yielded two classes of mutants: one that is defective both in insertion and excision and a second that is defective only in excision. We have described the second class below (see Xis protein). The insertion-defective or *int* mutants form a single complementation group. Most differed from wild type not only in prophage insertion and excision but also in phage crosses. They manifested reduced recombination in an interval containing *att*P (see Gottesman and Weisberg 1971). This established that *int* activity is not limited to the lysogenic pathway of phage development. Further studies established that recombination between any of the primary λ*att* sites and between *att*P and secondary sites requires *int* activity. A map of *int* and *xis* mutants, both point and deletion, is given in Figure 2. Mutations isolated by the abortive lysogeny test are interspersed with those isolated by the prophage curing test. Electron microscopy shows that the mutant sites lie in a region extending from within about 50 bp to about 1300 bp rightward of the crossover point in *att*P. Direct DNA sequencing shows that the region encoding the *int* polypeptide extends from 84 bp to 1151 bp rightward of the crossover point (see Appendix II; Davies 1980; Hoess et al. 1980).

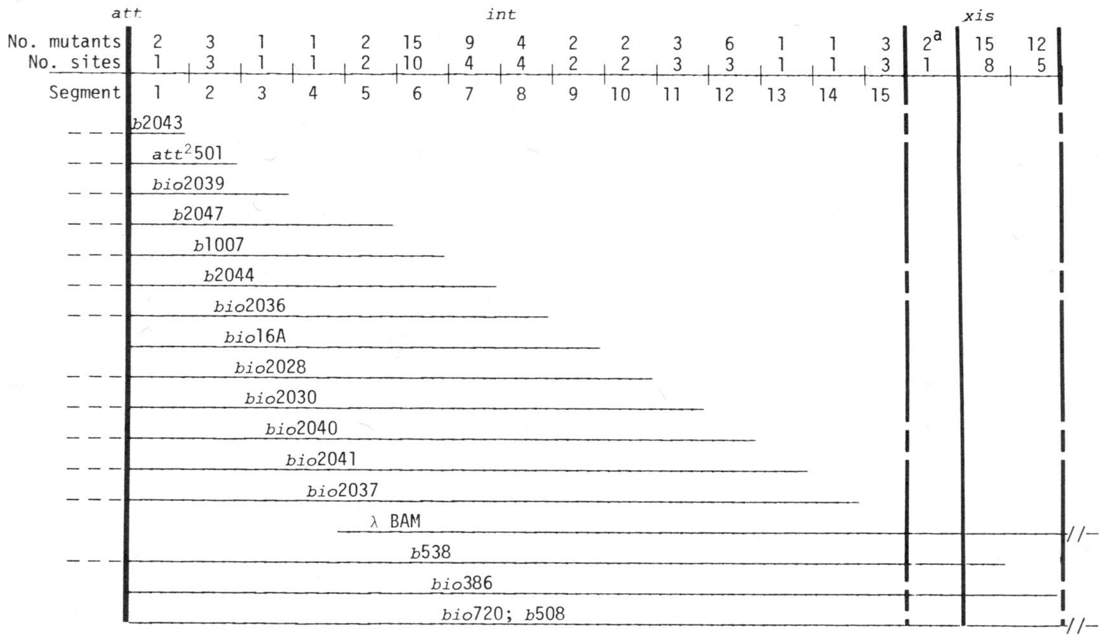

Figure 2 A genetic map of the *att-int-xis* region. Eighty-four independently isolated *int* and *xis* mutations were crossed one by one with a set of deletion mutations with endpoints in the *int-xis* region. The appearance of wild-type recombinants indicates that the wild-type allele of the mutation is not removed by the deletion. The results order the mutant sites and deletion endpoints along the genetic map. Occasionally, mutations in the same segment failed to give wild-type recombinants when crossed with one another; such groups of mutations are considered to define a single mutant site. Some deletions extend leftward beyond *att* and are indicated by dashed lines. The solid vertical lines indicate the boundaries of *int*, and the broken lines indicate the boundaries of *xis*. (Data from Enquist and Weisberg 1977a; R. Hoess, pers. comm.)

Two classes of *int* mutations were unusual in that the extent of their defect varied with the pair of *att* sites chosen to test them. The first and larger class contained mutants that were very defective in integrative (*att*P × *att*B) recombination but relatively proficient in excisive (*att*L × *att*R) recombination (Guarneros and Echols 1970; Enquist and Weisberg 1977b). These mutations probably reduce gp*int* activity to a level that is sufficient to promote *att*L × *att*R recombination but not *att*P × *att*B recombination. The assumption that more gp*int* activity is required for integrative than for excisive recombination is supported by experiments in which the level of wild-type gp*int* was limited by reducing *int* transcription or translation or by segregating a fixed quantity of protein among increasing numbers of cells. In all cases, integrative recombination was affected more severely or earlier than excisive recombination (Enquist et al. 1979a).

A second kind of unusual *int* mutation promotes *att*P × *att*B recombination more efficiently than any other combination tested (Enquist and Weisberg 1977a, and in prep.). Mutations that enlarge the range of gp*int* activity have also been found (see below, IHF; Xis Protein).

The purification of gp*int* awaited the development of methods to measure its function in vitro. One such assay measures integrative recombination (Nash 1975a; Mizuuchi and Nash 1976; Kikuchi and Nash 1978), and a second measures specific binding of gp*int* to DNA containing *att*P (Kamp 1973; Kotewicz et al. 1977; Kikuchi and Nash 1978). After purification, Int protein was shown to have a third activity: It promotes the relaxation of supercoiled DNA by breaking and religating polynucleotide chains and is therefore a topoisomerase (Kikuchi and Nash 1979b; see below).

The purified product migrates in SDS-gel electrophoresis as a single polypeptide of approximately 40,000 M_r (Kikuchi and Nash 1979a; Kotewicz et al. 1980). This is in good agreement with the molecular weight predicted from the sequence of the *int* gene (M_r = 40,330). The presumptive formylmethionine of gp*int* has been removed to leave serine as the amino terminus (Nash 1981a). Of the 356 amino acids, 69 are basic and 46 are acidic (Davies 1980; Hoess et al. 1980). Like other DNA-binding proteins, such as the *lac* and λ repressors, the aminoterminal portion is especially basic (35% of the first 20 amino acids are basic). At high ionic strength Int protein sediments as a monomer; however, electron microscopy indicates that, in the presence of *att*P-containing DNA, gp*int* forms multimers consisting of 4–8 monomers bound to *att*P (Hamilton et al. 1981; Better et al. 1982). The monomer appears to be rather asymmetric, and one of the purification protocols (Kotewicz et al. 1980) may yield an even more asymmetric or partially unfolded molecule.

Because purified Int protein binds to non-*att* DNA as well as to *att*-containing DNA, additional evidence was required to show sequence specificity. It was found that only those DNAs containing the P′ arm of the phage *att* site (i.e., *att*P or *att*L; see Fig. 1) form complexes that are stable in the presence of the polyanion heparin (Kikuchi and Nash 1978). We have further considered this special property of the P′ arm, along with other aspects of gp*int*-DNA recognition, in Attachment Sites, below.

Under optimal conditions, in vitro recombination with purified components is rapid. The intramolecular reaction, with *att*P and *att*B on the same supercoiled molecule, reaches completion in 3–5 minutes (10–30% of substrate converted to recombinant); the intermolecular reaction between a supercoiled molecule containing *att*P and a linear molecule containing *att*B requires approximately 45 minutes for completion (Nash and Robertson 1981). These experiments also confirmed the earlier suggestions that gp*int*-dependent recombination does not require an external energy source or any low-molecular-weight cofactors. The reaction is stimulated by 1 mM spermidine. $MgCl_2$ or high salt (165 mM KCl) can substitute, but the efficiency never exceeds one-half that found for optimum spermidine conditions. It is interesting that spermidine is not required for, and in fact inhibits, both the topoisomerase activity of gp*int* and the resolution of *att*-site Holliday structures (see below).

gp*int* acts catalytically as a topoisomerase, but the stoichiometry of the recombination reaction is complex and not well understood. The quantity of gp*int* required to produce a given amount of recombinant product depends upon the amount of substrate DNA. In a typical reaction mixture with DNA at 10 μg/ml, the ratio of gp*int* per recombinant is approximately 40, but a tenfold increase in the substrate DNA requires ten times as much gp*int* to produce the same absolute amount of recombinant (Nash and Robertson 1981). These findings raise the possibility that gp*int* acts noncatalytically in the recombination reaction.

The topoisomerase activity of gp*int* changes the linking number of supercoiled DNA in steps of one (rather than two), and it is therefore presumed to nick and reseal only one strand of the double helix at a time (Nash et al. 1981). Like recombination, nicking and resealing are independent of ATP and other cofactors. In contrast to recombination, topoisomerization is inhibited by spermidine. Int protein becomes attached to DNA via a protein-phosphodiester bond when it nicks DNA (N. Craig and H. Nash, pers. comm.). The complex probably conserves the energy of the bond for use in the religation step. The enzyme-DNA linkage is through a 3′-phosphate group. In this regard gp*int* resembles mammalian topoisomerase I (Champoux 1977; Halligan et al. 1982). However, it differs from topoisomerase I of *E. coli* (Tse et al. 1980) and from the resolvase protein of the transposon γδ (see below, Site-specific Recombination in Other Prokaryotes), which become linked to DNA through a 5′ phosphate. Int protein relaxes some supercoiled molecules that do not have a primary *att* site. Recent studies show that there are several sequences in the molecules used that closely resemble the gp*int* recognition sequences in primary *att* sites (see below, Attachment Sites).

Although the topoisomerase active site of gp*int* is almost certainly required for recombination, the protein must have at least one other functional domain. First, treatment of Int protein with *N*-ethylmaleimide abolishes its activity for recombination but does not affect the topoisomerase function. Second, an Int protein fragment that has an M_r of approximately 35,000 (instead of 40,000) has been generated by partial proteolytic digestion; although this fragment promotes relaxation of supercoiled DNA and resolution of Holliday structures involving the *att* sites (see below, Reaction Mechanisms), it is defective in the overall recombination reaction (N. Hasan and A. Landy, unpubl.).

IHF

It was discovered that integrative recombination in vitro requires a factor, now called IHF, that could be provided by an extract of nonlysogenic cells (Nash et al. 1977b; Kikuchi and Nash 1978). Shortly thereafter, selection schemes for isolating mutations in the gene(s) for IHF were devised (Miller and Friedman 1977, 1980; Miller et al. 1979; A. Kikuchi and R. Weisberg, in prep.). Although the schemes worked, the first well-characterized IHF mutation was isolated serendipitously by Williams et al. (1977). This mutation failed to support integrative recombination of λ, and extracts contained little or no active IHF. The mutation responsible for the phenotype is located near minute 38 on the *E. coli* chromosome. Miller and Friedman (1977, 1980) isolated and characterized several more mutations in this gene, which they named *himA*. *himA* mutations are recessive to *himA*$^+$ and, like the prototype, contain little or no active IHF (Miller et al. 1979). These workers also isolated mutations in two other host genes, *himB* and *himC*. The *himC* gene has not been characterized, but *himB* is probably identical to *gyrB*, the gene that encodes a subunit of DNA gyrase. This conclusion is based on genetic linkage between *himB* and *gyrB* and on analysis of a *himB* mutation that is temperature-sensitive for site-specific recombination and for DNA gyrase activity (L. Plantefaber et al., pers. comm.). The involvement of DNA gyrase in site-specific recombination probably reflects the requirement for supercoiled substrate (Mizuuchi and Nash 1976; see below).

A second selection scheme led to the isolation of mutations in a fourth gene, called *hip*. The best analyzed mutation is recessive to *hip*$^+$ and is located near minute 20 on the *E. coli* chromosome (Miller et al. 1979; A. Kikuchi and R. Weisberg, in prep.). Miller et al. (1981) have isolated mutations called *himD*, which are located near *hip* and confer a similar phenotype. Since *hip* and *himD* mutations fail to complement each other, and since both are complemented by a specialized transducing phage that carries a segment of bacterial DNA of approximately 4 kb from the *hip* region (E. Flamm and R. Weisberg, unpubl.), it is very likely that the mutations are located in the same gene. Crude extracts of a *hip* mutant contained little or no active IHF, but a mixture of crude extracts from *himA* and *hip* mutants was active. This suggests that IHF activity is the result of an interaction between gp*hip* and gp*himA*.

IHF has a native M_r of approximately 20,000, as calculated by sedimentation and gel permeation in solutions of high ionic strength (Nash and Robertson 1981). SDS-gel electrophoresis under denaturing conditions revealed two polypeptides in equimolar ratios: the α peptide, which is gp*himA*, has an M_r of approximately 10,500 (Miller and Nash 1981), and the β peptide, which is probably gp*hip* (Miller et al. 1981), has an M_r of approximately 9,500. Purified IHF and Int protein are sufficient to carry out efficient integrative recombination with supercoiled substrate, but DNA gyrase is also required for recombination of relaxed DNA in certain conditions. Indeed, DNA gyrase was discovered because of its involvement in integrative recombination (Mizuuchi et al. 1978). Excisive recombination in vitro also requires IHF (Abremski and Gottesman 1982).

IHF binds to several specific sites within *att*P, as discussed in the section on *att*-site structure. It also binds to non-*att* DNA, but the physiological role of such binding remains to be determined.

IHF facilitates several gp*int* functions. At limiting concentrations of Int protein, IHF stimulates (1) formation of heparin-resistant Int-protein complexes at the P′-arm site(s) (Nash and Robertson 1981), (2) binding by gp*int* at the *att*P core (N. Craig and H. Nash, pers. comm.), and (3) resolution by gp*int* of Holliday structures (Hsu and Landy, in prep.; see below, Reaction Mechanisms). It is not clear whether these facilitations of gp*int* functions are the result of IHF interaction with DNA, with gp*int*, or with both. IHF (like gp*int*) does not appear to act catalytically in recombination, since approximately 70 molecules are required for each recombinant produced (Nash and Robertson 1981). However, in contrast to gp*int*, this ratio is independent of the amount of parental DNA (or gp*int*) in the reaction; this same independence of DNA and gp*int* quantities is also observed for IHF stimulation of the formation of heparin-resistant Int-protein complexes at the P′ arm (see below).

Mutations in the *himA* or *hip* genes reduce integrative recombination in vivo two orders of magnitude or more (Miller and Friedman 1980; A. Kikuchi and R. Weisberg, in prep.). *att*P × *att*P and *att*P × *att*L recombination are also reduced about 100-fold in a *hip* mutant (Enquist et al. 1979b). In contrast, excisive recombination is less severely affected, especially when gp*int* is provided by infection with an *int*-constitutive (*int-c*) phage (Enquist et al. 1979a; Miller and Friedman 1980) or by derepression of a cro^- prophage (Gottesman and Abremski 1980). (The absence of gp*cro* is expected to increase transcription of *int* and *xis*.) In the second case, it has been proposed that gp*xis* compensates for the limitation in active IHF (Abremski and Gottesman 1981).

Two *int* mutants that partially relieve the host factor requirement for integrative and excisive recombination have been found; λ*int*-h3, which has been more fully characterized (Miller et al. 1980), and λ*xin* (C. Gritzmacher et al., unpubl.). The two mutations, whose location within *int* is not precisely known, may be independent isolates of the same mutation. Neither λ*int*-h3 nor λ*xin* requires a specific host factor mutant for its phenotype; thus, it is unlikely that they act by increasing the interaction of gp*int* with mutant IHF. B. Lange-Gustafson and H. Nash (pers. comm.) have shown that the *int*-h3 protein promotes a low level of integrative recombination in vitro in the absence of IHF. Both λ*xin* and λ*int*-h3 have other phenotypes. They can partially alleviate the *xis* requirement for excisive recombination (see below). In $himA^+hip^+$ hosts, both λ*int*-h3 and λ*xin* are able to recombine with greater than normal efficiency with secondary attachment sites. Several of these sites are identical with or are located very close to secondary sites used by wild-type λ (Miller et al. 1980; R. Hoess, pers. comm.; C. Gritzmacher et al., in prep.). Thus, the mutant *int* products have either the same specificity as gp*int*$^+$ or an overlapping specificity. Further characterization of the mutant proteins should shed light on the role of IHF in site-specific recombination.

Xis Protein

The first *xis* mutants were isolated by selecting λ variants that inserted their DNA into the bacterial chromosome with normal efficiency but formed lysogens that could not be cured by transient derepression (see Introduction; Guarneros and Echols 1970). Subsequently, more mutants were isolated by the red plaque test (see above, Int Protein; Enquist and Weisberg 1977a). The *xis* mutations form a single complementation group and are located immediately to the right of *int*. The DNA sequence shows that the coding region for gp*xis* extends from 1132 bp to 1347 bp (see Appendix II). According to this sequence, *xis*, with its carboxyl end at 1132 bp, overlaps *int*, with its amino end at 1152 bp (Hoess et al. 1980). Indeed, an unusual mutant, λ*int-xis*2266, which failed to complement both *int* and *xis* mutants (Enquist and Weisberg 1977a), has gained a base pair between positions 1138 and 1139 (R. Hoess, unpubl.).

xis function, together with that of *int*, is required for recombination in vivo between *att*L and *att*R. Both proteins are also required for prophage excision from secondary *att* sites (Shimada et al. 1972, 1973; R. Weisberg, unpubl.). In contrast, recombination between several other *att* sites is less strongly affected or is not affected at all by mutation in *xis* (Guarneros and Echols 1973). Most significantly, *att*P × *att*B recombination is *xis*-independent under standard conditions. Indeed, in some conditions, *xis* inhibits integrative recombination (Nash 1975b; Abremski and Gottesman 1982; J. Galindo and G. Guarneros, pers. comm.). The effect of *xis* on recombination between several *att*-site pairs is presented in Table 1. In addition to *att*P × *att*B, *att*L × *att*L recombination is also *xis*-independent. Most of the other *att* site pairs listed in Table 1 show some stimulation of recombination by gp*xis*, and this stimulation is sometimes more pronounced at 43°C than at 36°C (Guarneros and Echols 1973). It has recently been found that *att*R × *att*R recombination, unlike *att*L × *att*L, requires *xis* function (R. Weisberg, in prep.). In addition, the absence of *xis* function decreased the frequency of prophage curing from six secondary *att* sites by more than an order of magnitude (Shimada et al. 1972, 1973; R. Weisberg, unpubl.). In Excisive Recombination, below, we offer a hypothesis that ties some of these phenomena together (further suggestions are welcome).

The *xis* requirement even for *att*L × *att*R recombination is not absolute. *int-c* mutations, which synthesize large amounts of Int protein in the absence of the normal transcriptional activation (see Echols and Guarneros, this volume), promote a low level of excisive recombination in the absence of measurable *xis* ac-

Table 1 Effect of *xis* and high temperature on site-specific recombination

att pair	xis^+/xis^- (36°C)[a]		xis^+/xis^- (43°C)[a]	
P × B	25/18	1.4	0.7/1.1	0.63
L × R[b]	10/0.03	~333	13/n.t.[c]	—
L × L	14/10	1.4	12/13	0.92
L × P	11/4.1	2.7	11/0.4	28
R × P	5.7/0.4	14	0.07/0.04	~2
P × P	5.3/2.2	2.4	2.2/0.2	11

[a]Recombination frequency in a xis^+ infection/recombination frequency in a xis^- infection, followed by the ratio of the two.
[b]The L × R cross was carried out at 37°C instead of 36°C.
[c]n.t. indicates not tested. (Data from Echols 1970; Guarneros and Echols 1973.)

tivity (Shimada and Campbell 1974a,b; Abremski and Gottesman 1981). Two unusual *int* mutations, *int*-h3 and *xin*, in combination with *int-c*, further increase this level (Miller et al. 1980; C. Gritzmacher et al., unpubl.).

The nucleotide sequence of the *xis* gene indicates a protein product of 72 amino acids that is rich in basic amino acids (25% lysine and arginine). Preparations of gp*xis*, purified by different procedures, show a band corresponding to the expected 8630 M_r when electrophoresed in SDS gels under denaturing conditions (Abremski and Gottesman 1982; S. Wickner, pers. comm.; W. Bushman et al., in prep.). Additionally, Xis protein binds specifically to *att* DNA (see below, Primary Sites; S. Yin et al., in prep.).

gp*xis* promotes efficient recombination in vitro between *att*L and *att*R in the presence of gp*int* and IHF (Abremski and Gottesman 1982). Recombination requires Xis protein when the salt concentration is greater than 100 mM. It is accelerated by Xis protein in low salt, but the final extent of the low-salt reaction is independent of gp*xis*. Integrative recombination, in contrast, is insensitive to 100 mM NaCl and is inhibited by gp*xis*. Abremski and Gottesman (1982) have shown that a partially purified preparation of gp*xis* protects gp*int* from thermal inactivation. This observation, if confirmed with a pure preparation, would indicate a direct interaction between the two proteins. These workers have also shown that gp*xis*-dependent recombination does not require low-molecular-weight cofactors or an external energy source.

ATTACHMENT SITES

Primary Sites

Introduction

The attachment sites are regions of DNA that contain the structural determinants of site-specific recombination. The four primary *att* sites each consist of three elements: a crossover region that is common to all four, and two unique elements or arms that flank the crossover region. The symbol for the crossover region is O; P and P′ represent the left and right arms, respectively, of *att*P, and B and B′ represent the left and right arms, respectively, of *att*B. Recombination occurs by a reciprocal crossover. Thus, the insertion and excision reactions can be written formally in the following way (Guerrini 1969; Signer et al. 1969; Weisberg and Gottesman 1969; Gottesman and Weisberg 1971):

$$\text{POP}'(att\text{P}) + \text{BOB}'(att\text{B}) \underset{\text{gp}int\text{: IHF; gp}xis}{\overset{\text{gp}int\text{; IHF}}{\rightleftharpoons}} \text{BOP}'(att\text{L}) + \text{POB}'(att\text{R})$$

The existence of unique elements to the left and right of O is known from the observation that each *att* site differs from the others in recombinogenicity, requirement for gp*xis*, and temperature-sensitivity of the reaction (see Table 1). B and B′ are much smaller than P and P′ and probably consist of only a few nucleotide pairs each (see below).

The behavior of certain mutants defective in *att* function (Shulman and Gottesman 1973; Shulman et al. 1976) implied a crossover region rather than a crossover point. The best characterized mutations, *att*6 and *att*24, both had the following properties: (1) When present in one or both members of a pair of *att* sites, they depressed the frequency of recombination about 100-fold, and (2) the mutations could be moved by site-specific recombination from the original *att* sites in which they had been isolated to all of the others. The second property implies that the mutations are in a region common to all four *att* sites and suggests that points at which crossing over can occur are located both to the left and to the right of the sites changed by the mutations. Subsequent work, starting with the determination of the nucleotide sequence of the *att* sites, confirms these conclusions.

Sequence

The four primary *att* sites were isolated by identifying unique restriction endonuclease fragments whose sizes were altered by site-specific recombination (Marini and Landy 1977). The nucleotide sequence of the biologically important regions (see below) is shown in Figure 3 (Landy and Ross 1977). The sequence revealed a common region or "core" of 15 bp that is flanked by unique regions. The sequenced *att*L and *att*R sites were products of integrative recombination, and the sequenced *att*B site was a product of excisive recombination. Thus, recombination leaves the core unaltered, and this implies that the crossover must occur between identical bases within or immediately adjacent to the core. The properties of the *att* site mutations *att*6 and *att*24 suggested that they are located within the core between potential crossover points. Indeed, each lacks a T from the string of six located between positions −5 and 0 in the core (Ross et al. 1982). We have reviewed the evidence that confirms the location of the crossover points below.

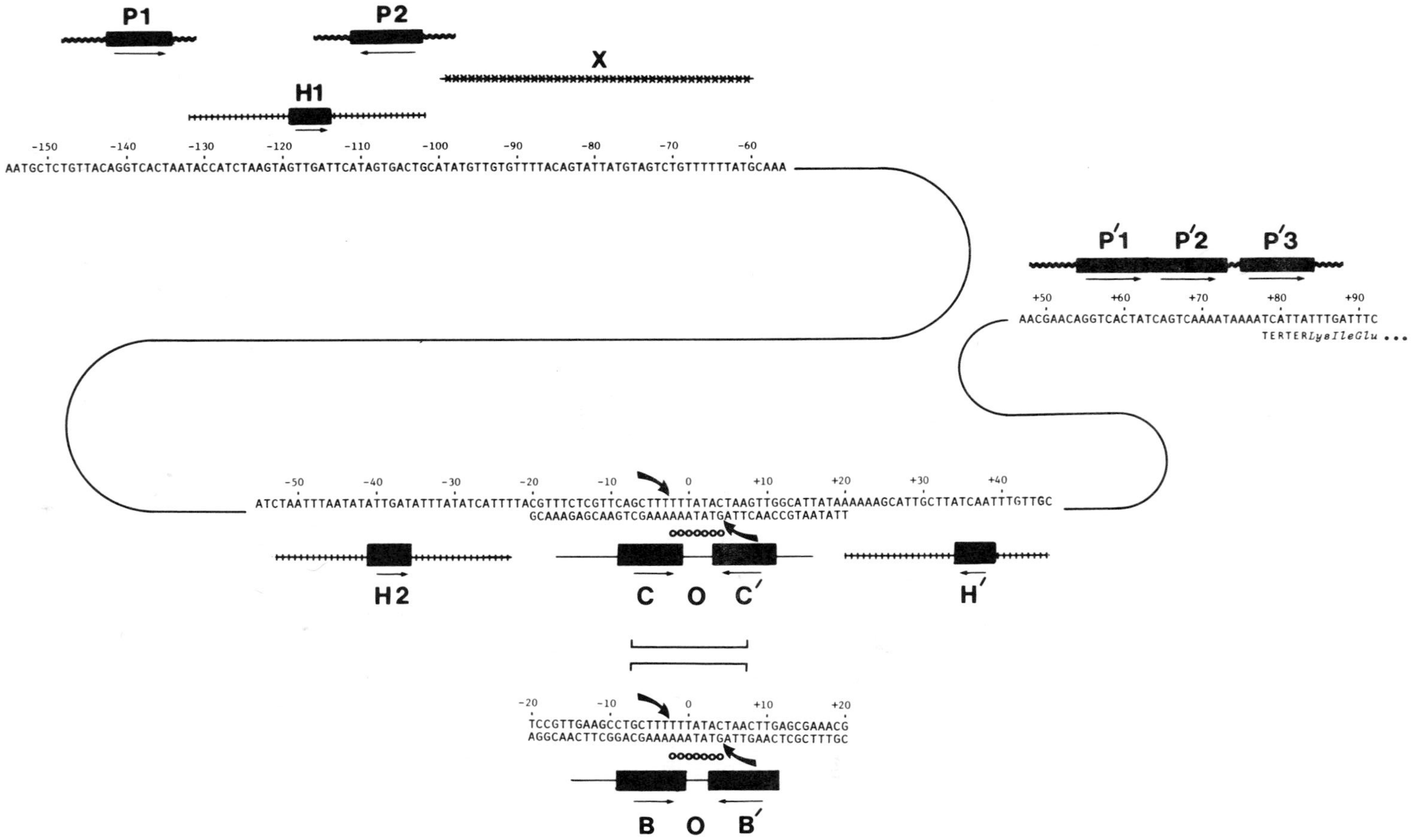

Figure 3 (See facing page for legend.)

The *att* sites manifest a strong bias in base composition. The core itself is 80% A + T, and in the phage *att* site the bias to A + T extends from position −200 to position +100 (average 70% A + T). The extremely high A + T content would favor local denaturation of the negatively supercoiled substrate (Brack et al. 1975; Botchan 1976), and this may be important for recombination.

Size

To determine the size of the *att* sites, deletion mutations extending toward the core or from the outside were constructed by recombinant DNA techniques and analyzed for function. It was found that the wild-type sequence between −152 and +82 was sufficient to confer full *att*P function (Hsu et al. 1980; Mizuuchi and Mizuuchi 1980). The right end of *att*P thus abuts the left end of the *int* structural sequence. In these experiments, the shortened *att*P sites were inserted in a supercoiled plasmid and were tested for their ability to recombine with a bacterial *att* site on a linear molecule. The inclusion of appropriate controls established that this assay registered only phage *att*-site function on the supercoiled molecule and not prophage or bacterial *att*-site function.

The size of *attB*, determined by the same kind of experiment, is about one-tenth that of *att*P; it consists of the core plus about 5 bp on either side (Mizuuchi et al. 1980). As shown below, the size difference is related to differences in the number and kinds of protein-binding sites that are required for biological function.

Figure 3 Functionally important regions in the sequences of *att*P and *att*B. The DNA sequences (Landy and Ross 1977) are shown double-stranded from −20 to +20, and only the *top* strand (5′ on the left) is shown for the remainder of *att*P (Hsu et al. 1980; Mizuuchi and Mizuuchi 1980). *att*B (*bottom*) is slightly shorter than the sequence shown (Mizuuchi and Mizuuchi 1980). Each of the features indicated is discussed in the text: the 15-bp common core region that is identical in *att*P and *att*B (⌐═┘); the location of the staggered cut sites (curved solid arrow) used for strand exchange and the overlap region between them (**oooo**), designated O (Mizuuchi et al. 1981); the junction-type binding sites for Int protein (—■—), designated C and C′ in *att*P and B and B′ in *att*B (Ross et al. 1979; Ross and Landy 1983); the IHF binding sites in *att*P (┼┼┼■┼┼┼), designated H1 and H2 in the P arm and H′ in the P′ arm (N. Craig and H. Nash, pers. comm.); the arm-type binding sites for Int protein (~~■~~), designated P1 and P2 in the P arm and P′1, P′2, and P′3 in the P′ arm (Davies et al. 1979; Ross et al. 1979; Hsu et al. 1980; Ross and Landy 1982); and the binding site for Xis protein (**xxxx**), designated X (S. Yin et al., in prep.). DNA protected against nuclease digestion by binding each of the proteins is indicated by the extent of their respective symbols. Within each protected region the consensus recognition sequence is indicated (■) along with the relative polarity (→) of each recognition sequence. The amino acid sequence and the two terminator codons correspond to the carboxyl terminus of the *int* gene as deduced from the DNA sequence.

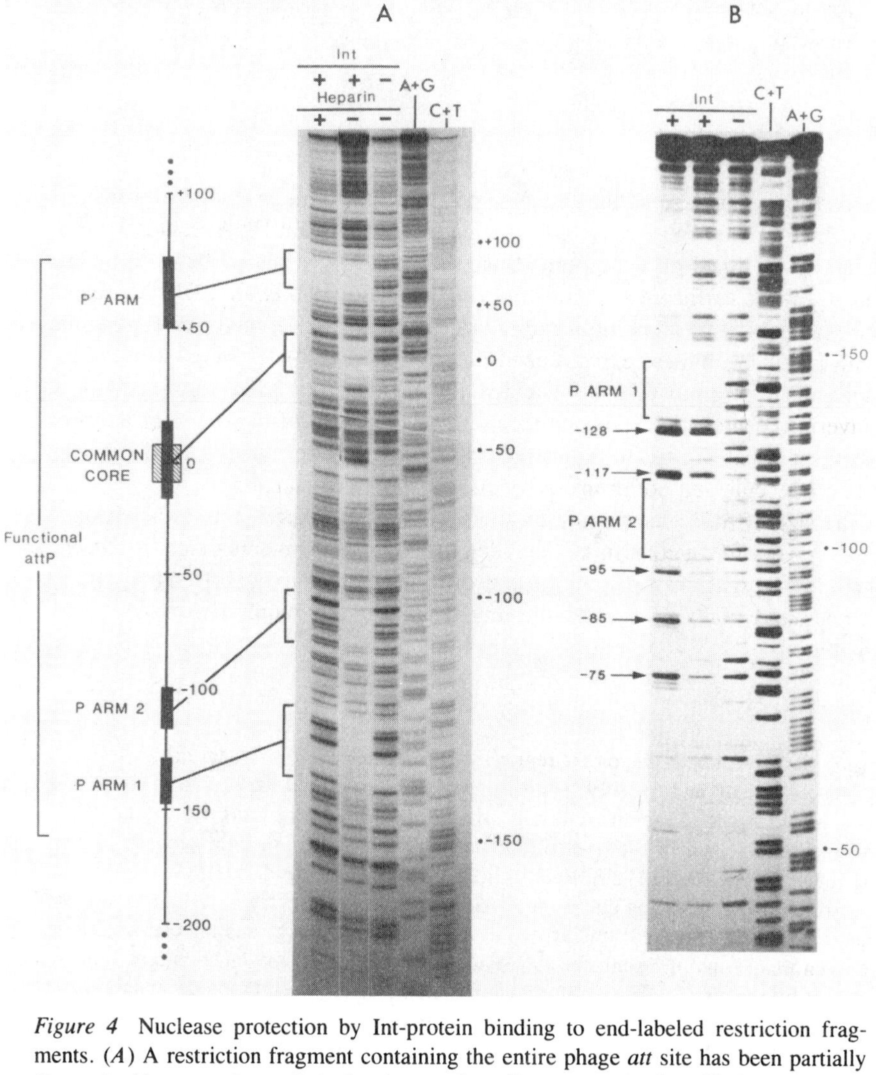

Figure 4 Nuclease protection by Int-protein binding to end-labeled restriction fragments. (*A*) A restriction fragment containing the entire phage *att* site has been partially digested with neocarzinostatin in the absence (lane 3) or presence of purified Int protein at 10 μg/ml (lane 2) or 20 μg/ml (lane 1). In lane 1 the gp*int*-DNA complex was challenged with heparin before neocarzinostatin digestion. A linear map of the minimal phage *att* site depicts the relative sizes and positions of the four gp*int*-protected sequences (■). (*B*) A restriction fragment containing the P arm and common core (run off the gel) that has been partially digested with DNase I in the absence (lane 3) and presence of purified Int protein at 2.5 μg/ml (lane 2) or 5 μg/ml (lane 1) exhibits a pattern of enhanced cutting at 10-bp intervals. (Reprinted, with permission, from Hsu et al. 1980.)

Int-protein-binding Sites in att*P and* att*B*

Figure 4 shows that Int protein specifically protects four regions of *att*P from digestion with neocarzinostatin, an agent that attacks DNA at A:T base pairs (Hsu et al. 1980). This and earlier experiments (Davies et al. 1979; Ross et al. 1979) indicate that gp*int* specifically interacts with nucleotides in these regions. Comparison of gp*int*-protected regions within the *att* sites and within non-*att* DNA reveals two distinct classes of recognition sites: the "junction-type sites," which are located at the core-arm junctions of both *att*P and *att*B, and the "arm-type sites," which occur in the arms of *att*P (Ross and Landy 1982). We consider first the junction-type sites.

The consensus data indicate that one recognition "domain" of gp*int* interacts with the sequence CAACTTNNT (where N is any base) (Ross and Landy 1983). This sequence or a variant of it is found at each of the four core-arm junctions. The left and right junction sites are designated C and C′, respectively, in *att*P, and B and B′, respectively, in *att*B (see Figs. 3 and 5). When the cores of *att*P and *att*B are aligned, C and B are in parallel orientation relative to each other and inverted relative to C′ and B′. Only C′ is a precise match to CAACTTNNT, and it binds Int protein well even when the C site is removed by deletion. In contrast, the C site binds less well when C′ is absent, a result suggesting that gp*int* binding to the core region is cooperative. Cooperativity might also promote binding to *att*B and to a secondary site (*int*155) that has two variant junction-type sites with the same configuration and spacing as *att*B (see Fig. 5) (Ross and Landy 1983; C. Gritzmacher and R. Weisberg, unpubl.).

The nucleotides in the *att*P and *att*B junction sites that come in close contact with bound Int protein have been determined by methylation protection experiments (Ross and Landy 1983). All but one of the contact positions are located along a single face of the double helix in both the major and minor grooves (Fig. 5). The position of the points of crossing over (Mizuuchi et al. 1980; see below, Analysis of *att*B) are situated in close proximity across the central major groove on the same face. Although several of the interactions are the same in all of the sites, others are not. The latter presumably reflect the apparent differences in binding strengths, which have been observed in nuclease protection experiments. These results, together with the gp*int*-binding studies on non-*att* DNA, point to a hierarchy of junction binding sites: C′ appears to be the strongest, and B′, the weakest. The relative strengths of the intermediate sites B and C have not been determined.

The consensus sequence of the arm-type Int-protein recognition site, C/AAGTCACTAT, has also been determined by the analysis of binding sites in *att*-containing DNA and non-*att* DNA (Fig. 6) (Ross and Landy 1982). Further support for some of the bases in this consensus sequence has been obtained from methylation protection experiments with *att*P (W. Ross and A. Landy, unpubl.). There are five such gp*int*-protected sequences in *att*P: two in the P arm (P1 and P2) and three in the P′ arm (P′1, P′2, and P′3). When considered in its entirety, the arm consensus sequence is asymmetric; however, it does have a

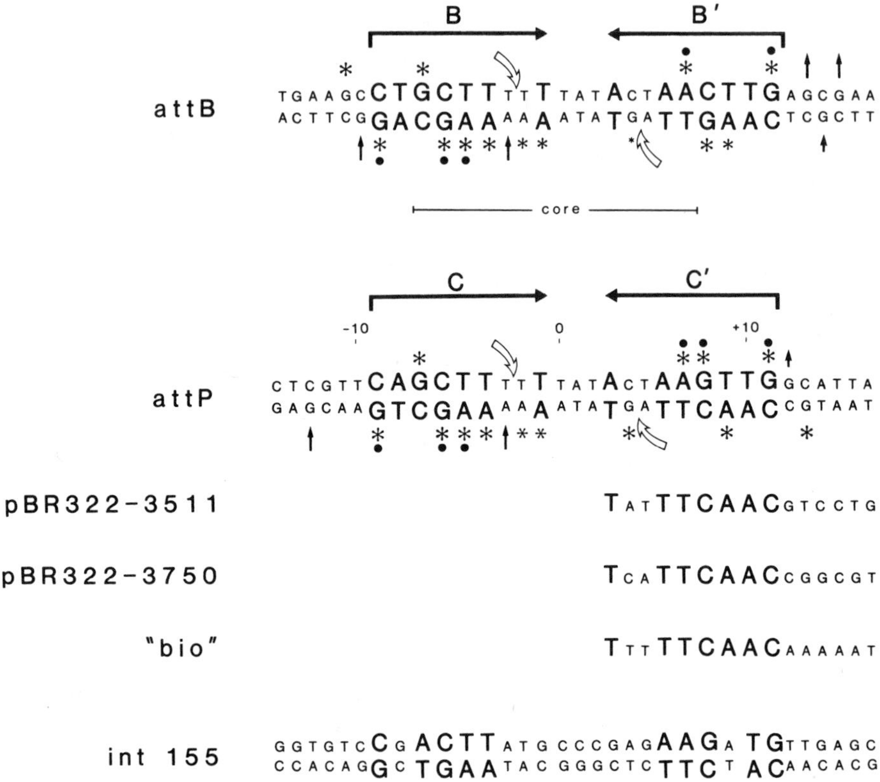

Figure 5 Junction-type Int-protein recognition sequences. Bases corresponding to the consensus sequence CAACTTNNT are in boldface type. Within the *att*P (C and C′) and *att*B (B and B′) recognition sites are marked those positions cut by Int during recombination (curved open arrow) and those bases for which bound Int protein affords protection against (*) or enhancement of (↑) methylation by dimethylsulfate. Non-*att* DNA sequences protected from nuclease digestion by gp*int* are aligned according to the consensus sequence (pBR322-3511, pBR322-3750, and *bio*). The *int*155 site (see text) is shown in the two bottom lines. (Reprinted, with permission, from Ross and Landy 1983.)

symmetrical component: AGTCACT. From the available data, it is not possible to say if the site is functionally symmetric. A symmetrical recognition sequence would suggest an Int-protein dimer as the functional binding unit. An asymmetric recognition sequence would confer polarity on gp*int* interaction at each site. According to this view, Int protein would bind with opposite polarity to the P1 and P2 arm sites (each of which gives its own distinct region of nuclease protection) and with the same polarity to each of the three adjacent consensus sites that constitute a single large nuclease-protected region in the P′ arm.

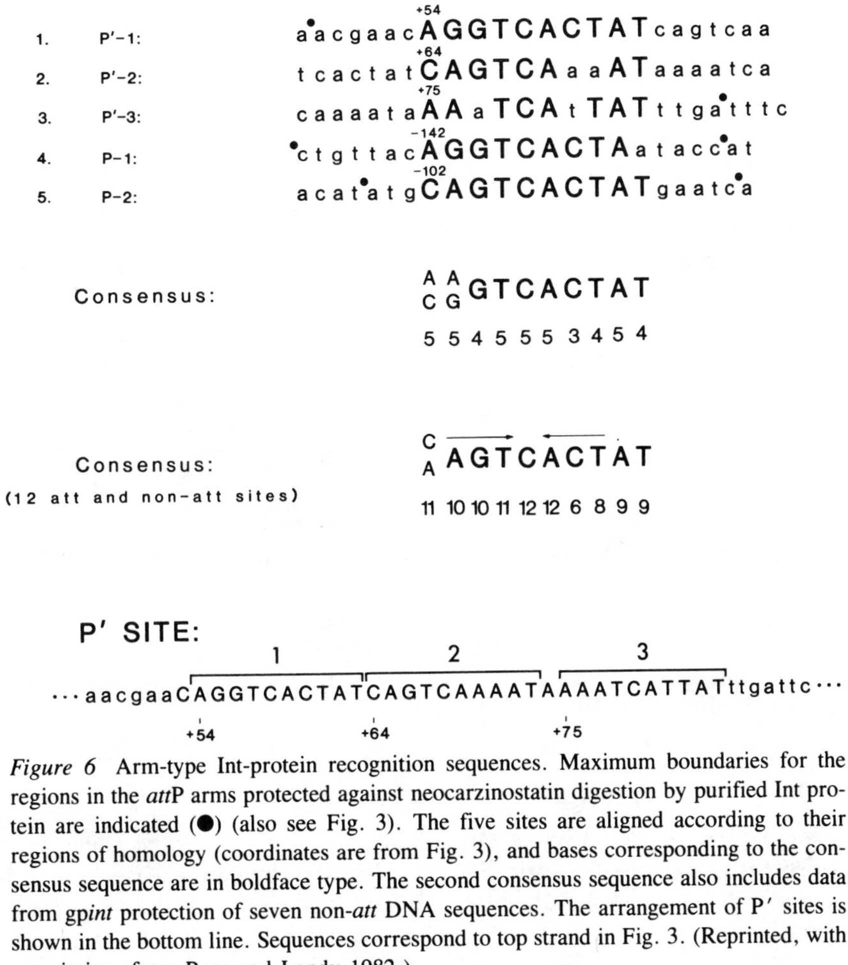

Figure 6 Arm-type Int-protein recognition sequences. Maximum boundaries for the regions in the *att*P arms protected against neocarzinostatin digestion by purified Int protein are indicated (●) (also see Fig. 3). The five sites are aligned according to their regions of homology (coordinates are from Fig. 3), and bases corresponding to the consensus sequence are in boldface type. The second consensus sequence also includes data from gp*int* protection of seven non-*att* DNA sequences. The arrangement of P′ sites is shown in the bottom line. Sequences correspond to top strand in Fig. 3. (Reprinted, with permission, from Ross and Landy 1982.)

It is possible that the unique configuration of recognition sequences in the P′ arm accounts for the heparin- and salt-resistant gp*int* binding to this region of *att*P. Alternatively, the distinctive properties of the P′ site may be a consequence of the unusual composition of this region; the 30-bp sequence extending rightward from the middle of P′2 is 90% A:T, although the overall composition of *att*P is approximately 70% A:T (see Fig. 6). Binding of Int protein is an intrinsic feature of the individual DNA sequences; gp*int* protection is seen for the P1 and P′ arm sites when present singly (P2 was not tested alone) and for the junction sites in the absence of any arm-type sites. However, the suggestion of cooperativity in binding at the junction sites has already been noted, and additional evidence for cooperativity is provided by electron micrographs of gp*int*-

att complexes (see below; Better et al. 1982). Possible roles for gp*int* binding to the arms will be considered below.

An additional feature of the gp*int*-DNA interaction is discernible in DNase I protection experiments. At Int protein concentrations higher than the minimum required to observe specific protection, there is a series of strongly enhanced bands at 10-bp intervals in the regions surrounding the P2, P′, and junction sites (Fig. 4) (Hsu et al. 1980; Ross and Landy 1982). These are to be distinguished from, but may be related to, a similar period observed within the region of P′ protected from exonuclease digestion (Davies et al. 1979). By analogy with studies on nucleosomes and DNA gyrase complexes, these 10-bp patterns of enhanced cutting suggest that DNA wraps around or lies on the surface of a higher order oligomer of gp*int*. Such a structure would be consistent with the observation by electron microscopy of gp*int* complexes interacting with the phage *att* site (Hamilton et al. 1981). The geometry and size of such complexes suggests that the wrapping involves about 230 bp of DNA and 4–8 gp*int* monomers (Better et al. 1982). Recent experiments of T. Pollack and H. Nash (unpubl.) suggest that the wrapping is centered around the *att*P core. This is inferred from the high frequency of knots formed when *att*P recombines with an *att*B site located on the same molecule but in inverted orientation. It is likely that both the left and right gp*int*-arm-binding sites are essential for wrapping, since analogous studies with an *att*L-*att*R substrate did not disclose high frequencies of knots. (*att*L and *att*R lack the P and P′ gp*int*-binding sites, respectively [see below, Analysis of *att*B].)

IHF-binding Sites

The phage *att* site interacts with IHF. There are three 30-bp regions in the phage *att* site that are protected against nuclease digestion by purified IHF, and each contains the sequence 5′-AANNNPuTTGAT (Fig. 3) (N. Craig and H. Nash, pers. comm.). This sequence, or a closely related variant, is also contained within several other IHF-protected regions that are found outside of *att*. The close proximity of the gp*int*- and IHF-protected regions within the phage *att* site suggests some degree of protein-protein interaction. Indeed, nuclease digestion in the presence of Int protein and IHF reveals protection of almost the entire 240-bp *att* site.

gpxis*-binding Site*

Nuclease protection experiments with highly purified gp*xis* establish that it is a specific DNA-binding protein. On the phage *att* site, gp*xis* protects DNA from −60 to −100 in the P arm (S. Yin et al., in prep.). As will be noted from the above discussion and Figure 3, the assignment of this region to gp*xis* interaction means that the entire 240 bp of *att*P can be accounted for in interactions with one of the three recombination proteins. The way in which gp*xis* affects recombination directionality remains unknown.

Analysis of att*B*

The *att*B site extends no more than 5 bp to either side of the common core (Mizuuchi et al. 1981), and it lacks IHF and arm-type gp*int*-binding sites as determined by nuclease protection experiments and by sequence inspection (Ross et al. 1979; Ross and Landy 1983; N. Craig and H. Nash, pers. comm.). *att*B and *att*P also differ in their topological requirements: The phage *att* site works best when it is on a supercoiled molecule, whereas the bacterial *att* site works best when it is on a linear molecule (in the standard in vitro reaction conditions). Like *att*P, *att*B contains a paired arrangement of a stronger and a weaker gp*int*-binding site around the 15-bp core.

The above comparisons suggest that functional differences between *att*P and *att*B are a result of the IHF- and arm gp*int*-binding sites in P and P′. This inference is confirmed by analysis of *att*P deletions that replace the IHF- and arm gp*int*-binding sites with heterologous DNA but leave unaltered the core and immediately adjacent regions. Such deletion mutants function as *att*B sites (Hsu et al. 1980; Mizuuchi and Mizuuchi 1980). Thus, a major feature of integrative site-specific recombination is the interaction between two unequal partners which, for convenience, have been referred to as "donor" and "recipient" *att* sites. The donor, or phage, *att* site is large (240 bp), has complex interactions with Int protein and IHF in both arms, and works best on a supercoiled molecule. A donor *att* site recombines efficiently with a much simpler recipient such as the bacterial *att* site. We propose that a recipient site is any short DNA sequence that provides the following: (1) homology with the core (see below), (2) gp*int* recognition sequences at the core boundaries, and (3) the absence of arm-type-binding sites in the flanking regions.

Location of the Crossover Points

The first evidence that recombination involves staggered (nonopposed) cuts in the two DNA strands came from the segregation patterns of core mutations such as *att*24 (see above) (Shulman and Gottesman 1973). *int*-promoted recombination between λ*att*24 and λatt^+ occasionally produced heterozygotes—phage particles that segregated both *att*24 and att^+ progeny upon further growth. Such heterozygosity suggests that recombination can produce a region of DNA around the *att*24 site in which the two polynucleotide chains are of different parentage. A simple way of accomplishing this is to exchange one DNA strand to the left and the other to the right of the *att*24 site. An experiment by K. Mizuuchi and M. Mizuuchi confirms this inference. They determined the segregation pattern of phosphorous atoms during in vitro recombination (Mizuuchi et al. 1981). The parents in this cross were an unlabeled *att*P-containing plasmid and an internally ^{32}P-labeled *att*B-containing fragment. The points at which the DNA strands are cut and exchanged determines the distribution of ^{32}P atoms to the *att*L and *att*R products. In these experiments ^{32}P was used not only as a tracer but also to generate cleavages by radioactive decay. By comparing the pattern of decay-induced cleavages in the *att*L and *att*R products with a sequencing ladder of the analogous DNA fragments, the junction (crossover point) between ^{32}P contributed from the *att*B partner and ^{31}P contributed from the *att*P partner was determined. The results show that in each strand the crossover occurs between two unique phosphorous atoms. If both cuts leave a 3′ phosphate and a 5′ hydroxyl, then they must occur between positions −3 and −2 in the top and +4 and +5 in the bottom strand, resulting in a 7-bp stagger. On the other hand, if both cuts leave a 3′ hydroxyl and a 5′ phosphate, then they must occur at positions −2/−1 in the top and +3/+4 in the bottom strand, resulting in a 5-bp stagger. Two kinds of evidence show that the first alternative is correct. The first evidence is genetic and is described below in Crossover Points in Secondary Sites. The second comes from the observation that Int protein can specifically cleave *att*-site DNA without the accompanying exchange and rejoining that normally occur during recombination. The sites of cleavage are between positions −3 and −2 in the top and +4 and +5 in the bottom strands (N. Craig and H. Nash, pers. comm.).

The location of the crossover points is not symmetrical within the core but is symmetrically placed relative to the two junction binding sites. Each crossover point is located between positions 7 and 8 of the 9-bp consensus binding sequence. The base pairs at these two positions are not conserved, although methylation protection experiments indicate that either one or both are in close contact with bound gp*int* in all of the sites. It therefore appears that cutting does not require a specific interaction of gp*int* with the immediately adjacent base pairs, and this might reflect a need for base movement during the reaction (Ross and Landy 1983).

The existence of staggered cuts within the core highlights the potential role of DNA-DNA homologies within the 7-bp heteroduplex region. This question arises when we consider recombination of secondary *att* sites that lack the canonical *att* sequence between the staggered cut sites, a segment that we propose to call the overlap region.

Secondary Sites

Comparison of Secondary and Primary Sites

When λ infects a host that is deleted for *att*B, the frequency of insertion is reduced 100- to 1000-fold. The rare insertions that do occur resemble λ insertion at *att*B in the following respects (Shimada et al. 1972; Hoess and Landy 1978; Landy et al. 1979): (1) They require gp*int*, (2) the integrative crossover is within the *att*P core, and (3) prophage excision requires gp*int* and gp*xis*. The sequences into which insertion occurs are called secondary *att* sites with the general designation ΔΔ′. Δ represents sequences to the left of the crossover locus, and Δ′, sequences to the right. Secondary sites have also been found in the λ chromosome and in foreign DNA that had been inserted in a λ vector (Parkinson 1971; Struhl 1981).

The secondary sites are functionally heterogenous as judged both by the frequency of recombination with *att*P and by the frequency of prophage excision (Shimada et al. 1972, 1973, 1975). Insertion of λ into a secondary site generates a pair of secondary prophage sites, ΔP′ and PΔ′, that have a family resemblance to *att*L and *att*R, respectively (Shimada et al. 1975). This resemblance and the simplicity and small size of *att*B suggest that secondary sites should be considered as variant recipient *att* sites.

Figure 7 and Table 2 show a comparison of *att*B and the core region of *att*P to the sequenced secondary *att* sites (Ross and Landy 1983). For those secondary *att* sites near the top of Figure 7, the homology with recipient *att*-site features is quite obvious, different features being matched to different degrees in each site. For example, the *proA/B* sequence has one rather good junction-type binding site and a high degree of homology to wild type in the overlap region; *galT* has two modest gp*int*-binding sites with perfect spacing and a moderate overlap region homology; and the *int*155 site has virtually no overlap region homology but does have two excellent junction binding sites with perfect spacing. It seems reasonable that each of these features will contribute to biological function, but we lack information to make quantitative predictions. Nevertheless, it is consistent with the implications of this ranking that several of the sequences in Figure 7, such as *proA/B* and *b*2, are considered to be relatively more efficient than most of the others by virtue of their independent recurrence in searches for secondary *att* sites.

Because the genetic methods for detecting secondary *att* sites are sensitive, even poor ones have been isolated. The ratio of the efficiency of the best to that of the worst is greater than 10^6. A more systematic analysis of the most frequently used sites should test the model that secondary *att* sites comprise a collection of recipient *att* sites of varying efficiency. Among the questions to be addressed are the relative contributions of the overlap homology, gp*int* recognition, and canonical spacings.

Crossover Points in Secondary Sites

Comparison of the nucleotide sequences of secondary prophage sites with those of *att*P and the corresponding secondary sites provides an independent method of determining the region or point at which crossing over occurred. Two points of crossing over, one between −3 and −2 and the other between +4 and +5 in the *att*P core can account for all of the insertions into secondary sites (Fig. 7). For two sites, *int*155 (Ross et al. 1982; Ross and Landy 1983; C. Gritzmacher and R. Weisberg, unpubl.) and *galT* (Weisberg et al. 1983), the sequencing of more than one insertion revealed that both crossover points can be used for each site (see Fig. 8). In addition, Int protein cleaves the *int*155 secondary site twice,

	INT	OVERLAP (↓ … ↑)	INT
	-10 -9 -8 -7 -6 -5 -4 -3	-2 -1 0 +1 +2 +3 +4	+5 +6 +7 +8 +9 +10 +11 +12
attB	AAGC**CTGCTT**T	T**T**TAT**A**C	T**AACTTG**AGCG
attP	CGTT**CAGCTT**T	T**T**TAT**A**C	T**AAGTTG**GCAT
galT-saf	CCGCGCCT**TT**G	T**T**TAT**A**C	A**AACCTG**CACC
b511	TGACT**AACTT**T	T**T**TAT**A**C	C**AA**TG**T**TTAAC
proA/B	AGTTGC**GCT**AA	T**T**TAT**A**C	G**A**G**GCT**TGCTT
trpC	GAAG**C**GTAA**T**G	T**T**TAT**A**A	ATG**GC**G**G**CAAT
b2	GGTAT**TGCTT**A	T**T**TATCG	A**AA**A**C**G**G**ACAG
SC 2669	AATCAC**A**T**T**AC	T**T**TAT**A**T	A**A**T**GT**ATAATT
galT	CCGCGCCT**TT**G	T**T**TTC**A**A	A**AACCTG**CACC
thrA	GTGCGG**GCTT**T	T**T**TCTGT	GTTT**C**CTGTAC
int 155	TGTC**C**G**ACTT**A	TGCCCGA	G**AAG**A**TG**TTGA
SC 2639	TGCTC**A**T**TT**T	T**T**AGT**A**T	ATT**CTT**CGAAG
rrnB	AGGTTG**GCT**AT	T**T**TACCA	CG**ACT**GTC
b522	TGCTTCTGC**T**T	TCGCC**A**C	C**A**T**C**A**T**TTCCA
rbt	GGACGC**G**G**TT**T	T**T**TCG**A**T	T**A**TTGACGCCG
SC 2667	AAAT**CA**CA**TT**A	C**T**TTATA	T**AA**TG**T**ATAAT
b508	TATCAGCAA**T**G	T**T**TATGT	A**AAG**AAACAGT
SC 2666	CAT**TACTT**T	A**T**ATA**A**T	GT**A**TAATTC

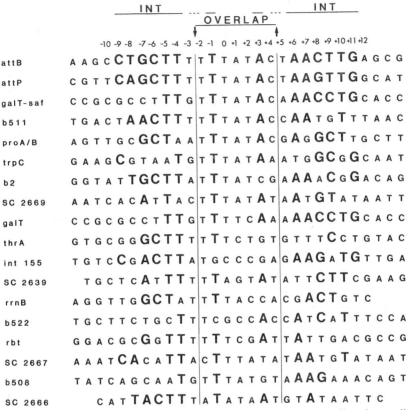

Figure 7 Comparison of secondary *att*-site sequences. Sequences are aligned according to cut positions, as judged by sequence analysis of the secondary prophage sites (for references, see Table 2; sites are presented in the same order of measured or predicted efficiency as those used in Table 2). In many instances, the secondary sites contain several bases of continuous homology with the *att*P core, which may include one or both of the cut positions; in these cases, the cuts in the secondary sites are assumed to occur in the same relative positions as in the *att*P and *att*B cores. The "overlap" region comprises the 7 bp between the cut positions in the top (↓) and bottom (↑) strands. Sequences correspond to the top strand in Fig. 3. The Int-protein junction-type recognition sequences in *att*P and *att*B are in boldface type (see Fig. 5), and any bases conserved in the secondary *att* sites are also bold. In some instances, better sequence matches with the recognition site are found spaced ±1 bp from the standard spacing (see Table 2) but are not indicated here. (Reprinted, with permission, from Ross and Landy 1983.)[1]

once in each strand. One cleavage is at a crossover point; the location of the other has not been precisely determined (N. Craig, pers. comm.). These results suggest that insertion into secondary sites, like insertion into *att*B, proceeds via staggered single-strand cuts at positions −3/−2 and +4/+5 (Fig. 5). Although

[1]Since the previous edition, the sequences for *proA/B* and SC 2639 have been corrected by the original authors.

Table 2 Conserved positions of secondary *att* sites

Secondary *att* site	Efficiency[a]	Matches (of seven): Int site I[b]	Int site II[c]	overlap region	Reference
galT-saf	10^{-4}	6	3, +5	7	Weisberg et al. (1983)
*b*511	—	6	4	7	Landy et al. (1979)
proA/B	10^{-4}	5	4	7	Pinkham et al. (1980)
trpC	10^{-5}	3	4	6	Christie and Platt (1979)
*b*2	—	5	4	5	Landy et al. (1979)
SC2669	—	4	3, +5	6	Struhl (1981)
galT	10^{-7}	6	3, +5	4	Bidwell and Landy (1979)
thrA	10^{-7}	5	+3	4	Chapman and Gardner (1981)
*int*155	—	5		1	Ross and Landy (1983); C. Gritzmacher and R. Weisberg (unpubl.)
SC2639	—	4	3	4	Struhl (1981)
rrnB	10^{-10}	4	−3	5	Csordas-Toth et al. (1979)
*b*522	—	3	−5	3	Landy et al. (1979)
rbt	—	4	2	4	Loviny et al. (1981)
SC2667	—	5	3	2	Struhl (1981)
*b*508	—	2	3	5	Landy et al. (1979)
SC2666	—	5	2	2	Struhl (1981)

[a]The sites are presented in an order reflecting known efficiencies where available (summarized in Pinkham et al. [1980] or, for *galT-saf*, in Weisberg et al. [1983]). When efficiencies were not available, they were predicted on the degree of conservation in the three regions. Data are taken from the sequence comparisons shown in Fig. 7. In each column the maximum number of matches is seven, either to the gp*int*-binding sequence $\mathrm{C}^{\mathrm{AA}}_{\mathrm{GT}}\mathrm{CTNNT}$ or to the overlap sequence TTTATAC.

[b]The best gp*int*-binding site with the correct position relative to one cut site.

[c]The best match on the opposite side of the overlap region. A potential gp*int*-binding site that differs in the canonical spacing by ±1 bp may be functionally significant, and such alignments are designated + or −. (The *att*P24 mutation, in which a 1-bp central core deletion brings the two binding sites closer together [Ross et al. 1982], recombines with wild-type *att*B ten times more efficiently than the best secondary sites [J. Auerbach and M. Gottesman, pers. comm.].) (Reprinted, with permission, from Ross and Landy 1983.)

this is an economical explanation of the data, other evidence suggests that insertion in *galT* frequently occurs by a flush cut at position −3/−2. (Weisberg et al. 1983). Further work is required to confirm this suggestion.

Excision from Secondary Sites and the Generation of saf *Mutations*

Does prophage excision from secondary sites proceed by crossing over at the same points as those used for insertions? If so, a cycle of insertion and excision should regenerate the original phage and host *att* sites. In fact, functionally wild-type *att*P and host sites can be recovered after an insertion-excision cycle in *att*B and in all secondary sites tested to date. For two sites, one in *galT* and one in *trpC*, excision can restore the original nucleotide sequences of the host sites

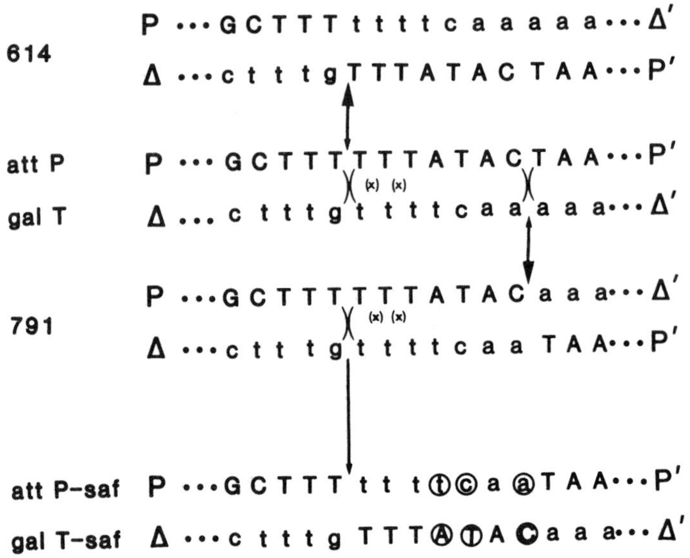

Figure 8 Prophage insertion and excision at a secondary *att* site in the *galT* gene of *E. coli*. Rows 3 and 4 show the top strands of the phage and host-core sequences. The two sequences are positioned so that the crossover points, determined by sequencing the prophage *att* sites, are aligned. Recombination at the left gives lysogen 614 (rows 1 and 2), and at the right, lysogen 791 (rows 5 and 6). The PΔ′ sites (rows 1 and 5) lie at the right prophage termini (with the map as conventionally drawn), and the ΔP′ sites (rows 2 and 6) lie at the left. Prophage excision can either regenerate the two original sites if the original crossover points are used (double-headed arrows) or generate two *saf* mutations if the alternative crossover point is used (single-headed arrow; rows 7 and 8). The altered bases are circled. Note that the left-hand crossover could have occurred between any of three adjacent Ts; however, the one indicated fits best with other results (see text). (Reprinted, with permission, from Mizuuchi et al. 1981; Weisberg et al. 1983.)

(Bidwell and Landy 1979; Christie and Platt 1979). However, if both of the insertion crossover points are also used for excision, then excision from certain secondary sites will generate mutant as well as wild-type products. This could occur as shown in Figure 8. The mutants result from interchanging the overlap regions of *att*P and the secondary site. Hence, they will occur only at those secondary sites where the two regions differ. In fact, such mutations have been found and are known as *saf* (*s*ite *af*finity) for reasons that are explained below.

The existence of *saf* mutations strongly suggests that excision, like insertion, can proceed by DNA exchange at two points: positions −3/−2 and +4/+5 of the core. As in the case of insertion into secondary *att* sites, we are uncertain if these points are the loci of single- or double-strand breaks in the DNA, and we refer the reader to the article by Weisberg et al. (1983) for a more complete discussion.

Role of Homology in Site-specific Recombination

Campbell (1962) originally proposed that crossing over between phage and host occurs within a small region of homology. We now know that this region is 15 bp long. Contrary to the original proposal, enlarging the region of homology does not necessarily increase the frequency of recombination. For example, recombination between BOP′ and BOB′ is considerably less efficient than recombination between POP′ and BOB′ (Weisberg and Gottesman 1969). Thus, regions outside of the core do not contribute to site-specific recombination in the same way as they would to general recombination. But how does the core itself contribute to recombination? The properties of *saf* mutations provide a partial answer to this question.

The best characterized *saf* mutation (λ*safG*) was isolated by prophage excision from the *galT* site as described above (see Fig. 4). It has base substitutions at positions +1, +2, and +4 of the core (Weisberg et al. 1983). The mutation has been crossed from *att*P to *att*B by a cycle of insertion and excision. *safG* reduced integrative recombination approximately 100-fold when present in either *att*P or *att*B. In contrast, recombination was nearly normal when both partners were mutant. Crosses between *att*L and *att*R gave similar results. Therefore, the match between the two recombining sites at positions +1, +2, and +4 is more important for recombination than the match to the primary site sequence at these positions. We suggest that the two DNA segments must interact with each other directly in this region and that this interaction is promoted by homology.

Although λ*att*P-λ*safG* inserts its DNA with low efficiency at *att*B, it inserts with greater than wild-type efficiency at other sites in the *E. coli* chromosome. Analysis by Southern hybridization shows that these sites differ from wild-type λ secondary sites (Weisberg et al. 1983). Although the sites preferred by λ*safG* have not yet been sequenced, it is tempting to speculate that they will have overlap regions similar or identical to the *galT* overlap region.

The *att*24 mutation, which behaves in crosses as if it were within the overlap region, deletes a T from the string of six between −5 and 0. The deletion does not inactivate the gp*int*-binding site at the left core-arm junction of *att*P. However, analysis of gp*int* binding to the λ*att*P24 core region suggests that the mutation shortens the distance between the left and right junction sites (Ross et al. 1982). The phenotype of *att*24 differs from that of *safG* in an important way. Recombination between *att*L24 and *att*R24, which occurs at about 1% of the fully wild-type frequency, is no more efficient than recombination between mutant and wild-type sites. This might reflect the effect of a shortened overlap region on DNA pairing, on gp*int* binding to the core-arm junction sites, or both.

The analysis of secondary sites and *saf* mutations provides strong evidence that the core has two roles. First, the bases of the overlap region participate in a DNA-DNA interaction. This interaction could be the complementary base-pairing needed to form the recombination joint, the pairing of base pairs needed to form the four-stranded structure proposed as an intermediate in recombination by Kikuchi and Nash (1979b; see below, A Model for Synapsis and Strand Exchange), or both. Second, core nucleotides outside of the overlap region, and probably some within it as well, form part of the core-arm junction Int-protein-binding sites. Secondary *att* sites are defective to varying degrees in one or both of these functions (see Table 2).

REACTION MECHANISMS

For purposes of discussion we have divided site-specific recombination into four phases: (1) formation of protein-DNA complexes, (2) association and juxtaposition of recombining *att* sites (synapsis), (3) crossing over of the first pair of DNA strands (formation of a crossed-strand exchange or Holliday structure) (Holliday 1964), and (4) crossing over of the second pair of DNA strands (resolution of the Holliday structure). It should be emphasized that this division is made in the absence of critical biochemical information, especially for the early steps where parts of phase one could be contemporaneous with, inseparable from, or even preceded by phase two.

Protein Binding and Supercoiling

Although we know a good deal about protein–*att*-site recognition, we know little about the kinetics or association constants of the interactions at each of the 12 or 13 individual recognition sites (Int protein, Xis protein, and IHF) in the two recombining partners. Similarly, very little is known about possible protein-protein interactions, especially those that might involve gp*xis*. Another area of ignorance is the role of DNA supercoiling. Efficient integrative recombination in vivo requires a supercoiled substrate, as evidenced by a strong inhibition in the presence of the gyrase inhibitor coumermycin (Kikuchi and Nash 1979a). A similar requirement for supercoiling is observed in vitro in the presence of KCl at concentrations greater than 50 mM (Mizuuchi et al. 1978; Pollack and Abremski 1979). In this case, however, it is possible to show that only *att*P DNA need be supercoiled and that, under certain conditions, recombination can be most efficient when the *att*B partner is linear (Mizuuchi and Mizuuchi 1979). Supercoiling also stimulates excisive recombination in vivo, although the requirement appears to be less stringent than for integrative recombination (Abremski and Gottesman 1979). Supercoiling strongly stimulates excisive recombination in vitro in the presence of KCl at concentrations greater than 100 mM (Pollock and Abremski 1979).

Supercoiling might enhance the binding of gp*int*, IHF, or both to attachment sites. Indeed, Better et al. (1982) have shown that a supercoiled molecule containing *att*P binds Int protein differently than its linear counterpart. Enhanced binding might result from localized melting of DNA. A precedent is furnished by the enhanced binding of RNA polymerase to partially melted regions of promoters in supercoiled DNA (Saucier and Wang 1972; Wang et al. 1977). The analogy is tempting because the phage *att* site has such a high A:T composition (70%) and because the in vitro condition permissive for recombination of non-supercoiled DNA (at a fivefold to tenfold reduced rate) is low ionic strength (Pollock and Abremski 1979). Both of these factors favor melting (Brack et al. 1975; Botchan 1976). Since there is some evidence to suggest that DNA is wrapped along the surface of gp*int* or some higher order multimer (see above, Int-protein-binding Sites in *att*P and *att*B), there might also be a protein-DNA configuration that is facilitated by supercoiling.

A Model for Synapsis and Strand Exchange

Eighty percent or more of the initial supercoiling in the parental DNA molecules is retained in the recombination products (Mizuuchi et al. 1980). The simplest explanation for the constraint that prevents free rotation with consequent loss of supercoiling is that rapid joining of the cut DNA to its new partner is assured by prior synapsis of the two DNA duplexes. Kikuchi and Nash (1979b) have proposed a mechanism for synapsis and strand exchange that commences with the pairing of two double helices within their core regions to yield a four-stranded helix (Fig. 9a). The proposal draws upon a stereochemically feasible model of four-stranded DNA in which the two double helices are interwrapped around their major groove surfaces and connected to one another by hydrogen bonds between identical base pairs (McGavin 1971). In an alternative version of the model, the two parental double helices interwrap around their minor groove surfaces (Wilson 1979; Nash 1981b; Nash et al. 1981). If synapsis occurs only between the 7-bp overlap regions, then four-stranded DNA comprises less than one turn of the helix. Two strands of the same polarity are then cut at the edge of the four-stranded region, and the broken ends are realigned with their new partners by rotation (in either direction) around the uncut strands (Fig. 9b,c). From the results described above, the 5′-hydroxyl side of the cut would remain stationary while rotation is accomplished by an Int-protein molecule that is covalently linked to DNA through a 3′-PO_4 side. Transfer by gp*int* of the 3′-PO_4 to the 5′-hydroxyl side of the new partner forms an intermediate that is topologically equivalent to a Holliday structure (Wilson 1979; Nash et al. 1981). To complete the crossover, the remaining two DNA strands are cut at the other edge of the overlap region, swiveled, and resealed in an analogous process (Fig. 9d,e). Finally, the hydrogen bonds between the two parental overlap regions are disrupted, the recombinant molecules separate, and the overlap regions of the two recombinant sites become base-paired by a switch in hydrogen bonding (Fig. 9f,g) (Wilson 1979). In the alternative version of the model (minor-groove interwrapping), recombination would occur in a similar way except that the switch in hydrogen bonding occurs before rather than after strand exchange (Wilson 1979).

The Kikuchi-Nash model is attractive because it explains in an economical way how the topoisomerase activity of gp*int* is used in recombination. The

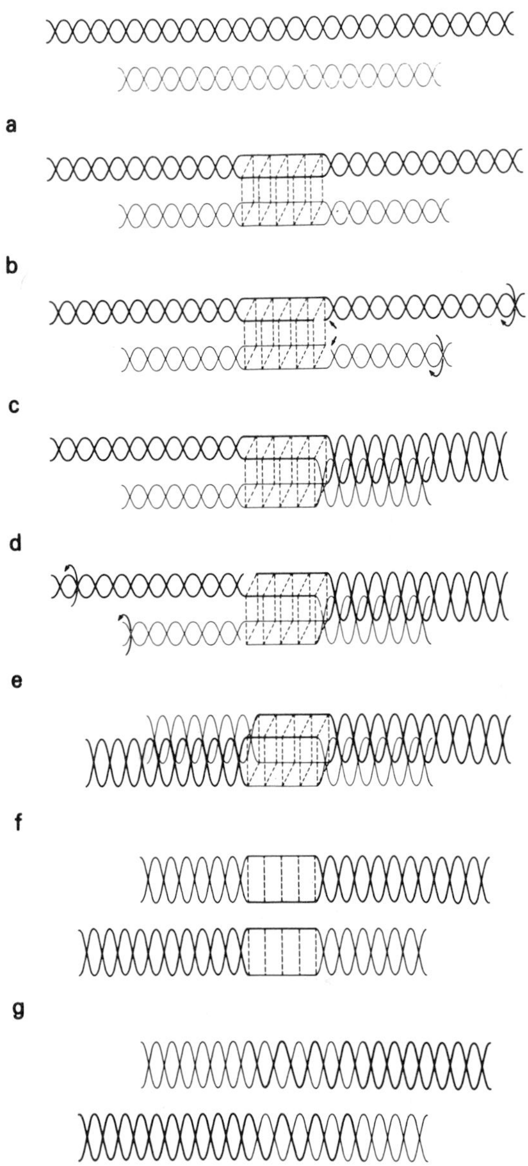

Figure 9 A model for integrative recombination. The details of the model are described in the text. For clarity, the four-stranded helix (*a*) has been straightened. Horizontal dashed lines represent Watson-Crick hydrogen bonds connecting the two strands of a single parent, and vertical dashed lines represent hydrogen bonds between the two parents. (Reprinted, with permission, from Nash 1981b.)

model predicts that recombination changes the number of superhelical turns by a specific amount (two units for inversion), and this prediction has recently been verified (H. Nash and T. Pollock, in prep.). Such a change is too small to have been detected by previous experiments (Mizuuchi et al. 1980). We have already pointed out that this model can account for the phenotype of *saf* mutations by assuming that they interfere with synapsis. Finally, the prediction that Holliday structures are intermediates in site-specific recombination is strongly supported by experiments described below.

Four-stranded regions in DNA have not been detected. If they exist they may be short or short-lived. Perhaps synapsed *att* sites are stabilized by gp*int* or IHF. Indeed, synapsis could be accomplished entirely by DNA-protein rather than by DNA-DNA interaction. This suggestion departs completely from the Kikuchi-Nash model since, if it is correct, the geometry of the synapsed *att* sites need not resemble that of four-stranded DNA. Acceptance of the Kikuchi-Nash model clearly requires direct evidence of the existence of four-stranded DNA in synapsis.

If recombination occurs by exchange of two strands at a time, Holliday structures will be intermediates. This will not be so if all four strands are exchanged simultaneously. Attempts at biochemical detection of Holliday structures have been unsuccessful, but a genetic approach has been fruitful. A Holliday structure can move with respect to a fixed point in the molecule by branch migration (Broker and Lehman 1971; Kim et al. 1972; Warner et al. 1979). Branch migration occurs only over regions of DNA homology. In such crosses as *att*P × *att*P or *att*P × *att*L, homology between the recombining sites is not limited to the core, and therefore migration of Holliday structures formed at the core might be extensive. Resolution of a migrated Holliday structure can produce a crossover at a distant point, which can be detected with appropriate genetic markers. Indeed, a low but significant frequency of *int*- and IHF-dependent recombination was found in a region that is close to but does not contain *att* (Echols and Green 1979; Enquist et al. 1979a). This suggests that recombination can proceed via a Holliday structure intermediate. The low apparent frequency of migrated structures might result from efficient resolution of single-strand exchanges within the attachment site, an explanation that is consistent with the efficiency of the in vitro reaction described next.

Resolution

To study the resolution phase, Hsu and Landy (in prep.) have made "synthetic" Holliday structures (Bell and Byers 1979) that have all the properties expected for a recombination intermediate between the primary *att* sites. Purified Int protein resolves these Holliday structures by a pair of cuts and religations that generate either *att*P and *att*B or *att*L and *att*R (depending on which pair of strands is cut; see Fig. 10). Control Holliday structures made from non-*att* DNA were not cut by gp*int*. Additional specificity for the recombination pathway is

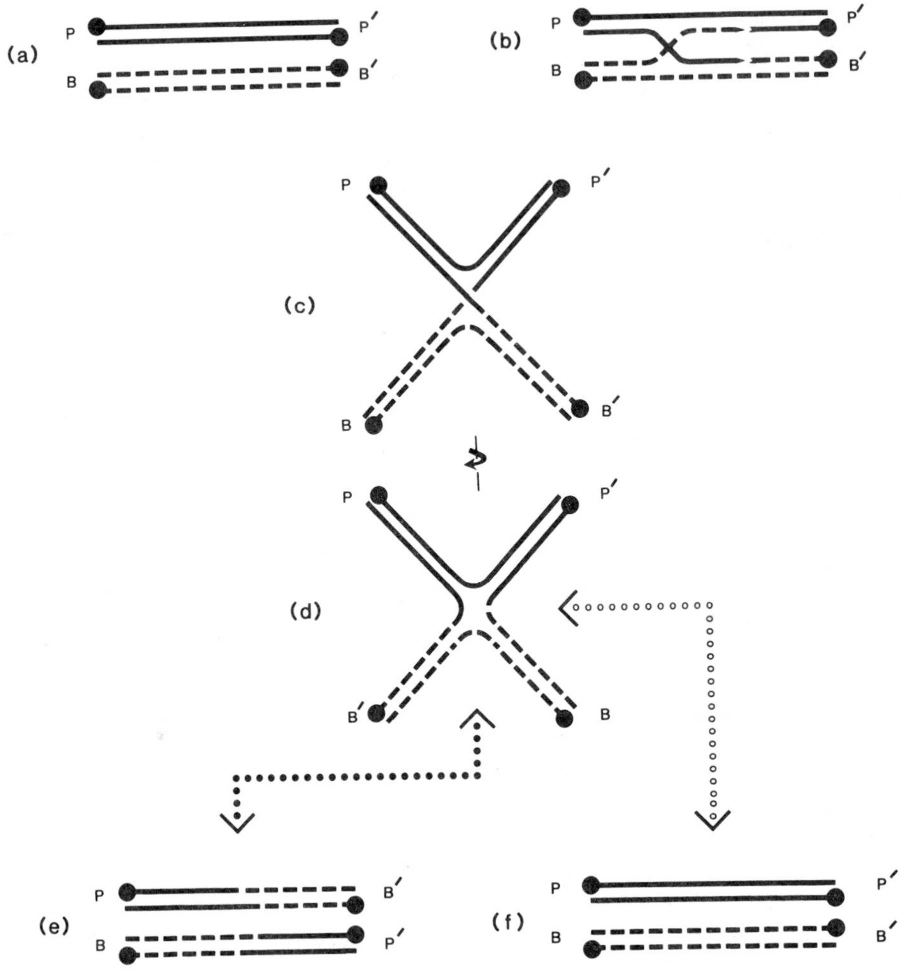

Figure 10 Formation and resolution of a Holliday structure intermediate by two cycles of reciprocal single-strand exchanges. Synapsis of the *att*P and *att*B parents (*a*) is followed by a reciprocal single-strand exchange (*b*) to generate a Holliday structure intermediate (*c*) that can also be drawn as *d* by simple rotation of the bottom legs around the crossed single strands. Resolution by cutting and ligating the second pair of DNA strands (the vertical cut) produces *att*L and *att*R (*e*), and resolution by recutting and ligating the same pair of DNA strands (the horizontal cut) regenerates the parental *att*P and *att*B. (Reprinted, with permission, from Hsu and Landy, in prep.)

indicated by the failure of other topoisomerases tested under these conditions (which are suitable for their topoisomerase activity) to cut either the *att* or non-*att* Holliday structures.

Although IHF stimulates resolution at low gp*int* concentrations, it is not required, and 100% of the Holliday structures can be resolved by gp*int* alone. The efficiency of the reaction supports the hypothesis that these structures represent true recombination intermediates. If so, IHF is required only for the earlier part of the reaction.

Synthetic Holliday structures have been useful in further dissection of the functional domains of the *att* sites (Hsu and Landy, in prep.). For example, Holliday structures that lack one or more of the gp*int*-binding sites in the P and P′ arms were still resolved. Hence, the role of gp*int*-binding sites in the arms must be for synapsis (phase 2) and/or the first crossover (phase 3). Additional manipulations of the Holliday structures may shed light on the cutting and ligation reactions and the necessary protein-DNA interactions at the core.

Excisive Recombination

Our understanding of excisive recombination is more limited than that of integrative recombination. Differences between the two reactions in their requirement for gp*xis* and gp*int* and in their resistance to high temperature show that excisive recombination is not a precise chemical reversal of integrative recombination (see also Echols and Guarneros, this volume). Nevertheless, we think that the mechanisms are similar: Most *int* mutants are defective in both reactions, and the analysis of excision from secondary sites suggests that the crossover points are the same for both reactions (see above, Excision from Secondary Sites and the Generation of *saf* Mutations).

What is the role of gp*xis* in excisive recombination and why is it dispensable for integrative recombination? Recent work suggests that gp*xis* interacts directly with gp*int* and binds specifically to the P arm (see above, Xis Protein; gp*xis*-binding Site). Consistent with this, Better et al. (1983) have found that gp*xis* promotes stable gp*int* binding to *att*R (POB′) but not to *att*B (BOB′). Previous results of Better et al. (1982) show that stable gp*int* binding to the P arm of *att*P does not require gp*xis*, probably because cooperative interactions among the 6–12 monomers of gp*int* bound to the numerous sites in *att*P stabilize the complex. Thus, gp*xis* might be essential for *att*R but not for *att*P function. Although this hypothesis is speculative, and additional assumptions are required to explain the large body of recombination data summarized in Table 1 above and Table 2 of Gottesman and Weisberg (1971), it nevertheless appears that we are close to understanding the action of Xis protein (see Nash 1981b).

SITE-SPECIFIC RECOMBINATION IN OTHER PROKARYOTES

Many temperate bacteriophages, including several that are not thought to be related to λ, form lysogens by prophage insertion. However, site-specific recombination has other roles as well. In the temperate phage P1, recombination at the *lox* site reduces the frequency of prophage loss (see below; Sternberg et al.

1980; Austin et al. 1981). A second P1-encoded pathway, also found in phage Mu, inverts a specific segment of phage DNA (Lee et al. 1974; Chow and Bukhari 1976; Yun and Vapnek 1977). Inversion expands the host range of Mu (Bukhari and Ambrosio 1978; Kamp et al. 1978; Van de Putte et al. 1980) and perhaps of P1 as well. These Mu and P1 pathways share their substrate specificity with a third pathway found in certain strains of *Salmonella* in which a DNA inversion allows the organism to alternate its flagellar proteins between two antigenic types, thereby escaping the immune defense of the host (Zieg et al. 1977; Iino and Kutsukake 1981). Yet another site-specific recombination system, this one encoded by transposons Tn*3* and $\gamma\delta$, resolves transposition intermediates into final products (Gill et al. 1978; Arthur and Sherratt 1979; Kostriken et al. 1981; Reed 1981a).

Many temperate phages form lysogens by prophage insertion at specific sites in the host chromosome. The mechanism of insertion of phages P2, P22, 21, and ϕ80 appears to differ from the λ*int* pathway mainly in the substrate specificity of the reaction. *int* mutants of P2, P22, and ϕ80 have been isolated as have *xis*$^-$ derivatives of P22 (Smith and Levine 1967; Choe 1969; Lindahl 1969; Y. Nishimune, cited in Sato 1970; H. Smith, cited in Susskind and Botstein 1978). Neither ϕ80 nor 21, which are both λ-related, can complement λ*int* or λ*xis* mutants (Kaiser and Masuda 1970; L. Enquist and R. Weisberg, unpubl.). Thus, it appears that the Int and Xis proteins of each phage are adapted to their cognate *att* site.

Phage P2, like λ, can insert at several different host sites, each of which has its own affinity for the prophage (Bertani and Six 1958; Six 1961; Kelly 1963). P2 *saf* mutations can be isolated by a cycle of insertion and excision at a less favored site (Six 1963, 1966). Upon relysogenization, P2 *saf* favors the site from which it came. Thus, a portion of the P2 phage *att* site may interact homologously with a bacterial *att* site.

The P1 *lox* and $\gamma\delta$ resolvase pathways are sufficiently well understood to merit a more detailed comparison with the λ*int* pathway. In both cases, the proteins and substrates have been characterized by in vitro recombination (Reed 1981b; K. Abremski et al., pers. comm.). The biologically significant reaction appears to be recombination between identical, directly repeated sites on a circular molecule (Sternberg et al. 1980; Reed 1981a). For P1, this reduces dimers of the prophage chromosome to monomers, thereby promoting accurate segregation of the plasmid prophage into daughter cells. The $\gamma\delta$-encoded pathway resolves a "cointegrate"—a transposition intermediate that consists of two fused chromosomes with a directly repeated copy of the transposon at each fusion joint. The products of recombination are two DNA circles.

A comparison of the λ*int*, $\gamma\delta$ resolvase, and P1 *lox* pathways is presented in Table 3. Of the many differences among these reactions, two are particularly intriguing. Resolvase, in contrast to the λ and P1 proteins, does not promote efficient recombination when the two sites are in inverted order. In addition, resolvase does not efficiently promote intermolecular recombination. Thus, the

Table 3 Comparison of conservative site-specific recombination pathways

Pathway	Stimulation of recombination by underwinding	Substantial production of uncatenated products	Enzyme-DNA linkage	Polarity of staggered cuts	Promotion of intermolecular recombination and inversion	Production of Holliday structures	Host-factor requirement
λ*int*	yes	no	3′	5′	yes	yes	yes
P1 *lox*	no	yes	?	?	yes	yes	no
γδ resolvase	yes	no	5′	3′	no	?	no

In vitro recombination of P1 *lox* sites has been studied by Hoess et al. (1982) and by K. Abremski and N. Sternberg (pers. comm.), who kindly made their results available before publication. The results suggesting that Holliday structures are intermediates in this pathway are described by Sternberg (1981). The properties of resolvase-promoted recombination have been described by Reed (1981b), Reed and Grindley (1981), and N. Grindley et al. (in prep.).

way in which the two *res* sites are connected to each other governs the extent or the direction of the reaction. The mechanism of this control is unknown. The second intriguing difference among the three reactions is in the catenation of the products. Intramolecular *int*-promoted recombination between directly repeated *att* sites on a supercoiled DNA molecule leads to catenated products (i.e., circular products that are connected as are adjacent links of a chain) (Nash et al. 1977a; Mizuuchi et al. 1980). This is an accidental consequence of the DNA writhe induced by underwinding. The inefficient recombination of relaxed DNA frequently produces uncatenated products (Pollock and Nash 1980). Resolvase-promoted recombinants are also catenated, presumably by the same mechanism. In contrast, intramolecular recombination of directly repeated *lox* sites on a supercoiled DNA molecule frequently produces uncatenated products even though superhelicity is conserved (K. Abremski et al., pers. comm.). It is not known how catenation is avoided in this pathway.

The λ*int*, P1 *lox*, and $\gamma\delta$ resolvase recombination pathways resemble one another in requiring neither DNA synthesis nor an external energy source, and in conserving superhelicity. Campbell (1980) has classified them as conservative reactions. They differ from one another both in specificity and in details of the reaction mechanism. The differences in specificity are almost certainly the consequence of differences in protein-DNA and perhaps also in DNA-DNA interactions. The significance of the other differences remains to be seen.

A second group of site-specific recombination pathways, classified by Campbell (1980) as duplicative, includes bacterial transposition. Current models of transposition postulate DNA breakage, single-strand exchange, and rejoining, followed by replication of the transposable element (see Kleckner 1981). The key difference between these models and our view of λ site-specific recombination is that the initial single-strand exchange is not thought to be reciprocal in transposition. Nonreciprocal exchange leaves a free 3′ end that serves as a primer for strand elongation. In the Kikuchi-Nash model of λ recombination, reciprocality of single-strand exchange might be favored by the approximate symmetry between the four core-junction Int-protein binding sites of *att*P and *att*B and by the overall stability of the paired *att* sites. Int protein bound at the two pairs of equivalent sites would cut, exchange, and rejoin the two pairs of strands before the complex falls apart. Even if a pair of strand ends failed to rejoin, Int protein attached to the 3′ end might prevent it from serving as a primer for strand elongation. In spite of these apparent differences, we would like to point out that the duplicative and conservative recombination pathways might be closely related to one another. Only slight rearrangements in the normal configuration of λ*att* sites might produce a duplicative transposition.

ACKNOWLEDGMENTS

We are grateful to Drs. Harrison Echols, Howard Nash, and Wilma Ross for their critical comments and to Ms. Terri Broderick for carefully typing the manuscript. We also thank our many colleagues, cited in the text, who have furnished us with unpublished information. A.L. acknowledges support from the National Institutes of Health (grant AI-13544) and the National Foundation March of Dimes (grant 1-543).

REFERENCES

Abremski, K. and S. Gottesman. 1979. The form of the DNA substrate required for excisive recombination of bacteriophage λ. *J. Mol. Biol.* **131:** 637.

——. 1981. Site-specific recombination: *xis*-independent excisive recombination of bacteriophage λ. *J. Mol. Biol.* **153:** 67.

——. 1982. Purification of the bacteriophage λ *xis* gene product required for λ excisive recombination. *J. Biol. Chem.* **257:** 9658.

Arthur, A. and D. Sherratt. 1979. Dissection of the transposition process: A transposon-encoded site-specific recombination system. *Mol. Gen. Genet.* **175:** 267.

Austin, S., M. Ziese, and N. Sternberg. 1981. A novel role for site-specific recombination in maintenance of bacterial replicons. *Cell* **25:** 729.

Bell, L. and B. Byers. 1979. Occurrence of crossed-strand exchange forms in yeast DNA during meiosis. *Proc. Natl. Acad. Sci.* **76:** 3445.

Bertani, G. and E. Six. 1958. Inheritance of prophage P2 in bacterial crosses. *Virology* **6:** 357.

Better, M., L. Chi, R.C. Williams, and H. Echols. 1982. Site-specific DNA condensation and pairing mediated by the Int protein of bacteriophage λ. *Proc. Natl. Acad. Sci.* **79:** 5837.

Better, M., S. Wickner, J. Auerbach, R. Williams, and H. Echols. 1983. Role of the Xis protein of bacteriophage λ in a specific reactive complex at the *att*R prophage attachment site. *Cell* **32:** 161.

Bidwell, K. and A. Landy. 1979. Structural features of λ site-specific recombination at a secondary *att* in *gal*T. *Cell* **16:** 397.

Botchan, P. 1976. An electron microscopic comparison of transcription on linear and superhelical DNA. *J. Mol. Biol.* **105:** 161.

Brack, C., T.A. Bickle, and R. Yuan. 1975. The relation of single-stranded regions in bacteriophage PM2 supercoiled DNA to the early melting sequences. *J. Mol. Biol.* **96:** 693.

Broker, T.R. and I.R. Lehman. 1971. Branched DNA molecules: Intermediates in T4 recombination. *J. Mol. Biol.* **60:** 131.

Bukhari, A. and L. Ambrosio. 1978. The invertible segment of bacteriophage Mu DNA determines the adsorption properties of Mu particles. *Nature* **271:** 575.

Campbell, A.M. 1962. Episomes. *Adv. Genet.* **11:** 101.

——. 1981. Some general questions about movable elements and their implications. *Cold Spring Harbor Symp. Quant. Biol.* **45:** 1.

Champoux, J.J. 1977. Strand breakage by the DNA untwisting enzyme results in covalent attachment of the enzyme to DNA. *Proc. Natl. Acad. Sci.* **74:** 3800.

Chapman, J. and J.F. Gardner. 1981. Secondary λ attachment site in the threonine operon attenuator of *Escherichia coli*. *J. Bacteriol.* **146:** 1046.

Choe, B. 1969. Integration-defective mutants of bacteriophage P2. *Mol. Gen. Genet.* **105:** 275.

Chow, L. and A. Bukhari. 1976. The invertible DNA segments of coliphages Mu and P1 are identical. *Virology* **74:** 242.

Christie, G.E. and T. Platt. 1979. A secondary attachment site for bacteriophage λ in *trp*C of *E. coli*. *Cell* **16:** 407.

Csordas-Toth, E., I. Boros, and P. Venetianer. 1979. Nucleotide sequence of a secondary attachment site for bacteriophage λ on *Escherichia coli* chromosome. *Nucleic Acids Res.* **7:** 1335.

Davies, R.W. 1980. DNA sequence of the *int-xis* P_I region of the bacteriophage attachment site for λ; overlap of the *int* and *xis* genes. *Nucleic Acids Res.* **8:** 1765.

Davies, R.W., P.H. Schreier, M.L. Kotewicz, and H. Echols. 1979. Studies on the binding of λ *int* protein to attachment site DNA: Identification of a tight binding site in the P′ region. *Nucleic Acids Res.* **7:** 2255.

Echols, H. 1970. Integrative and excisive recombination by bacteriophage λ: Evidence for an excision-specific recombination protein. *J. Mol. Biol.* **47:** 575.

Echols, H. and L. Green. 1979. Some properties of site-specific and general recombination inferred from *int*-initiated exchanges by bacteriophage λ. *Genetics* **93:** 297.

Enquist, L.W. and R.A. Weisberg. 1976. The red plaque test: A rapid method for identification of excision defective variants of bacteriophage λ. *Virology* **72:** 147.

——. 1977a. A genetic analysis of the *att-int-xis* region of *coli*phage λ. *J. Mol. Biol.* **111:** 97.

——. 1977b. Flexibility in attachment site recognition by λ integrase. In *DNA insertion elements, plasmids, and episomes* (ed. A. Bukhari et al.), p. 343. Cold Spring Harbor Laboratory, Cold Spring Harbor, New York.

Enquist, L.W., A. Kikuchi, and R.A. Weisberg. 1979a. The role of λ integrase in integration and excision. *Cold Spring Harbor Symp. Quant. Biol.* **43:** 1115.

Enquist, L.W., H. Nash, and R.A. Weisberg. 1979b. Strand exchange in site-specific recombination. *Proc. Natl. Acad. Sci.* **76:** 1363.

Gill, R., F. Heffron, C. Dougan, and S. Falkow. 1978. Analysis of sequences transposed by complementation of two classes of transposition-deficient mutants of Tn3. *J. Bacteriol.* **136:** 742.

Gingery, R. and H. Echols. 1967. Mutants of bacteriophage λ unable to integrate into the host chromosome. *Proc. Natl. Acad. Sci.* **58:** 1507.

Gottesman, M.E. and R.A. Weisberg. 1971. Prophage insertion and excision. In *The bacteriophage lambda* (ed. A.D. Hershey), p. 113. Cold Spring Harbor Laboratory, Cold Spring Harbor, New York.

Gottesman, M.E. and M. Yarmolinsky. 1968. Integration negative mutants of bacteriophage lambda. *J. Mol. Biol.* **31:** 487.

Gottesman, S. and K. Abremski. 1980. The role of *him*A and *xis* in λ site-specific recombination. *J. Mol. Biol.* **138:** 503.

Guarneros, G. and H. Echols. 1970. New mutants of bacteriophage λ with a specific defect in excision from the host chromosome. *J. Mol. Biol.* **47:** 565.

——. 1973. Thermal asymmetry of site-specific recombination by bacteriophage λ. *Virology* **52:** 30.

Guerrini, F. 1969. On the asymmetry of λ integration sites. *J. Mol. Biol.* **46:** 523.

Halligan, B., J. David, K. Edwards, and L. Liu. 1982. Intra- and intermolecular strand transfer by HeLa DNA topoisomerase I. *J. Biol. Chem.* **257:** 3995.

Hamilton, P., R. Yuan, and Y. Kikuchi. 1981. The nature of the complexes formed between the *int* protein and DNA. *J. Mol. Biol.* **152:** 163.

Hoess, R. and A. Landy. 1978. Structure of the λ *att* sites generated by *int*-dependent deletions. *Proc. Natl. Acad. Sci.* **75:** 5437.

Hoess, R.H., M. Ziese, and N. Sternberg. 1982. P1 site-specific recombination: Nucleotide sequence of the recombining sites. *Proc. Natl. Acad. Sci.* **79:** 3398.

Hoess, R.H., C. Foeller, K. Bidwell, and A. Landy. 1980. Site-specific recombination functions of bacteriophage λ. DNA sequence of regulatory regions and overlapping structural genes for Int and Xis. *Proc. Natl. Acad. Sci.* **77:** 2482.

Holliday, R. 1964. A mechanism for gene conversion in fungi. *Genet. Res.* **5:** 282.

Hsu, P.L., W. Ross, and A. Landy. 1980. The λ phage *att* site: Functional limits and interaction with *int* protein. *Nature* **285:** 85.

Iino, T. and K. Kutsukake. 1981. Trans-acting genes of bacteriophage P1 and Mu mediate inversion of a specific DNA segment involved in flagellar phase variation of *Salmonella*. *Cold Spring Harbor Symp. Quant. Biol.* **45:** 11.

Kaiser, A. and T. Masuda. 1970. Specificity in curing by heteroimmune superinfection. *Virology* **40:** 522.

Kamp, D. 1973. "*In vitro* Untersuchungen der Integrase des Bakterio-phagen λ." Ph.D. thesis, University of Cologne, Federal Republic of Germany.

Kamp, D., R. Kahman, D. Zipser, T. Broker, and L. Chow. 1978. Inversion of the G DNA segment of phage Mu controls phage infectivity. *Nature* **271:** 577.

Kelly, B. 1963. Localization of P2 prophage in two strains of *Escherichia coli*. *Virology* **19:** 32.

Kikuchi, Y. and H.A. Nash. 1978. The bacteriophage λ *int* gene product. A filter assay for genetic recombination, purification of Int and specific binding to DNA. *J. Biol. Chem.* **253:** 7149.

——. 1979a. Integrative recombination of bacteriophage λ: Requirement for supertwisted DNA *in vivo* and characterization of Int. *Cold Spring Harbor Symp. Quant. Biol.* **43:** 1099.

——. 1979b. Nicking-closing activity associated with bacteriophage λ *int* gene product. *Proc. Natl. Acad. Sci.* **76:** 3760.

Kim, J., P. Sharp, and N. Davidson. 1972. Electron microscope studies of heteroduplex DNA from a deletion mutant of bacteriophage Phi-X-174. *Proc. Natl. Acad. Sci.* **69:** 1948.

Kleckner, N. 1981. Transposable elements in prokaryotes. *Annu. Rev. Genet.* **15:** 341.

Kostriken, R., C. Morita, and F. Heffron. 1981. Transposon Tn3 encodes a site-specific recombination system: Identification of essential sequences, genes and actual sites of recombination. *Proc. Natl. Acad. Sci.* **78:** 4041.

Kotewicz, M., S. Chung, Y. Takeda, and H. Echols. 1977. Characterization of the integration protein of bacteriophage λ as a site-specific DNA-binding protein. *Proc. Natl. Acad. Sci.* **74:** 1511.

Kotewicz, M., E. Grzesiuk, W. Courschesne, R. Fischer, and H. Echols. 1980. Purification and characterization of the integration protein specified by bacteriophage λ. *J. Biol. Chem.* **255:** 2433.

Landy, A. and W. Ross. 1977. Viral integration and excision. Structure of the lambda *att* sites. *Science* **197:** 1147.

Landy, A., R.H. Hoess, K. Bidwell, and W. Ross. 1979. Site-specific recombination in bacteriophage λ structural features of recombining sites. *Cold Spring Harbor Symp. Quant. Biol.* **43:** 1089.

Lee, H., E. Ohtsubo, R. Deonier, and N. Davidson. 1974. Electron microscopic heteroduplex studies of sequence relations among plasmids of *Escherichia coli*. V. *ilv*$^+$ deletion mutants of F14. *J. Mol. Biol.* **89:** 585.

Lindahl, G. 1969. Multiple recombination mechanisms in bacteriophage P2. *Virology* **39:** 1861.

Loviny, T., M.S. Neuberger, and B.S. Hartley. 1981. Sequence of a secondary phage λ attachment site located between the penitol operons of *Klebsiella aerogenes*. *Biochem. J.* **193:** 631.

Marini, J.C, and A. Landy. 1977. The isolation of restriction fragments containing the *att* site of bacteriophage λ. *Virology* **76:** 196.

McGavin, S. 1971. Models of specifically paired like (homologous) nucleic acid structures. *J. Mol. Biol.* **55:** 293.

Miller, H.I. and D.I. Friedman. 1977. Isolation of *E. coli* mutants unable to support lambda integrative recombination. In *DNA insertion elements, plasmids, and episomes* (ed. A. Bukhari et al.), p. 349. Cold Spring Harbor Laboratory, Cold Spring Harbor, New York.

——. 1980. An *E. coli* gene product required for λ site-specific recombination. *Cell* **20:** 711.

Miller, H.I. and H.A. Nash. 1981. Direct role of the *him*A gene product in phage λ integration. *Nature* **290:** 523.

Miller, H.I., M. Kirk, and H. Echols. 1981. SOS induction and autoregulation of the *him*A gene for site-specific recombination in *Escherichia coli*. *Proc. Natl. Acad. Sci.* **78:** 6754.

Miller, H.I., M.M. Mozola, and D.I. Friedman. 1980. *int*-h: An *int* mutation of phage λ that enhances site-specific recombination. *Cell* **20:** 721.

Miller, H.I., A. Kikuchi, H.A. Nash, R.A. Weisberg, and D.I. Friedman.1979. Site-specific recombination of bacteriophage λ: The role of host gene products. *Cold Spring Harbor Symp. Quant. Biol.* **43:** 1121.

Mizuuchi, K. and M. Mizuuchi. 1979. Integrative recombination of bacteriophage λ: *In vitro* study of the intermolecular reaction. *Cold Spring Harbor Symp. Quant. Biol.* **43:** 1111.

Mizuuchi, K. and H.A. Nash. 1976. Restriction assay for integrative recombination of bacteriophage λ DNA *in vitro*. Requirement for closed circular DNA substrate. *Proc. Natl. Acad. Sci.* **73:** 3524.

Mizuuchi, K., M. Gellert, and H. Nash. 1978. Involvement of super-twisted DNA in integrative recombination of bacteriophage lambda. *J. Mol. Biol.* **121:** 375.

Mizuuchi, K., M. Gellert, R. Weisberg, and H. Nash. 1980. Catenation and supercoiling in the products of bacteriophage λ integrative recombination *in vitro. J. Mol. Biol.* **141:** 485.

Mizuuchi, K., R. Weisberg, L. Enquist, M. Mizuuchi, M. Buraczynska, C. Foeller, P.-L. Hsu, W. Ross, and A. Landy. 1981. Structure and function of the phage λ *att* site: Size, Int-binding sites and location of the crossover point. *Cold Spring Harbor Symp. Quant. Biol.* **45:** 429.

Mizuuchi, M. and K. Mizuuchi. 1980. Integrative recombination of bacteriophage λ: Extent of the DNA sequence involved in attachment site function. *Proc. Natl. Acad. Sci.* **77:** 3220.

Nash, H.A. 1975a. Integrative recombination of bacteriophage lambda DNA *in vitro. Proc. Natl. Acad. Sci.* **72:** 1072.

——. 1975b. Integrative recombination in bacteriophage lambda: Anlaysis of recombinant DNA. *J. Mol. Biol.* **91:** 501.

——. 1981a. Site-specific recombination protein of phage λ. In *The Enzymes* (ed. P. Boyer), p. 471. Academic Press, New York.

——. 1981b. Integration and excision of bacteriophage λ: The mechanism of conservative site-specific recombination. *Annu. Rev. Genet.* **15:** 143.

Nash, H.A. and C.A. Robertson. 1981. Purification and properties of the *E. coli* protein factor required for λ integrative recombination. *J. Biol. Chem.* **256:** 9246.

Nash, H.A., L.W. Enquist, and R.A. Weisberg. 1977a. On the role of the bacteriophage λ *int* gene product in site-specific recombination. *J. Mol. Biol.* **116:** 627.

Nash, H.A., K. Mizuuchi, L.W. Enquist, and R.A. Weisberg. 1981. Strand exchange in λ integrative recombination: Genetics, biochemistry, and models. *Cold Spring Harbor Symp. Quant. Biol.* **45:** 417.

Nash, H.A., K. Mizucchi, R. Weisberg, Y. Kikuchi, and M. Gellert. 1977b. Integrative recombination of bacteriophage λ—The biochemical approach to DNA insertions. In *DNA insertion elements, plasmids, and episomes* (ed. A. Bukhari et al.), p. 363. Cold Spring Harbor Laboratory, Cold Spring Harbor, New York.

Parkinson, J.S. 1971. Deletion mutants of bacteriophage lambda. II. Genetic properties of *att*-defective mutants. *J. Mol. Biol.* **56:** 385.

Pinkham, J.L., T. Platt, L.W. Enquist, and R.A. Weisberg. 1980. The secondary attachment site for bacteriophage λ in the *pro*A/B gene of *Escherichia coli. J. Mol. Biol.* **144:** 587.

Pollock, T.J. and K. Abremski. 1979. DNA without supertwists can be an *in vitro* substrate for site-specific recombination of bacteriophage λ. *J. Mol. Biol.* **131:** 651.

Pollock, T.J., and H. Nash. 1980. Catenation of the products of integrative recombination: Comparison of supertwisted and nonsupertwisted substrates. In *Mechanistic studies of DNA replication* (ed. B. Alberts), p. 953. Academic Press, New York.

Reed, R. 1981a. Resolution of cointegrates between transposons γδ and Tn3 defines the recombination site. *Proc. Natl. Acad. Sci.* **78:** 3428.

——. 1981b. Transposon-mediated site-specific recombination: A defined *in vitro* system. *Cell* **25:** 713.

Reed, R. and N. Grindley. 1981. Transposon-mediated site-specific recombination *in vitro*: DNA cleavage and protein-DNA linkage at the recombination site. *Cell* **25:** 721.

Ross, W. and A. Landy. 1982. Int recognizes two sequences in the phage *att* site: Characterization of arm-type sites. *Proc. Natl. Acad. Sci.* **79:** 7724.

——. 1983. Patterns of λ Int recognition in the regions of strand exchange. *Cell* **33:** 261.

Ross, W., M. Schulman, and A. Landy. 1982. Biochemical analysis of att-defective mutants of the phage λ site-specific recombination system. *J. Mol. Biol.* **156:** 505.

Ross, W., A. Landy, Y. Kikuchi, and H. Nash. 1979. Interaction of *Int* protein with specific sites on λ *att* DNA. *Cell* **18:** 297.

Sato, K. 1970. Genetic map of bacteriophage φ80: Genes on the right arm. *Virology* **40:** 1067.

Saucier, J.-M. and J.C. Wang. 1972. Angular alteration of the DNA helix by *E. coli* RNA polymerase. *Nat. New Biol.* **239:** 157.

Shimada, K. and A. Campbell. 1974a. *int*-constitutive mutants of bacteriophage lambda. *Proc. Natl. Acad. Sci.* **71:** 237.

——. 1974b. Lysogenization and curing by *int*-constitutive mutants of phage λ. *Virology* **60:** 157.

Shimada, K., R.A. Weisberg, and M.E. Gottesman. 1972. Prophage lambda at unusual chromosomal locations. I. Location of the secondary attachment sites and the properties of the lysogens. *J. Mol. Biol.* **63:** 483.

——. 1973. Prophage lambda at unusual chromosomal locations. II. Mutations induced by bacteriophage lambda in *Escherichia coli* K12. *J. Mol. Biol.* **80:** 297.

——. 1975. Prophage lambda at unusual chromosomal locations. III. The components of the secondary attachment sites. *J. Mol. Biol.* **93:** 415.

Shulman, M. and M. Gottesman. 1973. Attachment site mutants of bacteriophage lambda. *J. Mol. Biol.* **81:** 461.

Shulman, M., K. Mizuuchi, and M. Gottesman. 1976. New *att* mutants of phage λ. *Virology* **72:** 13.

Signer, E.R., J. Weil, and P.C. Kimball. 1969. Recombination in bacteriophage λ. III. Studies on the nature of the prophage attachment region. *J. Mol. Biol.* **46:** 543.

Six, E. 1961. Inheritance of prophage P2 in superinfection experiments. *Virology* **14:** 220.

——. 1963. Affinity of P2 *rd*1 for prophage sites on the chromosome of *Escherichia coli* strain C. *Virology* **10:** 375.

——. 1966. Specificity of P2 for prophage site I on the chromosome of *Escherichia coli* strain C. *Virology* **29:** 106.

Smith, H. and M. Levine. 1967. A phage P22 gene controlling the integration of prophage. *Virology* **31:** 207.

Sternberg, N. 1981. Bacteriophage P1 site-specific recombination. III. Strand exchange during recombination at *lox* sites. *J. Mol. Biol.* **150:** 603.

Sternberg, N., D. Hamilton, S. Austin, M. Yarmolinsky, and R. Hoess. 1980. Site-specific recombination and its role in the life cycle of bacteriophage P1. *Cold Spring Harbor Symp. Quant. Biol.* **44:** 297.

Struhl, K. 1981. Deletion, recombination and gene expression involving the bacteriophage λ attachment site. *J. Mol. Biol.* **152:** 517.

Susskind, M. and D. Botstein. 1978. Molecular genetics of bacteriophage P22. *Microbiol. Rev.* **42:** 385.

Tse, Y.-C., K. Kirkegaard, and J.C. Wang. 1980. Covalent bonds between protein and DNA. Formation of phosphotyrosine linkage between certain DNA topoisomerases and DNA. *J. Biol. Chem.* **255:** 5560.

Van de Putte, P., S. Cramer, and M. Giphart-Gassler. 1980. Invertible DNA determines host range specificity of bacteriophage Mu. *Nature* **286:** 218.

Wang, J.C., J.H. Jacobsen, and J.-M. Saucier. 1977. Physicochemical studies on interactions between DNA and RNA polymerase. Unwinding of the DNA helix by *Escherichia coli* RNA polymerase. *Nucleic Acids Res.* **4:** 1225.

Warner, R.C., R.A. Fischel, and F.C. Wheeler. 1979. Branch migration in recombination. *Cold Spring Harbor Symp. Quant. Biol.* **43:** 1089.

Weisberg, R. and M.E. Gottesman. 1969. The integration and excision defect of bacteriophage λdg. *J. Mol. Biol.* **46:** 565.

Weisberg, R., L. Enquist, C. Foeller, and A. Landy. 1983. A role for DNA homology in site-specific recombination: The isolation and characterization of a site-affinity mutant of coliphage λ. *J. Mol. Biol.* (in press).

Williams, J., D.L. Wulff, and H.A. Nash. 1977. A mutant of *Escherichia coli* deficient in a host function required for phage lambda integration and excision. In *DNA insertion elements, plasmids, and episomes* (ed. A. Bukhari et al.), p. 357. Cold Spring Harbor Laboratory, Cold Spring Harbor, New York.

Wilson, J.H. 1979. Nick-free formation of reciprocal heteroduplexes. A simple solution to the topological winding problem. *Proc. Natl. Acad. Sci.* **76:** 3641.

Yun, T. and D. Vapnek. 1972. Electron microscopic analyses of bacteriophages P1, P1CM, and P7. *Virology* **77:** 376.

Zieg, J., M. Silverman, M. Hilmen, and M. Simon. 1977. Recombinational switch for gene expression. *Science* **196:** 170.

Zissler, J. 1967. Integration negative (*int*) mutants of phage λ. *Virology* **31:** 189.

Phage Lambda's Accessory Genes

Donald Court*
Laboratory of Molecular Biology
National Institutes of Health
National Cancer Institute
Bethesda, Maryland 20205

Amos B. Oppenheim
Department of Microbiological Chemistry
The Hebrew University
Hadassah Medical School
Jerusalem, Israel

INTRODUCTION

Phage λ contains some 50 kbp of DNA, and approximately 50 genes have been revealed by analysis of mutants and phage proteins. Most of these genes are essential for perpetuation of the virus in either the lytic or the lysogenic mode of growth. For example, 27 genes are required for the phage to produce mature progeny and form plaques. Five other genes are required for repression, integration, and excision. Several other genes, which form the subject of this paper, are not absolutely required for either lysis or lysogeny. These genes are distributed in several places along the λ genome, and the regions carrying them can be deleted with minimal effects on phage development. Because they are not essential for "normal" phage development in standard hosts and under standard laboratory conditions these genes have been collectively designated as nonessential or dispensable. However, it is becoming clear that these genes are not functionless but, rather, they often have interesting and even essential roles in certain circumstances. Consequently, we prefer to call them accessory genes. In this paper, we review λ's accessory genes and discuss their role with respect to both λ and *Escherichia coli*.

Since these genes are dispensable for "normal" λ development, they have not been studied extensively, and our understanding of their expression and function is limited. Nevertheless, we can make a few general statements about their roles. Some appear to be nonessential for phage growth because an analogous or compensating function exists in the host. Others are essential to the phage when special conditions are encountered, such as a different host or environment. Still others appear to alter macromolecular synthesis of the host or the host's susceptibility to infection by viruses other than λ. These genes may function to ensure that λ survives the many different habitats it encounters in nature as opposed to the specialized conditions it finds in the laboratory. λ has maintained these genes

*Present address: Laboratory of Molecular Oncology, National Institutes of Health, National Cancer Institute, Bethesda, Maryland 20205.

in the face of selection; we believe that they, like other λ genes, will provide new and interesting control systems to study.

Genes of related function are clustered along the λ chromosome according to function (Fig. 1). The largest tract of dispensable DNA, between gene *J* and gene *N*, encompasses 33% of the λ genome. It includes the *b* genes to the left of *att* and the p_L operon genes to the right of *att* (see Fig. 1). The immunity region, composed of the *rex* (*r*II *ex*clusion) and *c*I genes, is also not required for plaque formation. The p_R operon is to the right of the immunity region; genes for transcriptional control (*cro*, *c*II, and *Q*) and essential replication genes (*O* and *P*) are located within this operon. Distal to *O* and *P* is a genetically silent region of phage λ for which no gene product has been identified. The deletion *nin*5 removes a 2.8-kb segment of this region without measurably affecting phage development. The right end of the λ genome contains gene *Q* and the genes involved in cell lysis. The lysis genes and the viral morphogenesis genes are part of the p_R' operon, which is positively controlled by gp*Q*. The last 2 kb at the right end of λ are also silent and are not known to contain any genes. A small deletion within this region is present in some λ derivatives (Fiandt et al. 1971). We discuss the *b* region, the p_L operon genes, the immunity region, and the *P-Q* region in greater detail in the following sections.

b REGION

The central region of the λ genome, bounded by the phage tail gene *J* and by *att*, includes a number of genes. Yet the existence of a set of plaque-forming deletion mutants (Kellenberger et al. 1961; Davis and Parkinson 1971) and the *lac*5 substitution (Ippen et al. 1971; Malamy et al. 1972) demonstrates that there are no "essential genes" in this region (Fig. 2).

Several proteins whose synthesis is dependent on the presence of the *b* region are resolved by polyacrylamide gel electrophoresis following infection of UV-irradiated cells (Hendrix 1971; Murialdo and Siminovitch 1972; Oppenheim et al. 1977; C. Epp and M.L. Pearson, in prep.). These proteins can be divided into two groups: those made early after infection, and those made late. The synthesis of at least four early proteins of 59 kD, 47 kD, 31 kD, and 24 kD is directed by the *b* region. Late after infection three additional proteins of 48 kD, 21 kD (Hendrix 1971), and 20.5 kD (Reeve and Shaw 1979) have been detected. The first two have been resolved from total cell extracts by gel electrophoresis, whereas the latter, *lom*, has been isolated as a component of the *E. coli* outer membrane following infection of minicells. The location of each of the early and late gene products of the *b* region has been determined by their presence or absence after infection by different λ*b* deletion mutants. The organization of genes in the *b* region that encode these proteins is illustrated in Figure 2 and Table 1. The DNA sequence of this region (see Appendix II) indicates three open reading frames in the right half of *b* capable of encoding the 59-kD, 47-kD,

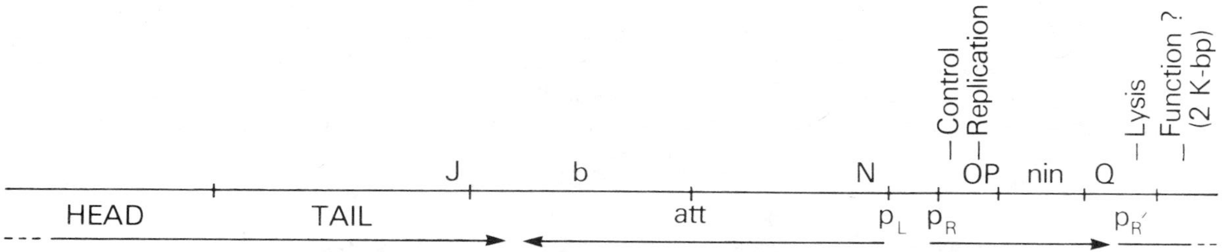

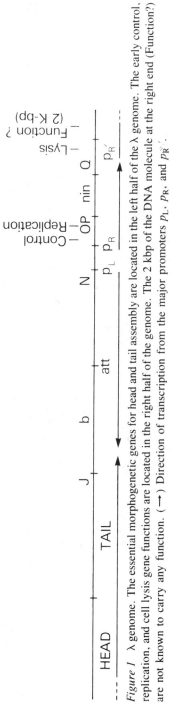

Figure 1 λ genome. The essential morphogenetic genes for head and tail assembly are located in the left half of the λ genome. The early control, replication, and cell lysis gene functions are located in the right half of the genome. The 2 kbp of the DNA molecule at the right end (Function?) are not known to carry any function. (→) Direction of transcription from the major promoters p_L, p_R, and p_R'.

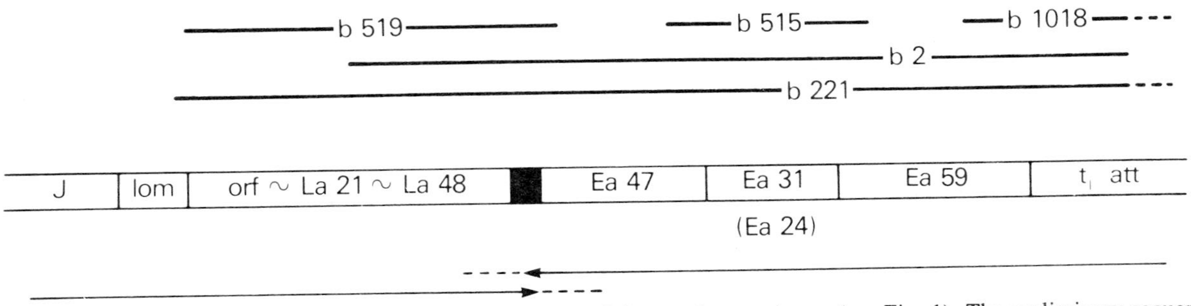

Figure 2 The *b* region of λ is located between tail gene *J* and the attachment site *att* (see Fig. 1). The preliminary sequence of this region (see Appendix II) shows 8500 bp of DNA between *J* and *att*. The early and late transcripts of this region are indicated, respectively, by arrows pointing left and right; (----) extent unknown. The t_l transcription terminator is shown between Ea59 and *att* (see text). Deletions used to locate the genes in this region are shown above the map as solid bars (see Table 1 for mapping: Szybalski and Szybalski 1979). (■) The intercistronic region between the early and late operons. The three genes for Ea47, Ea31, and Ea59 appear to saturate the early region sequence. The Ea24 protein appears to derive from Ea31 perhaps as a degradation product; alternatively, Ea24 synthesis may be controlled by Ea31, in which case the Ea24 gene may not even be located in *b*. The *lom* gene and genes for La21 and La48 only partly fill the late region. The sequence shows four large open reading frames (orf). Therefore, three genes (including those for La21 and La48) may be present to the right of *lom*. Their order is still in question.

Table 1 Genes in the *b* region

Deletion	Protein						
	gp*lom*	La21	La48	Ea47	Ea31	Ea24	Ea59(gp*ben*)
*b*221	–	–	–	–	–	–	–
*b*2	+	–	–	–	–	–	–
*b*515	+	+	+	–	–	–	–
*b*1018	+	+	+	+	+	+	–
*b*519	+	–	–	–	+	+	+

+ indicates that the radioactive protein can be observed on polyacrylamide gels after infection of a UV-irradiated host by a particular *b deletion*. λ^+ makes all of these products (Hendrix 1971; Reeve and Shaw 1979; C. Epp and M.L. Pearson, in prep.). Note that La21 and La48 are not mapped by this matrix nor are Ea31 and Ea24. Hendrix (1971) places La21 to the left of La48 because another deletion, *b*506 (not shown), makes La21 but not La48 and Ea47. However, there is confusion on this point because *b*2 does not delete as far to the left as *b*506 (Szybalski and Szybalski 1979), yet *b*2 is defective for both late proteins. Four genes large enough to encode the four early proteins will not fit in the DNA sequence where they map. C. Epp and M.L. Pearson (in prep.) suggested that the gene for Ea24 was nested within the gene for Ea59. The sequence of this region (see Appendix II) reveals an open reading frame for Ea59 in the correct position but none for Ea24 (see text for discussion).

and 31-kD proteins. The 24-kD protein may be a degradation product (Hendrix 1971) of the 31-kD protein (see legend to Fig. 2).

The 59-kD protein has been purified (S. Benchimol et al., pers. comm.). This protein is an endodeoxyribonuclease that is specific for supercoiled DNA and requires single-stranded DNA, ATP, and Mg ion as cofactors. The protein also has an ATPase activity that depends on single-stranded DNA. Two other DNA endonucleases that have been detected following λ infection are also seen only if the phage has an intact *b* region. Neither requires single-stranded DNA or ATP as a cofactor (Chowdhury et al. 1972; Rhodes and Meselson 1973). The function of these nucleases in the λ life cycle is unclear.

Those *b*-region proteins that are expressed soon after infection require an active p_L promoter and gp*N* (see below, p_L Operon). Their expression is subject to both *c*I and *cro* repression (Hendrix 1971; Oppenheim et al. 1977; Oppenheim and Oppenheim 1978). These results suggest that early *b*-gene transcription is initiated at p_L. Consistent with this idea, transcription of the *b* region early after infection was found to occur in the same direction as that from p_L (Bøvre and Szybalski 1969; Nijkamp et al. 1971). In addition, recent studies indicate that the p_L transcript continues beyond the *int* gene and through the t_I terminator into the *b* region (U. Schmeissner et al., unpubl.). This transcription is presumed to pass far into the *b* region and stop at a termination site that is resistant to the action of gp*N* (Gottesman et al. 1980; Honigman 1981). This *N*-resistant terminator may be essential for preventing RNA polymerase, transcribing the early p_L operon, from continuing into the virion assembly genes of the left arm and transcribing them in the antisense direction.

Late *b*-gene proteins are probably transcribed from p_R' by extension of the transcript of the late λ structural genes (Hendrix 1971). Transcription from p_R'

is antiterminated by gp*Q* (Grayhack and Roberts 1982; Forbes and Herskowitz 1983). Krell et al. (1972) showed that this transcript, like the early p_L transcript, terminates within the *b* segment. In fact, the early and late transcripts were found to overlap for a part of this segment (Bøvre and Szybalski 1969).

Thus, early gene expression of the *b* region is from the p_L promoter, whereas late gene expression is from the $p_R{}'$ promoter. However, let us point out that in vitro analyses reveal the presence of several promoters in the *b* segment itself (Botchan 1976; Vollenweider and Szybalski 1978; Kravchenko et al. 1979; Rosenvold et al. 1980). These promoters are oriented so that some transcribe in the same direction as p_L, and others, in the opposite direction. The function of these promoters is unclear. However, a recent observation by J.E. Shaw et al. (in prep.) indicates that they may be involved in expression of *b* proteins. These authors found that after λ infection of minicells, the synthesis of most λ proteins was governed by the same rules observed after infection of normal cells. Surprisingly, the synthesis of *b* proteins was no longer completely dependent upon transcription from p_L or upon gp*N*. This suggested that promoters in *b* might act in vivo to express *b* proteins under this set of experimental conditions. A *cis-trans* test reveals that expression of the Ea59 product is dependent on the p_L promoter (Hendrix 1970). However, similar experiments have not been done for other early *b*-region genes. Thus, the true location of the promoter(s) responsible for expression of *b* genes in normal cells has not been determined unambiguously. Three possible explanations may be considered. (1) the *b* promoters alone are responsible and depend upon gp*N* or a function under its control for activation. (2) the p_L promoter alone is responsible, and in normal conditions in *E. coli* cells the *b* promoters are inactive. (3) Both p_L and *b* promoters are active. Each of these possibilities raises new questions. If the first or third is correct, then what function activates the *b* promoters? If the second is correct, then what prevents *b*-promoter expression, and are there conditions other than minicells in which the *b* promoters are active? Thus, we find that questions remain about the function of the *b* proteins as well as about their mode of regulation.

p_L OPERON

Several genes are transcribed as a unit from the λ p_L promoter. Transcription of this operon is positively controlled by the *N*-gene product; gp*N* causes RNA polymerases initiating at p_L to transcribe past transcription stop signals located throughout the operon (Adhya et al. 1976; Friedman and Gottesman, this volume). The p_L promoter is regulated negatively by two different λ proteins, the products of the *c*I and *cro* genes. The *c*I repressor blocks initiation from p_L and maintains the prophage in a dormant state during lysogeny. After infection, in the absence of *c*I repressor, the *cro* protein depresses p_L transcription to one-fourth the initial rate by 8 minutes postinfection (Adhya et al. 1976). This reduction is important for a normal lytic cycle and phage burst (Eisen et al. 1975). The individual genes in the p_L operon have been identified primarily by their protein products on polyacrylamide gels. None except *N* is required for plaque formation in wild-type *E. coli* strains. Deletion and point mutants have been used to locate each gene in the operon, and with the availability of DNA sequence information, an accurate map has been constructed (see Fig. 3). Still some regions remain uncharted; namely, the segment between *xis* and *redA* (*exo*) and the order of *c*III and *kil*. We discuss each gene by its position along the p_L operon.

ral

λ phages that express *ral* (*r*estriction *al*leviation) are able to modulate the host restriction-modification system. The *ral* function is specific for type-I *Eco*K and *Eco*B systems. It does not affect the *Eco*PI system or other type-II enzymes (Zabeau et al. 1980). The gene is located near *N*, but its product has not been identified. However, from *bio* deletion mapping and sequence determinations, it is thought to have a molecular weight of 7600 (Fig. 3; Debrouwere et al. 1980a; Zabeau et al. 1980; Ineichen et al. 1981).

Host-controlled restriction modification is a means by which a cell can distinguish between its DNA and foreign DNA. Foreign DNA lacking proper modification signals (usually methyl groups) is subjected to endonuclease digestion (restriction). The host system includes three genes, *hsdS, hsdM* (modification), and *hsdR* (restriction), whose products form a multisubunit protein with both endonuclease and methylase functions (Vovis et al. 1974). *hsdS* encodes a specificity determinant (Boyer and Roulland-Dussoix 1969). When λ is grown on *E. coli* K, it is methylated by K and becomes resistant to K restriction. However, if λ·K infects *E. coli* B, which has a different specificity, λ·K DNA is restricted and degraded (Dussoix and Arber 1962).

Normal infection by unmodified λ results in restriction; only rarely (10^{-4}) do λ phage escape restriction. Those that do are modified and are no longer subject to restriction (Dussoix and Arber 1962). The presence or absence of *ral* has no effect on the frequency (10^{-4}) of escape from restriction by unmodified phage. Apparently the *ral* function is not made soon enough to protect its own DNA (Zabeau et al. 1980). However, there are two conditions in which the effect of *ral* is measurable. In one situation, restriction of unmodified phage DNA is reduced by coinfection with modified λ*ral*$^+$ helper; λ*ral* mutant helper is unable to protect unmodified phage. Protection does not require recombination between the helper and unmodified phage DNA (Zabeau et al. 1980; Toothman 1981). In a different set of experiments, modification of λ DNA is measured in a restriction-defective host, *hsdR*$^-$. Under these conditions, unmodified λ*ral*$^+$ phage DNA is completely modified after infection, whereas λ*ral*$^-$ DNA is only partially modified (Zabeau et al. 1980; A. Abeles, pers. comm.). Thus, the *ral* product stimulates complete methylation of unmodified DNA. Normally, the methylation enzyme is proficient only when one of the two DNA strands is already methylated (Vovis et al. 1974). The mechanism by which gp*ral* enhances methylase activity is not known. A direct interaction between gp*ral*

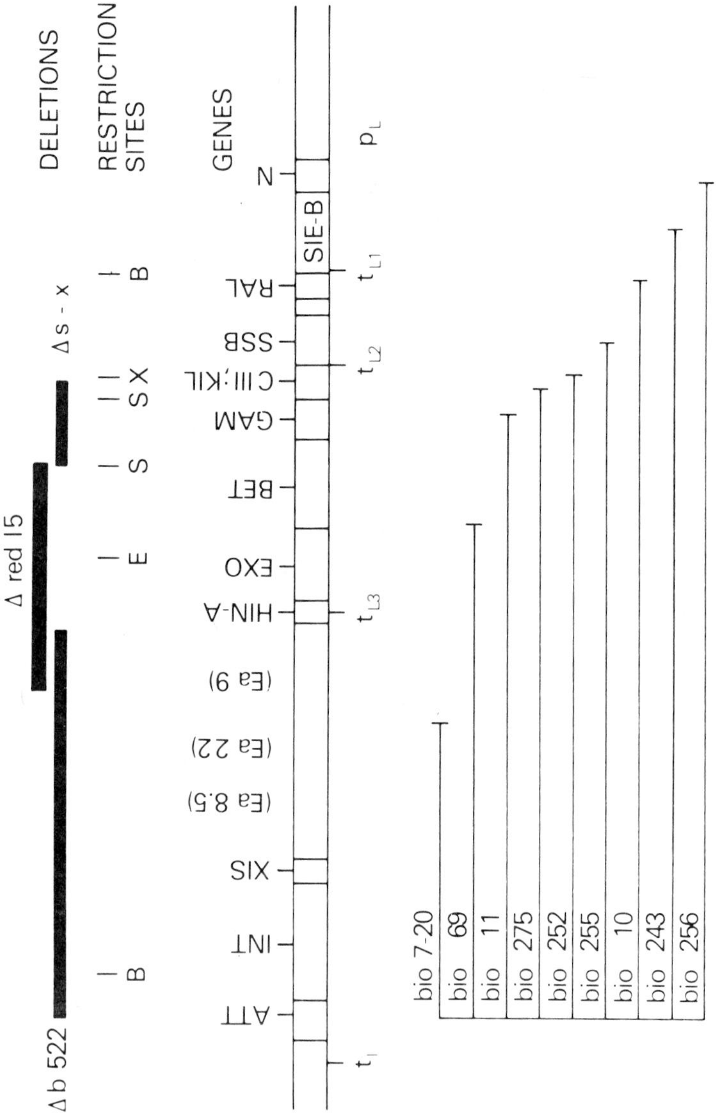

Figure 3 (See facing page for legend.)

and the multisubunit protein would be a simple explanation of the gp*ral* effect, but because of other changes in the cell caused by *ral* expression, an indirect effect may also be postulated (see below).

Debrouwere et al. (1980b) have proposed that gp*ral* affects other components of the cell, namely transcription termination and recombination. Polarity of nonsense mutations in the *lac* operon is reduced in strains expressing *ral*. Since such polarity in *lac* is caused by transcription termination by the host ρ factor, an effect upon ρ factor was postulated. Debrouwere et al. (1980b) have shown that recombination in certain regions of λ DNA appears to be reduced in ral^+ as opposed to ral^- phage crosses. An effect upon the host *recBC* recombination system was suggested by these investigators.

E. coli, mutant in ρ factor, shows severe pleiotropic alterations (Das et al. 1979). Debrouwere et al. (1980b) have proposed that a primary effect of *ral* upon ρ could cause all other observed effects including restriction alleviation. At present, however, this proposal is based upon rather indirect evidence. One property of ρ mutants is an inability to grow on succinate as a carbon source. Strain M5219 carrying a λ prophage with part of the p_L operon deleted can be made constitutive for *ral*; in this condition, the strain is also unable to utilize succinate (Debrouwere et al. 1980b). However, this effect is not caused by the expression of gp*ral*; strain M5219 is also constitutive for an adjacent gene product,

Figure 3 The p_L operon. The DNA region *att* through *N* of λ is represented. It contains some 8000 bp of the λ genome. The promoter p_L is used to transcribe this region from right to left. The genes in this operon are listed adjacent to open rectangles, which are drawn to scale to represent the sizes of these genes. Size has been determined in two ways: (1) The protein products of all of these genes except *xis, hinA*, cIII, *kil*, and *ral* have been identified by gel electrophoresis; (2) the gene has been identified for each from the DNA sequence for all except *ea*8.5, *ea*22, and *ea*9 (Hoess et al. 1980; Ineichen et al. 1981; Appendix II). The gene *sieB*, between *ral* and *N*, is encoded from the opposite DNA strand and is not made in the p_L transcript (see Fig. 4). Four transcription terminators are shown. The ρ-dependent terminator t_{L1} is located before the *ral* gene (Zabeau et al. 1980). The first independent terminator t_{L2} is defined genetically between *bio*11 and *bio*10 (Gottesman et al. 1982), probably between the cIII and *ssb* genes (Salstrom and Szybalski 1978b; Drahos and Szybalski 1981; A. Das, pers. comm.). A second independent terminator is located beyond *exo* (Salstrom and Szybalski 1978b; Luk and Szybalski 1982) in the proposed *hinA* gene (D. Court et al., unpubl.). Another terminator t_I is 200 bp beyond *att* (Oppenheim et al. 1982; U. Schmeissner et al., pers. comm). Several *bio* substitution deletions are represented below the map. The deletions extend from *att* various distances (represented by horizontal lines) to the right and have been used to map the p_L operon (Szybalski and Szybalski 1979). Restriction enzyme sites are indicated: (E) *Eco*RI, (S) *Sal*I, (X) *Xho*I, (B) *Bam*HI. Deletion mutants are shown as solid bars above the gene map: $\Delta b522$ begins at *att* and is $hinA^+$ (see text); ΔS-X removes DNA between the *Sal* and *Xho* sites (S. Rodgers, pers. comm.) and is $exo^+bet^-gam^-$cIII$^-$ kil^-ssb^+ (D. Court and M. Gottesman, unpubl.); $\Delta red15$ is $ea22^+ea9^-exo^-bet^-gam^+$ (C. Epp and M.L. Pearson, in prep.).

SSB (single-stranded binding protein); (see below, *ssb*) (Georgiou et al. 1979). Apparently the *ssb* product is the causal agent in blocking growth on succinate, since a strain, M5221, with a deletion of *ssb* but not *ral*, grows normally on succinate medium (D. Court, unpubl.). Thus, the effect of gene expression in this region is complex and interesting, and more information will be required to understand the full effect of *ral* expression.

ssb

The *ssb* (or *ea*10) gene was first identified by its protein product on acrylamide gels (Hendrix 1971). Recently, J. Kochan and H. Murialdo (pers. comm.) have identified gp*ssb* on two-dimensional O'Farrell gels and determined it to be 16 kD. Its location between the *ral* gene and *c*III was determined by deletion mapping. The deletion-substitution mutant *bio*252 is ssb^+, and *bio*255 (and *bio*214) are ssb^- (see Fig. 2; Hendrix 1971; Epp 1978). The *ssb* protein is the most abundant early λ protein made after infection (Hendrix 1971). It has now been purified and shown to be a single-stranded DNA binding protein that exists as an oligomer of 60 kD (R.W. Hendrix, pers. comm.). The amino acid composition of the pure protein compares well with that of a protein predicted from an open reading frame in the DNA of this region (Ineichen et al. 1981; R.W. Hendrix and W.E. Brown, pers. comm.).

General effects of *ssb* expression on host physiology have been observed following infection by λ. The most severe effects are found after infection by λ*cro* mutants where *ssb* is overexpressed. Under these conditions, phage and cell growth comes to a stop within 20 minutes; this has been called the Tro phenotype (Eisen et al. 1975). Total RNA and protein syntheses are reduced severely; DNA replication is also affected. Even though most cellular and phage protein levels were reduced during Tro inhibition, others, including gp*ssb* and the host proteins gp*groE* and gp*dnaK*, were synthesized at high levels. Although the Tro effect is relieved in part by deletions and point mutations in the *ssb* gene, other unidentified λ elements are also involved (Georgiou et al. 1979). In λ mutants defective in replication, no reduction in general host synthesis is caused by *ssb* even when the phage is cro^- (Georgiou et al. 1979; Court et al. 1980a). Replication may be involved directly or through a gene dosage effect. In normal λ^+ infections, the severe effects of λcro^- infection on total host macromolecular syntheses are not found, but more subtle changes in individual host proteins are observed (see below).

Drahos and Hendrix (1982) have measured the rates of synthesis of several cellular proteins. Proteins were isolated from two-dimensional O'Farrell gels, and the level of radioactive [^{35}S]methionine incorporated into each was measured during a short labeling period after infection by λ or λ deletion mutants. Drahos and Hendrix found that the synthesis of two host proteins gp*groE* and gp*dnaK*, which are important for phage λ growth and response to heat shock (see Furth and Wickner; Georgopoulos et al.; both this volume) were increased after infection of *bio*252 mutants (*ssb*$^+$*ral*$^+$) but not after infection with *bio*10 mutants (*ssb*$^-$*ral*$^-$). Most other proteins were unaffected. These results suggested an effect of *ssb*- and/or *ral*-gene products. Either of these two functions could act as regulatory proteins. The T4 gene *32* protein, like *ssb*, has a single-strand DNA binding activity and is known to regulate its own expression (Gold et al. 1976).

*c*III Gene

The product of the *c*III gene stimulates the lysogenic pathway of phage development. λ*c*III mutants form clear plaques as opposed to the normal turbid plaques made by λ$^+$. This is because fewer lysogenic bacteria survive in the center of *c*III plaques as compared to λ$^+$. Since λ*bio*275 forms turbid plaques and λ*bio*252 forms clear plaques (see Fig. 3), these two adjacent deletions define the *c*III-gene position. No protein has been identified as being the *c*III product. Deletion analysis and sequence information suggests that *c*III may overlap the *kil* gene (Greer 1975a; Honigman et al. 1981; Ineichen et al. 1981; D. Court and M. Gottesman, unpubl.; see legend to Fig. 3). The mechanism by which *c*III may affect lysogeny is discussed by Wulff and Rosenberg (this volume).

kil Gene

The *kil* (host *kil*ling by an induced λ prophage) gene product is expressed after prophage induction and causes cell death. Cell death occurs in induced prophage strains in which only the p_L operon is active. In this operon, the gene responsible for killing is defined by *bio* substitutions *bio*11 (*kil*$^+$) and *bio*275 (*kil*Δ) (see Fig. 3; Greer 1975a).

Expression of *kil* has several effects upon the host (Greer 1975a). After induction of *kil*$^+$ prophages, cell death is determined by 45 minutes, but protein and RNA synthesis continue normally for up to 3½ hours (Greer 1975a; Court et al. 1980a). DNA synthesis also continues for 2 hours at the same rate in *kil*$^+$ and *kil*Δ, but beyond that time DNA synthesis in the *kil*$^+$ decreases dramatically, and the *kil*$^+$ cells elongate to form filaments. Septation can be restored by the drug pantoyl lactone; however, cell death occurs even in the presence of the drug. The onset of death is also predetermined before DNA replication decreases. Thus, both of these effects, filamentation and reduced DNA synthesis, appear to be secondary effects of *kil* (Greer 1975a).

The primary effect of *kil* is not known; however, some host mutants, resistant to cell death when *kil* is expressed, have alterations in the cell envelope (Greer 1975b). Other mutants are cold-sensitive for growth. The cold-sensitive mutants are defective in the host *nusB* gene (C. Georgopoulos, pers. comm.). The product of *nusB* is required for the *N*-gene product to antiterminate transcription from p_L (see Friedman and Gottesman, this volume) and thereby express *kil*.

The *nusB*-host mutants are thus likely to affect expression of the *kil* gene rather than to affect the action of its product.

Point mutations have been isolated in the prophage that eliminate the killing effect. Many are pleiotropic *N* mutants. Of those that are N^+, six of seven map near *c*III presumably in *kil*; the other one maps to the right of *c*III (Greer 1975a) and is probably defective in the *N* *ut*ilization site (*nut*L). A *nut*L mutant, by preventing *N* antitermination, would also be Kil^- (Salstrom and Szybalski 1978a).

The *kil*-gene product has not yet been isolated or even identified on gels. A first identificaion of *kil* as a 16.5-kD protein appears to be wrong. This same protein proved to be gp*ssb* from analysis on two-dimensional O'Farrell gels (Greer and Ausubel 1979; J. Kochan and H. Murialdo, pers. comm.). The *kil*-mutant phage used for the original identification (Greer and Ausubel 1979) may have had a defect in both the *kil* and *ssb* genes. Another possibility is that the Kil phenotype requires both *kil* and *ssb* products.

The exact location of *kil* with respect to neighboring genes, *c*III and *gam,* is unclear. From the sequence analysis and deletion mapping of these three genes (Greer 1975a; Honigman et al. 1981; Ineichen et al. 1981; D. Court and M. Gottesman, unpubl.), it is likely that *kil* and *c*III overlap (Fig. 3).

red and *gam* Genes

Three genes that affect homologous recombination and phage replication are located together in the p_L operon, distal to the *kil* and *c*III genes. The *gam* gene has been well characterized. Its product inhibits several activities of the host *recBC* nuclease (Unger and Clark 1972; Sakaki et al. 1973; Karu et al. 1974). The *gam* subunit is 16 kD (Karu et al. 1974; C. Epp and M.L. Pearson, in prep.), consistent with the 414-bp open reading frame in this region (Ineichen et al. 1981).

The *redB* (or *bet*)-gene product promotes homologous recombination (Shulman et al. 1970). The product of this gene has been identified and found to be 30 kD (Hendrix 1971; Radding et al. 1971). This size agrees with the coding region of 771 bp determined from the DNA sequence of this region (Ineichen et al. 1981). Its function in recombination is yet to be determined.

The *redA* (or *exo*)-gene product also promotes recombination (Shulman et al. 1970). The purified product has a 5′-exodeoxyribonuclease activity (Little et al. 1967; Radding et al. 1971). The DNA sequence of this gene (see Appendix II) agrees with the 26 kD of the protein subunit.

A more detailed description of the functions of these three genes is given by Smith (this volume). Here we only summarize the numerous phenotypes that have been associated with these genes: Spi, the sensitivity of λ^+ to P2 interference; λ*red gam* double mutants, but not λ*red,* λ*gam,* or λ^+, can form plaques on a P2 lysogen (Zissler et al. 1971b); Fec, the inability of *red gam* double mutants to grow on *recA* strains (Zissler et al. 1971b); Feb, the inability of *red* or *gam* single mutants to grow on *polA* or *lig* strains (Zissler et al. 1971a); Ren, the inability of *red* mutants to grow on strains containing gp*rex* function (see below, Exclusion Systems of the λ Prophage); Suf, the addition of a sulfur group to a specific *E. coli* tRNA is dependent upon *red* (S. Adhya et al., pers. comm.).

S. Adhya et al. (unpubl.) have shown that expressions of the *red* genes and another as yet unidentified gene between the *bio*7-20 and *b*522 endpoints (Fig. 3) cause a specific change in a tRNA. Point mutations in *exo* or *bet* eliminate the tRNA effect, as does the *b*522 deletion. The *b*522 deletion still retains full recombination activity and the other phenotypes associated with red^{+} phage. Mapping and functional analyses indicate that tRNA modification is not dependent upon functions outside the p_L operon. An additional, perhaps unrelated, effect of *red* on tRNAs has been reported by Brégégère (1974, 1976).

Hin

λ Hin (*h*ost *in*hibition) function exerts broad pleiotropic effects upon the cell. (1) It reduces the uptake of several precursors (e.g., uridine) for macromolecular synthesis. (2) It reduces cAMP levels in the cell (Wu et al. 1971) and thereby affects expression of genes under the control of cAMP and its receptor protein, Crp. (3) It causes a small increase in ppGpp levels (twofold to threefold). (4) It enhances preferentially the stability of the p_L transcript. No evidence is yet available for the mechanism of action of the Hin function; however, all of the observed effects might be explained by an interaction with the cell membrane (see Court et al. 1980a,b).

Analysis of several deletions in the p_L operon defines two regions of the operon involved in the Hin phenotype (Court et al. 1980a, and unpubl.). Region A is defined by deletions *bio*7-20 and *bio*69 (Court et al. 1980a). Region B is between the *xis* gene and Δ*red*15 (Fig. 3; D. Court et al., unpubl.). The two regions can function independently, i.e., mutations in both *hinA* and *hinB* are required to observe the Hin^{-} phenotype. Both regions are relatively barren of genetic markers.

Region A is known to contain the *redA* gene for λ exonuclease and a small gene encoding an 8.5-kD protein of unknown function (C. Epp and M.L. Pearson, in prep.). Amber mutations in the *redA* gene have no effect on the Hin phenotype (Court et al. 1980a). Likewise, deletion of the 8.5-kD-encoding gene with the *b*522 deletion does not eliminate the Hin phenotype (D. Court et al., unpubl.). The *b*522 deletion leaves the *redA* gene intact, as well as 300 bp of DNA beyond *redA* (Luk and Szybalski 1982). In this interval, an open reading frame coding for 62 amino acids is present. The t_{L3} transcription terminator (Luk and Szybalski 1982) is located within this coding sequence. The *hinA* gene is likely to be the coding sequence between *b*522 and *redA*. However, as *cis-trans* tests have not been done, transcription termination at t_{L3} may be responsible for the Hin phenotype. Simple point mutations must be isolated before an unambiguous identification of the gene(s) can be made.

Two proteins of 22 kD and 8.5 kD are encoded by region B (Ausubel et al. 1971; Hendrix 1971; C. Epp and M.L. Pearson, in prep.). However, it is not known which, if either, is responsible for Hin.

Earlier work on Hin (Cohen and Chang 1970, 1971) had been interpreted to mean that p_L functions (namely *redA, redB,* and *gam*) were required for the Hin phenotype and that the primary effect was on general macromolecular synthesis. Now, neither of these interpretations appears correct. More definitive mapping (see below) indicates that these genes are not involved, and the apparent reduction in general synthesis observed is, in fact, a reduction in precursor uptake by the cell.

Also, Cohen and Chang (1971) had studied λ infection of *E. coli* P2 lysogens. Just as in infection of nonlysogens, there is an effect on β-galactosidase synthesis and a general host inhibition; however, in P2 lysogens the effect is magnified. They concluded that the phenomena of Hin and P2 interference (Spi) were interrelated. This interpretation is probably wrong, since P2 interference is almost certainly dependent upon a complex interaction between the products of the *red, gam,* and replication genes of λ with the *old*-gene product of P2 (Lindahl et al. 1970; Zissler et al. 1971b; Brégégère 1974), whereas the Hin phenotype is caused by genes located between *red* and *xis* (Fig. 3; Court et al. 1980a,b).

int and *xis* Genes

The genes *int* and *xis* are the last two genes in the p_L operon before *att* and the *b* region. They are both involved in site-specific recombination between phage and host DNA. These two functions are discussed in detail by Weisberg and Landy (this volume).

EXCLUSION SYSTEMS OF THE λ PROPHAGE

In the lysogenic state, a λ prophage expresses at least three different functions involved in exclusion of superinfecting phages. One of these, the λ repressor, is specific for λ; it prevents vegetative development of the prophage as well as all infecting homologous phages. Two other systems, *rex* (Howard 1967) and *sieB* (*s*uper*i*nfection *e*xclusion) (Susskind and Botstein 1980), express functions that exclude several unrelated phages, including lambdoid phages. Whereas the mechanism of action of the λcI repressor has been studied exhaustively and is well understood (see Gussin et al., this volume), our knowledge of how *rex* and *sieB* exclude superinfecting phage is rudimentary.

rex

E. coli lysogenic for λ prevents the growth of *r*II mutations of T4 phage (Benzer 1955). It does not block wild-type T4 growth. Several *rex*⁻ mutant prophages have now been isolated that lack exclusion (Howard 1967; Gussin and Peterson

1972). The mutations map in the immunity region next to the *c*I gene (Gussin and Peterson 1972). Recently, an analysis of the DNA sequence of the *rex* region (Landsmann et al. 1982) and complementation studies between *rex* mutations (K. Matz et al., in prep.) demonstrated that *rex* is composed of two adjacent cistrons, *rexA* and *rexB* (see Fig. 4). Phage 434 can also exclude heterologous phage. This exclusion is mediated by the gene *hex*, located in the immunity region. It is analogous to but does not complement either of the λ*rex* genes (R. Yocum and M. Susskind, pers. comm.; K. Matz et al., in prep.).

Several phages are excluded by rex^+ prophages; these include ϕ80 (Toothman and Herskowitz 1980a) in *E. coli*, as well as P22 and L in *Salmonella* (Susskind and Botstein 1980). Other phages, including λ itself, are resistant to exclusion only because they carry protective genes. When these genes are lost by mutation, the phage becomes sensitive; i.e., T4 *r*II (Benzer 1955), T5 *lr* (Jacquemin-Sablon and Lanni 1973), T7 *rbl* or gene *20* (Pao and Speyer 1975), and λ*red* or λ*ren* (Toothman and Herskowitz 1980b) mutations are all excluded by gp*rex*. In certain instances increased levels of gp*rex* are required to exclude (Toothman and Herskowitz 1980a,b).

Although several phages and mutants have been studied that are sensitive to gp*rex*, there is little known about the molecular mechanism of exclusion. The most extensively studied examples of gp*rex* exclusion involve mutants of T4, defective in the *r*II genes, and mutants of λ, defective in the *red* and *ren* genes. In both of these phages, the character of the mutations that cause sensitivity and the physiology of the abortive infections provide information about the exclusion system.

After infection of cells containing gp*rex*, normal development of the T4 and λ mutants continues for 20 minutes; after this time, the symptoms of exclusion become evident. Total macromolecular synthesis in the cell is reduced (protein, RNA, and DNA), and cellular respiration is inhibited. Inhibition of phage macromolecular synthesis is particularly severe and, unlike host syntheses, it does not recover during the abortive infection (Sauerbier et al. 1969; Toothman and Herskowitz 1980c). These symptoms do not occur if the infecting phage is defective in its own DNA replication (Toothman and Herskowitz 1980c) or if the host is defective in a membrane component, *tolC* (Campbell and Rolfe 1975).

Other evidence also suggests that Rex-dependent exclusion is involved with or triggered by DNA replication. This comes primarily from the isolation and analysis of λ mutants sensitive to Rex. The mutants fall into two groups: *red* and *ren*. λ*red* mutants, defective in general recombination, show reduced rates of DNA synthesis in wild-type hosts (Skalka 1974). In this regard, Red function may protect λ from Rex-dependent exclusion by removing intermediates sensitive to Rex or repairing DNA damaged by Rex. Another class of mutants, *ren*, which are represented by both point and deletion mutations (see below, *P* to *Q* Interval), have no known defect other than sensitivity to gp*rex*. A link between *ren* and DNA replication is again hinted at by the map position of *ren*, which is adjacent to the λ replication segment. Both *red* and *ren* mutations can be com-

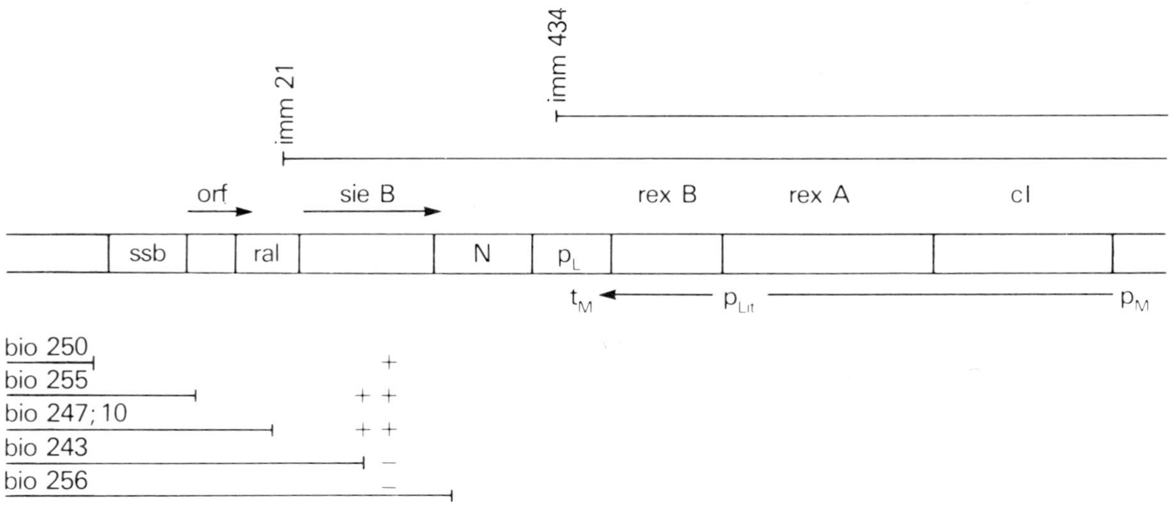

Figure 4 Exclusion genes. The λ exclusion genes are clustered in and near the immunity region of the phage. These genes are indicated (***sieB***, ***rexB***, ***rexA***, *c*I). The direction of their transcripts is shown by arrows. No promoter for the ***sieB*** gene is known. The terminator, t_M, and promoters p_M and p_{LIT} are in the immunity region (*imm*). The genes *ssb*, *ral*, and *N* are part of the p_L operon (see Fig. 3). Phages 21 and 434 recombine and form hybrids with λ. Hybrid phages with immunity (*imm*) of 21 or 434 have DNA nonhomologies in the region indicated by the horizontal lines above the gene map. The exact left end of imm^{21} is not known; it recombines with *bio*10 but not *bio*243. Different *bio* substitutions and their deletion extents are shown below the gene map (also see Fig. 3). The SieB phenotype is exhibited by λ prophages in *Salmonella*, including some *bio* substitution mutants. (+) Exclusion of temperate phage L; (+ +) Exclusion of both L and P22 phages; (−) $SieB^-$ where neither phage is excluded (Susskind and Botstein 1980). An open reading frame (orf) of 354 bp exists near the ends of the *bio*250 and *bio*255 deletions (Ineichen et al. 1981). Perhaps this orf is a gene involved in ***sieB*** expression, since *bio*250 and shorter deletions exclude only phage L; whereas *bio*255, *bio*247, and *bio*10 exclude L and P22 phages.

plemented by wild type for growth in the presence of gp*rex*; this suggests that these genes code for diffusible elements (Toothman and Herskowitz 1980b). Since the defect of a deletion mutant in either gene is less severe than a combined double mutant, the two functions may affect independent pathways of protection against gp*rex*.

Phage P22 is naturally sensitive to Rex, whereas λ is resistant (Susskind and Botstein 1980; Toothman and Herskowitz 1980a). Hybrid recombinants between λ and P22 exist (Botstein and Herskowitz 1974); some are sensitive to exclusion. Sensitive ones contain the *P-Q* region of P22 (genes *18* and *12*) but are complemented for growth by λ*ren*$^{+}$ helper phage (Toothman and Herskowitz 1980b).

The *c*I-*rex* region is subject to complex regulation. Transcription from two promoters, p_E and p_M, ensures coordinate expression of *c*I and *rex*. Establishment of *c*I and *rex* expression following infection requires the activation of p_E by the *c*II-gene product (Astrachan and Miller 1972; Belfort 1978). Maintenance of *c*I and *rex* expression in the immune lysogen requires activation of the p_M promoter by *c*I repressor (Yen and Gussin 1973; Lieb and Talland 1981). Detailed descriptions of these processes are given by Wulff and Rosenberg and Gussin et al. (both this volume). A third promoter, called p_{LIT}, is located at the end of the *rexA* gene (Pirrotta et al. 1980; U. Schmeissner, pers. comm.). Transcription from p_{LIT} should allow expression of only *rexB* (Hayes and Szybalski 1973; Landsman et al. 1982). Auxiliary factors may be required to activate this promoter, since RNA polymerase binds but does not initiate RNA synthesis here (Pirrotta et al. 1980). Hayes and Sybalski (1973) speculated that a transcription termination site was present beyond *rex*. This was deduced from "hybridization mapping" the end of the in vivo transcript. A sequence containing a U-rich region beyond a stem and loop is found just beyond the *rexB* gene (Landsman et al. 1982). This is the standard structure of a ρ-independent terminator (Rosenberg and Court 1979).

The purpose of three promoters for the *rex* genes is not understood. Preliminary data suggest that the p_{LIT} promoter is active and gp*rexB* is made in a repressed prophage defective in the p_M promoter (K. Matz et al., in prep.). The p_{LIT} promoter is also active during lytic development after prophage induction (Hayes and Szybalski 1973). This raises the possibility that *rexB* is transcribed from p_{LIT} but not from p_E and p_M. Mutations in the p_{LIT} promoter would allow resolution of this question.

sieB

The *sieB* function resembles *rex* exclusion in that it affects general macromolecular synthesis. This exclusion is observed only in *S. typhimurium* hosts lysogenic for either λ or P22 phages. Such lysogens exclude the development of several temperate *S. typhimurium* phages (Susskind et al. 1974; Susskind and Botstein

1980). On the basis of deletion analysis, the *sieB* gene is located between *ral* and *N* of λ. A similar location is found for *sieB* in P22 prophages (Susskind and Botstein 1980). The nucleotide sequence of this region reveals an open reading frame able to produce a protein of 21 kD and encompassing nearly the entire segment between *ral* and *N* (Ineichen et al. 1981). Interestingly, for this gene to be expressed, transcription must be rightward on λ, which is opposite to that of the p_L transcript. Since *sieB* is expressed in the prophage state and p_L is repressed, a promoter must be active in the repressed prophage to the left of *sieB*. In this regard some regulatory site or function appears to be located to the left of *sieB*, since deletions of this region affect exclusion (Susskind and Botstein 1980). More experiments will be required to understand the regulation of this gene and whether it is expressed in *E. coli*.

P TO *Q* INTERVAL

Approximately 7% of the λ genome lies between genes *P* and *Q*. Although no essential genes are found here, mutations in this region have revealed four mutant phenotypes (see Table 2). We discuss separately each of these phenotypes and the mutations that confer them.

N-independence (Nin)

Normally the *N* function is required for λ to grow lytically. In the absence of *N*, transcription termination occurs at different terminators. One of these, t_{R2}, is

Table 2 Phenotypes associated with the *P-Q* region

Mutation	Nin (*N* independence)[a]	Puq (*Q* overproduction)[b]	Ren (*rex* exclusion)[c]	Pas (*polA* suppression)[d]
*nin*5	+	(+)	+	+
*byp*1	(+)	+	−	−
*puq*3	−	+	0	+
*ren*20	−	0	+	−
*pasB*2	0	0	0	+

+ indicates that the mutant phenotype is observed with the mutation; − indicates that it is not observed; (+) indicates an intermediate effect; 0 indicates that the experiment has not yet been done.

[a]The ability of λ to form plaques in the absence of the *N*-gene product (Court and Sato 1969; Butler and Echols 1970; Hopkins 1970).

[b]A *Q* amber mutation cannot form plaques on a host with a weak suppressor, i.e., an ochre suppressor. Increasing the level of *Q* permits plaque formation on such strains (Sternberg and Enquist 1979).

[c]λ lysogens produce *c*I repressor and *rex*. Normal heteroimmune phage form plaques on λ lysogens, but certain mutants in the *P-Q* region do not plaque because of *rex* (Susskind and Botstein 1980; Toothman and Herskowitz 1980b).

[d]Some λ*bio* substitution mutations cannot form plaques on *pol*A hosts because the phage lack the *red* and *gam* genes. The *pas* mutations suppress this defect and allow plaque formation (Shizuya et al. 1974).

located between *P* and *Q*. In the absence of *N*, it blocks *Q* transcription. The deletion mutation, *nin*5, which removes t_{R2} as judged by in vitro transcription (Roberts 1975), allows *N* mutations to form plaques (Court and Sato 1969). A point mutation, *byp*, is situated between *nin*5 and *Q* (Sternberg and Enquist 1979). It also allows *N*-independent transcription beyond the *P-Q* region (Butler and Echols 1970; Hopkins 1970), but unlike *nin*5 it does not permit plaque formation in the absence of *N*. This suggests that the two mutations may affect transcription differently. Presumably, *byp* does not eliminate the t_{R2} terminator, since *byp* is located beyond *nin* and, presumably, t_{R2}. How then does *byp* increase *Q* transcription in the absence of *N*? Several possibilities exist: *byp* may inactivate a second terminator beyond t_{R2}; the mutation may create a new promoter that initiates transcription into the *Q* gene (although no constitutive expression of *Q* has been detected from λ*byp* mutations [Butler and Echols 1970; Sternberg and Enquist 1979]); *byp* may inactivate a promoter that transcribes toward p_R, thereby reducing convergent transcription and allowing increased rightward transcription through the *P-Q* region; finally, *byp* may increase the rate of *Q* translation from the p_R transcript. As yet, in vitro experiments to test transcription termination on *byp* DNA have not been reported, nor has the sequence change of the mutation been determined.

Q Overproduction (Puq)

The *Q*-gene product is required to activate transcription of all λ late genes (Dove 1966). Two results suggest that during normal λN^+ infections, gp*Q* levels in the cell are limiting for late gene expression: (1) Amber mutations in *Q* are not suppressed by ochre suppressors (Thomas 1967), and (2) the *Q* product does not function well in *trans* to activate expression of the late genes (Echols et al. 1976). Point mutations have been isolated that enhance the level of gp*Q* expression (Sternberg and Enquist 1979). These mutations, called *puq*, are located in the 2.8-kb DNA segment removed by the *nin*5 deletion. In addition, both *nin*5 and *byp* exhibit the Puq phenotype. The *puq* mutations, unlike *nin*5 and *byp*, do not show *N*-independent expression of *Q* (Table 2). With respect to the Puq phenotype, the *byp* mutant is stronger than *puq* mutations, which are stronger than the *nin*5 deletion (Sternberg and Enquist 1979). Thus, there are at least three different types of mutation to enhance gp*Q* levels. The mechanism by which the *puq* mutations increase late gene expression is not understood. Any explanation of the effect of *puq* must take into account the fact that gp*N* is required to observe this effect.

polA Suppression (PasB)

λ*red* and λ*gam* mutations are unable to form plaques on strains with defects in the *polA* gene. (Zissler et al. 1971a). λ*bio*-transducing phages, deleted for both *red* and *gam*, also fail to plaque on *polA*. Selection of pseudorevertants on *polA* strains yields phages that carry secondary mutations (*pasB*) located in the *P-Q* region (Shizuya et al. 1974). The *nin*5 deletion and the *puq* mutations confer a PasB phenotype to λ (Sternberg and Enquist 1979; Toothman and Herskowitz 1980b). Examination of published results (Table 2) reveals that *puq* and *pasB* mutations are thus far not distinguishable and may be the same mutation.

A Chi site appears to be required for the PasB suppression. This is inferred from the finding that *nin*5 and *puq* can suppress a *bio* substitution mutation, which happens to contain a Chi site, but not *red gam* point mutations, which lack such a site. Chi sites enhance *recBC*-dependent general recombination between λ DNA molecules (Henderson and Weil 1975; Malone et al. 1978; Smith, this volume).

The products of the genes *red* and *gam* are each thought to have a role in the formation of λ DNA concatemers, which are essential in the maturation and packaging of λ DNA. The *red* system may generate multimer length DNA directly by recombination or by generation of new replication forks (Skalka 1974). The *gam*-gene product has an indirect effect on concatemer formation. Its primary action is to inhibit the host *recBC* nuclease, which is known to degrade λ concatemers. The degradation of λ DNA by *recBC* nuclease is enhanced in *polA*-defective strains. It is believed that *polA* product repairs nuclease-sensitive intermediates in the normal replication of λ concatemers (Skalka 1974). The λ*bio* mutations, deleted for *red* and *gam*, are thus very sensitive to nuclease digestion in *polA*-defective strains. How might the *pasB* mutation decrease this sensitivity? The *P-Q* region is involved in DNA replication and, possibly, the switch of λ replication from a monomer circle to a rolling circle. In this light the *pasB* mutation (as well as *puq* or *nin*5) may alter a site for rolling-circle replication that is particularly sensitive to RecBC nuclease degradation. The requirement for Chi then is to generate new types of concatemers by using the same RecBC nuclease to generate recombinant dimers. However, several experiments remain to be done before accurate predictions can be made as to how *pasB* works. For example, a simple *cis-trans* test would determine whether *pasB* encodes a diffusible function or is a site.

rex Exclusion (Ren)

A gene product in the *P-Q* region protects λ phage from growth inhibition by gp*rex* (see above, Exclusion Systems of the λ Prophage). Mutations in this gene, *ren*, map between the *P* gene and the *nin*5 deletion (Toothman and Herskowitz 1980b). Part of the *ren* gene may be deleted in *nin*5, since *nin* is also Ren$^-$ (Table 2; Fig. 5). Complementation tests suggest that gp*ren* is a diffusible function (Toothman and Herskowitz 1980b). The DNA sequence immediately to the right of *P* contains an open reading frame of 288 bp (see Appendix II). The left end of the *nin*5 deletion is also located within this DNA segment (Kröger and Hobom 1982)

The function of *ren* is unclear; however, it may be linked to phage DNA

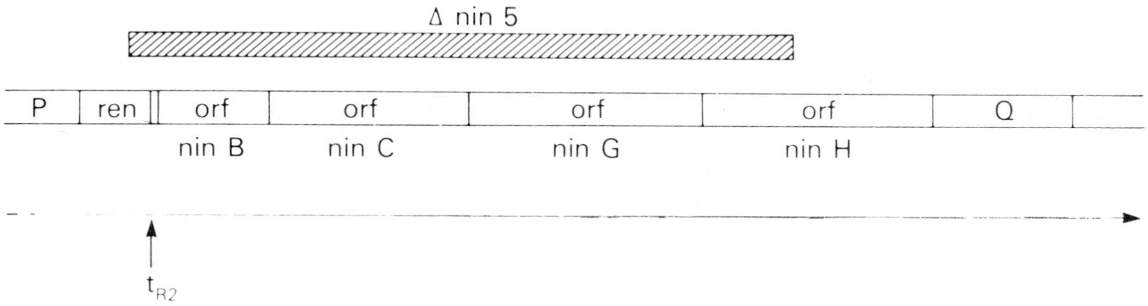

Figure 5 *P-Q* Region. The DNA between the *P* and *Q* genes of λ contains 3600 bp and five open reading frames (orf) able to encode proteins of at least 90 amino acids (Kröger and Hobom 1982) (see Appendix II). Only one functional gene, *ren*, is known to be located between *P* and *Q*, and it is the orf farthest to the left next to gene *P* (see text). The *nin*5 is a deletion mutation (Court and Sato 1969) caused by recombination between two 12-bp repeats separated by 2800 bp (Kröger and Hobom 1982). The arrow indicates direction of transcription from the p_R promoter (not shown). A ρ-independent termination site, t_{R2}, is just beyond *ren* (Roberts 1975; Kröger and Hobom 1982). Other terminators may also be located under the *nin* deletion. Potential genes for the orfs have been designated *nin* and are shown below the map (Kröger and Hobom 1982).

metabolism. Whereas λ*ren* mutants fail to form plaques on cells containing gp*rex*, certain *O* and *P* mutations suppress the defect of *ren* (Toothman and Herskowitz 1980b). The *O* and *P* genes function in DNA replication and are adjacent to *ren*. These three genes may be part of a replication cluster. In general, *ren* product may work by either inhibiting gp*rex* or by altering the replication of λ DNA so as to make it resistant to gp*rex* effects. Toothman and Herskowitz (1980b) have speculated that gp*ren* may be involved in the switch from early monomer to rolling-circle replication.

In summary, it appears that in addition to the *O* and *P* genes that are essential for λ DNA replication, auxiliary functions in the *P-Q* interval may be involved. In the pathway to mature virus formation, λ replication, late gene expression, and DNA packaging are closely coupled. In this respect mutations that increase DNA replication and, in particular, DNA concatemer formation may in turn cause elevated late gene expression and DNA packaging. Conversely, mutations that enhance late gene expression may salvage more efficiently a limited source of DNA concatemers. Mutations such as *nin*5 and *puq*3 affect late gene expression (Puq) and replication (PasB), whereas the *byp* mutation confers only the Puq phenotype on late gene expression (Table 2). These observations suggest that if these mutations have primary effects on only one function, i.e., replication or late gene expression, then it is replication that has the pleiotropic effect upon the other.

In conclusion, we still know very little about the *P-Q* interval. Recent sequence information (see Appendix II; Kröger and Hobom 1982) shows that there are several open reading frames (potential genes) in this segment of DNA. Obviously, more investigation will be required to fully understand the functions of each of these genes.

ACKNOWLEDGMENTS

We thank our colleagues who have sent us their unpublished results. We extend special thanks to I. Herskowitz and R. Hendrix, who took the time to read and comment on the manuscript during its preparation. Grants from the United States-Israel Binational Science Foundation (BSF) and the Israeli Academy of Sciences helped to support this effort.

REFERENCES

Adhya, S., M. Gottesman, B. de Crombrugghe, and D. Court. 1976. Transcription termination regulates gene expression. In *RNA polymerase* (ed. R. Losick and M. Chamberlin), p. 719. Cold Spring Harbor Laboratory, Cold Spring Harbor, New York.

Astrachan, L. and J.F. Miller. Regulation of λ*rex* expression after infection of *Escherichia coli* K by lambda bacteriophage. *J. Virol.* **9:** 510.

Ausubel, F., P. Voynow, E. Signer, and J. Mistry. 1971. Purification of proteins determined by two nonessential genes in lambda. In *The bacteriophage lambda* (ed. A.D. Hershey), p. 395. Cold Spring Harbor Laboratory, Cold Spring Harbor, New York.

Belfort, M. 1978. Anomalous behaviour of bacteriophage λ polypeptides in polyacrylamide gels: Resolution, identification, and control of the *rex* gene product. *J. Virol.* **28:** 270.

Benzer, S. 1955. Fine structure of a genetic region in bacteriophage. *Proc. Natl. Acad. Sci.* **41:** 344.

Botchan, P. 1976. An electron microscopic comparison of transcription on linear and superhelical DNA. *J. Mol. Biol.* **105:** 161.

Botstein, D. and I. Herskowitz. 1974. Properties of hybrids between Salmonella phage P22 and coliphage λ. *Nature* **251:** 584.

Bøvre, K. and W. Szybalski. 1969. Patterns of convergent and overlapping transcription in the *b*2 region of coliphage. *Virology* **38:** 614.

Boyer, H.W. and D. Roulland-Dussoix. 1969. A complementation analysis of the restriction and modification of DNA in *Escherichia coli. J. Mol. Biol.* **41:** 459.

Brégégère, F. 1974. Bacteriophage P2-λ interference: Inhibition of protein synthesis involves transfer RNA inactivation. *J. Mol. Biol.* **90:** 459.

——. 1976. Bacteriophage P2-lambda interference. II. Effects on the host under the control of lambda genes O and P. *J. Mol. Biol.* **104:** 411.

Butler, B. and H. Echols. 1970. Regulation of bacteriophage λ development by gene N: Properties of a mutation that bypasses N control of late protein synthesis. *Virology* **40:** 212.

Campbell, J.H. and B.G. Rolfe. 1975. Evidence for a dual control of the initiation of host-cell lysis caused by phage lambda. *Mol. Gen. Genet.* **139:** 1.

Chowdhury, M.R., S. Dunbar, and A. Becher. 1972. Induction of an endonuclease by some substitution and deletion variants of phage λ. *Virology* **49:** 314.

Cohen, S.N. and A.C.Y. Chang. 1970. Genetic expression in bacteriophage λ. I. Inhibition of *Escherichia coli* nucleic acid and protein synthesis during λ development. *J. Mol. Biol.* **49:** 557.

——. 1971. Genetic expression in bacteriophage λ. IV. Effects of P2 prophage on λ inhibition of host synthesis and λ gene expression. *Virology* **46:** 397.

Court, D. and K. Sato. 1969. Studies of novel transducing variants of lambda: Dispensability of genes N and Q. *Virology* **39:** 348.

Court, D., M. Gottesman, and M. Gallo. 1980a. Bacteriophage lambda Hin function. I. Pleiotropic alteration in host physiology. *J. Mol. Biol.* **138:** 715.

Court, D., B. de Crombrugghe, S. Adhya, and M. Gottesman. 1980b. Bacteriophage lambda Hin function. II. Enhanced stability of lambda messenger RNA. *J. Mol. Biol.* **138:** 731.

Das, A., D. Court, and S. Adhya. 1979. Pleiotropic effect of rho mutation in *Escherichia coli.* In Molecular basis of host virus interaction (ed. M. Chakravorty), p. 459. Science Press, Princeton.

Davis, R.W. and J.S. Parkinson. 1971. Deletion mutants of bacteriophage lambda. III. Physical structure of *att. J. Mol. Biol.* **56:** 403.

Debrouwere, L., M. Van Montagu, and J. Schell. 1980a. The ral gene of phage λ. III. Interference with *E. coli* ATP dependent functions. *Mol. Gen. Genet.* **179:** 81.

Debrouwere, L., M. Zabeau, M. Van Montagu, and J. Schell. 1980b. The ral gene of phage λ. II. Isolation and characterization of ral-deficient mutants. *Mol. Gen. Genet.* **179:** 75.

Dove, W.F. 1966. Action of the λ chromosome. I. Control of functions late in bacteriophage development. *J. Mol. Biol.* **19:** 187.

Drahos, D. and R. Hendrix. 1982. Effect of bacteriophage λ infection on synthesis of *gro*E protein and other *Escherichia coli* proteins. *J. Bacteriol.* **149:** 1050.

Drahos, D. and W. Szybalski. 1981. Antitermination and termination functions of the cloned *nut*L, *N*, and t_{L1} modules of coliphage lambda. *Gene* **16:** 261.

Dussoix, D. and W. Arber. 1962. Host specificity of DNA produced by *E. coli.* II. Control over acceptance of DNA from infecting phage lambda. *J. Mol. Biol.* **5:** 37.

Echols, H., D. Court, and L. Green. 1976. On the nature of *cis*-acting regulatory proteins and genetic organization in bacteriophage: The example of gene *Q* of bacteriophage λ. *Genetics* **83:** 5.

Eisen, H., M. Georgiou, C.P. Georgopoulos, S. Selzer, G. Gussin, and I. Herskowitz. 1975. The role of gene *cro* in phage λ development *Virology* **68:** 266.

Epp, C. 1978. "Early protein synthesis and its control in bacteriophage lambda." Ph.D. thesis, University of Toronto, Canada.

Fiandt, M., Z. Hradecna, H.A. Lozeron, and W. Szybalski. 1971. Electron micrographic mapping of deletions, insertions, inversions, and homologies in the DNAs of coliphages lambda and phi 80. In *The bacteriophage lambda* (ed. A.D. Hershey), p. 329. Cold Spring Harbor Laboratory, Cold Spring Harbor, New York.

Forbes, D. and I. Herskowitz. 1983. Polarity suppression by the *Q* gene product. *J. Mol. Biol.* **160:** 549.

Georgiou, M., H. Eisen, and C.P. Georgopoulos. 1979. An analysis of the Tro phenotype of bacteriophage λ. *Virology* **94:** 38.

Gold, L., P.Z. O'Farrell, and M. Russel. 1976. Regulation of gene 32 expression during bacteriophage T4 infection of *Escherichia coli*. *J. Biol. Chem.* **251:** 7251.

Gottesman, M.E., S.L. Adhya, and A. Das. 1980. Transcription antitermination by bacteriophage lambda *N* gene product. *J. Mol. Biol.* **140:** 57.

Grayhack, E.J. and J.W. Roberts. 1982. The phage λ *Q* gene product: Activity of a transcription antiterminator *in vitro*. *Cell* **30:** 637.

Greer, H. 1975a. The *kil* gene of bacteriophage lambda. *Virology* **66:** 589.

——. 1975b. Host mutants resistant to phage lambda killing. *Virology* **66:** 605.

Greer, H. and F.M. Ausubel. 1979. Radiochemical identification of the *kil* gene product of bacteriophage λ. *Virology* **95:** 577.

Gussin, G.N. and V. Peterson. 1972. Isolation and properties of rex⁻ mutants of bacteriophage lambda. *J. Virol.* **10:** 760.

Hayes, S. and W. Szybalski. 1973. Control of short leftward transcripts from the immunity and *ori* regions in induced coliphage lambda. *Mol. Gen. Genet.* **126:** 275.

Henderson, D. and J. Weil. 1975. Recombination-deficient deletions in bacteriophage λ and their interaction with *chi* mutations. *Genetics* **79:** 143.

Hendrix, R.W. 1970. "Proteins of bacteriophage lambda." Ph.D. thesis, Harvard University, Cambridge, Massachussetts.

——. 1971. Identification of proteins coded by phage lambda. In *The bacteriophage lambda* (ed. A.D. Hershey), p. 355. Cold Spring Harbor Laboratory, Cold Spring Harbor, New York.

Hoess, R.H., C. Foeller, K. Bidwell, and A. Landy. 1980. Site-specific recombination function of bacteriophage λ: DNA sequence of regulatory regions and overlapping structural genes for Int and Xis. *Proc. Natl. Acad. Sci.* **77:** 2482.

Honigman, A. 1981. Cloning and characterization of a transcription on termination signal in bacteriophage λ unresponsive to the *N* gene product. *Gene* **13:** 299.

Honigman, A., A.B. Oppenheim, B. Hohn, and T. Hohn. 1981. Plasmid vectors for positive selection of DNA inserts controlled by the λ p_L promoter, repressor, and antitermination function. *Gene* **13:** 289.

Hopkins, N. 1970. Bypassing a positive regulator: Isolation of a λ mutant that does not require N product to grow. *Virology* **40:** 223.

Howard, B.D. 1967. Phage lambda mutants deficient in *r*II exclusion. *Science* **158:** 1588.

Ineichen, K., J.C.W. Shepherd, and T.A. Bickle. 1981. The DNA sequence of the phage lambda genome between p_L and the gene *bet*. *Nucleic Acids Res.* **9:** 4639.

Ippen, K., J.A. Shapiro, and J.R. Beckwith. 1971. Transposition of the *lac* region of the *Escherichia coli* chromosome: Isolation of λ*lac* transducing bacteriophages. *J. Bacteriol.* **108:** 5.

Jacquemin-Sablon, A. and Y.T. Lanni. 1973. Lambda-repressed mutants of bacteriophage T5. I. Isolation and genetical characterization. *Virology* **56:** 230.

Karu, A., Y. Sakaki, H. Echols, and S. Linn. 1974. *In vitro* studies of the *gam* gene product of bacteriophage λ. In *Mechanisms in recombination* (ed. R.F. Grell), p. 95. Plenum Press, New York.

Kellenberger, G., M.L. Zichichi, and J. Weigle. 1961. A mutation affecting the DNA content of bacteriophage lambda and its lysogenizing properties. *J. Mol. Biol.* **3:** 399.

Kravchenko, V.V., S.K. Vassilanko, and M.A. Grachev. 1979. A rightward promoter to the left of the *att* site of λ phage DNA: Possible participant in site specific recombination. *Gene* **7:** 181.

Krell, K., M.E. Gottesman, J.S. Parks, and M.A. Eisenberg. 1972. Escape synthesis of the biotin operon in induced λ*b2* lysogens. *J. Mol. Biol.* **68:** 69.

Kröger, M. and G. Hobom. 1982. The *nin* region of bacteriophage lambda: A chain of 9 interlinked genes. *Gene* **20:** 25.

Landsmann, J., M. Kröger, and G. Hobom. 1982. The *rex* region of bacteriophage lambda: Two genes under threefold regulatory control. *Gene* **20:** 11.

Lieb, N. and C. Talland. 1981. IS elements in bacteriophage lambda gene *c*I prevent expression of gene *rex*. *J. Virol.* **38:** 789.

Lindahl, G., H. Sironi, H. Bialy, R. Calender. 1970. Bacteriophage λ abortive infection of bacteria lysogenic for phage P2. *Proc. Natl. Acad. Sci.* **66:** 587.

Little, J.W., I.R. Lehman, and A.D. Kaiser. 1967. An exonuclease induced by bacteriophage λ. I. Preparation of the crystalline enzyme. *J. Biol. Chem.* **242:** 672.

Luk, K.-C. and W. Szybalski. 1982. Transcription termination: Sequence and function of the rho-independent t_{L3} terminator in the major leftward operon of bacteriophage lambda. *Gene* **17:** 247.

Malamy, M.H., M. Fiandt, and W. Szybalski. 1972. Electron microscopy of polar insertions in the *lac* operon of *Escherichia coli*. *Mol. Gen. Genet.* **119:** 207.

Malone, R.E., D.K. Chattoraj, D.H. Faulds, M.M. Stahl, and F.W. Stahl. 1978. Hotspots for generalized recombination in the *Escherichia coli* chromosome. *J. Mol. Biol.* **121:** 473.

Murialdo, H. and L. Siminovitch. 1972. The morphogenesis of bacteriophage lambda. IV. Identification of gene products and control of the expression of morphogenetic information. *Virology* **48:** 785.

Nijkamp, H.J.J., K. BØvre, and W. Szybalski. 1971. Regulation of leftward transcription in the *J-b2-att* region of coliphage lambda. *Mol. Gen. Genet.* **111:** 11.

Oppenheim, A. and A.B. Oppenheim. 1978. Regulation of the *int* gene of bacteriophage: Activation by the *c*III gene products and the role of the p_I and p_L promoters. *Mol. Gen. Genet.* **165:** 39.

Oppenheim, A., S. Gottesman, and M. Gottesman. 1982. Regulation of *int* gene expression. *J. Mol. Biol.* **158:** 327.

Oppenheim, A., M. Belfort, N. Katzir, N. Kass, and A.B. Oppenheim. 1977. Interaction of *c*II, *c*III, and *cro* gene products in the regulation of early and late functions of phage. *Virology* **79:** 426.

Pao, C.C. and J.F. Speyer. 1975. Mutants of T7 bacteriophage inhibited by lambda prophage. *Proc. Natl. Acad. Sci.* **72:** 3642.

Pirrotta, V., K. Ineichen, and A. Walz. 1980. An unusual RNA polymerase binding site in the immunity region of phage lambda. *Mol. Gen. Genet.* **180:** 360.

Radding, C.M., J. Rosenzweig, F. Richards, and E. Cassuto. 1971. Separation and characterization of exonuclease, β protein, and a complex of both. *J. Biol. Chem.* **246:** 2510.

Reeve, J.N. and J.E. Shaw. 1979. Lambda encodes an outer membrane protein: The *lom* gene. *Mol. Gen. Genet.* **172:** 243.

Rhodes, M. and M. Meselson. 1973. An endonuclease induced by bacteriophage λ. *J. Biol. Chem.* **248:** 521.

Roberts, J.W. 1975. Transcription termination and late control in phage lambda. *Proc. Natl. Acad. Sci.* **72:** 330.

Rosenberg, M. and D. Court. 1979. Regulatory sequences involved in the promotion and termination of RNA transcription. *Annu. Rev. Genet.* **13:** 319.

Rosenvold, E.C., E. Calva, R.R. Burgess, and W. Szybalski. 1980. *In vitro* transcription from the *b2* region of bacteriophage λ. *Virology* **107:** 476.

Sakaki, Y., A.E. Karu, S. Linn, and H. Echols. 1973. Purification and properties of the γ-protein specified by bacteriophage λ: An inhibitor of the host *recBC* recombination enzyme. *Proc. Natl. Acad. Sci.* **70:** 2215.

Salstrom, J.S. and W. Szybalski. 1978a. Coliphage λ*nut*L⁻: A unique class of mutants defective in the site of gene N product utilization for antitermination of leftward transcription. *J. Mol. Biol.* **124:** 195.

——. 1978b. Transcription termination sites in the major leftward operon of coliphage lambda. *Virology* **88:** 252.

Sauerbier, W., S.M. Puck, A.R. Brautigam, and M. Hirsch-Kauffman. 1969. Control of gene function in bacteriophage T4. I. Ribonucleic acids and deoxynucleic acid metabolism in T4rII-infected lambda-lysogenic hosts. *J. Virol.* **4:** 742.

Shizuya, H., D. Brown, and A. Campbell. 1974. DNA polymerase I-dependent mutants of coliphage lambda. *J. Virol.* **13:** 947.

Shulman, M.J., L.M. Hallick, H. Echols, and E.R. Signer. 1970. Properties of recombination-deficient mutants of bacteriophage lambda. *J. Mol. Biol.* **52:** 501.

Skalka, A.M. 1974. A replicator's view of recombination (and repair). In *Mechanisms in recombination* (ed. R.F. Grell), p. 421. Plenum Press, New York.

Sternberg, N. and L. Enquist. 1979. Analysis of coliphage λ mutations that affect Q gene activity: Puq, byp, and nin5. *J. Virol.* **30:** 1.

Susskind, M.M. and D. Botstein. 1980. Superinfection exclusion by λ prophage in lysogens of *Salmonella typhimurium. Virology* **100:** 212.

Susskind, M.M., A. Wright, and D. Botstein. 1974. Superinfection exclusion by P22 prophage in lysogens of *Salmonella typhimurium.* IV. Genetics and physiology of sieB exclusion. *Virology* **62:** 367.

Szybalski, E.H. and W. Szybalski. 1979. A comprehensive molecular map of bacteriophage lambda. *Gene* **7:** 217.

Thomas, R., C. Leurs, C. Dambly, D. Parmentier, L. Lambert, P. Brachet, N. Lefebvre, S. Mousset, J. Porcheret, J. Szpirer, and D. Wauters. 1967. Isolation and characterization of new (sus) amber mutants of bacteriophage λ. *Mutat. Res.* **4:** 735.

Toothman, P. 1981. Restriction alleviation by bacteriophage lambda and lambda reverse. *J. Virol.* **38:** 621.

Toothman, P. and I. Herskowitz. 1980a. Rex-dependent exclusion of lambdoid phages. I. Prophage requirements for exclusion. *Virology* **102:** 133.

——. 1980b. Rex-dependent exclusion of lambdoid phages. II. Determinants of sensitivity to exclusion. *Virology* **102:** 147.

——. 1980c. Rex-dependent exclusion of lambdoid phages. III. Physiology of the abortive infection. *Virology* **102:** 161.

Unger, R.C. and A.J. Clark. 1972. Interaction of the recombination pathways of bacteriophage λ and its host *Escherichia coli* K12: Effects on exonuclease V activity. *J. Mol. Biol.* **70:** 539.

Vollenweider, H.J. and W. Szybalski. 1978. Electron microscopic mapping of RNA polymerase binding to coliphage lambda DNA. *J. Mol. Biol.* **123:** 485.

Vovis, G.F., K. Horiuchi, and N.D. Zinder. 1974. Kinetics of methylation of DNA by a restriction endonuclease from *Escherichia coli* B. *Proc. Natl. Acad. Sci.* **71:** 3810.

Wu, A.M., S. Ghosh, M. Willard, J. Davison, and H. Echols. 1971. Negative regulation by lambda: Repression of lambda RNA synthesis in vitro and host enzyme synthesis in vivo. In *The bacteriophage lambda* (ed. A.D. Hershey), p. 589. Cold Spring Harbor Laboratory, Cold Spring Harbor, New York.

Yen, K. and G. Gussin. 1973. Genetic characterization of a *prm*⁻ mutant of bacteriophage λ. *Virology* **56:** 300.

Zabeau, M., S. Freidman, M. Van Montagu, and J. Schell. 1980. The *ral* gene of phage λ. I. Identification of a non-essential gene that modulates restriction and modification in *E. coli. Mol. Gen. Genet.* **179:** 63.

Zissler, J., E. Signer, and F. Schaefer. 1971a. The role of recombination in growth of bacteriophage lambda. I. The gamma gene. In *The bacteriphage lambda* (ed. A.D. Hershey), p. 455. Cold Spring Harbor Laboratory, Cold Spring Harbor, New York.

——. 1971b. The role of recombination in growth of bacteriophage lambda. II. Inhibition of growth by prophage P2. In *The bacteriophage lambda* (ed. A.D. Hershey), p. 469. Cold Spring Harbor Laboratory, Cold Spring Harbor, New York.

Lambdoid Phage Head Assembly

Costa Georgopoulos, Kit Tilly,* and Sherwood Casjens
Department of Cellular, Viral, and Molecular Biology
University of Utah College of Medicine
Salt Lake City, Utah 84132

. . . the order and size of all the planets and spheres and heaven itself are so linked together that in no portion of it can anything be shifted without disrupting the remaining parts and the universe as a whole.
Copernicus: *De Revolutionibus*

INTRODUCTION

In bacteriophage morphogenesis, macromolecules are assembled into complex, elegant structures. The architecture of the finished virion is determined both by the characteristics of the structural proteins themselves and by the influence of other viral and host proteins that participate transiently in the assembly process. The interactions among these components can be dissected by physical and chemical analysis of purified viral proteins, assembly intermediates, and completed virus particles, as well as through genetic studies. Bacteriophage λ morphogenesis is a model for macromolecular assembly in general, because the virion is a relatively simple, well-defined structure that consists of a limited number of components and because the λ-*Escherichia coli* system is well understood and easily manipulated. An additional positive feature is the existence of related lambdoid phages (including *Salmonella* phage P22), whose morphogenetic pathways can be compared with that of λ to help derive information about evolutionarily conserved and variable elements in morphogenesis.

Much progress has been made since Kellenberger and Edgar (1971) discussed λ assembly in *The Bacteriophage Lambda*. We now know that the steps in bacteriophage λ head morphogenesis generally proceed in an obligate order, so that phage-coded proteins (with the help of the bacterial *groEL*- and *groES*-gene products) assemble into small aggregates, form large precursor structures called proheads, and finally combine with λ DNA and additional phage proteins to yield complete heads. We discuss the head assembly process in detail below, with the specific exclusion of the DNA packaging reaction, which has been reviewed elsewhere (Earnshaw and Casjens 1980; Feiss and Becker, this volume). The general topic of phage assembly has been ably reviewed recently by Wood and King (1979).

*Present address: Department of Biochemistry and Molecular Biology, Harvard University, Cambridge, Massachusetts 02138.

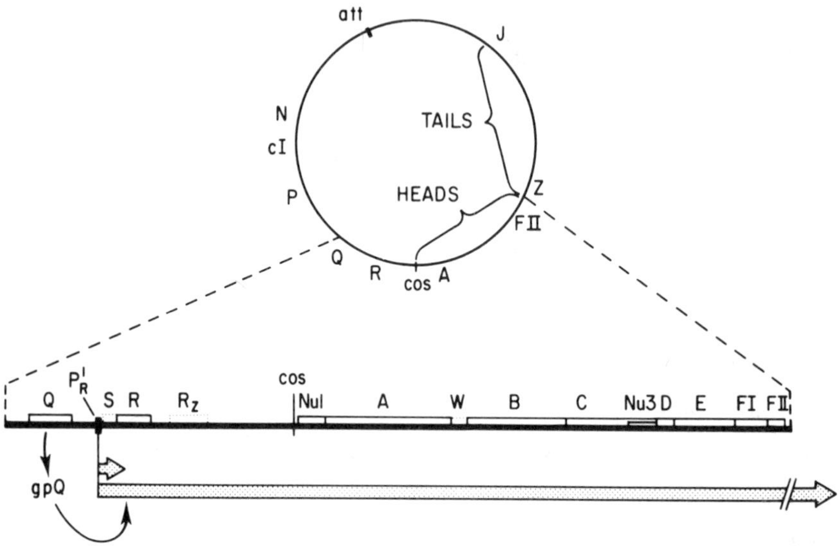

Figure 1 Physical map and transcription of the λ head gene region. Shown is a circular map of phage λ with a few genes for orientation. The heavy horizontal line represents the DNA of the expanded region (~13,000 bp; Daniels et al. 1980). Above it are the names, positions, and sizes of the known coding regions in this portion of the genome, indicated by thin solid lines or dotted lines for proteins not yet physically identified (Echols and Murialdo 1978; Young et al. 1979). DNA sequences outside the *cos* region are not yet available in this part of the chromosome, so some alterations in the exact size and position of genes are likely. Genes to the right of the *cos* site are discussed extensively in the text and by Feiss and Becker (this volume). Genes to the left are *S*, *R*, and R_z, which are required for cell lysis (Young et al. 1979), and *Q*, which is thought to positively regulate the expression of the late operon by stopping RNA polymerase termination 198 bp beyond the promoter for the late operon p_R' (Forbes and Herskowitz 1982).

PHAGE λ HEAD STRUCTURE

Genes and Proteins

The assembly of the λ head requires the products of at least ten phage genes (see Fig. 1; Weigle 1966; Parkinson 1968; Murialdo and Siminovitch 1972b; Boklage et al. 1973) and two host genes (*groEL* and *groES*; Georgopoulos et al. 1973; Tilly et al. 1981), although the finished head contains amino acid sequences coded by only six of these genes (*W, B, C, D, E,* and *FII*) and eight types of polypeptides (called gp*E*,[1] gp*D*, gp*B*, gp*W*, gp*FII*, p*B**, p*X1*, and p*X2*). Structures that accumulate after infection by mutants defective in each of the 12 genes have been analyzed (summarized in Tables 1 and 2). Phage with

[1]Our nomenclature for polypeptides follows that of Casjens and King (1975). gp refers to the primary protein gene product of a particular cistron. Thus, gp*E* refers to the entire polypeptide coded by gene *E*. p refers to a protein that has been modified or is of unknown genetic origin, and an asterisk indicates that the protein has been proteolytically cleaved. Therefore, p*B** refers to the cleaved form of the gene-*B* protein.

Table 1 Characteristics of λ head proteins

Gene product	Molecular weight (× 10^{-3})	Copies/mature prohead	Copies/phage	Head-related structures accumulated by mutant	Role
gp*Nu1*	21.5	<5	<5	mature proheads	DNA packaging
gp*A*	79	<1	<1	mature proheads	DNA packaging
gp*W*	?	—	+	incomplete, full heads	head completion
gp*B*	62	2–4(~12)[a]	2–4	aberrant proheads	head-tail connector, prohead assembly
p*B**	56	10–12	10–12	monsters, polyheads	head-tail connector
gp*C*	58	<1(~12)[a]	<1	aberrant proheads, monsters	prohead assembly
p*X1*	31	~6	~6		?
p*X2*	29	~6	~6		?
gp*Nu3*	19	<5(200)[a]	<10	aberrant proheads, monsters	prohead core
gp*D*	10	<10	420	aberrant (?) empty, expanded heads	DNA packaging, protein shell stabilization
gp*E*	38	420	420	no structures	protein shell, structural unit
gp*FI*	17	<5	<5	mature proheads	DNA packaging
gp*FII*	11	<1	5–6	incomplete full heads	head completion, head-tail connector
gp*groEL*	65	<1	<1	aberrant proheads	prohead assembly
gp*groES*	15	<5	<5	monsters, polyheads	

The references for these points are too numerous to delineate here. They can be found in the text and in Feiss and Becker (this volume).
[a]The numbers in parentheses are the approximate values for aberrant proheads in which the appropriate cleavages do not happen (C^- or *groE*$^-$).

Table 2 Physical characteristics of λ head-related structures

Structure	Protein composition	Sedimentation coefficient	Protein core	Radius (nm)	Shell thickness (nm)
head	gp*E*, gp*D*, gp*B*, p*B** p*X1*, p*X2*, gp*W*, gp*FII*	500S	no	32	0.4–0.5
mature prohead	gp*E*, gp*B*, p*B**, p*X1*, p*X2*	140S	no	25	0.9
putative immature prohead	(gp*E*, gp*Nu3*, gp*B*, gp*C*)[a]	(175S)	yes	(25)[a]	(0.9)[a]
B⁻ aberrant prohead	gp*E*, p*X1*, p*X2*	140S	no	25	0.9
C⁻ aberrant prohead	gp*E*, gp*Nu3*, gp*B*	175S	yes	25	0.9
Nu3⁻ aberrant prohead	gp*E*	175S	no	25	0.9
groEL⁻ or *groES*⁻ aberrant prohead	gp*E*, gp*Nu3*, gp*C*, (gp*B* variable, usually low)	175S	yes	25	0.9

Data from Hendrix and Casjens (1975); Hohn et al. (1975); Ray and Murialdo (1975); Zachary et al. (1976); Künzler and Hohn (1978); Earnshaw et al. (1979); Tilly et al. (1981).

[a]Immature wild-type proheads have not been analyzed directly in detail. The parentheses indicate probable values from analogies with related *C*⁻ and *groE*⁻ proheads.

mutations inactivating proteins that act at or after the DNA packaging step usually cause accumulation of structures that are normal assembly intermediates, whereas mutational inactivation of proteins involved in prohead assembly results in the accumulation of aberrant proheadlike structures (see below), as well as some "monster" assemblies termed spirals and polyheads, in which the coat protein shell is not closed properly. In infections lacking the building block of the protein shell, gp*E*, the other head proteins fail to assemble into large structures.

Structure of the Finished Head

Electron microscopy of negatively stained λ phage shows that the head is a fairly smooth-surfaced structure that is often hexagonal in outline (Fig. 2), but freeze-fracture-etched and negatively stained heads do show some substructure on the surface of the protein shell. Bayer and Bocharov (1973), using the former technique, found that there is a regular array of outward protrusions of "morphological units" on the surface that could be interpreted to be a triangulation number (T) = 21 *dextro* icosahedral surface lattice of hexamer-pentamer-clustered morphological units. (For a complete discussion of icosahedral virus structure and triangulation numbers, see Caspar and Klug [1963].) Since the number of capsid protein subunits in an icosahedral shell must be 60 T, subsequent determination that there were about 420 gp*E* and 420 gp*D* molecules per head made this structure seem unlikely for the protein shell (Casjens and Hendrix 1974a). Williams and Richards (1974) first recognized that two superimposed T = 7 *laevo* lattices (one trimer-clustered and one hexamer-pentamer-clustered) give an apparent T = 21 *dextro* hexamer-pentamer lattice if the two types of morphological units are indistinguishable (see Fig. 3). Hence, they proposed that one of the shell proteins was trimer-clustered, and the other, hexamer-pentamer-clustered, both in T = 7 *laevo* lattices. Imber et al. (1980) have shown that on polyheads (presumably a reasonable model for phage heads) gp*D* is trimer-clustered and gp*E* is hexamer-clustered.

If phage are disrupted so that the DNA is released, the head appears to be an easily deformed shell that has a stain-excluding knob at the proximal vertex (the vertex to which tails join) (Harrison et al. 1973). It is interesting to note that this structure is assembled as part of the head, but when virions are disrupted under certain conditions it remains bound to the tail rather than the head shell. Tsui and Hendrix (1980) showed that gp*B*, p*B**, and gp*FII* form this structure, which they called the "head-tail connector." It seems likely that gp*W* is also in the connector, but the gene-*W* polypeptide has not yet been physically identified. Its presence in the finished virion has only been detected by in vitro complementa-

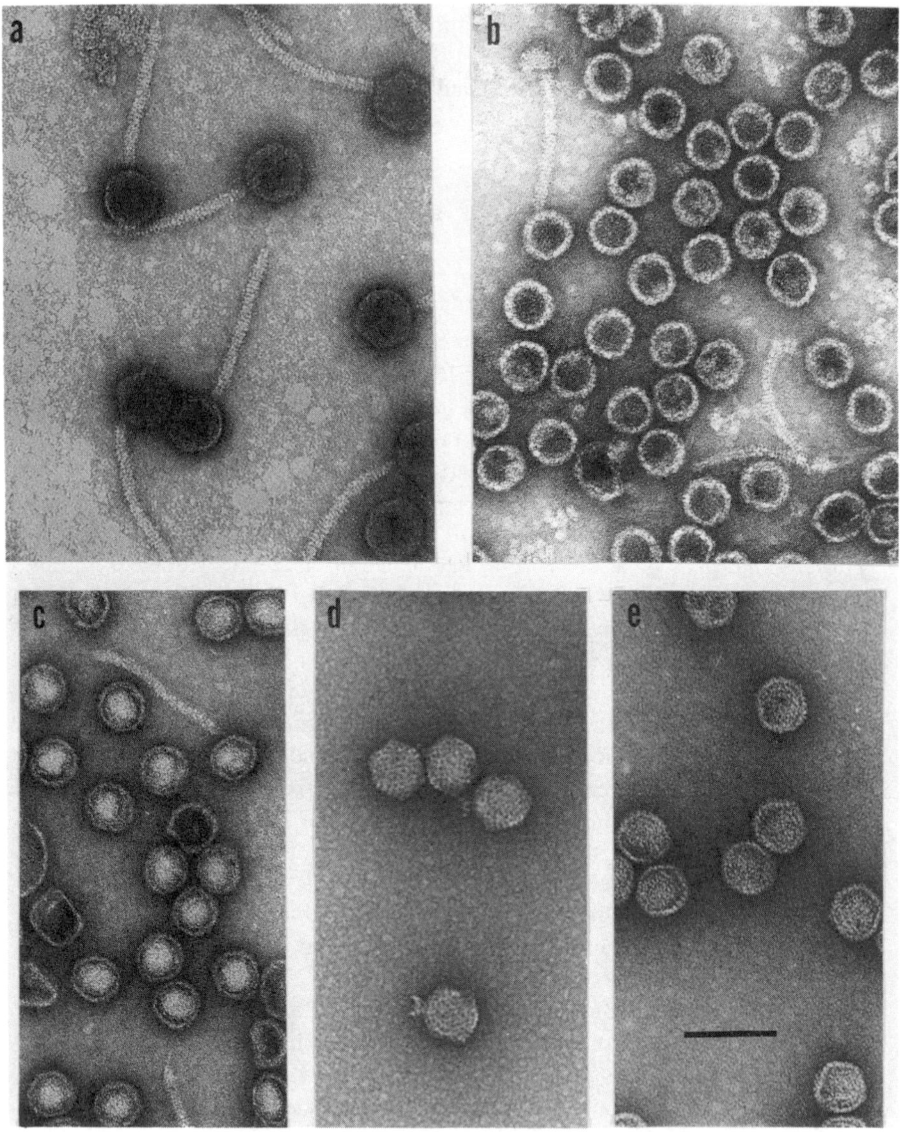

Figure 2 Negatively stained λ and P22 phage and proheads. (*a*) phage λ; (*b*) λ proheads produced in A^- infections; (*c*) core-containing λ proheads found in $groEL^-$ infections; (*d*) phage P22; (*e*) P22 proheads. Bar (*e*) represents 100 nm. Specimens were negatively stained with uranyl acetate. *a*, *b*, and *c* were kindly provided by Roger Hendrix (University of Pittsburgh, Pennsylvania).

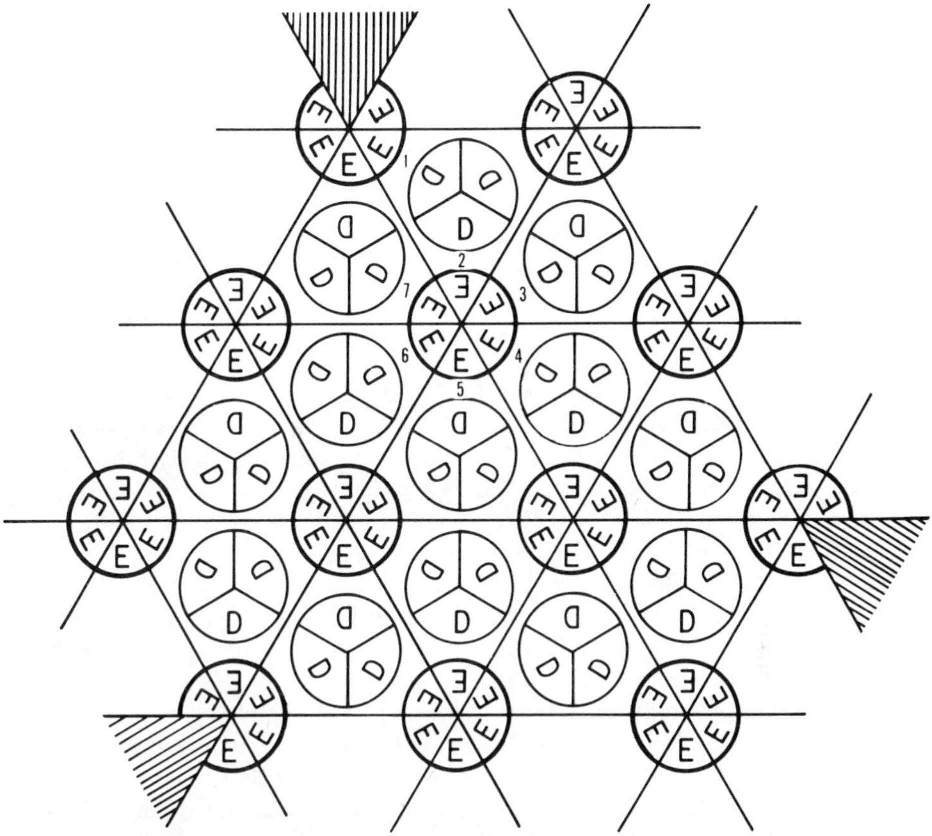

Figure 3 Arrangement of gp*E* and gp*D* polypeptides on the surface of the phage λ head. Each gp*E* has a portion of its polypeptide chain protruding from the surface near its adjacent local sixfold or fivefold axis of rotational symmetry, and each gp*D* molecule protrudes near its adjacent local threefold axis, thus forming morphological units or "clusters," which stand above the surface of the shell at these points. The trimers of gp*D* and hexamers and pentamers of gp*E* (shown diagrammatically as circles) represent these morphological units seen in the electron microscope. The actual arrangement of the polypeptide backbones is unknown; however, it is likely that the inner portion of the shell is composed of only gp*E*, with no large channels between individual molecules, and gp*D* molecules (only about one quarter of the size of gp*E*) occupy parts of "valleys" on the surface of the gp*E* shell. The three corners of one face of the icosahedral head (head vertices) are shown with one facet hatched. If the hatched area were cut out and the resulting edges folded and joined, the gp*E* pentamer at a vertex would be generated. The edges of one of the head's 20 faces are the lines joining the three vertices shown. Note that a gp*E*-gp*D* heterodimer is formally the protomer (smallest protein unit) in this structure, and that these protomers occupy seven different environments in the structure (indicated by numbers 1–7). These can be separated geometrically by their inability to be rotated into one another by rotation about the true symmetry axes of the icosahedral shell or visually by their different spatial relationships to a vertex or the center of a face.

tion tests (Katsura and Tsugita 1977). It has been speculated that the remaining head proteins, p*X1* and p*X2* (which contain gp*E* and gp*C* sequences; see below), are also present at the proximal vertex and form an adapter that joins the head shell and the connector (Murialdo and Ray 1975) or that they may be located at the 12 vertices of the head (Hendrix and Casjens 1975; Hohn et al. 1975; Zachary et al. 1976).

Prohead Structure

The structure of assembly intermediates and aberrant assemblies of gp*E* have also been studied in some detail. Karamata et al. (1962) first observed small, empty, spherical structures in productive λ infections. They showed that the structures were head-related and called them *petit* λ (Fig. 2b). Later, Hohn and Hohn (1974) and Kaiser et al. (1974) showed that these particles are direct precursors to the complete head, in that they are capable of encapsidating a DNA molecule in vitro without major protein reassortment. In accordance with this, the protein composition of these particles (called proheads) was shown to be a subset of that of the completed head, consisting of gp*E*, gp*B*, p*B**, p*X1*, and p*X2* (Table 1; Hendrix and Casjens 1975; Hohn et al. 1975). The gp*E* molecules are probably arranged in the same type of surface lattice on proheads as on phage (Hohn et al. 1976), but their conformation must be somewhat different, since the shell is thicker, the particle diameter is 22% smaller than that of heads (Künzler and Hohn 1978; Earnshaw et al. 1979), and the particles are circular, rather than hexagonal in outline. These particles appear empty by electron microscopy and low-angle X-ray scattering. Proheads, like finished heads, contain a single knob of negative-stain-excluding material associated with the coat protein shell, suggesting that the minor proteins could be clustered at the proximal vertex. Murialdo and Ray (1975) found that this knob correlated with the presence of polypeptides coded by genes *B* and *C* and proposed that they constitute the knob.

The aberrant proheadlike structures formed in B^-, C^-, $Nu3^-$, and $groE^-$ infections are also of interest in determining the roles of the proteins involved (Tables 1 and 2). These structures (historically also called *petit* λ) are not true assembly intermediates but are incorrect products formed when one of the above proteins is missing (Murialdo and Becker 1978b). Of particular interest are C^- and $groE^-$ proheads (Fig. 2c), which contain a large number of gp*Nu3* molecules (~200; Künzler and Hohn 1978; Earnshaw et al. 1979). These form an easily lost core or scaffold in the interior of the gp*E* shell (Hohn and Katsura 1977; Earnshaw et al. 1979). A small number of proheads that contain gp*Nu3* are also present in wild-type-infected cells (Hohn et al. 1975). The detailed arrangement of these proteins in the core is unknown.

λ HEAD ASSEMBLY PATHWAY

The overall stages in the assembly of the λ head from monomeric or oligomeric protein precursors have been elucidated by genetic and biochemical experiments and observations using electron microscopy (for reviews, see Hohn and Katsura 1977; Murialdo and Becker 1978a; Wood and King 1979). The in vitro assembly of proheads from soluble components (Murialdo and Becker 1977), the in vitro encapsidation of phage DNA by proheads (Kaiser and Masuda 1973), and the successful in vitro joining of heads to tails to give rise to infectious particles (Weigle 1966) have prepared the way for a more detailed analysis of the head assembly process. Head assembly proceeds through the orderly formation of discrete intermediates and can be conveniently divided into five stages (Fig. 4). Stage I is the formation of an as yet ill-defined small "initiator" structure, possibly consisting of gp*B*, gp*C*, and gp*Nu3*. One or both of the *groE* proteins may also be present. Stage II is the assembly of the coat protein (gp*E*) shell and gp*Nu3* on the initiator. Stage III consists of the conversion of this "immature" prohead to a "mature" prohead by the elimination of gp*Nu3*, the conversion of gp*B* to p*B**, and the conversion of gp*C* and some gp*E* to p*X1* and p*X2*. Stage IV is the encapsidation of the DNA molecule by the mature prohead, aided by the action of gp*Nu1*, gp*A*, and gp*FI*, with concomitant shell expansion and addition of gp*D* to the head shell. Stage V, completion, is the stabilization of the packaged DNA within the gp*E* shell and the preparation of the structure for tail joining by the addition of gp*W* and gp*FII* to the structure. The exact order of the individual steps within each stage has not always been conclusively demonstrated, since it has been difficult to detect partially assembled forms and, as mentioned above, several mutants do not accumulate true intermediate structures. With these difficulties in mind, we describe the current ideas about λ head assembly.

Stage I

Most of our limited knowledge about the possible sequence of events during this stage derives from the work of Murialdo and Becker (1977, 1978b) and Murialdo (1979). These investigators developed an in vitro system for the assembly and detection of functional proheads in crude extracts. For this reaction to occur, a source of unpolymerized gp*E*, gp*B*, gp*C*, and gp*Nu3*, as well as the *groE* proteins, must be present. Only $Nu3^-$ extracts can donate active unassembled gp*E*, presumably because it is sequestered in aberrant structures or proheads in other mutant-infected cells. Although there are aberrant structures in $Nu3^-$ extracts, they are fewer than in other mutant extracts. gp*B*, gp*C*, and gp*Nu3* can be donated by an E^- extract, in which they are not assembled into any prohead-type structures. The E^- extract is postulated to donate gp*B* and gp*Nu3* in the form of a complex, since mixed E^-B^- and E^-Nu3^- extracts cannot donate these proteins unless the concentration is increased threefold to fivefold (Fer-

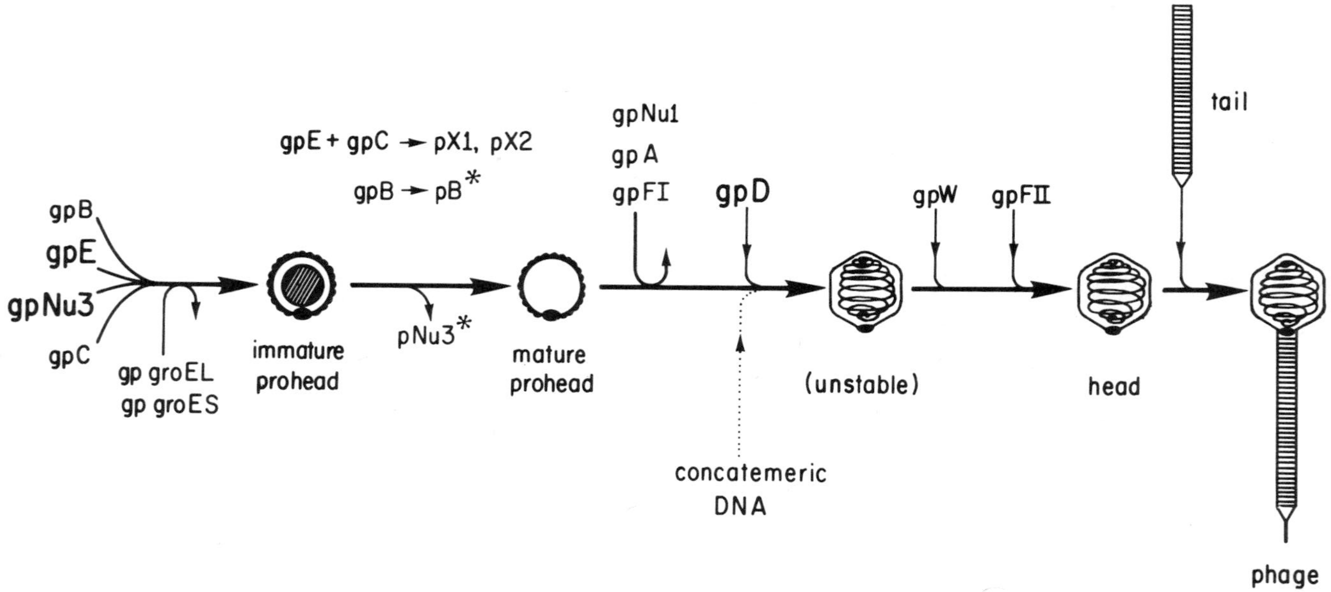

Figure 4 The phage λ head assembly pathway. Phage-coded proteins must enter and travel through the pathway as indicated for proper assembly to occur. The tail is assembled by an independent pathway (see Katsura, this volume) and joins spontaneously to completed heads. gp*E*, gp*D*, and gp*Nu3* are denoted by larger characters to indicate that large numbers of molecules of these proteins are required for head assembly.

rucci and Murialdo 1981), but no direct biochemical evidence for this gp*B*-gp*Nu3* complex exists. Formation of this biologically active complex requires the presence of at least gp*groEL*. In accordance with this requirement there is biochemical evidence for a gp*groEL*-gp*B* complex (Murialdo 1979). Advantage was taken of the fact that gp*groEL* is a decatetramer that sediments rapidly (Hendrix 1979; Hohn et al. 1979; Murialdo 1979). It was found that in *E*⁻ infections, a fraction of gp*B* cosedimented with gp*groEL*. This cosedimentation is probably not fortuitous because (1) a *B* amber fragment also cosediments, (2) functional gp*Nu3* is required for cosedimentation, and (3) genetic evidence suggests that gp*groEL* and gp*B* interact in vivo (Georgopoulos et al. 1973; Sternberg 1973). However, it should be noted that some of this complex does form in the *groEL*⁻ mutants that block head assembly (Murialdo 1979). Finally, Murialdo (1979) has also presented biochemical evidence that gp*B* can exist as a complex with various cleaved forms of gp*C* (p*C*′′, p*C*′′′, and so on), whose formation requires active gp*Nu3* and gp*groEL*; but this has not been shown to be an actual assembly intermediate. Thus, gp*B*-gp*Nu3*, gp*B*-gp*groEL*, and gp*B*-gp*C* associations have been seen or inferred, but it is important to stress again that biochemical evidence for the existence of the first and evidence that the second two are, in fact, intermediates is still lacking. It is possible to integrate these interactions into reaction pathways, which account for the contribution of each component to prohead assembly. The simplest of these, substantially that proposed by Ray and Murialdo (1975) and Murialdo (1979), is as follows:

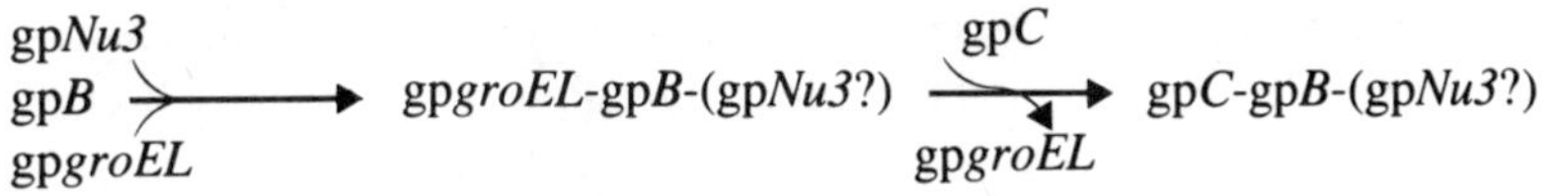

One problem with this hypothesis is the failure to detect biochemically any association between gp*Nu3* and the complexes. This could mean either that gp*Nu3* naturally dissociates from the complex and has a second assembly role during shell construction or that it artifactually dissociates during isolation. Hohn et al. (1979) have reported the association of six copies of gp*groEL* monomer with immature proheads derived from a specific *groEL*⁻ host. Whether this indicates a possible early specific association of gp*groEL* with immature proheads or an aberrant aggregation is not known. One possibility is that in this particular *groE*⁻ mutant, the mutated form of gp*groEL* binds more tightly to gp*B* and so is trapped within the immature prohead. The role(s) of gp*groES* is not yet defined in terms of these complexes, primarily because its existence has only recently been established (Tilly et al. 1981).

Recently, Kochan (1983) purified and characterized a "preconnector" particle consisting of 12 subunits of gp*B*. The particle has a donutlike structure when viewed in an electron microscope from the top, with an overall diameter of 15 nm and a 2-nm central hole. When viewed from the side, the particle resembles an asymmetrical dumbbell with a height of 13 nm. Functional gp*Nu3*, gp*groEL*, and, probably, gp*groES* are required for the production of the preconnector, which sediments at 25S. The preconnector, probably related to the head-tail connector structure observed by Tsui and Hendrix (1980), has been shown to act as an early intermediate in the correct assembly of the λ prohead (Kochan 1983).

Stage II

The process of gp*E* shell assembly is not well understood. Murialdo and Becker (1978a) have suggested that in all double-stranded DNA phages correct shell assembly begins on an initiator complex of minor proteins. The hypothetical gp*C*-gp*B*-(gp*Nu3*?) complex is certainly a candidate for such an initiator. Under normal conditions of assembly, gp*E* and gp*Nu3* are thought to polymerize on the putative initiation complex to form an immature prohead (containing gp*E*, gp*C*, gp*B*, and gp*Nu3*). Direct proof that such an intermediate actually exists has been elusive, but the possibility is supported by the following observations: (1) A small number of the proheads in wild-type-infected cells contain gp*Nu3* in large amounts, gp*C*, and gp*B* (Hohn et al. 1975; C. Georgopoulos, unpubl.); (2) Proheads containing a large amount of gp*Nu3* accumulate in C^- and *groE*$^-$ infections and in some B^- infections (Georgopoulos et al. 1979). Hohn et al. (1975) found that a small percentage of *groE*$^-$ proheads may be able to be processed and to package DNA; and (3) Zachary et al. (1976) showed that, during λ infection, core-containing proheads can be seen by electron microscopy early in infection and that these are replaced at later times by empty proheads. Such core-containing proheads accumulate in C^- or *groE*$^-$ infections.

It is not known how the gp*E* shell dimensions are determined, but since mutants blocked in each of the genes required to make proheads (except, of course, the *E* gene) assemble some coat protein shells of the correct dimensions and since the size of the chromosome does not affect shell dimensions (Earnshaw et al. 1979), most of the information for such assembly must lie within the coat protein itself. This notion is supported by the isolation of missense mutants of gene *E*, which give rise to abnormal head shell structures (Katsura 1980). If gp*B*, gp*groEL*, or gp*groES* is missing, some gp*E* assembles into spiral and tubular polyhead structures, indicating that the missing proteins are in some way required for proper shell construction (Mount et al. 1968; Murialdo and Siminovitch 1972b; Georgopoulos et al. 1973). If gp*Nu3* is absent, gp*E* assembly is slowed and fewer properly sized shells are made (Hohn et al. 1975; Ray and Murialdo 1975; Zachary et al. 1976). These *Nu3*$^-$ aberrant proheads contain only gp*E*, suggesting that gp*Nu3* makes prohead assembly more efficient in terms of shell construction and is required for incorporation of gp*B* and gp*C* into the particle. The protein composition of the aberrant proheads made during B^-, C^-, *Nu3*$^-$, or *groE*$^-$ infections, as well as the assembly of gp*E* into spirals and polyheads, presumably results from the polymerization of gp*E* on improper initiators or without an initiator complex (Table 2; Murialdo 1979).

Stage III

The maturation of proheads does not involve the action of any known phage gene products that are not already in the structure. There are three known changes during this conversion: (1) About three quarters of the gp*B* molecules are cleaved ed from 61,000 M_r to 56,000 M_r (Hendrix and Casjens 1975); (2) all of the gp*Nu3* is degraded and lost from the structure (Ray and Murialdo 1975); (3) each gp*C* molecule participates in a fusion/cleavage reaction with a gp*E* molecule to form the hybrid proteins p*X1* and p*X2* (Hendrix and Casjens 1974; Ray and Murialdo 1975), whose exact structures and mechanisms of formation are unknown. It is possible to determine which of the three reactions proceeds in various mutant infections. The results of such experiments (Hendrix and Casjens 1975; Hohn et al. 1975; Ray and Murialdo 1975; C. Georgopoulos, unpubl.) can be summarized as follows (see also Table 2): Cleavage of gp*B* does not occur if gp*C* or either of the *groE* proteins is missing. In the absence of gp*Nu3*, gp*B* is cleaved but is not found in the particles assembled under these conditions. Degradation of gp*Nu3* does not occur if gp*C* or either of the *groE* proteins is missing, but it does occur in the absence of gp*B*. None of the conversions occurs unless gp*E* is functional. A conservative interpretation of these results is that each of these processes is dependent upon proper assembly. The enzyme(s) that carries out these reactions has not yet been identified.

Stage IV

The DNA packaging reaction, in which mature proheads each package and cut a virion chromosome from the concatemeric replicating DNA with the aid of the gene-*Nu1*, -*A*, and -*FI* proteins, is being dealt with in detail by Feiss and Becker (this volume). It is during this stage that the gp*E* shell expands and gp*D* is added to the surface lattice. gp*D* adds in vitro only to expanded gp*E* structures (Imber et al. 1980).

Stage V

The phage head containing the newly packaged DNA molecule is quite unstable and cannot join to tails. The products of genes *W* and *FII* add to the incomplete head in that order to stabilize the head and construct the site that is recognized by tails (Casjens et al. 1972; Casjens 1974; Katsura and Tsugita 1977), which then bind spontaneously to form infectious phage particles (Weigle 1966, 1968).

P22 ASSEMBLY

Phage P22 is often considered to be a lambdoid phage by virtue of its having a λ-type early region (Susskind and Botstein 1978). The late region is regulated in a manner similar to that of λ, and there are genes analogous to *R* and *S* (and perhaps *A* and *Nu1*) in the late operon. The structural genes, however, have no obvious direct homology to those of λ. There are 11 phage gene products required under normal growth conditions for P22 head assembly, and all of them have been identified by gel electrophoresis (Botstein et al. 1973; Casjens and King 1974; Poteete and King 1977; Youderian and Susskind 1980). The pathway for P22 head assembly has overall similarities to the λ pathway (and to those of other phages; see Wood and King 1979; Earnshaw and Casjens 1980) and several differences that may be useful in distinguishing generalities from variations in the basic principles of phage head assembly. As with λ, P22 proheads and DNA packaging proteins encapsidate DNA from a concatemeric substrate, and several additional proteins are added that stabilize the head and prepare it for tail joining. Unlike λ, no proteolytic cleavages or covalent modifications of proteins are known to accompany P22 assembly (Casjens and King 1974).

P22 Head Structure

An electron micrograph of P22 is shown in Figure 2d. The head is quite similar in size and shape to that of λ, but the tail has a very different structure. The surface of the head has icosahedral symmetry and shows a T = 21 *dextro* lattice of morphological units (Casjens 1979), an arrangement similar to the surface of phage λ (see Fig. 3). The P22 head shell, however, contains only one species of major coat protein, gp*5* (Botstein et al. 1973). Physical measurements have shown that there are about 420 molecules of gp*5* per head, so the structure must be T = 7. Since superimposed T = 7 *laevo* trimer-clustered and hexamer-pentamer-clustered morphological unit lattices give an apparent T = 21 lattice (see above, Structure of the Finished Head), it seems likely that in the P22 head shell each gp*5* molecule has outward projections in positions equivalent to gp*D* and gp*E* in the λ head shell (i.e., local threefold, fivefold, and sixfold symmetry axes) (Casjens 1979). It is interesting to note that the molecular weight of gp*5* is very close to the sum of those for gp*E* and gp*D*; perhaps gp*5* fulfills the structural roles of both of these proteins.

The P22 prohead (Fig. 2e) contains gp*5* molecules in a basically similar arrangement to that found in the head, although the diameter is about 11% smaller (Earnshaw et al. 1976; Casjens 1979; Earnshaw 1979). About 220 molecules of gp*8* (called scaffolding protein) are also found in the prohead. These are thought to be inside the coat protein shell (Casjens and King 1974; Earnshaw et al. 1976), but their detailed spatial arrangement is unknown. Unlike λ, only one minor protein, gp*1*, must be present for the assembly of proheads that are competent to package DNA (Botstein et al. 1973; King et al. 1973). Only gp*5* and gp*8* must combine with gp*1* in the assembly of such proheads.

Including gp*1*, there are seven proteins found in small amounts (<30 molecules per phage) in finished phage heads. Four of these are found in normal proheads as well. Of these four, gp*1* is required for DNA packaging (and, perhaps, proper prohead assembly; see below), and gp*7*, gp*16*, and gp*20* are required only for successful DNA injection in the subsequent round of infection (Susskind and Botstein 1978). The last three, gp*4*, gp*10*, and gp*26*, are added after DNA packaging and stabilize the head and prepare it for tail joining. None of these minor proteins has been localized in the prohead or phage particle. Murialdo and Becker (1978a), however, have suggested that, by analogy with other phages, gp*1* is likely to lie at the proximal vertex.

P22 Assembly Pathway

Little is known about how proheads assemble from gp*1*, coat, and scaffolding proteins. gp*1* is a plausible candidate for the building block of the prohead assembly "initiator structure" (Murialdo and Becker 1978a; see above), but nonfunctional proheadlike structures do accumulate in 1^--infected cells. Fuller and King (1980) have purified coat and scaffolding proteins from dissaggregated proheads and found that they can form proheadlike structures in vitro, but neither protein assembles by itself. Similarly, in vivo, scaffolding protein alone does not appear to assemble into prohead-core structures, and coat protein alone polymerizes into aberrant structures more slowly than it normally assembles into proheads (Casjens and King 1974; Lenk et al. 1975). These data have been interpreted to mean that gp*5* and gp*8* coassemble to form a prohead, but an oligomeric association between the two proteins has not been demonstrated (Fuller and King 1980). The "injection proteins," gp*7*, gp*16*, and gp*20*, apparently must also be added during the formation of the prohead, since it has not been possible to add them to preformed proheads that lack them (Poteete et al. 1979). It is of interest to note that these three proteins are not added in obligate assembly steps (i.e., if they are absent, the assembly process continues and otherwise normal phage are made that lack these proteins). No minor protein or combination of minor proteins is required for the assembly of coat protein shells of the correct dimensions when scaffolding protein is present in normal amounts (S. Casjens, unpubl.), and scaffolding protein-defective infections produce some coat protein shells of the correct size (Earnshaw and King 1978). This suggests that coat protein alone has the information for the construction of such shells, and scaffolding protein may function to direct this information into more efficient or more accurate assembly. It is not known whether P22 assembly requires the *groEL*- or *groES*-gene products, since no *groE*$^-$ mutants of *Salmonella* exist. (*Salmonella* does have both *groE* genes; K. Tilly et al., unpubl.).

Figure 5 shows the remainder of the assembly pathway. Complete proheads package DNA in a complex reaction in which gp*2* and gp*3* must act, the coat protein shell expands 11% to phage diameter, and all of the scaffolding protein molecules exit intact. P22 does not have an isolatable analog to the mature prohead that lacks scaffolding protein; scaffolding protein exit has remained

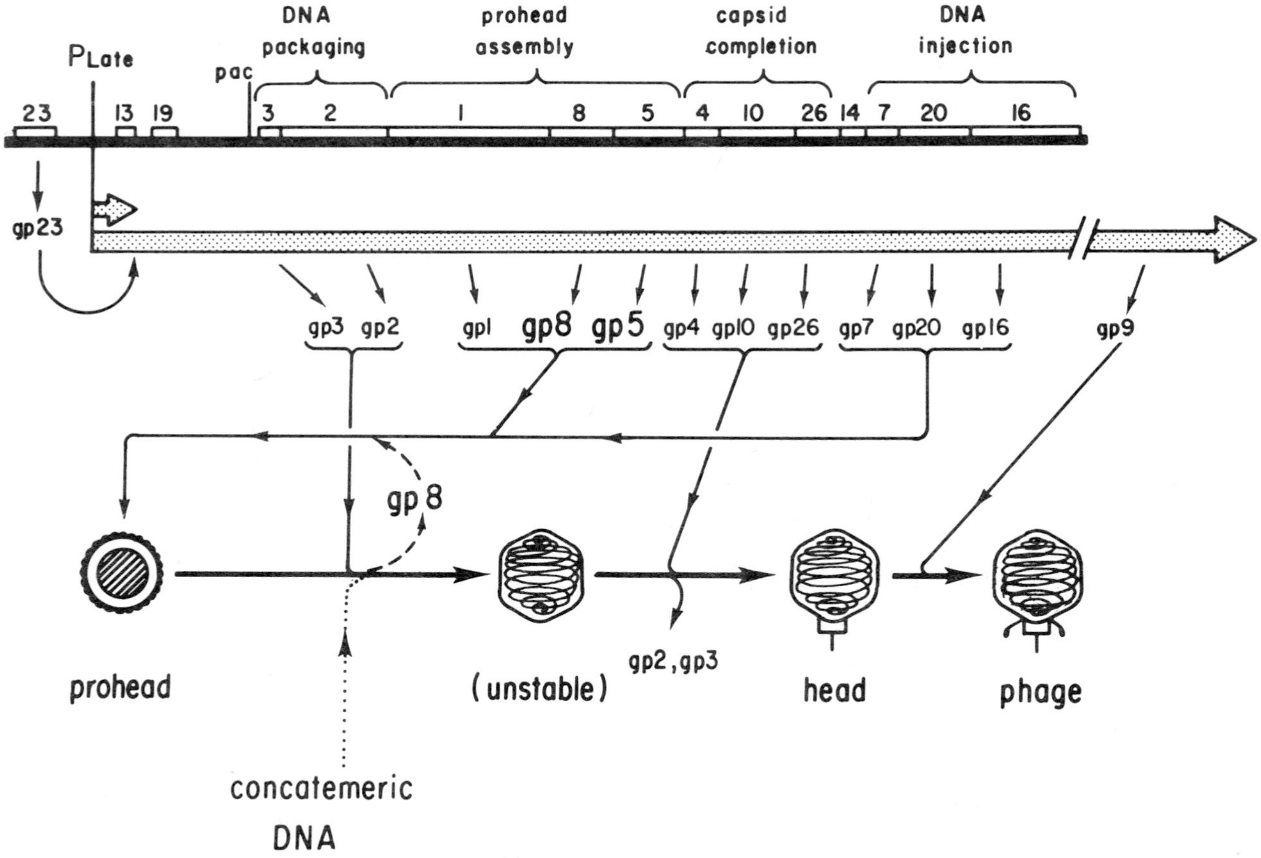

Figure 5 (See facing page for legend.)

coupled to DNA packaging in all experiments to date. It has been suggested, however, by analogy with other phage systems, that its exit may not be mechanistically involved in the DNA condensation reaction (Earnshaw and Casjens 1980). Shell expansion appears to be a property of coat protein alone (Earnshaw and King 1978), and a model for this expansion utilizing a hinge in the coat protein has been presented (Casjens 1979). Scaffolding protein can recycle to participate in further rounds of prohead assembly and thus acts catalytically in the prohead assembly and DNA packaging stages of phage assembly (King and Casjens 1974). The head thus formed is unstable, and addition of gp*4*, gp*10*, and gp*26* in subsequent steps stabilizes it (King et al. 1973; Lenk et al. 1975). Finally, gp*9*, the tail protein, is added to form a functional phage particle (Israel et al. 1967; Botstein et al. 1973).

MORPHOGENETIC PROTEIN GENES AND THEIR EXPRESSION

It is obvious that the morphogenetic genes of λ (and other phages) are not randomly arranged on the chromosome. First, the head and tail genes are present in nonoverlapping clusters (Weigle 1966). Furthermore, they seem to be arranged in such a manner that gene products that interact most intimately are coded by adjacent genes (Casjens and Hendrix 1974b). This arrangement may have evolved to conserve groups of genes that interact by minimizing the possibility of recombination within such groups (Stahl and Murray 1966). This idea is supported by the existence of the same gene order for analogous genes in other lambdoid phages and the divergence of lambdoid phage proteins in terms of molecular weight and functional specificity (Deeb 1970; Youderian 1978; K. Tilly, unpubl.; M. Feiss, pers. comm.). The result is the formation of groups or "modules" (Echols and Murialdo 1978) of genes coding for functional units such as prohead assembly, DNA packaging, head shell, and so on, that can be

Figure 5 Physical map and transcription of the P22 head gene region and P22 head assembly pathway. The heavy top line is an approximate physical map of the P22 late region covering about 18,000 bp (Chisholm et al. 1980; Weinstock and Botstein 1980). Immediately above this are the names, positions, and sizes of the coding regions for the known genes in the region (Chisholm et al. 1980; Youderian and Susskind 1980), and above these data are the general functions of the head genes. Other genes indicated are *13* and *19*, which are required for cell lysis, *14*, which is required for proper head assembly only at elevated temperatures (Youderian and Susskind 1980), and *23*, whose product is thought to be analogous to the λ gp*Q* (Weinstock et al. 1980). The mRNAs initiated in this region are shown as stippled arrows below the physical map. The proteins coded by the late mRNA are shown directly beneath it, and at the bottom the assembly pathway for phage P22 is given. Each protein must enter the pathway as indicated to be incorporated into the phage particle. gp*5* and gp*8* are denoted by larger characters to indicate that large numbers of these molecules are utilized in head assembly.

exchanged among related phages and then be allowed to diverge. The isolation by Feiss et al. (1981) of a viable λ phage that has acquired only the phage 21 head genes suggests that such recombinational events occur in nature.

Phage assembly appears, in general, to be a "steady-state situation," in which precursor polypeptides are continuously synthesized and assembled by the pathways discussed above. Assembly does not seem to be regulated by time of synthesis of the components. Many phages, however, coordinately express all of their morphogenetic genes only at late times after infection. Clustering of the morphogenetic genes facilitates this coordinate regulation. They are all transcribed from a single promoter in λ (Herskowitz and Signer 1971) and P22 (Weinstock et al. 1980; O.S. Casjens and M. Adams, unpubl.) and are therefore expressed together. It has been noted that the phage morphogenetic proteins are made in approximately the ratios needed during assembly (Brégégère and Gros 1971). How is this mediated if a single promoter is used? Ray and Pearson (1974, 1975) and O.S. Casjens and M. Adams (unpubl.) found, for λ and P22, respectively, that differential expression of the late proteins is most probably attained by differential translation initiation of the morphogenetic cistrons. This regulation is no doubt important, since in several instances it is known that artificially altering the ratios of morphogenetic proteins causes improper assembly (Floor 1970; Showe and Onorato 1978). Several additional mechanisms have been found for regulation of the levels of phage morphogenetic proteins. King et al. (1978) and O.S. Casjens and M. Adams (unpubl.) have found that the P22 scaffolding protein turns off its own synthesis at a posttranscriptional level when it is not assembled into proheads and that the rate of P22 coat protein synthesis increases (relative to the other head proteins) with time after infection. The former mechanism presumably allows the phage to make sufficient scaffolding protein and no more, whereas the second may allow synthesis of the minor proteins gp*1*, gp*7*, gp*16*, and gp*20* in excess early in infection to ensure their incorporation into proheads. Finally, we would like to speculate on another type of regulation that may occur. Parkinson (1968) and Murialdo and Siminovitch (1972a) observed genetic polarity among the λ head genes promoter-distal to some amber mutations. This was interpreted as classical "amber polarity" (Imamoto and Yanofsky 1967). However, Forbes and Herskowitz (1982) recently found that the gp*Q*-modified RNA polymerase is not susceptible to this type of polarity, making this explanation unlikely. Min Jou et al. (1972) and Oppenheim and Yanofsky (1980) have described an alternate type of polarity, which they call "translational coupling," in which translation of a gene is coupled to translation of the nearest upstream gene. Mechanisms that they proposed for this were (1) exposing the ribosomal binding site of the distal gene by translation of the proximal one, (2) translation of the two genes by the same ribosome, or (3) protection of the distal gene message from degradation by translation of the proximal one. This type of translational coupling is an attractive way of ensuring the synthesis of equal numbers of two polypeptides and could explain the λ polarity results.

Although gene clustering may be important in an evolutionary sense to the lambdoid phages, the genetic arrangement of the overlapping *C* and *Nu3* genes (Fig. 1) may have profound implications for prohead assembly. Translation of the *Nu3* gene begins at an initiation codon within the *C* gene and continues in phase with the *C* gene to the termination codon (Shaw and Murialdo 1980). This means that gp*Nu3* has the same amino acid sequence as the carboxyterminal portion of gp*C*. This identity may play a role in prohead assembly, allowing equivalent interactions between different proteins.

OTHER USES FOR PHAGE ASSEMBLY SYSTEMS

The phage assembly pathways clearly stand as model systems for the assembly of other supramolecular structures in nature, but in addition to this general use, the P22 and λ assembly pathways have been used to test several genetic methods of determining protein-protein interactions and the order of steps in a pathway. Georgopoulos et al. (1973), Jarvik and Botstein (1975), and Sternberg (1973, 1976) looked for and found mutations in other proteins that would correct a defect in a given protein. In general, such "second-site revertants" were located in genes whose protein products interact with the gene product carrying the first mutation, showing the efficacy of the general method. Jarvik and Botstein (1973) used independently isolated temperature-sensitive and cold-sensitive mutations in P22 genes to test a method for determining the order of events in a biological pathway. If a cold-sensitive mutation is placed in one gene that acts in a pathway and a temperature-sensitive mutation is placed in another, a shift from low nonpermissive temperature to high nonpermissive temperature would result in escape of intermediates through the entire pathway if the protein carrying the cold-sensitive mutation acts after the protein carrying the temperature-sensitive mutation (and functional intermediates accumulate at the first nonpermissive temperature), whereas no escape through the pathway would occur if their order of action were reversed. They were able to show that this method is of general applicability by using the previously characterized P22 head assembly pathway. Katsura (1981) has used phage assembly as a system in which to analyze the theory of genetic complementation between two related species, and Katsura (1980) and King and co-workers (Smith et al. 1980; Goldenberg and King 1981; Smith and King 1981) have used the λ gene *E* protein and the P22 gene *9* protein, respectively, to ask questions about the ways in which missense mutations can affect a protein's function and perhaps the pathway by which it folds to become a functional unit. Finally, phage and related structures (notably polyheads) have been useful in the development and testing of methods for obtaining three-dimensional structural information from electron micrographs (for review, see Steven 1981).

CURRENT ASPECTS OF PHAGE HEAD ASSEMBLY

The analysis of λ and P22, along with the other phages under study, can be used to formulate generalities about the nature of phage head assembly pathways. A phage is faced with the problem of constructing a complex supramolecular structure with only a limited genome to encode its components. They have solved the problem by using one protein, which we call coat protein (gp*E* and gp*5* in λ and P22, respectively), as the basic building block of the nucleic acid container. The coat protein carries most of the information for the assembly of the protein shell in its amino acid sequence and is able to assemble into a structure with icosahedral symmetry. The so-called minor proteins appear to be present at the proximal vertex. They contribute little to the maintenance of the integrity of the shell but, instead, are involved in controlling assembly of the coat protein, in attaching a tail (and, presumably, injection machinery) to that vertex, and probably in the DNA packaging reaction. Separation of these functions has usually not been possible, and it is quite likely that individual minor proteins perform more than one of these roles. In each double-stranded DNA phage under study, the structural proteins required for DNA packaging are assembled into a prohead (preformed DNA container) that is of smaller diameter than the finished phage head. Since no particles of intermediate size have been observed, expansion probably occurs by the rearrangement of the major capsid protein from one conformation to a second. All immature proheads contain a core of scaffolding protein present in about one-half the number of coat protein subunits (for review, see Earnshaw and Casjens 1980). Why do all phages have shell expansion and a scaffolding protein? The answer to this question is not known, but several hypotheses have been presented. For expansion of the coat protein shell, the hypotheses are the following: (1) Expansion causes a pressure differential that draws the DNA into the prohead (Hohn et al. 1974; Serwer 1976); (2) it helps create the macromolecule-free interior space in which to package DNA (Earnshaw and Casjens 1980); and (3) it is a coincidental result of a requirement that coat protein be in different DNA binding states before and after DNA packaging (see Earnshaw and Casjens 1980). The role of the scaffolding protein may be (1) to help guide the coat protein into a proper assembly (King et al. 1973; Showe and Black 1973), (2) to exclude host macromolecules from the interior of the coat protein shell (Earnshaw and Casjens 1980), or (3) to allow coassembly of the coat protein with the minor proteins (Murialdo and Ray 1975; Earnshaw and King 1978). None of these hypotheses has been rigorously excluded, nor are they, in general, mutually exclusive.

A number of additional problems remain unsolved. The positions of the minor proteins and arrangement of DNA within the finished head must be determined in more detail before a complete understanding of the roles of these proteins can be attained. Although some progress has been made in each of these directions (Gillin and Bode 1971, 1972; Earnshaw and Harrison 1977; Tsui and Hendrix 1980), it appears that technical advances will be required in order to gain a significantly more detailed picture of prohead and head structure. Such a picture will also be required to understand the nature of the extremely strong interprotein noncovalent bonds that hold the phage particles together. It is presumed that these will be similar to the nonpolar amino acid side-chain-patch, hydrogen-bond, and salt-bridge networks that bind the subunits of multimeric enzymes (Chothia and Janin 1975) and the coat protein subunits of tomato bushy stunt virus (Harrison 1980) and tobacco mosaic virus (Bloomer et al. 1978) together. X-ray diffraction analysis of tomato bushy stunt virus crystals has shown that the three coat protein states required by the T = 3 structure are attained by the movement of two relatively rigid folded domains with respect to one another (Harrison et al. 1978). Has a similar structure evolved for the more complex T = 7 phage heads, in which the proteins must be present in seven different states (Fig. 3), yet arranged in a perfectly ordered fashion? A final interesting question that only detailed structural analysis can solve is the following: Are the antigenically unrelated coat proteins of the various phage heads highly diverged from a single ancestor, in which case the peptide backbone might be expected to retain common folding characteristics, or are they the result of convergent independent solutions to a common biological problem?

The problems of initiation of prohead assembly and shell construction also require much clarification. It appears from the mutant studies that if the proper minor proteins are not present when shell assembly begins, proheadlike structures lacking various minor proteins are built, and these defects cannot be corrected (i.e., the particles are dead-end assemblies). Analysis of this problem is hampered by the apparent instability of the "initiator" complex and its precursors and the difficulty of studying small multicomponent structures that make up only a small fraction of the total cellular protein. It seems possible that use of the in vitro prohead assembly systems will aid in the purification of the components involved, the ultimate determination of the required order of action of the various components, and characterization of the various intermediates. While studying the early stages of prohead assembly, it will also be necessary to address the problem of the role of the host in the process. Although we do know that gp*groEL* and gp*groES* interact in vivo (Tilly and Georgopoulos 1982), that λ proheads made in *groEL*⁻ and *groES*⁻ bacteria have the same protein composition, and that gp*groEL* is required for the formation of the putative initiator complex (gp*groES* has not yet been tested in this regard), we still do not know what the *groE* proteins do in phage assembly or in the host, or why the only distantly related processes of head assembly in such diverse phages as λ and T4 and tail assembly in phage T5 (Zweig and Cummings 1973) require the same host protein (gp*groEL*) even though the phage proteins appear to be essentially unrelated. The maturation of λ proheads is also quite mysterious. What is the enzyme(s) that performs the protein modifications, and what are the functions of the modifications? Although Hohn et al. (1979) observed proteolysis of unassembled gp*Nu3* in fairly crude extracts in a reaction requiring active gp*groEL* and gp*groES*, the protease activity sedimented more slowly than the gp*groEL* decatetramer. It is not clear how this relates to Murialdo's (1979) observation that gp*groEL* apparently must function before any of the protein modifications can occur.

The use of genetics has been invaluable in the characterization of the proteins and assembly steps discussed above. Most of these studies have used nonsense mutations so that the null phenotype could be studied. This is no doubt the most easily interpretable use of mutants to analyze assembly. Some of the proteins involved in head assembly are likely to have several sequential functions in the pathway. (For a complete discussion of this point, see King [1980].) For example, in P22 assembly, gp*1* may first bind to gp*5* and later could serve as the binding site for, or be part of, the DNA packaging machinery. Use of nonsense 1^- mutations would only give information about the first of these two roles. The use of missense mutations (which unfortunately may give rise to experimental results that are more difficult to interpret) can in theory give information about the secondary functions if the first is unaffected by the mutation, and a few studies (Jarvik and Botstein 1973; Georgopoulos et al. 1979; Katsura 1980) have successfully found such mutations. We feel that this approach will be necessary in the future analysis of the prohead assembly and maturation and DNA packaging steps in the assembly pathway.

Although much has been learned about phage assembly in general terms, the physicochemical details of the mechanisms involved remain almost universally obscure. We feel that this will remain a fertile area of research for some time to come, particularly in the areas of detailed structural analysis of supramolecular structures, protein chemistry and genetics of interacting systems, and use of these as model systems for general genetic, chemical, and physical methodologies.

ACKNOWLEDGMENTS

We thank Helios Murialdo, Roger Hendrix, and Michael Feiss for communication of unpublished results and discussion of their ideas about head assembly, and Jerri Cohenour for preparation of the manuscript.

REFERENCES

Bayer, M.E. and A.F. Bocharov. 1973. The capsid structure of bacteriophage lambda. *Virology* **54:** 465.

Bloomer, A., J. Champness, G. Bricogne, R. Staden, and A. Klug. 1978. Protein disk of tobacco mosaic virus at 2.8 Å resolution showing the interactions within and between subunits. *Nature* **276:** 362.

Boklage, C., E. Wong, and V. Bode. 1973. The lambda *F* mutants belong to two cistrons. *Genetics* **75:** 221.

Botstein, D., C.H. Waddell, and J. King. 1973. Mechanism of head assembly and DNA encapsulation in *Salmonella* phage P22. I. Genes, proteins, structures and DNA maturation. *J. Mol. Biol.* **80:** 669.

Brégégère, F. and F. Gros. 1971. Migration électrophorétique et synthèse des protéines de maturation de bactériophage lambda. *Biochimie* **53:** 679.

Casjens, S.R. 1974. Bacteriophage lambda *FII* gene protein: Role in head assembly. *J. Mol. Biol.* **90:** 1.

——. 1979. Molecular organization of the bacteriophage P22 coat protein shell. *J. Mol. Biol.* **131:** 1.

Casjens, S.R. and R.W. Hendrix. 1974a. Location and amounts of the major structural proteins in bacteriophage lambda. *J. Mol. Biol.* **88:** 535.

——. 1974b. Comments on the arrangement of the morphogenetic genes of bacteriophage lambda. *J. Mol. Biol.* **90:** 20.

Casjens, S.R. and J. King. 1974. P22 Morphogenesis. I. Catalytic scaffolding protein in capsid assembly. *J. Supramol. Struct.* **2:** 202.

——. 1975. Virus assembly. *Annu. Rev. Biochem.* **44:** 555.

Casjens, S.R., T. Hohn, and A.D. Kaiser. 1972. Head assembly steps controlled by genes *F* and *W* in bacteriophage λ. *J. Mol. Biol.* **64:** 551.

Caspar, D.L.D. and A. Klug. 1963. Physical principles in the construction of regular viruses. *Cold Spring Harbor Symp. Quant. Biol.* **27:** 1.

Chisholm, R., R. Deans, E. Jackson, D. Jackson, and J. Rutila. 1980. A physical gene map of the bacteriophage P22 late region: Genetic analysis of cloned fragments of P22 DNA. *Virology* **102:** 172.

Chothia, C. and J. Janin. 1975. Principles of protein-protein recognition. *Nature* **256:** 705.

Daniels, D., J. DeWet, and F. Blattner. 1980. A new map of bacteriophage lambda DNA. *J. Virol.* **33:** 390.

Deeb, S.S. 1970. Studies on the *in vitro* assembly of bacteriophage φ80 and φ80-lambda hybrids. *J. Virol.* **9:** 174.

Earnshaw, W. 1979. Modeling of the small angle X-ray diffraction arising from the surface lattices of phages lambda and P22. *J. Mol. Biol.* **131:** 14.

Earnshaw, W.C. and S.R. Casjens. 1980. DNA packaging by the double-stranded DNA bacteriophages. *Cell* **21:** 319.

Earnshaw, W.C. and S.C. Harrison. 1977. DNA arrangement in isometric phage heads. *Nature* **268:** 598.

Earnshaw, W.C. and J. King. 1978. Structure of phage P22 protein aggregates formed in the absence of the scaffolding protein. *J. Mol. Biol.* **126:** 721.

Earnshaw, W., S. Casjens, and S.C. Harrison. 1976. Assembly of the head of bacteriophage P22: X-ray diffraction from heads, proheads, and related structures. *J. Mol. Biol.* **104:** 387.

Earnshaw, W.C., R.W. Hendrix, and J. King. 1979. Structural studies of bacteriophage lambda heads and proheads by small angle X-ray diffraction. *J. Mol. Biol.* **134:** 575.

Echols, H. and H. Murialdo. 1978. Genetic map of bacteriophage lambda. *Microbiol. Rev.* **42:** 577.

Feiss, M., R.A. Fisher, D. Siegele, and W. Widner. 1981. Bacteriophage λ and 21 packaging specificities. In *Bacteriophage assembly* (ed. M. DuBow), p. 213. Alan R. Liss, New York.

Ferrucci, F. and H. Murialdo. 1981. Bacteriophage λ prohead assembly: Assembly of biologically active precursors *in vitro*. In *Bacteriophage assembly* (ed. M. DuBow), p. 193. Alan R. Liss, New York.

Floor, E. 1970. Interaction of morphogenetic genes of bacteriophage T4. *J. Mol. Biol.* **47:** 293.

Forbes, D.D. and I. Herskowitz. 1982. Polarity suppression by the *Q* gene product of bacteriophage λ. *J. Mol. Biol.* **160:** 549.

Fuller, M. and J. King. 1980. Regulation of coat protein polymerization by the scaffolding protein of bacteriophage P22. *Biophys. J.* **32:** 381.

Georgopoulos, C.P., R.W. Hendrix, S.R. Casjens, and A.D. Kaiser. 1973. Host participation in bacteriophage lambda head assembly. *J. Mol. Biol.* **76:** 45.

Georgopoulos, C.P., R. Bisig, M. Magazin, H., Eisen, and D. Court. 1979. Novel bacteriophage lambda mutation affecting head assembly. *J. Virol.* **29:** 782.

Gillin, F.D. and V.C. Bode. 1971. The arrangement of DNA in lambda phage heads. II. λ DNA after exposure to micrococcal nuclease at the site of head-tail joining. *J. Mol. Biol.* **62:** 503.

——. 1972. Micrococcal nuclease treatment of bacteriophage heads alters the right-hand cohesive end of lambda deoxyribonucleic acid. *J. Virol.* **10:** 863.

Goldenberg, D. and J. King. 1981. Temperature-sensitive mutants blocked in the folding or subunit assembly of the bacteriophage P22 tail spike protein. II. Active mutant proteins matured at 30°C. *J. Mol. Biol.* **145:** 633.

Harrison, D.P., D.T. Brown, and V.C. Bode. 1973. The lambda head-tail joining reaction: Purification, properties, and structure of biologically active heads and tails. *J. Mol. Biol.* **79:** 437.

Harrison, S. 1980. Protein interfaces and intersubunit bonding. *Biophys. J.* **32:** 139.

Harrison, S., A. Olson, C. Schutt, F. Winkler, and G. Bricogne. 1978. Tomato bushy stunt virus at 2.9 Å resolution. *Nature* **276:** 368.

Hendrix, R.W. 1979. Purification and properties of *groE*, a host protein involved in bacteriophage assembly. *J. Mol. Biol.* **129:** 375.

Hendrix, R.W. and S.R. Casjens. 1974. Protein fusion: A novel reaction in bacteriophage λ head assembly. *Proc. Natl. Acad. Sci.* **71:** 1451.

——. 1975. Assembly of bacteriophage lambda heads: Protein processing and its genetic control in petit λ assembly. *J. Mol. Biol.* **91:** 187.

Herskowitz, I. and E. Signer. 1971. Control of transcription from the *r* strand of bacteriophage lambda. *Cold Spring Harbor Symp. Quant. Biol.* **35:** 355.

Hohn, B. and T. Hohn. 1974. Activity of empty, head-like particles for packaging of DNA of bacteriophage λ *in vitro*. *Proc. Natl. Acad. Sci.* **71:** 2372.

Hohn, T. and I. Katsura. 1977. Structure and assembly of bacteriophage lambda. *Curr. Top. Microbiol. Immunol.* **78:** 69.

Hohn, T., H. Flick, and B. Hohn. 1975. Petit λ, a family of particles from coliphage lambda infected cells. *J. Mol. Biol.* **98:** 107.

Hohn, T., M. Wurtz, and B. Hohn. 1976. Capsid transformation during packaging of bacteriophage λ DNA. *Philos. Trans. R. Soc. Lond. B.* **276:** 51.

Hohn, T., B. Hohn, A. Engel, M. Wurtz, and P.R. Smith. 1979. Isolation and characterization of the host protein *groE* involved in bacteriophage lambda assembly. *J. Mol. Biol.* **129:** 359.

Hohn, B., M. Wurtz, B. Klein, A. Lustig, and T. Hohn. 1974. Phage lambda DNA packaging *in vitro*. *J. Supramol. Struct.* **2:** 302.

Imamoto, T. and C. Yanofsky. 1967. Transcription of the tryptophan operon in polarity mutants of *Escherichia coli*. II. Evidence for normal production of tryp-mRNA molecules and for premature termination of transcription. *J. Mol. Biol.* **28:** 25.

Imber, R., A. Tsugita, M. Wurtz, and T. Hohn. 1980. The outer surface protein of bacteriophage lambda. *J. Mol. Biol.* **139:** 277.

Israel, V., T. Anderson, and M. Levine. 1967. *In vitro* morphogenesis of phage P22 from heads and base-plate parts. *Proc. Natl. Acad. Sci.* **57:** 284.

Jarvik, J. and D. Botstein. 1973. A genetic method for determining the order of events in a biological pathway. *Proc. Natl. Acad. Sci.* **70:** 2046.

——. 1975. Conditional lethal mutations that suppress genetic defects in morphogenesis by altering structural proteins. *Proc. Natl. Acad. Sci.* **72:** 2738.

Kaiser, A.D. and T. Masuda. 1973. *In vitro* assembly of bacteriophage lambda heads. *Proc. Natl. Acad. Sci.* **70:** 260.

Kaiser, A.D., M. Syvanen, and T. Masuda. 1974. Processing and assembly of the head of bacteriophage lambda. *J. Supramol. Struct.* **2:** 318.

Karamata, D., E. Kellenberger, G. Kellenberger, and M. Terzi. 1962. Sur une particule accompagnant le développement du coliphage λ. *Pathol. Microbiol.* **25:** 575.

Katsura, I. 1980. Structure and inherent properties of the bacteriophage lambda head shell. II. Isolation and initial characterization of prophage mutants defective in gene *E*. *J. Mol. Biol.* **142:** 387.

——. 1981. Assembly systems in molecular biology: A graph theory of genetic complementation between two related species. *J. Theor. Biol.* **88:** 503.

Katsura, I. and A. Tsugita. 1977. Purification and characterization of the major protein and the terminator protein of bacteriophage lambda tail. *Virology* **76:** 129.

Kellenberger, E. and R. Edgar. 1971. Structure and assembly of phage particles. In *The bacteriophage lambda* (ed. A.D. Hershey), p. 271. Cold Spring Harbor Laboratory, Cold Spring Harbor, New York.

King, J. 1980. Regulation of structural protein interactions as revealed in phage morphogenesis. In *Biological regulation and development* (ed. R. Goldberger), vol. 2, p. 101. Plenum Press, New York.

King, J. and S.R. Casjens. 1974. Catalytic head assembling protein in virus morphogenesis. *Nature* **251:** 112.

King, J., C. Hall, and S.R. Casjens. 1978. Control of the synthesis of phage P22 scaffolding protein is coupled to capsid assembly. *Cell* **15:** 551.

King, J., V. Lenk, and D. Botstein. 1973. Mechanism of head assembly and DNA encapsidation in *Salmonella* phage P22. II. Morphogenetic pathway. *J. Mol. Biol.* **80:** 697.

Kochan, J.P. 1983. "Identification and characterization of biologically active intermediates in bacteriophage lambda prohead assembly." Ph.D. thesis, University of Toronto, Toronto, Canada.

Künzler, P. and T. Hohn. 1978. Stages of bacteriophage lambda head morphogenesis: Physical analysis of particles in solution. *J. Mol. Biol.* **122:** 191.

Lenk, E., S.R. Casjens, J. Weeks, and J. King. 1975. Intracellular visualization of precursor capsids in phage P22 mutant infected cells. *Virology* **68:** 182.

Min Jou, W., G. Haegeman, M. Ysebaert, and W. Fiers. 1972. Nucleotide sequence of the gene coding for bacteriophage MS2 coat protein. *Nature* **237:** 82.

Mount, D.W.A., A.W. Harris, C.R. Fuerst, and L. Siminovitch. 1968. Mutations in bacteriophage lambda affecting particle morphogenesis. *Virology* **35:** 134.

Murialdo, H. 1979. Early intermediates in bacteriophage lambda prohead assembly. *Virology* **96:** 411.

Murialdo, H. and A. Becker. 1977. Assembly of biologically active proheads of bacteriophage lambda *in vitro. Proc. Natl. Acad. Sci.* **74:** 906.

——. 1978a. Head morphogenesis of complex double-stranded deoxyribonucleic acid bacteriophages. *Microbiol. Rev.* **42:** 529.

——. 1978b. A genetic analysis of bacteriophage lambda prohead assembly *in vitro. J. Mol. Biol.* **125:** 57.

Murialdo, H. and P. Ray. 1975. Model for arrangement of minor structural proteins in head of bacteriophage λ. *Nature* **257:** 815.

Murialdo, H. and L. Siminovitch. 1972a. The morphogenesis of bacteriophage lambda. IV. Identification of gene products and control of the expression of the morphogenetic information. *Virology* **48:** 785.

——. 1972b. The morphogenesis of phage lambda. V. Form-determining function of the genes required for the assembly of the head. *Virology* **48:** 824.

Oppenheim, D.S. and C. Yanofsky. 1980. Translation coupling during expression of the tryptophan operon of *Escherichia coli. Genetics* **95:** 785.

Parkinson, J.S. 1968. Genetics of the left arm of the chromosome of bacteriophage lambda. *Genetics* **59:** 311.

Poteete, A.R. and J. King. 1977. Functions of two new genes in *Salmonella* phage P22 assembly. *Virology* **76:** 725.

Poteete, A.R., V. Jarvik, and D. Botstein. 1979. Encapsulation of P22 DNA *in vitro. Virology* **95:** 550.

Ray, P. and H. Murialdo. 1975. The role of gene *Nu3* in bacteriophage lambda head morphogenesis. *Virology* **64:** 247.

Ray, P.N. and M.L. Pearson. 1974. Evidence for posttranscriptional control of the morphogenetic genes of bacteriophage lambda. *J. Mol. Biol.* **85:** 163.

——. 1975. Functional inactivation of bacteriophage λ morphogenetic gene mRNA. *Nature* **253:** 647.

Serwer, P. 1976. Internal proteins of bacteriophage T7. *J. Mol. Biol.* **107:** 271.

Shaw, J.E. and H. Murialdo. 1980. Morphogenetic genes *C* and *Nu3* overlap in bacteriophage λ. *Nature* **283:** 30.

Showe, M.K. and L.W. Black. 1973. Assembly core of bacteriophage T4: An intermediate in head formation. *Nat. New Biol.* **242:** 70.

Showe, M.K. and L. Onorato. 1978. Kinetic factors and form determination of the head of bacteriophage T4. *Proc. Natl. Acad. Sci.* **75:** 4165.

Smith, D. and J. King. 1981. Temperature-sensitive mutants blocked in folding or subunit assembly of the bacteriophage P22 tail spike protein. II. Inactive polypeptide chains synthesized at 39°C. *J. Mol. Biol.* **145:** 653.

Smith, D., P. Berget, and J. King. 1980. Temperature-sensitive mutants blocked in the folding or subunit assembly of the bacteriophage P22 tail-spike protein. I. Fine-structure mapping. *Genetics* **96:** 331.

Stahl, F.W. and N.E. Murray. 1966. The evolution of gene clusters and genetic circularity in microorganisms. *Genetics* **53:** 569.

Sternberg, N. 1973. Properties of a mutant of *Escherichia coli* defective in bacteriophage λ head formation (*groE*). II. The propagation of phage λ. *J. Mol. Biol.* **76:** 25.

——. 1976. A genetic analysis of bacteriophage λ head assembly. *Virology* **71:** 568.

Steven, A. 1981. Visualization of virus structure in three dimensions. *Methods Cell Biol.* **22:** 297.

Susskind, M. and D. Botstein. 1978. Molecular genetics of bacteriophage P22. *Microbiol. Rev.* **42:** 385.

Tilly, K. and C. Georgopoulos. 1982. Evidence that the two *Escherichia coli groE* morphogenetic gene products interact *in vivo. J. Bacteriol.* **149:** 1082.

Tilly, K., H. Murialdo, and C. Georgopoulos. 1981. Identification of a second *Escherichia coli groE* gene whose product is necessary for bacteriophage morphogenesis. *Proc. Natl. Acad. Sci.* **78:** 1629.

Tsui, L. and R.W. Hendrix. 1980. Head-tail connector of bacteriophage lambda. *J. Mol. Biol.* **142:** 419.

Weigle, J. 1966. Assembly of phage λ *in vitro. Proc. Natl. Acad. Sci.* **55:** 1462.

——. 1968. Studies on head-tail union in bacteriophage lambda. *J. Mol. Biol.* **33:** 483.

Weinstock, G. and D. Botstein. 1980. Genetics of bacteriophage P22. IV. Correlation of genetic and physical map using translocatable drug-resistance elements. *Virology* **106:** 92.

Weinstock, G., P. Riggs, and D. Botstein. 1980. Genetics of bacteriophage P22. III. The late operon. *Virology* **106:** 82.

Williams, R. and K. Richards. 1974. Capsid structure of bacteriophage lambda. *J. Mol. Biol.* **88:** 547.

Wood, W.B. and J. King. 1979. Genetic control of complex bacteriophage assembly. *Comp. Virol.* **13:** 581.

Youderian, P. 1978. "Genetic control of the length of lambdoid phage tails." Ph.D. thesis, Massachusetts Institute of Technology, Cambridge, Massachusetts.

Youderian, P. and M.M. Susskind. 1980. Identification of the products of bacteriophage P22 genes, including a new late gene. *Virology* **107:** 258.

Young, R., J. Way, S. Way, J. Yin, and M. Syvanen. 1979. Transposition mutagenesis of bacteriophage lambda. A new gene affecting cell lysis. *J. Mol. Biol.* **132:** 307.

Zachary, A., L.O. Simon, and S. Litwin. 1976. Lambda head morphogenesis as seen in the electron microscope. *Virology* **72:** 429.

Zweig, M. and D.J. Cummings. 1973. Cleavage of head and tail proteins during bacteriophage T5 assembly: Selective host involvement in the cleavage of a tail protein. *J. Mol. Biol.* **80:** 505.

DNA Packaging and Cutting

Michael Feiss
Department of Microbiology
College of Medicine
University of Iowa
Iowa City, Iowa 52242

Andrew Becker
Department of Medical Genetics
University of Toronto
Toronto, Ontario, Canada
M5S 1A8

INTRODUCTION

The head of bacteriophage λ is assembled when the products of two separate synthetic pathways come together. One, the DNA replication pathway, produces a linear polymer of λ chromosomes called a concatemer. In the other pathway, λ head proteins assemble to produce the empty capsid shell known as the prohead (Georgopoulos et al., this volume). DNA and proheads interact, and a chromosome from the concatemer is inserted into the cavity of the prohead and is separated from the unencapsidated remainder of the DNA by endonucleolytic cutting.

Knowledge of λ head assembly was presented by Yarmolinsky (1971) and by Kellenberger and Edgar (1971) in *The Bacteriophage Lambda*. The review of Hohn and Katsura (1977) offers a more recent summary of the subject. Casjens and King (1975), Murialdo and Becker (1978b), and Wood and King (1979) have written comparative reviews on head morphogenesis and DNA packaging. Most recently, the problem of DNA packaging in the double-stranded DNA phages has been updated in an elegant synthesis by Earnshaw and Casjens (1980). Because of space limitations in this paper, we restrict discussion of other viruses to results not emphasized in other reviews.

CHROMOSOME STRUCTURE AND DNA PACKAGING

The linear λ chromosome is injected into the cell, and annealing and ligation of the cohesive ends rapidly cyclize the molecule. Early in the growth cycle, bidirectional replication results in progeny rings. Later, rolling-circle replication produces the concatemers that are the substrates for DNA packaging (Skalka 1977; Furth and Wickner, this volume). DNA packaging generates mature DNA molecules through the introduction of staggered nicks at the sites of the annealed cohesive ends by the phage enzyme terminase. The site of action of terminase is called the cohesive end site, or *cos* (Emmons 1974; Feiss and Campbell 1974).

Model for Chromosome Packaging

Emmons (1974) and Feiss and Campbell (1974) examined the chromosomes packaged during infections by *cos* duplication mutants of λ. The manner in which *cos* duplications present in replicative concatemers were segregated at packaging led Emmons to propose a model for packaging and *cos* cutting in λ. The essential features of this model were (1) random selection of a *cos* site for packaging initiation, (2) processive packaging of two to three chromosomes in series from a concatemer, (3) dependence of *cos* cutting on packaging of a minimal length of DNA, and (4) a polarity of chromosome entry into the capsid from the *Nu1-A* end to the *R* end. Implicit in our résumé of the model (also to be found in Emmons [1974] and in Feiss and Campbell [1974]) is the idea that prohead-*cos* interactions underlie packaging specificity and that this specificity is mediated by terminase, the enzyme that also catalyzes *cos* cleavage. Ample evidence for this idea appears throughout this review.

Subsequent experiments with *cos* duplication phages, in which the packaging sequence length was limited to one, gave results supporting random initiation and processive and polarized packaging (Feiss and Bublitz 1975; Feiss et al. 1977). Dependence of *cos* cleavage on chromosome length (and hence on packaging) has been observed for phages with short chromosomes (Henderson and Weil 1977; Feiss and Siegele 1979). Feiss and Siegele found inefficient cleavage (75% cleavage) of terminal *cos* sites (the second *cos* cleavage needed to generate a mature chromosome) when the chromosome length was 77% of wild type. Efficiency of cleavage rose sharply to approximately 100% for a chromosome length of 90%. The basis of this dependence is obscure (see below). The length dependence of terminal *cos* cleavage plays only a minor role in the efficiency of packaging of chromosomes with lengths between 78% and 105%; within this range, efficiency is roughly constant (Feiss et al. 1977).

A wealth of additional support for polarized packaging includes the study of *doc* particles by Sternberg and Weisberg (1977a). *doc* particles result from packaging prophage DNA: *doc*L particles carry a DNA segment from *cos* rightward through the late genes into bacterial DNA, including *bio*; *doc*R DNA extends leftward from *cos* through *gal*. *doc*L particles are about 100-fold more abundant, presumably reflecting packaging asymmetry.

An in vitro demonstration of polarity has been obtained by B. Hohn and T. Bickle (cited in Hohn and Katsura 1977). Only the terminal fragment from the *Nu1-A* end of the chromosome is packaged when restriction fragments of λ DNA are used.

Structure of the DNA Substrate

Many DNA structures have been examined for packageability, and the results lead to the generalization that λ chromosomes are normally generated by cleavage of two *cos* sites. In addition to concatemeric DNA, a λ chromosome can be generated by cutting the *cos* sites of (1) tandem double prophages (Mousset and Thomas 1969) and (2) a repressed circular *cos* duplication phage chromosome (Feiss and Margulies 1973). The requirement for two *cos* sites led to models in which aligned *cos* sites were required for *cos* cleavage (Wang and Brezinski 1973; Freifelder et al. 1974). These alignment models have been ruled out by subsequent study of *doc*L particles (Sternberg and Weisberg 1977a). The *doc*L particles, headlike structures that carry a DNA segment from the left cohesive end through the late genes and into the bacterial DNA including *bio*, are produced in an amount roughly equivalent to the amounts of λ chromosomes produced when two prophages are present in tandem. The *doc*L particles would be much less abundant if *cos* alignment were necessary; *doc*L particles carry a DNA segment that protrudes from the head because of the absence of a terminal *cos* site for cleavage.

A long-standing mystery is the failure of monomer circles to be packaged (Szpirer and Brachet 1970; Stahl et al. 1972; Enquist and Skalka 1973). Models to explain the failure propose that there is a configuration for the DNA : terminase : capsid complex that is necessary for packaging and cleavage and that cannot be attained with circular monomers. Two models proposed so far (Sternberg and Weisberg 1977a) imagine a structural difficulty late in packaging: (1) a requirement for a segment of unpackaged DNA protruding from the head that monomers would not provide because they would be completely enveloped; (2) packaging of circular monomer DNA might leave a constrained loop of DNA outside the capsid that would block further packaging. With either model, the circular monomer : capsid complex must be unstable, because filled heads are rare in sections of infected bacteria when the DNA is mainly monomeric circles (Dawson et al. 1975), and the empty head structures found under these conditions are functional proheads, which means that the unstable circular monomer : capsid complexes would have to dissociate before capsid expansion (Dawson et al. 1976). The proposal of the packaging of a complete chromosome without expansion of the prohead must be considered problematical (see the discussion on the role of gp*D* below). We really do not know why monomeric circles are not packaged and would appreciate suggestions.

Although there are a number of examples of packaging of concatemeric DNA (λ, P22, T4, T7), phage P2 and its satellite P4 are exceptions in that the packaging substrate is monomeric circular DNA (Pruss et al. 1975).

TERMINASE

The production of the cohesive ends of λ DNA has been measured experimentally in three ways. One sensitive bioassay is the helper phage-dependent transfection system of Kaiser and Hogness (1960), where the uptake of the transforming λ DNA by helper-infected cells depends specifically upon the presence of cohesive ends (Kaiser and Inman 1965) homologous to those of the helper (Mandel 1967; Miller and Feiss 1981). Cohesive end production can also be measured by physical assay involving restriction enzyme analysis. The presence of free cohesive ends is indicated within the restriction pattern by a cut in the fragment spanning *cos*. A third, more inferential way of measuring *cos*-site cleavage in vitro is to ask that exogenously added concatemers of λ DNA be packaged into viable phage. Proper cleavage at the *cos* sites in this pathway is inferred from the generation of plaque-forming units.

Using the helper-dependent transfection system, Wang and Kaiser (1973) first observed terminase activity catalyzed by λ-infected cell extracts operating on covalently closed circular forms of λ DNA. Cohesive end production took place when the extracts were made from λA^+, but not from λA^-, phage-infected cells, implicating gp*A* as an essential phage-specific component of the terminase system.

In vivo evidence for interactions between the *A* product and specific DNA sequences during packaging was presented by Sternberg and Weisberg (1977a). The rare packaging of bacterial DNA was studied relative to the packaging of phage DNA. The ratio was changed for two missense mutations in *A* but not for several other missense mutations affecting prohead structural proteins. These results indicated that specific interactions required for packaging were affected by *A* missense mutations and, together with the results of Wang and Kaiser given above, reinforced the idea that *cos* cutting and DNA selection for packaging were under the control of the same protein, gp*A*.

Subsequent genetic studies (Weisberg et al. 1979), as well as biochemical work based on the establishment of cell-free systems for the packaging of λ DNA into phage (Kaiser and Masuda 1973; Hohn and Hohn 1974; Becker and Gold 1975; Kaiser et al. 1975), identified a second λ gene, *Nu1*, as essential for the postulated activities of terminase (Becker et al. 1977b; Sumner-Smith and Becker 1982).

Nu1 and *A* are the two leftmost genes on the λ chromosome, and mutations in either of them give identical phenotypes, namely, the accumulation of λ DNA concatemers as well as that of proheads that are wild type in their structure (Murialdo and Siminovitch 1972b) and are competent to package DNA in vitro (Becker et al. 1977b). This syndrome is taken as evidence of a failure to bring the DNA and proheads together for packaging and for *cos*-site cleavage, the two functions ascribed to terminase.

A single protein, now designated terminase, has been purified extensively on the basis of its ability to complement λA^- phage-infected cell extracts for packaging of λ DNA into infectious phage (Becker and Gold 1978; M. Gold and A. Becker, in prep.). The purified protein also complements λ$Nu1^-$ phage-infected cell extracts for in vitro packaging (Becker et al. 1977a; Sumner-Smith et al. 1982a). This result suggests that the terminase preparation is providing both polypeptides, gp*Nu1* and gp*A*, in active form to the system.

Indeed, during chromatography of terminase, two different polypeptides invariably copurified and were exactly coincident with the profile of terminase biological activity (M. Gold and A. Becker, in prep.). This was taken to mean that the active terminase is a heterooligomer of two nonidentical polypeptides. One of these is most certainly gp*A* (74–79 kD) (Murialdo and Siminovitch 1972a), on the basis of its comigration in denaturing gels with authentic *A* product. The other polypeptide is smaller (21.5 kD), and a variety of indirect evidence suggests that it is the product of the *Nu1* gene (Sumner-Smith and Becker 1982; Sumner-Smith et al. 1982a).

The retrieval of terminase as a gp*Nu1* and gp*A* heterooligomer may be the result of a selection imposed by the assay used in purification with a resultant discarding of other forms of these proteins, if they exist. It is possible that, in vivo, at least some gp*Nu1* and/or gp*A* molecules exist free of one another, entering the packaging pathway independently, presumably at adjacent points. Such a mechanism could ultimately bring the two proteins into a functional, physically stable association. At least during ϕ80 infection, gp*1* (homologous to gp*Nu1*) exists in several hundredfold excess over gp*2* (homologous to gp*A*) (Sumner-Smith and Becker 1982), so that its presence in vivo as a species free of gp*2* (though not necessarily its action as such) can be inferred with some certainty. Other phages produce terminase comprised of two polypeptides. For P2, genetic and biochemical evidence indicate that the terminase contains gp*M* and gp*P* (Bowden and Calendar 1979; Bowden and Modrich 1981); for P22, terminase comprises gp*2* and gp*3* (Raj et al. 1974; Schmieger and Backhaus 1975; Tye 1976; Poteete and Botstein 1979; Jackson et al. 1982; Lanski and Jackson 1982). By inference based on phenotypes, the corresponding genes in T7 would be *18* and *19* (Hausmann and LaRue 1969; Kerr and Sadowski 1975), and in T4 genes, *16* and *17* (for summary, see Black 1981; Black et al. 1981).

Physical measurements on terminase give 117 kD for the native promoter, the smallest form of the enzyme that shows all of the partial reactions so far detected. This mass can be accommodated in a protein made up of one gp*A* (75-kD) subunit and two gp*Nu1* (21-kD) subunits. Not all of the purified enzyme, however, is present in this protomeric form; an apparent dimer, as well as larger aggregates of the enzyme, have been observed.

Purified terminase is a multifunctional protein for which a number of partial reactions have been deduced. These activities of terminase, to be examined below from the viewpoint of DNA packaging and maturation, are as follows: (1) binding to the *cos* site, (2) cutting at *cos* to generate the cohesive ends, (3) prohead binding, and (4) hydrolysis of ATP to ADP and inorganic phosphate in the presence of DNA. More speculatively, it has been suggested that terminase plays a structural or catalytic role in the reaction of DNA entry into the capsid.

COHESIVE END SITE

The cohesive end site (*cos*) is defined as the site of action of terminase. This is a functional definition, and *cos* therefore includes all of the base pairs with which terminase has specific interactions that are required for function. Terminase performs several functions: recognition of the λ DNA by binding to *cos*, prohead binding, and *cos* cleavage. Hence, the interactions between terminase and *cos* may be complex. We describe here structural studies on the location, size, and sequence of *cos*.

Studies of transducing variants of λ, insertion mutations, and cloned DNA segments show that chromosomes can be packaged if they contain a small DNA segment that includes the annealed cohesive ends (for summary, see Syzbalski and Szybalski 1979). As of this writing, *cos* is confined to a 300-bp segment. The pk25 insertion of Tn*903* is about 100 bp to the *R* gene side of the annealed cohesive ends, and the t16 insertion is 200 bp to the *Nu1-A* side. Neither insertion disrupts *cos* function. Also, a cloned 400-bp *Hin*cII fragment with the cohesive end sequence centrally located is fully functional (Meyerowitz et al. 1980). *cos* is probably much smaller than 300 bp; recent trimming experiments in the laboratory of one of us (M.F.) show that a *cos* site retaining about 100 bp to the *Nu1-A* side is functional.

Sequencing studies have proceeded from the cohesive ends of λ into the duplex DNA at the mature DNA termini (Wu and Taylor 1971; Weigel et al. 1973; Nichols and Donelson 1978). The available sequence information, presented in Figure 1, reveals several features. The first feature is a segment of hyphenated rotational symmetry that includes the cohesive end sequence. The symmetry segment is 22 bp long; it contains 10 bp that are symmetrical and an additional 10 bp that are symmetrical with respect to purines and pyrimidines. The sites at which nicks are introduced by terminase to generate cohesive ends are symmetrically disposed. The conventional interpretation of twofold symmetry in a DNA sequence is that the bases interact with the symmetrically arranged subunits of a multimeric protein (Bernardi 1968; Murray et al. 1977). Such an interpretation has recently received experimental support from the determination of the structure of the *cro* repressor protein as a dimer that is symmetrically complementary to the symmetric bases of the operators to which it binds (Anderson et al. 1981). The high-order dependence of terminase activity on concentration suggests that the active form of terminase is a multimer (Hohn 1975; Becker et al. 1977a), an observation consistent with the symmetry hypothesis.

Comparative studies show the cohesive end sequence to be strongly conserved among lambdoid phages. The cohesive ends of phages λ, 21, ϕ80, 82, and 424 are identical (Murray and Murray 1973). The entire 22-bp symmetry segments of ϕ80 and λ-21 hybrid 19 (see below) have been sequenced and are identical to that of λ (Nichols 1978). One exception has been found: ϕD326 cohesive ends have a 2-base difference from λ as shown in Figure 1 (Murray et al. 1975). The two changes are transversions at positions showing purine : pyrimidine symmetry in λ. Phages ϕ80, 82, 424, 434, and ϕD326 are all able to package λ chromosomes and so have the same terminase specificity (Mousset and Thomas 1969; Kayajanian 1971; Murray et al. 1975). The sequence exception, ϕD326, is not different in terminase specificity, so the significance of the purine-pyrimidine symmetry at these two positions in λ is doubtful. Phage 21 (and the related λ-21 hybrid 19) packages λ chromosomes poorly and has a different ter-

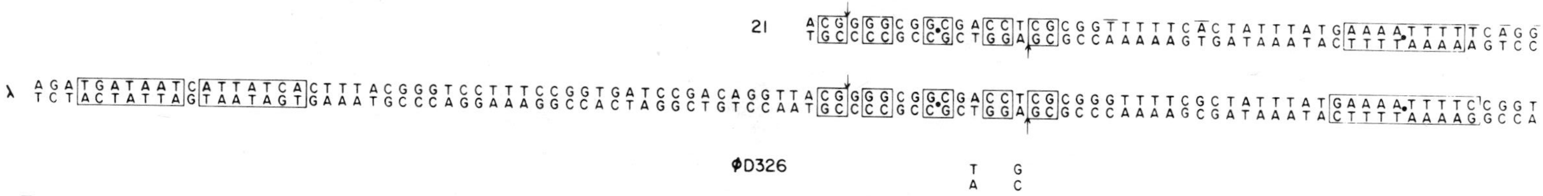

Figure 1 Sequences at *cos*. Arrows mark sites of introduction of nicks by terminase to generate cohesive ends. Boxed base pairs show rotational symmetry around axes marked by dots. The λ sequence is taken from Nichols and Donelson (1978). Nichols and Donelson noted considerable sequence variation in the segment to the left of the cohesive end sequence for two strains of λ*c*I857*Sam7*; only one sequence is shown here. The 21 sequence is from Nichols (1978). Base pairs for which λ and 21 differ are overlined in the 21 sequence. Base pairs in the cohesive ends that are different in ϕD326 are also shown (Murray et al. 1975).

minase specificity (Hohn 1975; Feiss et al. 1979). However, 21 terminase can cut a λ*cos* site under particular conditions (Feiss et al. 1979), as discussed below. For the lambdoid phages, then, the cohesive end symmetry segment sequence and terminase specificity are highly conserved. In sum, the comparative sequence information is consistent with but does not prove the view that the cohesive end symmetry segment interacts with symmetrically arranged terminase subunits.

To the *Nu1* side of the annealed cohesive ends there is another symmetry segment (Fig. 1) of unknown significance, though a role in terminase binding is an attractive guess. This symmetry segment, with its axis of rotation 31 bp away from the axis of the G + C-rich cohesive end segment, is very rich in A + T base pairs. The first ATG sequence that might mark the beginning of the *Nu1* gene is at the beginning of this symmetry element. Comparative information exists for ϕ80 and 21 (see Fig. 1) for this region (Nichols 1978). The ATG sequence is conserved for ϕ80 and 21; the symmetry segment is 2 bp larger for ϕ80 and 2 bp smaller for 21 when compared with λ. At about 50 bp to the *R* gene side of the cohesive end sequence there is another A + T-rich symmetry segment of unknown significance; this element is conserved in ϕ80 (Nichols 1978). One possibility for the role of this symmetry segment is interaction with the tail in completed phages (Thomas et al. 1978).

λ DNA PACKAGING

Site-specific Interaction of Terminase with λ DNA: Complex I

Biochemical Studies

Terminase : DNA packaging intermediates (called complex-I structures [Hohn and Katsura 1977]) have been identified in extracts of λ-infected cells (Hohn 1975; Hohn et al. 1976). Purified terminase also binds to λ DNA concatemers in vitro, and this binary complex, which can be isolated by simple physical techniques, can serve as an intermediate in head assembly (Becker et al. 1977a). Since terminase initiates λ DNA packaging from *cos* or its vicinity, specific binding of the enzyme to the DNA at this position was anticipated. Biochemical support for this supposition, however, was only provided by an unexpected property of the in vitro binding reaction; virtually all of the *cos* sites were cleaved, even in the absence of proheads (Becker and Gold 1978), indicating that, indeed, the formation of this packaging intermediate includes a contact between terminase and *cos* (Becker and Gold 1978).

When λ DNA concatemers are used as the substrate to be packaged in vitro, the formation of stable, biologically active binary complexes requires the addition of Mg^{++}, spermidine, ATP, and a basic (pI = 10) 22-kD *Escherichia coli* protein (Becker et al. 1977a; W. Parris and M. Gold, pers. comm.). These requirements are similar, if not identical, to those for the *cos* site cleavage that takes place at this time (Becker and Gold 1978). Thus, the binding and *cos* cutting reactions are not clearly separable in this system. It is of interest that ATP is needed to detect any biologically productive complex of sufficient stability to survive the dilution that takes place during isolation procedures.

It should be noted here that terminase catalyzes the hydrolysis of ATP to ADP and inorganic phosphate in the presence of DNA (M. Gold and A. Becker, in prep.). The ability to interact with or hydrolyze ATP may be a general property of the terminases of the double-stranded DNA phages (Poteete and Botstein 1979; D. Bowden, pers. comm.). The role of ATP in complex-I formation remains a matter for speculation. ATP hydrolysis could, e.g., be coupled to the movement of terminase along DNA in the enzyme's "search" for its specific site of binding (Yuan et al. 1980).

The order of the reaction between terminase and chromosomes has been estimated from kinetic and quasi-equilibrium studies. The data suggest that at least three, and possibly four, terminase protomers are required per complex formed when concatemers of λ DNA are used, and half that number are required when the substrate is linear monomeric λ DNA (Hohn 1975; Becker et al. 1977a, and unpubl.). Thus, the molar requirement for terminase in the "cut-plus-package" case is apparently twice that of the "package-only" case.

Genetic Studies

Terminase interaction(s) with *cos* may be complex because of the multiple functions of terminase: *cos* binding, prohead binding, and *cos* cutting. Although a physical description of these interactions is not yet available, evidence for such complexity is accumulating through the study of *cos* variants.

The *cos*1 mutation is a 400-bp deletion extending from *cos* (the deletion is presumed to lack the cohesive end sequence) into *Nu1* (Fisher et al. 1980). This mutation inactivates *cos* as a site for both the initiation and termination of packaging.

Phage 21 has a different packaging specificity from that of λ (Hohn 1975), as mentioned above. Each packages the other's DNA with an efficiency of about 1%. A hybrid phage, λ-21 hybrid 19, having the left cohesive end and head region from 21 and the rest of the chromosome from λ, was shown to have phage 21 packaging specificity (Feiss et al. 1979).

Packaging specificity of λ and λ-21 hybrid 19 is determined by the initial *cos* site: λ will not package a chromosome when the λ-21 hybrid *cos* site is at the site of initiation but will package a chromosome when the λ-21 *cos* site (called cos^{21}) is at the terminal position. The same conclusion is found for the specificity of the phage 21 packaging system: specificity for the initial *cos* site but not the terminal *cos* site (Feiss et al. 1979). The basis for λ and 21 specificities has been shown to involve a terminase : *cos* interaction (Feiss et al. 1981). The 21 *cos* site is cleaved about 1% as well as cos^{λ} by λ terminase in vitro. The failure to cut cos^{21} is due to a failure of λ terminase to bind to cos^{21}, a conclusion obtained from competition studies with the in vitro cleavage reaction (Feiss and Widner 1982). The λ and 21 packaging specificities can be explained by divergence in the interactions involved in binding of *cos* by terminase. This result implies that the cohesive end symmetry region is not the major determinant of terminase binding, because λ and λ-21 hybrid 19 are identical for the cohesive end symmetry sequence yet differ in terminase binding. Where might sequences for terminase binding lie? Hohn (1975) has argued that there might be a terminase binding site at the left chromosome end near to but separate from the cohesive end segment. She proposed this because (1) mature DNA can be packaged in vitro even though it does not have a complete *cos* segment, (2) packaging proceeds in a *Nu1-A* to *R* direction, and (3) λ and 21 have identical cohesive ends yet differ in packaging specificity. The newer results fit her proposal well, as do other results presented below. Terminase of λ indeed does not bind cos^{21}, and the base pairs involved must be at the left chromosome terminus outside the cohesive end symmetry region. The sequence at the left terminus of λ-21 hybrid 19 is presented in Figure 1, and it reveals many base-pair differences, some of which are likely involved in terminase-binding specificity.

Additional support for the concept of separate terminase-binding and cutting sites comes from study of the *cos*2 mutation, a deletion of the cohesive end sequence. A *cos*2 competitor plasmid binds λ terminase nearly as well as a cos^{+} plasmid (Feiss et al. 1983). Deletion of the cohesive end sequence does not alter terminase binding significantly, though cutting is abolished. Thus, *cos*2 is defective in cutting but not binding, and cos^{21} is defective in binding λ terminase but not cutting.

A second conclusion can be drawn from considering the question, How can λ terminase be unable to bind cos^{21} and still be able to cut cos^{21} in the terminal position? The obvious conclusion is that the terminase that cuts the terminal *cos* site binds at the initial *cos* site and is brought in contact with the terminal *cos* site by the packaging process. The implications of this conclusion in packaging are considered below. As far as *cos* structure is concerned, the sequences required at initial and terminal *cos* sites are obviously different. The terminase binding site is not required for cleavage of the terminal *cos* site. Separate binding and cutting sites have also been suggested for the enzymes involved in phage P22 (Casjens and Huang 1982) and Mu chromosome maturation (George and Bukhari 1981). A binding site necessary for packaging of adenovirus DNA is located near one end of the chromosome (Hammarskjöld and Winberg 1980).

Prohead Binding to Complex I: Complex II, A Prohead : Terminase : DNA Intermediate

Biochemical Studies

Prohead-DNA packaging intermediates have been demonstrated in λ-infected cell lysates both by function (Hohn et al. 1976) and by electron microscopy (Syvanen and Yin 1978; Yamagishi and Okamoto 1978; Okamoto and Yamagishi 1980). In addition, it has been possible to form such complexes in vitro (Becker et al. 1977a; Syvanen and Yin 1978) by allowing λ DNA, terminase, and proheads to interact. The intermediate formed, a prohead : terminase : DNA ternary complex (complex II), can be isolated in preparative velocity sedimentation gradients. It is defined by its convertibility into phage upon the addition of a source of gp*D*, gp*W*, gp*FII*, and tails but without the need for a further addition of proheads and terminase (Becker et al. 1977a).

The cofactor requirements for ternary complex formation are similar if not identical to those for the formation of the terminase : DNA binary complex : Mg^{++}, a polyamine, ATP, and one or more host-coded factors. Since attempts to form and isolate biologically productive binary terminase : prohead or DNA : prohead intermediates have not met with success, it appears that ternary complex formation proceeds, first, by the formation of the terminase : DNA complex, which is then followed by prohead binding.

Genetic Studies

A study by Okamoto and Yamagishi (1980) indicates that gp*Nu1* alone can bind proheads and DNA in a ternary complex. These authors examined, in the electron microscope, intermediates in λ DNA packaging that accumulated in vivo as a result of mutational blocks in certain head genes. In the case where gp*Nu1* was present but gp*A* was inactivated by an amber mutation, they observed empty proheads bound at *cos* on the DNA thread. These results suggest that gp*Nu1* is capable of an independent action in which it promotes the binding of proheads to DNA, perhaps at the proper site for the initiation of packaging. After this binding has taken place, gp*A* may enter the pathway, and DNA packaging and *cos* cutting take place. Whether or not such a sequential action by gp*Nu1* and gp*A* is operative in a wild-type infection is not known. On the surface, the findings of Okamoto and Yamagishi on the role of gp*Nu1* seem at variance with two sets of genetic data.

First, Sternberg and Weisberg (1977a) observed that missense mutations in the *A* gene altered the proportion of λ, λ*doc*L, λ*doc*R, and generalized transducing particles made, implying that it is the role of gp*A* to select which DNA can enter λ capsids. This apparent paradox can be resolved by postulating mechanisms in which a strong interaction between gp*Nu1* and gp*A* takes place as an early step and that both components, acting in concert, ultimately govern the specificity of the initiation step.

A separate study indicates that proheads also bind to the gp*A* component of terminase to form complex II (M. Feiss et al., in prep.). λ terminase is specific for cos^{λ} and λ proheads; 21 terminase is similarly specific. The λ-21 hybrid 33 terminase is specific for cos^{21} and λ proheads, and this hybrid specificity is due to a hybrid terminase. The phage-21-derived DNA of λ-21 hybrid 33 is the left 5% of the chromosome, a substitution that apparently results in a hybrid gp*A* in which the carboxyl terminus is of λ origin. Because the terminases of 21 and λ-21 hybrid 33 differ only in the carboxyl terminus of gp*A* yet differ in prohead specificity, it is concluded that the carboxyl terminus of gp*A* determines prohead specificity. The apparent conflict in the results of Okamoto and Yamagishi (1980) and of M. Feiss et al. (in prep.) could be resolved if gp*Nu1* can bind proheads to *cos*, but a productive complex leading to packaging requires a second specific interaction between the carboxyl terminus of gp*A* and the prohead.

Cohesive End Site Cleavage and DNA Packaging

It has been known for a long time that, in vivo, the generation of DNA molecules with cohesive ends depends not only on active terminase but also on functional proheads (Dove 1966; MacKinlay and Kaiser 1969; Wake et al. 1972; McClure and Gold 1973). The simple interpretation is that the appearance of molecules cut at *cos* depends on packaging and that terminase is not active to cut until triggered by packaging. Nevertheless, using extracts of cells infected by various λ head-gene mutants, Wang and Kaiser (1973) first showed that *cos* cutting in a cell-free system could take place in the absence of functional proheads, i.e., it could be uncoupled from packaging. This observation has been verified by showing that *cos* sites in λ DNA concatemers, as well as those in various cosmid DNAs (Becker and Gold 1978; Feiss et al. 1981; M. Gold et al., unpubl.), are cleaved in vitro by the action of isolated terminase.

Because in vivo terminase-dependent cleavage at *cos* requires proheads and may in fact depend upon extensive packaging of the DNA into proheads, why terminase should operate in vitro as a free nuclease is a puzzle (Becker and Gold 1978). The issue is particularly intriguing in the light of an observation by Hohn (1975), who showed that the λ DNA molecules accumulating during a $Nu1^{+}A^{+}prohead^{-}$ infection, which are known to be uncut, carry potentially active terminase molecules bound to them. It is possible that terminase is not juxtaposed to *cos* here, but the mechanism by which the enzyme is restrained, so to speak, in this case is not known. So far, *cos* cutting by isolated terminase has only been observed with purified DNA substrates.

Another possibility is that terminase does cleave *cos* sites without packaging in vivo, but the cohesive ends rapidly reanneal when packaging does not occur. One experiment asking about cutting without packaging was to search for end-to-end reassortment of λ chromosomes (Emmons 1974; Feiss and Campbell 1974). The negative results obtained indicate that if cutting-rejoining does occur, the same cohesive ends generated must reanneal together. The failure of monomeric circles to be packaged in vivo leads to similar notions.

The answer to whether *cos* cleavage depends on packaging may not turn out to be a strict yes or no. It may be that terminase in vivo has a low prohead-independent activity. Such an activity might make the first *cos* cleavage to initiate the processive packaging of a series of two or three chromosomes from a concatemer. Should packaging not quickly follow this cleavage, rejoining of the cohesive ends might ensue.

The apparent inability to cut chromosomes in the absence of packaging in vivo is typical for the double-stranded DNA phages. An exception is phage P22, where one terminase-dependent cleavage can take place in the absence of packaging (Jackson et al. 1982). This cleavage is at the *pac* site, the site of initiation of processive packaging of a sequence of chromosomes. The further cleavages of the sequence of packagings depend on headful packaging.

Role of gp*FI*

The role of the *FI*-gene product (Benchimol and Becker 1982) in λ head assembly is uncertain. The approximately 20-kD protein is made in large amounts (~ 10^5 copies) in a λ-infected cell but is not a structural component of the capsid.

The phenotype of *FI* mutations resembles that of *Nu1* and *A* mutations in that there is an accumulation of immature uncut DNA and of proheads that are competent for, but not engaged in, packaging (Murialdo and Siminovitch 1972b; Boklage et al. 1973; Hohn et al. 1975; Becker et al. 1977b). *FIam* mutations, however, are leaky (Dove 1966; Wiegle 1966; Parkinson 1968; MacKinlay and Kaiser 1969; Murialdo et al. 1981), as are double *FIam* mutations, suggesting that gp*FI* serves an ancillary role in the packaging reaction and that λ development is partially independent of *FI* function in vivo. Furthermore, packaging of λ DNA in vitro proceeds with equal efficiency in the presence or absence of gp*FI* (Becker et al. 1977b; Benchimol et al. 1978).

Pseudorevertants of *FIam* mutations have been isolated (Murialdo et al. 1981), and the mutations, called *fin*, map in the terminase genes. Some but not all of the *fin* mutations result in increased levels of gp*A*. Presumably, all of the *fin* mutations affect terminase activity, suggesting that an interplay exists between *FI* and terminase functions.

In vitro DNA packaging requires gp*FI* only when single-stranded DNA is present in the packaging reaction. gp*FI* prevents the action of an endonuclease that is activated by single-stranded DNA. The endonuclease, gp*ben*, is encoded by the *b* region of λ and, surprisingly, the endonuclease activity is not inhibited by gp*FI*. Whatever the significance of the gp*FI*-gp*ben* interaction, gp*FI* must have another function(s) because λFI^-ben^- does not grow (Benchimol et al. 1982; Sumner-Smith et al. 1982b).

DNA Entry and Condensation

In being packaged, the DNA is concentrated several hundredfold in the face of energetically unfavorable factors. These factors include polyelectrolyte repulsion from the charges present on the phosphodiester backbone, superhelical bending of the stiff duplex polymer, and conformational restriction of the DNA upon compaction (Riemer and Bloomfield 1980). The more recent speculations on mechanism benefit greatly from a better knowledge of the physical state and geometry of the packaged DNA derived principally from X-ray diffraction studies on DNA-filled heads. For six double-stranded DNA phages examined to date, the architecture of the packaged DNA is strikingly similar, suggesting a unitary underlying mechanism of DNA packaging.

Speculation has focused on connecting DNA entry with one or more capsid-related events known to occur at the time of packaging. If the search is to be for a mechanism that is common to all of the double-stranded DNA phages, it is reasonable to exclude from discussion those events that are peculiar to only some of the phages. The discussion will therefore not deal with events involving the accessory proteins that reside in proheads of certain phages (but not in λ)—neither their possible roles as spools or DNA-interacting lattices, nor their purported functions in promoting DNA collapse. The factors that are common to all of the phages include (1) introduction of phage DNA to the prohead by terminase, (2) presence of a specialized structure (the connector) at one corner of the icosahedral prohead, which can act as orifice for DNA entry, (3) interaction of the DNA with the inner surface of the capsid shell, (4) prohead expansion, (5) involvement of polyamines, and (6) utilization of ATP during packaging.

The interactions between λ DNA and terminase and of this binary complex with proheads have been discussed above. It is proposed that this *cos* DNA : terminase complex makes specific contact with a ringlike receptor structure, called the connector, present at the proximal prohead vertex (evidence for a connector-terminase interaction is available only in the case of phage T4 [Hsiao and Black 1977; Black et al. 1981]). The λ connector, which is constructed from four molecules of gp*B* and about eight molecules of p*B** (a cleavage product of gp*B*), serves as the primer for prohead assembly (Hendrix and Casjens 1975; Hohn et al. 1975; Murialdo and Becker 1978a; Tsui and Hendrix 1980). In addition, by virtue of being incorporated into the prohead, it provides a structure through which the DNA can pass during packaging and through which it can exit at injection. The connector has been purified recently on the basis of its ability to prime the assembly of biologically active proheads in vitro (J. Kochan and H. Murialdo, pers. comm.). In the electron microscope, these connectors appear as disks with radial 12-fold symmetry and an axial hole of about 20 Å diameter. It is assumed that the DNA, led by the leftmost chromosomal sequence, is passed into the prohead through this structure. It is not known whether the first DNA to enter penetrates as a free left end or if the double helix is looped in past the DNA of the left chromosomal extremity, which somehow remains fixed at the prohead orifice. For the free end to enter, *cos*-site cleavage must take place at the start of packaging, but the converse need not follow; *cos*-site cleavage at this time does not guarantee free-end entry nor exclude the possibility of a loop-in mechanism. Electron micrographs of the prohead-DNA complexes present in a λ D^- infection show predominantly capsid structures situated terminally on a DNA end, suggesting that the initiator *cos* site may be cut before packaging is completed (Okamoto and Yamagishi 1980).

Speculation on the fate of the DNA as it enters the prohead cavity must take account of the architecture of packaged DNA as deduced from physical studies. The DNA is B form, tightly packed to a density similar to hexagonal single crystals of DNA. The DNA appears bent smoothly in the head, forming concentric shells or layers in which adjacent helical segments run parallel, like thread on a spool. Several bacteriophages and reovirus (Harvey et al. 1981) give remarkably similar X-ray diffraction patterns, indicating structural similarity. The packed DNA can be described as an ordered, multilayered toroidal superhelix or solenoid. X-rays scattered from DNA on opposite sides of the head interfere, suggesting that these antipodal DNA segments are fixed in position with respect to one another and that this relationship is the same in all particles. Such a long-range regularity could arise if the outer-layer DNA segments interacted in a specific manner with the protein subunits in the lattice of the capsid shell (Earnshaw et al. 1976). Two types of argument favor the idea that the assembly of the DNA solenoid proceeds sequentially from outside inward. One states that the first DNA to enter the empty shell, because of its stiffness and charged nature, will seek a maximum radius of curvature and inevitably establish contact with the wall. Another argument states that to achieve an uncomplicated mode of exit upon injection, unwinding of the solenoid should be from the inside out, with the DNA nearest the axis of the spool the first to exit; hence, the assembly of the spool should proceed radially, from the outer DNA shell inward. Thus, the ordered pattern of the outer DNA shell may be interpreted as evidence of an ordered protein-DNA interaction taking place early in packaging. Drawing the analogy to crystallization (Earnshaw and Casjens 1980), this interaction has been viewed as a primary heterologous (DNA-protein) nucleation step for the DNA-DNA packing arrangement that is to follow (Richards et al. 1973; Earnshaw et al. 1976, 1978; Earnshaw and Harrison 1977). The major capsid protein, gp*E*, is known to bind to DNA (Brody 1973) and, in the capsid, the positively charged DNA binding sites are likely available on the inner surface of the shell (Hohn and Katsura 1977).

At some point after DNA packaging has begun, the prohead expands. Prohead expansion at packaging is a feature common to all of the double-stranded DNA phages. It is the result of a transition in the membrane lattice that involves a conformational change of the capsid protein subunits rather than a disassembly-reassembly process (Hohn et al. 1974; Kaiser et al. 1975; Laemmli et al. 1976; Lickfeld et al. 1976; Steven et al. 1976; Wurtz et al. 1976; Zachery and Simon 1977).

In λ, the capsid diameter expands 20%, corresponding to a 100% increase in capacity. B. Hohn (cited in Earnshaw and Casjens 1980) has examined the relationship between the amount of DNA packaged and prohead expansion. She observed that when the leftmost *Bam*HI fragment (11.4% of the λ chromosomal length) is packaged in vitro, proheads do not expand, whereas the packaging of a fragment 44.5% of λ DNA is accompanied by expansion. It may be calculated that the fraction of DNA that comes to occupy the outer solenoidal shell and hence to contact the prohead wall lies between these two values, but a cause-effect relationship is not established and the molecular trigger for expansion remains unknown.

It has been proposed (Hohn et al. 1974; Serwer 1975) that prohead expansion may be the mechanism that propels the DNA into the capsid. In these models, the increase in volume of a solvent-impermeable container leads to a pressure differential across the envelope that draws the DNA in. Per se, the mechanism cannot account for the packaging of the whole of the phage chromosome (Earnshaw and Casjens 1980) and may not be used at all for the entry of the first 10% or more of λ DNA. Apparently, capsid expansion is a consequence rather than a cause of DNA entry. Although of unquestionable importance (if for no other reason than to allow the accommodation of the entire chromosome in the head), the full significance of expansion does not yet seem appreciated.

After prohead expansion has taken place, gp*D* enters the capsid membrane in numbers equivalent to the major coat protein gp*E* (Casjens and Hendrix 1974; Imber et al. 1980). The entry of a normal-size chromosome requires that gp*D* be incorporated into the membrane lattice, otherwise the packaging apparatus disrupts (Sternberg and Weisberg 1977b). Neither the mechanism of this disruption nor the manner of the stabilization by gp*D* is clear. Some speculation on this issue may be possible on the basis of observed differences between the packing density for λ chromosomes of wild-type length versus that for deletion mutants of λ (Earnshaw and Harrison 1977).

According to X-ray diffraction data, λ chromosomes that are shorter than normal are packed more loosely, i.e., the double helices are farther apart. When does this loosening take place? One possibility is that packing proceeds according to the tight format throughout the DNA-entry cycle and that, for shortened chromosomes, loosening takes place after DNA entry is complete. Alternatively, the packing density may increase continuously during DNA entry, and the looser packing seen for shortened chromosomes is a record of the situation as it exists when DNA entry is terminated prematurely, so to speak. The second model is favored since it is not obvious why forces, initially holding the DNA tightly packed, are ultimately relaxed. For a packing density that increases continuously as more and more DNA is screwed in, shearing forces would come into play between apposed DNA segments and between the outer-shell DNA and the head membrane. Some readjustment in DNA-membrane contacts or reinforcement of the shell would seem in order, especially toward the end of DNA filling. gp*D* incorporation into the shell lattice may provide this function.

Intracapsid DNA is packed under pressure, with polyelectrolyte repulsion being the major energetic factor tending to destabilize the solenoid and, probably, the major force that drives the DNA from the head at injection (Earnshaw and Harrison 1977; Riemer and Bloomfield 1980). This thermodynamically unfavorable process may be compensated by binding multivalent cations to the DNA. Polyamines can, by themselves, induce the collapse of DNA to packing densities comparable to that of intracapsid DNA (Leng and Felsenfeld 1966;

Haynes et al. 1970; Laemmli 1975; Gosule and Schellman 1976; Chattoraj et al. 1978; Wilson and Bloomfield 1979), and crude in vitro packaging systems all require that polyamines be added. Earnshaw and Casjens (1980) have summarized the arguments, suggesting that the function of polyamines in DNA packaging is to stabilize the DNA-filled head rather than promote the packaging process directly.

Just as prohead expansion may be viewed as the result rather than the cause of DNA entry, so the DNA binding to the inner capsid wall may be seen as the consequence of the buildup in DNA concentration, rather than as a mechanism for pulling in the DNA (Laemmli 1970). Having taken these views, one is forced to seek another mechanism by which the DNA is transferred into the membrane cavity.

Currently under scrutiny are active transport models where the energy derived from ATP hydrolysis is used to ''pump'' the DNA into the head. A specific category of these models has terminase, confined at the orifice of the capsid membrane, catalyzing triphosphate cleavage and the translocation of the DNA into the prohead (Becker et al. 1977a; Black and Silverman 1978; Hendrix 1978; Murialdo and Becker 1978b; Poteete and Botstein 1978; Earnshaw and Casjens 1980). The terminase-catalyzed packaging models are based on the premises that (1) all of the in vitro DNA packaging reactions studied so far require ATP as a cofactor, (2) purified terminases catalyze ATP hydrolysis in the presence of DNA, and (3) precedents exist for enzyme translating along double-stranded DNA using the free energy derived from ATP cleavage; such, e.g., is the action of the restriction endonuclease *Eco*K (Yuan et al. 1980). To date, however, evidence is meager to indicate that terminase is directly involved in the mechanism of DNA passage into the prohead. Since several steps of the λ DNA packaging reaction can be studied as isolated events in vitro, some effort has been made to deal head-on with this question. Thus, it has been possible to show that both ATP and terminase are required for functions beyond those of complex I/II formation and of *cos*-site cleavage (Becker et al. 1977a), but what these functions are has not been established. Equally attractive models of DNA entry, which invoke ATP hydrolysis but not necessarily a translocase function for terminase, have been proposed.

Among these models, the one put forward by Hendrix (1978) is the most explicit. The Hendrix model is based on a mismatch in rotational symmetries between the proximal vertex of an icosahedral shell (fivefold) and the apposed connector ring (the gp*B*-derived connector in λ) with sixfold rotational symmetry. Because of the mismatch, the bonding between the capsid membrane and the connector ring is weak, so that rotation of one with respect to the other is possible. Packaging takes place when the DNA, driven by an ATP-coupled mechanism, is translated through the ring that rotates relative to the capsid. In one version of the model, the energy is applied at the DNA-connector interface, as would be the case if terminase, following the screw of the double helix, were acting as a DNA translocase at this position. Here, intracapsid DNA, restricted in its movement by bonding to the capsid wall, would accumulate supertwists, one for every turn of the double helix passed in. Passive rotation of the capsid membrane relative to the ring would relieve the tortional stresses imposed on the DNA. In a second model, the energy is applied between the ring and the prohead membrane, eliciting an active rotation of one with respect to the other. The DNA, like a threaded rod in a nut, is passively screwed in. This mechanism offers a model for a biological motor based on a mismatch of rotational symmetries. One full rotation is achieved in 30 discrete steps that correspond to the equivalent positions between a hexagon and a pentagon rotating 360 degrees relative to each other. Each step utilizes a relatively small packet of energy, as might be derived from the hydrolysis of an ATP molecule.

Role of gp*D*

The product of the *D* gene is a small protein (M_r = 11,000) that is found in phage heads in amounts equimolar to the major head protein gp*E* (Imber et al. 1980). The *D* protein acts late in packaging, adding to sites on the capsid generated by transformation of the prohead during packaging (Wurtz et al. 1976). The *D* protein acts after gp*A* in the packaging of λ^+ chromosomes (Kaiser et al. 1975). In cells infected with λD^-, a low but significant level of cohesive ends can be detected (Wake et al. 1972; McClure and Gold 1973; Wang and Kaiser 1973), and in sections, normal size (expanded) heads are seen that are partially filled with DNA (Lickfeld et al. 1976). Sternberg and Weisberg (1977b) speculated that gp*D* might stabilize λ heads and asked whether gp*D* might be dispensable for strains with deleted chromosomes. Parkinson and Huskey (1971) had previously shown that the stability of λ particles to chelating agents increases with decreasing DNA content. Sternberg and Weisberg discovered that λ deletion mutants with chromosomes less than 82% the size of λ^+ could grow without gp*D*. These phages lacking gp*D* have heads of normal size that contain chromosomes cut at the usual sites; they are very sensitive to chelating agents but can be stabilized by gp*D* addition in vitro. It was found that gp*D* is not involved in changing the capacity of the head, because there was no evidence that the size of the DNA packaged into *doc*L particles was reduced under D^- conditions. In the model favored by Sternberg and Weisberg, gp*D* acts to stabilize the head during packaging. There is a strong analogy between the roles of gp*D* and the T4 *soc* protein (Ishii and Yanagida 1975) in stabilizing heads. They propose that on infection by λD^- with a full-length chromosome, the D^- head becomes destabilized when filled with more than 82% of the chromosome and dissociates. Thus, gp*D* acts to stabilize the head and is not directly involved in prohead transformation or terminase action.

Cutting of the Terminal *cos* Site

Cutting of the terminal *cos* site does depend on packaging. Several lines of evidence support this assertion. Becker et al. (1977a) found that active ter-

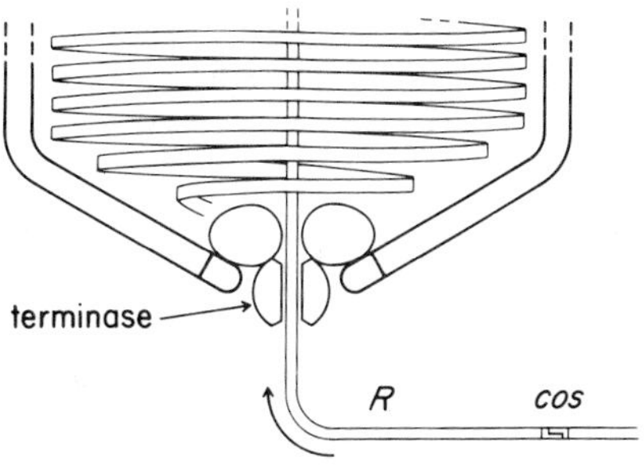

Figure 2 Scanning for the terminal *cos* site, showing part of a nearly filled head with terminase positioned at the connector. The DNA strand being packaged is "scanned" by terminase until the cohesive end symmetry segment of the terminal *cos* site is recognized and cut. Other aspects of packaging, such as the terminase-connector interaction, are purely hypothetical.

minase was required at a post-complex-II step in packaging. Kaiser et al. (1975) found a step requiring terminase that occurred after DNA packaging and gp*D* addition. Work cited above demonstrates that the probability of terminal *cos* cleavage depends on chromosome length and that terminase is brought into contact with the terminal *cos* site by packaging. The latter conclusion comes from the ability of λ terminase to cut a terminal *cos* site of phage-21 specificity, even though λ terminase cannot bind to a *cos* site of 21 specificity. The terminase molecule(s) that bind and cleave the initial *cos* site must also cleave the terminal *cos* site and be brought into contact with the terminal *cos* site by the packaging process. Figure 2 presents this model for scanning. What happens when a terminal *cos* site already has terminase bound to it? Packaging of the previous chromosome might pause until packaging of the next chromosome results in *cos* cleavage, or subunit dissociation and exchange might occur to result in the proper multimeric form of terminase. A logical consequence of the model is that all the terminase molecule(s) needed to package a chromosome bind at the initial *cos* site; the situation of failing to cut a terminal *cos* site because a required terminase protein has failed to bind it does not arise. Feiss et al. (1979) constructed lysogens of tandem prophages in which the initial *cos* site was cos^{λ} and the terminal *cos* site was of λ or 21 specificity. These lysogens were infected with a heteroimmune derivative of λ, and the yields of packaged prophages were measured. The yields of packaged prophages were about the same regardless of the specificity of the terminal *cos* site. In the helper packaging situation, therefore, chromosomes are packaged with equal efficiency whether or not they have a terminase-binding site at the terminal *cos* site. It follows that the terminase molecules that bind to the initial *cos* site are capable of cleaving both the initial and terminal *cos* sites. What fraction of terminal *cos* sites have already bound a terminase before the terminase bound at the initial *cos* site is brought into contact with the terminal *cos* site by packaging? Should terminase binding to *cos* be rate-limiting, the fraction might be very low.

These results, indicating that terminase is brought into contact with the terminal *cos* site through DNA packaging, correlate well with the results cited above, showing that the probability of cleavage of the terminal *cos* site depends on chromosome length. DNA packaging is the only identified assembly step in which chromosome length might be of functional significance. One can only speculate on the molecular explanation of the dependence of terminal *cos* cleavage on DNA length. One idea is that DNA packaging triggers prohead expansion, which might in turn activate terminase to recognize and cut. For short chromosomes the terminal *cos* site might sometimes be packaged before terminase activation. Another idea is that the velocity of DNA entry may be length-dependent—fast at first and then slowing as the head fills. The ability of terminase to cut the terminal *cos* site might then depend inversely on the packaging rate. To scan for the terminal *cos* implies that terminase is in contact with the DNA molecule throughout packaging; this is consistent with the proposal that terminase is a translocase. The fate of terminase following *cos* cleavage is unknown. Processive packaging could be a consequence of terminase remaining bound to the *Nu1-A* end of the next chromosome of the concatemer.

CONCLUSION

Much progress has been made in the last decade in understanding λ DNA packaging. A great deal has been gained from the establishment of cell-free packaging systems, which offer the opportunity to study the partial reactions of this complex pathway as isolated biochemical events, promoted by purified, well-characterized components. Yet λ research continues to enjoy the predictive force of detailed genetics. A central example is the terminase story. Although the enzyme is available in purified form and performs well its DNA binding, *cos*-cutting, and prohead-capture functions, genetics reminds us of the incompleteness of the biochemical treatment so far. Thus, the molecular basis for the control on the endonucleolytic activity of this enzyme remains to be determined, and an invitation is extended to describe the various interactive centers or domains of this protein. The structure of the enzyme's DNA target, *cos*, reflects this complexity, with separate sites for terminase binding and cutting.

Numerous other questions remain, and some are of central importance. Two such problems are the mechanism of DNA transport into the head and of the DNA condensation that takes place upon entry. An understanding of these problems is of interest not only to virus biology but could also provide a deeper insight into phenomena involving chromosome movement and folding in general.

ACKNOWLEDGMENTS

We thank Costa Georgopoulos, Helios Murialdo, Marvin Gold, and our colleagues and co-workers for criticism and discussion. During the writing of this review we were supported in part by grants from the National Institutes of Health and the Medical Research Council of Canada.

Note Added in Proof

The points at which terminase makes contact with *cos* DNA are being examined by DNase protection methods (A. Becker et al., in prep.). So far, information is available for the *r* strand. On a 220-bp *cos*-containing restriction fragment, the footprint (Galas and Schmitz 1978) is complex and extensive, with points of enhanced DNase cleavage, as well as protected and less well-protected segments alternating. DNA between positions of approximately −45 to +95, and perhaps beyond, is involved in this pattern (where −1 and +1 are the base pairs to the left and right, respectively, of the center of the cohesive end symmetry segment). Qualitatively, the footprint is similar to that induced by *E. coli* DNA gyrase about its sites of cutting, where points of enhanced DNase cleavage are interpreted as being "exposed" on the surface of the protein due to the wrapping of the DNA (Liu and Wang 1978; Fisher et al. 1981; Morrison and Cozzarelli 1981). The range of DNA involvement raises the question of which sequences are biologically essential or specific and which are involved merely because of contiguity or because terminase can interact with DNA less specifically. Information on this point comes from experiments where *cos*-containing DNA is resected and tested for terminase-dependent functions. Such studies (M. Feiss et al., in prep.) show *cos* to contain two separate sites, one where nicks are introduced, *cosN*, and a terminase binding site, *cosB*. *cosB* is located between +51 and a point between +93 and +190. DNA molecules carrying these base pairs compete for terminase as well as cos^{+} DNA in competition experiments (see text). *cosN* is confined to a segment from −22 to +24; DNA molecules carrying this segment and a functional *cosB* are cleaved normally in vitro. In vivo studies are in progress to ask whether *cosN* and *cosB* are sufficient for packaging or whether additional requirements exist, such as the *cosN-cosB* spacing or other recognition sites.

Related studies by Miwa and Matsubara (1983) yield a similar picture. They find that the sequences necessary for packaging into infectious phage extend from −40 to +120. They further show, in vitro, that there is a segment, called by them the minimal region for cos^{λ} cutting, that extends from −22 to +38. This site corresponds to *cosN* in the work discussed above. They also define an enhancing region, necessary for terminase binding and high efficiency *cos* cutting, that extends from +46 to +120. This segment corresponds to *cosB*.

REFERENCES

Anderson, W.F., D.H. Ohlendorf, Y. Takeda, and B.S. Matthews. 1981. Structure of the *cos* repressor from bacteriophage λ and its interaction with DNA. *Nature* **290:** 754.

Becker, A. and M. Gold. 1975. Isolation of the bacteriophage lambda *A*-gene protein. *Proc. Natl. Acad. Sci.* **72:** 581.

——. 1978. Enzymatic breakage of the cohesive end site of phage λ DNA: terminase (ter) reaction. *Proc. Natl. Acad. Sci.* **75:** 4199.

Becker, A., M. Marko, and M. Gold. 1977a. Early events in the *in vitro* packaging of bacteriophage λ DNA. *Virology* **78:** 291.

Becker, A., H. Murialdo, and M. Gold. 1977b. Studies on an *in vitro* system for the packaging and maturation of phage λ DNA. *Virology* **78:** 277.

Benchimol, S. and A. Becker. 1982. The *FI*-gene product of bacteriophage λ. Purification and properties. *Eur. J. Biochem.* **126:** 227.

Benchimol, S., H. Lucko, and A. Becker. 1982. Bacteriophage λ DNA packaging in vitro. The involvement of the *FI* gene product, single-stranded DNA, and a novel λ-directed protein in the packaging reaction. *J. Biol. Chem.* **257:** 5201

Benchimol, S., A. Becker, H. Murialdo, and M. Gold. 1978. The role of the bacteriophage lambda *FI*-gene product during phage head assembly. *Virology* **91:** 201.

Bernardi, G. 1968. Mechanism of action and structure of acid deoxyribonuclease. *Adv. Enzymol.* **31:** 1.

Black, L. 1981. The mechanism of bacteriophage DNA encapsidation. In *Bacteriophage assembly* (ed. M. Dubow), p. 97. Alan R. Liss, New York.

Black, L. and D. Silverman. 1978. Model for DNA packaging into bacteriophage T4 heads. *J. Virol.* **28:** 643.

Black, L.W., A.L. Zachary, and V. Manne. 1981. Studies of the mechanism of bacteriophage T4 DNA encapsidation. In *Bacteriophage assembly* (ed. M. DuBow), p. 111. Alan R. Liss, New York.

Boklage, C.E., E. C.-T. Wong, and V.C. Bode. 1973. The lambda *F* mutants belong to two cistrons. *Genetics* **75:** 221.

Bowden, D. and R. Calendar. 1979. Maturation of bacteriophage P2 DNA *in vitro*: A complex, site-specific system for DNA cleavage. *J. Mol. Biol.* **129:** 1.

Bowden, D.W. and P. Modrich. 1981. *In vitro* studies on the bacteriophage P2 terminase system. In *Bacteriophage assembly* (ed. M. DuBow), p. 223. Alan R. Liss, New York.

Brody, T. 1973. A DNA-binding form of the main structural protein of lambda heads. *Virology* **54:** 441.

Casjens, S.R. and R.W. Hendrix. 1974. Location and amounts of the major structural proteins in bacteriophage lambda. *J. Mol. Biol.* **88:** 535.

Casjens, S.R. and J. King. 1975. Virus assembly. *Annu. Rev. Biochem.* **44:** 555.

Chattoraj, D., L. Gosule, and J. Schellman. 1978. DNA condensation with polyamines. II. Electron microscopic studies. *J. Mol. Biol.* **121:** 327.

Dawson, P., B. Hohn, and A. Skalka. 1976. Functional empty capsid precursors produced by a lambda mutant defective for late λ DNA replication. *J. Virol.* **17:** 576.

Dawson, P., A. Skalka, and L. Simon. 1975. Bacteriophage lambda head morphogenesis: Studies on the role of DNA. *J. Mol. Biol.* **93:** 167.

Dove, W.F. 1966. Action of the lambda chromosome. I. The control of functions late in bacteriophage development. *J. Mol. Biol.* **19:** 187.

Earnshaw, W.C. and S.R. Casjens. 1980. DNA packaging by the double-stranded DNA bacteriophages. *Cell* **21:** 319.

Earnshaw, W.C. and S.C. Harrison. 1977. DNA arrangement in isometric phage heads. *Nature* **268:** 598.

Earnshaw, W., S. Casjens, and S.C. Harrison. 1976. Assembly of the head of bacteriophage P22: X-ray diffraction from heads, proheads, and related structures. *J. Mol. Biol.* **104:** 387.

Earnshaw, W.C., J. King, S.C. Harrison, and F.A. Eiserling. 1978. The structural organization of DNA packaged within the heads of T4 wild-type, isometric and giant bacteriophages. *Cell* **14:** 559.

Emmons, S.W. 1974. Bacteriophage lambda derivatives carrying two copies of the cohesive end site. *J. Mol. Biol.* **83:** 511.

Enquist, L.W. and A. Skalka. 1973. Replication of bacteriophage λ dependent on the function of host and viral genes. I. Interaction of *red*, *gam* and *rec*. *J. Mol. Biol.* **75:** 185.

Feiss, M. and A. Bublitz. 1975. Polarized packaging of bacteriophage lambda chromosomes. *J. Mol. Biol.* **94:** 583.

Feiss, M. and A. Campbell. 1974. Duplication of the bacteriophage lambda cohesive end site: Genetic studies. *J. Mol. Biol.* **83:** 527.

Feiss, M. and T. Margulies. 1973. On packaging of the bacteriophage lambda chromosome. *Mol. Gen. Genet.* **127:** 285.

Feiss, M. and D.A. Siegele. 1979. Packaging of the bacteriophage lambda chromosome: Dependence of *cos* cleavage on chromosome length. *Virology* **92:** 190.

Feiss, M. and W. Widner. 1982. Bacteriophage λ DNA packaging: Scanning for the terminal cohesive end site during packaging. *Proc. Natl. Acad. Sci.* **79:** 3498.

Feiss, M., I. Kobayashi, and W. Widner. 1982. Separate sites for binding and nicking of bacteriophage λ DNA by terminase. *Proc. Natl. Acad. Sci.* **80:** 955.

Feiss, M., R.A. Fisher, M.A. Crayton, and C. Egner. 1977. Packaging of the bacteriophage lambda chromosome: Effect of chromosome length. *Virology* **77:** 281.

Feiss, M., R.A. Fisher, D.A. Siegele, and W. Widner. 1981. Bacteriophage λ and 21 packaging specificities. In *Bacteriophage assembly* (ed. M. DuBow), p. 213. Alan R. Liss, New York.

Feiss, M., R.A. Fisher, D.A. Siegele, B.P. Nichols, and J.E. Donelson. 1979. Packaging of the bacteriophage lambda chromosome: A role for base sequence outside *cos*. *Virology* **92:** 56.

Fisher, R., Krizsanovich-Williams, and M. Feiss. 1980. Construction and characterization of a cohesive end site mutant of bacteriophage λ. *Virology* **107:** 144.

Fisher, L.M., K. Mizuuchi, M. O'Dea, H. Ohmori, and M. Gellert. 1981. Site-specific interaction of DNA gyrase with DNA. *Proc. Natl. Acad. Sci.* **78:** 4165.

Freifelder, D., L. Chud, and E. Levine. 1974. Requirement for maturation of *E. coli* bacteriophage lambda. *J. Mol. Biol.* **83:** 503.

Galas, D.J. and A. Schmitz. 1978. DNase footprinting: A simple method for the detection of protein-DNA binding specificity. *Nucleic Acids Res.* **5:** 3157.

George, M. and A.I. Bukhari. 1981. Heterogeneous host DNA attached to the left end of mature bacteriophage Mu DNA. *Nature* **292:** 175.

Gosule, L. and J. Schellman. 1976. Compact form of DNA induced by spermidine. *Nature* **259:** 333.

Hammarskjöld, M.L. and G. Winberg. 1980. Encapsidation of adenovirus 16 DNA is directed by a small DNA sequence at the left end of the genome. *Cell* **20:** 787.

Harvey, J.D., A.R. Bellamy, W.C. Earnshaw, and C. Schutt. 1981. Biophysical studies of reovirus type 3. Low-angle X-ray diffraction studies. *Virology* **112:** 240.

Hausmann, R. and K. LaRue. 1969. Variations in sedimentation patterns among deoxyribonucleic acids synthesized after infection of *Escherichia coli* by different amber mutants of bacteriophage T7. *J. Virol.* **3:** 278.

Haynes, M., R. Garrett, and W. Gratzer. 1970. Structure of nucleic acid-polybase complexes. *Biochemistry* **9:** 4410.

Henderson, D. and J. Weil. 1977. Morphogenesis of bacteriophage lambda deletion mutants: I. Abnormal head-related structures produced in normal *Escherichia coli*. *J. Mol. Biol.* **113:** 43.

Hendrix, R.W. 1978. Symmetry mismatch and DNA packaging in large bacteriophages. *Proc. Natl. Acad. Sci.* **75:** 4779.

Hendrix, R.W. and S.R. Casjens. 1975. Assembly of bacteriophage lambda heads: Protein processing and its genetic control in petit λ assembly. *J. Mol. Biol.* **91:** 187.

Hohn, B. 1975. DNA as a substrate for packaging into bacteriophage lambda *in vitro*. *J. Mol. Biol.* **98:** 93.

Hohn, B. and T. Hohn. 1974. Activity of empty, headlike particles for packaging of DNA of bacteriophage λ *in vitro*. *Proc. Natl. Acad. Sci.* **71:** 2372.

Hohn, B., M. Wurtz, B. Klein, A. Lustig, and T. Hohn. 1974. Phage lambda DNA packaging *in vitro*. *J. Supramol. Struct.* **2:** 302.

Hohn, T. and I. Katsura. 1977. Structure and assembly of bacteriophage lambda. *Curr. Top. Microbiol. Immunol.* **78:** 69.

Hohn, T., H. Flick, and B. Hohn. 1975. Petit λ, a family of particles from coliphage lambda infected cells. *J. Mol. Biol.* **98:** 107.

Hohn, T., M. Wurtz, and B. Hohn. 1976. Capsid transformation during packaging of bacteriophage λ DNA. *Philos. Trans. R. Soc. London B* **276:** 51.

Hsiao, C.L. and L.W. Black. 1977. DNA packaging and the pathway of bacteriophage T4 head assembly. *Proc. Natl. Acad. Sci.* **74:** 3652.

Imber, R., A. Tsugita, M. Wurtz, and T. Hohn. 1980. Outer surface protein of bacteriophage lambda. *J. Mol. Biol.* **139:** 277.

Ishii, T. and M. Yanagida. 1975. Molecular organization of the shell of the T even bacteriophage head. *J. Mol. Biol.* **97:** 655.

Jackson, E.N., F. Lanski, and C. Andres. 1982. Bacteriophage P22 mutants that alter the specificity of DNA packaging. *J. Mol. Biol.* **154:** 551.

Kaiser, A.D. and D.S. Hogness. 1960. The transformation of *Escherichia coli* with deoxyribonucleic acid isolated from bacteriophage λdg. *J. Mol. Biol.* **2:** 392.

Kaiser, A.D. and R.B. Inman. 1965. Cohesion and the biological activity of bacteriophage lambda DNA. *J. Mol. Biol.* **13:** 78.

Kaiser, D. and T. Masuda. 1973. *In vitro* assembly of bacteriophage lambda heads. *Proc. Natl. Acad. Sci.* **70:** 260.

Kaiser, A.D., M. Syvanen, and T. Masuda. 1975. DNA packaging steps in bacteriophage lambda head assembly. *J. Mol. Biol.* **91:** 175.

Kayajanian, G. 1971. Packaging of λ*dgal* DNA into the virions of other temperate phages. *Virology* **43:** 326.

Kellenberger, E. and R. Edgar. 1971. Structure and assembly of phage particles. In *The bacteriophage lambda* (ed. A.D. Hershey), p. 271. Cold Spring Harbor Laboratory, Cold Spring Harbor, New York.

Kerr, C. and P.D. Sadowski. 1975. Packaging and maturation of bacteriophage T7 DNA *in vitro*. *Proc. Natl. Acad. Sci.* **71:** 3545.

Laemmli, U.K. 1970. Cleavage of structural proteins during the assembly of the head of bacteriophage T4. *Nature* **227:** 680.

——. 1975. Characterization of DNA condensates induced by poly (ethylene oxide) and polylysine. *Proc. Natl. Acad. Sci.* **72:** 4288.

Laemmli, U.K., L.A. Amos, and A. Klug. 1976. Correlation between structural transformation and cleavage of the major head protein of T4 bacteriophage. *Cell* **7:** 191.

Lanski, F. and E.N. Jackson. 1982. Maturation cleavage of bacteriophage P22 DNA in the absence of DNA packaging. *J. Mol. Biol.* **154:** 565.

Leng, M. and G. Felsenfeld. 1966. The preferential interactions of polylysine and polyarginine with specific base sequences in DNA. *Proc. Natl. Acad. Sci.* **56:** 1325.

Lickfeld, K.G., B. Menge, B. Hohn, and T. Hohn. 1976. Morphogenesis of bacteriophage lambda: Electron microscopy of thin sections. *J. Mol. Biol.* **103:** 299.

Liu, L.F. and J.C. Wang. 1978. DNA-DNA gyrase complex:The wrapping of the DNA duplex outside the enzyme. *Cell* **15:** 979.

MacKinlay, A.G. and A.D. Kaiser. 1969. DNA replication in head mutants of bacteriophage λ. *J. Mol. Biol.* **39:** 679.

Mandel, M. 1969. Infectivity of phage P2 DNA in presence of helper phage. *Mol. Gen. Genet.* **99:** 88.

McClure, S.C.C. and M. Gold. 1973. A sedimentation analysis of DNA found in *Escherichia coli* infected with phage λ mutants. *Virology* **54:** 1.

Meyerowitz, E.M., G.M. Guild, L.S. Prestidge, and D.S. Hogness. 1980. A new high-capacity cosmid vector and its use. *Gene* **11:** 271.

Miller, G. and M. Feiss. 1981. Cohesive end annealing and the helper-mediated transformation system of phage λ. *Virology* **109:** 379.

Miwa, T. and K. Matsubara. 1982. Identification of sequences necessary for packaging DNA into lambda phage heads. *Gene* **20:** 267.

Morrison, A. and N. Cozzarelli. 1981. Contacts between DNA gyrase and its binding site on DNA: Features of symmetry and asymmetry revealed by protection from nucleases. *Proc. Natl. Acad. Sci.* **78:** 1416.

Mousset, S. and R. Thomas. 1969. Ter, a function which generates the ends of the mature λ chromosome. *Nature* **221:** 242.

Murialdo, H. and A. Becker. 1978a. A genetic analysis of bacteriophage lambda prohead assembly *in vitro. J. Mol. Biol.* **125:** 57.

——. 1978b. Head morphogenesis of complex double-stranded deoxyribonucleic acid bacteriophages. *Microbiol. Rev.* **42:** 529.

Murialdo, H. and L. Siminovitch. 1972a. The morphogenesis of bacteriophage λ. IV. Identification of gene products and control of the expression of the morphogenetic information. *Virology* **48:** 785.

——. 1972b. The morphogenesis of phage lambda. V. Form-determining function of the genes required for the assembly of the head. *Virology* **48:** 824.

Murialdo, H., W.L. Fife, A. Becker, M. Feiss, and J. Yochem. 1981. Bacteriophage lambda DNA maturation. The functional relationships among the products of genes *Nu1*, *A* and *FI*. *J. Mol. Biol.* **145:** 375.

Murray, K. and N. Murray. 1973. Terminal nucleotide sequences of DNA from temperate *coli* phages. *Nature* **243:** 134.

Murray, K., N.E. Murray, and G. Bertani. 1975. Base changes in the recognition site for *ter* functions in lambdoid phage DNA. *Nature* **254:** 262.

Murray, K., A.G. Isaksson-Forgen, M. Challberg, and P.T. England. 1977. Symmetrical nucleotide sequences in the recognition sites for the *ter* function of bacteriophages P2, 299 and 186. *J. Mol. Biol.* **112:** 471.

Nichols, B.P. 1978. "Deoxyribonucleotide sequence analysis around sites cleaved *in vivo* by endonucleases induced in bacteriophages T5, λ, φ80 and 21." Ph.D. thesis, University of Iowa, Iowa City.

Nichols, B.P. and J.E. Donelson. 1978. A 178 nucleotide sequence surrounding the *cos* site of λ DNA. *J. Virol.* **26:** 429.

Okamoto, M. and H. Yamagishi. 1980. Electron microscopic study of viral DNA packaging with lambda head mutants. *Int. J. Biol. Macromol.* **3:** 65.

Parkinson, J.S. 1968. Genetics of the left arm of the chromosome of bacteriophage lambda. *Genetics* **59:** 311.

Parkinson, J.S. and R.J. Huskey. 1971. Deletion mutants of bacteriophage lambda. I. Isolation and initial characterization. *J. Mol. Biol.* **56:** 369.

Poteete, A.R. and D. Botstein. 1979. Purification and properties of proteins essential to DNA encapsulation by phage P22. *Virology* **95:** 565.

Pruss, G.J., J.C. Wang, and R. Calendar. 1975. *In vitro* packaging of covalently closed circular monomers of bacteriophage DNA. *J. Mol. Biol.* **98:** 465.

Raj, A.S., A.Y. Raj, and H. Schmieger. 1974. Phage genes involved in the formation of generalized transducing particles in *Salmonella* phage P22. *Mol. Gen. Genet.* **135:** 175.

Richards, K.W., R.C. Williams, and R. Calendar. 1973. Mode of DNA packaging within bacteriophage heads. *J. Mol. Biol.* **78:** 255.

Riemer, S.C. and V.A Bloomfield. 1980. Packaging of DNA in bacteriophage heads: Some considerations on energetics. *Biopolymers* **17:** 784.

Schmieger, H. and H. Backhaus. 1976. Altered cotransduction frequencies exhibited by HT-mutants of *Salmonella* phage P22. *Mol. Gen. Genet.* **143:** 307.

Serwer, P. 1975. Buoyant density sedimentation of macromolecules in sodium iothalamate density gradients. *J. Mol. Biol.* **92:** 433.

Skalka, A.M. 1977. DNA replication-bacteriophage lambda. *Curr. Top. Microbiol. Immunol.* **78:** 201.

Stahl, F.W., K.D. McMilan, M.M. Stahl, R.E. Malone, Y. Nozv, and V.A. Russo. 1972. A role for recombination in the production of "free-loader" lambda bacteriophage particles. *J. Mol. Biol.* **68:** 57.

Sternberg, N. and R. Weisberg. 1977a. Packaging of coliphage λ DNA. I. The role of the cohesive end site and the *A* protein. *J. Mol. Biol.* **117:** 717.

——. 1977b. Packaging of coliphage lambda DNA. II. The role of the gene *D* protein. *J. Mol. Biol.* **117:** 733.

Steven, A.C., E. Couture, U. Aebi, and M.K. Showe. 1976. Structure of T4 polyhead transformation as a model for T4 capsid maturation. *J. Mol. Biol.* **106:** 187.

Sumner-Smith, M. and A. Becker. 1982. DNA packaging in the lambdoid phages: Identification of the products of ϕ80 genes *1* and *2*. *Virology* **111:** 629.

Sumner-Smith, M., A. Becker, and M. Gold. 1982a. DNA packaging in the lambdoid phages: The role of λ genes *Nu1* and *A*. *Virology* **111:** 642.

Sumner-Smith, M., S. Benchimol, H. Murialdo, and A. Becker. 1982b. The *ben* gene of bacteriophage λ. Mapping, identification and control of synthesis. *J. Mol. Biol.* **160:** 1.

Syvanen, M. and J. Yin. 1978. Studies of DNA packaging into the heads of bacteriophage lambda. *J. Mol. Biol.* **126:** 333.

Szpirer, J. and P. Brachet. 1970. Relations physiologiques entre les phages temperes λ et ϕ80. *Mol. Gen. Genet.* **108:** 78.

Szybalski, E.H. and W. Szybalski. 1979. A comprehensive molecular map of bacteriophage lambda. *Gene* **7:** 217.

Thomas, J., N. Sternberg, and R. Weisberg. 1978. Altered arrangement of the DNA in injection-defective lambda. *J. Mol. Biol.* **123:** 149.

Tsui, L. and R.W. Hendrix. 1980. Head-tail connector of bacteriophage lambda. *J. Mol. Biol.* **142:** 419.

Tye, B.-K. 1976. A mutant of phage P22 with randomly permuted DNA. *J. Mol. Biol.* **100:** 421.

Wake, R.G., A.D. Kaiser, and R.B. Inman. 1972. Isolation and structure of phage λ head-mutant DNA. *J. Mol. Biol.* **64:** 519.

Wang, J.C. and D.P. Brezinski. 1973. Alignment of two DNA helices: A model for recognition of DNA base sequences by the termini-generating enzymes of phage λ, 186 and P2. *Proc. Natl. Acad. Sci.* **70:** 2667.

Wang, J.C. and A.D. Kaiser. 1973. Evidence that the cohesive ends of mature λ DNA are generated by the gene *A* product. *Nat. New Biol.* **241:** 16.

Weigel, P.H., P.T. England, K. Murray, and R.W. Old. 1973. The 33 terminal nucleotide sequence of the cohesive ends of bacteriophage λ DNA. *J. Mol. Biol.* **57:** 491.

Weigle, J. 1966. Assembly of phage λ *in vitro*. *Proc. Natl. Acad. Sci.* **55:** 1462.

Weisberg, R.A., N. Sternberg, and E. Gallay. 1979. The *Nu1* gene of coliphage λ. *Virology* **95:** 99.

Wilson, R.W. and V. Bloomfield. 1979. Counterion-induced condensation of deoxyribonucleic acid. A light-scattering study. *Biochemistry* **18:** 2192.

Wood, W.B. and J. King. 1979. Genetic control of complex bacteriophage assembly. *Comp. Virol.* **13:** 581.

Wu, R. and E. Taylor. 1971. Nucleotide sequence analysis of DNA. II. Complete nucleotide sequence of the cohesive ends of bacteriophage λ DNA. *J. Mol. Biol.* **57:** 491.

Wurtz, M., J. Kistler, and T. Hohn. 1976. Surface structure of *in vitro* assembled bacteriophage lambda polyheads. *J. Mol. Biol.* **101:** 39.

Yamagishi, H. and M. Okamoto. 1978. Visualization of the intracellular development of bacteriophage λ with special reference to DNA packaging. *Proc. Natl. Acad. Sci.* **75:** 3706.

Yarmolinsky, M.B. 1971. Making and joining DNA ends. In *The bacteriophage lambda* (ed. A.D. Hershey), p. 97. Cold Spring Harbor Laboratory, Cold Spring Harbor, New York.

Yuan, R., D.L. Hamilton, and J. Burckhardt. 1980. DNA translocation by the restriction enzyme from *E. coli* K. *Cell* **20:** 237.

Zachary, A. and L.D. Simon. 1977. Size changes among λ capsid precursors. *Virology* **81:** 107.

Tail Assembly and Injection

Isao Katsura
Department of Biophysics and Biochemistry
Faculty of Science, University of Tokyo
Hongo, Tokyo, Japan

The tail of a bacteriophage is a supramolecular organelle with which the phage adsorbs to the host cell and injects its DNA into the cytoplasm. Only the relatively complex double-stranded DNA phages have a tail as a morphologically distinguishable, differentiated structure. The tails of these phages may be classified into three groups on the basis of morphology and function: long contractile (e.g., T4), long noncontractile (e.g., λ), and short (e.g., T7). In the long contractile and long noncontractile groups, tail assembly forms an independent branch in the morphogenetic pathway (Casjens and King 1975).

The assembly of the λ tail has been studied in detail. The results complement those of the T4 tail (King 1968) and give information useful to the studies of other tubular or helical structures, such as bacterial flagella (Iino 1977). Although the function of the λ tail itself may seem a very specialized field of research, problems contained in this study are more general: How do protein molecules function in assembled states where interaction between neighboring molecules is strong? How can we analyze a series of reactions occurring in such structures? The λ tail is a particularly advantageous model to use in attempting to understand the assembly and function of supramolecular structures due to the highly developed experimental approaches (genetic, structural, biochemical, immunological, and so on) available in this system.

STRUCTURE OF THE TAIL

The tail of phage λ is a thin flexible tube (135 nm in length), ending in a small conical part (15 nm in length) and a single tail fiber (23 nm in length) (Fig. 1) (Kellenberger and Edgar 1971; Hohn and Katsura 1977). Both the conical part and the tail tube have cross striations in electron micrographs and seem to consist of 3 disks (or 4) and 32 disks, respectively. By chemical treatments the tail tube can be dissociated into ringlike structures that correspond to single- or multi-layered disks (Bleviss and Easterbrook 1971; Katsura and Kühl 1974). An end view of the ringlike structure shows that an individual disk consists of an annular ring (''core'') 9 nm in diameter with a central hole 3 nm in diameter. In addition, there are probably six small knobs arranged around the core, giving an overall diameter to the structure of about 18 nm (Fig. 1). These knobs are thought to be responsible for the somewhat rough or fuzzy appearance of the tail tube. Various electron microscopy and biochemical data indicate that the disk is composed of six subunits of the major tail protein gp*V* (Buchwald et al. 1970b;

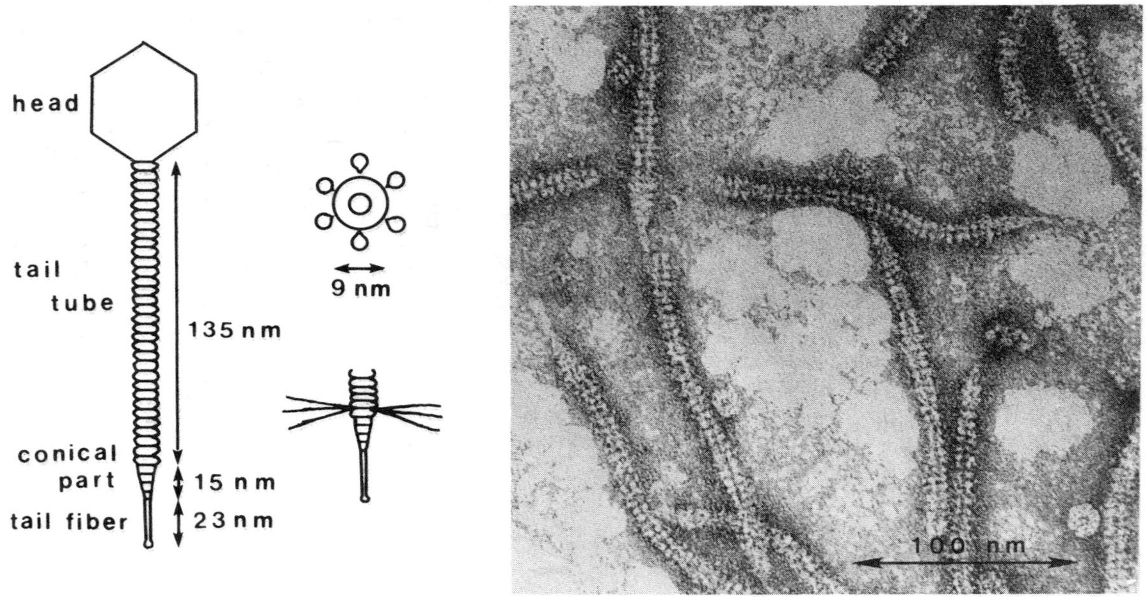

Figure 1 Structure of the λ tail. (*Left*) Schematic drawings of the λ phage particle, a disk (end view) composing the tail tube, and side fibers attached to the tail tip. (*Right*) Electron micrograph of free tails partially purified from a headless lysate. Negative staining with 1% uranyl acetate. The stain penetrates into the central hole of the free tail but not into that of the tail attached to the full head. The ringlike structures are *groE* particles (see Georgopoulos et al., this volume).

Casjens and Hendrix 1974a), each consisting of two separate folding domains. Surprisingly, mutants have been isolated in which as much as one third of the gp*V* molecule is absent (probably from the carboxyterminal end), yet the tails apparently function normally. In these mutants, the domain of gp*V* that constitutes the outer knob is missing, as shown by electron microscopy and hydrodynamic measurements (Katsura 1981).

Of the 32 disks constituting the tail tube, the double disks adjacent to the conical part often behave differently from the other 30 disks (Kühl and Katsura 1975; see also the penetration of negative stain in Fig. 1). However, there is so far no evidence that they are made of a protein different from gp*V*. Some micrographs show that there are several "side fibers" projecting from the double disks (Fig. 1; R. Hendrix, pers. comm.), whose origin and nature remain to be studied. In phage particles, the tail is joined by its head-proximal end to a structure called the knob or head-tail connector. However, this is considered to be a part of the head, since it is present in proheads and tailless heads but absent from free tails produced by mutants in head genes (Harrison et al. 1973; Tsui and Hendrix 1980).

TAIL GENES AND THEIR PRODUCTS

The genes whose products are directly required for the formation of the tail are called tail genes. The tail genes are located in a contiguous cluster on the λ genome, starting immediately to the right of the head-gene cluster. If one of the tail genes is defective, free heads and tail precursors are produced instead of infectious phage particles. These free heads are active; i.e., they can bind to free tails produced by head-gene mutants and yield infectious phage particles (Weigle 1966, 1968; Harrison et al. 1973). This shows that the head and the tail are formed independently and join at the last step of the assembly.

So far, 12 tail genes have been reported: *Z, U, V, G, T, H, "208," M, L, K, I,* and *J*, reading from left to right of the conventional representation of the genome (Campbell 1961; Thomas et al. 1967; Mount et al. 1968; Parkinson 1968). Of these, gene *T* and gene "*208*" have not been studied in detail and will be discussed very little. Most of the products of the tail genes have been detected in lysates of phage-infected cells by SDS-polyacrylamide gel electrophoresis (Table 1). The sum of their molecular weights accounts for the coding capacity of the tail region of the genome. This, together with the fact that several different mutant alleles have been isolated in most of the tail genes, argues that few, if any, tail genes remain to be discovered.

Most of the tail-gene products are found in finished phage particles, and their locations in the structure of the tail are known with varying degrees of certainty. The product of gene *J* is the part of the tail that interacts directly with the surface of the host cell, as judged by serological (Dove 1966; Buchwald and Siminovitch 1969) and genetic (Mount et al. 1968) evidence. This argues that gp*J* is at the tip of the tail and is generally taken to mean that gp*J* constitutes the tail fiber,

Table 1 Tail genes and their products

Gene	EM phenotype of amber mutants	Gene product: m.w. ($\times 10^{-3}$)	Gene product: presence (copies) in phage particle	Gene product: function
Z	defective particles and defective tails[7,10]	20[5]	+[5,6]	linkage of DNA end to the tail[17]
U	polytails[8,9]	14[1], 16[6]	+[5,6]	terminator[16] head-and-tail binding[15]
V	no tail-related structures have been detected by electron microscopy[8,10]	31[1], 32[3], 25[6]	170–210[2] 135–212[3]	major tail protein[19] injection[11,12]
G		33[1]	+[1]	initiator formation[18] injection[11]
T		(16)[1a]	(+)[1]	unknown
H		90→78[4 b] 87→79[1]	3–5[2 b] 6–7[3]	initiator formation[18] injection[11,12]
"208"	not checked	?	?	unknown
M	no tail-related structures have been detected by electron microscopy[8]	~10[6]	+[6]	initiator formation[18]
L		29[1]	+[1]	initiator formation[18]
K		27[1]	−?[1]	initiator formation[18]
I		?	?	initiator formation[18]
J		130[1], 140[3]	2–3[2] 3–4[3]	initiator formation[18] adsorption[8,11,13,14]

References: [1]Murialdo and Siminovitch (1972); [2]Murialdo and Siminovitch (1971); [3]Casjens and Hendrix (1974a); [4]Hendrix and Casjens (1974); [5]Katsura and Kühl (1974); [6]Katsura and Tsugita (1977); [7]Casjens and Hendrix (1974b); [8]Mount et al. (1968); [9]Kemp et al. (1968); [10]Kühl and Katsura (1975); [11]Katsura (1976b); [12]Scandella and Arber (1976); [13]Dove (1966); [14]Buchwald and Siminovitch (1969), Katsura and Kühl (1975a); [15]Katsura and Kühl (1975a); [16]Katsura (1976a); [17]Thomas et al. (1978); [18]Katsura and Kühl (1975b); [19]Buchwald et al. (1970a).

[a]The identification of this protein as gp*T* is tentative.

[b]Protein gp*H* is cleaved during the tail assembly. The molecular weights of both the uncleaved and the cleaved form are shown. Only the cleaved form (p*H**) is found in the finished phage particles.

although the identification of gp*J* with the fiber has never been made directly. The major tail protein gp*V* (Buchwald et al. 1970a) forms the tail tube: a stack of 32 hexameric rings (Casjens and Hendrix 1974a; Katsura and Tsugita 1977). Probably a hexamer of gp*U* is attached to the head-proximal end of the tail tube (Katsura and Tsugita 1977) and connects the tail to the head (Katsura and Kühl 1975a). The second most abundant component of the tail, p*H** (the cleavage product of gp*H*), is not present in polytails produced by mutants in gene *U* (Katsura 1976a). Since the polytails have both the conical part and the tail fiber, p*H** does not seem to constitute a major part of these structures. It may be located inside the tail tube, extending from the conical part. There is indirect evidence that gp*Z* may also be located inside the tail tube (Thomas et al. 1978). Most, if not all, of the other tail proteins (gp*G*, gp*M*, gp*L*, and 16K protein [gp*T*?]) are probably components of the tail tip (conical part plus tail fiber), because they are required for the formation of the initiator (see below), which contains gp*J* and therefore must lie at the distal end of the tail. The morphologically defined tail tip and the biochemically defined initiator are most likely the same or very similar structures, but this remains to be proved directly.

The electron microscopy phenotype of the tail-gene mutants is shown in Table 1. In addition to free heads, Z^- lysates accumulate noninfectious phagelike particles and inactive tails (Casjens and Hendrix 1974b; Kühl and Katsura 1975). U^- lysates contain aberrant polytails, having a tail tube of random length ranging up to many times normal length (Kemp et al. 1968; Mount et al. 1968). No tail-related structures have been observed by electron microscopy in the lysates of the other tail-gene mutants. The tail precursors, which are known on other grounds to accumulate in these lysates, have escaped microscopic detection, probably because they are very small or have shapes difficult to recognize as tail-related.

Two of the tail-gene products, gp*U* and gp*V*, have been purified and characterized with regard to molecular weight, amino acid composition, end groups, circular dichroism spectrum, and so on (Katsura and Tsugita 1977). Various mutations in gene *J* (Fuerst and Bingham 1978) and gene *V* (Katsura 1976b, 1981) have been located precisely within the genes, and the relationship between the phenotype and the location has been shown. Partial amino acid sequences in the aminoterminal regions of gp*V* and p*H** were determined by Walker et al. (1982).

ASSEMBLY

The assembly pathway of the λ tail, i.e., the order of action of the tail-gene products during the formation of the tail, has been established by detecting the in vitro complementation activity and serum-blocking power of tail precursors that are present in the lysates of tail-gene mutants (Fig. 2) (Katsura and Kühl 1975b; Katsura 1976a). As an example of this sort of experiment, consider a lysate of

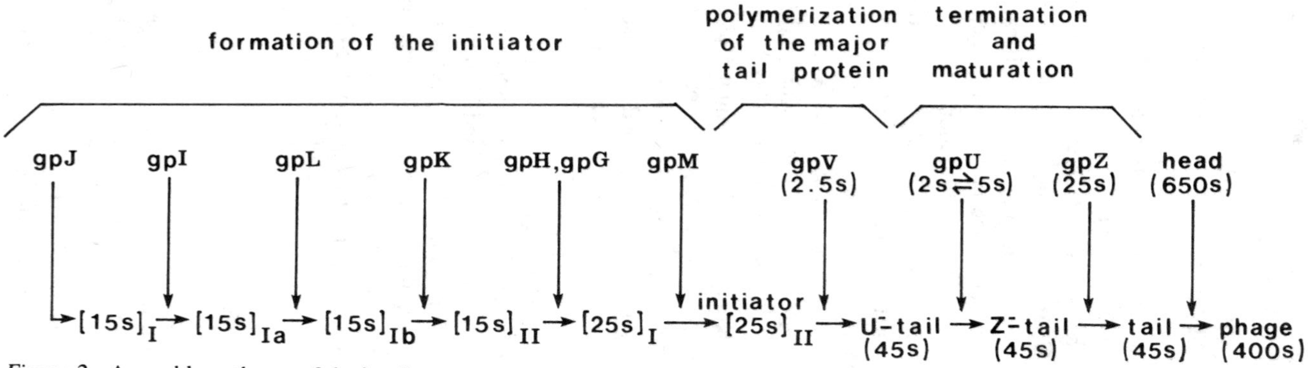

Figure 2 Assembly pathway of the λ tail.

cells infected with λV^- (a V^- lysate) and a similar J^- lysate. Neither lysate contains functional tails, but if they are mixed, tails assemble. These tails join to the heads that are already present in the lysates, and they are detected by plaque formation. If the J^- lysate is first fractionated by velocity sedimentation on a sucrose gradient and then complemented with the V^- lysate, the complementing material is found to sediment in a peak at 2.5S. This has been shown to be the unpolymerized form of gp*V*. Conversely, if the V^- lysate is first fractionated, the material that complements a J^- lysate can be seen to sediment at 25S. This is the "initiator," which is formed by the action of all of the tail proteins whose action precedes gp*V* in the assembly pathway (Fig. 2) and, in fact, it can also complement lysates defective in any of the latter proteins. Through an extensive series of such complementation experiments, the order of action of the various tail-gene products and something of the nature of the intermediate structures can be determined.

Tail assembly starts from gp*J*, the tail fiber located at the head-distal end in the finished phage particle. Then, the products of genes *I, L, K, H, (G)*, and *M* act in this order on the tail fiber. On the basis of in vitro complementation experiments, the position of the action of gp*G* in the pathway is ambiguous, because amber mutants in genes *G, H*, and *M* fail to complement one another in vitro. However, by examining the composition of gp*J*-containing structures in the lysates of tail mutants, Tsui and Hendrix (1983) concluded that the order is probably gp*H*, gp*G*, and gp*M*. The sedimentation coefficients of the assembly precursors are about 15S before the action of gp*H* and about 25S after that.

The interaction of at least seven gene products (from gp*J* to gp*M* in Fig. 2) yields the initiator or nucleus for the polymerization of the major tail protein gp*V*. The assembly of gp*V* on the initiator then proceeds to the correct tail length (32 disks) and pauses. At this point the product of gene *U* attaches to the proximal end and prevents further gp*V* polymerization. Already at this stage the precursor (Z^- tail) has the same shape as the finished tail and can join to the head. However, to yield an infectious phage particle, the Z^- tail has to be activated before joining to the head.

The second most abundant component of the λ tail (m.w. = 78,000–79,000) (Buchwald et al. 1970b) has been shown by peptide mapping to be a cleavage product of gp*H* (m.w. = 89,000–90,000) (Hendrix and Casjens 1974). The cleavage is coupled with assembly, because all of the gp*H* molecules in the finished phage particles are in the cleaved form, whereas a large amount of uncleaved gp*H* is present in cells irradiated with UV light before infection where assembly reactions occur very poorly (Murialdo and Siminovitch 1972). It has been shown, in fact, that the cleavage reaction takes place between the action of gp*U* and gp*Z* in the pathway (Tsui and Hendrix 1983). This is several steps after gp*H* has been incorporated into the initiator and after polymerization of gp*V* into the tail tube. The protease responsible for gp*H* cleavage has not been identified. The role of the cleavage of gp*H* in the assembly or function of the tail is not clear. It has been suggested that the cleavage may produce a high-energy configuration of p*H** in the tail, whose energy could be used to drive protein reorganization during DNA injection (Hendrix and Casjens 1974). Alternatively, if the model of length determination discussed below is correct, the end of gp*H* that is removed by cleavage may participate in that process and then be removed proteolytically after it has served its function.

Thus, the assembly pathway of the λ tail may be regarded as a series of three events: (1) formation of the initiator (*J, I, L, K, H, G, M*), (2) polymerization of the major tail protein (*V*), and (3) termination and maturation (*U*, gp*H* cleavage, *Z*). The order of action of the gene products is uniquely and strictly determined. In most cases, if the assembly is blocked by mutations, biologically active precursors accumulate in the cell. These include both assembly intermediates that accumulate behind the mutational block and unreacted proteins that would normally act later in the assembly pathway. However, for mutants in a few of the tail genes, assembly side reactions may take place and produce aberrant structures, such as inactive phagelike particles (Z^- particles) in Z^- lysates, polytails in U^- or U^-Z^- (double mutant) lysates, and polytubes in the lysates of double mutants U^-G^-, U^-H^-, U^-K^-, U^-I^-, or U^-J^- (Katsura 1976a; for discussion of polytails and polytubes, see the next section).

The order of action of the gene products during assembly (*J, I, L, K, H, G, M, V, U, Z*) is similar to the arrangement of the genes on the λ genome (*J, I, K, L, M, H, G, V, U, Z* from right to left). We argue (below) that the order of action is based on structural interactions between proteins involved in adjacent steps of the assembly pathway. Therefore, the similarity between order of functions and order of genes may have an evolutionary advantage, because it minimizes the number of new and possibly unproductive interactions that might arise by recombination in the tail region between λ and related phages having partially homologous chromosomes (Casjens and Hendrix 1974b).

REGULATION OF ASSEMBLY

Three aspects of regulation during assembly of the λ tail have been studied: (1) levels and kinetics of tail protein synthesis, (2) polymorphism in the assembly of the major tail protein due to defects in the initiation and termination, and (3) regulation of tail length.

The amounts of late proteins synthesized in λ-infected cells vary widely from gene to gene (Hendrix 1971; Murialdo and Siminovitch 1972). Of the tail proteins, gp*U*, gp*V*, and 16K protein (gp*T*?) are produced in large amounts (265–470 times as many copies as gp*A*, a head protein), gp*L* is made in intermediate amounts (44 times as many copies as gp*A*), and gp*G*, gp*H*, gp*K*, and gp*J* are synthesized in small amounts (12–18 times as many copies as gp*A*). It is reasonable that the cell produces a large amount of the major tail protein gp*V*, which forms more than 80% of the tail. Since abortive polytails are formed in the absence of gp*U* action, a large number of gp*U* molecules may be required to prevent this. The reason, if any, for the large production of 16K protein is not known. The control of the rate of protein synthesis seems to be made at the translational level; the amount of mRNA per unit length of DNA is constant within the experimental error throughout the region of structural genes (Ray and Pearson 1974), and selective inactivation of mRNA is not large enough to account for the difference in the amounts of various tail proteins (Ray and Pearson 1975). In fact, a major part, if not all, of the mRNA coding for the tail proteins is a long molecule starting from the late promoter p_R', going over the cohesive site, and transcribing the structural genes from left to right in the conventional representation (see Friedman and Gottesman, this volume).

The time when various tail proteins start to appear in λ-infected cells corresponds to the direction of the late mRNA synthesis, from gene *Z* to gene *J* (Murialdo and Siminovitch 1972). This means that at the initial stage of tail protein synthesis, proteins that act at late steps in assembly appear first and wait for proteins required for early steps. This may be reasonable, because assembly intermediates at early steps are not stable even in the host cell (Katsura 1976a) and because abortive polytails are formed if gp*J* to gp*V* in the assembly pathway are synthesized before the production of gp*U*. However, except for the very initial stage of tail protein synthesis, it can be said that all of the tail proteins are produced at the same time. Therefore, the order of action of proteins in the assembly pathway must be determined by the interactions between tail protein molecules themselves and not by the time course of their synthesis. That is, each step in assembly reactions should generate the reaction site for the next step, thus forming the sequential order of assembly pathway (Showe and Kellenberger 1975; Wood and King 1979). Although experimental proof has been given in few cases, possible mechanisms of the site generation have been discussed quite often. For instance, let us consider a case in which proteins *A* and *B* join first and then protein *C* interacts with the *A-B* complex. A possible explanation for this ordering may be that *C* has two binding sites, one for *A* and one for *B*, and that both sites have to be used at the same time for a stable interaction. An alternative mechanism would be that the site on *B* for interaction with *C* is available only after conformational change of *B* that is induced by binding of *A* to *B*.

The λ tail gives a good example of the regulation of protein polymerization both by initiation and by termination. The major tail protein gp*V* may polymerize into aberrant polytails and polytubes, as well as normal tails. Polytails and polytubes are distinguished by the fact that polytails have the conical part and the tail fiber, whereas polytubes do not. Both structures show wide length distributions, unlike normal tails. The average length of polytubes (several microns) is much larger than that of polytails (several hundred nanometers). Genetic evidence shows that polytubes in vivo are formed by aberrant initiation of polymerization on a pseudoinitiator made only of gp*L* and gp*M* instead of the $(25S)_{II}$ initiator (Fig. 2) of the normal pathway (Katsura 1976a). Structures indistinguishable from these polytubes are formed by dissociation of the tail tube into ringlike structures at pH 2 and subsequent assembly at neutral pH (Bleviss and Easterbrook 1971; Katsura and Kühl 1974). Whether there is any pseudoinitiator in this case is not clear.

In contrast to polytubes, polytails are formed by a defect in the termination of polymerization of gp*V* (Katsura 1976a). The polymerization stops for a while at the correct tail length (U^- tail in Fig. 2), even in the absence of the terminator protein gp*U*. This seems to be the essential step in length determination of the λ tail, but the mechanism of the transient pause is not known. In the normal pathway gp*U* and gp*Z* act on the U^- tail and produce the normal tail. However, in the absence of gp*U* the polymerization of gp*V* resumes, and the assembly reaction enters an aberrant pathway, yielding polytails. Since polytails contain neither gp*H* nor its cleaved form (Katsura 1976a), the resumption of polymerization of gp*V* may be triggered by the release of (probably uncleaved) gp*H* from the U^- tail. This gives rise to relaxation in another part of the regulation: gp*Z* can act on the polytail before the action of gp*U*, although the action of gp*U* on the U^- tail is prerequisite to the action of gp*Z* in the normal pathway (Katsura 1976a).

Perhaps the most interesting question about regulation of tail assembly, and one that has so far successfully resisted definitive solution, is the question of how correct tail length is determined. Length determination is apparently quite precise; any variation in length greater than about ±1–2 disks around the canonical 32 disks should have been detected in the measurements that have been made.

To determine what genes could be involved in the length determination, Youderian (1978) performed a series of genetic experiments, taking advantage of the fact that the tail tube of the lambdoid phage φ80 is about 10% longer than the λ tail tube. By measuring the length of the tails produced by complementation between λ and φ80 tail proteins, he proved that genes *Z, U, V, G,* and *K* do not determine the tail length. Furthermore, by constructing λ-φ80 hybrid genomes and measuring the tail length of their phage particles, he showed that the information for the length determination exists in the DNA region of genes *V, G, T, H, "208," M,* and *L*. Finally, he suggested that gp*H* may act as a template to determine the length of the tail, because he observed that the φ80 protein homologous to p*H** has a slightly higher molecular weight than the λ p*H**. This is in agreement with the recent observation that two other lambdoid phages, HK97 and HK022, which have tails 17% longer and 11% shorter than λ, produce correspondingly larger and smaller gp*H* homologs (R. Hendrix, pers. comm.).

Most of the available evidence is thus compatible with a model in which gp*H*, which is assembled as part of the initiator, becomes extended along the length of the tail during gp*V* polymerizaton and, by some unspecified mechanism, marks the stopping point at a place determined by the size of the gp*H* polypeptide. This would make λ tail length determination grossly similar to length determination in tobacco mosaic virus and filamentous bacteriophage, although in these cases the templates are not proteins but nucleic acid molecules. However, this simple view of length determination is somewhat clouded by the existence of viable λ mutants that carry small deletions in gene *H* and make slightly shortened versions of gp*H* and p*H**, yet seem to have normal tail length (I. Katsura, unpubl.).

FUNCTION OF THE TAIL: STEPS IN DNA INJECTION

The process of delivery of the λ DNA from the phage particle to the host cytoplasm has been analyzed into several successive steps. Two steps have been recognized for many years, namely "reversible" and "irreversible" adsorption (Lieb 1954) or "adsorption" and "injection" (Bode and Kaiser 1965). It is not clear whether these two classifications are the same, though they look very similar. Reversible adsorption occurs even at low temperature. After this step the phage sediments together with the bacterium but still can be inactivated by antiphage serum. In contrast to this, irreversible adsorption is dependent on temperature and the energy metabolism of the host cell. The phage-bacterium complex becomes resistant to treatment by a Waring blendor only after injection.

In the studies mentioned above, it was not clear whether there is a state in which the phage has adsorbed to the bacterium irreversibly but has not yet injected its DNA. This was clarified by MacKay and Bode (1976a), who defined the steps of "lag," "trigger," and "uptake" after adsorption. Although even the free tail can adsorb to the bacterium, the lag reaction requires that the head be attached to the tail. It occurs almost instantaneously at 37°C but takes a long time at 23°C. The lag function is followed by trigger or ejection of DNA, but if the condensed state of the DNA in the head is stabilized by putrescine (1.4-diaminobutane), the reaction stops between lag and trigger. In the normal pathway of DNA injection, the ejection of DNA is always accompanied by the uptake of DNA by the host cell. However, if the phage-bacterium complex is treated with chloroform before the lag function, the uptake does not occur and the DNA is found in the culture medium.

Analogous studies have been performed using isolated λ-receptor protein (Schwartz 1975; MacKay and Bode 1976b; Roa and Scandella 1976) or membranes (Zgaga et al. 1973) instead of the whole cell. The reactions between the phage and the isolated receptor show three steps: reversible interaction, irreversible binding (phage inactivation), and DNA ejection, which is temperature-dependent. Curiously, chloroform or ethanol is required for the second step, but this requirement is suppressed by mutations in either gp*J* of phage λ or the receptor of the host cell. The requirement of organic solvent probably corresponds to the finding in the studies using isolated membranes (Zgaga et al. 1973) that the outer membrane has to form a complex with the inner membrane to be adsorbed by λ phage irreversibly.

HOST AND PHAGE PROTEINS REQUIRED FOR ADSORPTION AND INJECTION

Host proteins that play essential roles in adsorption and injection have been recognized mainly by genetic studies. The *Escherichia coli* gene *lamB* codes for the λ-receptor protein located in the outer membrane (Thirion and Hofnung 1972; Randall-Hazelbauer and Schwartz 1973). Another *E. coli* gene, *ptsM*, which codes for an inner membrane protein, plays an essential role in the DNA injection of λ phage (Elliot and Arber 1978). The products of both genes have functions of sugar transport in addition to those in the infection of λ phage. The *E. coli* mutations in these genes that do not allow the infection of λ phage are called λ^r (lambda-resistant) and *pel*$^-$ (*pe*netration of *l*ambda) for genes *lamB* and *ptsM*, respectively. Phage mutations that overcome these host mutations have been isolated and named λ*h* (host-range) (Appleyard et al. 1956) and λ*hp* (host-range of *pel*$^-$ bacteria) (Scandella and Arber 1974). The λ*h* mutations cluster in the right end of gene *J* (Mount et al. 1968; Fuerst and Bingham 1978), and the λ*hp* mutations are located in gene *H* and gene *V* (Scandella and Arber 1976), which shows that the products of these tail genes function in adsorption and injection, respectively. A curious property of the *pel*$^-$ block to injection is that it is much more severe for phages with genomes that are smaller than wild type, and the block can be largely overcome by DNA insertions or duplications, which increase the length of the genome (Emmons et al. 1975).

Another class of phage mutations that produce normal-looking but noninfectious phage particles has been found among prophage mutations (Katsura 1976b). They map in genes *V, G, H,* and *J*. Amber mutations in gene *Z* also produce noninfectious phage particles. The nature of the defect has been clarified in this last case. The right-hand cohesive end of the DNA, which penetrates partway into the tail in normal phage particles, is not in contact with the tail in Z^- particles. The same defect is also found in phage particles containing DNA without the right-hand cohesive end (Thomas et al. 1978).

STRUCTURAL AND KINETIC ASPECTS OF THE INFECTIOUS PROCESS

The rate of adsorption of λ phage to *E. coli* cells has been measured and compared with the rate of collision calculated according to the diffusion theory of von Smoluchovsky (Schwartz 1976). Under the best conditions, as many as 40% of the collisions result in adsorption. The high efficiency is explained by the high mobility of the tail fiber due to bending of the flexible tail, as well as rotation of the phage particle. The tip of the tail fiber and the receptor molecule on the cell surface can attain with high probability the proper geometrical orientation for adsorption when the phage particle collides with the cell.

Unlike the case of phages such as T4, no change in the tail structure has been observed by electron microscopy during adsorption and injection of λ. However, it has been suggested that p*H** is probably ejected together with DNA (Roa 1981). It may be a pilot protein, i.e., a protein that enters the cell along with the DNA, though there is no direct evidence that it enters the bacterial cell. The DNA molecule is thought to be ejected right end first through the tail tube (Chattoraj and Inman 1974; Saigo and Uchida 1974; Thomas 1974); this is opposite to the direction of DNA movement during packaging (see Feiss and Becker, this volume). It is not known from which part of the tail the DNA molecule comes out—from the end of the tail fiber or somewhere else in the tail tip. Perhaps most plausible, though untested, is the idea that the tail tip undergoes a structural rearrangement before, or concomitant with, injection. The mechanism by which the DNA passes through the cell envelope is largely unknown. The fact that both of the host proteins known to be involved in adsorption and injection function in sugar transport suggests that the phage in some way parasitizes this transport mechanism. Bayer (1968) has argued that phage adsorption to *E. coli*, and therefore DNA injection, occurs at sites of adhesion between the inner and outer membrane.

CONCLUSION AND SCOPE OF FUTURE STUDIES

The studies reviewed in this paper show that the λ tail is a good model system to study the assembly and function of supramolecular structures, especially with respect to sequential ordering of assembly, regulation of protein polymerization, length determination, and probably conformational reorganization of assembled structures. The work on tail assembly has been typical of molecular biological experimentation, in the sense that it has relied heavily on the synergism between genetic and biochemical methods. As a whole, studies on injection are less advanced than studies on assembly, partly because there is so far no general method to obtain intermediates in injection by genetic means. To date, little is known about the nature of minor proteins, the arrangements of proteins in the tail tip, and the structures of the 15S and 25S assembly intermediates. However, considering the remarkable progress that has been made during the past 10 years, we may well expect successful studies on these subjects in the near future, which will certainly yield information essential to the problems of length determination, gp*H* cleavage, and protein reorganization during the infectious process.

ACKNOWLEDGMENTS

I thank Dr. H. Noda for reading the manuscript critically and for giving various useful suggestions to me. Thanks are also due to Drs. R. Hendrix, M. Roa, M. Schwartz, and P. Youderian for sending unpublished results.

REFERENCES

Appleyard, R.K., J.F. McGregor, and K.M. Baird. 1956. Mutations to extended host range and the occurrence of phenotypic mixing in the temperate coliphage λ. *Virology* **2:** 565.

Bayer, M.E. 1968. Adsorption of bacteriophages to adhesions between wall and membrane of *Escherichia coli. J. Virol.* **2:** 346.

Bleviss, M. and K.E. Easterbrook. 1971. Self-assembly of bacteriophage lambda tails. *Can. J. Microbiol.* **17:** 947.

Bode, V.C. and A.D. Kaiser. 1965. Changes in the structure and activity of λ DNA in a superinfected immune bacterium. *J. Mol. Biol.* **14:** 399.

Buchwald, M. and L. Siminovitch. 1969. Production of serum-blocking material by mutants of the left arm of the λ chromosome. *Virology* **38:** 1.

Buchwald, M., H. Murialdo, and L. Siminovitch. 1970a. The morphogenesis of bacteriophage lambda. II. Identification of principal structural proteins. *Virology* **42:** 390.

Buchwald, M., P. Steed-Glaister, and L. Siminovitch. 1970b. The morphogenesis of bacteriophage lambda. I. Purification and characterization of λ heads and λ tails. *Virology* **42:** 375.

Campbell, A. 1961. Sensitive mutants of bacteriophage λ. *Virology* **14:** 22.

Casjens, S.R. and R.W. Hendrix. 1974a. Location and amounts of the major structural proteins in bacteriophage lambda. *J. Mol. Biol.* **88:** 535.

——. 1974b. Comments on the arrangement of the morphogenetic genes of bacteriophage lambda. *J. Mol. Biol.* **90:** 20.

Casjens, S. and J. King. 1975. Virus assembly. *Annu. Rev. Biochem.* **44:** 555.

Chattoraj, D. and R.B. Inman. 1974. Location of DNA ends in P2, 186, P4 and lambda bacteriophage heads. *J. Mol. Biol.* **87:** 11.

Dove, W.F. 1966. Action of the lambda chromosome. I. Control of functions late in bacteriophage development. *J. Mol. Biol.* **19:** 187.

Elliot, J. and W. Arber. 1978. *E. coli* K-12 *pel* mutants, which block phage λ DNA injection, coincide with *ptsM*, which determines a component of a sugar transport system. *Mol. Gen. Genet.* **161:** 1.

Emmons, S.W., V. MacCosham, and R.L. Baldwin. 1975. Tandem genetic duplication in phage lambda. III. The frequency of duplication mutants in two derivatives of phage lambda is independent of known recombination systems. *J. Mol. Biol.* **91:** 133.

Fuerst, C.R. and H. Bingham. 1978. Genetic and physiological characterization of the J gene of bacteriophage lambda. *Virology* **87:** 437.

Harrison, D.P., D.T. Brown, and V.C. Bode. 1973. The lambda head-tail joining reaction: Purification, properties and structure of biologically active heads and tails. *J. Mol. Biol.* **79:** 437.

Hendrix, R.W. 1971. Identification of proteins coded in phage lambda. In *The bacteriophage lambda* (ed. A.D. Hershey), p. 355. Cold Spring Harbor Laboratory, Cold Spring Harbor, New York.

Hendrix, R.W. and S.R. Casjens. 1974. Protein cleavage in bacteriophage λ tail assembly. *Virology* **61:** 156

Hohn, T. and I. Katsura. 1977. Morphogenesis of bacteriophage lambda. *Curr. Top. Microbiol. Immunol.* **78:** 69.

Iino, T. 1977. Genetics of structure and function of bacterial flagella. *Annu. Rev. Genet.* **11:** 161.

Katsura, I. 1976a. Morphogenesis of bacteriophage lambda tail: Polymorphism in the assembly of the major tail protein. *J. Mol. Biol.* **107:** 307.

——. 1976b. Isolation of λ prophage mutants defective in structural genes: Their use for the study of bacteriophage morphogenesis. *Mol. Gen. Genet.* **148:** 31.

——. 1981. Structure and function of the major tail protein of bacteriophage lambda. Mutants having small major tail protein molecules in their virion. *J. Mol. Biol.* **146:** 493.

Katsura, I. and P.W. Kühl. 1974. A regulator protein for the length determination of bacteriophage lambda tail. *J. Supramol. Struct.* **2:** 239.

——. 1975a. Morphogenesis of the tail of bacteriophage λ. II. *In vitro* formation and properties of phage particles with extra long tails. *Virology* **63:** 238.

——. 1975b. Morphogenesis of the tail of bacteriophage λ. III. Morphogenetic pathway. *J. Mol. Biol.* **91:** 257.

Katsura, I. and A. Tsugita. 1977. Purification and characterization of the major protein and the terminator protein of bacteriophage lambda tail. *Virology* **76:** 129.

Kellenberger, E. and R.S. Edgar. 1971. Structure and assembly of phage particles. In *The bacteriophage lambda* (ed. A.D. Hershey), p. 271. Cold Spring Harbor Laboratory, Cold Spring Harbor, New York.

Kemp, C.L., A.F. Howatson, and L. Siminovitch. 1968. Electron microscopic studies of mutants of lambda bacteriophage. I. General description and quantitation of viral products. *Virology* **36:** 490.

King, J. 1968. Assembly of the tail of bacteriophage T4. *J. Mol. Biol.* **32:** 231.

Kühl, P.W. and I. Katsura. 1975. Morphogenesis of the tail of bacteriophage λ. I. *In vitro* intratail complementation. *Virology* **63:** 221.

Lieb, M. 1954. Studies on lysogenization in *Escherichia coli. Cold Spring Harbor Symp. Quant. Biol.* **18:** 71.

MacKay, D.J. and V.C. Bode. 1976a. Events in lambda injection between phage adsorption and DNA entry. *Virology* **72:** 154.

——. 1976b. Binding to isolated phage receptors and λ DNA release *in vitro. Virology* **72:** 167.

Mount, D.W.A., A.W. Harris, C.R. Fuerst, and L. Siminovitch. 1968. Mutation in bacteriophage lambda affecting particle morphogenesis. *Virology* **35:** 134.

Murialdo, H. and L. Siminovitch. 1971. The morphogenesis of bacteriophage lambda. III. Identification of genes specifying morphogenetic proteins. In *The bacteriophage lambda* (ed. A.D. Hershey), p. 711. Cold Spring Harbor Laboratory, Cold Spring Harbor, New York.

——. 1972. The morphogenesis of bacteriophage lambda. IV. Identification of gene products and control of the expression of the morphogenetic information. *Virology* **48:** 785.

Parkinson, J.S. 1968. Genetics of the left arm of the chromosome of bacteriophage lambda. *Genetics* **59:** 311.

Randall-Hazelbauer, L. and M. Schwartz. 1973. Isolation of the bacteriophage lambda receptor from *Escherichia coli. J. Bacteriol.* **116:** 1436.

Ray, P.N. and M.L. Pearson. 1974. Evidence for post-transcriptional control of the morphogenetic genes of bacteriophage lambda. *J. Mol. Biol.* **85:** 163.

——. 1975. Functional inactivation of bacteriophage λ morphogenetic gene mRNA. *Nature* **253:** 647.

Roa, M. 1981. Receptor-triggered ejection of DNA and protein in phage lambda. *FEMS Microbiol. Lett.* **11:** 257.

Roa, M. and D. Scandella. 1976. Multiple steps during the interaction between coliphage lambda and its receptor protein *in vitro. Virology* **72:** 182.

Saigo, K. and H. Uchida. 1974. Connection of the right-hand terminus of DNA to the proximal end of the tail in bacteriophage lambda. *Virology* **61:** 524.

Scandella, D. and W. Arber. 1974. An *Escherichia coli* mutant which inhibits the injection of phage λ DNA. *Virology* **58:** 504.

——. 1976. Phage λ DNA injection into *Escherichia coli pel*⁻ mutants is restored by mutations in phage genes *V* or *H. Virology* **69:** 206.

Schwartz, M. 1975. Reversible interaction between coliphage lambda and its receptor protein. *J. Mol. Biol.* **99:** 185.

——. 1976. The adsorption of coliphage lambda to its host: Effect of variations in the surface density of receptor and in phage-receptor affinity. *J. Mol. Biol.* **103:** 521.

Showe, M.K. and E.K. Kellenberger. 1975. Control mechanisms in virus assembly. In *Control process in viral multiplication* (ed. D.C. Burke and W.C. Russell), p. 407. Cambridge University Press, Cambridge, England.

Thirion, J.P. and M. Hofnung. 1972. On some aspects of phage λ resistance in *E. coli* K12. *Genetics* **71:** 207.

Thomas, J.O. 1974. Chemical linkage of the tail to the right-hand end of bacteriophage lambda DNA. *J. Mol. Biol.* **87:** 1.

Thomas, J.O., N. Sternberg, and R. Weisberg. 1978. Altered arrangement of the DNA in injection-defective lambda bacteriophage. *J. Mol. Biol.* **123:** 149.

Thomas, R., C. Leurs, C. Dambly, D. Parmenties, L. Lambert, P. Brachet, N. Lefebvre, S. Mousset, J. Porcheret, J. Szpirer, and D. Wauters. 1967. Isolation and characterization of new sus (amber) mutants of bacteriophage λ. *Mutat. Res.* **4:** 735.

Tsui, L. and R.W. Hendrix. 1980. Head-tail connector of bacteriophage lambda. *J. Mol. Biol.* **142:** 419.

——. 1983. Proteolytic processing of phage lambda tail protein gpH: Timing of the cleavage. *Virology* **125:** 257.

Walker, J.E., A.D. Auffret, A. Carne, A. Gurnett, P. Hanisch, D. Hill, and M. Saraste. 1982. Solid-phase sequence analysis of polypeptides eluted from polyacrylamide gels. An aid to interpretation of DNA sequences exemplified by the *Escherichia coli unc* operon and bacteriophage lambda. *Eur. J. Biochem.* **123:** 253.

Weigle, J. 1966. Assembly of phage lambda *in vitro. Proc. Natl. Acad. Sci.* **55:** 1462.

———. 1968. Studies on head-tail union in bacteriophage lambda. *J. Mol. Biol.* **33:** 483.

Wood, W.B. and J. King. 1979. Genetic control of complex bacteriophage assembly. *Comp. Virol.* **13:** 581.

Youderian, P.A. 1978. "Genetic control of the length of lambdoid phage tails." Ph.D. thesis, Massachusetts Institute of Technology, Cambridge.

Zgaga, V., M. Medić, E. Salaj-Šmic, D. Novak, and M. Wrischer. 1973. Infection of *Escherichia coli* envelope-membrane complex with lambda phage: Adsorption and penetration. *J. Mol. Biol.* **79:** 697.

Bacteriophage P22 Antirepressor and Its Control

Miriam M. Susskind and Philip Youderian
Department of Molecular Genetics and Microbiology
University of Massachusetts Medical School
Worcester, Massachusetts 01605

In the past 10 years, it has become increasingly clear that the genome of temperate *Salmonella* phage P22 is organized and regulated in much the same way as the genome of λ (or of other related lambdoid coliphages). Alignment of the P22 and λ prophage genetic maps (Fig. 1) reveals that the two genomes are, for the most part, composed of analogous clusters of functionally related genes arranged in the same order. This congruence of genetic structure and function is particularly striking in the left half of the prophage, where there is a one-to-one correspondence between many individual P22 and λ genes, including the genes encoding major regulatory proteins as well as their sites of action. Thus, the regulatory mechanisms responsible for temporal control of gene expression during lytic development and for establishment and maintenance of repression in P22 closely parallel the mechanisms that operate in λ.

P22 and λ not only share common features of genetic and physiological organization but also show a considerable degree of DNA sequence homology (Skalka and Hanson 1972). Furthermore, P22 and λ recombine in vivo to produce viable hybrids (Gemski et al. 1972; Botstein and Herskowitz 1974). For all of these reasons, P22 has properly come to be considered a member of the lambdoid family. However, P22 differs profoundly from λ (and other lambdoid coliphages) in many properties, most of which are determined by genes in the right half of the prophage map. The pathway of P22 capsid assembly differs substantially from the λ head assembly pathway (King et al. 1973); most notably, P22 DNA is packaged by a "sequential headful" mechanism that yields DNA molecules with circularly permuted and repetitious ends, rather than unique, cohesive ends (Tye et al. 1974a,b). This difference in packaging mechanism accounts for many features of P22 that are not shared by the lambdoid coliphages: the ability to mediate generalized transduction (Zinder and Lederberg 1952), the circularity of the vegetative genetic map (Gough and Levine 1968; Botstein et al. 1972), and the indispensable role of homologous recombination in circularization of the infecting phage DNA molecule and, hence, in phage growth (Botstein and Matz 1970). The most conspicuous morphological difference between the two phages is in the structure of the tail. P22 has a short, sixfold symmetric baseplate, formed by the product of a single tail gene (Israel et al. 1967; Botstein et al. 1973); in contrast, λ has a complex, elongated tail structure for which 11 gene products are required (Katsura, this volume).

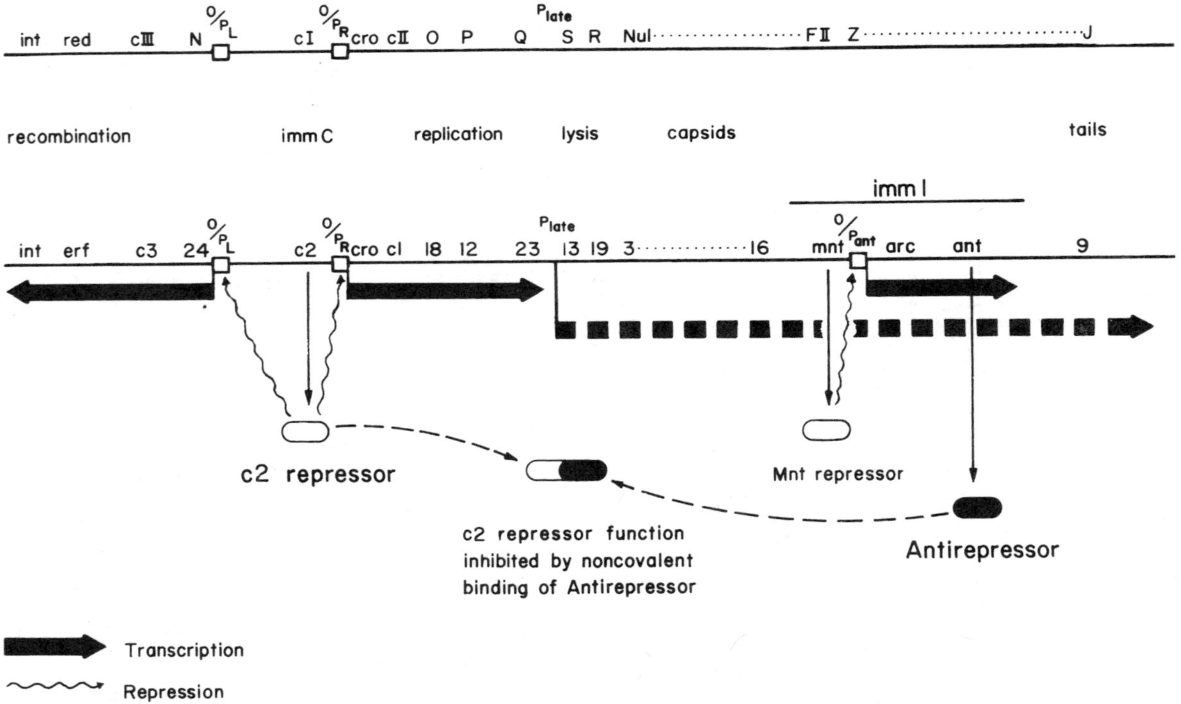

Figure 1 Comparison of the prophage maps of P22 and λ. (→) Transcription units; (----→) transcription dependent on gene-*23* function; (~~~→) negative control during lysogeny by repressors acting at operator/promoter sites (□).

Embedded within the P22 late operon, between the genes controlling head morphogenesis and the tail gene, lies a unique cluster of genes involved in prophage repression and superinfection immunity (Fig. 1). This region, *immI*, does not directly regulate the expression of essential phage genes; rather, it exerts its effect indirectly by interacting with the primary repression mechanism (*c*2 repressor, analogous to λ*c*I repressor) specified by the *immC* region (analogous to the λ immunity region). The key element of the *immI* region is the *ant* gene, which encodes an antirepressor protein that inhibits the activity of P22*c*2 repressor, as well as other primary lambdoid phage repressors. Expression of the *ant* gene is negatively regulated during lysogeny and during the lytic cycle of infection by the linked *mnt* (for *maint*enance of lysogeny) and *arc* (for *a*n*t*i*r*epressor *c*ontrol) genes, respectively. The purpose of this paper is to present the current state of knowledge concerning the regulatory mechanisms encoded by *immI*: antirepressor, the control of antirepressor synthesis, and the relationship of the *immI* region to the late operon.

LAMBDOID CONTROL CIRCUITS IN P22

Except for the additional control circuits specified by the *immI* region, the regulation of gene expression in P22 closely resembles that in phage λ. Most of the P22 genome, including all of the genes coding for essential vegetative functions, is organized into three operons: the p_L and p_R early operons and the p_{LATE} operon (Fig. 1). During the lytic cycle of phage growth, expression of the early operons is positively regulated by the product of gene *24* (analogous to λ*N*) and is later turned down by the product of the *cro* gene. Late gene expression is dependent on the positive regulator encoded by gene *23*, which is analogous to λ*Q*. During lysogeny, the *c*2-gene product (*c*2 repressor, analogous to λ*c*I repressor) represses the early operons and turns on expression of the *c*2 gene from a leftward promoter, p_{RM}, which overlaps p_R. During the establishment of lysogeny, high-level synthesis of repressor from the p_{RE} promoter requires the products of genes *c*1 and *c*3 (analogous to λ*c*II and λ*c*III, respectively).

The role of gene *24* in positive regulation of early gene expression (and, consequently, in late gene expression) is indicated by the pleiotropic effects of 24^- mutations. P22 24^- mutants have been shown to be defective in a variety of early and late phage functions (DNA synthesis, lysogenization, lysis; Hilliker and Botstein 1975), in synthesis of virtually all phage-coded proteins that have been identified by SDS-polyacrylamide gel electrophoresis (Lew and Casjens 1975; P. Youderian and M.M. Susskind, unpubl.), and in synthesis of total phage RNA (Pipas and Reeves 1977; E.N. Jackson, unpubl.). The analogy between *24* and *N* is further supported by the fact that *24* can substitute for *N* in λ/P22 hybrids (Botstein and Herskowitz 1974), can complement a λN^- defect in *trans* (Friedman and Ponce-Campos 1975; Hilliker and Botstein 1976), and can suppress polarity during p_L-promoted expression of the *Escherichia coli gal* operon

(Hilliker et al. 1978). Thus, the product of gene *24*, like the λ*N* protein, presumably acts to antiterminate transcription within the p_L and p_R operons.

The product of P22 gene *23* positively regulates late gene expression. During lytic infection with P22 *23*⁻ mutants, late proteins are not synthesized (Botstein et al. 1973; Lew and Casjens 1975; Youderian and Susskind 1980), and the late genes are not transcribed (E.N. Jackson, unpubl.). Gene *23* can substitute for the λ*Q* gene in λ/P22 hybrids and encodes an activity functionally indistinguishable from λ*Q* activity (Botstein and Herskowitz 1974). The product of gene *23*, like *Q* product, presumably acts to extend a small leader transcript (P22 transcript *a*, analogous to the λ 6S transcript) initiated at p_{LATE}, the promoter for late gene expression (Roberts et al. 1976).

The primary mechanism of repression in P22 is entirely analogous to that in λ. The P22*c2* gene encodes a repressor that is directly responsible for repressing transcription of the early operons. P22*c2* repressor has been purified and shown to bind specifically to two operators, o_L and o_R (Ballivet and Eisen 1978; Poteete et al. 1980). Each operator consists of three partially symmetrical, 18-bp repressor binding sites with different but related sequences (Poteete et al. 1980). P22*c2* repressor and λ repressor do not recognize each other's operators; the operators of the two phages, although similar in overall organization, show no sequence homology and differ in the size and relative spacing of the repressor binding sites. Despite these differences, the molecular details of the interaction of P22 repressor with o_R strikingly parallel the λ case; thus, *c2* repressor exhibits a similar order of affinity in binding site selection, binds in a pairwise cooperative manner, and activates P22 p_{RM} by binding to o_{R2} (Poteete and Ptashne 1982). P22*c2* repressor and its operators are nearly identical to the repressor and operators of homoimmune coliphage 21 (Botstein and Herskowitz 1974; Ballivet et al. 1977; Ballivet and Eisen 1978; Poteete et al. 1980).

In both P22 and λ, the first gene to the right of p_R, *cro*, encodes a second repressor that turns down expression of the early genes during the lytic cycle of phage infection (Winston and Botstein 1981). Although the biochemical mechanism of action of the P22*cro*-gene product has yet to be elucidated, it is expected to work in the same way as the λ Cro repressor, which binds to o_L and o_R and reduces transcription from p_L and p_R (Gussin et al., this volume). Thus, the P22*immC* region, like the λ immunity region, specifies a dual repressor system: Transcription from p_L and p_R is negatively regulated by the *c2* gene, located immediately to the left of p_R, during lysogeny, and by the *cro* gene, located immediately to the right of p_R, during lytic infection.

ANTIREPRESSOR

Early genetic studies showed that prophage P22, unlike prophage λ, specifies two different repressors that are necessary for maintenance of lysogeny and superinfection immunity. Thus, P22 prophages with temperature-sensitive mutations in either *c*2 (Levine and Smith 1964) or *mnt* (Gough 1968) are induced when heated to nonpermissive temperature, indicating that the products of both genes are necessary for repression of prophage P22. Both repressors are also required for superinfection immunity; lysogens carrying defective prophages deleted for either the *c*2 region (*immC*) or the *mnt* region (*immI*) support the growth of superinfecting P22 wild type (Chan and Botstein 1972). Furthermore, immunity specificity is determined by both *immC* and *immI*. In crosses between P22 and the related heteroimmune *Salmonella* phage L, two types of recombinants are formed that differ in immunity specificity from each other and from both parental types (Bezdek and Amati 1968). Another early observation that underscored the bipartite nature of P22 repression/immunity was the isolation of virulent mutations (i.e., mutations that allow superinfecting phage to grow in an immune lysogen) in both *immC* and *immI* (Bronson and Levine 1971).

In 1972, Levine proposed that the *immI* region encodes an antagonist of the primary *c*2 repressor and that this ''antirepressor'' is negatively regulated by Mnt repressor. This hypothesis was later substantiated by the isolation and characterization of *ant*$^-$ mutations (Botstein et al. 1975; Levine et al. 1975). P22*mnt-ts ant*$^-$ lysogens do not induce at high temperature, indicating that the sole function of Mnt repressor in the maintenance of lysogeny is to repress the synthesis of antirepressor. P22*c*2$^+$*mnt*$^-$ (*ant*$^-$) lysogens are immune to superinfecting P22*ant*$^-$ phage, indicating that the role of Mnt repressor in superinfection immunity is to prevent synthesis of antirepressor by the superinfecting phage. The virulent mutations in *immI* (called *virA* or *Vy*) confer the virulent phenotype only in *cis* to a functional *ant* gene, indicating that these mutations render the synthesis of antirepressor constitutive in the presence of Mnt repressor. The novel immunity specificities of P22/L hybrids are accounted for by the hypothesis that P22 and L have heterospecific *immC* repressors, both of which are sensitive to P22 antirepressor but that the *immI* region of phage L has neither an active *ant* nor *mnt* gene (Susskind and Botstein 1978).

Although P22*ant*$^-$ mutants are unable to grow in P22*c*2$^+$*mnt*$^-$ lysogens, *ant*$^-$ mutants are silent under most conditions. P22*ant*$^-$ mutants are indistinguishable from P22 wild type in their ability to grow lytically and establish lysogeny upon infection of a sensitive (nonlysogenic) host, indicating that antirepressor does not play a substantial role in the decision between lysis and lysogeny. The most likely explanation for this observation is that the small amount of antirepressor produced by wild-type phage, although sufficient to inactivate the maintenance level of *c*2 repressor in a P22*c*2$^+$*mnt*$^-$ lysogen, is not sufficient to inactivate the high level of *c*2 repressor produced during establishment (Gough 1977). Antirepressor influences the decision between lysis and lysogeny only under circumstances in which the control of antirepressor synthesis is deranged as a consequence of mutation (e.g., *mnt*$^-$, *Vy*, or *arc*$^-$).

As prophage, P22*ant*$^-$ mutants are also indistinguishable from P22 wild type in ability to undergo spontaneous induction or induction by agents such as UV light and mitomycin C (Botstein et al. 1975; Levine et al. 1975), indicating that antirepressor does not play a substantial role in these processes either. For both P22 and λ, this type of induction is mediated by cleavage of *c*2 and *c*I repressor by the host RecA protein (Roberts et al. 1978; Phizicky and Roberts 1980; Sauer et al. 1982). Antirepressor works by a different mechanism (see below), does not require the RecA function, and is active against mutant forms of λ and P22 repressor that are insensitive to RecA-mediated induction (Botstein et al. 1975; Susskind and Botstein 1975).

MECHANISM OF ACTION OF ANTIREPRESSOR

Antirepressor can be assayed in vitro as an inhibitor of the operator-specific DNA binding activity of λ*c*I repressor (Susskind and Botstein 1975) or P22*c*2 repressor (J. DeAnda and R.T. Sauer, unpubl.). Early experiments employing crude lysates as a source of antirepressor identified the repressor protein as the target of antirepressor and established that antirepressor does not inactivate repressor by covalent modification or cleavage, since repressor that has been inactivated by antirepressor can be recovered in active form by denaturation and selective renaturation (Susskind and Botstein 1975).

Using this assay, antirepressor has been purified to near homogeneity and shown to be the *ant*-gene product by determination of its aminoterminal amino acid sequence (R.T. Sauer et al., in prep.). The nucleotide sequence of the *ant* gene demonstrates that antirepressor has a subunit molecular weight of approximately 34,600 (300 amino acids), in good agreement with the molecular weight determined by gel electrophoresis (Susskind 1980).

More recent experiments using purified antirepressor show that the carboxyterminal domain of λ repressor is much more efficient than the aminoterminal domain in "sparing" P22 repressor from inhibition by antirepressor (J. DeAnda and R.T. Sauer, unpubl.). The carboxyterminal domains of both λ and P22 repressor mediate repressor dimerization as well as cooperative interactions between repressor dimers (Gussin et al., this volume). Thus, it seems likely that the mechanism of inactivation of repressor by antirepressor involves either cooligomerization of antirepressor with repressor or inhibition of repressor oligomerization.

Antirepressor binds with low affinity to DNA, apparently without sequence specificity. Furthermore, inhibition of repressor activity by antirepressor is more efficient in the presence of nonspecific DNA, suggesting that nonspecific DNA may be directly involved in antirepressor activity. One mechanism consistent with the observed data is that antirepressor cooligomerizes with repressor and, as a result, the repressor in this complex is "relocated" to nonspecific DNA binding sites (J. DeAnda and R.T. Sauer, unpubl.).

CONTROL OF ANTIREPRESSOR SYNTHESIS

Figure 2 summarizes our current picture of the regulatory circuits governing antirepressor synthesis. Rightward transcription of the antirepressor gene (*ant*)

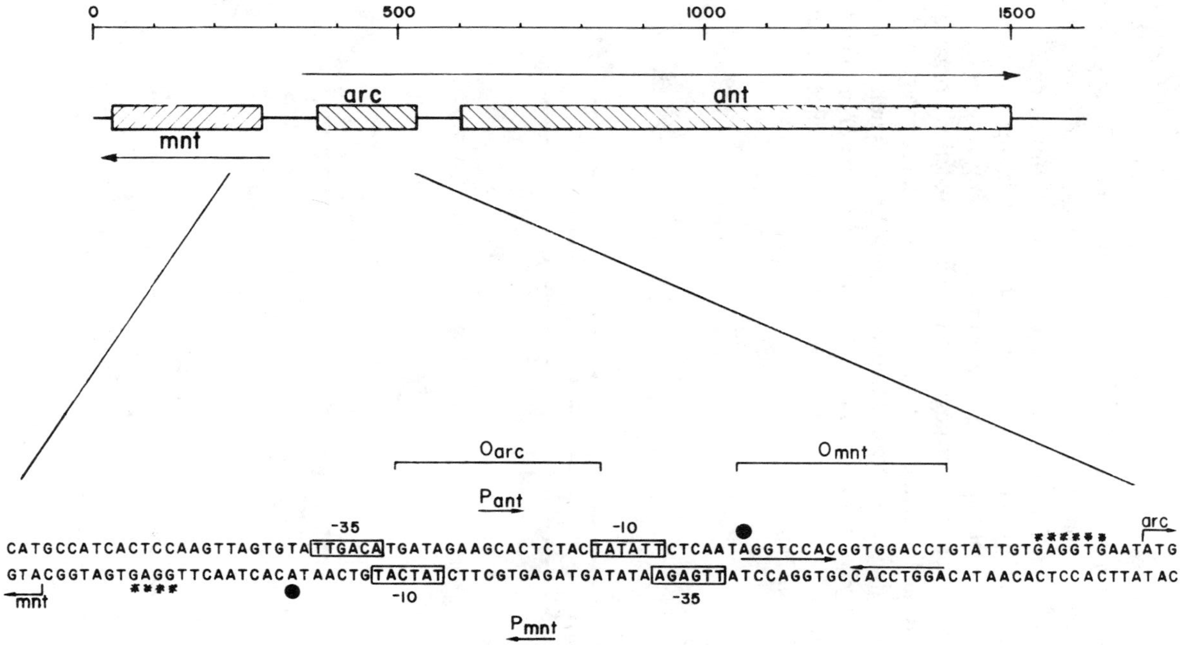

Figure 2 The *immI* region. The −10 hexamer and −35 hexamer of p_{ANT} and p_{MNT} are boxed. (●) Probable startpoints of transcription; (*) potential ribosome recognition sequences (R.T. Sauer et al., in prep.). The locations of o_{MNT} and o_{ARC} are deduced from the sequence analysis of operator-constitutive mutations (see text).

initiates at a promoter, p_{ANT}, located between the *mnt* and *arc* genes. Synthesis of antirepressor is repressed during lysogeny by the product of the *mnt* gene, which is transcribed leftward from a promoter, p_{MNT}, that overlaps p_{ANT}. The site of action of the Mnt repressor (o_{MNT}) is defined by mutations called *Vy*, which permit constitutive antirepressor synthesis in the presence of Mnt repressor. Synthesis of antirepressor is negatively regulated during lytic infection by the product of the *arc* gene, the first gene transcribed from p_{ANT}. The site of action of the Arc repressor (o_{ARC}) is defined by mutations that permit constitutive antirepressor synthesis in the presence of high levels of Arc repressor. Mnt and Arc are presumed to be DNA-binding proteins that prevent the initiation of transcription from p_{ANT}. Together, Mnt and Arc comprise a dual repressor system similar in several respects to the P22*immC* or λ dual repressor systems (Fig. 3). Thus, the *immI* region, which superimposes a second-order regulatory mechanism on the primary immunity system by encoding antirepressor, is somewhat analogous in genetic and functional organization to the P22*immC* and λ immunity regions. All three immunity regions include a pair of diverging promoters, a leftward promoter for synthesis of a repressor active during lysogeny (*mnt* in *immI*, *c2* in *immC*, and cI in λ) and a rightward promoter for synthesis of a repressor active during lytic infection (*arc* in *immI*, *cro* in P22*immC* and λ).

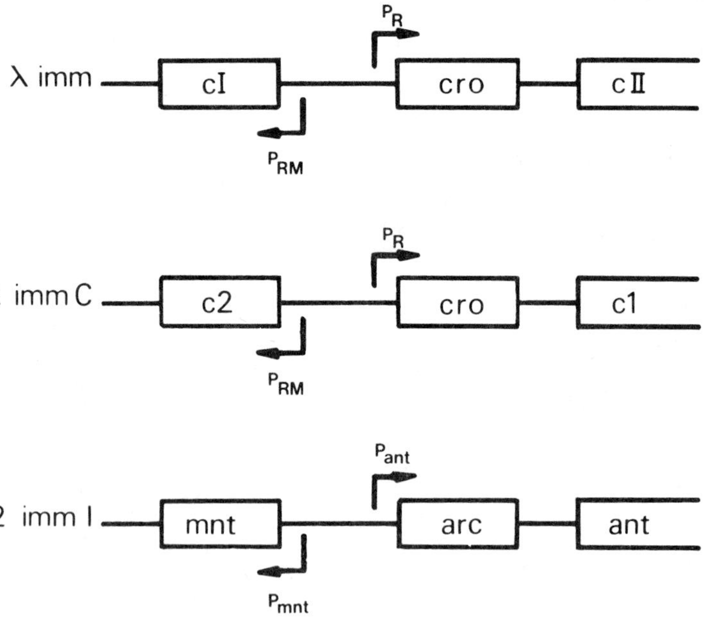

Figure 3 Analogy among the λ*imm*, P22*immC*, and P22*immI* regions.

ARC REPRESSOR

During lytic infection, synthesis of antirepressor is negatively regulated by the product of the first gene in the *ant* operon, *arc*. With P22 wild-type or *mnt*$^-$ phage, synthesis of antirepressor occurs at a low rate at early times and is shut off at late times. In contrast, P22*arc*$^-$ mutants dramatically overproduce antirepressor protein; antirepressor is the major protein (phage or host) produced during *arc*$^-$ infection (Susskind 1980).

The *arc*-gene product not only negatively regulates synthesis of antirepressor but negatively regulates its own synthesis as well. The product of the *arc* gene has been identified by gel electrophoresis as a small protein that is present in lysates of *sup*^{0} cells infected with P22*arc*$^+$ phage but absent in lysates of cells infected with P22*arc-am* phage (Youderian et al. 1982b). P22*arc-am* phage overproduce Arc protein as well as antirepressor during infection of a *sup*$^+$ host that does not restore Arc activity (Youderian et al. 1982b). The finding that Arc protein is overproduced under this Arc$^-$ condition indicates that *arc*-gene expression is autogenously regulated.

The *arc* and *ant* genes are expressed in vivo from the same promoter (p_{ANT}). Phages carrying a p_{ANT} point mutation or deletion synthesize neither Arc protein nor antirepressor (Youderian et al. 1982b). DNA sequence analysis confirms that the *arc* gene lies between p_{ANT} and *ant*, is transcribed rightward, and codes for a small protein (~ 50 amino acids; R.T. Sauer et al., in prep.). Thus, *arc* and *ant* form an operon. The fact that Arc protein regulates expression of both genes in the operon supports the idea that Arc protein works by binding to DNA and blocking initiation of transcription from p_{ANT}.

Arc$^-$ LETHAL PHENOTYPE

Failure of *arc*$^-$ phage to regulate antirepressor synthesis leads secondarily to the failure to produce progeny phage. Thus, amber mutations in *arc* confer a conditional-lethal phenotype. During infection of unirradiated *sup*^{0} cells, P22*arc-am* phage not only overproduce antirepressor but also show an aberrant pattern of production of other phage proteins (Susskind 1980). In particular, all of the phage late proteins, which are required for host cell lysis and assembly of the virion, are underproduced.

Evidence that the Arc$^-$ lethal phenotype is a secondary effect of the overproduction of antirepressor has been provided by the analysis of revertants of P22*arc-am* phage selected for ability to form plaques on a *sup*^{0} host (Susskind 1980; Susskind and Youderian 1982; Youderian et al. 1982a). The majority of these revertants retain the *arc-am* mutation and have acquired a second mutation that suppresses the Arc$^-$ lethal phenotype. Without exception, these suppressing mutations affect synthesis of antirepressor. Most of these pseudorevertants have the Ant$^-$ phenotype (inability to grow on P22*c2*$^+$*mnt*$^-$ lysogens) and produce little or no antirepressor protein. The suppressing mutations in these Ant$^-$ pseudorevertants are either in the *ant* structural gene or in the *ant* promoter. A substantial fraction of pseudorevertants of P22*arc-am* phage retain the Ant^+ phenotype and carry suppressing mutations that reduce the rate of antirepressor synthesis but do not abolish it. The level of antirepressor produced by these phages in sup^0 cells is less than the lethal level produced by their *arc-am* parent but more than the level produced by P22 arc^+ phage. Many of these Ant^+ pseudorevertants carry ''mild'' promoter-down mutations in p_{ANT}.

The lethal phenotype of *arc-am* mutations appears to be related to antirepressor synthesis rather than to antirepressor activity. The Arc^- lethal defect is suppressed by mutations in p_{ANT} that result in only a modest decrease in the rate of antirepressor synthesis (Youderian et al. 1982a) but is not suppressed by missense or carboxyterminal amber mutations in *ant* that completely inactivate antirepressor (Susskind 1980). Although it could be argued that Ant protein has another activity that is responsible for the lethal effect, no suppressor mutations that detoxify the protein (without affecting its rate of synthesis) have been found. Since all of the suppressing mutations in *arc-am* pseudorevertants affect the *synthesis* of antirepressor, we are drawn to the hypothesis that overproduction of antirepressor protein per se leads to the observed failure of P22 late gene expression under Arc^- conditions. That is, elevated translation of antirepressor apparently interferes with the expression of essential phage genes.

ant PROMOTER

The map position of *Vy* mutations suggested that the *ant* gene is transcribed rightward from a promoter between *mnt* and *arc* (Botstein et al. 1975; Levine et al. 1975; Susskind 1980). Many of the Ant^- mutations in pseudorevertants of *arc-am* phage were also found to map in this region, suggesting that these mutations inactivate the *ant* promoter (Susskind 1980). The identification and characterization of the p_{ANT} transcript and of these promoter-down mutations have confirmed these predictions (Susskind and Youderian 1982; Youderian et al. 1982a; R.T. Sauer et al., in prep.).

When P22 DNA is transcribed in vitro by *E. coli* RNA polymerase, several major transcripts are produced (Roberts et al. 1976), one of which has been identified as the p_{ANT} transcript (Susskind and Youderian 1982). The results of experiments using restriction-fragment templates show that this transcript (1.2 kb) is synthesized rightward and that the transcribed region includes the *arc* and *ant* genes. The p_{ANT} transcript is produced at a reduced rate from templates carrying mutations predicted to reduce the activity of the *ant* promoter. Reversion of an Ant^- phage carrying a severe promoter-down mutation by selection for the Ant^+ phenotype simultaneously restores synthesis of antirepressor protein in vivo and synthesis of the p_{ANT} transcript in vitro (Susskind and Youderian 1982).

Since the p_{ANT} transcript is one of the most abundant products synthesized in vitro from a wild-type P22 DNA template, transcription from p_{ANT} does not appear to depend on ancillary proteins. Thus, *ant* promoter-down mutations directly affect the interaction of RNA polymerase with p_{ANT}, rather than the interaction of the promoter with a positive activator. The isolation and DNA sequence analysis of a large number of p_{ANT} mutations has provided a simple genetic approach to determine which features of the promoter sequence are important for RNA polymerase recognition (Youderian et al. 1982a). This analysis is especially informative in view of the fact that the wild-type *ant* promoter has a ''near-consensus'' sequence (Fig. 4).

As shown in Figure 4, both severe and mild p_{ANT} promoter-down mutations cluster in two regions that correspond to the most conserved sequence homologies among wild-type prokaryotic promoters, the −35 hexamer and the −10 hexamer (for review, see Rosenberg and Court 1979). Most p_{ANT} promoter-down mutations reduce the homology of p_{ANT} with the consensus sequence; this result provides strong genetic support for the idea that the consensus homologies among wild-type prokaryotic promoters define the critical primary structural features of the promoter. Consistent with this idea, many of the p_{ANT} promoter-down mutations correspond to mutations that have been shown to decrease the activity of other promoters.

Mnt REPRESSOR

Levine et al. (1975) and Botstein et al. (1975) showed that the *mnt*-gene product is required for maintenance of lysogeny by P22*ant*⁺ prophage and is required for

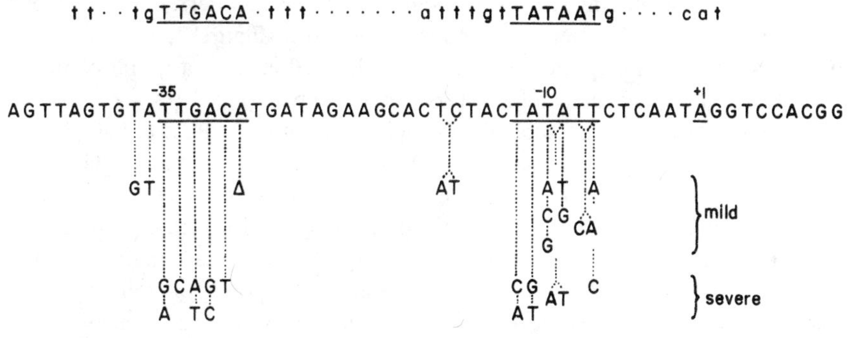

Figure 4 p_{ANT} promoter-down mutations. The top line shows the consensus promoter sequence (Rosenberg and Court 1979). Highly conserved positions (occurring in >50% of wild-type promoters) are indicated by uppercase letters, and weakly conserved positions (occurring in >33% of wild-type promoters) are indicated by lowercase letters. The sequence of the wild-type *ant* promoter is shown on the next line. The −35 hexamer and −10 hexamer and presumed transcription startpoint are underlined. Shown below are the sequence changes for 25 single and tandem double p_{ANT} promoter-down mutations (Youderian et al. 1982a). (Δ) A deleted base pair.

immunity of lysogens to superinfecting P22*ant*$^+$ phage. These findings indicated that *mnt* codes for a negative regulator of *ant*-gene expression. Furthermore, the demonstration that a *Vy* mutation confers virulence only if it lies *cis* to a functional *ant* gene indicates that the *Vy* mutation alters a regulatory site and thereby renders *ant*-gene expression insensitive to Mnt control. It was inferred that the *mnt*-gene product is a repressor that inhibits initiation of transcription from p_{ANT} by binding to an operator that is altered by the *Vy* mutations.

Since the *arc* gene is expressed from p_{ANT}, the hypothesis that Mnt acts at the level of transcription initiation predicts that Mnt represses synthesis of Arc protein as well as antirepressor. This prediction has been confirmed by the analysis of proteins synthesized by superinfecting phage in nonimmune ($c2^-$) *mnt*$^+$ and *mnt-am* lysogens. In the *mnt*$^+$ lysogen, synthesis of both Arc protein and antirepressor by superinfecting phage is repressed. Even *arc*$^-$ phage, which normally synthesize very high levels of antirepressor, produce no detectable Ant protein in the presence of Mnt (Youderian et al. 1982b).

The *mnt* gene has been identified in the DNA sequence of *immI* by analysis of *mnt* amber and missense mutations (R.T. Sauer et al., in prep.). The *mnt* gene is read leftward, as predicted by the analogy in Figure 3. However, the *mnt*-coding sequence is only 83 amino acids long; Mnt is thus only about one-third as large as λ*c*I or P22*c*2 repressor. It is noteworthy that the amino acid sequences of the Mnt and Arc proteins are clearly related; Mnt repressor is also closely related in sequence to the λ Cro protein.

ant OPERATORS

The location of the binding site for Mnt repressor has been inferred by the DNA sequence analysis of *Vy* mutations. As summarized in Figure 2, *Vy* mutations alter sites within a 17-bp palindrome (o_{MNT}) located at the startpoint of p_{ANT} transcription (P. Youderian and M.M. Susskind, unpubl.).

o_{MNT} does not closely resemble the o_L and o_R operators of P22, λ, and phage 21. The most striking difference is that the o_{MNT} palindrome is not reiterated; the other phage operators are composed of three repressor binding sites, each a partially symmetric sequence of 17 bp (λ) or 18 bp (P22, 21) (see Gussin et al., this volume). Furthermore, whereas most or all of the repressor binding sites in o_L and o_R are located 5′ relative to the p_L and p_R startpoints of transcription, most or all of the o_{MNT} palindrome lies 3′ relative to the startpoint of p_{ANT} transcription. Thus, o_{MNT} more closely resembles many bacterial operators (e.g., *E. coli trp* or *lac*) than other temperate phage operators.

The site of Mnt repressor binding (o_{MNT}) is not the site of Arc repressor binding (o_{ARC}). *Vy* mutations (including a deletion of the central 7 bp of the o_{MNT} palindrome) do not confer the Arc$^-$ phenotype (overproduction of antirepressor during infection), indicating that these mutations do not affect the binding site for Arc repressor. Furthermore, *Vy* mutants are not able to grow in a

P22*c*2$^+$*mnt*$^-$ lysogen carrying a plasmid that produces Arc repressor constitutively. In contrast, phages with mutations between the −10 hexamer and the −35 hexamer of p_{ANT}, which have been isolated by a variety of selections, are able to grow in this "Arc-immune" host (P. Youderian et al., unpubl.). (Since p_{ANT} is only partially active, these mutations do not exhibit the Arc$^-$ lethal phenotype.) As summarized in Figure 2, the Arc-insensitive phenotype of these mutations suggests that the binding site for Arc repressor is located within p_{ANT}, in a region of partial dyad symmetry.

REGULATION OF THE *mnt* GENE

Figure 2 shows that the p_{ANT} region has another axis of dyad symmetry located between the −35 hexamer and that the −10 hexamer, and that on the other DNA strand are two hexamers that define a leftward (opposing) promoter. Two genetic observations indicate that this is the promoter for expression of the *mnt* gene, which begins 28 bp in the 3′ direction from the leftward −10 hexamer. First, two of the mutations that reduce the activity of the *ant* promoter also confer an Mnt$^-$ phenotype. These include a severe p_{ANT} promoter-down mutation that alters a base pair common to both the −10 hexamer of p_{ANT} (TATATT → TATATC) and the −35 hexamer of the leftward promoter (TTGAGA → TTGAGG), and a mild p_{ANT} promoter-down mutation that deletes the last base of the p_{ANT} −35 hexamer (TTGACA → TTGACΔ) and the "invariant T" of the leftward −10 hexamer (TATCAT → TATCAΔ). Second, phages that carry deletions extending rightward from endpoints immediately to the right of o_{MNT} retain Mnt$^+$ activity (immunity); in contrast, a deletion that extends rightward from a site between the −35 hexamer and the −10 hexamer of p_{ANT} (and therefore between the −10 hexamer and the −35 hexamer of the leftward promoter) is phenotypically Mnt$^-$. It therefore appears that the p_{MNT} promoter is located directly opposite the p_{ANT} promoter (P. Youderian and M.M. Susskind, unpubl.).

At present, little is known about how the synthesis of Mnt repressor is regulated. The striking extent of overlap between p_{ANT} and p_{MNT} raises the possibility that differential expression from these promoters may be regulated by a simple competition mechanism. The results of in vitro promoter strength assays (McClure 1980) show that p_{ANT} and p_{MNT} compete for the mutually exclusive binding of RNA polymerase; *ant* promoter-down mutations indirectly activate p_{MNT} (W.R. McClure et al., unpubl.). This finding predicts that p_{MNT} may be positively regulated by Mnt (and possibly Arc) repressor by an indirect mechanism: the relief of promoter competition by repression of p_{ANT}. The position of o_{MNT} with respect to p_{MNT} also raises the possibility that Mnt repressor may directly activate p_{MNT}, in a manner analogous to the activation of the λ p_{RM} promoter by λ*c*I repressor bound at o_{R2} (Meyer and Ptashne 1980; Meyer et al. 1980; see Gussin et al., this volume).

ant TERMINATOR

Even without addition of ρ termination factor, the p_{ANT} transcript terminates in vitro between the end of *ant* and the beginning of gene *9*, which encodes the phage tail protein (Susskind and Youderian 1982). The existence of a transcription termination signal (t_{ANT}) in this region had been inferred from the evidence that gene *9* is not expressed from p_{ANT} in vivo. Although the exact 3′ end of the p_{ANT} transcript has yet to be determined, the results of transcription experiments using templates cleaved with restriction endonucleases indicate that t_{ANT} lies close to the 3′end of *ant*. This region includes a sequence that closely resembles known prokaryotic transcription termination signals: T_6A immediately preceded by a G-C-rich "stem-and-loop" sequence (R.T. Sauer et al., in prep.). Further evidence that this sequence is t_{ANT} has recently been provided by the characterization of a set of recombinant plasmids carrying P22 gene *9* (Berget et al. 1983). Gene *9* is expressed very inefficiently by a plasmid in which the *E. coli lac* UV5 promoter is positioned upstream from a fragment containing the end of *ant*, t_{ANT}, and all of gene *9*. Deletions extending from a site within *ant* into or beyond the t_{ANT} stem-and-loop sequence markedly increase p*9* synthesis in vivo and cause extension of the corresponding transcript in vitro. Neither effect is seen with deletions that leave the stem-and-loop sequence intact. These results not only confirm the identity of the t_{ANT} sequence but also demonstrate its ability to block transcription of gene *9*.

Although the p_{ANT} transcript terminates at t_{ANT}, the p_{LATE} transcript apparently extends through the entire *immI* region (i.e., through t_{ANT}) and continues into gene *9*. The product of gene *23* (*p23*), which is required for expression of all P22 late genes including gene *9*, is thought to activate transcription from p_{LATE} by antiterminating a small leader transcript (transcript *a*; Roberts et al. 1976). It is likely that *p23* acts a second time to antiterminate transcription at t_{ANT} to allow expression of gene *9* (Weinstock et al. 1980). The reason for postulating that gene *9* is expressed from p_{LATE}, rather than a *23*-dependent promoter between t_{ANT} and gene *9*, is the observation by Weinstock et al. (1980) that Tn*1* insertions in genes *20*, *16*, and *ant* are polar on expression of gene *9*.

Studies of P22 transcription in vivo support the idea that gene *9* is expressed from the p_{LATE} transcript. S.R. Casjens and M.B. Adams (unpubl.) measured the lag period before the onset of exponential decay of late gene mRNAs following rifampicin arrest of RNA initiation; for gene *9* as well as five head morphogenesis genes, the lag time varies proportionately with the distance between the gene and p_{LATE}. Furthermore, E.N. Jackson and T. Weighous (unpubl.) have shown that *23*-dependent rightward transcription of the *immI* region occurs late in infection.

Although there is good evidence that the *ant* gene is transcribed as part of the p_{LATE} operon, antirepressor protein is not synthesized from the p_{LATE} transcript. P22 phage carrying a p_{ANT} promoter-down point mutation produce *p9* in normal amounts late in infection but synthesize no detectable antirepressor protein

(Susskind and Youderian 1982). This is true even if the phage carries both *mnt*$^-$ and *arc*$^-$ mutations, indicating that the failure to synthesize antirepressor is not due to the action of either Mnt or Arc, the only known negative regulators of antirepressor synthesis.

The observation that the *ant* gene is expressed from the p_{ANT} transcript but not from the longer, antiterminated p_{LATE} transcript is reminiscent of the observation that the λ*int* gene is expressed efficiently during infection from the shorter p_{INT} transcript but not from the longer, antiterminated p_L transcript (Guarneros and Galindo 1979; Belfort 1980; Epp et al. 1981; Guarneros et al. 1982). The p_{INT} transcript terminates at t_{INT}, shortly before completion of an RNase III recognition site that is included on the p_L transcript. Cleavage of the p_L transcript by RNase III removes the t_{INT} stem-and-loop sequence and presumbly allows rapid exonucleolytic degradation of *int*-coding sequences (Rosenberg and Schmeissner 1982). It remains to be seen whether a similar mechanism prevents *ant*-gene expression from p_{LATE}.

Recently, we have demonstrated that several pseudorevertants of P22*arc-am* phage carry mutations in the t_{ANT} "stem" sequence (M.M. Susskind et al., unpubl.). Like all other mutations that suppress the Arc$^-$ lethal phenotype, these mutations reduce the rate of p_{ANT}-promoted synthesis of antirepressor. The λ*int* paradigm suggests two possible explanations for the effect of these t_{ANT} mutations on expression of the upstream *ant* gene. If the mutations efficiently prevent termination at t_{ANT}, extension of the p_{ANT} transcript may introduce a site for endonucleolytic attack; as noted above, this model would also explain why antirepressor is not synthesized from the antiterminated p_{LATE} transcript. Alternatively, termination may still occur at t_{ANT}, but destabilization of the terminator stem structure at the 3′ end of the transcript may increase its susceptibility to exonucleolytic attack (see Rosenberg and Schmeissner 1982).

REFERENCES

Ballivet, M. and H. Eisen. 1978. Purification and properties of phage P22 c2 repressor. *Eur. J. Biochem.* **82:** 175.

Ballivet, M., L.F. Reichardt, and H. Eisen. 1977. Purification and properties of coliphage 21 repressor. *Eur. J. Biochem.* **73:** 601.

Belfort, M. 1980. The cII-independent expression of the phage λ *int* gene in RNase III-defective *E. coli*. *Gene* **11:** 149.

Berget, P.B., A.R. Poteete, and R.T. Sauer. 1983. Control of phage P22 tail protein expression by transcription termination. *J. Mol. Biol.* **164:** 561.

Bezdek, M. and P. Amati. 1968. Evidence for two immunity regulator systems in temperate bacteriophages P22 and L. *Virology* **36:** 701.

Botstein, D. and I. Herskowitz. 1974. Properties of hybrids between *Salmonella* phage P22 and coliphage λ. *Nature* **251:** 584.

Botstein, D. and M.J. Matz. 1970. A recombination function essential to the growth of bacteriophage P22. *J. Mol. Biol.* **54:** 417.

Botstein, D., R.K. Chan, and C.H. Waddell. 1972. Genetics of bacteriophage P22. II. Gene order and gene function. *Virology* **49:** 268.

Botstein, D., C.H. Waddell, and J. King. 1973. Mechanism of head assembly and DNA encapsulation in *Salmonella* phage P22. I. Genes, proteins, structures and DNA maturation. *J. Mol. Biol.* **80:** 669.

Botstein, D., K.K. Lew, V. Jarvik, and C.A. Swanson, Jr. 1975. Role of antirepressor in the bipartite control of repression and immunity by bacteriophage P22. *J. Mol. Biol.* **91:** 439.

Bronson, M.J. and M. Levine. 1971. Virulent mutants of bacteriophage P22. I. Isolation and genetic analysis. *J. Virol.* **7:** 559.

Chan, R.K. and D. Botstein. 1972. Genetics of bacteriophage P22. I. Isolation of prophage deletions which affect immunity to superinfection. *Virology* **49:** 257.

Epp, C., M. Pearson, and L. Enquist. 1981. Downstream regulation of *int* gene expression by the *b2* region in phage lambda. *Gene* **13:** 327.

Friedman, D. and R. Ponce-Campos. 1975. Differential effect of phage regulator functions on transcription from various promoters: Evidence that the P22 gene *24* and the λ gene *N* products distinguish three classes of promoters. *J. Mol. Biol.* **98:** 537.

Gemski, P., Jr., L.S. Baron, and N. Yamamoto. 1972. Formation of hybrids between coliphage λ and *Salmonella* phage P22 with a *Salmonella typhimurium* hybrid sensitive to these phages. *Proc. Natl. Acad. Sci.* **69:** 3110.

Gough, M. 1968. Second locus of bacteriophage P22 necessary for the maintenance of lysogeny. *J. Virol.* **2:** 992.

——. 1977. Establishment mode repressor synthesis blunts phage P22 antirepressor activity. *J. Mol. Biol.* **111:** 55.

Gough, M. and M. Levine. 1968. The circularity of the phage P22 linkage map. *Genetics* **58:** 161.

Guarneros, G. and J.M. Galindo. 1979. The regulation of integrative recombination by the *b2* region and the *c*II gene of bacteriophage λ. *Virology* **95:** 119.

Guarneros, G., C. Montanez, T. Hernandez, and D. Court. 1982. Posttranscriptional control of bacteriophage λ *int* gene expression from a site distal to the gene. *Proc. Natl. Acad. Sci.* **79:** 238.

Hilliker, S. and D. Botstein. 1975. An early regulatory gene of *Salmonella* phage P22 analogous to gene *N* of coliphage λ. *Virology* **68:** 510.

——. 1976. Specificity of genetic elements controlling regulation of early functions in temperate bacteriophages. *J. Mol. Biol.* **106:** 537.

Hilliker, S., M. Gottesman, and S. Adhya. 1978. The activity of *Salmonella* phage P22 gene 24 product in *Escherichia coli*. *Virology* **86:** 37.

Israel, J.V., T.F. Anderson, and M. Levine. 1967. *In vitro* morphogenesis of phage P22 from heads and base-plate parts. *Proc. Natl. Acad. Sci.* **57:** 284.

King, J., E.V. Lenk, and D. Botstein. 1973. Mechanism of head assembly and DNA encapsulation in *Salmonella* phage P22. II. Morphogenetic pathway. *J. Mol. Biol.* **80:** 697.

Levine, M. 1972. Replication and lysogeny with phage P22 in *Salmonella typhimurium*. *Curr. Top. Microbiol. Immunol.* **58:** 135.

Levine, M. and H.O. Smith. 1964. Sequential gene action in the establishment of lysogeny. *Science* **146:** 1581.

Levine, M., S. Truesdell, T. Ramakrishnan, and M.J. Bronson. 1975. Dual control of lysogeny by bacteriophage P22: An antirepressor locus and its controlling elements. *J. Mol. Biol.* **91:** 421.

Lew, K. and S. Casjens. 1975. Identification of early proteins coded by bacteriophage P22. *Virology* **68:** 525.

McClure, W.R. 1980. Rate-limiting steps in RNA chain initiation. *Proc. Natl. Acad. Sci.* **77:** 5364.

Meyer, B.J. and M. Ptashne. 1980. Gene regulation at the right operator (O_R) of bacteriophage λ. III. λ repressor directly activates gene transcription. *J. Mol. Biol.* **139:** 195.

Meyer, B.J., R. Maurer, and M. Ptashne. 1980. Gene regulation at the right operator (O_R) of bacteriophage λ. II. O_R1, O_R2, and O_R3: Their roles in mediating the effects of repressor and cro. *J. Mol. Biol.* **139:** 163.

Phizicky, E.M. and J.W. Roberts. 1980. Kinetics of *RecA* protein-mediated inactivation of repressors of phage λ and phage P22. *J. Mol. Biol.* **139:** 319.

Pipas, J.M. and R.H. Reeves. 1977. Patterns of transcription in bacteriophage P22-infected *Salmonella typhimurium*. *J. Virol.* **21:** 825.

Poteete, A.R. and M. Ptashne. 1982. Control of transcription by the bacteriophage P22 repressor. *J. Mol. Biol.* **157:** 21.

Poteete, A.R., M. Ptashne, M. Ballivet, and H. Eisen. 1980. Operator sequences of bacteriophages P22 and 21. *J. Mol. Biol.* **137:** 81.

Roberts, J.W., C.W. Roberts, and N.L. Craig. 1978. *Escherichia coli recA* gene product inactivates phage λ repressor. *Proc. Natl. Acad. Sci.* **75:** 4714.

Roberts, J.W., C.W. Roberts, S. Hilliker, and D. Botstein. 1976. Transcription termination and regulation in bacteriophages P22 and lambda. In *RNA polymerase* (ed. R. Losick and M. Chamberlin), p. 707. Cold Spring Harbor Laboratory, Cold Spring Harbor, New York.

Rosenberg, M. and D. Court. 1979. Regulatory sequences involved in the promotion and termination of RNA transcription. *Annu. Rev. Genet.* **13:** 319.

Rosenberg, M. and U. Schmeissner. 1982. Regulation of gene expression by transcription termination and RNA processing. In *Interaction of translational and transcriptional controls in the regulation of gene expression* (ed. M. Grunberg-Manago and B. Safer), p. 1. Elsevier Science Publishing, New York.

Sauer, R.T., M.J. Ross, and M. Ptashne. 1982. Cleavage of the λ and P22 repressors by *recA* protein. *J. Biol. Chem.* **257:** 4458.

Skalka, A. and P. Hanson. 1972. Comparisons of the distributions of nucleotides and common sequences in deoxyribonucleic acid from selected bacteriophages. *J. Virol.* **9:** 583.

Susskind, M.M. 1980. A new gene of bacteriophage P22 which regulates synthesis of antirepressor. *J. Mol. Biol.* **138:** 685.

Susskind, M.M. and D. Botstein. 1975. Mechanism of action of *Salmonella* phage P22 antirepressor. *J. Mol. Biol.* **98:** 413.

———. 1978. Repression and immunity in *Salmonella* phages P22 and L: Phage L lacks a functional secondary immunity system. *Virology* **89:** 618.

Susskind, M.M. and P. Youderian. 1982. Transcription *in vitro* of the bacteriophage P22 antirepressor gene. *J. Mol. Biol.* **154:** 427.

Tye, B.-K., R.K. Chan, and D. Botstein. 1974a. Packaging of an oversize transducing genome by *Salmonella* phage P22. *J. Mol. Biol.* **85:** 485.

Tye, B.-K., J.A. Huberman, and D. Botstein. 1974b. Non-random circular permutation of phage P22 DNA. *J. Mol. Biol.* **85:** 501.

Weinstock, G.M., P.D. Riggs, and D. Botstein. 1980. Genetics of bacteriophage P22. III. The late operon. *Virology* **106:** 82.

Winston, F. and D. Botstein. 1981. Control of lysogenization by phage P22. I. The P22 *cro* gene. *J. Mol. Biol.* **152:** 209.

Youderian, P. and M.M. Susskind. 1980. Identification of the products of bacteriophage P22 genes, including a new late gene. *Virology* **107:** 258.

Youderian, P., S. Bouvier, and M.M. Susskind. 1982a. Sequence determinants of promoter activity. *Cell* **30:** 843.

Youderian, P., S.J. Chadwick, and M.M. Susskind. 1982b. Autogenous regulation by the bacteriophage P22 *arc* gene product. *J. Mol. Biol.* **154:** 449.

Zinder, N.D. and J. Lederberg. 1952. Genetic exchange in *Salmonella*. *J. Bacteriol.* **64:** 679.

Evolution of the Lambdoid Phages

Allan Campbell
Department of Biological Sciences
Stanford University
Stanford, California 94305

David Botstein
Department of Biology
Massachusetts Institute of Technology
Cambridge, Massachusetts 02139

The purpose of this paper is to attempt to explain the origin of bacteriophage λ. For such an explanation to make sense, we must first define what we mean by "lambda," for there is abundant experimental evidence that the organisms that we call by this name are members of a large family of similar temperate bacteriophages that share the same genome organization and the same life-style. Furthermore, members of the lambdoid family can still undergo exchanges of genes or groups of genes, producing clearly different but, nevertheless, perfectly functional lambdoid phages.

DEFINITION OF THE LAMBDOID FAMILY

A central problem then becomes the definition of the limits of the λ family. We suggest that the criterion for inclusion should be the ability to exchange genetic information with λ by homologous recombination with λ itself. If this definition is accepted, the family comprises not only the lambdoid phages of *Escherichia coli* but also includes phage P22, whose normal host is *Salmonella typhimurium*, as well as the large family of *Salmonella* phages related to P22. P22 has a grossly different virion morphology and lacks some features common to the lambdoid coliphages, including the cohesive DNA ends. When one takes this broad view of family membership, such striking similarities remain among all the phages in the family that abandonment of the traditional criteria for inclusion in the family (ability to grow on *E. coli* and cohesiveness of the DNA ends to the DNA ends of λ itself) is amply justified. Definition of the family simply by the ability to recombine with others in the family is consistent with classical definitions of species.

The common properties of the lambdoid family, defined solely by the ability to exchange information, can be summarized as follows. All members make a choice upon infection between lysis and lysogenization of the host. So far as is known, this decision is always made by circuits of similar design. All members of the family integrate the prophage DNA into the host chromosome via a circular DNA intermediate. All have a similar system of positive regulators, which exert transcriptional control by antitermination of transcripts of operons that are always organized in the same way. Finally, the early lytic functions are organized into divergently transcribed operons regulated negatively during lysogeny by a single repressor specified by a gene lying between the operons.

The variations among members of the lambdoid family include many differences in regulatory specificity. The repressor from one member may or may not recognize the operators of another; the positive regulators from one member may or may not be able to prevent termination of the transcripts of another. Yet all of the members have repressors, operators, and positive regulators that work in essentially the same way. Not only regulatory specificities differ, however, among members of the family. The integrase from one member may or may not recognize the attachment sequences of another; the adsorption apparatus of different members may allow adsorption to different hosts, resulting in host ranges that are very diverse. It is the common gene organization that allows the interchange of information, be it whole genes, groups of genes, or parts of genes to result in organisms that are fully functional and can express and regulate their newly acquired genes normally.

Despite these arguments that what has evolved is a continually interbreeding family and not really a single organism, it seems necessary to explain the origin of the first "primordial" lambdoid phage from something sufficiently different as to be unrecognizable as a member of the current family. This paper begins with an attempt at such an explanation, followed by speculation about how the family grew, and concludes with a short description of the kind of population dynamics that apparently govern further evolution of these phages. We have tried to test these ideas against what is known of the properties of lambdoid phages and of the mechanisms for generating diversity among them.

ORIGIN OF PRIMORDIAL λ

Possible Modes of Origin: Linear Evolution or Association of Preexisting Genetic Elements

Consideration of the possible ways in which λ might have evolved starts with the question of the origin of viruses in general. Luria et al. (1978) discuss two hypotheses: (1) Viruses might be degenerate descendants of cellular parasites, or (2) viral genomes might have evolved from genetic elements of the host. The second hypothesis is strongly favored.

More specific hypotheses for evolution of viral genomes from host elements can be classified according to whether the postulated cellular progenitors of a virus are single or multiple. Although the question has rarely been explicitly addressed, much of the available discussion on virus evolution has been couched in terms that imply a single progenitor. For a large and complex organism like λ, this idea means that all the various virus functions (such as replication, insertion, assembly, and lysis) must have developed by linear evolution from a single precursor. On the other hand, if multiple progenitors are envisioned, the viral genome is viewed as a chimeric structure comprising several segments, each of which is derived from a cellular counterpart. Replication genes and origins might have derived from those of the host or a plasmid; lysis genes, from the determinants of enzymes that are involved in cell envelope metabolism; and assembly genes, from host genes that code for subcellular structures.

Such a multiple origin, in which the viral genome appears as a composite of several evolutionarily and functionally distinct ''modules'' is very appealing. Like other living things, lambdoid phages must at some time have evolved toward increasing complexity. If we ask how λ acquired some new property, such as the ability to insert its DNA into the chromosome, e.g., it is easier to imagine that an ancestral, noninserting virus co-opted the insertion genes of some transposon than it is to believe that the virus evolved an integrase system by mutation of its own DNA. The arrangement of genes in λ also encourages the postulation of an origin from various modules that have remained largely distinct and separable from one another. Such modules might include sets of genes determining immunity, insertion, recombination, replication, lysis, the head structures, and the tail structures. Within each such module, evolution by gene duplication and diversification, along lines consonant with current concepts of cellular genes determining metabolic pathways, is plausible. Such evolution could have occurred both before and after the postulated recruitment of the module into the virus; it is even possible to imagine that an improved version of a module that appears in the host might replace, by recombination, a virus copy or vice versa.

By discussing the origin of primordial λ separately from development of the diverse lambdoid phages, we imply that all lambdoid phages indeed had a common ancestor. The alternative is that similar modules associated in parallel to produce several members of the family independently. Most of the known facts concerning comparative genome structure in the lambdoid family are easier to understand on the assumption of a common ancestral phage, although some limitations to this assumption will become clear as we proceed. Before going further, however, it is important to observe that the comparative genetics of the lambdoid phages have generated a concept of ''modular'' genome organization (Botstein and Herskowitz 1974; Susskind and Botstein 1978). There is no necessary logical connection between the modules of that concept (wherein the modules are thought to behave as units in the population biology of the lambdoid family) and the functional modules proposed as the original building blocks of primordial λ.

The problem can be illustrated by considering the results of heteroduplex analysis, in which lambdoid phages tested pairwise exhibit some segments of good homology, and some, of no homology, with sudden transitions between the two types of segment (Simon et al. 1971; Szybalski and Szybalski 1974). There is no reason to suppose that each heterologous segment bounded by homologous segments necessarily constitutes a module in the sense of a unit of function. Yet exchange, by homologous recombination, of the heterologous segment between the two phages is often possible. In the case of plasmids, interspersion of homologous with heterologous segments has sometimes been interpreted to imply that the junction points represent events of illegitimate site-specific recombination during plasmid evolution (Cohen 1976). Were we to make the same inference for λ, we would have to deny the descent of the lambdoids from a common ancestor and would have difficulty dealing with the observations that the genomic organization of the phages remains similar despite gross divergence in the regions specifying function.

Early physical studies of λ DNA revealed another type of interspersion. Successive segments of λ DNA differ in their base composition. The transitions between AT-rich and GC-rich segments are frequently abrupt. Before direct sequencing, the distribution of such segments was most accurately determined by electron microscopy of partially denatured molecules. Comparison of the partial denaturation maps of λ and the related phages 21 and 424 shows that, after allowance is made for some deletions or insertions that must be postulated in order to align the maps at all, the arrangement of AT-rich and GC-rich segments in these three phages is similar, even in those parts of the genome that are heterologous by heteroduplex or recombinational analysis (Highton and Whitfield 1975). This result indicates that the pattern of interspersion of segments of different base content is an ancient feature of the λ genome, which antedates the divergence of these three phages from their common ancestor.

Similarities between λ and Host Genes

If viral genomes originate from associations of diverse host modules, then segments of the viral genome might be more closely related to functionally analogous segments of the host DNA than they are to one another. The examples given below, which are suggestive of such a situation, should be interpreted cautiously in light of the following considerations. (1) Observed similarities could reflect convergent evolution favored by common functional constraint, rather than common ancestry. (2) Relatedness by itself provides no information on the direction of evolution. A host cell may as well have co-opted a virus module as the reverse. (3) Host modules may have undergone as extensive an evolution as the virus modules subsequent to their divergence from a common ancestor.

Replication Origins and Genes

Nucleotide sequences of DNA segments containing the origins of replication of many members of the lambdoid family have been determined and have been shown to be very similar to one another (Grosschedl and Hobom 1978; Moore et al. 1979). The origin of chromosomal replication from *E. coli* (*oriC*) has a nucleotide sequence that is also very similar (Meijer et al. 1979). Some of the proteins that must act, directly or indirectly, on these origins of replication are used by the phages as well as the host (e.g., the DnaG and DnaB proteins), whereas others are specific (Wickner 1979). For instance, *oriC* requires the DnaA and DnaC proteins, both of which are apparently dispensable for λ DNA replication. Instead, λ replication requires the *O* and *P* proteins. Different lambdoid origins require different *O* and *P* analogs: The system shows specificity. The λ*P* protein interacts with the DnaB function of the host (Georgopoulos and Herskowitz 1971; Wickner 1979), and it is known that the *O* protein has a complex specificity that involves recognition of both the origin of replication and the *P*-DnaB protein complex (Furth et al. 1979). The picture is fully consistent with an evolutionary relationship between the host origin of replication and the origins of replication of the lambdoid phages.

Insertion Sites and Genes

The 15-bp sequence common to *attPP′* and *attBB′* has been compared with the internal resolution site of transposon Tn*3* (identical to the insertion-sequence element γδ) and to the 14-bp termini of the flagellar antigen switch region in *S. typhimurium*. All three of these sequences contain crossover points for reciprocal site-specific recombination. When appropriately aligned, the flagellar switch matches λ in 9 of 14 bases, whereas the Tn*3* is dissimilar to the other two. Comparison of the amino acid sequence of λ integrase and the corresponding enzymes from the other systems (i.e., the TnpR and Hin proteins) shows significant homology between TnpR and Hin but none between Hin and Int (Simon et al. 1980).

EVOLUTION OF LAMBDOID PHAGES FROM PRIMORDIAL λ

Possible Mechanisms of Diversification

As we document later, existing λ-related phages have a common arrangement of functional modules (and generally of genes within modules), although the modules themselves may differ extensively from those of λ, both in base sequence and in specificity. Furthermore, natural λ-related phages exhibit various combinations of specificities. Phage λ and 434, e.g., differ in their *c*I and *cro* proteins and their cognate operator sites but have equivalent integrase proteins and *att* sites, whereas phages P22 and 21 have indistinguishable *c*I specificities but differ in many other functions, including integrase.

Two mechanisms that may underlie this diversification are (1) mutational divergence in different lines of descendants of primordial λ, followed by recombinational reshuffling of mutated segments and (2) replacement of λ modules by the recruitment of functionally equivalent modules from the host or from unrelated viruses. The latter mechanism would constitute a continuation of the accumulation of host modules that originated primordial λ, with the modification that new modules are replacing old ones rather than simply being added.

Both mechanisms may have contributed to λ evolution. We might ask what their relative contribution has been. We would certainly like to know which mechanism is responsible for any particular variation under discussion, such as the difference between the immunity regions of phages λ and 21. We must ask whether this historical question can in principle be answered by examining existing phages and hosts.

We are speaking here of segments of the two phage genomes whose relevant properties are the following: The λ immunity segment and the 21 immunity segment perform the same function in the life cycle of the phage. They are located at equivalent positions with respect to other genes, and their internal genetic structure (arrangement of *N*, *c*I, *cro*, *c*II, and operator promoter sites) is very similar. On the other hand, these two immunity segments are sufficiently dissimilar that (1) the DNAs show no signs of homology, either in heteroduplex analysis or in ability to generate identified viable recombinants, and (2) the specificities of all recognition proteins and their cognate recognition sites are distinct, with no detectable cross-reaction.

In principle, two types of information might comprise substantial evidence for mutational divergence from a common precursor. (1) Complete DNA sequencing of the two immunity regions might show sufficient sequence conservation to indicate a common origin. (2) Studies on a sufficient number of lambdoid phages might furnish the missing links required to demonstrate that the λ and 21 immunity segments must be related, because both are obviously related to the immunity region of a third phage. For example, a common relative of λ and 21 is phage P22, whose gene-*24* protein (equivalent to *N* of λ) can recognize the *nut* sites of both λ and 21 (Hilliker and Botstein 1976).

Evidence that might support recruitment from different sources would be identification and characterization of the sources themselves. For example, the idea that λ first acquired *att* and *int* by the insertion into an ancestral virus of a transposon whose internal resolution site could now be used to insert the phage is as good as any other. The insertion module of phage 21 might then be derived from some other transposon, rather than having diverged from that of λ. Identification of transposons with the appropriate sequence homologies or functional specificities would be helpful.

Mutational Divergence and Natural Selection

If we divide the λ genome into functional modules, then mutational divergence within modules followed by recombinational reshuffling of mutated modules might generate a family of viruses whose DNAs show patterns of interspersion of homology and nonhomology. Such patterns are rather generally observed in the DNAs of related organisms, not just in viruses. They may arise by various mechanisms, some of which are unrelated to recombination. The introns of eukaryotic genes, e.g., have frequently diverged more than their bracketing exons (Perler et al. 1980). That observation can be rationalized on the basis that sequences with no specific function diverge more rapidly than functional sequences, because random mutations are more frequently tolerated. A species in which functional modules have diverged but remain recombinable because of intervening homologous segments might suggest the opposite—that functional segments diverge more rapidly than spacer regions.

One point worth noting is that whatever relevance the concept of functional modules may have to the divergence of λ-related phages from primordial λ, the homologous segments shared by pairs of these phages are seldom functionally inert spacer regions. With the possible exceptions of parts of the *b*2 and *nin*5 regions, λ has little, if any, functionless DNA. For example, some hybrids between λ and phage φ80 specify a hybrid exonuclease, unlike that of either parent, because they arose by recombination within the *redX* gene (Szpirer et al. 1969); another example is the production by recombination of λ-φ80 hybrid *O* proteins with a hybrid specificity (Furth and Yates 1978). Therefore, the selective pressures for conservation of function may be as strong in the homologous regions as in the heterologous ones.

There may also be circumstances where natural selection favors divergence rather than conservation of functional specificities. One reason for the plethora of host range, immunity, replication, and insertion specificities among the lambdoid phages may well be the fact that any new rare type will have some selective advantage over the common types, simply because it is different from them. For example, a few individuals of phage 21 introduced into a community containing *E. coli* and λ should selectively increase relative to λ, because phage 21 can multiply on bacteria lysogenic for λ and, as long as phage 21 is rare, there will be few bacteria lysogenic for phage 21.

As in the somewhat analogous case of multiple mating types in higher fungi (Raper 1966), a selective advantage of rare types is adequate to explain why the population remains highly polymorphic. With a few reasonable accessory assumptions, it can explain not only the perpetuation of diversity but also a high rate of diversification. A new immunity specificity can hardly emerge full-blown from a single mutation. Repressor, Cro, and operator sites must all change to generate the new specificity. However, if mutants that have taken the first steps toward a new immunity (such as virulent mutants) are selectively enriched initially, then they may have the opportunity of undergoing extensive additional mutation, leading to the same kind of balanced state that λ itself exhibits. In this case, the selective pressures on such functional regions to change so that they reach a new adaptive peak may generate a rate of divergence far greater than the rate of random mutational drift.

Obviously, a combination of factors contributes to the rates of change of different segments of the genome. In addition to those already discussed, the possibility remains that recombinational reshuffling itself has been sufficiently important to the lambdoid phages, so that homologous segments have been conserved to facilitate that process. As the course of natural selection cannot be influenced by anticipated benefits, this would mean that homologous segments have been conserved in the successful recombinants of the past, which have become the prevalent varieties of the present.

COMPARATIVE BIOLOGY OF LAMBDOID PHAGES

All members of the family of lambdoid phages have a similar genetic map, in the sense that the order of genes relative to the functions specified by those genes is always the same. It is convenient to go through the genes in the prophage order, since some members of the family do not have defined vegetative DNA ends (e.g., P22), whereas all have their integration genes at the same relative position. Starting at the left end of the prophage, one finds the integration genes followed by the recombination genes, a *c*III-like gene and, finally, a positive regulator of early gene transcription (*N* in λ). All of these genes are transcribed leftward from p_L and are subject to positive regulation by the promoter-proximal *N*-like gene. All of the lambdoids have a similar immunity region, consisting of a gene specifying a repressor flanked by two operators in inverted orientation to one another, which control two divergent operons. Beyond the rightward early promoter (p_R) all of the phages have a *cro*-like gene followed by a *c*II-like gene, two DNA replication genes, and the origin of early DNA replication. Many of the phages have a relatively blank (few functions known) region followed by a positive regulator of late transcription (gene *Q* in λ). All the phages thus far investigated have a single late promoter that responds to the *Q*-like function. The order of functions in the late operon is lysis (two or three genes), followed by DNA maturation/head assembly, and finally tail assembly/adsorption and injection. Finally, the last segment in the prophage is a relatively blank region (*b*2 in λ) that sometimes contains prophage functions, such as antigen conversion (in P22).

This invariant order is retained even in cases where the actual functions encoded at each point are radically different. As indicated briefly above, heteroduplex analysis of different lambdoids pairwise yields a similar pattern in all pairs thus far examined. Regions of homology (in the same order and orientation) are interspersed with regions of nonhomology). In many cases, tests of function (mechanism, cross-complementation, or other tests of cross-specificity) have been possible; often these can be correlated to the heteroduplex observations.

It has frequently been observed that the lambdoid phages can recombine with one another quite readily, resulting in the exchange of regions of nonhomology via crossovers in the regions of homology. Many such crossovers result in viable, normally regulated temperate phages having specificities of function that are the composite of the two parental lambdoids. Even members of the family as divergent as λ and P22 can recombine, exchanging their recombination, immunity, replication, and late control genes (Botstein and Herskowitz 1974; Hilliker and Botstein 1976).

The regulatory organization of the family is an important feature that allows recombination to generate functional recombinants with different specificities. The most relevant feature is the proximity of regulatory genes to their sites of recognition (if not action). The principle is best exemplified by the functions of repressor (which acts only on flanking operators) and the late regulator, which apparently recognizes sites at or near the late promoter but has effects (via the antitermination mechanism) that can be many genes distant. The long operons can be controlled from sites near their beginning; thus, substitution of a foreign functional unit (e.g., tail genes) from another family member has no disruptive regulatory consequence. Similarly, exchange of regulatory specificity need only involve the exchange of a small region (i.e., immunity or Q-p_R').

Regions of function that include many gene products that must act together (e.g., head assembly) appear to have a subsegment organization that has the consequence of permitting exchanges without loss of function (Casjens and Hendrix 1974; Feiss et al. 1981; see also Georgopoulos et al., this volume). This organization of subfunctions is maintained even in P22, whose head structure and assembly pathway are so divergent that exchange of the subfunctions with λ seems no longer to be possible. The maintenance of this underlying organization could be argued to support a common origin for both assembly pathways. Similar subgroups can be discerned in the genes controlling lambdoid tail assembly (Youderian 1978); in this case, however, the P22 region consists of only a single gene, whereas λ tail assembly requires 12 (Katsura, this volume).

More detailed comparisons can be made by perusing the papers in this volume that describe the biology of each of the functions of the lambdoid phages. The overall picture of a series of regions of function that vary often in specificity and sometimes in mechanism among the members of the lambdoid family strongly supports the view that these phages show a modular structure that could reflect the architecture of a common progenitor and/or an ongoing evolution through the recruitment, diversification, and exchange of functional segments.

λ-RELATED DEFECTIVE PROPHAGES

Natural Examples

In nature, lambdoid phages are encountered as prophages, as well as free virions. λ-related DNA has also been observed in bacterial strains classified as nonlysogens on the basis of the failure to elicit the formation of viable phage after treatments that would induce a λ prophage. Using DNA hybridization, λ-related DNA has been found in laboratory strains of *E. coli*, as well as other enterobacterial species (Anilionis and Riley 1980; Riley and Anilionis 1980).

The strains that harbor such cryptic λ DNA include cured derivatives of *E. coli* K12 commonly used for the propagation of λ. The analysis of pseudorevertants of various λ mutants has made it clear that some of these segments can recombine with an infecting λ phage to generate recombinants in which certain genes have been replaced by functionally homologous DNA. The most thoroughly characterized of these λ-related segments is the Rac-defective prophage, which is located at 31 minutes on the K12 chromosome and has an estimated length of 26.5 kb (Kaiser and Murray 1979). The ability of λ Rac to recombine with λ came to light when lysates of λ*bio* phages deleted for *red* and *gam* were scored for pseudorevertants that plate on *recA*$^-$*recBC*$^+$ hosts (Zissler et al. 1971; Gottesman et al. 1974). In these phages (called λ reverse), λ DNA extending from coordinate 45.1 (in the *b*2 region) through 71.1 (around *c*III) has been replaced by nonhomologous DNA. They have regained recombination proficiency and encode exonuclease VIII (ExoVIII), which differs radically from λ exonuclease in its physical properties (Gillen et al. 1977).

The ExoVIII gene of this defective prophage is ordinarily not expressed from the bacterial chromosome. A possible explanation is that the defective prophage makes a repressor. This explanation is supported by the existence of recessive mutations (*sbcA*) that cause constitutive expression of ExoVIII. The *sbcA* mutations were selected from *recA*$^+$*recBC*$^-$ strains, in which they restore mitomycin resistance and recombination proficiency, as though ExoVIII could replace the *recBC*-coded enzyme ExoV. Certain strains of K12 (such as the standard multiply auxotrophic recipient AB1157) have been found to lack the Rac prophage altogether. This condition (designated *rac*$^-$) first manifested itself in a transient burst of recombination proficiency in crosses between *rac*$^+$ donors and *recBC*$^-$ derivatives of AB1157, which was attributed to zygotic induction of a defective prophage (Low 1973).

The Rac prophage also includes a replication origin (*oriJ*), which can be rescued as an autonomous plasmid similar to λdv (Diaz et al. 1979). If the analogy with λdv is apt, the plasmid should include not only *oriJ* but also replication genes, counterparts of *O* and *P*. The Rac prophage thus encodes several phage functions—recombination gene(s) analogous to *red*, replication determinants, and a specific repressor. There is also some evidence indicating the presence of a specific integration-excision system and of a restriction alleviation function.

Restriction mapping of the Rac prophage and heteroduplex studies on λ, λ reverse, and a λ derivative bearing a cloned segment of K12 DNA that includes the *recE* gene have located some of these determinants of the Rac prophage. The segment carried by λ reverse derives from the two ends of the Rac prophage, as expected for a segment spanning the attachment site. The positions of the *oriJ* segment and of a deletion producing an SbcA phenotype are both compatible with the hypothesis that the Rac prophage was derived from a typical lambdoid phage that became defective by deletion of most of the DNA in the *O-J* region.

The *E. coli* K12 chromosome also contains a recombinable segment that can supply functions equivalent to those of λ genes *Q, S, R,* presumably *Rz*, and *Nu1*, as well as the p_R' and *cos* sites (Strathern and Herskowitz 1975; Fisher and Feiss 1980). Rescue of this segment into λ requires the presence of a functional recombination system (*rec* or *red*). Therefore, rescue probably occurs by homologous crossing over. The *Q, S,* and *R* analogs in this segment have been named *q*, *s*, and *r*. Genes *qsr* are rescued as a unit (observed as a heterologous substitution extending from 84 to 95 in heteroduplex studies). The rescued *q* gene has a different specificity from that of λ*Q*. Thus, a phage (λ*qin*) bearing the *qsr* substitution will not turn on late transcription from $p_R{}'$ of λ. Likewise, λQ^+ will not activate late transcription from λ*qin*. The *qsr* substitution was first isolated as a small plaque mutation (p*4*) of wild-type λ (Jacob and Wollman 1954). Recurrences were later selected from phages bearing deletions or double mutations in *Q*, *S*, and *R* (Sato and Campbell 1970; Strathern and Herskowitz 1975).

Southern hybridizations show that *qsr* DNA is not present on F′123, which carries the Rac prophage. Therefore, it must come from a prophage fragment located elsewhere in the K12 genome. Strain AB1157, which lacks the Rac prophage, is also missing part of the *qsr* prophage. Rescue of *qsr* is not observed in AB1157. Rescue of *cos* and *Nu1* by a mutant (λ*cos*1) bearing a 400-bp deletion of these determinants occurs both in AB1157 and in other K12 strains. In most K12 strains, *cos* rescue is accompanied by *qsr* substitution about 1% of the time. This joint rescue indicates that the *qsr* prophage extends through *cos* and *Nu1*. The rescue of *cos* and *Nu1* from AB1157 indicates that either AB1157 retains part of the *qsr* prophage or K12 carries a third defective prophage that includes *cos* and *Nu1* but not *qsr* (Fisher and Feiss 1980).

Generation of Defective λ in the Laboratory

In the laboratory, defective prophages can arise from complete prophages by standard mutational mechanisms, such as point mutation or simple deletion, as well as by events more specifically related to the virus life cycle, such as aberrant excision and reinsertion (as for λd*gal*) or complex interactions with insertion-sequence elements (as for λ cryptic; Zissler et al. 1977). Presumably, at least some of the λ-related DNA encountered in natural strains has arisen from lambdoid prophages by similar degenerative changes.

Defective Prophages as Sources of Modules

The usual mechanism of gene rescue from naturally defective phages, exemplified by the *qsr* and *cos* pseudorevertants, is homologous recombination, dependent on the *red* or *rec* genes. This implies that rescuable heterologous segments are bracketed by DNA that is homologous to the DNA of the rescuing phage and positioned appropriately on the phage DNA. The presence of such homologies might indicate that the defective prophages are descendants of primordial λ.

Incorporation of the *qsr* segment into λ can proceed by a pathway that apparently involves both homologous and illegitimate recombination. The *qsr* substitution phages λ p*4* and λ*qin* are thought to arise by a double crossover between λ and the *qsr* prophage, with the crossover points lying in homologous DNA that surrounds the substitution. However, this pathway is not open to a recipient phage bearing the *nin*5 deletion, which removes the λ DNA to the left of the substituted region. Nevertheless, rescue of *cos* into a *nin*5*cos*1*Ram* phage is accomplished about 1% of the time by rescue of the *r* gene (Fisher and Feiss 1980). The resulting phages prove to carry additions, adjacent to *QSR*, of all or part of *qsr*, plus the adjoining *cos* region. Generation of the addition phages is *recA*-dependent, suggesting that crossing over within the homologous DNA around *cos* is a necessary step in the process. However, the addition phages also include a novel joint to the left of the added DNA, which must be created by illegitimate recombination. A 700-bp sequence within the *qsr* substitution is present in two copies in some of the addition phages. Whether this sequence plays any part in the generation of the novel joint is unknown.

In the context of our discussion here, the addition phages are noteworthy as functional phages that arise by illegitimate recombination between an infecting phage and a defective prophage in the host chromosome. Interspersion of homologous DNA may facilitate the interchange of nonhomologous segments but is clearly not an obligatory requirement.

Significance of Defective Prophages and Gene Rescue from Them

Some λ-related DNA is present in most bacterial strains likely to be infected by λ in nature. Generalizing from the K12 results, we presume that rescue of DNA segments from defective prophages into superinfecting lambdoid phages constitutes part of the natural population biology of lambdoid phages.

Two conceptually distinct interpretations of defective prophages are possible. In one view they represent "genetic debris" (Strathern and Herskowitz 1975) that has accumulated in cells from previous virus infections and is in the process of decay and loss. An alternative interpretation is that regardless of how they originally entered the cell, the rescuable segments are in fact serving some purpose to the host and therefore can be considered as virus modules that have been appropriated by the host.

It is hard to design a feasible experiment that could distinguish these alternatives with absolute rigor. Where the rescuable genes are never known to be expressed from the defective prophage and where their deletion has no observed effect on host phenotype, we may tentatively classify them as genetic debris. However, the *sbcA* mutations provide an instructive example of how conditions that might plausibly arise in nature (exposure of a *recB* mutant to agents such as mitomycin that damage DNA) can select for expression of prophage genes. It is unlikely that everything that starts out as genetic debris is discarded rather than co-opted.

Having gone that far, one must accept that such a virus module that can be appropriated by the host is difficult to distinguish, even conceptually, from a host module that was appropriated in the past by a virus. Thus, once we include defective prophages as part of the phage population (which seems a logical necessity), the line between reshuffling genes within the phage population and recapitulation of the kinds of recruitment events that generated primordial λ becomes indistinct. At present, dependence of rescue on *red* and *rec* provides the best presumptive evidence for reshuffling within a pool of genes that function primarily as phage rather than host genes.

POPULATION BIOLOGY OF LAMBDOID PHAGES

Four major themes run through the arguments and evidence summarized above: (1) The lambdoid phages have in common a genomic organization that reflects a modular construction; (2) many diverse specificities (or mechanisms) are represented among the extended lambdoid family for many of the phage functions; (3) the segments containing functional units can be reshuffled by recombination to produce viable hybrid phages; and (4) functional and sequence similarities exist, in at least some cases, to host genes, some of which may currently function in host metabolism, whereas others may represent defective prophages or even genetic debris.

Given these features, we must ask what kind of evolutionary mechanisms are now operating upon the lambdoid family. Several considerations seem evident. First, the evolution has not been, and cannot now be, a simple matter of mutation and selection through linear descent for the simple reason that any improvement that occurred in λ would soon find its way into other phages in the family by recombination. Should the improvement in function be substantial, then we could expect that the movement of the function through the family might be very rapid. Thus, the unit of selection is not the phage but some lesser functional unit that retains its ability to be recombined into many members of the lambdoid family. Elsewhere this unit has been called a module (Botstein and Herskowitz 1974; Susskind and Botstein 1978) but, as we discussed above, the modules defined in this way need not resemble the functional units that we imagine came together to form the hypothetical primordial λ.

In such a theory, the product of evolution is not a given phage but, instead, the family of phages with interchangeable genetic elements (modules), each of which carries out a particular biological function. Each phage encountered in nature is a favorable combination of modules (one for each function) that has been selected to work well together in filling a particular niche. Exchange of a given module for another with the same biological function occurs by recombination among members of the family via crossovers in regions of homology that are interspersed among regions of nonhomology. No assumption need be made about whether the homology is within regions of function or not, nor whether every pair of lambdoid phages have any homologous DNA between any two given modules; it suffices that there are regions of homology between an exchanging pair.

The only constraints upon members of the family are that (1) they retain the genome organization that allows newly acquired modules to be expressed and to function properly in their new genetic environment, and that (2) they retain regions of homology with some other members so that the modules they contain do not become isolated from the rest of the family. Phages in the same interbreeding population can differ widely in any characteristic (including DNA structure, host range, and morphology), since these are aspects of individual modules. Conversely, phages of the same family that contain essentially identical modules can be found as parasites of hosts, which themselves are only very distantly related.

Selection at the level of the module is for (1) good execution of function, (2) retention of flanking homology ensuring proper placement in the genome of the recipient phage during an exchange of modules, and (3) retention of functional compatibility with the maximum number of combinations of other functional units.

The sources of modules in such a theory need not be limited to other lambdoid phages. Defective prophages, or even host genes that retain homology with phage modules of common ancestry, might serve. One prediction of this view is that one might find in a phage an origin of DNA replication and analogs of λ genes *O* and *P* that are identical to the origin and replication genes of a bacterium.

This view of lambdoid phage evolution has the advantage of providing a rationale for the existence of the P22 antirepressor and its attendant regulation (see Susskind and Youderian, this volume). The role of antirepressor is to facilitate reassortment of the functional modular segments to fit changing environmental conditions. Antirepressor has a very broad specificity for lambdoid repressors (Susskind and Botstein 1975). By inducing any prophages present in a new bacterial host, antirepressor allows prophages to replicate and thereby stimulates recombination with the superinfecting phage. Furthermore, derepression of prophage functions might sometimes be essential when P22 cannot use its own function in the new host.

Finally, it should be noted that in such a scheme it is not the complete phages that are the products of evolution. The fittest modules and the diverse population of lambdoids is the product of the process of selection. Primordial λ gave the family its genome architecture; its descendants have only that architecture in common.

REFERENCES

Anilionis, A. and M. Riley. 1980. Conservation and variation of nucleotide sequences within related bacterial genomes: *Escherichia coli* strains. *J. Bacteriol.* **143:** 355.

Botstein, D. and I. Herskowitz. 1974. Properties of hybrids between *Salmonella* phage P22 and coliphage lambda. *Nature* **251:** 584.

Casjens, S.R. and R. Hendrix. 1974. Comments on the arrangement of morphogenetic genes of bacteriophage lambda. *J. Mol. Biol.* **90:** 20.

Cohen, S.N. 1976. Transposable genetic elements and plasmid evolution. *Nature* **263:** 731.

Diaz, R., P. Barnsley, and R.H. Pritchard. 1979. Location and characterization of a new replication origin in the *E. coli* K12 chromosome. *Mol. Gen. Genet.* **175:** 151.

Feiss, M., R.A. Fisher, D. Siegete, and W. Widner. 1981. Bacteriophage lambda and 21 packaging specificities. In *Bacteriophage assembly* (ed. M. Dubow), p. 213. Alan R. Liss, New York.

Fisher, R. and M. Feiss. 1980. Reversion of a cohesive end site mutant of bacteriophage lambda by recombination with a defective prophage. *Virology* **107:** 160.

Furth, M.E. and J.L. Yates. 1978. Specificity determinants for lambda DNA replication. II. Structure of O proteins of lambda ϕ82 hybrid phages and of a ϕ mutant defective in the origin of replication. *J. Mol. Biol.* **126:** 227.

Furth, M.E., J.L. Yates, and W.F. Dove. 1979. Positive and negative control of bacteriophage lambda DNA replication. *Cold Spring Harbor Symp. Quant. Biol.* **43:** 147.

Georgopoulos, C.P. and I. Herskowitz. 1971. *Escherichia coli* mutants blocked in lambda DNA synthesis. In *The bacteriophage lambda* (ed. A.D. Hershey), p. 553. Cold Spring Harbor Laboratory, Cold Spring Harbor, New York.

Gillen, J.R., A.E. Karu, H. Nagaishi, and A.J. Clark. 1977. Characterization of the deoxyribonuclease determined by lambda reverse as exonuclease VIII of *Escherichia coli. J. Mol. Biol.* **113:** 27.

Gottesman, M.M., M.E. Gottesman, S. Gottesman, and M. Gellert. 1974. Characterization of bacteriophage lambda reverse as an *Escherichia coli* phage carrying a unique set of host-derived recombination functions. *J. Mol. Biol.* **88:** 471.

Grosschedl, R. and G. Hobom. 1979. DNA sequences and structural homologies of the replication origins of lambdoid bacteriophages. *Nature* **277:** 621.

Highton, P.J. and M. Whitfield. 1975. Similarities between the DNA molecules of bacteriophages 424, lambda and 21, determined by denaturation and electron microscopy. *Virology* **63:** 438.

Hilliker, S. and D. Botstein. 1976. Specificity of genetic elements controlling regulation of early functions in temperate bacteriophage. *J. Mol. Biol.* **106:** 537.

Jacob, F. and E.L. Wollman. 1954. Étude génétique d'un bactériophage tempéré d'*Escherichia coli*. I. Le system génétique du bactériophage lambda. *Ann. Inst. Pasteur* **87:** 653.

Kaiser, K. and N.E. Murray. 1979. Physical characterization of the "Rac prophage" in *E. coli* K12. *Mol. Gen. Genet.* **175:** 159.

Low, K.B. 1973. Restoration by the *rac* locus of recombinant forming ability in $recB^-$ and $recC^-$ merozygotes of *Escherichia coli* K12. *Mol. Gen. Genet.* **122:** 119.

Luria, S.E., J.E. Darnell, Jr., D. Baltimore, and A. Campbell. 1978. *General virology*, 3rd edition. John Wiley, New York.

Meijer, M., E. Beck, F.G. Hansen, H.E.N. Bergmans, W. Messer, K. von Meyenburg, and H. Schaller. 1979. Nucleotide sequence of the origin of replication of the *Escherichia coli* K12 chromosome. *Proc. Natl. Acad. Sci.* **76:** 580.

Moore, D.D., K. Denniston-Thompson, K.E. Kruger, M.E. Furth, B.G. Williams, D.L. Daniels, and F.R. Blattner. 1979. Dissection and comparative anatomy of the origins of replication of lambdoid phages. *Cold Spring Harbor Symp. Quant. Biol.* **43:** 155.

Perler, F., A. Efstratiadis, R. Lomedico, W. Gilbert, R. Kolodner, and J. Dodgson. 1980. The evolution of genes: The chick preproinsulin gene. *Cell* **20:** 555.

Raper, J.R. 1966. *Genetics of sexuality in fungi*. Ronald Press, New York.

Riley, M. and A. Anilionis. 1980. Conservation and variation of nucleotide sequences within related bacterial genomes: Enterobacteria. *J. Bacteriol.* **143:** 366.

Sato, K. and A. Campbell. 1970. Specialized transduction by lambda phage from a deletion lysogen. *Virology* **41:** 474.

Simon, M.N., R.W. Davis, and N. Davidson. 1971. Heteroduplexes of DNA molecules of lambdoid phages: Physical mapping of their base sequence relationships by electron microscopy. In *The bacteriophage lambda* (ed. A.D. Hershey), p. 313. Cold Spring Harbor Laboratory, Cold Spring Harbor, New York.

Simon, M., J. Zieg, M. Silverman, G. Mandel, and R. Doolittle. 1980. Phage variation: Evolution of a controlling element. *Science* **209:** 1370.

Strathern, A. and I. Herskowitz. 1975. Defective prophage in *Escherichia coli* K12 strains. *Virology* **67:** 136.

Susskind, M.M. and D. Botstein. 1975. Mechanism of action of *Salmonella* phage P22 antirepressor. *J. Mol. Biol.* **98:** 413.

——. 1978. Molecular genetics of bacteriophage P22. *Microbiol. Rev.* **42:** 385.

Szpirer, J., R. Thomas, and C.M. Radding. 1969. Hybrids of bacteriophage lambda and ϕ80: A study of nonvegetative functions. *Virology* **37:** 585.

Szybalski, W. and E.H. Szybalski. 1974. Visualization of the evolution of viral genomes. In *Viruses, evolution and cancer*, (ed. E. Kurstak and K. Maramorosch), p. 563. Academic Press, New York.

Wickner, S.H. 1979. DNA replication proteins of *Escherichia coli* and phage λ. *Cold Spring Harbor Symp. Quant. Biol.* **43:** 303.

Youderian, P. 1978. "Genetic analysis of the length of the tails of lambdoid bacteriophages." Ph.D. thesis, Massachusetts Institute of Technology, Cambridge.

Zissler, J., E. Signer, and F. Schaeffer. 1971. The role of recombination in growth of bacteriophage lambda. I. The gamma gene. In *The bacteriophage lambda* (ed. A.D. Hershey), p. 455. Cold Spring Harbor, Cold Spring Harbor, New York.

Zissler, J., E. Mosharaffa, W. Pilacinski, and W. Szybalski. 1977. Position effects of insertion sequences IS2 near the gene for prophage λ insertion and excision. In *DNA insertion sequences, plasmids, and episones* (ed. A.I. Bukhari, et al.), p. 381. Cold Spring Harbor Laboratory, Cold Spring Harbor, New York.

A Beginner's Guide to Lambda Biology

Werner Arber
Department of Microbiology
Biozentrum, University of Basel
CH-4056 Basel, Switzerland

INTRODUCTION

Figure 1 is a portrait of bacteriophage λ. The phage particle is composed of roughly half protein and half double-stranded DNA. The size of the genome of wild-type λ phage is about 48 kb, corresponding to a mass of about 32×10^6 daltons.

Extensive genetic studies have been done with this bacteriophage (see Campbell 1971). Mutations with particular phenotypes, such as the formation of clear plaques (*c*) rather than turbid plaques (c^+), or other aspects of plaque appearance, as well as host-range mutations (*h*), played important roles in the early investigations. More recent work has made extensive use of conditionally lethal mutations, including, in particular, *sus* (*su*pressor *s*ensitive) mutations, also known as nonsense mutations. The mutational events leading to *sus* generate any of the three nonsense codons: UAG for the amber mutations, UAA for the ochre mutations, and UGA for the opal mutations. Most of the *sus* mutations in use are amber, and the symbol *am* is sometimes used for particular *sus* mutations. For work with *sus* mutations, two bacterial host strains are needed: one that does not allow the reproduction of the phage (nonpermissive strain), and a second in which the phage is able to reproduce (permissive strain or suppressor strain).

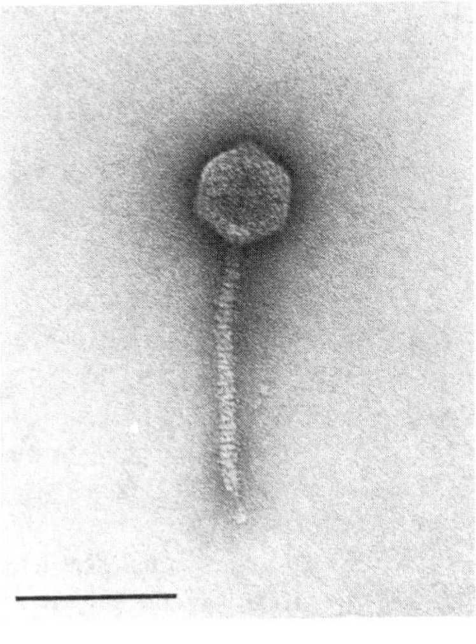

Figure 1 Electron micrograph of bacteriophage λ particle, negatively stained with uranyl acetate. Bar represents 0.1 μm. (Photo courtesy of M. Wurtz.)

Such suppressor strains usually contain a mutation in a tRNA gene, so that a new species of tRNA is produced that is able to give sense to an otherwise nonsense codon (i.e., to insert an amino acid at what would otherwise be a protein chain termination codon).

A second class of conditionally lethal mutations, the temperature-sensitive mutations, are often more difficult to work with than are *sus* mutations. A major problem is the difficulty of accurately controlling the temperature of cultures in petri dishes. Most *sus* and temperature-sensitive mutations are single-base substitutions. In addition to these, deletions and substitutions play important roles in genetic studies of bacteriophage λ.

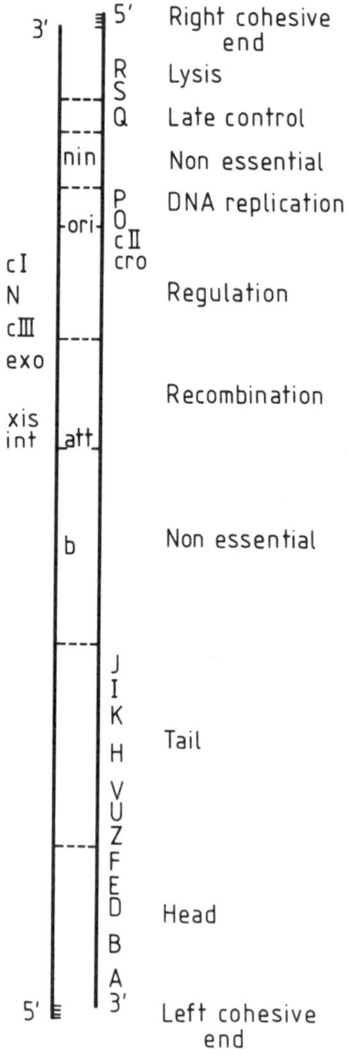

Figure 2 The λ genome map. Genetic symbols are explained in Appendix I; see also Szybalski and Szybalski (1979). Not all known genes are shown. The genes are labeled next to the information strand. Sites *att* and *ori* and the regions *b* and *nin* are indicated between the two strands. Map distances represent approximate physical distances. (*Right*) Functions of genome regions.

Genetic and physiological investigations have identified most of λ's essential genes as well as their functions (Szybalski and Szybalski 1979; also see Appendix I). As with other bacteriophages, functionally related genes are linked, as shown in Figure 2. Not only do the genes responsible for the head formation and those responsible for the tail formation form groups of linked genes, but genes expressed early in the phage infection are linked as are those expressed late.

VEGETATIVE REPRODUCTION AND PHAGE MORPHOGENESIS

The first step in the infection of a host cell by bacteriophage λ is its adsorption on the cell surface (Fig. 3). Adsorption involves a specific interaction between the tip of the phage tail (*J* protein) and a component of the outer cell membrane, the product of the bacterial gene *lamB* (Murialdo and Siminovitch 1972; Randall-

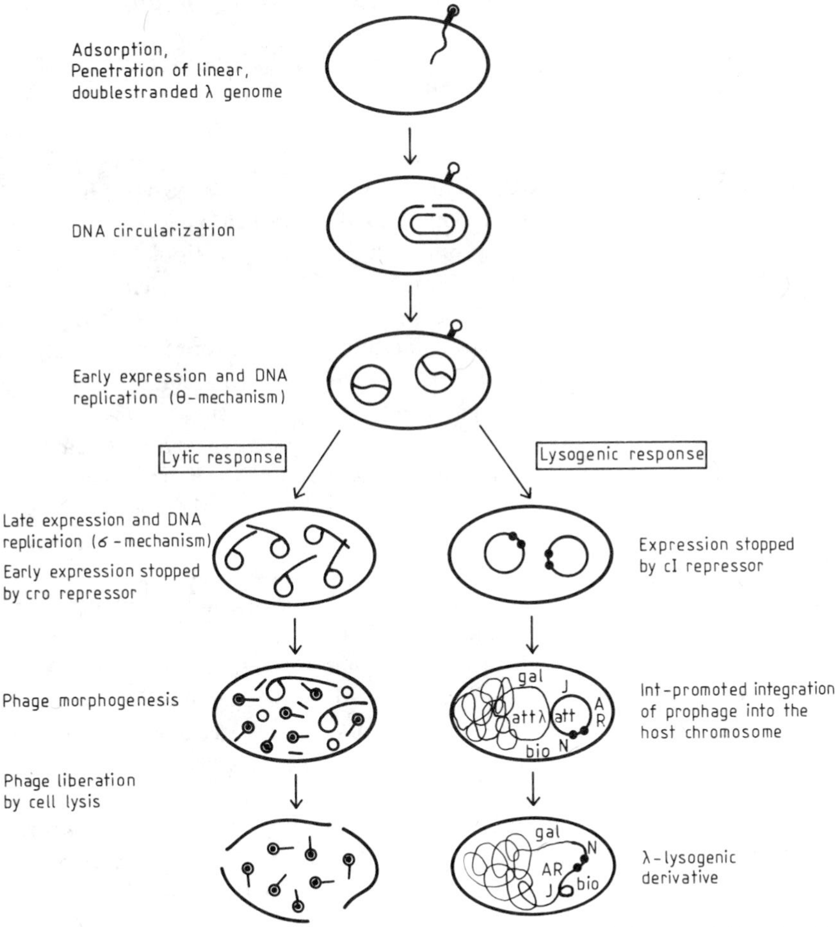

Figure 3 Schematic representation of responses obtained upon infection of *E. coli* with phage λ. For further explication, see text.

Hazelbauer and Schwartz 1973). The *lamB* product is a maltose transport protein, and its expression is induced by maltose. Therefore, λ adsorbs best to cells grown in the presence of maltose (but in the absence of glucose, to avoid catabolite repression).

Penetration of the phage DNA into the host cell is triggered by a specific interaction between components of the phage tail and a component of the inner cell membrane, which is most likely the product of gene *ptsM*, a component of a phosphotransferase system responsible for the transport of several sugars (Elliott and Arber 1978).

The genome of bacteriophage λ in the head of the phage particle is a linear double-stranded DNA molecule. Shortly after its penetration into the host cell, the DNA molecule circularizes. This is possible because the λ DNA molecule has complementary single-stranded ends 12 nucleotides long (see Yarmolinsky 1971; Feiss and Becker, this volume). The site of the joined ends, together with adjacent sequences required for DNA packaging, is generally called *cos* (*co*hesive end *s*ite).

DNA replication is directed by phage genes *O* and *P*, but several host functions are also required (Skalka 1977; Furth and Wickner, this volume). The replication occurs in two phases. First, the circular genome is replicated in the θ mode. Starting from the region of genes *c*II and *O*, replication forks move in both directions along the DNA until they meet each other on the other side of the circular DNA molecule. The products of this replication are two circular daughters. In a second phase, σ or rolling-circle replication produces concatemeric forms—long DNA molecules composed of several tandem λ genomes.

Simultaneously with DNA replication, λ genes are expressed according to a controlled scheme (see below, Regulation of Gene Expression). About halfway through the latent period, enough structural proteins have accumulated to allow the formation of progeny phage particles. The morphogenetic pathway (Hohn and Katsura 1977; Georgopoulos et al.; Katsura; both this volume) is composed of two independent branches: assembly of the head and assembly of the tail.

The head precursor, *petit* λ, is assembled from proteins coded for by the phage genes *E* (main capsid protein), *Nu3* (core protein), *C*, and *B*, and with the help of *groE* functions of the host. For the next step, DNA packaging, the products of phage genes *Nu1, A, FI,* and *D*, as well as host factors, are required. During packaging, DNA is cut out of concatemers to unit-size molecules with cohesive ends. The head filled with DNA then reacts with the products of λ genes *W* and *FII*, thereby becoming competent to attach a tail. Packaging of DNA from either concatemeric or linear monomeric precursors, with head completion and tail attachment, can be performed in vitro in suitable extracts.

Since DNA packaging depends on the specific recognition and cutting of two *cos* sites, virions of deletion strains of λ contain less DNA (have shorter genomes) than do reference-type virions. Similarly, insertion derivatives contain longer DNA molecules. Genome sizes varying from about 105% to about 78% of that of reference type λ are packaged with equally high efficiency. For genomes larger or smaller than these limits, the efficiency with which mature phage particles are obtained rapidly decreases. This could be due to either inefficient packaging or instability of the phage particles (Weil et al. 1975; Feiss et al. 1977).

Phage progeny particles are liberated by lysis of the host cell, which occurs spontaneously if phage genes *S*, *R*, and *Rz* are properly expressed.

LYSOGENIZATION

λ is a temperate bacteriophage. After infection, instead of reproducing vegetatively, its genome may become mostly silent and may be incorporated into the host chromosome (Fig. 3). This event is called lysogenization, and the integrated λ is said to be a prophage. During infection leading to lysogenization, only early phage functions are expressed. At some time after the infection, the phage genome comes under repression due to the product of gene *c*I. In the presence of the product of the early gene *int*, site-specific recombination incorporates the λ genome into the host chromosome. This recombination usually occurs between *att*P on the phage genome and *att*B on the *Escherichia coli* chromosome. These two sites share a 15-bp homology, but adjacent, nonhomologous nucleotide sequences are also essential for the reaction. This site-specific integration is not promoted by the general recombination system of *E. coli* or that of λ; only the bacteriophage *int*-gene product (together with some bacterial factors) is involved. Using purified *int* enzyme together with host extracts, the process can be studied in vitro (Nash 1977; Weisberg and Landy, this volume).

Lysogenic bacteria allow enough *c*I repressor to be produced (see below, Regulation of Gene Expression) to insure repression of all other phage functions (except *rex* [*r*II *ex*clusion], noted historically for its role in blocking the growth of T4*r*II mutants). Lysogenic cells are therefore also immune against superinfecting λ phage.

Phage λ can form double lysogens, most conveniently detected when the two components are distinguished by genetic markers. These double lysogens carry one copy of each prophage type, usually in tandem arrangement. They are relatively unstable and segregate single lysogens, which often carry a recombinant prophage. They also segregate genetically pure double lysogens as a result of recombinational interactions between the tandem prophages (homogenotization).

Occasionally, integration of the λ genome can occur at secondary attachment sites found at various places along the *E. coli* chromosome. These rare events are studied with *E. coli* mutants in which the normal *att*B site is deleted (Shimada et al. 1972).

Integration of the λ genome into the bacterial chromosome is reversible. Excision of the prophage requires two phage gene products besides bacterial factors, namely the product of gene *int* (responsible for integration) and the product of phage gene *xis*. The requirement for both *int*- and *xis*-gene products for excision is one piece of evidence that the attachment sites of the phage and of the host cell are not limited to their homologous segment. Otherwise, one might expect the integration and the excision processes to be equivalent.

Since phage genes *int* and *xis* are under repression in the lysogenic state, excision occurs only if repression is lifted. The lifting of repression is called induction; the vegetative cycle of bacteriophage growth is induced. Induction generally follows interference with bacterial DNA replication by mechanisms discussed by Roberts and Devoret (this volume). Convenient means of inducing a lysogenic culture are UV irradiation or treatment with mitomycin C. Alternatively, if gene *c*I, responsible for the production of the λ repressor, has a temperature-sensitive mutation, induction occurs upon incubation of lysogenic cells at a temperature that inactivates the repressor.

Rarely, and only with particular λ mutants or host strains, can bacteria be formed that carry λ as a plasmid rather than as a chromosomally integrated prophage. Relevant to the use of λ in recombinant DNA technology is the maintenance as plasmids of derivatives of λ with extensive deletions. The most extreme of these deletion mutants are known as λdv, and they consist of only those short DNA segments that specify the phage functions for DNA replication and its control (Berg 1974; Matsubara 1976).

RECOMBINATION

According to early determinations, the DNA molecules contained in the wild-type λ phage particles of a one-cycle growth have the probability of about 0.5 of participating in genetic exchange during that growth cycle (Wollman and Jacob 1954; Kaiser 1955; but see Stahl 1979). Extensive studies have shown that the genetic distances between particular genes of bacteriophage λ, established by genetic recombination between phage mutants, do not exactly correspond to physical distances as established by electron microscopy and restriction cleavage methods. This indicates that the chance of recombination is subject to local variations.

λ recombines in bacterial *recA* mutants, in which recombination between host chromosomal markers is practically zero. The residual λ recombination is explained by two facts. First, a pair of genes (*redα* and *redβ* = *exo* and *bet*, respectively) is responsible for a λ-specific enzyme system promoting general recombination. The contribution of this Red system in conditions of full expression is as important as that of the Rec system of general bacterial recombination. Second, the *int* enzyme, responsible for integration of the phage into the host chromosome, can also promote recombination between two λ genomes; such recombination occurs uniquely at the *att* site and is genetically manifest only for pairs of markers that straddle *att*. This recombination is conveniently studied with *red*$^-$ phage in *recA*$^-$ bacterial hosts. Alternatively, *int*$^-$ phage mutations permit study of recombination by the λ Red mechanism in *recA*$^-$ hosts (Gottesman and Weisberg 1971; Signer 1971). The Red system is involved in DNA replication, and under certain conditions recombination is required for chromosome packaging. These roles of recombination in the λ lytic cycle are detailed by Smith (this volume).

TRANSDUCTION

With a low probability, prophage excision occurs by an illegitimate exchange. These illegitimate acts of recombination result in hybrid structures between the phage genome and adjacent segments of the bacterial chromosome (Campbell 1962; Franklin 1971). In the resulting transducing phage, part of the phage genome is substituted by a segment of the bacterial chromosome. With lysogens carrying the prophage at the ordinary att^{λ} site, this process can produce either λ*gal* or λ*bio* phages. Bacteria carrying the λ prophage integrated at other sites produce other types of transducing phages by illegitimate excision. In this way genes from many parts of the *E. coli* chromosome can be integrated in vivo into the λ genome. The limitation is in obtaining a lysogen in which the prophage is near the gene to be picked up.

Because of their mode of formation, structures of the genomes of transducing phages obey the following rules: Substitution of phage genes by bacterial DNA starts at the *att* site, extends rightward or leftward, and ends at random. However, one does not find deletions of phage genes extending to *cos*, presumably because such deleted genomes do not get packaged into phage particles. Depending on the direction and extent of the deletion, transducing phages may be defective either for lytic growth or only for lysogenization, or they may be missing no functions required for either branch of their life cycle. However, because of the hybrid nature of their *att* site, the rate of lysogenization of transducing phages always differs from that of the corresponding wild type and is usually reduced. Transducing phages defective for lytic growth can still be propagated in the presence of helper phage genomes, which provide the missing functions. Helper phage also increases the lysogenization rate of transducing phage by facilitating their integration. Such lysogens are usually doubles, carrying the helper and the defective transducer in tandem.

NONESSENTIAL REGIONS OF THE PHAGE GENOME

Extensive study of transducing phages, as well as of other deletion derivatives of phage λ, showed that part of the wild-type genome is not essential. In particular, the *b* region, occupying about 18% of the length of the λ genome, is important neither for vegetative replication nor for lysogenization. (A more or less intact *att* region [at the right border of the *b* region] is of course essential for Int-mediated integration.) Another, smaller, nonessential region is known as *nin* (for *N* *in*dependence). For a more detailed discussion, see Court and Oppenheim (this volume).

PHAGES RELATED TO λ

Bacteriophages 434, 21, 82, and ϕ80 are related to phage λ. The DNAs of each of these so-called lambdoid phages show a partial homology with bacteriophage λ DNA, and they also show a partial homology with one another (Fiandt et al. 1971; Simon et al. 1971). Viable recombinants between any pair of these related phages can be isolated. Such hybrids can, e.g., have the tail specificity of phage 434 but the attachment site and the immunity characteristics of phage λ, or vice versa. Since the corresponding gene areas, e.g., the immunity regions, of the different lambdoid phages are not of identical length, hybrids can be obtained that have smaller genomes than that of bacteriophage λ. These small genomes made even smaller by deletions of nonessential regions are used in the construction of phage λ gene-cloning vectors, which permit the incorporation of relatively large fragments of foreign DNA. When working with hybrid phages, one should pay attention to their particular characteristics. It may be important, e.g., to know which adsorption characteristics a phage has, as well as which type of immunity it confers.

REGULATION OF GENE EXPRESSION

Figure 4 shows, in a simplified way, the sequence of events leading to gene expression after infection of a host cell with bacteriophage λ (see also Thomas 1971; Pirrotta 1976; Weisberg et al. 1977; Adhya and Gottesman 1978; Szybalski and Szybalski 1979; Ptashne et al. 1980; Friedman and Gottesman; Wulff and Rosenberg; Echols and Guarneros; all this volume).

In the first phase, soon after phage infection, *E. coli* RNA polymerase associates with the λ DNA and starts transcription at two promoters, p_L and p_R. Transcription goes left from p_L until it reaches t_{L1}, where it terminates with an efficiency of 80%, or t_{L2}. The resulting mRNA codes for *N*-gene product. At the same time, the transcription starting at p_R goes to the right. About half of these mRNA molecules stop at a terminator (t_{R1}) left of gene *c*II; the other half of the transcripts are continued until they reach a second terminator (t_{R2}) within the *nin* region. The common product of these two messengers is that of gene *cro*. The longer molecules of the mixed mRNA population also specify the products of genes *O* and *P*.

In the second phase, as soon as a certain level of *N*-gene product is available, this protein starts to act as an antiterminator. As a consequence of this activity, the transcripts starting either from p_L or p_R are now much longer than they were in the first phase. From the transcript starting at p_L, more *N* product is made, which increases the antitermination activity, and several additional gene products are also produced, particularly *c*III protein and the λ products involved in recombination and integration.

From the transcript starting at p_R, more *cro* product is made along with considerable amounts of the products of genes *c*II, *O*, *P*, and *Q*. The products of genes *O* and *P* are involved in the replication of the λ genome (see above, Vegetative Reproduction and Phage Morphogenesis). As soon as a sufficient

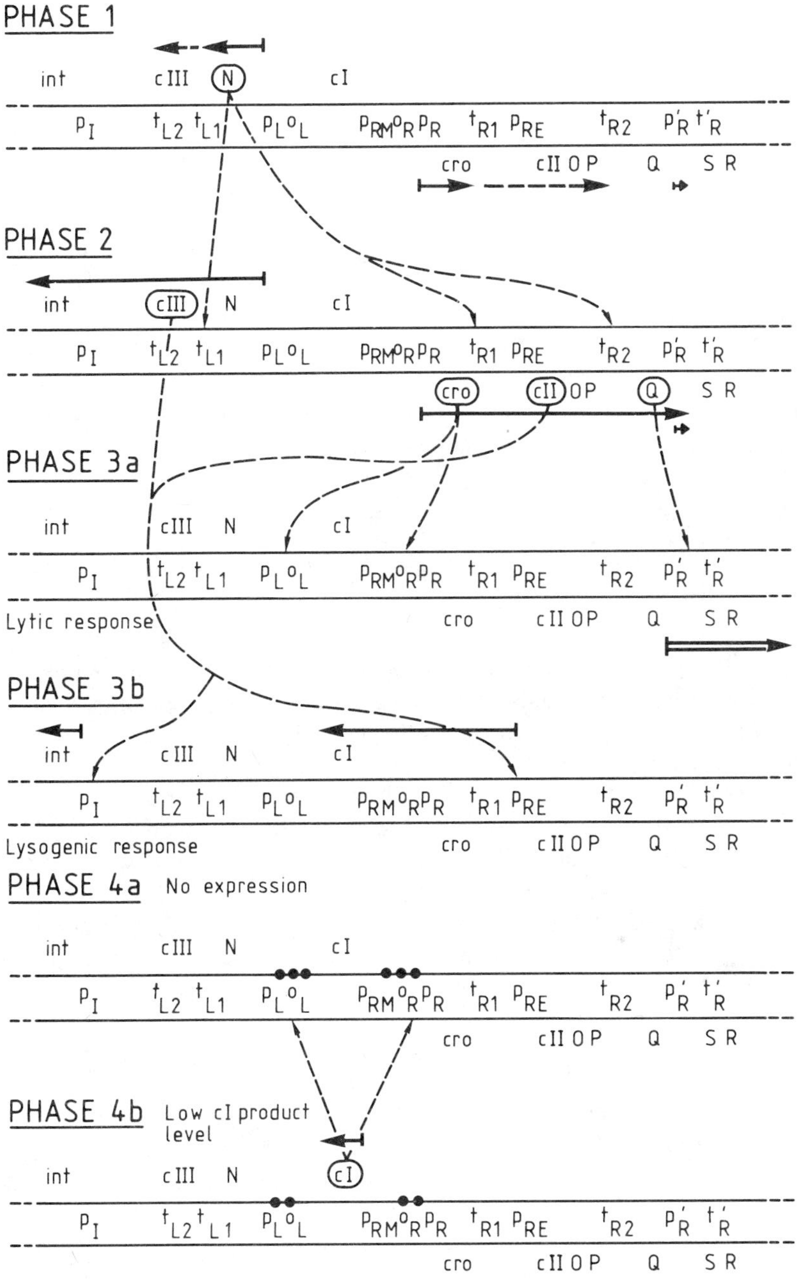

Figure 4 Schematic representation of gene expression in phage λ. Only the most relevant part of the genome is shown and, on this, only the relevant genes and sites. As in Fig. 2, genes are labeled next to the information strand, and relevant control sites are given between the two strands: (p) promoter; (o) operator; (t) terminator. (→) Transcripts; (●) *c*I repressor molecules (phase 4). Distances between sites and genes are not drawn to scale. For further explanation, see text.

amount of *cro* protein accumulates, it acts as a repressor both on the operator o_L and o_R. In so doing, it shuts off the early transcription starting at p_L and p_R (Gussin et al., this volume). This occurs about 15 minutes after infection.

In the third phase (as already mentioned), *Q* protein has been made. This acts as a positive regulator, probably by overcoming the action of a terminator, t_R', to the right of p_R', which otherwise gives rise to the synthesis of only a very short RNA molecule. This marks the beginning of the third phase. In this late phase, much mRNA is made from genes *S* and *R* and from all of the head and tail genes located beyond the *cos* region of the λ genome.

Throughout the second phase, the products of genes *c*II and *c*III have been accumulating. Their action can lead to an alternative program of gene expression in the third phase. Together, the *c*II and *c*III products can activate transcription from still two other promoter sites, p_I (gene *int*) and p_{RE} (*r*epression *e*stablishment). This is the critical event that determines whether the phage continues vegetative reproduction or lysogenizes the host. The transcription initiated at p_{RE}, going leftward, transcribes the antisense strand of the gene *cro* en route to transcribing gene *c*I. The product of *c*I is the λ repressor. Just like the product of gene *cro*, it has an affinity for both o_L and o_R. However, the *c*I protein seems to be metabolically more stable than *cro* protein and fills the repressor binding subsites within o_L and o_R in a different order than *cro* protein (see Gussin et al., this volume). When *c*I accumulates rapidly enough, it can block phage DNA replication and late gene expression by preventing expression of genes *O, P,* and *Q*. The cell will then survive infection, and the repressed λ genome can be inserted as prophage into the bacterial chromosome by integrase, the product of *int*, which is heavily expressed in phase 3b by leftward transcription from p_I. Since, at this time, both p_L and p_R are repressed, no new *c*III and *c*II products will be made. Consequently, transcription from p_{RE} is stopped and no new *c*I protein is made from messenger initiated there.

In the fourth phase (the lysogenic condition), the lysogenized bacterium survives the phage infection and continues its division. In so doing, it replicates the integrated prophage each time it replicates its own chromosome. The *c*I repressor originally made from p_{RE} will be diluted by cell division, but repressor made from a second promoter, p_{RM} (*r*epressor *m*aintenance), takes its place. The transcript from this promoter also goes left traversing gene *c*I (and gene *rex*). p_{RM} is activated by repressor binding to a specific subsite within o_R and, therefore, the product of the *c*I gene stimulates *c*I (and *rex*) transcription in a lysogen. The accumulation of repressor appears to be limited by inefficient translation of the p_{RM}-derived transcript (see Gussin et al., this volume).

Let us now examine the phenotypes of a few mutations. Mutational inactivation of gene *c*I results in failure to produce *c*I repressor. Thus, mutations in *c*I form clear plaques, since they are not able to lysogenize the host. Note that the *c*I mutants retain sensitivity to wild-type *c*I repressor and are therefore unable to grow in a lysogenic strain; i.e., K12(λ) is immune to superinfection with either λ or λ*c* mutants.

λ mutants able to grow in K12(λ) are called λ*vir*. These are usually double (or triple) mutants, with one mutation in o_L and a second in o_R. These mutations affect the operators such that they no longer bind *c*I repressor.

Nin deletions remove terminator t_{R2}, located between genes *P* and *Q*, allowing transcription of *Q* in the absence of *N* protein. The resulting *Q* protein allows expression of the late genes, so that vegetative growth of λ*nin* is independent of *N*. λ*N nin* phages do not grow normally, however, because of their failure to express the recombination functions Red and Gam, which are involved in late DNA replication.

*c*II or *c*III mutants rarely establish lysogeny unless the corresponding gene products are provided by a coinfecting "helper." Once lysogeny has been established, however, *c*II and *c*III mutants can be maintained in the prophage state, since the products of these two genes play no role in prophage maintenance.

As mentioned above, hybrids between λ and related phages may not be sensitive to λ repressor. The hybrid λi^{434}, e.g., sensitive to 434 repressor, carries all λ genes except for the region p_L to p_R, which is derived from phage 434. This block of genes tends to stay together (as do other blocks of cofunctional genes; see Cambell and Botstein, this volume) in crosses between lambdoid phages. Thus, the virulent phages that would result from recombination between *c*I and the sites of action of *c*I protein rarely arise. In the hybrid λi^{21}, the region from *N* through *c*II is derived from phage 21, so that this phage too is a proper temperate phage, sensitive to phage 21 repressor but not to λ repressor.

Finally, it should be mentioned that the bacterial genes are not fully shut off by λ infection. This is important to remember when studying the products of λ gene expression. A widely used trick is to expose the cells prior to infection to a dose of UV irradiation sufficient to severely impair the expression of the bacterial genes without destroying the ability of the bacterium to synthesize the proteins coded in the undamaged λ DNA.

A HISTORICAL VIEW OF THE CHILDHOOD OF PHAGE λ AND ITS FIRST STEPS INTO BIOLOGICAL RESEARCH

Around 1950, Joshua and Esther Lederberg conducted their experiments demonstrating bacterial conjugation. Much of this work was done with derivatives of strain K12 of *E. coli*. Among these were many mutants that had been induced by heavy UV irradiation. To their surprise, a few such bacterial mutants showed particularly strange behavior; although they grew well in culture, the mutants lysed when they were incubated in conjugation experiments together with normal donor cultures of nonmutant K12 strains. This puzzling situation found the following explanation. K12 and most of its early derivatives are naturally lysogenic for phage λ, but a few of the rare survivors from the heavy UV irradiation had been cured of the λ prophage. When such cured strains were incubated together with normal λ-lysogenic K12 strains, two things could happen. First, since the λ lysogens produce λ phage particles at a low rate by spontaneous derepression, these phages could infect the cured nonlysogenic bacterial mutants and eventually lyse them. Second, if the λ-lysogenic strain donated by conjugation a part of its chromosomal DNA containing the λ prophage to a nonlysogenic recipient cell, the λ genome could undergo its vegetative growth cycle in the repressorless recipient cell ("zygotic induction"). This process would also liberate free phage particles, which eventually could infect the remaining nonlysogenic bacteria. Phage λ was thus discovered indirectly.

Phage λ entered the scientific literature in 1951 in an abstract published by Esther Lederberg. It was given the name lambda (λ) in analogy to the name of the kappa (κ) factor carried by *Paramecium*. λ, the greek letter following κ, by chance corresponds to the first letter of the name of the phage's discoverer.

λ-sensitive K12 strains then came into the hands of other investigators, and basic research on λ took off relatively fast. The first physiological data on λ are contained in an article by Weigle and Delbrück (1951), devoted to the mutual exclusion between phages λ and T5. Soon after this, Bertani and Weigle (1953) found that strain *E. coli* C is also sensitive to λ, and this led them to discover host-controlled modification of λ and of phage P2. At the same time, studies by Lederberg and Lederberg (1953), as well as by Wollman (1953), showed the site of attachment of λ prophage to be near the *gal* markers of the *E. coli* chromosome.

The next 10 years—the youth of phage λ—brought a number of important contributions of general interest for biological research. Use of λ-lysogenic bacteria, which express λ gene *rex*, allowed Benzer (1961) to draw an extensive and precise fine-structure map of the *r*II genes of phage T4. Results from studies with λ control mutants were decisive for development of the operon concept (Jacob and Monod 1961). Investigations with λ prophage mutants affecting genes essential for vegetative phage growth, but not for prophage maintenance, opened new general approaches to the problem of identifying the functions of particular genes (Jacob et al. 1957; Arber and Kellenberger 1958). This approach was soon extended into a very general method by the use of conditionally lethal mutants (Campbell 1959). These mutants were originally called *hd* (*h*ost-*d*ependent) and were later identified as suppressor-sensitive mutants. Hand in hand with the studies of λ mutants in vegetative reproduction, the phenomenon of specialized transduction of *gal* markers by λ (Morse et al. 1956a,b) found its basic explanation (Campbell 1957; Arber 1958; Arber et al. 1960). As a direct consequence of this research, the concept of helper virus for the reproduction of viruses with defective genomes emerged, and our understanding of the integration mechanism of viruses in host chromosomes ripened (Campbell 1962). Phage λ served in the basic investigations of the molecular mechanisms of DNA restriction and modification (Arber and Dussoix 1962; Dussoix and Arber 1962), and the studies of its single-stranded cohesive ends (Hershey et al. 1963) revealed the first enzyme reaction to produce staggered breaks in double-stranded DNA molecules.

ACKNOWLEDGMENTS

A first version of this paper served as handout to the participants of a course held in Basel in 1976 (European Molecular Biology Organization). For its redaction, Noreen and Ken Murray and Barbara Hohn gave their valuable assistance.

REFERENCES

Adhya S. and M. Gottesmann. 1978. Control of transcription termination. *Annu. Rev. Biochem.* **47:** 967.

Arber, W. 1958. Transduction des caractères Gal par le bactefiophage lambda. *Arch. Sci. Genève* **11:** 259.

Arber, W. and D. Dussoix. 1962. Host specificity of DNA produced by *Escherichia coli*. I. Host controlled modification of bacteriophage λ. *J. Mol. Biol.* **5:** 18.

Arber, W. and G. Kellenberger. 1958. Study of the properties of seven defective-lysogenic strains derived from *Escherichia coli* K12(λ). *Virology* **5:** 458.

Arber, W., G. Kellenberger, and J. Weigle. 1960. The defectiveness of lambda transducing phage. In *Papers on bacterial genetics* (ed. E.A. Adelberg), p. 224. Little, Brown, Boston.

Benzer, S. 1961. On the topography of the genetic fine structure. *Proc. Natl. Acad. Sci.* **47:** 403.

Berg, D.E. 1974. Genes of phage λ essential for λdv plasmids. *Virology* **62:** 224.

Bertani, G. and J.J. Weigle. 1953. Host controlled variation in bacterial viruses. *J. Bacteriol.* **65:** 113.

Campbell, A. 1957. Transduction and segregation in *Escherichia coli* K12. *Virology* **4:** 366.

——. 1959. Ordering of genetic sites in bacteriophage λ by the use of galactose-transducing defective phages. *Virology* **9:** 293.

——. 1962. Episomes. *Adv. Genet.* **11:** 101.

——. 1971. Genetic structure. In *The bacteriophage lambda* (ed. A.D. Hershey), p. 13. Cold Spring Harbor Laboratory, Cold Spring Harbor, New York.

Dussoix, D. and W. Arber. 1962. Host specificity of DNA produced by *Escherichia coli*. II. Control over acceptance of DNA from infecting phage λ. *J. Mol. Biol.* **5:** 37.

Elliott, J. and W. Arber. 1978. *E. coli* K12 *pel*⁻ mutants, which block phage λ DNA injection, coincide with *ptsM*, which determines a component of a sugar transport system. *Mol. Gen. Genet.* **161:** 1.

Feiss, M., R.A. Fisher, M.A. Crayton, and C. Egner. 1977. Packaging of the bacteriophage λ chromosome: Effect of the chromosome length. *Virology* **77:** 281.

Fiandt, M., Z. Hradecna, H.A. Lozeron, and W. Szybalski. 1971. Electron micrographic mapping of deletions, insertions, inversions, and homologies in the DNAs of coliphages lambda and phi 80. In *The bacteiophage lambda* (ed. A.D. Hershey), p. 329. Cold Spring Harbor Laboratory, Cold Spring Harbor, New York.

Franklin, N.C. 1971. Illegitimate recombination. In *The bacteriophage lambda* (ed. A.D. Hershey), p. 175. Cold Spring Harbor Laboratory, Cold Spring Harbor, New York.

Gottesman, M.E. and R.A. Weisberg. 1971. Prophage insertion and excision. In *The bacteriophage lambda* (ed. A.D. Hershey), p. 113. Cold Spring Harbor Laboratory, Cold Spring Harbor, New York.

Hershey, A.D., E. Burgi, and L. Ingraham. 1963. Cohesion of DNA molecules isolated from phage lambda. *Proc. Natl. Acad. Sci.* **49:** 748.

Hohn, T. and I. Katsura. 1977. Structure and assembly of bacteriophage lambda. *Curr. Top. Microbiol. Immunol.* **78:** 69.

Jacob, F. and J. Monod. 1961. Genetic regulatory mechanisms in the synthesis of proteins. *J. Mol. Biol.* **3:** 318.

Jacob, F., C.R. Fuerst, and E.L. Wollman. 1957. Recherches sur les bactéries lysogènes défectives. II. Les types physiologiques liés aux mutations du prophage. *Ann. Inst. Pasteur* **93:** 724.

Kaiser, A.D. 1955. Genetic study of the temperate coliphage λ. *Virology* **1:** 424.

Lederberg, E.M. 1951. Lysogenicity in *E. coli* K12. *Genetics* **36:** 560.

Lederberg, E.M. and J. Lederberg. 1953. Genetic studies of lysogenicity in *Escherichia coli*. *Genetics* **38:** 51.

Matsubara, K. 1976. Genetic structure and regulation of a replicon of plasmid λdv. *J. Mol. Biol.* **102:** 427.

Morse, M.L., E.M. Lederberg, and J. Lederberg. 1956a. Transduction in *Escherichia coli* K12. *Genetics* **41:** 142.

——. 1956b. Transductional heterogenotes in *Escherichia coli*. *Genetics* **41:** 758.

Murialdo, H. and L. Siminovitch. 1972. The morphogenesis of bacteriophage lambda. IV. Identification of gene products and control of the expression of the morphogenetic information. *Virology* **48:** 785.

Nash, H.A. 1977. Integration and excision of bacteriophage lambda. *Curr. Top. Microbiol. Immunol.* **78:** 171.

Pirrotta, V. 1976. The λ repressor and its action. *Curr. Top. Microbiol. Immunol.* **74:** 21.

Ptashne, M., A. Jeffrey, A.D. Johnson, R. Maurer, B.J. Meyer, C.O. Pabo, T.M. Roberts, and R.T. Sauer. 1980. How the λ repressor and cro work. *Cell* **19:** 1.

Randall-Hazelbauer, L. and M. Schwartz. 1973. Isolation of the bacteriophage lambda receptor from *Escherichia coli*. *J. Bacteriol.* **116:** 1436.

Shimada, K., R.A. Weisberg, and M.E. Gottesman. 1972. Prophage lambda at unusual chromsomal locations. I. Location of the secondary attachment sites and the properties of the lysogens. *J. Mol. Biol.* **63:** 483.

Signer, E. 1971. General recombination. In *The bacteriophage lambda* (ed. A.D. Hershey), p. 139. Cold Spring Harbor Laboratory, Cold Spring Harbor, New York.

Simon, M.N., R.W. Davis, and N. Davidson. 1971. Heteroduplexes of DNA molecules of lambdoid phages: Physical mapping of their base sequence relationships by electron microscopy. In *The bacteriophage lambda* (ed. A.D. Hershey), p. 313. Cold Spring Harbor Laboratory, Cold Spring Harbor, New York.

Skalka, A.M. 1977. DNA replication—Bacteriophage lambda. *Curr. Top. Microbiol. Immunol.* **78:** 201.

Stahl, F.W. 1979. *Genetic recombination*. W.H. Freeman, San Francisco.

Szybalski, E.H. and W. Szybalski. 1979. A comprehensive molecular map of bacteriophage lambda. *Gene* **7:** 217.

Thomas, R. 1971. Regulation of gene expression in bacteriophage lambda. *Curr. Top. Microbiol. Immunol.* **56:** 13.

Weigle, J. and M. Delbrück. 1951. Mutual exclusion between an infecting phage and a carried phage. *J. Bacteriol.* **62:** 301.

Weil, J., N. DeWein, and A. Casale. 1975. Morphogenesis of λ with genomes containing excess DNA: Functional particles containing 12 and 15% excess DNA. *Virology* **63:** 352.

Weisberg, R.A., S. Gottesman, and M.E. Gottesman. 1977. Bacteriopohage λ: The lysogenic pathway. *Comp. Virol.* **8:** 197.

Wollman, E.L. 1953. Sur le déterminisme génétique de la lysogénie. *Ann. Inst. Pasteur* **84:** 281.

Wollman, E.L. and F. Jacob. 1954. Etude génétique d'un bactériophage tempéré. II. Mécanisme de la recombinaison génétique. *Ann. Inst. Pasteur* **87:** 674.

Yarmolinsky, M.B. 1971. Making and joining DNA ends. In *The bacteriophage lambda* (ed. A.D. Hershey), p. 97. Cold Spring Harbor Laboratory, Cold Spring Harbor, New York.

Phage Lambda and Molecular Cloning

Noreen E. Murray
Department of Molecular Biology
University of Edinburgh
Edinburgh EH9 3JR, Scotland

For more than 20 years, transducing derivatives of lambdoid phages have facilitated functional and structural analyses of bacterial genes. Restriction enzymes that make staggered cuts within specific DNA sequences (targets) produce discrete fragments with short cohesive ends (Hedgpeth et al. 1972; Mertz and Davis 1972; Bigger et al. 1973), and this discovery immediately indicated that fragments of DNA could be spliced into the severed arms of a λ vector molecule. The cohered fragments would then be joined covalently by DNA ligase (Gellert 1967), and the recombinant genomes, transducing phages, recovered by transfection of *Escherichia coli* (Mandel and Higa 1970). This goal was quickly realized (Murray and Murray 1974; Thomas et al. 1974), and detailed analyses of the type previously possible for only a few *E. coli* genes became generally applicable. The ability to insert DNA fragments from any source into plasmid and phage vectors, supplemented with the new techniques for rapid DNA sequence determination (Maxam and Gilbert 1977; Sanger et al. 1977), has already generated many exciting advances in diverse areas of molecular biology.

The development of the phage λ chromosome as a receptor for fragments of DNA was assured by the availability of well-characterized deletion mutants, by the means of selecting for the loss of restriction targets and by prior identification of the essential components of the phage transcriptional circuits and replication system. The precise knowledge attained by 1972 showed the genetic engineer which regions of the λ genome could be sacrificed and, also, which special features could be manipulated to advantage. Although this wealth of information is instructive to the student of λ, it is not necessarily so to the molecular biologist wishing to use a λ cloning vector. The main aim of this paper is to relate the biology and genetics of phage λ to the cloning and manipulation of foreign DNA sequences. When thinking about cloning strategies, however, it is important to keep in mind that cloning systems continue to evolve. In 1974 a plasmid vector would have been chosen for cloning ''large'' DNA fragments, whereas λ offered the advantage of a powerful, controllable promoter. Just a few years later, large fragments could be more readily cloned in λ, and powerful promoters had been incorporated into plasmid vectors. Novel host-vector systems, like any other new product, may bring new problems, many of which can be avoided by an understanding of the biology of vectors and their hosts.

BIOLOGY OF PHAGE λ

Both introductory (Arber, this volume) and specialized accounts of the biology of phage λ precede this paper, but summaries of those features most relevant to the use of λ as a cloning vector are nevertheless included here.

Encapsidation

Basic to the use of phage λ as a vector is the fact that more than 40% (20 kb) of the λ genome does not code for any function essential to productive infection. Deletion of this DNA, however, despite the retention of a functional set of essential genes, produces a noninfective phage. Infectivity is lost because encapsidation of the virion DNA has a minimum (Bellet et al. 1971) as well as a maximum (Weil et al. 1973) size requirement, but infectivity can be maintained by replacing nonessential λ genes with an alternative DNA sequence of appropriate size. The encapsidated λ chromosome is a linear duplex DNA molecule with self-complementary, and therefore cohesive, single-stranded projections at the 5′ ends, which allow the molecule to circularize upon infection (Strack and Kaiser 1965). Encapsidation not only requires molecules of appropriate size but, also, the DNA sequence (*cos*) that includes the cohesive ends (Feiss et al. 1979).

The effective packaging of only those molecules of appropriate length is fundamental to the design of λ vectors and to the use of encapsidation in vitro as an efficient means of recovering recombinant genomes. It is also of consequence that, despite the minimum and maximum size requirements, the degree of tolerance is such that there is normally little discrimination among molecules in the range of 80–100% (or even 105%) of the normal length. In general, phages with more than 25% of their genome deleted are noninfective, those with 22–25% deleted grow poorly and are easily outgrown by derivatives with an increased DNA content, and phages with more than a normal complement of DNA may be outgrown by deletion mutants.

The substrate for encapsidation in vivo (see Fig. 1) is multimeric DNA in which the *cos* sequences of λ are so located that cutting within them generates monomeric genomes of appropriate length (Feiss et al. 1977; see Feiss and Becker, this volume). Normally, these concatemers are generated by the rolling-circle mode of replication, but when this is blocked, recombination between monomeric circular genomes produces dimeric circles, an alternative concatemeric substrate. Linear, but not circular, monomers can be packaged in vitro, although these are not found in the bacterial cell.

Application of the technique of in vitro encapsidation (Hohn and Hohn 1974; Becker and Gold 1975) to the recovery of recombinant DNA molecules (Hohn and Murray 1977; Sternberg et al. 1977) has had a great impact on the use of λ as a cloning system. This technique recovers λ genomes with efficiencies as high as 10^{-2}, approximately 100 times that achieved by transfection of bacterial cells.

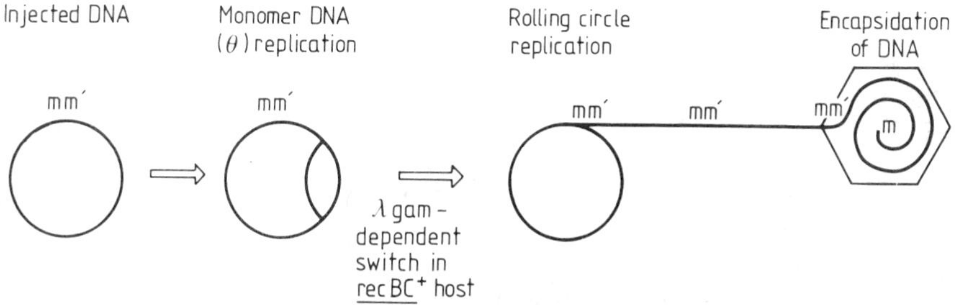

Figure 1 Infective pathway of bacteriophage λ. The phage DNA forms a covalently closed circle through pairing of the cohesive ends, m and m′, followed by ligation. Early DNA replication by the θ mode generates circular monomeric molecules. The switch to the rolling-circle mode of replication is dependent on the product of the *gam* gene, which inhibits the host *recBC* nuclease. The multimeric DNA is the substrate for encapsidation.

Gene Organization and Transcription

The genome of phage λ (Fig. 2) may be divided into three regions. The left-hand region includes all of the genes (from *A* through *J*) whose products are necessary to package phage DNA and produce an infectious virion. The central region (between genes *J* and *N*) contains no genes whose products are necessary for plaque-forming ability. The remaining portion of the genome, from *N* rightward, includes all the major control elements, the genes necessary for phage replication (*O* and *P*), and those for cell lysis (*S* and *R*).

All essential genes and many nonessential genes are transcribed from three major promoters: p_L, p_R, and $p_R{}'$ (see Fig. 2). Following infection of a sensitive host (see Arber; Friedman and Gottesman; Wulff and Rosenberg; all this volume), transcription proceeds leftward from p_L through gene *N* and rightward from p_R through gene *cro*. In the absence of the product of gene *N*, most transcripts terminate just beyond these genes at the sites t_L and t_{R1}. A second termination signal, t_{R2}, impedes further progress of those transcripts from p_R that evade termination at t_{R1}. *N*-gene product exerts its positive regulatory role by causing RNA polymerase to ignore these "stop" signals and continue leftward beyond t_L into the nonessential genes, rightward through the replication genes *O* and *P*, and then through t_{R2} and gene *Q*. The product of *Q* is essential for transcription of all the "late" genes of λ. Since the linear chromosome of the phage circularizes on infection, *Q*-dependent transcription from $p_R{}'$ (located between genes *Q* and *S*; see Fig. 2) continues through genes *S* and *R* into *A* and beyond *J*. It follows that the product of gene *N* is normally essential for the activation of gene *Q*. However, if t_{R2} is deleted, *N*-independent "leakage" of transcription through t_{R1} can suffice to provide *O, P,* and *Q* functions. Most λ vectors include a deletion (*nin*5), which removes t_{R2}. This makes *N* and the p_L promoter dispensable for phage growth (Court and Sato 1969) and extends the rightward limit of the nonessential central region up to, but excluding, p_R.

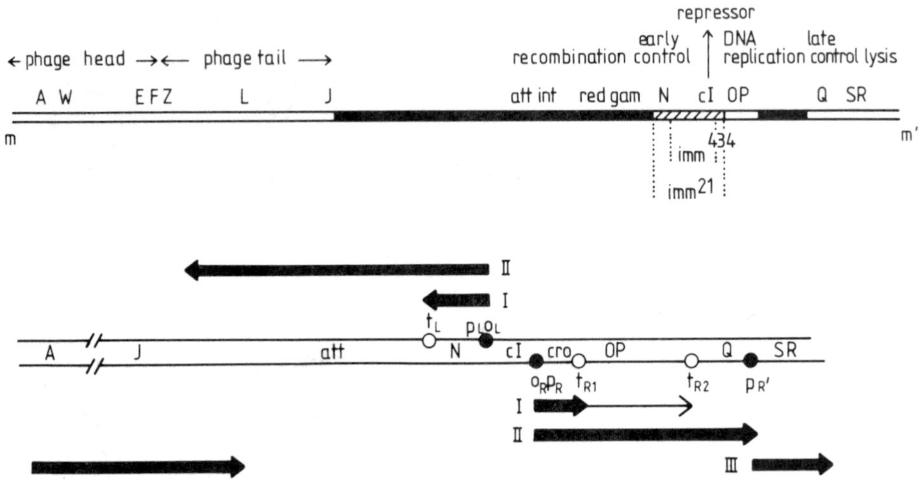

Figure 2 (*Top*) The genetic map includes some key markers and transcriptional circuits of bacteriophage λ. The black regions are nonessential, whereas the hatched region is nonessential only in the absence of the termination site t_{R2}. The dotted lines indicate the endpoints of the immunity regions of the λ*imm*^{434} and λ*imm*^{21} phages; o_L and o_R are the repressor binding sites. (*Bottom*) This map (not drawn to scale) indicates the organization of the phage control region, including the sites at which transcription is initiated and terminated. Heavy arrows represent the major transcripts. (●) Major promoters; (○) major termination signals. (*I*) The immediate early transcripts initiated at p_L or p_R and terminating at t_L and t_{R1}, respectively, in the absence of the product of gene *N*; some transcripts escape t_{R1} but terminate at t_{R2}; (*II*) In the presence of the *N* protein, early transcripts initiated at p_L continue through t_L, and those from p_R continue through t_{R1} and t_{R2}. (*III*) Late transcription of the circular genome, dependent on the *Q*-gene product, continues through genes *S, R, A* and, eventually, *J*.

Phenotypes of Deletion Derivatives

In any λ vector, at least part of the nonessential region of the genome is deleted to make space for donor DNA. The consequences of some of these deletions are summarized in Figure 3. Deletion of the DNA between genes *J* and *att* (the site at which a sequence-specific crossover integrates λ into the *E. coli* chromosome) has no detectable effect, although this region is not completely devoid of genes and does include some promoter and termination sequences. If the deletion extends further to the right (e.g., *b*189), so as to remove all or part of *att*, then normal integration of the phage into the *E. coli* chromosome and excision of the prophage from the chromosome (see Weisberg and Landy, this volume) are impaired. λ vectors designed to accommodate fragments larger than 12 kb normally lack the attachment site.

Deletion of the genes *int* and *xis*, the products of which mediate integrative and excisive recombination, is of no more consequence than loss of *att* itself, but

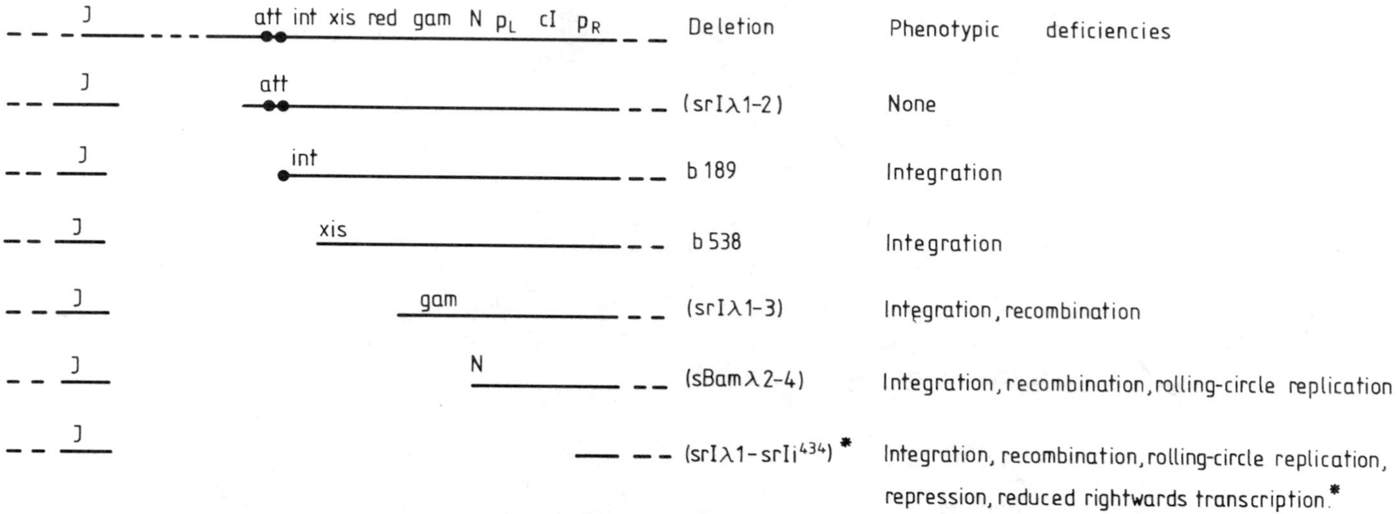

Figure 3 Phenotypic effects associated with the deletion of nonessential genes of phage λ. The nomenclature for restriction targets uses the abbreviation for the enzyme (Smith and Nathans 1973), followed by the genome and the number of the site within the genome. The targets of the λ genome are numbered from left to right; *sr*I*i*^{434} refers to an *Eco*RI target in the *imm*^{434} region. (*) A phage deleted for gene *N* forms a small plaque if t_{R2} is removed by *nin*5, a deletion located between genes *P* and *Q*.

deletion of the neighboring gene *redA* results in a phage defective in general recombination. Although the absence of recombination may increase the stability of incorporated DNA sequences, it decreases the rate of DNA replication. Consequently, the burst size of a *red*⁻ phage in a *rec*⁺ host is one-third that of a wild-type λ (Enquist and Skalka 1973; see Smith, this volume). This deficiency is aggravated in a *recA* host, but lysates with titers of around 10^{10} pfu/ml can still be obtained.

The *gam* gene of λ, located just to the right of the *red* genes, plays a crucial role in the normal infective program of the phage. The *gam* product inhibits the host *recBC* nuclease (Unger and Clark 1972), and only in the absence of this enzyme can the rolling-circle mode of DNA replication be initiated (Greenstein and Skalka 1975). In the absence of *gam* function, therefore, λ cannot generate linear concatemeric DNA, the normal substrate for encapsidation (see Fig. 1). Gam^- phage do make plaques, however, because the recombination systems Red and Rec generate dimers via circle-by-circle recombination, and these dimers are substrates for encapsidation. A DNA sequence designated Chi (see Smith, this volume) improves the growth of Gam^- phage in rec^+ bacteria by stimulating RecBC-dependent crossing over. In a rec^+ bacterium, phage that are both Red^- and Gam^- are totally dependent on Rec-mediated exchange. Such phage sorely need a Chi sequence, whose stimulation of dimer formation via recombination enables the phage to make a modest, rather than a vanishingly small, plaque. On $recBC^-$ hosts, red^-gam^- phages can form plaques, even in the absence of Chi, but they do not yield high-titer lysates. A more vigorous host for lysate production is an $sbcA^-recBC^-$ strain (M. Stahl, pers. comm.); the *sbcA* mutation activates the compensatory *recE* pathway (Barbour et al. 1970). A $recBC^+$ alternative harbors a plasmid carrying the *gam* gene of λ. Transcription of the *gam* gene in this plasmid is from the late λ promoter $p_R{}'$ and is activated in *trans* following infection by Q^+ λ (G. Crouse, pers. comm.). This plasmid alleviates the replication barrier encountered by red^-gam^- phage in a $recA^-$ host.

A phage deleted for gene *N*, even in the absence of t_{R2}, produces minute plaques, since transcription of genes *O, P,* and *Q* is reduced, and that of *red* and *gam*, should they still be present, is severely limited. In a $recBC^+$ host, this N^- phage is dependent on Chi-stimulated recombination for the generation of concatemeric DNA.

Recombinant plasmids are usually propagated in $recA^-$ hosts, since some sequences, particularly repetitive ones, are more stable in the absence of recombination functions. Homologous recombination between misaligned, repeated sequences in phage λ leads to duplication and deletion derivatives (Bellet et al. 1971), but some constraint on this process is imposed by the size of the resulting phage genomes. Few data are available on Rec-dependent instabilities in λ vectors (McClements et al. 1981), although *rec*-independent instabilities in both plasmid and phage systems are well documented (Brutlag et al. 1977; McClements et al. 1981; Chia et al. 1982). In practice, some workers have preferred to be cautious and suffer the lower titers obtained on propagating red^-

phage in *recA*⁻ hosts, others have compromised by using red^-gam^+ phage in rec^+ (phenotypically $RecBC^-$) hosts, and many have risked the use of red^-gam^- phages propagated in Rec^+ hosts to clone slightly larger fragments of DNA. RecBC strains may be a safer alternative, and recent experiments demonstate that palindromic sequences can be cloned in λ by using a *recBC sbcB* host. The palindromic sequences, however, remain prone to deletion (D.R.F. Leach, pers. comm.).

Transcription of Cloned Sequences

Most transducing phages made in vitro, like those made in vivo (see Campbell 1971), incorporate foreign DNA in the central nonessential region between genes *J* and *N*. In such cases, transcription of the incorporated genes from the major λ promoters, p_L or $p_R{}'$, is possible.

Extensive data on the expression of bacterial genes show that genes inserted downstream of p_L are transcribed from this promoter (e.g., Franklin 1971), even if they are to the left of the attachment site (Hopkins et al. 1976). Few incorporated DNA sequences are likely to impede transcription initiated from p_L in the presence of the product of gene *N* (Adhya et al. 1974; Franklin 1974; Segawa and Imamoto 1974). In many cases, genes inserted in the alternative orientation are effectively transcribed from $p_R{}'$ (Bøvre and Szybalski 1969; Wilson and Murray 1979; Burt and Brammar 1982a,b). Inserted DNA sequences, though transcribed, are not necessarily expressed efficiently, even if they are from prokaryotic genes (Mercereau-Puijalon and Kourilsky 1976) or from eukaryotic genes lacking intervening sequences (Mellado et al. 1981).

Transcription from p_L may well impede transcription from an incorporated promoter oriented in the opposing direction (Ward and Murray 1979). This effect is moderated early after infection when the activity of p_L is reduced (see Friedman and Gottesman; Gussin et al; both this volume). Conversely, transcription from a strong promoter, carried by an inserted DNA fragment that does not also include an appropriate terminator sequence, may have a deleterious effect on any host-vector system.

MAKING λ VECTORS

Manipulation of Restriction Targets

A derivative of λ that is to serve as a vector must provide space and an appropriate restriction target located in nonessential DNA, but none in essential regions of the genome. Restriction enzymes recognize specific, short nucleotide sequences. In DNA of uniform composition, a specific tetranucleotide would occur by chance once in every 256 bp, and a specific hexanucleotide, once in every 4016 bp. Since the λ genome is nearly 50 kb in length, it usually has more than one target sequence for any given endonuclease (see Fig. 4). Although vec-

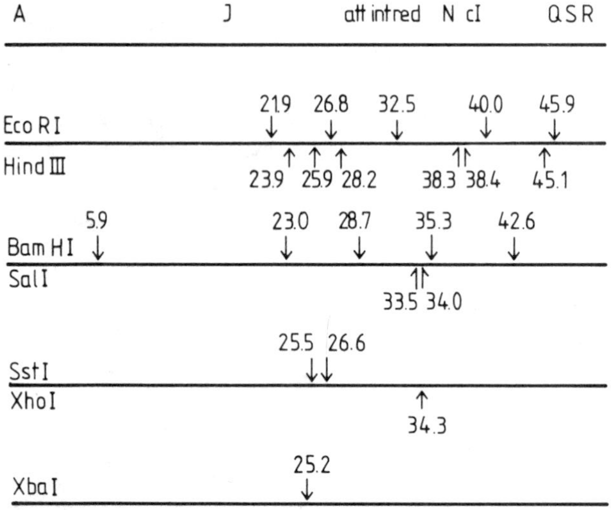

Figure 4 Restriction targets within the λ genome. (→) The positions of the targets for those enzymes for which λ vectors are available. Each of these enzymes recognizes a hexanucleotide sequence. The positions of the targets are taken from the information compiled by Williams and Blattner (1980) and are normalized with respect to the *Eco*RI site at 40.0 kb. The left-hand end of λ is then calculated to be at nucleotide 403, and the right-hand end, at nucleotide 49,500.

tors for DNA fragmented with *Sal*I or *Sst*I require trivial manipulation of restriction targets, those for DNA fragmented with other endonucleases have required the removal of as many as five targets.

The plasmid coding for *Eco*RI is resident in a λ-sensitive strain of *E. coli*. It was therefore possible to use this strain to select mutations leading to loss of *Eco*RI targets in the λ genome, since such mutations should confer preferential survival when unmodified phage enter a restricting host (Arber and Kuhnlein 1967; Murray et al. 1973). This strategy proved effective, and targets within essential genes could be mutated without any associated deleterious phenotypic effect (Murray and Murray 1974; Rambach and Tiollais 1974; Thomas et al. 1974).

An alternative route was obligatory for the manipulation of the targets for *Hin*dIII, an enzyme isolated from *Haemophilus influenzae*. Fortuitously, any of the six targets could be removed by the use of deletions or substitutions of the λ genome (Murray and Murray 1975).

An in vitro approach was necessary to remove one of the five targets for *Bam*HI. Four targets, located in nonessential regions of the λ genome (Perricaudet and Tiollais 1975; Haggerty and Schleif 1976), were removed by either deletions or substitutions. Phage retaining the one target located in an essential region of the genome were mutagenized. DNA isolated from these phage was treated with *Bam*HI endonuclease, and surviving molecules were recovered. After a series of cycles of enrichment, genomes lacking all targets for *Bam*HI were isolated (Klein and Murray 1979; Rimm et al. 1980). Their selection depended on in vitro encapsidation as an efficient means of recovering intact DNA molecules. The cohesive ends of the DNA fragments generated by *Bam*HI anneal with those of fragments generated by a number of other enzymes, e.g., *Bgl*II, *Mbo*I, and *Sau*3a, and a vector for *Bam*HI fragments therefore has a variety of uses. More general diversification of the role of λ as a vector, exemplified by Messing et al. (1981) in the construction of vectors from the single-stranded DNA phage M13, relies on the insertion of adapter (poly-linker) sequences. In this way, a vector for *Bam*HI fragments has been converted into a vector for use with *Bam*HI, *Eco*RI, or *Sal*I (A.-M. Frischauf et al., in prep.).

TYPES OF VECTORS

A λ genome with some nonessential DNA deleted, but retaining a single target for a restriction enzyme, is known as an insertion vector. Those in common use lack approximately 20 percent of their genome. If λ genomes are recovered by in vitro encapsidation, conditions may be chosen that favor recovery of longer (recombinant) molecules (Sternberg et al. 1977; Williams and Blattner 1980).

Replacement vectors are phages that retain two targets flanking a replaceable segment of DNA. Digestion with the restriction enzyme separates the central fragment from the two arms of the vector, which carry all the essential genes of the phage but comprise a genome length too short to be encapsidated. Plaque-forming phages must therefore include a DNA fragment of appropriate size between the two arms of the vector: either a donor fragment or the original central fragment. Efficient cutting of both restriction targets in the vector DNA molecules is therefore critical in the use of a replacement vector. This must be checked stringently by assaying the residual infectivity of the digested vector DNA when a reduction of three to five orders of magnitude should be observed. If a severalfold molar excess of donor fragments is used, the majority of the recovered phages are recombinant. However, an excess of donor DNA fragments increases the opportunity for their oligomerization, which tends to reduce the yield of plaque-forming phages.

The first replacement vector (Thomas et al. 1974) was *red*$^-$*gam*$^+$, with space to accommodate *Eco*RI fragments up to a length of approximately 15 kb. Alternative *red*$^-$*gam*$^+$ vectors for *Hin*dIII and *Eco*RI (Murray et al. 1977; also see Fig. 9) have the necessary space for fragments as large as 16 kb and 19 kb, respectively, whereas *gam*$^-$ variants for *Eco*RI, *Hin*dIII, or *Bam*HI may accommodate 20 kb or more of donor DNA (Murray and Murray 1975; Blattner et al. 1977; Loenen and Brammar 1980; also see Figs. 5 and 9). Replacement vectors for cloning much longer fragments (>25 kb) require deletion of essential λ genes (Perricaudet and Tiollais 1975) and could be used if the essential late gene products were provided from λ genes carried by a plasmid.

RECOGNITION AND SELECTION OF RECOMBINANTS

In a ligation reaction with fragments of donor DNA and cleaved λ vector, many infectious DNA molecules have merely restituted the parental vector chromosomes. Some λ vectors offer genetic approaches that permit the recognition or even the selection of recombinant genomes.

Genetic Tests and Screens

In the case of an insertion vector, recombinant phages may be recognized if insertion of a DNA fragment inactivates a nonessential gene or modifies some aspect of phage phenotype dependent on genome size. Insertion at the *Eco*RI target within a phage gene coding for a recombination function (*redA*) produces a recombination-deficient (Red⁻) phage (Murray and Murray 1974). Testing of individual plaques can reveal Red⁻ phages by their inability to form plaques on a ligase-deficient strain of *E. coli*. An alternative approach relies on the fact that gene-*D* function is essential only if the phage genome exceeds 82% of the wild-type length (Sternberg et al. 1977). Vectors containing an amber mutation in gene *D* can therefore be used to detect recombinants that have incorporated sufficient donor DNA to prevent plaque formation in the absence of an amber suppressor. For *Eco*RI insertion vectors, inclusion of a donor fragment increases the number of *Eco*RI targets, and this is correlated with a measurable decrease in the efficiency of plaque formation on an *Eco*RI-restricting host.

More generally useful screens are based on color tests or changes in plaque morphology and are applicable to as many as 10^3 plaques on a single plate. Insertion of DNA into the *lacZ* gene carried by a λ*lacZ* transducing phage inactivates the *lacZ* gene, and recombinant (*lacZ*⁻) phage are easily detected (Blattner et al. 1977; Pourcel and Tiollais 1977; Pourcel et al. 1979), since β-galactosidase, the product of the *lacZ* gene, hydrolyzes a colorless substrate to form a nondiffusible blue pigment (see Arber et al., this volume). Insertion at the *Bam*HI site within the *int* gene can be recognized by "the red plaque test" (Enquist and Weisberg 1976; Klein and Murray 1979; see Arber et al., this volume). A commonly used screen, which depends on the insertional inactivation of the gene coding for the phage repressor (Murray et al. 1977), also confers the advantage of selection (see below).

For replacement vectors, a change in phenotype associated with substitution of the central DNA fragment facilitates the recognition of recombinants and provides a convenient assay for monitoring the efficiency of recovery of recombinant phages. This genetic screen can be achieved if a gene in the central fragment imparts a readily recognizable phenotype, irrespective of its orientation. Screens using *lacZ* are the most common, but alternatives are reviewed by Williams and Blattner (1980). The *lacZ* gene in the central fragment of a vector is identified by blue plaques on a *lacZ*⁻ host (Blattner et al. 1977; Murray et al.

1977). Alternatively, the presence of the *lac* operator may be detected using *lacZ*⁺ bacteria, since phage replication produces sufficient copies of the *lac* operator to titrate the *lac* repressor and so induce the chromosomal *lacZ* gene (Blattner et al. 1977). A third variation relies on suppression of an amber mutation in the *lacZ* gene of the host by a suppressor gene within the central fragment of the vector (Murray et al. 1977). In this case, the medium includes inducer as well as the indicator substrate (see Arber et al., this volume).

Genetic Selection of Recombinants

The so-called "immunity" insertion vector has the advantage that recombinants may be selected. This selection relies on the use of restriction targets in the gene, *c*I, coding for the phage repressor protein. The vectors used have the immunity of phage 434, since the *c*I analog of this phage includes single targets for both *Eco*RI and *Hin*dIII (Murray et al. 1977). The insertion of donor DNA into the *c*I gene of λimm^{434} is recognized by a change from a turbid to a clear-plaque morphology. Not only do the recombinants grow more vigorously than the vector, but they may be selected by using as indicator a strain with a mutation in a gene whose product normally antagonizes the establishment of repression (Lathe and Lecocq 1977). This strain (RW2241), usually known as POP101, carries a mutation (*lyc*7) that is probably an allele at the *hfl* locus (Belfort and Wulff 1971). In the POP101 host, λimm^{434} (or λimm^{21}) phage, other than $c\text{I}^-$ and $c\text{II}^-$ derivatives, establish repression so effectively that they form plaques with an efficiency of about 10^{-4} (Lecocq and Gathoye 1973). Pirrotta and colleagues (e.g., Scherer et al. 1981), following the suggestion of R. Lathe, have successfully used POP101 to eliminate nonrecombinants. A restriction-deficient derivative of POP101 (NM514) is available.

Removal of the central fragment from the arms of a replacement vector prior to the addition of the donor DNA ensures that most of the infective genomes will be recombinant. Nevertheless, genetic selection confers a useful additional or alternative enrichment for recombinant genomes.

One selection system uses a derivative of the original replacement vector (Thomas et al. 1974), in which the replaceable DNA is a dimer of two small fragments of T5 DNA (Davison et al. 1979). Each fragment includes the T5 gene *A*3, in the presence of which phage λ, like T5, will not grow in an *E. coli* host carrying plasmid ColIb.

An alternative system, easily applicable to vectors designed to provide maximum space, makes use of the fact that $\lambda red^- gam^-$ phages are able to form plaques on a *rec*⁺ host lysogenic for phage P2 (Zissler et al. 1971). Such phages are said to have a Spi⁻ phenotype, since, unlike wild-type λ, they are no longer *s*ensitive to *P*2 *i*nterference (see Smith; Court and Oppenheim; both this volume). The selection of recombinants as Spi⁻ phages is discussed in detail later in the text.

Detection of Functional Cloned Genes

Compensation for a genetic lesion of *E. coli* by bacterial genes in a λ phage has traditionally provided an efficient means for the selection of transducing phages (see, e.g., Campbell 1971). This remains a basis for detecting recombinants, including bacterial (e.g., Murray and Murray 1974; Cameron et al. 1975; Borck et al. 1976; Silverman et al. 1976) and sometimes even eukaryotic genes (e.g, Struhl et al. 1976).

Phages compensating for a host defect may be selected as lysogenic transductants. If the λ vector is integration-deficient, integration may be achieved by coinfecting with a wild-type helper phage. Alternatively, *E. coli* DNA within a λ phage provides homology with the chromosome, and either this or a resident prophage can serve as a substrate for *rec*-dependent integration (see Auerbach and Howard-Flanders 1981). The expression of cloned genes may also be detected in the lytic phase. In this case, a gene acquired by the phage compensates for the host defect by providing the necessary enzyme for bacterial growth and the consequent phage and plaque production. This selection works efficiently for genes that must rely on the phage promoters (either p_L or p_R') for their transcription and obviates the need for stable lysogens. Many cloned *E. coli* genes (see, e.g., Borck et al. 1976; Guest and Stephens 1980) and some fungal genes (Struhl et al. 1976) were isolated in this way. The technical details of detecting plaques in the absence of a good bacterial lawn are discussed by Arber et al. (this volume).

Tests relying on the functional expression of incorporated genes are limited by the availability of suitably defective strains of *E. coli*, although in some cases the potential of a wild-type bacterium may be extended; e.g., derivatives of λ carrying the amidase genes of *Pseudomonas aeruginosa* confer the ability to use acetamide as nitrogen source (Drew et al. 1980). Enzyme assays linked to pH-dependent color changes can frequently be adapted for in situ screens and, in principle, a wide range of other tests based on histochemical stains may be applicable.

Detection by Immunoassay

Assays have been developed for detecting the translation products of cloned genes independent of the production of a functional enzyme. These methods based on specific cross-reaction with antibodies (Sanzey et al. 1976; Skalka and Shapiro 1977), can detect a portion of a polypeptide chain that includes an antigenic determinant in the appropriate configuration. The sensitivity of screens on the basis of precipitation of antibody in the neighborhood of a plaque can be increased by using colorimetric assays of enzyme-antibody conjugates. Immunoassays of this type, using a *Corynebacterium diphtheriae* phage, have permitted the screening of 10^6 plaques for the release of diphtheria toxin (Kaplan et al. 1981).

The usefulness of the solid-phase radioimmunoassay of Broome and Gilbert (1978) has been extended by Kemp and Cowman (1981), who detected antibody bound to antigen by its ability to bind ^{125}I-labeled staphylococcal A protein, a reagent that specifically reacts with the F2c part of most antibodies. This method, which has the advantage that the radioactively labeled reagent is always the same one, was used to detect immunoglobulin-C μ chains in *E. coli* colonies but is applicable to plaques (Wenman and Lovett 1982).

Immunoassays are applicable to lysogenic cells. An integration-proficient expression vector λgt11 (λ*lac*5cI857*nin*5*Sam*100) readily allows this approach (R.A. Young and R. Davis, pers. comm.). The foreign DNA, or cDNA, is inserted into the *Eco*RI site in the *lacZ* gene. Efficient lysogeny is effected by the use of an *hfl*⁻ host (see above, Genetic Selection of Recombinants), and a temperature shift induces the synthesis of a hybrid protein whose stability may be increased if the host is *lon*⁻ (see Mount 1980).

Screening by Nucleic Acid Hybridization

This technique is extremely sensitive, readily detecting sequences of 1 kb or less, and limited only by the availability of a radioactively labeled RNA or DNA probe. Phage plaques are a ready source of both free and packaged DNA. The simplest and quickest method of screening plaques by in situ hybridization (Benton and Davis 1977) follows the direct transfer of phage DNA from plaques to nitrocellulose filters simply by contact. The surface of a standard agar plate covered with 10^4 plaques can be screened in this way, and the scale can be increased by using larger plates or even cafeteria trays (Blattner et al. 1978).

Selection of Homologous Sequences via Recombination

As an alternative to screening by hybridization, B. Seed and colleagues (Maniatis et al. 1982) have devised a selection using a recombinational test for homologous DNA. For this purpose, they constructed a small plasmid (~1 kb), comprising a replication origin, a suppressor gene (*supF*), and restriction targets into which a DNA probe can be inserted. This plasmid is maintained in the host by demanding suppression of amber mutations in each of two genes conferring resistance to different antibiotics. Homology between the probe and a sequence cloned within λ is required for the plasmid to recombine with, and hence integrate into, the phage genome. Phages acquiring the plasmid may be selected as follows. A population of recombinants made in a vector such as Charon 4A or EMBL3A, which has amber mutations in the λ genes *A* and *B*, is propagated on a Rec⁺ host carrying the *supF* plasmid. Phages in the lysate that have incorporated the *supF* plasmid are then selected as plaques on a suppressor-free host.

Spi⁻ RECOMBINANTS

Spi$^-$ derivatives of phage λ, defined by their ability to form plaques on *E. coli* lysogenic for phage P2, were identified by Zissler et al. (1971) as *red$^-$gam$^-$* mutants. Subsequently, the inclusion in the λ genome of a nucleotide sequence (Chi) that stimulates Rec-dependent recombination was shown to enhance the plaque size of *red$^-$gam$^-$* phages (Lam et al. 1974; Henderson and Weil 1975). However, a *red$^+$gam$^-$* Chi phage itself gives minute plaques on a host lysogenic for P2 (S. Brenner and N.E. Murray, unpubl.) The selection of *red$^-$gam$^-$* (Spi$^-$) recombinants in the presence of a Chi sequence (Murray and Murray 1975) was first used effectively by D.F. Ward (see Murray 1977) for a *Bam*HI replacement vector despite the presence of an additional *Bam*HI target in the left arm of the vector (see Fig. 5, line 1). The removal of this extra *Bam*HI target (Klein and Murray 1979; Rimm et al. 1980) and the multipurpose potential of *Bam*HI vectors has led to the production of new families of vectors in which the *red* and *gam* genes of λ are carried in the central replaceable fragment of DNA. Selection against Spi$^+$ vector phages requires expression of *gam*, but preferably both *red* and *gam*. Where inversion of the central fragment of the vector does not sever *red* and *gam* from p_L, transcription of these genes from this promoter is assured. For vectors where *Bam*HI-mediated inversion separates *red* and *gam* from p_L, the phage remain sensitive to inhibition by P2, even though the *gam*-dependent switch to the normal means of replication may be delayed. In one case (Fig. 5, line 4), transcription of *gam* can be initiated from a weak constitutive promoter located to the left of the attachment site (Loenen and Brammar 1980), whereas in the remainder, transcription of *gam* from p_R' is implicated (N.E. Murray, unpubl.). Since intramolecular site-specific recombination can delete the *red* and *gam* genes, *int$^-$* phages (e.g., NM1122 and Charon 30 [see Fig. 5]; EMBL3 and EMBL4 [see Fig. 9]) show a lower background of Spi$^-$ derivatives.

Vector L47 (Loenen and Brammar 1980) can be used as a *Bam*HI, *Eco*RI, *Hin*dIII, *Sal*I, or *Xho*I vector. This vector (Fig. 5, line 3) includes ChiA, (Stahl et al. 1975), and recombinants generated with *Bam*HI, *Eco*RI, or *Hin*dIII may be selected as Spi$^-$ phages. Both the *Eco*RI and the *Hin*dIII recombinants are *N$^-$*, as well as *gam$^-$*. Charon 30 (Fig. 5, line 4) and Charon 28 (Rimm et al. 1980) can be used with the same spectrum of enzymes, but the Spi selection is only possible when used as *Bam*HI vectors. However, Rimm et al. do not suggest selecting recombinants on a P2 lysogen, since neither Charon 28 nor Charon 30 includes a Chi sequence. Under the severe selective conditions imposed by the P2 lysogen, a Spi$^-$ phage without a Chi sequence might never be detected. Derivatives of Charon 30 and Charon 28 are best recovered on a *recBC$^-$* host, where a recombinant lacking a Chi sequence will not be at a selective disadvantage.

An alternative phage, 1059 (Fig. 5, line 5), incorporating ChiC (Stahl et al. 1975), was designed specifically as a *Bam*HI vector (Karn et al. 1980) and offers more space than other *Bam*HI vectors. Its disadvantage is the presence of a

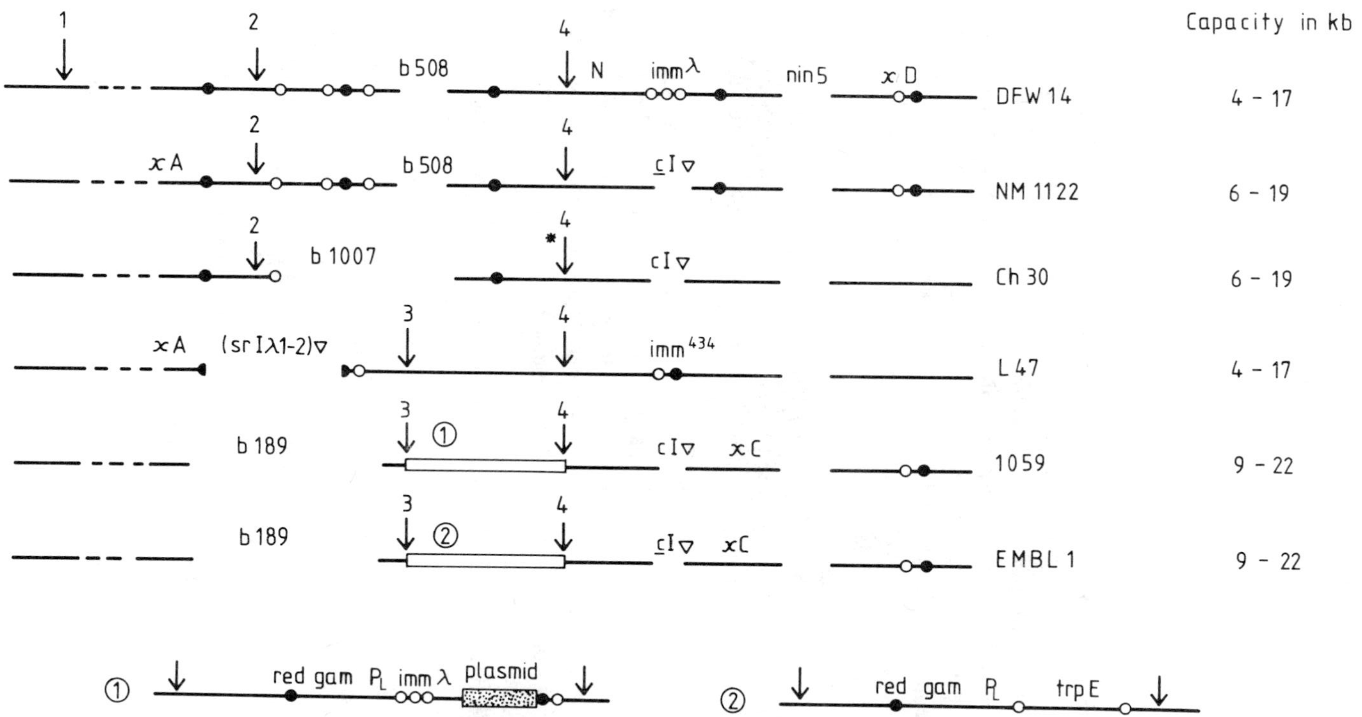

Figure 5 *Bam*HI vectors yielding Spi⁻ recombinants. Restriction targets for *Bam*HI (→), *Eco*RI (●), and *Hin*dIII (○). The bottom line shows some details of the central fragments from vectors 1059 and EMBL1. χA, χC, and χD are Chi sites; (---) omission of some DNA. The minimum and maximum capacities (as *Bam*HI vectors) are based on genome sizes of 38 kb and 51 kb, respectively (Williams and Blattner 1980). (*) The replaceable segment of Ch30 is a dimer of the fragment shown.

ColE1 plasmid sequence within the central fragment. Inevitably, some phages that retain plasmid DNA survive the selection procedure and are all too readily detected by their hybridization with a probe made from most of the commonly used plasmids. Modifications to vector 1059 have alleviated this problem (A.-M. Frischauf et al., in prep.); either the central fragment from the primitive *Bam*HI vector DFW14 (see Fig. 5, line 1) may be used, or the *Hin*dIII fragment containing the plasmid sequence may be replaced by a fragment of *E. coli* DNA (EMBL1; Fig. 5, line 6).

The range of vector EMBL1 has been extended. Mutations removing the *Eco*RI targets were selected, and the *Bam*HI targets were replaced by linkers carrying recognition sequences for the enzymes *Bam*HI, *Eco*RI, and *Sal*I (A.-M. Frischauf et al., in prep.). These derivatives (EMBL3 and EMBL4) are effective for the selective cloning of large DNA fragments generated with *Eco*RI and *Sal*I, as well as *Bam*HI. Each *Bam*HI target has an *Eco*RI target at one side and a *Sal*I target at the other. The outer flanking targets serve to separate donor DNA from vector where insertion at the *Bam*HI site alters the sequence, so that the *Bam*HI target is destroyed. This is commonly the case when fragments generated by an enzyme with a tetranucleotide recognition sequence (e.g., *Sau*3a) are cloned in a vector for fragments generated by an enzyme recognizing a hexanucleotide sequence (e.g., *Bam*HI). Cutting the inner flanking targets can be of general use, since this inactivates the central fragment. Extraction with isopropanol removes the short cohesive fragments that would otherwise participate in the ligation reaction.

Since the full Spi$^-$ phenotype is characteristic of phage that are Red$^-$ and Gam$^-$, alternative selection systems may be devised. For example, a Spi$^-$ recombinant results from insertional inactivation of the *red* gene of a *gam*^{am} Chi vector (Yankofsky et al. 1978). Similarly, recombinants derived from a *red*$^-$*gam*^{am} Chi replacement vector, in which the central fragment carries a suppressor, have a conditional Spi$^-$ phenotype (N.E. Murray, unpubl.). In both cases, the stringency of selection is impaired by the weak Spi$^-$ phenotype of the vector; nevertheless, these systems provide strong enrichment for conditional Spi$^-$ recombinants which, once isolated, may be propagated as Gam$^+$ phage in permissive Rec$^+$ or Rec$^-$ hosts.

It should be stressed that Spi$^-$ λ phages have a low burst size and form particularly small plaques on a P2 lysogen. Each Chi mutation increases the plaque size of these phages similarly. The medium used for the recovery of Spi$^-$ phages can affect the plaque size markedly; bacteria grow slowly on poor media, and the plaques are consequently larger. An *sbcA*$^-$*recBC*$^-$ host alleviates the replication problems associated with Spi$^-$ phages, thereby increasing plaque size (M. Stahl, pers. comm.), although not if the vector includes the mutation *Aam*32 present in Charon 4A and EMBL3A (A.-M. Frischauf et al., in prep.); whereas a *recA*$^-$ host carying a *gam*$^+$ plasmid permits propagation of *gam*$^-$ phages even in the absence of recombination pathways (G. Crouse, pers. comm.). The handling of *red*$^-$*gam*$^-$ phages is discussed by Arber et al. (this volume).

BIOLOGICALLY CONTAINED λ HOST SYSTEMS

Concern about potential hazards of novel recombinant molecules endowed with the ability to replicate in *E. coli* led to guidelines for the physical and biological containment of such molecules. Host-vector systems were tested before accreditation for a normal (EK1), or higher (EK2), level of biological containment (see, e.g., Blattner et al. 1977; Leder et al. 1977). Currently, the requirements for such strains are generally more relaxed, but they remain dependent on the relevant national advisory board.

STRATEGIES FOR CLONING IN PHAGE λ

The full impact of the recombinant DNA technology required the detection of single-copy DNA from a complex genome as one of a population (library) of recombinant molecules derived from fragmented but unfractionated donor DNA. The realization of this goal (e.g., Maniatis et al. 1978) depended on the combination of a few simple technical advances.

Libraries of Recombinants

In principle, a random population of DNA fragments, prepared either by nuclease digestion or mechanical shearing, can be recovered in a phage or plasmid vector. If sufficient recombinants can be recovered and screened, then, even for a complex eukaryote, any single gene sequence can be isolated, amplified, and purified. Given the mean length of the DNA fragments cloned and the size of the haploid genome, the number of independent recombinants needed to ensure that a single-copy sequence is present in the library with an arbitrary probability can be calculated (Clarke and Carbon 1976). Obviously, the size of the library required decreases as the size of the cloned DNA fragments increases. Fragments of 20–25 kb in length can be cloned in a λ vector. These are much smaller than the largest fragments (>40 kb) that have been cloned in plasmid vectors, but in practice, large fragments are recovered only inefficiently in plasmid vectors (see, e.g., Hohn and Murray 1977). In contrast, the recovery of recombinants via in vitro encapsidation of λ can reflect the distribution of fragment sizes of the donor DNA (Maniatis et al. 1978).

The construction of truly representative libraries of any genome requires randomly fragmented donor DNA, but initially it was not easy to recover such fragments efficiently in a λ vector. When plasmid vectors are used, the donor DNA can be sheared and provided with 3′-homopolymeric extensions before annealing to cleaved vector molecules having complementary 3′-homopolymeric extensions (Wensink et al. 1975). λ vectors are not ideal for this approach, since it is difficult to prepare long, linear DNA molecules free of nicks at which 3′-homopolymeric sequences would be added. Alternative ways of recovering randomly fragmented DNA in λ vectors are now available (see below), but even in their absence, donor DNA digested with *Eco*RI, *Hin*dIII, or *Bam*HI was often an effective source of a particular DNA sequence (e.g., Blattner et al. 1978), providing the DNA sequence could be identified by a labeled probe. The now classical technique of hybridization to fractionated digests of DNA (Southern 1975) indicates the sizes of the fragments to be cloned, thus permitting a judicious choice of enzymes and vectors. Any one restriction enzyme generates a population of donor DNA fragments of heterogeneous size, but the vector and conditions of encapsidation together impose a size fractionation and consequent enrichment for the recombinant. With donor fragments of heterogeneous sizes, sequences that are nonadjacent in the donor genome are occasionally cloned together, but this can be prevented by treatment with phosphatase before ligation to the vector (Ullrich et al. 1977).

The cloning of random DNA fragments in a λ vector was achieved by Maniatis et al. (1978) through the use of synthetic linkers bearing targets for *Eco*RI (Bahl et al. 1977; Scheller et al. 1977) as an alternative to homopolymeric extensions. The linkers were joined to blunt-ended donor DNA fragments by the action of T4 DNA ligase (Sgaramella 1972). The strategy used by Maniatis et al. (1978) is outlined in Figure 6. Donor DNA of high molecular weight was fragmented, either by shearing followed by treatment with a single-strand specific nuclease or by partial digestion with a combination of two restriction endonucleases, each of which recognized only a tetranucleotide sequence and generated molecules with blunt ends. The enzymes *Hae*III (recognition sequence GGCC) and *Alu*I (AGCT) were used with the aim of generating a near random collection of DNA fragments. Fragments of approximately 20 kb in length were isolated after centrifugation through a sucrose gradient. The *Eco*RI targets within these DNA fragments were modified with *Eco*RI methylase (Greene et al. 1975) before the covalent attachment of synthetic *Eco*RI linkers and generation of cohesive ends by digestion with *Eco*RI endonuclease. The resulting fragments were then joined to the purified arms of an *Eco*RI vector using a previously established optimal ratio of donor-to-vector DNA. Recombinant genomes were then recovered by in vitro encapsidation. This carefully executed series of reactions effectively produced libraries of recombinants.

More recent advances enable the less biochemically experienced to construct equally representative libraries. The simpler approach depends on the use of vectors that accommodate large fragments of DNA having the same cohesive ends (GATC) as those generated by *Bam*HI (Karn et al. 1980; Loenen and Brammar 1980). Of particular relevance are the enzymes *Mbo*I and *Sau*3a, which specifically recognize this tetranucleotide sequence. A procedure for the use of such *Bam*HI vectors is shown in Figure 7. Partial digestion of genomic DNA by *Sau*3a is an effective way of generating nearly random populations of high-molecular-weight DNA fragments. The tetranucleotide GATC should occur on average once every 256 bp in DNA of random sequence with 50% G + C, and hence only 1 in 80 of these sites should be cut to produce DNA fragments of approximately 20 kb in length. The composition of the recognition

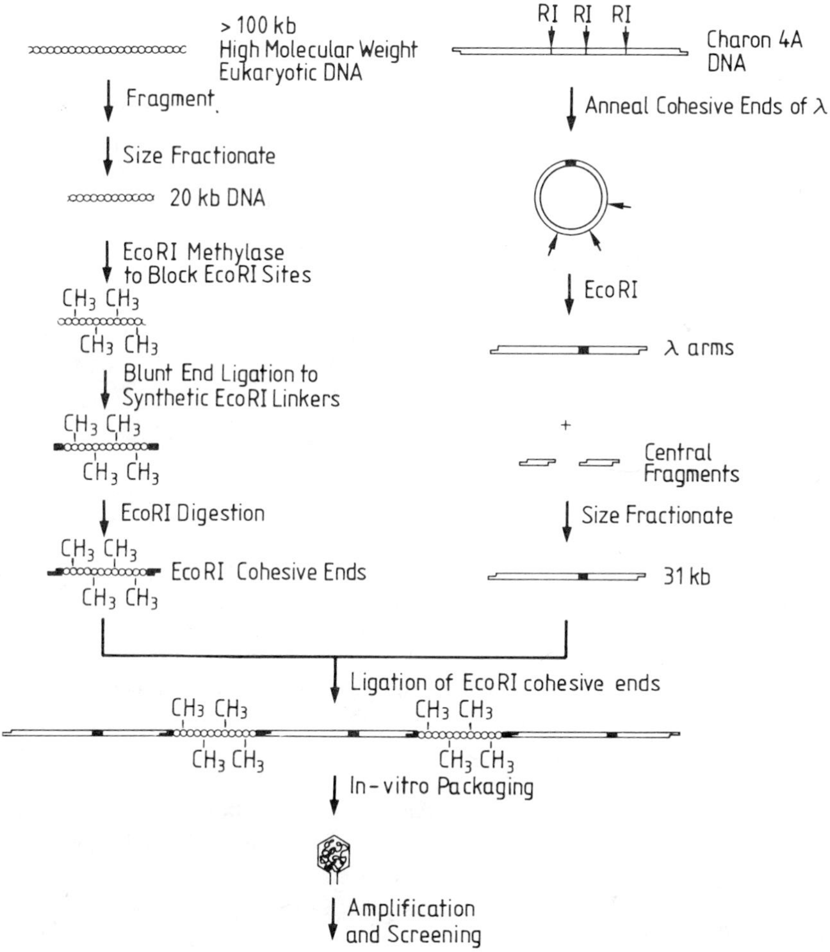

Figure 6 Schematic diagram illustrating a strategy used to construct libraries of random eukaryotic DNA fragments. (Derived from Maniatis et al. 1978.)

sequence is such that deviations in overall base composition have minimal effect on the distribution and frequency of targets. The advantage of making libraries using *Sau*3a and *Mbo*I is both the technical ease and the high efficiency of ligation using cohesive rather than blunt ends.

Recombinants Obtained with Limited Amounts of Donor DNA and cDNA

Detection of a particular DNA sequence among the plaques of a genomic library is usually dependent on a suitable probe with which to screen by hybridization, but for the vast majority of genes no such probe is available. For a gene whose

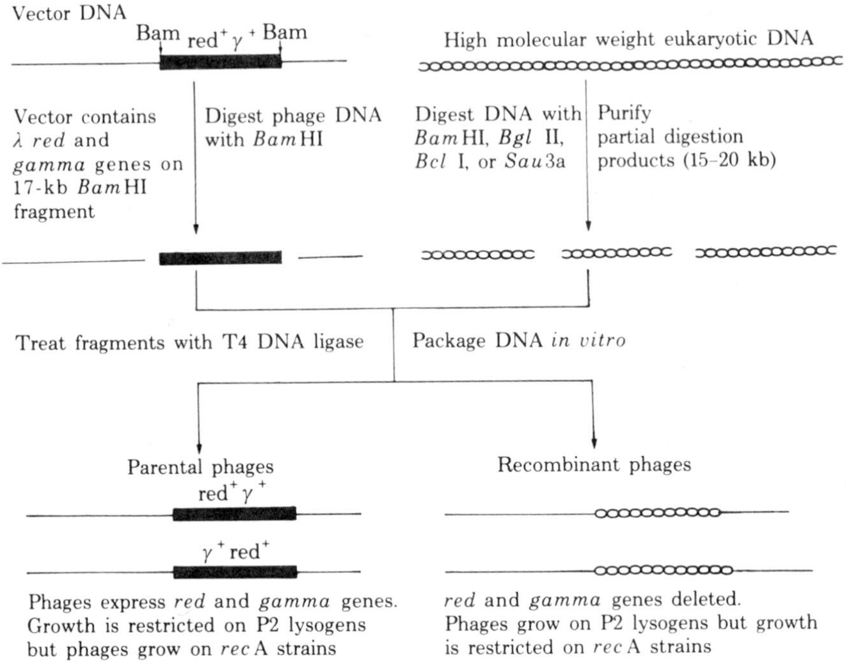

Figure 7 Scheme for the use of *Bam*HI vectors for constructing a genomic library. (Reprinted, with permission, from Karn et al. 1980.)

message or polypeptide product can be recognized, alternative screens can sometimes be developed. In special cases, genes of interest may be localized within a cytologically defined region of a chromosome, and the polytene chromosomes of *Drosophila* provide many such examples. Microdissection of *Drosophila* polytene chromosomes has been used to obtain DNA from defined regions as small as 200 kb (Scalenghe et al. 1981). Although only picogram quantities of DNA can be obtained, a combination of micromanipulative methods for the biochemical reactions and in vitro encapsidation of the DNA gave recombinant phages that could be identified with the correct polytene band by in situ hybridization. These methods increase the accessibility of cytologically defined genes in polytene chromosomes.

For their experiments, Scalenghe et al. (1981) chose a λ vector that would accept fragments of all sizes up to 11–12 kb and that would facilitate the recognition of recombinants as clear-plaque derivatives (Murray et al. 1977). Efficient ligation was obtained by keeping the reactions to nanoliter volumes so as to increase the reactant concentrations.

The microscale reactions described by Scalenghe et al. (1981) are clearly applicable whenever DNA for cloning is limiting, as it is from some cultured cells and viruses that are obtainable only in minute amounts.

Again, taking advantage of methods of efficient recovery and selection of recombinants, cDNA libraries of transcripts from *Drosophila* embryos have been isolated (Scherer et al. 1981). cDNA was made from poly(A)$^+$ RNA, the 3′ ends were made flush by the use of the Klenow fragment of DNA polymerase I, and synthetic *Eco*RI linkers attached covalently by T4 DNA ligase.

ANALYSIS OF RECOMBINANTS

Physical Analysis of DNA

DNA, free of RNA and host DNA, is readily prepared from phage λ (see Arber et al., this volume). A yield of 50–100 μg of DNA from a 100-ml culture should be anticipated, but more than half of this will be from the vector moiety of the DNA. Because of the complexity of the vector, detailed restriction mapping of the cloned sequence using enzymes that recognize a tetranucleotide sequence requires the subcloning of fragments in a small, well-characterized plasmid vector.

The linear duplex molecules of phage λ are excellent substrates for heteroduplex analysis (Davidson and Szybalski 1971; Davis et al. 1971). Deletion and substitution derivatives of phage λ provide excellent internal reference points. A pronounced AT-rich region within the right arm of the chromosome provides a striking marker for partial denaturation mapping (Schnös and Inman 1970).

The two strands of λ DNA are readily separated by equilibrium centrifugation in cesium chloride in the presence of poly[r(U,G)] (Hradecna and Szybalski 1967), because of the asymmetric distribution of poly[r(U,G)]-binding sites. Since the polarity of each strand has been defined, hybridization of mRNA to the separated strands of a recombinant genome simultaneously identifies both the coding strand and the polarity of a cloned gene.

Genetic Analyses

General

The ease with which phage may be tested on genetically different strains and the ability to reclone, in the classical sense, by plaque purification, expedite the characterization of genes and their mutant derivatives. Genes cloned in λ may be studied with the phage in either the lytic or the lysogenic mode. In the latter case, a single copy of a cloned gene per bacterial chromosome can be maintained, thereby minimizing some selective pressures. Occasionally, the cell can tolerate one copy of a cloned gene but not the many copies associated with a multicopy plasmid, and some genes whose products are deleterious to the cell can be propagated in the lytic mode only where infection is destined to be a terminal event (Mileham et al. 1980). Such genes usually have minimal effect on phage growth if they are dependent on the phage late promoter for their expression or if they are inserted within the immunity region oriented so that their transcription is opposed by that of the *c*I gene.

Recombination

Cloned segments of DNA within λ can be subjected to rigorous analysis by genetic recombination. Such analysis is most incisive when the functional expression of the cloned genes supplies a basis for the selection of recombinants. Alternatively, markers flanking a cloned sequence can be used to select exchanges in the interval between them. An example is the identification of a hotspot for genetic recombination within the genome of phage P1, following the transfer of fragments of P1 DNA to genetically marked, recombination-deficient λ vectors (Sternberg and Hamilton 1981). The most stringent selection results if the vectors are devoid of homology within the selected interval. This principle was used to orient fragments of the *trp* operon of *E. coli* (Hopkins et al. 1976), and a pair of λ cloning vectors that fulfill this requirement (Carroll et al. 1980) have been used to recover the entire coding sequence of the mouse d*hfr* gene free of introns (G.F. Crouse et al., pers. comm.).

The analysis of phage that include *E. coli* genes brings unique advantages and problems, since homologous sequences are present in the *E. coli* chromosome. Homology provides a substrate for recombination and can permit the integration of the phage into the *E. coli* chromosome or the transfer of genetic markers in either direction between phage and bacterial chromosomes. Integration via homology has useful genetic consequences. If the lysogen is made in an Hfr strain, previously unmapped genes can be positioned on the genetic map, since the site of integration can easily be located by zygotic induction (Wollman and Jacob 1954; Shimada et al. 1972). Induction of a prophage, irrespective of its location, may be followed by excision, and occasional errors in these events sequester additional bacterial DNA sequences within the phage genome. Aberrant excision can result in concomitant gain of bacterial DNA and loss of phage DNA. Phages with longer genomes may be detected by their increased buoyant density (Bellett et al. 1971) or by their greatly enhanced ability to form plaques on particular hosts (Emmons et al. 1975; Elliot and Arber 1978). If the original vector was *gam*$^+$, some aberrant excision events produce *red*$^-$*gam*$^-$ phages. These Spi$^-$ phages, selected on a P2 lysogen, have been used to extend a cloned chromosomal sequence and to orient it with respect to neighboring bacterial genes (Murray and Kelley 1979). Guest and Stephens (1980) isolated λ*nadC* and λ*lpd* phages from libraries of *gam*$^+$-transducing phages and made "Spi$^-$ extension derivatives" to clone and characterize the intervening interval of the *E. coli* chromosome (Fig. 8).

The high frequency of genetic transfer between a transducing phage and the bacterial chromosome has important technical consequences. It may often be prudent to propagate a mutant sequence in a host carrying either the same mutation or an extensive deletion of that region. Alternatively, a *recA*$^-$ host may be used for a *red*$^-$ phage. Deletions in *E. coli* DNA cloned in λ are readily isolated (see below) and transferred to the host chromosome, provided that sufficient homology remains on either side of the deletion. Even in the absence of selective

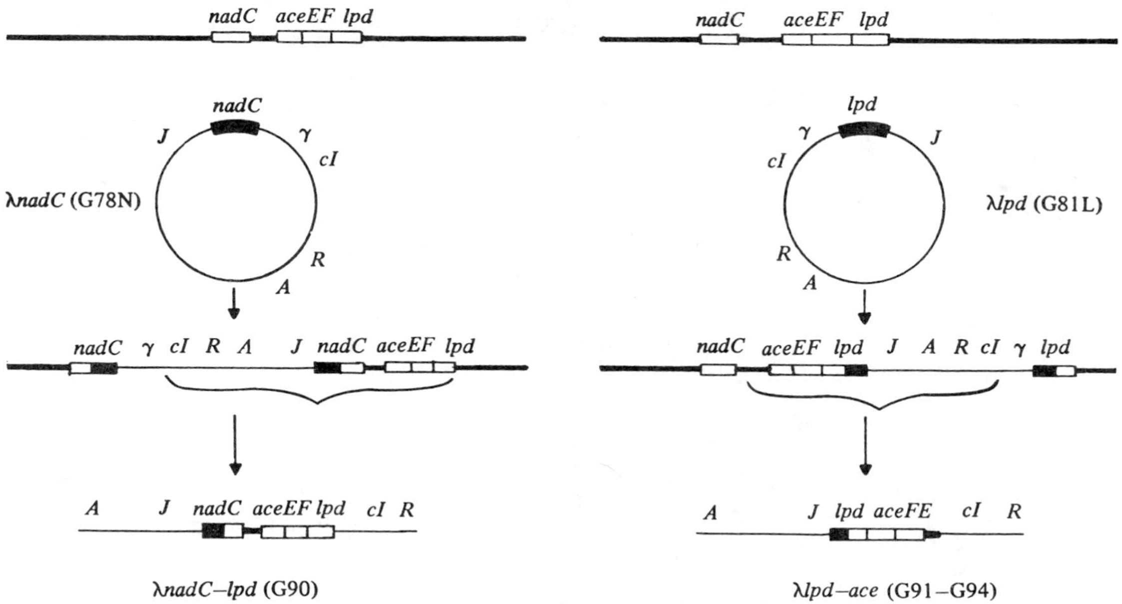

Figure 8 Diagrammatic representation of the extension of the transducing ranges of λ*nadC* and λ*lpd* phages by integration in the region of bacterial DNA homology and aberrant excision of the transducing phages. The orientation of the bacterial DNA within the phage dictates the orientation of the transducing phage. The detection of transducing phages including neighboring genes was facilitated by selecting for λ Spi^- derivatives. (Reprinted, with permission, from Guest and Stephens 1980.)

pressure, a nonlethal mutation may be transferred from phage to bacterium, or host to phage, following *rec*-dependent integration of the transducing phage. Induction of a lysogenic bacterium, heterozygous for a mutation within the duplicated region of homology, gives a lysate in which the proportion of phage that carry the mutation is determined by the relative lengths of flanking homology. If the mutation is in the center of the homologous region, 50% of the excised phage will be mutant and, likewise, 50% of the cured bacteria. This ready transfer of mutations can be applied to the construction of strains. Relevant examples include making restriction-deficient (*hsdR*⁻) and recombination-deficient (*recA*⁻) hosts.

Generation and Isolation of Mutants

Phage λ itself is a substrate for chemical mutagenesis in vitro, and growth of phages in mutator strains *mutD* or *mutT* (Cox 1976) effectively generates both transitions and transversions in vivo (Enquist and Weisberg 1977; see Arber et al., this volume).

Deletions may be made in vitro simply by the excision of the DNA between two restriction targets (Murray and Murray 1974). In most cases, this approach is not ideally suited to phage recombinants, since the long arms of the vector are apt to contain targets for the restriction enzyme. An alternative use of endonucleases for the isolation of deletion mutants relies on the fact that *E. coli* can join nonhomologous DNA sequences. The two fragments of a phage genome cut at a single restriction target can cohere by the long complementary sequences of their natural ends (*cos*), which are then covalently joined by the host ligase. Circularization by illegitimate recombination introduces deletions of variable size flanking the site previously defined by a restriction target. This type of mechanism has been demonstrated for *E. coli* (Murray and Murray 1974) but was used most effectively for SV40 in mammalian cells (Lai and Nathans 1974; Carbon et al. 1975).

Naturally occurring deletion mutants of phage λ and its derivatives are easily selected on plates containing chelating agents. This depends on the finding that deletion derivatives are less sensitive to chelating agents than is wild-type λ (Parkinson and Huskey 1971; see Arber et al., this volume). Deletions that extend from the cloned fragment into the *gam* gene of the phage can be isolated by selection for deletion mutants of Spi⁻ phenotype, and such deletions can orient the incorporated genes with respect to the vector chromosome. Should the vector include the phage attachment site and a functional *int* gene, then a major class of deletions results from intramolecular, *int*-mediated excision events between *att* and sequences that share homology with the core sequence of a normal attachment site (Davis and Parkinson 1971; see Weisberg and Landy, this volume). Deletion derivatives of recombinants isolated in a *Dam* vector are isolated among those phage forming plaques on a suppressor-free host (Sternberg et al. 1979; see Arber et al., this volume).

Expression

During lytic infection, the major λ promoters (p_L, p_R and p_R') are generally responsible for most transcription, including that of genes incorporated in the central region of the phage. In the prophage state, the major λ promoters are repressed, and failure to express the incorporated genes suggests that either the cloned genes have been severed from their promoter or their promoter is ineffective in the *E. coli* host used. Conversely, expression in the prophage state, irrespective of the orientation of the cloned fragment within the phage, implies transcription from an included promoter or promoterlike sequence, although the possibility of transcription from neighboring bacterial phage or even novel fusion sequences must always be borne in mind.

Cloned *E. coli* promoter sequences can be analyzed by their interaction with host control proteins and RNA polymerase. Operon fusions, in which a promoter is fused to the *lacZ* structural gene, facilitate refined genetic approaches (e.g., Ptashne et al. 1980). A number of in vivo and in vitro techniques exist for the construction of such fusions (Casadaban 1976; Windass and Brammar 1979; Holowchuck et al. 1980), and their genetic exploitation has been outlined recently by Beckwith (1981).

Gene as well as operon fusions can be used to study expression. Since the amino terminus of β-galactosidase is not essential for enzymatic activity (Muller-Hill and Kania 1974), fusions that replace the 5′-coding segment of the *lacZ* gene with that of another gene may produce a fusion polypeptide that retains β-galactosidase activity (e.g., Muller-Hill and Kania 1974; Bassford et al. 1979). Both transcription and translation of this fusion sequence can be regulated by the DNA sequence substituted for the control region of the *lacZ* gene. The biochemical resectioning necessary to make such fusions is generally easier to execute in plasmid recombinants (see Casadaban et al. 1980), but following cloning in a λ*lacZ* phage, the in-vivo-generated deletions so readily isolated in λ can be a useful alternative to in vitro approaches. Furthermore, λ*lacZ* vectors permitting fusions in any reading frame are available (Charnay et al. 1978). Fusion polypeptides should help to correlate functional roles with particular domains of a polypeptide (e.g., Kania and Muller-Hill 1977; Bassford et al. 1979).

Detection of Polypeptides and Determination of the Direction of Transcription

The polypeptides encoded by a cloned fragment of DNA are readily detected following infection of either UV-irradiated cells (Jaskunas et al. 1975) or minicells (Adler et al. 1966; Reeve 1979). In the former case, a dose of irradiation is chosen that eliminates most of the transcriptional potential of the *E. coli* chromosome, whereas minicells are devoid of a bacterial chromosome. Transcription from phage promoters can be avoided by infecting a λ lysogen, but this has the secondary consequence that phage DNA replication, and hence gene amplification, is prevented. Under these conditions, the polypeptides from cloned RNA polymerase genes (*rpoBC*) were easily detected, whereas that of the cloned DNA polymerase I (*polA*) gene, despite its large size, was not (Newman et al. 1979). This shows that a weak promoter may not be detected.

On infection of a UV-irradiated nonlysogen, most transcription is from the powerful λ promoters. Orientation-dependent timing of expression, driven by the λ promoters, can be used to identify the products of early transcription leftward (from p_L) or products of late transcription rightward (from p_R') (e.g., Wilson and Murray 1979). Alternatively, transcription from the late promoter p_R' can be distinguished from that of the early promoter p_L, because it is more susceptible to high levels of irradiation (Hendrix 1971).

Amplification of Gene Products

The expression of a prokaryotic gene, cloned in the central region of λ (between genes *J* and *N*; see Fig. 2) and transcribed from either its own or a λ promoter, can often be boosted so that the gene product comprises at least a few percent of the soluble cell proteins of the host. Technically, it is convenient to propagate a lysogen and induce prophage excision and replication by the inactivation of a thermosensitive repressor.

The yield of polypeptide encoded by a gene with a good promoter may be increased by delaying cell lysis, so that DNA replication can increase the number of gene copies. This was first done by the use of a mutation in gene *S* (Muller-Hill et al. 1968) that prevents cell lysis but permits DNA replication and protein synthesis to continue for some hours (Harris et al. 1967). Moir and Brammar (1976) obtained significantly better amplification of the products of the *trp* operon of a λ*trp* phage by using a mutation in gene *Q*. Under these conditions, expression of all late functions of λ is blocked; hence, lysis and the packaging of the replicated genomes are prevented. This approach has been used to amplify both *E. coli* ligase (Panasenko et al. 1977) and DNA polymerase I (Kelley et al. 1977). In both of these cases, it was not possible to clone the functional gene and its promoter in a multicopy plasmid vector, presumably because of the consequent overproduction of the gene products.

In the absence of an efficient promoter within the cloned sequence, it is possible to take advantage of transcription from a λ promoter. Moir and Brammar (1976) used λ*trp* phages lacking the *trp* promoter to investigate ways of optimizing gene expression initiated at p_L. This requires that moderation of transcription by the product of λ's *cro* gene be prevented. Hence, a *cro*$^-$ mutation was used, and cell lysis was inhibited by mutations in genes *Q* and *S*. Cells infected with λ*trp cro*$^-$*Q*$^-$*S*$^-$ phages, in which the *trp* genes were expressed exclusively from promoter p_L, contained up to 20% of their soluble protein as the five polypeptide products of the *trp* operon. However, these experiments relied on infection at relatively high multiplicity with phages in which the *trp* genes were in close proximity to p_L. Experiments depending on the induction of either a λ*trp* prophage (W.J. Brammar and N.E. Murray, unpubl.) or λ *polA* prophage, in which the *polA* gene is left of the attachment site, were less effective (Murray and Kelley 1979). This, at least in part, probably reflects the poor replication of phage DNA in the absence of the *cro* repressor (Folkmanis et al. 1977). The more successful use of *cro*⁻ derivatives following infection of *E. coli* is tempered by the fact that *cro*⁻ phages are difficult to propagate. A temperature-sensitive *cro* allele can be used, but it remains difficult to optimize conditions for derepressed transcription from p_L.

One simple alternative is to use p_R' under conditions in which both encapsidation of DNA and cell lysis are prevented. T4 DNA ligase can then comprise a few percent of the soluble cell protein following the induction of a prophage, in which the T4 DNA ligase gene is transcribed from the *Q*-dependent late promoter (Murray et al. 1979).

Other derivatives of λ are available in which genes may be cloned downstream of an efficient, cap-insensitive (UV5) *lac* promoter (Pourcel et al. 1979). For some λ*lac* expression vectors, the cloned gene may be inserted close to the start of the *lacZ* gene, so that a fusion product, dependent on both the transcription and translation signals of *lac*, results (Charnay et al. 1978, 1979). These systems have been used for the expression of eukaryotic genes (Charnay et al. 1980).

λ IN CONJUNCTION WITH PLASMIDS

λ Promoters in Plasmids

Transcription from the λ promoters p_L and p_R is efficient and can be controlled by the products of the λ genes *c*I and *cro*. As already discussed, it is hard to take full advantage of p_L in phage λ, since preventing the moderation of early transcription reduces DNA replication, but p_L in λ within a multicopy plasmid vector is an attractive alternative (Bernhard et al. 1979; Remaut et al. 1981). A gene whose product is to be amplified may be cloned downstream of the p_L promoter, transcription from which is prevented by the temperature-sensitive repressor encoded by a defective *cro*⁻ prophage. Following derepression, these hybrid plasmids can yield very high levels of the gene product (Remaut et al. 1981). Optimal amplification may, however, require termination signals beyond the cloned gene in order that high levels of transcription do not continue into vector sequences, thereby interfering with or enhancing transcription of the vector (Remaut et al. 1981).

In those cases where expression of the cloned gene in a multicopy plasmid is deleterious to the host, the normal promoter sequence can be replaced by a tightly controlled λ promoter. Such recombinant plasmids are readily recovered and propagated in a λ immune cell (M. Zabeau, pers. comm.).

Cosmids

A short region including the cohered ends, *cos*, of λ (see Feiss and Becker, this volume) is necessary for recognition by the λ-specific packaging proteins. The

DNA of special plasmids, termed cosmids (Collins and Hohn 1978), which include the *cos* site of λ, can be efficiently packaged into λ capsids providing the *cos* sequences are located in direct tandem arrangement, 37–52 kb apart. If, therefore, the cosmid vector is only 10 kb in length, encapsidation selects for recombinants, including donor DNA fragments of 30–40 kb.

Representative gene libraries in cosmids are consequently smaller than with λ vectors, but their screening by in situ DNA hybridization is less sensitive, in part, because of a decrease in copy number.

Phasmids

A third combination of plasmid and λ sequences has been devised to take advantage of attractive features of both phage and plasmid vector systems. A phasmid is a λ derivative containing one copy of a multicopy plasmid vector (Brenner et al. 1982). The plasmid component includes a λ attachment site that facilitates integration of the plasmid into the phage vector and also its release from the phage. Sequences cloned within this special plasmid may be transferred to any λ-sensitive host as part of an encapsidated phage genome and subjected to genetic analyses by conventional phage techniques. Alternatively, the plasmid component may be released, and DNA free of the long λ arms may be prepared from the plasmid.

USE AND CHOICE OF λ VECTORS

The vectors illustrated in Figures 5 and 9 preferentially include new ones, otherwise vectors were chosen either as representative examples or because they have been used extensively.

Figure 9 Some widely used and some new vectors (but see also Fig. 5). Vector DNA is shown by a single line; non-λ vector sequences, by a heavy line with the origin indicated above. Gaps represent deletions with the name of the deletion given below. Sequences that are replaced by donor DNA are shown as double lines, that for EMBL3 and EMBL4 as hatched areas (since part is λ and part is *E. coli*) and interrupted since it is not drawn to scale. The only targets shown are those that may be used for cloning. The cloning capacities indicated are based on a minimum size limit of 38 kb and a maximum of 51 kb (Williams and Blattner 1980). (*) The central fragment of EMBL3 and EMBL4 is as shown for EMBL1 in Fig. 5. For EMBL3, the *Sal*I targets are peripheral, whereas in EMBL4 the linkers are reversed and the *Eco*RI targets are outermost. In λgt-λB, λB is the fragment of λ DNA between *Eco*RI targets s*r*Iλ1 and s*r*Iλ2. The λB fragment of λgt-λB may be replaced by a genetically marked fragment, e.g., *supE* in NM781, *lac* in NM1039, and T5 gene *A3* in λgtWES, T5-622. The ChiC *gam*^{am} derivative of NM762 (NM1173) allows good enrichment for recombinants. NM1149 is a single alternative to the *Eco*RI (NM607) and *Hin*dIII (NM590) vectors, as is its *red* derivative NM1150 and the *red*$^-$*gam*$^-$ vector Charon 7. All of the vectors mentioned are listed in Appendix III.

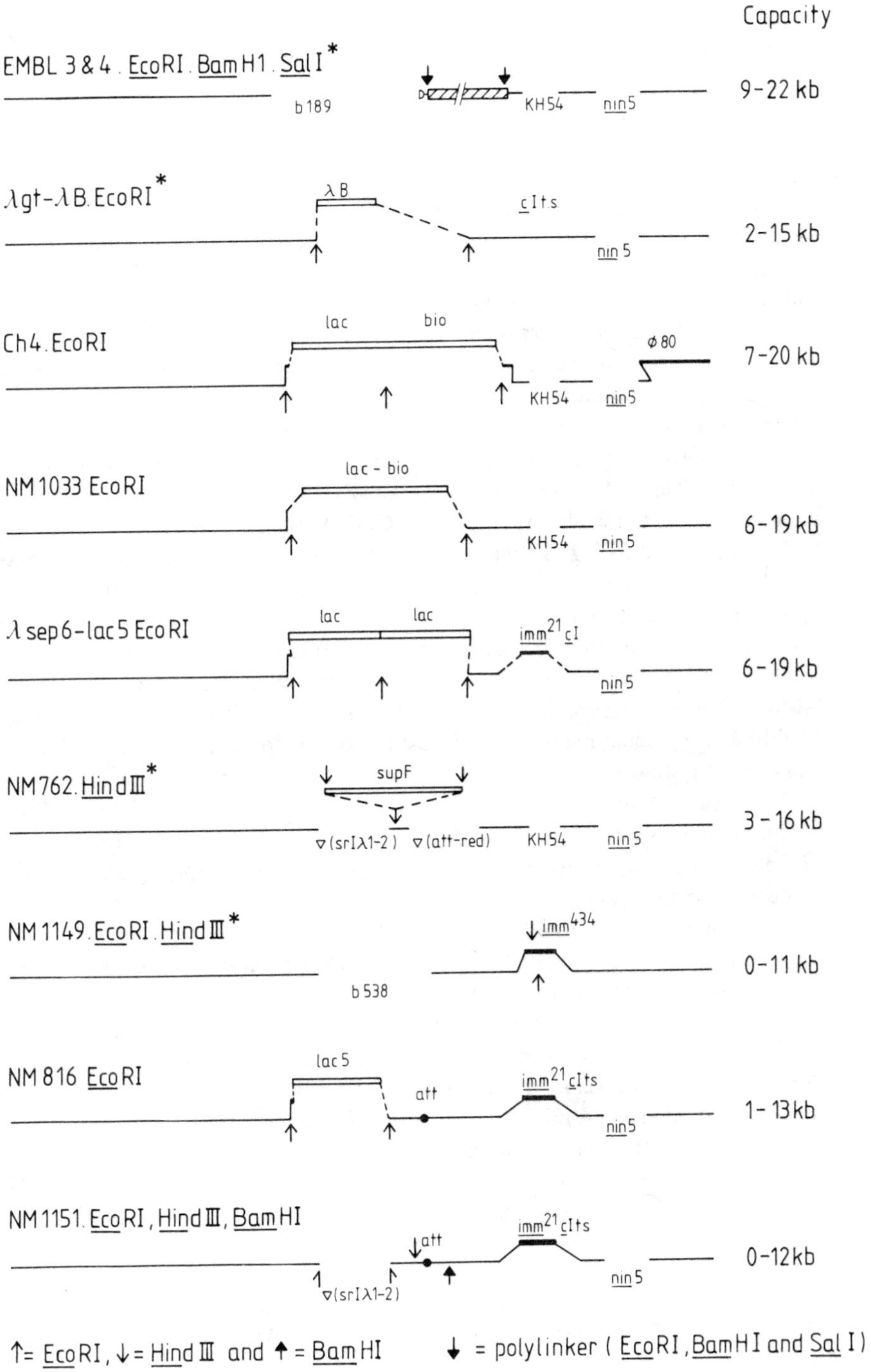
Capacity
EMBL 3 & 4 . EcoRI . BamH1 . SalI *
b189
KH54
nin5
9–22 kb
λgt–λB. EcoRI *
λB
cIts
nin 5
2–15 kb
Ch4. EcoRI
lac
bio
φ80
KH54
nin5
7–20 kb
NM1033 EcoRI
lac – bio
KH54
nin 5
6–19 kb
λ sep6–lac5 EcoRI
lac
lac
imm21 cI
nin 5
6–19 kb
NM762. HindIII *
supF
▽(srIλ1-2)
▽(att-red)
KH54
nin 5
3 – 16 kb
NM1149. EcoRI . HindIII *
imm434
b538
0–11 kb
NM 816 EcoRI
lac5
att
imm21 cIts
nin5
1– 13 kb
NM1151. EcoRI , HindIII, BamHI
att
imm21 cIts
▽(srIλ1-2)
nin5
0–12 kb
↑= EcoRI, ↓= HindIII and ⬆ = BamHI
⬇ = polylinker (EcoRI, BamHI and SalI)

Figure 9 (See facing page for legend.)

λ vectors are commonly used for making genomic libraries, since large DNA fragments are recovered efficiently, and plaques are a convenient and sensitive substrate for screening by hybridization. The simplest way to make representative libraries is to clone in a *Bam*HI vector DNA that has been partially digested with *Sau*3a. For a vector that generates Spi^- recombinants, either the restituted vector genomes should be eliminated by genetic selection on a P2 lysogen, or a *recBC*$^-$ strain should be used to minimize selection against Spi^- phages. The *Bam*HI targets in vectors EMBL3 and EMBL4 are within a polylinker sequence (Fig. 9, line 1); therefore, on digestion with *Bam*HI endonuclease, both the central fragment and the vector arms retain useful restriction targets. The targets at the ends of the central fragment can provide a means for its biochemical inactivation, whereas those in the arms serve to separate the vector moiety from *Sau*3a donor fragments cloned into the *Bam*HI targets.

An alternative way of cloning random fragments is to shear the donor DNA and attach linkers (Maniatis et al. 1978). *Eco*RI linkers are generally used, since the methylase is available to modify any *Eco*RI targets in the fragmented donor DNA. Vectors L47 (Fig. 5, line 3) or EMBL4 (Fig. 9, line 1) provide the maximum space, but the N^+gam^- recombinant derivatives of EMBL4 are the more vigorous. Alternatively, some gam^+ vectors, NM1033 (Fig. 9, line 4) and λ*sep6-lac5* (Fig. 9, line 5) (Meyerowitz and Hogness 1982), have space for DNA fragments of almost 20 kb, and in contrast to their predecessors (e.g., NM631, NM647), recombinant derivatives of these red^-gam^+ vectors grow as readily as the vector itself. For DNA digested by *Hin*dIII, L47 has space for fragments of almost 20 kb in length and permits the stringent selection of Spi^- (but N^-) recombinants. The more vigorous Gam^+N^+ alternatives, NM762 (Fig. 9, line 6) and its gam^{am} Chi derivative (NM1173), have space for fragments of 16 kb, and the latter allows enrichment for the conditional Spi^- recombinants.

Ideally, prefractionated donor fragments and the purified arms of a replacement vector are used in ligation reactions. Purified arms are particularly recommended if there is no genetic selection against vector genomes. The optimal concentrations of donor and vector DNA for ligation reactions depend on the sizes of both components and can be calculated by a simplification of the Jacobson-Stockmayer expression (see, e.g., Williams and Blattner 1980). Unless the donor fragments are small, concentrations of the vector are likely to be within the range of 60–300 μg/ml. Such concentrations also favor concatemer formation, the preferred substrate for encapsidation.

In all cases where libraries of recombinants are to be screened by hybridization with a sequence cloned in another vector as probe, it is essential to establish that there is no homology between the two vector systems. Obviously λ sequences are common to λ and cosmids, whereas λ, M13 (Messing et al. 1981), and plasmid vectors frequently include short sequences from the *lac* operon. DNAs from recombinant λ phage have been used as probes for screening a cosmid library following treatment of the phage DNA with BAL-31 nuclease to remove the *cos* region (K. Kaiser, pers. comm.).

Insertion vectors, accompanied by an efficient screen or means of selection, have proved useful in a variety of circumstances, most particularly when donor DNA or cDNA is limiting (Scalenghe et al. 1981; Scherer et al. 1981). The vector most commonly used has been the *red*⁻ immunity vector, NM641 (Murray et al. 1977), although for many purposes the higher titers provided by a *red*⁺ phage might be advantageous. One such *red*⁺ immunity vector (NM1149; Fig. 9, line 7) can be used for *Eco*RI, *Hin*dIII, or a combination of both, and a derivative with a deletion in the *redA* gene is also available (NM1150). The vector NM1151 (Fig. 9, line 9) is a *red*⁺ insertion vector for *Bam*HI fragments; recombinants are recognized by insertional inactivation of the *int* gene (Enquist and Weisberg 1976; Klein and Murray 1979).

It is difficult to generalize about the choice of λ vector for the cloning of a prokaryotic gene, where recognition of the recombinant usually relies on expression of the included gene. The selective system to be used may impose some features on the vector, but a phage able to establish repression is usually advantageous. Such phages, however, yield high-titer lysates less readily than *c*I⁻ phages do, and unless the genetic selection relies on compensation for a temperature-sensitive lethal mutation of the host, a phage coding for a heat-sensitive repressor is an obvious compromise.

Some experiments require a phage capable of integrating into the *E. coli* chromosome. Integration-proficient insertion vectors are available for *Hin*dIII and *Eco*RI fragments (see, e.g., Fig. 9, line 9), and there is an integration-proficient replacement vector for *Eco*RI fragments (see Fig. 9, line 8). This vector, though useful, is imperfect in that simple deletion of the central fragment leaves a phage capable of forming minute plaques (Wilson and Murray 1979). Extra capacity is achieved in both of these vectors by substituting imm^{21} for imm^{λ}. It may be useful to remember that recombinant derivatives of some Spi⁻ *Bam*HI vectors retain a functional attachment site. Loss of the *Bam*HI fragment between s*bam*λ3 (in *int*) and s*bam*λ4 (left of *N*) inactivates *int* but not the phage attachment site. L47 (Fig. 5, line 3) has a normal attachment site, and if the *int* product is provided in *trans*, derivatives of this phage can integrate at the chromosomal attachment site. A variant of L47, in which $imm^{434}cI^-$ is replaced by the immunity region of λ carrying a *c*I gene coding for a temperature-sensitive repressor (D. Bramhill and D.W. Burt, pers. comm.), allows the establishment of lysogeny.

The most commonly used λ vectors are for DNA fragmented by *Eco*RI, *Hin*dIII, or *Bam*HI. Convenient maps of λ vectors have been compiled by Williams and Blattner (1980). These investigators also tabulate the restriction targets of λ vectors (see also Appendix III) and their cloning capacities for DNA fragmented by *Eco*RI, *Hin*dIII, *Bam*HI, *Sal*I, *Xho*I, *Xba*I, *Sst*I, and combinations of these enzymes. These data are a useful guide to the appropriate vector.

ACKNOWLEDGMENTS

I am indebted to W.J. Brammar, D. Bramhill, G.F. Crouse, and H. Lehrach for providing unpublished information; to G. Cesareni, D.J. Finnegan, J.R. Guest,

K. Kaiser, H. Lehrach, R. Mellado, K. Murray, J.S. Parkinson, V. Pirrotta, D.F. Ward, and M. Zabeau for many constructive suggestions; to Helen Senior who did many of the unpublished experiments cited in this review; and to Wendy Moses for much patient typing. This article was written while I was on leave at the European Molecular Biology Laboratory (Heidelberg, Germany), the support of which is gratefully acknowledged.

REFERENCES

Adhya, S., M. Gottesman, and B. de Crombrugghe. 1974. Release of polarity in *Escherichia coli* by gene N of phage λ: Termination and antitermination of transcription. *Proc. Natl. Acad. Sci.* **71:** 2534.

Adler, H.I., W.D. Fisher, A. Cohen, and A.A. Hardigree. 1966. Miniature *Escherichia coli* cells deficient in DNA. *Proc. Natl. Acad. Sci.* **57:** 321.

Arber, W. and U. Kuhnlein. 1967. Mutationeller Verlust B-Spezifischer Restriction des Bakteriophagen fd. *Pathol. Microbiol.* **30:** 946.

Auerbach, J.I. and P. Howard-Flanders. 1981. Identification of wild-type or mutant alleles of bacterial genes cloned on a bacteriophage lambda vector: Isolation of uvrC(Am) and other mutants. *J. Bacteriol.* **146:** 713.

Bahl, C.P., K.J. Marians, R. Wu, J. Stawinsky, and S. Narang. 1977. A general method for inserting specific DNA sequences into cloning vehicles. *Gene* **1:** 81.

Barbour, S.D., H. Nagaishi, A. Templin, and A.J. Clark. 1970. Biochemical and genetic studies of recombination proficiency in *E. coli*. II. Rec^+ revertants caused by indirect suppression of Rec^- mutations. *Proc. Natl. Acad. Sci.* **67:** 128.

Bassford, P., T. Silhavy, and J. Beckwith. 1979. Use of gene fusion to study secretions of maltose-binding protein into *Escherichia coli* periplasm. *J. Bacteriol.* **139:** 19.

Becker, A. and M. Gold. 1975. Isolation of the bacteriophage lambda A protein. *Proc. Natl. Acad. Sci.* **72:** 581.

Beckwith, J. 1981. A genetic approach to characterizing complex promoters in *E. coli*. *Cell* **23:** 307.

Belfort, M. and D.L. Wulff. 1971. A mutant of *Escherichia coli* that is lysogenized with high frequency. In *The bacteriophage lambda* (ed. A.D. Hershey), p. 739. Cold Spring Harbor Laboratory, Cold Spring Harbor, New York.

Bellet, A.J.D., H.G. Busse, and R.L. Baldwin. 1971. Tandem genetic duplications in a derivative of phage lambda. In *The bacteriophage lambda* (ed. A.D. Hershey), p. 501. Cold Spring Harbor Laboratory, Cold Spring Harbor, New York.

Benton, W.D. and R.W. Davis. 1977. Screening λ gt recombinant clones by hybridization to single plaques *in situ*. *Science* **196:** 180.

Bernhard, H.U., E. Remaut, M.V. Hershfield, H.K. Das, D.R. Helinski, C. Yanofsky, and N. Franklin. 1979. Construction of plasmid cloning vehicles that promote gene expression from the bacteriophage λ p_L promoter. *Gene* **5:** 59.

Bigger, C.H., K. Murray, and N.E. Murray. 1973. Recognition sequence of a restriction enzyme. *Nat. New Biol.* **244:** 7.

Blattner, F.R., A.E. Blechl, K. Denniston-Thompson, H.E. Faber, J.E. Richards, J.L. Slightom, P.W. Tucker, and O. Smithies. 1978. Cloning fetal γ-globin and mouse α-type globin DNA: Preparation and screening of shotgun collections. *Science* **202:** 1279.

Blattner, F.R., B.G. Williams, A.E. Blechl, K. Denniston-Thompson, H.E. Faber, L. Furlong, J.D. Grunwald, D.O. Kiefer, D.D. Moore, J.W. Schumm, E.L. Sheldon, and O. Smithies. 1977. Charon phages: Safer derivatives of bacteriophage lambda for DNA cloning. *Science* **196:** 161.

Borck, K., J.D. Beggs, W.J. Brammar, A.S. Hopkins, and N.E. Murray. 1976. The construction *in vitro* of transducing derivatives of phage lambda. *Mol. Gen. Genet.* **146:** 199.

Bøvre, K. and W. Szybalski. 1969. Patterns of convergent and overlapping transcription within the *b2* region of coliphage λ. *Virology* **38:** 614.

Brenner, S., G. Cesareni, and J. Karn. 1982. Phasmids; hybrids between ColEl plasmids, *E. coli*, and bacteriophage lambda. *Gene* **17:** 27.

Broome, S. and W. Gilbert. 1978. Immunological screening method to detect specific translation products. *Proc. Natl. Acad. Sci.* **75:** 2746.

Brutlag, D.L., K. Fry, T. Nelson, and P. Huey. 1977. Synthesis of hybrid bacterial plasmids containing highly repeated satellite DNA. *Cell* **10:** 509.

Burt, D.W. and W.J. Brammar. 1982a. The *cis*-specificity of the *Q*-gene product of bacteriophage lambda. *Mol. Gen. Genet.* **185:** 468.

——. 1982b. Transcriptional termination sites in the *b2* region of bacteriophage lambda that are unresponsive to antitermination. *Mol. Gen. Genet.* **185:** 462.

Cameron, J.R., S.M. Panasenko, I.R. Lehman, and R.W. Davis. 1975. *In vitro* construction of bacteriophage λ carrying segments of the *Escherichia coli* chromosome, selection of hybrids containing the gene for DNA ligase. *Proc. Natl. Acad. Sci.* **72:** 3416.

Campbell, A. 1971. Genetic structure. In *The bacteriophage lambda* (ed. A.D. Hershey), p. 13. Cold Spring Harbor Laboratory, Cold Spring Harbor, New York.

Carbon, J., J.A. Shapiro, and P. Berg. 1975. Biochemical procedure for production of small deletions in simian virus 40 DNA. *Proc. Natl. Acad. Sci.* **72:** 1392.

Carroll, D., R.S. Ajioka, and C. Georgopoulos. 1980. Bacteriophage lambda cloning vehicles for studies of genetic recombination. *Gene* **10:** 261.

Casadaban, M.J. 1976. Transposition and fusion of the lac genes to selected promoters in *E. coli* using bacteriophage λ and Mu. *J. Mol. Biol.* **104:** 541.

Casadaban, M.J., J. Chou, and S.N. Cohen. 1980. *In vitro* gene fusions that join an enzymatically active β galactosidase segment to amino-terminal fragments of exogenous proteins: *E. coli* vectors for the detection and cloning of translation initiation signals. *J. Bacteriol.* **143:** 971.

Charnay, P., M. Perricaudet, F. Galibert, and P. Tiollais. 1978. Bacteriophage lambda and plasmid vectors, allowing fusion of cloned genes in each of the three translational phases. *Nucleic Acids Res.* **5:** 4479.

Charnay, P., M. Gervais, A. Louise, F. Galibert, and P. Tiollais. 1980. Biosynthesis of hepatitis B virus surface antigens in *Escherichia coli*. *Nature* **286:** 893.

Charnay, P., A. Louise, A. Fritsch, D. Perrin, and P. Tiollais. 1979. Bacteriophage lambda-*E. coli* K12 vector host system for gene cloning and expression under lactose promoter control. II. Fragment insertion at the vicinity of the *lac* UV5 promoter. *Mol. Gen. Genet.* **170:** 171.

Chia, W., M.R.D. Scott, and P.W.J. Rigby. 1982. The construction of cosmid libraries of eukaryotic DNA using the Homer series of vectors. *Nucleic Acids Res.* **10:** 2503.

Clarke, L. and J. Carbon. 1976. A colony bank containing synthetic ColEl hybrid plasmids representative of the entire *E. coli* genome. *Cell* **9:** 91.

Collins, J. and B. Hohn. 1978. Cosmids: A type of plasmid gene-cloning vector that is packageable *in vitro* in bacteriophage λ heads. *Proc. Natl. Acad. Sci.* **75:** 4242.

Court, D. and K. Sato. 1969. Studies of novel transducing variants of lambda, dispensability of genes N and Q. *Virology* **39:** 348.

Cox, E.C. 1976. Bacterial mutator genes and the control of spontaneous mutation. *Annu. Rev. Genet.* **10:** 135.

Davidson, N. and W. Szybalski. 1971. Physical and chemical characteristics of lambda DNA. In *The bacteriophage lambda* (ed. A.D. Hershey), p. 45. Cold Spring Harbor Laboratory, Cold Spring Harbor, New York.

Davis, R.H. and J.S. Parkinson. 1971. Deletion mutants of bacteriophage lambda. III. Physical structure of att^{λ}. *J. Mol. Biol.* **56:** 403.

Davis, R.W., M. Simon, and N. Davidson. 1971. Electron microscope heteroduplex methods for mapping regions of base sequence homology in nucleic acids. *Methods Enzymol.* **21:** 413.

Davison, J., F. Brunel, and M. Merchez. 1979. A new host-vector system allowing selection for foreign DNA inserts in bacteriophage λgtWES. *Gene* **8:** 69.

Drew, R.E., P.H. Clarke, and W.J. Brammar. 1980. The construction *in vitro* of derivatives of bacteriophage lambda carrying the amidase genes of *Pseudomonas aeruginosa. Mol. Gen. Genet.* **177:** 311.

Elliot, J. and W. Arber. 1978. *E. coli* K-12 *pel* mutants, which block phage λ DNA injection, coincide with *ptsM*, which determines a component of a sugar transport system. *Mol. Gen. Genet.* **161:** 1.

Emmons, S.W., V. MacCosham, and R.L. Baldwin. 1975. Tandem genetic duplications in phage lambda. III. The frequency of duplication mutants in two derivatives of phage lambda is independent of known recombination systems. *J. Mol. Biol.* **91:** 133.

Enquist, L. and A. Skalka. 1973. Replication of lambda DNA dependent on the function of host and viral genes. *J. Mol. Biol.* **75:** 185.

Enquist, L.W. and R.A. Weisberg. 1976. The red plaque test: A rapid method for identification of excision defective variants of bacteriophage lambda. *Virology* **72:** 147.

——. 1977. A genetic analysis of the *att-int-xis* region of coliphage lambda. *J. Mol. Biol.* **111:** 97.

Feiss, M., R.A. Fisher, M.A. Crayton, and C. Egner. 1977. Packaging of the bacteriophage λ chromosome: Effect of chromosome length. *Virology* **77:** 281.

Feiss, M., R.A. Fisher, D.A. Siegele, B.P. Nichols, and J.E. Donelson. 1979. Packaging of the bacteriophage chromosome: A role for base sequence outside *cos. Virology* **92:** 56.

Folkmanis, A., W. Maltzman, P. Mellon, A. Skalka, and H. Echols. 1977. The essential role of the *cro* gene in the lytic development of bacteriophage λ. *Virology* **81:** 352.

Franklin, N.C. 1971. The N operon of lambda: Extent and regulation as observed in fusions to the tryptophan operon of *Escherichia coli.* In *The bacteriophage lambda* (ed. A.D. Hershey), p. 621. Cold Spring Harbor Laboratory, Cold Spring Harbor, New York.

——. 1974. Altered reading of genetic signals fused to the N operon of bacteriophage λ: Genetic evidence for modification of RNA polymerase by the product of the N gene. *J. Mol. Biol.* **89:** 33.

Gellert, M. 1967. Formation of covalent circles of lambda DNA by *E. coli* extracts. *Proc. Natl. Acad. Sci.* **57:** 148.

Greene, P.J., M.S. Poonian, A.L. Nussbaum, L. Tobias, D.E. Garfen, H.W. Boyer, and H.M. Goodman. 1975. Restriction and modification of a self complementary octanucleotide containing the *Eco*RI substrate. *J. Mol. Biol.* **99:** 237.

Greenstein, M. and A. Skalka. 1975. Replication of bacteriophage lambda DNA: *In vivo* studies of the interaction between the viral *gamma* and the host *recBC* DNase. *J. Mol. Biol.* **97:** 543.

Guest, J.R. and P.E. Stephens. 1980. Molecular cloning of the pyruvate dehydrogenase complex genes of *Escherichia coli. J. Gen. Microbiol.* **121:** 277.

Haggerty, D.M. and R.F. Schleif. 1976. Location in bacteriophage lambda DNA of cleavage sites of the site specific endonuclease from *Bacillus amyloliquefaciens* H. *J. Virol.* **18:** 659.

Harris, A.W., D.W. Mount, C.R. Fuerst, and L. Siminovitch. 1967. Mutations in bacteriophage lambda affecting host cell lysis. *Virology* **32:** 553.

Hedgpeth, J., H.M. Goodman, and H.W. Boyer. 1972. DNA nucleotide sequences restricted by the RI endonuclease. *Proc. Natl. Acad. Sci.* **69:** 3448.

Henderson, D. and J. Weil. 1975. Recombination deficient deletions in bacteriophage lambda and their interaction with *chi* mutations. *Genetics* **79:** 143.

Hendrix, R.W. 1971. Identification of proteins coded in phage lambda. In *The bacteriophage lambda* (ed. A.D. Hershey), p. 355. Cold Spring Harbor Laboratory, Cold Spring Harbor, New York.

Hohn, B. and T. Hohn. 1974. Activity of empty, headlike particles for packaging of DNA of bacteriophage λ *in vitro. Proc. Natl. Acad. Sci.* **71:** 2372.

Hohn, B. and K. Murray. 1977. Packaging recombinant DNA molecules into bacteriophage particles *in vitro. Proc. Natl. Acad. Sci.* **74:** 3259.

Holowchuck, E.W., J.D. Freisen, and N.P. Fiil. 1980. Bacteriophage λ vehicle for the direct cloning of *E. coli* promoter DNA sequences: Feedback regulation of the *rplJL-rpoBC* operon. *Proc. Natl. Acad. Sci.* **77:** 2124.

Hopkins, A.S., N.E. Murray, and W.J. Brammar. 1976. Characterization of λ*trp*-transducing bacteriophage made *in vitro*. *J. Mol. Biol.* **107:** 549.

Hradecna, Z. and W. Szybalski. 1967. Fractionation of the complementary strands of coliphage λ DNA based on the asymmetric distribution of the poly(I,G)-binding sites. *Virology* **32:** 633.

Jaskunas, S.R., L. Lindahl, M. Nomura, and R.R. Burgess. 1975. Identification of two copies of the gene for the elongation factor EF-Tu in *E. coli*. *Nature* **257:** 458.

Kania, J. and B. Muller-Hill. 1977. Construction, isolation and implications of repressor-galactosidase: β-galactosidase hybrid molecules. *Eur. J. Biochem.* **79:** 381.

Kaplan, D.A., L. Naumovski, and R.J. Collier. 1981. Chromogenic detection of antigen in bacteriophage plaques: A microplaque method applicable to large-scale screening. *Gene* **13:** 211.

Karn, J., S. Brenner, L. Barnett, and G. Cesareni. 1980. Novel bacteriophage λ cloning vector. *Proc. Natl. Acad. Sci.* **77:** 5172.

Kelley, W.S., K. Chalmers, and N.E. Murray. 1977. Isolation and characterization of a λ*polA* transducing phage. *Proc. Natl. Acad. Sci.* **74:** 5632.

Kemp, D.J. and A.F. Cowman. 1981. Direct immunoassay for detecting *E. coli* colonies which contain polypeptides encoded by cloned DNA segments. *Proc. Natl. Acad. Sci.* **78:** 4520.

Klein, B. and K. Murray. 1979. Phage lambda receptor chromosomes for DNA fragments made with restriction endonuclease I of *Bacillus amyloliquefaciens* H. *J. Mol. Biol.* **133:** 289.

Lai, C.J. and D. Nathans. 1974. Deletion mutants of simian virus 40 generated by enzymatic excision of DNA segments from the viral genome. *J. Mol. Biol.* **89:** 179.

Lam, S.T., M.M. Stahl, K.D. McMilin, and F.W. Stahl. 1974. Rec-mediated recombinational hot spot activity in bacteriophage lambda. II. A mutation which causes hot spot activity. *Genetics* **77:** 425.

Lathe, R. and J.P. Lecocq. 1977. Overproduction of a viral protein during infection of a *lyc* mutant of *E. coli* with λ*imm*^{434}. *Virology* **83:** 204.

Lecocq, J.P. and A.M. Gathoye. 1973. Une mutation bacterienne augmentant la frequence de lysogenisation par le phage λ. *Arch. Int. Biochem. Physiol.* **81:** 803.

Leder, P., D. Tiemeier, and L. Enquist. 1977. EK2 derivatives of bacteriophage lambda useful in the cloning of DNA from higher organisms: The λgtWES system. *Science* **196:** 175.

Loenen, W.A.M. and W.J. Brammar. 1980. A bacteriophage lambda vector for cloning large DNA fragments made with several restriction enzymes. *Gene* **10:** 249.

Mandel, M. and A. Higa. 1970. Calcium-dependent bacteriophage DNA infection. *J. Mol. Biol.* **53:** 159.

Maniatis, T., E.F. Fritsch, and J. Sambrook. 1982. Molecular cloning. A laboratory manual. Cold Spring Harbor Laboratory, Cold Spring Harbor, New York.

Maniatis, T., R.L. Hardison, R. Lacy, J. Lauer, C. O'Connell, D. Quon, G.K. Sim, and A. Efstratiadis. 1978. Isolation of structural genes from libraries of eukaryotic DNA. *Cell* **15:** 687.

Maxam, A.M. and W. Gilbert. 1977. A new method for sequencing DNA. *Proc. Natl. Acad. Sci.* **74:** 560.

McClements, W.L., R. Dhar, D.G. Blair, L. Enquist, M. Oskarsson, and G.F. Vande Woude. 1981. The long terminal repeat of Moloney sarcoma provirus. *Cold Spring Harbor Symp. Quant. Biol.* **45:** 699.

Mellado, R.P., H. Delius, B. Klein, and K. Murray. 1981. Transcription of sea urchin histone genes in *Escherichia coli*. *Nucleic Acids Res.* **9:** 3889.

Mercereau-Puijalon, O. and P. Kourilsky. 1976. Escape synthesis of β-galactosidase under the control of bacteriophage lambda. *J. Mol. Biol.* **108:** 733.

Mertz, J.E. and R.W. Davis. 1972. Cleavage of DNA by RI restriction endonuclease generates cohesive ends. *Proc. Natl. Acad. Sci.* **69:** 3370.

Messing, J., R. Crea, and P.H. Seeburg. 1981. A system for shotgun DNA sequencing. *Nucleic Acids Res.* **9:** 309.

Meyerowitz, E. and D. Hogness. 1982. Molecular organization of a *Drosophila* puff site that responds to ecdysone. *Cell* **28:** 165.

Mileham, A.J., H.R. Revel, and N.E. Murray. 1980. Molecular cloning of the T4 genome: Organization and expression of the *frd*-DNA ligase region. *Mol. Gen. Genet.* **179:** 227.

Moir, A. and W.J. Brammar. 1976. The use of specialized transducing phages in the amplification of enzyme production. *Mol. Gen. Genet.* **149:** 87.

Mount, D.W. 1980. The genetics of protein degradation in bacteria. *Annu. Rev. Genet.* **14:** 279.

Muller-Hill, B. and J. Kania. 1974. Lac repressor can be fused to β-galactosidase. *Nature* **249:** 561.

Muller-Hill, B., L. Crapo, and W. Gilbert. 1968. Mutants that make more *lac* repressor. *Proc. Natl. Acad. Sci.* **59:** 1259.

Murray, K. 1977. Making use of coliphage lambda. In *Recombinant molecules: Impact on science and society* (ed. R.F. Beers, Jr. and E.G. Bassett), p. 249. Raven Press, New York.

Murray, K. and N.E. Murray. 1975. Phage lambda receptor chromosomes for DNA fragments made with restriction endonuclease III of *Haemophilus influenzae* and restriction endonuclease I of *Escherichia coli. J. Mol. Biol.* **98:** 551.

Murray, N.E. and W.S. Kelley. 1979. Characterization of λ *polA* transducing phages; effective expression of the *E. coli polA* gene. *Mol. Gen. Genet.* **175:** 77.

Murray, N.E. and K. Murray. 1974. Manipulation of restriction targets in phage λ to form receptor chromosomes for DNA fragments. *Nature* **251:** 331.

Murray, N.E., W.J. Brammar, and K. Murray. 1977. Lambdoid phages that simplify the recovery of *in vitro* recombinants. *Mol. Gen. Genet.* **150:** 53.

Murray, N.E., S. Bruce, and K. Murray. 1979. Molecular cloning of the DNA ligase gene from T4. II. Expression of the gene and purification of its product. *J. Mol. Biol.* **132:** 493.

Murray, N.E., P. Manduca de Ritis, and L.A. Foster. 1973. DNA targets for the *Escherichia coli* K restriction system analysed in recombinants between phage phi80 and lambda. *Mol. Gen. Genet.* **120:** 261.

Newman, A.J., T.G. Linn, and R.S. Hayward. 1979. Evidence for co-transcription of the RNA polymerase genes *rpoBC* with a ribosomal protein gene of *Escherichia coli. Mol. Gen. Genet.* **169:** 195.

Panasenko, S.N., J.R. Cameron, R.W. Davis, and I.R. Lehman. 1977. Five hundred fold overproduction of DNA ligase after induction of a hybrid lambda lysogen constructed *in vitro. Science* **196:** 188.

Parkinson, J.S. and R.J. Huskey. 1971. Deletion mutants of bacteriophage lambda. I. Isolation and initial characterization. *J. Mol. Biol.* **56:** 369.

Perricaudet, M. and P. Tiollais. 1975. Defective phage lambda chromosome, potential vector for DNA fragments obtained after cleavage by endonuclease *Bam*HI. *FEBS Lett.* **56:** 7.

Pourcel, C. and P. Tiollais. 1977. λp*lac*5 derivatives, potential vectors for DNA fragments cleaved by *Sst* I. *Gene* **1:** 281.

Pourcel, C., C. Marchal, A. Loise, A. Fritsch, and P. Tiollais. 1979. Bacteriophage lambda-*E. coli* K12 vector-host system for cloning and expression under lactose promoter control. I. DNA insertion at the *lacZ Eco*RI restriction site. *Mol. Gen. Genet.* **170:** 161.

Ptashne, M., A. Jeffrey, A.D. Johnson, R. Maurer, B.J. Meyer, C.O. Pabo, T.M. Roberts, and R.T. Sauer. 1980. How the λ repressor and Cro work. *Cell* **19:** 1.

Rambach, A. and P. Tiollais. 1974. Bacteriophage λ having *Eco*RI endonuclease sites only in the non-essential region of the genome. *Proc. Natl. Acad. Sci.* **71:** 3927.

Reeve, J.N. 1979. Use of minicells for bacteriophage-directed polypeptide synthesis. *Methods Enzymol.* **68:** 493.

Remaut, E., P. Stanssens, and W. Fiers. 1981. Plasmid vectors for high-efficiency expression controlled by the p_L promoter of coliphage λ. *Gene* **15:** 81.

Rimm, D.L., D. Horness, J. Kucera, and F.R. Blattner. 1980. Construction of coliphage lambda Charon vectors with *Bam*HI cloning sites. *Gene* **12:** 301.

Sanger, F., S. Nicklen, and A.R. Coulson. 1977. DNA sequencing with chain-terminating inhibitors. *Proc. Natl. Acad. Sci.* **74:** 5463.

Sanzey, B., O. Mercereau, T. Ternynzk, and P. Kourilsky. 1976. Methods for identification of recombinants of phage λ. *Proc. Natl. Acad. Sci.* **73:** 3394.

Scalenghe, F., E. Turco, J.E. Edstrom, V. Pirrotta, and M. Melli. 1981. Microdissection and cloning of DNA from a specific region of *Drosophila melanogaster* polytene chromosomes. *Chromosoma* **82:** 205.

Scheller, R.H., R.E. Dickerson, H.W. Boyer, A.D. Riggs, and K. Itakura. 1977. Chemical synthesis of restriction enzyme recognition sites useful for cloning. *Science* **196:** 177.

Scherer, G., J. Telford, C. Baldari, and V. Pirrotta. 1981. Isolation of cloned genes differentially expressed at early and late stages of Drosophila embryonic development. *Dev. Biol.* **86:** 438.

Schnös, M. and R. Inman. 1970. Position of branch points in replicating λ DNA. *J. Mol. Biol.* **51:** 61.

Segawa, T. and F. Imamoto. 1974. Diversity of regulation of genetic transcription. II. Specific relaxation of polarity read-through transcription of the translocated *trp* operon in bacteriophage lambda. *J. Mol. Biol.* **87:** 741.

Sgaramella, V. 1972. Enzymatic oligomerisation of bacteriophage P22 DNA and of linear simian virus 40 DNA. *Proc. Natl. Acad. Sci.* **69:** 3389.

Shimada, K., R.A. Weisberg, and M.E. Gottesman. 1972. Prophage lambda at unusual chromsomal locations. I. Location of the secondary attachment sites and the properties of the lysogens. *J. Mol. Biol.* **63:** 483.

Silverman, M., P. Matsumara, R. Draper, S. Edwards, and M. Simon. 1976. Expression of flagellar genes carried by bacteriophage lambda. *Nature* **261:** 248.

Skalka, A. and L. Shapiro. 1977. Screening for recombinant DNAs with *in vitro* immunoassays. *ICN-UCLA Symp. Mol. Cell. Biol.* **8:** 75.

Smith, H.O. and D. Nathans. 1973. Nomenclature for restriction enzymes. *J. Mol. Biol.* **81:** 419.

Southern, E.M. 1975. Detection of specific sequences among DNA fragments separated by gel electrophoresis. *J. Mol. Biol.* **98:** 503.

Stahl, F.W., J.M. Crasemann, and M.M. Stahl. 1975. Rec-mediated recombinational hot spot activity in bacteriophage lambda. III. Chi mutations are site mutations stimulating Rec-mediated recombination. *J. Mol. Biol.* **94:** 203.

Sternberg, N. and D. Hamilton. 1981. Bacteriophage P1 site-specific recombination. I. Recombination between *loxP* sites. *J. Mol. Biol.* **150:** 467.

Sternberg, N., D. Tiemeier, and L. Enquist. 1977. *In vitro* packaging of a λ *Dam* vector containing *Eco*RI DNA fragments of *Escherichia coli* and phage P1. *Gene* **1:** 255.

Sternberg, N., D. Hamilton, L. Enquist, and R. Weisberg. 1979. A simple technique for the isolation of deletion mutants of phage lambda. *Gene* **8:** 35.

Strack, H.B. and A.D. Kaiser. 1965. On the structure of the ends of lambda DNA. *J. Mol. Biol.* **12:** 36.

Struhl, K., J.R. Cameron, and R.W. Davis. 1976. Functional genetic expression of eukaryotic DNA in *E. coli. Proc. Natl. Acad. Sci.* **73:** 1471.

Thomas, M., J.R. Cameron, and R.W. Davis. 1974. Viable molecular hybrids of bacteriophage λ and eukaryotic DNA. *Proc. Natl. Acad. Sci.* **71:** 3927.

Ullrich, A., J. Shine, J. Chirgwin, R. Pictet, E. Tischer, W.J. Rutter, and H.M. Goodman. 1977. Rat insulin genes: Construction of plasmids containing the coding sequences. *Science* **196:** 1313.

Unger, R.C. and A.J. Clark. 1972. Interaction of the recombination pathways of bacteriophage λ and its host *Escherichia coli* K12. Effects on exonuclease activity. *J. Mol. Biol.* **70:** 539.

Ward, D.F. and N.E. Murray. 1979. Convergent transcription in bacteriophage λ: Interference with gene expression. *J. Mol. Biol.* **133:** 249.

Weil, J., R. Cunningham, R. Martin, B. Mitchell, and B. Bolling. 1973. Characteristics of λ p4, a λ derivative containing 9% excess DNA. *Virology* **50:** 373.

Wenman, W.M. and M.A. Lovett. 1982. Expression in *E. coli* of *Chlamydia trachomatis* antigen recognized during human infection. *Nature* **296:** 68.

Wensink, P.C., D.J. Finnegan, J.E. Donelson, and D.S. Hogness. 1975. A system for mapping DNA sequences in the chromosomes of *Drosophila melanogaster. Cell* **3:** 315.

Williams, B.G. and F.R. Blattner. 1980. Bacteriophage lambda vectors for DNA cloning. In *Genetic engineering* (ed. J.K. Setlow and A. Mullander), vol. 2, p. 201. Plenum Press, New York.

Wilson, G.G. and N.E. Murray. 1979. Molecular cloning of the DNA ligase gene from bacteriophage T4. I. Characterization of the recombinants. *J. Mol. Biol.* **132:** 471.

Windass, J.D. and W.J. Brammar. 1979. Aberrant immunity behaviour of hybrid λimm^{21} phages containing the DNA of ColEl-type plasmids. *Mol. Gen. Genet.* **172:** 329.

Wollman, E.L. and F. Jacob. 1954. Lysogenie et recombinaison genetique chez *Escherichia coli* K12. *C. R. Acad. Sci.* **239:** 455.

Yankovsky, N.K., B.A. Rebentish, V.G. Bogush, and V.N. Krylov. 1978. The creating and studying of a new type of λ DNA vector molecule. *Genetika* **14:** 1164.

Zissler, J., E.R. Signer, and F. Schaefer. 1971. The role of recombination in growth of bacteriophage lambda. In *The bacteriophage lambda* (ed. A.D. Hershey), p. 445. Cold Spring Harbor Laboratory, Cold Spring Harbor, New York.

Experimental Methods for Use with Lambda

Werner Arber*
Department of Microbiology
Biozentrum, University of Basel
Ch-4056 Basel, Switzerland

Lynn Enquist*
Molecular Genetics, Inc.
Minnetonka, Minnesota 55343

Barbara Hohn*
Friedrich Miescher Institut
Basel, Switzerland

Noreen E. Murray and Kenneth Murray*
Department of Molecular Biology
University of Edinburgh
Edinburgh EH9 3JR, Scotland

INTRODUCTION

This paper is addressed particularly (but not exclusively) to those who use λ as a vector for cloning genes of other organisms. In it we describe techniques for growing and purifying phage and offer a brief menu of tests that are useful for verification of genetic characters found in some widely used vectors. We also suggest methods for simple strain construction, mutagenesis, DNA purification, and in vitro DNA packaging. Alternative and overlapping procedures can also be found (Miller 1972; Davis et al. 1980; Maniatis et al. 1982; Berman et al. 1983). We emphasize that the choice of route is often a matter of individual preference and that methods can and should be varied to suit the particular phage strain, laboratory conditions, and the convenience of the investigator. We have indicated why certain manipulations are to be preferred or avoided, but we warn the reader that the reasons are often speculative and are presented to demystify our subject rather than to inhibit experimentation.

GROWTH AND STORAGE OF BACTERIAL HOST STRAINS

The commonly used hosts for the growth of phage λ are derivatives of *Escherichia coli* K12. Those few strains to which particular reference is made in this paper are given in Table 1. All known genes of *E. coli* K12 and their symbols are listed by Bachmann and Low (1980).

*Compiled from these laboratories and from those of Frank and Mary Stahl and Robert Weisberg.

Table 1 Some *E. coli* strains commonly used as hosts for λ

Strain	Partial genotype	References	Main features and use
C600[a]	*supE tonA lacY*	Appleyard (1954)	permissive indicator; host for lysates
5K[a]	*supE tonA lacY hsdR*	Hubacek and Glover (1970)	r^-m^+ permissive indicator; good transfection host
R594[a]	*sup*^{0}	Campbell (1965)	nonpermissive host for λ*am* phage
W3101[a]	*sup*^{0}	Bachmann (1972)	nonpermissive host for λ*am* phage
Y-mel[a]	*supF*	see Goldberg and Howe (1969)	indicator for λ*Sam*7
N1323(λ)[b]	*lop*8 (λ*c*I857*Sam*7)	see Gottesman et al. (1973) for *lop*8	source of λ DNA
WA803[a]	*supE hsdS*	Wood (1966)	r^-m^-; indicator and transfection host
ED8654[a]	*supE supF hsdR trpR lacY*	Borck et al. (1976)	r^-m^+; indicator and transfection host
NM531[b]	*supE supF hsdR trpR lacY recA*13	N.E. Murray (pers. comm.)	as ED8654 but Rec^-
ED8767[b]	*supE supF hsdS lacY recA*56	Murray et al. (1977)	r^-m^- Rec^- indicator; good for cosmids
NM538[b,c]	*supF hsdR trpR lacY*	A.-M. Frischauf et al. (in prep.)	r^-m^+ Su^+ host; good for *Sam*7 and *Aam Bam* phages
NM539[b,c]	*supF hsdR lacY* (P2cox3)	A.-M. Frischauf et al. (in prep.)	derivative of NM538; selection of λ Spi^-
SM32[g]	*lon*∇*galE sulA strA*	Mizusawa and Ward (1982)	for better growth of *c*I^+ phages
NM514[b]	*hsdR lyc*7 (*lyc*7 is an *hfl* allele)	for *lyc*7, see Lecocq and Gathoye (1973)	r^-m^+; selection of λ*imm*^{434} *c*I^- (see Plaque Morphology Mutants)

JC8679[a,b,c]	*recB*21*recC*22*sbcA*23	Gillen et al. (1981)	host for *red*⁻ and Spi⁻ Chi^0 phage
NM519[a,b,c]	*recB*21*recC*22*sbcA*23*hsdR*	N.E. Murray (pers. comm.)	r^-m^+; for recovery of Spi⁻ Chi^0 phage
GC507[h]	*supE supF hsdS trpR met recA*13/*pgam* ∇1 *bla*	G.F. Crouse (pers. comm.)	Rec⁻ host for Spi⁻ phage (see Recombination-deficient Mutants; Fec⁻ and Spi⁻ Phenotypes
K388[e]	*rpoB* (Snu1)	Baumann and Friedman (1976)	scoring *nin*5 and Chi
LE289[f]	(λ[*int*-FII]) *galT supF*	L. Enquist and R. Weisberg (1976 and pers. comm.)	scoring *xis* or *int*
WA2127[a]	*ptsM supE hsdS tonA lac*	Elliott and Arber (1978)	enriching for phage with longer genomes
LE30[f]	*mutD*5	Enquist and Weisberg (1977)	mutator host
BHB2688[d]	*recA* (λ*imm*^{434}c*I*ts *b*2*red*3*Eam*4*Sam*7)/λ	Hohn (1979)	in vitro packaging extracts
BHB2690[d]	*recA* (λ*imm*^{434}c*I*ts *b*2*red*3*Dam*15*Sam*7)/λ		

[a]From B. Bachmann (Yale University School of Medicine, New Haven, Connecticut.
[b]From N.E. Murray (University of Edinburgh, Scotland).
[c]Note that λ*Aam*32 phages and, hence, Charon 4A, grow poorly on *recBC supE* hosts and λ*Aam*32 *Bam*1 *sbam*1^0 phages grow poorly on *supE* hosts.
[d]From B. Hohn (Friedrich Miescher Institut, Basel, Switzerland).
[e]From D. Friedman (University of Michigan School of Medicine).
[f]From L. Enquist (Molecular Genetics, Inc., Minnetonka, Minnesota).
[g]From D.F. Ward (Uniform Services University of the Health Sciences, Bethesda, Maryland).
[h]From G.F. Crouse (National Cancer Institut, Frederich, Maryland).

For the titration and propagation of λ, bacteria are best grown in a rich medium, such as tryptone broth, with aeration at 37°C, and used when they have reached a concentration of no more than 5×10^8 cells/ml to 1×10^9 cells/ml. Saturated overnight cultures may also be used, but to avoid cell damage and death these should not be incubated for a long time in the stationary phase.

One of the following procedures can be used to ensure optimal adsorption of phage λ to the bacteria: (1) growth of the bacteria in the presence of 0.4% maltose but in the absence of glucose (see below, Preparation of Phage Stocks), or (2) centrifugation of the bacteria (normally grown in tryptone broth) and resuspension in 10 mM $MgSO_4$, followed by aeration at 37°C for 1 hour. These procedures are based on the specific interaction between bacteriophage tail components and the *lamB* receptor substance of the bacterial surface and can be applied whenever the phage (or packaged cosmids) have the tail fiber encoded by gene *J* of λ. They can also be used for λ derivatives that have the host range of φ80 (*h*80), although such phage do not adsorb as efficiently as *h*λ phage even at the preferred $MgSO_4$ concentration of 1 mM. Cultures starved in Mg^{++} can be stored at 4°C for several days and used as λ indicator without noticeable loss of the efficiency of plating of the phage. Cultures of recombination- or repair-deficient strains of bacteria die more quickly than other strains.

Bacteria may be stored at room temperature as stabs in vials of solidified medium (for recipe, see 3. Stab Agar in the Appendix to this paper). The vials must be free of detergent and should be tightly sealed to prevent desiccation (e.g., sealed with parafilm or wax). Alternatively, bacteria may be stored at −70°C in 12–15% glycerol. In practice, approximately 1 ml of a fresh bacterial culture is pipetted into 120–150 μl of glycerol. The vials may be opened for sampling without thawing and then returned to −70°C. Bacteria may also be stored as lyophilized cultures.

TITRATION OF PHAGE

The phage lysate is diluted serially in growth medium or Tris Mg gelatin (TMG) buffer (see 16a in the Appendix to this paper). An aliquot from each dilution is pipetted into 0.2 ml of indicator bacteria, prepared as described above, in Growth and Storage of Bacterial Host Strains. This preadsorption mixture is incubated for 10–15 minutes at 37°C. About 2.5 ml of soft agar is then added to the tube, and the total mixture is plated on tryptone, trypticase (BBL), or Luria agar (LA) plates (for recipes, see the Appendix to this paper). The choice of medium is influenced by the phage, the laboratory, and the experiment. The less rich trypticase agar is preferred when it is advisable to enhance the size of the plaques, whereas the richer tryptone or Luria agar can ease the distinction between clear and turbid plaques (see below, Plaque Morphology Mutants). After the top agar has solidified, the plates are inverted and incubated overnight at 37°C. λ plaques obtained by this method should be rather uniform in size (titration without preadsorption usually gives plaques of various sizes).

λ AMBER OR *sus* MUTATIONS

Amber mutations are available in most genes of phage λ. They are suppressor-sensitive (*sus*) and grow normally only in a "permissive host" in which the amber mutation is suppressed. The suppressor mutation in the permissive host can be in any of a number of genes, i.e., *supI*, *supII*, or *supIII*, now more frequently referred to, respectively, as *supD*, *supE*, and *tyrT* (*supF*). These genes code for tRNAs (suppressors are mutant tRNAs) that can translate chain-termination codons. The suppressor thus allows translation of a gene carrying a nonsense mutation to give a functional polypeptide.

λ*am* mutations in essential genes can readily be identified by spotting phage at a titer of 10^5 particles/ml onto a pair of tryptone plates seeded in a top layer of soft agar with either nonpermissive (*sup*^{0}) or permissive (e.g., *supE*, sometimes also designated *sup*$^+$ or Su$^+$, although they are mutants) indicators. Spots can be produced with a sterile wire loop or capillary tubes. Appropriate controls should always be included. After the spots have dried, the plates are incubated at 37°C.

Most λ*am* mutants were isolated by their ability to grow in the permissive indicator, C600 (*supE*), but not in the nonpermissive host, R594 or W3101 (*sup*^{0}). Although *supE* is often the most effective suppressor, λ*am* mutations are generally suppressed by *supD*, *supE*, or *supF*. Occasionally, an amber mutation is suppressed only by *supE* or *supD* (e.g., *Pam*3) or by *supF* (e.g., *Sam*7). The amber mutations in gene *S* provide a convenient method for preparing high-titer phage stocks (see sections below on Liquid Lysis and Induction Techniques). *S*-gene product is required for cell lysis; when λ*Sam* phage infect or are induced from within nonpermissive hosts, intracellular phage production continues for an abnormally long time. The cells containing large numbers of particles can then be concentrated by low-speed centrifugation prior to being lysed artificially. Since *Sam*7 requires suppression by *supF* for normal function, either a *sup*^{0} or a *supE* host can be used. The latter option is valuable when the phage to be prepared contains an amber mutation in another essential gene. Another mutation in gene *S*, *Sam*100, is suppressed, albeit weakly, by *supE*, as well as by *supF*; its usefulness in this method is therefore confined to preparation of stocks in *sup*^{0} strains.

OTHER RELEVANT MUTATIONS OF λ

Plaque Morphology Mutants

The lysogenic state of λ involves the propagation by the bacterial chromosome of a potentially lethal set of phage genes. To prevent the expression of these genes in a lysogen, λ synthesizes a repressor encoded by gene *c*I. Repressor acts by binding to operators o_L and o_R, thereby preventing transcription of certain key genes. As a result of repressor action, plaques formed by λ wild type contain many surviving cells ("turbid" plaques).

Mutations in gene *c*I lead to a defective repressor and plaques that are clear, since they contain no surviving cells. Because of repressor, a λ lysogenic cell is "immune" to superinfection with λ*c*I mutants, as well as λ wild type. λ*c*I857 has a mutation in the *c*I gene that leads to a temperature-sensitive repressor. Lysogens of this phage may be propagated at low temperatures (32–34°C), and the resident prophage induced by temperatures above 37°C.

Phage mutated in both operators o_L and o_R are virulent (λ*vir*) and form a clear plaque even on a strain lysogenic for phage λ, because the repressor is unable to bind at the operators. Similarly, λ wild type can form plaques on a heteroimmune lysogen carrying, e.g., phage 434, because the 434 repressor protein does not bind to the operators of phage λ.

Some other mutations influence the turbidity of λ plaques. These are chiefly mutations in the *c*II and *c*III genes, whose products stimulate lysogenization but are not required for maintenance of repression. The products of some host genes stimulate (e.g., *himA*; see Wulff and Rosenberg, this volume) or antagonize (e.g., *hfl*; see Wulff and Rosenberg; Murray; both this volume) the establishment of repression, and *hfl* mutants have been used to select λ*c*I$^-$ recombinants (see Murray, this volume).

Host-range Mutations

Failure of bacteria to absorb λ is a frequent cause of bacterial resistance to λ. λ*h* mutants, which grow both on the resistant strain and on the standard hosts for λ, can be selected for certain nonadsorbing bacterial strains (e.g., CR63). These host-range mutations map in phage gene *J*, which is responsible for the tip of the phage tail. Phage having mutations in gene *J* frequently lose infectivity more rapidly than λ$^+$.

Insertion-deficient Strains of λ and the Red Plaque Test

Derivatives of phage λ lacking either *att*, the site at which insertion of the prophage into the bacterial chromosome takes place, or the phage-coded function (*int*) necessary to catalyze this site-specific recombination event, are insertion-deficient. Although these phages give turbid plaques (because they are able to make *c*I-type repressor), the surviving cells within these turbid plaques are not stable lysogens. They are "abortive lysogens," in which the nonintegrated λ genomes are repressed. The stable lysogens characteristic of an insertion-proficient phage are readily distinguished from abortive lysogens, since descendants of the former remain immune to superinfection by challenging λ phage (Gottesman and Yarmolinsky 1968).

Red plaques, detected on special *E. coli* strains and indicator plates, are diagnostic of λ*int*$^+$ and *xis*$^+$ (Enquist and Weisberg 1976, 1977). Phage with a defect in either *int* or *xis*, or with another insertion specificity (e.g., ϕ80), make white plaques. Recombinant derivatives of λ vectors with cloning sites in *int*, or with a replaceable fragment containing *int* and *xis*, can thus be identified as white plaques.

The indicator bacterium (see Table 1) carries a λ prophage within a gene involved in galactose fermentation. A deletion within the prophage extends from the *int* gene through the *c*I gene, thereby rendering the cell sensitive to superinfection. Infection of this Gal⁻ strain by an *int*⁺*xis*⁺ λ phage results in excision of the prophage, restoring a functional *gal* operon. Tetrazolium agar with galactose as carbon source is the best indicator, although galactose MacConkey agar works well for spot tests (see 7. MacConkey Agar and 8. TTC Galactose Agar, in the Appendix to this paper). In either case, the lawn remains white and the centers of turbid plaques are red for *gal*⁺ (excision) or white for *gal*⁻ (no excision). The red color develops in 30–48 hours at 32°C and 12–24 hours at 37°C.

Recombination-deficient Mutants; Fec⁻ and Spi⁻ Phenotypes

Mutants of λ defective in recombination (*red*⁻) can be distinguished from wild type by their failure to form plaques on hosts deficient in polymerase I (*polA*) or ligase (*lig*) (see Smith, this volume). The latter condition is achieved when *E. coli*, having a temperature-sensitive mutation in the *lig* gene (e.g. *ligts*7), is propagated at 37°C, or even at 32°C, a temperature at which ligase is limiting for phage but not bacterial growth.

The products of the *red* genes are involved not only in recombination but also in DNA replication, possibly by initiating replication forks at sites of genetic exchange. λ *red*⁻ consequently grow less well than λ *red*⁺ (see Smith, this volume).

The product of the λ gene *gam* inactivates *E. coli* exonuclease V (ExoV, the *recBC* product). Since ExoV blocks the transition from early to late replication, *gam*⁻ phage also grow less well than λ*gam*⁺.

λ*red*⁻*gam*⁻ phages grow very poorly on wild-type bacteria and are unable to form plaques on a *recA*⁻*recBC*⁺ host (i.e., the Fec⁻ phenotype). The plaque size of *red*⁻*gam*⁻ phages in *recA*⁺ hosts is much enhanced by inclusion in the λ genome of a DNA sequence (Chi) that stimulates *rec*-mediated recombination (Smith, this volume). The host strain and media further influence the size of plaque achieved by *red*⁻*gam*⁻ phages. Fresh plates, nutritionally marginal media, and reduced bacterial inocula all increase the plaque size.

A *red*⁻*gam*⁻ Chi phage is recognized by its ability to form plaques on a host lysogenic for phage P2, i.e., by its Spi⁻ phenotype. A convenient way to grow *red*⁻*gam*⁻ phages (or the single mutants) is in a host that is *recB*⁻*sbcA*⁻. These bacteria are themselves vigorous, because the *sbcA* mutation compensates for the repair deficiencies conferred by the *recB* mutation. The *red*⁻*gam*⁻ λ grows well on such a host for two reasons: (1) The *recB* mutation prevents the production of ExoV, so that late λ replication is restored, and (2) the *sbcA* mutation derepresses a bacterial Red-like recombination pathway (RecE). A restriction-deficient derivative of a *recB*⁻*sbcA*⁻ strain is recommended for the recovery of *red*⁻*gam*⁻ recombinants from vectors such as Charon 4 and Charon 30. When it is desirable to minimize recombination, a *recA*⁻ strain carrying the *gam* gene of λ within a plasmid (see Table 1) is more vigorous than either a $recB^-$ or a $recA^-recB^-$ host. Transcription of the *gam* gene in this plasmid is from the late λ promoter $p_R{}'$ and is activated following infection by Q^+ λ. (G. Crouse, pers. comm.).

Deletion of t_{R2}

t_{R2} is a site of transcriptional termination located between genes *P* and *Q*. In wild-type λ, the product of gene *N* allows transcription to continue rightward past t_{R2} into gene *Q* and leftward past t_{L1} into genes *gam* and *red*. A commonly used deletion of t_{R2}, present in several cloning vectors, is *nin*5. This deletion may be detected because λ*nin*5, but not λ^+, strains will grow on bacteria (e.g., *groN* and super-Nus) in which the *N*-gene product is ineffective. In the absence of effective *N* function, even the growth of λ*nin*5 is poor, because the *red* and *gam* genes are not expressed. Therefore, the test for *nin*5 works best if 10^4–10^6 phage are spotted onto the lawn of super-Nus bacteria. For a λ*nin*5 phage the presence of a Chi sequence can be identified when the phage suspension is diluted further to distinguish between minute (Chi^0) and small (Chi^+) plaques (A. Anilionis and W.J. Brammar, pers. comm.). A *groN* host remains the preferred host for *imm*λ phages, but a super-Nus strain, e.g., K388 (Baumann and Friedman 1976) at 42°C should be used for λimm^{21} phages.

PREPARATION OF PHAGE STOCKS

Serial propagation of phage strains inevitably selects for faster growing variants. For example, other things being equal, wild-type phage will grow better than amber mutations even on suppressing hosts. Although the growth advantage may be small, its effect is cumulative and will therefore be substantial over many cycles of growth. To avoid such selection we strongly recommend that phage stocks be started from a single plaque whenever feasible. A rapid method for the isolation of single plaques is presented by Davis et al. (1980) and Enquist and Weisberg (1976). A single plaque will generally contain 10^5–10^7 phage, depending, among other things, on its size, the stability of the phage, and the freshness of the plate. If carry-over of the old indicator strain is not desired, the plate from which the plaque is picked may be inverted for 10–15 minutes over a glass dish containing $CHCl_3$. The opportunities for unwanted evolution are minimized by isolating a young (i.e., 6–8 hr) plaque. A single fresh plaque may be picked in a sterile capillary tube or pasteur pipette and, when dispersed in a small volume of buffer (e.g., 0.3 ml), may contain a suitable number of particles to inoculate one plate. For healthy phages, a fraction of the resuspended plaque should be adequate. For very sick phages, it may be necessary to replate a dilution of the young plaque and pick 10–20 young plaques to use as inoculum. A simpler system of amplifying inocula, when lysates are required from many clones and contamination with faster growing variants will not cause problems, is to spot samples of phage, each from a single purified plaque, onto a lawn of indicator bacteria. The area of lysis (a "macroplaque"), achieved after incubation for 6–8 hours, provides the inoculum for making a lysate either by confluent lysis on a plate or a liquid culture.

Indicator bacteria are prepared according to the methods in Growth and Storage of Bacterial Host Strains (above). Exponentially growing cells appear to give better yields than overnight cultures for phage that form turbid plaques. This may be due to reduced lysogenization in the first growth cycle. Note that bacteria grown in the presence of maltose possess much more λ-receptor protein (*lamB* product) than bacteria grown in the presence of glucose. It seems, therefore, good practice to grow phage stocks in media supplemented with glucose, since loss of phage by readsorption to cells and cell debris is then small, but adsorption is still sufficient to assure infection of practically all of the bacteria present. When λ is grown in maltose media, it is anticipated that an important fraction of phage can be lost by readsorption to cell debris. In fact, however, some experienced λ laboratories prefer to use maltose-grown cells.

Confluent Lysis

The medium used in the plates may be tryptone, trypticase, or Luria agar, but whichever medium is used, the plates should be thick and are best used as soon as they harden. Higher titers can be obtained by adding the following supplements (suggested by Dr. Yuzo Nozu) to the bottom agar: 0.3% glucose, 7.5×10^{-2} mM $CaCl_2$, 4×10^{-3} mM $FeCl_3$ or $FeSO_4$, and 2 mM $MgSO_4$. The purpose of the glucose is twofold: It represses the synthesis of λ receptors, thereby reducing adsorption of phage, and it also increases the proportion of infected cells that enter the lytic cycle. $CaCl_2$ reduces phage readsorption, and iron may increase respiration.

For a single plate, 10^5–10^6 phage (use the higher value for small plaque formers) are added to 0.2 ml of indicator bacteria. After 10–15 minutes at 37°C or at room temperature, to allow phage adsorption, molten soft agar is added and the mixture is poured onto the fresh plate. Some workers use 2.5 ml of soft agar, others use 1.5–2.5 ml of soft agar and 1.5–2.5 ml of warm broth. The resulting moist plates are best incubated right-side-up and at 37°C unless the phage bears a temperature-sensitive lethal mutation.

Lysates are ready for harvesting after 4–8 hours, depending on temperature, host cell, and phage genotype. At the time of harvesting, the plaques should touch one another. It is useful to prepare a plate containing no phage, only indicator bacteria. This will help in deciding whether a clear plate is due to slow growth of the host or to lysis. Extensive incubation of lysates, e.g., overnight, can provide extra opportunity for the enrichment of fast-growing mutants and, if the phage forms turbid plaques, for loss of titer by adsorption to immune cells.

Two harvesting methods are in use. Impatient investigators transfer the top layer to a tube containing about 0.2 ml of $CHCl_3$. This can be achieved with either a 10-ml pipette or a sterile microscope slide. Some investigators add a few milliliters of phage buffer to the tube. The contents of the tube are mixed well, left for 5 minutes at room temperature, and the lysate is clarified by centrifugation for 10 minutes at 5000–8000 rpm. The lysate is poured off into a detergent-free tube (either acid-cleaned or new), since the phage are particularly sensitive to trace amounts of detergent.

Alternatively, a few milliliters of phage buffer or broth can be layered onto the plate (after confluent lysis has been obtained), and the plates left at room temperature for at least 1 hour or refrigerated overnight. $CHCl_3$ may be added to the plate or after the phage suspension has been pipetted into a tube. Bacterial debris is removed by centrifugation.

Regardless of the method of harvesting, plate lysates normally yield titers between 10^{10} and 10^{11} pfu/ml. The first method leaves some agar in the lysate, but this stabilizes the phage (see below, To Preserve a λ Stock). However, in lysates intended for the isolation of DNA, the residual agar may interfere with digestion of the DNA by endonucleases. Agarose in place of agar in the top layer reduces this problem.

Liquid Lysis

Yields with this method are frequently lower than from plate lysates. However, it is simpler, easier to scale up, and avoids agar in the stock. Host bacteria may be grown according to the methods in Growth and Storage of Bacterial Host Strains (above) to reach a concentration of 10^9 cells/ml. They are then infected with 10^5 phage/ml to 10^8 phage/ml. Phage adsorption is allowed for 10–15 minutes at 37°C. The infection mixture is then diluted 10–100-fold into fresh growth medium, e.g., Luria broth (L-broth), made 1–10 mM in $MgSO_4$. Blattner et al. (1977) recommend a broth using N-Z-amine (Humko-Sheffield), and M. Melli (pers. comm.) recommend buffering Luria broth with 50 mM Tris-HCl (pH 7.5). The Nozu supplement described above, in Confluent Lysis, also tends to improve the results. The diluted culture is aerated at 37°C for several hours or even overnight. Good aeration is vital for high yields and is most readily achieved if the flask is no more than one-fifth full. The time of massive cell lysis is primarily a function of the amount of phage used as inoculum, but it also depends on the genotype of the phage, the host strain, and the growth medium. With λ*c*I$^-$, lysis should eventually occur even if the cell density becomes relatively high. However, prolonged incubation risks the enrichment of faster growing mutants. Of course, if lysis occurs early, i.e., while the cell density is still low, the phage yield will be low. This can be avoided by decreasing the initial inoculum of phage, increasing the initial inoculum of cells, or both. After lysis, which can be detected by a drop in the optical density of the culture or by inspection under the phase-contrast microscope, $CHCl_3$ is added and incubation is continued for 5–10 minutes. The phage stock is purified from bacterial debris by low-speed centrifugation. Typical titers are 1×10^{10} pfu/ml to 5×10^{10} pfu/ml.

Growth for only one lytic cycle may sometimes be indicated, e.g., to provide a phage population with a new type of modification (in this case, λ is grown on a strain that modifies but does not restrict DNA, i.e., r^-m^+), or if one desires a limited replication to maintain the proportion of various genotypes present in a mixed population of phage particles. Also, genetic crosses between two phage strains are usually made in one growth cycle. The bacteria to be infected are grown as described above, in Growth and Storage of Bacterial Host Strains. Infection is made at a multiplicity appropriate to the particular experiment. After 10–15 minutes of adsorption at 37°C without aeration, the infected bacteria can be washed if necessary by centrifugation (which removes most of the unadsorbed phage particles), resuspended in fresh medium at a concentration of no more than 3×10^8 cells/ml, and further incubated at 37°C with aeration. Incubation of more diluted cell suspensions may increase the phage yield per infected cell and it reduces the risk of phage loss by readsorption, but it decreases the final titer of the lysate. Lysis should be observed after 60–70 minutes of incubation. $CHCl_3$ is then added to complete the lysis and to kill the surviving bacteria. The debris is removed by low-speed centrifugation.

Depending on the phage used and on the growth conditions, phage yields varying between 10 pfu/infected cell and 200 pfu/infected cell can be expected. At cell densities above about 3×10^8/ml, the yield drops, so that it is not advantageous to increase the cell density to much over 5×10^8 cells/ml.

Growth for only one lytic cycle in nonpermissive bacteria yields high-titer lysates of *Sam*7 (or *Sam*100) mutants of λ. In this case, a multiplicity of infection of 2 and a period of incubation of 3–4 hours is suggested. The cells must be lysed with $CHCl_3$. For large volumes, and if a concentrated lysate of λ*Sam*7 is desired, a small sample can first be tested for lysis by addition of a drop of chloroform. If this test is positive, the bacteria are sedimented, resuspended in 1/100 volume of medium, lysed by the addition of chloroform, and, if desired, the lysate treated with DNase. After sedimenting the debris, the supernatant should contain about 10^{12} pfu/ml.

For phage strains that form turbid plaques, 5 μg/ml of mitomycin C can be added to the growth medium, but the growth tube should then be protected from light. This procedure inhibits lysogenization. A simpler alternative is to use a host in which transcription of the repressor gene is reduced. Mizusawa and Ward (1982) have used a *lon*⁻ strain (see Table 1).

Lysogenization by λ*cIts* phage is best prevented as follows: Adsorb the phage to the bacteria for 15 minutes at 32°C, and incubate the tube for an additional 15 minutes at 45°C. Then dilute into growth medium and incubate at 37°C until lysis.

INDUCTION OF LYSOGENS

Batch induction of lysogenic bacteria is a convenient method to produce liquid stocks of all those phage strains that can be kept as a prophage. This is a particularly good way of reducing the opportunity for selection of faster growing derivatives. The lysogenic bacteria, while still in the exponential phase of growth, are induced to phage production by one of the methods described here. Note, however, that methods 1, 2, and 3 will not induce prophages that carry the *ind*⁻ mutation. Many phage strains that carry the *cIts*857 mutation are also *ind*⁻. Likewise, hosts that carry a *recA* mutation are not inducible by these methods.

Induction Techniques

1. Induction with Mitomycin C. The lysogenic bacteria are grown in a common growth medium (e.g., tryptone, Luria broth, or M9a) to a concentration of about 2×10^8/ml. 5 μg/ml of mitomycin C is then added for induction of phage reproduction. The culture is protected from light and well aerated at 37°C until lysis occurs. Chloroform is then added, and the lysate is cleared from debris by centrifugation.

2. Induction by UV Irradiation. Care must be taken that the suspension medium in which the lysogenic bacteria are UV-irradiated does not itself absorb UV light (use M9a medium prepared with filtered casaminoacids [see 10 and 11b in the Appendix to this paper], or resuspend the bacteria in phosphate buffer [see 20 in the Appendix to this paper]). In addition, bacteria in the lower part of the suspension should not be shielded against the UV light by those above. About 10 ml of culture containing bacteria at 2×10^8/ml can be irradiated without noticeable shielding effect in a normal size petri dish. A germicidal lamp, with maximum output at a wavelength of 260 nm, should be used. At the optimal induction dose, up to 90% of the bacteria may be productive and, except for highly UV-sensitive mutants, such as many *uvr* and *rec* strains, usually one to a few percent of λ lysogenic bacteria survive. Exceeding the optimal dose of radiation reduces the number of phage liberated per cell. Therefore, the dependence of phage yield on the duration of irradiation should be determined in advance. After the UV irradiation, cultures in M9a medium are again incubated at 37°C with good aeration, whereas bacteria irradiated in the buffer must first be transferred to fresh growth medium. (This can be achieved by adding one-tenth volume of 10× growth medium to cells irradiated in buffer.) Photoreactivation is prevented by protecting the irradiated bacteria from visible light. After lysis, a few drops of chloroform are added, and the lysate is purified from bacterial debris by centrifugation.

3. Induction of thy⁻ *λ Lysogens by Thymidine Starvation.* This induction method, applicable to strains with a *thy*⁻ mutation (for the isolation of *thy*⁻ mutations; see Miller [1972]), is useful in the preparation of λ labeled in its DNA with radioactive thymidine. The bacteria are grown in M9a medium (see 11b in the Appendix to this paper) supplemented with 10 μg/ml of thymidine to about 2×10^8/ml, washed, resuspended in M9a medium without thymidine, and incubated at 37°C for 2 hours, at which time 10–25 μg/ml of thymidine is added to the culture. After an additional incubation at 37°C for 90–120 minutes, most of the cells will have lysed and produced phage. Chloroform is then added, and the lysate is purified from bacterial debris by centrifugation.

4. Induction of λc*I*ts *Lysogens.* Lysogens harboring a prophage with temperature-sensitive repressor (*cIts* mutation should be incubated at 42–45°C for induction. The lysogens are grown at 32°C until they reach 2 × 10^8/ml. The temperature of the culture is then shifted to 42–45°C, incubation is continued for 10–15 minutes at this temperature, and finally the culture is transferred to 37°C (the optimal temperature for phage growth) and further incubated until lysis occurs. Vigorous aeration should be continued throughout.

Notes on Induction Techniques

The latent period of λ induced by UV irradiation or thymidine starvation is longer than that measured after heat induction. Massive lysis within 70–120 minutes after induction and incubation at 37°C points to an efficient induction.

Foaming at the time of lysis can be prevented if aeration is done by shaking or rotating the culture, rather than by bubbling. Flasks should be filled to no more than one fifth (and preferably only one tenth) of their nominal capacity.

An induced, nonpermissive, λ*Sam* lysogen will not lyse before chloroform is added. Chloroform should be added to the cultures after 3 hours of aeration at 37°C to obtain optimal yield (see above, Liquid Lysis).

The extensive DNA replication associated with prolonged incubation following induction of λ*Sam* may lead to an increase in the number of single-strand breaks, apparently because the host cells' ligase activity becomes limiting. This can be avoided by using a ligase overproducing host, e.g., *lop*8 or *lop*11 (Gottesman et al. 1973).

PHAGE PURIFICATION

There are several methods for recovery and purification of phage from lysates: (1) differential sedimentation, (2) zone or equilibrium centrifugation in density gradients, (3) precipitation with polyethylene glycol, or (4) combinations of these. It may be useful to titer the phage preparation at each stage, even though this measure of only viable particles may underestimate the DNA yield by a factor of two or three. Phage suspensions should be handled gently and swirled slowly—never shaken. The largest loss of viable phage will usually be at the first sedimentation (up to 50%) but should not be great thereafter. Overall yields of over 30% (on the basis of titration of viable particles) from the initial lysate are not uncommon, even when several purification stages are involved. It is advisable to keep magnesium ions (usually supplied as $MgSO_4$) present at all stages.

Cellular DNA and RNA in phage lysates may be removed by treatment with 1–10 μg/ml deoxyribonuclease and ribonuclease at room temperature for at least 1 hour.

1. Differential Sedimentation. The lysate is clarified by centrifugation at 10,000 rpm for 10 minutes (e.g., in a 6 × 250-ml angle rotor), and the phage is then pelleted by centrifugation at 45,000*g* for 2½ hours at 4°C. The supernatant is decanted, and the pellets (which should contain 99% of the phage) are resuspended (~6 ml of TMG/100-ml centrifuge tube, or pro rata) by shaking gently (conveniently on a horizontal rotary shaker) overnight in the cold room. The phage suspension is again clarified by centrifugation (10,000 rpm for 10 min), and pancreatic DNase and RNase (pure preparations to avoid the risk of proteolysis) are added to 10 μg/ml. After 1–2 hours at room temperature (or 30 min at 37°C), the phage are again pelleted; it is convenient to use one centrifuge tube to make subsequent resuspension easier. The pellet of phage, which should now be quite clear, is resuspended as before, but in approximately one-tenth volume of TMG, and again clarified by centrifugation at 10,000 rpm for 10 minutes. If the sample is to be stored at this stage, add chloroform (see note on chloroform below, in To Preserve a λ Stock). For further purification, equilibrium centrifugation in cesium chloride (CsCl) or sedimentation through step gradients of CsCl solutions is used.

2. Precipitation with Polyethylene Glycol 6000. Precipitation of λ with polyethylene glycol (PEG) is not quantitative, but the method is simple and fast (Bertani and Bertani 1970; Yamamoto et al. 1970; Hohn et al. 1975). The lysate is made 0.5 M in NaCl, and PEG is added as powder, or preferentially from a 35% w/v stock solution, to a final concentration of 10% w/v PEG. The mixture is kept at 4°C for at least 1 hour, and the precipitated phage is recovered by low-speed centrifugation; the supernatant is decanted and may be assayed for residual phage activity to assess the efficiency of precipitation. The precipitate is dispersed in about 1/25 of the initial volume of TMG by gentle rotary shaking for several hours to extract the phage (residual PEG can be removed by shaking with an equal volume of chloroform), and the suspension is clarified by centrifugation (5000 rpm for 5 min). It may be desirable to extract the pellet a second time in the same way. The phage stock must then be purified further, e.g., by sedimentation through a CsCl step gradient.

3. Equilibrium Centrifugation. Equilibrium (isopycnic) centrifugation separates the components of a solution according to their buoyant densities. A density gradient is generally established under the force generated upon centrifugation within about 10 hours, and particles in the suspension slowly move to positions corresponding to their buoyant densities. Equilibrium is reached faster at higher speed, but since the steepness of the gradient increases with speed of centrifugation, lower speeds give better resolution. For preparations where sharp peaks are not essential, shorter times of centrifugation can be used than are customary for analytical experiments.

The buoyant density of a λ particle varies with the size of its genome; deletion derivatives are lighter and insertion derivatives, heavier than the wild type

(Weigle et al. 1959; Bellet et al. 1971; Davidson and Szybalski 1971). The centrifuge tubes are loaded with the phage particles suspended in a CsCl solution corresponding to the buoyant density of the phage. Upon centrifugation, phage will then form a band at about the middle of the tube.

The desired density of CsCl in the phage suspension is usually 41.5% w/w. This is readily achieved by weighing the phage suspension in a tared centrifuge tube and adding solid CsCl (41.5/58.5 × the weight of the phage suspension) or by mixing the suspension with an appropriate volume of a saturated, or nearly saturated, stock solution of CsCl in TMG. The density of a CsCl solution is conveniently determined by measuring its refractive index, $[n]^0_{20}$, which depends primarily upon the concentration of CsCl and, to a much lesser extent, upon the suspension medium. The relationship between density and the refractive index of a CsCl solution is given by Sober (1970). CsCl solutions (in 0.1 M Tris-HCl at pH 7.5, 0.5% w/v NaCl, and 0.1% w/v NH_4Cl) with a refractive index $[n]^0_{20} = 1.3815$ have been used routinely for wild-type λ (genome size 100%) and $[n]^0_{20} = 1.3805$ for λ*b*2 (genome size 88%). If necessary, the suspension or solution of phage in CsCl may be clarified by centrifugation (15,000 rpm for 1 hr).

After centrifugation to equilibrium (12–16 hr in a 3 × 5-ml swinging-bucket rotor [22,000–30,000 rpm, 20°C] or longer with larger tubes, but much shorter times with a rotor for vertical tubes), the phage will usually be visible as an opalescent band if at least 10^{11} particles are present. The band is best seen by illuminating the tube from above or from one side with a strong light and viewing it from the front against a dark background. Other components of the solution may also form bands, but phage can easily be monitored by their infectivity. The phage are collected by puncturing the base of the tube and allowing the solution to drip out slowly, by inserting a syringe through the side of the tube just below the phage band or from the top by withdrawing the phage band with a pasteur pipette (or with a suitable peristaltic pump).

If the centrifugation is to be repeated, the phage are diluted to a convenient volume with a solution of CsCl of appropriate density (i.e., that of the phage), and the equilibrium run is carried out as before. The phage should then be collected from the opposite direction to that used after the first centrifugation, because nucleic acid contaminants sediment during the equilibrium run, whereas protein floats to the surface.

Purified phage in CsCl solution can be stored in sterile tubes in the refrigerator. CsCl is removed by dialysis against TMG (to maintain infective particles) or 10 mM Tris-HCl (pH 8), 1 mM EDTA prior to the extraction of DNA.

4. CsCl Step Gradients. The phage preparation is placed on top of two or more layers of CsCl solutions of different density, and the phage are sedimented so that they form a band at the interface of two of the CsCl layers. Preferably, the suspension of phage is first partly purified by differential centrifugation (see method 1) or precipitation with PEG (method 2). CsCl solutions with densities of, e.g., 1.3 g/cm^3, 1.5 g/cm^3, and 1.7 g/cm^3 are prepared in phage buffer. A formula relating percentage of w/w to density, ρ, of the solution is

$$\% w/w = 137.48 - \frac{138.11}{\rho}$$

Equal volumes of the three solutions are pipetted sequentially (in the order 1.3 g/cm^3, 1.5 g/cm^3, and 1.7 g/cm^3) into a centrifuge tube (e.g., 1.2 ml each in a 5-ml tube), and the phage solution ($\sim$1 ml at 10^{11} phage/ml or more in this example) is layered on top from a pipette. A phage band should be visible about halfway down the tube after centrifugation for about 1 hour at 30,000 rpm (3 × 5-ml swinging-bucket rotor), and the phage are then collected as described in method 3).

The step gradient procedure is quicker than equilibrium centrifugation, but the purification is less complete, particularly with large volumes. The concentration steps may be adjusted to suit the individual experiment by varying the number, volume, and density of the CsCl layers and the volume of the λ suspension. Speed and time of centrifugation should be adjusted so that most of the phage form a band between a layer of CsCl at a density somewhat lower than that of λ and another layer of CsCl at a density somewhat higher than that of λ.

A final modification is a reverse step gradient in which the density of the phage suspension is adjusted to 1.7 g/ml with CsCl and the suspension overlaid with CsCl steps of lower density, so that the phage float to a density interface. This method is efficient for removal of contaminating DNA and RNA (Davis et al. 1980) and may be used following a descending step gradient.

Phage recovered by centrifugation through CsCl step gradients are often taken through an equilibrium centrifugation cycle as a final stage of purification.

5. *Glycerol Step Gradients.* Phage can be recovered quickly from a lysate by pelleting through a glycerol step gradient. There are two steps in the gradient: 40% v/v and 5% v/v glycerol in 50 mM Tris-HCl (pH 7.5), 10 mM $MgSO_4$. Phage particles are pelleted quantitatively, whereas ribosomes band in the 40% glycerol. DNA, RNA, and other small or light contaminants do not sediment during centrifugation. The method may be used for as little as 1 ml of clarified lysate or larger volumes where, e.g., the phage have been concentrated by PEG precipitation.

The step gradient is prepared in a polyallomer centrifuge tube. A convenient rotor is the Ti50 or SW41. For this, 3 ml of 40% v/v glycerol is overlaid carefully with 3 ml of 5% v/v glycerol, over which the clarified phage lysate is carefully layered and, if necessary, the tube filled with TMG. After centrifugation at 35,000 rpm for 1 hour at 4°C, the supernatant is decanted and the pellet resuspended in 1 ml or less of 50 mM Tris-HCl (pH 7.5), 10 mM $MgSO_4$ in readiness for DNA extraction. Centrifugation through a CsCl step gradient using the flotation method from a solution of high density (1.7 g/ml) provides a convenient means of further purification before preparation of DNA from the phage (method 4).

TO PRESERVE A λ STOCK

Under appropriate conditions, phage λ kept in suspension at 4°C are stable over periods of several years. However, "appropriate conditions" are difficult to define.

Lysates should be freed from bacterial debris and nonlysed bacteria to prevent adsorption of the phage. Low-speed centrifugation with recovery of the supernatant is recommended as a first stage, though it does not remove all of the bacteria. Since ultrafiltration frequently causes λ to lose a considerable part of its infectivity, lysates are usually sterilized by the addition of a few drops of chloroform.

Phage λ is relatively insensitive to chloroform, although this depends on the batch of chloroform used. Chloroform undergoes photolysis and so should be extracted with anhydrous sodium bicarbonate before use. Alternatively, one may use freshly distilled chloroform. Chloroform can exert a long-term inactivation; although little inactivation of λ is obtained upon adding chloroform, additional inactivation may occur upon storing, even if the excess chloroform is removed.

Another way to keep contaminating bacteria from proliferation in phage stocks is to add 10 mM Na_3N (sodium azide).

The presence of 0.01% gelatin or small amounts of agar make stocks more stable; unpurified plate stocks which, of course, contain agar are in fact usually relatively stable. In contrast, highly purified phage are often inactivated quickly, the precise rate depending somewhat on the suspension medium. To prevent fast inactivation, the suspension medium should contain about 1 mM Mg^{++}. Addition of gelatin and 1% glycerol may also help. Recommended suspension media are tryptone broth, SMC (see 14, in the Appendix to this paper), and Tris buffers (e.g., 10 mM Tris with 1–10 mM $MgSO_4$), which can also be supplemented with NaCl. pH should range between 7 and 8.5.

Phage λ survives freezing and thawing if appropriate media are used, e.g., 3XD (see 13 in the Appendix to this paper) or 15% glycerol. Phage stocks can therefore be deep frozen, although they are usually stored at 4°C for greater convenience. Screw-cap vials with tinfoil lids are recommended, and they should be well sealed to avoid evaporation of the suspension medium.

Note that heavy shaking or strong fluorescent light tend to inactivate λ; stocks should thus be stored in the dark.

The stability of a phage is also a function of its genotype. In particular, deletion and insertion derivatives may behave differently from their parental strains. It may be possible to recover phage from a nonviable lysate by transfection (see below).

CONSTRUCTION OF λ LYSOGENS

Spotting Phage on a Lawn of Permissive Bacteria. 0.2 ml of the bacteria to be lysogenized are plated in 2.5 ml of soft agar on a tryptone agar plate; a drop of the phage suspension (concentration between 10^4 phage/ml and 10^8 phage/ml) is deposited, left to dry, and incubated overnight at 37°C (or at 32°C in case of a λ*c*I*ts* phage). Bacteria are picked from the turbid zone of lysis with a sterile wire or toothpick, suspended in a small volume of sterile buffer or tryptone broth, and streaked out on a tryptone plate so as to obtain single colonies upon incubation. These colonies are tested as described below, in To Test Cultures for Lysogeny.

Infection at Known Multiplicity in Liquid. The bacteria are grown in tryptone broth supplemented with 0.4% maltose until near the end of the exponential phase of growth (~10^9 bacteria/ml). The bacteria are then infected with a known multiplicity of λ. Usually, the higher the multiplicity is, the higher the proportion of lysogens. A multiplicity of about five phages per cell is usually used. After 10–15 minutes of adsorption at 37°C, the bacteria are serially diluted, and samples of the dilutions are spread on tryptone agar plates so as to obtain between 20 colonies/plate and 500 colonies/plate (or they are streaked out with a platinum loop to obtain single colonies). The plates are incubated overnight at 37°C (or 32°C for λ*c*I*ts*). The resulting colonies are analyzed as described below, in To Test Cultures for Lysogeny. This method is frequently satisfactory when the host is nonpermissive (e.g., lysogenization of *sup*^{0} bacteria with λ*c*I857*Sam*7), provided that the cells are sensitive to phage adsorption.

If a low efficiency of lysogenization is expected, the infected bacteria are spread on a plate that has been seeded with about 10^9 homoimmune phage of strain λ*c*I$^-$, λ*c*I$^-$*h*80, or both. Individual colonies grown on these plates are purified and tested for lysogeny. Special care has to be taken to check for phage-resistant colonies (see below) or bacteria lysogenic for a recombinant of the desired phage with the phage used for the selection.

TO TEST CULTURES FOR LYSOGENY

Two properties of lysogenic bacteria are easily tested: the release of λ upon induction, and their immunity against superinfection by λ. Bacteria from each colony to be tested are suspended in a small volume of sterile medium and, if necessary, incubated for some time to reach a higher cell density (cell densities of about 10^8/ml are convenient for these tests).

To test for phage production, about 10^8 indicator bacteria are plated in 2.5 ml of soft agar on a predried tryptone agar plate. Small drops of the bacteria to be tested are then deposited on top of the plate, left to dry, and the plate incubated overnight at 37°C. Phage production by the lysogenic bacteria results in the lysis of the indicator.

To test for immunity, about 10^6 λ*c*I⁻ phages are streaked with a platinum loop across a predried tryptone agar plate. After this streak is dry, individual streaks of each of the bacterial cultures to be tested are laid in one direction across the λ*c*I⁻ streak (do not move back and forth). The plates are incubated at 37°C (or 32°C for *c*I*ts* mutations) overnight. After incubation, lysogens will have grown over the entire length of the streak, whereas nonlysogens will be lysed from the phage streak onward, sometimes to the end, depending on the amount of phage deposited. Cells that are lysogenic for a prophage with immunity other than λ, e.g., imm^{434}, must be tested with a *c*I mutation of the corresponding immunity, e.g., λimm^{434}. When immunity is not specified, it is usually assumed to be λ.

In an analogous cross-streak test, the strains can be checked for resistance to adsorption of phage λ: Instead of λ*c*I⁻, use a λ*vir* mutation, which can grow both in lysogenic and nonlysogenic bacteria as long as it is able to adsorb to the cells. Frequently, bacteria are cross-checked against both λ*c*I⁻ and λ*vir* on the same plate. The two phages are placed in parallel streaks, about 2 cm apart, and the bacteria to be tested are streaked across both phages, in the order λ*c*I⁻-λ*vir*.

Where large numbers of colonies must be screened, replica-plating techniques (see Miller 1972) may be used. The colonies are replicated to plates having indicator bacteria in hard agar, usually 4–5 ml of molten tryptone agar containing 1.3% agar. After incubation, the imprint of lysogenic colonies appears surrounded by a halo of lysis due to spontaneous phage production. More pronounced phage production is obtained by irradiating the replicated plate before incubation with an optimal induction dose of UV (see above, Induction of Lysogens, technique 2).

To test for λ immunity, about 10^7 λ*c*I⁻ phage are spread on top of a second predried tryptone agar plate and used for replication. After replication, the plate should be irradiated with one third of its normal inducing dose of UV light. λ immune, i.e., lysogenic, colonies grown for 8–16 hours are fat, whereas the colonies formed by sensitive bacteria are lean because of attack by the λ*c*I⁻ phage. UV irradiation, for unknown reasons, improves the distinction.

Note that *recA*⁻ lysogens, λ*ind*⁻ lysogens, and defective lysogens (e.g., a nonpermissive bacterial strain carrying a prophage with an amber mutation) may not respond positively in the phage production test; however, they will be immune against superinfecting λ. The lysogens used to make extracts for in vitro packaging cannot be checked by either of these tests, but they carry a *c*I*ts* prophage and are killed by temperatures that induce the prophage.

To screen for temperature-sensitive lysogens, it is usually sufficient to touch each colony to be tested with a toothpick and make short streaks in corresponding positions on two plates. One plate is incubated at 32°C, and the other, at 42°C. Lysogens will show no, or very little, growth at the higher temperature.

WORK WITH DELETION AND INSERTION DERIVATIVES OF PHAGE λ

Because the DNA of λ particles has unique ends, deletion and insertion mutants have chromosomes that are of abnormal length. Phage particles containing between about 75% and 108% of the reference type DNA (mass = 31.9 × 10^6) are reasonably stable and, unless they lack essential genes, are infective (Feiss et al. 1977).

Equilibrium centrifugation in a CsCl solution (see above, Phage Purification, method 3) provides a convenient method to determine the DNA content of a phage particle. Enrichment for phages with either larger or smaller genomes can be effected biologically as follows.

1. Enrichment of Insertion Mutants. Some bacterial *ptsM* mutations block penetration of λ DNA after phage adsorption, to a degree depending upon the DNA content of the phage particle and on the growth medium. On tryptone agar plates made up with Bacto-agar (Difco), λ reference type grows with an efficiency of about 0.2–0.5, but the λ*b*221 deletion mutant (78% DNA), with an efficiency of only about 10^{-6}. The efficiency of plating of lesser deletions is intermediate. Thus, these *ptsM*⁻ bacteria allow efficient enrichment of insertion derivatives in a population of λ deletion mutants. Enrichment of insertions is also possible in one cycle of growth, but the enrichment factors obtained are considerably smaller than the reciprocal efficiency of plating values. Note also that the largest insertions give the highest degree of enrichment (Emmons et al. 1975; Elliott and Arber 1978).

2. Enrichment and Selection of Deletion Mutants. Magnesium ions are necessary to maintain the structural integrity of the phage head. Chelating agents, such as EDTA or pyrophosphate, which remove magnesium ions, result in the bursting of any λ head with a DNA content in excess of 96% of wild type. Consequently, naturally occurring deletion derivatives of λ can be selected by exposure to chelating agents (Parkinson and Huskey 1971). Phenotypic variants of λ (λ*), resistant to chelating agents for reasons other than a lowered DNA content, exist in any population of λ phages. Multiple cycles of selection eliminate λ* phages. A simple way of achieving multiple rounds of selection is by demanding plaque formation on media containing chelating agents, either sodium pyrophosphate (1–20 mM) or EDTA (~0.3–1 mM).

The concentration of chelating agent needed varies not only depending on the DNA content of the phage and the size of the desired deletion but also on the batch of tryptone, or trypticase, agar used; it is determined empirically and must be high enough to eliminate a background of parental phage. Molten trypticase or tryptone agar is supplemented with EDTA from a stock solution (200 mM at pH 8) and then poured into level plates. These plates should be used within a few days, since evaporation will change the EDTA concentration. Confusingly, this change will vary from plate to plate. Some investigators add chelating agent to the top agar, others do not; Mg^{++}, however, should be avoided.

The isolation of independent deletions is assured by the following procedure. Single plaques are amplified by stabbing with a toothpick to a fresh lawn. After incubation, each area of lysis is harvested in a pasteur pipette and transferred to 0.5 ml of broth (no Mg^{++}) containing one drop of $CHCl_3$. These phage suspensions should have titers of 10^7–10^8 pfu/ml. 0.1-ml samples are plated with indicator bacteria on plates supplemented with the appropriate concentration of EDTA. After overnight incubation, one EDTA-resistant plaque is picked from each plate and repurified on EDTA plates. The resulting plaques will be pure clones of EDTA-resistant phage.

An alternative way of isolating deletion derivatives of λ phage that carry an amber mutation in gene *D* is described by Sternberg and Weisberg (1977) and Sternberg et al. (1979).

MUTAGENESIS

λ is a good substrate for mutagenesis either in vitro or in vivo, and the frequency of mutation to clear-plaque derivatives is a convenient monitor for the efficiency of mutagenic treatment. Spontaneous mutations are produced by a number of different mechanisms including, e.g., the insertion of a transposable genetic element. Therefore, the use of mutants induced by a specific mutagen may often be more advisable than that of spontaneous mutants.

Use of Mutagens

Chemical mutagenesis with hydroxylamine is simple and has the virtue of being highly specific (G-to-A transitions), efficient, and is effective in vitro (see Davis et al. 1980). A less specific alternative is based on mutagenic repair of UV-damaged phage (Weigle 1953). Phage stocks are irradiated with UV and used to infect cells whose SOS pathway is then induced by a brief exposure to UV (Enquist and Weisberg 1977).

2 ml of phage stock (10^9 pfu/ml) are exposed to UV (about 4000 erg/mm^2) in TMG as a thin film in a plastic petri dish. This reduces the number of plaque-forming units by about five orders of magnitude. The irradiated phage stock can be stored at 4°C. Infectious phage are recovered in the following way, avoiding exposure to bright light. 0.1 ml of cells and 0.1 ml of lysate are mixed as a droplet in a plastic petri dish and incubated for 5 minutes at room temperature. The infected cells are exposed to UV (300 erg/mm^2). 10-μl samples are dispensed to foil-covered tubes containing 1 ml of Luria broth and 10 mM $MgSO_4$. The samples are agitated at 38°C for 2 hours, and 2 drops of $CHCl_3$ are added. Each tube should contain 500–1000 pfu, and 2% of these should be clear-plaque mutations.

Use of Mutator Strains

Mutations in some bacterial genes (*mut* loci) endow the cell with an elevated frequency of mutation. The mutator allele *mutD5* (Fowler et al. 1974) can increase mutation frequencies by three to four orders of magnitude, if the cells are grown in rich broth. *mutD5* promotes all possible base-pair substitutions (transitions occur more frequently than transversions) and frameshift mutations. *mutT* (Treffer's mutator) is also very efficient, and it primarily causes A-to-C transversions. The following procedures have been used for the *mutD*5 strain (Enquist and Weisberg 1977).

1. To make 2-ml minilysates of mutagenized phage, infect 2×10^8 to 5×10^8 *mutD*5 cells in 10 mM $MgSO_4$ at a multiplicity of five phage per cell. Incubate 10 minutes for adsorption and then dilute 10-μl samples to each of 10–20 screw-cap tubes containing 2 ml of Luria broth and 10 mM $MgSO_4$. Shake at 38°C for 2 hours and then add one drop of $CHCl_3$ and titer. The lysates are stable at 4°C for at least 1 year.
2. To mutagenize single plaques, plate for single plaques at 38°C on *mutD*5 indicator bacteria on tryptone (or Luria broth) plates. After incubation, pick single plaques to tubes with 0.2 ml of TMG plus one drop of $CHCl_3$. Plate out phage to select or screen for mutants, and then pick, redilute, and plate to reclone the mutant phage.

DETECTION OF PARTICULAR TRANSDUCING PHAGES

Phage lines that carry inserts of non-λ DNA can sometimes be detected by functional tests, particularly if the function can compensate for a genetic defect in an *E. coli* strain. Such complementation tests can be made using either lysogens or lytic infection.

1. Detection as Lysogens. Lysogens of recombinant phages are made in the defective *E. coli* host, and either the phenotypes of selected lysogens are tested or lysogens of the appropriate phenotype are selected. Since many λ vectors lack genes or the attachment site necessary for lysogenization, a helper phage is used, and some dilysogens will include the hybrid phage. If the helper is heteroimmune, selection can be made for the immunity of the recombinant phage.

2. Detection by Lytic Infection. This method has the advantage that no helper phage is needed. It relies on the inability of λ phage to multiply in a host whose growth is blocked by a genetic defect. Should an infecting phage carry a gene whose product substitutes for the host deficiency, the cell grows, supports growth of the infecting phage, and is lysed. Multiple rounds of infection lead to plaque formation. Transcription of the cloned gene has, in some cases, been shown to rely solely on the phage promoters. Plaques dependent upon complementation take from 10 hours to a few days to develop when selected as follows.

The phage suspension is diluted in buffer, and no more than 10^6 phage are adsorbed to the appropriate indicator bacteria suspended in 10 mM $MgSO_4$. The infected bacteria are then plated on minimal agar plates in 2.5–3 ml of top agar (in water or minimal medium) supplemented with 10 mM $MgSO_4$. The plaques selected are of variable morphology, and quite turbid plaques have been seen from a λ*c*I$^-$ transducing phage. However, this is not always the case, and two alternative approaches ensure a visible lawn and, consequently, the detection of lysis. One (see Davis et al. 1980) uses sufficient indicator bacteria (~ 10^9/plate) to give a slightly turbid lawn in the absence of cell growth. Davis et al. (1980) also suggest long-wavelength UV irradiation after the addition of 10 μg of ethidium bromide as a means of "staining" areas of lysis (plaques). An alternative method (e.g., Borck et al. 1976) is to use fewer cells (e.g., 2×10^8) but sufficient supplement (e.g., one drop of broth to the top agar) so that the cells grow, but very poorly. Plaques are then readily distinguished from revertant colonies (they frequently do not look like classical plaques but have prominent fluffy or granular turbid centers).

Phage from a selected plaque can be purified nonselectively, but it is important to confirm the complementation response of phage recovered in this way.

3. Detection of LacZ or Suppressors. The screening for recombinant derivatives of many λ vectors relies on the color tests available for the detection of β-galactosidase. Lac^+ plaques are detected on either lactose MacConkey agar or on medium (e.g., tryptone) containing 5-bromo-4-chloro-3-indolyl-β-D-galactoside (Xgal) as substrate and 10 mM isopropyl-β-D-thiogalactoside (IPTG) as inducer (see 9 in the Appendix to this paper). Phages including the entire *lacZ* gene, or even the *Eco*RI fragment containing most of this gene, are detected by complementation with an appropriate *lacZ* host, e.g., *lac*ZM15, which has a small deletion in the *lacZ* gene but produces a polypeptide able to serve as an "omega-donor" in allelic complementation tests. Lac^+ plaques are red on lactose-MacConkey agar and blue on Xgal plates. In the former test, immune Lac^+ cells in the center of a plaque become red; in the latter, the substituted galactoside is hydrolyzed in the agar to give a nondiffusing blue pigment. This second test is equally applicable to clear and turbid plaques. An alternative test is for the repressor binding sequence of the *lac* operator. The binding of the *lac* repressor to the replicating phage genomes leads to derepression of the chromosomal *lacZ* gene. This test therefore uses only indicator substrate Xgal (no inducer) and a *lacZ*$^+$ host. Phages carrying amber suppressors are detected as Lac^+ plaques when the indicator cells have an amber mutation in the *lacZ* gene; inducer is added to the Xgal medium. Recombination with the host chromosome can replace the suppressor, *supF*, carried by a λ vector with its nonsuppressing allele, if the vector is grown in other than a *supF* host.

EXTRACTION OF λ DNA

The mass of λ^+ DNA is 3.2×10^7; i.e., 3×10^7 g of DNA contain 6×10^{23} molecules; 1 μg of λ DNA contains 2×10^{10} molecules, or 2×10^{10} phage particles contain 1 μg of DNA. DNA concentration is conveniently measured by adsorption of a solution at 259 nm; a solution having $A_{259} = 1.0$ has a concentration of 50 μg of DNA (double-stranded)/ml. Convenient concentrations of lambdoid phage DNA are 100–300 μg/ml, or A_{259} of 2–6, which would result from deproteinization of a phage suspension at a concentration of about 2×10^{12} phage/ml to 6×10^{12} phage/ml.

DNA preparation usually starts with a phage suspension after purification by equilibrium centrifugation in CsCl solution. The concentration may be between 10^{12} pfu/ml and 10^{14} pfu/ml, but this underestimates, often by a factor of two to three, the number of phage particles and, hence, DNA molecules. The phage sample should be diluted as necessary with 10 mM Tris-HCl (pH 8.0), 1 mM EDTA and dialyzed against 100 volumes of this buffer in the cold room for about 2 hours. The dialyzed solution is mixed with an equal volume of phenol (see below) and rolled gently at room temperature for about 1 minute. The liquid phases are allowed to separate, and the phenol phase (lower) is removed with a sterile pasteur pipette. The aqueous phase is extracted twice more with phenol in this way and then dialyzed against 10 mM Tris-HCl (pH 8.0), 1 mM EDTA (at least 100 volumes), the buffer being changed four times in the course of about 20 hours. Phenol may also be removed by extraction (three times) with chloroform (see note above, in To Preserve a λ Stock) or ether, the residual solvent being removed by evaporation in a stream of nitrogen. The DNA solution is then transferred to a sterile tube, the volume being most conveniently measured by weighing, and its absorbance at 235 nm, 259 nm, and 280 nm, measured against the dialysis buffer as blank in the spectrophotometer. The ratios of absorbance at 259/235 and 259/280 should be about 2.0. Dilute solutions of DNA may be concentrated by precipitation with ethanol (2–2.5 volumes) after making the solution 0.3 M in sodium acetate at pH 7. After several hours at −20°C, or about an hour at −70°C, the DNA is pelleted in the centrifuge (e.g., 10,000 rpm for 10 min), allowed to drain dry or dried briefly under vacuum, and dissolved in the requisite volume of 10 mM Tris-HCl (pH 8.0), 1 mM EDTA.

Preparation of the phenol is important. It can be distilled in a stream of nitrogen and stored frozen, or collected and stored for limited periods (at 4°C) under air-free distilled water and extracted immediately before use with an equal volume of 0.5 M Tris-HCl (pH 8.5) (the aqueous phase being discarded). Do not use phenol that has been exposed to air for long periods. A simpler alternative, preferred by some workers, is a mixture of phenol, *m*-cresol and 8-hydroxyquinoline (see 19. Preparation of Phenol without Distillation, in the Appendix to this paper); this reagent is stable for some weeks at room temperature (Blin and Stafford 1976). Dialysis tubing should be thoroughly washed with water and boiled before use for 10–15 minutes in 2 mM EDTA, and gloves should always be worn for preparing and handling DNA solutions (to protect the DNA, not your hands).

TRANSFECTION

For transfection, a modification of the $CaCl_2$ starvation procedure (Mandel and Higa 1970), in which the cells are grown in Luria broth (Lederberg and Cohen 1974), is used. An overnight culture is grown from a single colony, diluted

1:50 in fresh Luria broth, and grown to an OD_{650} of 0.5–0.6 (i.e., 2 × 10^8 cells/ml to 3 × 10^8 cells/ml). The cells are cooled on ice for 20 minutes or spun down directly in the cold (4000 rpm for 5–10 min). The cells are resuspended in 1/2 initial volume of ice-cold 0.1 M $CaCl_2$ and then repelleted immediately and resuspended in 1/20 initial volume of ice-cold 0.1 M $CaCl_2$. The cells are left on ice for at least 30 minutes (they will stay competent for some hours and probably a few days).

The DNA to be recovered is diluted freshly, e.g., in a mixture of saline sodium citrate (SSC) (see 15 in the Appendix to this paper), and 0.1 M $CaCl_2$. Use 0.1 ml of diluted DNA (never more than 100 ng) to 0.1 ml of competent cells. Leave on ice for 10 minutes. Heat shock at 37°C for 5 minutes (this is best done in thin-walled tubes). Plate in 2.5 ml of molten top agar (45°C) supplemented with 10 mM $MgSO_4$. If the plaques are to be transferred to nitrocellulose for hybridization, use agarose rather than agar in the top layer. Beware of the fact that agarose sets more quickly than agar. Good competent cells should give 10^6 pfu/μg of λ DNA.

IN VITRO PACKAGING OF λ AND RECOMBINANT DNA

In vitro packaging of λ DNA was first described in 1974 (Hohn and Hohn 1974). The procedure given here is a modification of one of the original methods (Becker and Gold 1975). Both donor strains are *recA*⁻ and λ-resistant; the prophage in BHB2690 is λ*imm*^{434}*cIts b2red3Dam15Sam7*, and that in BHB2688 is λ*imm*^{434}*cIts b2red3Eam4Sam7*.

Preparation of Packaging Extracts

1. Sonicated Extract from Induced Prehead Donor BHB2690. Streak out strain BHB2690 on a Luria plate and incubate overnight at 32°C. A duplicate plate incubated at 42°C should not show growth. Inoculate 500 ml of N-Z broth (see 6 in the Appendix to this paper) prewarmed to 32°C, with enough bacteria to give an OD_{600} of 0.08 to 0.1. Incubate with aeration at 32°C until an OD of 0.3 is reached. Induce by shaking the flask slowly in a waterbath at 45°C for 15 minutes. Incubate at 37°C for 3 hours with vigorous aeration. Check for successful induction by adding a drop of $CHCl_3$ to a small sample of the culture, which should clear in a few minutes. Spin down the cells (5000 rpm for 10 min), and remove all extra liquid. Resuspend the pellet in 2.5 ml of cold sonication buffer (see 17 in the Appendix to this paper). Sonicate with short blasts until the solution clears and is no longer viscous. Avoid foaming, and keep the tube on ice so that the temperature of the solution does not exceed 4°C. Pellet for 10 minutes at 10,000 rpm. To the supernatant (2 ml) add an equal volume of cold sonication buffer and one-sixth volume of packaging buffer (see 18 in the Appendix to this paper). Dispense 15-μl samples or multiples in precooled Eppendorf tubes and freeze the open tubes in liquid N_2. Store at −70°C.

2. *Freeze-thaw Lysate from Induced Packaging Protein Donor BH2688.* Inoculate three flasks containing 500 ml of prewarmed broth; otherwise, proceed exactly as with the prehead donor strain (BHB2690) up to the centrifugation of the induced cells. Remove all extra liquid and resuspend the pellets in a total of 2 ml of cold sucrose solution (10% w/v sucrose, 50 mM Tris-HCl at pH 7.5). Transfer the suspension to a precooled centrifuge tube, add 100 μl of fresh lysozyme solution (2 mg/ml in 0.25 M Tris-HCl at pH 7.5), and mix gently. Freeze in liquid N_2. Thaw in ice, add 100 μl of packaging buffer, and mix thoroughly. Centrifuge at 35,000 rpm for 1 hour at 4°C, collect the supernatant, distribute 10-μl portions into precooled Eppendorf tubes, and plunge the open tubes in liquid N_2. Store at −70°C.

In Vitro Packaging

Tubes are removed from the deep freezer (or the liquid N_2, if packaging is to be performed immediately after production of the extracts), and thawed on ice. 15 μl of the sonicated extract is added to one tube containing freeze-thaw extract, and the DNA to be packaged is added (up to a few micrograms in not more than 2 μl). The solution is mixed and incubated for 1 hour at room temperature. It is then diluted with 0.5 ml of TMG buffer and adsorbed to appropriate indicator bacteria (grown in Luria broth + 0.4% maltose to late exponential phase and transferred to 10 mM $MgSO_4$). The mixture is plated in top agar + 10 mM $MgSO_4$ on tryptone or trypticase plates (for cosmids, the appropriate selective drug is added). λ wild-type particles may be recovered with efficiencies up to (and infrequently more than) 2×10^8 pfu/μg of DNA.

APPENDIX
MEDIA, BUFFERS, AND REAGENTS

1. *Tryptone Broth Media*
   a. Tryptone Broth:

      10 g Bacto-tryptone (Difco)
      5 g NaCl
      1 liter H_2O
      pH 7.2

   b. Tryptone Agar:

      10 g Bacto-tryptone (Difco)
      5 g NaCl
      13 g Bacto-agar (Difco)
      1 liter H_2O
      pH 7.2

      The agar concentration can be varied between 10 g/liter and 15 g/liter. For many purposes, Bacto-agar (Difco) can be replaced by less well-purified, but cheaper, brands of agar. We recommend choosing an agar concentration giving a hardness comparable to that of 1.3% Bacto-agar (Difco). Supplementing tryptone or trypticase agar with 10 mM $MgSO_4$ has been shown to enhance the efficiency with which phage λ forms plaques. Petri dishes (15 × 100-mm) on a level surface should be filled one-half to two-thirds full of molten agar and not moved until the agar has solidified. For phage assays, the plates should be allowed to dry overnight (at least) before use.

2. *Soft Agar*
   In principle, one can prepare for each solid medium used, a corresponding soft agar (also called top agar), using 6.5 g of Bacto-agar (Difco)/liter to 8 g of Bacto-agar/liter. Often the exact content of the soft agar is unimportant, but care should be taken that one does not use a rich soft agar for plating on poor media. In principle, water soft agar could be used for plating on most solid media. Note that plates with soft agar top layer are difficult to replicate. For use as master plates in replication, the agar content should be increased (e.g., at least 8 g/liter).

   a. Tryptone Soft Agar:

      10 g Bacto-tryptone (Difco)
      5 g NaCl
      7 g Bacto-agar (Difco)
      1 liter H_2O
      pH 7.2

b. Water Soft Agar:

6.5 g Bacto-agar (Difco)
1 liter H_2O

3. *Stab agar*
A soft tryptone agar (frequently supplemented with thymidine) may be used for stabs, e.g.,

a. Soft Tryptone Agar:

9 g Bacto-tryptone (Difco)
7.5 g Bacto-agar (Difco)
5 g NaCl
0.1 g thymidine
1 liter H_2O
pH 7.2

b. Alternative Stab Agar:

8 g nutrient broth (Difco)
7.6 g Bacto-agar (Difco)
0.1 g thymidine (if required)
1 liter H_2O
pH 7.2

4. *L-broth Media*
A large variety of recipes are found in the literature under the names of LB or Luria broth and LA or Luria agar. These media all contain tryptone (usually 10 g/liter) and yeast extract (usually 5 g/liter). They also contain variable amounts of NaCl and are sometimes supplemented with glucose, Mg^{++}, and Ca^{++}. Be aware of this variety and do not blindly copy at random one of several recipes you may find. As an example, Miller (1972) lists in his excellent guide to experiments in molecular genetics two quite different recipes, one for LB and the other for Luria broth.

a. LB (Luria Broth):

10 g Bacto-tryptone (Difco)
5 g yeast extract (Difco)
10 g NaCl (some people use 5 g NaCl, as originally given by Lennox 1955)
1 liter H_2O
pH 7.2

If desired, complete before use with 10–30 ml of 10% glucose/liter and with other supplements such as Ca^{++} and Mg^{++}.

b. LA (Luria Agar):

10 g Bacto-tryptone (Difco)
5 g yeast extract (Difco)
10 g NaCl
10 g Bacto-agar (Difco) (some people use 15 g Bacto-agar)
1 liter H_2O
pH 7.2

If desired, add after sterilization 10–30 ml of 10% glucose/liter and other supplements, such as Ca^{++} and Mg^{++}.

5. *BBL Trypticase Agar*
10 g trypticase (Baltimore Biological Laboratories)
5 g NaCl
10 g Bacto-agar (Difco)
1 liter H_2O
pH 7.2

6. *N-Z Broth*
10 g N-Z-amine (Humko-Sheffield, Linnhurst, N.J.)
5 g NaCl
2 g $MgCl_2.6H_2O$
1 liter H_2O
pH 7.5

7. *MacConkey Agar*
35–40 g MacConkey Difco base
1 liter H_2O
Add desired sugar after sterilization
lactose (10 g/liter)
galactose (10 g/liter)

8. *Triphenyl-tetrazolium-chloride (TTC) Galactose Agar*
Supplement 1 liter of sterilized (molten) tryptone agar (see 1. Tryptone Broth Media) with

50 ml 20% galactose
2.5 ml TTC solution
TCC agar plates should be stored in the dark.

a. TTC Solution:

1 g 2,3,5-triphenyl-tetrazolium-chloride in
100 ml H_2O

Dissolve and boil 5 min (do not autoclave). Store in the dark at 4°C.

9. *Xgal;* lac *Indicator Plates (expensive)*
5-Bromo-4-chloro-3-indolyl-β-D-galactoside (Xgal; 40 mg/liter) is used with various media, such as BBL trypticase or tryptone agar. Isopropyl-β-D-thiogalactoside (IPTG; 1 mM) may be used as inducer. Alternatively, stock solutions of Xgal (20–50 μl) and IPTG (20 μl) may be added to the top agar instead of to the plates. The stock solutions, Xgal (20 mg/ml in dimethylformamide) and IPTG (20 mg/ml) should be stored at −20°C.

10. *Filtered Amino Acids*
Casaminio acids, technical grade (Difco): Add 30 g activated charcoal to 1 liter of 20% solution; leave 5 min, boil, and filter through paper to remove the charcoal.

11. *M9 Synthetic Media*
Prepare the following stock solutions and sterilize.

Salt Mixture:

70 g Na_2HPO_4
30 g KH_2PO_4
5 g NaCl
10 g NH_4Cl
1 liter H_2O

Glucose (4%)

Calcium Chloride:

10 mM: 1.5 g $CaCl_2.2H_2O$; 1 liter H_2O

Magnesium Sulphate:

100 mM: 2.5 g $MgSO_4.7H_2O$; 100 ml H_2O

Ferric Citrate:

0.3%: 0.3 g $FeC_6H_5O_7.3H_2O$; 100 ml H_2O

a. *M9 Medium*
Mix as follows:

780 ml sterile H_2O
100 ml salt mixture
100 ml glucose (4%)
10 ml $CaCl_2$ (10 mM)
10 ml $MgSO_4$ (100 mM)
0.2 ml ferric citrate solution

b. *M9a Medium*
M9 supplemented with 20% casamino acids; 5 ml/100 ml

12. *Nozu Supplements*
Make medium:

0.3% in glucose
7.5×10^{-2} mM in $CaCl_2$
2 mM $MgSO_4$
4×10^{-3} mM in $FeCl_3$ or $FeSO_4$

13. *3XD Medium*
4.5 g KH_2PO_4
10.5 g Na_2HPO_4
3 g NH_4Cl
882 ml H_2O

After sterilization, add:

12 ml 100 mM $MgSO_4$
3 ml 100 mM $CaCl_2$
30 ml glycerol
6 ml gelatin (0.5%)
67 ml casamino acids (20%)

14. *SMC (Salt Mg Ca)*
100 ml salt mixture (same as for M9; see 11).
10 ml 100 mM $MgSO_4$
10 ml 10 mM $CaCl_2$
880 ml H_2O

15. *SSC (Saline Sodium Citrate)*
150 mM NaCl
15 mM Na citrate

16. *Tris Buffers*
Tris buffers are used in a large variety of recipes. The following are two examples.

a. TMG (Tris Mg Gelatin). Dilution Buffer for Phage:

10 mM Tris base
10 mM $MgSO_4$
0.01% gelatin
Adjust to pH 7.4 with HCl

b. Tris Salts:

100 mM Tris base
0.5% NaCl
0.1% NH_4Cl
Adjust to pH 7.5 with HCl

17. *Sonication Buffer (for in Vitro Packaging Extracts)*
    20 mM Tris-HCl (pH 8.0)
    1 mM EDTA
    3 mM $MgCl_2$
    5 mM β-mercaptoethanol

18. *Packaging Buffer*
    6 mM Tris-HCl (pH 8.0)
    50 mM spermidine
    50 mM putrescine
    20 mM $MgCl_2$
    30 mM ATP
    30 mM β-mercaptoethanol

19. *Preparation of Phenol without Distillation*
    1 lb. bottle Mallinkrodt phenol
    100 ml 2 M Tris-HCl (pH 7.8)
    130 ml H_2O

    Heat (37°C or 45°C) with shaking until dissolved (there should be barely an aqueous phase)

    Add:

    25 ml *m*-cresol
    1 ml 2-mercaptoethanol
    500 mg 8-hydroxyquinoline

20. *Phosphate Buffer*
    7 g Na_2HPO_4
    4 g NaCl
    3 g KH_2PO_4
    1 liter H_2O

    For use, make 1 mM in $MgSO_4$

REFERENCES

Appleyard, R.K. 1954. Segregation of new lysogenic types during growth of a doubly lysogenic strain derived from *E. coli* K12. *Genetics* **39:** 440.

Bachmann, B.J. 1972. Pedigrees of some mutant strains of *Escherichia coli* K12. *Bacteriol. Rev.* **36:** 525.

Bachmann, B.J. and K.B. Low. 1980. Linkage map of *Escherichia coli* K12. *Microbiol. Rev.* **44:** 1.

Baumann, M.F. and D.I. Friedman. 1976. Cooperative effects of bacterial mutations affecting *N* expression. II. Isolation and characterization of mutations in the *rif* region. *Virology* **73:** 128.

Becker, A. and M. Gold. 1975. Isolation of the bacteriophage lambda A protein. *Proc. Natl. Acad. Sci.* **72:** 581.

Bellet, A.J.D., H.G. Busse, and R.L. Baldwin. 1971. Tandem genetic duplications in a derivative of phage lambda. In *The bacteriophage lambda* (ed. A.D. Hershey), p. 501. Cold Spring Harbor Laboratory, Cold Spring Harbor, New York.

Berman, M., L. Enquist, and T. Silhavy. 1983. *Experiments with Gene Fusions.* Cold Spring Harbor Laboratory, Cold Spring Harbor, New York. (In press.)

Bertani, L.E. and G. Bertani. 1970. Preparation and characterization of temperate, non-inducible bacteriophage P2. *J. Gen. Virol.* **6:** 201.

Blattner, F.R., B.G. Williams, A.E. Blechl, K. Denniston-Thompson, H.E. Faber, L. Furlong, D.S. Grunwald, D.O. Kiefer, D.D. Moore, J.W. Schumm, E.L. Sheldon, and O. Smithies. 1977. Charon phages: Safer derivatives of bacteriophage lambda for DNA cloning. *Science* **196:** 161.

Blin, N. and D.W. Stafford. 1976. A general method for isolation of high molecular weight DNA from eukaryotes. *Nucleic Acids Res.* **3:** 2303.

Borck, K., J.D. Beggs, W.J. Brammar, A.S. Hopkins, and N.E. Murray. 1976. The construction *in vitro* of transducing derivatives of phage lambda. *Mol. Gen. Genet.* **146:** 199.

Campbell, A. 1965. The steric effect in lysogenization by bacteriophage lambda. I. Lysogenization by a partially diploid strain of *E. coli* K12. *Virology* **27:** 329.

Davidson, N. and W. Szybalski. 1971. Physical and chemical characterization of lambda DNA. In *The bacteriophage lambda* (ed. A.D. Hershey), p. 45. Cold Spring Harbor Laboratory, Cold Spring Harbor, New York.

Davis, R.W., D. Botstein, and J.R. Roth. 1980. *A manual for genetic engineering. Advanced bacterial genetics.* Cold Spring Harbor Laboratory, Cold Spring Harbor, New York.

Elliott, J. and W. Arber. 1978. *E. coli* K12 *pel* mutants, which block phage λ DNA injection, coincide with *ptsM*, which determines a component of a sugar transport system. *Mol. Gen. Genet.* **161:** 1.

Emmons, S.W., V. MacCosham, and R.L. Baldwin. 1975. Tandem genetic duplications in phage lambda. III. The frequency of duplication mutants in two derivatives of phage lambda is independent of known recombination systems. *J. Mol. Biol.* **91:** 133.

Enquist, L.W. and R.A. Weisberg. 1976. The red plaque test: A rapid method for identification of excision defective variants of bacteriophage lambda. *Virology* **72:** 147.

——. 1977. A genetic analysis of the *att-int-xis* region of coliphage lambda. *J. Mol. Biol.* **111:** 97.

Feiss, M., R.A. Fisher, M.A. Crayton, and C. Egner. 1977. Packaging of the bacteriophage λ chromosome: Effect of chromosome length. *Virology* **77:** 281.

Fowler, R.G., G.E. Degner, and E.C. Cox. 1974. Mutational specificity of a conditional *Escherichia coli* mutator, *mutD5*. *Mol. Gen. Genet.* **133:** 179.

Gillen, J.R., D.K. Willis, and A.J. Clark. 1981. Analysis of the RecE pathway of genetic recombination in *Escherichia coli* K12. *J. Bacteriol.* **145:** 521.

Goldberg, A.R. and M. Howe. 1969. New mutations in the *S* cistron of bacteriophage λ affecting host cell lysis. *Virology* **38:** 200.

Gottesman, M.E. and M.B. Yarmolinsky. 1968. Integration-negative mutants of bacteriophage λ. *J. Mol. Biol.* **31:** 487.

Gottesman, M.M., M.L. Hicks, and M. Gellert. 1973. Genetics and function of DNA ligase in *Escherichia coli. J. Mol. Biol.* **77:** 531.

Hohn, B. 1979. *In vitro* packaging of λ and cosmid DNA. *Methods Enzymol.* **68:** 299.

Hohn, B. and T. Hohn. 1974. Activity of empty, headlike particles for packaging of DNA of bacteriophage λ *in vitro*. *Proc. Natl. Acad. Sci.* **71:** 2372.

Hohn, T., H. Flick, and B. Hohn. 1975. Petit λ, a family of particles from coliphage lambda infected cells. *J. Mol. Biol.* **98:** 107.

Hubacek, J. and S.W. Glover. 1970. Complementation analysis of temperature sensitive host specificity mutations in *Escherichia coli. J. Mol. Biol.* **50:** 111.

Lecocq, J.P. and A.M. Gathoye. 1973. Une mutation bacterienne augmentant la frequence de lysogenisation par le phage λ. *Arch. Int. Biochem. Physiol.* **81:** 803.

Lederberg, E.M. and G.N. Cohen. 1974. Transformation of *Salmonella typhimurium* by plasmid deoxyribonucleic acid. *J. Bacteriol.* **119:** 1072.

Lennox, E.S. 1955. Transduction of linked genetic characters of the host by bacteriophage P1. *Virology* **1:** 190.

Mandel, M. and A. Higa. 1970. Calcium-dependent bacteriophage DNA infections. *J. Mol. Biol.* **53:** 159.

Maniatis, T., E.F. Fritsch, and J. Sambrook. 1982. *Molecular Cloning. A Laboratory Manual.* Cold Spring Harbor Laboratory, Cold Spring Harbor, New York.

Miller, J.H. 1972. *Experiments in molecular genetics.* Cold Spring Harbor Laboratory, Cold Spring Harbor, New York.

Mizusawa, S. and D.F. Ward. 1982. A bacteriophage lambda vector for cloning with *Bam*HI and *Sau*3A. *Gene* **20:** 317.

Murray, N.E., W.J. Brammar, and K. Murray. 1977. Lambdoid phages that simplify the recovery of *in vitro* recombinants. *Mol. Gen. Genet.* **150:** 53.

Parkinson, J.S. and R.J. Huskey. 1971. Deletion mutants of bacteriophage lambda. Isolation and initial characterisation. *J. Mol. Biol.* **56:** 369.

Sober, H.A., ed. 1970. *The handbook of biochemistry and molecular biology*, p. J-292. Chemical Rubber, Cleveland Ohio.

Sternberg, N. and R. Weisberg. 1977. Packaging of coliphage lambda DNA. II. The role of the gene D protein. *J. Mol. Biol.* **117:** 733.

Sternberg, N., D. Hamilton, L. Enquist, and R. Weisberg. 1979. A simple technique for the isolation of deletion mutants of phage lambda. *Gene* **8:** 35.

Weigle, J.J. 1953. Induction of mutations in a bacterial virus. *Proc. Natl. Acad. Sci.* **39:** 628.

Weigle, J., M. Meselson, and K. Paigen. 1959. Density alterations associated with transducing ability in bacteriophage lambda. *J. Mol. Biol.* **1:** 379.

Wood, W.B. 1966. Host specificity of DNA produced by *E. coli*: Bacterial mutations affecting the restriction and modification of DNA. *J. Mol. Biol.* **16:** 118.

Yamamoto, K.R., B.M. Alberts, R. Benzinger, L. Lawthorne, and G. Treiber. 1970. Rapid bacteriophage sedimentation in the presence of polyethylene glycol and its application to large-scale virus purification. *Virology* **40:** 734.

Appendices

APPENDIX I

A Molecular Map of Coliphage Lambda

Compiled by Donna L. Daniels, John L. Schroeder,* Waclaw Szybalski,† Fred Sanger,‡ and Frederick R. Blattner**
**Laboratory of Genetics and †McArdle Laboratory*
University of Wisconsin, Madison, Wisconsin 53706, and
‡MRC Laboratory of Molecular Biology, Cambridge, England

The molecular map of λ is presented in the following sections:

Figure 1 is a scale drawing of the λ map.

Table 1 presents our tabulation of all the sites on the λ map. Listed are the names of all mapped sites (columns A and E), their base-pair coordinates on the 5′-3′ *l* strand of λ DNA (column C), four-character symbols used by the computer for sequence annotation in Appendix II (column D), a description of the site (columns F and H), and the method of localization on the map (column G). Restriction sites that have been mapped experimentally are shown in column A. All other sites are specified in column E.

Table 2 lists the published sources of λ DNA sequence.

Table 3 shows the sources of mapping information for sites other than genes.

Table 4 shows the sources of mapping information for protein-coding genes and open reading frames (orfs).

Appendix II presents the complete sequence of λ with all mapping information.

SOURCES

The sources of information used to compile the molecular map of phage λ DNA are

1. The complete nucleotide sequence of λ. Various groups have sequenced regions of different strains of λ. This information has been collated, and sequences have been compared from all sources to determine overlaps and discrepancies. This information forms eight "sequence blocks," which are indicated in Table 1 (column B) by ditto marks enclosed by the symbols ▽ and △. Within these blocks are only four unresolved discrepancies (single base-pair changes). Sanger et al. (1982) presented the complete sequence of λ*c*I*ind*1*ts*857*S*7 based on sequence determined in their laboratory and on compilation and comparison with published λ sequences (see Table 2). The sequence that has been annotated is that compiled by Sanger et al. (1982) changed to "wild type." These changes are *ind*1 to *ind*⁺ at 37589; *c*I*ts*857 to

*c*I⁺ at 37742; *Sam*7 to *S*⁺ at 45352; and C to T correction of a typographical error in Sanger et al. (1982) at 44374.

2. A map of 171 restriction sites. This map of 20 enzymes was determined by gel electrophoresis and is presented in column A of Table 1. It was constructed by merging the published 67-site map for 15 enzymes (Daniels and Blattner 1982) with the individual enzyme maps of *Pvu*II of L. MacHattie (pers. comm.), *Hin*dII (*Hin*cII) of Robinson and Landy (1977a,b), *Pst*I (Smith et al. 1976; Legerski et al. 1978), *Sph*I, *Mst*II (D.L. Daniels, unpubl.), and *Cla*I (Mayer et al. 1981; D.L. Daniels, unpubl.). The restriction map predicted from the DNA sequence agrees with the experimental map, confirming both.
3. Compilation of electron micrographic mapping information by Szybalski and Szybalski (1979).
4. The λ gene map by Echols and Murialdo (1978).
5. Sequence changes of a large number of mutations, both from published sequences of λ variants and personal communications from λ workers (see Table 3).

NUMBERING SYSTEM

The λ sequence as presented includes both of the sticky ends. The numbering system is as follows: Base 1 is the first (5′) base of the left sticky end of the *l* strand of the λ DNA molecule as isolated from the capsid. This base is denoted LEND (left end). The first base of the double-stranded portion (denoted CLND [complementary strand left end]) is 13; the last base of the double-stranded portion (denoted REND [right end]) is 48502; the last base of the right sticky end (denoted CRND [complementary strand right end]) is 48514. The length of the circular form of λ DNA is 48502 bp and runs from LEND to REND.

ANNOTATION CONVENTIONS

Annotations may refer to one of three features of sequence: a particular base, an internucleotide space, or a map interval.

1. The particular base form is used for point mutations and a few other sites. The number refers to the affected base.
2. Internucleotide spaces are annotated for restriction cut sites, hybrid (hy) phage junctions, and insertion mutations (i). The space is located between the numbered base and the one following it. Restriction sites were obtained by computer search for the recognition sites of the enzymes tabulated in Appendix II and labeled at the cut site on the *l* strand. Predicted lengths of *l* strand fragments can be obtained by subtraction. Lengths of *r* strand fragments may need to be corrected depending on the position of the 3′-cut site for the particular enzymes used.

3. Map intervals are annotated for genes, open reading frames, deletions, substitutions, sequence blocks, λdv plasmids, and transcriptional units. These annotations occur in left-right pairs. The numbers refer to the first and last bases of the interval, respectively. For genes, these are the first base of the ATG or GTC start codon and the third base of the last amino acid codon, not including the termination codon. For deletions and substitutions the numbers refer to the first and last bases of λ that are deleted. Sequence block markers refer to the first and last bases sequenced in the particular reference. For λdv plasmids the annotated interval is the region of λ that is present in the plasmid (not the region of λ deleted to create the plasmid). For transcription units the two ends of the RNA molecule are denoted.

ACCURACY

Most sites have been mapped precisely by DNA sequencing. A few amber mutations could be assigned exactly by fine-structure recombinational mapping combined with patterns of suppression.

Sites that have been mapped less precisely are also indicated. Deletion, substitution, and insertion endpoints determined by heteroduplex analysis were translated from the %λ map of Szybalski and Szybalski (1979) to the base-pair coordinate system by linear interpolation between landmark sites, as described by Daniels et al. (1980). Coordinates for such sites are preceded by a tilde (~). The method used to determine each site is indicated in Table 1 (column G).

Gene endpoints are indicated in italics in Table 1 (column D). Their most probable locations were determined from the sequence based on the presence of an open reading frame starting with an AUG or a GUG, the presence of a possible ribosome binding site (RBS; Shine and Dalgarno 1974; Stormo 1982a,b), similarity to the codon usage of other λ genes (Staden and McLachlan 1982), and consistency with the %λ gene maps of Echols and Murialdo (1978) and Szybalski and Szybalski (1979). In several cases, the gene assignments were confirmed by determination of the first few aminoterminal amino acids (see Table 4).

LAMBDA GENES

Head Assembly (see p. 281)	*Nu1*, *A*, *FI* DNA packaging. *Nu1* and *A* proteins are components of terminase, which cleaves at *cos*. *W*, *FII* Head completion. *FII* protein is needed for head-tail joining. *B*, *C* Initiation of head assembly. *B* protein forms first intermediate; also involved in head-tail joining. *Nu3* Prohead core; a scaffolding protein. *E* Major head structural protein. *D* Structural protein involved in stabilization of head.
Tail Assembly (see p. 334)	*Z*, *U* Tail completion. *U* is also involved in head-tail joining. *V* Major tail structural protein. *G*, *H*, *M*, *L*, *K*, *I*, *J* Tail initiation. *J* determines host range and presumably codes for the tail fiber; *H* determines tail length. *T* Role in tail assembly unknown.
Recombination	*int*, *xis* Components of site-specific recombination system for prophage integration and excision; catalyzes recombination between *att* sites. *exo*, *bet* General recombination system (Red system); (*exo*) exonuclease; (*bet*) promotes DNA renaturation. *gam* Binds to and inhibits *recBC* enzyme.
Exclusion	*ral* Restriction alleviation; protects λ from host *Eco*B and *Eco*K systems. *sieB* Superinfection exclusion by λ prophage in *S. typhimurium*; *sieB* is probably *git*. *rexB*, *rexA* Exclusion system of a λ prophage; excludes T4 *r*II mutants and other phages. *ren* Protects λ from Rex-dependent exclusion.
Regulation	*c*I "λ repressor"; maintains prophage state by repressing transcription from p_L and p_R and by activating transcription from p_M; binds to operators o_L and o_R. *c*II Establishment of lysogeny; activates transcription of *int* (from p_I) and of *c*I (from p_E). *c*III Establishment of lysogeny; regulates stability of *c*II protein. *cro* Repressor that acts during lytic growth to turn off transcription from p_L, p_R, and p_M. *N* Transcription antiterminator; allows transcription initiated at p_L and p_R to continue into the delayed early genes. *Q* Transcription antiterminator; allows transcription initiated at p'_R to continue into the late genes.
Replication	*O*, *P* DNA replication initiation. *O* protein binds to *ori*; *P* protein binds to *O* and host proteins.
Lysis	*S*, *R*, *Rz* Cellular lysis. *S* disrupts inner membrane function; *R*, λ endolysin, degrades the cell wall.
Other	*lom* λ outer membrane protein. *kil* Killing of host upon prophage induction. *ben* A DNase; Ea59 protein. *ssb* Single-stranded DNA-binding protein; Ea10 protein.

Genes are listed in left-to-right map order. Sites are described on page 473. Prepared by Ira Herskowitz.

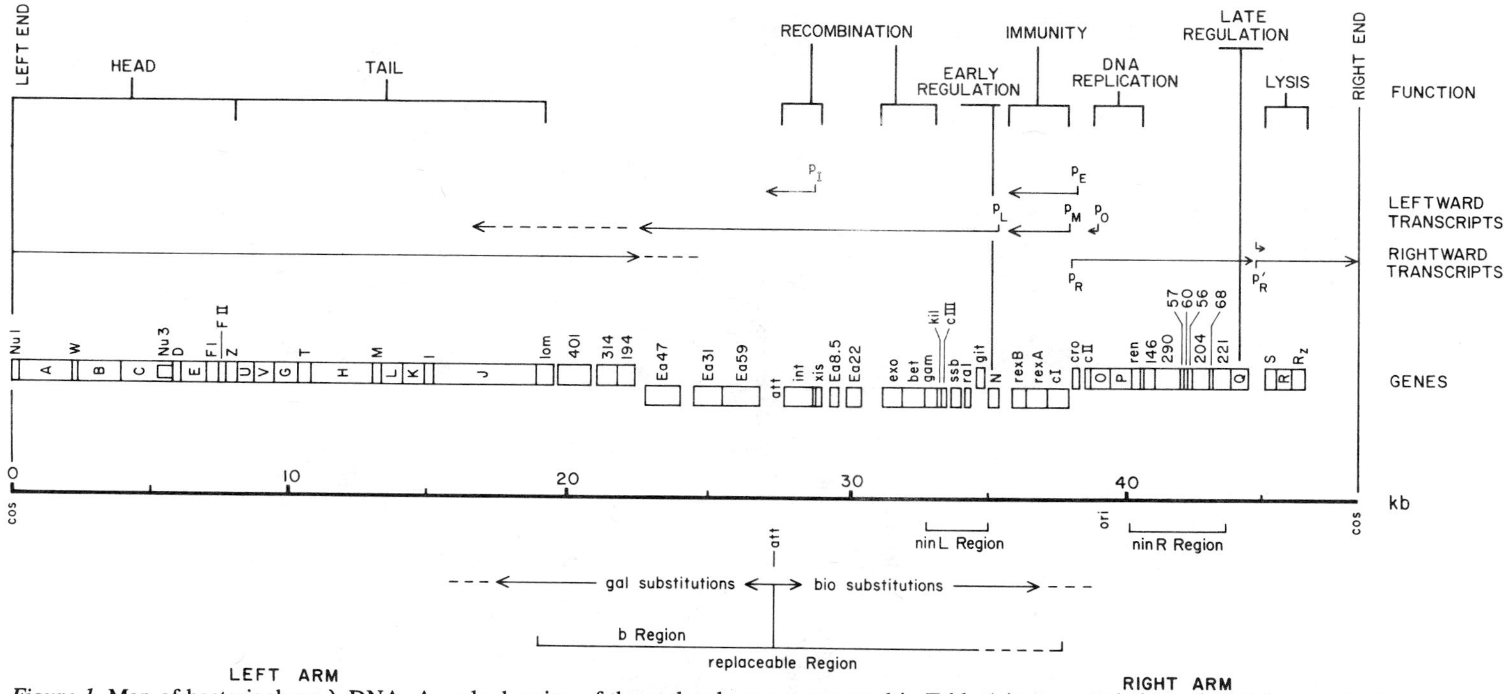

Figure 1. Map of bacteriophage λ DNA. A scale drawing of the molecular map presented in Table 1 is presented above the kilobase scale, beginning and ending at the cohesive end site (*cos*). Gene clusters with related functions are indicated above the brackets, and the regulatory genes, *N* (involved in early regulation) and *Q* (involved in late regulation) are indicated by vertical lines. Known promoters are denoted by *p* with a subscript to indicate their unique points of origin. (p_I) *int* protein promoter; (p_E) establishment promoter for *c*I; (p_M) maintenance promoter for *c*I: (p_L) major leftward promoter; (p_R) major rightward promoter; (p_O) *oop* promoter; (p_R') late promoter. (→) Extent and direction of transcription; (----) readthrough. Map positions of genes as determined by analysis of open reading frames (ORFs) in the DNA sequence are indicated. Known genes (identified either genetically by mutation analysis or functionally by SDS-gel electrophoresis of protein product, or both) are indicated with letter names, whereas ORFs, presumed from the sequence to code for protein products but with no previously known gene assignments, are given numbers corresponding to the coding capacity of the ORF. Major areas of substitution and deletion mutations are indicated below the scale. (*att*) attachment site; (*ori*) origin of replication.

KEY TO TABLE 1

Columns:

(A)	Restriction site
(B)	Region sequenced by more than one laboratory (see Table 2) (between symbols ▽ and △)
(C)	Base-pair coordinates (if the number is preceded by a tilde [~] the site is approximate)
(D)	Gene or site symbol for computerized sequence annotation in Appendix II (gene endpoints are in italics)
(E)	Name of the site
(F)	Site description (see below)
(G)	Method of mapping (see below)
(H)	Comments

Site Descriptions (F):

dl	Left end of a deletion
dr	Right end of a deletion
sl	Left end of a substitution
sr	Right end of a substitution
dl-*att*	Left end of a deletion whose right end is at *att*
att-dr	Right end of a deletion whose left end is at *att*
sl-*att*	Left end of a substitution whose right end is at *att*
att-sr	Right end of a substitution whose left end is at *att*
i	Insertion
5′	First base (startpoint) of an RNA transcript
3′	Last base (endpoint) of an RNA transcript
bs	Protein binding site
pm	Point mutation
M-NH3	Amino terminus of an orf beginning with AUG
V-NH3	Amino terminus of an orf beginning with GUG
COOH	Carboxyl terminus of an orf; corresponds to the last base of the coding sequence
hy	λ::lambdoid junction in hybrid phages; the genotype of the hybrid and the particular novel junction mapped (::) are specified in Comments
plasmid-*l*	Left endpoint on the λ map for the λdv plasmid
plasmid-*r*	Right endpoint on the λ map for the λdv plasmid
f(*l*)	Left end of a site possibly significant for biological function (see f in Methods of Mapping)
f(c)	Center of a site possibly significant for biological function (see f in Methods of Mapping)

Methods of Mapping (G):

sq	Wild-type or mutant site sequenced
RNA	RNA end determined by sequencing or fine-structure S1 mapping of the RNA, as well as promoterlike or terminatorlike characteristics in the sequence
AA	Amino acids at amino terminus of a protein were sequenced
G	Genetic analysis: gene assigned to a particular orf or startpoint based on genetic evidence that the gene spans a sequenced site (e.g., restriction site, point mutation, insertion site, and so on); genetic analysis of point mutations by fine-structure mapping and suppression phenotypes, which allows accurate assignment to the sequence; genetic mapping of a biologically significant site to within a limited region on the sequence
RES	Restriction mapping of mutations, which modify the restriction pattern
EM	Mapping by electron microscopy of heteroduplexes
f	Site assumed from features of the sequence, such as homology to other such sites and/or mapping of a biological function to the vicinity of the sequence feature; position of the presumed left end (*l*) or center (c) of the sequence is specified

TABLE 1 A MOLECULAR MAP OF BACTERIOPHAGE LAMBDA

A	B	C	D	E	F	G	H
ZEND		0	ZEND	Zero END			used for fragment endpoints
	▽	1	SB1L	Sequence Block 1L	start		
	"	1	LEND	Left END		sq	first base of sticky end (l-strand)
	"	13	CLND	3' end of r-strand			first base of double stranded DNA
	△	72	SB1L	Sequence Block 1L	end		
		191	*NU1*	Nul (orf-181)	M-NH3	G	
		~194	NU16	Nul t16	i	EM	
HIN2		199					
PVU2		211					
BGL2		415					
		711	*A*	A (orf-641)	V-NH3		
		733	*NU1*	Nul	COOH		
HIN2		734					
HPA1		734					
		~919	AROE	λaroE	sl	EM	
		~919	TRKA	λtrkA	sl	EM	
		~919	SPC2	λspc2	sl	EM	
		~919	TRKA	λtrkA	sl	EM	
		~919	SPC2	λspc2	sl	EM	
		~919	FUS2	λfus2	sl	EM	
		~919	SPC1	λspc1	sl	EM	
		~968	TP2	Tp2	i(Tn402)	EM	
		~1305	G258	λgal258	sl-att	EM	
		~1451	G914	λgalA914	sl-att	EM	
		~1838	GAL3	λgal3	sl-att	EM	
		~1886	G102	λgalT-N102	sl-att	EM	
PVU2		1919					
SPH1		2216					
PVU2		2387					
PVU2		2528					
PST1		2560					
		2633	*W*	W (orf-68)	M-NH3		
		2633	*A*	A	COOH		
		2642	WAM	Wam403	pm	sq	C to T
PST1		2824					
		2836	*B*	B (orf-533)	M-NH3		
		2836	*W*	W	COOH		
		2902	B*	B*		AA	start processed protein of capsid
		~2950	RF18	λrif-18	sl-att	EM	
		~2950	RF20	λrif-20	sl-att	EM	
PVU2		3060					

PST1	3629					
PVU2	3639					
PST1	3644					
	~3675	G24	λgal24	sl-att	EM	
PST1	3860					
	~4111	S104	λgalS104	sl-att	EM	
CLA1	4199					
	~4255	S101	λgalS101	sl-att	EM	
PST1	4374					
	4418	*C*	C (orf-439)	V-NH3		
	4434	*B*	B	COOH		
	~4544	GALC	λgalC	sl-att	EM	
PST1	4713					
AVA1	4720					
PST1	4913					
PST1	5124					
	5132	*NU3*	Nu3 (orf-201)	M-NH3	AA	
PST1	5218					
HIN2	5269					
HPA1	5269					
BAM1	5505					
PST1	5686					
HIN2	5710					
HPA1	5710					
	5734	*NU3*	Nu3	COOH		
	5734	*C*	C	COOH		
	5747	*D*	D (orf-110)	M-NH3	AA	
	6076	*D*	D	COOH		
	6135	*E*	E (orf-341)	M-NH3		
	~6287	KN11	kan11	i	EM	
	7157	*E*	E	COOH		
	7202	*FI*	FI (orf-132)	M-NH3		
	~7497	G730	λgalF730	sl-att	EM	
	7597	*FI*	FI	COOH		
	7612	*FII*	FII (orf-117)	V-NH3	AA	
PVU2	7833					
HIN2	7950					
HPA1	7950					
	7962	*FII*	FII	COOH		
	7977	*Z*	Z (orf-192)	M-NH3		
	~8125	GALZ	λgalZ	sl-att	EM	
HIN2	8201					
HPA1	8201					
PST1	8524					
	8552	*U*	U (orf-131)	M-NH3		
	8552	*Z*	Z	COOH		
BCL1	8844					masked by dam methylation

TABLE 1 (continued)

A	B	C	D	E	F	G	H
		8944	*U*	U	COOH		
		8955	*V*	V (orf-246)	M-NH3	AA	
HIN2		9056					
		~9335	GALG	λgalG	sl-att	EM	
BCL1		9361					masked by dam methylation
		~9577	TRKA	λtrkA	sr	EM	
PST1		9617					
HIN2		9626					
		9692	*V*	V	COOH		
		9711	*G*	G (orf-140)	M-NH3		
PST1		9781					
		10115	*T*	T (orf-144)	V-NH3		
		10130	*G*	G	COOH		
		10542	*H*	H (orf-853)	M-NH3		
		10546	*T*	T	COOH		
		~10883	H706	λgalH706	sl-att	EM	
HIN2		11585					
HPA1		11585					
PST1		11767					
PST1		11839					
PVU1		11937					
SPH1		12006					
PVU2		12101					
PVU2		12164					
		13100	*M*	M (orf-109)	M-NH3		
		13100	*H*	H	COOH		
		13426	*M*	M	COOH		
		13429	*L*	L (orf-232)	M-NH3		
		~13743	GALK	λgalK	sl-att	EM	
HIN2		13785					
BCL1		13820					masked by dam methylation
		14124	*L*	L	COOH		
		14276	*K*	K (orf 199)	M-NH3		
PST1		14298					
PST1		14385					
		14773	*I*	I (orf-223)	M-NH3		protein not identified
	▽	14869	SB2	Sequence Block 2	start		
	"	14872	*K*	K	COOH		
HIN2	"	14993					masked by EcoK modification
HPA1	"	14993					masked by EcoK modification
	"	15441	*I*	I	COOH		

	"	15505	*J*	J (orf-1132)	M-NH3		
CLA1	"	15584					masked by dam methylation
PVU2	"	16080					
PST1	"	16085					
CLA1	"	16121					
PST1	"	16235					
KPN1	"	17057					
HIN2	"	17076					
PST1	"	17394					
	"	18482	T'J4	t'j4 terminator	3'	RNA	terminates leftward transcription
KPN1	"	18560					
	"	18597	T'J3	t'j3	3'	RNA	terminates leftward transcription
	"	~18630	H434	h434	sl	EM	
	"	18637	T'J2	t'j2	3'	RNA	terminates leftward transcription
	"	18671	T'J1	t'j1	3'	RNA	terminates leftward transcription
HIN2	"	18756					
	"	18900	*J*	J	COOH		
	"	18965	*LOM*	lom (orf-206)	M-NH3		
	Δ	18969	SB2	Sequence Block 2	end		
		~19066	H434	h434	sr	EM	
		19369	LAC5	lac5	sl	sq	
AVA1		19397					
SMA1		19399					
		19582	*LOM*	lom	COOH		
		~19646	B221	b221	dl	EM	
		~19646	B519	b519	dl	EM	
		19650	*ORF*	orf-401	M-NH3		
		~19700	B189	b189	dl-att	RES	
PVU2		19718					
PST1		19837					
HIN2		19841					
		~19985	PGL1	λpgl1	sl-att	EM	
PVU2		20061					
PST1		20285					
SST2		20323					
		~20420	B536	b536	dl	EM	
		~20469	G130	λgal130	sl-att	EM	
SST2		20533					
HIN2		20569					
PVU2		20697					
		~20809	B538	b538	dl	EM	
		20852	*ORF*	orf-401	COOH		
AVA1		20999					
		21029	*ORF*	orf-314	M-NH3		
ECR1		21226					
		~21292	B506	b506	dl	EM	
SST2		21609					

TABLE 1 (continued)

A	B	C	D	E	F	G	H
	▽	21661	SB3	Sequence Block 3	start		
	‖	21714	DIFF	strain difference		sq	A in cb2 G in cI857S7
	△	21737	SB3	Sequence Block 3	end		
		21738	B2	b2	dl-att	sq	
		~21825	REV	λrev	sl	EM	
HIN2		21904					
HPA1		21904					
		21970	*ORF*	orf-314	COOH		
		21973	*ORF*	orf-194	M-NH3		
BAM1		22346					
BGL2		22425					
PST1		22425					
		~22530	C105	λcam105	i(Tn9)	RES	dl for deletions pMS370;968;969;972
		22554	*ORF*	orf-194	COOH		
		22689	*EA47*	Ea47	COOH		
		~22840	B536	b536	dr	EM	
PVU2		22993					
		~23033	B519	b519	dr	EM	
		~23082	AROE	λaroE	sr	EM	
HIN3		23130					
	▽	23131	SB4	Sequence Block 4	start		
HIN2	‖	23147					
	‖	23231	PBL	pBL	5'	RNA	leftward promoter in b region
	△	23269	SB4	Sequence Block 4	end		
		~23278	LAC5	lac5	sr	EM	
		~23472	SPC2	λspc2	sr	EM	
	▽	23495	SB5	Sequence Block 5	start		
	△	23548	SB5	Sequence Block 5	end		
		23549	B511	b511	dl-att	sq	
		~23616	1007	b1007	dl	EM	
		23918	*EA47*	Ea47 (orf-410)	M-NH3		
SPH1		23946					
		~24003	B515	b515	dl	EM	
		~24020	C107	cam107	i(Tn9)	EM RES	dl for deletions pXJS; pQLC 137-144
		~24197	B506	b506	dr	EM	
SPH1		24375					
		~24390	HYF	hyF	hy	EM	λ::att80imm80 OP80 QSR80
		~24390	HY42	hy42	hy	EM	λ::att80imm80 O80 PQSRλ
		~24390	HYM	hyM	hy	EM	λ::att80immλ OPλ QSR80
		~24390	T60	λtrp60-3	sl	EM	
XBA1		24508					partially blocked by dam methylation

		24512	*EA31*	Ea31	COOH			
SST1		24776						
		~24924	CAM3	cam3	i(Tn9)	EM		
HIN3		25157						
		25399	*EA31*	Ea31 (orf-296)	M-NH3			
		25399	*EA59*	bEa59	COOH			
		~25843	B515	b515	dr	EM		
SST1		25881						
		~25891	CAM1	cam1	i(Tn9)	EM		
ECR1		26104						
PVU1		26258						
CLA1		26617						
		~26714	1018	b1018	dl	EM		
MST2		26718						
HIN2		26744						
PST1		26932						
		26973	*EA59*	bEa59 (orf-525)	M-NH3			
HIN2		27318						
HPA1		27318						
SPH1		27378						
PVU2		27414						
	▽	27479	SB6	Sequence Block 6	start			
HIN3	″	27479						
	″	~27488	SD	sd	i	EM		
	″	~27488	SPC1	λspc1	sr	EM		
	″	27514	D106	Δ106	dr	sq		BAL31 from Hin3; is Sib+
	″	27534	D119	Δ119	dr	sq		BAL31 from Hin3; is Sib+
	″	27537	SIB	sib3	pm	sq		C to T
	″	27538	TI	tI terminator	3'	RNA		end of int message
	″	27546	D125	Δ125	dr	sq		BAL31 from Hin3; is Sib-
	″	27547	HEF	hef13	pm	sq		G to A
	″	27562	D112	Δ112	dr	sq		BAL31 from Hin3; is Sib-
	″	27568	SIB	sib2	pm	sq		C to T
	″	27570	PH53	PHΔ53;54	dr	sq		ExoIII-S1 from Hin3; is Sib- Att+
	″	27573	SIB	sib1	pm	sq		G to T
	″	27583	INT1	P arm 1	bs			int binds (DNase protection 20bp)
	″	27615	PH55	PHΔ55	dr	sq		ExoIII-S1 from Hin3; is Att-
	″	27615	INT2	P arm 2	bs			int binds (DNase protection 20bp)
	″	27714	INT3	core site	bs			int binds (DNase protection 34bp)
	″	27727	ATT-	att 2; 6; 24	pm	sq		att minus pm deletes T
	″	27731	ATT	att	f(c)	sq	f	center of common core
	″	27772	B936	bio936	att-sr	sq		
	″	27779	INT4	P' arm	bs			int binds (DNase protection 41bp)
	″	27815	*INT*	int	COOH			
AVA1	″	27887						
	″	27896	501	att501	dr	sq		att- int-
BAM1	″	27972						

TABLE 1 (continued)

A	B	C	D	E	F	G	H
	"	~28312	1007	b1007	dr	EM	
	"	~28506	B16A	bio16A	att-sr	EM	
	"	~28603	RD10	Δ red10	dl	EM	
	"	~28640	MGB	λ MGB	i(Tn9)	RES	dl for deletions pQLC 95-118
	"	~28845	2034	bio2034	att-sr	EM	
	"	28863	*XIS*	xis	COOH		
	"	28882	*INT*	int (orf-356)	M-NH3		
	"	~28894	2037	bio2037	att-sr	EM	
HIN2	"	28928					
	"	28975	T841	trp Δ 841	sl	sq	
	"	~28991	I548	int-c548	i(IS2)	EM	
	"	29063	XIS6	xis am6	pm	sq	G to A
	"	29065	SI	sI pI	5'	RNA	start of int message sq
	"	29076	INTC	int-c226;262;518	pm	sq	C to T
	"	29078	*XIS*	xis (orf-72)	M-NH3		
	"	~29088	IC57	int-c57	i(IS2)	EM	
	"	~29088	I60	int-c60	i(IS2)	EM	
	"	~29088	I508	int-c508	i(IS2)	EM	
	"	29091	C2	cII site	bs	f	cII binds -35 region of pI
	"	~29093	B538	b538	dr	EM	
	"	29094	T303	trp Δ 303	sl	sq	
	"	29126	T29	trp Δ 29	sl	sq	
	"	~29185	B386	bio386	att-dr	EM	
	"	29377	*E8.5*	Ea8.5	COOH		
	"	~29379	C842	crg 842	i(IS2)	EM	
	"	~29427	RED1	Δ red1	dl	EM	
	"	29655	*E8.5*	Ea8.5 (orf-93)	M-NH3		
	"	29711	B508	b508	att-dr	sq	
	Δ	29811	SB6	Sequence Block 6	end		
		~29815	BI2	bi2	i(IS2)	EM	
		29850	*EA22*	Ea22	COOH		
CLA1		30290					
		~30348	7-20	bio7-20	att-sr	EM	
		30395	*EA22*	Ea22 (orf-182)	V-NH3		
		~30397	B122	bio122	dr	EM	
		~30445	B221	b221	dr	EM	
	▽	30493	SB7	Sequence Block 7	start		
	"	~30520	RD15	Δ red15	dl	EM G	
	"	30529	CHIB	chiB	pm	sq	121 delete C; 131 C to G
	Δ	30569	SB7	Sequence Block 7	end		
	▽	31043	SB8	Sequence Block 8	start		

	"	31043	B522	b522	att-dr	sq	
	"	31262	TL3	tL3 terminator	3'	RNA	
	"	~31267	HY1	Hy1	hy	EM	λ::imm80 OP 80 QSR80
	"	~31267	HY5	Hy5	hy	EM	80::immλ OPλ QSR80
	"	~31267	HYM	HyM	hy	EM	λatt80::immλ OPλ QSR80
	"	~31267	RED1	Δ red1	dr	EM	
	"	31351	*EXO*	exo	COOH		
	"	~31412	1018	b1018	dr	EM	
	"	~31412	1451	b1451	dl	EM	
	"	~31461	KAN3	kan3	i(Tn6)	EM	
AVA1	"	31617					
SMA1	"	31619					
ECR1	"	31747					
HIN2	"	31809					
HPA1	"	31809					
	"	~31897	1451	b1451	dr	EM	
CLA1	"	31991					masked by dam methylation
	"	~31994	B72	bio72	att-sr	EM	
	"	~31994	B69	bio69	att-sr	EM	
PST1	"	32009					
	"	32028	*EXO*	exo (orf-226)	M-NH3		
	"	32028	*BET*	bet	COOH		
	"	~32042	B169	bio169	att-sr	EM	
HIN2	"	32219					
HPA1	"	32219					
	"	~32236	RD10	Δ red10	dr	EM	
PST1	"	32256					
	"	~32527	B267	bio267	att-sr	EM	
	"	~32624	B74	bio74	att-sr	EM	
BCL1	"	32729					masked by dam methylation
SAL1	"	32745					
HIN2	"	32747					
	"	32810	*BET*	bet (orf-261)	M-NH3		
	"	~32818	RD15	Δ red15	dr	EM	
	"	32819	*GAM*	gam	COOH		
	"	~32867	NL4	ninL4	dl	EM	
	"	~32915	1453	b1453	att-dr	EM	
CLA1	"	32964					masked by dam methylation
	"	~32964	B11	bio11	att-sr	EM	
	"	~33012	BIO1	bio1	att-sr	EM	
	"	33035	DIFF	difference		sq	Ineichen et al. (1981) G(not C)
	"	33100	TL2D	tL2d terminator	3'	RNA	
	"	~33109	NL30	ninL30	dl	EM	
	"	33141	TL2C	tL2c terminator	3'	RNA	RNA ends at this T and at next A
	"	33190	*KIL*	kil	COOH		
	"	33232	*GAM*	gam (orf-138)	M-NH3	G	
SAL1	"	33244					

TABLE 1 (continued)

A	B	C	D	E	F	G	H
HIN2	"	33246					
	"	33302	*CIII*	cIII	COOH		
	"	~33303	B275	bio275	att-sr	EM	
	"	~33303	1319	b1319	att-dr	EM	
	"	~33303	NL63	ninL63	dl	EM	
	"	33330	*KIL*	orf-47 (kil?)	M-NH3		protein not identified
	"	33441	C611	cIIIam611	pm	sq	C to T
	"	~33449	NL20	ninL20	dl	EM	
	"	~33449	NL44	ninL44	dl	EM	
	"	~33449	B252	bio252	att-sr	EM	
	"	33463	*CIII*	cIII (orf-54)	M-NH3	G	
	"	33494	TL2B	tL2b terminator	3'	RNA	
	"	~33497	B250	bio250	att-sr	EM	
AVA1	"	33498					
XHO1	"	33498					
	"	33539	*EA10*	Ea10	COOH		
	"	~33546	V203	dv203	plasmid-1	EM	
CLA1	"	33585					
	"	~33739	NL8	ninL8	dl	EM	
	"	~33885	NL88	ninL88	dl	EM	
	"	33904	*EA10*	Ea10 (orf-122)	M-NH3		
	"	33930	TL2A	tL2a terminator	3'	RNA	
	"	~34055	B214	bio214	att-sr	EM	
	"	34090	*RAL*	ral	COOH		
	"	34104	TP44	trp44	dr	sq	
	"	~34127	IMML	immL	sl	EM	
	"	~34224	KAN1	kan 1	i(Tn5)	EM	
	"	~34249	B10	bio10	att-sr	EM	
	"	34287	*RAL*	ral (orf-66)	M-NH3		
MST2	"	34319					
	"	~34370	REV	λrev	sr	EM	
	"	34379	I21	imm21	sl	sq	
	"	34497	*GIT*	git sieB	M-NH3		
BAM1	"	34499					
	"	34560	TL1	tL1 terminator	3'	RNA	
	"	34640	NL99	ninL99	dl	sq	N independent to the left
	"	~34661	B243	bio243	att-sr	EM	
CLA1	"	34697					
	"	~34758	B233	bio233	att-sr	EM	
	"	~34758	T60	trp60-3	sr	EM	
	"	~34855	B232	bio232	att-sr	EM	

	"	~34855	NL63	ninL63	dr	EM	
	"	~34952	NL4	ninL4	dr	EM	
	"	34969	TP48	trp48	sl	sq	
	"	35035	*GIT*	git sieB	COOH		
	"	35040	*N*	N	COOH		
	"	~35049	NL44	ninL44	dr	EM	
CLA1	"	35051					
	"	35060	NL99	ninL99	dr	sq	
	"	~35097	V154	dv 154	plasmid-1	EM	
	"	~35100	B256	bio256	att-sr	EM	
	"	~35100	Q115	pλQLC115	dr	RES	
	"	35116	AM22	Nam22	pm	sq	C to T
	"	~35120	Q108	pλQLC108	dr	RES	
	"	~35130	Q118	pλQLC118	dr	RES	
	"	~35140	Q109	pλQLC109	dr	RES	
	"	~35150	Q105	pλQLC105	dr	RES	
	"	~35160	Q101	pλQLC101	dr	RES	
	"	~35160	Q143	pλQLC143	dr	RES	
	"	35165	AM7	Nam7	pm	sq	GT to AA
	"	~35180	1974	pλXJS1974	dr	RES	
	"	35191	MAR2	Mar2	pm	sq	G to A
	"	~35194	NL20	ninL20	dr	EM	
	"	~35194	NL8	ninL8	dr	EM	
	"	~35200	Q117	pλQLC117	dr	RES	
	"	~35200	2300	pλXJS2300	dr	RES	
	"	~35210	M972	pλMS972	dr	RES	
	"	~35210	QL98	pλQLC98	dr	RES	
	"	~35220	Q104	pλQLC104	dr	RES	
	"	35245	2521	Nam2521	pm	sq	G to T
HIN2	"	35261					
HPA1	"	35261					
	"	~35270	M370	pλMS370	dr	RES	same as pλCM1
	"	35287	AM53	Nam53	pm	sq	A to T
	"	~35290	1976	pλXJS1076	dr	RES	
	"	~35291	NL88	n nL88	dr	EM	
	"	~35330	M968	pλMS968	dr	RES	
	"	35310	2524	Nam2524	pm	sq	C to T
	"	35324	219	Nam219	pm	sq	C to A
	"	~35340	N2-1	bioN2-1	att-sr	EM	
	"	~35350	Q102	pλQLC102	dr	RES	
	"	35360	*N*	N (orf-107)	M-NH3		
	"	~35420	Q144	pλQLC144	dr	RES	
	"	~35440	QL95	pλQLC95	dr	RES	
	"	35465	DV1	dv1	plasmid-1	sq	
	"	~35485	NL32	ninL32	dr	EM	
	"	35518	NUTL	nutL	f(1)	f G	N utilization (17bp stem-loop)
	"	35528	NUT-	nutL63;96;18	pm	sq	63 C to A; 96 C to G; 18 C to T

TABLE 1 (continued)

A	B	C	D	E	F	G	H
	"	~35530	Q138	pλQLC138	dr	RES	
	"	35530	NUT-	nutL3	pm	sq	delete G
	"	~35570	Q103	pλQLC103	dr	RES	
	"	35576	3H-1	bio3H-1	att-sr	sq	
	"	35582	SL	sL pL	5'	RNA	
	"	35584	I4	imm434	sl	sq	
	"	35591	OL1	oL1	f(1) bs	f	17-bp diad symmetry cI binds
	"	35596	V2	vir2;v003	pm	sq	vir2 C to A; v003 C to T
	"	35606	V101	vir101	pm	sq	T to C
	"	35613	SEX1	sex1	pm	sq	G to A
HIN2	"	35615					
	"	35615	OL2	oL2	f(1) bs	f	17 bp cI binds cro binds
	"	35619	2668	vL2668	pm	sq	
	"	35620	V169	v169	pm	sq	
	"	35622	V305	v305	pm	sq	G to T
	"	35635	OL3	oL3	f(1) bs	f	17 bp cro binds
	"	35639	L668	1668	pm	sq	
BGL2	"	35711					
PVU1	"	35791					
	"	35804	TIMM	tIMM terminator	3'	f RNA	presumed 3' from pRE pRM plit
	"	35828	*REXB*	rexB	COOH		
	"	35848	KH54	KH54	dl	sq	
	"	35887	DIFF	difference		sq	Ineichen et al. (1981) delete A
	"	~35915	NL30	ninL30	dr	EM	
	"	35940	REX-	rex209	pm	sq	G to A
	"	35947	REX-	rex111	pm	sq	G to A
	"	~36010	M969	pλ969	dr	RES	
	"	~36010	24-5	bio24-5	att-sr	EM	
	"	~36153	KH67	KH67	dl	EM	
	"	~36153	16-3	bio16-3	att-sr	EM	
	"	36259	*REXB*	rexB (orf-144)	M-NH3		
	"	~36270	Q139	pλQLC139	dr	RES	
	"	36278	*REXA*	rexA	COOH		
	"	~36320	Q137	pλQLC137	dr	RES	
	"	36322	LIT	s lit p lit	5'	f RNA	start of lit (mis) RNA
	"	~36390	T75	bio t75	att-sr	EM	
	"	36398	KH70	KH70	dl	sq	
	"	~36520	Q111	pλQLC111	dr	RES	
	"	~36560	Q141	pλQLC141	dr	RES	
	"	~36670	Q142	pλ142	dr	RES	
	"	~36770	173	tet 173	i(Tn10)	EM	

	■	~36770	174	tet 174	i(Tn10)	EM	
	■	36817	PC1	pc 1	pm	sq	
	■	~36818	KAN2	kan2	i(Tn5)	EM	
	■	~36890	2299	pλXJS2299	dr	RES	
HIN3	■	36895					
CLA1	■	36966					
PST1	■	37005					
	■	~37010	1975	pλXJS1975	dr	RES	
	■	~37100	Q112	pλQLC112	dr	RES	
	■	~37110	1973	pλXJS1973	dr	RES	
	■	37114	*REXA*	rexA (orf-279)	M-NH3		
	■	~37150	Q106	pλQLC106	dr	RES	
	■	~37160	Q110	pλQLC110	dr	RES	
	■	37182	V021	dv021	plasmid-r	sq	
	■	~37200	Q107	pλQLC107	dr	RES	
	■	37230	*CI*	cI	COOH		
	■	37255	UA72	cI UA72	pm	sq	C to A
	■	37272	IC11	cI IC11	pm	sq	insert C
	■	37281	SP31	cI SP31	i (IS1)	sq	
	■	37287	AM14	cIam14	pm	G	C to A
	■	37293	NT1	cI NT1	pm	sq	G to T
	■	37308	504	cIam504	pm	G	G to C
	■	37309	BP83	cI BP83	pm	sq	T to C
	■	37313	505	cIam505	pm	G	G to A
	■	37320	K100	KH100	i(IS5)	sq	
	■	37330	UV59	cI UV59	pm	sq	A to G
	■	37333	UV63	cI UV63	pm	sq	delete A
	■	37334	IC28	cI IC28	pm	sq	insert A
BCL1	■	37352					masked by dam methylation
	■	37376	ET26	cI ET26	pm	sq	C to A
	■	37384	AA21	cI AA21	pm	sq	insert C
	■	37385	AA2	cI AA2	pm	sq	delete C
	■	37387	NT16	cI NT16	pm	sq	insert A
	■	37396	UV76	cI UV76	pm	sq	A to T
	■	37398	SP1	cI SP1	i (IS1)	sq	
	■	37412	NT15	cI NT15	pm	sq	delete G
	■	37422	TN10	171;172;473;474	i(Tn10)	sq	
	■	37423	13	cI 13	pm	sq	delete T
HIN2	■	37433					
	■	37442	UV64	cI UV64	pm	sq	GAAT to AAA
HIN3	■	37459					
	■	37531	UV73	cI UV73	pm	sq	GC to T
	■	37538	BL46	cI BL46	pm	sq	T to C
	■	~37579	KH67	KH67	dr	EM	
	■	~37579	30-7	bio 30-7	att-sr	EM	
(HIN3)		37584					created by ind1;as in most cI857
	■	37589	IND1	ind1	pm	sq	C to T lysogen not UV inducible

TABLE 1 (continued)

A	B	C	D	E	F	G	H
	‖	37604	PF17	cI PF17	pm	sq	insert C
	‖	37623	AA3	cI AA3	pm	sq	insert A
	‖	37624	SP48	cI SP48	pm	sq	delete A
	‖	~37627	V266	dv266	plasmid-1	EM	
	‖	37629	ET39	cI ET39	pm	sq	G to T
	‖	37629	499	cIam499	pm	G	G to C
	‖	37630	B138	cI BP138	pm	sq	T to C
	‖	37635	212	cIam212	pm	G	A to C
	‖	37651	ET3	cI ET3	pm	sq	G to C
	‖	37680	AM34	cIam34	pm	G	C to A
	‖	37682	E86	cIamE86	pm	sq	C to A
	‖	37687	54-1	cI54-1	pm	sq	A to C
	‖	37690	UV52	cI UV52	pm	sq	T to C
	‖	37691	ET15	cI ET15	pm	sq	C to A
	‖	37694	R82	cIamR82	pm	sq	T to A
	‖	37702	UV96	CI UV96	pm	sq	G to A
	‖	37705	UV97	cI UV97	pm	sq	G to A
	‖	37706	UV79	cI uv79	pm	sq	G to A
	‖	37707	PF12	cI PF12;cI AA6	pm	sq	insert G; delete G
	‖	37708	11-3	cI11-3;cICP175	pm	sq	11-3 C to T;CP175 C to A
	‖	37709	60-1	cI60-1	pm	sq	T to G
	‖	37710	UV47	cI UV47	pm	sq	A to T
	‖	37711	5-3	cI5-3	pm	sq	A to G
	‖	37718	U116	cI UV116;cI UV74	pm	sq	UV116 C to T;UV77 C to A
	‖	37720	BP84	cI BP84	pm	sq	A to G
	‖	37726	BU30	cI BU30	pm	sq	A to G
	‖	37727	UV66	cI UV66	pm	sq	AC to TT
	‖	37727	KH70	KH70	dr	sq	
	‖	37733	SP44	cI SP44;cI SP37	pm	sq	SP44 G to A;SP37 G to T
	‖	~37740	Q113	pλQLC113	dr	RES	
	‖	37742	857	cI857	pm	sq	C to T cIts
	‖	37744	63-1	cI63-1	pm	sq	A to C
	‖	37746	ET7	cI ET7	pm	sq	C to A
	‖	37756	BL71	cI BL71	pm	sq	T to C
	‖	37762	UA8	cI UA8	pm	sq	G to T
	‖	37765	UV61	cI UV61	pm	sq	T to A
	‖	37768	DV93	dvh93	plasmid-1	sq	
	‖	37768	UV62	cI UV62	pm	sq	A to G
	‖	37768	L57	cIamL57	pm	sq	A to C
	‖	37772	UV41	cI UV41	pm	sq	C to T
	‖	37773	34-1	cI34-1	pm	sq	A to C

"	37775	UV44	cI UV44	pm	sq	T to C
"	37775	U117	cI UV117	pm	sq	T to G
"	37780	NT17	cI NT17	pm	sq	C to A
"	37780	26-2	cI26-2	pm	sq	C to T
"	37781	UV91	cI UV91	pm	sq	CC to TT
"	37781	UA33	cI UA33;cI SP43	pm	sq	UA33 C to A;SP43 C to T
"	37781	CP32	cICP32	pm	sq	C to A
"	37784	UV46	cI UV46	pm	sq	T to C
"	37784	53-1	cI53-1	pm	sq	T to C
"	37787	U108	cI UV108	pm	sq	A to T
"	37789	UV37	cI UV37	pm	sq	A to G
"	37789	L50	cIamL50	pm	sq	A to C
"	37791	NT3	cI NT3	pm	sq	deltete A
"	37792	CP-9	cICP-9;cI29-6	pm	sq	CP9 G to T; 29-6 G to A
"	37798	BL42	cI BL42	pm	sq	A to G
"	37801	SP38	cI SP38	pm	sq	C to T
"	37804	37-1	cI37-1	pm	sq	G to A
"	37805	B102	cI BP102	pm	sq	A to G
"	37806	TN10	475	i(Tn10)	sq	
"	37806	UA68	cI UA68	pm	sq	C to A
"	37807	BL70	cI BL70	pm	sq	T to C
"	37807	40-3	cI40-3	pm	sq	T to A
"	37808	282	cIam282	pm	G	G to A
"	37808	NT25	cI NT25;cI UA77	pm	sq	NT25 G to A;UA77 G to T
"	37808	Q44	cIamQ44	pm	sq	G to A
"	37810	ET22	cI ET22	pm	sq	C to G
"	37810	38-2	cI38-2	pm	sq	C to T
"	37811	SP61	cI SP61	pm	sq	C to T
"	37812	PF2	cI PF2;cI AA13	pm	sq	PF2 insert C;AA13 delete C
"	37813	SP28	cI SP28	pm	sq	A to G
"	37813	9-1	cI9-1	pm	sq	A to G
"	37814	B120	cI BU120	pm	sq	T to C
"	37816	SP41	cI SP41	pm	sq	C to A
"	~37817	T124	bio t124	att-sr	EM	
"	37817	UV86	cI UV86	pm	sq	C to T
"	37818	AA13	cI AA13	pm	sq	delete C
"	37819	CP7	cICP7	pm	sq	A to T
"	37826	PC2	pc 2	pm	sq	
"	37828	SP5	cI SP5	pm	sq	G to A
"	37831	BU39	cI BU39	pm	sq	A to G
"	37835	12-1	cI12-1	pm	sq	A to G
"	37838	PC3	pc 3	pm	sq	
"	37840	BL51	cI BL51	pm	sq	T to C
"	37841	SP3	cI SP3	pm	sq	G to A
"	37841	Q33	cIamQ33	pm	sq	G to A
"	37844	UA6	cI UA6;cI BL89	pm	sq	UA6 A to Y;BL89 A to G
"	37844	47-1	cI47-1	pm	sq	A to G

TABLE 1 (continued)

A	B	C	D	E	F	G	H
	"	37846	TN10	408	i(Tn10)	sq	
	"	37846	UV14	cI UV14	pm	sq	A to G
	"	37846	L31	cIamL31	pm	sq	A to C
	"	37846	4-3	cI4-3	pm	sq	A to G
	"	37852	BL84	cI BL84	pm	sq	A to G
	"	37852	1-1	cI1-1	pm	sq	A to G
	"	37860	SP42	cI SP42	pm	sq	delete T
	"	37861	UV80	cI UV60	pm	sq	insert TT
	"	37862	UV57	cI UV57	pm	sq	T to C
	"	37863	UV88	cI UV88	pm	sq	C to TTT
	"	37865	ET28	cI ET28	pm	sq	T to G
	"	37870	ET13	cI ET13	pm	sq	insert T
	"	37872	302	cIam302	pm	G	A to C
	"	37872	56-1	cI56-1;cI5-1	pm	sq	56-1 A to C;5-1 A to G
	"	37873	32-1	cI32-1;cI29-7	pm	sq	32-1 T to G;29-7 T to C
	"	37882	UV85	cI UV85	pm	sq	T to C
	"	37883	SP39	cI SP39;cI BL80	pm	sq	SP39 T to G;BL80 T to C
	"	37886	U112	cI UV112;cI UV93	pm	sq	UV112 G to A;UV93 G to Y
	"	37886	36-1	cI36-1	pm	sq	G to A
	"	37894	BL97	cI BL97	pm	sq	G to A
	"	37894	SP52	cI SP52	i(IS1)	sq	
	"	37894	SP60	cI SP60	i(IS1)	sq	
	"	37894	177	cI CP177	pm	sq	G to T
	"	37896	SP51	cI SP51	pm	sq	G to C
	"	37897	BP80	cI BP80	pm	sq	T to C
	"	37898	UV23	cI UV23	pm	sq	C to T
	"	37901	UV6	cI UV6	pm	sq	C to T
	"	37901	E13	cIamE13	pm	sq	C to A
	"	37903	BL60	cI BL60	pm	sq	A to G
	"	37903	29-2	cI29-2	pm	sq	A to G
	"	37904	SP53	cI SP53	i(IS5)	sq	
	"	37905	UA54	cI UA54	pm	sq	C to A
	"	37907	B100	cI BL100	pm	sq	G to A
	"	37910	UV34	cI UV34	pm	sq	C to T
	"	37915	SP27	cI SP27	pm	sq	G to A
	"	37917	UV90	cI UV90	pm	sq	T to A
	"	37918	L7	cIamL7	pm	sq	A to C
	"	37919	SP14	cI SP14	i(IS5)	sq	
	"	37919	SP35	cI SP35	i(IS5)	sq	
	"	37922	UV12	cI UV12	pm	sq	insert R
	"	37925	KH54	KH54	dr	sq	

	"	37926	SP57	cI SP57	pm	sq	C to A
	"	37928	57-1	cI57-1	pm	sq	T to G
	"	37930	SP62	cI SP62	pm	sq	T to A
	"	37932	UA60	cI UA60	pm	sq	insert T
	"	37935	UA9	cI UA9	pm	sq	G to T
	"	37938	UV28	cI UV28	pm	sq	C to T
	"	37939	BP81	cI BP81	pm	sq	A to G
	"	37940	*CI*	cI (orf-237)	M-NH3	AA	
	"	37940	SM	sM pM	5'	RNA	formerly called pRM
	"	37941	BP86	cI BP86	pm	sq	A to G
	"	37947	SP2	cI SP2	i(IS5)	sq	
	"	37950	SP46	cI SP46	i(IS1)	sq	
	"	37951	OR3	oR3	f(1) bs	f	17 bp sequence cI binds
	"	37954	E37	pRM-E37	pm	sq	C to T
	"	37955	VC3	vC3	pm	sq	A to G
	"	37957	3R1	OR3-r1	pm	sq	C to T
	"	37958	3R2	OR3-r2; OR3-r3	pm	sq	-r2 G to T; -r3 G to A
	"	37965	C12	OR3-c12	pm	sq	A to G
	"	37966	C10	OR3-c10	pm	sq	T to C
	"	37971	UP1	prm up-1	pm	sq	A to G
	"	37973	M104	pRM-M104;116;U31	pm	sq	C to T
	"	37973	UV26	cI UV26	pm	sq	C to T
	"	37974	OR2	oR2	f(1) bs	f	cI binds (17bp)
	"	37975	BU20	cI BU20	pm	sq	A to G
	"	37978	E104	pRM-E104;v3c	pm	sq	E104 A to C; v3c A to G
	"	37979	V1	v1;vir146;pRM-E93	pm	sq	v1 C to A; E93 C to T
	"	37980	UV51	cI UV51	pm	sq	CC to TT
	"	~37983	S274	spi 274	att-sr	EM	
	"	37984	BR1	BR1	pm	sq	C to T
	"	37985	VN	vN	pm	sq	G to T
	"	37986	MAC1	MAC1;MAC41	pm	sq	MAC1 T to A; MAC41 T to C
	"	37987	MCC8	MCC8;MCH31;AAH8	pm	sq	MCC8 G to C; MCH31 G to A; AAH8 G to T
HIN2	"	37989					
	"	37989	MAH4	MAH4;AAH2	pm	sq	MAH4 delete T;AAH2 T to A
	"	37990	CH20	MCH20; MCH9	pm	sq	MCH20 G to C;MCH9 delete G
	"	37991	X3	pR-x3	pm	sq	A to G
	"	37992	MCH8	MCH8	pm	sq	C to T
	"	37998	OR1	oR1	f(1) bs	f	cI binds (17bp)
	"	37999	2668	vR2668;virR18	pm	sq	A to G
	"	38000	668	VR668	pm	sq	insert C
	"	38003	V326	vs326; BC1	pm	sq	vs326 C to A;BC1 C to T
	"	38006	U119	UV119	pm	sq	delete G (clear)
	"	38007	UV8	pRM-UV8;v3	pm	sq	UV8 C to T; v3 C to A
	"	38008	U93	pRM-U93;M36	pm	sq	U93 G to A
	"	38009	V387	vs387;vC1;NR5	pm	sq	vs387 G to C; vC1 G to T
	"	38009	UV11	UV11	pm	sq	G to A (clear)
	"	38010	B143	BU143	pm	sq	T to C (clear)

TABLE 1 (continued)

A	B	C	D	E	F	G	H
	"	38023	SR	sR pR	5'	RNA	
	"	~38031	CRO2	Δ cro2	dl	EM	
	"	38041	*CRO*	cro (orf-66)	M-NH3	AA	
BGL2	"	38103					
	"	38135	TRO	tRO terminator	3'	RNA	
	"	~38197	CRO2	Δ cro2	dr	EM	
AVA1	"	38214					
	"	38238	*CRO*	cro	COOH		
	"	38245	I4	imm434	sr	sq	
	"	38264	NUTR	NutR	f(1)	f	N utilization (homology to nutL)
	"	38302	CIN1	cin-1	pm	sq	G to A
	"	38306	CNC1	CNC1	pm	sq	T to C
	"	38307	CNC8	CNC8	PM	sq	A to G
	"	38334	TR1	tR1 terminator	3'	RNA	
	"	38339	DYA3	dya3	pm	sq	T to C
	"	38341	C17	c17		sq	9-bp duplication
	"	38343	SE	sE pE	5'	RNA	formerly called pRE
	"	38344	DYA2	dya2	pm	sq	A to G
	"	38350	R32	r32	i(IS2)	sq	
	"	38350	3048	cY3048	pm	sq	A to G
	"	38354	3071	cY3071;c3073a	pm	sq	cY3071 T to C; c3073a T to A
	"	38354	2001	cY2001	pm	sq	T to G
	"	38356	3088	cII3088	pm	sq	A to G
	"	38357	3019	cY3019	pm	sq	C to T
	"	38359	CIR5	cir5	pm	sq	T to C
	"	38360	*CII*	cII (orf-97)	M-NH3	AA	
	"	38360	3086	cII3086	pm	sq	A to G
	"	38361	3067	cII3067	pm	sq	T to C
	"	38362	3105	cII3105	pm	sq	G to A
	"	38363	3073	c3073b	pm	sq	G to C
	"	38364	3059	cII3059	pm	sq	T to A
	"	38364	CAN1	can1	pm	sq	T to C
	"	38368	CTR2	ctr2;ctr3	pm	sq	T to C
	"	38369	CTR1	ctr1	pm	sq	G to A
	"	38369	C2	cII site	bs		cII binds -35 region of pRE
	"	38370	3003	cY3003	pm	sq	C to T
	"	38371	CY42	cY42	pm	sq	A to T
	"	38372	3077	cY3077	pm	sq	A to G
	"	38375	3114	cII3114;cII3109	pm	sq	cII3114 A to G;cII3109 A to T
	"	38376	844	cY844	pm	sq	A to G
	"	38377	3075	cY3075	pm	sq	AC to TT

	"	38378	3107	cY3107	pm	sq	C to T
	"	38379	3008	cY3008	pm	sq	G to A
	"	38380	3001	cY3001	pm	sq	C to T
	"	38380	3095	cY3095	pm	sq	C to A
	"	38381	3072	cY3072;cY3078	pm	sq	cY3072 A to C;cY3078 A to G
	"	38382	3098	cY3098	pm	sq	A to G
	"	38384	3623	cII3623	pm	sq	G to T
	"	38385	3085	cII3085	pm	sq	A to G
	"	38387	3104	cII3104	pm	sq	G to A
	"	38388	3091	cII3091	pm	sq	C to T
	"	38391	3056	cII3056	pm	sq	T to C
	"	38393	3099	cII3099	pm	sq	C to T
	"	38396	3081	cII3081	pm	sq	A to T
	"	38399	3097	cII3097	pm	sq	G to A
	"	38400	3111	cII3111	pm	sq	AG to T
	"	38403	3092	cII3092	pm	sq	G to A
	"	38430	2002	cII2002	pm	sq	T to C
	"	38453	3641	cII3641	pm	sq	G to T
	"	38472	3062	cII3062	pm	sq	C to T
	"	38474	AM60	cIIam60	pm	sq	C to T
	"	38476	3283	cII3283a	pm	sq	C to A
	"	38478	3283	cII3283b	pm	sq	T to A
	"	38483	CHIC	chiC	pm	sq	A to T
	"	38488	3302	cII3302	pm	sq	G to A
	"	38498	68	cII68	pm	sq	T to A
	"	38523	3258	cII3258a	pm	sq	T to G
	"	38538	AM41	cIIam41	pm	sq	G to A
	"	38543	ICE	replication ice	f(1)	f G	5bp stem;5bp loop;proposed inceptor
HIN2	"	38548					
	"	38549	PK35	pk35	i(Tn903)	sq	
	"	38564	3258	cII3258b	pm	sq	C to T
	"	38571	3468	cII3468	pm	sq	C to T
	"	38582	3386	cII3386	pm	sq	G to A
	"	38592	3139	cII3139	pm	sq	T to C
	"	38599	OOP	oop terminator	3'	RNA	
	"	38617	I21	imm21	sr	sq	
	"	38634	3639	cII3639	pm	sq	A to G
	"	38642	3638	cII3638	pm	sq	A to G
	"	38650	*CII*	cII	COOH		
	"	38651	3520	cII3520	pm	sq	T to C
	"	38675	OOP	oop start point	5'	RNA	
	"	38686	*O*	O (orf-299)	M-NH3	AA	
BGL2	"	38754					
BGL2	"	38814					
	"	39034	ITN1	ori iteron 1	f(1) bs	f	18 bp repeated seq O binds
	"	39054	ITN2	ori iteron 2	f(1) bs	f	18 bp O binds-DNase protection
	"	39078	ITN3	ori iteron 3	f(1)bs	f	18 bp O binds-DNase protection

TABLE 1 (continued)

A	B	C	D	E	F	G	H
	"	39092	R93	r93	dl	sq	ori- mutation
	"	39101	ITN4	ori iteron 4	f(l)bs	f	18 bp O binds-DNase protection
	"	39115	R93	r93	dr	sq	ori- mutation
	"	39120	R99	r99	dl	sq	ori- mutation
	"	39122	TI12	ti12(tiny 12)	pm	sq	C to A; cis replication defective
	"	39131	R99	r99	dr	sq	ori- mutation
	"	39138	R96	r96	dl	sq	ori- mutation
	"	39152	R96	r96	dr	sq	ori- mutation
	"	39158	HY42	Hy42	hy	sq	λatt80imm80 080::PQSRλ
	"	39165	RP82	rep82	hy	sq	λ rep82::PQSRλ
ECR1	"	39168					
	"	39268	RIC	ric5b	pm	sq	C to T
	"	39292	RIC	ric5b	pm	sq	G to A
	"	~39373	A-20	bioE5a-20	att-sr	EM	
SPH1	"	39422					
	"	39582	*P*	P (orf-233)	M-NH3		
	"	39582	*O*	O	COOH		
HIN2	"	39608					
HPA1	"	39608					
HIN2	"	39836					
HPA1	"	39836					
AVA1	"	39888					
SMA1	"	39890					
	"	40091	DV93	dvh93	plasmid-r	sq	
	"	40280	*REN*	ren (orf-96)	M-NH3		
	"	40280	*P*	P	COOH		
	"	40335	V021	dv021	plasmid-r	sq	
	"	~40354	PQ4	PQ Δ 4	dl	EM	
	"	~40354	CF1	cf1	dl	EM	
SST2	"	40389					
	"	~40404	PQ9	PQ Δ 9	dl	EM	
	"	~40502	P4	P4	sl	EM	
	"	~40502	Q3	qin3	sl	EM	
	"	~40502	A3	qinA3	sl	EM	
	"	~40502	M3	λgalM3	sl	EM	
	"	~40502	C3	qinC3	sl	EM	
	"	40502	NIN5	nin5	dl	sq	N independent growth
	"	~40551	V203	dv203	plasmid-r	EM	
	"	~40551	PQ1	PQΔ 1	dl	EM	
	"	40567	*REN*	ren	COOH		
	"	~40600	NI24	nin24	dl	EM	

	"	40624	TR2	tR2 terminator	3'	f	presumed genetically defined tR2
	"	40644	*ORF*	nin orf-146	M-NH3		
	"	~40796	PQ8A	PQ Δ 8a	dl	EM	
HIN2	"	40942					
	"	41011	TN10	627	i(Tn10)	sq	
	"	41081	*ORF*	nin orf-290	M-NH3		
	"	41081	*ORF*	nin orf-146	COOH		
	"	~41090	D86	Δ 86	dl	EM	
	"	~41139	B1	b1	i(IS2)	EM	
	"	~41287	3000	b3000	dl	EM	
CLA1	"	41364					
	"	~41679	IMML	immL	sr	EM	
BAM1	"	41732					
	"	41781	TN10	623	i(Tn10)	sq	
	"	~41924	PQ9	PQ Δ 9	dr	EM	
	"	41950	*ORF*	nin orf-57	M-NH3		
	"	41950	*ORF*	nin orf-290	COOH		
	"	~41973	3000	b3000	dr	EM	
CLA1	"	42021					
	"	42090	*ORF*	nin orf-60	V-NH3		
	"	42120	*ORF*	nin orf-57	COOH		
	"	42269	*ORF*	nin orf-56	V-NH3		
	"	42269	*ORF*	nin orf-60	COOH		
	"	~42317	D86	Δ 86	dr	EM	
	"	42429	*ORF*	nin orf-204	M-NH3		
	"	42436	*ORF*	nin orf-56	COOH		
	"	42645	DV1	dv1	plasmid-r	sq	
	"	~42660	PQ8A	PQ Δ 8a	dr	EM	
	"	~43004	NI24	nin24	dr	EM	
	"	43040	*ORF*	nin orf-68	M-NH3		
	"	43040	*ORF*	nin orf-204	COOH		
	"	43082	DIFF	difference		sq	Kroger and Hobom (1982);G (not A)
HIN2	"	43183					
	"	43224	*ORF*	nin orf-221	M-NH3		
	"	43243	*ORF*	nin orf-68	COOH		
	"	43307	NIN5	nin5	dr	sq	
	"	~43346	CF1	cf1	sr	EM	
	"	~43394	PQ9	PQ Δ 9	dr	EM	
	"	~43443	PQ1	PQ Δ 1	dr	EM	
BCL1	"	43682					masked by dam methylation
	"	~43688	V266	dv266	plasmid-r	EM	
CLA1	"	43825					
	"	~43836	PQ4	PQΔ 4	dr	EM	
	"	~43885	HYM	HyM	hy	EM	λatt80immOPλ::QSR80
	"	~43885	HY5	Hy5	hy	EM	80immλ OPλ::QSR80
	"	43886	*Q*	Q (orf-207)	M-NH3		
	"	43886	*ORF*	nin orf-221	COOH		

TABLE 1 (continued)

A	B	C	D	E	F	G	H
	=	~44081	TP1	Tp1	i(Tn402)	EM	
HIN3	=	44141					
	=	~44326	V154	dv154	plasmid-r	EM	
	=	~44424	M3	λgalM3	sr	EM	
	=	44498	DIFF	difference			Daniels (1982) -G; Petrov (1981) -A
	=	44506	*Q*	Q	COOH		
	=	44587	6S	p6S pR'	5'	RNA	start of 6S RNA; λ late RNA
	=	~44600	QUT	qut	f(c)	G	(maps between 44587 and 44610)
	=	~44606	PK3	pk3	i(Tn903)	EM	
	=	44621	*ORF*	orf-64	M-NH3		
	=	44780	T6S	t6S	3'	RNA	end of 6S RNA
	=	44812	*ORF*	orf-64	COOH		
ECR1	=	44972					
	=	45027	CHID	chiD	pm	sq	
	=	45348	S9B	Sts9B	pm	sq	G to A
	=	45186	*S*	S (orf-107)	M-NH3		protein not identified
	=	~45336	422	422	i(Tn5)	EM;RES	
	=	45352	S7	Sam7	pm	sq	G to A
	=	45493	*R*	R (orf-158)	M-NH3	AA	
	=	45506	*S*	S	COOH		
	=	45853	296	λ296	sl	sq	deletes 309 bp;inserts pBR322
	=	45966	*RZ*	Rz (orf-153)	M-NH3	G	
	=	45966	*R*	R	COOH		
	=	~45994	FUS2	λfus2	sr	EM	
	=	46008	DK23	dK23	i(TN903)	sq	
	=	~46043	Q3	qin3	sr	EM	
	=	~46043	A3	qinA3	sr	EM	
	=	~46043	P4	P4	sr	EM	
	=	~46043	C3	qinC3	sr	EM	
	=	~46141	DK6	dk 6	i(Tn903)	EM;RES	
	=	46161	296	λ296	sl	sq	deletes 309 bp;inserts pBR322
BCL1	=	46366					masked by dam methylation
	=	46423	*RZ*	Rz	COOH		
CLA1	=	46439					
	Δ	46467	SB8	Sequence Block 8	end		
		~47471	PK26	pk26	i(Tn903)	EM	
		~47520	PK22	pk22	i(Tn903)	EM	
		~47520	PK24	pk24	i(Tn903)	EM	
		~47618	DA3	Δa3	dl	EM	
HIN2		47938					
BCL1		47942					masked by dam methylation

		~48158	PK21	pk21	i(Tn903)	EM	
		~48207	DA3	Δa3	dr	EM	
HIN2		48298					
	▽	48391	SB1R	Sequence Block 1R	start		
	"	~48403	PK25	pk25	i(Tn903)	EM	
REND	"	48502		Right END			used for fragment endpoints
	"	48514	CRND	5' end of r-strand			last base of sticky end
	Δ	48514	SB1R	Sequence Block 1R	end		

TABLE 2 SOURCES OF AVAILABLE λ DNA SEQUENCES

Gene Region	Source	Coordinates
λ*c*I857*S*7	Complete Sequence	
	Sanger et al. (1982)	1-28461
		28808-35245
		35280-37709
		38039-38197
		38499-38680
		38757-39657
		39700-44169
		44501-48514
	Sequence Block 1L [1-72]	
	Wu and Taylor 1971	1-12
cos	Nichols and Donelson (1978)	13-72
	Sequence Block 2 [14869-18969]	
J	F. Sanger, A. R. Coulson, and G.-F. Hong, (1980) as cited by Daniels et al. (1982)	
	Sequence Block 3 [21661-21737]	
b-region	Hoess and Landy (1978)	
	Sequence Block 4 [23129-23495]	
*p*bL	Rosenvold et al. (1980) (sequence inaccurate)	23131-23248
	K.-C. Luk and W. Szybalski (unpubl.) (sequence accurate with one typographical error at 23143 in Abstracts of Bacteriophage Meetings, Cold Spring Harbor Laboratory, 1982)	23129-23269
	Sequence Block 5 [23495-23548]	
b-region	Hoess and Landy (1978)	
	Sequence Block 6 [27479-29811]	
att	Hsu et al. (1980)	27479-27633
att	Landy and Ross (1977)	27617-27934
att	Davies et al. (1977)	27616-28935
int-xis	Hoess et al. (1980)	27725-29275
	Davies (1980)	27724-29525
	Abraham et al. (1980)	28929-29198
	R. H. Hoess, K. Bidwell and A. Landy, pers. comm. (1980)	29065-29811
xis-red	Davies et al. (1978)	29711-29811
xis-red	Hoess and Landy (1978)	29710-29778
	Sequence Block 7 [30493-30569]	
*chi*B	Smith et al. (1980)	
	Sequence Block 8 [31042-46467]	
xis-red	Davies et al. (1978)	31043-31058
xis-red	Hoess and Landy (1978)	31042-31129
*t*L3	Luk and Szybalski (1982a, b)	30299-31408
bet-exo	F. Sanger, A. R. Coulson, and G.-F. Hong, (1980) as cited by Daniels et al (1982)	31270-32891
bet-N	Ineichen et al. (1981)	32503-35905
*N-nut*L	Franklin and Bennett (1979)	34956-35615
*p*L mRNA start	Dalhberg and Blattner (1975)	35434-35618
*p*L*o*L	Ptashne et al. (1976)	35578-35667
	Kleid et al. (1975)	35583-35600
*o*L	Horn and Wells (1981)	35468-35819
rex	Landsmann et al. (1982)	35437-37348
*c*I	Sauer (1978)	37225-37936
*p*R*o*R	Ptashne et al. (1976)	37903-38027
*p*M(prm)	Walz et al. (1976)	37905-37989
cro-O	Schwarz et al. (1978)	37990-38982

TABLE 2 (Continued)

Gene Region	Source	Coordinates
cro	Roberts et al. (1977)	38041-38241
oop	Dalhberg and Blattner (1973)	38597-38672
O	Scherer (1978)	38597-39688
P	Schwarz et al. (1980)	37768-40293
ori	Denniston-Thompson et al. (1977)	39062-39170
ori-O	Moore et al. (1979)	38008-39328
nin	Kröger and Hobom (1982)	40218-43972
Q-p'_R-*qut*-6S	Daniels and Blattner (1982)	43681-45218
Q	Petrov et al. (1981)	43860-45001
Q-S	Daniels (1981)	43681-45634
6S region	Sklar (1977)	44467-44807
6S RNA	Lebowitz et al. (1971)	44588-44780
chiD	Smith et al. (1981)	44972-45057
R protein	Imada and Tsugita (1971)	45493-45963
R	D. L. Daniels (unpubl.)	45635-45977
Rz	H. Ikeda, pers. comm. (1980);	46064-46443
S-R-Rz	F. Sanger, A. R. Coulson and G.-F. Hong, (1980) as cited by Daniels et al. (1982)	44467-46467
	Sequence Block 1R [48391-48514]	
cos	Nichols and Donelson (1978)	48391-48502
	Wu and Taylor (1971)	48503-48514

TABLE 3 SOURCES OF MAPPING INFORMATION FOR SITES ASSIGNED TO EXACT POSITION IN SEQUENCE

Site name	Reference
left end	Wu and Taylor (1971); Nichols and Donelson (1978)
*Wam*403 (as in λ*WES*)	G.-F. Hong and F. Sanger, unpubl. (1983)
*B**	Walker et al. (1982)
*Hpa*I, *Hinc*II sites overlap mK site	Daniels et al. (1980); A. Glasgow and M. Howe, pers. comm. (1983)
t'J4 terminator	Luk and Szybalski (1983b)
t'J3 terminator	---ibid.
t'J2 terminator	---ibid.
t'J1 terminator	---ibid.
*b*189	A. Frischauf, H. Lehrach, A. Poustka, N. Murray, pers. comm. (1983)
*lac*5 left	J. S. Salstrom and D. Newman, pers. comm. (1982)
A in cb2; *G* in cI857S7	R. Hoess and A. Landy, pers. comm. (1983)
*b*2	Hoess and Landy (1978)
λcam105	Muster et al. (1983)
*pb*L	Rosenvold et al. (1980); K.-C. Luk and W. Szybalski, unpubl. (1982)
*b*511	Hoess and Landy (1978)
λcam107	Muster et al. (1983)
Δ106	Court et al. (1983)
Δ119	---ibid.
*sib*3	Guarneros et al. (1982)
Δ125	Court et al. (1983)
*t*I terminator	Luk et al. (1982); U. Schmeissner, pers. comm. (1981);
*hef*13	Guarneros et al. (1982)
Δ112	Court et al. (1983).
*sib*2	Guarneros et al. (1982)
pHΔ53	Hsu et al. (1980)
*sib*1	Guarneros et al. (1982)
INT 1 binding site	Hsu et al. (1980).
INT 2 binding site	---ibid.
pHΔ55	---ibid.
INT 3 binding site	---ibid.
att core	Davis et al. (1977); Landy and Ross (1977)
att 2; 6; 24	Ross et al. (1982)
*bio*936	R. Weisberg, pers. comm. (1982)
INT 4 binding site	Hsu et al. (1980)
*att*501	Ross et. al. (1982)
λMGB22	L. MacHattie, pers. comm. (1983)
*trp*Δ841	Abraham et al. (1980)
xis am6	Abraham et al. (1980); Hoess et al. (1980);
*s*I *p*I	Schmeissner et al. (1981)
int-c226;262;518	Abraham et al. (1980); Hoess et al. (1980).
cII binding site	Ho and Rosenberg (1982)
*trp*Δ303	Abraham et al. (1980)
*trp*Δ29	---ibid.
*b*508	Davies et al. (1978); Hoess and Landy (1978)
Δ*red* 15	G. Smith, pers. comm. (1983)
*chi*B	Smith et al. (1980)
*b*522	Davies et al. (1978); Hoess and Landy (1978)
*t*L3 terminator	Luk et al. (1982)
G instead of C at 33035	Ineichen et al. (1982)
*t*L2d terminator	K.-C. Luk and W. Szybalski (1983a)
*t*L2c terminator	---ibid.
*c*IIIam611	Knight and Echols (1983)
*t*L2b terminator	K.-C. Luk and W. Szybalski (1983a)
*t*L2a terminator	---ibid.
*nin*L99	Somasekhar et al. (1982)
*trp*44 right	N. Franklin, pers. comm. (1983)
*imm*21 left	---ibid.
*t*L1 terminator	Drahos and Szybalski (1981)
*trp*48	Franklin and Bennett (1979)
pλQLC115	L. MacHattie, pers. comm. (1983)
*N*am22	Franklin and Bennett (1979).

TABLE 3 (Continued)

Site name	Reference
pλQLC108	L. MacHattie, pers. comm. (1983)
pλQLC118	---ibid.
pλQLC109	---ibid.
pλQLC105	---ibid.
pλQLC101	---ibid.
pλQLC143	---ibid.
*N*am7	Franklin and Bennett (1979)
pλXJS1974	L. MacHattie, pers. comm. (1983)
*N*mar2	Franklin and Bennett (1979)
pλQLC117	L. MacHattie, pers. comm. (1983)
pλXJS2300	---ibid.
pλMS972	---ibid.
pλQLC98	---ibid.
pλQLC104	---ibid.
*N*am2521	Franklin and Bennett (1979)
pλMS370	Muster et al. (1983)
*N*am53	Franklin and Bennett (1979)
pλXJS1076	L. MacHattie, pers. comm. (1983)
pλMS968	---ibid.
*N*am2524	Franklin and Bennett (1979)
*N*am219	---ibid.
pλQLC102	L. MacHattie, pers. comm. (1983)
pλQLC144	---ibid.
pλQLC95	---ibid.
dvl left	Dahlberg and Blattner (1975); Landsmann et al. (1982)
*nut*L	Salstrom and Szybalski (1978); Drahos and Szybalski (1981); Drahos et al. (1982); Somasekhar et al. (1982)
pλQLC138	L. MacHattie, pers. comm. (1983)
pλQLC103	---ibid.
*bio*3h-1	G. Somasekhar and W. Szybalski, unpubl. (1982)
*s*L *p*L	Blattner et al. (1972)
*imm*434 left	Landsmann et al. (1982); R. Pastrana, pers. comm. (1982)
*o*L1	Ptashne et al. (1976)
*vir*2,v003	Dahlberg and Blattner (1975); Flashman (1978)
*vir*101	Dahlberg and Blattner (1975)
*p*L *sex*1	Kleid et al. (1976)
*o*L2	Ptashne et al. (1976)
*v*L2668	Bailone and Devoret (1978)
*v*169	---ibid.
*v*305	Flashman (1978)
*o*L3	Ptashne et al. (1976)
1 668	Bailone and Devoret (1982)
*t*IMM terminator	Landsmann et al. (1982); K.-C. Luk and W. Szybalski, unpubl. (1983)
*KH*54	K.-C. Luk and W. Szybalski, unpubl. (1983)
delete A at 35887	Ineichen et al. (1981)
*rex*209	Landsmann et al. (1982)
*rex*111	---ibid.
pλ969	L. MacHattie, pers. comm. (1983)
pλQLC139	---ibid.
pλQLC137	---ibid.
s lit *p* lit	Landsmann et al. (1982); K.-C. Luk and W. Szybalski, in prep. (1983)
*KH*70	K.-C. Luk and W. Szybalski, unpubl. (1983)
pλQLC111	L. MacHattie, pers. comm. (1983)
pλQLC141	---ibid.
pλ142	---ibid.
pc 1	Guarente et al. (1982)
pλXJS2299	L. MacHattie, pers. comm. (1983)
pλXJS1975	---ibid.
pλQLC112	---ibid.
pλXJS1973	---ibid
pλQLC106	---ibid.

TABLE 3 (Continued)

Site name	Reference
pλQLC110	L. MacHattie, pers. comm. (1983)
pλQLC107	---ibid.
dv021 left	Landsman et al. (1982)
*c*I UA72	[a] F.Hutchinson, T.R. Skopek, R.D. Wood, pers. comm. (1983)
*c*I IC11	---ibid.
*c*I SP31	---ibid.
*c*Iam14	Lieb (1981)
*c*I NT1	F.Hutchinson, T.R. Skopek, R.D. Wood, pers. comm. (1983)
*c*Iam504	Lieb (1981)
*c*I BP83	F.Hutchinson, T.R. Skopek, R.D. Wood, pers. comm. (1983)
*c*Iam505	Lieb (1981)
KH100	Landsmann et al. (1982)
*c*I UV59	F.Hutchinson, T.R. Skopek, R.D. Wood, pers. comm. (1983)
*c*I UV63	---ibid.
*c*I IC28	---ibid.
*c*I ET26	---ibid.
*c*I AA21	---ibid.
*c*I AA2	---ibid.
*c*I NT16	---ibid.
*c*I UV76	---ibid.
*c*I SP1	---ibid.
*c*I NT15	---ibid.
*c*I171;172;473;474::Tn*10*	Kleckner (1979)
*c*I 13	F. Hutchinson, T.R. Skopek, R.D. Wood, pers. comm. (1983)
*c*I UV64	---ibid.
*c*I UV73	---ibid.
*c*I BL46	---ibid.
ind1	Sanger et al. (1982); R.T. Sauer, pers. comm. (1980)
*c*I PF17	F. Hutchinson, T.R. Skopek, R.D. Wood, pers comm. (1983)
*c*I AA3	---ibid.
*c*I SP48	---ibid.
*c*I ET39	---ibid.
*c*Iam499	Lieb (1981).
*c*I BP138	F. Hutchinson, T.R. Skopek, R.D. Wood, pers. comm. (1983)
*c*Iam212	Lieb (1981)
*c*I ET3	F.Hutchinson, T.R. Skopek, R.D. Wood, pers. comm. (1983)
*c*Iam34	Lieb (1981)
*c*IamE86	Hecht et al. (1983)
*c*I54-1	---ibid.
*c*IamR82	---ibid.
*c*I UV52	F.Hutchinson, T.R. Skopek, R.D. Wood, pers. comm. (1983)
*c*I ET15	---ibid.
*c*I UV96	---ibid.
*c*I UV97	---ibid.
*c*I UV79	---ibid.
*c*I PF12;*c*I AA6	---ibid.
*c*I11-3;*c*ICP175	Hecht et al. (1983)
*c*I60-1	---ibid.
*c*I UV47	F.Hutchinson, T.R. Skopek, R.D. Wood, pers. comm. (1883)
*c*I5-3	Hecht et al. (1983)
*c*I UV116;*c*I UV74	F.Hutchinson, T.R. Skopek, R.D. Wood, pers. comm. (1983)
*c*I BP84	---ibid.
*c*I BU30	---ibid.
*c*I UV66	---ibid.
*c*I SP44;*c*I SP37	---ibid.

TABLE 3 (Continued)

Site name	Reference
pλQLC113	L. MacHattie, pers. comm. (1983)
*c*I857	Lieb (1981); R. T. Sauer, pers. comm. (1980)
*KH*70	K.-C. Luk and W. Szybalski (unpubl.)
*c*I63-1	Hecht et al. (1983)
*c*I ET7	F.Hutchinson, T.R. Skopek, R.D. Wood, pers. comm. (1983)
*c*I BL71	---ibid.
*c*I UA8	---ibid.
*c*I UV61	---ibid.
dvh93 left	Schwarz et al. (1980)
*c*I UV62	F.Hutchinson, T.R. Skopek, R.D. Wood, pers. comm. (1983)
*c*IamL57	Hecht et al. (1983)
*c*I UV41	F.Hutchinson, T.R. Skopek, R.D. Wood, pers. comm. (1983)
*c*I34-1	Hecht et al. (1983)
*c*I UV117	F.Hutchinson, T.R. Skopek, R.D. Wood, pers. comm. (1983)
*c*I UV44	---ibid.
*c*I NT17	---ibid.
*c*I26-2	Hecht et al. (1983)
*c*I UA33;*c*I SP43	F.Hutchinson, T.R. Skopek, R.D. Wood, pers. comm. (1983)
*c*I UV91	---ibid.
*c*ICP32	Hecht et al. (1983)
*c*I UV46	F.Hutchinson, T.R. Skopek, R.D. Wood, pers. comm. (1983)
*c*I53-1	Hecht et al. (1983)
*c*I UV108	F.Hutchinson, T.R. Skopek, R.D. Wood, pers. comm. (1983)
*c*I UV37	---ibid.
*c*IamL50	Hecht et al. (1983)
*c*I NT3	F.Hutchinson, T.R. Skopek, R.D. Wood, pers. comm. (1983)
*c*ICP-9;*c*I29-6	Hecht et al. (1983)
*c*I BL42	F.Hutchinson, T.R. Skopek, R.D. Wood, pers. comm. (1983)
*c*I SP38	---ibid.
*c*I37-1	Hecht et al. (1983)
*c*I BP102	F.Hutchinson, T.R. Skopek, R.D. Wood, pers. comm. (1983)
*c*I475::Tn*10*	Kleckner (1979)
*c*I UA68	F.Hutchinson, T.R. Skopek, R.D. Wood, pers. comm. (1983)
*c*I BL70	---ibid.
*c*I40-3	Hecht et al. (1983)
*c*Iam282	Lieb (1981)
*c*I NT25;*c*I UA77	F.Hutchinson, T.R. Skopek, R.D. Wood, pers. comm. (1983)
*c*IamQ44	Hecht et al. (1983)
*c*I ET22	F.Hutchinson, T.R. Skopek, R.D. Wood, pers. comm. (1983)
*c*I38-2	Hecht et al. (1983)
*c*I SP61	F.Hutchinson, T.R. Skopek, R.D. Wood, pers. comm. (1983)
*c*I PF2;*c*I AA13	---ibid.
*c*I SP28	---ibid.
*c*I9-1	Hecht et al. (1983)
*c*I BU120	F.Hutchinson, T.R. Skopek, R.D. Wood, pers. comm. (1983)
*c*I SP41	---ibid.
*c*I UV86	---ibid.
*c*I AA13	---ibid.
*c*ICP7	Hecht et al. (1983)
pc 2	A. Hochschild, pers. comm. (1982)
*c*I SP5	F.Hutchinson, T.R. Skopek, R.D. Wood, pers. comm. (1983)

TABLE 3 (Continued)

Site name	Reference
*c*I BU39	F.Hutchinson, T.R. Skopek, R.D. Wood, pers. comm. (1983)
*c*I12-1	Hecht et al. (1983)
pc 3	A. Hochschild, pers. comm. (1982)
*c*I BL51	F.Hutchinson, T.R. Skopek, R.D. Wood, pers. comm. (1983)
*c*I SP3	---ibid.
*c*IamQ33	Hecht et al. (1983)
*c*I UA6;*c*I BL89	F.Hutchinson, T.R. Skopek, R.D. Wood, pers. comm. (1983)
*c*I47-1	Hecht et al. (1983)
*c*I408::Tn*10*	Kleckner (1979)
*c*I UV14	F.Hutchinson, T.R. Skopek, R.D. Wood, pers. comm. (1983)
*c*IamL31	Hecht et al. (1983)
*c*I4-3	---ibid.
*c*I BL84	F.Hutchinson, T.R. Skopek, R.D. Wood, pers. comm. (1983)
*c*I1-1	Hecht et al. (1983)
*c*I SP42	F.Hutchinson, T.R. Skopek, R.D. Wood, pers. comm. (1983)
*c*I UV80	---ibid.
*c*I UV57	---ibid.
*c*I UV88	---ibid.
*c*I ET28	---ibid.
*c*I ET13	---ibid.
*c*Iam302	Lieb (1981)
*c*I56-1;*c*I5-1	Hecht et al. (1983)
*c*I32-1;*c*I29-7	---ibid.
*c*I UV85	F.Hutchinson, T.R. Skopek, R.D. Wood, pers. comm. (1983)
*c*I SP39;*c*I BL80	---ibid.
*c*I UV112;*c*I UV93	---ibid.
*c*I36-1	Hecht et al. (1983)
*c*I SP60	F.Hutchinson, T.R. Skopek, R.D. Wood, pers. comm. (1983)
*c*I SP52	---ibid.
*c*I BL97	---ibid.
*c*I CP177	Hecht et al. (1983)
*c*I SP51	F.Hutchinson, T.R. Skopek, R.D. Wood, pers. comm. (1983)
*c*I BP80	---ibid.
*c*I UV23	---ibid.
*c*I UV6	---ibid.
*c*IamE13	Hecht et al. (1983)
*c*I BL60	F.Hutchinson, T.R. Skopek, R.D. Wood, pers. comm. (1983)
*c*I29-2	Hecht et al. (1983)
*c*I SP53	F.Hutchinson, T.R. Skopek, R.D. Wood, pers. comm. (1983)
*c*I UA54	---ibid.
*c*I BL100	---ibid.
*c*I UV34	---ibid.
*c*I SP27	---ibid.
*c*I UV90	---ibid.
*c*IamL7	Hecht et al. (1983)
*c*I SP35	F.Hutchinson, T.R. Skopek, R.D. Wood, pers. comm. (1983)
*c*I SP14	---ibid.
*c*I UV12	---ibid.
*KH*54	K.-C. Luk and W. Szybalski (unpubl.)
*c*I SP57	F.Hutchinson, T.R. Skopek, R.D. Wood, pers. comm. (1983)
*c*I57-1	Hecht et al. (1983)
*c*I SP62	F.Hutchinson, T.R. Skopek, R.D. Wood, pers. comm. (1983)
*c*I UA60	---ibid.

TABLE 3 (Continued)

Site name	Reference
*c*I UA9	F.Hutchinson, T.R. Skopek, R.D. Wood, pers. comm. (1983)
*c*I UV28	---ibid.
*c*I BP81	---ibid.
*s*M,*p*M	Walz et al. (1976)
*c*I BP86	F.Hutchinson, T.R. Skopek, R.D. Wood, pers. comm. (1983)
*c*I SP2	---ibid.
*c*I SP46	---ibid.
*o*R3	Ptashne et al. (1976)
prm-E37	Rosen et al. (1980)
*v*C3	Flashman (1976); Meyer et al. (1980)
oR3-r1	Meyer et al. (1980)
oR3-r2 oR3-r3	---ibid.
oR3-c12	---ibid.
oR3-c10	---ibid.
prm up-1	---ibid.
prm-M104;116;U31	Meyer et al. (1975); Kleid et al. (1976); Rosen et al. (1980)
*c*I BU20	F.Hutchinson, T.R. Skopek, R.D. Wood, pers. comm. (1983)
*o*R2	Kleid et al. (1976); Ptashne et al. (1976)
*c*I UV26	F.Hutchinson, T.R. Skopek, R.D. Wood, pers. comm. (1983)
prm-E104	G. Gussin, pers. comm. (1982)
v3c	Flashman (1976); Flashman (1978)
v1;*vir*146	Maniatis et al. (1975); Schwarz et al. (1980)
*c*I UV51	F.Hutchinson, T.R. Skopek, R.D. Wood, pers. comm. (1983)
prm-E93	Rosen et al. (1980)
BR1	J. Eliason, pers. comm. (1982)
vN	Maniatis et al. (1975); Flashman (1978)
MAC1 MAC41	J. Eliason, pers. comm. (1982)
MCC8,MCH31,AAH8	---ibid.
MAH4,AAH2	---ibid.
MCH20 MCH9	---ibid.
*p*R-x3	Hawley (1982)
MCH8	J. Eliason, pers. comm. (1982)
*o*R1	Ptashne et al. (1976)
VR2668	Bailone and Devoret (1978); R. Devoret, pers. comm. (1982)
*vir*R18	Flashman (1978)
VR668	Bailone and Devoret (1978); R. Devoret, pers. comm (1982)
vs326	Maniatis et al. (1975)
BC1	J. Eliason, pers. comm (1982)
prm-UV8	Rosen et al. (1980)
v3	Maniatis et al. (1975); Schwarz et al. (1980)
prm-UV93;M36	Rosen et al. (1980)
vs387 *v*C1 NR5	Meyer et al. (1975); J. Eliason, pers. comm. (1982)
UV119	F.Hutchinson, T.R. Skopek, R.D. Wood, pers. comm. (1983)
UV11	---ibid.
BU143	---ibid.
*s*R *p*R	Blattner and Dahlberg (1972)
*t*R0 terminator	Calva and Burgess (1980)
*imm*434 right	Rosenberg et al. (1978); Schwarz et al. (1978); Grosschedl and Schwarz (1979)
*nut*R	Rosenberg et al. (1978)
Cin-1	Wulff et al. (1980)
CNC1	---ibid
CNC8	---ibid
*t*R1 terminator	Rosenberg et al. (1978); Calva and Burgess (1980).
dya3	D.L. Wulff and M.E. Mahoney, pers. comm. (1983)

TABLE 3 (Continued)

Site name	Reference
*c*17	Rosenberg et al. (1978)
*s*E,*p*E	Schmeissner et al. (1980)
dya2	D.L. Wulff and M.E. Mahoney, pers. comm. (1983)
*r*32	Rosenberg et al. (1978)
*cy*3048	Wulff et al. (1980)
*cy*3071;*cy*3073a	D.L. Wulff and M.E. Mahoney, pers. comm. (1983)
*cy*2001	Rosenberg et al. (1978); Schmeissner et al. (1980); Wulff et al. (1980)
*c*II3088	D.L. Wulff and M.E. Mahoney, pers. comm. (1983)
*cy*3019	Wulff et al. (1980)
cir5	D.L. Wulff and M.E. Mahoney, pers. comm. (1983)
*c*II3086	---ibid.
*c*II3067	---ibid.
*c*II3105	---ibid.
*cy*3073b	---ibid.
*c*II3059	---ibid.
*can*1	Wulff et al. (1980)
ctr2;ctr3	D.L. Wulff and M.E. Mahoney, pers. comm. (1983)
ctr1	---ibid.
*c*II binding site	Ho and Rosenberg (1982)
*cy*3003	Wulff et al. (1980)
*cy*42	Schwarz et al. (1978); Wulff et al. (1980)
*cy*3077	D.L. Wulff and M.E. Mahoney, pers. comm. (1983)
*c*II3114;*c*II3109	---ibid.
*cy*844	Wulff et al. (1980)
*cy*3075	D.L. Wulff and M.E. Mahoney, pers. comm. (1983)
*cy*3107	---ibid.
*cy*3008	Wulff et al. (1980)
*cy*3001	---ibid
*cy*3095	D.L. Wulff and M.E. Mahoney, pers. comm. (1983)
*cy*3072;*cy*3078	---ibid.
*cy*3098	---ibid.
*c*II3623	---ibid.
*c*II3085	---ibid.
*c*II3104	---ibid.
*c*II3091	---ibid.
*c*II3056	---ibid.
*c*II3099	---ibid.
*c*II3081	---ibid.
*c*II3097	---ibid.
*c*II3111	---ibid.
*c*II3092	---ibid.
*c*II2002	Wulff et al. (1980)
*c*II3641	D.L. Wulff and M.E. Mahoney, pers. comm. (1983)
*c*II3062	---ibid.
*c*IIam60	Schwarz et al. (1978)
*c*II3283a	D.L. Wulff and M.E. Mahoney, pers. comm. (1983)
*c*II3283b	---ibid.
*c*II3302	---ibid.
*chi*C	Sprague et al. (1978)
*c*II68	Schwarz et al. (1978)
*c*II2358a	D.L. Wulff and M.E. Mahoney, pers. comm. (1983)
*c*IIam41	Schwarz et al. (1978)
ice	Hobom et al. (1978)
pk35	Moore et al. (1979)
*c*II3258b	D.L. Wulff and M.E. Mahoney, pers. comm. (1983)
*c*II3468	---ibid.
*c*II3386	---ibid.
*c*II3139	---ibid.
to (*oop* terminator)	Dahlberg and Blattner (1973)
*imm*21 right	Schwarz et al. (1978); Rosenberg et al. (1980)
*c*II3639	D.L. Wulff and M.E. Mahoney, pers. comm. (1983)
*c*II3638	---ibid.
*c*II3520	---ibid.
so (*oop* start point)	Blattner and Dahlberg (1972); Dahlberg and Blattner (1973)

TABLE 3 (Continued)

Site name	Reference
ori iteron 1	Denniston-Thompson et al. (1977); Hobom et al. (1979); Moore et al. (1981); Tsurimoto and Matsubara (1981); K. Zahn, pers. comm. (1982)
ori iteron 2	---ibid.
ori iteron 3	---ibid.
r93	Denniston-Thompson et al. (1977)
ori iteron 4	Hobom et al. (1979); Matsubara et al. (1981); Moore et al. (1981); K. Zahn, pers comm. (1982)
r99	Denniston-Thompson et al. (1977)
tiny 12	Hobom et al. (1979); Moore et al. (1979, 1981)
r96	Denniston-Thompson et al. (1977)
hy42	Moore et al. (1979; 1981)
rep82	Moore et al. (1981)
ri[c]5b	Moore and Blattner (1982)
λdvh93	Schwartz et al. (1980)
dv021 right	Kröger and Hobom (1982)
*nin*5 left	---ibid.
*t*R2 terminator	---ibid.
627	Halling et al. (1982)
623	---ibid.
dv1 right	Kröger and Hobom (1982)
G instead of A at 43082	---ibid.
*nin*5 right	---ibid.
A instead of C at 44498	Petrov et al. (1981) - A; Daniels and Blattner (1982) - G
*p*6S *pR′*	Lebowitz et al. (1971); Blattner and Dahlberg (1972); Calva and Burgess (1980); Grayhack and Roberts (1982); D. L. Daniels, H. Lozeron, and F. R. Blattner, in prep. (1983)
qut	Somasekhar and Szybalski (1983); E. Grayhack, C. Hart, X. Yang and J. Roberts, pers. comm. (1982)
6S RNA	Lebowitz et al. (1971); Calva and Burgess (1980)
*chi*D	Smith et al. (1981)
λ422	Daniels and Blattner (1982)
*S*ts9B	R. Grimaila, G. Neal, R. Young, pers. comm. (1983)
*S*am7	Daniels (1981); Sanger et al. (1982)
λ296	Ikeda et al. (1982)
dK23	Benedik, M., R. Young, A. Taylor and A. Campbell. pers. comm. (1982)
dk6	---ibid.
right end	Wu and Taylor (1971); Nichols and Donelson (1978)

The mapping information that assigns these sites exactly in the sequence is referenced. For references to restriction sites, see text. For most of the other sites see Szybalski and Szybalski (1979).

[a] For mutants isolated by R. Hutchinson, T.R. Shopek, and R. D. Wood (pers. comm.) symbols BL, BB, or BV refer to mutagenesis by 5-bromouracil; AL or AA, to 9-aminoacridine; PF, to proflavine; IC, to ICR141; SP, to spontaneous mutants; UV, to 254 nm UV light; ET or VA, to acetophenone + 313 mm UV light; NT, to only the host irradiated with 254 nm UV light.

TABLE 4 METHOD OF ASSIGNMENT OF START CODONS OF λ PROTEINS

Gene	Predicted traits[a]	Method of Assignment
Nu1	191-733 orf-181 20.4	gene *Nu1* inactivated by insertion *Nu1t16* mapped by restriction enzymes (M. Gould, A. Becker, and H. Murialdo pers. comm. 1982)
A	711-2633 orf-641 73.3	uncertain; probable start based on RBS; several possible other starts in same reading frames
W	2633-2836 orf-68 7.6	probable start based on RBS is established as correct by amber mutation at 2,642 (Sanger et al. 1982)
B	2836-4434 orf-533 59.5	probable start based on RBS: one of two possible before B*
B*		start of processed *B*-coded protein in capsid, first few amino acids determined (Sanger et al. 1982; Walker et al. 1982)
C	4418-5734 orf-439 45.9	probable start based on RBS, best of six possibilities: genes *C* and *Nu3* overlap in same reading frame (Shaw and Murialdo 1980)
Nu3	5132-5734 orf-201 20.8	based on measured size of *Nu3* protein and on RBS, several other possibilities
D	5747-6076 orf-110 11.6	first few amino acids sequenced (Walker et al. 1982)
E	6135-7157 orf-341 38.2	first few amino acids sequenced (Walker et al. 1982)
FI	7202-7597 orf-132 14.3	probable start of two possible based on RBS
FII	7612-7962 orf-117 12.7	probable start of two possible based on RBS
Z	7977-8552 orf-192 21.6	probable start of three possible based on RBS
U	8552-8944 orf-131 14.6	only probable start based on RBS
V	8955-9692 orf-246 25.8	first few amino acids sequenced (Walker et al. 1982)
G	9711-10130 orf-140 15.6	most probable of two possible starts; good sequence in region of this open-reading frame. Reasonable codon usage and RBS. *V* and *T* predicted sizes fit well with physiochemical data (see Szybalski and Szybalski 1979), but *G* predicted M_r is about half its measured size.
T	10115-10546 orf-144 16.1	uncertain; probable start out of several possible based on RBS

TABLE 4 (Continued)

Gene	Predicted traits	Method of Assignment
H	10542-13100 orf-853 92.3	first few amino acids sequenced (Walker et al. 1982)
M	13100-13426 orf-109 12.5	probable start of five possible based on RBS
L	13429-14124 orf-232 25.7	probable start out of several possible based on RBS
K	14276-14872 orf-199 23.0	probable start based on RBS and size of K protein
I	14773-15441 orf-223 23.1	probable start out of three possible based on RBS, protein not identified on gels so M_r information not available
J	15505-18900 orf-1132 124.4	probable start based on RBS
lom	18965-19582 orf-206 21.9	open reading frame is probably coding for product based on codon usage, RBS. The product is *lom* based on genetic map, hydrophobicity (*lom* is a membrane protein), and M_r
orf-401 19650-20852 39.9 *orf*-314 21029-21970 32.0 *orf*-194 21973-22554 21.6		three orf's probably code for protein products based on codon usage. Orf 401 and 194 have good RBS. orf 314 poor RBS. A single base phase shift could merge orf 314 with a short orf preceding it yielding an orf 335 with a good RBS. This suggests the possibility that there is a mutation in the sequenced strain, λ*c*I857*S*7
b*Ea*47	23918-22689 48.1 orf-410	based on codon usage and RBS (Epp 1978; Szybalski and Szybalski 1979; Sanger et al. 1982; Court and Oppenheim, this volume)
b*Ea*31	25399-34512 orf-296 34.6	probable start (among two possible ones) based on codon usage and RBS (Epp 1978; Szybalski and Szybalski 1979; Court and Oppenheim, this volume)
b*Ea*59	26973-25399 orf-525 59.5	only probable start based on codon usage and RBS (Epp 1978; Court and Oppenheim, this volume)
int	28882-27815 orf-356 40.3	Davies 1980; Hoess et al. 1980
xis	29078-28863 orf-72 8.6	Davies 1980; Hoess et al. 1980
*Ea*8.5	29655-29377 orf-93 10.8	only probable start based on codon usage and RBS

TABLE 4 (Continued)

Gene	Predicted traits	Method of Assignment
Ea22	30395-29850 orf-182 20.9	only probable start based on codon usage and RBS
exo	32028-31351 orf-226 25.9	most probable start based on codon usage and RBS
bet	32810-32028 orf-261 29.7	only probable start based on RBS (Ineichen et al. 1981)
gam	33232-32819 orf-138 16.3	start uncertain (Ineichen et al. 1981; Sanger et al. 1982; Zagursky and Hays 1983; S. Friedman and J. Hays . pers. comm. 1983) but to left of *Sal*I @ 33244 (W. Loenen and F.R. Blattner, unpubl. 1982)
orf-47	33330-33190 (kil?) 5.5	This open reading frame has reasonable codon usage and RBS, protein not firmly identified. (Ineichen et al. 1981; Sanger et al. 1982; Court and Oppenheim, this volume)
*c*III	33463-33302 orf-54 6.0	This open reading frame is identified by cIIIam611. It has reasonable codon usage and RBS (Ineichen et al. 1981; Sanger et al. 1982; Knight and Echols 1983; Court and Oppenheim, this volume)
*Ea*10	33904-33539 orf-122 13.8	has been shown to be single-stranded binding protein and in some maps is called ssb (Ineichen et al 1981; Court and Oppenheim, this volume)
ral	34287-34090 orf-66 7.6	(Ineichen et al. 1981)
git	34497-35033 orf-179 21.0	probably *sieB* (Ineichen et al. 1981; Court and Oppenheim, this volume)
N	35360-35040 orf-107 12.3	(Franklin and Bennett 1979; N. Franklin, pers. comm. 1983)
rexB	36259-35828 orf-144 16.0	probable start of two possible based on RBS (Landsmann et al. 1982)
rexA	37114-36278 orf-279 31.3	only possible start based on RBS (Landsmann et al. 1982; Sanger et al. 1982)
*c*I	37940-37230 orf-237 26.2	amino acid sequence (Ptashne et al. 1976; Sauer 1978)
cro	38041-38238 orf-66 7.4	amino acid sequence (Hsiang et al. 1977; Roberts et al. 1977; Schwarz et al. 1978
*c*II	38360-38650 orf-97 11.1	first few amino acids sequenced (Roberts et al. 1977; Schwarz et al. 1978; Ho et al. 1982)

TABLE 4 (Continued)

Gene	Predicted traits	Method of Assignment
O	38686-39582 orf-299 33.9	(Roberts et al. 1977; Schwarz et al. 1978)
P	39582-40280 orf-233 26.5	(Schwarz et al. 1980)
ren	40280-40567 orf-96 10.6	probable start based on RBS; (Toothman and Herskowitz 1979); showed *ren* point mutations map to the left of *nin*5 deletions and that *nin*-deleted phage were *ren*-. This reading frame spans the *nin*5 endpoint (Kröger and Hobom 1982)
orf-146	40644-41081 ninB 16.6	(Kröger and Hobom 1982; Sanger et al. 1982)
orf-290	41081-41950 ninC 33.6	(Kröger and Hobom 1982; Sanger et al. 1982)
orf-57	41950-42120 ninD 7.0	(Kröger and Hobom 1982; Sanger et al. 1982)
orf-60	42090-42269 ninE 7.4	(Kröger and Hobom 1982; Sanger et al. 1982)
orf-56	42269-42436 ninF 6.3	(Kröger and Hobom 1982; Sanger et al. 1982)
orf-204	42429-43040 ninG 24.1	(Kröger and Hobom 1982; Sanger et al. 1982)
orf-68	43040-43243 ninH 7.9	(Kröger and Hobom 1982; Sanger et al. 1982)
orf-221	43224-43886 ninI 25.2	(Kröger and Hobom 1982; Sanger et al. 1982)
Q	43886-44506 orf-207 22.5	(Petrov et al. 1981; Daniels and Blattner 1982; Kröger and Hobom 1982; Sanger et al. 1982)
orf-64	44621-44812 7.1	(Daniels and Blattner 1982) possible function based on similar orf's in analogous control regions of P22, 82 (J. Roberts, pers. comm.)
S	45186-45506 orf-107 11.5	(Daniels and Blattner 1982) probable start based on RBS, reading frame assigned based on sequence of λSam7 (Daniels 1981; Sanger et al. 1982)
R	45493-45966 orf-158 17.8	amino acid sequence, (Imada and Tsugita 1971; Bienkowska-Szewczyk et al 1981; Daniels 1981; Sanger et al. 1982)

TABLE 4 (Continued)

Gene	Predicted traits	Method of Assignment
Rz	45966-46423 orf-153 17.2	most probable start based on RBS, also site of *Rz* insertion mutation dK23 has been sequenced, and the BclI site at 46366 has been shown to be within *Rz* (Garrett and Young, 1982; Sanger et al 1982; Ikeda, 1981, pers. comm.)

Genes were assigned to open reading frames, as described in the text and in Sanger et al. (1982) and Daniels et al. (1983). The assignment of genetically defined genes to open reading frames is probably correct except for kil, which is uncertain. The choice of start point is often fairly arbitrary and may, in fact, be incorrect. It is not known whether the three new open reading frames in the *b* region (401, 314 and 194) or the eight new open reading frames in the *nin* region (146, 290, 57, 60, 56, 204, 68, 221), or the open reading frame in the *Q-S* intergenic region (orf 64), actually code for products or not. For further discussion of gene assignments see Sanger et al. (1982).

Abbreviations: ORF, open reading frame: RBS, ribosome-binding site.

[a]Predicted traits are the coordinates of the gene endpoints, length in amino acids and molecular weight in kilodaltons.

ACKNOWLEDGMENTS

This is paper 2665 of the Laboratory of Genetics, University of Wisconsin, Madison, supported by National Institutes of Health grants AI-18214, GM-21812, and GM-28252 (to. F.R.B.) and a special grant from Cold Spring Harbor Laboratory.

REFERENCES

Abraham, J., D. Mascarenhas, R. Fischer, M. Benedik, A. Campbell, and H. Echols. 1980. DNA sequence of regulatory region for integration gene of bacteriophage λ. *Proc. Natl. Acad. Sci.* **77:** 2477.

Bailone, A. and R. Devoret. 1978. Isolation of ultravirulent mutants of phage λ. *Virology* **84:** 547.

Bieńkowska-Szewczyk, K., B. Lipińska, and A. Taylor. 1981. The *R* gene product of bacteriophage λ is the murein transglycosylase. *Mol. Gen. Genet.* **184:** 111.

Blattner, F.R. and J.E. Dahlberg. 1972. RNA synthesis startpoints in bacteriophage lambda: Are the promoter and operator transcribed? *Nat. New Biol.* **237:** 227.

Calva, E. and R.R. Burgess. 1980. Characterization of a ρ-dependent termination site within the *cro* gene of bacteriophage λ. *J. Biol. Chem.* **225:** 11017.

Court, D., U. Schmeissner, M. Rosenberg, A. Oppenheim, G. Guarneros, and C. Montanez. 1983. Processing of λ *int* RNA: A mechanism for gene control. *J. Mol. Biol.* (in press).

Dahlberg, J.E. and F.R. Blattner. 1973. In vitro transcription products of lambda DNA: Nucleotide sequences and regulatory sites. In *Virus Research. Proceedings of the ICN-UCLA Symposium* (ed. C.F. Fox and W.S. Robinson), p. 533. Academic Press, New York.

——. 1975. Sequence of the promoter-operator proximal region of the major leftward RNA of bacteriophage lambda. *Nucleic Acids Res.* **2:** 1441.

Daniels, D.L. 1981. "Control of late transcription in bacteriophage lambda." Ph.D. thesis, University of Wisconsin, Madison.

Daniels, D.L. and F.R. Blattner. 1982. Nucleotide sequence of the *Q* gene and the *Q* to *S* intergenic region of bacteriophage lambda. *Virology* **117:** 81.

Daniels, D.L., J.R. deWet, and F.R. Blattner. 1980. A new map of bacteriophage lambda DNA. *J. Virol.* **33:** 390.

Daniels, D.L., F. Sanger, and A.R. Coulson. 1983. Features of bacteriophage lambda—Analysis of the complete nucleotide sequence. *Cold Spring Harbor Symp. Quant. Biol.* **47:** 1009.

Daniels, D.L., J.L. Schroeder, W. Szybalski, and F.R. Blattner. 1982. A molecular map of bacteriophage lambda. In *Genetic maps* (ed. S.J. O'Brien), Vol. 2, p. 1. Cold Spring Harbor Laboratory, New York.

Davies, R.W. 1980. DNA sequence of the *int-xis* p_I region of the bacteriophage lambda; overlap of the *int* and *xis* genes. *Nucleic Acids Res.* **8:** 1765.

Davies, R.W., P.H. Schreier, and D.E. Büchel. 1977. Nucleotide sequence of the attachment site of coliphage lambda. *Nature* **270:** 757.

——. 1978. Determination of the endpoints of partial deletion mutants of the attachment site of bacteriophage lambda by DNA sequencing. *Nucleic Acids Res.* **5:** 3209.

Denniston-Thompson, K., D.D. Moore, K.E. Kruger, M.E. Furth, and F.R. Blattner. 1977. Physical structure of the replication origin of bacteriophage lambda. *Science* **198:** 1051.

Drahos, D. and W. Szybalski. 1981. Antitermination and termination functions of the cloned *nut*L, *N* and t_{L1} modules of coliphage lambda. *Gene* **16:** 261.

Drahos, D., G.R. Galluppi, M. Caruthers, and W. Szybalski. 1982. Synthesis of the *nut*L DNA segments and analysis of antitermination and termination functions in coliphage lambda. *Gene* **18:** 343.

Echols, H. and H. Murialdo. 1978. Genetic map of bacteriophage lambda. *Microbiol. Rev.* **42:** 577.

Epp, C. 1978. "Early protein synthesis and its control in bacteriophage lambda." Ph.D. thesis, University of Toronto, Canada.

Flashman, S.M. 1976. "The relation between structure and function in the operators of bacteriophage lambda: An analysis using o^c mutations." Ph.D. thesis, Harvard University, Cambridge, Massachusetts.

——. 1978. Mutational analysis of the operators of bacteriophage lambda. *Mol. Gen. Genet.* **166:** 61.

Franklin, N.C. and G.N. Bennett. 1979. The *N* protein of bacteriophage lambda, defined by its DNA sequence, is highly basic. *Gene* **8:** 107.

Garrett, J.M. and R. Young. 1982. Lethal action of bacteriophage λ *S* gene. *J. Virol.* **44:** 886.

Grayhack, E.J. and J.W. Roberts. 1982. The phage lambda *Q* gene product: Activity of a transcription antiterminator in vitro. *Cell* **30:** 637.

Grosschedl, R. and E. Schwarz. 1979. Nucleotide sequence of the *cro-c*II-*oop* region of bacteriophage 434 DNA. *Nucleic Acids Res.* **6:** 867.

Guarente, L., J.S. Nye, A. Hochschild, and M. Ptashne. 1982. Mutant λ phage repressor with a specific defect in its positive control function. *Proc. Natl. Acad. Sci.* **79:** 2236.

Guarneros, G., C. Montanez, T. Hernandez, and D. Court. 1982. Posttranscriptional control of bacteriophage λ *int* gene expression from a site distal to the gene. *Proc. Natl. Acad. Sci.* **79:** 238.

Halling, S.M., R.W. Simons, J.C. Way, R.B. Walsh, and N. Kleckner. 1982. DNA sequence organization of IS*10*-right of Tn*10* and comparison with IS*10*-left. *Proc. Natl. Acad. Sci.* **79:** 2608.

Hawley, D.K. 1982. "Control of transcription initiation frequency of p_R and p_{RM} promoters of bacteriophage lambda." Ph.D. thesis, Harvard University, Cambridge, Massachusetts.

Hecht, M.H., H.C.M. Nelson, and R.T. Sauer. 1982. Mutations in λ repressor's amino-terminal domain: Implications for protein stability and DNA binding. *Proc. Natl. Acad. Sci.* **80:** 2676.

Ho, Y.-s., and M. Rosenberg. 1982. Characterization of the phage lambda regulatory protein cII. *Ann. Microbiol. (Paris)* **133A:** 215.

Ho, Y.-s., M. Lewis, and M. Rosenberg. 1982. Purification and properties of a transcriptional activator. The cII protein of phage λ. *J. Biol. Chem.* **257:** 9128.

Hobom, G., R. Grosschedl, M. Lusky, G. Scherer, E. Schwarz, and H. Kössel. 1979. Functional analysis of the replicator structure of lambdoid bacteriophage DNAs. *Cold Spring Harbor Symp. Quant. Biol.* **43:** 165.

Hoess, R.H. and A. Landy. 1978. Structure of the lambda *att* sites generated by *int*-dependent deletions. *Proc. Natl. Acad. Sci.* **75:** 5437.

Hoess, R.H., C. Foeller, K. Bidwell, and A. Landy. 1980. Site-specific recombination functions of bacteriophage λ: DNA sequence of regulatory regions and overlapping structural genes for Int and Xis. *Proc. Natl. Acad. Sci.* **77:** 2482.

Horn, G.T. and R.D. Wells. 1981. The leftward promoter of bacteriophage λ. *J. Biol. Chem.* **256:** 1998.

Hsiang, M.W., R.D. Cole, Y. Takeda, and H. Echols. 1977. Amino acid sequence of Cro regulatory protein of bacteriophage lambda. *Nature* **270:** 275.

Hsu, P.-L., W. Ross, and A. Landy. 1980. The λ phage *att* site: Functional limits and interaction with Int protein. *Nature* **285:** 85.

Ikeda, H., K. Aoki, and A. Naito. 1982. Illegitimate recombination mediated *in vitro* by DNA gyrase of *Escherichia coli:* Structure of recombinant DNA molecules. *Proc. Natl. Acad. Sci.* **79:** 3724.

Imada, M. and A. Tsugita. 1971. Amino acid sequence of λ phage endolysin. *Nat. New Biol.* **233:** 230.

Ineichen, K., J.C.W. Shepherd, and T.A. Bickle. 1981. The DNA sequence of the phage lambda genome between p_L and the gene *bet*. *Nucleic Acids Res.* **9:** 4639.

Kleckner, N. 1979. DNA sequence analysis of Tn*10* insertions: Origin and role of 9 bp flanking repetitions during Tn*10* translocation. *Cell* **16:** 711.

Kleid, D.G., K.L. Agarwal, and H.G. Khorana. 1975. The nucleotide sequence in the promoter region of the gene *N* in bacteriophage λ. *J. Biol. Chem.* **250:** 5574.

Kleid, D., Z. Humayun, A. Jeffrey, and M. Ptashne. 1976. Novel properties of a restriction endonuclease isolated from *Haemophilus parahaemolyticus*. *Proc. Natl. Acad. Sci.* **73:** 293.

Knight, D.M. and H. Echols. 1983. The *c*III gene and protein of bacteriophage λ. *J. Mol. Biol.* **163:** 505.

Kröger, M. and G. Hobom. 1982. A chain of interlinked genes in the *nin*R region of bacteriophage lambda. *Gene* **20:** 25.

Landsmann, J., M. Kröger, and G. Hobom. 1982. The *rex* region of bacteriophage lambda: Two genes under three-way control. *Gene* **20:** 11.

Landy, A. and W. Ross. 1977. Viral integration and excision: Structure of the lambda *att* sites. *Science* **197:** 1147.

Lebowitz, P., S.M. Weissman, and C.M. Radding. 1971. Nucleotide sequence of a ribonucleic acid transcribed in vitro from λ phage deoxyribonucleic acid. *J. Biol. Chem.* **246:** 5120.

Legerski, R.J., J.L. Hodnett, and H.B. Gray, Jr. 1978. Extracellular nucleases of *Pseudomonas* BAL31. III. Use of the double-stranded deoxyribonuclease activity as the basis of a convenient method for the mapping of fragments of DNA produced by cleavage with restriction enzymes. *Nucleic Acids Res.* **5:** 1445.

Lieb, M. 1981. A fine structure map of spontaneous and induced mutations in the lambda repressor gene, including insertions of IS elements. *Mol. Gen. Genet.* **184:** 364.

Luk, K.-C. and W. Szybalski. 1982a. Transcription termination: Sequence and function of the rho-independent t_{L3} terminator in the major leftward operon of bacteriophage lambda. *Gene* **17:** 247.

——. 1982b. Characterization of the cloned terminators t_{R1}, t_{L3} and t_I, and the *nut*R antitermination site of coliphage lambda. *Gene* **20:** 127.

——. 1983a. The t_{L2} cluster of transcription termination sites between genes *bet* and *ral* of coliphage lambda. *Virology* **125:** 403.

——. 1983b. A cluster of leftward, rho-dependent t'_J terminators in the *J* gene of coliphage lambda. *Gene* **21:** 175.

Luk, K.-C., P. Dobrzański, and W. Szybalski. 1982. Cloning and characterization of the termination site t_I for the gene *int* transcript in coliphage lambda. *Gene* **17:** 259.

Maniatis, T., M. Ptashne, K. Backman, D. Kleid, S. Flashman, A. Jeffrey, and R. Maurer. 1975. Recognition sequences of repressor and polymerase in the operators of bacteriophage lambda. *Cell* **5:** 109.

Mayer, H., R. Grosschedl, H. Schütte, and G. Hobom. 1981. *Cla*I, a new restriction endonuclease from *Caryophanon latum* L. *Nucleic Acids Res.* **9:** 4833.

Meyer, B.J., D.G. Kleid, and M. Ptashne. 1975. Repressor turns off transcription of its own gene. *Proc. Natl. Acad. Sci.* **72:** 4785.

Meyer, B.J., R. Maurer, and M. Ptashne. 1980. Gene regulation at the right operator (o_R) of bacteriophage λ. II. o_R1, o_R2, and o_R3: Their roles in mediating the effects of repressor and cro. *J. Mol. Biol.* **139:** 163.

Moore, D.D. and F.R. Blattner. 1982. Appendix: Sequence of λ*ri*^{c}5b. *J. Mol. Biol.* **154:** 81.

Moore, D.D., K.J. Denniston, and F.R. Blattner. 1981. Sequence organization of the origins of DNA replication in lambdoid coliphages. *Gene* **14:** 91.

Moore, D.D., K. Denniston-Thompson, K.E. Kruger, M.E. Furth, B.G. Williams, D.L. Daniels, and F.R. Blattner. 1979. Dissection and comparative anatomy of the origins of replication of lambdoid phages. *Cold Spring Harbor Symp. Quant. Biol.* **43:** 155.

Muster, C.J., L.A. MacHattie, and J.A. Shapiro. 1983. *p*λCM system: Observations on the roles of transposable elements in formation and breakdown of plasmids derived from bacteriophage lambda replicons. *J. Bacteriol.* **153:** 976.

Nichols, B.P. and J.E. Donelson. 1978. 178-Nucleotide sequence surrounding the *cos* site of bacteriophage lambda DNA. *J. Virol.* **26:** 429.

Petrov, N.A., V.A. Karginov, N.N. Mikriukov, O.I. Serpinski, and V.V. Kravchenko. 1981. Complete nucleotide sequence of the bacteriophage λ DNA region containing gene *Q* and promoter p'_R. *FEBS Lett.* **133:** 316.

Ptashne, M., K. Bachman, M.Z. Humayun, A. Jeffrey, R. Maurer, B. Meyer, and R.T. Sauer. 1976. Autoregulation and function of a repressor in bacteriophage lambda. *Science* **194:** 156.

Roberts, T.M., H. Shimatake, C. Brady, and M. Rosenberg. 1977. Sequence of *cro* gene of bacteriophage lambda. *Nature* **270:** 274.

Robinson, L.H. and A. Landy. 1977a. *Hin*dII, *Hin*dIII, and *Hpa*I restriction fragment maps of bacteriophage λ DNA. *Gene* **2:** 1.

——. 1977b. *Hin*dII, *Hin*dIII, and *Hpa*I restriction fragment maps of the left arm of bacteriophage λ DNA. *Gene* **2:** 33.

Rosen, E.D., J.L. Hartley, K. Matz, B.P. Nichols, K.M. Young, J.E. Donelson, and G.N. Gussin. 1980. DNA sequence analysis of *prm*⁻ mutations of coliphage lambda. *Gene* **11:** 197.

Rosenberg, M., D. Court, H. Shimatake, C. Brady, and D.L. Wulff. 1978. The relationship between function and DNA sequence in an intercistronic regulatory region in phage λ. *Nature* **272:** 414.

——. 1980. Structure and function of an intercistronic regulatory region in bacteriophage lambda. In *The operon*, (ed. J.H. Miller and W.S. Reznikoff), p. 345. Cold Spring Harbor Laboratory, Cold Spring Harbor, New York.

Rosenvold, E.C., E. Calva, R.R. Burgess, and W. Szybalski. 1980. *In vitro* transcription from the *b2* region of bacteriophage λ. *Virology* **107:** 476.

Ross, W., M. Shulman, and A. Landy. 1982. Biochemical analysis of *att*-defective mutants of the phage lambda site-specific recombination system. *J. Mol. Biol.* **156:** 505.

Salstrom, J.S. and W. Szybalski. 1978. Coliphage λ*nut*L⁻: A unique class of mutants defective in the site of gene *N* product utilization for antitermination of leftward transcription. *J. Mol. Biol.* **124:** 195.

Sanger, F., A.R. Coulson, G.F. Hong, D.F. Hill, and G.B. Petersen. 1982. Nucleotide sequence of bacteriophage λ DNA. *J. Mol. Biol.* **162:** 729.

Sauer, R.T. 1978. DNA sequence of the bacteriophage λ *c*I gene. *Nature* **276:** 301.

Scherer, G. 1978. Nucleotide sequence of the *O* gene and of the origin of replication in bacteriophage lambda DNA. *Nucleic Acids Res.* **5:** 3141.

Schmeissner, U., D. Court, K. McKenney, and M. Rosenberg. 1981. Positively activated transcription of λ integrase gene initiates with UTP *in vivo. Nature* **292:** 173.

Schmeissner, U., D. Court, H. Shimatake, and M. Rosenberg. 1980. Promoter for the establishment of repressor synthesis in bacteriophage λ. *Proc. Natl. Acad. Sci.* **77:** 3191.

Schwarz, E., G. Scherer, G. Hobom, and H. Kössel. 1978. Nucleotide sequence of *cro*, *c*II and part of the *O* gene in phage λ DNA. *Nature* **272:** 410.

——. 1980. The primary structure of the phage λ *P* gene completes the nucleotide sequence of the plasmid λdvh93. *Biochem. Int.* **1:** 386.

Shaw, J.E. and H. Murialdo. 1980. Morphogenetic genes *C* and *Nu*3 overlap in bacteriophage λ. *Nature* **283:** 30.

Shine, J. and L. Dalgarno. 1974. The 3′-terminal sequence of *Escherichia coli* 16S ribosomal RNA: Complementarity to nonsense triplets and ribosome binding sites. *Proc. Natl. Acad. Sci.* **71:** 1342.

Sklar, J.L. 1977. "Structure and function of two regions of DNA controlling the synthesis of prokaryotic RNAs." Ph.D. thesis, Yale University, New Haven, Connecticut.

Smith, D.I., F.R. Blattner, and J. Davies. 1976. The isolation and partial characterization of a new restriction endonuclease from *Providencia stuarti. Nucleic Acids Res.* **3:** 343.

Smith, G.R., D.W. Schultz, and J.M. Crasemann. 1980. Generalized recombination: Nucleotide sequence homology between Chi recombinational hotspots. *Cell* **19:** 785.

Smith, G.R., M. Comb, D.W. Schultz, D.L. Daniels, and F.R. Blattner. 1981. Nucleotide sequence of the Chi recombinational hot spot χ^+D in bacteriophage lambda. *J. Virol.* **37:** 336.

Somasekhar, G. and W. Szybalski. 1983. The *qut* site required for transcription antitermination by gene *Q* product of coliphage lambda. *Fed. Proc.* **42:** 2038.

Somasekhar, G., D. Drahos, J.S. Salstrom, and W. Szybalski. 1982. Sequence changes in coliphage lambda mutants affecting the *nut*L antitermination site and termination by t_{L1} and t_{L2}. *Gene* **20:** 477.

Sprague, K.U., D.H. Faulds, and G.R. Smith. 1978. A single base-pair change creates a Chi recombinational hot spot in bacteriophage λ. *Proc. Natl. Acad. Sci.* **75:** 6182.

Staden, R. and A.D. McLachlan. 1982. Codon preference and its use in identifying protein coding regions in long DNA sequences. *Nucleic Acids Res.* **10:** 141.

Stormo, G.D., T.D. Schneider, and L.M. Gold. 1982a. Characterization of translational initiation sites in *E. coli. Nucleic Acids. Res.* **10:** 2971.

Stormo, G.D., T.D. Schneider, L. Gold, and A. Ehrenfeucht. 1982b. Use of the "Perceptron" algorithm to distinguish translation initiation sites in *E. coli. Nucleic Acids Res.* **10:** 2997.

Szybalski, E.H. and W. Szybalski. 1979. A comprehensive molecular map of bacteriophage lambda. *Gene* **7:** 217.

Toothman, P. and I. Herskowitz. 1979. *Rex*-dependent exclusion of lambdoid phages. I. Prophage requirements for exclusion. *Virology* **102:** 133.

Tsurimoto, T. and K. Matsubara. 981. Purified bacteriophage lambda *O* protein binds to four repeating sequences at the λ replication origin. *Nucleic Acids Res.* **9:** 1789.

Walker, J.E., A.D. Auffret, A. Carne, A. Gurnett, P. Hansch, D. Hill, and M. Saraste. 1982. Solid-phase sequence analysis of polypeptides eluted from polyacrylamide gels. An aid to interpretation of DNA sequences exemplified by the *Escherichia coli unc* operon and bacteriophage lambda. *Eur. J. Biochem.* **123:** 253.

Walz, A., V. Pirrotta, and K. Ineichen. 1976. λ repressor regulates the switch between p_R and p_{rm} promoters. *Nature* **262:** 665.

Wu, R. and E. Taylor. 1971. Nucleotide sequence analysis of DNA. II. Complete nucleotide sequence of the cohesive ends of bacteriophage λ DNA. *J. Mol. Biol.* **57:** 491.

Wulff, D.L., M. Beher, S. Izumi, J. Beck, M. Mahoney, H. Shimatake, C. Brady, D. Court, and M. Rosenberg. 1980. Structure and function of the *cy* control region of bacteriophage lambda. *J. Mol. Biol.* **138:** 209.

Zagursky, R.J. and J.B. Hays. 1983. Expression of the phage λ recombination genes *exo* and *bet* under *lacPO* control on a multi-copy plasmid. *Gene* (in press).

APPENDIX II

Complete Annotated Lambda Sequence

Compiled by Donna L. Daniels, John L. Schroeder,* Waclaw Szybalski,† Fred Sanger,‡ Alan R. Coulson,‡ Guo F. Hong,‡ Diane F. Hill,‡ George B. Petersen,‡ and Frederick R. Blattner**

**Laboratory of Genetics and †McArdle Laboratory*
University of Wisconsin, Madison, Wisconsin 53706, and
‡MRC Laboratory of Molecular Biology, Cambridge, England

Here we present the completely annotated λ DNA sequence. The *l* strand of the λ "wild-type" DNA[1] is presented above the ruler line in the 5′–3′ direction, and the *r* strand, below the ruler in the 3′–5′ direction. Translation in all six frames is presented below the DNA sequence in the single-letter amino acid code. Those translation frames believed to actually code for proteins are capitalized. The single-letter amino acid code is as follows:

A = Ala	G = Gly	M = Met	S = Ser
C = Cys	H = His	N = Asn	T = Thr
D = Asp	I = Ile	P = Pro	V = Val
E = Glu	K = Lys	Q = Gln	W = Trp
F = Phe	L = Leu	R = Arg	Y = Tyr

All restriction cut sites predicted from their recognition sequences on the *l* strand are indicated above the base to the left of the internucleotide site of cutting. All mapped sites presented in Appendix I are presented here as annotations to the sequence. The four-character designation is indicated above the numbered base, and the complete annotation is repeated below the line of sequence. For annotation numbering conventions, see Appendix I.

Following the sequence are two explanatory tables.

Table 1 lists the restriction enzyme recognition sequences for which the sequence was searched.

Table 2 summarizes all restriction sites predicted by the λ DNA sequence.[1] For a summary of all of the experimentally mapped restriction sites and the annotations, see Appendix I.

[1]Various groups have sequenced different strains of λ. Sanger et al. (1982) presented the complete sequence of λc*I**ind*1 *ts*857*S*7. The sequence presented here and in Appendix I is that sequence changed at four positions to λ wild type. These changes are *ind*1 to *ind*$^+$ at 37589; c*I* *ts*857 to c*I*$^+$ at 37742; *Sam*7 to *S*$^+$ at 45352; and C to T correction of a typographical error in Sanger et al. (1982) at 44374.

```
THIS IS  MAP LAMBDA15 WITH LAM15 + ENZ + PRO;


SL F        CT      M                X    H          M   M              S                   A
BE N        LH      N                M    P          B   B              B                   H
1N U        NA      L                N    A          0   0              1                   A
LD H        D1      1                1    2          2   2              L                   3
/
GGGCGGCGACCTCGCGGGTTTTCGCTATTTATGAAAATTTTCCGGTTTAAGGCGTTTCCGTTCTTCTTCGTCATAACTTAATGTTTTTATTTAAAATACC
----.----+----.----+----.----+----.----+----.----+----.----+----.----+----.----+----.----+----.----+  1ØØ
            GCGCCCAAAAGCGATAAATACTTTTAAAAGGCCAAATTCCGCAAAGGCAAGAAGAAGCAGTATTGAATTACAAAAATAAATTTTATGG

g  r  r  p  r  g  f  s  l  f  m  k  i  f  r  f  k  a  f  p  f  f  f  v  i  t  .  c  f  y  l  k  y  p
 g  g  d  l  a  g  f  r  y  l  .  k  f  s  g  l  r  r  f  r  s  s  s  s  .  l  n  v  f  i  .  n  t
  a  a  t  s  r  v  f  a  i  y  e  n  f  p  v  .  g  v  s  v  l  l  r  h  n  l  m  f  l  f  k  i  p
----.----+----.----+----.----+----.----+----.----+----.----+----.----+----.----+----.----+----.----+
    r  r  g  r  p  n  e  s  n  i  f  i  k  r  n  l  a  n  g  n  k  k  t  m  v  .  h  k  .  k  f  y
  p  p  s  r  a  p  k  r  .  k  h  f  n  e  p  k  l  r  k  r  e  e  e  d  y  s  l  t  k  i  .  f  v
   a  a  v  e  r  t  k  a  i  .  s  f  k  g  t  .  p  t  e  t  r  r  r  .  l  k  i  n  k  n  l  i  g

SB1L        1         Sequence Block 1L    start
LEND        1         Left END                          sq      first base of sticky end
CLND        13        3' end of r-strand                        first base of double stranded DNA
SB1L        72        Sequence Block 1L    end


         M                 M                   HH         M                                N  N     H*
         N                 N                   AA         N                                U  U     IØ
         L                 L                   EE         L                                1  1     NØ
         1                 1                   13         1                                   6     21
                                               /                                                     /
CTCTGAAAAGAAAGGAAACGACAGGTGCTGAAAGCGAGGCTTTTTGGCCTCTGTCGTTTCCTTTCTCTGTTTTTGTCCGTGGAATGAACAATGGAAGTCA
----.----+----.----+----.----+----.----+----.----+----.----+----.----+----.----+----.----+----.----+  2ØØ
GAGACTTTTCTTTCCTTTGCTGTCCACGACTTTCGCTCCGAAAAACCGGAGACAGCAAAGGAAAGAGACAAAAACAGGCACCTTACTTGTTACCTTCAGT

  l  k  r  k  e  t  t  g  a  e  s  e  a  f  w  p  l  s  f  p  f  s  v  f  v  r  g  m  n  n  g  s  q
l  .  k  e  r  k  r  q  v  l  k  a  r  l  f  g  l  c  r  f  l  s  l  f  l  s  v  e  .  t  M  E  V  N
 s  e  k  k  g  n  d  r  c  .  k  r  g  f  l  a  s  v  v  s  f  l  c  f  c  p  w  n  e  q  w  k  s
----.----+----.----+----.----+----.----+----.----+----.----+----.----+----.----+----.----+----.----+
g  r  f  l  f  s  v  v  p  a  s  l  s  a  k  q  g  r  d  n  g  k  e  t  k  t  r  p  i  f  l  p  l  .
 r  q  f  s  l  f  r  c  t  s  f  a  l  s  k  p  r  q  r  k  r  e  r  n  k  d  t  s  h  v  i  s  t
  e  s  f  f  p  f  s  l  h  q  f  r  p  k  k  a  e  t  t  e  k  r  q  k  q  g  h  f  s  c  h  f  d

NU1       191       Nu1 (orf-181)        M-NH3       G
NU16      194       Nu1 t16              i           EM
*ØØ1      199      Mapped site HIN2


       NF PA*        B                    R                                  M                   M
       SN VLØ        B                    S                                  N                   N
       PU UUØ        V                    A                                  L                   L
       BH 212        1                    1                                  1                   1
            //
ACAAAAAGCAGCTGGCTGACATTTTCGGTGCGAGTATCCGTACCATTCAGAACTGGCAGGAACAGGGAATGCCCGTTCTGCGAGGCGGTGGCAAGGGTAA
----.----+----.----+----.----+----.----+----.----+----.----+----.----+----.----+----.----+----.----+  3ØØ
TGTTTTTCGTCGACCGACTGTAAAAGCCACGCTCATAGGCATGGTAAGTCTTGACCGTCCTTGTCCCTTACGGGCAAGACGCTCCGCCACCGTTCCCATT

 q  k  a  a  g  .  h  f  r  c  e  y  p  y  h  s  e  l  a  g  t  g  n  a  r  s  a  r  r  w  q  g  .
  K  K  Q  L  A  D  I  F  G  A  S  I  R  T  I  Q  N  W  Q  E  Q  G  M  P  V  L  R  G  G  G  K  G  N
t  k  s  s  w  l  t  f  s  v  r  v  s  v  p  f  r  t  g  r  n  r  e  c  p  f  c  e  a  v  a  r  v  m
----.----+----.----+----.----+----.----+----.----+----.----+----.----+----.----+----.----+----.----+
  c  f  a  a  p  q  c  k  r  h  s  y  g  y  w  e  s  s  a  p  v  p  f  a  r  e  a  l  r  h  c  p  y
l  l  f  c  s  a  s  m  k  p  a  l  i  r  v  m  .  f  q  c  s  c  p  i  g  t  r  r  p  p  p  l  p  l
 v  f  l  l  q  s  v  n  e  t  r  t  d  t  g  n  l  v  p  l  f  l  s  h  g  n  q  s  a  t  a  l  t

*ØØ2      211      Mapped site PVU2
NU1       ...      CONTINUES IN PHASE TWO
```

```
        H   F               S     H         B F       AFM HHNS          F
        N   N               F     G         B O       LNN HPCC          N
        F   U               A     U         V K       UUL AAIR          U
        1   H               N      2         1 1       1H1 1211          H
                                                             /
TGAGGTGCTTTATGACTCTGCCGCCGTCATAAAATGGTATGCCGAAAGGGATGCTGAAATTGAGAACGAAAAGCTGCGCCGGGAGGTTGAAGAACTGCGG
---.----+----.----+----.----+----.----+----.----+----.----+----.----+----.----+----.----+----.----+ 400
ACTCCACGAAATACTGAGACGGCGGCAGTATTTTACCATACGGCTTTCCCTACGACTTTAACTCTTGCTTTTCGACGCGGCCCTCCAACTTCTTGACGCC

 . g a l . l c r r h k m v c r k g c . n . e r k a a p g g . r t a a
 E V L Y D S A A V I K W Y A E R D A E I E N E K L R R E V E E L R
  r c f m t l p p s . n g m p k g m l k l r t k s c a g r l k n c g
---.----+----.----+----.----+----.----+----.----+----.----+----.----+----.----+----.----+----.----+
 h p a k h s q r r . l i t h r f p h q f q s r f a a g p p q l v a
  s t s . s e a a t m f h y a s l s a s i s f s f s r r s t s s s r
 i l h k i v r g g d y f p i g f p i s f n l v f l q a p l n f f q p

NU1     ...     CONTINUES IN PHASE TWO

MM HH   B   *BXM   E S           R          T     AM T   MH  H         H              MH
BN AA   G   0GHB   C C           S          A     FL H   SH  A         G              BN
OL EE   L   0L00   R R           A          Q     LU A   TA  E         A              OF
21 13   1   3221   2 1           1          1     31 1   11  3         1              21
    /       ///
CAGGCCAGCGAGGCAGATCTCCAGCCAGGAACTATTGAGTACGAACGCCATCGACTTACGCGTGCGCAGGCCGACGCACAGGAACTGAAGAATGCCAGAG
---.----+----.----+----.----+----.----+----.----+----.----+----.----+----.----+----.----+----.----+ 500
GTCCGGTCGCTCCGTCTAGAGGTCGGTCCTTGATAACTCATGCTTGCGGTAGCTGAATGCGCACGCGTCCGGCTGCGTGTCCTTGACTTCTTACGGTCTC

  g q r g r s p a r n y . v r t p s t y a c a g r r t g t e e c q r
 Q A S E A D L Q P G T I E Y E R H R L T R A Q A D A Q E L K N A R D
 r p a r q i s s q e l l s t n a i d l r v r r p t h r n . r m p e
---.----+----.----+----.----+----.----+----.----+----.----+----.----+----.----+----.----+----.----+
 a p w r p l d g a l f . q t r v g d v . a h a p r r v p v s s h w l
  c a l s a s r w g p v i s y s r w r s v r a c a s a c s s f f a l
   l g a l c i e l w s s n l v f a m s k r t r l g v c l f q l i g s

*003    415     Mapped site BGL2
NU1     ...     CONTINUES IN PHASE TWO

 N                       R           TBM           H     T  S   BH        NF
 S                       S           HIB           P     A  D   VG        SN
 P                       A           ANO           H     Q  U   UI        PU
 B                       1           111           1     1  1   1J        BH
                                                                 /
ACTCCGCTGAAGTGGTGGAAACCGCATTCTGTACTTTCGTGCTGTCGCGGATCGCAGGTGAAATTGCCAGTATTCTCGACGGGCTCCCCCTGTCGGTGCA
---.----+----.----+----.----+----.----+----.----+----.----+----.----+----.----+----.----+----.----+ 600
TGAGGCGACTTCACCACCTTTGGCGTAAGACATGAAAGCACGACAGCGCCTAGCGTCCACTTTAACGGTCATAAGAGCTGCCCGAGGGGGACAGCCACGT

 l r . s g g n r i l y f r a v a d r r . n c q y s r r a p p v g a
 S A E V V E T A F C T F V L S R I A G E I A S I L D G L P L S V Q
 t p l k w w k p h s v l s c c r g s q v k l p v f s t g s p c r c s
---.----+----.----+----.----+----.----+----.----+----.----+----.----+----.----+----.----+----.----+
 s r q l p p f r m r y k r a t a s r l h f q w y e r r a g g t p a
 s e a s t t s v a n q v k t s d r i a p s i a l i r s p s g r d t c
 v g s f h h f g c e t s e h q r p d c t f n g t n e v p e g g q r h

NU1     ...     CONTINUES IN PHASE TWO

 F    BH       A  N              E                F F T H H B        FNHS
 N    BP       F  S              C                N N H H G B        OCPC
 U    VA       L  P              R                U U A A U V        KIAR
 H    12       3  C              V                H H 1 1 2 1        1121
                                                                       //
GCGGCGTTTTCCGGAACTGGAAAACCGACATGTTGATTTCCTGAAACGGGATATCATCAAAGCCATGAACAAAGCAGCCGCGCTGGATGAACTGATACCG
---.----+----.----+----.----+----.----+----.----+----.----+----.----+----.----+----.----+----.----+ 700
CGCCGCAAAAGGCCTTGACCTTTTGGCTGTACAACTAAAGGACTTTGCCCTATAGTAGTTTCGGTACTTGTTTCGTCGGCGCGACCTACTTGACTATGGC

 a a f s g t g k p t c . f p e t g y h q s h e q s s r a g . t d t g
 R R F P E L E N R H V D F L K R D I I K A M N K A A A L D E L I P
 g v f r n w k t d m l i s . n g i s s k p . t k q p r w m n . y r
---.----+----.----+----.----+----.----+----.----+----.----+----.----+----.----+----.----+----.----+
 a a n e p v p f g v h q n g s v p y . . l w s c l l r a p h v s v
 r r k g s s s f r c t s k r f r s i m l a m f l a a a s s s i g
 l p t k r f q f v s m n i e q f p i d d f g h v f c g r q i f q y r

NU1     ...     CONTINUES IN PHASE TWO
```

```
       D   A        T      B    NHH**   F   F           T HHNS                    HH         E
       D            A      B    UPI00   N   N           H HPCC                    AP         C
       E            Q      V    1AN00   U   U           A AAIR                    EA         P
       1            1      1     1245   H   H           1 1211                    32         1
                                 ///                         /
GGGTTGCTGAGTGAATATATCGAACAGTCAGGTTAACAGGCTGCGGCATTTTGTCCGCGCCGGGCTTCGCTCACTGTTCAGGCCGGAGCCACAGACCGCC
----.----+----.----+----.----+----.----+----.----+----.----+----.----+----.----+----.----+----.----+ 800
CCCAACGACTCACTTATATAGCTTGTCAGTCCAATTGTCCGACGCCGTAAAACAGGCGCGGCCCGAAGCGAGTGACAAGTCCGGCCTCGGTGTCTGGCGG

 v  a  e  .  i  y  r  t  v  r  l  t  g  c  g  i  l  s  a  p  g  f  a  h  c  s  g  r  s  h  r  p  p
G  L  L  S  E  Y  I  E  Q  S  G  .  q  a  a  a  f  c  p  r  r  a  s  l  t  v  q  a  g  a  t  d  r  r
 g  c  .  V  N  I  S  N  S  Q  V  N  R  L  R  H  F  V  R  A  G  L  R  S  L  F  R  P  E  P  Q  T  A
----.----+----.----+----.----+----.----+----.----+----.----+----.----+----.----+----.----+----.----+
p  t  a  s  h  i  y  r  v  t  l  n  v  p  q  p  m  k  d  a  g  p  k  a  .  q  e  p  r  l  w  l  g  g
 p  n  s  l  s  y  i  s  c  d  p  .  c  a  a  a  n  q  g  r  r  a  e  s  v  t  .  a  p  a  v  s  r
  p  q  q  t  f  i  d  f  l  .  t  l  l  s  r  c  k  t  r  a  p  s  r  e  s  n  l  g  s  g  c  v  a

A        711     A (orf-641)       V-NH3
NU1      733     Nu1               COOH
*004     734     Mapped site HIN2
*005     734     Mapped site HPA1
```

```
  S       H             F            H         E S        HH                N    A H             F
  F       G             0            N         C C        HA                S    S A             0
  A       U             K            F         R R        AE                P    U E             K
  N       2             1            1         2 1        12                B    1 3             1

GTTGAATGGGCGGATGCTAATTACTATCTCCCGAAAGAATCCGCATACCAGGAAGGGCGCTGGGAAACACTGCCCTTTCAGCGGGCCATCATGAATGCGA
----.----+----.----+----.----+----.----+----.----+----.----+----.----+----.----+----.----+----.----+ 900
CAACTTACCCGCCTACGATTAATGATAGAGGGCTTTCTTAGGCGTATGGTCCTTCCCGCGACCCTTTGTGACGGGAAAGTCGCCCGGTAGTACTTACGCT

 l  n  g  r  m  l  i  t  i  s  r  k  n  p  h  t  r  k  g  a  g  k  h  c  p  f  s  g  p  s  .  m  r
  .  m  g  g  c  .  l  l  s  p  e  r  i  r  i  p  g  r  a  l  g  n  t  a  l  s  a  g  h  h  e  c  d
V  E  W  A  D  A  N  Y  Y  L  P  K  E  S  A  Y  Q  E  G  R  W  E  T  L  P  F  Q  R  A  I  M  N  A  M
----.----+----.----+----.----+----.----+----.----+----.----+----.----+----.----+----.----+----.----+
 n  f  p  r  i  s  i  v  i  e  r  f  f  g  c  v  l  f  p  a  p  f  c  q  g  k  l  p  g  d  h  i  r
r  q  i  p  p  h  .  n  s  d  g  s  l  i  r  m  g  p  l  a  s  p  f  v  a  r  e  a  p  w  .  s  h  s
 t  s  h  a  s  a  l  .  .  r  g  f  s  d  a  y  w  s  p  r  q  s  v  s  g  k  .  r  a  m  m  f  a

A        ...     CONTINUES IN PHASE THREE
```

```
    F       M  B  ATSTSFS         H        H       B           EF T
    N       N  B  RRPRPUP         P        P       B           CN P
    U       L  V  OKCKCSC         H        H       V           1U 2
    H       1  1  EA2A221         1        1       1           5H
                   //////
TGGGCAGCGACTACATCCGTGAGGTGAATGTGGTGAAGTCTGCCCGTGTCGGTTATTCCAAAATGCTGCTGGGTGTTTATGCCTACTTTATAGAGCATAA
----.----+----.----+----.----+----.----+----.----+----.----+----.----+----.----+----.----+----.----+ 1000
ACCCGTCGCTGATGTAGGCACTCCACTTACACCACTTCAGACGGGCACAGCCAATAAGGTTTTACGACGACCCACAAATACGGATGAAATATCTCGTATT

w  a  a  t  t  s  v  r  .  m  w  .  s  l  p  v  s  v  i  p  k  c  c  w  v  f  m  p  t  l  .  s  i  s
 g  q  r  l  h  p  .  g  e  c  g  e  v  c  p  c  r  l  f  q  n  a  a  g  c  l  c  l  l  y  r  a  .
  G  S  D  Y  I  R  E  V  N  V  V  K  S  A  R  V  G  Y  S  K  M  L  L  G  V  Y  A  Y  F  I  E  H  K
----.----+----.----+----.----+----.----+----.----+----.----+----.----+----.----+----.----+----.----+
 h  a  a  v  v  d  t  l  h  i  h  h  l  r  g  t  d  t  i  g  f  h  q  q  t  n  i  g  v  k  y  l  m
  p  c  r  s  c  g  h  p  s  h  p  s  t  q  g  h  r  n  n  w  f  a  a  p  h  k  h  r  s  .  l  a  y
i  p  l  s  .  m  r  s  t  f  t  t  f  d  a  r  t  p  .  e  l  i  s  s  p  t  .  a  .  k  i  s  c  l

AROE     919     lambdaaroE        sl            EM
TRKA     919     lambdatrkA        sl            EM
SPC2     919     lambdaspc2        sl            EM
TRKA     919     lambdatrkA        sl            EM
SPC2     919     lambdaspc2        sl            EM
FUS2     919     lambdafus2        sl            EM
SPC1     919     lambdaspc1        sl            EM
TP2      968     Tp2               i(Tn402)      EM
A        ...     CONTINUES IN PHASE THREE
```

```
F   H     B               S H         F  H                         H    B        EF
N   H     B               F G         O  P                         N    B        CN
U   A     V               A U         K  H                         F    V        1U
H   1     1               N 2         1  1                         3    1        5H

GCAGCGCAACACCCTTATCTGGTTGCCGACGGATGGTGATGCCGAGAACTTTATGAAAACCCACGTTGAGCCGACTATTCGTGATATTCCGTCGCTGCTG
----.----+----.----+----.----+----.----+----.----+----.----+----.----+----.----+----.----+----.----+ 1100
CGTCGCGTTGTGGGAATAGACCAACGGCTGCCTACCACTACGGCTCTTGAAATACTTTTGGGTGCAACTCGGCTGATAAGCACTATAAGGCAGCGACGAC

  s  a  t  p  l  s  g  c  r  r  m  v  m  p  r  t  l  .  k  p  t  l  s  r  l  f  v  i  f  r  r  c  w
 a  a  q  h  p  y  l  v  a  d  g  w  .  c  r  e  l  y  e  n  p  r  .  a  d  y  s  .  y  s  v  a  a  g
  Q  R  N  T  I  T  W  L  P  T  D  G  D  A  F  N  F  M  K  T  H  V  F  P  T  I  R  D  I  P  S  L  I
----.----+----.----+----.----+----.----+----.----+----.----+----.----+----.----+----.----+----.----+
 l  l  a  v  g  k  d  p  q  r  r  i  t  i  g  l  v  k  h  f  g  v  n  l  r  s  n  t  i  n  r  r  q  q
  a  a  c  c  g  .  r  t  a  s  p  h  h  h  r  s  s  .  s  f  g  r  q  a  s  .  e  h  y  e  t  a  a
  c  r  l  v  r  i  q  n  g  v  s  p  s  a  s  f  k  i  f  v  w  t  s  g  v  i  r  s  i  g  d  s  s

A       ...      CONTINUES IN PHASE THREE
```

```
 HH AH                  HNS  H                X                        H  F S        F
 HA SA                  PCC  P                M                        G  C C        N
 AE UE                  AIR  H                N                        I  R R        U
 12 13                  211  1                1                        C  2 1        H
                         /
GCGCTGGCCCCGTGGTATGGCAAAAAGCACCGGGATAACACGCTCACCATGAAGCGTTTCACTAATGGGCGTGGCTTCTGGTGCCTGGGCGGTAAAGCGG
----.----+----.----+----.----+----.----+----.----+----.----+----.----+----.----+----.----+----.----+ 1200
CGCGACCGGGGCACCATACCGTTTTTCGTGGCCCTATTGTGCGAGTGGTACTTCGCAAAGTGATTACCCGCACCGAAGACCACGGACCCGCCATTTCGCC

  r  w  p  r  g  m  a  k  s  t  g  i  t  r  s  p  .  s  v  s  l  m  g  v  a  s  g  a  w  a  v  k  r
   a  g  p  v  v  w  q  k  a  p  g  .  h  a  h  h  e  a  f  h  .  w  a  w  l  l  v  p  g  r  .  s  g
 A  L  A  P  W  Y  G  K  K  H  R  D  N  T  L  T  M  K  R  F  T  N  G  R  G  F  W  C  L  G  G  K  A  A
----.----+----.----+----.----+----.----+----.----+----.----+----.----+----.----+----.----+----.----+
  r  q  g  r  p  i  a  f  l  v  p  i  v  r  e  g  h  l  t  e  s  i  p  t  a  e  p  a  q  a  t  f  r
 p  a  p  g  t  t  h  c  f  a  g  p  y  c  a  .  w  s  a  n  .  .  h  a  h  s  r  t  g  p  r  y  l  p
  a  s  a  g  h  y  p  l  f  c  r  s  l  v  s  v  m  f  r  k  v  l  p  r  p  k  q  h  r  p  p  l  a

A       ...      CONTINUES IN PHASE THREE
```

```
                  H         BF        F                                            F S
                  G         BO        N                                            C C
                  U         VK        U                                            R R
                  2         11        H                                            2 1

CAAAAAACTACCGTGAAAAGTCGGTGGATGTGGCGGGTTATGATGAACTTGCTGCTTTTGATGATGATATTGAACAGGAAGGCTCTCCGACGTTCCTGGG
----.----+----.----+----.----+----.----+----.----+----.----+----.----+----.----+----.----+----.----+ 1300
GTTTTTTGATGGCACTTTTCAGCCACCTACACCGCCCAATACTACTTGAACGACGAAAACTACTACTATAACTTGTCCTTCCGAGAGGCTGCAAGGACCC

 q  k  t  t  v  k  s  r  w  m  w  r  v  m  m  n  l  l  l  l  m  m  i  l  n  r  k  a  l  r  r  s  w  v
  k  k  l  p  .  k  v  g  g  c  g  g  l  .  .  t  c  c  f  .  .  .  y  .  t  g  r  l  s  d  v  p  g
   K  N  Y  R  E  K  S  V  D  V  A  G  Y  D  E  L  A  A  F  D  D  D  T  F  Q  F  G  S  P  T  F  L  G
----.----+----.----+----.----+----.----+----.----+----.----+----.----+----.----+----.----+----.----+
  c  f  v  v  t  f  l  r  h  i  h  r  t  i  i  f  k  s  s  k  i  i  i  n  f  l  f  a  r  r  r  e  q
   l  f  s  g  h  f  t  p  p  h  p  p  n  h  h  v  q  q  k  q  h  h  y  q  v  p  l  s  e  s  t  g  p
 a  f  f  .  r  s  f  d  t  s  t  a  p  .  s  s  s  a  a  k  s  s  s  i  s  c  s  p  e  g  v  n  r  p

A       ...      CONTINUES IN PHASE THREE
```

```
   G   H       F F  C BHH                     M      H                     F     H  B
   2   P       C O  F AAA                     N      G                     N     N  B
   5   H       P K  R LEE                     L      I                     U     F  V
   8   1       1 1  1 113                     1      C                     H     1  1
                       //
TGACAAGCGTATTGAAGGCTCGGTCTGGCCAAAGTCCATCCGTGGCTCCACGCCAAAAGTGAGAGGCACCTGTCAGATTGAGCGTGCAGCCAGTGAATCC
----.----+----.----+----.----+----.----+----.----+----.----+----.----+----.----+----.----+----.----+ 1400
ACTGTTCGCATAACTTCCGAGCCAGACCGGTTTCAGGTAGGCACCGAGGTGCGGTTTTCACTCTCCGTGGACAGTCTAACTCGCACGTCGGTCACTTAGG

  t  s  v  l  k  a  r  s  g  q  s  p  s  v  a  p  r  q  k  .  e  a  p  v  r  l  s  v  q  p  v  n  p
 .  q  a  y  .  r  l  g  l  a  k  v  h  p  w  l  h  a  k  s  e  r  h  l  s  d  .  a  c  s  q  .  i  p
  D  K  R  I  E  G  S  V  W  P  K  S  I  R  G  S  T  P  K  V  R  G  T  C  Q  I  F  R  A  A  S  F  S
----.----+----.----+----.----+----.----+----.----+----.----+----.----+----.----+----.----+----.----+
 t  v  l  t  n  f  a  r  d  p  w  l  g  d  t  a  g  r  w  f  h  s  a  g  t  l  n  l  t  c  g  t  f  g
  h  c  a  y  q  l  s  p  r  a  l  t  w  g  h  s  w  a  l  l  s  l  c  r  d  s  q  a  h  l  w  h  i
  s  l  r  i  s  p  e  t  q  g  f  d  m  r  p  e  v  g  f  t  l  p  v  q  .  i  s  r  a  a  l  s  d

G258    1305      lambdagal258       sl-att      EM
A       ...      CONTINUES IN PHASE THREE
```

```
                                  M               G                        A       H HH         MA H
                                  N               9                        C       G AA         NC P
                                  l               1                        Y       A EF         LY A
                                  1               4                        1       1 13         11 2
                                                                                      /          /
CCGCATTTTATGCGTTTTCATGTTGCCTGCCCGCATTGCGGGGAGGAGCAGTATCTTAAATTTGGCGACAAAGAGACGCCGTTTGGCCTCAAATGGACGC
----.----+----.----+----.----+----.----+----.----+----.----+----.----+----.----+----.----+----.----+ 1500
GGCGTAAAATACGCAAAAGTACAACGGACGGGCGTAACGCCCCTCCTCGTCATAGAATTTAAACCGCTGTTTCTCTGCGGCAAACCGGAGTTTACCTGCG

 r  i  l  c  v  f  m  l  p  a  r  i  a  g  r  s  s  i  l  n  l  a  t  k  r  r  r  l  a  s  n  g  r
  a  f  y  a  f  s  c  c  l  p  a  l  r  g  g  a  v  s  .  i  w  r  q  r  d  a  v  w  p  q  m  d  a
P  H  F  M  R  F  H  V  A  C  P  H  C  G  E  E  Q  Y  L  K  F  G  D  K  E  T  P  F  G  L  K  W  T  P
----.----+----.----+----.----+----.----+----.----+----.----+----.----+----.----+----.----+----.----+
  r  m  k  h  t  k  m  n  g  a  r  m  a  p  l  l  l  i  k  f  k  a  v  f  l  r  r  k  a  e  f  p  r
g  a  n  .  a  n  e  h  q  r  g  a  n  r  p  p  a  t  d  .  i  q  r  c  l  s  a  t  q  g  .  i  s  a
 g  c  k  i  r  k  .  t  a  q  g  c  q  p  s  s  c  y  r  l  n  p  s  l  s  v  g  n  p  r  l  h  v

G914     1451        lambdagalA914       sl-att       EM
A        ...       CONTINUES IN PHASE THREE


H   H         F    M              HF                   F         A  S                             F
G   G         O    N              GO                   C         L  F                             C
U   A         K    L              AK                   1         U  A                             P
2   1         1    1              11                   5         1  N                             1

CGGATGACCCCTCCAGCGTGTTTTATCTCTGCGAGCATAATGCCTGCGTCATCCGCCAGCAGGAGCTGGACTTTACTGATGCCCGTTATATCTGCGAAAA
----.----+----.----+----.----+----.----+----.----+----.----+----.----+----.----+----.----+----.----+ 1600
GCCTACTGGGGAGGTCGCACAAAATAGAGACGCTCGTATTACGGACGCAGTAGGCGGTCGTCCTCGACCTGAAATGACTACGGGCAATATAGACGCTTTT

r  m  t  p  p  a  c  f  i  s  a  s  i  m  p  a  s  s  a  s  r  s  w  t  l  l  m  p  v  i  s  a  k  r
 g  .  p  l  q  r  v  l  s  l  r  a  .  c  l  r  h  p  p  a  g  a  g  l  y  .  c  p  l  y  l  r  k
  D  D  P  S  S  V  F  Y  L  C  E  H  N  A  C  V  I  R  Q  Q  E  L  D  F  T  D  A  R  Y  I  C  E  K
----.----+----.----+----.----+----.----+----.----+----.----+----.----+----.----+----.----+----.----+
 r  i  v  g  g  a  h  k  i  e  a  l  m  i  g  a  d  d  a  l  l  l  q  v  k  s  i  g  t  i  d  a  f
  p  h  g  r  w  r  t  k  d  r  r  a  y  h  r  r  .  g  g  a  p  a  p  s  .  q  h  g  n  y  r  r  f
g  s  s  g  e  l  t  n  .  r  q  s  c  l  a  q  t  m  r  w  c  s  s  s  k  v  s  a  r  .  i  q  s  f

A        ...       CONTINUES IN PHASE THREE


  HNSXBM   AA              F                H             H  M                                   R H
  PCCHIB   SV              O                P             P  B                                   S P
  AIRONO   UA              K                A             H  O                                   A H
  211211   12              1                2             1  2                                   1 1
    / //    /
GACCGGGATCTGGACCCGTGATGGCATTCTCTGGTTTTCGTCATCCGGTGAAGAGATTGAGCCACCTGACAGTGTGACCTTTCACATCTGGACAGCGTAC
----.----+----.----+----.----+----.----+----.----+----.----+----.----+----.----+----.----+----.----+ 1700
CTGGCCCTAGACCTGGGCACTACCGTAAGAGACCAAAAGCAGTAGGCCACTTCTCTAACTCGGTGGACTGTCACACTGGAAAGTGTAGACCTGTCGCATG

  p  g  s  g  p  v  m  a  f  s  g  f  r  h  p  v  k  r  l  s  h  l  t  v  .  p  f  t  s  g  q  r  t
d  r  d  l  d  p  .  w  h  s  l  v  f  v  i  r  .  r  d  .  a  t  .  q  c  d  l  s  h  l  d  s  v  q
 T  G  I  W  T  R  D  G  I  L  W  F  S  S  S  G  E  E  I  E  P  P  D  S  V  T  F  H  I  W  T  A  Y
----.----+----.----+----.----+----.----+----.----+----.----+----.----+----.----+----.----+----.----+
l  g  p  d  p  g  t  i  a  n  e  p  k  r  .  g  t  f  l  n  l  w  r  v  t  h  g  k  v  d  p  c  r  v
 s  r  s  r  s  g  h  h  c  e  r  t  k  t  m  r  h  l  s  q  a  v  q  c  h  s  r  e  c  r  s  l  t
  v  p  i  q  v  r  s  p  m  r  q  n  e  d  d  p  s  s  i  s  g  g  s  l  t  v  k  .  m  q  v  a  y

A        ...       CONTINUES IN PHASE THREE


            E S                     H             F                                H             M
            C C                     G             O                                G             N
            R R                     U             K                                E             L
            2 1                     2             1                                2             1

AGCCCGTTCACCACCTGGGTGCAGATTGTCAAAGACTGGATGAAAACGAAAGGGGATACGGGAAAACGTAAAACCTTCGTAAACACCACGCTCGGTGAGA
----.----+----.----+----.----+----.----+----.----+----.----+----.----+----.----+----.----+----.----+ 1800
TCGGGCAAGTGGTGGACCCACGTCTAACAGTTTCTGACCTACTTTTGCTTTCCCCTATGCCCTTTTGCATTTTGGAAGCATTTGTGGTGCGAGCCACTCT

  a  r  s  p  p  g  c  r  l  s  k  t  g  .  k  r  k  g  i  r  e  n  v  k  p  s  .  t  p  r  s  v  r
   p  v  h  h  l  g  a  d  c  q  r  l  d  e  n  e  r  g  y  g  k  t  .  n  l  r  k  h  h  a  r  .  d
S  P  F  T  T  W  V  Q  I  V  K  D  W  M  K  T  K  G  D  T  G  K  R  K  T  F  V  N  T  T  L  G  E  T
----.----+----.----+----.----+----.----+----.----+----.----+----.----+----.----+----.----+----.----+
  a  r  e  g  g  p  h  l  n  d  f  v  p  h  f  r  f  p  i  r  s  f  t  f  g  e  y  v  g  r  e  t  l
c  g  t  .  w  r  p  a  s  q  .  l  s  s  s  f  s  l  p  y  p  f  v  y  f  r  r  l  c  w  a  r  h  s
 l  g  n  v  v  q  t  c  i  t  l  s  q  i  f  v  f  p  s  v  p  f  r  l  v  k  t  f  v  v  s  p  s

A        ...       CONTINUES IN PHASE THREE
```

```
   H          S     HH     G  F                                 HH        G
   P          F     PG     A  0                                 HA        1
   H          A     AU     L  K                                 AF        Ø
   1          N     22     3  1                                 12        2

CGTGGGAGGCGAAAATTGGCGAACGTCCGGATGCTGAAGTGATGGCAGAGCGGAAAGAGCATTATTCAGCGCCCGTTCCTGACCGTGTGGCTTACCTGAC
----.----+----.----+----.----+----.----+----.----+----.----+----.----+----.----+----.----+----.----+  19ØØ
GCACCCTCCGCTTTTAACCGCTTGCAGGCCTACGACTTCACTACCGTCTCGCCTTTCTCGTAATAAGTCGCGGGCAAGGACTGGCACACCGAATGGACTG

r  g  r  r  k  l  a  n  v  r  m  l  k  .  w  q  s  g  k  s  i  i  q  r  p  f  l  t  v  w  l  t  .  p
 v  g  g  e  n  w  r  t  s  g  c  .  s  d  g  r  a  e  r  a  l  f  s  a  r  s  .  p  c  g  l  p  d
  W  E  A  K  T  G  F  R  P  D  A  E  V  M  A  E  R  K  E  H  Y  S  A  P  V  P  D  R  V  A  Y  L  T
----.----+----.----+----.----+----.----+----.----+----.----+----.----+----.----+----.----+----.----+
r  p  l  r  f  n  a  f  t  r  i  s  f  h  h  c  l  p  f  l  m  i  .  r  g  n  r  v  t  h  s  v  q
 t  p  p  s  f  q  r  v  d  p  h  q  l  s  p  l  a  s  l  a  n  n  l  a  r  e  q  g  h  p  k  g  s
v  h  s  a  f  i  p  s  r  g  s  a  s  t  i  a  s  r  f  s  c  .  e  a  g  t  g  s  r  t  a  .  r  v

GAL3    1838       lambdagal3             sl-att       EM
G1Ø2    1886       lambdagalT-N1Ø2        sl-att       EM
Ø       ...      CONTINUES IN PHASE THREE


  H     T H    N  PA*AA              HT     H       AMHHNS F      A H       H           A HF
  P     A N    S  VLØSV              HH     G       SNAPCC 0      L P       P           S AN
  A     Q F    P  UUØUA              AA     U       ULEAIR K      U H       A           U EU
  2     1 1    B  21612              11     2       113211 1      1 1       2           1 3H
                   /// /                /               /
CGCCGGTATCGACTCCCAGCTGGACCGCTACGAAATGCGCGTATGGGGATGGGGGCCGGGTGAGGAAAGCTGGCTGATTGACCGGCAGATTATTATGGGC
----.----+----.----+----.----+----.----+----.----+----.----+----.----+----.----+----.----+----.----+  2ØØØ
GCGGCCATAGCTGAGGGTCGACCTGGCGATGCTTTACGCGCATACCCCTACCCCCGGCCCACTCCTTTCGACCGACTAACTGGCCGTCTAATAATACCCG

 p  v  s  t  p  s  w  t  a  t  k  c  a  y  g  d  g  g  r  v  r  k  a  g  .  l  t  g  r  l  l  w  a
r  r  y  r  l  p  a  g  p  l  r  n  a  r  m  g  m  g  a  g  .  g  k  l  a  d  .  p  a  d  y  y  g  p
 A  G  I  D  S  Q  L  D  R  Y  E  M  R  V  W  G  W  G  P  G  F  E  S  W  L  I  D  R  Q  I  I  M  G
----.----+----.----+----.----+----.----+----.----+----.----+----.----+----.----+----.----+----.----+
g  g  t  d  v  g  l  q  v  a  v  f  h  a  y  p  s  p  p  r  t  l  f  a  p  q  n  v  p  l  n  n  h  a
 r  r  y  r  s  g  a  p  g  s  r  f  a  r  i  p  i  p  a  p  h  p  f  s  a  s  q  g  a  s  .  .  p
  a  p  i  s  e  w  s  s  r  .  s  i  r  t  h  p  h  p  g  p  s  s  l  q  s  i  s  r  c  i  i  i  p

*ØØ6    1919       Mapped site PVU2
Ø       ...        CONTINUES IN PHASE THREE


      B  EB       EF  F H M  H       HH    F               H             T      E      E
      B  CB       CN  N G N  G       AA    0               P             A      C      C
      V  0V       1U  U A L  U       EE    K               A             Q      R      1
      1  B1       5H  H 1 1  2       13    1               2             1      V      5
          /                           /
CGCCACGACGATGAACAGACGCTGCTGCGTGTGGATGAGGCCATCAATAAAACCTATACCCGCCGGAATGGTGCAGAAATGTCGATATCCCGTATCTGCT
----.----+----.----+----.----+----.----+----.----+----.----+----.----+----.----+----.----+----.----+  21ØØ
GCGGTGCTGCTACTTGTCTGCGACGACGCACACCTACTCCGGTAGTTATTTTGGATATGGGCGGCCTTACCACGTCTTTACAGCTATAGGGCATAGACGA

 a  t  t  m  n  r  r  c  c  v  w  m  r  p  s  i  k  p  i  p  a  g  m  v  q  k  c  r  y  p  v  s  a
  p  r  r  .  t  d  a  a  a  c  g  .  g  h  q  .  n  l  y  p  p  e  w  c  r  n  v  d  i  p  y  l  l
R  H  D  D  E  Q  T  I  L  R  V  D  E  A  T  N  K  T  Y  T  R  R  N  G  A  E  M  S  I  S  R  I  C  W
----.----+----.----+----.----+----.----+----.----+----.----+----.----+----.----+----.----+----.----+
 a  v  v  i  f  l  r  q  q  t  h  i  l  g  d  i  f  g  i  g  a  p  i  t  c  f  h  r  y  g  t  d  a
g  g  r  r  h  v  s  a  a  a  h  p  h  p  w  .  y  f  r  y  g  g  s  h  h  l  f  t  s  i  g  y  r  s
 r  w  s  s  s  c  v  s  s  r  t  s  s  a  m  l  l  v  .  v  r  r  f  p  a  s  i  d  i  d  r  i  q

Ø         ...      CONTINUES IN PHASE THREE


                                      T               HNS  MB  F     H          A S
                                      A               PCC  BT  0     P          C F
                                      Q               AIR  ON  K     H          C A
                                      1               211  11  1     1          1 N
                                                        /   /
GGGATACTGGCGGGATTGACCCGACCATTGTGTATGAACGCTCGAAAAAACATGGGCTGTTCCGGGTGATCCCCATTAAAGGGGCATCCGTCTACGGAAA
----.----+----.----+----.----+----.----+----.----+----.----+----.----+----.----+----.----+----.----+  22ØØ
CCCTATGACCGCCCTAACTGGGCTGGTAACACATACTTGCGAGCTTTTTTGTACCCGACAAGGCCCACTAGGGGTAATTTCCCCGTAGGCAGATGCCTTT

g  i  l  a  g  l  t  r  p  l  c  m  n  a  r  k  n  m  g  c  s  g  .  s  p  l  k  g  h  p  s  t  e  s
 g  y  w  r  d  .  p  d  h  c  v  .  t  l  e  k  t  w  a  v  p  g  d  p  h  .  r  g  i  r  l  r  k
  D  T  G  G  I  D  P  T  I  V  Y  E  R  S  K  K  H  G  L  F  R  V  I  P  I  K  G  A  S  V  Y  G  K
----.----+----.----+----.----+----.----+----.----+----.----+----.----+----.----+----.----+----.----+
p  i  s  a  p  n  v  r  g  n  h  i  f  a  r  f  f  m  p  q  e  p  h  d  g  n  f  p  c  g  d  v  s
 p  y  q  r  s  q  q  s  w  q  t  h  v  s  s  f  v  h  a  t  g  p  s  g  w  .  l  p  m  r  r  r  f
q  s  v  p  p  i  s  g  v  m  t  y  s  r  e  f  f  c  p  s  n  r  t  i  g  m  l  p  a  d  t  .  p  f
```

```
A       ...       CONTINUES IN PHASE THREE

 H   C BHH     SN*                              H           R         T
 P   F AAA     PS0                              G           S         H
 A   R LFF     HP0                              E           A         A
 2   1 113     1C7                              2           1         1
        //      //
GCCGGTGGCCAGCATGCCACGTAAGCGAAACAAAAACGGGGTTTACCTTACCGAAATCGGTACGGATACCGCGAAAGAGCAGATTTATAACCGCTTCACA
----.----+----.----+----.----+----.----+----.----+----.----+----.----+----.----+----.----+----.----+ 2300
CGGCCACCGGTCGTACGGTGCATTCGCTTTGTTTTTGCCCCAAATGGAATGGCTTTAGCCATGCCTATGGCGCTTTCTCGTCTAAATATTGGCGAAGTGT

  r  w  p  a  c  h  v  s  e  t  k  t  g  f  t  l  p  k  s  v  r  i  p  r  k  s  r  f  i  t  a  s  h
 a  g  g  q  h  a  t  .  a  k  q  k  r  g  l  p  y  r  n  r  y  g  y  r  e  r  a  d  l  .  p  l  h  t
   P  V  A  S  M  P  R  K  R  N  K  N  G  V  Y  L  T  E  I  G  T  D  T  A  K  E  Q  T  Y  N  R  F  T
----.----+----.----+----.----+----.----+----.----+----.----+----.----+----.----+----.----+----.----+
 l  r  h  g  a  h  w  t  l  s  v  f  v  p  n  v  k  g  f  d  t  r  i  g  r  f  l  l  n  i  v  a  e  c
  a  p  p  w  c  a  v  y  a  f  c  f  r  p  k  g  .  r  f  r  y  p  y  r  s  l  a  s  k  y  g  s  .
   g  t  a  l  m  g  r  l  r  f  l  f  p  t  .  r  v  s  i  p  v  s  v  a  f  s  c  i  .  l  r  k  v

*007      2216       Mapped site SPHI
A         ...        CONTINUES IN PHASF THREF

   A  H    H H        X    FNSHH             H      NSH         M              HFENF PA* F  B  B
   C  P    G G        M    OCCPG             N      CCP         B              HNCSN VL0 C  B  B
   Y  A    A U        N    KTRAI             F      IRA         0              AU1PU UU0 1  V  V
   1  2    1 2        1    1112C             3      112         1              1H5BH 218 5  1  1
                             ///                       //                        /    //
CTGACGCCGGAAGGGGATGAACCGCTTCCCGGTGCCGTTCACTTCCCGAATAACCCGGATATTTTTGATCTGACCGAAGCGCAGCAGCTGACTGCTGAAG
----.----+----.----+----.----+----.----+----.----+----.----+----.----+----.----+----.----+----.----+ 2400
GACTGCGGCCTTCCCCTACTTGGCGAAGGGCCACGGCAAGTGAAGGGCTTATTGGGCCTATAAAAACTAGACTGGCTTCGCGTCGTCGACTGACGACTTC

  .  r  r  k  g  m  n  r  f  p  v  p  f  t  s  r  i  t  r  i  f  l  i  .  p  k  r  s  s  .  l  l  k
   d  a  g  r  g  .  t  a  s  r  c  r  s  l  p  e  .  p  g  y  f  .  s  d  r  s  a  a  a  d  c  .  r
 L  T  P  F  G  D  F  P  L  P  G  A  V  H  F  P  N  N  P  D  I  F  D  L  T  F  A  Q  Q  L  T  A  E  F
----.----+----.----+----.----+----.----+----.----+----.----+----.----+----.----+----.----+----.----+
  q  r  r  f  p  i  f  r  k  g  t  g  n  v  e  r  i  v  r  i  n  k  i  q  g  f  r  l  l  q  s  s  f
 v  s  a  p  l  p  h  v  a  e  r  h  r  e  s  g  s  y  g  p  y  k  q  d  s  r  l  a  a  a  s  q  q  l
 s  v  g  s  p  s  s  g  s  g  p  a  t  .  k  g  f  l  g  s  i  k  s  r  v  s  a  c  c  s  v  a  s

*008      2387       Mapped site PVU2
A         ...        CONTINUES IN PHASE THREE

      T M          H              F                        M        H     T                B   H    F
      A B          G              0                        N        G     A                B   H    N
      Q 0          U              K                        L        A     Q                V   A    U
      1 2          2              1                        1        1     1                1   1    H

AGCAGGTCGAAAAATGGGTGGATGGCAGGAAAAAAATACTGTGGGACAGCAAAAAGCGACGCAATGAGGCACTCGACTGCTTCGTTTATGCGCTGGCGGC
----.----+----.----+----.----+----.----+----.----+----.----+----.----+----.----+----.----+----.----+ 2500
TCGTCCAGCTTTTTACCCACCTACCGTCCTTTTTTTATGACACCCTGTCGTTTTTCGCTGCGTTACTCCGTGAGCTGACGAAGCAAATACGCGACCGCCG

 s  r  s  k  n  g  w  m  a  g  k  k  y  c  g  t  a  k  s  d  a  m  r  h  s  t  a  s  f  m  r  w  r  r
  a  g  r  k  m  g  g  w  q  e  k  n  t  v  g  q  q  k  a  t  q  .  g  t  r  l  l  r  l  c  a  g  g
   Q  V  E  K  W  V  D  G  R  K  K  I  L  W  D  S  K  K  R  R  N  E  A  L  D  C  F  V  Y  A  I  A  A
----.----+----.----+----.----+----.----+----.----+----.----+----.----+----.----+----.----+----.----+
 l  l  d  f  f  p  h  i  a  p  f  f  y  q  p  v  a  f  l  s  a  i  l  c  e  v  a  e  n  i  r  q  r
  a  p  r  f  i  p  p  h  c  s  f  f  v  t  p  c  c  f  a  v  c  h  p  v  r  s  s  r  k  h  a  p  p
 s  c  t  s  f  h  t  s  p  l  f  f  i  s  h  s  l  l  f  r  r  l  s  a  s  s  q  k  t  .  a  s  a  a

A         ...        CONTINUES IN PHASE THREE

HHF MH        S  N      NF PA*XBMBD B    HFE            M  P*     H      MF     F      B
HAN SH        F  S      SN VL0HIBBD B    HNC            N  S0     G      BN     0      B
AEU TA        A  P      PU UU0ONOVE V    AU1            L  T1     U      OU     K      V
12H 11        N  B      BH 21921111 1    1H5            1  10     2      2H     1      1
  /                         // ///          /               /
GCTGCGCATCAGTATTTCCCGCTGGCAGCTGGATCTCAGTGCGCTGCTGGCGAGCCTGCAGGAAGAGGATGGTGCAGCAACCAACAAGAAAACACTGGCA
----.----+----.----+----.----+----.----+----.----+----.----+----.----+----.----+----.----+----.----+ 2600
CGACGCGTAGTCATAAAGGGCGACCGTCGACCTAGAGTCACGCGACGACCGCTCGGACGTCCTTCTCCTACCACGTCGTTGGTTGTTCTTTTGTGACCGT

  c  a  s  v  f  p  a  g  s  w  i  s  v  r  c  w  r  a  c  r  k  r  m  v  q  q  p  t  r  k  h  w  q
 a  a  h  q  y  f  p  l  a  a  g  s  q  c  a  a  g  e  p  a  g  r  g  w  c  s  n  q  q  e  n  t  g  r
  L  R  I  S  I  S  R  W  Q  L  D  L  S  A  L  L  A  S  L  Q  E  E  D  G  A  A  T  N  K  K  T  L  A
----.----+----.----+----.----+----.----+----.----+----.----+----.----+----.----+----.----+----.----+
 r  q  a  d  t  n  g  a  p  l  q  i  e  t  r  q  q  r  a  q  l  f  l  i  t  c  c  g  v  l  f  c  q  c
  a  a  c  .  y  k  g  s  a  a  p  d  .  h  a  a  p  s  g  a  p  l  p  h  h  l  l  w  c  s  f  v  p
  s  r  m  l  i  e  r  q  c  s  s  r  l  a  s  s  a  l  r  c  s  s  s  p  a  a  v  l  l  f  v  s  a
```

```
*009    2528    Mapped site PVU2
*010    2560    Mapped site PST1
A       ...     CONTINUES IN PHASE THREE
```

```
                M  H     H      WA   T F W HB        NF MF       BF
                N  P     G            H O A GB        SN BN       GN
                L  A     U            A K M AV        PU OU       LU
                1  2     2            1 1   11        BH 2H       1H
                                 /          /
GATTACGCCCGTGCCTTATCCGGAGAGGATGAATGACGCGACAGGAAGAACTTGCCGCTGCCCGTGCGGCACTGCATGACCTGATGACAGGTAAACGGGT
----.----+----.----+----.----+----.----+----.----+----.----+----.----+----.----+----.----+----.----+ 2700
CTAATGCGGGCACGGAATAGGCCTCTCCTACTTACTGCGCTGTCCTTCTTGAACGGCGACGGGCACGCCGTGACGTACTGGACTACTGTCCATTTGCCCA

 i  t  p  v  p  y  p  e  r  m  n  d  a  t  g  r  t  c  r  c  p  c  g  t  a  .  p  d  d  r  .  t  g
  l  r  p  c  l  i  r  r  g  .  M  T  R  Q  E  E  L  A  A  A  R  A  A  L  H  D  L  M  T  G  K  R  V
D  Y  A  R  A  L  S  G  E  D  E  .  r  d  r  k  n  l  p  l  p  v  r  h  c  m  t  .  .  q  v  n  g  w
----.----+----.----+----.----+----.----+----.----+----.----+----.----+----.----+----.----+----.----+
 i  v  g  t  g  .  g  s  l  i  f  s  a  v  p  l  v  q  r  q  g  h  p  v  a  h  g  s  s  l  y  v  p
l  n  r  g  h  r  i  r  l  p  h  i  v  r  c  s  s  s  a  a  a  r  a  a  s  c  s  r  i  v  p  l  r  t
  s  .  a  r  a  k  d  p  s  s  s  h  r  s  l  f  f  k  g  s  g  t  r  c  q  m  v  q  h  c  t  f  p

W       2633    W (orf-68)      M-NH3
A       2633    A               COOH
WAM     2642    Wam403          pm        sq     C to T
```

```
        R                             C H G                                   A          E   H
        S                             F A D                                   L          C   P
        A                             R F I                                   U          P   A
        1                             1 3 2                                   1          1   2

GGCAACAGTACAGAAAGACGGACGAAGGGTGGAGTTTACGGCCACTTCCGTGTCTGACCTGAAAAAATATATTGCAGAGCTGGAAGTGCAGACCGGCATG
----.----+----.----+----.----+----.----+----.----+----.----+----.----+----.----+----.----+----.----+ 2800
CCGTTGTCATGTCTTTCTGCCTGCTTCCCACCTCAAATGCCGGTGAAGGCACAGACTGGACTTTTTTATATAACGTCTCGACCTTCACGTCTGGCCGTAC

g  n  s  t  e  r  r  t  k  g  g  v  y  g  h  f  r  v  .  p  e  k  i  y  c  r  a  g  s  a  d  r  h  d
 A  T  V  Q  K  D  G  R  R  V  E  F  T  A  T  S  V  S  D  L  K  K  Y  I  A  F  L  E  V  Q  T  G  M
  q  q  y  r  k  t  d  e  g  w  s  l  r  p  l  p  c  l  t  .  k  n  i  l  q  s  w  k  c  r  p  a  .
----.----+----.----+----.----+----.----+----.----+----.----+----.----+----.----+----.----+----.----+
p  l  l  v  s  l  r  v  f  p  p  t  .  p  w  k  r  t  q  g  s  f  i  y  q  l  a  p  l  a  s  r  c
 a  v  t  c  f  s  p  r  l  t  s  n  v  a  v  e  t  d  s  r  f  f  y  i  a  s  s  s  t  c  v  p  m
h  c  c  y  l  f  v  s  s  p  h  l  k  r  g  s  g  h  r  v  q  f  f  i  n  c  l  q  f  h  l  g  a  h

W       ...     CONTINUES IN PHASE TWO
```

```
               AHA   P*          BW                         B   A HH   B             F   HTH      H
               SGV   S0                                     S   S AP   B             N   HHN      P
               UAA   T1                                     T   U EA   V             U   AAF      A
               112   11                                     X   1 32   1             H   113      2
                 /      /           /                                                      //
ACACAGCGACGCAGGGGACCTGCAGGATTTTATGTATGAAAACGCCCACCATTCCCACCCTTCTGGGGCCGGACGGCATGACATCGCTGCGCGAATATGC
----.----+----.----+----.----+----.----+----.----+----.----+----.----+----.----+----.----+----.----+ 2900
TGTGTCGCTGCGTCCCCTGGACGTCCTAAAATACATACTTTTGCGGGTGGTAAGGGTGGGAAGACCCCGGCCTGCCGTACTGTAGCGACGCGCTTATACG

 t  a  t  q  g  t  c  r  i  l  c  M  K  T  P  T  I  P  T  L  L  G  P  D  G  M  T  S  L  R  E  Y  A
T  Q  R  R  R  G  P  A  G  F  Y  V  .  k  r  p  p  f  p  p  f  w  g  r  t  a  .  h  r  c  a  n  m  p
  h  s  d  a  g  d  l  q  d  f  m  y  e  n  a  h  h  s  h  p  s  g  a  g  r  h  d  i  a  a  r  i  c
----.----+----.----+----.----+----.----+----.----+----.----+----.----+----.----+----.----+----.----+
s  v  a  v  c  p  v  q  l  i  k  h  i  f  v  g  v  m  g  v  r  r  p  g  s  p  m  v  d  s  r  s  y  a
 v  c  r  r  l  p  g  a  p  n  .  t  h  f  r  g  g  n  g  g  k  q  p  r  v  a  h  c  r  q  a  f  i
  c  l  s  a  p  s  r  c  s  k  i  y  s  f  a  w  w  e  w  g  e  p  a  p  r  c  s  m  a  a  r  i  h

*011    2824    Mapped site PST1
B       2836    B (orf-533)     M-NH3
W       2836    W               COOH
```

```
 B               NF  M       B                   RR         S       H     F        F B
 *               SN  N       B                   FF         F       G     N        0 B
                 PU  L       V                   12         A       U     U        K V
                 BH  1       1                   80         N       2     H        1 1
                                                   /
CGGTTATCACGGCGGTGGCAGCGGATTTGGAGGGCAGTTGCGGTCGTGGAACCCACCGAGTGAAAGTGTGGATGCAGCCCTGTTGCCCAACTTTACCCGT
----.----+----.----+----.----+----.----+----.----+----.----+----.----+----.----+----.----+----.----+ 3000
GCCAATAGTGCCGCCACCGTCGCCTAAACCTCCCGTCAACGCCAGCACCTTGGGTGGCTCACTTTCACACCTACGTCGGGACAACGGGTTGAAATGGGCA

 G  Y  H  G  G  G  S  G  F  G  G  Q  L  R  S  W  N  P  P  S  E  S  V  D  A  A  L  L  P  N  F  T  R
  v  i  t  a  v  a  a  d  l  e  g  s  c  g  r  g  t  h  r  v  k  v  w  m  q  p  c  c  p  t  l  p  v
r  l  s  r  r  w  q  r  i  w  r  a  v  a  v  v  e  p  t  e  .  k  c  g  c  s  p  v  a  q  l  y  p  w
----.----+----.----+----.----+----.----+----.----+----.----+----.----+----.----+----.----+----.----+
  p  .  .  p  p  p  l  p  n  p  p  c  n  r  d  h  f  g  g  l  s  l  t  s  a  a  r  n  g  l  k  v  r
g  t  i  v  a  t  a  a  s  k  s  p  l  q  p  r  p  v  w  r  t  f  t  h  i  c  g  q  q  g  v  k  g  t
 r  n  d  r  r  h  c  r  i  q  l  a  t  a  t  t  s  g  v  s  h  f  h  p  h  l  g  t  a  w  s  .  g

B*      2902      B*                               AA     start processed protein of capsid
RE18    2950      lambdarif-18         sl-att      EM
RE20    2950      lambdarif-20         sl-att      EM
B       ...       CONTINUES IN PHASE ONE


           TH    M       R             F   F   B        N  PA*F    BM  S         E          H   D
           HH    B       S             0   N   B        S  VL0N    IB  F         C          P   D
           AA    0       A             K   U   V        P  UU1U    NO  A         P          A   E
           11    1       1             1   H   1        B  212H    11  N         1          2   1
                                                            ///
GGCAATGCCCGCGCAGACGATCTGGTACGCAATAACGGCTATGCCGCCAACGCCATCCAGCTGCATCAGGATCATATCGTCGGGTCTTTTTTCCGGCTCA
----.----+----.----+----.----+----.----+----.----+----.----+----.----+----.----+----.----+----.----+ 3100
CCGTTACGGGCGCGTCTGCTAGACCATGCGTTATTGCCGATACGGCGGTTGCGGTAGGTCGACGTAGTCCTAGTATAGCAGCCCAGAAAAAAGGCCGAGT

G  N  A  R  A  D  D  L  V  R  N  N  G  Y  A  A  N  A  I  Q  L  H  Q  D  H  I  V  G  S  F  F  R  L  S
 a  m  p  a  q  t  i  w  y  a  i  t  a  m  p  p  t  p  s  s  c  i  r  i  i  s  s  g  l  f  s  g  s
  q  c  p  r  r  r  s  g  t  q  .  r  l  c  r  q  r  h  p  a  a  s  g  s  y  r  r  v  f  f  p  a  q
----.----+----.----+----.----+----.----+----.----+----.----+----.----+----.----+----.----+----.----+
 p  l  a  r  a  s  s  r  t  r  l  l  p  .  a  a  l  a  m  w  s  c  .  s  .  i  t  p  d  k  k  r  s
  a  i  g  a  c  v  i  q  y  a  i  v  a  i  g  g  v  g  d  l  q  m  l  i  m  d  d  p  r  k  e  p  e
h  c  h  g  r  l  r  d  p  v  c  y  r  s  h  r  w  r  w  g  a  a  d  p  d  y  r  r  t  k  k  g  a  .

*012    3060      Mapped site PVU2
B       ...       CONTINUES IN PHASE ONE


            A     HH       M        S             M    M    T           F           M     B BH      E
            L     HA       N        F             B    N    H           N           N     B BG      C
            U     AF       L        A             0    L    A           U           L     V VU      1
            1     12       1        N             2    1    1           H           1     1 12      5

GTCATCGCCCAAGCTGGCGCTATCTGGGCATCGGGGAGGAAGAAGCCCGTGCCTTTTCCCGCGAGGTTGAAGCGGCATGGAAAGAGTTTGCCGAGGATGA
----.----+----.----+----.----+----.----+----.----+----.----+----.----+----.----+----.----+----.----+ 3200
CAGTAGCGGGTTCGACCGCGATAGACCCGTAGCCCCTCCTTCTTCGGGCACGGAAAAGGGCGCTCCAACTTCGCCGTACCTTTCTCAAACGGCTCCTACT

  H  R  P  S  W  R  Y  L  G  I  G  F  E  F  A  R  A  F  S  R  F  V  E  A  A  W  K  E  F  A  E  D  D
v  i  a  q  a  g  a  i  w  a  s  g  r  k  k  p  v  p  f  p  a  r  l  k  r  h  g  k  s  l  p  r  m  t
 s  s  p  k  l  a  l  s  g  h  r  g  g  r  s  p  c  l  f  p  r  g  .  s  g  m  e  r  v  c  r  g  .
----.----+----.----+----.----+----.----+----.----+----.----+----.----+----.----+----.----+----.----+
l  .  r  g  l  q  r  .  r  p  m  p  s  s  s  a  r  a  k  e  r  s  t  s  a  a  h  f  s  n  a  s  s  s
 t  m  a  w  a  p  a  i  q  a  d  p  l  f  f  g  t  g  k  g  a  l  n  f  r  c  p  f  l  k  g  l  i
  d  d  g  l  s  a  s  d  p  c  r  p  p  l  l  g  h  r  k  g  r  p  q  l  p  m  s  l  t  q  r  p  h

B       ...       CONTINUES IN PHASE ONE
```

```
EF FF                                          HH           C BHH                         H    HH
CN NO                                          NN           F AAA                         P    AA
1U UK                                          FF           R LEE                         H    EE
5H H1                                          13           1 113                         1    13
                                               /             //                                /
CTGCTGCTGCATTGACGTTGAGCGAAAACGCACGTTTACCATGATGATTCGGGAAGGTGTGGCCATGCACGCCTTTAACGGTGAACTGTTCGTTCAGGCC
----.----+----.----+----.----+----.----+----.----+----.----+----.----+----.----+----.----+----.----+ 3300
GACGACGACGTAACTGCAACTCGCTTTTGCGTGCAAATGGTACTACTAAGCCCTTCCACACCGGTACGTGCGGAAATTGCCACTTGACAAGCAAGTCCGG

 C  C  C  T  D  V  E  R  K  R  T  F  T  M  M  I  R  E  G  V  A  M  H  A  F  N  G  E  L  F  V  Q  A
  a  a  a  l  t  l  s  e  n  a  r  l  p  .  .  f  g  k  v  w  p  c  t  p  l  t  v  n  c  s  f  r  p
l  l  l  h  .  r  .  a  k  t  h  v  y  h  d  d  s  g  r  c  g  h  a  r  l  .  r  .  t  v  r  s  g  h
----.----+----.----+----.----+----.----+----.----+----.----+----.----+----.----+----.----+----.----+
 q  q  q  m  s  t  s  r  f  r  v  n  v  m  i  i  r  s  p  t  a  m  c  a  k  l  p  s  s  n  t  .  a
v  a  a  a  n  v  n  l  s  f  a  r  k  g  h  h  n  p  f  t  h  g  h  v  g  k  v  t  f  q  e  n  l  g
 s  s  s  c  q  r  q  a  f  v  c  t  .  w  s  s  e  p  l  h  p  w  a  r  r  .  r  h  v  t  r  e  p

B       ...       CONTINUES IN PHASE ONE
```

```
E S                    TF    H        HH          F   H     S       H      H
C C                    HN    P        PG          O   H     F       P      P
R R                    AU    A        AU          K   A     A       A      A
2 1                    1H    2        22          1   1     N       2      2

ACCTGGGATACCAGTTCGTCGCGGCTTTTCCGGACACAGTTCCGGATGGTCAGCCCGAAGCGCATCAGCAACCCGAACAATACCGGCGACAGCCGGAACT
----.----+----.----+----.----+----.----+----.----+----.----+----.----+----.----+----.----+----.----+ 3400
TGGACCCTATGGTCAAGCAGCGCCGAAAAGGCCTGTGTCAAGGCCTACCAGTCGGGCTTCGCGTAGTCGTTGGGCTTGTTATGGCCGCTGTCGGCCTTGA

T  W  D  T  S  S  S  R  L  F  R  T  Q  F  R  M  V  S  P  K  R  I  S  N  P  N  N  T  G  D  S  R  N  C
 p  g  i  p  v  r  r  g  f  s  g  h  s  s  g  w  s  a  r  s  a  s  a  t  r  t  i  p  a  t  a  g  t
  l  g  y  q  f  v  a  a  f  p  d  t  v  p  d  g  q  p  e  a  h  q  q  p  e  q  y  r  r  q  p  e  l
----.----+----.----+----.----+----.----+----.----+----.----+----.----+----.----+----.----+----.----+
v  q  s  v  l  e  d  r  s  k  r  v  c  n  r  i  t  l  g  f  r  m  l  l  g  f  l  v  p  s  l  r  f
 g  p  i  g  t  r  r  p  k  e  p  c  l  e  p  h  d  a  r  l  a  d  a  v  r  v  i  g  a  v  a  p  v
w  r  p  y  w  n  t  a  a  k  g  s  v  t  g  s  p  .  g  s  a  c  .  c  g  s  c  y  r  r  c  g  s  s

B       ...       CONTINUES IN PHASE ONE
```

```
    H                 N       F  HH       M             SE S   H    F    F
    P                 S       N  HA       N             FC C   G    N    O
    A                 P       U  AF       L             AR R   U    U    K
    2                 B       H  12       1             N2 1   2    H    1

GCCGTGCCGGTGTGCAGATTAATGACAGCGGTGCGGCGCTGGGATATTACGTCAGCGAGGACGGGTATCCTGGCTGGATGCCGCAGAAATGGACATGGAT
----.----+----.----+----.----+----.----+----.----+----.----+----.----+----.----+----.----+----.----+ 3500
CGGCACGGCCACACGTCTAATTACTGTCGCCACGCCGCGACCCTATAATGCAGTCGCTCCTGCCCATAGGACCGACCTACGGCGTCTTTACCTGTACCTA

  R  A  G  V  Q  I  N  D  S  G  A  A  L  G  Y  Y  V  S  E  D  G  Y  P  G  W  M  P  Q  K  W  T  W  I
a  v  p  v  c  r  l  m  t  a  v  r  r  w  d  i  t  s  a  r  t  g  i  l  a  g  c  r  r  n  g  h  g  y
 p  c  r  c  a  d  .  .  q  r  c  g  a  g  i  l  r  q  r  g  r  v  s  w  l  d  a  a  e  m  d  m  d
----.----+----.----+----.----+----.----+----.----+----.----+----.----+----.----+----.----+----.----+
q  r  a  p  t  c  i  l  s  l  p  a  a  s  p  y  .  t  l  s  s  p  y  g  p  q  i  g  c  f  h  v  h  i
 a  t  g  t  h  l  n  i  v  a  t  r  r  q  s  i  v  d  a  l  v  p  i  r  a  p  h  r  l  f  p  c  p
  g  h  r  h  a  s  .  h  c  r  h  p  a  p  i  n  r  .  r  p  r  t  d  q  s  s  a  a  s  i  s  m  s

B       ...       CONTINUES IN PHASE ONE
```

```
               NSH   B HTH         M          M                H  T
               CCP   S HHH         N          N                N  H
               IRA   S AAA         L          L                F  A
               112   2 111         1          1                1  1
                //      /
ACCCCGTGAGTTACCCGGCGGGCGCCTCGTTCATTCACGTTTTTGAACCCGTGGAGGACGGGCAGACTCGCGGTGCAAATGTGTTTTACAGCGTGATG
----.----+----.----+----.----+----.----+----.----+----.----+----.----+----.----+----.----+----.----+ 3600
TGGGGCACTCAATGGGCCGCCCGCGCGGAGCAAGTAAGTGCAAAAACTTGGGCACCTCCTGCCCGTCTGAGCGCCACGTTTACACAAAATGTCGCACTAC

 P  R  E  L  P  G  G  R  A  S  F  I  H  V  F  E  P  V  E  D  G  Q  T  R  G  A  N  V  F  Y  S  V  M
  p  v  s  y  p  a  g  a  p  r  s  f  t  f  l  n  p  w  r  t  g  r  l  a  v  q  m  c  f  t  a  .  w
t  p  .  v  t  r  r  a  r  l  v  h  s  r  f  .  t  r  g  g  r  a  d  s  r  c  k  c  v  l  q  r  d  g
----.----+----.----+----.----+----.----+----.----+----.----+----.----+----.----+----.----+----.----+
 g  r  s  n  g  p  p  r  a  e  n  m  .  t  k  s  g  t  s  s  p  c  v  r  p  a  f  t  n  .  l  t  i
y  g  t  l  .  g  a  p  a  g  r  e  n  v  n  k  f  g  h  l  v  p  l  s  a  t  c  i  h  k  v  a  h  h
 v  g  h  t  v  r  r  a  r  r  t  .  e  r  k  q  v  r  p  p  r  a  s  e  r  h  l  h  t  k  c  r  s

B       ...       CONTINUES IN PHASE ONE
```

```
S       B     T   M   FB  P*      NF  PA*F P*  BHH                           F   G          A
F       B     A   B   NB  S0      SN  VL0N S0  BHA                           N   2          L
A       V     Q   0   UV  T1      PU  UU1U T1  VAF                           U   4          U
N       1     1   2   H1  13      BH  214H 15  112                           H              1
                          /           ///  /
GAGCAGATGAAGATGCTCGACACGCTGCAGAACACGCAGCTGCAGAGCGCCATTGTGAAGGCGATGTATGCCGCCACCATTGAGAGTGAGCTGGATACGC
----.----+----.----+----.----+----.----+----.----+----.----+----.----+----.----+----.----+----.----+ 3700
CTCGTCTACTTCTACGAGCTGTGCGACGTCTTGTGCGTCGACGTCTCGCGGTAACACTTCCGCTACATACGGCGGTGGTAACTCTCACTCGACCTATGCG

E  Q  M  K  M  L  D  T  L  Q  N  T  Q  L  Q  S  A  I  V  K  A  M  Y  A  A  T  I  E  S  E  L  D  T  Q
s  r  .  r  c  s  t  r  c  r  t  r  s  c  r  a  p  l  .  r  r  c  m  p  p  p  l  r  v  s  w  i  r
 a  d  e  d  a  r  h  a  a  e  h  a  a  a  e  r  h  c  e  g  d  v  c  r  h  h  .  e  .  a  g  y  a
----.----+----.----+----.----+----.----+----.----+----.----+----.----+----.----+----.----+----.----+
s  c  i  f  i  s  s  v  s  c  f  v  c  s  c  l  a  m  t  f  a  i  y  a  a  v  m  s  l  s  s  s  v
 l  l  h  l  h  e  v  r  q  l  v  r  l  q  l  a  g  n  h  l  r  h  i  g  g  g  n  l  t  l  q  i  r
p  a  s  s  s  a  r  c  a  a  s  c  a  a  a  s  r  w  q  s  p  s  t  h  r  w  w  q  s  h  a  p  y  a

*013    3629     Mapped site PST1
*014    3639     Mapped site PVU2
*015    3644     Mapped site PST1
G24     3675      lambdaga124           sl-att       EM
B        ...     CONTINUES IN PHASE ONE
```

```
                          HT            NF          B     H                            F HT         F  F   HHH
                          HH            SN          B     P                            N PH         N  N   HAP
                          AA            PU          V     A                            U HA         U  U   AEA
                          11            BH          1     2                            H 11         H  H   122
                           /                                                              /                 /
AGTCAGCGATGGATTTTATTCTGGGCGCGAACAGTCAGGAGCAGCGGGAAAGGCTGACCGGCTGGATTGGTGAAATTGCCGCGTATTACGCCGCAGCGCC
----.----+----.----+----.----+----.----+----.----+----.----+----.----+----.----+----.----+----.----+ 3800
TCAGTCGCTACCTAAAATAAGACCCGCGCTTGTCAGTCCTCGTCGCCCTTTCCGACTGGCCGACCTAACCACTTTAACGGCGCATAATGCGGCGTCGCGG

 S  A  M  D  F  I  L  G  A  N  S  Q  F  Q  R  E  R  L  T  G  W  I  G  E  I  A  A  Y  Y  A  A  A  P
s  q  r  w  i  l  f  w  a  r  t  v  r  s  s  g  k  g  .  p  a  g  l  v  k  l  p  r  i  t  p  q  r  r
 v  s  d  g  f  y  s  g  r  e  q  s  g  a  a  g  k  a  d  r  l  d  w  .  n  c  r  v  l  r  r  s  a
----.----+----.----+----.----+----.----+----.----+----.----+----.----+----.----+----.----+----.----+
c  d  a  i  s  k  i  r  p  a  f  l  .  s  c  r  s  l  s  v  p  q  i  p  s  i  a  a  y  .  a  a  a  g
 l  .  r  h  i  k  n  q  a  r  v  t  l  l  l  p  f  p  q  g  a  p  n  t  f  n  g  r  i  v  g  c  r
  t  l  s  p  n  .  e  p  r  s  c  d  p  a  a  p  f  a  s  r  s  s  q  h  f  q  r  t  n  r  r  l  a

B        ...     CONTINUES IN PHASE ONE
```

```
AA BBHM      H       RS            HNS  H        H     P*   D              T             FB
SV GBPN      H       SF            PCC  N        P     S0   D              T             CB
UA LVAL      A       AA            AIR  F        H     T1   E              H             OV
12 1121      1       1N            211  1        1     16   1              2             B1
 /  //                /             /                   /                                 /
GGTCCGGCTGGGAGGCGCAAAAGTACCGCACCTGATGCCGGGTGACTCACTGAACCTGCAGACGGCTCAGGATACGGATAACGGCTACTCCGTGTTTGAG
----.----+----.----+----.----+----.----+----.----+----.----+----.----+----.----+----.----+----.----+ 3900
CCAGGCCGACCCTCCGCGTTTTCATGGCGTGGACTACGGCCCACTGAGTGACTTGGACGTCTGCCGAGTCCTATGCCTATTGCCGATGAGGCACAAACTC

 V  R  L  G  G  A  K  V  P  H  L  M  P  G  D  S  L  N  L  Q  T  A  Q  D  T  D  N  G  Y  S  V  F  E
  s  g  w  e  a  q  k  y  r  t  .  c  r  v  t  h  .  t  c  r  r  l  r  i  r  i  t  a  t  p  c  l  s
g  p  a  g  r  r  k  s  t  a  p  d  a  g  .  l  t  e  p  a  d  g  s  g  y  g  .  r  l  l  r  v  .  a
----.----+----.----+----.----+----.----+----.----+----.----+----.----+----.----+----.----+----.----+
  t  r  s  p  p  a  f  t  g  c  r  i  g  p  s  e  s  f  r  c  v  a  .  s  v  s  l  p  .  e  t  n  s
r  d  p  q  s  a  c  f  y  r  v  q  h  r  t  v  .  q  v  q  l  r  s  l  i  r  i  v  a  v  g  h  k  l
 p  g  a  p  l  r  l  l  v  a  g  s  a  p  h  s  v  s  g  a  s  p  e  p  y  p  y  r  s  s  r  t  q

*016    3860     Mapped site PST1
B        ...     CONTINUES IN PHASE ONE
```

```
E BF        F HNS             F A    NSHB            A          A H
C BN        N PCC             N L    CCPB            L          S A
1 VU        U AIR             U U    IRAV            U          U E
5 1H        H 211             H 1    1121            1          1 3
                /                     //
CAGTCACTGCTGCGGTATATCGCTGCCGGGCTGGGTGTCTCGTATGAGCAGCTTTCCCGGAATTACGCCCAGATGAGCTACTCCACGGCACGGGCCAGTG
---.----+----.----+----.----+----.----+----.----+----.----+----.----+----.----+----.----+----.----+ 4000
GTCAGTGACGACGCCATATAGCGACGGCCCGACCCACAGAGCATACTCGTCGAAAGGGCCTTAATGCGGGTCTACTCGATGAGGTGCCGTGCCCGGTCAC

Q S L L R Y I A A G L G V S Y E Q L S R N Y A Q M S Y S T A R A S A
 s h c c g i s l p g w v s r m s s f p g i t p r . a t p r h g p v
 v t a a v y r c r a g c l v . a a f p e l r p d e l l h g t g q c
---.----+----.----+----.----+----.----+----.----+----.----+----.----+----.----+----.----+----.----+
 c d s s r y i a a p s p t e y s c s e r f . a w i l . e v a r a l
 l . q q p i d s g p q t d r i l l k g p i v g l h a v g r c p g t
a t v a a t y r q r a p h r t h a a k g s n r g s s s w p v p w h

B    ...    CONTINUES IN PHASE ONE
```

```
H         R         F F  H             S                      M      HH  M
N         S         N O  N             F                      N      AA  B
F         A         U K  F             A                      L      EE  0
1         1         H 1  3             N                      1      13  2
                                                                     /
CGAACGAGTCGTGGGCGTACTTTATGGGGCGGCGAAAATTCGTCGCATCCCGTCAGGCGAGCCAGATGTTTCTGTGCTGGCTGGAAGAGGCCATCGTTCG
---.----+----.----+----.----+----.----+----.----+----.----+----.----+----.----+----.----+----.----+ 4100
GCTTGCTCAGCACCCGCATGAAATACCCCGCCGCTTTTAAGCAGCGTAGGGCAGTCCGCTCGGTCTACAAAGACACGACCGACCTTCTCCGGTAGCAAGC

 N E S W A Y F M G R R K F V A S R Q A S Q M F L C W L E E A I V R
r t s r g r t l w g g e n s s h p v r r a r c f c a g w k r p s f a
 e r v v g v l y g a a k i r r i p s g e p d v s v l a g r g h r s
---.----+----.----+----.----+----.----+----.----+----.----+----.----+----.----+----.----+----.----+
a f s d h a y k i p r r f n t a d r . a l w i n r h q s s s a m t r
 r v l r p r v k h p p s f e d c g t l r a l h k q a p q f l g d n
 s r t t p t s . p a a f i r r m g d p s g s t e t s a p l p w r e

B    ...    CONTINUES IN PHASE ONE
```

```
F T     S     H     B HTH                 E S                        H      C BHH C*T
N H     1     P     S HHH                 C C                        P      F AAA L0A
U A     0     H     S AAA                 R R                        A      R LEE A1Q
H 1     4     1     2 111                 2 1                        2      1 113 171
                      /                                                       //  //
CCGCGTGGTGACGTTACCTTCAAAAGCGCGCTTCAGTTTTCAGGAAGCCCGCAGTGCCTGGGGGAACTGCGACTGGATAGGCTCCGGTCGTATGGCCATC
---.----+----.----+----.----+----.----+----.----+----.----+----.----+----.----+----.----+----.----+ 4200
GGCGCACCACTGCAATGGAAGTTTTCGCGCGAAGTCAAAAGTCCTTCGGGCGTCACGGACCCCCTTGACGCTGACCTATCCGAGGCCAGCATACCGGTAG

 R V V T L P S K A R F S F Q E A R S A W G N C D W I G S G R M A I
  a w . r y l q k r a s v f r k p a v p g g t a t g . a p v v w p s
 p r g d v t f k s a l q f s g s p q c l g e l r l d r l r s y g h r
---.----+----.----+----.----+----.----+----.----+----.----+----.----+----.----+----.----+----.----+
  r t t v n g e f a r k l k . s a r l a q p f q s q i p e p r i a m
 a a h h r . r . f r a e t k l f g a t g p p v a v p y a g t t h g d
 g r p s t v k l l a s . n e p l g c h r p s s r s s l s r d y p w

S104    4111    lambdagalS104    sl-att    EM
*017    4199    Mapped site CLA1
B       ...     CONTINUES IN PHASE ONE
```

```
E           S              H  H  D  RS             MH   T       H
C           F              P  P  D  S1             SH   H       P
P           A              H  A  E  A0             TA   A       H
1           N              1  2  1  11             11   1       1

GATGGTCTGAAAGAAGTTCAGGAAGCGGTGATGCTGATAGAAGCCGGACTGAGTACCTACGAGAAAGAGTGCGCAAAACGCGGTGACGACTATCAGGAAA
---.----+----.----+----.----+----.----+----.----+----.----+----.----+----.----+----.----+----.----+ 4300
CTACCAGACTTTCTTCAAGTCCTTCGCCACTACGACTATCTTCGGCCTGACTCATGGATGCTCTTTCTCACGCGTTTTGCGCCACTGCTGATAGTCCTTT

D G L K E V Q E A V M L I E A G L S T Y E K E C A K R G D D Y Q E I
 m v . k k f r k r . c . . k p d . v p t r k s a q n a v t t i r k
 w s e r s s g s g d a d r s r t e y l r e r v r k t r . r l s g n
---.----+----.----+----.----+----.----+----.----+----.----+----.----+----.----+----.----+----.----+
 s p r f s t . s a t i s i s a p s l v . s f s h a f r p s s . . s
  i t q f f n l f r h h q y f g s q t g v l f l a c f a t v v i l f
r h d s l l e p l p s a s l l r v s y r r s l t r l v r h r s d p f
```

```
S101     4255      lambdagalS101        sl-att       EM
B        ...       CONTINUES IN PHASE ONE


        E     AA                  HH   F  HE      B    B   E S   BF  F  FP*B    H  BHNS  F     M  M
        C     SV                  HA   N  PC      B    B   C C   GN  N  NS0B    N  BPCC  N     N  N
        1     UA                  AE   U  AP      V    V   R R   LU  U  UT1V    F  VAIR  U     L  L
        5     12                  12   H  21      1    1   2 1   1H  H  H181    1  1211  H     1  1
              /                                                            /           /
TTTTTGCCCAGCAGGTCCGTGAAACGATGGAGCGCCGTGCAGCCGGTCTTAAACCGCCCGCCTGGGCGGCTGCAGCATTTGAATCCGGGCTGCGACAATC
----.----+----.----+----.----+----.----+----.----+----.----+----.----+----.----+----.----+----.----+ 4400
AAAAACGGGTCGTCCAGGCACTTTGCTACCTCGCGGCACGTCGGCCAGAATTTGGCGGGCGGACCCGCCGACGTCGTAAACTTAGGCCCGACGCTGTTAG

 F  A  Q  Q  V  R  E  T  M  E  R  R  A  A  G  L  K  P  P  A  W  A  A  A  A  F  E  S  G  L  R  Q  S
f  l  p  s  r  s  v  k  r  w  s  a  v  q  p  v  l  n  r  p  p  g  r  l  q  h  l  n  p  g  c  d  n  q
 f  c  p  a  g  p  .  n  d  g  a  p  c  s  r  s  .  t  a  r  l  g  g  c  s  i  .  i  r  a  a  t  i
----.----+----.----+----.----+----.----+----.----+----.----+----.----+----.----+----.----+----.----+
i  k  a  w  c  t  r  s  v  i  s  r  r  a  a  p  r  l  g  g  a  q  a  a  a  a  n  s  d  p  s  r  c  d
 n  k  g  l  l  d  t  f  r  h  l  a  t  c  g  t  k  f  r  g  g  p  r  s  c  c  k  f  g  p  q  s  l
  k  q  g  a  p  g  h  f  s  p  a  g  h  l  r  d  .  v  a  r  r  p  p  q  l  ■  q  i  r  a  a  v  i

*018     4374      Mapped site PSTI
B        ...       CONTINUES IN PHASE ONE


                B C E   M    AF   B                      B  HH      EST NF                          HT
                B   C   B    LN                          G  AA      CFT SN                          HH
                V   1   0    UU                          L  EE      0AH PU                          AA
                1   5   2    1H                          1  13      BN2 BH                          11
                                                             /       /                               /
AACAGAGGAGGAGAAGAGTGACAGCAGAGCTGCGTAATCTCCCGCATATTGCCAGCATGGCCTTTAATGAGCCGCTGATGCTTGAACCCGCCTATGCGCG
----.----+----.----+----.----+----.----+----.----+----.----+----.----+----.----+----.----+----.----+ 4500
TTGTCTCCTCCTCTTCTCACTGTCGTCTCGACGCATTAGAGGGCGTATAACGGTCGTACCGGAAATTACTCGGCGACTACGAACTTGGGCGGATACGCGC

 T  E  E  E  K  S  D  S  R  A  A  .  s  p  a  y  c  q  h  g  l  .  .  a  a  d  a  .  t  r  l  c  a
  q  r  r  r  r  V  T  A  E  L  R  N  L  P  H  I  A  S  M  A  F  N  E  P  L  M  L  F  P  A  Y  A  R
n  r  g  g  e  e  .  q  q  s  c  v  i  s  r  i  l  p  a  w  p  l  ■  s  r  .  c  l  n  p  p  ■  r  g
----.----+----.----+----.----+----.----+----.----+----.----+----.----+----.----+----.----+----.----+
 v  s  s  s  f  l  s  l  l  a  a  y  d  g  a  y  q  w  c  p  r  .  h  a  a  s  a  q  v  r  r  h  a
.  c  l  l  l  l  t  v  a  s  s  r  l  r  g  c  i  a  l  ■  a  k  l  s  g  s  i  s  s  g  a  .  a  r
 l  l  p  p  s  s  h  c  c  l  q  t  i  e  r  ■  n  g  a  h  g  k  i  l  r  q  h  k  f  g  g  i  r

C        4418      C (orf-439)          V-NH3
B        4434      B                    COOH


               H       HH  A    BM F    FS G   H  B        H F            M  E S B           NT
               H       AA  L    IB C    NF A   G  B        P O            N  C C G           RH
               A       EE  U    N0 1    UA L   U  V        A K            L  R R L           UA
               1       13  1    11 5    HN C   2  1        2 1            1  2 1 1           11
                        /               /                                                    /
GGTTTTCTTTTGTGCGCTTGCAGGCCAGCTTGGGATCAGCAGCCTGACGGATGCGGTGTCCGGCGACAGCCTGACTGCCCAGGAGGCACTCGCGACGCTG
----.----+----.----+----.----+----.----+----.----+----.----+----.----+----.----+----.----+----.----+ 4600
CCAAAAGAAAACACGCGAACGTCCGGTCGAACCCTAGTCGTCGGACTGCCTACGCCACAGGCCGCTGTCGGACTGACGGGTCCTCCGTGAGCGCTGCGAC

g  f  l  l  c  a  c  r  p  a  w  d  q  q  p  d  g  c  g  v  r  r  q  p  d  c  p  g  g  t  r  d  a  g
 V  F  F  C  A  L  A  G  Q  L  G  I  S  S  L  T  D  A  V  S  G  D  S  L  T  A  Q  E  A  L  A  T  L
  f  s  f  v  r  l  q  a  s  l  g  s  a  a  .  r  ■  r  c  p  a  t  a  .  l  p  r  r  h  s  r  r  w
----.----+----.----+----.----+----.----+----.----+----.----+----.----+----.----+----.----+----.----+
 p  k  r  k  h  a  q  l  g  a  q  s  .  c  g  s  p  h  p  t  r  r  c  g  s  q  g  p  p  v  r  s  a
  t  k  k  q  a  s  a  p  w  s  p  i  l  l  r  v  s  a  t  d  p  s  l  r  v  a  w  s  a  s  a  v  s
p  n  e  k  t  r  k  c  a  l  k  p  d  a  a  q  r  i  r  h  g  a  v  a  q  s  g  l  l  c  e  r  r  q

GALC     4544      lambdagalC           sl-att       EM
C        ...       CONTINUES IN PHASE TWO
```

```
  H    H            HAA        AH                      B           S F  H      H              BH
  G    P            PSV        SA                      B           F N  P      P              BP
  A    A            HUA        UE                      V           A U  A      A              VA
  1    2            112        13                      1           N H  2      2              12
                     //
GCATTATCCGGTGATGATGACGGACCACGACAGGCCCGCAGTTATCAGGTCATGAACGGCATCGCCGTGCTGCCGGTGTCCGGCACGCTGGTCAGCCGGA
----.----+----.----+----.----+----.----+----.----+----.----+----.----+----.----+----.----+----.----+ 4700
CGTAATAGGCCACTACTACTGCCTGGTGCTGTCCGGGCGTCAATAGTCCAGTACTTGCCGTAGCGGCACGACGGCCACAGGCCGTGCGACCAGTCGGCCT

 i  i  r  .  .  .  r  t  t  t  g  p  q  l  s  g  h  e  r  h  r  r  a  a  g  v  r  h  a  g  q  p  d
A  L  S  G  D  D  D  G  P  R  Q  A  R  S  Y  Q  V  M  N  G  I  A  V  L  P  V  S  G  T  L  V  S  R  T
  h  y  p  v  ■  ■  t  d  h  d  r  p  a  v  i  r  s  .  t  a  s  p  c  c  r  c  p  a  r  w  s  a  g
----.----+----.----+----.----+----.----+----.----+----.----+----.----+----.----+----.----+----.----+
p  ■  i  r  h  h  h  r  v  v  v  p  g  c  n  d  p  .  s  r  c  r  r  a  a  p  t  r  c  a  p  .  g  s
 a  n  d  p  s  s  s  p  g  r  c  a  r  l  .  .  t  ■  f  p  ■  a  t  s  g  t  d  p  v  s  t  l  r
  c  .  g  t  i  i  v  s  w  s  l  g  a  t  i  l  d  h  v  a  d  g  h  q  r  h  g  a  r  q  d  a  p

C       ...     CONTINUES IN PHASE TWO


 T     HHHFFP*   R *A BH      H      F              B             F      MB                          T
 H     GAHNNS0   S 0V BG      P      0              B             N      BI                          A
 A     AEAUUT1   A 2A VU      A      K              V             U      ON                          Q
 1     121HH19   1 01 12      2      1              1             H      11                          1
          // /       /                                                    /
CGCGGGCGCTGCAGCCGTACTCGGGGATGACCGGTTACAACGGCATTATCGCCCGTCTGCAACAGGCTGCCAGCGATCCGATGGTGGACGGCATTCTGCT
----.----+----.----+----.----+----.----+----.----+----.----+----.----+----.----+----.----+----.----+ 4800
GCGCCCGCGACGTCGGCATGAGCCCCTACTGGCCAATGTTGCCGTAATAGCGGGCAGACGTTGTCCGACGGTCGCTAGGCTACCACCTGCCGTAAGACGA

 a  g  a  a  a  v  l  g  d  d  r  l  q  r  h  y  r  p  s  a  t  g  c  q  r  s  d  g  g  r  h  s  a
  R  A  L  Q  P  Y  S  G  M  T  G  Y  N  G  I  I  A  R  L  Q  Q  A  A  S  D  P  M  V  D  G  I  L  L
r  g  r  c  s  r  t  r  g  .  p  v  t  t  a  l  s  p  v  c  n  .  l  p  a  i  r  w  w  t  a  f  c  s
----.----+----.----+----.----+----.----+----.----+----.----+----.----+----.----+----.----+----.----+
  a  p  a  a  a  t  s  p  s  s  r  n  c  r  c  .  r  g  d  a  v  p  q  w  r  d  s  p  p  r  c  e  a
v  r  a  s  c  g  y  e  p  i  v  p  .  l  p  ■  i  a  r  r  c  c  a  a  l  s  g  i  t  s  p  ■  r  s
 r  p  r  q  l  r  v  r  p  h  g  t  v  v  a  n  d  g  t  q  l  l  s  g  a  i  r  h  h  v  a  n  q

*019    4713    Mapped site PST1
*020    4720    Mapped site AVA1
C       ...     CONTINUES IN PHASE TWO


              NSH  H             F            H                                  H          HH
              CCP  G             0            H                                  P          HA
              IRA  U             K            A                                  A          AE
              112  2             1            1                                  2          12
                //
CGATATGGACACGCCCGGCGGGATGGTGGCGGGGGCATTTGACTGCGCTGACATCATCGCCCGTGTGCGTGACATAAAACCGGTATGGGCGCTTGCCAAC
----.----+----.----+----.----+----.----+----.----+----.----+----.----+----.----+----.----+----.----+ 4900
GCTATACCTGTGCGGGCCGCCCTACCACCGCCCCCGTAAACTGACGCGACTGTAGTAGCGGGCACACGCACTGTATTTTGGCCATACCCGCGAACGGTTG

r  y  g  h  a  r  r  d  g  g  g  g  i  .  l  r  .  h  h  r  p  c  a  .  h  k  t  g  ■  g  a  c  q  r
 D  M  D  T  P  G  G  M  V  A  G  A  F  D  C  A  D  I  I  A  R  V  R  D  I  K  P  V  W  A  L  A  N
  i  w  t  r  p  a  g  w  w  r  g  h  l  t  a  l  t  s  s  p  v  c  v  t  .  n  r  y  g  r  l  p  t
----.----+----.----+----.----+----.----+----.----+----.----+----.----+----.----+----.----+----.----+
 r  y  p  c  a  r  r  s  p  p  p  p  ■  q  s  r  q  c  .  r  g  h  a  h  c  l  v  p  i  p  a  q  w
  s  i  s  v  g  p  p  i  t  a  p  a  n  s  q  a  s  ■  ■  a  r  t  r  s  ■  f  g  t  h  a  s  a  l
e  i  h  v  r  g  a  p  h  h  r  p  c  k  v  a  s  v  d  d  g  t  h  t  v  y  f  r  y  p  r  k  g  v

C       ...     CONTINUES IN PHASE TWO


            P* T                    HF      NSHA   M         E      NSH    H          A
            S0 T                    GN      CCPC   N         C      CCP    G          C
            T2 H                    AU      IRAY   L         P      IRA    A          Y
            11 2                    1H      1121   1         1      112    1          1
             /                                //                      //
GACATGAACTGCAGTGCAGGTCAGTTGCTTGCCAGTGCCGCCTCCCGGCGTCTGGTCACGCAGACCGCCCGGACAGGCTCCATCGGCGTCATGATGGCTC
----.----+----.----+----.----+----.----+----.----+----.----+----.----+----.----+----.----+----.----+ 5000
CTGTACTTGACGTCACGTCCAGTCAACGAACGGTCACGGCGGAGGGCCGCAGACCAGTGCGTCTGGCGGGCCTGTCCGAGGTAGCCGCAGTACTACCGAG

 h  e  l  q  c  r  s  v  a  c  q  c  r  l  p  a  s  g  h  a  d  r  p  d  r  l  h  r  r  h  d  g  s
D  M  N  C  S  A  G  Q  L  L  A  S  A  A  S  R  R  L  V  T  Q  T  A  R  T  G  S  I  G  V  M  M  A  H
  t  .  t  a  v  q  v  s  c  l  p  v  p  p  p  g  v  w  s  r  r  p  p  g  q  a  p  s  a  s  .  w  l
----.----+----.----+----.----+----.----+----.----+----.----+----.----+----.----+----.----+----.----+
r  c  s  s  c  h  l  d  t  a  q  w  h  r  r  g  a  d  p  .  a  s  r  g  s  l  s  w  r  r  .  s  p  e
 s  ■  f  q  l  a  p  .  n  s  a  l  a  a  e  r  r  r  t  v  c  v  a  r  v  p  e  ■  p  t  ■  i  a
  v  h  v  a  t  c  t  l  q  k  g  t  g  g  g  p  t  q  d  r  l  g  g  p  c  a  g  d  a  d  h  h  s
```

```
*021    4913      Mapped site PST1
C       ...       CONTINUES IN PHASE TWO
```

```
B             F   H                                  N   F  F           BH            FM           HH
B             N   H                                  S   N  N           BG            OB           PG
V             U   A                                  P   U  U           VU            KO           AU
1             H   1                                  B   H  H           12            12           22
                                                                        /             /
ACAGTAATTACGGTGCTGCGCTGGAGAAACAGGGTGTGGAAATCACGCTGATTTACAGCGGCAGCCATAAGGTGGATGGCAACCCCTACAGCCATCTTCC
----.----+----.----+----.----+----.----+----.----+----.----+----.----+----.----+----.----+----.----+ 5100
TGTCATTAATGCCACGACGCGACCTCTTTGTCCCACACCTTTAGTGCGACTAAATGTCGCCGTCGGTATTCCACCTACCGTTGGGGATGTCGGTAGAAGG

q  .  l  r  c  c  a  g  e  t  g  c  g  n  h  a  d  l  q  r  q  p  .  g  g  w  q  p  l  q  p  s  s
 S  N  Y  G  A  A  L  E  K  Q  G  V  E  I  T  L  I  Y  S  G  S  H  K  V  D  G  N  P  Y  S  H  L  P
t  v  i  t  v  l  r  w  r  n  r  v  w  k  s  r  .  f  t  a  a  a  i  r  w  ■  a  t  p  t  a  i  f  r
----.----+----.----+----.----+----.----+----.----+----.----+----.----+----.----+----.----+----.----+
 c  y  n  r  h  q  a  p  s  v  p  h  p  f  .  a  s  k  c  r  c  g  y  p  p  h  c  g  r  c  g  d  e
.  l  l  .  p  a  a  s  s  f  c  p  t  s  i  v  s  i  .  l  p  l  w  l  t  s  p  l  g  .  l  w  r  g
 v  t  i  v  t  s  r  q  l  f  l  t  h  f  d  r  q  n  v  a  a  a  ■  l  h  i  a  v  g  v  a  ■  k
```

```
C       ...       CONTINUFS IN PHASE TWO
```

```
     A  AHNS F         P* NSHHN       T FH          MH                   H  H                  B
     C  APCC 0         S0 CCPGU       T 0G          SH                   P  A                  B
     Y  TAIR K         T2 IRAU3       H KA          TA                   A  E                  V
     1  2211 1         12 1122        2 11          11                   2  3                  1
           /            /  ///
GGATGACGTCCGGGAGACACTGCAGTCCCGGATGGACGCAACCCGCCAGATGTTTGCGCAGAAGGTGTCGGCATATACCGGCCTGTCCGTGCAGGTTGTG
----.----+----.----+----.----+----.----+----.----+----.----+----.----+----.----+----.----+----.----+ 5200
CCTACTGCAGGCCCTCTGTGACGTCAGGGCCTACCTGCGTTGGGCGGTCTACAAACGCGTCTTCCACAGCCGTATATGGCCGGACAGGCACGTCCAACAC

g  .  r  p  g  d  t  a  v  p  d  g  r  n  p  p  d  v  c  a  e  g  v  g  i  y  r  p  v  r  a  g  c  a
 D  D  V  R  E  T  L  Q  S  R  M  D  A  T  R  Q  M  F  A  Q  K  V  S  A  Y  T  G  L  S  V  Q  V  V
 ■  t  s  g  r  h  c  s  p  g  w  t  q  p  a  r  c  l  r  r  r  c  r  h  i  p  a  c  p  c  r  l  c
----.----+----.----+----.----+----.----+----.----+----.----+----.----+----.----+----.----+----.----+
p  h  r  g  p  s  v  a  t  g  s  p  r  l  g  g  s  t  q  a  s  p  t  p  ■  y  r  g  t  r  a  p  q
 s  s  t  r  s  v  s  c  d  r  i  s  a  v  r  w  i  n  a  c  f  t  d  a  y  v  p  r  d  t  c  t  t
r  i  v  d  p  l  c  q  l  g  p  h  v  c  g  a  l  h  k  r  l  l  h  r  c  i  g  a  q  g  h  l  n  h
```

```
*022    5124      Mapped site PST1
NU3     5132        Nu3 (orf-201)        M-NH3        AA
C       ...       CONTINUES IN PHASE TWO
```

```
 M           F   P*  RN   M      S   HH         H    B               HH**S     H    M      S        S
 N           N   S0  SS   N      F   AA         P    G               PI00F     P    B      F        F
 L           U   T2  AP   L      A   EE         A    L               AN22A     H    0      A        A
 1           H   13  1B   1      N   13         2    1               1245N     1    1      N        N
                  /                   /                               ////
CTGGATACCGAGGCTGCAGTGTACAGCGGTCAGGAGGCCATTGATGCCGGACTGGCTGATGAACTTGTTAACAGCACCGATGCGATCACCGTCATGCGTG
----.----+----.----+----.----+----.----+----.----+----.----+----.----+----.----+----.----+----.----+ 5300
GACCTATGGCTCCGACGTCACATGTCGCCAGTCCTCCGGTAACTACGGCCTGACCGACTACTTGAACAATTGTCGTGGCTACGCTAGTGGCAGTACGCAC

 g  y  r  g  c  s  v  q  r  s  g  g  h  .  c  r  t  g  .  .  t  c  .  q  h  r  c  d  h  r  h  a  .
L  D  T  E  A  A  V  Y  S  G  Q  E  A  I  D  A  G  L  A  D  E  L  V  N  S  T  D  A  I  T  V  M  R  D
 w  i  p  r  l  q  c  t  a  v  r  r  p  l  ■  p  d  w  l  ■  n  l  l  t  a  p  ■  r  s  p  s  c  v
----.----+----.----+----.----+----.----+----.----+----.----+----.----+----.----+----.----+----.----+
a  p  y  r  p  q  l  t  c  r  d  p  p  w  q  h  r  v  p  q  h  v  q  .  c  c  r  h  s  .  r  .  a  h
 s  s  v  s  a  a  t  y  l  p  .  s  a  ■  s  a  p  s  a  s  s  s  t  l  l  v  s  a  i  v  t  ■  r
  q  i  g  l  s  c  h  v  a  t  l  l  g  n  i  g  s  q  s  i  f  k  n  v  a  g  i  r  d  g  d  h  t
```

```
*023    5218      Mapped site PST1
*024    5269      Mapped site HIN2
*025    5269      Mapped site HPA1
C       ...       CONTINUES IN PHASE TWO
NU3     ...       CONTINUES IN PHASE TWO
```

```
      H             F    M  D        H              H
      G             O    N  D        N              N
      U             K    L  E        F              F
      2             1    1  1        3              1

ATGCACTGGATGCACGTAAATCCCGTCTCTCAGGAGGGCGAATGACCAAAGAGACTCAATCAACAACTGTTTCAGCCACTGCTTCGCAGGCTGACGTTAC
---------+---------+---------+---------+---------+---------+---------+---------+---------+---------+ 5400
TACGTGACCTACGTGCATTTAGGGCAGAGAGTCCTCCCGCTTACTGGTTTCTCTGAGTTAGTTGTTGACAAAGTCGGTGACGAAGCGTCCGACTGCAATG

  c  t  g  c  t  .  i  p  s  l  r  r  a  n  d  q  r  d  s  i  n  n  c  f  s  h  c  f  a  g  .  r  y
   A  L  D  A  R  K  S  R  L  S  G  G  R  M  T  K  E  T  Q  S  T  T  V  S  A  T  A  S  Q  A  D  V  T
 m  h  w  m  h  v  n  p  v  s  q  e  g  e  .  p  k  r  l  n  q  q  l  f  q  p  l  l  r  r  l  t  l  l
---------+---------+---------+---------+---------+---------+---------+---------+---------+---------+
  h  v  p  h  v  y  i  g  d  r  l  l  a  f  s  w  l  s  e  i  l  l  q  k  l  w  q  k  a  p  q  r  .
 s  a  s  s  a  r  l  d  r  r  e  p  p  r  i  v  l  s  v  .  d  v  v  t  e  a  v  a  e  c  a  s  t  v
   i  c  q  i  c  t  f  g  t  e  .  s  p  s  h  g  f  l  s  l  .  c  s  n  .  g  s  s  r  l  s  v  n

C          ...       CONTINUES IN PHASE TWO
NU3        ...       CONTINUES IN PHASE TWO
```

```
      H    M                        BHTF HF   H       B HT H  M      NF        F  B          F
      G    N                        GHHN HN   P       B PH H  B      SN        N  B          N
      I    L                        LAAU AU   A       V HA A  O      PU        U  V          U
      C    1                        111H 1H   2       1 11 1  1      BH        H  1          H
                                     ///
TGACGTGGTGCCAGCGACGGAGGGCGAGAACGCCAGCGCGGCGCAGCCGGACGTGAACGCGCAGATCACCGCAGCGGTTGCGGCAGAAAACAGCCGCATT
---------+---------+---------+---------+---------+---------+---------+---------+---------+---------+ 5500
ACTGCACCACGGTCGCTGCCTCCCGCTCTTGCGGTCGCGCCGCGTCGGCCTGCACTTGCGCGTCTAGTGGCGTCGCCAACGCCGTCTTTTGTCGGCGTAA

 .  r  g  a  s  d  g  g  r  e  r  q  r  g  a  a  g  r  e  r  a  d  h  r  s  g  c  g  r  k  q  p  h  y
  D  V  V  P  A  T  E  G  E  N  A  S  A  A  Q  P  D  V  N  A  Q  I  T  A  A  V  A  A  E  N  S  R  I
   t  w  c  q  r  r  r  a  r  t  p  a  r  r  s  r  t  .  t  r  r  s  p  q  r  l  r  q  k  t  a  a  l
---------+---------+---------+---------+---------+---------+---------+---------+---------+---------+
  q  r  p  a  l  s  p  p  r  s  r  w  r  p  a  a  p  r  s  r  a  s  .  r  l  p  q  p  l  f  c  g  c
   s  t  t  g  a  v  s  p  s  f  a  l  a  a  c  g  s  t  f  a  c  i  v  a  a  t  a  a  s  f  l  r  m
 s  v  h  h  w  r  r  l  a  l  v  g  a  r  r  l  r  v  h  v  r  l  d  g  c  r  n  r  c  f  v  a  a  n

C          ...       CONTINUES IN PHASE TWO
NU3        ...       CONTINUES IN PHASE TWO
```

```
   B*BXMB M  M    M               T     H     AMMT                NSH                 AH    F
   I0AHBT N  N    N               H     G     FLBH                CCP                 SA    N
   N2MOON l  l    L               A     A     LUOA                IRA                 UE    U
   161211 1  1    1               1     1     3121                112                 13    H
    ////                                                           //
ATGGGGATCCTCAACTGTGAGGAGGCTCACGGACGCGAAGAACAGGCACGCGTGCTGGCAGAAACCCCCGGTATGACCGTGAAAACGGCCCGCCGCATTC
---------+---------+---------+---------+---------+---------+---------+---------+---------+---------+ 5600
TACCCCTAGGAGTTGACACTCCTCCGAGTGCCTGCGCTTCTTGTCCGTGCGCACGACCGTCTTTGGGGGCCATACTGGCACTTTTGCCGGGCGGCGTAAG

  g  d  p  q  l  .  g  g  s  r  t  r  r  t  g  t  r  a  g  r  n  p  r  y  d  r  e  n  g  p  p  h  s
 M  G  I  L  N  C  E  E  A  H  G  R  E  E  Q  A  R  V  L  A  E  T  P  G  M  T  V  K  T  A  R  R  I  L
  w  g  s  s  t  v  r  r  l  t  d  a  k  n  r  h  a  c  w  q  k  p  p  v  .  p  .  k  r  p  a  a  f
---------+---------+---------+---------+---------+---------+---------+---------+---------+---------+
 .  p  s  g  .  s  h  p  p  e  r  v  r  l  v  p  v  r  a  p  l  f  g  r  y  s  r  s  f  p  g  g  c  e
  i  p  i  r  l  q  s  s  a  .  p  r  s  s  c  a  r  t  s  a  s  v  g  p  i  v  t  f  v  a  r  r  m
   h  p  d  e  v  t  l  l  s  v  s  a  f  f  l  c  a  h  q  c  f  g  g  t  h  g  h  f  r  g  a  a  n

*026      5505      Mapped site BAM1
C          ...      CONTINUES IN PHASE TWO
NU3        ...      CONTINUES IN PHASE TWO
```

```
GCHF F          BS  H  B HTH          H BMS         S H  BH H  N      F  P*B  SNSH
DFAN N          BD  G  S HHH          H IBF         D G  BP G  S      N  S0S  FCCP
IREU U          VU  I  S AAA          A NOA         U I  VA I  P      U  T2T  AIRA
213H H          11  A  2 111          1 11N         1 C  12 C  B      H  17E  N112
/               /      /                 /                                 //   //
TGGCCGCAGCACCACAGAGTGCACAGGCGCGCAGTGACACTGCGCTGGATCGTCTGATGCAGGGGGCACCGGCACCGCTGGCTGCAGGTAACCCGGCATC
----.----+----.----+----.----+----.----+----.----+----.----+----.----+----.----+----.----+----.----+ 5700
ACCGGCGTCGTGGTGTCTCACGTGTCCGCGCGTCACTGTGACGCGACCTAGCAGACTACGTCCCCCGTGGCCGTGGCGACCGACGTCCATTGGGCCGTAG

 g  r  s  t  t  e  c  t  g  a  q  .  h  c  a  g  s  s  d  a  g  g  t  g  t  a  g  c  r  .  p  g  i
 A  A  A  P  Q  S  A  Q  A  R  S  D  T  A  L  D  R  L  M  Q  G  A  P  A  P  L  A  A  G  N  P  A  S
w  p  q  h  h  r  v  h  r  r  a  v  t  l  r  w  i  v  .  c  r  g  h  r  h  r  w  l  q  v  t  r  h  l
----.----+----.----+----.----+----.----+----.----+----.----+----.----+----.----+----.----+----.----+
 p  r  l  v  v  s  h  v  p  a  c  h  c  q  a  p  d  d  s  a  p  p  v  p  v  a  p  q  l  y  g  p  ■
r  a  a  a  g  c  l  a  c  a  r  l  s  v  a  s  s  r  r  i  c  p  a  g  a  g  s  a  a  p  l  g  a  d
 q  g  c  c  w  l  t  c  l  r  a  t  v  s  r  q  i  t  q  h  l  p  c  r  c  r  q  s  c  t  v  r  c

*027    5686    Mapped site PST1
C       ...     CONTINUES IN PHASE TWO
NU3     ...     CONTINUES IN PHASE TWO
```

```
  S   HH**             NC  H      D    F                          F                  NSH
  F   PI00             U   G           O                          N                  CCP
  A   AN22             3   U           K                          U                  IRA
  N   1289                 2           1                          H                  112
      ///             /                                                               //
TGATGCCGTTAACGATTTGCTGAACACACCAGTGTAAGGGATGTTTATGACGAGCAAAGAAACCTTTACCCATTACCAGCCGCAGGGCAACAGTGACCCG
----.----+----.----+----.----+----.----+----.----+----.----+----.----+----.----+----.----+----.----+ 5800
ACTACGGCAATTGCTAAACGACTTGTGTGGTCACATTCCCTACAAATACTGCTCGTTTCTTTGGAAATGGGTAATGGTCGGCGTCCCGTTGTCACTGGGC

.  c  r  .  r  f  a  e  h  t  s  v  r  d  v  y  d  e  q  r  n  l  y  p  l  p  a  a  g  q  q  .  p  g
 D  A  V  N  D  L  L  N  T  P  V  .  g  ■  f  M  T  S  K  E  T  F  T  H  Y  Q  P  Q  G  N  S  D  P
 ■  p  l  t  i  c  .  t  h  q  c  k  g  c  l  .  r  a  k  k  p  l  p  i  t  s  r  r  a  t  v  t  r
----.----+----.----+----.----+----.----+----.----+----.----+----.----+----.----+----.----+----.----+
q  h  r  .  r  n  a  s  c  v  l  t  l  s  t  .  s  s  c  l  f  r  .  g  n  g  a  a  p  c  c  h  g
 s  a  t  l  s  k  s  f  v  g  t  y  p  i  n  i  v  l  l  s  v  k  v  w  .  w  g  c  p  l  l  s  g
r  i  g  n  v  i  q  q  v  c  w  h  l  p  h  k  h  r  a  f  f  g  k  g  ■  v  l  r  l  a  v  t  v  r

*028    5710    Mapped site HIN2
*029    5710    Mapped site HPA1
NU3     5734    Nu3              COOH
C       5734    C                COOH
D       5747    D (orf-110)      M-NH3     AA
```

```
         T H NSH              HH     S   N                          M  A          H   H
         H H CCP              HA     F   S                          N  L          G   G
         A A IRA              AE     A   P                          L  U          U   I
         1 1 112              12     N   B                          1  1          2   C
              //
GCTCATACCGCAACCGCGCCCGGCGGATTGAGTGCGAAAGCGCCTGCAATGACCCCGCTGATGCTGGACACCTCCAGCCGTAAGCTGGTTGCGTGGGATG
----.----+----.----+----.----+----.----+----.----+----.----+----.----+----.----+----.----+----.----+ 5900
CGAGTATGGCGTTGGCGCGGGCCGCCTAACTCACGCTTTCGCGGACGTTACTGGGGCGACTACGACCTGTGGAGGTCGGCATTCGACCAACGCACCCTAC

 s  y  r  n  r  a  r  r  i  e  c  e  s  a  c  n  d  p  a  d  a  g  h  l  q  p  .  a  g  c  v  g  w
A  H  T  A  T  A  P  G  G  L  S  A  K  A  P  A  M  T  P  L  M  L  D  T  S  S  R  K  L  V  A  W  D  G
 l  i  p  q  p  r  p  a  d  .  v  r  k  r  l  q  .  p  r  .  c  w  t  p  p  a  v  s  w  l  r  g  ■
----.----+----.----+----.----+----.----+----.----+----.----+----.----+----.----+----.----+----.----+
p  e  y  r  l  r  a  r  r  i  s  h  s  l  a  q  l  s  g  a  s  a  p  c  r  w  g  y  a  p  q  t  p  h
 a  .  v  a  v  a  g  p  p  n  l  a  f  a  g  a  i  v  g  s  i  s  s  v  e  l  r  l  s  t  a  h  s
 s  ■  g  c  g  r  g  a  s  q  t  r  f  r  r  c  h  g  r  q  h  q  v  g  g  a  t  l  q  n  r  p  i

D       ...     CONTINUES IN PHASE TWO
```

```
HB    F     F        B        EF    E                                    H       M       H
GB    0     N        B        CN    C                                    P       N       G
EV    K     U        V        1U    P                                    A       L       U
21    1     H        1        5H    1                                    2       1       2
/
GCACCACCGACGGTGCTGCCGTTGGCATTCTTGCGGTTGCTGCTGACCAGACCAGCACCACGCTGACGTTCTACAAGTCCGGCACGTTCCGTTATGAGGA
----.----+----.----+----.----+----.----+----.----+----.----+----.----+----.----+----.----+----.----+ 6000
CGTGGTGGCTGCCACGACGGCAACCGTAAGAACGCCAACGACGACTGGTCTGGTCGTGGTGCGACTGCAAGATGTTCAGGCCGTGCAAGGCAATACTCCT

 h  h  r  r  c  c  r  w  h  s  c  g  c  c  .  p  d  q  h  h  a  d  v  l  q  v  r  h  v  p  l  .  g
  T  T  D  G  A  A  V  G  I  L  A  V  A  A  D  Q  T  S  T  T  L  T  F  Y  K  S  G  T  F  R  Y  E  D
a  p  p  t  v  l  p  l  a  f  l  r  l  l  l  t  r  p  a  p  r  .  r  s  t  s  p  a  r  s  v  m  r  m
----.----+----.----+----.----+----.----+----.----+----.----+----.----+----.----+----.----+----.----+
  c  w  r  r  h  q  r  q  c  e  q  p  q  q  q  g  s  w  c  w  a  s  t  r  c  t  r  c  t  g  n  h  p
p  v  v  s  p  a  a  t  p  m  r  a  t  a  a  s  w  v  l  v  v  s  v  n  .  l  d  p  v  n  r  .  s  s
 a  g  g  v  t  s  g  n  a  n  k  r  n  s  s  v  l  g  a  g  r  q  r  e  v  l  g  a  r  e  t  i  l

D        ...        CONTINUES IN PHASE TWO


S  B HMGCHFH     F                              AA   T     H     B                 D S
D  B GNDFAOP     N                              SV   H     P     G                   F
U  V ILIREKA     U                              UA   A     A     L                   A
1  1 A121312     H                              12   1     2     1                   N
      / / /                                     /
TGTGCTCTGGCCGGAGGCTGCCAGCGACGAGACGAAAAAACGGACCGCGTTTGCCGGAACGGCAATCAGCATCGTTTAACTTTACCCTTCATCACTAAAG
----.----+----.----+----.----+----.----+----.----+----.----+----.----+----.----+----.----+----.----+ 6100
ACACGAGACCGGCCTCCGACGGTCGCTGCTCTGCTTTTTTGCCTGGCGCAAACGGCCTTGCCGTTAGTCGTAGCAAATTGAAATGGGAAGTAGTGATTTC

c  a  l  a  g  g  c  q  r  r  d  e  k  t  d  r  v  c  r  n  g  n  q  h  r  l  t  l  p  f  i  t  k  g
 V  L  W  P  E  A  A  S  D  E  T  K  K  R  T  A  F  A  G  T  A  I  S  I  V  .  l  y  p  s  s  l  k
  c  s  g  r  r  l  p  a  t  r  r  k  n  g  p  r  l  p  e  r  q  s  a  s  f  n  f  t  l  h  h  .  r
----.----+----.----+----.----+----.----+----.----+----.----+----.----+----.----+----.----+----.----+
 h  a  r  a  p  p  q  w  r  r  s  s  f  v  s  r  t  q  r  f  p  l  .  c  r  k  v  k  g  k  m  v  l
  t  s  q  g  s  a  a  l  s  s  v  f  f  r  v  a  n  a  p  v  a  i  l  m  t  .  s  .  g  e  d  s  f
i  h  e  p  r  l  s  g  a  v  l  r  f  f  p  g  r  k  g  s  r  c  d  a  d  n  l  k  v  r  .  .  .  l

D       6076       D                 COOH


HF         BF                            E  T    R             E     F                   MB NH
AN         GN                               A    S             C     N                   BI SG
EU         LU                               Q    A             1     U                   ON PA
3H         1H                               1    1             5     H                   11 B1
                                                                                          /
GCCGCCTGTGCGGCTTTTTTTACGGGATTTTTTTATGTCGATGTACACAACCGCCCAACTGCTGGCGGCAAATGAGCAGAAATTTAAGTTTGATCCGCTG
----.----+----.----+----.----+----.----+----.----+----.----+----.----+----.----+----.----+----.----+ 6200
CGGCGGACACGCCGAAAAAAATGCCCTAAAAAAATACAGCTACATGTGTTGGCGGGTTGACGACCGCCGTTTACTCGTCTTTAAATTCAAACTAGGCGAC

  r  l  c  g  f  f  y  g  i  f  l  c  r  c  t  q  p  p  n  c  w  r  q  m  s  r  n  l  s  l  i  r  c
 a  a  c  a  a  f  f  t  g  f  f  y  v  d  v  h  n  r  p  t  a  g  g  k  .  a  e  i  .  v  .  s  a  v
  p  p  v  r  l  f  l  r  d  f  f  M  S  M  Y  T  T  A  Q  L  L  A  A  N  E  Q  K  F  K  F  D  P  L
----.----+----.----+----.----+----.----+----.----+----.----+----.----+----.----+----.----+----.----+
 p  r  r  h  p  k  k  .  p  i  k  k  h  r  h  v  c  g  g  l  q  q  r  c  i  l  l  f  k  l  k  i  r  q
  a  a  q  a  a  k  k  v  p  n  k  .  t  s  t  c  l  r  g  v  a  p  p  l  h  a  s  i  .  t  q  d  a
   g  g  t  r  s  k  k  r  s  k  k  i  d  i  y  v  v  a  w  s  s  a  a  f  s  c  f  n  l  n  s  g  s

E       6135        F (orf-341)          M-NH3


                                 AH                                    HNS          HHK R
                                 LP                                    PCC          HAN S
                                 UH                                    AIR          AE1 A
                                 11                                    211          121 1
                                                                        /            /
TTTCTGCGTCTCTTTTTCCGTGAGAGCTATCCCTTCACCACGGAGAAAGTCTATCTCTCACAAATTCCGGGACTGGTAAACATGGCGCTGTACGTTTCGC
----.----+----.----+----.----+----.----+----.----+----.----+----.----+----.----+----.----+----.----+ 6300
AAAGACGCAGAGAAAAAGGCACTCTCGATAGGGAAGTGGTGCCTCTTTCAGATAGAGAGTGTTTAAGGCCCTGACCATTTGTACCGCGACATGCAAAGCG

  f  c  v  s  f  s  v  r  a  i  p  s  p  r  r  k  s  i  s  h  k  f  r  d  w  .  t  w  r  c  t  f  r
   s  a  s  l  f  p  .  e  l  s  l  h  h  g  e  s  l  s  l  t  n  s  g  t  g  k  h  g  a  v  r  f  a
 F  L  R  L  F  F  R  E  S  Y  P  F  T  T  E  K  V  Y  L  S  Q  I  P  G  L  V  N  M  A  L  Y  V  S  P
----.----+----.----+----.----+----.----+----.----+----.----+----.----+----.----+----.----+----.----+
  k  q  t  e  k  e  t  l  a  i  g  e  g  r  l  f  d  i  e  .  l  n  r  s  q  y  v  h  r  q  v  n  r
 t  e  a  d  r  k  g  h  s  s  d  r  .  w  p  s  l  r  d  r  v  f  e  p  v  p  l  c  p  a  t  r  k  a
 n  r  r  r  k  k  r  s  l  .  g  k  v  v  s  f  t  .  r  e  c  i  g  p  s  t  f  m  a  s  y  t  e

KN11    6287        kan11              i             EM
E        ...        CONTINUES IN PHASE THREE
```

```
  M H           H         F             M  HNS                 H    H
  N P           P         N             N  PCC                 N    G
  L A           H         U             L  AIR                 F    A
  1 2           1         H             1  211                 1    1
                                            /
CGATTGTTTCCGGTGAGGTTATCCGTTCCCGTGGCGGCTCCACCTCTGAATTTACGCCGGGATATGTCAAGCCGAAGCATGAAGTGAATCCGCAGATGAC
----.----+----.----+----.----+----.----+----.----+----.----+----.----+----.----+----.----+----.----+ 6400
GCTAACAAAGGCCACTCCAATAGGCAAGGGCACCGCCGAGGTGGAGACTTAAATGCGGCCCTATACAGTTCGGCTTCGTACTTCACTTAGGCGTCTACTG

r  l  f  p  v  r  l  s  v  p  v  a  a  p  p  l  n  l  r  r  d  ■  s  s  r  s  ■  k  .  i  r  r  .
 d  c  f  r  .  g  y  p  f  p  w  r  l  h  l  .  i  y  a  g  i  c  q  a  e  a  .  s  e  s  a  d  d  p
  I  V  S  G  E  V  I  R  S  R  G  G  S  T  S  E  F  T  P  G  Y  V  K  P  K  H  E  V  N  P  Q  M  T
----.----+----.----+----.----+----.----+----.----+----.----+----.----+----.----+----.----+----.----+
 r  n  n  g  t  l  n  d  t  g  t  a  a  g  g  r  f  k  r  r  s  i  d  l  r  l  ■  f  h  i  r  l  h
  s  q  k  r  h  p  .  g  n  g  h  r  s  w  r  q  i  .  a  p  i  h  .  a  s  a  h  l  s  d  a  s  s
g  i  t  e  p  s  t  i  r  e  r  p  p  e  v  e  s  n  v  g  p  y  t  l  g  f  c  s  t  f  g  c  i  v
E       ...      CONTINUES IN PHASE THREE
```

```
          HH     XMB   F HM     AA  NSH       F            S          N         A C BHH
          PG     HBI   O NB     SV  CCP       N            F          S         L F AAA
          AU     OON   K FO     UA  IRA       U            A          P         U R LEE
          22     211   1 12     12  112       H            N          C         1 1 113
                 //             /   //                                                //
CCTGCGTCGCCTGCCGGATGAAGATCCGCAGAATCTGGCGGACCCGGCTTACCGCCGCCGTCGCATCATCATGCAGAACATGCGTGACGAAGAGCTGGCC
----.----+----.----+----.----+----.----+----.----+----.----+----.----+----.----+----.----+----.----+ 6500
GGACGCAGCGGACGGCCTACTTCTAGGCGTCTTAGACCGCCTGGGCCGAATGGCGGCGGCAGCGTAGTAGTACGTCTTGTACGCACTGCTTCTCGACCGG

  c  v  a  c  r  ■  k  i  r  r  i  w  r  t  r  l  t  a  a  v  a  s  s  c  r  t  c  v  t  k  s  w  p
p  a  s  p  a  g  .  r  s  a  e  s  g  g  p  g  l  p  p  p  s  h  h  h  a  e  h  a  .  r  r  a  g  h
 L  R  R  I  P  D  E  D  P  Q  N  L  A  D  P  A  Y  R  R  R  R  I  I  M  Q  N  M  R  D  E  E  L  A
----.----+----.----+----.----+----.----+----.----+----.----+----.----+----.----+----.----+----.----+
g  q  t  a  q  r  i  f  i  r  l  i  q  r  v  r  s  v  a  a  t  a  d  d  h  l  v  h  t  v  f  l  q  g
 g  a  d  g  a  p  h  l  d  a  s  d  p  p  g  p  k  g  g  g  d  c  .  .  a  s  c  a  h  r  l  a  p
  r  r  r  r  g  s  s  s  g  c  f  r  a  s  g  a  .  r  r  r  r  ■  ■  ■  c  f  ■  r  s  s  s  s  a
E       ...      CONTINUES IN PHASF THREE
```

```
M  D S T            M             A          H       H       TMHBMH       A HFM
B  D F A            B             F          G       P       ABPINP       S ANN
O  E A Q            O             L          E       A       QOHNLA       U FUL
2  1 N 1            2             2          2       2       111112       1 3H1
                                                               //            /
ATTGCTCAGGTCGAAGAGATGCAGGCAGTTTCTGCCGTGCTTAAGGGCAAATACACCATGACCGGTGAAGCCTTCGATCCGGTTGAGGTGGATATGGGCC
----.----+----.----+----.----+----.----+----.----+----.----+----.----+----.----+----.----+----.----+ 6600
TAACGAGTCCAGCTTCTCTACGTCCGTCAAAGACGGCACGAATTCCCGTTTATGTGGTACTGGCCACTTCGGAAGCTAGGCCAACTCCACCTATACCCGG

  l  l  r  s  k  r  c  r  q  f  l  p  c  l  r  a  n  t  p  .  p  v  k  p  s  i  r  l  r  w  i  w  a
  c  s  g  r  r  d  a  g  s  f  c  r  a  .  g  q  i  h  h  d  r  .  s  l  r  s  g  .  g  g  y  g  p
I  A  Q  V  E  E  M  Q  A  V  S  A  V  L  K  G  K  Y  T  M  T  G  E  A  F  D  P  V  E  V  D  M  G  R
----.----+----.----+----.----+----.----+----.----+----.----+----.----+----.----+----.----+----.----+
  n  s  l  d  f  l  h  l  c  n  r  g  h  k  l  a  f  v  g  h  g  t  f  g  e  i  r  n  l  h  i  h  a
w  q  e  p  r  l  s  a  p  l  k  q  r  a  .  p  c  i  c  w  s  r  h  l  r  r  d  p  q  p  p  y  p  g
■  a  .  t  s  s  i  c  a  t  e  a  t  s  l  p  l  y  v  ■  v  p  s  a  k  s  g  t  s  t  s  i  p
E       ...      CONTINUES IN PHASE THREE
```

```
                    H  F                                                      ET        T H
                    P  N                                                      CA        H H
                    A  U                                                      RQ        A A
                    2  H                                                      V1        1 1
                                                                              /
GCAGTGAGGAGAATAACATCACGCAGTCCGGCGGCACGGAGTGGAGCAAGCGTGACAAGTCCACGTATGACCCGACCGACGATATCGAAGCCTACGCGCT
----.----+----.----+----.----+----.----+----.----+----.----+----.----+----.----+----.----+----.----+ 6700
CGTCACTCCTCTTATTGTAGTGCGTCAGGCCGCCGTGCCTCACCTCGTTCGCACTGTTCAGGTGCATACTGGGCTGGCTGCTATAGCTTCGGATGCGCGA

a  v  r  r  i  t  s  r  s  p  a  a  r  s  g  a  s  v  t  s  p  r  ■  t  r  p  t  i  s  k  p  t  r  .
 q  .  g  e  .  h  h  a  v  r  r  h  g  v  e  q  a  .  q  v  h  v  .  p  d  r  r  y  r  s  l  r  a
  S  E  E  N  N  I  T  Q  S  G  G  T  E  W  S  K  R  D  K  S  T  Y  D  P  T  D  D  I  E  A  Y  A  L
----.----+----.----+----.----+----.----+----.----+----.----+----.----+----.----+----.----+----.----+
 a  t  l  l  i  v  d  r  l  g  a  a  r  l  p  a  l  t  v  l  g  r  i  v  r  g  v  i  d  f  g  v  r
  c  h  p  s  y  c  .  a  t  r  r  c  p  t  s  c  a  h  c  t  w  t  h  g  s  r  r  y  r  l  r  r  a
r  l  s  s  f  l  ■  v  c  d  p  p  v  s  h  l  l  r  s  l  d  v  y  s  g  v  s  s  i  s  a  .  a  s
E       ...      CONTINUES IN PHASE THREE
```

```
 N   B       H   TMB       HH            A
 S   S       P   ABI       HA            L
 P   T       H   QON       AE            U
 B   X       1   111       12            1
                  /
GAACGCCAGCGGTGTGGTGAATATCATCGTGTTCGATCCGAAAGGCTGGGCGCTGTTCCGTTCCTTCAAAGCCGTCAAGGAGAAGCTGGATACCCGTCGT 6800
---.---+---.---+---.---+---.---+---.---+---.---+---.---+---.---+---.---+---.---+
CTTGCGGTCGCCACACCACTTATAGTAGCACAAGCTAGGCTTTCCGACCCGCGACAAGGCAAGGAAGTTTCGGCAGTTCCTCTTCGACCTATGGGCAGCA

 t p a v w . i s s c s i r k a g r c s v p s k p s r r s w i p v v
e r q r c g e y h r v r s e r l g a v p f l q s r q g e a g y p s w
 N A S G V V N I I V F D P K G W A L F R S F K A V K E K L D T R R
---.---+---.---+---.---+---.---+---.---+---.---+---.---+---.---+---.---+---.---+
q v g a t h h i d d h e i r f a p r q e t g e f g d l l l q i g t t
 s r w r h p s y . r t r d s l s p a t g n r . l r . p s a p y g d
  f a l p t t f i ■ t n s g f p q a s n r e k l a t l s f s s v r r

E    ...    CONTINUES IN PHASE THREE
```

```
     A    N      E E S H            H           F C BHH         H
     L    S      C C C P            G           O F AAA         P
     U    P      P R R H            U           K R LEE         A
     1    B      1 2 1 1            2           1 1 113         2
                                                      //
GGCTCTAATTCCGAGCTGGAGACAGCGGTGAAAGACCTGGGCAAAGCGGTGTCCTATAAGGGGATGTATGGCGATGTGGCCATCGTCGTGTATTCCGGAC 6900
---.---+---.---+---.---+---.---+---.---+---.---+---.---+---.---+---.---+---.---+
CCGAGATTAAGGCTCGACCTCTGTCGCCACTTTCTGGACCCGTTTCGCCACAGGATATTCCCCTACATACCGCTACACCGGTAGCAGCACATAAGGCCTG

 a l i p s w r q r . k t w a k r c p i r g c ■ a ■ w p s s c i p d
  l . f r a g d s g e r p g q s g v l . g d v w r c g h r r v f r t
G S N S E L E T A V K D L G K A V S Y K G M Y G D V A I V V Y S G Q
---.---+---.---+---.---+---.---+---.---+---.---+---.---+---.---+---.---+---.---+
 a r i g l q l c r h f v q a f r h g i l p h i a i h g d d h i g s
h s . n r a p s l p s l g p c l p t r y p s t h r h p w r r t n r v
 p e l e s s s v a t f s r p l a t d . l p i y p s t a ■ t t y e p

E    ...    CONTINUES IN PHASE THREE
```

```
RH        A          H   E          D      TE   B  MH       FS   H
SG        C          P   C          D      HC   B  SH       NF   G
AA        Y          A   O          E      AP   V  TA       UA   U
11        1          2   K          1      11   1  11       HN   2
                                                             /
AGTACGTGGAAAAACGGCGTCAAAAAGAACTTCCTGCCGGACAACACGATGGTGCTGGGGAACACTCAGGCACGCGGTCTGCGCACCTATGGCTGCATTCA 7000
---.---+---.---+---.---+---.---+---.---+---.---+---.---+---.---+---.---+---.---+
TCATGCACCTTTTTGCCGCAGTTTTTCTTGAAGGACGGCCTGTTGTGCTACCACGACCCCTTGTGAGTCCGTGCGCCAGACGCGTGGATACCGACGTAAGT

s t w k t a s k r t s c r t t r w c w g t l r h a v c a p ■ a a f r
 v r g k r r q k e l p a g g q h d g a g e h s g t r s a h l w l h s
  Y V E N G V K K N F L P D N T M V L G N T Q A R G L R T Y G C I Q
---.---+---.---+---.---+---.---+---.---+---.---+---.---+---.---+---.---+---.---+
l v h f v a d f l v e q r v v r h h q p v s l c a t q a g i a a n
 t r p f r r . f s s g a p c c s p a p s c e p v r d a c r h s c e
c y t s f p t l f f k r g s l v i t s p f v . a r p r r v . p q ■ .

E    ...    CONTINUES IN PHASE THREE
```

```
     F HHT            M       B    H HMB H   HTH        H
     O GHH            N       S    P PBI P   HHP        N
     K AAA            L       T    A HON A   AAH        F
     1 111            1       E    2 111 2   111        1
         /                             //     //
GGATGCGGACGCACAGCGCGAAGGCATTAACGCCTCTGCCCGTTACCCGAAAAACTGGGTGACCACCGGCGATCCGGCGCGTGAGTTCACCATGATTCAG 7100
---.---+---.---+---.---+---.---+---.---+---.---+---.---+---.---+---.---+---.---+
CCTACGCCTGCGTGTCGCGCTTCCGTAATTGCGGAGACGGGCAATGGGCTTTTTGACCCACTGGTGGCCGCTAGGCCGCGCACTCAAGTGGTACTAAGTC

 ■ r t h s a k a l t p l p v t r k t g . p p a i r r v s s p . f s
g c g r t a r r h . r l c p l p e k l g d h r r s g a . v h h d s v
 D A D A Q R E G I N A S A R Y P K N W V T T G D P A R E F T M I Q
---.---+---.---+---.---+---.---+---.---+---.---+---.---+---.---+---.---+---.---+
l i r v c l a f a n v g r g t v r f v p h g g a i r r t l e g h n l
 p h p r v a r l c . r r q g n g s f s p s w r r d p a h t . w s e
  s a s a c r s p ■ l a e a r . g f f q t v v p s g a r s n v ■ i .

E    ...    CONTINUES IN PHASE THREE
```

```
SB N      EF                              R      E     AH    A H    M      H
FB S      CN                              S            SA    S A    N      N
AV P      1U                              A            UE    U E    L      F
N1 B      5H                              1            13    1 3    1      1
/
TCAGCACCGCTGATGCTGCTGGCTGACCCTGATGAGTTCGTGTCCGTACAACTGGCGTAATCATGGCCCTTCGGGGCCATTGTTTCTCTGTGGAGGAGTC
----.----+----.----+----.----+----.----+----.----+----.----+----.----+----.----+----.----+----.----+ 7200
AGTCGTGGCGACTACGACGACCGACTGGGACTACTCAAGCACAGGCATGTTGACCGCATTAGTACCGGGAAGCCCCGGTAACAAAGAGACACCTCCTCAG

 q  h  r  .  c  c  w  l  t  l  m  s  s  c  p  y  n  w  r  n  h  g  p  s  g  p  l  f  l  c  g  g  v
  s  t  a  d  a  a  g  .  p  .  .  v  r  v  r  t  t  g  v  i  m  a  l  r  g  h  c  f  s  v  e  e  s
S  A  P  L  M  L  L  A  D  P  D  E  F  V  S  V  Q  L  A  .  s  w  p  f  g  a  i  v  s  l  w  r  s  p
----.----+----.----+----.----+----.----+----.----+----.----+----.----+----.----+----.----+----.----+
 .  c  r  q  h  q  q  s  v  r  i  l  e  h  g  y  l  q  r  l  .  p  g  e  p  g  n  n  r  q  p  p  t
t  l  v  a  s  a  a  p  q  g  q  h  t  r  t  r  v  v  p  t  i  m  a  r  r  p  w  q  k  e  t  s  s  d
 d  a  g  s  i  s  s  a  s  g  s  s  n  t  d  t  c  s  a  y  d  h  g  k  p  a  m  t  e  r  h  l  l

E     7157        E             COOH


F                                                      H                                    HMH
I                                                      P                                    HBA
                                                       H                                    AOE
                                                       1                                    122
                                                                                              /
CATGACGAAAGATGAACTGATTGCCCGTCTCCGCTCGCTGGGTGAACAACTGAACCGTGATGTCAGCCTGACGGGGACGAAAGAAGAACTGGCGCTCCGT
----.----+----.----+----.----+----.----+----.----+----.----+----.----+----.----+----.----+----.----+ 7300
GTACTGCTTTCTACTTGACTAACGGGCAGAGGCGAGCGACCCACTTGTTGACTTGGCACTACAGTCGGACTGCCCCTGCTTTCTTCTTGACCGCGAGGCA

h  d  e  r  .  t  d  c  p  s  p  l  a  g  .  t  t  e  p  .  c  q  p  d  g  d  e  r  r  t  g  a  p  c
 M  T  K  D  E  L  I  A  R  L  R  S  L  G  E  Q  L  N  R  D  V  S  L  T  G  T  K  E  E  L  A  L  R
  .  r  k  m  n  .  l  p  v  s  a  r  w  v  n  n  .  t  v  m  s  a  .  r  g  r  k  k  n  w  r  s  v
----.----+----.----+----.----+----.----+----.----+----.----+----.----+----.----+----.----+----.----+
w  s  s  l  h  v  s  q  g  d  g  s  a  p  h  v  v  s  g  h  h  .  g  s  p  s  s  l  l  v  p  a  g
 m  v  f  s  s  s  i  a  r  r  r  e  s  p  s  c  s  f  r  s  t  l  r  v  p  v  f  s  s  s  a  s  r
g  h  r  f  i  f  q  n  g  t  e  a  r  q  t  f  l  q  v  t  i  d  a  q  r  p  r  f  f  f  q  r  e  t

EI    7202        FI (orf-132)        M-NH3


      MA      A       H       H F         D   HNMS        H   M
      NL      L       G       P O         D   PCNC        P   N
      LU      U       U       A K         E   AILR        A   L
      11      1       2       2 1         1   2111        2   1
                                                 /
GTGGCAGAGCTGAAAGAGGAGCTTGATGACACGGATGAAACTGCCGGTCAGGACACCCCTCTCAGCCGGGAAAATGTGCTGACCGGACATGAAAATGAGG
----.----+----.----+----.----+----.----+----.----+----.----+----.----+----.----+----.----+----.----+ 7400
CACCGTCTCGACTTTCTCCTCGAACTACTGTGCCTACTTTGACGGCCAGTCCTGTGGGGAGAGTCGGCCCTTTTACACGACTGGCCTGTACTTTTACTCC

 g  r  a  e  r  g  a  .  .  h  g  .  n  c  r  s  g  h  p  s  q  p  g  k  c  a  d  r  t  .  k  .  g
V  A  E  L  K  E  E  L  D  D  T  D  E  T  A  G  Q  D  T  P  L  S  R  E  N  V  L  T  G  H  E  N  E  V
 w  q  s  .  k  r  s  l  m  t  r  m  k  l  p  v  r  t  p  l  s  a  g  k  m  c  .  p  d  m  k  m  r
----.----+----.----+----.----+----.----+----.----+----.----+----.----+----.----+----.----+----.----+
h  p  l  a  s  l  p  a  q  h  c  p  h  f  q  r  d  p  c  g  e  .  g  p  f  h  a  s  r  v  h  f  h  p
 t  a  s  s  f  s  s  s  s  s  v  s  s  v  a  p  .  s  v  g  r  l  r  s  f  t  s  v  p  c  s  f  s
  h  c  l  q  f  l  l  k  i  v  r  i  f  s  g  t  l  v  g  r  e  a  p  f  i  h  q  g  s  m  f  i  l

EI      ...      CONTINUES IN PHASE TWO


BM      HF H      B H                              B        AFS  H            GT H
IB      HN P      B N                              B        LNF  P            7H G
NO      AU A      V F                              V        UUA  H            3A U
11      1H 2      1 1                              1        1HN  1            01 2
                                                             /
TGGGATCAGCGCAGCCGGATACCGTGATTCTGGATACGTCTGAACTGGTCACGGTCGTGGCACTGGTGAAGCTGCATACTGATGCACTTCACGCCACGCG
----.----+----.----+----.----+----.----+----.----+----.----+----.----+----.----+----.----+----.----+ 7500
ACCCTAGTCGCGTCGGCCTATGGCACTAAGACCTATGCAGACTTGACCAGTGCCAGCACCGTGACCACTTCGACGTATGACTACGTGAAGTGCGGTGCGC

 g  i  s  a  a  g  y  r  d  s  g  y  v  .  t  g  h  g  r  g  t  g  e  a  a  y  .  c  t  s  r  h  a
  G  S  A  Q  P  D  T  V  I  L  D  T  S  E  L  V  T  V  V  A  L  V  K  L  H  T  D  A  L  H  A  T  R
w  d  q  r  s  r  i  p  .  f  w  i  r  l  n  w  s  r  s  w  h  w  .  s  c  i  l  m  h  f  t  p  r  g
----.----+----.----+----.----+----.----+----.----+----.----+----.----+----.----+----.----+----.----+
 p  i  l  a  a  p  y  r  s  e  p  y  t  q  v  p  .  p  r  p  v  p  s  a  a  y  q  h  v  e  r  w  a
t  p  d  a  c  g  s  v  t  i  r  s  v  d  s  s  t  v  t  t  a  s  t  f  s  c  v  s  a  s  .  a  v  r
 h  s  .  r  l  r  i  g  h  n  q  i  r  r  f  q  d  r  d  h  c  q  h  l  q  m  s  i  c  k  v  g  r

G730    7497      lambdagalF730       sl-att      EM
EI      ...      CONTINUES IN PHASE TWO
```

```
        B    F        F  HNS                        H    B   F           B      HTFHESC BHH        MF
        B    O        N  PCC                        P    G   N           B      HHNACCF AAA        NI
        V    K        U  AIR                        A    L   U           V      AAUERRR LEE        L
        1    1        H  211                        2    1   H           1      11H3211 113        1
                          /                                                      /// /   //
GGATGAACCTGTGGCATTTGTGCTGCCGGGAACGGCGTTTCGTGTCTCTGCCGGTGTGGCAGCCGAAATGACAGAGCGCGGCCTGGCCAGAATGCAATAA
----.----+----.----+----.----+----.----+----.----+----.----+----.----+----.----+----.----+----.----+ 7600
CCTACTTGGACACCGTAAACACGACGGCCCTTGCCGCAAAGCACAGAGACGGCCACACCGTCGGCTTTACTGTCTCGCGCCGGACCGGTCTTACGTTATT

g  .  t  c  g  i  c  a  a  g  n  g  v  s  c  l  c  r  c  g  s  r  n  d  r  a  r  p  g  q  n  a  i  t
 D  E  P  V  A  F  V  L  P  G  T  A  F  R  V  S  A  G  V  A  A  E  M  T  E  R  G  L  A  R  M  Q  .
  ■  n  l  w  h  l  c  c  r  e  r  r  f  v  s  l  p  v  w  q  p  k  .  q  s  a  a  w  p  e  c  n  n
----.----+----.----+----.----+----.----+----.----+----.----+----.----+----.----+----.----+----.----+
 p  h  v  q  p  ■  q  a  a  p  f  p  t  e  h  r  q  r  h  p  l  r  f  s  l  a  r  g  p  w  f  a  i
  s  s  g  t  a  n  t  s  g  p  v  a  n  r  t  e  a  p  t  a  a  s  i  v  s  r  p  r  a  l  i  c  y
p  i  f  r  h  c  k  h  q  r  s  r  r  k  t  d  r  g  t  h  c  g  f  h  c  l  a  a  q  g  s  h  l  l

EI     7597        FI                 COOH
```

```
        HH F      T  SB     T    F          T H          T   R        F
        HA I      A  FB     A    N          H H          H   S        O
        AE I      Q  AV     Q    U          A A          A   A        K
        12        1  N1     1    H          1 1          1   1        1
                      /
CGGGAGGCGCTGTGGCTGATTTCGATAACCTGTTCGATGCTGCCATTGCCCGCGCCGATGAAACGATACGCGGGTACATGGGAACGTCAGCCACCATTAC
----.----+----.----+----.----+----.----+----.----+----.----+----.----+----.----+----.----+----.----+ 7700
GCCCTCCGCGACACCGACTAAAGCTATTGGACAAGCTACGACGGTAACGGGCGCGGCTACTTTGCTATGCGCCCATGTACCCTTGCAGTCGGTGGTAATG

  g  g  a  V  A  D  F  D  N  I  F  D  A  A  I  A  R  A  D  E  T  I  R  G  Y  M  G  T  S  A  T  I  T
r  e  a  l  w  l  i  s  i  t  c  s  ■  l  p  l  p  a  p  ■  k  r  y  a  g  t  w  e  r  q  p  p  l  h
 g  r  r  c  g  .  f  r  .  p  v  r  c  c  h  c  p  r  r  .  n  d  t  r  v  h  g  n  v  s  h  h  y
----.----+----.----+----.----+----.----+----.----+----.----+----.----+----.----+----.----+----.----+
v  p  p  a  t  a  s  k  s  l  r  n  s  a  a  ■  a  r  a  s  s  v  i  r  p  y  ■  p  v  d  a  v  ■  v
 r  s  a  s  h  s  i  e  i  v  q  e  i  s  g  n  g  a  g  i  f  r  y  a  p  v  h  s  r  .  g  g  n
  p  l  r  q  p  q  n  r  y  g  t  r  h  q  w  q  g  r  r  h  f  s  v  r  t  c  p  f  t  l  w  w  .

EII    7612        FII (orf-117)      V-NH3      AA
```

```
H            H             H                    A    H          HT         T
P            P             P                    L    P          HH         T
A            H             H                    U    A          AA         H
2            1             1                    1    2          11         2
                                                                 /
ATCCGGTGAGCAGTCAGGTGCGGTGATACGTGGTGTTTTTGATGACCCTGAAAATATCAGCTATGCCGGACAGGGCGTGCGCGTTGAAGGCTCCAGCCCG
----.----+----.----+----.----+----.----+----.----+----.----+----.----+----.----+----.----+----.----+ 7800
TAGGCCACTCGTCAGTCCACGCCACTATGCACCACAAAAACTACTGGGACTTTTATAGTCGATACGGCCTGTCCCGCACGCGCAACTTCCGAGGTCGGGC

S  G  E  Q  S  G  A  V  I  R  G  V  F  D  D  P  E  N  I  S  Y  A  G  Q  G  V  R  V  E  G  S  S  P
 p  v  s  s  q  v  r  .  y  v  v  f  l  ■  t  l  k  i  s  a  ■  p  d  r  a  c  a  l  k  a  p  a  r
i  r  .  a  v  r  c  g  d  t  w  c  f  .  .  p  .  k  y  q  l  c  r  t  g  r  a  r  .  r  l  q  p  v
----.----+----.----+----.----+----.----+----.----+----.----+----.----+----.----+----.----+----.----+
 d  p  s  c  d  p  a  t  i  r  p  t  k  s  s  g  s  f  i  l  .  a  p  c  p  t  r  t  s  p  e  l  g
c  g  t  l  l  .  t  r  h  y  t  t  n  k  i  v  r  f  i  d  a  i  g  s  l  a  h  a  n  f  a  g  a  r
■  r  h  a  t  l  h  p  s  v  h  h  k  q  h  g  q  f  y  .  s  h  r  v  p  r  a  r  q  l  s  w  g

EII     ...        CONTINUES IN PHASE ONE
```

```
          H M    B     F NF PA*FF   B            M            H    M      HH
          P N    B     N SN VL0NN   B            N            P    B      PG
          A L    V     U PU UU3UU   V            L            H    0      AU
          2 1    1     H BH 210HH   1            1            1    1      22
                              ///
TCCCTGTTTGTCCGGACTGATGAGGTGCGGCAGCTGCGGCGTGGAGACACGCTGACCATCGGTGAGGAAAATTTCTGGGTAGATCGGGTTTCGCCGGATG
----.----+----.----+----.----+----.----+----.----+----.----+----.----+----.----+----.----+----.----+ 7900
AGGGACAAACAGGCCTGACTACTCCACGCCGTCGACGCCGCACCTCTGTGCGACTGGTAGCCACTCCTTTTAAAGACCCATCTAGCCCAAAGCGGCCTAC

S  L  F  V  R  T  D  E  V  R  Q  L  R  R  G  D  T  L  T  I  G  E  E  N  F  W  V  D  R  V  S  P  D  D
 p  c  l  s  g  l  ■  r  c  g  s  c  g  v  e  t  r  .  p  s  v  r  k  i  s  g  .  i  g  f  r  r  ■
  p  v  c  p  d  .  .  g  a  a  a  a  a  w  r  h  a  d  h  r  .  g  k  f  l  g  r  s  g  f  a  g  .
----.----+----.----+----.----+----.----+----.----+----.----+----.----+----.----+----.----+----.----+
 d  r  n  t  r  v  s  s  t  r  c  s  r  r  p  s  v  s  v  ■  p  s  s  f  k  q  t  s  r  t  e  g  s
  g  q  k  d  p  s  i  l  h  p  l  q  p  t  s  v  r  q  g  d  t  l  f  i  e  p  y  i  p  n  r  r  i
t  g  t  q  g  s  q  h  p  a  a  a  a  a  h  l  c  a  s  w  r  h  p  f  n  r  p  l  d  p  k  a  p  h

*030    7833       Mapped site PVU2
EII     ...        CONTINUES IN PHASE ONE
```

```
        F                           R           HH**     NF  F       H      ZC BHH F   E
        O                           S           PI00     SN  I       G       F AAA O   C
        K                           A           AN33     PU  I       U      R  LEE K   P
        1                           1           1212     BH          2      1  113 1   1
                                                  ///                           //
ATGGCGGAAGTTGTCATCTCTGGCTTGGACGGGGCGTACCGCCTGCCGTTAACCGTCGCCGCTGAAAGGGGGATGTATGGCCATAAAAGGTCTTGAGCAG
----.----+----.----+----.----+----.----+----.----+----.----+----.----+----.----+----.----+----.----+ 8000
TACCGCCTTCAACAGTAGAGACCGAACCTGCCCCGCATGGCGGACGGCAATTGGCAGCGGCGACTTTCCCCCTACATACCGGTATTTTCCAGAACTCGTC

 G G S C H L W L G R G V P P A V N R R R . k g d v w p . k v l s r
■ a e v v i s g l d g a y r l p l t v a a e r g ■ y g h k r s . a g
 w r k l s s l a w t g r t a c r . p s p l k g g c M A I K G L E Q
----.----+----.----+----.----+----.----+----.----+----.----+----.----+----.----+----.----+----.----+
s p p l q . r q s p r p t g g a t l r r r q f p s t h g y f t k l l
 i a s t t ■ e p k s p a y r r g n v t a a s l p i y p w l l d q a
 h r f n d d r a q v p r v a q r . g d g s f p p h i a ■ f p r s c

*031    7950    Mapped site HIN2
*032    7950    Mapped site HPA1
FII     7962      FII                  COOH
Z       7977      Z (orf-192)          M-NH3


H           D        M             H E S H    F  F    BC BHH  F     T            T   H E         A
A           D        N             G C C G    N  N    GF AAA  O     H            H   G C         C
E           E        L             I R R I    U  U    LR LEE  K     A            A   A R         Y
3           1        1             C 2 1 C    H  H    11 113  1     1            1   1 V         1
                                                          //
GCCGTTGAAAACCTCAGCCGTATCAGCAAAACGGCGGTGCCTGGTGCCGCCGCAATGGCCATTAACCGCGTTGCTTCATCCGCGATATCGCAGTCGGCGT
----.----+----.----+----.----+----.----+----.----+----.----+----.----+----.----+----.----+----.----+ 8100
CGGCAACTTTTGGAGTCGGCATAGTCGTTTTGCCGCCACGGACCACGGCGGCGTTACCGGTAATTGGCGCAACGAAGTAGGCGCTATAGCGTCAGCCGCA

 p l k t s a v s a k r r c l v p p q w p l t a l l h p r y r s r r
  r . k p q p y q q n g g a w c r r n g h . p r c f i r d i a v g v
A V E N L S R I S K T A V P G A A A M A I N R V A S S A I S Q S A S
----.----+----.----+----.----+----.----+----.----+----.----+----.----+----.----+----.----+----.----+
 g n f v e a t d a f r r h r t g g c h g n v a n s . g r y r l r r
p r q f g . g y . c f p p a q h r r l p w . g r q k ■ r s i a t p t
 a t s f r l r i l l v a t g p a a a i a ■ l r t a e d a i d c d a

Z         ...       CONTINUES IN PHASE THREE


                        RG  H                   A HES          A H                    HH  H
                        SA  P                   S ACC          S A                    AA  N
                        AL  A                   U ERR          U E                    EE  F
                        1Z  2                   1 321          1 3                    13  1
                         /                         /                                   /
CACAGGTTGCCCGTGAGACAAAGGTACGCCGGAAACTGGTAAAGGAAAGGGCCAGGCTGAAAAGGGCCACGGTCAAAAATCCGCAGGCCAGAATCAAAGT
----.----+----.----+----.----+----.----+----.----+----.----+----.----+----.----+----.----+----.----+ 8200
GTGTCCAACGGGCACTCTGTTTCCATGCGGCCTTTGACCATTTCCTTTCCCGGTCCGACTTTTCCCGGTGCCAGTTTTTAGGCGTCCGGTCTTAGTTTCA

h r l p v r q r y a g n w . r k g p g . k g p r s k i r r p e s k l
 t g c p . d k g t p e t g k g k g q a e k g h g q k s a g q n q s
 Q V A R E T K V R R K L V K E R A R L K R A T V K N P Q A R I K V
----.----+----.----+----.----+----.----+----.----+----.----+----.----+----.----+----.----+----.----+
 . l n g t l c l y a p f q y l f p g p q f p g r d f i r l g s d f
 v p q g h s l p v g s v p l p f p w a s f p w p . f d a p w f . l
d c t a r s v f t r r f s t f s l a l s f l a v t l f g c a l i l t

GALZ    8125      lambdagalZ           sl-att      EM
Z         ...     CONTINUES IN PHASE THREE
```

```
HH**NHS                    A          HT            H T HF     A        F      F          B
PIØØCPC                    L          HH            G H HN     C        O      N          B
AN33IAR                    U          AA            A A AU     Y        K      U          V
1234121                    1          11            1 1 1H     1        1      H          1
 /// //                                 /
TAACCGGGGGGATTTGCCCGTAATCAAGCTGGGTAATGCGCGGGTTGTCCTTTCGCGCCGCAGGCGTCGTAAAAAGGGGCAGCGTTCATCCCTGAAAGGT
----.----+----.----+----.----+----.----+----.----+----.----+----.----+----.----+----.----+----.----+ 8300
ATTGGCCCCCCTAAACGGGCATTAGTTCGACCCATTACGCGCCCAACAGGAAAGCGCGGCGTCCGCAGCATTTTTCCCCGTCGCAAGTAGGGACTTTCCA

 t  g  g  i  c  p  .  s  s  w  v  m  r  g  l  s  f  r  a  a  g  v  v  k  r  g  s  v  h  p  .  k  v
.  p  g  g  f  a  r  n  q  a  g  .  c  a  g  c  p  f  a  p  q  a  s  .  k  g  a  a  f  i  p  e  r  w
 N  R  G  D  L  P  V  I  K  L  G  N  A  R  V  V  L  S  R  R  R  R  R  K  K  G  Q  R  S  S  L  K  G
----.----+----.----+----.----+----.----+----.----+----.----+----.----+----.----+----.----+----.----+
n  v  p  p  i  q  g  y  d  l  q  t  i  r  p  n  d  k  r  a  a  p  t  t  f  l  p  l  t  .  g  q  f  t
 .  g  p  p  n  a  r  l  .  a  p  y  h  a  p  q  g  k  a  g  c  a  d  y  f  p  a  a  n  m  g  s  l
  l  r  p  s  k  g  t  i  l  s  p  l  a  r  t  t  r  e  r  r  l  r  r  l  f  p  c  r  e  d  r  f  p

*Ø33    82Ø1      Mapped site HIN2
*Ø34    82Ø1      Mapped site HPA1
Z        ...      CONTINUES IN PHASE THREE
```

```
T F  F           B    B              NSH  HT                    GCHH         N      F           B
T N  N           B    S              CCP  HH                    DFAP         S      N           B
H U  U           V    T              IRA  AA                    IREA         P      U           V
2 H  H           1    E              112  11                    2132         C      H           1
                                       //   /                     /
GGCGGCAGCGTGCTTGTGGTGGGTAACCGTCGTATTCCCGGCGCGTTTATTCAGCAACTGAAAAATGGCCGGTGGCATGTCATGCAGCGTGTGGCTGGGA
----.----+----.----+----.----+----.----+----.----+----.----+----.----+----.----+----.----+----.----+ 8400
CCGCCGTCGCACGAACACCACCCATTGGCAGCATAAGGGCCGCGCAAATAAGTCGTTGACTTTTTACCGGCCACCGTACAGTACGTCGCACACCGACCCT

 a  a  a  c  l  w  w  v  t  v  v  f  p  a  r  l  f  s  n  .  k  m  a  g  g  m  s  c  s  v  w  l  g
  r  q  r  a  c  g  g  .  p  s  y  s  r  r  v  y  s  a  t  e  k  w  p  v  a  c  h  a  a  c  g  w  e
G  G  S  V  L  V  V  G  N  R  R  I  P  G  A  F  I  Q  Q  L  K  N  G  R  W  H  V  M  Q  R  V  A  G  K
----.----+----.----+----.----+----.----+----.----+----.----+----.----+----.----+----.----+----.----+
 a  a  a  h  k  h  h  t  v  t  t  n  g  a  r  k  n  l  l  q  f  i  a  p  p  m  d  h  l  t  h  s  p
h  r  c  r  a  q  p  p  y  g  d  y  e  r  r  t  .  e  a  v  s  f  h  g  t  a  h  .  a  a  h  p  q  s
 p  p  l  t  s  t  t  p  l  r  r  i  g  p  a  n  i  .  c  s  f  f  p  r  h  c  t  m  c  r  t  a  p

Z        ...      CONTINUES IN PHASE THREE
```

```
                   B              H     H   NF               A                        M      X
                   S              P     G   SN               H                        B      M
                   T              H     I   PU               A                        O      N
                   X              1     C   BH               3                        2      1

AAAACCGTTACCCCATTGATGTGGTGAAAATCCCGATGGCGGTGCCGCTGACCACGGCGTTTAAACAAAATATTGAGCGGATACGGCGTGAACGTCTTCC
----.----+----.----+----.----+----.----+----.----+----.----+----.----+----.----+----.----+----.----+ 8500
TTTTGGCAATGGGGTAACTACACCACTTTTAGGGCTACCGCCACGGCGACTGGTGCCGCAAATTTGTTTTATAACTCGCCTATGCCGCACTTGCAGAAGG

k  t  v  t  p  l  m  w  .  k  s  r  w  r  c  r  .  p  r  r  l  n  k  i  l  s  g  y  g  v  n  v  f  r
 k  p  l  p  h  .  c  g  e  n  p  d  g  g  a  a  d  h  g  v  .  t  k  y  .  a  d  t  a  .  t  s  s
  N  R  Y  P  I  D  V  V  K  I  P  M  A  V  P  L  T  T  A  F  K  Q  N  I  E  R  I  R  R  E  R  L  P
----.----+----.----+----.----+----.----+----.----+----.----+----.----+----.----+----.----+----.----+
 f  v  t  v  g  n  i  h  h  f  d  r  h  r  h  r  q  g  r  r  k  f  l  i  n  l  p  y  p  t  f  t  k
  f  g  n  g  w  q  h  p  s  f  g  s  p  p  a  a  s  w  p  t  .  v  f  y  q  a  s  v  a  h  v  d  e
f  f  r  .  g  m  s  t  t  f  i  g  i  a  t  g  s  v  v  a  n  l  c  f  i  s  r  i  r  r  s  r  r  g

Z        ...      CONTINUES IN PHASE THREE
```

```
     BA          HF  FP*M     D BSH          F  UZ                    FS    R  H B           F
     BL          HN  NSØN     D BFG          O                        NF    S  G B           O
     VU          AU  UT3L     E VAU          K                        UA    A  U V           K
     11          1H  H151     1 1N2          1                        HN    1  2 1           1
                        /         //            /
GAAAGAGCTGGGCTATGCGCTGCAGCATCAACTGAGGATGGTAATAAAGCGATGAAACATACTGAACTCCGTGCAGCCGTACTGGATGCACTGGAGAAGC
----.----+----.----+----.----+----.----+----.----+----.----+----.----+----.----+----.----+----.----+ 8600
CTTTCTCGACCCGATACGCGACGTCGTAGTTGACTCCTACCATTATTTCGCTACTTTGTATGACTTGAGGCACGTCGGCATGACCTACGTGACCTCTTCG

 k  s  w  a  m  r  c  s  i  n  .  g  w  .  .  s  d  e  t  y  .  t  p  c  s  r  t  g  c  t  g  e  a
e  r  a  g  l  c  a  a  a  s  t  e  d  g  n  k  a  M  K  H  T  E  L  R  A  A  V  L  D  A  L  E  K  H
 K  E  L  G  Y  A  L  Q  H  Q  L  R  M  V  I  K  R  .  n  i  l  n  s  v  q  p  y  w  m  h  w  r  s
----.----+----.----+----.----+----.----+----.----+----.----+----.----+----.----+----.----+----.----+
r  f  l  q  a  i  r  q  l  m  l  q  p  h  y  y  l  s  s  v  y  q  v  g  h  l  r  v  p  h  v  p  s  a
 s  l  a  p  s  h  a  a  a  d  v  s  s  p  l  l  a  i  f  c  v  s  s  r  a  a  t  s  s  a  s  s  f
  f  s  s  p  .  a  s  c  c  .  s  l  i  t  i  f  r  h  f  m  s  f  e  t  c  g  y  q  i  c  q  l  l
```

```
*035      8524      Mapped site PST1
U         8552        U (orf-131)          M-NH3
Z         8552        Z                    COOH


      HNS                       N      M                   H          H          H    HH
      PCC                       S      N                   P          P          P    HA
      AIR                       P      L                   A          H          A    AE
      211                       B      1                   2          1          2    12
        /
ATGACACCGGGGCGACGTTTTTTGATGGTCGCCCCGCTGTTTTTGATGAGGCGGATTTTCCGGCAGTTGCCGTTATCTCACCGGCGCTGAATACACGGG
----.----+----.----+----.----+----.----+----.----+----.----+----.----+----.----+----.----+----.----+  8700
TACTGTGGCCCCGCTGCAAAAAACTACCAGCGGGGCGACAAAAACTACTCCGCCTAAAAGGCCGTCAACGGCAAATAGAGTGGCCGCGACTTATGTGCCC

 .  h  r  g  d  v  f  .  w  s  p  r  c  f  .  .  g  g  f  s  g  s  c  r  l  s  h  r  r  .  i  h  g
  D  T  G  A  T  F  F  D  G  R  P  A  V  F  D  E  A  D  F  P  A  V  A  V  Y  L  T  G  A  F  Y  T  G
*  t  p  g  r  r  f  l  *  v  a  p  l  f  l  *  r  r  i  f  r  q  l  p  f  i  s  p  a  l  n  t  r  a
----.----+----.----+----.----+----.----+----.----+----.----+----.----+----.----+----.----+----.----+
  h  c  r  p  s  t  k  q  h  d  g  r  q  k  q  h  p  p  n  e  p  l  q  r  k  d  .  r  r  q  i  c  p
c  s  v  p  a  v  n  k  s  p  r  g  a  t  k  s  s  a  s  k  g  a  t  a  t  .  r  v  p  a  s  y  v  p
 *  v  g  p  r  r  k  k  i  t  a  g  s  n  k  i  l  r  i  k  r  c  n  g  n  i  e  g  a  s  f  v  r

U         ...       CONTINUES TN PHASF TWO


      A      M     EBS           AF      T                 D   H   H  H S     A H       H     FH   NSH
      L      B     CBC           LN      A                 D   G   P  N F     L G       G     ON   CCP
      U      0     RVR           UU      Q                 E   I   A  F A     U U       U     KF   IRA
      1      2     211           1H      1                 1   C   2  1 N     1 2       2     11   112
                                                                                               /    //
CGAAGAGCTGGACAGCGATACCTGGCAGGCGGAGCTGCATATCGAAGTTTTCCTGCCTGCTCAGGTGCCGGATTCAGAGCTGGATGCGTGGATGGAGTCC
----.----+----.----+----.----+----.----+----.----+----.----+----.----+----.----+----.----+----.----+  8800
GCTTCTCGACCTGTCGCTATGGACCGTCCGCCTCGACGTATAGCTTCAAAAGGACGGACGAGTCCACGGCCTAAGTCTCGACCTACGCACCTACCTCAGG

r  r  a  g  q  r  y  l  a  g  g  a  a  y  r  s  f  p  a  c  s  g  a  g  f  r  a  g  c  v  d  g  v  p
 E  E  L  D  S  D  T  W  Q  A  F  I  H  I  E  V  F  L  P  A  Q  V  P  D  S  E  L  D  A  W  M  E  S
  k  s  w  t  a  i  p  g  r  r  s  c  i  s  k  f  s  c  l  l  r  c  r  i  q  s  w  *  r  g  w  s  p
----.----+----.----+----.----+----.----+----.----+----.----+----.----+----.----+----.----+----.----+
r  l  a  p  c  r  y  r  a  p  p  a  a  y  r  l  k  g  a  q  e  p  a  p  n  l  a  p  h  t  s  p  t
 s  s  s  s  l  s  v  q  c  a  s  s  c  i  s  t  k  r  g  a  .  t  g  s  e  s  s  s  a  h  i  s  d
a  f  l  q  v  a  i  g  p  l  r  l  q  *  d  f  n  e  q  r  s  l  h  r  i  .  l  q  i  r  p  h  l  g

U         ...       CONTINUES IN PHASE TWO


  F      H             H E NSH       H     *BM          B C BHHN F           H  S HT
  0      P             P C CCP       P     0CB          S F AAAS N           P  F HH
  K      A             H R IRA       H     310          T R LEEP U           A  A AA
  1      2             1 V 112       1     611          X 1 113B H           2  N 11
                            //               //              ///                   /
CGGATTTATCCGGTGATGAGCGATATCCCGGCACTGTCAGATTTGATCACCAGTATGGTGGCCAGCGGCTATGACTACCGGCGCGACGATGATGCGGGCT
----.----+----.----+----.----+----.----+----.----+----.----+----.----+----.----+----.----+----.----+  8900
GCCTAAATAGGCCACTACTCGCTATAGGGCCGTGACAGTCTAAACTAGTGGTCATACCACCGGTCGCCGATACTGATGGCCGCGCTGCTACTACGCCCGA

  d  l  s  g  d  e  r  y  p  g  t  v  r  f  d  h  q  y  g  g  q  r  l  .  l  p  a  r  r  .  c  g  l
R  I  Y  P  V  M  S  D  I  P  A  I  S  D  I  I  T  S  M  V  A  S  G  Y  D  Y  R  R  D  D  D  A  G  L
g  f  i  r  .  .  a  i  s  r  h  c  q  i  .  s  p  v  w  w  p  a  a  *  t  t  g  a  t  *  *  r  a
----.----+----.----+----.----+----.----+----.----+----.----+----.----+----.----+----.----+----.----+
g  s  k  d  p  s  s  r  y  g  p  v  t  l  n  s  .  w  y  p  p  w  r  s  h  s  g  a  r  r  h  h  p  s
 r  i  .  g  t  i  l  s  i  g  a  s  d  s  k  i  v  l  i  t  a  l  p  .  s  .  r  r  s  s  s  a  p
 p  n  i  r  h  h  a  i  d  r  c  q  .  i  q  d  g  t  h  h  g  a  a  i  v  v  p  a  v  i  i  r  a

*036      8844      Mapped site BCl1
U         ...       CONTINUES IN PHASE TWO
```

```
             M                       M     U          V  H   R                H        H   HHNSAA
             B                       N                   G   S                P        G   PPCCSV
             0                       L                   A   A                A        I   AHIRUA
             1                       1                   1   1                2        C   211112
                                                                                              ////
TGTGGAGTTCAGCCGATCTGACTTATGTCATTACCTATGAAATGTGAGGACGCTATGCCTGTACCAAATCCTACAATGCCGGTGAAAGGTGCCGGGACCA
----.----+----.----+----.----+----.----+----.----+----.----+----.----+----.----+----.----+----.----+ 9000
ACACCTCAAGTCGGCTAGACTGAATACAGTAATGGATACTTTACACTCCTGCGATACGGACATGGTTTAGGATGTTACGGCCACTTTCCACGGCCCTGGT

 v  e  f  s  r  s  d  l  c  h  y  l  .  n  v  r  t  l  c  l  y  q  i  l  q  c  r  .  k  v  p  g  p
  W  S  S  A  D  L  T  Y  V  I  T  Y  E  M  .  g  r  y  a  c  t  k  s  y  n  a  g  e  r  c  r  d  h
c  g  v  q  p  i  .  l  m  s  l  p  m  k  c  e  d  a  M  P  V  P  N  P  T  M  P  V  K  G  A  G  T  T
----.----+----.----+----.----+----.----+----.----+----.----+----.----+----.----+----.----+----.----+
  t  s  n  l  r  d  s  k  h  .  .  r  h  f  t  l  v  s  h  r  y  w  i  r  c  h  r  h  f  t  g  p  g
k  h  l  e  a  s  r  v  .  t  m  v  .  s  i  h  p  r  .  a  q  v  l  d  .  l  a  p  s  l  h  r  s  w
 q  p  t  .  g  i  q  s  i  d  n  g  i  f  h  s  s  a  i  g  t  g  f  g  v  i  g  t  f  p  a  p  v

U       8944       U                  COOH
V '     8955       V (orf-246)        M-NH3       AA
```

```
                        B          HTHH                H H*        T                A       A   NSH H
                        S          PHNN                G I0        H                C       C   CCP G
                        T          HAFF                A N3        A                P       Y   IRA A
                        E          1131                1 27        1                1       1   112 1
                                    ///                   /                                      //
CCCTGTGGGTTTATAAGGGGAGCGGTGACCCTTACGCGAATCCGCTTTCAGACGTTGACTGGTCGCGTCTGGCAAAAGTTAAAGACCTGACGCCCGGCGA
----.----+----.----+----.----+----.----+----.----+----.----+----.----+----.----+----.----+----.----+ 9100
GGGACACCCAAATATTCCCCTCGCCACTGGGAATGCGCTTAGGCGAAAGTCTGCAACTGACCAGCGCAGACCGTTTTCAATTTCTGGACTGCGGGCCGCT

p  c  g  f  i  r  g  a  v  t  l  t  r  i  r  f  q  t  l  t  g  r  v  w  q  k  l  k  t  .  r  p  a  n
 p  v  g  l  .  g  e  r  .  p  l  r  e  s  a  f  r  r  .  l  v  a  s  g  k  s  .  r  p  d  a  r  r
  L  W  V  Y  K  G  S  G  D  P  Y  A  N  P  L  S  D  V  D  W  S  R  L  A  K  V  K  D  L  T  P  G  E
----.----+----.----+----.----+----.----+----.----+----.----+----.----+----.----+----.----+----.----+
 g  q  p  n  i  l  p  a  t  v  r  v  r  i  r  k  .  v  n  v  p  r  t  q  c  f  n  f  v  q  r  g  a
  g  t  p  k  y  p  s  r  h  g  k  r  s  d  a  k  l  r  q  s  t  a  d  p  l  l  .  l  g  s  a  r  r
v  r  h  t  .  l  p  l  p  s  g  .  a  f  g  s  e  s  t  s  q  d  r  r  a  f  t  l  s  r  v  g  p  s

*037    9056       Mapped site HIN2
V        ...       CONTINUES IN PHASE THREE
```

```
    N   D H              A     TS                  M         HNS                  H           A
    S   D N              L     AF                  B         PCC                  P           L
    P   E F              U     QA                  0         AIR                  A           U
    B   1 1              1     1N                  2         211                  2           1
                                                               /
ACTGACCGCTGAGTCCTATGACGACAGCTATCTCGATGATGAAGATGCAGACTGGACTGCGACCGGGCAGGGGCAGAAATCTGCCGGAGATACCAGCTTC
----.----+----.----+----.----+----.----+----.----+----.----+----.----+----.----+----.----+----.----+ 9200
TGACTGGCGACTCAGGATACTGCTGTCGATAGAGCTACTACTTCTACGTCTGACCTGACGCTGGCCCGTCCCCGTCTTTAGACGGCCTCTATGGTCGAAG

  .  p  l  s  p  m  t  t  a  i  s  m  m  k  m  q  t  g  l  r  p  g  r  g  r  n  l  p  e  i  p  a  s
t  d  r  .  v  l  .  r  q  l  s  r  .  .  r  c  r  l  d  c  d  r  a  g  a  e  i  c  r  r  y  q  l  h
 L  T  A  E  S  Y  D  D  S  Y  L  D  D  E  D  A  D  W  T  A  T  G  Q  G  Q  K  S  A  G  D  T  S  F
----.----+----.----+----.----+----.----+----.----+----.----+----.----+----.----+----.----+----.----+
f  q  g  s  l  g  i  v  v  a  i  e  i  i  f  i  c  v  p  s  r  g  p  l  p  l  f  r  g  s  i  g  a  e
 v  s  r  q  t  r  h  r  c  s  d  r  h  h  l  h  l  s  s  q  s  r  a  p  a  s  i  q  r  l  y  w  s
  s  v  a  s  d  .  s  s  l  .  r  s  s  s  s  a  s  q  v  a  v  p  c  p  c  f  d  a  p  s  v  l  k

V        ...       CONTINUFS IN PHASE THREE
```

```
  S       H      NSH    F B   EF        HHFBE
  F       G      CCP    0 B   CN        HANBC
  A       U      IRA    K V   1U        AEUV1
  N       2      112    1 1   5H        12H15
                  //                      / /
ACGCTGGCGTGGATGCCCGGAGAGCAGGGGCAGCAGGCGCTGCTGGCGTGGTTTAATGAAGGCGATACCCGTGCCTATAAAATCCGCTTCCCGAACGGCA
----.----+----.----+----.----+----.----+----.----+----.----+----.----+----.----+----.----+----.----+ 930
TGCGACCGCACCTACGGGCCTCTCGTCCCCGTCGTCCGCGACGACCGCACCAAATTACTTCCGCTATGGGCACGGATATTTTAGGCGAAGGGCTTGCCGT

 r  w  r  g  c  p  e  s  r  g  s  r  r  c  w  r  g  l  m  k  a  i  p  v  p  i  k  s  a  s  r  t  a
  a  g  v  d  a  r  r  a  g  a  a  g  a  a  g  v  v  .  .  r  r  y  p  c  l  .  n  p  l  p  e  r  h
T  L  A  W  M  P  G  E  Q  G  Q  Q  A  L  L  A  W  F  N  E  G  D  T  R  A  Y  K  I  R  F  P  N  G  T
----.----+----.----+----.----+----.----+----.----+----.----+----.----+----.----+----.----+----.----+
 r  q  r  p  h  g  s  l  l  p  l  l  r  q  q  r  p  k  i  f  a  i  g  t  g  i  f  d  a  e  r  v  a
.  a  p  t  s  a  r  l  a  p  a  a  p  a  a  p  t  t  .  h  l  r  y  g  h  r  y  f  g  s  g  s  r  c
 v  s  a  h  i  g  p  s  c  p  c  c  a  s  s  a  h  n  l  s  p  s  v  r  a  .  l  i  r  k  g  f  p

V        ...       CONTINUES IN PHASE THREE
```

```
   T                    E         G                    HH   *BM         H          H        A  A
   A                    C         A                    PP   0CB         P          P        C  A
   Q                    1         L                    HH   3L0         H          H        Y  T
   1                    5         G                    11   811         1          1        1  2
                                                        /    //
CGGTCGATGTGTTCCGTGGCTGGGTCAGCAGTATCGGTAAGGCGGTGACGGCGAAGGAAGTGATCACCCGCACGGTGAAAGTCACCAATGTGGGACGTCC
----.----+----.----+----.----+----.----+----.----+----.----+----.----+----.----+----.----+----.----+ 9400
GCCAGCTACACAAGGCACCGACCCAGTCGTCATAGCCATTCCGCCACTGCCGCTTCCTTCACTAGTGGGCGTGCCACTTTCAGTGGTTACACCCTGCAGG

r  s  m  c  s  v  a  g  s  a  v  s  v  r  r  .  r  r  r  k  .  s  p  a  r  .  k  s  p  m  w  d  v  r
 g  r  c  v  p  w  l  g  q  q  y  r  .  g  g  d  g  e  g  s  d  h  p  h  g  e  s  h  q  c  g  t  s
  V  D  V  F  R  G  W  V  S  S  I  G  K  A  V  T  A  K  E  V  I  T  R  T  V  K  V  T  N  V  G  R  P
----.----+----.----+----.----+----.----+----.----+----.----+----.----+----.----+----.----+----.----+
 r  d  i  h  e  t  a  p  d  a  t  d  t  l  r  h  r  r  l  f  h  d  g  a  r  h  f  d  g  i  h  s  t
  p  r  h  t  g  h  s  p  .  c  y  r  y  p  p  s  p  s  p  l  s  .  g  c  p  s  l  .  w  h  p  v  d
v  t  s  t  n  r  p  q  t  l  l  i  p  l  a  t  v  a  f  s  t  i  v  r  v  t  f  t  v  l  t  p  r  g

GALG     9335        lambdagalG            sl-att      EM
*038     9361       Mapped site BCL1
V         ...       CONTINUES IN PHASE THREE
```

```
 T          M     F   M     NB  F     H            A       H              M       SH  H
 A          B     N   B     SB  N     P            C       G              N       DP  G
 Q          0     U   0     PV  U     A            Y       A              L       UH  I
 1          1     H   2     B1  H     2            1       1              1       11  A

GTCGATGGCAGAAGATCGCAGCACGGTAACAGCGGCAACCGGCATGACCGTGACGCCTGCCAGCACCTCGGTGGTGAAAGGGCAGAGCACCACGCTGACC
----.----+----.----+----.----+----.----+----.----+----.----+----.----+----.----+----.----+----.----+ 9500
CAGCTACCGTCTTCTAGCGTCGTGCCATTGTCGCCGTTGGCCGTACTGGCACTGCGGACGGTCGTGGAGCCACCACTTTCCCGTCTCGTGGTGCGACTGG

 r  w  q  k  i  a  a  r  .  q  r  q  p  a  .  p  .  r  l  p  a  p  r  w  .  k  g  r  a  p  r  .  p
v  d  g  r  r  s  q  h  g  n  s  g  n  r  h  d  r  d  a  c  q  h  l  g  g  e  r  a  e  h  h  a  d  r
 S  M  A  E  D  R  S  T  V  T  A  A  T  G  M  T  V  T  P  A  S  T  S  V  V  K  G  Q  S  T  T  L  T
----.----+----.----+----.----+----.----+----.----+----.----+----.----+----.----+----.----+----.----+
r  r  h  c  f  i  a  a  r  y  c  r  c  g  a  h  g  h  r  r  g  a  g  r  h  h  f  p  l  a  g  r  q  g
 t  s  p  l  l  d  c  c  p  l  l  p  l  r  c  s  r  s  a  q  w  c  r  p  p  s  l  a  s  c  w  a  s
  d  i  a  s  s  r  l  v  t  v  a  a  v  p  m  v  t  v  g  a  l  v  e  t  t  f  p  c  l  v  v  s  v

V         ...       CONTINUES IN PHASE THREE
```

```
   HH  M   H                     A                                         T         H            B
   AA  N   P                     L                                         R         P            B
   EE  L   A                     U                                         K         H            V
   13  1   2                     1                                         A         1            1
    /
GTGGCCTTCCAGCCGGAGGGCGTAACCGACAAGAGCTTTCGTGCGGTGTCTGCGGATAAAACAAAAGCCACCGTGTCGGTCAGTGGTATGACCATCACCG
----.----+----.----+----.----+----.----+----.----+----.----+----.----+----.----+----.----+----.----+ 9600
CACCGGAAGGTCGGCCTCCCGCATTGGCTGTTCTCGAAAGCACGCCACAGACGCCTATTTTGTTTTCGGTGGCACAGCCAGTCACCATACTGGTAGTGGC

 w  p  s  s  r  r  a  .  p  t  r  a  f  v  r  c  l  r  i  k  q  k  p  p  c  r  s  v  v  .  p  s  p
  g  l  p  a  g  g  r  n  r  q  e  l  s  c  g  v  c  g  .  n  k  s  h  r  v  g  q  w  y  d  h  h  r
V  A  F  Q  P  E  G  V  T  D  K  S  F  R  A  V  S  A  D  K  T  K  A  T  V  S  V  S  G  M  T  I  T  V
----.----+----.----+----.----+----.----+----.----+----.----+----.----+----.----+----.----+----.----+
 h  g  e  l  r  l  a  y  g  v  l  a  k  t  r  h  r  r  i  f  c  f  g  g  h  r  d  t  t  h  g  d  g
r  p  r  g  a  p  p  r  l  r  c  s  s  e  h  p  t  q  p  y  f  l  l  w  r  t  p  .  h  y  s  w  .  r
  t  a  k  w  g  s  p  t  v  s  l  l  k  r  a  t  d  a  s  l  v  f  a  v  t  d  t  l  p  i  v  m  v

TRKA     9577        lambdatrkA            sr          EM
V         ...       CONTINUES IN PHASE THREE
```

```
            F   P*       H*      H         H   B             F H           H                 V    H
            N   S0       I0      P         P   B             N P           P                      P
            U   T3       N4      A         A   V             U H           H                      A
            H   19       20      2         2   1             H 1           1                      2
                 /        /
TGAACGGCGTTGCTGCAGGCAAGGTCAACATTCCGGTTGTATCCGGTAATGGTGAGTTTGCTGCGGTTGCAGAAATTACCGTCACCGCCAGTTAATCCGG
----.----+----.----+----.----+----.----+----.----+----.----+----.----+----.----+----.----+----.----+ 9700
ACTTGCCGCAACGACGTCCGTTCCAGTTGTAAGGCCAACATAGGCCATTACCACTCAAACGACGCCAACGTCTTTAATGGCAGTGGCGGTCAATTAGGCC

.  t  a  l  l  q  a  r  s  t  f  r  l  y  p  v  m  v  s  l  l  r  l  q  k  l  p  s  p  p  v  n  p  e
 e  r  r  c  c  r  q  g  q  h  s  g  c  i  r  .  w  .  v  c  c  g  c  r  n  y  r  h  r  q  l  i  r
  N  G  V  A  A  G  K  V  N  I  P  V  V  S  G  N  G  E  F  A  A  V  A  E  I  T  V  T  A  S  .  s  g
----.----+----.----+----.----+----.----+----.----+----.----+----.----+----.----+----.----+----.----+
 h  v  a  n  s  c  a  l  d  v  n  r  n  y  g  t  i  t  l  k  s  r  n  c  f  n  g  d  g  g  t  l  g
  s  r  r  q  q  l  c  p  .  c  e  p  q  i  r  y  h  h  t  q  q  p  q  l  f  .  r  .  r  w  n  i  r
t  f  p  t  a  a  p  l  t  l  m  g  t  t  d  p  l  p  s  n  a  a  t  a  s  i  v  t  v  a  l  .  d  p
```

```
*039     9617      Mapped site PST1
*040     9626      Mapped site HIN2
V        9692      V                          COOH


H        G           H H                                                       FP* H       B       S
N                    N N                                                       NS0 H       B       F
F                    F F                                                       UT4 A       V       A
1                    3 1                                                       H11 1       1       N
                                                                                /
AGAGTCAGCGATGTTCCTGAAAACCGAATCATTTGAACATAACGGTGTGACCGTCACGCTTTCTGAACTGTCAGCCCTGCAGCGCATTGAGCATCTCGCC
----.----+----.----+----.----+----.----+----.----+----.----+----.----+----.----+----.----+----.----+ 9800
TCTCAGTCGCTACAAGGACTTTTGGCTTAGTAAACTTGTATTGCCACACTGGCAGTGCGAAAGACTTGACAGTCGGGACGTCGCGTAACTCGTAGAGCGG

  s  q  r  c  s  .  k  p  n  h  l  n  i  t  v  .  p  s  r  f  l  n  c  q  p  c  s  a  l  s  i  s  p
r  v  s  d  v  p  e  n  r  i  i  .  t  .  r  c  d  r  h  a  f  .  t  v  s  p  a  a  h  .  a  s  r  p
 e  s  a  M  F  L  K  T  E  S  F  E  H  N  G  V  T  V  T  L  S  E  L  S  A  L  Q  R  I  E  H  L  A
----.----+----.----+----.----+----.----+----.----+----.----+----.----+----.----+----.----+----.----+
s  l  .  r  h  e  q  f  g  f  .  k  f  ■  v  t  h  g  d  r  k  r  f  q  .  g  q  l  a  n  l  ■  e  g
 l  t  l  s  t  g  s  f  r  i  ■  q  v  y  r  h  s  r  .  a  k  q  v  t  l  g  a  a  c  q  a  d  r
  s  d  a  i  n  r  f  v  s  d  n  s  c  l  p  t  v  t  v  s  e  s  s  d  a  r  c  r  ■  s  c  r  a

G        9711      G (orf-140)                M-NH3
*041     9781      Mapped site PST1


                           H             H                   A       MH  H     HT                 F
                           N             P                   C       BG  P     HH                 O
                           F             A                   Y       OA  A     AA                 K
                           1             2                   1       21  2     11                 1
                                                                      /        /
CTGATGAAACGGCAGGCAGAACAGGCGGAGTCAGACAGCAACCGGAAGTTTACTGTGGAAGACGCCATCAGAACCGGCGCGTTTCTGGTGGCGATGTCCC
----.----+----.----+----.----+----.----+----.----+----.----+----.----+----.----+----.----+----.----+ 9900
GACTACTTTGCCGTCCGTCTTGTCCGCCTCAGTCTGTCGTTGGCCTTCAAATGACACCTTCTGCGGTAGTCTTGGCCGCGCAAAGACCACCGCTACAGGG

  .  .  n  g  r  q  n  r  r  s  q  t  a  t  g  s  l  l  w  k  t  p  s  e  p  a  r  f  w  w  r  c  p
 d  e  t  a  g  r  t  g  g  v  r  q  q  p  e  v  y  c  g  r  r  h  q  n  r  r  v  s  g  g  d  v  p
L  M  K  R  Q  A  E  Q  A  E  S  D  S  N  R  K  F  T  V  E  D  A  I  R  T  G  A  F  L  V  A  M  S  L
----.----+----.----+----.----+----.----+----.----+----.----+----.----+----.----+----.----+----.----+
  q  h  f  p  l  c  f  l  r  l  .  v  a  v  p  l  k  s  h  f  v  g  d  s  g  a  r  k  q  h  r  h  g
g  s  s  v  a  p  l  v  p  p  t  l  c  c  g  s  t  .  q  p  l  r  w  .  f  r  r  t  e  p  p  s  t  g
 r  i  f  r  c  a  s  c  a  s  d  s  l  l  r  f  n  v  t  s  s  a  ■  l  v  p  a  n  r  t  a  i  d

G        ...       CONTINUES IN PHASE THREE


                  S            MH                           E                   E S AHM
                  F            BG                           C                   C C SAN
                  A            OA                           O                   R R UEL
                  N            21                           B                   2 1 131
                                /                                                     /
TGTGGCATAACCATCCGCAGAAGACGCAGATGCCGTCCATGAATGAAGCCGTTAAACAGATTGAGCAGGAAGTGCTTACCACCTGGCCCACGGAGGCAAT
----.----+----.----+----.----+----.----+----.----+----.----+----.----+----.----+----.----+----.----+ 10000
ACACCGTATTGGTAGGCGTCTTCTGCGTCTACGGCAGGTACTTACTTCGGCAATTTGTCTAACTCGTCCTTCACGAATGGTGGACCGGGTGCCTCCGTTA

c  g  i  t  i  r  r  r  r  r  c  r  p  .  ■  k  p  l  n  r  l  s  r  k  c  l  p  p  g  p  r  r  q  f
 v  a  .  p  s  a  e  d  a  d  a  v  h  e  .  s  r  .  t  d  .  a  g  s  a  y  h  l  a  h  g  g  n
  W  H  N  H  P  Q  K  T  Q  M  P  S  M  N  E  A  V  K  Q  I  E  Q  E  V  L  T  T  W  P  T  E  A  I
----.----+----.----+----.----+----.----+----.----+----.----+----.----+----.----+----.----+----.----+
 q  p  ■  v  ■  r  l  l  r  l  h  r  g  h  i  f  g  n  f  l  n  l  l  f  h  k  g  g  p  g  r  l  c
  t  a  y  g  d  a  s  s  a  s  a  t  w  s  h  l  r  .  v  s  q  a  p  l  a  .  w  r  a  w  p  p  l
r  h  c  l  w  g  c  f  v  c  i  g  d  ■  f  s  a  t  l  c  i  s  c  s  t  s  v  v  q  g  v  s  a  i

G        ...       CONTINUES IN PHASE THREE
```

```
       RH                                          H  M        A HNSASHBAHHA
       SP                                          P  N        C PCCSDGUPGAS
       AA                                          H  L        Y AIRUUAUAIFU
       12                                          1  1        1 21111111J31
                                                                  /// ////
TTCTCATGCTGAAAACGTGGTGTACCGGCTGTCTGGTATGTATGAGTTTGTGGTGAATAATGCCCCTGAACAGACAGAGGACGCCGGGCCCGCAGAGCCT
----.----+----.----+----.----+----.----+----.----+----.----+----.----+----.----+----.----+----.----+ 10100
AAGAGTACGACTTTTGCACCACATGGCCGACAGACCATACATACTCAAACACCACTTATTACGGGGACTTGTCTGTCTCCTGCGGCCCGGGCGTCTCGGA

 l m l k t w c t g c l v c m s l w . i m p l n r q r t p g p q s l
f s c . k r g v p a v w y v . v c g e . c p . t d r g r r a r r a c
S H A E N V V Y R L S G M Y E F V V N N A P E Q T E D A G P A E P
----.----+----.----+----.----+----.----+----.----+----.----+----.----+----.----+----.----+----.----+
n r m s f v h h v p q r t h i l k h h i i g r f l c l v g p g c l r
 k e h q f r p t g a t q y t h t q p s y h g q v s l p r r a r l a
  e . a s f t t y r s d p i y s n t t f l a g s c v s s a p g a s g

G    ...    CONTINUES IN PHASE THREE
```

```
     XT T     AGD H         HT          T     F  HNSH
     M  A     LD  P         HH          T     O  PCCG
     N  Q     UE  H         AA          H     K  AIRU
     1  1     11  1         11          2     1  2112
      /        /             /                    //
GTTTCTGCGGGAAAGTGTTCGACGGTGAGCTGAGTTTTGCCCTGAAACTGGCGCGTGAGATGGGGCGACCCGACTGGCGTGCCATGCTTGCCGGGATGTC
----.----+----.----+----.----+----.----+----.----+----.----+----.----+----.----+----.----+----.----+ 10200
CAAAGACGCCCTTTCACAAGCTGCCACTCGACTCAAAACGGGACTTTGACCGCGCACTCTACCCCGCTGGGCTGACCGCACGGTACGAACGGCCCTACAG

f l r e s v r r . a e f c p e t g a . d g a t r l a c h a c r d v
 f c g k V F D G E L S F A L K L A R E M G R P D W R A M L A G M S
V S A G K C S T V S . v l p . n w r v r w g d p t g v p c l p g c h
----.----+----.----+----.----+----.----+----.----+----.----+----.----+----.----+----.----+----.----+
 n r r s l t r r h a s n q g s v p a h s p a v r s a h w a q r s t
q k q p f t n s p s s l k a r f s a r s i p r g s q r a m s a p i d
 t e a p f h e v t l q t k g q f q r t l h p s g v p t g h k g p h

I    10115    T (orf-144)    V-NH3
G    10130    G              COOH
```

```
   F      H      R         E          HNS   R  S  DH
   O      G      S         C          PCC   S  D  DG
   K      I      A         1          AIR   A  U  EI
   1      C      1         5          211   1  1  1A
                                        /
ATCCACGGAGTATGCCGACTGGCACCGCTTTTACAGTACCCATTATTTTCATGATGTTCTGCTGGATATGCACTTTTCCGGGCTGACGTACACCGTGCTC
----.----+----.----+----.----+----.----+----.----+----.----+----.----+----.----+----.----+----.----+ 10300
TAGGTGCCTCATACGGCTGACCGTGGCGAAAATGTCATGGGTAATAAAAGTACTACAAGACGACCTATACGTGAAAAGGCCCGACTGCATGTGGCACGAG

i h g v c r l a p l l q y p l f s . c s a g y a l . r a d v h r a q
 S T E Y A D W H R F Y S T H Y F H D V L L D M H F S G L T Y T V L
  p r s m p t g t a f t v p i i f m m f c w i c t f p g . r t p c s
----.----+----.----+----.----+----.----+----.----+----.----+----.----+----.----+----.----+----.----+
m w p t h r s a g s k c y g n n e h h e a p y a s k r a s t c r a
 d v s y a s q c r k . l v w . k . s t r s s i c k e p s v y v t s
. g r l i g v p v a k v t g m i k m i n q q i h v k g p q r v g h e

I    ...    CONTINUES IN PHASE TWO
```

```
     FMB H  A  N  S    E    MH HT      E  MS    M
     OBI P  V  S  F    C    NP HH      C  BF    B
     KON A  A  P  A    1    LA AA      O  OA    O
     111 2  3  B  N    5    12 11      B  2N    2
      /                        /
AGCCTGTTTTTCAGCGATCCGGATATGCATCCGCTGGATTTCAGTCTGCTGAACCGGCGCGAGGCTGACGAAGAGCCTGAAGATGATGTGCTGATGCAGA
----.----+----.----+----.----+----.----+----.----+----.----+----.----+----.----+----.----+----.----+ 10400
TCGGACAAAAAGTCGCTAGGCCTATACGTAGGCGACCTAAAGTCAGACGACTTGGCCGCGCTCCGACTGCTTCTCGGACTTCTACTACACGACTACGTCT

 p v f q r s g y a s a g f q s a e p a r g . r r a . r . c a d a e
S L F F S D P D M H P L D F S L L N R R E A D E E P E D D V L M Q K
 a c f s a i r i c i r w i s v c . t g a r l t k s l k m m c . c r
----.----+----.----+----.----+----.----+----.----+----.----+----.----+----.----+----.----+----.----+
. g t k . r d p y a d a p n . d a s g a r p q r l a q l h h a s a s
 l r n k l s g s i c g s s k l r s f r r s a s s s g s s s t s i c
  a q k e a i r i h m r q i e t q q v p a l s v f l r f i i h q h l

I    ...    CONTINUES IN PHASE TWO
```

```
  F      M    H                AH NSH                         NSHH         M F       H
  N      N    P                SA CCP                         CCPG         N O       G
  U      L    A                UE IRA                         IRAU         L K       U
  H      1    2                13 112                         1122         1 1       2
                                   //                              ///
AAGCGGCAGGGCTTGCCGGAGGTGTCCGCTTTGGCCCGGACGGGAATGAAGTTATCCCCGCTTCCCCGGATGTGGCGGACATGACGGAGGATGACGTAAT
----.----+----.----+----.----+----.----+----.----+----.----+----.----+----.----+----.----+----.----+ 10500
TTCGCCGTCCCGAACGGCCTCCACAGGCGAAACCGGGCCTGCCCTTACTTCAATAGGGGCGAAGGGGCCTACACCGCCTGTACTGCCTCCTACTGCATTA

 s  g  r  a  c  r  r  c  p  l  w  p  g  r  e  .  s  y  p  r  f  p  g  c  g  g  h  d  g  g  .  r  n
  A  A  G  L  A  G  G  V  R  F  G  P  D  G  N  E  V  I  P  A  S  P  D  V  A  D  M  T  E  D  D  V  M
k  r  q  g  l  p  e  v  s  a  l  a  r  t  g  m  k  l  s  p  l  p  r  m  w  r  t  .  r  r  m  t  .  c
----.----+----.----+----.----+----.----+----.----+----.----+----.----+----.----+----.----+----.----+
  l  p  l  a  q  r  l  h  g  s  q  g  p  r  s  h  l  .  g  r  k  g  p  h  p  p  c  s  p  p  h  r  l
f  a  a  p  s  a  p  p  t  r  k  p  g  s  p  f  s  t  i  g  a  e  g  s  t  a  s  m  v  s  s  s  t  i
 f  r  c  p  k  g  s  t  d  a  k  a  r  v  p  i  f  n  d  g  s  g  r  i  h  r  v  h  r  l  i  v  y

I       ...     CONTINUFS IN PHASE TWO
```

```
  F               BMM          H   H    H   T    H       M              S  H    H     FCH G   F
  O               IBN          N   P             P       B              F  N    G     NFA D   O
  K               NOL          F   A             A       O              A  F    U     URE I   K
  1               111          1   2             2       1              N  1    2     H13 2   1
                                                                                       /
GCTGATGACAGTATCAGAAGGGATCGCAGGAGGAGTCCGGTATGGCTGAACCGGTAGGCGATCTGGTCGTTGATTTGAGTCTGGATGCGGCCAGATTTGA
----.----+----.----+----.----+----.----+----.----+----.----+----.----+----.----+----.----+----.----+ 10600
CGACTACTGTCATAGTCTTCCCTAGCGTCCTCCTCAGGCCATACCGACTTGGCCATCCGCTAGACCAGCAACTAAACTCAGACCTACGCCGGTCTAAACT

a  d  d  s  i  r  r  d  r  r  r  s  p  v  w  l  n  r  .  a  i  w  s  l  i  .  v  w  m  r  p  d  l  t
 L  M  T  V  S  E  G  I  A  G  G  V  R  Y  G  .  t  g  r  r  s  g  r  .  f  e  s  g  c  g  q  i  .
  .  .  q  y  q  k  g  s  q  e  e  s  g  M  A  E  P  V  G  D  L  V  V  D  L  S  L  D  A  A  R  F  D
----.----+----.----+----.----+----.----+----.----+----.----+----.----+----.----+----.----+----.----+
 a  s  s  l  i  l  l  s  r  l  l  l  g  t  h  s  f  r  y  a  i  q  d  n  i  q  t  q  i  r  g  s  k
  s  i  v  t  d  s  p  i  a  p  p  t  r  y  p  q  v  p  l  r  d  p  r  q  n  s  d  p  h  p  w  i  q
h  q  h  c  y  .  f  p  d  c  s  s  d  p  i  a  s  g  t  p  s  r  t  t  s  k  l  r  s  a  a  l  n  s

H     10542     H (orf-853)     M-NH3
I     10546     T               COOH
```

```
         C BHHHH       A             RS                     N   F                   D    B        HH
         F AAAGN       C             SF                     S   N                   D    B        HA
         R LEEAF       Y             AA                     P   U                   E    V        AE
         1 11311       1             1N                     B   H                   1    1        12
            ///                      /
CGAGCAGATGGCCAGAGTCAGGCGTCATTTTTCTGGTACGGAAAGTGATGCGAAAAAAACAGCGGCAGTCGTTGAACAGTCGCTGAGCCGACAGGCGCTG
----.----+----.----+----.----+----.----+----.----+----.----+----.----+----.----+----.----+----.----+ 10700
GCTCGTCTACCGGTCTCAGTCCGCAGTAAAAAGACCATGCCTTTCACTACGCTTTTTTTGTCGCCGTCAGCAACTTGTCAGCGACTCGGCTGTCCGCGAC

  s  r  w  p  e  s  g  v  i  f  l  v  r  k  v  m  r  k  k  q  r  q  s  l  n  s  r  .  a  d  r  r  w
r  a  d  g  q  s  q  a  s  f  f  w  y  g  k  .  c  e  k  n  s  g  s  r  .  t  v  a  e  p  t  g  a  g
 E  Q  M  A  R  V  R  R  H  F  S  G  T  E  S  D  A  K  K  T  A  A  V  V  E  Q  S  L  S  R  Q  A  L
----.----+----.----+----.----+----.----+----.----+----.----+----.----+----.----+----.----+----.----+
v  l  l  h  g  s  d  p  t  m  k  r  t  r  f  t  i  r  f  f  c  r  c  d  n  f  l  r  q  a  s  l  r  q
 r  a  s  p  w  l  .  a  d  n  k  q  y  p  f  h  h  s  f  f  l  p  l  r  q  v  t  a  s  g  v  p  a
  s  c  i  a  l  t  l  r  .  k  e  p  v  s  l  s  a  f  f  v  a  a  t  t  s  c  d  s  l  r  c  a  s

H       ...     CONTINUES IN PHASE THREE
```

```
  F                                       FB             F    H                 C BHH   F A         B
  N                                       NB             N    P                 F AAA   N L         B
  U                                       UV             U    H                 R LEE   U U         V
  H                                       H1             H    1                 1 113   H 1         1
                                         /                                        //
GCTGCACAGAAAGCGGGGATTTCCGTCGGGCAGTATAAAGCCGCCATGCGTATGCTGCCTGCACAGTTCACCGACGTGGCCACGCAGCTTGCAGGCGGGC
----.----+----.----+----.----+----.----+----.----+----.----+----.----+----.----+----.----+----.----+ 10800
CGACGTGTCTTTCGCCCCTAAAGGCAGCCCGTCATATTTCGGCGGTACGCATACGACGGACGTGTCAAGTGGCTGCACCGGTGCGTCGAACGTCCGCCCG

  l  h  r  k  r  g  f  p  s  g  s  i  k  p  p  c  v  c  c  l  h  s  s  p  t  w  p  r  s  l  q  a  g
   c  t  e  s  g  d  f  r  r  a  v  .  s  r  h  a  y  a  a  c  t  v  h  r  r  g  h  a  a  c  r  r  a
A  A  Q  K  A  G  I  S  V  G  Q  Y  K  A  A  M  R  M  L  P  A  Q  F  T  D  V  A  T  Q  L  A  G  G  Q
----.----+----.----+----.----+----.----+----.----+----.----+----.----+----.----+----.----+----.----+
  s  c  l  f  r  p  n  g  d  p  l  i  f  g  g  h  t  h  q  r  c  l  e  g  v  h  g  r  l  k  c  a  p
p  q  v  s  l  p  s  k  r  r  a  t  y  l  r  w  a  y  a  a  q  v  t  .  r  r  p  w  a  a  q  l  r  a
 a  a  c  f  a  p  i  e  t  p  c  y  l  a  a  m  r  i  s  g  a  c  n  v  s  t  a  v  c  s  a  p  p

H       ...     CONTINUES IN PHASE THREE
```

```
      B     MB  E   F                       H    H      H   MB        F           H  HH   M
      B     BI  C   N                       N    P      G   BI        O           7  PP   B
      V     ON  1   U                       F    H      U   ON        K           0  AH   0
      1     11  5   H                       1    1      2   11        1           6  21   1
             /                                               /
AAAGTCCGTGGCTGATCCTGCTGCAACAGGGGGGGCAGGTGAAGGACTCCTTCGGCGGGATGATCCCCATGTTCAGGGGGCTTGCCGGTGCGATCACCCT
----.----+----.----+----.----+----.----+----.----+----.----+----.----+----.----+----.----+----.----+ 10900
TTTCAGGCACCGACTAGGACGACGTTGTCCCCCCCGTCCACTTCCTGAGGAAGCCGCCCTACTAGGGGTACAAGTCCCCCGAACGGCCACGCTAGTGGGA

k  v  r  g  .  s  c  c  n  r  g  g  r  .  r  t  p  s  a  g  .  s  p  c  s  g  g  l  p  v  r  s  p  c
 k  s  v  a  d  p  a  a  t  g  g  a  g  e  g  l  l  r  r  d  d  p  h  v  q  g  a  c  r  c  d  h  p
  S  P  W  L  I  L  L  Q  Q  G  G  Q  V  K  D  S  F  G  G  M  I  P  M  F  R  G  L  A  G  A  I  T  L
----.----+----.----+----.----+----.----+----.----+----.----+----.----+----.----+----.----+----.----+
 l  t  r  p  q  d  q  q  l  l  p  p  l  h  l  v  g  e  a  p  h  d  g  h  e  p  p  k  g  t  r  d  g
  f  d  t  a  s  g  a  a  v  p  p  a  p  s  p  s  r  r  r  s  s  g  w  t  .  p  a  q  r  h  s  .  g
c  l  g  h  s  i  r  s  c  c  p  p  c  t  f  s  e  k  p  p  i  i  g  m  n  l  p  s  a  p  a  i  v  r

H706  10883        lambdagalH706      sl-att       EM
H      ...        CONTINUES IN PHASE THREE
```

```
           A H       B     M       H      H         E S                                           AA
           S A       S     N       P      H         C C                                           SV
           U E       T     L       A      A         R R                                           UA
           1 3       X     1       2      1         2 1                                           12
                                                                                                   /
GCCGATGGTGGGGGCCACCTCGCTGGCGGTGGCGACCGGTGCGCTGGCGTATGCCTGGTATCAGGGCAACTCAACCCTGTCCGATTTCAACAAAACGCTG
----.----+----.----+----.----+----.----+----.----+----.----+----.----+----.----+----.----+----.----+ 11000
CGGCTACCACCCCCGGTGGAGCGACCGCCACCGCTGGCCACGCGACCGCATACGGACCATAGTCCCGTTGAGTTGGGACAGGCTAAAGTTGTTTTGCGAC

  r  w  w  g  p  p  r  w  r  w  r  p  v  r  w  r  m  p  g  i  r  a  t  q  p  c  p  i  s  t  k  r  w
a  d  g  g  g  h  l  a  g  g  g  d  r  c  a  g  v  c  l  v  s  g  q  l  n  p  v  r  f  q  q  n  a  g
 P  M  V  G  A  T  S  L  A  V  A  T  G  A  L  A  Y  A  W  Y  Q  G  N  S  T  L  S  D  F  N  K  T  L
----.----+----.----+----.----+----.----+----.----+----.----+----.----+----.----+----.----+----.----+
q  r  h  h  p  g  g  r  q  r  h  r  g  t  r  q  r  i  g  p  i  l  a  v  .  g  q  g  i  e  v  f  r  q
 a  s  p  p  p  w  r  a  p  p  p  s  r  h  a  p  t  h  r  t  d  p  c  s  l  g  t  r  n  .  c  f  a
  g  i  t  p  a  v  e  s  a  t  a  v  p  a  s  a  y  a  q  y  .  p  l  e  v  r  d  s  k  l  l  v  s

H      ...        CONTINUES IN PHASE THREE
```

```
       H                        M           AA           HNS  B   F                   E         H
       P                        B           SV           PCC  G   N                   C         N
       A                        0           UA           AIR  L   U                   P         F
       2                        1           12           211  1   H                   1         1
                                             /             /
GTCCTTTCCGGCAATCAGGCGGGACTGACGGCAGATCGTATGCTGGTCCTGTCCAGAGCCGGGCAGGCGGCAGGGCTGACGTTTAACCAGACCAGCGAGT
----.----+----.----+----.----+----.----+----.----+----.----+----.----+----.----+----.----+----.----+ 11100
CAGGAAAGGCCGTTAGTCCGCCCTGACTGCCGTCTAGCATACGACCAGGACAGGTCTCGGCCCGTCCGCCGTCCCGACTGCAAATTGGTCTGGTCGCTCA

 s  f  p  a  i  r  r  d  .  r  q  i  v  c  w  s  c  p  e  p  g  r  r  q  g  .  r  l  t  r  p  a  s
  p  f  r  q  s  g  g  t  d  g  r  s  y  a  g  p  v  q  s  r  a  g  g  r  a  d  v  .  p  d  q  r  v
V  L  S  G  N  Q  A  G  L  T  A  D  R  M  L  V  L  S  R  A  G  Q  A  A  G  L  T  F  N  Q  T  S  E  S
----.----+----.----+----.----+----.----+----.----+----.----+----.----+----.----+----.----+----.----+
  d  k  g  a  i  l  r  s  q  r  c  i  t  h  q  d  q  g  s  g  p  l  r  c  p  q  r  k  v  l  g  a  l
p  g  k  r  c  d  p  p  v  s  p  l  d  y  a  p  g  t  w  l  r  a  p  p  l  a  s  t  .  g  s  w  r  t
 t  r  e  p  l  .  a  p  s  v  a  s  r  i  s  t  r  d  l  a  p  c  a  a  p  s  v  n  l  w  v  l  s

H      ...        CONTINUES IN PHASE THREE
```

```
  D     H                  M        H  D    H                         HTF              HMM  S
  D     H                  N        G  D    P                         HHO              PNN  F
  E     A                  L        A  E    H                         AAK              ALL  A
  1     1                  1        1  1    1                         111              211  N
                                                                       //                /
CACTCAGCGCACTGGTTAAGGCGGGGGTAAGCGGTGAGGCTCAGATTGCGTCCATCAGCCAGAGTGTGGCGCGTTTCTCCTCTGCATCCGGCGTGGAGGT
----.----+----.----+----.----+----.----+----.----+----.----+----.----+----.----+----.----+----.----+ 11200
GTGAGTCGCGTGACCAATTCCGCCCCCATTCGCCACTCCGAGTCTAACGCAGGTAGTCGGTCTCACACCGCGCAAAGAGGAGACGTAGGCCGCACCTCCA

h  s  a  h  w  l  r  r  g  .  a  v  r  l  r  l  r  p  s  a  r  v  w  r  v  s  p  l  h  p  a  w  r  w
 t  q  r  t  g  .  g  g  g  k  r  .  g  s  d  c  v  h  q  p  e  c  g  a  f  l  l  c  i  r  r  g  g
  L  S  A  L  V  K  A  G  V  S  G  E  A  Q  I  A  S  I  S  Q  S  V  A  R  F  S  S  A  S  G  V  F  V
----.----+----.----+----.----+----.----+----.----+----.----+----.----+----.----+----.----+----.----+
 .  e  a  c  q  n  l  r  p  y  a  t  l  s  l  n  r  g  d  a  l  t  h  r  t  e  g  r  c  g  a  h  l
  v  .  r  v  p  .  p  p  p  l  r  h  p  e  s  q  t  w  .  g  s  h  p  a  n  r  r  q  m  r  r  p  p
d  s  l  a  s  t  l  a  p  t  l  p  s  a  .  i  a  d  m  l  w  l  t  a  r  k  e  e  a  d  p  t  s  t

H      ...        CONTINUES IN PHASE THREE
```

```
    T                      A        E      A  A                                A
    T                      L        C      C  A                                F
    H                      U        P      Y  T                                L
    1                      1        1      1  2                                3
GGACAAGGTCGCTGAAGCCTTCGGGAAGCTGACCACAGACCCGACGTCGGGGCTGACGGCGATGGCTCGCCAGTTCCATAACGTGTCGGCGGAGCAGATT
----.----+----.----+----.----+----.----+----.----+----.----+----.----+----.----+----.----+----.----+ 11300
CCTGTTCCAGCGACTTCGGAAGCCCTTCGACTGGTGTCTGGGCTGCAGCCCCGACTGCCGCTACCGAGCGGTCAAGGTATTGCACAGCCGCCTCGTCTAA

  t  r  s  l  k  p  s  g  s  .  p  q  t  r  r  r  g  .  r  r  w  l  a  s  s  i  t  c  r  r  s  r  l
 g  q  g  r  .  s  l  r  e  a  d  h  r  p  d  v  g  a  d  g  d  g  s  p  v  p  .  r  v  g  g  a  d  c
  D  K  V  A  E  A  F  G  K  L  T  T  D  P  T  S  G  L  T  A  M  A  R  Q  F  H  N  V  S  A  E  Q  I
----.----+----.----+----.----+----.----+----.----+----.----+----.----+----.----+----.----+----.----+
 h  v  l  d  s  f  g  e  p  l  q  g  c  v  r  r  r  p  q  r  r  h  s  a  l  e  m  v  h  r  r  l  l  n
  p  c  p  r  q  l  r  r  s  a  s  w  l  g  s  t  p  a  s  p  s  p  e  g  t  g  y  r  t  p  p  a  s
   s  l  t  a  s  a  k  p  f  s  v  v  s  g  v  d  p  s  v  a  i  a  r  w  n  w  l  t  d  a  s  c  i

H      ...      CONTINUES IN PHASE THREE


           D      F      H   B       HNS            FM         HF                      E       F
           D      N      P   B       PCC            NN         AN                      C       N
           E      U      A   V       AIR            UL         EU                      P       U
           1      H      2   1       211            H1         3H                      1       H
                                       /             /
GCGTATGTTGCTCAGTTGCAGCGTTCCGGCGATGAAGCCGGGGCATTGCAGGCGGCGAACGAGGCCGCAACGAAAGGGTTTGATGACCAGACCCGCCGCC
----.----+----.----+----.----+----.----+----.----+----.----+----.----+----.----+----.----+----.----+ 11400
CGCATACAACGAGTCAACGTCGCAAGGCCGCTACTTCGGCCCCGTAACGTCCGCCGCTTGCTCCGGCGTTGCTTTCCCAAACTACTGGTCTGGGCGGCGG

  r  m  l  l  s  c  s  v  p  a  m  k  p  g  h  c  r  r  r  t  r  p  q  r  k  g  l  m  t  r  p  a  a
   v  c  c  s  v  a  a  f  r  r  .  s  r  g  i  a  g  g  e  r  g  r  n  e  r  v  .  .  p  d  p  p  p
 A  Y  V  A  Q  L  Q  R  S  G  D  E  A  G  A  L  Q  A  A  N  E  A  A  T  K  G  F  D  D  Q  T  R  R  L
----.----+----.----+----.----+----.----+----.----+----.----+----.----+----.----+----.----+----.----+
  r  i  n  s  l  q  l  t  g  a  i  f  g  p  c  q  l  r  r  v  l  g  c  r  f  p  k  i  v  l  g  a  a
 q  t  h  q  e  t  a  a  n  r  r  h  l  r  p  m  a  p  p  s  r  p  r  l  s  l  t  q  h  g  s  g  g  g
  a  y  t  a  .  n  c  r  e  p  s  s  a  p  a  n  c  a  a  f  s  a  a  v  f  p  n  s  s  w  v  r  r

H      ...      CONTINUES IN PHASE THREE


            S          E  E S                 HT          S       H             F                  S
            D          C  C C                 HH          F       G             O                  F
            U          P  R R                 AA          A       U             K                  A
            1          1  2 1                 11          N       2             1                  N
                                               /
TGAAAGAGAACATGGGCACGCTGGAGACCTGGGCAGACAGGACTGCGCGGGCATTCAAATCCATGTGGGATGCGGTGCTGGATATTGGTCGTCCTGATAC
----.----+----.----+----.----+----.----+----.----+----.----+----.----+----.----+----.----+----.----+ 11500
ACTTTCTCTTGTACCCGTGCGACCTCTGGACCCGTCTGTCCTGACGCGCCCGTAAGTTTAGGTACACCCTACGCCACGACCTATAACCAGCAGGACTATG

 .  k  r  t  w  a  r  w  r  p  g  q  t  g  l  r  g  h  s  n  p  c  g  m  r  c  w  i  l  v  v  l  i  p
  e  r  e  h  g  h  a  g  d  l  g  r  q  d  c  a  g  i  q  i  h  v  g  c  g  a  g  y  w  s  s  .  y
  K  E  N  M  G  T  L  E  T  W  A  D  R  T  A  R  A  F  K  S  M  W  D  A  V  L  D  I  G  R  P  D  T
----.----+----.----+----.----+----.----+----.----+----.----+----.----+----.----+----.----+----.----+
  q  f  l  v  h  a  r  q  l  g  p  c  v  p  s  r  p  c  e  f  g  h  p  i  r  h  q  i  n  t  t  r  i
   s  l  s  c  p  c  a  p  s  r  p  l  c  s  q  a  p  m  .  i  w  t  p  h  p  a  p  y  q  d  d  q  y
 r  f  s  f  m  p  v  s  s  v  q  a  s  l  v  a  r  a  n  l  d  m  h  s  a  t  s  s  i  p  r  g  s  v

H      ...      CONTINUES IN PHASE THREE


  T H           B M          F                           H      MH  H             F HH**        HT
  H H           B N          N                           N      SH  G             O PI00        HH
  A A           V L          U                           F      TA  U             K AN44        AA
  1 1           1 1          H                           1      11  2             1 1223        11
                                                                                     ///         /
CGCGCAGGAGATGCTGATTAAGGCAGAGGCTGCGTATAAGAAAGCAGACGACATCTGGAATCTGCGCAAGGATGATTATTTTGTTAACGATGAAGCGCGG
----.----+----.----+----.----+----.----+----.----+----.----+----.----+----.----+----.----+----.----+ 11600
GCGCGTCCTCTACGACTAATTCCGTCTCCGACGCATATTCTTTCGTCTGCTGTAGACCTTAGACGCGTTCCTACTAATAAAACAATTGCTACTTCGCGCC

  r  r  r  c  .  l  r  q  r  l  r  i  r  k  q  t  t  s  g  i  c  a  r  m  i  i  l  l  t  m  k  r  g
 r  a  g  d  a  d  .  g  r  g  c  v  .  e  s  r  r  h  l  e  s  a  q  g  .  l  f  c  .  r  .  s  a  g
  A  Q  E  M  L  I  K  A  E  A  A  Y  K  K  A  D  D  I  W  N  L  R  K  D  D  Y  F  V  N  D  E  A  R
----.----+----.----+----.----+----.----+----.----+----.----+----.----+----.----+----.----+----.----+
 g  r  l  l  h  q  n  l  c  l  s  r  i  l  f  c  v  v  d  p  i  q  a  l  i  i  i  k  n  v  i  f  r  p
  r  a  p  s  a  s  .  p  l  p  q  t  y  s  l  l  r  c  r  s  d  a  c  p  h  n  n  q  .  r  h  l  a
   a  c  s  i  s  i  l  a  s  a  a  y  l  f  a  s  s  m  q  f  r  r  l  s  s  .  k  t  l  s  s  a  r

*042   11585      Mapped site HIN2
*043   11585      Mapped site HPA1
H      ...      CONTINUES IN PHASE THREE
```

```
  HT      H   M         F AH         H       F              D   EF    H     B            MHFEEF
  HH      G   B         O SA         H       N              D   CN    N     B            SHNCCN
  AA      U   O         K UE         A       U              E   1U    F     V            TAU11U
  11      2   1         1 13         1       H              1   5H    1     1            11H55H
  /                                                                                         /
GCGCGTTACTGGGATGATCGTGAAAAGGCCCGTCTTGCGCTTGAAGCCGCCCGAAAGAAGGCTGAGCAGCAGACTCAACAGGACAAAAATGCGCAGCAGC
----.----+----.----+----.----+----.----+----.----+----.----+----.----+----.----+----.----+----.----+ 11700
CGCGCAATGACCCTACTAGCACTTTTCCGGGCAGAACGCGAACTTCGGCGGGCTTTCTTCCGACTCGTCGTCTGAGTTGTCCTGTTTTTACGCGTCGTCG

 r  v  t  g  m  i  v  k  r  p  v  l  r  l  k  p  p  e  r  r  l  s  s  r  l  n  r  t  k  m  r  s  s
  a  l  l  g  .  s  .  k  g  p  s  c  a  .  s  r  p  k  e  g  .  a  a  d  s  t  g  q  k  c  a  a  a
A  R  Y  W  D  D  R  E  K  A  R  L  A  L  E  A  A  R  K  K  A  E  Q  Q  T  Q  Q  D  K  N  A  Q  Q  Q
----.----+----.----+----.----+----.----+----.----+----.----+----.----+----.----+----.----+----.----+
 r  t  v  p  i  i  t  f  l  g  t  k  r  k  f  g  g  s  l  l  s  l  l  l  s  l  l  v  f  i  r  l  l
p  a  n  s  p  h  d  h  f  p  g  d  q  a  q  l  r  g  f  s  p  q  a  a  s  e  v  p  c  f  h  a  a  a
 a  r  .  q  s  s  r  s  f  a  r  r  a  s  s  a  a  r  f  f  a  s  c  c  v  .  c  s  l  f  a  c  c

H       ...      CONTINUES IN PHASE THREE
```

```
   HB B                        M          H  MB            F   P*A NF    H
   GB B                        N          H  BB            N   S0C SN    G
   AV V                        L          A  OV            U   T4Y PU    A
   11 1                        1          1  21            H   141 BH    1
                                                                /
AGAGCGATACCGAAGCGTCACGGCTGAAATATACCGAAGAGGCGCAGAAGGCTTACGAACGGCTGCAGACGCCGCTGGAGAAATATACCGCCCGTCAGGA
----.----+----.----+----.----+----.----+----.----+----.----+----.----+----.----+----.----+----.----+ 11800
TCTCGCTATGGCTTCGCAGTGCCGACTTTATATGGCTTCTCCGCGTCTTCCGAATGCTTGCCGACGTCTGCGGCGACCTCTTTATATGGCGGGCAGTCCT

r  a  i  p  k  r  h  g  .  n  i  p  k  r  r  r  r  l  t  n  g  c  r  r  r  w  r  n  i  p  p  v  r  k
 e  r  y  r  s  v  t  a  e  i  y  r  r  g  a  e  g  l  r  t  a  a  d  a  a  g  e  i  y  r  p  s  g
  S  D  T  E  A  S  R  L  K  Y  T  E  E  A  Q  K  A  Y  E  R  L  Q  T  P  L  E  K  Y  T  A  R  Q  E
----.----+----.----+----.----+----.----+----.----+----.----+----.----+----.----+----.----+----.----+
 l  a  i  g  f  r  .  p  q  f  i  g  f  l  r  l  l  s  v  f  p  q  l  r  r  q  l  f  i  g  g  t  l
  s  r  y  r  l  t  v  a  s  i  y  r  l  p  a  s  p  k  r  v  a  a  s  a  a  p  s  i  y  r  g  d  p
c  l  s  v  s  a  d  r  s  f  y  v  s  s  a  c  f  a  .  s  r  s  c  v  g  s  s  f  y  v  a  r  .  s

*044   11767     Mapped site PST1
H       ...      CONTINUES IN PHASE THREE
```

```
          M                          P*                       F F                                 H
          B                          S0                       N N                                 G
          O                          T4                       U U                                 A
          2                          15                       H H                                 1
                                     /
AGAACTGAACAAGGCACTGAAAGACGGGAAAATCCTGCAGGCGGATTACAACACGCTGATGGCGGCGGCGAAAAAGGATTATGAAGCGACGCTGAAAAAG
----.----+----.----+----.----+----.----+----.----+----.----+----.----+----.----+----.----+----.----+ 11900
TCTTGACTTGTTCCGTGACTTTCTGCCCTTTTAGGACGTCCGCCTAATGTTGTGCGACTACCGCCGCCGCTTTTTCCTAATACTTCGCTGCGACTTTTTC

  n  .  t  r  h  .  k  t  g  k  s  c  r  r  i  t  t  r  .  w  r  r  r  k  r  i  m  k  r  r  .  k  s
r  t  e  q  g  t  e  r  r  e  n  p  a  g  g  l  q  h  a  d  g  g  g  e  k  g  l  .  s  d  a  e  k  a
 F  L  N  K  A  L  K  D  G  K  I  L  Q  A  D  Y  N  T  L  M  A  A  A  K  K  D  Y  F  A  T  I  K  K
----.----+----.----+----.----+----.----+----.----+----.----+----.----+----.----+----.----+----.----+
f  f  q  v  l  c  q  f  v  p  f  d  q  l  r  i  v  v  r  q  h  r  r  r  f  l  i  i  f  r  r  q  f  l
 l  v  s  c  p  v  s  l  r  s  f  g  a  p  p  n  c  c  a  s  p  p  p  s  f  p  n  h  l  s  a  s  f
  s  s  f  l  a  s  f  s  p  f  i  r  c  a  s  .  l  v  s  i  a  a  a  f  f  s  .  s  a  v  s  f  f

*045   11839     Mapped site PST1
H       ...      CONTINUES IN PHASE THREE
```

```
                                 M P*     B  S   HM F  E         H     H       H
                                 B V0     B  D   GB N  C         G     P       G
                                 O U4     V  U   IO U  1         A     A       A
                                 1 16     1  1   A2 H  5         1     2       1

CCGAAACAGTCCAGCGTGAAGGTGTCTGCGGGCGATCGTCAGGAAGACAGTGCTCATGCTGCCCTGCTGACGCTTCAGGCAGAACTCCGGACGCTGGAGA
----.----+----.----+----.----+----.----+----.----+----.----+----.----+----.----+----.----+----.----+ 12000
GGCTTTGTCAGGTCGCACTTCCACAGACGCCCGCTAGCAGTCCTTCTGTCACGAGTACGACGGGACGACTGCGAAGTCCGTCTTGAGGCCTGCGACCTCT

 r  n  s  p  a  .  r  c  l  r  a  i  v  r  k  t  v  l  m  l  p  c  .  r  f  r  q  n  s  g  r  w  r
  e  t  v  q  r  e  g  v  c  g  r  s  s  g  r  q  c  s  c  c  p  a  d  a  s  g  r  t  p  d  a  g  e
P  K  Q  S  S  V  K  V  S  A  G  D  R  Q  E  D  S  A  H  A  A  L  L  T  L  Q  A  E  L  R  T  I  E  K
----.----+----.----+----.----+----.----+----.----+----.----+----.----+----.----+----.----+----.----+
 r  f  l  g  a  h  l  h  r  r  a  i  t  l  f  v  t  s  m  s  g  q  q  r  k  l  c  f  e  p  r  q  l
a  s  v  t  w  r  s  p  t  q  p  r  d  d  p  l  c  h  e  h  q  g  a  s  a  e  p  l  v  g  s  a  p  s
 g  f  c  d  l  t  f  t  d  a  p  s  r  .  s  s  l  a  .  a  a  r  s  v  s  .  a  s  s  r  v  s  s
```

```
*046   11937    Mapped site PVU1
H       ...     CONTINUES IN PHASE THREE


     SN*H                   E   F   HHNHS  B              H         TM  RM          F   H        N
     PS0P                   C   N   HACPC  B              N         HN  SN          N   H        S
     HP4A                   1   U   AEIAR  V              F         AL  AL          U   A        P
     1C72                   5   H   12121  1              1         11  11          H   1        B
      ///                              //                                /
AGCATGCCGGAGCAAATGAGAAAATCAGCCAGCAGCGCCGGGATTTGTGGAAGGCGGAGAGTCAGTTCGCGGTACTGGAGGAGGCGGCGCAACGTCGCCA
----.----+----.----+----.----+----.----+----.----+----.----+----.----+----.----+----.----+----.----+ 12100
TCGTACGGCCTCGTTTACTCTTTTAGTCGGTCGTCGCGGCCCTAAACACCTTCCGCCTCTCAGTCAAGCGCCATGACCTCCTCCGCCGCGTTGCAGCGGT

s  m  p  e  q  m  r  k  s  a  s  s  a  g  i  c  g  r  r  r  v  s  s  r  y  w  r  r  r  r  n  v  a  s
 a  c  r  s  k  .  e  n  q  p  a  a  p  g  f  v  e  g  g  e  s  v  r  g  t  g  g  g  g  a  t  s  p
  H  A  G  A  N  E  K  I  S  Q  Q  R  R  D  L  W  K  A  E  S  Q  F  A  V  L  E  F  A  A  Q  R  R  Q
----.----+----.----+----.----+----.----+----.----+----.----+----.----+----.----+----.----+----.----+
l  m  g  s  c  i  l  f  d  a  l  l  a  p  i  q  p  l  r  l  t  l  e  r  y  q  l  l  r  r  l  t  a
 a  h  r  l  l  h  s  f  .  g  a  a  g  p  n  t  s  p  p  s  d  t  r  p  v  p  p  p  p  a  v  d  g
f  c  a  p  a  f  s  f  i  l  w  c  r  r  s  k  h  f  a  s  l  .  n  a  t  s  s  s  a  a  c  r  r  w

*047   12006    Mapped site SPH1
H       ...     CONTINUES IN PHASE THREE


PA*                  E        H                    HR B     N  PA*  F
VL0                  C        H                    GS B     S  VL0  N
UU4                  1        A                    AA V     P  UU4  U
218                  5        1                    11 1     B  219  H
 //                                                              //
GCTGTCTGCACAGGAGAAATCCCTGCTGGCGCATAAAGATGAGACGCTGGAGTACAAACGCCAGCTGGCTGCACTTGGCGACAAGGTTACGTATCAGGAG
----.----+----.----+----.----+----.----+----.----+----.----+----.----+----.----+----.----+----.----+ 12200
CGACAGACGTGTCCTCTTTAGGGACGACCGCGTATTTCTACTCTGCGACCTCATGTTTGCGGTCGACCGACGTGAACCGCTGTTCCAATGCATAGTCCTC

 c  l  h  r  r  n  p  c  w  r  i  k  m  r  r  w  s  t  n  a  s  w  l  h  l  a  t  r  l  r  i  r  s
a  v  c  t  g  e  i  p  a  g  a  .  r  .  d  a  g  v  q  t  p  a  g  c  t  w  r  q  g  y  v  s  g  a
 L  S  A  Q  E  K  S  L  L  A  H  K  D  E  T  L  E  Y  K  R  Q  L  A  A  L  G  D  K  V  T  Y  Q  E
----.----+----.----+----.----+----.----+----.----+----.----+----.----+----.----+----.----+----.----+
l  q  r  c  l  l  f  g  q  q  r  m  f  i  l  r  q  l  v  f  a  l  q  s  c  k  a  v  l  n  r  i  l  l
 a  t  q  v  p  s  i  g  a  p  a  y  l  h  s  a  p  t  c  v  g  a  p  q  v  q  r  c  p  .  t  d  p
 s  d  a  c  s  f  d  r  s  a  c  l  s  s  v  s  s  y  l  r  w  s  a  a  s  p  s  l  t  v  y  .  s

*048   12101    Mapped site PVU2
*049   12164    Mapped site PVU2
H       ...     CONTINUES IN PHASE THREE


 HH      T H     HFE         B H      E   F          B    ASHF                  HNS            H
 HA      H H     HNC         B N      C   N          B    SFAN                  PCC            P
 AE      A A     AU1         V F      1   U          V    UAEU                  AIR            A
 12      1 1     1H5         1 3      5   H          1    1N3H                  211            2
                   /                                                              /
CGCCTGAACGCGCTGGCGCAGCAGGCGGATAAATTCGCACAGCAGCAACGGGCAAAACGGGCCGCCATTGATGCGAAAAGCCGGGGGCTGACTGACCGGC
----.----+----.----+----.----+----.----+----.----+----.----+----.----+----.----+----.----+----.----+ 12300
GCGGACTTGCGCGACCGCGTCGTCCGCCTATTTAAGCGTGTCGTCGTTGCCCGTTTTGCCCGGCGGTAACTACGCTTTTCGGCCCCCGACTGACTGGCCG

 a  .  t  r  w  r  s  r  r  i  n  s  h  s  s  n  g  q  n  g  p  p  l  m  r  k  a  g  g  .  l  t  g
  p  e  r  a  g  a  a  g  g  .  i  r  t  a  a  t  g  k  t  g  r  h  .  c  e  k  p  g  a  d  .  p  a
R  L  N  A  L  A  Q  Q  A  D  K  F  A  Q  Q  Q  R  A  K  R  A  A  I  D  A  K  S  R  G  L  T  D  R  Q
----.----+----.----+----.----+----.----+----.----+----.----+----.----+----.----+----.----+----.----+
 a  q  v  r  q  r  l  l  r  i  f  e  c  l  l  l  p  c  f  p  g  g  n  i  r  f  a  p  p  q  s  v  p
a  g  s  r  a  p  a  a  p  p  y  i  r  v  a  a  v  p  l  v  p  r  w  q  h  s  f  g  p  a  s  q  g  a
 r  r  f  a  s  a  c  c  a  s  l  n  a  c  c  c  r  a  f  r  a  a  m  s  a  f  l  r  p  s  v  s  r

H       ...     CONTINUES IN PHASE THREE
```

```
                           HH                      N         HH                         F  ES      F
                           HA                      S         HA                         C  CC      N
                           AE                      P         AE                         P  RR      U
                           12                      B         12                         1  21      H

AGGCAGAACGGGAAGCCACGGAACAGCGCCTGAAGGAACAGTATGGCGATAATCCGCTGGCGCTGAATAACGTCATGTCAGAGCAGAAAAAGACCTGGGC
----.----+----.----+----.----+----.----+----.----+----.----+----.----+----.----+----.----+----.----+ 12400
TCCGTCTTGCCCTTCGGTGCCTTGTCGCGGACTTCCTTGTCATACCGCTATTAGGCGACCGCGACTTATTGCAGTACAGTCTCGTCTTTTTCTGGACCCG

r  q  n  g  k  p  r  n  s  a  .  r  n  s  ■  a  i  i  r  w  r  .  i  t  s  c  q  s  r  k  r  p  g  r
 g  r  t  g  s  h  g  t  a  p  e  g  t  v  w  r  .  s  a  g  a  e  .  r  h  v  r  a  e  k  d  l  g
  A  F  R  E  A  T  E  Q  R  L  K  E  Q  Y  G  D  N  P  L  A  L  N  N  V  M  S  E  Q  K  K  T  W  A
----.----+----.----+----.----+----.----+----.----+----.----+----.----+----.----+----.----+----.----+
l  c  f  p  f  g  r  f  l  a  q  i  f  l  i  a  i  i  r  q  r  q  i  v  d  h  .  l  l  f  l  g  p
 p  l  v  p  l  w  p  v  a  g  s  p  v  t  h  r  y  d  a  p  a  s  y  r  .  t  l  a  s  f  s  r  p
c  a  s  r  s  a  v  s  c  r  r  f  s  c  y  p  s  l  g  s  a  s  f  l  t  ■  d  s  c  f  f  v  q  a

H          ...        CONTTNUES TN PHASE THREE
```

```
     E      A   MT       H         SHH F    H                        HH M
     C      L   BH       G         TAA 0    P                        HA B
     P      U   0A       U         UEE K    A                        AE 0
     1      1   21       2         113 1    2                        12 2
                                     //
GGCTGAAGACCAGCTTCGCGGGAACTGGATGGCAGGCCTGAAGTCCGGCTGGAGTGAGTGGGAAGAGAGCGCCACGGACAGTATGTCGCAGGTAAAAAGT
----.----+----.----+----.----+----.----+----.----+----.----+----.----+----.----+----.----+----.----+ 12500
CCGACTTCTGGTCGAAGCGCCCTTGACCTACCGTCCGGACTTCAGGCCGACCTCACTCACCCTTCTCTCGCGGTGCCTGTCATACAGCGTCCATTTTTCA

 l  k  t  s  f  a  g  t  g  w  q  a  .  s  p  a  g  v  s  g  k  r  a  p  r  t  v  c  r  r  .  k  v
g  .  r  p  a  s  r  e  l  d  g  r  p  e  v  r  l  e  .  v  g  r  e  r  h  g  q  y  v  a  g  k  k  c
 A  E  D  Q  L  R  G  N  W  M  A  G  L  K  S  G  W  S  E  W  E  E  S  A  T  D  S  M  S  Q  V  K  S
----.----+----.----+----.----+----.----+----.----+----.----+----.----+----.----+----.----+----.----+
r  s  f  v  l  k  a  p  v  p  h  c  a  q  l  g  a  p  t  l  p  f  l  a  g  r  v  t  h  r  l  y  f  t
 p  q  l  g  a  e  r  s  s  s  p  l  g  s  t  r  s  s  h  t  p  l  s  r  w  p  c  y  t  a  p  l  f
  a  s  s  w  s  r  p  f  q  i  a  p  r  f  d  p  q  l  s  h  s  s  l  a  v  s  l  i  d  c  t  f  l

H          ...        CONTINUES IN PHASF THREF
```

```
  F       E HB                      S     F           H                   H HF A         B
  N       C GB                      F     N           P                   P HN L         B
  U       P EV                      A     U           A                   H AU U         V
  H       1 21                      N     H           2                   1 1H 1         1

GCAGCCACGCAGACCTTTGATGGTATTGCACAGAATATGGCGGCGATGCTGACCGGCAGTGAGCAGAACTGGCGCAGCTTCACCCGTTCCGTGCTGTCCA
----.----+----.----+----.----+----.----+----.----+----.----+----.----+----.----+----.----+----.----+ 12600
CGTCGGTGCGTCTGGAAACTACCATAACGTGTCTTATACCGCCGCTACGACTGGCCGTCACTCGTCTTGACCGCGTCGAAGTGGGCAAGGCACGACAGGT

 q  p  r  r  p  l  ■  v  l  h  r  i  w  r  r  c  .  p  a  v  s  r  t  g  a  a  s  p  v  p  c  c  p
  s  h  a  d  l  .  w  y  c  t  e  y  g  g  d  a  d  r  q  .  a  e  l  a  q  l  h  p  f  r  a  v  h
A  A  T  Q  T  F  D  G  I  A  Q  N  M  A  A  M  L  T  G  S  E  Q  N  W  R  S  F  T  R  S  V  L  S  M
----.----+----.----+----.----+----.----+----.----+----.----+----.----+----.----+----.----+----.----+
 c  g  r  l  g  k  i  t  n  c  l  i  h  r  r  h  q  g  a  t  l  l  v  p  a  a  e  g  t  g  h  q  g
h  l  w  a  s  r  q  h  y  q  v  s  y  p  p  s  a  s  r  c  h  a  s  s  a  c  s  .  g  n  r  a  t  w
 a  a  v  c  v  k  s  p  i  a  c  f  i  a  a  i  s  v  p  l  s  c  f  q  r  l  k  v  r  e  t  s  d

H          ...        CONTINUES IN PHASE THREE
```

```
                  A                                     F    HH     B      F      HF    H     T S
                  F                                     N    HA     B      0      GN    H     H F
                  L                                     U    AE     V      K      AU    A     A A
                  2                                     H    12     1      1      1H    1     1 N

TGATGACAGAAATTCTGCTTAAGCAGGCAATGGTGGGGATTGTCGGGAGTATCGGCAGCGCCATTGGCGGGGCTGTTGGTGGCGGCGCATCCGCGTCAGG
----.----+----.----+----.----+----.----+----.----+----.----+----.----+----.----+----.----+----.----+ 12700
ACTACTGTCTTTAAGACGAATTCGTCCGTTACCACCCCTAACAGCCCTCATAGCCGTCGCGGTAACCGCCCCGACAACCACCGCCGCGTAGGCGCAGTCC

.  .  q  k  f  c  l  s  r  q  w  w  g  l  s  g  v  s  a  a  p  l  a  g  l  l  v  a  a  h  p  r  q  a
 d  d  r  n  s  a  .  a  g  n  g  g  d  c  r  e  y  r  q  r  h  w  r  g  c  w  w  r  r  i  r  v  r
  M  T  E  I  L  L  K  Q  A  M  V  G  I  V  G  S  I  G  S  A  I  G  G  A  V  G  G  G  A  S  A  S  G
----.----+----.----+----.----+----.----+----.----+----.----+----.----+----.----+----.----+----.----+
 h  h  c  f  n  q  k  l  l  c  h  h  p  n  d  p  t  d  a  a  g  n  a  p  s  n  t  a  a  c  g  r  .
  s  s  l  f  e  a  .  a  p  l  p  p  s  q  r  s  y  r  c  r  w  q  r  p  q  q  h  r  r  ■  r  t  l
■  i  v  s  i  r  s  l  c  a  i  t  p  i  t  p  l  i  p  l  a  ■  p  p  a  t  p  p  p  a  d  a  d  p

H          ...        CONTINUES IN PHASE THREE
```

```
  R B     B HFN F BF           M  H           H  F      N  H
  S B     G ANS N GN           N  P           P  N      S  P
  A V     L EUP U LU           L  A           A  U      P  H
  1 1     1 3HB H 1H           1  2           2  H      B  1
             /
CGGTACAGCCATTCAGGCCGCTGCGGCGAAATTCCATTTTGCAACCGGAGGATTTACGGGAACCGGCGGCAAATATGAGCCAGCGGGGATTGTTCACCGT
----.----+----.----+----.----+----.----+----.----+----.----+----.----+----.----+----.----+----.----+ 12800
GCCATGTCGGTAAGTCCGGCGACGCCGCTTTAAGGTAAAACGTTGGCCTCCTAAATGCCCTTGGCCGCCGTTTATACTCGGTCGCCCCTAACAAGTGGCA

  v  q  p  f  r  p  l  r  r  n  s  i  l  q  p  e  d  l  r  e  p  a  a  n  m  s  q  r  g  l  f  t  v
 r  y  s  h  s  g  r  c  g  e  i  p  f  c  n  r  r  i  y  g  n  r  r  q  i  .  a  s  g  d  c  s  p  w
G  T  A  I  Q  A  A  A  A  K  F  H  F  A  T  G  G  F  T  G  T  G  G  K  Y  E  P  A  G  I  V  H  R
----.----+----.----+----.----+----.----+----.----+----.----+----.----+----.----+----.----+----.----+
a  t  c  g  n  l  g  s  r  r  f  e  m  k  c  g  s  s  k  r  s  g  a  a  f  i  l  w  r  p  n  n  v  t
 r  y  l  w  e  p  r  q  p  s  i  g  n  q  l  r  l  i  .  p  f  r  r  c  i  h  a  l  p  s  q  e  g
  p  v  a  m  .  a  a  a  a  f  n  w  k  a  v  p  p  n  v  p  v  p  p  l  y  s  g  a  p  i  t  .  r

H      ...     CONTINUES IN PHASE THREE


  M       HM                  H  B        H  S  H        HTF       H           R H
  B       PN                  P  G        N  F  P        HHN       P           S P
  0       HL                  A  L        F  A  A        AAU       A           A A
  2       11                  2  1        1  N  2        11H       2           1 2
                                                         //
GGTGAGTTTGTCTTCACGAAGGAGGCAACCAGCCGGATTGGCGTGGGGAATCTTTACCGGCTGATGCGCGGCTATGCCACCGGCGGTTATGTCGGTACAC
----.----+----.----+----.----+----.----+----.----+----.----+----.----+----.----+----.----+----.----+ 12900
CCACTCAAACAGAAGTGCTTCCTCCGTTGGTCGGCCTAACCGCACCCCTTAGAAATGGCCGACTACGCGCCGATACGGTGGCCGCCAATACAGCCATGTG

  v  s  l  s  s  r  r  r  q  p  a  g  l  a  w  g  i  f  t  g  .  c  a  a  m  p  p  a  v  m  s  v  h
 .  v  c  l  h  e  g  g  n  q  p  d  w  r  g  e  s  l  p  a  d  a  r  l  c  h  r  r  l  c  r  y  t
G  E  F  V  F  T  K  F  A  T  S  R  I  G  V  G  N  L  Y  R  L  M  R  G  Y  A  T  G  G  Y  V  G  T  P
----.----+----.----+----.----+----.----+----.----+----.----+----.----+----.----+----.----+----.----+
  t  l  k  d  e  r  l  l  c  g  a  p  n  a  h  p  i  k  v  p  q  h  a  a  i  g  g  a  t  i  d  t  c
h  h  t  q  r  .  s  p  p  l  w  g  s  q  r  p  s  d  k  g  a  s  a  r  s  h  w  r  r  n  h  r  y  v
 p  s  n  t  k  v  f  s  a  v  l  r  i  p  t  p  f  r  .  r  s  i  r  p  .  a  v  p  p  .  t  p  v

H      ...     CONTINUES IN PHASF THREE


NS F        B HH      A  HNS                              H            B    AA H
CC N        B GP      C  PCC                              P            B    SV P
IR U        V AA      Y  AIR                              H            V    UA A
11 H        1 12      1  211                              1            1    12 2
/                         /                                                  /
CGGGCAGCATGGCAGACAGCCGGTCGCAGGCGTCCGGGACGTTTGAGCAGAATAACCATGTGGTGATTAACAACGACGGCACGAACGGGCAGATAGGTCC
----.----+----.----+----.----+----.----+----.----+----.----+----.----+----.----+----.----+----.----+ 13000
GCCCGTCGTACCGTCTGTCGGCCAGCGTCCGCAGGCCCTGCAAACTCGTCTTATTGGTACACCACTAATTGTTGCTGCCGTGCTTGCCCGTCTATCCAGG

r  a  a  w  q  t  a  g  r  r  r  p  g  r  l  s  r  i  t  m  w  .  l  t  t  t  a  r  t  g  r  .  v  r
 g  q  h  g  r  q  p  v  a  g  v  r  d  v  .  a  e  .  p  c  g  d  .  q  r  r  h  e  r  a  d  r  s
  G  S  M  A  D  S  R  S  Q  A  S  G  T  F  E  Q  N  N  H  V  V  I  N  N  D  G  T  N  G  Q  I  G  P
----.----+----.----+----.----+----.----+----.----+----.----+----.----+----.----+----.----+----.----+
 r  a  a  h  c  v  a  p  r  l  r  g  p  r  k  l  l  i  v  m  h  h  n  v  v  v  a  r  v  p  l  y  t
  p  c  c  p  l  c  g  t  a  p  t  r  s  t  q  a  s  y  g  h  p  s  .  c  r  r  c  s  r  a  s  l  d
g  p  l  m  a  s  l  r  d  c  a  d  p  v  n  s  c  f  l  w  t  t  i  l  l  s  p  v  f  p  c  i  p  g

H      ...     CONTINUES IN PHASE THREE


  F                  AH      HS            S                     HH M  H          MH
  N                  SA      GD            F                     AA N  P
  U                  UE      IU            A                     EE L  A
  H                  13      C1            N                     13 1  2
                              /                                  /                 /
GGCTGCTCTGAAGGCGGTGTATGACATGGCCCGCAAGGGTGCCCGTGATGAAATTCAGACACAGATGCGTGATGGTGGCCTGTTCTCCGGAGGTGGACGA
----.----+----.----+----.----+----.----+----.----+----.----+----.----+----.----+----.----+----.----+ 13100
CCGACGAGACTTCCGCCACATACTGTACCGGGCGTTCCCACGGGCACTACTTTAAGTCTGTGTCTACGCACTACCACCGGACAAGAGGCCTCCACCTGCT

  l  l  .  r  r  c  m  t  w  p  a  r  v  p  v  m  k  f  r  h  r  c  v  m  v  a  c  s  p  e  v  d  d
g  c  s  e  g  g  v  .  h  g  p  q  g  c  p  .  .  n  s  d  t  d  a  .  w  w  p  v  l  r  r  w  t  M
 A  A  L  K  A  V  Y  D  M  A  R  K  G  A  R  D  E  I  Q  T  Q  M  R  D  G  G  L  F  S  G  G  G  R
----.----+----.----+----.----+----.----+----.----+----.----+----.----+----.----+----.----+----.----+
r  s  s  q  l  r  h  i  v  h  g  a  l  t  g  t  i  f  n  l  c  l  h  t  i  t  a  q  e  g  s  t  s  s
 p  q  e  s  p  p  t  h  c  p  g  c  p  h  g  h  h  f  e  s  v  s  a  h  h  h  g  t  r  r  l  h  v
  a  a  r  f  a  t  y  s  m  a  r  l  p  a  r  s  s  i  .  v  c  i  r  s  p  p  r  n  e  p  p  p  r

M      13100     M (orf-109)     M-NH3
H      13100     H               COOH
```

```
 E X  N  M          NSH H          AFA               H            HD     HH
 C M  S  B          CCP G          SOV               H            PD     HA
 P N  P  O          IRA U          UKA               A            HE     AE
 1 1  B  2          112 2          112               1            11     12
                    //              /
TGAAGACCTTCCGCTGGAAAGTGAAACCCGGTATGGATGTGGCTTCGGTCCCTTCTGTAAGAAAGGTGCGCTTTGGTGATGGCTATTCTCAGCGAGCGCC
----.----+----.----+----.----+----.----+----.----+----.----+----.----+----.----+----.----+----.----+ 13200
ACTTCTGGAAGGCGACCTTTCACTTTGGGCCATACCTACACCGAAGCCAGGGAAGACATTCTTTCCACGCGAAACCACTACCGATAAGAGTCGCTCGCGG

 e d l p l e s e t r y g c g f g p f c k k g a l w . w l f s a s a
  K T F R W K V K P G M D V A S V P S V R K V R F G D G Y S Q R A P
 . r p s a g k . n p v w m w l r s l l . e r c a l v m a i l s e r l
----.----+----.----+----.----+----.----+----.----+----.----+----.----+----.----+----.----+----.----+
 s s r g s s l s v r y p h p k p g k q l f p a s q h h s n e a l a
 i f v k r q f t f g p i s t a e t g e t l f t r k p s p . e . r a g
 h l g e a p f h f g t h i h s r d r r y s l h a k t i a i r l s r

M      ...     CONTINUES IN PHASE TWO
```

```
 HBNS               R              H M  M        HH    RB   H          S    H    M
 PGCC               S              G N  N        AA    SS   N          D    G    B
 ALIR               A              A L  L        EE    AT   F          U    I    O
 2111               1              1 1  1        13    1X   1          1    A    2
  //                                              /
TGCCGGGCTGAATGCCAACCTGAAAAACGTACAGCGTGACGCTTTCTGTCCCCCGTGAGGAGGCCACGGTACTGGAGTCGTTTCTGGAAGAGCACGGGGGC
----.----+----.----+----.----+----.----+----.----+----.----+----.----+----.----+----.----+----.----+ 13300
ACGGCCCGACTTACGGTTGGACTTTTTGCATGTCGCACTGCGAAAGACAGGGGGCACTCCTCCGGTGCCATGACCTCAGCAAAGACCTTCTCGTGCCCCCG

 c r a e c q p e n v q r d a f c p p . g g h g t g v v s g r a r g l
  A G L N A N L K T Y S V T L S V P R E E A T V L E S F L E E H G G
  p g . m p t . k r t a . r f l s p v r r p r y w s r f w k s t g a
----.----+----.----+----.----+----.----+----.----+----.----+----.----+----.----+----.----+----.----+
 q r a s h w g s f t c r s a k q g g h p p w p v p t t e p l a r p
  a p s f a l r f v y l t v s e t g r s s a v t s s d n r s s c p p
 r g p q i g v q f r v a h r k r d g t l l g r y q l r k q f l v p a

M      ...     CONTINUES IN PHASE TWO
```

```
                 A F   H        F        B      MH H        BT          F
                 C N   G        N        S      SH P        BH          N
                 Y U   A        U        T      TA H        VA          U
                 1 H   1        H        E      11 1        11          H

TGGAAATCCTTTCTGTGGACGCCGCCTTATGAGTGGCGGCAGATAAAGGTGACCTGCGCAAAATGGTCGTCGCGGGTCAGTATGCTGCGTGTTGAGTTCA
----.----+----.----+----.----+----.----+----.----+----.----+----.----+----.----+----.----+----.----+ 13400
ACCTTTAGGAAAGACACCTGCGGCGGAATACTCACCGCCGTCTATTTCCACTGGACGCGTTTTACCAGCAGCGCCCAGTCATACGACGCACAACTCAAGT

  e i l s v d a a l . v a a d k g d l r k m v v a g q y a a c . v q
 W K S F L W T P P Y E W R Q I K V T C A K W S S R V S M L R V E F S
  g n p f c g r r l m s g g r . r . p a q n g r r g s v c c v l s s
----.----+----.----+----.----+----.----+----.----+----.----+----.----+----.----+----.----+----.----+
 s s i r e t s a a k h t a a s l p s r r l i t t a p . y a a h q t .
  q f d k r h v g g . s h r c i f t v q a f h d d r t l i s r t s n
  p f g k q p r r r i l p p l y l h g a c f p r r p d t h q t n l e

M      ...     CONTINUES IN PHASE TWO
```

```
 H          S     M L H     FH                                   C H G   S    H
 H          F         P     CP                                   F A D   D    G
 A          A         H     RA                                   R E I   U    I
 1          N         1     V2                                   1 3 2   1    A

GCGCAGAGTTTGAACAGGTGGTGAACTGATGCAGGATATCCGGCAGGAAACACTGAATGAATGCACCCGTGCGGAGCAGTCGGCCAGCGTGGTGCTCTGG
----.----+----.----+----.----+----.----+----.----+----.----+----.----+----.----+----.----+----.----+ 13500
CGCGTCTCAAACTTGTCCACCACTTGACTACGTCCTATAGGCCGTCCTTTGTGACTTACTTACGTGGGCACGCCTCGTCAGCCGGTCGCACCACGAGACC

 r r v . t g g e l M Q D I R Q E T L N E C T R A E Q S A S V V L W
  A E F E Q V V N . c r i s g r k h . m n a p v r s s r p a w c s g
 a q s l n r w . t d a g y p a g n t e . m h p c g a v g q r g a l g
----.----+----.----+----.----+----.----+----.----+----.----+----.----+----.----+----.----+----.----+
  r l t q v p p s s i c s i r c s v s f s h v r a s c d a l t t s q
 l a s n s c t t f q h l i d p l f c q i f a g t r l l r g a h h e p
  a c l k f l h h v s a p y g a p f v s h i c g h p a t p w r p a r

M      13426      M                 COOH
L      13429      L (orf-232)       M-NH3
```

```
    T  M                                                         H   HB    HES
    A  N                                                         P   PS    PCC
    Q  L                                                         H   AT    HRR
    1  1                                                         1   2E    121
                                                                           /
GAAATCGACCTGACAGAGGTCGGTGGAGAACGTTATTTTTTCTGTAATGAGCAGAACGAAAAAGGTGAGCCGGTCACCTGGCAGGGGCGACAGTATCAGC
----.----+----.----+----.----+----.----+----.----+----.----+----.----+----.----+----.----+----.----+ 13600
CTTTAGCTGGACTGTCTCCAGCCACCTCTTGCAATAAAAAAGACATTACTCGTCTTGCTTTTTCCACTCGGCCAGTGGACCGTCCCCGCTGTCATAGTCG

E  I  D  L  T  E  V  G  G  E  R  Y  F  F  C  N  E  Q  N  E  K  G  E  P  V  T  W  Q  G  R  Q  Y  Q  P
 k  s  t  .  q  r  s  v  e  n  v  i  f  s  v  m  s  r  t  k  k  v  s  r  s  p  g  r  g  d  s  i  s
  n  r  p  d  r  g  r  w  r  t  l  f  f  l  .  .  a  e  r  k  r  .  a  g  h  l  a  g  a  t  v  s  a
----.----+----.----+----.----+----.----+----.----+----.----+----.----+----.----+----.----+----.----+
s  i  s  r  v  s  t  p  p  s  r  .  k  k  q  l  s  c  f  s  f  p  s  g  t  v  q  c  p  r  c  y  .
 f  d  v  q  c  l  d  t  s  f  t  i  k  e  t  i  l  l  v  f  f  t  l  r  d  g  p  l  p  s  l  i  l
p  f  r  g  s  l  p  r  h  l  v  n  n  k  r  y  h  a  s  r  f  l  h  a  p  .  r  a  p  a  v  t  d  a

L      ...      CONTINUES IN PHASE ONE


                                         H       R  TH                        RH     B     HNSH
                                         G       S  HH                        SP     S     PCCG
                                         I       A  AA                        AH     T     AIRU
                                         C       1  11                        11     E     2112
                                                                                             //
CGTATCCCATTCAGGGGAGCGGTTTTGAACTGAATGGCAAAGGCACCAGTACGCGCCCCACGCTGACGGTTTCTAACCTGTACGGTATGGTCACCGGGAT
----.----+----.----+----.----+----.----+----.----+----.----+----.----+----.----+----.----+----.----+ 13700
GCATAGGGTAAGTCCCCTCGCCAAAACTTGACTTACCGTTTCCGTGGTCATGCGCGGGGTGCGACTGCCAAAGATTGGACATGCCATACCAGTGGCCCTA

 Y  P  I  Q  G  S  G  F  E  L  N  G  K  G  T  S  T  R  P  T  L  T  V  S  N  L  Y  G  M  V  T  G  M
r  i  p  f  r  g  a  v  l  n  .  m  a  k  a  p  v  r  a  p  r  .  r  f  l  t  c  t  v  w  s  p  g  w
 v  s  h  s  g  e  r  f  .  t  e  w  q  r  h  q  y  a  p  h  a  d  g  f  .  p  v  r  y  g  h  r  d
----.----+----.----+----.----+----.----+----.----+----.----+----.----+----.----+----.----+----.----+
g  y  g  m  .  p  l  p  k  s  s  f  p  l  p  v  l  v  r  g  v  s  v  t  e  l  r  y  p  i  t  v  p  i
 r  i  g  n  l  p  a  t  k  f  q  i  a  f  a  g  t  r  a  g  r  q  r  n  r  v  q  v  t  h  d  g  p
  t  d  w  e  p  s  r  n  q  v  s  h  c  l  c  w  y  a  g  w  a  s  p  k  .  g  t  r  y  p  .  r  s

L      ...      CONTINUES IN PHASE ONE


         F     HM                   AA H G              S       H             F    HH*
         O     NB                   SV P A              F       G             O    PI0
         K     FO                   UA A L              A       U             K    HN5
         1     12                   12 2 K              N       2             1    120
                                     /                                              //
GGCGGAAGATATGCAGAGTCTGGTCGGCGGAACGGTGGTCCGGCGTAAGGTTTACGCCCGTTTTCTGGATGCGGTGAACTTCGTCAACGGAAACAGTTAC
----.----+----.----+----.----+----.----+----.----+----.----+----.----+----.----+----.----+----.----+ 13800
CCGCCTTCTATACGTCTCAGACCAGCCGCCTTGCCACCAGGCCGCATTCCAAATGCGGGCAAAAGACCTACGCCACTTGAAGCAGTTGCCTTTGTCAATG

 A  E  D  M  Q  S  L  V  G  G  T  V  V  R  R  K  V  Y  A  R  F  L  D  A  V  N  F  V  N  G  N  S  Y
  r  k  i  c  r  v  w  s  a  e  r  w  s  g  v  r  f  t  p  v  f  w  m  r  .  t  s  s  t  e  t  v  t
g  g  r  y  a  e  s  g  r  r  n  g  g  p  a  .  g  l  r  p  f  s  g  c  g  e  l  r  q  r  k  q  l  r
----.----+----.----+----.----+----.----+----.----+----.----+----.----+----.----+----.----+----.----+
 a  s  s  i  c  l  r  t  p  p  v  t  t  r  r  l  t  .  a  r  k  r  s  a  t  f  k  t  l  p  f  l  .
h  r  f  i  h  l  t  q  d  a  s  r  h  d  p  t  l  n  v  g  t  k  q  i  r  h  v  e  d  v  s  v  t  v
p  p  l  y  a  s  d  p  r  r  f  p  p  g  a  y  p  k  r  g  n  e  p  h  p  s  s  r  .  r  f  c  n

GALK   13743      lambdagalK          sl-att      EM
*050   13785      Mapped site HIN2
L      ...        CONTINUES IN PHASE ONE


   MB  HM E         *BM  NF  H   H            F      D   B HT            H    RM
   BI  PN C         0CB  SN  P   H            N      D   B HH            P    SN
   ON  AL O         5LO  PU  H   A            U      E   V AA            H    AL
   11  21 B         111  BH  1   1            H      1   1 11            1    11
   /                 //                                    /
GCCGATCCGGAGCAGGAGGTGATCAGCCGCTGGCGCATTGAGCAGTGCAGCGAACTGAGCGCGGTGAGTGCCTCCTTTGTACTGTCCACGCCGACGGAAA
----.----+----.----+----.----+----.----+----.----+----.----+----.----+----.----+----.----+----.----+ 13900
CGGCTAGGCCTCGTCCTCCACTAGTCGGCGACCGCGTAACTCGTCACGTCGCTTGACTCGCGCCACTCACGGAGGAAACATGACAGGTGCGGCTGCCTTT

A  D  P  E  Q  E  V  I  S  R  W  R  I  E  Q  C  S  E  L  S  A  V  S  A  S  F  V  L  S  T  P  T  E  T
 p  i  r  s  r  r  .  s  a  a  g  a  l  s  s  a  a  n  .  a  r  .  v  p  p  l  y  c  p  r  r  r  k
  r  s  g  a  g  g  d  q  p  l  a  h  .  a  v  q  r  t  e  r  g  e  c  l  l  c  t  v  h  a  d  g  n
----.----+----.----+----.----+----.----+----.----+----.----+----.----+----.----+----.----+----.----+
 a  s  g  s  c  s  t  i  l  r  q  r  m  s  c  h  l  s  s  l  a  t  l  a  e  k  t  s  d  v  g  v  s
  g  i  r  l  l  l  h  d  a  a  p  a  n  l  l  a  a  f  q  a  r  h  t  g  g  k  y  q  g  r  r  r  f
r  r  d  p  a  p  p  s  .  g  s  a  c  q  a  t  c  r  v  s  r  p  s  h  r  r  q  v  t  w  a  s  p  f
```

```
*051    13820      Mapped site BCL1
L        ...       CONTINUES IN PHASE ONE


H       HH    F  HNS             C BHH         E S AAH      T             H        AA H        TH
G       HA    O  PCC             F AAA         C C SVG      H             P        SV P        HG
U       AE    K  AIR             R LFE         R R UAE      A             H        UA A        AU
2       12    1  211             1 113         2 1 122      1             1        12 2        12
                   /                //              //                              /
CGGATGGCGCTGTTTTTCCGGGACGTATCATGCTGGCCAACACCTGCACCTGGACCTATCGCGGTGACGAGTGCGGTTATAGCGGTCCGGCTGTCGCGGA
----.----+----.----+----.----+----.----+----.----+----.----+----.----+----.----+----.----+----.----+  14000
GCCTACCGCGACAAAAAGGCCCTGCATAGTACGACCGGTTGTGGACGTGGACCTGGATAGCGCCACTGCTCACGCCAATATCGCCAGGCCGACAGCGCCT

  D  G  A  V  F  P  G  R  I  M  L  A  N  T  C  T  W  T  Y  R  G  D  E  C  G  Y  S  G  P  A  V  A  D
r  ■  a  l  f  f  r  d  v  s  c  w  p  t  p  a  p  g  p  i  a  v  t  s  a  v  i  a  v  r  l  s  r  ■
 g  w  r  c  f  s  g  t  y  h  a  g  q  h  l  h  l  d  l  s  r  .  r  v  r  l  .  r  s  g  c  r  g
----.----+----.----+----.----+----.----+----.----+----.----+----.----+----.----+----.----+----.----+
v  s  p  a  t  k  g  p  r  i  ■  s  a  l  v  q  v  q  v  .  r  p  s  s  h  p  .  l  p  g  a  t  a  s
 r  i  a  s  n  k  r  s  t  d  h  q  g  v  g  a  g  p  g  i  a  t  v  l  a  t  i  a  t  r  s  d  r
  p  h  r  q  k  e  p  v  y  .  a  p  w  c  r  c  r  s  r  d  r  h  r  t  r  n  y  r  d  p  q  r  p

L        ...       CONTINUES IN PHASE ONE


          F               E               F         DB                                       F
          O               C               N         DB                                       N
          K               R               U         EV                                       U
          1               V               H         11                                       H

TGAATATGACCAGCCAACGTCCGATATCACGAAGGATAAATGCAGCAAATGCCTGAGCGGTTGTAAGTTCCGCAATAACGTCGGCAACTTTGGCGGCTTC
----.----+----.----+----.----+----.----+----.----+----.----+----.----+----.----+----.----+----.----+  14100
ACTTATACTGGTCGGTTGCAGGCTATAGTGCTTCCTATTTACGTCGTTTACGGACTCGCCAACATTCAAGGCGTTATTGCAGCCGTTGAAACCGCCGAAG

 E  Y  D  Q  P  T  S  D  I  T  K  D  K  C  S  K  C  L  S  G  C  K  F  R  N  N  V  G  N  F  G  G  F
  n  ■  t  s  q  r  p  i  s  r  r  i  n  a  a  n  a  .  a  v  v  s  s  a  i  t  s  a  t  l  a  a  s
.  i  .  p  a  n  v  r  y  h  e  g  .  ■  q  q  ■  p  e  r  l  .  v  p  q  .  r  r  q  l  w  r  l  p
----.----+----.----+----.----+----.----+----.----+----.----+----.----+----.----+----.----+----.----+
  s  y  s  w  g  v  d  s  i  v  f  s  l  h  l  l  h  r  l  p  q  l  n  r  l  l  t  p  l  k  p  p  k
i  f  i  v  l  w  r  g  i  d  r  l  i  f  a  a  f  a  q  a  t  t  l  e  a  i  v  d  a  v  k  a  a  e
 h  i  h  g  a  l  t  r  y  .  s  p  y  i  c  c  i  g  s  r  n  y  t  g  c  y  r  r  c  s  q  r  s

L        ...       CONTINUES IN PHASE ONE


                       L                     H      H        H     NSH         HN     H       F
                                             N      N        H     CCP         HS     N       N
                                             F      F        A     IRA         AP     F       U
                                             1      1        1     112         1B     1       H
                                                                     //
CTTTCCATTAACAAACTTTCGCAGTAAATCCCATGACACAGACAGAATCAGCGATTCTGGCGCACGCCCGGCGATGTGCGCCAGCGGAGTCGTGCGGCTT
----.----+----.----+----.----+----.----+----.----+----.----+----.----+----.----+----.----+----.----+  14200
GAAAGGTAATTGTTTGAAAGCGTCATTTAGGGTACTGTGTCTGTCTTAGTCGCTAAGACCGCGTGCGGGCCGCTACACGCGGTCGCCTCAGCACGCCGAA

L  S  I  N  K  L  S  Q  .  i  p  .  h  r  q  n  q  r  f  w  r  t  p  g  d  v  r  q  r  s  r  a  a  s
 f  p  l  t  n  f  r  s  k  s  h  d  t  d  r  i  s  d  s  g  a  r  p  a  ■  c  a  s  g  v  v  r  l
  f  h  .  q  t  f  a  v  n  p  ■  t  q  t  e  s  a  i  l  a  h  a  r  r  c  a  p  a  e  s  c  g  f
----.----+----.----+----.----+----.----+----.----+----.----+----.----+----.----+----.----+----.----+
 r  e  ■  l  l  s  e  c  y  i  g  h  c  l  c  f  .  r  n  q  r  v  g  p  s  t  r  w  r  l  r  a  a
  k  g  n  v  f  k  r  l  l  d  w  s  v  s  l  i  l  s  e  p  a  r  g  a  i  h  a  l  p  t  t  r  s
g  k  w  .  c  v  k  a  t  f  g  ■  v  c  v  s  d  a  i  r  a  c  a  r  r  h  a  g  a  s  d  h  p  k

L      14124      L                    COOH


        M    H                                    H  M    H     H          K   B H           F  MP*
        N    P                                    P  N    P     P              B P           N  BS0
        L    A                                    A  L    A     H              V A           U  0T5
        1    2                                    2  1    2     1              1 2           H  212
                                                                                                  /
CGTGGTAAGCACGCCGGAGGGGGAAAGATATTTCCCCTGCGTGAATATCTCCGGTGAGCCGGAGGCTATTTCCGTATGTCGCCGGAAGACTGGCTGCAGG
----.----+----.----+----.----+----.----+----.----+----.----+----.----+----.----+----.----+----.----+  14300
GCACCATTCGTGCGGCCTCCCCCTTTCTATAAAGGGGACGCACTTATAGAGGCCACTCGGCCTCCGATAAAGGCATACAGCGGCCTTCTGACCGACGTCC

  w  .  a  r  r  r  g  k  d  i  s  p  a  .  i  s  p  v  s  r  r  l  f  p  y  v  a  g  r  l  a  a  g
r  g  k  h  a  g  g  g  k  i  f  p  l  r  e  y  l  r  .  a  g  g  y  f  r  M  S  P  E  D  W  L  Q  A
 v  v  s  t  p  e  g  e  r  y  f  p  c  v  n  i  s  g  e  p  e  a  i  s  v  c  r  r  k  t  g  c  r
----.----+----.----+----.----+----.----+----.----+----.----+----.----+----.----+----.----+----.----+
e  h  y  a  r  r  l  p  f  s  i  e  g  a  h  i  d  g  t  l  r  l  s  n  g  y  t  a  p  l  s  a  a  p
 r  p  l  c  a  p  p  p  f  i  n  g  r  r  s  y  r  r  h  a  p  p  .  k  r  i  d  g  s  s  q  s  c
  t  t  l  v  g  s  p  s  l  y  k  g  q  t  f  i  e  p  s  g  s  a  i  e  t  h  r  r  f  v  p  q  l
```

```
K       14276     K (orf-199)         M-NH3
*052    14298     Mapped site PST1

                         H HH AA            NSBHE     E S M D     B H    H   F  F   P*
                         P HA SV            CCSPC     C C N D     B A    P   N  N   S0
                         H AE UA            IRTAP     R R L E     V E    A   U  U   T5
                         1 12 12            11X21     2 1 1 1     1 3    2   H  H   13
                             /               / /                                     /
CAGAAATGCAGGGTGAGATTGTGGCGCTGGTCCACAGCCACCCCGGTGGTCTGCCCTGGCTGAGTGAGGCCGACCGGCGGCTGCAGGTGCAGAGTGATTT
----.----+----.----+----.----+----.----+----.----+----.----+----.----+----.----+----.----+----.----+ 14400
GTCTTTACGTCCCACTCTAACACCGCGACCAGGTGTCGGTGGGGCCACCAGACGGGACCGACTCACTCCGGCTGGCCGCCGACGTCCACGTCTCACTAAA

 r  n  a  g  .  d  c  g  a  g  p  q  p  p  r  w  s  a  l  a  e  .  g  r  p  a  a  a  g  a  e  .  f
  F  M  Q  G  E  I  V  A  L  V  H  S  H  P  G  G  I  P  W  L  S  F  A  D  R  R  L  Q  V  Q  S  D  L
q  k  c  r  v  r  l  w  r  w  s  t  a  t  p  v  v  c  p  g  .  v  r  p  t  g  g  c  r  c  r  v  i  c
----.----+----.----+----.----+----.----+----.----+----.----+----.----+----.----+----.----+----.----+
 l  f  a  p  h  s  q  p  a  p  g  c  g  g  r  h  d  a  r  a  s  h  p  r  g  a  a  a  p  a  s  h  n
a  s  i  c  p  s  i  t  a  s  t  w  l  w  g  p  p  r  g  g  q  s  l  s  a  s  r  r  s  c  t  c  l  s  k
 c  f  h  l  t  l  n  h  r  q  d  v  a  v  g  t  t  q  g  p  q  t  l  g  v  p  p  q  l  h  l  t  i

*053    14385     Mapped site PST1
K       ...       CONTINUFS IN PHASE TWO

    B   E      HNS     H         N        HF        HSNS F  HH S   H                 S
    G   C      PCC     N         S        PN        PFCC N  HA D   G                 F
    L   P      AIR     F         P        HU        AAIR U  AF U   I                 A
    1   1      211     1         B        1H        2N11 H  12 1   A                 N
                /                                      //
GCCGTGGTGGCTGGTCTGCCGGGGGACGATTCATAAGTTCCGCTGTGTGCCGCATCTCACCGGGCGGCGCTTTGAGCACGGTGTGACGGACTGTTACACA
----.----+----.----+----.----+----.----+----.----+----.----+----.----+----.----+----.----+----.----+ 14500
CGGCACCACCGACCAGACGGCCCCCTGCTAAGTATTCAAGGCGACACACGGCGTAGAGTGGCCCGCCGCGAAACTCGTGCCACACTGCCTGACAATGTGT

a  v  v  a  g  l  p  g  d  d  s  .  v  p  l  c  a  a  s  h  r  a  a  l  .  a  r  c  d  g  l  l  h  t
 P  W  W  L  V  C  R  G  T  I  H  K  F  R  C  V  P  H  L  T  G  R  R  F  F  H  G  V  T  D  C  Y  T
  r  g  g  w  s  a  g  g  r  f  i  s  s  a  v  c  r  i  s  p  g  g  a  l  s  t  v  .  r  t  v  t  h
----.----+----.----+----.----+----.----+----.----+----.----+----.----+----.----+----.----+----.----+
a  t  t  a  p  r  g  p  s  s  e  y  t  g  s  h  a  a  d  .  r  a  a  s  q  a  r  h  s  p  s  n  c
 g  h  h  s  t  q  r  p  v  i  .  l  n  r  q  t  g  c  r  v  p  r  r  k  s  c  p  t  v  s  q  .  v
q  r  p  p  q  d  a  p  p  r  n  m  l  e  a  t  h  r  m  e  g  p  p  a  k  l  v  t  h  r  v  t  v  c

K       ...       CONTINUFS IN PHASE TWO

    HNSH          F  S         H    M     H          F  C H G H            M
    PCCG          O  F         P    N     G          O  F A D N            N
    AIRU          K  A         A    L     U          K  R E I F            L
    2112          1  N         2    1     2          1  1 3 2 1            1
     //
CTGTTCCGGGATGCTTATCATCTGGCGGGGATTGAGATGCCGGACTTTCATCGTGAGGATGACTGGTGGCGTAACGGCCAGAATCTCTATCTGGATAATC
----.----+----.----+----.----+----.----+----.----+----.----+----.----+----.----+----.----+----.----+ 14600
GACAAGGCCCTACGAATAGTAGACCGCCCCTAACTCTACGGCCTGAAAGTAGCACTCCTACTGACCACCGCATTGCCGGTCTTAGAGATAGACCTATTAG

  v  p  g  c  l  s  s  g  g  d  .  d  a  g  l  s  s  .  g  .  l  v  a  .  r  p  e  s  l  s  g  .  s
L  F  R  D  A  Y  H  L  A  G  I  E  M  P  D  F  H  R  E  D  D  W  W  R  N  G  Q  N  L  Y  L  D  N  L
 c  s  g  m  l  i  i  w  r  g  l  r  c  r  t  f  i  v  r  m  t  g  g  v  t  a  r  i  s  i  w  i  i
----.----+----.----+----.----+----.----+----.----+----.----+----.----+----.----+----.----+----.----+
v  t  g  p  h  k  d  d  p  p  s  q  s  a  p  s  e  d  h  p  h  s  t  a  y  r  g  s  d  r  d  p  y  d
 s  n  r  s  a  .  .  r  a  p  i  s  i  g  s  k  .  r  s  s  s  q  h  r  l  p  w  f  r  .  r  s  l
  q  e  p  i  s  i  m  q  r  p  n  l  h  r  v  k  m  t  l  i  v  p  p  t  v  a  l  i  e  i  q  i  i

K       ...       CONTINUFS IN PHASE TWO

                H       N  F   B HNS      EF                      H H   F
                G       S  N   B PCC      CN                      N N   N
                I       P  U   V AIR      1U                      F F   U
                C       B  H   1 211      5H                      3 1   H
                                  /
TGGAGGCGACGGGGCTGTATCAGGTGCCGTTGTCAGCGGCACAGCCGGGCGATGTGCTGCTGTGCTGTTTTGGTTCATCAGTGCCGAATCACGCCGCAAT
----.----+----.----+----.----+----.----+----.----+----.----+----.----+----.----+----.----+----.----+ 14700
ACCTCCGCTGCCCCGACATAGTCCACGGCAACAGTCGCCGTGTCGGCCCGCTACACGACGACACGACAAAACCAAGTAGTCACGGCTTAGTGCGGCGTTA

 g  g  d  g  a  v  s  g  a  v  v  s  g  t  a  g  r  c  a  a  v  l  f  w  f  i  s  a  e  s  r  r  n
  E  A  T  G  L  Y  Q  V  P  L  S  A  A  Q  P  G  D  V  L  L  C  C  F  G  S  S  V  P  N  H  A  A  I
w  r  r  r  g  c  i  r  c  r  c  q  r  h  s  r  a  m  c  c  c  a  v  l  v  h  q  c  r  i  t  p  q  f
----.----+----.----+----.----+----.----+----.----+----.----+----.----+----.----+----.----+----.----+
 p  p  s  p  a  t  d  p  a  t  t  l  p  v  a  p  r  h  a  a  t  s  n  q  n  m  l  a  s  d  r  r  l
r  s  a  v  p  s  y  .  t  g  n  d  a  a  c  g  p  s  t  s  s  h  q  k  p  e  d  t  g  f  .  a  a  i
 q  l  r  r  p  q  i  l  h  r  q  .  r  c  l  r  a  i  h  q  q  a  t  k  t  .  .  h  r  i  v  g  c
```

```
K       ...       CONTINUES IN PHASE TWO


   BF B     AFE F             D   M       R       I  F      BHH   H  A
   BN B     LNC N             D   N       S          N      BGG   P  C
   VU V     UU1 U             E   L       A          U      VAA   H  Y
   1H 1     1H5 H             1   1       1          H      111   1  1
              /
TTACTGCGGCGACGGCGAGCTGCTGCACCATATTCCTGAACAACTGAGCAAACGAGAGAGGTACACCGACAAATGGCAGCGACGCACACACTCCCTCTGG
----.----+----.----+----.----+----.----+----.----+----.----+----.----+  14800
AATGACGCCGCTGCCGCTCGACGACGTGGTATAAGGACTTGTTGACTCGTTTGCTCTCTCCATGTGGCTGTTTACCGTCGCTGCGTGTGTGAGGGAGACC

l l r r r r a a a p y s . t t e q t r e v h r q M A A T H T L P L A
 Y C G D G E L L H H I P E Q L S K R E R Y T D K W Q R R T H S L W
  t a a t a s c c t i f l n n . a n e r g t p t n g s d a h t p s g
----.----+----.----+----.----+----.----+----.----+----.----+----.----+
 k s r r r r a a a g y e q v v s c v l s t c r c i a a v c v s g r
  . q p s p s s s c w i g s c s l l r s l y v s l h c r r v c e r q
n v a a v a l q q v * n r f l q a f s l p v g v f p l s a c v g e p

I       14773     I (orf-223)        M-NH3              protein not identified
K       ...       CONTINUES IN PHASE TWO


  M HNS  B HTH     S                  F   T  S S  K   F          HBNSS
  N PCC  S HHH     F                  N   A  F B      O          PGCCD
  l AIR  S AAA     A                  U   Q  A 2      K          ALIRU
  1 211  2 111     N                  H   1  N        1          21111
     /      /                                                      ///
CGTCACCGGGCATGGCGCGCATCTGCCTTTACGGGGATTTACAACGATTTGGTCGCCGCATCGACCTTCGTGTGAAAACGGGGGCTGAAGCCATCCGGGC
----.----+----.----+----.----+----.----+----.----+----.----+----.----+  14900
GCAGTGGCCCGTACCGCGCGTAGACGGAAATGCCCCTAAATGTTGCTAAACCAGCGGCGTAGCTGGAAGCACACTTTTGCCCCCGACTTCGGTAGGCCCG

  S P G M A R I C L Y G D I Q R F G R R I D L R V K T G A E A I R A
R H R A W R A S A F T G I Y N D I V A A S T F V . k r g l k p s g h
 v t g h g a h l p l r g f t t i w s p h r p s c e n g g . s h p g
----.----+----.----+----.----+----.----+----.----+----.----+----.----+
a d g p * a r m q r . p s k c r n p r r * s r r t f v p a s a * r a
 r . r a h r a d a k v p i . l s k t a a d v k t h f r p s f g d p
  t v p c p a c r g k r p n v v i q d g c r g e h s f p p q l w g p

SB2     14869     Sequence Block 2   start
K       14872     K                  COOH
I       ...       CONTINUES IN PHASE ONE


 C BHH  A  NSH          D              R     HNS   A  A E   HNS HH** H
 F AAA  L  CCP          D              S     PCC   C  A C   PCC PI00 H
 R LFE  U  IRA          E              A     AIR   Y  T O   AIR AN55 A
 1 113  1  112          1              1     211   1  2 K   211 1245 1
   //      //                                  /              /   ///
ACTGGCCACACAGCTCCCGGCGTTTCGTCAGAAACTGAGCGACGGCTGGTATCAGGTACGGATTGCCGGGCGGGACGTCAGCACGTCCGGGTTAACGGCG
----.----+----.----+----.----+----.----+----.----+----.----+----.----+  15000
TGACCGGTGTGTCGAGGGCCGCAAAGCAGTCTTTGACTCGCTGCCGACCATAGTCCATGCCTAACGGCCCGCCCTGCAGTCGTGCAGGCCCAATTGCCGC

 L A T Q L P A F R Q K L S D G W Y Q V R I A G R D V S T S G L T A
  w p h s s r r f v r n . a t a g i r y g l p g g t s a r p g . r r
t g h t a p g v s s e t e r r l v s g t d c r a g r q h v r v n g a
----.----+----.----+----.----+----.----+----.----+----.----+----.----+
 s a v c s g a n r . f s l s p q y . t r i a p r s t l v d p n v a
c q g c l e r r k t l f q a v a p i l y p n g p p v d a r g p . r r
 v p w v a g p t e d s v s r r s t d p v s q r a p r . c t r t l p

*054    14993     Mapped site HIN2
*055    14993     Mapped site HPA1
I       ...       CONTINUES IN PHASE ONE
```

```
    H               HH                H     HNSA H                B    E S     F
    N               HA                N     PCCS A                B    C C     N
    F               AE                F     AIRU E                V    R R     U
    1               12                1     2111 3                1    2 1     H
                                              /
CAGTTACATGAGACTCTGCCTGATGGCGCTGTAATTCATATTGTTCCCAGAGTCGCCGGGGCCAAGTCAGGTGGCGTATTCCAGATTGTCCTGGGGGCTG
---------+---------+---------+---------+---------+---------+---------+---------+---------+---------+ 15100
GTCAATGTACTCTGAGACGGACTACCGCGACATTAAGTATAACAAGGGTCTCAGCGGCCCCGGTTCAGTCCACCGCATAAGGTCTAACAGGACCCCCGAC

Q  L  H  E  T  L  P  D  G  A  V  I  H  I  V  P  R  V  A  G  A  K  S  G  G  V  F  Q  I  V  L  G  A  A
 s  y  m  r  l  c  l  m  a  l  .  f  i  l  f  p  e  s  p  g  p  s  q  v  a  y  s  r  l  s  w  g  l
  v  t  .  d  s  a  .  w  r  c  n  s  y  c  s  q  s  r  r  g  q  v  r  w  r  i  p  d  c  p  g  g  c
---------+---------+---------+---------+---------+---------+---------+---------+---------+---------+
 c  n  c  s  v  r  g  s  p  a  t  i  .  i  t  g  l  t  a  p  a  i  d  p  p  t  n  w  i  t  r  p  a
  l  .  m  l  s  q  r  i  a  s  y  n  m  n  n  g  s  d  g  p  g  l  .  t  a  y  e  l  n  d  q  p  s
a  t  v  h  s  e  a  q  h  r  q  l  e  y  q  e  w  l  r  r  p  w  t  l  h  r  i  g  s  q  g  p  p  q

I      ...      CONTINUES IN PHASE ONE
```

```
F       HBM        H        F        BF      BA BHHF      H        S     H
N       PIB        P        N        BN      GS BAPO      P        F     G
U       ANO        A        U        VU      LU VEAK      A        A     I
H       211        2        H        1H      11 1321      2        N     C
                                                //
CCGCCATTGCCGGATCATTCTTTACCGCCGGAGCCACCCTTGCAGCATGGGGGGCAGCCATTGGGGCCGGTGGTATGACCGGCATCCTGTTTTCTCTCGG
---------+---------+---------+---------+---------+---------+---------+---------+---------+---------+ 15200
GGCGGTAACGGCCTAGTAAGAAATGGCGGCCTCGGTGGGAACGTCGTACCCCCCGTCGGTAACCCCGGCCACCATACTGGCCGTAGGACAAAAGAGAGCC

  A  I  A  G  S  F  F  T  A  G  A  T  L  A  A  W  G  A  A  I  G  A  G  G  M  T  G  I  L  F  S  L  G
p  p  l  p  d  h  s  l  p  p  e  p  p  l  q  h  g  g  q  p  l  g  p  v  v  .  p  a  s  c  f  l  s  v
 r  h  c  r  i  i  l  y  r  r  s  h  p  c  s  m  g  g  s  h  w  g  r  w  y  d  r  h  p  v  f  s  r
---------+---------+---------+---------+---------+---------+---------+---------+---------+---------+
a  a  m  a  p  d  n  k  v  a  p  a  v  r  a  a  h  p  a  a  m  p  a  p  p  i  v  p  m  r  n  e  r  p
 g  g  n  g  s  .  e  k  g  g  s  g  g  k  c  c  p  p  c  g  n  p  g  t  t  h  g  a  d  q  k  r  e
  r  w  q  r  i  m  r  .  r  r  l  w  g  q  l  m  p  p  l  w  q  p  r  h  y  s  r  c  g  t  k  e  r

I      ...      CONTINUES IN PHASE ONE
```

```
    S   H   S    H     H             S A
    D   G   F    H     G             N C
    U   I   A    A     I             A C
    1   A   N    1     C             1 1

TGCCAGTATGGTGCTCGGTGGTGTGGCGCAGATGCTGGCACCGAAAGCCAGAACTCCCCGTATACAGACAACGGATAACGGTAAGCAGAACACCTATTTC
---------+---------+---------+---------+---------+---------+---------+---------+---------+---------+ 15300
ACGGTCATACCACGAGCCACCACACCGCGTCTACGACCGTGGCTTTCGGTCTTGAGGGGCATATGTCTGTTGCCTATTGCCATTCGTCTTGTGGATAAAG

  A  S  M  V  L  G  G  V  A  Q  M  L  A  P  K  A  R  T  P  R  I  Q  T  T  D  N  G  K  Q  N  T  Y  F
   p  v  w  c  s  v  v  w  r  r  c  w  h  r  k  p  e  l  p  v  y  r  q  r  i  t  v  s  r  t  p  i  s
c  q  y  g  a  r  w  c  g  a  d  a  g  t  e  s  q  n  s  p  y  t  d  n  g  .  r  .  a  e  h  l  f  l
---------+---------+---------+---------+---------+---------+---------+---------+---------+---------+
 a  l  i  t  s  p  p  t  a  c  i  s  a  g  f  a  l  v  g  r  i  c  v  v  s  l  p  l  c  f  v  .  k
t  g  t  h  h  e  t  t  h  r  l  h  q  c  r  f  g  s  s  g  t  y  l  c  r  i  v  t  l  l  v  g  i  e
 h  w  y  p  a  r  h  h  p  a  s  a  p  v  s  l  w  f  e  g  y  v  s  l  p  y  r  y  a  s  c  r  n

I      ...      CONTINUES IN PHASE ONE
```

```
    M       E S               R        HT     AM T    D   M
    N       C C               S        HH     FL H    D   B
    L       R R               A        AA     LU A    E   O
    1       2 1               1        11     31 1    1   1
                                       /
TCCTCACTGGATAACATGGTTGCCCAGGGCAATGTTCTGCCTGTTCTGTACGGGGAAATGCGCGTGGGGTCACGCGTGGTTTCTCAGGAGATCAGCACGG
---------+---------+---------+---------+---------+---------+---------+---------+---------+---------+ 15400
AGGAGTGACCTATTGTACCAACGGGTCCCGTTACAAGACGGACAAGACATGCCCCTTTACGCGCACCCCAGTGCGCACCAAAGAGTCCTCTAGTCGTGCC

S  S  L  D  N  M  V  A  Q  G  N  V  L  P  V  L  Y  G  E  M  R  V  G  S  R  V  V  S  Q  E  I  S  T  A
 p  h  w  i  t  w  l  p  r  a  m  f  c  l  f  c  t  g  k  c  a  w  g  h  a  w  f  l  r  r  s  a  r
  l  t  g  .  h  g  c  p  g  q  c  s  a  c  s  v  r  g  n  a  r  g  v  t  r  g  f  s  g  d  q  h  g
---------+---------+---------+---------+---------+---------+---------+---------+---------+---------+
 e  e  s  s  l  m  t  a  w  p  l  t  r  g  t  r  y  p  s  i  r  t  p  d  r  t  t  e  .  s  i  l  v
  g  .  q  i  v  h  n  g  l  a  i  n  q  r  n  q  v  p  f  h  a  h  p  .  a  h  n  r  l  l  d  a  r
r  r  v  p  y  c  p  q  g  p  c  h  e  a  q  e  t  r  p  f  a  r  p  t  v  r  p  k  e  p  s  .  c  p

I      ...      CONTINUES IN PHASE ONE
```

```
                                S       HI                                                 M
                                F       P                                                  N
                                A       H                                                  L
                                N       1                                                  1
                                         /
CAGACGAAGGGGACGGTGGTCAGGTTGTGGTGATTGGTCGCTGATGCAAAATGTTTTATGTGAAACCGCCTGCGGGCGGTTTTGTCATTTATGGAGCGTG
----.----+----.----+----.----+----.----+----.----+----.----+----.----+----.----+----.----+----.----+ 15500
GTCTGCTTCCCCTGCCACCAGTCCAACACCACTAACCAGCGACTACGTTTTACAAAATACACTTTGGCGGACGCCCGCCAAAACAGTAAATACCTCGCAC

  D  E  G  D  G  G  Q  V  V  V  I  G  R  .  c  k  m  f  y  v  k  p  p  a  g  g  f  v  i  y  g  a  .
q  t  k  g  t  v  v  r  l  w  .  l  v  a  d  a  k  c  f  m  .  n  r  l  r  a  v  l  s  f  m  e  r  e
 r  r  r  g  r  w  s  g  c  g  d  w  s  l  m  q  n  v  l  c  e  t  a  c  g  r  f  c  h  l  w  s  v
----.----+----.----+----.----+----.----+----.----+----.----+----.----+----.----+----.----+----.----+
a  s  s  p  s  p  p  .  t  t  t  i  p  r  q  h  l  i  n  .  t  f  g  g  a  p  p  k  t  m  .  p  a  h
 c  v  f  p  v  t  t  l  n  h  h  n  t  a  s  a  f  h  k  i  h  f  r  r  r  a  t  k  d  n  i  s  r
  l  r  l  p  r  h  d  p  q  p  s  q  d  s  i  c  f  t  k  h  s  v  a  q  p  r  n  q  .  k  h  l  t

I      15441         I                COOH


     J                               T HT                                T S     M  C*T
                                     H HH                                D F     B  L0A
                                     A AA                                E A     0  A5Q
                                     1 11                                1 N     1  161
                                        /                                           //
AGGAATGGGTAAAGGAAGCAGTAAGGGGCATACCCCGCGCGAAGCGAAGGACAACCTGAAGTCCACGCAGTTGCTGAGTGTGATCGATGCCATCAGCGAA
----.----+----.----+----.----+----.----+----.----+----.----+----.----+----.----+----.----+----.----+ 15600
TCCTTACCCATTTCCTTCGTCATTCCCCGTATGGGGCGCGCTTCGCTTCCTGTTGGACTTCAGGTGCGTCAACGACTCACACTAGCTACGGTAGTCGCTT

 g  M  G  K  G  S  S  K  G  H  T  P  R  E  A  K  D  N  L  K  S  T  Q  L  L  S  V  I  D  A  I  S  E
  e  w  v  k  e  a  v  r  g  i  p  r  a  k  r  r  t  t  .  s  p  r  s  c  .  v  .  s  m  p  s  a  k
r  n  g  .  r  k  q  .  g  a  y  p  a  r  s  e  g  q  p  e  v  h  a  v  a  e  c  d  r  c  h  q  r  r
----.----+----.----+----.----+----.----+----.----+----.----+----.----+----.----+----.----+----.----+
  p  i  p  l  p  l  l  l  p  c  v  g  r  s  a  f  s  l  r  f  d  v  c  n  s  l  t  i  s  a  m  l  s
s  s  h  t  f  s  a  t  l  p  m  g  r  a  f  r  l  v  v  q  l  g  r  l  q  q  t  h  d  i  g  d  a  f
 l  f  p  y  l  f  c  y  p  a  y  g  a  r  l  s  p  c  g  s  t  w  a  t  a  s  h  s  r  h  w  .  r

J       15505       J (orf-1132)       M-NH3
*056    15584       Mapped site CLA1


A H          AA H  H     B       F   EF           R   H      M     D                    H
S A          SV P  G     B       O   CN           S   P      N     D                    P
U E          UA A  U     V       K   1U           A   A      L     E                    A
1 3          12 2  2     1       1   5H           1   2      1     1                    2
              /
GGGCCGATTGAAGGTCCGGTGGATGGCTTAAAAAGCGTGCTGCTGAACAGTACGCCGGTGCTGGACACTGAGGGGAATACCAACATATCCGGTGTCACGG
----.----+----.----+----.----+----.----+----.----+----.----+----.----+----.----+----.----+----.----+ 15700
CCCGGCTAACTTCCAGGCCACCTACCGAATTTTTCGCACGACGACTTGTCATGCGGCCACGACCTGTGACTCCCCTTATGGTTGTATAGGCCACAGTGCC

G  P  I  E  G  P  V  D  G  L  K  S  V  L  L  N  S  T  P  V  L  D  T  E  G  N  T  N  I  S  G  V  T  V
 g  r  l  k  v  r  w  m  a  .  k  a  c  c  .  t  v  r  r  c  w  t  l  r  g  i  p  t  y  p  v  s  r
  a  d  .  r  s  g  g  w  l  k  k  r  a  a  e  q  y  a  g  a  g  h  .  g  e  y  q  h  i  r  c  h  g
----.----+----.----+----.----+----.----+----.----+----.----+----.----+----.----+----.----+----.----+
 p  g  i  s  p  g  t  s  p  k  f  l  t  s  s  f  l  v  g  t  s  s  v  s  p  f  v  l  m  d  p  t  v
  p  r  n  f  t  r  h  i  a  .  f  a  h  q  q  v  t  r  r  h  q  v  s  l  p  i  g  v  y  g  t  d  r
l  a  s  q  l  d  p  p  h  s  l  f  r  a  a  s  c  y  a  p  a  p  c  q  p  s  y  w  c  i  r  h  .  p

J       ...       CONTINUES IN PHASE ONE


        HNS                HHM  H           H      H      M              R               H     M
        PCC                PNN  P           N      P      N              S               P     B
        AIR                HFL  A           F      A      L              A               H     0
        211                111  2           1      2      1              1               1     1
          /
TGGTGTTCCGGGCTGGTGAGCAGGAGCAGACTCCGCCGGAGGGATTTGAATCCTCCGGCTCCGAGACGGTGCTGGGTACGGAAGTGAAATATGACACGCC
----.----+----.----+----.----+----.----+----.----+----.----+----.----+----.----+----.----+----.----+ 15800
ACCACAAGGCCCGACCACTCGTCCTCGTCTGAGGCGGCCTCCCTAAACTTAGGAGGCCGAGGCTCTGCCACGACCCATGCCTTCACTTTATACTGTGCGG

  V  F  R  A  G  F  Q  F  Q  T  P  P  E  G  F  E  S  S  G  S  E  T  V  L  G  T  E  V  K  Y  D  T  P
w  c  s  g  l  v  s  r  s  r  l  r  r  r  d  l  n  p  p  a  p  r  r  c  w  v  r  k  .  n  m  t  r  r
 g  v  p  g  w  .  a  g  a  d  s  a  g  g  i  .  i  l  r  l  r  d  g  a  g  y  g  s  e  i  .  h  a
----.----+----.----+----.----+----.----+----.----+----.----+----.----+----.----+----.----+----.----+
t  t  n  r  a  p  s  c  s  c  v  g  g  s  p  n  s  d  e  p  e  s  v  t  s  p  v  s  t  f  y  s  v  g
 h  h  e  p  s  t  l  l  l  l  s  r  r  l  s  k  f  g  g  a  g  l  r  h  q  t  r  f  h  f  i  v  r
  p  t  g  p  q  h  a  p  a  s  e  a  p  p  i  q  i  r  r  s  r  s  p  a  p  y  p  l  s  i  h  c  a

J       ...       CONTINUES IN PHASE ONE
```

```
                             T         T H            R                  H           M     H   H
                             A         T H            S                  G           N     N   P
                             Q         H A            A                  E           L     F   H
                             1         2 1            1                  2           1     1   1

GATCACCCGCACCATTACGTCTGCAAACATCGACCGTCTGCGCTTTACCTTCGGTGTACAGGCACTGGTGGAAACCACCTCAAAGGGTGACAGGAATCCG
----.----+----.----+----.----+----.----+----.----+----.----+----.----+----.----+----.----+----.----+ 15900
CTAGTGGGCGTGGTAATGCAGACGTTTGTAGCTGGCAGACGCGAAATGGAAGCCACATGTCCGTGACCACCTTTGGTGGAGTTTCCCACTGTCCTTAGGC

 I  T  R  T  I  T  S  A  N  I  D  R  L  R  F  T  F  G  V  Q  A  L  V  E  T  T  S  K  G  D  R  N  P
  s  p  a  p  l  r  l  q  t  s  t  v  c  a  l  p  s  v  y  r  h  w  w  k  p  p  q  r  v  t  g  i  r
d  h  p  h  h  y  v  c  k  h  r  p  s  a  l  y  l  r  c  t  g  t  g  g  n  h  l  k  g  .  q  e  s  v
----.----+----.----+----.----+----.----+----.----+----.----+----.----+----.----+----.----+----.----+
  i  v  r  v  m  v  d  a  f  m  s  r  r  r  k  v  k  p  t  c  a  s  t  s  v  v  e  f  p  s  l  f  g
r  d  g  a  g  n  r  r  c  v  d  v  t  q  a  k  g  e  t  y  l  c  q  h  f  g  g  .  l  t  v  p  i  r
 s  .  g  c  w  .  t  q  l  c  r  g  d  a  s  .  r  r  h  v  p  v  p  p  f  w  r  l  p  h  c  s  d

J       ...      CONTINUES IN PHASE ONE


           E                                         H   H                                  M    HH
           C                                         P   P                                  N    AA
           1                                         H   H                                  L    EE
           5                                         1   1                                  1    13
                                                                                                  /
TCGGAAGTCCGCCTGCTGGTTCAGATACAACGTAACGGTGGCTGGGTGACGGAAAAAGACATCACCATTAAGGGCAAAACCACCTCGCAGTATCTGGCCT
----.----+----.----+----.----+----.----+----.----+----.----+----.----+----.----+----.----+----.----+ 16000
AGCCTTCAGGCGGACGACCAAGTCTATGTTGCATTGCCACCGACCCACTGCCTTTTTCTGTAGTGGTAATTCCCGTTTTGGTGGAGCGTCATAGACCGGA

S  E  V  R  L  L  V  Q  I  Q  R  N  G  G  W  V  T  F  K  D  I  T  I  K  G  K  T  T  S  Q  Y  L  A  S
 r  k  s  a  c  w  f  r  y  n  v  t  v  a  g  .  r  k  k  t  s  p  l  r  a  k  p  p  r  s  i  w  p
  g  s  p  p  a  g  s  d  t  t  .  r  w  l  g  d  g  k  r  h  h  h  .  g  q  n  h  l  a  v  s  g  l
----.----+----.----+----.----+----.----+----.----+----.----+----.----+----.----+----.----+----.----+
 d  s  t  r  r  s  t  .  i  c  r  l  p  p  q  t  v  s  f  s  m  v  m  l  p  l  v  v  e  c  y  r  a
  r  f  d  a  q  q  n  l  y  l  t  v  t  a  p  h  r  f  f  v  d  g  n  l  a  f  g  g  r  l  i  q  g
t  p  l  g  g  a  p  e  s  v  v  y  r  h  s  p  s  p  f  l  c  .  w  .  p  c  f  w  r  a  t  d  p  r

J       ...      CONTINUES IN PHASE ONE


       M   B    H   F  F T H     S     HH     MH H    FA  H    FHB    E   N  PA*F P*
       N   S    P   N  N H H     F     PG     SH G    OC  P    OGB    C   S  VL0N S0
       L   T    H   U  U A A     A     AU     TA U    KY  A    KAV    P   P  UU5U T5
       1   F    1   H  H 1 1     N     22     11 2    11  2    111    1   B  217H 18
                                                                /              /// /
CGGTGGTGATGGGTAACCTGCCGCCGCGCCCGTTTAATATCCGGATGCGCAGGATGACGCCGGACAGCACCACAGACCAGCTGCAGAACAAAACGCTCTG
----.----+----.----+----.----+----.----+----.----+----.----+----.----+----.----+----.----+----.----+ 16100
GCCACCACTACCCATTGGACGGCGGCGCGGGCAAATTATAGGCCTACGCGTCCTACTGCGGCCTGTCGTGGTGTCTGGTCGACGTCTTGTTTTGCGAGAC

 V  V  M  G  N  L  P  P  R  P  F  N  I  R  M  R  R  M  T  P  D  S  T  T  D  Q  L  Q  N  K  T  L  W
r  w  .  w  v  t  c  r  r  a  r  l  i  s  g  c  a  g  .  r  r  t  a  p  q  t  s  c  r  t  k  r  s  g
 g  g  d  g  .  p  a  a  a  p  v  .  y  p  d  a  q  d  d  a  g  q  h  h  r  p  a  a  e  q  n  a  l
----.----+----.----+----.----+----.----+----.----+----.----+----.----+----.----+----.----+----.----+
e  t  t  i  p  l  r  g  g  r  g  n  l  i  r  i  r  l  i  v  g  s  l  v  v  s  w  s  c  f  l  v  s  q
 r  h  h  h  t  v  q  r  r  a  r  k  i  d  p  h  a  p  h  r  r  v  a  g  c  v  l  q  l  v  f  r  e
  p  p  s  p  y  g  a  a  a  g  t  .  y  g  s  a  c  s  s  a  p  c  c  w  l  g  a  a  s  c  f  a  r

*057    16080    Mapped site PVU2
*058    16085    Mapped site PST1
J       ...      CONTINUES IN PHASE ONE


                    C*T                                                  H                 F  E       B
                    L0A                                                  N                 N  C       B
                    A5Q                                                  F                 U  1       V
                    191                                                  1                 H  5       1
                     //
GTCGTCATACACTGAAATCATCGATGTGAAACAGTGCTACCCGAACACGGCACTGGTCGGCGTGCAGGTGGACTCGGAGCAGTTCGGCAGCCAGCAGGTG
----.----+----.----+----.----+----.----+----.----+----.----+----.----+----.----+----.----+----.----+ 16200
CAGCAGTATGTGACTTTAGTAGCTACACTTTGTCACGATGGGCTTGTGCCGTGACCAGCCGCACGTCCACCTGAGCCTCGTCAAGCCGTCGGTCGTCCAC

 S  S  Y  T  E  I  I  D  V  K  Q  C  Y  P  N  T  A  L  V  G  V  Q  V  D  S  E  Q  F  G  S  Q  Q  V
  r  h  t  l  k  s  s  m  .  n  s  a  t  r  t  r  h  w  s  a  c  r  w  t  r  s  s  s  a  a  s  r  .
v  v  i  h  .  n  h  r  c  e  t  v  l  p  e  h  g  t  g  r  r  a  g  g  l  g  a  v  r  q  p  a  g  e
----.----+----.----+----.----+----.----+----.----+----.----+----.----+----.----+----.----+----.----+
 d  d  y  v  s  i  m  s  t  f  c  h  .  g  f  v  a  s  t  p  t  c  t  s  e  s  c  n  p  l  w  c  t
p  r  .  v  s  f  d  d  i  h  f  l  a  v  r  v  r  c  q  d  a  h  l  h  v  r  l  l  e  a  a  l  l  h
 t  t  m  c  q  f  .  r  h  s  v  t  s  g  s  c  p  v  p  r  r  a  p  p  s  p  a  t  r  c  g  a  p
```

```
*059  16121      Mapped site CLA1
J      ...       CONTINUES IN PHASE ONE

        H           HT            P*H     T                   TF     H N                        A   H
        P           HH            S0G     A                   HN     G S                        H   P
        H           AA            T6I     Q                   AU     A P                        A   A
        1           11            10C     1                   1H     1 B                        3   2
                     /             //
AGCCGTAATTATCATCTGCGCGGGCGTATTCTGCAGGTGCCGTCGAACTATAACCCGCAGACGCGGCAATACAGCGGTATCTGGGACGGAACGTTTAAAC
----.----+----.----+----.----+----.----+----.----+----.----+----.----+----.----+----.----+----.----+ 16300
TCGGCATTAATAGTAGACGCGCCCGCATAAGACGTCCACGGCAGCTTGATATTGGGCGTCTGCGCCGTTATGTCGCCATAGACCCTGCCTTGCAAATTTG

S  R  N  Y  H  L  R  G  R  I  L  Q  V  P  S  N  Y  N  P  Q  T  R  Q  Y  S  G  I  W  D  G  T  F  K  P
 a  v  i  i  i  c  a  g  v  f  c  r  c  r  r  t  i  t  r  r  r  g  n  t  a  v  s  g  t  e  r  l  n
  p  .  l  s  s  a  r  a  y  s  a  g  a  v  e  l  .  p  a  d  a  a  i  q  r  y  l  g  r  n  v  .  t
----.----+----.----+----.----+----.----+----.----+----.----+----.----+----.----+----.----+----.----+
 l  r  l  .  .  r  r  p  r  i  r  c  t  g  d  f  .  l  g  c  v  r  c  y  l  p  i  q  s  p  v  n  l
  a  t  i  i  ■  q  a  p  t  n  q  l  h  r  r  v  i  v  r  l  r  p  l  v  a  t  d  p  v  s  r  k  f
s  g  y  n  d  d  a  r  a  y  e  a  p  a  t  s  s  y  g  a  s  a  a  i  c  r  y  r  p  r  f  t  .  v

*060  16235      Mapped site PST1
J      ...       CONTINUES IN PHASE ONE

                    HHES          F                  T H           E            F  H              F
                    AACC          0                  H H           C            N  G              0
                    EERR          K                  A A           0            U  U              K
                    1321          1                  1 1           K            H  2              1
                     ///
CGGCATACAGCAACAACATGGCCTGGTGTCTGTGGGATATGCTGACCCATCCGCGCTACGGCATGGGGAAACGTCTTGGTGCGGCGGATGTGGATAAATG
----.----+----.----+----.----+----.----+----.----+----.----+----.----+----.----+----.----+----.----+ 16400
GCCGTATGTCGTTGTTGTACCGGACCACAGACACCCTATACGACTGGGTAGGCGCGATGCCGTACCCCTTTGCAGAACCACGCCGCCTACACCTATTTAC

 A  Y  S  N  N  M  A  W  C  L  W  D  M  L  T  H  P  R  Y  G  M  G  K  R  L  G  A  A  D  V  D  K  W
r  h  t  a  t  t  w  p  g  v  c  g  i  c  .  p  i  r  a  t  a  w  g  n  v  l  v  r  r  ■  w  i  n  g
 g  i  q  q  q  h  g  l  v  s  v  g  y  a  d  p  s  a  l  r  h  g  e  t  s  w  c  g  g  c  g  .  ■
----.----+----.----+----.----+----.----+----.----+----.----+----.----+----.----+----.----+----.----+
g  a  y  l  l  l  ■  a  q  h  r  h  s  i  s  v  w  g  r  .  p  ■  p  f  r  r  p  a  a  s  t  s  l  h
 r  c  v  a  v  v  h  g  p  t  q  p  i  h  q  g  ■  r  a  v  a  h  p  f  t  k  t  r  r  i  h  i  f
  p  ■  c  c  c  c  p  r  t  d  t  p  y  a  s  g  d  a  s  r  c  p  s  v  d  q  h  p  p  h  p  y  i

J      ...       CONTINUES IN PHASE ONE

   HH           C H GS RR                H             F        HF T H        S       R
   HA           F A DC RS                P             N        PN H H        F       S
   AE           R E IA UA                A             U        HU A A        A       A
   12           1 3 21 11                2             H        1H 1 1        N       1
                     /  /
GGCGCTGTATGTCATCGGCCAGTACTGCGACCAGTCAGTGCCGGACGGCTTTGGCGGCACGGAGCCGCGCATCACCTGTAATGCGTACCTGACCACACAG
----.----+----.----+----.----+----.----+----.----+----.----+----.----+----.----+----.----+----.----+ 16500
CCGCGACATACAGTAGCCGGTCATGACGCTGGTCAGTCACGGCCTGCCGAAACCGCCGTGCCTCGGCGCGTAGTGGACATTACGCATGGACTGGTGTGTC

 A  L  Y  V  I  G  Q  Y  C  D  Q  S  V  P  D  G  F  G  G  T  E  P  R  I  T  C  N  A  Y  L  T  T  Q
  r  c  ■  s  s  a  s  t  a  t  s  q  c  r  t  a  l  a  a  r  s  r  a  s  p  v  ■  r  t  .  p  h  s
g  a  v  c  h  r  p  v  l  r  p  v  s  a  g  r  l  w  r  h  g  a  a  h  h  l  .  c  v  p  d  h  t  a
----.----+----.----+----.----+----.----+----.----+----.----+----.----+----.----+----.----+----.----+
 a  s  y  t  ■  p  w  y  q  s  w  d  t  g  s  p  k  p  p  v  s  g  r  ■  v  q  l  a  y  r  v  v  c
p  r  q  i  d  d  a  l  v  a  v  l  .  h  r  v  a  k  a  a  r  l  r  a  d  g  t  i  r  v  q  g  c  l
 p  a  t  h  .  r  g  t  s  r  g  t  l  a  p  r  s  q  r  c  p  a  a  c  .  r  y  h  t  g  s  w  v

J      ...       CONTINUES IN PHASE ONE

          H   S   DH    F   S              H       H                         H         AA
          G   D   DG    0   F              H       P                         G         SV
          U   U   EI    K   A              A       A                         A         UA
          2   1   1A    1   N              1       2                         1         12
                                                                                         /
CGTAAGGCGTGGGATGTGCTCAGCGATTTCTGCTCGGCGATGCGCTGTATGCCGGTATGGAACGGGCAGACGCTGACGTTCGTGCAGGACCGACCGTCGG
----.----+----.----+----.----+----.----+----.----+----.----+----.----+----.----+----.----+----.----+ 16600
GCATTCCGCACCCTACACGAGTCGCTAAAGACGAGCCGCTACGCGACATACGGCCATACCTTGCCCGTCTGCGACTGCAAGCACGTCCTGGCTGGCAGCC

R  K  A  W  D  V  L  S  D  F  C  S  A  M  R  C  M  P  V  W  N  G  Q  T  L  T  F  V  Q  D  R  P  S  D
 v  r  r  g  ■  c  s  a  i  s  a  r  r  c  a  v  c  r  y  g  t  g  r  r  .  r  s  c  r  t  d  r  r
  .  g  v  g  c  a  q  r  f  l  l  g  d  a  l  y  a  g  ■  e  r  a  d  a  d  v  r  a  g  p  t  v  g
----.----+----.----+----.----+----.----+----.----+----.----+----.----+----.----+----.----+----.----+
 r  l  a  h  s  t  s  l  s  k  q  e  a  i  r  q  i  g  t  h  f  p  c  v  s  v  n  t  c  s  r  g  d
  t  l  r  p  i  h  e  a  i  e  a  r  r  h  a  t  h  r  y  p  v  p  l  r  q  r  e  h  l  v  s  r  r
a  y  p  t  p  h  a  .  r  n  r  s  p  s  a  s  y  a  p  i  s  r  a  s  a  s  t  r  a  p  g  v  t  p
```

```
J       ...       CONTINUES IN PHASE ONE

          AA            S             HH   H    B HTHF              A        HH       AA         M
          SV            F             PG   P    S HHHO              L        HA       SV         N
          UA            A             AU   H    S AAAK              U        AE       UA         L
          12            N             22   1    2 1111              1        12       12         1
           /                                       /                                /
ATAAGACGTGGACCTATAACCGCAGTAATGTGGTGATGCCGGATGATGGCGCGCCGTTCCGCTACAGCTTCAGCGCCCTGAAGGACCGCCATAATGCCGT
----.----+----.----+----.----+----.----+----.----+----.----+----.----+----.----+----.----+----.----+ 16700
TATTCTGCACCTGGATATTGGCGTCATTACACCACTACGGCCTACTACCGCGCGGCAAGGCGATGTCGAAGTCGCGGGACTTCCTGGCGGTATTACGGCA

  K  T  W  T  Y  N  R  S  N  V  V  M  P  D  D  G  A  P  F  R  Y  S  F  S  A  L  K  D  R  H  N  A  V
 i  r  r  g  p  i  t  a  v  m  w  .  c  r  m  m  a  r  r  s  a  t  a  s  a  p  .  r  t  a  i  m  p  l
 .  d  v  d  l  .  p  q  .  c  g  d  a  g  .  w  r  a  v  p  l  q  l  q  r  p  e  g  p  p  .  c  r
----.----+----.----+----.----+----.----+----.----+----.----+----.----+----.----+----.----+----.----+
 s  l  v  h  v  .  l  r  l  l  t  t  i  g  s  s  p  a  g  n  r  .  l  k  l  a  r  f  s  r  w  l  a  t
  i  l  r  p  g  i  v  a  t  i  h  h  h  r  i  i  a  r  r  e  a  v  a  e  a  g  q  l  v  a  m  i  g
   y  s  t  s  r  y  g  c  y  h  p  s  a  p  h  h  r  a  t  g  s  c  s  .  r  g  s  p  g  g  y  h  r

J       ...       CONTINUES IN PHASE ONE

               H                                H                   MHH                         S
               P                                L                   BAA                         F
               H                                U                   OEE                         A
               1                                1                   213                         N
                                                                     //
TGAGGTGAACTGGATTGACCCGAACAACGGCTGGGAGACGGCGACAGAGCTTGTTGAAGATACGCAGGCCATTGCCCGTTACGGTCGTAATGTTACGAAG
----.----+----.----+----.----+----.----+----.----+----.----+----.----+----.----+----.----+----.----+ 16800
ACTCCACTTGACCTAACTGGGCTTGTTGCCGACCCTCTGCCGCTGTCTCGAACAACTTCTATGCGTCCGGTAACGGGCAATGCCAGCATTACAATGCTTC

  E  V  N  W  I  D  P  N  N  G  W  E  T  A  T  E  L  V  E  D  T  Q  A  I  A  R  Y  G  R  N  V  T  K
   r  .  t  g  l  t  r  t  t  a  g  r  r  r  q  s  l  l  k  i  r  r  p  l  p  v  t  v  v  m  l  r  r
 .  g  e  l  d  .  p  e  q  r  l  g  d  g  d  r  a  c  .  r  y  a  g  h  c  p  l  r  s  .  c  y  e  d
----.----+----.----+----.----+----.----+----.----+----.----+----.----+----.----+----.----+----.----+
   s  t  f  q  i  s  g  f  l  p  q  s  v  a  v  s  s  t  s  s  v  c  a  m  a  r  .  p  r  l  t  v  f
 n  l  h  v  p  n  v  r  v  v  a  p  l  r  r  c  l  k  n  f  i  r  l  g  n  g  t  v  t  t  i  n  r  l
 q  p  s  s  s  q  g  s  c  r  s  p  s  p  s  l  a  q  q  l  y  a  p  w  q  g  n  r  d  y  h  .  s

J       ...       CONTINUES IN PHASE ONE

 H      M      F R      HNS             T HHNS                      E            E    H
 G      B      O S      PCC             H HPCC                      C            C    G
 U      O      K A      AIR             A AAIR                      1            P    A
 2      2      1 1      211             1 1211                      5            1    1
                          /                 /
ATGGATGCCTTTGGCTGTACCAGCCGGGGGCAGGCACACCGCGCCGGGCTGTGGCTGATTAAAACAGAACTGCTGGAAACGCAGACCGTGGATTTCAGCG
----.----+----.----+----.----+----.----+----.----+----.----+----.----+----.----+----.----+----.----+ 16900
TACCTACGGAAACCGACATGGTCGGCCCCCGTCCGTGTGGCGCGGCCCGACACCGACTAATTTTGTCTTGACGACCTTTGCGTCTGGCACCTAAAGTCGC

 M  D  A  F  G  C  T  S  R  G  Q  A  H  R  A  G  L  W  L  I  K  T  E  L  L  E  T  Q  T  V  D  F  S  V
 w  m  p  l  a  v  p  a  g  g  r  h  t  a  p  g  c  g  .  l  k  q  n  c  w  k  r  r  p  w  i  s  a
  g  c  l  w  l  y  q  p  g  a  g  t  p  r  r  a  v  a  d  .  n  r  t  a  g  n  a  d  r  g  f  q  r
----.----+----.----+----.----+----.----+----.----+----.----+----.----+----.----+----.----+----.----+
 i  s  a  k  p  q  v  l  r  p  c  a  c  r  a  p  s  h  s  i  l  v  s  s  s  s  v  c  v  t  s  k  l
  h  i  g  k  a  t  g  a  p  p  l  c  v  a  g  p  q  p  q  n  f  c  f  q  q  f  r  l  g  h  i  e  a
 s  p  h  r  q  s  y  w  g  p  a  p  v  g  r  r  a  t  a  s  .  f  l  v  a  p  f  a  s  r  p  n  .  r

J       ...       CONTINUES IN PHASE ONE

     H      X           R HNS                                  H           H
     H      M           S PCC                                  P           P
     A      N           A AIR                                  A           A
     1      1           1 211                                  2           2
                           /
TCGGCGCAGAAGGGCTTCGCCATGTACCGGGCGATGTTATTGAAATCTGCGATGATGACTATGCCGGTATCAGCACCGGTGGTCGTGTGCTGGCGGTGAA
----.----+----.----+----.----+----.----+----.----+----.----+----.----+----.----+----.----+----.----+ 17000
AGCCGCGTCTTCCCGAAGCGGTACATGGCCCGCTACAATAACTTTAGACGCTACTACTGATACGGCCATAGTCGTGGCCACCAGCACACGACCGCCACTT

   G  A  E  G  L  R  H  V  P  G  D  V  I  E  I  C  D  D  D  Y  A  G  I  S  T  G  G  R  V  L  A  V  N
 s  a  q  k  g  f  a  m  y  r  a  m  l  l  k  s  a  m  m  t  m  p  v  s  a  p  v  v  v  c  w  r  .  t
  r  r  r  r  a  s  p  c  t  g  r  c  y  .  n  l  r  .  .  l  c  r  y  q  h  r  w  s  c  a  g  g  e
----.----+----.----+----.----+----.----+----.----+----.----+----.----+----.----+----.----+----.----+
 t  p  a  s  p  s  r  w  t  g  p  s  t  i  s  i  q  s  s  s  .  a  p  i  l  v  p  p  r  t  s  a  t  f
  d  a  c  f  p  k  a  m  y  r  a  i  n  n  f  d  a  i  i  v  i  g  t  d  a  g  t  t  t  h  q  r  h
   r  r  l  l  a  e  g  h  v  p  r  h  .  q  f  r  r  h  h  s  h  r  y  .  c  r  h  d  h  a  p  p  s

J       ...       CONTINUES IN PHASE ONE
```

```
    E H  NSH         H T  BH  F         F         H H R KM*TH       E S    H*              H   M
    C P  CCP         G A  BG  0         N         P G S PN0HH       C C    I0              P   N
    P H  IRA         A Q  VA  K         U         A I A NL6AA       R R    N6              A   L
    1 1  112         1 1  11  1         H         2 C 1 11111       2 1    22              2   1
          //                                               //               /
CAGCCAGACCCGGACGCTGACGCTCGACCGTGAAATCACGCTGCCATCCTCCGGTACCGCGCTGATAAGCCTGGTTGACGGAAGTGGCAATCCGGTCAGC
----.----+----.----+----.----+----.----+----.----+----.----+----.----+----.----+----.----+----.----+ 17100
GTCGGTCTGGGCCTGCGACTGCGAGCTGGCACTTTAGTGCGACGGTAGGAGGCCATGGCGCGACTATTCGGACCAACTGCCTTCACCGTTAGGCCAGTCG

 S  Q  T  R  T  L  T  L  D  R  E  I  T  L  P  S  S  G  T  A  L  I  S  L  V  D  G  S  G  N  P  V  S
  a  r  p  g  r  .  r  s  t  v  k  s  r  c  h  p  p  v  p  r  .  .  a  w  l  t  e  v  a  i  r  s  a
q  p  d  p  d  a  d  a  r  p  .  n  h  a  a  i  l  r  y  r  a  d  k  p  g  .  r  k  w  q  s  g  q  r
----.----+----.----+----.----+----.----+----.----+----.----+----.----+----.----+----.----+----.----+
  l  w  v  r  v  s  v  s  s  r  s  i  v  s  g  d  e  p  v  a  s  i  l  r  t  s  p  l  p  l  g  t  l
v  a  l  g  p  r  q  r  e  v  t  f  d  r  q  w  g  g  t  g  r  q  y  a  q  n  v  s  t  a  i  r  d  a
 c  g  s  g  s  a  s  a  r  g  h  f  .  a  a  m  r  r  y  r  a  s  l  g  p  q  r  f  h  c  d  p  .

*061    17057      Mapped site KPN1
*062    17076      Mapped site HIN2
J        ...       CONTINUES IN PHASE ONE
```

```
        H                                                                    B      A B  AF        F
        P                                                                    B      L B  LN        N
        H                                                                    V      U V  UU        U
        1                                                                    1      1 1  1H        H

GTGGAGGTTCAGTCCGTCACCGACGGCGTGAAGGTAAAAGTGAGCCGTGTTCCTGACGGTGTTGCTGAATACAGCGTATGGGAGCTGAAGCTGCCGACGC
----.----+----.----+----.----+----.----+----.----+----.----+----.----+----.----+----.----+----.----+ 17200
CACCTCCAAGTCAGGCAGTGGCTGCCGCACTTCCATTTTCACTCGGCACAAGGACTGCCACAACGACTTATGTCGCATACCCTCGACTTCGACGGCTGCG

V  E  V  Q  S  V  T  D  G  V  K  V  K  V  S  R  V  P  D  G  V  A  E  Y  S  V  W  E  L  K  L  P  T  L
 w  r  f  s  p  s  p  t  a  .  r  .  k  .  a  v  f  l  t  v  l  l  n  t  a  y  g  s  .  s  c  r  r
  g  g  s  v  r  h  r  r  r  e  g  k  s  e  p  c  s  .  r  c  c  .  i  q  r  m  g  a  e  a  a  d  a
----.----+----.----+----.----+----.----+----.----+----.----+----.----+----.----+----.----+----.----+
 t  s  t  .  d  t  v  s  p  t  f  t  f  t  l  r  t  g  s  p  t  a  s  y  l  t  h  s  s  f  s  g  v
  h  l  n  l  g  d  g  v  a  h  l  y  f  h  a  t  n  r  v  t  n  s  f  v  a  y  p  l  q  l  q  r  r
r  p  p  e  t  r  .  r  r  r  s  p  l  l  s  g  h  e  q  r  h  q  q  i  c  r  i  p  a  s  a  a  s  a

J        ...       CONTINUES IN PHASE ONE
```

```
   HHB         N   F                                H                  F     N  HMB         HH
   HGB         S   N                                P                  N     S  PNB         AA
   AAV         P   U                                H                  U     P  ALV         EE
   111         B   H                                1                  H     C  211         13
                                                                                             /
TGCGCCAGCGACTGTTCCGCTGCGTGAGTATCCGTGAGAACGACGACGGCACGTATGCCATCACCGCCGTGCAGCATGTGCCGGAAAAAGAGGCCATCGT
----.----+----.----+----.----+----.----+----.----+----.----+----.----+----.----+----.----+----.----+ 17300
ACGCGGTCGCTGACAAGGCGACGCACTCATAGGCACTCTTGCTGCTGCCGTGCATACGGTAGTGGCGGCACGTCGTACACGGCCTTTTTCTCCGGTAGCA

  R  Q  R  L  F  R  C  V  S  I  R  E  N  D  D  G  T  Y  A  I  T  A  V  Q  H  V  P  E  K  E  A  I  V
c  a  s  d  c  s  a  a  .  v  s  v  r  t  t  t  a  r  m  p  s  p  p  c  s  m  c  r  k  k  r  p  s  w
 a  p  a  t  v  p  l  r  e  y  p  .  e  r  r  r  h  v  c  h  h  r  r  a  a  c  a  g  k  r  g  h  r
----.----+----.----+----.----+----.----+----.----+----.----+----.----+----.----+----.----+----.----+
s  r  w  r  s  n  r  q  t  l  i  r  s  f  s  s  p  v  y  a  m  v  a  t  c  c  t  g  s  f  s  a  m  t
 q  a  l  s  q  e  a  a  h  t  d  t  l  v  v  v  a  r  i  g  d  g  g  h  l  m  h  r  f  f  l  g  d
  a  g  a  v  t  g  s  r  s  y  g  h  s  r  r  r  c  t  h  w  .  r  r  a  a  h  a  p  f  l  p  w  r

J        ...       CONTINUES IN PHASE ONE
```

```
            H                                      H   F  N        F          B               P*N
            H                                      P   N  S        N          B               S0S
            A                                      H   U  P        U          V               T6P
            1                                      1   H  B        H          1               13B
                                                                                               /
GGATAACGGGGCGCACTTTGACGGCGAACAGAGTGGCACGGTGAATGGTGTCACGCCGCCAGCGGTGCAGCACCTGACCGCAGAAGTCACTGCAGACAGC
----.----+----.----+----.----+----.----+----.----+----.----+----.----+----.----+----.----+----.----+ 17400
CCTATTGCCCCGCGTGAAACTGCCGCTTGTCTCACCGTGCCACTTACCACAGTGCGGCGGTCGCCACGTCGTGGACTGGCGTCTTCAGTGACGTCTGTCG

 D  N  G  A  H  F  D  G  E  Q  S  G  T  V  N  G  V  T  P  P  A  V  Q  H  L  T  A  E  V  T  A  D  S
  i  t  g  r  t  l  t  a  n  r  v  a  r  .  m  v  s  r  r  q  r  c  s  t  .  p  q  k  s  l  q  t  a
g  .  r  g  a  l  .  r  r  t  e  w  h  g  e  w  c  h  a  a  s  g  a  a  p  d  r  r  s  h  c  r  q  r
----.----+----.----+----.----+----.----+----.----+----.----+----.----+----.----+----.----+----.----+
  s  l  p  a  c  k  s  p  s  c  l  p  v  t  f  p  t  v  g  g  a  t  c  c  r  v  a  s  t  v  a  s  l
h  i  v  p  r  v  k  v  a  f  l  t  a  r  h  i  t  d  r  r  w  r  h  l  v  q  g  c  f  d  s  c  v  a
 p  y  r  p  a  s  q  r  r  v  s  h  c  p  s  h  h  .  a  a  l  p  a  a  g  s  r  l  l  .  q  l  c
```

```
*063   17394     Mapped site PST1
J        ...     CONTINUES IN PHASE ONE

               HT      H             H             N          F
               HH      G             P             S          N
               AA      E             H             P          U
               11      2             1             B          H
                /
GGGGAATATCAGGTGCTGGCGCGATGGGACACACCGAAGGTGGTGAAGGGCGTGAGTTTCCTGCTCCGTCTGACCGTAACAGCGGACGACGGCAGTGAGC
----.----+----.----+----.----+----.----+----.----+----.----+----.----+----.----+----.----+----.----+ 17500
CCCCTTATAGTCCACGACCGCGCTACCCTGTGTGGCTTCCACCACTTCCCGCACTCAAAGGACGAGGCAGACTGGCATTGTCGCCTGCTGCCGTCACTCG

G E Y Q V L A R W D T P K V V K G V S F L L R L T V T A D D G S E R
g n i r c w r d g t h r r w . r a . v s c s v . p . q r t t a v s
 g i s g a g a ■ g h t e g g e g r e f p a p s d r n s g r r q . a
----.----+----.----+----.----+----.----+----.----+----.----+----.----+----.----+----.----+----.----+
p s y . t s a r h s v g f t t f p t l k r s r r v t v a s s p l s
 p f i l h q r s p v c r l h h l a h t e q e t q g y c r v v a t l
r p i d p a p a i p c v s p p s p r s n g a g d s r l l p r r c h a

J        ...     CONTINUES IN PHASE ONE

        AH NSH                           HH            HNS
        SA CCP                           HA            PCC
        UE IRA                           AE            AIR
        13 112                           12            211
           //                                            /
GGCTGGTCAGCACGGCCCGGACGACGGAAACCACATACCGCTTCACGCAACTGGCGCTGGGGAACTACAGGCTGACAGTCCGGGCGGTAAATGCGTGGGG
----.----+----.----+----.----+----.----+----.----+----.----+----.----+----.----+----.----+----.----+ 17600
CCGACCAGTCGTGCCGGGCCTGCTGCCTTTGGTGTATGGCGAAGTGCGTTGACCGCGACCCCTTGATGTCCGACTGTCAGGCCCGCCATTTACGCACCCC

 L V S T A R T T E T T Y R F T Q L A L G N Y R L T V R A V N A W G
g w s a r p g r r k p h t a s r n w r w g t t g . q s g r . ■ r g g
 a g q h g p d d g n h i p l h a t g a g e l q a d s p g g k c v g
----.----+----.----+----.----+----.----+----.----+----.----+----.----+----.----+----.----+----.----+
r s t l v a r v v s v v y r k v c s a s p f . l s v t r a t f a h p
 p q d a r g p r r f g c v a e r l q r q p v v p q c d p r y i r p
  a p . c p g s s p f w ■ g s . a v p a p s s c a s l g p p l h t p

J        ...     CONTINUES IN PHASE ONE

FF  H   MB BH  A        H     F   BH  F M    T  B    A   A  HNS  H
CN  G   BT BP  C        P     N   GP  N N    A  B    L   C  PCC  G
1U  A   ON VA  Y        A     U   LA  U L    Q  V    U   Y  AIR  A
5H  1   11 12  1        2     H   12  H 1    1  1    1   1  211  1
         /                         /                         /
GCAGCAGGGCGATCCGGCGTCGGTATCGTTCCGGATTGCCGCACCGGCAGCACCGTCGAGGATTGAGCTGACGCCGGGCTATTTTCAGATAACCGCCACG
----.----+----.----+----.----+----.----+----.----+----.----+----.----+----.----+----.----+----.----+ 17700
CGTCGTCCCGCTAGGCCGCAGCCATAGCAAGGCCTAACGGCGTGGCCGTCGTGGCAGCTCCTAACTCGACTGCGGCCCGATAAAAGTCTATTGGCGGTGC

 Q Q G D P A S V S F R I A A P A A P S R I E L T P G Y F Q I T A T
  s r a i r r r y r s g l p h r q h r r g l s . r r a i f r . p p r
a a g r s g v g i v p d c r t g s t v e d . a d a g l f s d n r h a
----.----+----.----+----.----+----.----+----.----+----.----+----.----+----.----+----.----+----.----+
 c c p s g a d t d n r i a a g a a g d l i s s v g p . k . i v a v
p l l a i r r r y r e p n g c r c c r r p n l q r r a i k l y g g r
 a a p r d p t p i t g s q r v p l v t s s q a s a p s n e s l r w

J        ...     CONTINUES IN PHASE ONE

F     S           R                                E          AM T
N     F           S                                C          FL H
U     A           A                                R          LU A
H     N           1                                V          31 1

CCGCATCTTGCCGTTTATGACCCGACGGTACAGTTTGAGTTCTGGTTCTCGGAAAAGCAGATTGCGGATATCAGACAGGTTGAAACCAGCACGCGTTATC
----.----+----.----+----.----+----.----+----.----+----.----+----.----+----.----+----.----+----.----+ 17800
GGCGTAGAACGGCAAATACTGGGCTGCCATGTCAAACTCAAGACCAAGAGCCTTTTCGTCTAACGCCTATAGTCTGTCCAACTTTGGTCGTGCGCAATAG

P H L A V Y D P T V Q F E F W F S E K Q I A D I R Q V E T S T R Y L
 r i l p f ■ t r r y s l s s g s r k s r l r i s d r l k p a r v i
  a s c r l . p d g t v . v l v l g k a d c g y q t g . n q h a l s
----.----+----.----+----.----+----.----+----.----+----.----+----.----+----.----+----.----+----.----+
 g c r a t . s g v t c n s n q n e s f c i a s i l c t s v l v r .
  r ■ k g n i v r r y l k l e p e r f l l n r i d s l n f g a r t i
a a d q r k h g s p v t q t r t r p f a s q p y . v p q f w c a n d

J        ...     CONTINUES IN PHASE ONE
```

```
  R    HH R      F                HNSAH                                            HM
  S    HA S      N                PCCSA                                            NN
  A    AE A      U                AIRUE                                            FL
  1    12 1      H                21113                                            31
                                   //                                               /
TTGGTACGGCGCTGTACTGGATAGCCGCCAGTATCAATATCAAACCGGGCCATGATTATTACTTTTATATCCGCAGTGTGAACACCGTTGGCAAATCGGC
----.----+----.----+----.----+----.----+----.----+----.----+----.----+----.----+----.----+----.----+ 17900
AACCATGCCGCGACATGACCTATCGGCGGTCATAGTTATAGTTTGGCCCGGTACTAATAATGAAAATATAGGCGTCACACTTGTGGCAACCGTTTAGCCG

 G  T  A  L  Y  W  I  A  A  S  I  N  I  K  P  G  H  D  Y  Y  F  Y  I  R  S  V  N  T  V  G  K  S  A
l  v  r  r  c  t  g  .  p  p  v  s  i  s  n  r  a  m  i  i  t  f  i  s  a  v  .  t  p  l  a  n  r  h
 w  y  g  a  v  l  d  s  r  q  y  q  y  q  t  g  p  .  l  l  l  l  y  p  q  c  e  h  r  w  q  i  g
----.----+----.----+----.----+----.----+----.----+----.----+----.----+----.----+----.----+----.----+
r  p  v  a  s  y  q  i  a  a  l  i  l  i  l  g  p  w  s  .  .  k  .  i  r  l  t  f  v  t  p  l  d  a
 k  t  r  r  q  v  p  y  g  g  t  d  i  d  f  r  a  m  i  i  v  k  i  d  a  t  h  v  g  n  a  f  r
  q  y  p  a  t  s  s  l  r  w  y  .  y  .  v  p  g  h  n  n  s  k  y  g  c  h  s  c  r  q  c  i  p

J      ...     CONTINUFS IN PHASE ONE
```

```
      H          S           B  E S                            H H     B           AFE
      A          F           S  C C                            N N     B           LNC
      E          A           T  R R                            F F     V           UU1
      3          N           E  2 1                            3 1     1           1H5
                                                                                    /
ATTCGTGGAGGCCGTCGGTCGGGCGAGCGATGATGCGGAAGGTTACCTGGATTTTTTCAAAGGCAAGATAACCGAATCCCATCTCGGCAAGGAGCTGCTG
----.----+----.----+----.----+----.----+----.----+----.----+----.----+----.----+----.----+----.----+ 18000
TAAGCACCTCCGGCAGCCAGCCCGCTCGCTACTACGCCTTCCAATGGACCTAAAAAAGTTTCCGTTCTATTGGCTTAGGGTAGAGCCGTTCCTCGACGAC

F  V  F  A  V  G  R  A  S  D  D  A  E  G  Y  L  D  F  F  K  G  K  I  T  E  S  H  L  G  K  E  L  L
 s  w  r  p  s  v  g  r  a  m  m  r  k  v  t  w  i  f  s  k  a  r  .  p  n  p  i  s  a  r  s  c  w
i  r  g  g  r  r  s  g  e  r  .  c  g  r  l  p  g  f  f  q  r  q  d  n  r  i  p  s  r  q  g  a  a  g
----.----+----.----+----.----+----.----+----.----+----.----+----.----+----.----+----.----+----.----+
 n  t  s  a  t  p  r  a  l  s  s  a  s  p  .  r  s  k  k  l  p  l  i  v  s  d  w  r  p  l  s  s  s
c  e  h  l  g  d  t  p  r  a  i  i  r  f  t  v  q  i  k  e  f  a  l  y  g  f  g  m  e  a  l  l  q  q
 m  r  p  p  r  r  d  p  s  r  h  h  p  l  n  g  p  n  k  .  l  c  s  l  r  i  g  d  r  c  p  a  a

J      ...     CONTINUES IN PHASE ONE
```

```
     T MA                 E  M  B        AT  S      H           F            B
     A NL                 C  N  S        SA  F      G           O            G
     Q LU                 1  L  T        UQ  A      U           K            L
     1 11                 5  1  X        21  N      2           1            1
                                          /
GAAAAAGTCGAGCTGACGGAGGATAACGCCAGCAGACTGGAGGAGTTTTCGAAAGAGTGGAAGGATGCCAGTGATAAGTGGAATGCCATGTGGGCTGTCA
----.----+----.----+----.----+----.----+----.----+----.----+----.----+----.----+----.----+----.----+ 18100
CTTTTTCAGCTCGACTGCCTCCTATTGCGGTCGTCTGACCTCCTCAAAAGCTTTCTCACCTTCCTACGGTCACTATTCACCTTACGGTACACCCGACAGT

E  K  V  E  L  T  E  D  N  A  S  R  L  E  E  F  S  K  E  W  K  D  A  S  D  K  W  N  A  M  W  A  V  K
 k  k  s  s  .  r  r  i  t  p  a  d  w  r  s  f  r  k  s  g  r  m  p  v  i  s  g  m  p  c  g  l  s
  k  s  r  a  d  g  g  .  r  q  q  t  g  g  v  f  e  r  v  e  g  c  q  .  .  v  e  c  h  v  g  c  q
----.----+----.----+----.----+----.----+----.----+----.----+----.----+----.----+----.----+----.----+
 s  f  t  s  s  v  s  s  l  a  l  l  s  s  s  n  e  f  s  h  f  s  a  l  s  l  h  f  a  m  h  a  t
  f  f  d  l  q  r  l  i  v  g  a  s  q  l  l  k  r  f  l  p  l  i  g  t  i  l  p  i  g  h  p  s  d
p  f  l  r  a  s  p  p  y  r  w  c  v  p  p  t  k  s  l  t  s  p  h  w  h  y  t  s  h  w  t  p  q  .

J      ...     CONTINUES IN PHASE ONE
```

```
      E                      T T     HHDM     MM                D               F
      C                      H T     AADN     NN                D               N
      P                      A H     EEEL     LL                E               U
      1                      1 2     1311     11                1               H
                                      /
AAATTGAGCAGACCAAAGACGGCAAACATTATGTCGCGGGTATTGGCCTCAGCATGGAGGACACGGAGGAAGGCAAACTGAGCCAGTTTCTGGTTGCCGC
----.----+----.----+----.----+----.----+----.----+----.----+----.----+----.----+----.----+----.----+ 18200
TTTAACTCGTCTGGTTTCTGCCGTTTGTAATACAGCGCCCATAACCGGAGTCGTACCTCCTGTGCCTCCTTCCGTTTGACTCGGTCAAAGACCAACGGCG

  I  E  Q  T  K  D  G  K  H  Y  V  A  G  I  G  L  S  M  E  D  T  E  F  G  K  L  S  Q  F  L  V  A  A
k  l  s  r  p  k  t  a  n  i  m  s  r  v  l  a  s  a  w  r  t  r  r  k  a  n  .  a  s  f  w  l  p  p
 n  .  a  d  q  r  r  q  t  l  c  r  g  y  w  p  q  h  g  g  h  g  g  r  q  t  e  p  v  s  g  c  r
----.----+----.----+----.----+----.----+----.----+----.----+----.----+----.----+----.----+----.----+
l  i  s  c  v  l  s  p  l  c  .  t  a  p  i  p  r  l  m  s  s  v  s  s  p  l  s  l  w  n  r  t  a  a
 f  n  l  l  g  f  v  a  f  m  i  d  r  t  n  a  e  a  h  l  v  r  l  f  a  f  q  a  l  k  q  n  g
  f  q  a  s  w  l  r  c  v  n  h  r  p  y  q  g  .  c  p  p  c  p  p  l  c  v  s  g  t  e  p  q  r

J      ...     CONTINUES IN PHASF ONE
```

```
                        NSH       T             H               A       HH
                        CCP       T             H               F       HA
                        IRA       H             A               L       AE
                        112       2             1               3       12
                         //
CAATCGTATCGCATTTATTGACCCGGCAAACGGGAATGAAACGCCGATGTTTGTGGCGCAGGGCAACCAGATATTCATGAACGACGTGTTCCTGAAGCGC
----.----+----.----+----.----+----.----+----.----+----.----+----.----+----.----+----.----+----.----+  18300
GTTAGCATAGCGTAAATAACTGGGCCGTTTGCCCTTACTTTGCGGCTACAAACACCGCGTCCCGTTGGTCTATAAGTACTTGCTGCACAAGGACTTCGCG

 N  R  I  A  F  I  D  P  A  N  G  N  E  T  P  M  F  V  A  Q  G  N  Q  I  F  M  N  D  V  F  L  K  R
  i  v  s  h  l  l  t  r  q  t  g  m  k  r  r  c  l  w  r  r  a  t  r  y  s  .  t  t  c  s  .  s  a
q  s  y  r  i  y  .  p  g  k  r  e  .  n  a  d  v  c  g  a  g  q  p  d  i  h  e  r  r  v  p  e  a  p
----.----+----.----+----.----+----.----+----.----+----.----+----.----+----.----+----.----+----.----+
 l  r  i  a  n  i  s  g  a  f  p  f  s  v  g  i  n  t  a  c  p  l  w  i  n  m  f  s  t  n  r  f  r
g  i  t  d  c  k  n  v  r  c  v  p  i  f  r  r  h  k  h  r  l  a  v  l  y  e  h  v  v  h  e  q  l  a
w  d  y  r  m  .  q  g  p  l  r  s  h  f  a  s  t  q  p  a  p  c  g  s  i  .  s  r  r  t  g  s  a

J     ...     CONTINUES IN PHASE ONE


   AH          N  F  F        H  H  M          H          A                E
   SA          S  N  N        P  A  N          P          L                C
   UE          P  U  U        A  E  L          A          U                R
   13          B  H  H        2  3  1          2          1                V

CTGACGGCCCCCACCATTACCAGCGGCGGCAATCCTCCGGCCTTTTCCCTGACACCGGACGGAAAGCTGACCGCTAAAAATGCGGATATCAGTGGCAGTG
----.----+----.----+----.----+----.----+----.----+----.----+----.----+----.----+----.----+----.----+  18400
GACTGCCGGGGGTGGTAATGGTCGCCGCCGTTAGGAGGCCGGAAAAGGGACTGTGGCCTGCCTTTCGACTGGCGATTTTTACGCCTATAGTCACCGTCAC

L  T  A  P  T  I  T  S  G  G  N  P  P  A  F  S  L  T  P  D  G  K  L  T  A  K  N  A  D  I  S  G  S  V
 .  r  p  p  p  l  p  a  a  a  i  l  r  p  f  p  .  h  r  t  e  s  .  p  l  k  m  r  i  s  v  a  v
  d  g  p  h  h  y  q  r  r  q  s  s  g  l  f  p  d  t  g  r  k  a  d  r  .  k  c  g  y  q  w  q  c
----.----+----.----+----.----+----.----+----.----+----.----+----.----+----.----+----.----+----.----+
 r  v  a  g  v  m  v  l  p  p  l  g  g  a  k  e  r  v  g  s  p  f  s  v  a  l  f  a  s  i  l  p  l
  q  r  g  g  g  n  g  a  a  a  i  r  r  g  k  g  q  c  r  v  s  l  q  g  s  f  i  r  i  d  t  a  t
g  s  p  g  w  w  .  w  r  r  c  d  e  p  r  k  g  s  v  p  r  f  a  s  r  .  f  h  p  y  .  h  c  h

J     ...     CONTINUES IN PHASE ONE


              HNS   D  H          A       R      MR D          T
              PCC   D  G          L       S      NS D          '
              AIR   E  A          U       A      LA E          J
              211   1  1          1       1      11 1          4
               /
TGAATGCGAACTCCGGGACGCTCAGTAATGTGACGATAGCTGAAAACTGTACGATAAACGGTACGCTGAGGGCGGAAAAAATCGTCGGGGACATTGTAAA
----.----+----.----+----.----+----.----+----.----+----.----+----.----+----.----+----.----+----.----+  18500
ACTTACGCTTGAGGCCCTGCGAGTCATTACACTGCTATCGACTTTTGACATGCTATTTGCCATGCGACTCCCGCCTTTTTTAGCAGCCCCTGTAACATTT

  N  A  N  S  G  T  L  S  N  V  T  I  A  E  N  C  T  I  N  G  T  L  R  A  E  K  I  V  G  D  I  V  K
 .  m  r  t  p  g  r  s  v  m  .  r  .  l  k  t  v  r  .  t  v  r  .  g  r  k  k  s  s  g  t  l  .  r
 e  c  e  l  r  d  a  q  .  c  d  d  s  .  k  l  y  d  k  r  y  a  e  g  g  k  n  r  r  g  h  c  k
----.----+----.----+----.----+----.----+----.----+----.----+----.----+----.----+----.----+----.----+
 t  f  a  f  e  p  v  s  i  l  l  t  v  i  a  s  f  q  v  i  f  p  v  s  l  a  s  f  i  t  p  s  m  t  f
  h  i  r  v  g  p  r  e  t  i  h  r  y  s  f  v  t  r  y  v  t  r  q  p  r  f  f  d  d  p  v  n  y
   s  h  s  s  r  s  a  .  y  h  s  s  l  q  f  s  y  s  l  r  y  a  s  p  p  f  f  r  r  p  c  q  l

T'J4   18482       t'J4 terminator     3'      RNA     terminates leftward transcription
J      ...        CONTINUES IN PHASE ONE


  F       HTF       T H                   GCH      H R K*H R       F                 M  T
  N       HHN       H H                   DFA      G S P0P S       O                 B  '
  U       AAU       A A                   IRE      I A N6H A       K                 O  J
  H       11H       1 1                   213      C 1 141 1       1                 1  3
           //                              /           //
GGCGGCGAGCGCGGCTTTTCCGCGCCAGCGTGAAAGCAGTGTGGACTGGCCGTCAGGTACCCGTACTGTCACCGTGACCGATGACCATCCTTTTGATCGC
----.----+----.----+----.----+----.----+----.----+----.----+----.----+----.----+----.----+----.----+  18600
CCGCCGCTCGCGCCGAAAAGGCGCGGTCGCACTTTCGTCACACCTGACCGGCAGTCCATGGGCATGACAGTGGCACTGGCTACTGGTAGGAAAACTAGCG

 A  A  S  A  A  F  P  R  Q  R  E  S  S  V  D  W  P  S  G  T  R  T  V  T  V  T  D  D  H  P  F  D  R
  r  r  a  r  l  f  r  a  s  v  k  a  v  w  t  g  r  q  v  p  v  l  s  p  .  p  m  t  i  l  l  i  a
g  g  e  r  g  f  s  a  p  a  .  k  q  c  g  l  a  v  r  y  p  y  c  h  r  d  r  .  p  s  f  .  s  p
----.----+----.----+----.----+----.----+----.----+----.----+----.----+----.----+----.----+----.----+
 a  a  l  a  a  k  g  r  w  r  s  l  l  t  s  q  g  g  d  p  v  r  v  t  v  t  v  s  s  w  g  k  s  r
 l  r  r  a  r  s  k  r  a  l  t  f  a  t  h  v  p  r  .  t  g  t  s  d  g  h  g  i  v  m  r  k  i  a
  p  p  s  r  p  k  e  a  g  a  h  f  c  h  p  s  a  t  l  y  g  y  q  .  r  s  r  h  g  d  k  q  d
```

```
*064   18560      Mapped site KPN1
T'J3   18597        t'j3                   3'           RNA     terminates leftward transcription
J       ...       CONTINUES IN PHASE ONE


               N            TH      T     R    N   F            H  T   T           S  RR
               S            H4      '     S    S   N            N  A   '           C  RS
               P            A3      J     A    P   U            F  Q   J           A  UA
               B            14      2     1    B   H            3  1   1           1  11
                                                                                       /
CAGATAGTGGTGCTTCCGCTGACGTTTCGCGGAAGTAAGCGTACTGTCAGCGGCAGGACAACGTATTCGATGTGTTATCTGAAAGTACTGATGAACGGTG
----.----+----.----+----.----+----.----+----.----+----.----+----.----+----.----+----.----+----.----+ 18700
GTCTATCACCACGAAGGCGACTGCAAAGCGCCTTCATTCGCATGACAGTCGCCGTCCTGTTGCATAAGCTACACAATAGACTTTCATGACTACTTGCCAC

Q  I  V  V  L  P  L  T  F  R  G  S  K  R  T  V  S  G  R  T  T  Y  S  M  C  Y  L  K  V  L  M  N  G  A
 r  .  w  c  f  r  .  r  f  a  e  v  s  v  l  s  a  a  g  q  r  i  r  c  v  i  .  k  y  .  .  t  v
  d  s  g  a  s  a  d  v  s  r  k  .  a  y  c  q  r  q  d  n  v  f  d  v  l  s  e  s  t  d  e  r  c
----.----+----.----+----.----+----.----+----.----+----.----+----.----+----.----+----.----+----.----+
 w  i  t  t  s  g  s  v  n  r  p  l  l  r  v  t  l  p  l  v  v  y  e  i  h  .  r  f  t  s  i  f  p
  l  y  h  h  k  r  q  r  k  a  s  t  l  l  t  s  d  a  a  p  c  r  i  r  h  t  i  q  f  y  q  h  v  t
g  s  l  p  a  e  a  s  t  e  r  f  y  a  y  q  .  r  c  s  l  t  n  s  t  n  d  s  l  v  s  s  r  h

H434   18630        h434                   sl           EM
T'J2   18637        t'j2                   3'           RNA     terminates leftward transcription
T'J1   18671        t'j1                   3'           RNA     terminates leftward transcription
J       ...       CONTINUES IN PHASE ONE


             H   HTFM           R                     H*    NN                  MB
             P   HHNN           S                     I0    SS                  BI
             H   AAUL           A                     N6    PP                  ON
             1   11H1           1                     25    CB                  11
                    ///                                 /                          /
CGGTGATTTATGATGGCGCGGCGAACGAGGCGGTACAGGTGTTCTCCCGTATTGTTGACATGCCAGCGGGTCGGGGAAACGTGATCCTGACGTTCACGCT
----.----+----.----+----.----+----.----+----.----+----.----+----.----+----.----+----.----+----.----+ 18800
GCCACTAAATACTACCGCGCCGCTTGCTCCGCCATGTCCACAAGAGGGCATAACAACTGTACGGTCGCCCAGCCCCTTTGCACTAGGACTGCAAGTGCGA

  V  I  Y  D  G  A  A  N  E  A  V  Q  V  F  S  R  I  V  D  M  P  A  G  R  G  N  V  I  L  T  F  T  L
r  .  f  ■  ■  a  r  r  t  r  r  y  r  c  s  p  v  l  l  t  c  q  r  v  g  e  t  .  s  .  r  s  r  l
 g  d  l  .  w  r  g  e  r  g  g  t  g  v  l  p  y  c  .  h  a  s  g  s  g  k  r  d  p  d  v  h  a
----.----+----.----+----.----+----.----+----.----+----.----+----.----+----.----+----.----+----.----+
a  t  i  .  s  p  a  a  f  s  a  t  c  t  n  e  r  i  t  s  ■  g  a  p  r  p  f  t  i  r  v  n  v  s
 r  h  n  i  i  a  r  r  v  l  r  y  l  h  e  g  t  n  n  v  h  w  r  t  p  s  v  h  d  q  r  e  r
  p  s  k  h  h  r  p  s  r  p  p  v  p  t  r  g  y  q  q  c  a  l  p  d  p  f  r  s  g  s  t  .  a

*065   18756      Mapped site HIN2
J       ...       CONTINUES IN PHASE ONE


             H                  S A                                       H      HH             SE JD
             N                  N C                                       P      HA             FC  D
             F                  A C                                       H      AE             AP  E
             3                  1 1                                       1      12             N1  1
                                                                                                 /  /
TACGTCCACACGGCATTCGGCAGATATTCCGCCGTATACGTTTGCCAGCGATGTGCAGGTTATGGTGATTAAGAAACAGGCGCTGGGCATCAGCGTGGTC+ 18900
ATGCAGGTGTGCCGTAAGCCGTCTATAAGGCGGCATATGCAAACGGTCGCTACACGTCCAATACCACTAATTCTTTGTCCGCGACCCGTAGTCGCACCAG

 T  S  T  R  H  S  A  D  I  P  P  Y  T  F  A  S  D  V  Q  V  M  V  I  K  K  Q  A  L  G  I  S  V  V
  r  p  h  g  i  r  q  i  f  r  r  i  r  l  p  a  ■  c  r  l  w  .  l  r  n  r  r  w  a  s  a  w  s
y  v  h  t  a  f  g  r  y  s  a  v  y  v  c  q  r  c  a  g  y  g  d  .  e  t  g  a  g  h  q  r  g  l
----.----+----.----+----.----+----.----+----.----+----.----+----.----+----.----+----.----+----.----+
  v  d  v  r  c  e  a  s  i  g  g  y  v  n  a  l  s  t  c  t  i  t  i  l  f  c  a  s  p  ■  l  t  t
k  r  g  c  p  ■  r  c  i  n  r  r  i  r  k  g  a  i  h  l  n  h  h  n  l  f  l  r  q  a  d  a  h  d
 .  t  w  v  a  n  p  l  y  e  a  t  y  t  q  w  r  h  a  p  .  p  s  .  s  v  p  a  p  c  .  r  p

J      18900        J                      COOH
```

```
     M                HNS                     M      S          L   HS                            F
     N                PCC                     N      F          O   PB                            N
     L                AIR                     L      A          M   H2                            U
     1                211                     1      N              1                             H
                        /                                            /
TGAGTGTGTTACAGAGGTTCGTCCGGGAACGGGCGTTTTATTATAAAACAGTGAGAGGTGAACGATGCGTAATGTGTGTATTGCCGTTGCTGTCTTTGCC
----.----+----.----+----.----+----.----+----.----+----.----+----.----+----.----+----.----+----.----+ 19000
ACTCACACAATGTCTCCAAGCAGGCCCTTGCCCGCAAAATAATATTTTGTCACTCTCCACTTGCTACGCATTACACACATAACGGCAACGACAGAAACGG

.  v  c  y  r  g  s  s  g  n  g  r  f  i  i  k  q  .  e  v  n  d  a  .  c  v  y  c  r  c  c  l  c  r
 e  c  v  t  e  v  r  p  g  t  g  v  l  l  .  n  s  e  r  .  t  M  R  N  V  C  I  A  V  A  V  F  A
  s  v  l  q  r  f  v  r  e  r  a  f  y  y  k  t  v  r  g  e  r  c  v  ■  c  v  l  p  l  l  s  l  p
----.----+----.----+----.----+----.----+----.----+----.----+----.----+----.----+----.----+----.----+
 q  t  h  .  l  p  e  d  p  f  p  r  k  i  i  f  c  h  s  t  f  s  a  y  h  t  y  q  r  q  q  r  q
  s  h  t  v  s  t  r  g  p  v  p  t  k  n  y  f  l  s  l  h  v  i  r  l  t  h  i  a  t  a  t  k  a
r  l  t  n  c  l  n  t  r  s  r  a  n  .  .  l  v  t  l  p  s  r  h  t  i  h  t  n  g  n  s  d  k  g

LOM    18965     lom (orf-206)     M-NH3
SB2    18969     Sequence Block 2  end


                    HH AH                      R                 H            HNS   R
                    PP SA                      S                 4            PCC   S
                    HA UE                      A                 3            AIR   A
                    12 13                      1                 4            211   1
                                                                                /
GCACTTGCGGTGACAGTCACTCCGGCCCGTGCGGAAGGTGGACATGGTACGTTTACGGTGGGCTATTTTCAAGTGAAACCGGGTACATTGCCGTCGTTGT
----.----+----.----+----.----+----.----+----.----+----.----+----.----+----.----+----.----+----.----+ 19100
CGTGAACGCCACTGTCAGTGAGGCCGGGCACGCCTTCCACCTGTACCATGCAAATGCCACCCGATAAAAGTTCACTTTGGCCCATGTAACGGCAGCAACA

  t  c  g  d  s  h  s  g  p  c  g  r  w  t  w  y  v  y  g  g  l  f  s  s  e  t  g  y  i  a  v  v  v
A  L  A  V  T  V  T  P  A  R  A  E  G  G  H  G  T  F  T  V  G  Y  F  Q  V  K  P  G  T  L  P  S  L  S
 h  l  r  .  q  s  l  r  p  v  r  k  v  d  ■  v  r  l  r  w  a  i  f  k  .  n  r  v  h  c  r  r  c
----.----+----.----+----.----+----.----+----.----+----.----+----.----+----.----+----.----+----.----+
r  v  q  p  s  l  .  e  p  g  h  p  l  h  v  h  y  t  .  p  p  s  n  e  l  s  v  p  y  ■  a  t  t  t
 a  s  a  t  v  t  v  g  a  r  a  s  p  p  c  p  v  n  v  t  p  .  k  .  t  f  g  p  v  n  g  d  n
  c  k  r  h  c  d  s  r  g  t  r  f  t  s  ■  t  r  k  r  h  a  i  k  l  h  f  r  t  c  q  r  r  q

H434   19066     h434              sr            EM
LOM    ...       CONTINUES IN PHASE TWO


            H      H                           R          A                            EHSH      F
            P      N                           S          L                            CPCG      N
            A      F                           A          U                            RHRA      U
            2      1                           1          1                            2111      H

CGGGCGGGGATACCGGTGTGAGTCATCTGAAAGGGATTAACGTGAAGTACCGTTATGAGCTGACGGACAGTGTGGGGGTGATGGCTTCCCTGGGGTTCGC
----.----+----.----+----.----+----.----+----.----+----.----+----.----+----.----+----.----+----.----+ 19200
GCCCGCCCCTATGGCCACACTCAGTAGACTTTCCCTAATTGCACTTCATGGCAATACTCGACTGCCTGTCACACCCCCACTACCGAAGGGACCCCAAGCG

 g  r  g  y  r  c  e  s  s  e  r  d  .  r  e  v  p  l  .  a  d  g  q  c  g  g  d  g  f  p  g  v  r
  G  G  D  T  G  V  S  H  L  K  G  I  N  V  K  Y  R  Y  E  L  T  D  S  V  G  V  M  A  S  L  G  F  A
r  a  g  i  p  v  .  v  i  .  k  g  l  t  .  s  t  v  ■  s  .  r  t  v  w  g  .  w  l  p  w  g  s  p
----.----+----.----+----.----+----.----+----.----+----.----+----.----+----.----+----.----+----.----+
  p  r  p  y  r  h  s  d  d  s  l  s  .  r  s  t  g  n  h  a  s  p  c  h  p  p  s  p  k  g  p  t  r
d  p  p  s  v  p  t  l  .  r  f  p  i  l  t  f  y  r  .  s  s  v  s  l  t  p  t  i  a  e  r  p  n  a
 r  a  p  i  g  t  h  t  ■  q  f  p  n  v  h  l  v  t  i  l  q  r  v  t  h  p  h  h  s  g  q  p  e

LOM    ...       CONTINUES IN PHASE TWO


 T  T          F         BM HNS                                                 GCHH AA H
 H  A          N         BN PCC                                                 DFAP SV P
 A  Q          U         VL AIR                                                 IREA UA A
 1  1          H         11 211                                                 2132 12 2
                          /   /                                                  /      /
CGCGTCGAAAAAGAGCAGCACAGTGATGACCGGGGAGGATACGTTTCACTATGAGAGCCTGCGTGGACGTTATGTGAGCGTGATGGCCGGACCGGTTTTA
----.----+----.----+----.----+----.----+----.----+----.----+----.----+----.----+----.----+----.----+ 19300
GCGCAGCTTTTTCTCGTCGTGTCACTACTGGCCCCTCCTATGCAAAGTGATACTCTCGGACGCACCTGCAATACACTCGCACTACCGGCCTGGCCAAAAT

r  v  e  k  e  q  h  s  d  d  r  g  g  y  v  s  l  .  e  p  a  w  t  l  c  e  r  d  g  r  t  g  f  t
 A  S  K  K  S  S  T  V  M  T  G  E  D  T  F  H  Y  E  S  L  R  G  R  Y  V  S  V  M  A  G  P  V  L
  r  r  k  r  a  a  q  .  .  p  g  r  i  r  f  t  ■  r  a  c  v  d  v  ■  .  a  .  w  p  d  r  f  y
----.----+----.----+----.----+----.----+----.----+----.----+----.----+----.----+----.----+----.----+
 r  t  s  f  s  c  c  l  s  s  r  p  p  y  t  e  s  h  s  g  a  h  v  n  h  s  r  s  p  r  v  p  k
  a  d  f  f  l  l  v  t  i  v  p  s  s  v  n  .  .  s  l  r  r  p  r  .  t  l  t  i  a  p  g  t  k
g  r  r  f  l  a  a  c  h  h  g  p  l  i  r  k  v  i  l  a  q  t  s  t  i  h  a  h  h  g  s  r  n  .

LOM    ...       CONTINUES IN PHASE TWO
```

```
                        R   N  GCHH    B               AA H      R  L                            *NXASNHS*
                        S   C  DFAP    G               SV P      S  A                            0CMVCCPC0
                        A   0  IREA    L               UA A      A  C                            6IAARIAR6
                        1   1  2132    1               12 2      1  5                            611111217
                                /                        /                                       ///////
CAAATCAGTAAGCAGGTCAGTGCGTACGCCATGGCCGGAGTGGCTCACAGTCGGTGGTCCGGCAGTACAATGGATTACCGTAAGACGGAAATCACTCCCG
----.----+----.----+----.----+----.----+----.----+----.----+----.----+----.----+----.----+----.----+ 19400
GTTTAGTCATTCGTCCAGTCACGCATGCGGTACCGGCCTCACCGAGTGTCAGCCACCAGGCCGTCATGTTACCTAATGGCATTCTGCCTTTAGTGAGGGC

  n  q  .  a  g  q  c  v  r  h  g  r  s  g  s  q  s  v  v  r  q  y  n  g  l  p  .  d  g  n  h  s  r
Q  I  S  K  Q  V  S  A  Y  A  M  A  G  V  A  H  S  R  W  S  G  S  T  M  D  Y  R  K  T  E  I  T  P  G
 k  s  v  s  r  s  v  r  t  p  w  p  e  w  l  t  v  g  g  p  a  v  q  w  i  t  v  r  r  k  s  l  p
----.----+----.----+----.----+----.----+----.----+----.----+----.----+----.----+----.----+----.----+
v  f  .  y  a  p  .  h  t  r  w  p  r  l  p  e  c  d  t  t  r  c  y  l  p  n  g  y  s  p  f  .  e  r
 c  i  l  l  c  t  l  a  y  a  m  a  p  t  a  .  l  r  h  d  p  l  v  i  s  .  r  l  v  s  i  v  g
  l  d  t  l  l  d  t  r  v  g  h  g  s  h  s  v  t  p  p  g  a  t  c  h  i  v  t  l  r  f  d  s  g

LAC5   19369     lac5                   sl          sq
*066   19397     Mapped site AVA1
*067   19399     Mapped site SMA1
LOM     ...      CONTINUES IN PHASE TWO
```

```
                       E S                  F       D        M        S A        H    H   F
                       C C                  N       D        N        N C        G    P   N
                       R R                  U       E        L        A C        A    A   U
                       2 1                  H       1        1        1 1        1    2   H

GGTATATGAAAGAGACGACCACTGCCAGGGACGAAAGTGCAATGCGGCATACCTCAGTGGCGTGGAGTGCAGGTATACAGATTAATCCGGCAGCGTCCGT
----.----+----.----+----.----+----.----+----.----+----.----+----.----+----.----+----.----+----.----+ 19500
CCATATACTTTCTCTGCTGGTGACGGTCCCTGCTTTCACGTTACGCCGTATGGAGTCACCGCACCTCACGTCCATATGTCTAATTAGGCCGTCGCAGGCA

 v  y  e  r  d  d  h  c  q  g  r  k  c  n  a  a  y  l  s  g  v  e  c  r  y  t  d  .  s  g  s  v  r
  Y  M  K  E  T  T  T  A  R  D  E  S  A  M  R  H  T  S  V  A  W  S  A  G  I  Q  I  N  P  A  A  S  V
g  i  .  k  r  r  p  l  p  g  t  k  v  q  c  g  i  p  q  w  r  g  v  q  v  y  r  l  i  r  q  r  p  s
----.----+----.----+----.----+----.----+----.----+----.----+----.----+----.----+----.----+----.----+
  t  y  s  l  s  s  w  q  w  p  r  f  h  l  a  a  y  r  l  p  t  s  h  l  y  v  s  .  d  p  l  t  r
p  y  i  f  s  v  v  v  a  l  s  s  l  a  i  r  c  v  e  t  a  h  l  a  p  i  c  i  l  g  a  a  d  t
 p  i  h  f  l  r  g  s  g  p  v  f  t  c  h  p  m  g  .  h  r  p  t  c  t  y  l  n  i  r  c  r  g

LOM     ...      CONTINUES IN PHASE TWO
```

```
  B                        H                  R       H                          L      E S
  B                        P                  S       N                          0      C C
  V                        A                  A       F                          M      R R
  1                        2                  1       1                                 2 1

CGTTGTTGATATTGCTTATGAAGGCTCCGGCAGTGGCGACTGGCGTACTGACGGATTCATCGTTGGGGTCGGTTATAAATTCTGATTAGCCAGGTAACAC
----.----+----.----+----.----+----.----+----.----+----.----+----.----+----.----+----.----+----.----+ 19600
GCAACAACTATAACGAATACTTCCGAGGCCGTCACCGCTGACCGCATGACTGCCTAAGTAGCAACCCCAGCCAATATTTAAGACTAATCGGTCCATTGTG

r  c  .  y  c  l  .  r  l  r  q  w  r  l  a  y  .  r  i  h  r  w  g  r  l  .  i  l  i  s  q  v  t  q
 V  V  D  I  A  Y  E  G  S  G  S  G  D  W  R  T  D  G  F  I  V  G  V  G  Y  K  F  .  l  a  r  .  h
  l  l  i  l  l  m  k  a  p  a  v  a  t  g  v  l  t  d  s  s  l  g  s  v  i  n  s  d  .  p  g  n  t
----.----+----.----+----.----+----.----+----.----+----.----+----.----+----.----+----.----+----.----+
 r  q  q  y  q  k  h  l  s  r  c  h  r  s  a  y  q  r  i  .  r  q  p  r  n  y  i  r  i  l  w  t  v
  t  t  s  i  a     s  p  e  p  l  p  s  q  r  v  s  p  n  m  t  p  t  p  .  l  n  q  n  a  l  y  c
d  n  n  i  n  s  i  f  a  g  a  t  a  v  p  t  s  v  s  e  d  n  p  d  t  i  f  e  s  .  g  p  l  v

LOM    19582     lom                  COOH
```

```
                  H     H                    BB  O    H             H                          H  RB
                  P     P                    25  R    P             N                          P  S1
                  A     A                    21  F    H             F                          A  A8
                  2     2                    19       1             1                          2  19
                                             /
AGTGTTATGACAGCCCGCCGGAACCGGTGGGCTTTTTTGTGGGGTGAATATGGCAGTAAAGATTTCAGGAGTCCTGAAAGACGGCACAGGAAAACCGGTA
----.----+----.----+----.----+----.----+----.----+----.----+----.----+----.----+----.----+----.----+ 19700
TCACAATACTGTCGGGCGGCCTTGGCCACCCGAAAAAACACCCCACTTATACCGTCATTTCTAAAGTCCTCAGGACTTTCTGCCGTGTCCTTTTGGCCAT

  c  y  d  s  p  p  e  p  v  g  f  f  v  g  .  i  w  q  .  r  f  q  e  s  .  k  t  a  q  e  n  r  y
s  v  m  t  a  r  r  n  r  w  a  f  l  w  g  e  y  g  s  k  d  f  r  s  p  e  r  r  h  r  k  t  g  t
 v  l  .  q  p  a  g  t  g  g  l  f  c  g  v  n  M  A  V  K  I  S  G  V  L  K  D  G  T  G  K  P  V
----.----+----.----+----.----+----.----+----.----+----.----+----.----+----.----+----.----+----.----+
c  h  .  s  l  g  g  s  g  t  p  k  k  t  p  h  i  h  c  y  l  n  .  s  d  q  f  v  a  c  s  f  r  y
 l  t  i  v  a  r  r  f  r  h  a  k  k  h  p  s  y  p  l  l  s  k  l  l  g  s  l  r  c  l  f  v  p
  t  n  h  c  q  a  p  v  p  p  s  k  q  p  t  f  i  a  t  f  i  e  p  t  r  f  s  p  v  p  f  g  t
```

```
B221   19646     b221          dl          EM
B519   19646     b519          dl          EM
ORF    19650     orf-401       M-NH3
B189   19700     b189          dl-att      RFS
```

```
              N  PA*                           B             SH  DBH   H   HH        HNS  F
              S  VL0                           S             DP  DVG   N   PG        PCC  0
              P  UU6                           T             UH  EUI   F   AU        AIR  K
              B  218                           X             11  11J   1   22        211  1
                  //                                               /                   /
CAGAACTGCACCATTCAGCTGAAAGCCAGACGTAACAGCACCACGGTGGTGGTGAACACGGTGGGCTCAGAGAATCCGGATGAAGCCGGGCGTTACAGCA
----.----+----.----+----.----+----.----+----.----+----.----+----.----+----.----+----.----+----.----+ 19800
GTCTTGACGTGGTAAGTCGACTTTCGGTCTGCATTGTCGTGGTGCCACCACCACTTGTGCCACCCGAGTCTCTTAGGCCTACTTCGGCCCGCAATGTCGT

 r  t  a  p  f  s  .  k  p  d  v  t  a  p  r  w  w  .  t  r  w  a  q  r  i  r  m  k  p  g  v  t  a
  e  l  h  h  s  a  e  s  q  t  .  q  h  h  g  g  g  e  h  g  g  l  r  e  s  g  .  s  r  a  l  q  h
Q  N  C  T  I  Q  L  K  A  R  R  N  S  T  T  V  V  V  N  T  V  G  S  E  N  P  D  E  A  G  R  Y  S  M
----.----+----.----+----.----+----.----+----.----+----.----+----.----+----.----+----.----+----.----+
 l  v  a  g  n  l  q  f  g  s  t  v  a  g  r  h  h  h  v  r  h  a  .  l  i  r  i  f  g  p  t  v  a
v  s  s  c  w  e  a  s  l  w  v  y  c  c  w  p  p  p  s  c  p  p  s  l  s  d  p  h  l  r  a  n  c  c
  c  f  q  v  m  .  s  f  a  l  r  l  l  v  v  t  t  t  f  v  t  p  e  s  f  g  s  s  a  p  r  .  l

*068   19718     Mapped site PVU2
ORF    ...       CONTINUES IN PHASE THREE
```

```
H          R  FF    R               P*  H*                     HNSAAH                  H       MNHS
G          S  00    S               S0  I0                     PCCSVP                  N       BCPC
U          A  KK    A               T6  N7                     AIRUAH                  F       OIAR
2          1  11    1               19  20                     211121                  1       2121
               /                     /   /                       / //                            //
TGGATGTGGAGTACGGTCAGTACAGTGTCATCCTGCAGGTTGACGGTTTTCCACCATCGCACGCCGGGACCATCACCGTGTATGAAGATTCACAACCGGG
----.----+----.----+----.----+----.----+----.----+----.----+----.----+----.----+----.----+----.----+ 19900
ACCTACACCTCATGCCAGTCATGTCACAGTAGGACGTCCAACTGCCAAAAGGTGGTAGCGTGCGGCCCTGGTAGTGGCACATACTTCTAAGTGTTGGCCC

w  m  w  s  t  v  s  t  v  s  s  c  r  l  t  v  f  h  h  r  t  p  g  p  s  p  c  m  k  i  h  n  r  g
 g  c  g  v  r  s  v  q  c  h  p  a  g  .  r  f  s  t  i  a  r  r  d  h  h  r  v  .  r  f  t  t  g
  D  V  E  Y  G  Q  Y  S  V  I  L  Q  V  D  G  F  P  P  S  H  A  G  T  I  T  V  Y  E  D  S  Q  P  G
----.----+----.----+----.----+----.----+----.----+----.----+----.----+----.----+----.----+----.----+
 h  i  h  l  v  t  l  v  t  d  d  q  l  n  v  t  k  w  w  r  v  g  p  g  d  g  h  i  f  i  .  l  r
  p  h  p  t  r  d  t  c  h  .  g  a  p  q  r  n  e  v  m  a  r  r  s  w  .  r  t  h  l  n  v  v  p
m  s  t  s  y  p  .  y  l  t  m  r  c  t  s  p  k  g  g  d  c  a  p  v  m  v  t  y  s  s  e  c  g  p

*069   19837     Mapped site PST1
*070   19841     Mapped site HIN2
ORF    ...       CONTINUES IN PHASE THREE
```

```
          H             M   S    H       MNSHXFGGCHHBH F                  M         P H   HMT AM T
          G             N   F    G       NCCPMODDFAPBG N                  N         G G   HBH FL H
          A             L   A    U       LIRAAKIIREAVA U                  L         L A   AOA LU A
          1             1   N    2       1112312213211 H                  1         1 1   121 31 1
                                            /// //////                                      /
GACGCTGAATGATTTTCTCTGTGCCATGACGGAGGATGATGCCCGGCCGGAGGTGCTGCGTCGTCTTGAACTGATGGTGGAAGAGGTGGCGCGTAACGCG
----.----+----.----+----.----+----.----+----.----+----.----+----.----+----.----+----.----+----.----+ 20000
CTGCGACTTACTAAAAGAGACACGGTACTGCCTCCTACTACGGGCCGGCCTCCACGACGCAGCAGAACTTGACTACCACCTTCTCCACCGCGCATTGCGC

 r  .  m  i  f  s  v  p  .  r  r  m  m  p  g  r  r  c  c  v  v  l  n  .  w  w  k  r  w  r  v  t  r
d  a  e  .  f  s  l  c  h  d  g  g  .  c  p  a  g  g  a  a  s  s  .  t  d  g  g  r  g  g  a  .  r  v
  T  L  N  D  F  L  C  A  M  T  E  D  D  A  R  P  E  V  L  R  R  L  E  L  M  V  E  E  V  A  R  N  A
----.----+----.----+----.----+----.----+----.----+----.----+----.----+----.----+----.----+----.----+
p  r  q  i  i  k  e  t  g  h  r  l  i  i  g  p  r  l  h  q  t  t  k  f  q  h  h  f  l  h  r  t  v  r
 s  a  s  h  n  e  r  h  w  s  p  p  h  h  g  a  p  p  a  a  d  d  q  v  s  p  p  l  p  p  a  y  r
  v  s  f  s  k  r  q  a  m  v  s  s  s  a  r  g  s  t  s  r  r  r  s  s  i  t  s  s  t  a  r  l  a

PGL1   19985        lambdapgl1          sl-att      EM
ORF    ...       CONTINUES IN PHASE THREE
```

```
                 R         T     H S   HMN     B         N PA*SFD       TFAH S                  H D
                 S         H     G F   PBA     B         S VLØFND       HNSA F                  N D
                 A         A     A A   AOE     V         P UU7AUF       AUUF A                  F E
                 1         1     1 N   221     1         B 211NH1       1H13 N                  1 1
                                          /                  // //
TCCGTGGTGGCACAGAGTACGGCAGACGCGAAGAAATCAGCCGGCGATGCCAGTGCATCAGCTGCTCAGGTCGCGGCCCTTGTGACTGATGCAACTGACT
----.----+----.----+----.----+----.----+----.----+----.----+----.----+----.----+----.----+----.----+  20100
AGGCACCACCGTGTCTCATGCCGTCTGCGCTTCTTTAGTCGGCCGCTACGGTCACGTAGTCGACGAGTCCAGCGCCGGGAACACTGACTACGTTGACTGA

 p  w  w  h  r  v  r  q  t  r  r  n  q  p  a  m  p  v  h  q  l  l  r  s  r  p  l  .  l  m  q  l  t
  r  g  g  t  e  y  g  r  r  e  e  i  s  r  r  c  q  c  i  s  c  s  g  r  g  p  c  d  .  c  n  .  l
S  V  V  A  Q  S  T  A  D  A  K  K  S  A  G  D  A  S  A  S  A  A  Q  V  A  A  L  V  T  D  A  T  D  S
----.----+----.----+----.----+----.----+----.----+----.----+----.----+----.----+----.----+----.----+
  g  h  h  c  l  t  r  c  v  r  l  f  .  g  a  i  g  t  c  .  s  s  l  d  r  g  k  h  s  i  c  s  v
t  r  p  p  v  s  y  p  l  r  s  s  i  l  r  r  h  w  h  m  l  q  e  p  r  p  g  q  s  q  h  l  q  s
 d  t  t  a  c  l  v  a  s  a  f  f  d  a  p  s  a  l  a  d  a  a  .  t  a  a  r  t  v  s  a  v  s

*Ø71  20061     Mapped site PVU2
ORF    ...      CONTINUES IN PHASE THREE
```

```
      T HF         B    H    B   F        H SAD             H    H M    F          S  HH
      H HN         B    P    G   N        G FLD             P    H N    N          F  AA
      A AU         V'   A    L   U        A AUE             A    A L    U          A  EE
      1 1H         1    2    1   H        1 N11             2    1 1    H          N  13
                                             //                                       /
CAGCACGCGCCGCCAGCACGTCCGCCGGACAGGCTGCATCGTCAGCTCAGGAAGCGTCCTCCGGCGCAGAAGCGGCATCAGCAAAGGCCACTGAAGCGGA
----.----+----.----+----.----+----.----+----.----+----.----+----.----+----.----+----.----+----.----+  20200
GTCGTGCGCGGCGGTCGTGCAGGCGGCCTGTCCGACGTAGCAGTCGAGTCCTTCGCAGGAGGCCGCGTCTTCGCCGTAGTCGTTTCCGGTGACTTCGCCT

q  h  a  p  p  a  r  p  p  d  r  l  h  r  q  l  r  k  r  p  p  a  q  k  r  h  q  q  r  p  l  k  r  k
 s  t  r  r  q  h  v  r  r  t  g  c  i  v  s  s  g  s  v  l  r  r  r  s  g  i  s  k  g  h  .  s  g
  A  R  A  A  S  T  S  A  G  Q  A  A  S  S  A  Q  E  A  S  S  G  A  E  A  A  S  A  K  A  T  E  A  E
----.----+----.----+----.----+----.----+----.----+----.----+----.----+----.----+----.----+----.----+
.  c  a  g  g  a  r  g  g  s  l  s  c  r  .  s  l  f  r  g  g  a  c  f  r  c  .  c  l  g  s  f  r
 l  v  r  r  w  c  t  r  r  v  p  q  m  t  l  e  p  l  t  r  r  r  l  l  p  m  l  l  p  w  q  l  p
e  a  r  a  a  l  v  d  a  p  c  a  a  d  d  a  .  s  a  d  e  p  a  s  a  a  d  a  f  a  v  s  a  s

ORF    ...      CONTINUES IN PHASE THREE
```

```
        F  F  F    H  B          MTF  FCH GH       H   BF          B     HH      F FP*         B
        N  N  N    N  B          NHN  NFA DG       P   GN          B     NG      N NSØ         B
        U  U  U    F  V          LAU  URE IE       A   LU          V     FA      U UT7         V
        H  H  H    1  1          11H  H13 22       2   1H          1     31      H H12         1
                                        /  /                                         /
AAAAAGTGCCGCAGCCGCAGAGTCCTCAAAAAACGCGGCGGCCACCAGTGCCGGTGCGGCGAAAACGTCAGAAACGAATGCTGCAGCGTCACAACAATCA
----.----+----.----+----.----+----.----+----.----+----.----+----.----+----.----+----.----+----.----+  20300
TTTTTCACGGCGTCGGCGTCTCAGGAGTTTTTTGCGCCGCCGGTGGTCACGGCCACGCCGCTTTTGCAGTCTTTGCTTACGACGTCGCAGTGTTGTTAGT

 k  v  p  q  p  q  s  p  q  k  t  r  r  p  p  v  p  v  r  r  k  r  q  k  r  m  l  q  r  h  n  n  q
k  k  c  r  s  r  r  v  l  k  k  r  g  g  h  q  c  r  c  g  e  n  v  r  n  e  c  c  s  v  t  t  i  s
 K  S  A  A  A  A  E  S  S  K  N  A  A  A  T  S  A  G  A  A  K  T  S  E  T  N  A  A  A  S  Q  Q  S
----.----+----.----+----.----+----.----+----.----+----.----+----.----+----.----+----.----+----.----+
f  f  t  g  c  g  c  l  g  .  f  v  r  r  g  g  t  g  t  r  r  f  r  .  f  r  i  s  c  r  .  l  l  .
 f  f  h  r  l  r  l  t  r  l  f  r  p  p  w  w  h  r  h  p  s  f  t  l  f  s  h  q  l  t  v  v  i
  f  l  a  a  a  a  s  d  e  f  f  a  a  a  v  l  a  p  a  a  f  v  d  s  v  f  a  a  a  d  c  c  d

*Ø72  20285     Mapped site PST1
ORF    ...      CONTINUES IN PHASE THREE
```

```
 F                 N TSMF*GCHH M          HF      S                  HHM        M F           B
 N                 S HSNNØDFAG N          AN      F                  AAN        N N           B
 U                 P ATLU7IREA L          EU      A                  EEL        L U           V
 H                 B 121H32131 1          3H      N                  131        1 H           1
                       // ///                                         //
GCCGCCACGTCTGCCTCCACCGCGGCCACGAAAGCGTCAGAGGCCGCCACTTCAGCACGAGATGCGGTGGCCTCAAAAGAGGCAGCAAAATCATCAGAAA
----.----+----.----+----.----+----.----+----.----+----.----+----.----+----.----+----.----+----.----+  20400
CGGCGGTGCAGACGGAGGTGGCGCCGGTGCTTTCGCAGTCTCCGGCGGTGAAGTCGTGCTCTACGCCACCGGAGTTTTCTCCGTCGTTTTAGTAGTCTTT

 p  p  r  l  p  p  p  r  p  r  k  r  q  r  p  p  l  q  h  e  m  r  w  p  q  k  r  q  q  n  h  q  k
  r  h  v  c  l  h  r  g  h  e  s  v  r  g  r  h  f  s  t  r  c  g  g  l  k  r  g  s  k  i  i  r  n
A  A  T  S  A  S  T  A  A  T  K  A  S  E  A  A  T  S  A  R  D  A  V  A  S  K  E  A  A  K  S  S  E  T
----.----+----.----+----.----+----.----+----.----+----.----+----.----+----.----+----.----+----.----+
  g  g  r  r  g  g  g  r  g  r  f  r  .  l  g  g  s  .  c  s  i  r  h  g  .  f  l  c  c  f  .  .  f
l  r  w  t  q  r  w  r  p  w  s  l  t  l  p  r  w  k  l  v  l  h  p  p  r  l  l  p  l  l  i  m  l  f
 a  a  v  d  a  e  v  a  a  v  f  a  d  s  a  a  v  e  a  r  s  a  t  a  e  f  s  a  a  f  d  d  s
```

```
*073   20323     Mapped site SST2
ORF      ...     CONTINUES IN PHASE THREE
```

```
                S  HB    F A     B   M F        E S  BF G       H   E S
                F  P5    N L     B   N N        C C  GN 1       N   C C
                A  A3    U U     V   L U        R R  LU 3       F   R R
                N  26    H 1     1   1 H        2 1  1H 0       3   2 1

CGAACGCATCATCAAGTGCCGGTCGTGCAGCTTCCTCGGCAACGGCGGCAGAAAATTCTGCCAGGGCGGCAAAAACGTCCGAGACGAATGCCAGGTCATC
----.----+----.----+----.----+----.----+----.----+----.----+----.----+----.----+----.----+----.----+  20500
GCTTGCGTAGTAGTTCACGGCCAGCACGTCGAAGGAGCCGTTGCCGCCGTCTTTTAAGACGGTCCCGCCGTTTTTGCAGGCTCTGCTTACGGTCCAGTAG

r  t  h  h  q  v  p  v  v  q  l  p  r  q  r  r  q  k  i  l  p  g  r  q  k  r  p  r  r  *  p  g  h  l
 e  r  i  i  k  c  r  s  c  s  f  l  g  n  g  g  r  k  f  c  q  g  g  k  n  v  r  d  e  c  q  v  i
  N  A  S  S  S  A  G  R  A  A  S  S  A  T  A  A  E  N  S  A  R  A  A  K  T  S  E  T  N  A  R  S  S
----.----+----.----+----.----+----.----+----.----+----.----+----.----+----.----+----.----+----.----+
 r  v  c  .  .  t  g  t  t  c  s  g  r  c  r  r  c  f  i  r  g  p  r  c  f  r  g  l  r  i  g  p  .
  s  r  *  *  l  h  r  d  h  l  k  r  p  l  p  p  l  f  n  q  w  p  p  l  f  t  r  s  s  h  w  t  *
v  f  a  d  d  l  a  p  r  a  a  e  e  a  v  a  a  s  f  e  a  l  a  a  f  v  d  s  v  f  a  l  d  d
```

```
B536   20420      b536                dl        EM
G130   20469       lambdagal130       sl - att  EM
ORF      ...      CONTINUES IN PHASF THREE
```

```
   E  NF           B  HH    NF TSMF*      HN  F  FH    F     H*        BSM      F
   C  SN           B  HA    SN HSNN0      GS  N  NG    0     I0        BFN      N
   1  PU           V  AE    PU ATLU7      AP  U  UA    K     N7        VAL      U
   5  BH           1  12    BH 121H4      1B  H  H1    1     25        1N1      H
                                  //      /     /          /
TGAAACAGCAGCGGAACGGAGCGCCTCTGCCGCGGCAGACGCAAAAACAGCGGCGGCGGGGAGTGCGTCAACGGCATCCACGAAGGCGACAGAGGCTGCG
----.----+----.----+----.----+----.----+----.----+----.----+----.----+----.----+----.----+----.----+  20600
ACTTTGTCGTCGCCTTGCCTCGCGGAGACGGCGCCGTCTGCGTTTTTGTCGCCGCCGCCCCTCACGCAGTTGCCGTAGGTGCTTCCGCTGTCTCCGACGC

 k  q  q  r  n  g  a  p  l  p  r  q  t  q  k  q  r  r  r  g  v  r  q  r  h  p  r  r  r  q  r  l  r
.  n  s  s  g  t  e  r  l  c  r  g  r  r  k  n  s  g  g  g  e  c  v  n  g  i  h  e  g  d  r  g  c  g
 E  T  A  A  E  R  S  A  S  A  A  A  D  A  K  T  A  A  A  G  S  A  S  T  A  S  T  K  A  T  E  A  A
----.----+----.----+----.----+----.----+----.----+----.----+----.----+----.----+----.----+----.----+
r  f  c  c  r  f  p  a  g  r  g  r  c  v  c  f  c  r  r  r  p  t  r  .  r  c  g  r  l  r  c  l  s  r
 q  f  l  l  p  v  s  r  r  q  r  p  l  r  l  f  l  p  p  p  s  h  t  l  p  *  w  s  p  s  l  p  q
  s  v  a  a  s  r  l  a  e  a  a  a  s  a  f  v  a  a  a  p  l  a  d  v  a  d  v  f  a  v  s  a  a
```

```
*074   20533      Mapped site SST2
*075   20569      Mapped site HIN2
ORF      ...      CONTINUES IN PHASE THREE
```

```
                  S        F      F F          H                   A  MN PA*
                  F        N      N N          N                   L  BS VL0
                  A        U      U U          F                   U  OP UU7
                  N        H      H H          3                   1  2B 216
                                                                       /  //
GGAAGTGCGGTATCAGCATCGCAGAGCAAAAGTGCGGCAGAAGCGGCGGCAATACGTGCAAAAAATTCGGCAAAACGTGCAGAAGATATAGCTTCAGCTG
----.----+----.----+----.----+----.----+----.----+----.----+----.----+----.----+----.----+----.----+  20700
CCTTCACGCCATAGTCGTAGCGTCTCGTTTTCACGCCGTCTTCGCCGCCGTTATGCACGTTTTTTAAGCCGTTTTGCACGTCTTCTATATCGAAGTCGAC

 e  v  r  y  q  h  r  r  a  k  v  r  q  k  r  r  q  y  v  q  k  i  r  q  n  v  q  k  i  .  l  q  l
  k  c  g  i  s  i  a  e  q  k  c  g  r  s  g  g  n  t  c  k  k  f  g  k  t  c  r  r  y  s  f  s  c
G  S  A  V  S  A  S  Q  S  K  S  A  A  E  A  A  A  I  R  A  K  N  S  A  K  R  A  E  D  I  A  S  A  V
----.----+----.----+----.----+----.----+----.----+----.----+----.----+----.----+----.----+----.----+
 s  t  r  y  .  c  r  l  a  f  t  r  c  f  r  r  c  y  t  c  f  i  r  c  f  t  c  f  i  y  s  .  s
p  f  h  p  i  l  *  a  s  c  f  h  p  l  l  p  p  l  v  h  l  f  n  p  l  v  h  l  l  y  l  k  l  q
 p  l  a  t  d  a  d  c  l  l  l  a  a  s  a  a  a  i  r  a  f  f  e  a  f  r  a  s  s  i  a  e  a
```

```
*076   20697      Mapped site PVU2
ORF      ...      CONTINUES IN PHASE THREE
```

```
MST H    H            F                F ADE      B            B          F
NFH H    G            O                N LDC      B            B          N
LAA A    U            K                U UE1      V            V          U
1N1 1    2            1                H 115      1            1          H

TCGCGCTTGAGGATGCGGACACAACGAGAAAGGGGATAGTGCAGCTCAGCAGTGCAACCAACAGCACGTCTGAAACGCTTGCTGCAACGCCAAAGGCGGT
----.----+----.----+----.----+----.----+----.----+----.----+----.----+----.----+----.----+----.----+  20800
AGCGCGAACTCCTACGCCTGTGTTGCTCTTTCCCCTATCACGTCGAGTCGTCACGTTGGTTGTCGTGCAGACTTTGCGAACGACGTTGCGGTTTCCGCCA

s  r  l  r  *  r  t  q  r  e  r  g  .  c  s  s  a  v  q  p  t  a  r  l  k  r  l  l  q  r  q  r  r  l
 r  a  .  g  c  g  h  n  e  k  g  d  s  a  a  q  q  c  n  q  q  h  v  .  n  a  c  c  n  a  k  g  g
  A  L  E  D  A  D  T  T  R  K  G  I  V  Q  L  S  S  A  T  N  S  T  S  E  T  L  A  A  T  P  K  A  V
----.----+----.----+----.----+----.----+----.----+----.----+----.----+----.----+----.----+----.----+
 d  r  k  l  i  r  v  c  r  s  l  p  y  h  l  e  a  t  c  g  v  a  r  r  f  r  k  s  c  r  w  l  r
  r  a  q  p  h  p  c  l  s  f  p  s  l  a  a  .  c  h  l  w  c  c  t  q  f  a  q  q  l  a  l  p  p
t  a  s  s  s  a  s  v  v  l  f  p  i  t  c  s  l  l  a  v  l  l  v  d  s  v  s  a  a  v  g  f  a  t

ORF      ...      CONTINUES IN PHASE THREE
```

```
     B H          F               H   O  H             T HD
     5 G          O               P   R  P             H HD
     3 U          K               A   F  A             A AE
     8 2          1               2      2             1 11

TAAGGTGGTAATGGATGAAACGAACAGAAAAGCCCACTGGACAGTCCGGCACTGACCGGAACGCCAACAGCACCAACCGCGCTCAGGGGAACAAACAATA
----.----+----.----+----.----+----.----+----.----+----.----+----.----+----.----+----.----+----.----+  20900
ATTCCACCATTACCTACTTTGCTTGTCTTTTCGGGTGACCTGTCAGGCCGTGACTGGCCTTGCGGTTGTCGTGGTTGGCGCGAGTCCCCTTGTTTGTTAT

  r  w  .  w  *  k  r  t  e  k  p  t  g  q  s  g  t  d  r  n  a  n  s  t  n  r  a  q  g  n  k  q  y
 .  g  g  n  g  .  n  e  q  k  s  p  l  d  s  p  a  l  t  g  t  p  t  a  p  t  a  l  r  g  t  n  n  t
 K  V  V  M  D  E  T  N  R  K  A  H  W  T  V  R  H  .  p  e  r  q  q  h  q  p  r  s  g  e  q  t  i
----.----+----.----+----.----+----.----+----.----+----.----+----.----+----.----+----.----+----.----+
n  l  h  y  h  i  f  r  v  s  f  g  v  p  c  d  p  v  s  r  f  a  l  l  v  l  r  a  .  p  f  l  c  y
 .  p  p  l  p  h  f  s  c  f  l  g  s  s  l  g  a  s  v  p  v  g  v  a  g  v  a  s  l  p  v  f  l
  l  t  t  i  s  s  v  f  l  f  a  w  q  v  t  r  c  q  g  s  r  w  c  c  w  g  r  e  p  s  c  v  i

B538   20809       b538                 dl        EM
ORF    20852       orf-401              COOH
```

```
     T                 R  GCHF T       H     THAM T    H          H              GCHF  F   HH*A
     T                 S  DFAN H       G     APFL H    G          G              DFAN  N   HA0V
     H                 A  IREU A       A     QHLU A    A          A              IREU  U   AE7A
     2                 1  213H 1       1     1131 1    1          1              213H  H   1271
                            /                  /                                   /       //
CCCAGATTGCGAACACCGCTTTTGTACTGGCCGCGATTGCAGATGTTATCGACGCGTCACCTGACGCACTGAATACGCTGAATGAACTGGCCGCAGCGCT
----.----+----.----+----.----+----.----+----.----+----.----+----.----+----.----+----.----+----.----+  21000
GGGTCTAACGCTTGTGGCGAAAACATGACCGGCGCTAACGTCTACAATAGCTGCGCAGTGGACTGCGTGACTTATGCGACTTACTTGACCGGCGTCGCGA

 p  d  c  e  h  r  f  c  t  g  r  d  c  r  c  y  r  r  v  t  .  r  t  e  y  a  e  .  t  g  r  s  a
  q  i  a  n  t  a  f  v  l  a  a  i  a  d  v  i  d  a  s  p  d  a  l  n  t  l  n  e  l  a  a  a  l
p  r  l  r  t  p  l  l  y  w  p  r  l  q  *  l  s  t  r  h  l  t  h  .  i  r  .  *  n  w  p  q  r  s
----.----+----.----+----.----+----.----+----.----+----.----+----.----+----.----+----.----+----.----+
 g  s  q  s  c  r  k  q  v  p  r  s  q  l  h  .  r  r  t  v  q  r  v  s  y  a  s  h  v  p  r  l  a
v  w  i  a  f  v  a  k  t  s  a  a  i  a  s  t  i  s  a  d  g  s  a  s  f  v  s  f  s  s  a  a  a  s
 g  l  n  r  v  g  s  k  y  q  g  r  n  c  i  n  d  v  r  .  r  v  c  q  i  r  q  i  f  q  g  c  r

*077   20999     Mapped site AVA1
```

```
   B MB                O      T H                          M      HH
   B BI                R      H H                          B      HA
   V ON                F      A A                          0      AE
   1 11                       1 1                          2      12
      /
CGGGAATGATCCAGATTTTGCTACCACCATGACTAACGCGCTTGCGGGTAAACAACCGAAGAATGCGACACTGACGGCGCTGGCAGGGCTTTCCACGGCG
----.----+----.----+----.----+----.----+----.----+----.----+----.----+----.----+----.----+----.----+  21100
GCCCTTACTAGGTCTAAAACGATGGTGGTACTGATTGCGCGAACGCCCATTTGTTGGCTTCTTACGCTGTGACTGCCGCGACCGTCCCGAAAGGTGCCGC

r  e  .  s  r  f  c  y  h  h  d  .  r  a  c  g  .  t  t  e  e  c  d  t  d  g  a  g  r  a  f  h  g  e
 g  n  d  p  d  f  a  t  t  M  T  N  A  L  A  G  K  Q  P  K  N  A  T  L  T  A  L  A  G  L  S  T  A
  g  *  i  q  i  l  l  p  p  .  l  t  r  l  r  v  n  n  r  r  *  r  h  .  r  r  w  q  g  f  p  r  r
----.----+----.----+----.----+----.----+----.----+----.----+----.----+----.----+----.----+----.----+
 r  s  h  d  l  n  q  .  w  w  s  .  r  a  q  p  y  v  v  s  s  h  s  v  s  p  a  p  l  a  k  w  p
  p  f  s  g  s  k  a  v  v  *  v  l  a  s  a  p  l  c  g  f  f  a  v  s  v  a  s  a  p  s  e  v  a
e  p  i  i  w  i  k  s  g  g  h  s  v  r  k  r  t  f  l  r  l  i  r  c  q  r  r  q  c  p  k  g  r  r

ORF    21029       orf-314              M-NH3
```

```
                    S             F                  H D
                    F             N                  N D
                    A             U                  F E
                    N             H                  1 1

AAAAATAAATTACCGTATTTTGCGGAAAAATGATGCCGCCAGCCTGACTGAACTGACTCAGGTTGGCAGGGATATTCTGGCAAAAAATTCCGTTGCAGATG
----.----+----.----+----.----+----.----+----.----+----.----+----.----+----.----+----.----+----.----+ 21200
TTTTTATTTAATGGCATAAAACGCCTTTTTACTACGGCGGTCGGACTGACTTGACTGAGTCCAACCGTCCCTATAAGACCGTTTTTTAAGGCAACGTCTAC

  k  .  i  t  v  f  c  g  k  .  c  r  q  p  d  .  t  d  s  g  w  q  g  y  s  g  k  k  f  r  c  r  c
K  N  K  L  P  Y  F  A  E  N  D  A  A  S  L  T  E  L  T  Q  V  G  R  D  I  L  A  K  N  S  V  A  D  V
 k  i  n  y  r  i  l  r  k  ■  ■  p  p  a  .  l  n  .  l  r  l  a  g  i  f  w  q  k  i  p  l  q  ■
----.----+----.----+----.----+----.----+----.----+----.----+----.----+----.----+----.----+----.----+
s  f  y  i  v  t  n  q  p  f  h  h  r  w  g  s  q  v  s  e  p  q  c  p  y  e  p  l  f  n  r  q  l  h
 f  f  l  n  g  y  k  a  s  f  s  a  a  l  r  v  s  s  v  .  t  p  l  s  i  r  a  f  f  e  t  a  s
  f  i  f  .  r  i  k  r  f  i  i  g  g  a  q  s  f  q  s  l  n  a  p  i  n  q  c  f  i  g  n  c  i

ORF      ...      CONTINUES IN PHASE TWO
```

```
               A HH      *EH    HH    BH          H MB       C BHH        E           S      B
               S AP      ØCN    PA    GP          H BI       F AAA        C           F      5
               U EA      7RF    HE    LA          A ON       R LEE        R           A      Ø
               1 32      813    13    12          1 11       1 113        V           N      6
                          //     /     /              /         //
TTCTTGAATACCTTGGGGCCGGTGAGAATTCGGCCTTTCCGGCAGGTGCGCCGATCCCGTGGCCATCAGATATCGTTCCGTCTGGCTACGTCCTGATGCA
----.----+----.----+----.----+----.----+----.----+----.----+----.----+----.----+----.----+----.----+ 21300
AAGAACTTATGGAACCCCGGCCACTCTTAAGCCGGAAAGGCCGTCCACGCGGCTAGGGCACCGGTAGTCTATAGCAAGGCAGACCGATGCAGGACTACGT

 s  .  i  p  w  g  r  .  e  f  g  l  s  g  r  c  a  d  p  v  a  i  r  y  r  s  v  w  l  r  p  d  a
  L  E  Y  L  G  A  G  E  N  S  A  F  P  A  G  A  P  I  P  W  P  S  D  I  V  P  S  G  Y  V  L  M  Q
f  l  n  t  l  g  p  v  r  i  r  p  f  r  q  v  r  r  s  r  g  h  q  i  s  f  r  l  a  t  s  .  c  r
----.----+----.----+----.----+----.----+----.----+----.----+----.----+----.----+----.----+----.----+
  e  q  i  g  q  p  r  h  s  n  p  r  e  p  l  h  a  s  g  t  a  ■  l  y  r  e  t  q  s  r  g  s  a
t  r  s  y  r  p  a  p  s  f  e  a  k  g  a  p  a  g  i  g  h  g  d  s  i  t  g  d  p  .  t  r  i  c
 n  k  f  v  k  p  g  t  l  i  r  g  k  r  c  t  r  r  d  r  p  w  .  i  d  n  r  r  a  v  d  q  h

*Ø78     21226      Mapped site ECR1
B5Ø6     21292        B5Ø6                  dl           EM
ORF      ...        CONTINUES IN PHASE TWO
```

```
                                           T                      M
                                           H                      N
                                           A                      L
                                           1                      1

GGGGCAGGCGTTTGACAAATCAGCCTACCCAAAACTTGCTGTCGCGTATCCATCGGGTGTGCTTCCTGATATGCGAGGCTGGACAATCAAGGGGAAACCC
----.----+----.----+----.----+----.----+----.----+----.----+----.----+----.----+----.----+----.----+ 21400
CCCCGTCCGCAAACTGTTTAGTCGGATGGGTTTTGAACGACAGCGCATAGGTAGCCCACACGAAGGACTATACGCTCCGACCTGTTAGTTCCCCTTTGGG

g  a  g  v  .  q  i  s  l  p  k  t  c  c  r  v  s  i  g  c  a  s  .  y  a  r  l  d  n  q  g  e  t  r
 G  Q  A  F  D  K  S  A  Y  P  K  L  A  V  A  Y  P  S  G  V  L  P  D  M  R  G  W  T  I  K  G  K  P
  g  r  r  l  t  n  q  p  t  q  n  l  l  s  r  i  h  r  v  c  f  l  i  c  e  a  g  q  s  r  g  n  p
----.----+----.----+----.----+----.----+----.----+----.----+----.----+----.----+----.----+----.----+
 p  a  p  t  q  c  i  l  r  g  l  v  q  q  r  t  d  ■  p  h  a  e  q  y  a  l  s  s  l  .  p  s  v
  p  c  a  n  s  l  d  a  .  g  f  s  a  t  a  y  g  d  p  t  s  g  s  i  r  p  q  v  i  l  p  f  g
l  p  l  r  k  v  f  .  g  v  w  f  k  s  d  r  i  w  r  t  h  k  r  i  h  s  a  p  c  d  l  p  f  g

ORF      ...      CONTINUES IN PHASE TWO
```

```
 N                    D        H             F       F                H   RS
 S                    D        G             O       O                P   SF
 P                    E        U             K       K                A   AA
 B                    1        2             1       1                2   1N

GCCAGCGGTCGTGCTGTATTGTCTCAGGAACAGGATGGAATTAAGTCGCACACCCACAGTGCCAGTGCATCCGGTACGGATTTGGGGACGAAAACCACAT
----.----+----.----+----.----+----.----+----.----+----.----+----.----+----.----+----.----+----.----+ 21500
CGGTCGCCAGCACGACATAACAGAGTCCTTGTCCTACCTTAATTCAGCGTGTGGGTGTCACGGTCACGTAGGCCATGCCTAAACCCCTGCTTTTGGTGTA

  q  r  s  c  c  i  v  s  g  t  g  w  n  .  v  a  h  p  q  c  q  c  i  r  y  g  f  g  d  e  n  h  i
A  S  G  R  A  V  L  S  Q  E  Q  D  G  I  K  S  H  T  H  S  A  S  A  S  G  T  D  L  G  T  K  T  T  S
 p  a  v  v  l  y  c  l  r  n  r  ■  e  l  s  r  t  p  t  v  p  v  h  p  v  r  i  w  g  r  k  p  h
----.----+----.----+----.----+----.----+----.----+----.----+----.----+----.----+----.----+----.----+
r  w  r  d  h  q  i  t  e  p  v  p  h  f  .  t  a  c  g  c  h  w  h  ■  r  y  p  n  p  s  s  f  w  ■
 a  l  p  r  a  t  n  d  .  s  c  s  p  i  l  d  c  v  w  l  a  l  a  d  p  v  s  k  p  v  f  v  v
  g  a  t  t  s  y  q  r  l  f  l  i  s  n  l  r  v  g  v  t  g  t  c  g  t  r  i  q  p  r  f  g  c

ORF      ...      CONTINUES IN PHASE TWO
```

```
                                    T       H        T  H           S    BH           D
                                    A       G        A  N           D    VG           D
                                    Q       I        Q  F           U    UI           E
                                    1       C        1  3           1    1J           1
                                                                          /
CGTCGTTTGATTACGGGACGAAAACAACAGGCAGTTTCGATTACGGCACCAAATCGACGAATAACACGGGGGCTCATGCTCACAGTCTGAGCGGTTCAAC
----.----+----.----+----.----+----.----+----.----+----.----+----.----+----.----+----.----+----.----+ 21600
GCAGCAAACTAATGCCCTGCTTTTGTTGTCCGTCAAAGCTAATGCCGTGGTTTAGCTGCTTATTGTGCCCCCGAGTACGAGTGTCAGACTCGCCAAGTTG

 v  v  .  l  r  d  e  n  n  r  q  f  r  l  r  h  q  i  d  e  .  h  g  g  s  c  s  q  s  e  r  f  n
  S  F  D  Y  G  T  K  T  T  G  S  F  D  Y  G  T  K  S  T  N  N  T  G  A  H  A  H  S  L  S  G  S  T
r  r  l  i  t  g  r  k  q  q  a  v  s  i  t  a  p  n  r  r  i  t  r  g  l  ■  l  t  v  .  a  v  q  q
----.----+----.----+----.----+----.----+----.----+----.----+----.----+----.----+----.----+----.----+
  t  t  q  n  r  s  s  f  l  l  c  n  r  n  r  c  w  i  s  s  y  c  p  p  e  h  e  c  d  s  r  n  l
d  d  n  s  .  p  v  f  v  v  p  l  k  s  .  p  v  l  d  v  f  l  v  p  a  .  a  .  l  r  l  p  e  v
 r  r  k  i  v  p  r  f  c  c  a  t  e  i  v  a  g  f  r  r  i  v  r  p  s  ■  s  v  t  q  a  t  .
```

ORF ... CONTINUES IN PHASE TWO

```
  A HFNTS*S    H            B        H             F     H  S
  S ANSHSØD    G            S        G             O     N  B
  U EUPAT7U    I            T        U             K     F  3
  1 3HB1291    A            X        2             1     1
      /  /
AGGGGCCGCGGGTGCTCATGCCCACACAAGTGGTTTAAGGATGAACAGTTCTGGCTGGAGTCAGTATGGAACAGCAACCATTACAGGAAGTTTATCCACA
----.----+----.----+----.----+----.----+----.----+----.----+----.----+----.----+----.----+----.----+ 21700
TCCCCGGCGCCCACGAGTACGGGTGTGTTCACCAAATTCCTACTTGTCAAGACCGACCTCAGTCATACCTTGTCGTTGGTAATGTCCTTCAAATAGGTGT

r  g  r  g  c  s  c  p  h  k  w  f  k  d  e  q  f  w  l  e  s  v  w  n  s  n  h  y  r  k  f  i  h  s
 G  A  A  G  A  H  A  H  T  S  G  L  R  M  N  S  S  G  W  S  Q  Y  G  T  A  T  I  T  G  S  L  S  T
  g  p  r  v  l  ■  p  t  q  v  v  .  g  .  t  v  l  a  g  v  s  ■  e  q  q  p  l  q  e  v  y  p  q
----.----+----.----+----.----+----.----+----.----+----.----+----.----+----.----+----.----+----.----+
 l  p  r  p  h  e  h  g  c  l  h  n  l  s  s  c  n  q  s  s  d  t  h  f  l  l  w  .  l  f  n  i  w
  p  a  a  p  a  .  a  w  v  l  p  k  l  i  f  l  e  p  q  l  .  y  p  v  a  v  ■  v  p  l  k  d  v
c  p  g  r  t  s  ■  g  v  c  t  t  .  p  h  v  t  r  a  p  t  l  i  s  c  c  g  n  c  s  t  .  g  c
```

*Ø79 21609 Mapped site SST2
SB3 21661 Sequence Block 3 start
ORF ... CONTINUES IN PHASE TWO

```
             D                      STB                F          B        H   R             H S
             I                      BA2                N          B        P   S             P D
             F                      3Q                 U          V        A   A             A U
             F                        1                H          1        2   1             2 1
                                     //
GTTAAAGGAACCAGCACACAGGGTATTGCTTATTTATCGAAAACGGACAGTCAGGGCAGCCACAGTCACTCATTGTCCGGTACAGCCGTGAGTGCCGGTG
----.----+----.----+----.----+----.----+----.----+----.----+----.----+----.----+----.----+----.----+ 21800
CAATTTCCTTGGTCGTGTGTCCCATAACGAATAAATAGCTTTTGCCTGTCAGTCCCGTCGGTGTCAGTGAGTAACAGGCCATGTCGGCACTCACGGCCAC

  .  r  n  q  h  t  g  y  c  l  f  i  e  n  g  q  s  g  q  p  q  s  l  i  v  r  y  s  r  e  c  r  c
V  K  G  T  S  T  Q  G  I  A  Y  L  S  K  T  D  S  Q  G  S  H  S  H  S  L  S  G  T  A  V  S  A  G  A
 l  k  e  p  a  h  r  v  l  l  i  y  r  k  r  t  v  r  a  a  t  v  t  h  c  p  v  q  p  .  v  p  v
----.----+----.----+----.----+----.----+----.----+----.----+----.----+----.----+----.----+----.----+
l  .  l  f  w  c  v  p  y  q  k  n  i  s  f  p  c  d  p  c  g  c  d  s  ■  t  r  y  l  r  s  h  r  h
 t  l  p  v  l  v  c  p  i  a  .  k  d  f  v  s  l  .  p  l  w  l  .  e  n  d  p  v  a  t  l  a  p
  n  f  s  g  a  c  l  t  n  s  i  .  r  f  r  v  t  l  a  a  v  t  v  .  q  g  t  c  g  h  t  g  t
```

DIFF 21714 strain difference sq A in cb2 G in cI857S7
SB3 21737 Sequence Block 3 end
B2 21738 b2 dl-att sq
ORF ... CONTINUES IN PHASE TWO

```
H    NMH              F  R  MH         H    S      S    H                                H     B
G    SSH              O  E  SH         P    F      D    G                                P     B
I    PTA              K  V  TA         A    A      U    I                                H     V
A    C11              1     11         2    N      1    A                                1     1

CACATGCGCATACAGTTGGTATTGGTGCGCACCAGCATCCGGTTGTTATCGGTGCTCATGCCCATTCTTTCAGTATTGGTTCACACGGACACACCATCAC
----.----+----.----+----.----+----.----+----.----+----.----+----.----+----.----+----.----+----.----+ 21900
GTGTACGCGTATGTCAACCATAACCACGCGTGGTCGTAGGCCAACAATAGCCACGAGTACGGGTAAGAAAGTCATAACCAAGTGTGCCTGTGTGGTAGTG

 t  c  a  y  s  w  y  w  c  a  p  a  s  g  c  y  r  c  s  c  p  f  f  q  y  w  f  t  r  t  h  h  h
  H  A  H  T  V  G  I  G  A  H  Q  H  P  V  V  I  G  A  H  A  H  S  F  S  I  G  S  H  G  H  T  I  T
h  m  r  i  q  l  v  l  v  r  t  s  i  r  l  l  s  v  l  m  p  i  l  s  v  l  v  h  t  d  t  p  s  p
----.----+----.----+----.----+----.----+----.----+----.----+----.----+----.----+----.----+----.----+
 v  h  a  y  l  q  y  q  h  a  g  a  d  p  q  .  r  h  e  h  g  n  k  .  y  q  n  v  r  v  c  w  .
a  c  a  c  v  t  p  i  p  a  c  w  c  g  t  t  i  p  a  .  a  w  e  k  l  i  p  e  c  p  c  v  m  v
 c  m  r  m  c  n  t  n  t  r  v  l  m  r  n  n  d  t  s  m  g  m  r  e  t  n  t  .  v  s  v  g  d

REV     21825       lambdarev           sl          FM
ORF      ...      CONTINUES IN PHASE TWO


   HH** F             T                                M                O  O
   PI00 N             H                                N                R  R
   AN88 U             A                                L                F  F
   1201 H             1                                1
    ///
CGTTAACGCTGCGGGTAACGCGGAAAACACCGTCAAAAACATTGCATTTAACTATATTGTGAGGCTTGCATAATGGCATTCAGAATGAGTGAACAACCAC
----.----+----.----+----.----+----.----+----.----+----.----+----.----+----.----+----.----+----.----+ 22000
GCAATTGCGACGCCCATTGCGCCTTTTGTGGCAGTTTTTGTAACGTAAATTGATATAACACTCCGAACGTATTACCGTAAGTCTTACTCACTTGTTGGTG

r  .  r  c  g  .  r  g  k  h  r  q  k  h  c  i  .  l  y  c  e  a  c  i  M  A  F  R  M  S  E  Q  P  R
 V  N  A  A  G  N  A  E  N  T  V  K  N  I  A  F  N  Y  I  V  R  L  A  .  w  h  s  e  .  v  n  n  h
  l  t  l  r  v  t  r  k  t  p  s  k  t  l  h  l  t  i  l  .  g  l  h  n  g  i  q  n  e  .  t  t  t
----.----+----.----+----.----+----.----+----.----+----.----+----.----+----.----+----.----+----.----+
 r  .  r  q  p  y  r  p  f  c  r  .  f  c  q  m  .  s  y  q  s  a  q  m  i  a  n  l  i  l  s  c  g
  t  l  a  a  p  l  a  s  f  v  t  l  f  m  a  n  l  .  i  t  l  s  a  y  h  c  e  s  h  t  f  l  w
g  n  v  s  r  t  v  r  f  v  g  d  f  v  n  c  k  v  i  n  h  p  k  c  l  p  m  .  f  s  h  v  v  v

*080    21904       Mapped site HIN2
*081    21904       Mapped site HPA1
ORF     21970         orf-314               COOH
ORF     21973         orf-194               M-NH3


AA                 E    GCHH                                   H     HH           HE M                R
SV                 C    DFAP                                   P     PG           PC N                S
UA                 1    IREA                                   H     HA           AP L                A
12                 5    2132                                   1     11           21 1                1
 /                       /                                         /
GGACCATAAAAATTTATAATCTGCTGGCCGGAACTAATGAATTTATTGGTGAAGGTGACGCATATATTCCGCCTCATACCGGTCTGCCTGCAAACAGTAC
----.----+----.----+----.----+----.----+----.----+----.----+----.----+----.----+----.----+----.----+ 22100
CCTGGTATTTTTAAATATTAGACGACCGGCCTTGATTACTTAAATAACCACTTCCACTGCGTATATAAGGCGGAGTATGGCCAGACGGACGTTTGTCATG

 T  I  K  I  Y  N  L  L  A  G  T  N  E  F  I  G  E  G  D  A  Y  I  P  P  H  T  G  L  P  A  N  S  T
g  p  .  k  f  i  i  c  w  p  e  l  m  n  l  l  v  k  v  t  h  i  f  r  l  i  p  v  c  l  q  t  v  p
  d  h  k  n  l  .  s  a  g  r  n  .  .  i  y  w  .  r  .  r  i  y  s  a  s  y  r  s  a  c  k  q  y
----.----+----.----+----.----+----.----+----.----+----.----+----.----+----.----+----.----+----.----+
r  v  m  f  i  .  l  r  s  a  p  v  l  s  n  i  p  s  p  s  a  y  i  g  g  .  v  p  r  g  a  f  l  v
 p  g  y  f  n  i  i  q  q  g  s  s  i  f  k  n  t  f  t  v  c  i  n  r  r  m  g  t  q  r  c  v  t
  s  w  l  f  k  y  d  a  p  r  f  .  h  i  .  q  h  l  h  r  m  y  e  a  e  y  r  d  a  q  l  c  y

ORF      ...      CONTINUFS IN PHASF ONE
```

```
       T              H                        M                 S      S E        M
       T              P                        N                 F      F C        B
       H              A                        L                 A      A P        0
       2              2                        1                 N      N 1        2

CGATATTGCACCGCCAGATATTCCGGCTGGCTTTGTGGCTGTTTTCAACAGTGATGAGGCATCGTGGCATCTCGTTGAAGACCATCGGGGTAAAACCGTC
----.----+----.----+----.----+----.----+----.----+----.----+----.----+----.----+----.----+----.----+  22200
GCTATAACGTGGCGGTCTATAAGGCCGACCGAAACACCGACAAAAGTTGTCACTACTCCGTAGCACCGTAGAGCAACTTCTGGTAGCCCCATTTTGGCAG

 D  I  A  P  P  D  I  P  A  G  F  V  A  V  F  N  S  D  E  A  S  W  H  L  V  E  D  H  R  G  K  T  V
  i  l  h  r  q  i  f  r  l  a  l  w  l  f  s  t  v  *  r  h  r  g  i  s  l  k  t  i  g  v  k  p  s
r  y  c  t  a  r  y  s  g  w  l  c  g  c  f  q  q  .  .  g  i  v  a  s  r  .  r  p  s  g  .  n  r  l
----.----+----.----+----.----+----.----+----.----+----.----+----.----+----.----+----.----+----.----+
 s  i  a  g  g  s  i  g  a  p  k  t  a  t  k  l  l  s  s  a  d  h  c  r  t  s  s  w  r  p  l  v  t
g  i  n  c  r  w  i  n  r  s  a  k  h  s  n  e  v  t  i  l  c  r  p  *  e  n  f  v  *  p  t  f  g  d
 r  y  q  v  a  l  y  e  p  q  s  q  p  q  k  .  c  h  h  p  *  t  a  d  r  q  l  g  d  p  y  f  r

ORF       ...      CONTINUES IN PHASE ONE


          H   AM T    H            AA      H         E S    M   HNS
          P   FL H    G            SV      P         C C    N   PCC
          A   LU A    A            UA      A         R R    L   AIR
          2   31 1    1            12      2         2 1    1   211
                                    /                             /
TATGACGTGGCTTCCGGCGACGCGTTATTTATTTCTGAACTCGGTCCGTTACCGGAAAATTTTACCTGGTTATCGCCGGGAGGGGAATATCAGAAGTGGA
----.----+----.----+----.----+----.----+----.----+----.----+----.----+----.----+----.----+----.----+  22300
ATACTGCACCGAAGGCCGCTGCGCAATAAATAAAGACTTGAGCCAGGCAATGGCCTTTTAAAATGGACCAATAGCGGCCCTCCCCTTATAGTCTTCACCT

Y  D  V  A  S  G  D  A  L  F  I  S  E  L  G  P  L  P  E  N  F  T  W  L  S  P  G  G  E  Y  Q  K  W  N
 *  t  w  l  p  a  t  r  y  l  f  l  n  s  v  r  y  r  k  i  l  p  g  y  r  r  e  g  n  i  r  s  g
  .  r  g  f  r  r  r  v  i  y  f  .  t  r  s  v  t  g  k  f  y  l  v  i  a  g  r  g  i  s  e  v  e
----.----+----.----+----.----+----.----+----.----+----.----+----.----+----.----+----.----+----.----+
  .  s  t  a  e  p  s  a  n  n  i  e  s  s  p  g  n  g  s  f  k  v  q  n  d  g  p  p  s  y  .  f  h
 i  v  h  s  g  a  v  r  .  k  n  r  f  e  t  r  .  r  f  i  k  g  p  .  r  r  s  p  f  i  l  l  p
r  h  r  p  k  r  r  r  t  i  .  k  q  v  r  d  t  v  p  f  n  .  r  t  i  a  p  l  p  i  d  s  t  s

ORF       ...      CONTINUES IN PHASE ONE


        E S             H                 HB*BXMBNHSM               S M
        C C             P                 PI0AHBICPCN               F B
        R R             H                 AN0MOONIARL               A 0
        2 1             1                 21212111211               N 2
                                           //// ///
ACGGCACAGCCTGGGTGAAGGATACGGAAGCAGAAAAACTGTTCCGGATCCGGGAGGCGGAAGAAACAAAAAAAAGCCTGATGCAGGTAGCCAGTGAGCA
----.----+----.----+----.----+----.----+----.----+----.----+----.----+----.----+----.----+----.----+  22400
TGCCGTGTCGGACCCACTTCCTATGCCTTCGTCTTTTTGACAAGGCCTAGGCCCTCCGCCTTCTTTGTTTTTTTTCGGACTACGTCCATCGGTCACTCGT

  G  T  A  W  V  K  D  T  E  A  E  K  L  F  R  I  R  E  A  E  E  T  K  K  S  L  M  Q  V  A  S  E  H
t  a  q  p  g  .  r  i  r  k  q  k  n  c  s  g  s  g  r  r  k  k  q  k  k  a  .  c  r  .  p  v  s  i
 r  h  s  l  g  e  g  y  g  s  r  k  t  v  p  d  p  g  g  g  r  n  k  k  k  p  d  a  g  s  q  .  a
----.----+----.----+----.----+----.----+----.----+----.----+----.----+----.----+----.----+----.----+
f  p  v  a  q  t  f  s  v  s  a  s  f  s  n  r  i  r  s  a  s  s  v  f  f  l  r  i  c  t  a  l  s  c
 v  a  c  g  p  h  l  i  r  f  c  f  f  q  e  p  d  p  l  r  f  f  c  f  f  a  q  h  l  y  g  t  l
  r  c  l  r  p  s  p  y  p  l  l  f  v  t  g  s  g  p  p  p  l  f  l  f  f  g  s  a  p  l  w  h  a

*082    22346      Mapped site BAM1
ORF       ...      CONTINUES IN PHASE ONE


     HFSB   H     F    P**BFXM                         M     M    E S                M
     HNFB   G     N    S00GOHB                         B     N    C C                B
     AUAV   U     U    T08LKOO                         0     L    R R                0
     1HN1   2     H    1342121                         2     1    2 1                2
      //               /// //
TATTGCGCCGCTTCAGGATGCTGCAGATCTGGAAATTGCAACGAAGGAAGAAACCTCGTTGCTGGAAGCCTGGAAGAAGTATCGGGTGTTGCTGAACCGT
----.----+----.----+----.----+----.----+----.----+----.----+----.----+----.----+----.----+----.----+  22500
ATAACGCGGCGAAGTCCTACGACGTCTAGACCTTTAACGTTGCTTCCTTCTTTGGAGCAACGACCTTCGGACCTTCTTCATAGCCCACAACGACTTGGCA

 I  A  P  L  Q  D  A  A  D  L  E  I  A  T  K  E  E  T  S  L  L  E  A  W  K  K  Y  R  V  L  L  N  R
  l  r  r  f  r  *  l  q  i  w  k  l  q  r  r  k  k  p  r  c  w  k  p  g  r  s  i  g  c  c  .  t  v
y  c  a  a  s  g  c  c  r  s  g  n  c  n  e  g  r  n  l  v  a  g  s  l  e  e  v  s  g  v  a  e  p  c
----.----+----.----+----.----+----.----+----.----+----.----+----.----+----.----+----.----+----.----+
 i  a  g  s  .  s  a  a  s  r  s  i  a  v  f  s  s  v  e  n  s  s  a  q  f  f  y  r  t  n  s  f  r
*  n  r  r  k  l  i  s  c  i  q  f  n  c  r  l  f  f  g  r  q  q  f  g  p  l  l  i  p  h  q  q  v  t
 y  q  a  a  e  p  h  q  l  d  p  f  q  l  s  p  l  f  r  t  a  p  l  r  s  s  t  d  p  t  a  s  g
```

```
*083  22425     Mapped site BGL2
*084  22425     Mapped site PST1
ORF   ...       CONTINUES IN PHASE ONE


                             C HHE                   O                  F
                             1 AAC                   R                  N
                             0 EE1                   F                  U
                             5 135                                      H
                                //
GTTGATACATCAACTGCACCTGATATTGAGTGGCCTGCTGTCCCTGTTATGGAGTAATCGTTTTGTGATATGCCGCAGAAACGTTGTATGAAATAACGTT
----.----+----.----+----.----+----.----+----.----+----.----+----.----+----.----+----.----+----.----+ 22600
CAACTATGTAGTTGACGTGGACTATAACTCACCGGACGACAGGGACAATACCTCATTAGCAAAACACTATACGGCGTCTTTGCAACATACTTTATTGCAA

V  D  T  S  T  A  P  D  I  E  W  P  A  V  P  V  M  E  .  s  f  c  d  m  p  q  k  r  c  m  k  .  r  s
 l  i  h  q  l  h  l  i  l  s  g  l  l  s  l  l  w  s  n  r  f  v  i  c  r  r  n  v  v  .  n  n  v
  .  y  i  n  c  t  .  y  .  v  a  c  c  p  c  y  g  v  i  v  l  .  y  a  a  e  t  l  y  e  i  t  f
----.----+----.----+----.----+----.----+----.----+----.----+----.----+----.----+----.----+----.----+
 t  s  v  d  v  a  g  s  i  s  h  g  a  t  g  t  i  s  y  d  n  q  s  i  g  c  f  r  q  i  f  y  r
  n  i  c  .  s  c  r  i  n  l  p  r  s  d  r  n  h  l  l  r  k  t  i  h  r  l  f  t  t  h  f  l  t
h  q  y  m  l  q  v  q  y  q  t  a  q  q  g  q  .  p  t  i  t  k  h  y  a  a  s  v  n  y  s  i  v  n

C105  22530     lambdacam105      i(Tn9)     RES     dl for deletions pMS370;968;969;97
ORF   22554     orf-194           COOH


                         AD                    M                                  H     E          H
                         LD                    B                                  N     A          P
                         UE                    0                                  F     4          H
                         11                    2                                  1     7          1

CTGCGGTTAGTTAGTATATTGTAAAGCTGAGTATTGGTTTATTTGGCGATTATTATCTTCAGGAGAATAATGGAAGTTCTATGACTCAATTGTTCATAGT
----.----+----.----+----.----+----.----+----.----+----.----+----.----+----.----+----.----+----.----+ 22700
GACGCCAATCAATCATATAACATTTCGACTCATAACCAAATAAACCGCTAATAATAGAAGTCCTCTTATTACCTTCAAGATACTGAGTTAACAAGTATCA

  a  v  s  .  y  i  v  k  l  s  i  g  l  f  g  d  y  y  l  q  e  n  n  g  s  s  m  t  q  l  f  i  v
l  r  l  v  s  i  l  .  s  .  v  l  v  y  l  a  i  i  i  f  r  r  i  m  e  v  l  .  l  n  c  s  .  c
 c  g  .  l  v  y  c  k  a  e  y  w  f  i  w  r  l  l  s  s  g  e  .  w  k  f  y  d  s  i  v  h  s
----.----+----.----+----.----+----.----+----.----+----.----+----.----+----.----+----.----+----.----+
e  a  t  l  .  y  i  t  f  s  l  i  p  k  n  p  s  .  .  r  .  s  f  l  p  l  e  i  v  .  N  N  M  T
 r  r  n  t  l  i  n  y  l  q  t  n  t  .  k  a  i  i  i  k  l  l  i  i  s  t  r  h  s  l  q  e  y
  q  p  .  n  t  y  q  l  a  s  y  q  n  i  q  r  n  n  d  e  p  s  y  h  f  n  .  s  e  i  t  .  l

EA47  22689     Ea47              COOH


                                                                H E                         M
                                                                N C                         B
                                                                F 1                         0
                                                                1 5                         2

GTTTACATCACCGCCAATTGCTTTTAAGACTGAACGCATGAAATATGGTTTTTCGTCATGTTTTGAGTCTGCTGTTGATATTTCTAAAGTCGGTTTTTTT
----.----+----.----+----.----+----.----+----.----+----.----+----.----+----.----+----.----+----.----+ 22800
CAAATGTAGTGGCGGTTAACGAAAATTCTGACTTGCGTACTTTATACCAAAAAGCAGTACAAAACTCAGACGACAACTATAAAGATTTCAGCCAAAAAAA

 f  t  s  p  p  i  a  f  k  t  e  r  m  k  y  g  f  s  s  c  f  e  s  a  v  d  i  s  k  v  g  f  f
  l  h  h  r  q  l  l  l  r  l  n  a  .  n  m  v  f  r  h  v  l  s  l  l  l  i  f  l  k  s  v  f  f
v  y  i  t  a  n  c  f  .  d  .  t  h  e  i  w  f  f  v  m  f  .  v  c  c  .  y  f  .  s  r  f  f  f
----.----+----.----+----.----+----.----+----.----+----.----+----.----+----.----+----.----+----.----+
  N  V  D  G  G  I  A  K  L  V  S  R  M  F  Y  P  K  E  D  H  K  S  D  A  T  S  I  E  L  T  P  K  K
h  k  c  .  r  w  n  s  k  l  s  f  a  h  f  i  t  k  r  .  t  k  l  r  s  n  i  n  r  f  d  t  k  k
 t  .  m  v  a  l  q  k  .  s  q  v  c  s  i  h  n  k  t  m  n  q  t  q  q  q  y  k  .  l  r  n  k

EA47  ...       CONTINUES IN PHASE FOUR
```

```
                                       B           H   X                  X
                                       5           N   M                  M
                                       3           F   N                  N
                                       6           1   1                  1

TCTTCGTTTTCTCTAACTATTTTCCATGAAATACATTTTTGATTATTATTTGAATCAATTCCAATTACCTGAAGTCTTTCATCTATAATTGGCATTGTAT
----.----+----.----+----.----+----.----+----.----+----.----+----.----+----.----+----.----+----.----+ 22900
AGAAGCAAAAGAGATTGATAAAAGGTACTTTATGTAAAAACTAATAATAAACTTAGTTAAGGTTAATGGACTTCAGAAAGTAGATATTAACCGTAACATA

s s f s l t i f h e i h f . l l f e s i p i t . s l s s i i g i v c
 l r f l . l f s m k y i f d y y l n q f q l p e v f h l . l a l y
  f v f s n y f p . n t f l i i i . i n s n y l k s f i y n w h c m
----.----+----.----+----.----+----.----+----.----+----.----+----.----+----.----+----.----+----.----+
 E E N E R V I K W S I C K Q N N N S D I G I V Q L R E D I I P M T
  r r k r . s n e m f y m k s . . k f . n w n g s t k . r y n a n y
k k t k e l . k g h f v n k i i i q i l e l . r f d k m . l q c q i

B536   22840      b536                 dr           EM
EA47   ...        CONTINUES IN PHASE FOUR
```

```
          S T                     D         A        E                           B          N PA*F
          F T                     D         L        C                           B          S VL0N
          A H                     E         U        R                           V          P UU8U
          N 2                     1         1        V                           1          B 215H
                                                                                               ///
GTATTGGTTTATTGGAGTAGATGCTTGCTTTTCTGAGCCATAGCTCTGATATCCAAATGAAGCCATAGGCATTTGTTATTTTGGCTCTGTCAGCTGCATA
----.----+----.----+----.----+----.----+----.----+----.----+----.----+----.----+----.----+----.----+ 23000
CATAACCAAATAACCTCATCTACGAACGAAAAGACTCGGTATCGAGACTATAGGTTTACTTCGGTATCCGTAAACAATAAAACCGAGACAGTCGACGTAT

  i g l l e . m l a f l s h s s d i q m k p . a f v i l a l s a a .
v l v y w s r c l l f . a i a l i s k . s h r h l l f w l c q l h n
 y w f i g v d a c f s e p . l . y p n e a i g i c y f g s v s c i
----.----+----.----+----.----+----.----+----.----+----.----+----.----+----.----+----.----+----.----+
H I P K N S Y I S A K R L W L E S I W I F G Y A N T I K A R D A A Y
 t n t . q l l h k s k q a m a r i d l h l w l c k n n q s q . s c
  y q n i p t s a q k e s g y s q y g f s a m p m q . k p e t l q m

*085   22993      Mapped site PVU2
EA47   ...        CONTINUES IN PHASE FOUR
```

```
           T        M    M      B                        H              F     T   A
           T        B    B      5                        G              0     T   R
           H        0    0      1                        U              K     H   0
           2        2    1      9                        2              1     2   E

ACGCCAAAAAATATATTTATCTGCTTGATCTTCAAATGTTGTATTGATTAAATCAATTGGATGGAATTGTTTATCATAAAAAATTAATGTTTGAATGTGA
----.----+----.----+----.----+----.----+----.----+----.----+----.----+----.----+----.----+----.----+ 23100
TGCGGTTTTTTATATAAATAGACGAACTAGAAGTTTACAACATAACTAATTTAGTTAACCTACCTTAACAAATAGTATTTTTTAATTACAAACTTACACT

 r q k i y l s a . s s n v v l i k s i g w n c l s . k i n v . m .
  a k k y i y l l d l q m l y . l n q l d g i v y h k k l m f e c d
t p k n i f i c l i f k c c i d . i n w m e l f i i k n . c l n v i
----.----+----.----+----.----+----.----+----.----+----.----+----.----+----.----+----.----+----.----+
 R W F I Y K D A Q D E F T T N I L D I P H F Q K D Y F I L T Q I H
l a l f y i . r s s r . i n y q n f . n s p i t . . l f n i n s h s
 v g f f i n i q k i k l h q i s . i l q i s n n i m f f . h k f t

B519   23033      b519                 dr           EM
AROE   23082      lambdaaroE           sr           EM
EA47   ...        CONTINUES IN PHASE FOUR
```

```
        A                   *HSA              H*M            T
        H                   0IBL              I0B            A
        A                   8N4U              N8O            Q
        3                   63 1              272            1
                             ///               //
TAACCGTCCTTTAAAAAAGTCGTTTCTGCAAGCTTGGCTGTATAGTCAACTAACTCTTCTGTCGAAGTGATATTTTTAGGCTTATCTACCAGTTTTAGAC
----.----+----.----+----.----+----.----+----.----+----.----+----.----+----.----+----.----+----.----+ 23200
ATTGGCAGGAAATTTTTTCAGCAAAGACGTTCGAACCGACATATCAGTTGATTGAGAAGACAGCTTCACTATAAAAATCCGAATAGATGGTCAAAATCTG

 .  p  s  f  k  k  v  v  s  a  s  l  a  v  .  s  t  n  s  s  v  e  v  i  f  l  g  l  s  t  s  f  r  r
  n  r  p  l  k  k  s  f  l  q  a  w  l  y  s  q  l  t  l  l  s  k  .  y  f  .  a  y  l  p  v  l  d
   t  v  l  .  k  s  r  f  c  k  l  g  c  i  v  n  .  l  f  c  r  s  d  i  f  r  l  i  y  q  f  .  t
----.----+----.----+----.----+----.----+----.----+----.----+----.----+----.----+----.----+----.----+
 Y  G  D  K  L  F  T  T  E  A  L  K  A  T  Y  D  V  L  E  E  T  S  T  I  N  K  P  K  D  V  L  K  L
  l  r  g  k  f  f  d  n  r  c  a  q  s  y  l  .  s  v  r  r  d  f  h  y  k  .  a  .  r  g  t  k  s
 i  v  t  r  .  f  l  r  k  q  l  s  p  q  i  t  l  .  s  k  q  r  l  s  i  k  l  s  i  .  w  n  .  v

*086   23130     Mapped site HIN3
SB4    23131       Sequence Block 4     start
*087   23147     Mapped site HIN2
EA47    ...      CONTINUES IN PHASE FOUR
```

```
  M H                        P                            S        L       A             E
  B G                        B                            B        A       H             C
  0 A                        L                            4        C       A             R
  2 1                                                              5       3             2

GCTCTTTAATATCTTCAGGAATTATTTTATTGTCATATTGTATCATGCTAAATGACAATTTGCTTATGGAGTAATCTTTTAATTTTAAATAAGTTATTCT
----.----+----.----+----.----+----.----+----.----+----.----+----.----+----.----+----.----+----.----+ 23300
CGAGAAATTATAGAAGTCCTTAATAAAATAACAGTATAACATAGTACGATTTACTGTTAAACGAATACCTCATTAGAAAATTAAAATTTATTCAATAAGA

   s  l  i  s  s  g  i  i  l  l  s  y  c  i  m  l  n  d  n  l  l  m  e  .  s  f  n  f  k  .  v  i  l
 a  l  .  y  l  q  e  l  f  y  c  h  i  v  s  c  .  m  t  i  c  l  w  s  n  l  l  i  l  n  k  l  f  s
  l  f  n  i  f  r  n  y  f  i  v  i  l  y  h  a  k  .  q  f  a  y  g  v  i  f  .  f  .  i  s  y  s
----.----+----.----+----.----+----.----+----.----+----.----+----.----+----.----+----.----+----.----+
R  E  K  I  D  F  P  I  I  K  N  D  Y  Q  I  M  S  F  S  L  K  S  I  S  Y  D  K  L  K  L  Y  T  I  R
 a  r  .  y  r  .  s  n  n  .  q  .  i  t  d  h  .  i  v  i  q  k  h  l  l  r  k  i  k  f  l  n  n
  s  k  l  i  k  l  f  .  k  i  t  m  n  y  .  a  l  h  c  n  a  .  p  t  i  k  .  n  .  i  l  .  e

PBL    23231       pBL                  5'        RNA      leftward promoter in b region
SB4    23269       Sequence Block 4     end
LAC5   23278        lac5                sr        EM
EA47    ...       CONTINUES IN PHASE FOUR
```

```
 S                H  TH                H                                            E
 C                N  AN                P                                            C
 R                F  QF                H                                            1
 1                1  13                1                                            5
                      /
CCTGGCTTCATCAAATAAAGAGTCGAATGATGTTGGCGAAATCACATCGTCACCCATTGGATTGTTTATTTGTATGCCAAGAGAGTTACAGCAGTTATAC
----.----+----.----+----.----+----.----+----.----+----.----+----.----+----.----+----.----+----.----+ 23400
GGACCGAAGTAGTTTATTTCTCAGCTTACTACAACCGCTTTAGTGTAGCAGTGGGTAACCTAACAAATAAACATACGGTTCTCTCAATGTCGTCAATATG

  l  a  s  s  n  k  e  s  n  d  v  g  e  i  t  s  s  p  i  g  l  f  i  c  m  p  r  e  l  q  q  l  y
   w  l  h  q  i  k  s  r  m  m  l  a  k  s  h  r  h  p  l  d  c  l  f  v  c  q  e  s  y  s  s  y  t
 p  g  f  i  k  .  r  v  e  .  c  w  r  n  h  i  v  t  h  w  i  v  y  l  y  a  k  r  v  t  a  v  i  h
----.----+----.----+----.----+----.----+----.----+----.----+----.----+----.----+----.----+----.----+
  R  A  E  D  F  L  S  D  F  S  T  P  S  I  V  D  D  G  M  P  N  N  I  Q  I  G  L  S  N  C  C  N  Y
 e  q  s  .  .  i  f  l  r  i  i  n  a  f  d  c  r  .  g  n  s  q  k  n  t  h  w  s  l  .  l  l  .  v
  g  p  k  m  l  y  l  t  s  h  h  q  r  f  .  m  t  v  w  q  i  t  .  k  y  a  l  l  t  v  a  t  i

EA47    ...       CONTINUES IN PHASE FOUR
```

```
                  AD        N    H                                     S                  H   S
                  LD        S    N                                     P                  N   B
                  UE        P    F                                     C                  F   5
                  11        C    3                                     2                  1

ATTCTGCCATAGATTATAGCTAAGGCATGTAATAATTCGTAATCTTTTAGCGTATTAGCGACCCATCGTCTTTCTGATTTAATAATAGATGATTCAGTTA
----.----+----.----+----.----+----.----+----.----+----.----+----.----+----.----+----.----+----.----+ 23500
TAAGACGGTATCTAATATCGATTCCGTACATTATTAAGCATTAGAAAATCGCATAATCGCTGGGTAGCAGAAAGACTAAATTATTATCTACTAAGTCAAT

i  l  p  .  i  i  a  k  a  c  n  n  s  .  s  f  s  v  l  a  t  h  r  l  s  d  l  i  i  d  d  s  v  k
 f  c  h  r  l  .  l  r  h  v  i  i  r  n  l  l  a  y  .  r  p  i  v  f  l  i  .  .  .  ■  i  q  l
  s  a  i  d  y  s  .  g  ■  .  .  f  v  i  f  .  r  i  s  d  p  s  s  f  .  f  n  n  r  .  f  s  .
----.----+----.----+----.----+----.----+----.----+----.----+----.----+----.----+----.----+----.----+
 M  R  G  Y  I  I  A  L  A  H  L  L  E  Y  D  K  L  T  N  A  V  W  R  R  F  S  K  I  I  S  S  E  T
  n  q  w  l  n  y  s  l  c  t  i  i  r  l  r  k  a  y  .  r  g  ■  t  k  r  i  .  y  y  i  i  .  n
c  e  a  ■  s  .  l  .  p  ■  y  y  n  t  i  k  .  r  i  l  s  g  d  d  k  q  n  l  l  l  h  n  l  .
```

SPC2 23472 lambdaspc2 sr EM
SB5 23495 Sequence Block 5 start
EA47 ... CONTINUES IN PHASE FOUR

```
                                               SB                                   M
                                               B5                                   B
                                               51                                   0
                                                1                                   2

AATATGAAGGTAATTTCTTTTGTGCAAGTCTGACTAACTTTTTTATACCAATGTTTAACATACTTTCATTTGTAATAAACTCAATGTCATTTTCTTCAAT
----.----+----.----+----.----+----.----+----.----+----.----+----.----+----.----+----.----+----.----+ 23600
TTATACTTCCATTAAAGAAAACACGTTCAGACTGATTGAAAAAATATGGTTACAAATTGTATGAAAGTAAACATTATTTGAGTTACAGTAAAAGAAGTTA

 y  e  g  n  f  f  c  a  s  l  t  n  f  f  i  p  ■  f  n  i  l  s  f  v  i  n  s  ■  s  f  s  s  ■
n  ■  k  v  i  s  f  v  q  v  .  l  t  f  l  y  q  c  l  t  y  f  h  l  .  .  t  q  c  h  f  l  q  c
  i  .  r  .  f  l  l  c  k  s  d  .  l  f  y  t  n  v  .  h  t  f  i  c  n  k  l  n  v  i  f  f  n
----.----+----.----+----.----+----.----+----.----+----.----+----.----+----.----+----.----+----.----+
L  Y  S  P  L  K  K  Q  A  L  R  V  L  K  K  I  G  I  N  L  M  S  E  N  T  I  F  E  I  D  N  E  E  I
 f  i  f  t  i  e  k  t  c  t  q  s  v  k  k  y  w  h  k  v  y  k  .  k  y  y  v  .  h  .  k  r  .
  i  h  l  y  n  r  k  h  l  d  s  .  s  k  .  v  l  t  .  c  v  k  ■  q  l  l  s  l  t  ■  k  k  l
```

SB5 23548 Sequence Block 5 end
B511 23549 b511 dl-att sq
EA47 ... CONTINUES IN PHASE FOUR

```
               1                    M                                                           BM
               0                    N                                                           IB
               0                    L                                                           NO
               7                    1                                                           11

GTAAGATGAAATAAGAGTAGCCTTTGCCTCGCTATACATTTCTAAATCGCCTTGTTTTTCTATCGTATTGCGAGAATTTTTAGCCCAAGCCATTAATGGA
----.----+----.----+----.----+----.----+----.----+----.----+----.----+----.----+----.----+----.----+ 23700
CATTCTACTTTATTCTCATCGGAAACGGAGCGATATGTAAAGATTTAGCGGAACAAAAAGATAGCATAACGCTCTTAAAAATCGGGTTCGGTAATTACCT

 .  d  e  i  r  v  a  f  a  s  l  y  i  s  k  s  p  c  f  s  i  v  l  r  e  f  l  a  q  a  i  n  g
  k  ■  k  .  e  .  p  l  p  r  y  t  f  l  n  r  l  v  f  l  s  y  c  e  n  f  .  p  k  p  l  ■  d
v  r  .  n  k  s  s  l  c  l  a  i  h  f  .  i  a  l  f  f  y  r  i  a  r  i  f  s  p  s  h  .  w  i
----.----+----.----+----.----+----.----+----.----+----.----+----.----+----.----+----.----+----.----+
  Y  S  S  I  L  T  A  K  A  E  S  Y  M  E  L  D  G  Q  K  E  I  T  N  R  S  N  K  A  W  A  M  L  P
h  l  i  f  y  s  y  g  k  g  r  .  v  n  r  f  r  r  t  k  r  d  y  q  s  f  k  .  g  l  g  n  i  s
 t  l  h  f  l  l  l  r  q  r  a  i  c  k  .  i  a  k  n  k  .  r  i  a  l  i  k  l  g  l  w  .  h
```

1007 23616 b1007 dl EM
EA47 ... CONTINUFS IN PHASE FOUR

```
                                                                                                M
                                                                                                N
                                                                                                L
                                                                                                1

TCATTTTTCCATTTTTCAATAACATTATTGTTATACCAAATGTCATATCCTATAATCTGGTTTTTGTTTTTTTGAATAATAAATGTTACTGTTCTTGCGG
----.----+----.----+----.----+----.----+----.----+----.----+----.----+----.----+----.----+----.----+ 23800
AGTAAAAAGGTAAAAAGTTATTGTAATAACAATATGGTTTACAGTATAGGATATTAGACCAAAAACAAAAAAACTTATTATTTACAATGACAAGAACGCC

s  f  f  h  f  s  i  t  l  l  l  y  q  ■  s  y  p  i  i  w  f  l  f  f  .  i  i  n  v  t  v  l  a  v
 h  f  s  i  f  q  .  h  y  c  y  t  k  c  h  i  l  .  s  g  f  c  f  f  e  .  .  ■  l  l  f  l  r
  i  f  p  f  f  n  n  i  i  v  i  p  n  v  i  s  y  n  l  v  f  v  f  l  n  n  k  c  y  c  s  c  g
----.----+----.----+----.----+----.----+----.----+----.----+----.----+----.----+----.----+----.----+
  D  N  K  W  K  E  I  V  N  N  N  Y  W  I  D  Y  G  I  I  Q  N  K  N  K  Q  I  I  F  T  V  T  R  A
 .  k  e  ■  k  .  y  c  .  q  .  v  l  h  .  i  r  y  d  p  k  q  k  k  s  y  y  i  n  s  n  k  r
i  ■  k  g  n  k  l  l  ■  i  t  i  g  f  t  ■  d  .  l  r  t  k  t  k  k  f  l  l  h  .  q  e  q  p

EA47    ...     CONTINUES IN PHASE FOUR


            XH                 X                           F          B            M
            MN                 M                           N          B            B
            NF                 N                           U          V            0
            11                 1                           H          1            2

TTTGGAGGAATTGATTCAAATTCAAGCGAAATAATTCAGGGTCAAAATATGTATCAATGCAGCATTTGAGCAAGTGCGATAAATCTTTAAGTCTTCTTTC
----.----+----.----+----.----+----.----+----.----+----.----+----.----+----.----+----.----+----.----+ 23900
AAACCTCCTTAACTAAGTTTAAGTTCGCTTTATTAAGTCCCAGTTTTATACATAGTTACGTCGTAAACTCGTTCACGCTATTTAGAAATTCAGAAGAAAG

  w  r  n  .  f  k  f  k  r  n  n  s  g  s  k  y  v  s  ■  q  h  l  s  k  c  d  k  s  l  s  l  l  s
f  g  g  i  d  s  n  s  s  e  i  i  q  g  q  n  ■  y  q  c  s  i  .  a  s  a  i  n  l  .  v  f  f  p
 l  e  e  l  i  q  i  q  a  k  .  f  r  v  k  i  c  i  n  a  a  f  e  q  v  r  .  i  f  k  s  s  f
----.----+----.----+----.----+----.----+----.----+----.----+----.----+----.----+----.----+----.----+
T  Q  L  F  Q  N  L  N  L  R  F  L  E  P  D  F  Y  T  D  I  C  C  K  L  L  H  S  L  D  K  L  R  R  E
 n  p  p  i  s  e  f  e  l  s  i  i  .  p  .  f  i  y  .  h  l  ■  q  a  l  a  i  f  r  .  t  k  k
  k  s  s  n  i  .  i  .  a  f  y  n  l  t  l  i  h  i  l  a  a  n  s  c  t  r  y  i  k  l  d  e  k

EA47    ...     CONTINUES IN PHASE FOUR


N                E                     S      SN*         M                   H                     T
C                A                     F      PS0         N                   P                     T
0                4                     A      HP8         L                   H                     H
1                7                     N      1C8         1                   1                     2
                                               //
CCATGGTTTTTTAGTCATAAAACTCTCCATTTTGATAGGTTGCATGCTAGATGCTGATATATTTTAGAGGTGATAAAATTAACTGCTTAACTGTCAATGT
----.----+----.----+----.----+----.----+----.----+----.----+----.----+----.----+----.----+----.----+ 24000
GGTACCAAAAAATCAGTATTTTGAGAGGTAAAACTATCCAACGTACGATCTACGACTATATAAAATCTCCACTATTTTAATTGACGAATTGACAGTTACA

 h  g  f  l  v  i  k  l  s  i  l  i  g  c  ■  l  d  a  d  i  f  .  r  .  .  n  .  l  l  n  c  q  c
  ■  v  f  .  s  .  n  s  p  f  .  .  v  a  c  .  ■  l  i  y  f  r  g  d  k  i  n  c  l  t  v  n  v
p  w  f  f  s  h  k  t  l  h  f  d  r  l  h  a  r  c  .  y  i  l  e  v  i  k  l  t  a  .  l  s  ■  .
----.----+----.----+----.----+----.----+----.----+----.----+----.----+----.----+----.----+----.----+
  W  P  K  K  T  M  f  s  e  ■  k  i  p  q  ■  s  s  a  s  i  n  .  l  h  y  f  .  s  s  l  q  .  h
g  ■  t  k  .  d  y  f  e  g  n  q  y  t  a  h  .  i  s  i  y  k  l  p  s  l  i  l  q  k  v  t  l  t
 g  h  n  k  l  .  l  v  r  w  k  s  l  n  c  a  l  h  q  y  i  k  s  t  i  f  n  v  a  .  s  d  i

EA47    23918      Ea47 (orf-410)       M-NH3
*088    23946      Mapped site SPH1


   B         M      C     H
   5         B      1     N
   1         0      0     F
   5         1      7     1

AATACAAGTTGTTTGATCTTTGCAATGATTCTTATCAGAAACCATATAGTAAATTAGTTACACAGGAAATTTTTAATATTATTATTATCATTCATTATGT
----.----+----.----+----.----+----.----+----.----+----.----+----.----+----.----+----.----+----.----+ 24100
TTATGTTCAACAAACTAGAAACGTTACTAAGAATAGTCTTTGGTATATCATTTAATCAATGTGTCCTTTAAAAATTATAATAATAATAGTAAGTAATACA

n  t  s  c  l  i  f  a  ■  i  l  i  r  n  h  i  v  n  .  l  h  r  k  f  l  i  l  l  l  s  f  i  ■  y
 i  q  v  v  .  s  l  q  .  f  l  s  e  t  i  .  .  i  s  y  t  g  n  f  .  y  y  y  y  h  s  l  c
  y  k  l  f  d  l  c  n  d  s  y  q  k  p  y  s  k  l  v  t  q  e  i  f  n  i  i  i  i  i  h  y  v
----.----+----.----+----.----+----.----+----.----+----.----+----.----+----.----+----.----+----.----+
 l  v  l  q  k  i  k  a  i  i  r  i  l  f  w  i  t  f  .  n  c  l  f  n  k  i  n  n  n  d  n  ■  i
  i  c  t  t  q  d  k  c  h  n  k  d  s  v  ■  y  y  i  l  .  v  p  f  k  .  y  .  .  .  .  e  n  h
y  y  l  n  n  s  r  q  l  s  e  .  .  f  g  y  l  l  n  t  v  c  s  i  k  l  i  i  i  i  ■  .  .  t

B515    24003      b515            dl         EM
C107    24020      cam107          i(Tn9)     EM RES  dl for deletions pXJS; pQLC 137-14
```

```
                               A                             F    A        B             F  F   B
                               F                             N    F        B             N  N   5
                               L                             U    L        V'            U  U   0
                               3                             H    3        1             H  H   6

ATTAAAATTAGAGTTGTGGCTTGGCTCTGCTAACACGTTGCTCATAGGAGATATGGTAGAGCCGCAGACACGTCGTATGCAGGAACGTGCTGCGGCTGGC
----.----+----.----+----.----+----.----+----.----+----.----+----.----+----.----+----.----+----.----+ 24200
TAATTTTAATCTCAACACCGAACCGAGACGATTGTGCAACGAGTATCCTCTATACCATCTCGGCGTCTGTGCAGCATACGTCCTTGCACGACGCCGACCG

 .  n  .  s  c  g  l  a  l  l  t  r  c  s  .  e  i  w  .  s  r  r  h  v  v  c  r  n  v  l  r  l  a
i  k  i  r  v  v  a  w  l  c  .  h  v  a  h  r  r  y  g  r  a  a  d  t  s  y  a  g  t  c  c  g  w  l
 l  k  l  e  l  w  l  g  s  a  n  t  l  l  i  g  d  m  v  e  p  q  t  r  r  m  q  e  r  a  a  a  g
----.----+----.----+----.----+----.----+----.----+----.----+----.----+----.----+----.----+----.----+
y  .  f  .  l  q  p  k  a  r  s  v  r  q  e  y  s  i  h  y  l  r  l  c  t  t  h  l  f  t  s  r  s  a
 i  l  i  l  t  t  a  q  s  q  .  c  t  a  .  l  l  y  p  l  a  a  s  v  d  y  a  p  v  h  q  p  q
  n  f  n  s  n  h  s  p  e  a  l  v  n  s  m  p  s  i  t  s  g  c  v  r  r  i  c  s  r  a  a  a  p

B506   24197       b506           dr         EM


             H               E X                                                 R              M
             P               C M                                                 S              N
             H               O N                                                 A              L
             1               B 1                                                 1              1

TGGTGAACTTCCGATAGTGCGGGTGTTGAATGATTTCCAGTTGCTACCGATTTTACATATTTTTTGCATGAGAGAATTTGTACCACCTCCCACCGACCAT
----.----+----.----+----.----+----.----+----.----+----.----+----.----+----.----+----.----+----.----+ 24300
ACCACTTGAAGGCTATCACGCCCACAACTTACTAAAGGTCAACGATGGCTAAAATGTATAAAAAACGTACTCTCTTAAACATGGTGGAGGGTGGCTGGTA

 g  e  l  p  i  v  r  v  l  n  d  f  q  l  l  p  i  l  h  i  f  c  m  r  e  f  v  p  p  p  t  d  h
  v  n  f  r  .  c  g  c  .  m  i  s  s  c  y  r  f  y  i  f  f  a  .  e  n  l  y  h  l  p  p  t  i
w  .  t  s  d  s  a  g  v  e  .  f  p  v  a  t  d  f  t  y  f  l  h  e  r  i  c  t  t  s  h  r  p  s
----.----+----.----+----.----+----.----+----.----+----.----+----.----+----.----+----.----+----.----+
  p  s  s  g  i  t  r  t  n  f  s  k  w  n  s  g  i  k  c  i  k  q  m  l  s  n  t  g  g  g  v  s  w
s  t  f  k  r  y  h  p  h  q  i  i  e  l  q  .  r  n  .  m  n  k  a  h  s  f  k  y  w  r  g  g  v  m
 q  h  v  e  s  l  a  p  t  s  h  n  g  t  a  v  s  k  v  y  k  k  c  s  l  i  q  v  v  e  w  r  g


          R          A                 H                                   SN*           HHHT A
          S          V                 P                                   PS0           YYY6 V
          A          R                 A                                   HP8           F4M0 R
          1          2                 2                                   1C9             2  2
                                                                            //            ///
CTATGACTGTACGCCACTGTCCCTAGGACTGCTATGTGCCGGAGCGGACATTACAAACGTCCTTCTCGGTGCATGCCACTGTTGCCAATGACCTGCCTAG
----.----+----.----+----.----+----.----+----.----+----.----+----.----+----.----+----.----+----.----+ 24400
GATACTGACATGCGGTGACAGGGATCCTGACGATACACGGCCTCGCCTGTAATGTTTGCAGGAAGAGCCACGTACGGTGACAACGGTTACTGGACGGATC

l  .  l  y  a  t  v  p  r  t  a  m  c  r  s  g  h  y  k  r  p  s  r  c  m  p  l  l  p  m  t  c  l  g
 y  d  c  t  p  l  s  l  g  l  l  c  a  g  a  d  i  t  n  v  l  l  g  a  c  h  c  c  q  .  p  a  .
  m  t  v  r  h  c  p  .  d  c  y  v  p  e  r  t  l  q  t  s  f  s  v  h  a  t  v  a  n  d  l  p  r
----.----+----.----+----.----+----.----+----.----+----.----+----.----+----.----+----.----+----.----+
 r  h  s  y  a  v  t  g  l  v  a  i  h  r  l  p  c  .  l  r  g  e  r  h  m  g  s  n  g  i  v  q  r
  .  s  q  v  g  s  d  r  p  s  s  h  a  p  a  s  m  v  f  t  r  r  p  a  h  w  q  q  w  h  g  a  .
d  i  v  t  r  w  q  g  .  s  q  .  t  g  s  r  v  n  c  v  d  k  e  t  c  a  v  t  a  l  s  r  g  l

*089   24375     Mapped site SPH1
HYF    24390      hyF                  hy         EM      lambda::att80imm80 OP80 QSR80
HY42   24390      hy42                 hy         EM      lambda::att80imm80 080 PQSRlambda
HYM    24390      hyM                  hy         EM      lambda::att80immlambdaOPlambdaQSR8
I60    24390        lambdatrp60-3      sl         EM


                      H                             M  M   H        H
                      P                             N  N   N        G
                      A                             L  L   F        A
                      2                             1  1   3        1

GAATTGGTTAGCAAGTTACTACCGGATTTTGTAAAAACAGCCCTCCTCATATAAAAAGTATTCGTTCACTTCCGATAAGCGTCGTAATTTTCTATCTTTC
----.----+----.----+----.----+----.----+----.----+----.----+----.----+----.----+----.----+----.----+ 24500
CTTAACCAATCGTTCAATGATGGCCTAAAACATTTTTGTCGGGAGGAGTATATTTTTCATAAGCAAGTGAAGGCTATTCGCAGCATTAAAAGATAGAAAG

  i  g  .  q  v  t  t  g  f  c  k  n  s  p  p  h  i  k  s  i  r  s  l  p  i  s  v  v  i  f  y  l  s
e  l  v  s  k  l  l  p  d  f  v  k  t  a  l  l  i  .  k  v  f  v  h  f  r  .  a  s  .  f  s  i  f  h
 n  w  l  a  s  y  y  r  i  l  .  k  q  p  s  s  y  k  k  y  s  f  t  s  d  k  r  r  n  f  l  s  f
----.----+----.----+----.----+----.----+----.----+----.----+----.----+----.----+----.----+----.----+
p  i  p  .  c  t  v  v  p  n  q  l  f  l  g  g  .  i  f  l  i  r  e  s  g  i  l  t  t  i  k  .  r  e
 s  n  t  l  l  n  s  g  s  k  t  f  v  a  r  r  m  y  f  t  n  t  .  k  r  y  a  d  y  n  e  i  k
  f  q  n  a  l  .  .  r  i  k  y  f  c  g  e  e  y  l  f  y  e  n  v  e  s  l  r  r  l  k  r  d  k
```

```
       *X XEMB      M     M                                               X
       0B HABI      B     N                                               M
       9A 030N      0     L                                               N
       01 2111      2     1                                               1
        /   //
ATCATATTCTAGATCCCTCTGAAAAAATCTTCCGAGTTTGCTAGGCACTGATACATAACTCTTTTCCAATAATTGGGGAAGTCATTCAAATCTATAATAG
----.----+----.----+----.----+----.----+----.----+----.----+----.----+----.----+----.----+----.----+ 24600
TAGTATAAGATCTAGGGAGACTTTTTTAGAAGGCTCAAACGATCCGTGACTATGTATTGAGAAAAGGTTATTAACCCCTTCAGTAAGTTTAGATATTATC

 s  y  s  r  s  l  .  k  n  l  p  s  l  l  g  t  d  t  .  l  f  s  n  n  w  g  s  h  s  n  l  .  .
  h  i  l  d  p  s  e  k  i  f  r  v  c  .  a  l  i  h  n  s  f  p  i  i  g  e  v  i  q  i  y  n  r
i  i  f  .  i  p  l  k  k  s  s  e  f  a  r  h  .  y  i  t  l  f  q  .  l  g  k  s  f  k  s  i  i  g
----.----+----.----+----.----+----.----+----.----+----.----+----.----+----.----+----.----+----.----+
  d  y  e  l  d  r  q  f  f  r  g  l  k  s  p  v  s  v  y  s  k  e  l  l  q  p  l  .  e  f  r  y  y
.  .  i  r  s  g  e  s  f  i  k  r  t  q  .  a  s  i  c  l  e  k  g  i  i  p  s  t  ■  .  i  .  l  l
 ■  ■  n  .  I  G  R  F  F  D  E  S  N  A  L  C  Q  Y  M  V  R  K  W  Y  N  P  F  D  N  L  D  I  I

*090   24508      Mapped site XBA1
EA31   24512        Ea31                  COOH
```

```
                    B            AFE
                    B            LNC
                    V            UU1
                    1            1H5
                                  /
GTTTCAGATTTGCTTCAATAAATTCTGACTGTAGCTGCTGAAACGTTGCGGTTGAACTATATTTCCTTATAACTTTTACGAAAGAGTTTCTTTGAGTAAT
----.----+----.----+----.----+----.----+----.----+----.----+----.----+----.----+----.----+----.----+ 24700
CAAAGTCTAAACGAAGTTATTTAAGACTGACATCGACGACTTTGCAACGCCAACTTGATATAAAGGAATATTGAAAATGCTTTCTCAAAGAAACTCATTA

v  s  d  l  l  q  .  i  l  t  v  a  a  e  t  l  r  l  n  y  i  s  l  .  l  l  r  k  s  f  f  e  .  s
 f  q  i  c  f  n  k  f  .  l  .  l  l  k  r  c  g  .  t  i  f  p  y  n  f  y  e  r  v  s  l  s  n
  f  r  f  a  s  i  n  s  d  c  s  c  .  n  v  a  v  e  l  y  f  l  i  t  f  t  k  e  f  l  .  v  i
----.----+----.----+----.----+----.----+----.----+----.----+----.----+----.----+----.----+----.----+
 t  e  s  k  s  .  y  i  r  v  t  a  a  s  v  n  r  n  f  .  i  e  k  y  s  k  r  f  l  k  k  s  y
  n  .  i  q  k  l  l  n  q  s  y  s  s  f  r  q  p  q  v  i  n  g  .  l  k  .  s  l  t  e  k  l  l
P  K  L  N  A  E  I  F  E  S  Q  L  Q  Q  F  T  A  T  S  S  Y  K  R  I  V  K  V  F  S  N  R  Q  T  I

EA31    ...      CONTINUES IN PHASE SIX
```

```
                                  M                      A            TS  A SHBH*MM
                                  N                      L            TD  L SGVG0BB
                                  L                      U            HU  U TIUI900
                                  1                      1            21  1 1A1J122
                                                                             //// /
CACTTCACTCAAGTGCTTCCCTGCCTCCAAACGATACCTGTTAGCAATATTTAATAGCTTGAAATGATGAAGAGCTCTGTGTTTGTCTTCCTGCCTCCAG
----.----+----.----+----.----+----.----+----.----+----.----+----.----+----.----+----.----+----.----+ 24800
GTGAAGTGAGTTCACGAAGGGACGGAGGTTTGCTATGGACAATCGTTATAAATTATCGAACTTTACTACTTCTCGAGACACAAACAGAAGGACGGAGGTC

  l  h  s  s  a  s  l  p  p  n  d  t  c  .  q  y  l  i  a  .  n  d  e  e  l  c  v  c  l  p  a  s  s
h  f  t  q  v  l  p  c  l  q  t  i  p  v  s  n  i  .  .  l  e  ■  ■  k  s  s  v  f  v  f  l  p  p  v
 t  s  l  k  c  f  p  a  s  k  r  y  l  l  a  i  f  n  s  l  k  .  .  r  a  l  c  l  s  s  c  l  q
----.----+----.----+----.----+----.----+----.----+----.----+----.----+----.----+----.----+----.----+
d  s  .  e  l  a  e  r  g  g  f  s  v  q  .  c  y  k  i  a  q  f  s  s  s  s  q  t  q  r  g  a  e  l
 .  k  v  .  t  s  g  q  r  w  v  i  g  t  l  l  i  .  y  s  s  i  i  f  l  e  t  n  t  k  r  g  g
  V  E  S  L  H  K  G  A  E  L  R  Y  R  N  A  I  N  L  L  K  F  H  H  L  A  R  H  K  D  E  Q  R  W

*091   24776      Mapped site SST1
EA31    ...       CONTINUES IN PHASE SIX
```

```
  MHNS                            NSH   H                                                      T
  NPCC                            CCP   P                                                      A
  LAIR                            IRA   A                                                      Q
  1211                            112   2                                                      1
     /                             //
TTCGCCGGGCATTCAACATAAAAACTGATAGCACCCGGAGTTCCGGAAACGAAATTTGCATATACCCATTGCTCACGAAAAAAAATGTCCTTGTCGATAT
----.----+----.----+----.----+----.----+----.----+----.----+----.----+----.----+----.----+----.----+ 24900
AAGCGGCCCGTAAGTTGTATTTTTGACTATCGTGGGCCTCAAGGCCTTTGCTTTAAACGTATATGGGTAACGAGTGCTTTTTTTTACAGGAACAGCTATA

 s  p  g  i  q  h  k  n  .  .  h  p  e  f  r  k  r  n  l  h  i  p  i  a  h  e  k  k  c  p  c  r  y
  r  r  a  f  n  i  k  t  d  s  t  r  s  s  g  n  e  i  c  i  y  p  l  l  t  k  k  n  v  l  v  d  i
f  a  g  h  s  t  .  k  l  i  a  p  g  v  p  e  t  k  f  a  y  t  h  c  s  r  k  k  ■  s  l  s  i  .
----.----+----.----+----.----+----.----+----.----+----.----+----.----+----.----+----.----+----.----+
  e  g  p  ■  .  c  l  f  q  y  c  g  s  n  r  f  r  f  k  c  i  g  ■  a  .  s  f  f  h  g  q  r  y
t  r  r  a  n  l  ■  f  v  s  l  v  r  l  e  p  f  s  i  q  ■  y  g  n  s  v  f  f  f  t  r  t  s  i
 N  A  P  C  E  V  Y  F  S  I  A  G  P  T  G  S  V  F  N  A  Y  V  W  Q  E  R  F  F  I  D  K  D  I

EA31    ...       CONTINUES IN PHASE SIX
```

```
 H    H        F   R   C       M                                   TF                   H
 G    N        O   S   A       N                                   HN                   P
 U    F        K   A   M       L                                   AU                   H
 2    1        1   1   3       1                                   1H                   1

AGGGATGAATCGCTTGGTGTACCTCATCTACTGCGAAAACTTGACCTTTCTCTCCCATATTGCAGTCGCGGCACGATGGAACTAAATTAATAGGCATCAC
----.----+----.----+----.----+----.----+----.----+----.----+----.----+----.----+----.----+----.----+ 25000
TCCCTACTTAGCGAACCACATGGAGTAGATGACGCTTTTGAACTGGAAAGAGAGGGTATAACGTCAGCGCCGTGCTACCTTGATTTAATTATCCGTAGTG

r  d  e  s  l  g  v  p  h  l  l  r  k  l  d  l  s  l  p  y  c  s  r  g  t  m  e  l  n  .  .  a  s  p
 g  m  n  r  l  v  y  l  i  y  c  e  n  l  t  f  l  s  h  i  a  v  a  a  r  w  n  .  i  n  r  h  h
  g  .  i  a  w  c  t  s  s  t  a  k  t  .  p  f  s  p  i  l  q  s  r  h  d  g  t  k  l  i  g  i  t
----.----+----.----+----.----+----.----+----.----+----.----+----.----+----.----+----.----+----.----+
 l  s  s  d  s  p  t  g  .  r  s  r  f  s  s  r  e  r  g  y  q  l  r  p  v  i  s  s  f  .  y  a  d
  p  i  f  r  k  t  y  r  m  .  q  s  f  k  v  k  r  e  w  i  a  t  a  a  r  h  f  .  i  l  l  c  .
Y  P  H  I  A  Q  H  V  E  D  V  A  F  V  Q  G  K  E  G  M  N  C  D  R  C  S  P  V  L  N  I  P  M  V

CAM3   24924      cam3                i(Tn9)     EM
EA31    ...       CONTINUES IN PHASE SIX
```

```
  S                              M   M              H                    H
  F                              B   B              P                    P
  A                              0   0              H                    H
  N                              1   2              1                    1

CGAAAATTCAGGATAATGTGCAATAGGAAGAAAATGATCTATATTTTTTGTCTGTCCTATATCACCACAAAATGGACATTTTTCACCTGATGAAACAAGC
----.----+----.----+----.----+----.----+----.----+----.----+----.----+----.----+----.----+----.----+ 25100
GCTTTTAAGTCCTATTACACGTTATCCTTCTTTTACTAGATATAAAAAACAGACAGGATATAGTGGTGTTTTACCTGTAAAAAGTGGACTACTTTGTTCG

  k  i  q  d  n  v  q  .  e  e  n  d  l  y  f  l  s  v  l  y  h  h  k  m  d  i  f  h  l  m  k  q  a
r  k  f  r  i  m  c  n  r  k  k  m  i  y  i  f  c  l  s  y  i  t  t  k  w  t  f  f  t  .  .  n  k  h
 e  n  s  g  .  c  a  i  g  r  k  .  s  i  f  f  v  c  p  i  s  p  q  n  g  h  f  s  p  d  e  t  s
----.----+----.----+----.----+----.----+----.----+----.----+----.----+----.----+----.----+----.----+
g  f  i  .  s  l  t  c  y  s  s  f  s  r  y  k  k  d  t  r  y  .  w  l  i  s  m  k  .  r  i  f  c  a
 r  f  n  l  i  i  h  l  l  f  f  i  i  .  i  k  q  r  d  .  i  v  v  f  h  v  n  k  v  q  h  f  l
  S  F  E  P  Y  H  A  I  P  L  F  H  D  I  N  K  T  Q  G  I  D  G  C  F  P  C  K  E  G  S  S  V  L

EA31    ...       CONTINUES IN PHASE SIX
```

```
  N        T                                            *HA                       B           D
  S        T                                            0IL                       S           D
  P        H                                            9NU                       T           E
  C        2                                            231                       E           1
                                                         /
ATGTCATCGTAATATGTTCTAGCGGGTTTGTTTTTATCTCGGAGATTATTTTCATAAAGCTTTTCTAATTTAACCTTTGTCAGGTTACCAACTACTAAGG
----.----+----.----+----.----+----.----+----.----+----.----+----.----+----.----+----.----+----.----+ 25200
TACAGTAGCATTATACAAGATCGCCCAAACAAAAATAGAGCCTCTAATAAAAGTATTTCGAAAAGATTAAATTGGAAACAGTCCAATGGTTGATGATTCC

 c  h  r  n  m  f  .  r  v  c  f  y  l  g  d  y  f  h  k  a  f  l  i  .  p  l  s  g  y  q  l  l  r
  v  i  v  i  c  s  s  g  f  v  f  i  s  e  i  i  f  i  k  l  f  .  f  n  l  c  q  v  t  n  y  .  g
m  s  s  .  y  v  l  a  g  l  f  l  s  r  r  l  f  s  .  s  f  s  n  l  t  f  v  r  l  p  t  t  k  v
----.----+----.----+----.----+----.----+----.----+----.----+----.----+----.----+----.----+----.----+
  h  .  r  l  i  n  .  r  t  q  k  .  r  p  s  .  k  .  l  a  k  r  i  .  g  k  d  p  .  w  s  s  l
c  t  m  t  i  h  e  l  p  n  t  k  i  e  s  i  i  k  m  f  s  k  .  n  l  r  q  .  t  v  l  .  .  p
 M  D  D  Y  Y  T  R  A  P  K  N  K  D  R  L  N  N  E  Y  L  K  E  L  K  V  K  T  L  N  G  V  V  L

*092   25157      Mapped site HIN3
EA31    ...       CONTINUES IN PHASE SIX
```

```
     M                                          T   A
     N                                          A   L
     L                                          Q   U
     1                                          1   1

TTGTAGGCTCAAGAGGGTGTGTCCTGTCGTAGGTAAATAACTGACCTGTCGAGCTTAATATTCTATATTGTTGTTCTTTCTGCAAAAAAGTGGGGAAGTG
----.----+----.----+----.----+----.----+----.----+----.----+----.----+----.----+----.----+----.----+ 25300
AACATCCGAGTTCTCCCACACAGGACAGCATCCATTTATTGACTGGACAGCTCGAATTATAAGATATAACAACAAGAAAGACGTTTTTTCACCCCTTCAC

l  .  a  q  e  g  v  s  c  r  r  .  i  t  d  l  s  s  l  i  f  y  i  v  v  l  s  a  k  k  w  g  s  e
 c  r  l  k  r  v  c  p  v  v  g  k  .  l  t  c  r  a  .  y  s  i  l  l  f  f  l  q  k  s  g  e  v
  v  g  s  r  g  c  v  l  s  .  v  n  n  .  p  v  e  l  n  i  l  y  c  c  s  f  c  k  k  v  g  k  .
----.----+----.----+----.----+----.----+----.----+----.----+----.----+----.----+----.----+----.----+
 n  y  a  .  s  p  t  d  q  r  l  y  i  v  s  r  d  l  k  i  n  .  i  t  t  r  e  a  f  f  h  p  l
  q  l  s  l  l  t  h  g  t  t  p  l  y  s  v  q  r  a  .  y  e  i  n  n  n  k  r  c  f  l  p  s  t
T  T  P  E  L  P  H  T  R  D  Y  T  F  L  Q  G  T  S  S  L  I  R  Y  Q  Q  E  K  Q  L  F  T  P  F  H

EA31    ...       CONTINUES IN PHASE SIX
```

```
                                      S                                   M                M     EE
                                      F                                   N                B     AA
                                      A                                   L                0     35
                                      N                                   1                2     19
                                                                                                  /
AGTAATGAAATTATTTCTAACATTTATCTGCATCATACCTTCCGAGCATTTATTAAGCATTTCGCTATAAGTTCTCGCTGGAAGAGGTAGTTTTTTCATT
----.----+----.----+----.----+----.----+----.----+----.----+----.----+----.----+----.----+----.----+ 25400
TCATTACTTTAATAAAGATTGTAAATAGACGTAGTATGGAAGGCTCGTAAATAATTCGTAAAGCGATATTCAAGAGCGACCTTCTCCATCAAAAAAGTAA

  .  .  n  y  f  .  h  l  s  a  s  y  l  p  s  i  y  .  a  f  r  y  k  f  s  l  e  e  v  v  f  s  l
s  n  e  i  i  s  n  i  y  l  h  h  t  f  r  a  f  i  k  h  f  a  i  s  s  r  w  k  r  .  f  f  h  c
 v  m  k  l  f  l  t  f  i  c  i  i  p  s  e  h  l  l  s  i  s  l  .  v  l  a  g  r  g  s  f  f  i
----.----+----.----+----.----+----.----+----.----+----.----+----.----+----.----+----.----+----.----+
s  y  h  f  .  k  .  c  k  d  a  d  y  r  g  l  m  .  .  a  n  r  .  l  n  e  s  s  s  t  t  k  e  n
 l  l  s  i  i  e  l  m  .  r  c  .  v  k  r  a  n  i  l  c  k  a  i  l  e  r  q  f  l  y  n  k  .
  T  I  F  N  N  R  V  N  I  Q  M  M  G  E  S  C  K  N  L  M  E  S  Y  T  R  A  P  L  P  L  K  K  M

EA31  25399      Ea31 (orf-296)      M-NH3
EA59  25399      bEa59               COOH
```

```
 R                                   A        T                                         X
 S                                   H        A                                         M
 A                                   A        Q                                         N
 1                                   3        1                                         1

GTACTTTACCTTCATCTCTGTTCATTATCATCGCTTTTAAAACGGTTCGACCTTCTAATCCTATCTGACCATTATAATTTTTTAGAATGGTTTCATAAGA
----.----+----.----+----.----+----.----+----.----+----.----+----.----+----.----+----.----+----.----+ 25500
CATGAAATGGAAGTAGAGACAAGTAATAGTAGCGAAAATTTTGCCAAGCTGGAAGATTAGGATAGACTGGTAATATTAAAAAATCTTACCAAAGTATTCT

 y  f  t  f  i  s  v  h  y  h  r  f  .  n  g  s  t  f  .  s  y  l  t  i  i  i  f  .  n  g  f  i  r
  t  l  p  s  s  l  f  i  i  i  a  f  k  t  v  r  p  s  n  p  i  .  p  l  .  f  f  r  m  v  s  .  e
v  l  y  l  h  l  c  s  l  s  s  l  l  k  r  f  d  l  l  i  l  s  d  h  y  n  f  l  e  w  f  h  k  k
----.----+----.----+----.----+----.----+----.----+----.----+----.----+----.----+----.----+----.----+
 y  k  v  k  m  e  t  .  .  .  r  k  .  f  p  e  v  k  .  d  .  r  v  m  i  i  k  .  f  p  k  m  l
Q  V  K  G  E  D  R  N  M  I  M  A  K  L  V  T  R  G  E  L  G  I  Q  G  N  Y  N  K  L  I  T  E  Y  S
  t  s  .  r  .  r  q  e  n  d  d  s  k  f  r  n  s  r  r  i  r  d  s  w  .  l  k  k  s  h  n  .  l

EA59    ...      CONTINUES IN PHASE FIVE
```

```
   A    H                                                                     M       DH         H
   L    N                                                                     N       DP         N
   U    F                                                                     L       EH         F
   1    1                                                                     1       11         3
                                                                                       /
AAGCTCTGAATCAACGGACTGCGATAATAAGTGGTGGTATCCAGAATTTGTCACTTCAAGTAAAAACACCTCACGAGTTAAAACACCTAAGTTCTCACCG
----.----+----.----+----.----+----.----+----.----+----.----+----.----+----.----+----.----+----.----+ 25600
TTCGAGACTTAGTTGCCTGACGCTATTATTCACCACCATAGGTCTTAAACAGTGAAGTTCATTTTTGTGGAGTGCTCAATTTTGTGGATTCAAGAGTGGC

k  l  .  i  n  g  l  r  .  .  v  v  v  s  r  i  c  h  f  k  .  k  h  l  t  s  .  n  t  .  v  l  t  e
 s  s  e  s  t  d  c  d  n  k  w  w  y  p  e  f  v  t  s  s  k  n  t  s  r  v  k  t  p  k  f  s  p
  a  l  n  q  r  t  a  i  i  s  g  g  i  q  n  l  s  l  q  v  k  t  p  h  e  l  k  h  l  s  s  h  r
----.----+----.----+----.----+----.----+----.----+----.----+----.----+----.----+----.----+----.----+
 f  s  q  i  l  p  s  r  y  y  t  t  t  d  l  i  q  .  k  l  y  f  c  r  v  l  .  f  v  .  t  r  v
  L  E  S  D  V  S  Q  S  L  L  H  H  Y  G  S  N  T  V  E  L  L  F  V  E  R  T  L  V  G  L  N  E  G
f  a  r  f  .  r  v  a  i  i  l  p  p  i  w  f  k  d  s  .  t  f  v  g  .  s  n  f  c  r  l  e  .  r

EA59    ...      CONTINUES IN PHASE FIVE
```

```
              H                                               N                   S MRR
              P                                               S                   C NRS
              A                                               P                   A LUA
              2                                               C                   1 111
                                                                                     //
AATGTCTCAATATCCGGACGGATAATATTTATTGCTTCTCTTGACCGTAGGACTTTCCACATGCAGGATTTTGGAACCTCTTGCAGTACTACTGGGGAAT
----.----+----.----+----.----+----.----+----.----+----.----+----.----+----.----+----.----+----.----+ 25700
TTACAGAGTTATAGGCCTGCCTATTATAAATAACGAAGAGAACTGGCATCCTGAAAGGTGTACGTCCTAAAACCTTGGAGAACGTCATGATGACCCCTTA

 c  l  n  i  r  t  d  n  i  y  c  f  s  .  p  .  d  f  p  h  a  g  f  w  n  l  l  q  y  y  w  g  m
n  v  s  i  s  g  r  i  i  f  i  a  s  l  d  r  r  t  f  h  m  q  d  f  g  t  s  c  s  t  t  g  e  .
 m  s  q  y  p  d  g  .  y  l  l  l  l  l  t  v  g  l  s  t  c  r  i  l  e  p  l  a  v  l  l  g  n
----.----+----.----+----.----+----.----+----.----+----.----+----.----+----.----+----.----+----.----+
s  h  r  l  i  r  v  s  l  i  .  q  k  e  q  g  y  s  k  g  c  a  p  n  q  f  r  k  c  y  .  q  p  i
 F  T  F  I  D  P  R  I  I  N  I  A  E  R  S  R  L  V  K  W  M  C  S  K  P  V  E  Q  L  V  V  P  S
  i  d  .  y  g  s  p  y  y  k  n  s  r  k  v  t  p  s  e  v  h  l  i  k  s  g  r  a  t  s  s  p  f

EA59    ...      CONTINUES IN PHASE FIVE
```

```
                                   T  S                                    S    H       F
                                   A  F                                    F    G       O
                                   Q  A                                    A    U       K
                                   1  N                                    N    2       1

GAGTTGCAATTATTGCTACACCATTGCGTGCATCGAGTAAGTCGCTTAATGTTCGTAAAAAAGCAGAGAGCAAAGGTGGATGCAGATGAACCTCTGGTTC
----.----+----.----+----.----+----.----+----.----+----.----+----.----+----.----+----.----+----.----+ 25800
CTCAACGTTAATAACGATGTGGTAACGCACGTAGCTCATTCAGCGAATTACAAGCATTTTTTCGTCTCTCGTTTCCACCTACGTCTACTTGGAGACCAAG

 s  c  n  y  c  y  t  i  a  c  i  e  .  v  a  .  c  s  .  k  s  r  e  q  r  w  ■  q  ■  n  l  w  f
  v  a  i  i  a  t  p  l  r  a  s  s  k  s  l  n  v  r  k  k  a  e  s  k  g  g  c  r  .  t  s  g  s
e  l  q  l  l  l  h  h  c  v  h  r  v  s  r  l  ■  f  v  k  k  q  r  a  k  v  d  a  d  e  p  l  v  h
----.----+----.----+----.----+----.----+----.----+----.----+----.----+----.----+----.----+----.----+
  l  q  l  .  q  .  v  ■  a  h  ■  s  y  t  a  .  h  e  y  f  l  l  s  c  l  h  i  c  i  f  r  q  n
H  T  A  I  I  A  V  G  N  R  A  D  L  L  D  S  L  T  R  L  F  A  S  L  L  P  P  H  L  H  V  F  P  E
 s  n  c  n  n  s  c  w  q  t  c  r  t  l  r  k  i  n  t  f  f  c  l  a  f  t  s  a  s  s  g  r  t

EA59     ...     CONTINUES IN PHASE FIVE
```

```
MTH                                             B                                N   S  A SHBH*HAT  C
NAN                                             5                                S   D  L SGVG0NSA  A
LQF                                             1                                P   U  U TIUT9FUQ  M
113                                             5                                C   1  1 1A1J3321  1
 /                                                                                        ///////
ATCGAATAAAACTAATGACTTTTCGCCAACGACATCTACTAATCTTGTGATAGTAAATAAAACAATTGCATGTCCAGAGCTCATTCGAAGCAGATATTTC
----.----+----.----+----.----+----.----+----.----+----.----+----.----+----.----+----.----+----.----+ 25900
TAGCTTATTTTGATTACTGAAAAGCGGTTGCTGTAGATGATTAGAACACTATCATTTATTTTGTTAACGTACAGGTCTCGAGTAAGCTTCGTCTATAAAG

i  e  .  n  .  .  l  f  a  n  d  i  y  .  s  c  d  s  k  .  n  n  c  ■  s  r  a  h  s  k  q  i  f  l
 s  n  k  t  n  d  f  s  p  t  t  s  t  n  l  v  i  v  n  k  t  i  a  c  p  e  l  i  r  s  r  y  f
  r  i  k  l  ■  t  f  r  q  r  h  l  l  i  l  .  .  .  i  k  q  l  h  v  q  s  s  f  e  a  d  i  s
----.----+----.----+----.----+----.----+----.----+----.----+----.----+----.----+----.----+----.----+
■  s  y  f  .  h  s  k  a  l  s  ■  .  .  d  q  s  l  l  y  f  l  q  ■  d  l  a  .  e  f  c  i  n
 D  F  L  V  L  S  K  E  G  V  V  D  V  L  R  T  I  T  F  L  V  I  A  H  G  S  S  M  R  L  L  Y  K
.  r  i  f  s  i  v  k  r  w  r  c  r  s  i  k  h  y  y  i  f  c  n  c  t  w  l  e  n  s  a  s  i  e

B515   25843      b515               dr          EM
*093   25881      Mapped site SST1
CAM1   25891        caml             i(Tn9)      EM
EA59    ...       CONTINUES IN PHASE FIVE
```

```
                                                 H                 M                M
                                                 N                 B                N
                                                 F                 O                L
                                                 1                 2                1

TGGATATTGTCATAAAACAATTTAGTGAATTTATCATCGTCCACTTGAATCTGTGGTTCATTACGTCTTAACTCTTCATATTTAGAAATGAGGCTGATGA
----.----+----.----+----.----+----.----+----.----+----.----+----.----+----.----+----.----+----.----+ 26000
ACCTATAACAGTATTTTGTTAAATCACTTAAATAGTAGCAGGTGAACTTAGACACCAAGTAATGCAGAATTGAGAAGTATAAATCTTTACTCCGACTACT

  d  i  v  i  k  q  f  s  e  f  i  i  v  h  l  n  l  w  f  i  t  s  .  l  f  i  f  r  n  e  a  d  e
w  i  l  s  .  n  n  l  v  n  l  s  s  s  t  .  i  c  g  s  l  r  l  n  s  s  y  l  e  ■  r  l  ■  s
 g  y  c  h  k  t  i  .  .  i  y  h  r  p  l  e  s  v  v  h  y  v  l  t  l  h  i  .  k  .  g  .  .
----.----+----.----+----.----+----.----+----.----+----.----+----.----+----.----+----.----+----.----+
r  s  i  t  ■  f  c  n  l  s  n  i  ■  t  w  k  f  r  h  n  ■  v  d  .  s  k  ■  n  l  f  s  a  s  s
 Q  I  N  D  Y  F  L  K  T  F  K  D  D  D  V  Q  I  Q  P  E  N  R  R  L  E  E  Y  K  S  I  L  S  I
  p  y  q  .  l  v  i  .  h  i  .  .  r  g  s  s  d  t  t  .  .  t  k  v  r  .  i  .  f  h  p  q  h

EA59     ...     CONTINUES IN PHASE FIVE
```

```
                               D              A
                               D              L
                               E              U
                               1              1

GTTCCATATTTGAAAAGTTTTCATCACTACTTAGTTTTTTGATAGCTTCAAGCCAGAGTTGTCTTTTTCTATCTACTCTCATACAACCAATAAATGCTGA
----.----+----.----+----.----+----.----+----.----+----.----+----.----+----.----+----.----+----.----+ 26100
CAAGGTATAAACTTTTCAAAAGTAGTGATGAATCAAAAAACTATCGAAGTTCGGTCTCAACAGAAAAAGATAGATGAGAGTATGTTGGTTATTTACGACT

 f  h  i  .  k  v  f  i  t  t  .  f  f  d  s  f  k  p  e  l  s  f  s  i  y  s  h  t  t  n  k  c  .
  s  i  f  e  k  f  s  s  l  l  s  f  l  i  a  s  s  q  s  c  l  f  l  s  t  l  i  q  p  i  n  a  e
v  p  y  l  k  s  f  h  h  y  l  v  f  .  .  l  q  a  r  v  v  f  f  y  l  l  s  y  n  q  .  ■  l  k
----.----+----.----+----.----+----.----+----.----+----.----+----.----+----.----+----.----+----.----+
  n  w  i  q  f  t  k  ■  v  v  .  n  k  s  l  k  l  g  s  n  d  k  e  i  .  e  .  v  v  l  l  h  q
L  E  M  N  S  F  N  E  D  S  S  L  K  K  I  A  E  L  W  L  Q  R  K  R  D  V  R  M  C  G  I  F  A  S
 t  g  y  k  f  l  k  .  .  .  k  t  k  q  y  s  .  a  l  t  t  k  k  .  r  s  e  y  l  w  y  i  s
```

```
EA59      ...       CONTINUES IN PHASE FIVE

   *E  D       M                A             F        HB                      R
   0C  D       B                H             N        NB                      S
   9R  E       0                A             U        FV                      A
   41  1       1                3             H        11                      1
    /
AATGAATTCTAAGCGGAGATCGCCTAGTGATTTTAAACTATTGCTGGCAGCATTCTTGAGTCCAATATAAAAGTATTGTGTACCTTTTGCTGGGTCAGGT
----.----+----.----+----.----+----.----+----.----+----.----+----.----+----.----+----.----+----.----+  26200
TTACTTAAGATTCGCCTCTAGCGGATCACTAAAATTTGATAACGACCGTCGTAAGAACTCAGGTTATATTTTCATAACACATGGAAAACGACCCAGTCCA

n  e  f  .  a  e  i  a  .  .  f  .  t  i  a  g  s  i  l  e  s  n  i  k  v  l  c  t  f  c  w  v  r  l
 m  n  s  k  r  r  s  p  s  d  f  k  l  l  l  a  a  f  l  s  p  i  .  k  y  c  v  p  f  a  g  s  g
  .  i  l  s  g  d  r  l  v  i  l  n  y  c  w  q  h  s  .  v  q  y  k  s  i  v  y  l  l  l  g  q  v
----.----+----.----+----.----+----.----+----.----+----.----+----.----+----.----+----.----+----.----+
 f  s  n  .  a  s  i  a  .  h  n  .  v  i  a  p  l  m  r  s  d  l  i  f  t  n  h  v  k  q  q  t  l
  I  F  E  L  R  L  D  G  L  S  K  L  S  N  S  A  A  N  K  L  G  I  Y  F  Y  Q  T  G  K  A  P  D  P
f  h  i  r  l  p  s  r  r  t  i  k  f  .  q  q  c  c  e  q  t  w  y  l  l  i  t  y  r  k  s  p  .  t

*094    26104      Mapped site ECR1
EA59      ...      CONTINUES IN PHASE FIVE

   M                 BM                               M  PT*               H      H  T
   N                 IB                               B  VA0               N      N  A
   L                 NO                               0  UQ9               F      F  Q
   1                 11                               1  115               1      1  1
                                                          //
TGTTCTTTAGGAGGAGTAAAAGGATCAAATGCACTAAACGAAACTGAAACAAGCGATCGAAAATATCCCTTTGGGATTCTTGACTCGATAAGTCTATTAT
----.----+----.----+----.----+----.----+----.----+----.----+----.----+----.----+----.----+----.----+  26300
ACAAGAAATCCTCCTCATTTTCCTAGTTTACGTGATTTGCTTTGACTTTGTTCGCTAGCTTTTATAGGGAAACCCTAAGAACTGAGCTATTCAGATAATA

  f  f  r  r  s  k  r  i  k  c  t  k  r  n  .  n  k  r  s  k  i  s  l  w  d  s  .  l  d  k  s  i  i
c  s  l  g  g  v  k  g  s  n  a  l  n  e  t  e  t  s  d  r  k  y  p  f  g  i  l  d  s  i  s  l  l  f
 v  l  .  e  e  .  k  d  q  m  h  .  t  k  l  k  q  a  i  e  n  i  p  l  g  f  l  t  r  .  v  y  y
----.----+----.----+----.----+----.----+----.----+----.----+----.----+----.----+----.----+----.----+
n  n  k  l  l  l  l  l  i  l  h  v  l  r  f  q  f  l  r  d  f  i  d  r  q  s  e  q  s  s  l  d  i  i
 Q  E  K  P  P  T  F  P  D  F  A  S  F  S  V  S  V  L  S  R  F  Y  G  K  P  I  R  S  E  I  L  R  N
  t  r  .  s  s  y  f  s  .  i  c  .  v  f  s  f  c  a  i  s  f  i  g  k  p  n  k  v  r  y  t  .  .

*095    26258      Mapped site PVU1
EA59      ...      CONTINUFS IN PHASF FIVE

                                              H
                                              P
                                              H
                                              1

TTTCAGAGAAAAAATATTCATTGTTTTCTGGGTTGGTGATTGCACCAATCATTCCATTCAAAATTGTTGTTTTACCACACCCATTCCGCCCGATAAAAGC
----.----+----.----+----.----+----.----+----.----+----.----+----.----+----.----+----.----+----.----+  26400
AAAGTCTCTTTTTTATAAGTAACAAAAGACCCAACCACTAACGTGGTTAGTAAGGTAAGTTTTAACAACAAAATGGTGTGGGTAAGGCGGGCTATTTTCG

 f  r  e  k  i  f  i  v  f  w  v  g  d  c  t  n  h  s  i  q  n  c  c  f  t  t  p  i  p  p  d  k  s
  s  e  k  k  y  s  l  f  s  g  l  v  i  a  p  i  i  p  f  k  i  v  v  l  p  h  p  f  r  p  i  k  a
f  q  r  k  n  i  h  c  f  l  g  w  .  l  h  q  s  f  h  s  k  l  l  f  y  h  t  h  s  a  r  .  k  h
----.----+----.----+----.----+----.----+----.----+----.----+----.----+----.----+----.----+----.----+
 k  l  s  f  i  n  m  t  k  q  t  p  s  q  v  l  .  e  m  .  f  q  q  k  v  v  g  m  g  g  s  l  l
N  E  S  F  F  Y  E  N  N  E  P  N  T  I  A  G  I  M  G  N  L  I  T  T  K  G  C  G  N  R  G  I  F  A
 k  .  l  f  f  i  .  q  k  r  p  q  h  n  c  w  d  n  w  e  f  n  n  n  .  w  v  w  e  a  r  y  f

EA59      ...       CONTINUES IN PHASE FIVF

                        H        H           M              H   HNSS    H
                        P        G           N              N   PCCD    G
                        H        E           L              F   AIRU    I
                        1        2           1              1   2111    A
                                                                  /
ATGAATGTTCGTGCTGGGCATAGAATTAACCGTCACCTCAAAAGGTATAGTTAAATCACTGAATCCGGGAGCACTTTTTCTATTAAATGAAAAGTGGAAA
----.----+----.----+----.----+----.----+----.----+----.----+----.----+----.----+----.----+----.----+  26500
TACTTACAAGCACGACCCGTATCTTAATTGGCAGTGGAGTTTTCCATATCAATTTAGTGACTTAGGCCCTCGTGAAAAAGATAATTTACTTTTCACCTTT

m  n  v  r  a  g  h  r  i  n  r  h  l  k  r  y  s  .  i  t  e  s  g  s  t  f  s  i  k  .  k  v  e  i
 .  m  f  v  l  g  i  e  l  t  v  t  s  k  g  i  v  k  s  l  n  p  g  a  l  f  l  l  n  e  k  w  k
  e  c  s  c  w  a  .  n  .  p  s  p  q  k  v  .  l  n  h  .  i  r  e  h  f  f  y  .  m  k  s  g  n
----.----+----.----+----.----+----.----+----.----+----.----+----.----+----.----+----.----+----.----+
 m  f  t  r  a  p  c  l  i  l  r  .  r  l  l  y  l  .  i  v  s  d  p  l  v  k  e  i  l  h  f  t  s
  H  I  N  T  S  P  M  S  N  V  T  V  E  F  P  I  T  L  D  S  F  G  P  A  S  K  R  N  F  S  F  H  F
c  s  h  e  h  q  a  y  f  .  g  d  g  .  f  t  y  n  f  .  q  i  r  s  c  k  k  .  .  i  f  l  p  f
```

EA59 ... CONTINUES IN PHASE FIVE

```
                              A                                   DM    M                 H
                              F                                   DN    N                 G
                              L                                   EL    L                 A
                              3                                   11    1                 1

TCTGACAATTCTGGCAAACCATTTAACACACGTGCGAACTGTCCATGAATTTCTGAAAGAGTTACCCCTCTAAGTAATGAGGTGTTAAGGACGCTTTCAT
----.----+----.----+----.----+----.----+----.----+----.----+----.----+----.----+----.----+----.----+  26600
AGACTGTTAAGACCGTTTGGTAAATTGTGTGCACGCTTGACAGGTACTTAAAGACTTTCTCAATGGGGAGATTCATTACTCCACAATTCCTGCGAAAGTA

  .  q  f  w  q  t  i  .  h  t  c  e  l  s  m  n  f  .  k  s  y  p  s  k  .  .  g  v  k  d  a  f  i
 s  d  n  s  g  k  p  f  n  t  r  a  n  c  p  .  i  s  e  r  v  t  p  l  s  n  e  v  l  r  t  l  s  f
l  t  i  l  a  n  h  l  t  h  v  r  t  v  h  e  f  l  k  e  l  p  l  .  v  m  r  c  .  g  r  f  h
----.----+----.----+----.----+----.----+----.----+----.----+----.----+----.----+----.----+----.----+
 i  q  c  n  q  c  v  m  .  c  v  h  s  s  d  m  f  k  q  f  l  .  g  e  l  y  h  p  t  l  s  a  k  m
 D  S  L  E  P  L  G  N  L  V  R  A  F  Q  G  H  I  E  S  L  T  V  G  R  L  L  S  T  N  L  V  S  E
  r  v  i  r  a  f  w  k  v  c  t  r  v  t  w  s  n  r  f  s  n  g  r  .  t  i  l  h  .  p  r  k  .
```

EA59 ... CONTINUES IN PHASE FIVE

```
              C*T  C BHH                     A                A                D
              L0A  F AAA                     L                H                D
              A9Q  R LEE                     U                A                E
              161  1 113                     1                3                1
               //     //
TTTCAATGTCGGCTAATCGATTTGGCCATACTACTAAATCCTGAATAGCTTTAAGAAGGTTATGTTTAAAACCATCGCTTAATTTGCTGAGATTAACATA
----.----+----.----+----.----+----.----+----.----+----.----+----.----+----.----+----.----+----.----+  26700
AAAGTTACAGCCGATTAGCTAAACCGGTATGATGATTTAGGACTTATCGAAATTCTTCCAATACAAATTTTGGTAGCGAATTAAACGACTCTAATTGTAT

 f  n  v  g  .  s  i  w  p  y  y  .  i  l  n  s  f  k  k  v  m  f  k  t  i  a  .  f  a  e  i  n  i
  s  m  s  a  n  r  f  g  h  t  t  k  s  .  i  a  l  r  r  l  c  l  k  p  s  l  n  l  l  r  l  t  .
f  q  c  r  l  i  d  l  a  i  l  l  n  p  e  .  l  .  e  g  y  v  .  n  h  r  l  i  c  .  d  .  h  s
----.----+----.----+----.----+----.----+----.----+----.----+----.----+----.----+----.----+----.----+
  k  l  t  p  .  d  i  q  g  y  .  .  i  r  f  l  k  l  f  t  i  n  l  v  m  a  .  n  a  s  i  l  m
N  E  I  D  A  L  R  N  P  W  V  V  L  D  Q  I  A  K  L  L  N  H  K  F  G  D  S  L  K  S  L  N  V  Y
  k  .  h  r  s  i  s  k  a  m  s  s  f  g  s  y  s  .  s  p  .  t  .  f  w  r  k  i  q  q  s  .  c
```

*096 26617 Mapped site CLA1
EA59 ... CONTINUES IN PHASE FIVE

```
     H     1 ' M*D                     H*                   M
     P     0   S0D                     I0                   B
     H     1   T9E                     N9                   0
     1     8   271                     28                   2
                //                      /
GTAGTCAATGCTTTCACCTAAGGAAAAAAACATTTCAGGGAGTTGACTGAATTTTTTATCTATTAATGAATAAGTGCTTACTTCTTCTTTTTGACCTACA
----.----+----.----+----.----+----.----+----.----+----.----+----.----+----.----+----.----+----.----+  26800
CATCAGTTACGAAAGTGGATTCCTTTTTTTGTAAAGTCCCTCAACTGACTTAAAAAATAGATAATTACTTATTCACGAATGAAGAAGAAAAACTGGATGT

v  v  n  a  f  t  .  g  k  k  h  f  r  e  l  t  e  f  f  i  y  .  .  i  s  a  y  f  f  f  l  t  y  k
  .  s  m  l  s  p  k  e  k  n  i  s  g  s  .  l  n  f  l  s  i  n  e  .  v  l  t  s  s  f  .  p  t
  s  q  c  f  h  l  r  k  k  t  f  q  g  v  d  .  i  f  y  l  l  m  n  k  c  l  l  l  l  f  d  l  q
----.----+----.----+----.----+----.----+----.----+----.----+----.----+----.----+----.----+----.----+
 t  t  l  a  k  v  .  p  f  f  c  k  l  s  n  v  s  n  k  i  .  .  h  i  l  a  .  k  k  k  k  v  .
  Y  D  I  S  E  G  L  S  F  F  M  E  P  L  Q  S  F  K  K  D  I  L  S  Y  T  S  V  E  E  K  Q  G  V
l  l  .  h  k  .  r  l  f  f  v  n  .  p  t  s  q  i  k  .  r  n  i  f  l  h  k  s  r  r  k  s  r  c
```

1018 26714 b1018 dl EM
*097 26718 Mapped site MST2
*098 26744 Mapped site HIN2
EA59 ... CONTINUES IN PHASE FIVE

```
                             EH
                             CP
                             RH
                             V1

AAACCAATTTTAACATTTCCGATATCGCATTTTTCACCATGCTCATCAAAGACAGTAAGATAAAACATTGTAACAAAGGAATAGTCATTCCAACCATCTG
----.----+----.----+----.----+----.----+----.----+----.----+----.----+----.----+----.----+----.----+ 26900
TTTGGTTAAAATTGTAAAGGCTATAGCGTAAAAAGTGGTACGAGTAGT   TGTCATTCTATTTTGTAACATTGTTTCCTTATCAGTAAGGTTGGTAGAC

 t  n  f  n  i  s  d  i  a  f  f  t  m  l  i  k  d  s  k  i  k  h  c  n  k  g  i  v  i  p  t  i  c
k  p  i  l  t  f  p  i  s  h  f  s  p  c  s  s  k  t  v  r  .  n  i  v  t  k  e  .  s  f  q  p  s  a
 n  q  f  .  h  f  r  y  r  i  f  h  h  a  h  q  r  q  .  d  k  t  l  .  q  r  n  s  h  s  n  h  l
----.----+----.----+----.----+----.----+----.----+----.----+----.----+----.----+----.----+----.----+
l  v  l  k  l  m  e  s  i  a  n  k  v  m  s  m  l  s  l  l  i  f  c  q  l  l  p  i  t  m  g  v  m  q
 F  G  I  K  V  N  G  I  D  C  K  E  G  H  E  D  F  V  T  L  Y  F  M  T  V  F  S  Y  D  N  W  G  D
 f  w  n  .  c  k  r  y  r  m  k  .  w  a  .  .  l  c  y  s  l  v  n  y  c  l  f  l  .  e  l  w  r

EA59     ...      CONTINUES IN PHASE FIVE


                                     P*                M                E
                                     S0                N                A
                                     T9                L                5
                                     19                1                9
                                     /
CTCGTAGGAATGCCTTATTTTTTTCTACTGCAGGAATATACCCGCCTCTTTCAATAACACTAAACTCCAACATATAGTAACCCTTAATTTTATTAAAATA
----.----+----.----+----.----+----.----+----.----+----.----+----.----+----.----+----.----+----.----+ 27000
GAGCATCCTTACGGAATAAAAAAAGATGACGTCCTTATATGGGCGGAGAAAGTTATTGTGATTTGAGGTTGTATATCATTGGGAATTAAAATAATTTTAT

 s  .  e  c  l  i  f  f  y  c  r  n  i  p  a  s  f  n  n  t  k  l  q  h  i  v  t  l  n  f  i  k  i
  r  r  n  a  l  f  f  s  t  a  g  i  y  p  p  l  s  i  t  l  n  s  n  i  .  .  p  l  i  l  l  k  .
l  v  g  m  p  y  f  f  l  l  q  e  y  t  r  l  f  q  .  h  .  t  p  t  y  s  n  p  .  f  y  .  n  n
----.----+----.----+----.----+----.----+----.----+----.----+----.----+----.----+----.----+----.----+
 e  y  s  h  r  i  k  k  .  q  l  f  i  g  a  e  k  l  l  v  l  s  w  c  i  t  v  r  l  k  i  l  i
A  R  L  F  A  K  N  K  E  V  A  P  I  Y  G  G  R  E  I  V  S  F  E  L  M  y  y  g  k  i  k  n  f  y
 s  t  p  i  g  .  k  k  r  s  c  s  y  v  r  r  k  .  y  c  .  v  g  v  y  l  l  g  .  n  .  .  f

*099   26932      Mapped site PST1
EA59   26973      bEa59 (orf-525)     M-NH3


                   F        BXM
                   N        IHB
                   U        NO0
                   H        121
                            /
ACCGCAATTTATTTGGCGGCAACACAGGATCTCTCTTTTAAGTTACTCTCTATTACATACGTTTTCCATCTAAAAATTAGTAGTATTGAACTTAACGGGG
----.----+----.----+----.----+----.----+----.----+----.----+----.----+----.----+----.----+----.----+ 27100
TGGCGTTAAATAAACCGCCGTTGTGTCCTAGAGAGAAAATTCAATGAGAGATAATGTATGCAAAAGGTAGATTTTTAATCATCATAACTTGAATTGCCCC

t  a  i  y  l  a  a  t  q  d  l  s  f  k  l  l  s  i  t  y  v  f  h  l  k  i  s  s  i  e  l  n  g  a
 p  q  f  i  w  r  q  h  r  i  s  l  l  s  y  s  l  l  h  t  f  s  i  .  k  l  v  v  l  n  l  t  g
  r  n  l  f  g  g  n  t  g  s  l  f  .  v  t  l  y  y  i  r  f  p  s  k  n  .  .  y  .  t  .  r  g
----.----+----.----+----.----+----.----+----.----+----.----+----.----+----.----+----.----+----.----+
 v  a  i  .  k  a  a  v  c  s  r  e  k  l  n  s  e  i  v  y  t  k  w  r  f  i  l  l  i  s  s  l  p
  g  c  n  i  q  r  c  c  l  i  e  r  k  l  .  e  r  n  c  v  n  e  m  .  f  n  t  t  n  f  k  v  p
l  r  l  k  n  p  p  l  v  p  d  r  k  .  t  v  r  .  .  m  r  k  g  d  l  f  .  y  y  q  v  .  r  p


          S                 A                                              S    H
          F                 L                                              D    G
          A                 U                                              U    I
          N                 1                                              1    A

CATCGTATTGTAGTTTTCCATATTTAGCTTTCTGCTTCCTTTTGGATAACCCACTGTTATTCATGTTGCATGGTGCACTGTTTATACCAACGATATAGTC
----.----+----.----+----.----+----.----+----.----+----.----+----.----+----.----+----.----+----.----+ 27200
GTAGCATAACATCAAAAGGTATAAATCGAAAGACGAAGGAAAACCTATTGGGTGACAATAAGTACAACGTACCACGTGACAAATATGGTTGCTATATCAG

  s  y  c  s  f  p  y  l  a  f  c  f  l  l  d  n  p  l  l  f  m  l  h  g  a  l  f  i  p  t  i  .  s
h  r  i  v  v  f  h  i  .  l  s  a  s  f  w  i  t  h  c  y  s  c  c  m  v  h  c  l  y  q  r  y  s  l
 i  v  l  .  f  s  i  f  s  f  l  l  p  f  g  .  p  t  v  i  h  v  a  w  c  t  v  y  t  n  d  i  v
----.----+----.----+----.----+----.----+----.----+----.----+----.----+----.----+----.----+----.----+
a  d  y  q  l  k  g  y  k  a  k  q  k  r  k  s  l  g  s  n  n  m  n  c  p  a  s  n  i  g  v  i  y  d
 c  r  i  t  t  k  w  i  .  s  e  a  e  k  q  i  v  w  q  .  e  h  q  m  t  c  q  k  y  w  r  y  l
 m  t  n  y  n  e  m  n  l  k  r  s  g  k  p  y  g  v  t  i  .  t  a  h  h  v  t  .  v  l  s  i  t
```

```
    A                      M     A                     X    MS  RR
    V                      B     L                     M    BC  RS
    A                      0     U                     N    0A  UA
    3                      2     1                     1    21  11
                                                                 /
TATTAATGCATATATAGTATCGCCGAACGATTAGCTCTTCAGGCTTCTGAAGAAGCGTTTCAAGTACTAATAAGCCGATAGATAGCCACGGACTTCGTAG
----.----+----.----+----.----+----.----+----.----+----.----+----.----+----.----+----.----+----.----+  27300
ATAATTACGTATATATCATAGCGGCTTGCTAATCGAGAAGTCCGAAGACTTCTTCGCAAAGTTCATGATTATTCGGCTATCTATCGGTGCCTGAAGCATC

 i  n  a  y  i  v  s  p  n  d  .  l  f  r  l  l  k  k  r  f  k  y  .  .  a  d  r  .  p  r  t  s  .
  l  m  h  i  .  y  r  r  t  i  s  s  s  g  f  .  r  s  v  s  s  t  n  k  p  i  d  s  h  g  l  r  s
y  .  c  i  y  s  i  a  e  r  l  a  l  q  a  s  e  e  a  f  q  v  l  i  s  r  .  i  a  t  d  f  v  a
----.----+----.----+----.----+----.----+----.----+----.----+----.----+----.----+----.----+----.----+
 i  l  a  y  i  t  d  g  f  s  .  s  k  l  s  r  f  f  r  k  l  y  .  y  a  s  l  y  g  r  v  e  y
r  n  i  c  i  y  y  r  r  v  i  l  e  e  p  k  q  l  l  t  e  l  v  l  l  g  i  s  l  w  p  s  r  l
  .  .  h  m  y  l  i  a  s  r  n  a  r  .  a  e  s  s  a  n  .  t  s  i  l  r  y  i  a  v  s  k  t


                 HH**                 M                               A      SN*
                 PI11                 N                               V      PS1
                 AN00                 L                               A      HP0
                 1201                 1                               3      1C2
                   ///                                                         //
CCATTTTTCATAAGTGTTAACTTCCGCTCCTCGCTCATAACAGACATTCACTACAGTTATGGCGGAAAGGTATGCATGCTGGGTGTGGGGAAGTCGTGAA
----.----+----.----+----.----+----.----+----.----+----.----+----.----+----.----+----.----+----.----+  27400
GGTAAAAAGTATTCACAATTGAAGGCGAGGAGCGAGTATTGTCTGTAAGTGATGTCAATACCGCCTTTCCATACGTACGACCCACACCCCTTCAGCACTT

p  f  f  i  s  v  n  f  r  s  s  l  i  t  d  i  h  y  s  y  g  g  k  v  c  m  l  g  v  g  k  s  .  k
 h  f  s  .  v  l  t  s  a  p  r  s  .  q  t  f  t  t  v  m  a  e  r  y  a  c  w  v  w  g  s  r  e
  i  f  h  k  c  .  l  p  l  l  a  h  n  r  h  s  l  q  l  w  r  k  g  m  h  a  g  c  g  e  v  v  k
----.----+----.----+----.----+----.----+----.----+----.----+----.----+----.----+----.----+----.----+
 g  n  k  m  l  t  l  k  r  e  e  s  m  v  s  m  .  .  l  .  p  p  f  t  h  m  s  p  t  p  f  d  h
  w  k  e  y  t  n  v  e  a  g  r  e  y  c  v  n  v  v  t  i  a  s  l  y  a  h  q  t  h  p  l  r  s
a  m  k  .  l  h  .  s  g  s  r  a  .  l  l  c  e  s  c  n  h  r  f  p  i  c  a  p  h  p  s  t  t  f

*100   27318     Mapped site HIN2
*101   27318     Mapped site HPA1
*102   27378     Mapped site SPH1


B    H    N PA*F             M                                            S*HA     SSM
B    G    S VL1N             B                                            B1IL     DPN
V    A    P UU0U             0                                            60NU      CL
1    1    B 213H             2                                             431      11
             ///                                                           ///      /
AGAAAAGAAGTCAGCTGCGTCGTTTGACATCACTGCTATCTTCTTACTGGTTATGCAGGTCGTAGTGGGTGGCACACAAAGCTTTGCACTGGATTGCGAG
----.----+----.----+----.----+----.----+----.----+----.----+----.----+----.----+----.----+----.----+  27500
TCTTTTCTTCAGTCGACGCAGCAAACTGTAGTGACGATAGAAGAATGACCAATACGTCCAGCATCACCCACCGTGTGTTTCGAAACGTGACCTAACGCTC

  k  r  s  q  l  r  r  l  t  s  l  l  s  s  y  w  l  c  r  s  .  w  v  a  h  k  a  l  h  w  i  a  r
r  k  e  v  s  c  v  v  .  h  h  c  y  l  l  t  g  y  a  g  r  s  g  w  h  t  k  l  c  t  g  l  r  g
 e  k  k  s  a  a  s  f  d  i  t  a  i  f  l  l  v  m  q  v  v  v  g  g  t  q  s  f  a  l  d  c  e
----.----+----.----+----.----+----.----+----.----+----.----+----.----+----.----+----.----+----.----+
f  f  l  l  .  s  r  r  k  v  d  s  s  d  e  .  q  n  h  l  d  y  h  t  a  c  l  a  k  c  q  i  a  l
 l  f  s  t  l  q  t  t  q  c  .  q  .  r  r  v  p  .  a  p  r  l  p  h  c  v  f  s  q  v  p  n  r
  s  f  f  d  a  a  d  n  s  m  v  a  i  k  k  s  t  i  c  t  t  t  p  p  v  c  l  k  a  s  s  q  s

*103   27414     Mapped site PVU2
SB6    27479       Sequence Block 6       start
*104   27479     Mapped site HIN3
SD     27488       sd                     i          EM
SPC1   27488        lambdaspc1            sr         EM
```

```
             D                   D  ST       DH H      H     D  M  S P  HS        I
             1                   1  II       1E H      P     1  B  I H  HI        N
             0                   1  B        2F A      H     1  0  B 5  AB        T
             6                   9           5  1      1     2  2    3  1         1
                                                                         /
GCTTTGTGCTTCTCTGGAGTGCGACAGGTTTGATGACAAAAAATTAGCGCAAGAAGACAAAAATCACCTTGCGCTAATGCTCTGTTACAGGTCACTAATA
----.----+----.----+----.----+----.----+----.----+----.----+----.----+----.----+----.----+----.----+ 27600
CGAAACACGAAGAGACCTCACGCTGTCCAAACTACTGTTTTTTAATCGCGTTCTTCTGTTTTTAGTGGAACGCGATTACGAGACAATGTCCAGTGATTAT

 l  c  a  s  l  e  c  d  r  f  d  d  k  k  l  a  q  e  d  k  n  h  l  a  l  m  l  c  y  r  s  l  i
  f  v  l  l  w  s  a  t  g  l  m  t  k  n  .  r  k  k  t  k  i  t  l  r  .  c  s  v  t  g  h  .  y
a  l  c  f  s  g  v  r  q  v  .  .  q  k  i  s  a  r  r  q  k  s  p  c  a  n  a  l  l  q  v  t  n  t
----.----+----.----+----.----+----.----+----.----+----.----+----.----+----.----+----.----+----.----+
  s  q  a  e  r  s  h  s  l  n  s  s  l  f  n  a  c  s  s  l  f  .  r  a  s  i  s  q  .  l  d  s  i
p  k  t  s  r  q  l  a  v  p  k  i  v  f  f  .  r  l  f  v  f  i  v  k  r  .  h  e  t  v  p  .  .  y
 a  k  h  k  e  p  t  r  c  t  q  h  c  f  i  l  a  l  l  c  f  d  g  q  a  l  a  r  n  c  t  v  l
```

D106	27514	delta106	dr	sq		BAL31 from Hin3; is Sib+
D119	27534	delta119	dr	sq		BAL31 from Hin3; is Sib+
SIB	27537	sib3	pm	sq		C to T
TI	27538	tI terminator	3'	RNA		end of int message
D125	27546	delta125	dr	sq		BAL31 from Hin3; is Sib-
HEF	27547	hef13	pm	sq		G to A
D112	27562	delta112	dr	sq		BAL31 from Hin3; is Sib-
SIB	27568	sib2	pm	sq		C to T
PH53	27570	PHdelta53;54	dr	sq		ExoIII-S1 from Hin3; is Sib- Att+
SIB	27573	sib1	pm	sq		G to T
INT1	27583	P arm 1	bs			int binds (DNase protection 20bp)

```
    D         PIH             N
    D         HNN             D
    E         5TF             E
    1         521             1
               //
CCATCTAAGTAGTTGATTCATAGTGACTGCATATGTTGTGTTTTACAGTATTATGTAGTCTGTTTTTTATGCAAAATCTAATTTAATATATTGATATTTA
----.----+----.----+----.----+----.----+----.----+----.----+----.----+----.----+----.----+----.----+ 27700
GGTAGATTCATCAACTAAGTATCACTGACGTATACAACACAAAATGTCATAATACATCAGACAAAAAATACGTTTTAGATTAAATTATATAACTATAAAT

p  s  k  .  l  i  h  s  d  c  i  c  c  v  l  q  y  y  v  v  c  f  l  c  k  i  .  f  n  i  l  i  f  i
 h  l  s  s  .  f  i  v  t  a  y  v  v  f  y  s  i  m  .  s  v  f  y  a  k  s  n  l  i  y  .  y  l
  i  .  v  v  d  s  .  .  l  h  m  l  c  f  t  v  l  c  s  l  f  f  m  q  n  l  i  .  y  i  d  i  y
----.----+----.----+----.----+----.----+----.----+----.----+----.----+----.----+----.----+----.----+
 g  d  l  y  n  i  .  l  s  q  m  h  q  t  k  c  y  .  t  t  q  k  k  h  l  i  .  n  l  i  n  i  n
  w  r  l  l  q  n  m  t  v  a  y  t  t  n  .  l  i  i  y  d  t  k  .  a  f  d  l  k  i  y  q  y  k
v  m  .  t  t  s  e  y  h  s  c  i  n  h  k  v  t  n  h  l  r  n  k  i  c  f  r  i  .  y  i  s  i  .
```

PH55	27615	PHdelta55	dr	sq		ExoIII-S1 from Hin3; is Att-
INT2	27615	P arm 2	bs			int binds (DNase protection 20bp)

```
              I        A  A   A   D                                    B      I
              N        L  T   T   D                                    9      N
              T        U  T   T   E                                    3      T
              3        1  -       1                                    6      4

TATCATTTTACGTTTCTCGTTCAGCTTTTTATACTAAGTTGGCATTATAAAAAAGCATTGCTTATCAATTTGTTGCAACGAACAGGTCACTATCAGTCA
----.----+----.----+----.----+----.----+----.----+----.----+----.----+----.----+----.----+----.----+ 27800
ATAGTAAAATGCAAAGAGCAAGTCGAAAAAATATGATTCAACCGTAATATTTTTTCGTAACGAATAGTTAAACAACGTTGCTTGTCCAGTGATAGTCAGT

  s  f  y  v  s  r  s  a  f  l  y  .  v  g  i  i  k  k  h  c  l  s  i  c  c  n  e  q  v  t  i  s  q
y  h  f  t  f  l  v  q  l  f  y  t  k  l  a  l  .  k  s  i  a  y  q  f  v  a  t  n  r  s  l  s  v  k
 i  i  l  r  f  s  f  s  f  f  i  l  s  w  h  y  k  k  a  l  l  i  n  l  l  q  r  t  g  h  y  q  s
----.----+----.----+----.----+----.----+----.----+----.----+----.----+----.----+----.----+----.----+
i  d  n  .  t  e  r  e  a  k  k  y  .  t  p  m  i  f  f  c  q  k  d  i  q  q  l  s  c  t  v  i  l  .
 y  .  k  v  n  r  t  .  s  k  .  v  l  n  a  n  y  f  l  m  a  .  .  n  t  a  v  f  l  d  s  d  t
  i  m  k  r  k  e  n  l  k  k  i  s  l  q  c  .  l  f  a  n  s  i  l  k  n  c  r  v  p  .  .  .  d
```

INT3	27714	core site	bs			int binds (DNase protection 34bp)
ATT-	27727	att 2; 6; 24	pm	sq		att minus pm deletes T
ATT	27731	att	f(c)	sq	f	center of common core
B936	27772	bio936	att-sr	sq		
INT4	27779	P' arm	bs			int binds (DNase protection 41bp)

```
             I                                      MS            N             *A     5
             N                                      NF            C             1V     0
             T                                      LA            0             0A     1
                                                    1N            1             51
                                                                                 /
AAATAAAATCATTATTTGATTTCAATTTTGTCCCACTCCCTGCCTCTGTCATCACGATACTGTGATGCCATGGTGTCCGACTTATGCCCGAGAAGATGTT
----.----+----.----+----.----+----.----+----.----+----.----+----.----+----.----+----.----+----.----+ 27900
TTTATTTTAGTAATAAACTAAAGTTAAAACAGGGTGAGGGACGGAGACAGTAGTGCTATGACACTACGGTACCACAGGCTGAATACGGGCTCTTCTACAA

 n  k  i  i  i  .  f  q  f  c  p  t  p  c  l  c  h  h  d  t  v  m  p  w  c  p  t  y  a  r  e  d  v
  i  k  s  l  f  d  f  n  f  v  p  l  p  a  s  v  i  t  i  l  .  c  h  g  v  r  l  m  p  e  k  m  l
k  .  n  h  y  l  i  s  i  l  s  h  s  l  p  l  s  s  r  y  c  d  a  m  v  s  d  l  c  p  r  r  c  .
----.----+----.----+----.----+----.----+----.----+----.----+----.----+----.----+----.----+----.----+
  f  l  i  m  i  q  n  .  n  q  g  v  g  q  r  q  .  .  s  v  t  i  g  h  h  g  v  .  a  r  s  s  t
 l  i  f  d  n  n  s  k  l  k  t  g  s  g  a  e  t  m  v  i  s  h  h  w  p  t  r  s  i  g  s  f  i  n
 f  y  f  .  .  K  I  E  I  K  D  W  E  R  G  R  D  D  R  Y  Q  S  A  M  T  D  S  K  H  G  L  L  H

INT    27815     int                  COOH
*105   27887     Mapped site AVA1
501    27896     att501               dr          sq        att- int-
```

```
  M                             H              MH              B*BXMB   AT   S  E
  B                             N              SH              I1AHBI   SA   F  C
  0                             F              TA              N0MOON   UQ   A  P
  2                             1              11              161211   21   N  1
                                                                ////     /
GAGCAAACTTATCGCTTATCTGCTTCTCATAGAGTCTTGCAGACAAACTGCGCAACTCGTGAAAGGTAGGCGGATCCCCTTCGAAGGAAAGACCTGATGC
----.----+----.----+----.----+----.----+----.----+----.----+----.----+----.----+----.----+----.----+ 28000
CTCGTTTGAATAGCGAATAGACGAAGAGTATCTCAGAACGTCTGTTTGACGCGTTGAGCACTTTCCATCCGCCTAGGGGAAGCTTCCTTTCTGGACTACG

e  q  t  y  r  l  s  a  s  h  r  v  l  q  t  n  c  a  t  r  e  r  .  a  d  p  l  r  r  k  d  l  m  l
 s  k  l  i  a  y  l  l  l  i  e  s  c  r  q  t  a  q  l  v  k  g  r  r  i  p  f  e  g  k  t  .  c
  a  n  l  s  l  i  c  f  s  .  s  l  a  d  k  l  r  n  s  .  k  v  g  g  s  p  s  k  e  r  p  d  a
----.----+----.----+----.----+----.----+----.----+----.----+----.----+----.----+----.----+----.----+
 s  c  v  .  r  k  d  a  e  .  l  t  k  c  v  f  q  a  v  r  s  l  y  a  s  g  r  r  l  f  s  r  i
  l  l  s  i  a  .  r  s  r  m  s  d  q  l  c  v  a  c  s  t  f  p  l  r  i  g  k  s  p  f  v  q  h
Q  A  F  K  D  S  I  Q  K  E  Y  L  R  A  S  L  S  R  L  E  H  F  T  P  P  D  G  E  F  S  L  G  S  A

*106   27972     Mapped site BAM1
INT    ...       CONTINUES IN PHASE SIX
```

```
      B HTH                  HH            F NTS                                    H
      S HHH                  PG            0 RHF                                    N
      S AAA                  AU            K UAA                                    F
      2 111                  22            1 11N                                    1
         /                                    //
TTTTCGTGCGCGCATAAAATACCTTGATACTGTGCCGGATGAAAGCGGTTCGCGACGAGTAGATGCAATTATGGTTTCTCCGCCAAGAATCTCTTTGCAT
----.----+----.----+----.----+----.----+----.----+----.----+----.----+----.----+----.----+----.----+ 28100
AAAAGCACGCGCGTATTTTATGGAACTATGACACGGCCTACTTTCGCCAAGCGCTGCTCATCTACGTTAATACCAAAGAGGCGGTTCTTAGAGAAACGTA

  f  v  r  a  .  n  t  l  i  l  c  r  m  k  a  v  r  d  e  .  m  q  l  w  f  l  r  q  e  s  l  c  i
 f  s  c  a  h  k  i  p  .  y  c  a  g  .  k  r  f  a  t  s  r  c  n  y  g  f  s  a  k  n  l  f  a  f
  f  r  a  r  i  k  y  l  d  t  v  p  d  e  s  g  s  r  r  v  d  a  i  m  v  s  p  p  r  i  s  l  h
----.----+----.----+----.----+----.----+----.----+----.----+----.----+----.----+----.----+----.----+
 s  k  t  r  a  y  f  v  k  i  s  h  r  i  f  a  t  r  s  s  y  i  c  n  h  n  r  r  w  s  d  r  q  m
 k  e  h  a  c  l  i  g  q  y  q  a  p  h  f  r  n  a  v  l  l  h  l  .  p  k  e  a  l  f  r  k  a
  K  R  A  R  M  F  Y  R  S  V  T  G  S  S  L  P  E  R  R  T  S  A  I  I  T  E  G  G  L  I  E  K  C

INT    ...       CONTINUES IN PHASE SIX
```

```
                                         S         H          F               T
                                         F         G          0               A
                                         A         U          K               Q
                                         N         2          1               1

TTATCAAGTGTTTCCTTCATTGATATTCCGAGAGCATCAATATGCAATGCTGTTGGGATGGCAATTTTTACGCCTGTTTTGCTTTGCTCGACATAAAGAT
----.----+----.----+----.----+----.----+----.----+----.----+----.----+----.----+----.----+----.----+ 28200
AATAGTTCACAAAGGAAGTAACTATAAGGCTCTCGTAGTTATACGTTACGACAACCCTACCGTTAAAAATGCGGACAAAACGAAACGAGCTGTATTTCTA

 y  q  v  f  p  s  l  i  f  r  e  h  q  y  a  m  l  l  g  w  q  f  l  r  l  f  c  f  a  r  h  k  d
  i  k  c  f  l  h  .  y  s  e  s  i  n  m  q  c  c  w  d  g  n  f  y  a  c  f  a  l  l  d  i  k  i
l  s  s  v  s  f  i  d  i  p  r  a  s  i  c  n  a  v  g  m  a  i  f  t  p  v  l  l  c  s  t  .  r  y
----.----+----.----+----.----+----.----+----.----+----.----+----.----+----.----+----.----+----.----+
 .  .  t  n  g  e  n  i  n  r  s  c  .  y  a  i  s  n  p  h  c  n  k  r  r  n  q  k  a  r  c  l  s
n  i  l  h  k  r  .  q  y  e  s  l  m  l  i  c  h  q  q  s  p  l  k  .  a  q  k  a  k  s  s  m  f  i
 K  D  L  T  E  K  M  S  I  G  L  A  D  I  H  L  A  T  P  I  A  I  K  V  G  T  K  S  Q  E  V  Y  L
```

```
INI      ...      CONTINUES IN PHASE SIX

E           EE          H                  NSH                      D        B  T
C           CC          P                  CCP                      D        B  T
R           RP          H                  IRA                      E        V  H
V           V1          1                  112                      1        1  2
                                           //
ATCCATCTACGATATCAGACCACTTCATTTCGCATAAATCACCAACTCGTTGCCCGGTAACAACAGCCAGTTCCATTGCAAGTCTGAGCCAACATGGTGA
----.----+----.----+----.----+----.----+----.----+----.----+----.----+----.----+----.----+----.----+ 28300
TAGGTAGATGCTATAGTCTGGTGAAGTAAAGCGTATTTAGTGGTTGAGCAACGGGCCATTGTTGTCGGTCAAGGTAACGTTCAGACTCGGTTGTACCACT

i  h  l  r  y  q  t  t  s  f  r  i  n  h  q  l  v  a  r  .  q  q  p  v  p  l  q  v  .  a  n  m  v  m
 s  i  y  d  i  r  p  l  h  f  a  .  i  t  n  s  l  p  g  n  n  s  q  f  h  c  k  s  e  p  t  w  .
  p  s  t  i  s  d  h  f  i  s  h  k  s  p  t  r  c  p  v  t  t  a  s  s  i  a  s  l  s  q  h  g  d
----.----+----.----+----.----+----.----+----.----+----.----+----.----+----.----+----.----+----.----+
 i  w  r  r  y  .  v  v  e  n  r  m  f  .  w  s  t  a  r  y  c  c  g  t  g  n  c  t  q  a  l  m  t
  d  m  .  s  i  l  g  s  .  k  a  y  i  v  l  e  n  g  p  l  l  l  w  n  w  q  l  d  s  g  v  h  h
Y  G  D  V  I  D  S  W  K  M  F  C  L  D  G  V  R  Q  G  T  V  V  A  L  E  M  A  L  R  L  W  C  P  S

INI      ...      CONTINUES IN PHASE SIX

 H E HF 1              H               M     B         MF  HT    F        B
 N C PN 0              N               B     B         NN  HH    N        B
 F 1 HU 0              F               0     V         LU  AA    U        V
 1 5 1H 7              3               1     1         1H  11    H        1
                                                            /
TGATTCTGCTGCTTGATAAATTTTCAGGTATTCGTCAGCCGTAAGTCTTGATCTCCTTACCTCTGATTTTGCTGCGCGAGTGGCAGCGACATGGTTTGTT
----.----+----.----+----.----+----.----+----.----+----.----+----.----+----.----+----.----+----.----+ 28400
ACTAAGACGACGAACTATTTAAAAGTCCATAAGCAGTCGGCATTCAGAACTAGAGGAATGGAGACTAAAACGACGCGCTCACCGTCGCTGTACCAAACAA

  i  l  l  l  d  k  f  s  g  i  r  q  p  .  v  l  i  s  l  p  l  i  l  l  r  e  w  q  r  h  g  l  l
 .  f  c  c  l  i  n  f  q  v  f  v  s  r  k  s  .  s  p  y  l  .  f  c  c  a  s  g  s  d  m  v  c  c
 d  s  a  a  .  .  i  f  r  y  s  s  a  v  s  l  d  l  l  t  s  d  f  a  a  r  v  a  a  t  w  f  v
----.----+----.----+----.----+----.----+----.----+----.----+----.----+----.----+----.----+----.----+
i  i  r  s  s  s  l  n  e  p  i  r  .  g  y  t  k  i  e  k  g  r  i  k  s  r  s  h  c  r  c  p  k  n
 h  n  q  q  k  i  f  k  .  t  n  t  l  r  l  d  q  d  g  .  r  q  n  q  q  a  l  p  l  s  m  t  q
  S  E  A  A  Q  Y  I  K  L  Y  E  D  A  T  L  R  S  R  R  V  E  S  K  A  A  R  T  A  A  V  H  N  T

1007   28312       b1007             dr           EM
INI      ...      CONTINUES IN PHASE SIX

      HH   A            AM    D S   M                  A F  H          M
      AA   L            VN    D F   B                  C N  G          N
      EE   U            AL    E A   0                  Y U  A          L
      13   1            31    1 N   1                  1 H  1          1
      /
GTTATATGGCCTTCAGCTATTGCCTCTCGGAATGCATCGCTCAGTGTTGATCTGATTAACTTGGCTGACGCCGCCTTGCCCTCGTCTATGTATCCATTGA
----.----+----.----+----.----+----.----+----.----+----.----+----.----+----.----+----.----+----.----+ 28500
CAATATACCGGAAGTCGATAACGGAGAGCCTTACGTAGCGAGTCACAACTAGACTAATTGAACCGACTGCGGCGGAACGGGAGCAGATACATAGGTAACT

  l  y  g  l  q  l  l  p  l  g  m  h  r  s  v  l  i  .  l  t  w  l  t  p  p  c  p  r  l  c  i  h  .
  y  m  a  f  s  y  c  l  s  e  c  i  a  q  c  .  s  d  .  l  g  .  r  r  l  a  l  v  y  v  s  i  e
v  i  w  p  s  a  i  a  s  r  n  a  s  l  s  v  d  l  i  n  l  a  d  a  a  l  p  s  s  m  y  p  l  s
----.----+----.----+----.----+----.----+----.----+----.----+----.----+----.----+----.----+----.----+
  n  y  p  r  .  s  n  g  r  p  i  c  r  e  t  n  i  q  n  v  q  s  v  g  g  q  g  r  r  h  i  w  q
q  .  i  a  k  l  .  q  r  e  s  h  m  a  .  h  q  d  s  .  s  p  q  r  r  r  a  r  t  .  t  d  m  s
 T  I  H  G  E  A  I  A  E  R  F  A  D  S  L  T  S  R  I  L  K  A  S  A  A  K  G  E  D  I  Y  G  N

INI      ...      CONTINUES IN PHASE SIX

   BF         M        H          ES         M                ME T
   1N         B        P          CF         N                BC T
   6U         0        H          PA         L                00 H
   AH         2        1          1N         1                2B 2

GCATTGCCGCAATTTCTTTTGTGGTGATGTCTTCAAGTGGAGCATCAGGCAGACCCCTCCTTATTGCTTTAATTTTGCTCATGTAATTTATGAGTGTCTT
----.----+----.----+----.----+----.----+----.----+----.----+----.----+----.----+----.----+----.----+ 28600
CGTAACGGCGTTAAAGAAAACACCACTACAGAAGTTCACCTCGTAGTCCGTCTGGGGAGGAATAACGAAATTAAAACGAGTACATTAAATACTCACAGAA

a  l  p  q  f  l  l  w  .  c  l  q  v  e  h  q  a  d  p  s  l  l  l  .  f  c  s  c  n  l  .  v  s  s
 h  c  r  n  f  f  c  g  d  v  f  k  w  s  i  r  q  t  p  p  y  c  f  n  f  a  h  v  i  y  e  c  l
  i  a  a  i  s  f  v  v  m  s  s  s  g  a  s  g  r  p  l  l  i  a  l  i  l  l  m  .  f  m  s  v  f
----.----+----.----+----.----+----.----+----.----+----.----+----.----+----.----+----.----+----.----+
 a  n  g  c  n  r  k  h  h  h  r  .  t  s  c  .  a  s  g  e  k  n  s  .  n  q  e  h  l  k  h  t  d
  c  q  r  l  k  k  q  p  s  t  k  l  h  l  m  l  c  v  g  g  .  q  k  l  k  a  .  t  i  .  s  h  r
L  M  A  A  I  E  K  T  T  I  D  E  L  P  A  D  P  L  G  R  R  I  A  K  I  K  S  M  Y  N  I  L  T  K
```

```
B16A   28506       bio16A            att-sr      EM
INT      ...      CONTINUES IN PHASE SIX


  R   H     E    C BMHHES            M M                                      H     M     T
  D   N     C    F ANAACC            B G                                      N     N     T
  1   F     1    R LLEERR            0 B                                      F     L     H
  0   1     5    1 111321            1                                        1     1     2
                     ////
CTGCTTGATTCCTCTGCTGGCCAGGATTTTTTCGTAGCGATCAAGCCATGAATGTAACGTAACGGAATTATCACTGTTGATTCTCGCTGTCAGAGGCTTG
----.----+----.----+----.----+----.----+----.----+----.----+----.----+----.----+----.----+----.----+ 28700
GACGAACTAAGGAGACGACCGGTCCTAAAAAAGCATCGCTAGTTCGGTACTTACATTGCATTGCCTTAATAGTGACAACTAAGAGCGACAGTCTCCGAAC

  a  .  f  l  c  w  p  g  f  f  r  s  d  q  a  m  n  v  t  .  r  n  y  h  c  .  f  s  l  s  e  a  c
 l  l  d  s  s  a  g  q  d  f  f  v  a  i  k  p  .  m  .  r  n  g  i  i  t  v  d  s  r  c  q  r  l  v
c  l  i  p  l  l  a  r  i  f  s  .  r  s  s  h  e  c  n  v  t  e  l  s  l  l  i  l  a  v  r  g  l
----.----+----.----+----.----+----.----+----.----+----.----+----.----+----.----+----.----+----.----+
e  a  q  n  r  q  q  g  p  n  k  r  l  s  .  a  m  f  t  v  y  r  f  .  .  q  q  n  e  s  d  s  a  q
 r  s  s  e  e  a  p  w  s  k  k  t  a  i  l  g  h  i  y  r  l  p  i  i  v  t  s  e  r  q  .  l  s
  Q  K  I  G  R  S  A  L  I  K  E  Y  R  D  L  W  S  H  L  T  V  S  N  D  S  N  I  R  A  T  L  P  K

RD10   28603        delta red10        dl        EM
MGB    28640        lambda MGB         i(Tn9)    RES     dl for deletions pQLC 95-118
INT      ...       CONTINUES IN PHASF SIX


                               HH      A              HH                                           AA
                               AA      L              NN                                           SV
                               EE      U              FF                                           UA
                               13      1              13                                           12
                               /                      /                                             /
TGTTTGTGTCCTGAAAATAACTCAATGTTGGCCTGTATAGCTTCAGTGATTGCGATTCGCCTGTCTCTGCCTAATCCAAACTCTTTACCCGTCCTTGGGT
----.----+----.----+----.----+----.----+----.----+----.----+----.----+----.----+----.----+----.----+ 28800
ACAAACACAGGACTTTTATTGAGTTACAACCGGACATATCGAAGTCACTAACGCTAAGCGGACAGAGACGGATTAGGTTTGAGAAATGGGCAGGAACCCA

 v  c  v  l  k  i  t  q  c  w  p  v  .  l  q  .  l  r  f  a  c  l  c  l  i  q  t  l  y  p  s  l  g
  f  v  s  .  k  .  l  n  v  g  l  y  s  f  s  d  c  d  s  p  v  s  a  .  s  k  l  f  t  r  p  w  v
c  l  c  p  e  n  n  s  m  l  a  c  i  a  s  v  i  a  i  r  l  s  l  p  n  p  n  s  l  p  v  l  g  s
----.----+----.----+----.----+----.----+----.----+----.----+----.----+----.----+----.----+----.----+
  t  q  t  r  f  i  v  .  h  q  g  t  y  s  .  h  n  r  n  a  q  r  q  r  i  w  v  r  .  g  d  k  p
t  n  t  d  q  f  y  s  l  t  p  r  y  l  k  l  s  q  s  e  g  t  e  a  .  d  l  s  k  v  r  g  q  t
 H  K  H  G  S  F  L  E  I  N  A  Q  I  A  E  T  I  A  I  R  R  D  R  G  L  G  F  E  K  G  T  R  P

INT      ...       CONTINUES IN PHASF SIX


                                            2       NSH  HH   X  M               I M MB      2     M
                                            0       CCP  HA   I  B               N B BI      0     N
                                            3       IRA  AE   S  0               T 0 ON      3     L
                                            4       112  12      2                 2 11      7     1
                                                     //                                /
CCCTGTAGCAGTAATATCCATTGTTTCTTATATAAAGGTTAGGGGGTAAATCCCGGCGCTCATGACTTCGCCTTCTTCCCATTTCTGATCCTCTTCAAAA
----.----+----.----+----.----+----.----+----.----+----.----+----.----+----.----+----.----+----.----+ 28900
GGGACATCGTCATTATAGGTAACAAAGAATATATTTCCAATCCCCCATTTAGGGCCGCGAGTACTGAAGCGGAAGAAGGGTAAAGACTAGGAGAAGTTTT

p  c  s  s  n  i  h  c  f  l  y  k  g  .  g  v  n  p  g  a  h  d  f  a  f  f  p  f  l  i  l  f  k  r
 p  v  a  v  i  s  i  v  s  y  i  k  v  r  g  .  i  p  a  l  m  t  s  p  s  s  h  f  .  s  s  s  k
  l  .  q  .  y  p  l  f  l  i  .  r  l  g  g  k  s  r  r  s  .  l  r  l  l  p  i  s  d  p  l  q  k
----.----+----.----+----.----+----.----+----.----+----.----+----.----+----.----+----.----+----.----+
 g  q  l  l  l  i  w  q  k  k  y  l  p  .  p  t  f  g  p  a  .  S  K  A  K  K  G  N  R  I  R  K  L
  g  t  a  t  i  d  m  t  e  .  i  f  t  l  p  y  i  g  a  s  m  v  e  g  e  e  w  k  q  d  e  e  f
D  R  Y  C  Y  Y  G  N  N  R  I  Y  L  N  P  P  L  D  R  R  E  H  S  R  R  R  G  M  e  s  g  r  .  f

2034   28845        bio2034           att-sr      EM
XIS    28863        xis               COOH
INT    28882        int (orf-356)     M-NH3
2037   28894        bio2037           att-sr      EM
```

```
HH           T         H*      N    HH         FM             M  TH           I
AA           A         I1      S    NN         OB             N  8P           5
EE           Q         N0      P    FF         KO             L  4A           4
13           1         27      B    13         12             1  12           8
 /                      /            /
GGCCACCTGTTACTGGTCGATTTAAGTCAACCTTTACCGCTGATTCGTGGAACAGATACTCTCTTCCATCCTTAACCGGAGGTGGGAATATCCTGCATTC
----.----+----.----+----.----+----.----+----.----+----.----+----.----+----.----+----.----+----.----+ 29000
CCGGTGGACAATGACCAGCTAAATTCAGTTGGAAATGGCGACTAAGCACCTTGTCTATGAGAGAAGGTAGGAATTGGCCTCCACCCTTATAGGACGTAAG

  p  p  v  t  g  r  f  k  s  t  f  t  a  d  s  w  n  r  y  s  l  p  s  l  t  g  g  g  n  i  l  h  s
 g  h  l  l  l  v  d  l  s  q  p  l  p  l  i  r  g  t  d  t  l  f  h  p  .  p  e  v  g  i  s  c  i  p
  a  t  c  y  w  s  i  .  v  n  l  y  r  .  f  v  e  q  i  l  s  s  i  l  n  r  r  w  e  y  p  a  f
----.----+----.----+----.----+----.----+----.----+----.----+----.----+----.----+----.----+----.----+
L  G  G  T  V  P  R  N  L  D  V  K  V  A  S  E  H  F  L  Y  E  R  G  D  K  V  P  P  P  F  I  R  C  E
 p  w  r  n  s  t  s  k  l  .  g  k  g  s  i  r  p  v  s  v  r  k  w  g  .  g  s  t  p  i  d  q  m
  a  v  q  .  q  d  i  .  t  l  r  .  r  q  n  t  s  c  i  s  e  e  m  r  l  r  l  h  s  y  g  a  n
```

*107	28928	Mapped site HIN2			
I841	28975	trp delta 841	sl	sq	
I548	28991	int-c548	i(IS2)	EM	
XIS	...	CONTINUES IN PHASE FOUR			

```
     T      X            A  A                  X S      R I X      IIIC BT
     A      M            C  A                  I I      S N I      C652 53
     Q      N            Y  T                  S        A T S      500  30
     1      1            1  2                  6        1 C        7 8  83
                                                                    //
CCGAACCCATCGACGAACTGTTTCAAGGCTTCTTGGACGTCGCTGGCGTGCGTTCCACTCCTGAAGTGTCAAGTACATCGCAAAGTCTCCGCAATTACAC
----.----+----.----+----.----+----.----+----.----+----.----+----.----+----.----+----.----+----.----+ 29100
GGCTTGGGTAGCTGCTTGACAAAGTTCCGAAGAACCTGCAGCGACCGCACGCAAGGTGAGGACTTCACAGTTCATGTAGCGTTTCAGAGGCGTTAATGTG

  r  t  h  r  r  t  v  s  r  l  l  g  r  r  w  r  a  f  h  s  .  s  v  k  y  i  a  k  s  p  q  l  h
   e  p  i  d  e  l  f  q  g  f  l  d  v  a  g  v  r  s  t  p  e  v  s  s  t  s  q  s  l  r  n  y  t
 p  n  p  s  t  n  c  f  k  a  s  w  t  s  l  a  c  v  p  l  l  k  c  q  v  h  r  k  v  s  a  i  t  r
----.----+----.----+----.----+----.----+----.----+----.----+----.----+----.----+----.----+----.----+
  R  V  W  R  R  V  T  E  L  S  R  P  R  R  Q  R  A  N  W  E  Q  L  T  L  Y  M  a  f  d  g  c  n  c
 g  s  g  m  s  s  s  n  .  p  k  k  s  t  a  p  t  r  e  v  g  s  t  d  l  v  d  c  l  r  r  l  .  v
  g  f  g  d  v  f  q  k  l  a  e  q  v  d  s  a  h  t  g  s  r  f  h  .  t  c  r  l  t  e  a  i  v
```

XIS6	29063	xis am6	pm	sq	G to A
SI	29065	sI pI	5'	RNA	start of int message sq
INTC	29076	int-c226;262;518	pm	sq	C to T
XIS	29078	xis (orf-72)	M-NH3		
IC57	29088	int-c57	i(IS2)	EM	
I60	29088	int-c60	i(IS2)	EM	
I508	29088	int-c508	i(IS2)	EM	
C2	29091	cII site	bs	f	cII binds -35 region of pI
B538	29093	b538	dr	EM	
I303	29094	trp delta 303	sl	sq	

```
                    F T       M         H ATH                N    F    HB
                    N 2       B         N SAN                S    0    H3
                    U 9       0         F UQF                P    K    A8
                    H         2         3 213                C    1    16
                                           //                           /
GCAAGAAAAAACCGCCATCAGGCGGCTTGGTGTTCTTTCAGTTCTTCAATTCGAATATTGGTTACGTCTGCATGTGCTATCTGCGCCCATATCATCCAGT
----.----+----.----+----.----+----.----+----.----+----.----+----.----+----.----+----.----+----.----+ 29200
CGTTCTTTTTTGGCGGTAGTCCGCCGAACCACAAGAAAGTCAAGAAGTTAAGCTTATAACCAATGCAGACGTACACGATAGACGCGGGTATAGTAGGTCA

 a  r  k  n  r  h  q  a  a  w  c  s  f  s  s  s  i  r  i  l  v  t  s  a  c  a  i  c  a  h  i  i  q  w
  q  e  k  t  a  i  r  r  l  g  v  l  s  v  l  q  f  e  y  w  l  r  l  h  v  l  s  a  p  i  s  s  s
  k  k  k  p  p  s  g  g  l  v  f  f  q  f  f  n  s  n  i  g  y  v  c  m  c  y  l  r  p  y  h  p  v
----.----+----.----+----.----+----.----+----.----+----.----+----.----+----.----+----.----+----.----+
  a  l  f  f  r  w  .  a  a  q  h  e  k  l  e  e  i  r  i  n  t  v  d  a  h  a  i  q  a  w  i  m  w
   c  s  f  v  a  m  l  r  s  p  t  r  e  t  r  .  n  s  y  q  n  r  r  c  t  s  d  a  g  m  d  d  l
 r  l  f  f  g  g  d  p  p  k  t  n  k  .  n  k  l  e  f  i  p  .  t  q  m  h  .  r  r  g  y  .  g  t
```

I29	29126	trp delta 29	sl	sq
B386	29185	bio386	att-dr	EM

```
                    T                                       H
                    A                                       N
                    Q                                       F
                    1                                       1

GGTCGTAGCAGTCGTTGATGTTCTCCGCTTCGATAACTCTGTTGAATGGCTCTCCATTCCATTCTCCTGTGACTCGGAAGTGCATTTATCATCTCCATAA
----.----+----.----+----.----+----.----+----.----+----.----+----.----+----.----+----.----+----.----+ 29300
CCAGCATCGTCAGCAACTACAAGAGGCGAAGCTATTGAGACAACTTACCGAGAGGTAAGGTAAGAGGACACTGAGCCTTCACGTAAATAGTAGAGGTATT

 s . q s l ■ f s a s i t l l n g s p f h s p v t r k c i y h l h k
g r s s r . c s p l r . l c . ■ a l h s i l l . l g s a f i i s i k
 v v a v v d v l r f d n s v e w l s i p f s c d s e v h l s s p .
----.----+----.----+----.----+----.----+----.----+----.----+----.----+----.----+----.----+----.----+
h d y c d n i n e a e i v r n f p e g n w e g t v r f h ■ . . r w l
 p r l l r q h e g s r y s q q i a r w e ■ r r h s p l a n i ■ e ■
  t t a t t s t r r k s l e t s h s e ■ g n e q s e s t c k d d g y
```

```
                                           TM                               E C       M R
                                           HN                               8 8       N S
                                           AL                               . 4       L A
                                           11                               5 2       1 1
                                           /
AACAAAACCCGCCGTAGCGAGTTCAGATAAAATAAATCCCCGCGAGTGCGAGGATTGTTATGTAATATTGGGTTTAATCATCTATATGTTTTGTACAGAG
----.----+----.----+----.----+----.----+----.----+----.----+----.----+----.----+----.----+----.----+ 29400
TTGTTTTGGGCGGCATCGCTCAAGTCTATTTTATTTAGGGGCGCTCACGCTCCTAACAATACATTATAACCCAAATTAGTAGATATACAAAACATGTCTC

 t k p a v a s s d k i n p r e c e d c y v i l g l i i y ■ f c t e
  q n p p . r v q i k . i p a s a r i v ■ . y w v . s s i c f v q r
n k t r r s e f r . n k s p r v r g l l c n i g f n h l y v l y r e
----.----+----.----+----.----+----.----+----.----+----.----+----.----+----.----+----.----+----.----+
 v f g a t a l e s l i f g r s h s s q . t i n p k i ■ . i n q v s
f c f g g y r t . i f y i g a l a l i t i y y q t . D D I H K T C L
 f l v r r l s n l y f l d g r t r p n n h l i p n l . r y t k y l
```

E8.5 29377 Ea8.5 COOH

C842 29379 crg 842 i(IS2) EM

```
                      R  R                                              M
                      S  E                                              B
                      A  D                                              0
                      1  1                                              2

AGGGCAAGTATCGTTTCCACCGTACTCGTGATAATAATTTTGCACGGTATCAGTCATTTCTCGCACATTGCAGAATGGGGATTTGTCTTCATTAGACTTA
----.----+----.----+----.----+----.----+----.----+----.----+----.----+----.----+----.----+----.----+ 29500
TCCCGTTCATAGCAAAGGTGGCATGAGCACTATTATTAAAACGTGCCATAGTCAGTAAAGAGCGTGTAACGTCTTACCCCTAAACAGAAGTAATCTGAAT

r a s i v s t v l v i i i l h g i s h f s h i a e w g f v f i r l i
 g q v s f p p y s . . . f c t v s v i s r t l q n g d l s s l d l
  g k y r f h r t r d n n f a r y q s f l a h c r ■ g i c l h . t y
----.----+----.----+----.----+----.----+----.----+----.----+----.----+----.----+----.----+----.----+
 l a l i t e v t s t i i i k c p i l . k e c ■ a s h p n t k ■ l s
  P C T D N G G Y E H Y Y N Q V T D T M E R V N C F P S K D E N S K
s p l y r k w r v r s l l k a r y . d n r a c q l i p i q r . . v .
```

RED1 29427 delta red1 dl EM

E8.5 ... CONTINUES IN PHASE FIVE

```
                      H                                                               HBXM
                      N                                                               PIHB
                      F                                                               ANOO
                      1                                                               2121
                                                                                         /
TAAACCTTCATGGAATATTTGTATGCCGACTCTATATCTATACCTTCATCTACATAAACACCTTCGTGATGTCTGCATGGAGACAAGACACCGGATCTGC
----.----+----.----+----.----+----.----+----.----+----.----+----.----+----.----+----.----+----.----+ 29600
ATTTGGAAGTACCTTATAAACATACGGCTGAGATATAGATATGGAAGTAGATGTATTTGTGGAAGCACTACAGACGTACCTCTGTTCTGTGGCCTAGACG

  n l h g i f v c r l y i y t f i y i n t f v ■ s a w r q d t g s a
. t f ■ e y l y a d s i s i p s s t . t p s . c l h g d k t p d l h
 k p s w n i c ■ p t l y l y l h l h k h l r d v c ■ e t r h r i c
----.----+----.----+----.----+----.----+----.----+----.----+----.----+----.----+----.----+----.----+
i f r . p i n t h r s . i . v k ■ . ■ f v k t i d a h l c s v p d a
 Y V K M S Y K Y A S E I D I G E D V Y V G E H H R C P S L V G S R
  l g e h f i q i g v r y r y r . r c l c r r s t q ■ s v l c r i q
```

E8.5 ... CONTINUES IN PHASE FIVE

```
            DDE1 HNF1              E8.5                    ALU1

ACAACATTGATAACGCCCAATCTTTTTGCTCAGACTCTAACTCATTGATACTCATTTATAAACTCCTTGCAATGTATGTCGTTTCAGCTAAACGGTATCA
----.----+----.----+----.----+----.----+----.----+----.----+----.----+----.----+----.----+----.----+ 29700
TGTTGTAACTATTGCGGGTTAGAAAAACGAGTCTGAGATTGAGTAACTATGAGTAAATATTTGAGGAACGTTACATACAGCAAAGTCGATTTGCCATAGT

 q h . . r p i f l l r l . l i d t h l . t p c n v c r f s . t v s
  n i d n a q s f c s d s n s l i l i y k l l a m y v v s a k r y q
 t t l i t p n l f a q t l t h . y s f i n s l q c m s f q l n g i s
----.----+----.----+----.----+----.----+----.----+----.----+----.----+----.----+----.----+----.----+
  c c q y r g i k k s l s . s m s v . k y v g q l t h r k l . v t d
 C L M S L A W D K Q E S E L E N I S M . l s r a i y t t e a l r y .
 v v n i v g l r k a . v r v . q y e n i f e k c h i d n . s f p i

E8.5   29655        Ea8.5 (orf-93)      M-NH3


        B508               TTH2            RSA1              HNF1        SFAN

GCAATGTTTATGTAAAGAAACAGTAAGATAATACTCAACCCGATGTTTGAGTACGGTCATCATCTGACACTACAGACTCTGGCATCGCTGTGAAGACGAC
----.----+----.----+----.----+----.----+----.----+----.----+----.----+----.----+----.----+----.----+ 29800
CGTTACAAATACATTTCTTTGTCATTCTATTATGAGTTGGGCTACAAACTCATGCCAGTAGTAGACTGTGATGTCTGAGACCGTAGCGACACTTCTGCTG

 a m f m . r n s k i i l n p m f e y g h h l t l q t l a s l . r r r
  q c l c k e t v r . y s t r c l s t v i i . h y r l w h r c e d d
  n v y v k k q . d n t q p d v . v r s s s d t t d s g i a v k t t
----.----+----.----+----.----+----.----+----.----+----.----+----.----+----.----+----.----+----.----+
 a i n i y l f l l i i s l g i n s y p . . r v s c v r a d s h l r
  c h k h l s v t l y y e v r h k l v t m m q c . l s q c r q s s s
 l l t . t f f c y s l v . g s t q t r d d d s v v s e p m a t f v v

B508   29711        b508                att-dr     sq


THA1 MBO2 HGA1 SB6 BI2        MBO1   SFAN2 EFA2   HGAF1 HNF3       NDE1  HPHF1 FOK1

GCGAAATTCAGCATTTTCACAAGCGTTATCTTTTACAAAACCGATCTCACTCTCCTTTGATGCGAATGCCAGCGTCAGACATCATATGCAGATACTCACC
----.----+----.----+----.----+----.----+----.----+----.----+----.----+----.----+----.----+----.----+ 29900
CGCTTTAAGTCGTAAAAGTGTTCGCAATAGAAAATGTTTTGGCTAGAGTGAGAGGAAACTACGCTTACGGTCGCAGTCTGTAGTATACGTCTATGAGTGG

  e i q h f h k r y l l q n r s h s p l m r m p a s d i i c r y s p
 a k f s i f t s v i f y k t d l t l l . c e c q r q t s y a d t h l
 r n s a f s q a l s f t k p i s l s f d a n a s v r h h m q i l t
----.----+----.----+----.----+----.----+----.----+----.----+----.----+----.----+----.----+----.----+
 r s i . c k . l r . r k c f r d . E G K I R I G A D S M M H L Y E G
 a f n l m k v l t i k . l v s r v r r q h s h w r . v d y a s v .
  r f e a n e c a n d k v f g i e s e k s a f a l t l c . i c i s v

SB6    29811        Sequence Block 6    end
BI2    29815        bi2                 i(IS2)     EM
EA22   29850        Ea22                COOH


        SFAN           SFAN1 MNL1           TAQ1            ECP1     TAQ1     FNUH MBO2  XMN1 HPH1

TGCATCCTGAACCCATTGACCTCCAACCCCGTAATAGCGATGCGTAATGATGTCGATAGTTACTAACGGGTCTTGTTCGATTAACTGCCGCAGAAACTCT
----.----+----.----+----.----+----.----+----.----+----.----+----.----+----.----+----.----+----.----+ 30000
ACGTAGGACTTGGGTAACTGGAGGTTGGGGCATTATCGCTACGCATTACTACAGCTATCAATGATTGCCCAGAACAAGCTAATTGACGGCGTCTTTGAGA

 a s . t h . p p t p . . r c v m m s i v t n g s c s i n c r r n s
  h p e p i d l q p r n s d a . . c r . l l t g l v r l t a a e t l
 c i l n p l t s n p v i a m r n d v d s y . r v l f d . l p q k l f
----.----+----.----+----.----+----.----+----.----+----.----+----.----+----.----+----.----+----.----+
  A D Q V W Q G G V G Y Y R H T I I D I T V L P D Q E I L Q R L F E
 r c g s g m s r w g r l l s a y h h r y n s v p r t r n v a a s v r
 q m r f g n v e l g t i a i r l s t s l . . r t k n s . s g c f s

EA22   ...    CONTINUES IN PHASE FOUR
```

```
E S B   T                 M    H      E SH             F        M
C C S   T                 B    N      C CG             0        B
R R T   H                 0    F      R RU             K        0
2 1 E   2                 2    1      2 12             1        2

TCCAGGTCACCAGTGCAGTGCTTGATAACAGGAGTCTTCCCAGGATGGCGAACAACAAGAAACTGGTTTCCGTCTTCACGGACTTCGTTGCTTTCCAGTT
----.----+----.----+----.----+----.----+----.----+----.----+----.----+----.----+----.----+----.----+ 30100
AGGTCCAGTGGTCACGTCACGAACTATTGTCCTCAGAAGGGTCCTACCGCTTGTTGTTCTTTGACCAAAGGCAGAAGTGCCTGAAGCAACGAAAGGTCAA

s  r  s  p  v  q  c  l  i  t  g  v  f  p  g  w  r  t  t  r  n  w  f  p  s  s  r  t  s  l  l  s  s  l
 p  g  h  q  c  s  a  .  .  q  e  s  s  q  d  g  e  q  q  e  t  g  f  r  l  h  g  l  r  c  f  p  v
  q  v  t  s  a  v  l  d  n  r  s  l  p  r  m  a  n  n  k  k  l  v  s  v  f  t  d  f  v  a  f  q  f
----.----+----.----+----.----+----.----+----.----+----.----+----.----+----.----+----.----+----.----+
 E  L  D  G  T  C  H  K  I  V  P  T  K  G  P  H  R  V  V  L  F  Q  N  G  D  E  R  V  E  N  S  E  L
  g  p  .  w  h  l  a  q  y  c  s  d  e  w  s  p  s  c  c  s  v  p  k  r  r  .  p  s  r  q  k  g  t
k  w  t  v  l  a  t  s  s  l  l  l  r  g  l  i  a  f  l  l  f  s  t  e  t  k  v  s  k  t  a  k  w  n

EA22    ...    CONTINUES IN PHASE FOUR
```

```
    F                              B             F                               A
    0                              B             N                               L
    K                              V             U                               U
    1                              1             H                               1

TAGCAATACGCTTACTCCCATCCGAGATAACACCTTCGTAATACTCACGCTGCTCGTTGAGTTTTGATTTTGCTGTTTCAAGCTCAACACGCAGTTTCCC
----.----+----.----+----.----+----.----+----.----+----.----+----.----+----.----+----.----+----.----+ 30200
ATCGTTATGCGAATGAGGGTAGGCTCTATTGTGGAAGCATTATGAGTGCGACGAGCAACTCAAAACTAAAACGACAAAGTTCGAGTTGTGCGTCAAAGGG

  a  i  r  l  l  p  s  e  i  t  p  s  .  y  s  r  c  s  l  s  f  d  f  a  v  s  s  s  t  r  s  f  p
.  q  y  a  y  s  h  p  r  .  h  l  r  n  t  h  a  a  r  .  v  l  i  l  l  f  q  a  q  h  a  v  s  l
 s  n  t  l  t  p  i  r  d  n  t  f  v  i  l  t  l  l  v  e  f  .  f  c  c  f  k  l  n  t  q  f  p
----.----+----.----+----.----+----.----+----.----+----.----+----.----+----.----+----.----+----.----+
K  A  I  R  K  S  G  D  S  I  V  G  E  Y  Y  E  R  Q  E  N  L  K  S  K  A  T  E  L  E  V  R  L  K  G
 .  c  y  a  .  e  w  g  l  y  c  r  r  l  v  .  a  a  r  q  t  k  i  k  s  n  .  a  .  c  a  t  e
  l  l  v  s  v  g  m  r  s  l  v  k  t  i  s  v  s  s  t  s  n  q  n  q  q  k  l  s  l  v  c  n  g

EA22    ...    CONTINUES IN PHASE FOUR
```

```
          H              E MS    TF                   F                E                 C*T
          H              C NC    HN                   0                C                 L1A
          A              R LR    AU                   K                1                 A0Q
          1              2 11    1H                   1                5                 181
                            /                                                             //
TACTGTTAGCGCAATATCCTCGTTCTCCTGGTCGCGGCGTTTGATGTATTGCTGGTTTCTTTCCCGTTCATCCAGCAGTTCCAGCACAATCGATGGTGTT
----.----+----.----+----.----+----.----+----.----+----.----+----.----+----.----+----.----+----.----+ 30300
ATGACAATCGCGTTATAGGAGCAAGAGGACCAGCGCCGCAAACTACATAACGACCAAAGAAAGGGCAAGTAGGTCGTCAAGGTCGTGTTAGCTACCACAA

 t  v  s  a  i  s  s  f  s  w  s  r  r  l  m  y  c  w  f  l  s  r  s  s  s  s  s  s  t  i  d  g  v
  l  l  a  q  y  p  r  s  p  g  r  g  v  .  c  i  a  g  f  f  p  v  h  p  a  v  p  a  q  s  m  v  l
y  c  .  r  n  i  l  v  l  l  v  a  a  f  d  v  l  l  v  s  f  p  f  i  q  q  f  q  h  n  r  w  c  y
----.----+----.----+----.----+----.----+----.----+----.----+----.----+----.----+----.----+----.----+
 V  T  L  A  I  D  E  N  E  Q  D  R  R  K  I  Y  Q  Q  N  R  E  R  E  D  L  L  E  L  V  I  S  P  T
r  s  n  a  c  y  g  r  e  g  p  r  p  t  q  h  i  a  p  k  k  g  t  .  g  a  t  g  a  c  d  i  t  n
  .  q  .  r  l  i  r  t  r  r  t  a  a  n  s  t  n  s  t  e  k  g  n  m  w  c  n  w  c  l  r  h  h

*108    30290    Mapped site CLA1
EA22    ...      CONTINUES IN PHASE FOUR
```

```
          H    E                        A      7          F               D                   E ABT
          G    C                        V      -          N               D                   A S1A
          A    P                        A      2          U               E                   2 U2Q
          1    1                        3      0          H               1                   2 221
                                                                                                 //
ACCAATTCATGGAAAAGGTCTGCGTCAAATCCCCAGTCGTCATGCATTGCCTGCTCTGCCGCTTCACGCAGTGCCTGAGAGTTAATTTCGCTCACTTCGA
----.----+----.----+----.----+----.----+----.----+----.----+----.----+----.----+----.----+----.----+ 30400
TGGTTAAGTACCTTTTCCAGACGCAGTTTAGGGGTCAGCAGTACGTAACGGACGAGACGGCGAAGTGCGTCACGGACTCTCAATTAAAGCGAGTGAAGCT

t  n  s  w  k  r  s  a  s  n  p  q  s  s  c  i  a  c  s  a  a  s  r  s  a  .  e  l  i  s  l  t  s  n
 p  i  h  g  k  g  l  r  q  i  p  s  r  h  a  l  p  a  l  p  l  h  a  v  p  e  s  .  f  r  s  l  r
  q  f  m  e  k  v  c  v  k  s  p  v  v  m  h  c  l  l  c  r  f  t  q  c  l  r  v  n  f  a  h  f  e
----.----+----.----+----.----+----.----+----.----+----.----+----.----+----.----+----.----+----.----+
 V  L  E  H  F  L  D  A  D  F  G  W  D  D  H  M  A  Q  E  A  A  E  R  L  A  Q  S  N  I  E  S  V  e
  g  i  .  p  f  p  r  r  .  i  g  l  r  .  a  n  g  a  r  g  s  .  a  t  g  s  l  .  n  r  e  s  r
.  w  n  m  s  f  t  q  t  l  d  g  t  t  m  c  q  r  s  q  r  k  v  c  h  r  l  t  l  k  a  .  k  s
```

```
7-20    30348     bio7-20             att-sr      EM
EA22    30395     Ea22 (orf-182)      V-NH3
B122    30397     bio122              dr          EM


           M             XMB   E S     M    B                H     EMS  A                 BMS
           N             HBI   C C     N    2                G     CBC  C                 IBB
           L             OON   R R     L    2                A     ROR  Y                 NO7
           1             211   2 1     1    1                1     221  1                 11
                          //                                        /
ACCTCTCTGTTTACTGATAAGTTCCAGATCCTCCTGGCAACTTGCACAAGTCCGACAACCCTGAACGACCAGGCGTCTTCGTTCATCTATCGGATCGCCA
----.----+----.----+----.----+----.----+----.----+----.----+----.----+----.----+----.----+----.----+  30500
TGGAGAGACAAATGACTATTCAAGGTCTAGGAGGACCGTTGAACGTGTTCAGGCTGTTGGGACTTGCTGGTCCGCAGAAGCAAGTAGATAGCCTAGCGGT

  l  s  v  y  .  .  v  p  d  p  p  g  n  l  h  k  s  d  n  p  e  r  p  g  v  f  v  h  l  s  d  r  h
t  s  l  f  t  d  k  f  q  i  l  l  a  t  c  t  s  p  t  t  l  n  d  q  a  s  s  f  i  y  r  i  a  t
 p  l  c  l  l  i  s  s  r  s  s  w  q  l  a  q  v  r  q  p  .  t  t  r  r  l  r  s  s  i  g  s  p
----.----+----.----+----.----+----.----+----.----+----.----+----.----+----.----+----.----+----.----+
f  r  e  t  .  q  y  t  g  s  g  g  p  l  k  c  l  d  s  l  g  s  r  g  p  t  k  t  .  r  d  s  r  w
 v  e  r  n  v  s  l  n  w  i  r  r  a  v  q  v  l  g  v  v  r  f  s  w  a  d  e  n  ■  .  r  i  a
  g  r  q  k  s  i  l  e  l  d  e  q  c  s  a  c  t  r  c  g  q  v  v  l  r  r  r  e  d  i  p  d  g

B221    30445     b221                dr          EM
SB7     30493     Sequence Block 7    start


                   R       ECS            F   H                    HSM       S               H
                   D       CHC            N   H                    HBB       F               N
                   1       RIR            U   A                    A70       A               F
                   5       2B1            H   1                    1 2       N               1
                                                                     /
CACTCACAACAATGAGTGGCAGATATAGCCTGGTGGTTCAGGCGGCGCATTTTTATTGCTGTGTTGCGCTGTAATTCTTCTATTTCTGATGCTGAATCAA
----.----+----.----+----.----+----.----+----.----+----.----+----.----+----.----+----.----+----.----+  30600
GTGAGTGTTGTTACTCACCGTCTATATCGGACCACCAAGTCCGCCGCGTAAAAATAACGACACAACGCGACATTAAGAAGATAAAGACTACGACTTAGTT

 t  h  n  n  e  w  q  i  .  p  g  g  s  g  g  a  f  l  l  l  c  c  a  v  i  l  l  f  l  ■  l  n  q
  l  t  t  ■  s  g  r  y  s  l  v  v  q  a  a  h  f  y  c  c  v  a  l  .  f  f  y  f  .  c  .  i  n
h  s  q  q  .  v  a  d  i  a  w  w  f  r  r  r  i  f  i  a  v  l  r  c  n  s  s  i  s  d  a  e  s  ■
----.----+----.----+----.----+----.----+----.----+----.----+----.----+----.----+----.----+----.----+
  v  .  l  l  s  h  c  i  y  g  p  p  e  p  p  a  n  k  n  s  h  q  a  t  i  r  r  n  r  i  s  f  .
v  s  v  v  i  l  p  l  y  l  r  t  t  .  a  a  c  k  .  q  q  t  a  s  y  n  k  .  k  q  h  q  i  l
 c  e  c  c  h  t  a  s  i  a  q  h  n  l  r  r  ■  k  i  a  t  n  r  q  l  e  e  i  e  s  a  s  d

RD15    30520     delta red15         dl          EM G
CHIB    30529     chiB                pm          sq       121 delete C; 131 C to G
SB7     30569     Sequence Block 7    end


                                         M                 H    H                   A
                                         N                 N    P                   L
                                         L                 F    H                   U
                                         1                 3    1                   1

TGATGTCTGCCATCTTTCATTAATCCCTGAACTGTTGGTTAATACGCTTGAGGGTGAATGCGAATAATAAAAAAGGAGCCTGTAGCTCCCTGATGATTTT
----.----+----.----+----.----+----.----+----.----+----.----+----.----+----.----+----.----+----.----+  30700
ACTACAGACGGTAGAAAGTAATTAGGGACTTGACAACCAATTATGCGAACTCCCACTTACGCTTATTATTTTTTCCTCGGACATCGAGGGACTACTAAAA

.  c  l  p  s  f  i  n  p  .  t  v  g  .  y  a  .  g  .  ■  r  i  i  k  k  e  p  v  a  p  .  .  f  c
 d  v  c  h  l  s  l  i  p  e  l  l  v  n  t  l  e  g  e  c  e  .  .  k  r  s  l  .  l  p  d  d  f
  ■  s  a  i  f  h  .  s  l  n  c  w  l  i  r  l  r  v  n  a  n  n  k  k  g  a  c  s  s  l  ■  i  l
----.----+----.----+----.----+----.----+----.----+----.----+----.----+----.----+----.----+----.----+
 h  h  r  g  d  k  ■  l  g  q  v  t  p  .  y  a  q  p  h  i  r  i  i  f  f  s  g  t  a  g  q  h  n
  s  t  q  w  r  e  n  i  g  s  s  n  t  l  v  s  s  p  s  h  s  y  y  f  l  l  r  y  s  g  s  s  k
i  i  d  a  ■  k  .  .  d  r  f  q  q  n  i  r  k  l  t  f  a  f  l  l  f  p  a  q  l  e  r  i  i  k
```

```
                             A   H   N   E S      H     H            TM
                             C   G   S   C C      N     N            HB
                             Y   A   P   R R      F     F            AO
                             1   1   C   2 1      1     1            12
                                                                      /
GCTTTTCATGTTCATCGTTCCTTAAAGACGCCGTTTAACATGCCGATTGCCAGGCTTAAATGAGTCGGTGTGAATCCCATCAGCGTTACCGTTTCGCGGT
----.----+----.----+----.----+----.----+----.----+----.----+----.----+----.----+----.----+----.----+ 30800
CGAAAAGTACAAGTAGCAAGGAATTTCTGCGGCAAATTGTACGGCTAACGGTCCGAATTTACTCAGCCACACTTAGGGTAGTCGCAATGGCAAAGCGCCA

  f  s  c  s  s  f  l  k  d  a  v  .  h  a  d  c  q  a  .  m  s  r  c  e  s  h  q  r  y  r  f  a  v
 a  f  h  v  h  r  s  l  k  t  p  f  n  m  p  i  a  r  l  k  .  v  g  v  n  p  i  s  v  t  v  s  r  c
l  f  m  f  i  v  p  .  r  r  r  l  t  c  r  l  p  g  l  n  e  s  v  .  i  p  s  a  l  p  f  r  g
----.----+----.----+----.----+----.----+----.----+----.----+----.----+----.----+----.----+----.----+
 q  k  e  h  e  d  n  r  l  s  a  t  .  c  a  s  q  w  a  .  i  l  r  h  s  d  w  .  r  .  r  k  a  t
  a  k  .  t  .  r  e  k  f  v  g  n  l  m  g  i  a  l  s  l  h  t  p  t  f  g  m  l  t  v  t  e  r
   s  k  m  n  m  t  g  .  l  r  r  k  v  h  r  n  g  p  k  f  s  d  t  h  i  g  d  a  n  g  n  r  p


     R               T         H   E                             B   B       F
     S               A         P   C                             B   G       N
     A               Q         A   1                             V   L       U
     1               1         2   5                             1   1       H

GCTTCTTCAGTACGCTACGGCAAATGTCATCGACGTTTTTATCCGGAAACTGCTGTCTGGCTTTTTTTGATTTCAGAATTAGCCTGACGGGCAATGCTGC
----.----+----.----+----.----+----.----+----.----+----.----+----.----+----.----+----.----+----.----+ 30900
CGAAGAAGTCATGCGATGCCGTTTACAGTAGCTGCAAAAATAGGCCTTTGACGACAGACCGAAAAAAACTAAAGTCTTAATCGGACTGCCCGTTACGACG

  l  l  q  y  a  t  a  n  v  i  d  v  f  i  r  k  l  l  s  g  f  f  .  f  q  n  .  p  d  g  q  c  c
   f  f  s  t  l  r  q  m  s  s  t  f  l  s  g  n  c  c  l  a  f  f  d  f  r  i  s  l  t  g  n  a  a
 a  s  s  v  r  y  g  k  c  h  r  r  f  y  p  e  t  a  v  w  l  f  l  i  s  e  l  a  .  r  a  m  l  r
----.----+----.----+----.----+----.----+----.----+----.----+----.----+----.----+----.----+----.----+
  s  r  .  y  a  v  a  f  t  m  s  t  k  i  r  f  s  s  d  p  k  k  q  n  .  f  .  g  s  p  c  h  q
 h  k  k  l  v  s  r  c  i  d  d  v  n  k  d  p  f  q  q  r  a  k  k  s  k  l  i  l  r  v  p  l  a  a
  a  e  e  t  r  .  p  l  h  .  r  r  k  .  g  s  v  a  t  q  s  k  k  i  e  s  n  a  q  r  a  i  s


     M E   D                                                TB     N  F        A
     N C   D                                                TB     S  N        V
     L 1   E                                                HV     P  U        A
     1 5   1                                                21     B  H        3
                                                             /
GAAGGGCGTTTTCCTGCTGAGGTGTCATTGAACAAGTCCCATGTCGGCAAGCATAAGCACACAGAATATGAAGCCCGCTGCCAGAAAAATGCATTCCGTG
----.----+----.----+----.----+----.----+----.----+----.----+----.----+----.----+----.----+----.----+ 31000
CTTCCCGCAAAAGGACGACTCCACAGTAACTTGTTCAGGGTACAGCCGTTCGTATTCGTGTGTCTTATACTTCGGGCGACGGTCTTTTTACGTAAGGCAC

 e  g  r  f  p  a  e  v  s  l  n  k  s  h  v  g  k  h  k  h  t  e  y  e  a  r  c  q  k  n  a  f  r  g
  k  g  v  f  l  l  r  c  h  .  t  s  p  m  s  a  s  i  s  t  q  n  m  k  p  a  a  r  k  m  h  s  v
   r  a  f  s  c  .  g  v  i  e  q  v  p  c  r  q  a  .  a  h  r  i  .  s  p  l  p  e  k  c  i  p  w
----.----+----.----+----.----+----.----+----.----+----.----+----.----+----.----+----.----+----.----+
  s  p  r  k  g  a  s  t  d  n  f  l  d  w  t  p  l  c  l  c  v  s  y  s  a  r  q  w  f  f  a  n  r
   f  p  t  k  r  s  l  h  .  q  v  l  g  m  d  a  l  m  l  v  c  f  i  f  g  a  a  l  f  i  c  e  t
 r  l  a  n  e  q  q  p  t  m  s  c  t  g  h  r  c  a  y  a  c  l  i  h  l  g  s  g  s  f  h  m  g  h


     E S                                 SB      A  S     H    F    FM    X
     C C                                 B5      L  F     G    N    OB    M
     R R                                 82      U  A     U    U    KO    N
     2 1                                  2      1  N     2    H    12    1
                                         /
GTTGTCATACCTGGTTTCTCTCATCTGCTTCTGCTTTCGCCACCATCATTTCCAGCTTTTGTGAAAGGGATGCGGCTAACGTATGAAATTCTTCGTCTGT
----.----+----.----+----.----+----.----+----.----+----.----+----.----+----.----+----.----+----.----+ 31100
CAACAGTATGGACCAAAGAGAGTAGACGAAGACGAAAGCGGTGGTAGTAAAGGTCGAAAACACTTTCCCTACGCCGATTGCATACTTTAAGAAGCAGACA

   c  h  t  w  f  l  s  s  a  s  a  f  a  t  i  i  s  s  f  c  e  r  d  a  a  n  v  .  n  s  s  s  v
 v  v  i  p  g  f  s  h  l  l  l  l  s  p  p  s  f  p  a  f  v  k  g  m  r  l  t  y  e  i  l  r  l  f
  l  s  y  l  v  s  l  i  c  f  c  f  r  h  h  h  f  q  l  l  .  k  g  c  g  .  r  m  k  f  f  v  c
----.----+----.----+----.----+----.----+----.----+----.----+----.----+----.----+----.----+----.----+
 p  q  .  v  q  n  r  e  d  a  e  a  k  a  v  m  m  e  l  k  q  s  l  s  a  a  l  t  h  f  e  e  d  t
  t  t  m  g  p  k  e  .  r  s  r  s  e  g  g  d  n  g  a  k  t  f  p  i  r  s  v  y  s  i  r  r  r
   n  d  y  r  t  e  r  m  q  k  q  k  r  w  w  .  k  w  s  k  h  f  p  h  p  .  r  i  f  n  k  t  q

SB8    31043    Sequence Block 8    start
B522   31043    b522                att-dr     sq
```

```
                         H                             T           S            M
                         N                             A           F            B
                         F                             Q           A            0
                         1                             1           N            2

TTCTACTGGTATTGGCACAAACCTGATTCCAATTTGAGCAAGGCTATGTGCCATCTCGATACTCGTTCTTAACTCAACAGAAGATGCTTTGTGCATACAG
----.----+----.----+----.----+----.----+----.----+----.----+----.----+----.----+----.----+----.----+ 31200
AAGATGACCATAACCGTGTTTGGACTAAGGTTAAACTCGTTCCGATACACGGTAGAGCTATGAGCAAGAATTGAGTTGTCTTCTACGAAACACGTATGTC

 s  t  g  i  g  t  n  l  i  p  i  .  a  r  l  c  a  i  s  i  l  v  l  n  s  t  e  d  a  l  c  i  q
  l  l  v  l  a  q  t  .  f  q  f  e  q  g  y  v  p  s  r  y  s  f  l  t  q  q  k  m  l  c  a  y  s
f  y  w  y  w  h  k  p  d  s  n  l  s  k  a  m  c  h  l  d  t  r  s  .  l  n  r  r  c  f  v  h  t  a
----.----+----.----+----.----+----.----+----.----+----.----+----.----+----.----+----.----+----.----+
  e  v  p  i  p  v  f  r  i  g  i  q  a  l  s  h  a  m  e  i  s  t  r  l  e  v  s  s  a  k  h  m  c
n  r  s  t  n  a  c  v  q  n  w  n  s  c  p  .  t  g  d  r  y  e  n  k  v  .  c  f  i  s  q  a  y  l
 k  .  q  y  q  c  l  g  s  e  l  k  l  l  a  i  h  w  r  s  v  r  e  .  s  l  l  l  h  k  t  c  v


         M     D     NMF                      T   HHHRHNHS             M
         N     D     SNN                      L   YYYEACPC             N
         L     E     PLU                      3   15MDEIAR             L
         1     1     B1H                             13121             1
                       /                          ///////
CCCCTCGTTTATTATTTATCTCCTCAGCCAGCCGCTGTGCTTTCAGTGGATTTCGGATAACAGAAAGGCCGGGAAATACCCAGCCTCGCTTTGTAACGGA
----.----+----.----+----.----+----.----+----.----+----.----+----.----+----.----+----.----+----.----+ 31300
GGGGAGCAAATAATAAATAGAGGAGTCGGTCGGCGACACGAAAGTCACCTAAAGCCTATTGTCTTTCCGGCCCTTTATGGGTCGGAGCGAAACATTGCCT

p  l  v  y  y  l  s  p  q  p  a  a  v  l  s  v  d  f  g  .  q  k  g  r  e  i  p  s  l  a  l  .  r  s
 p  s  f  i  i  y  l  l  s  q  p  l  c  f  q  w  i  s  d  n  r  k  a  g  k  y  p  a  s  l  c  n  g
  p  r  l  l  f  i  s  s  a  s  r  c  a  f  s  g  f  r  i  t  e  r  p  g  n  t  q  p  r  f  v  t  e
----.----+----.----+----.----+----.----+----.----+----.----+----.----+----.----+----.----+----.----+
 g  r  t  .  .  k  d  g  .  g  a  a  t  s  e  t  s  k  p  y  c  f  p  r  s  i  g  l  r  a  k  y  r
  g  e  n  i  i  .  r  r  l  w  g  s  h  k  .  h  i  e  s  l  l  f  a  p  f  y  g  a  e  s  q  l  p
a  g  r  k  n  n  i  e  e  a  l  r  q  a  k  l  p  n  r  i  v  s  l  g  p  f  v  w  g  r  k  t  v  s

TL3    31262      tL3 terminator     3'       RNA
HY1    31267      Hy1                hy       EM      lambda::imm80 OP 80 QSR80
HY5    31267      Hy5                hy       EM      80::immlambda OPlambda QSR80
HYM    31267      HyM                hy       EM      lambdaatt80::immlambdaOPlambdaQSR8
RED1   31267      delta red1         dr       EM


 A            H     NSH   MS H     H          E F
 C            H     CCP   NF G     G          X 0
 C            A     IRA   LA A     U          0 K
 1            1     112   1N 1     2            1
                    //
GTAGACGAAAGTGATTGCGCCTACCCGGATATTATCGTGAGGATGCGTCATCGCCATTGCTCCCCAAATACAAAACCAATTTCAGCCAGTGCCTCGTCCA
----.----+----.----+----.----+----.----+----.----+----.----+----.----+----.----+----.----+----.----+ 31400
CATCTGCTTTCACTAACGCGGATGGGCCTATAATAGCACTCCTACGCAGTAGCGGTAACGAGGGGTTTATGTTTTGGTTAAAGTCGGTCACGGAGCAGGT

  r  r  k  .  l  r  l  p  g  y  y  r  e  d  a  s  s  p  l  l  p  k  y  k  t  n  f  s  q  c  l  v  h
v  d  e  s  d  c  a  y  p  d  i  i  v  r  m  r  h  r  h  c  s  p  n  t  k  p  i  s  a  s  a  s  s  i
 .  t  k  v  i  a  p  t  r  i  l  s  .  g  c  v  i  a  i  a  p  q  i  q  n  q  f  q  p  v  p  r  p
----.----+----.----+----.----+----.----+----.----+----.----+----.----+----.----+----.----+----.----+
l  l  r  f  h  n  r  r  g  p  y  .  r  s  s  a  d  d  g  n  s  g  l  y  l  v  l  k  l  w  h  r  t  w
 t  s  s  l  s  q  a  .  g  s  i  i  t  l  i  r  .  R  W  Q  E  G  F  V  F  G  I  E  A  L  A  E  D
  y  v  f  t  i  a  g  v  r  i  n  d  h  p  h  t  m  a  m  a  g  w  i  c  f  w  n  .  g  t  g  r  g

EXO    31351      exo                COOH


 M  T     11 H    M             F     R           K               SHH          T
 N  A     04 P    B             0     S           A               TAA          H
 L  Q     15 A    0             K     A           N               UEE          A
 1  1     81 2    1             1     1           3               113          1
           /                                                       //
TTTTTTCGATGAACTCCGGCACGATCTCGTCAAAACTCGCCATGTACTTTTCATCCCGCTCAATCACGACATAATGCAGGCCTTCACGCTTCATACGCGG
----.----+----.----+----.----+----.----+----.----+----.----+----.----+----.----+----.----+----.----+ 31500
AAAAAAGCTACTTGAGGCCGTGCTAGAGCAGTTTTGAGCGGTACATGAAAAGTAGGGCGAGTTAGTGCTGTATTACGTCCGGAAGTGCGAAGTATGCGCC

 f  f  d  e  l  r  h  d  l  v  k  t  r  h  v  l  f  i  p  l  n  h  d  i  m  q  a  f  t  l  h  t  r
  f  s  m  n  s  g  t  i  s  s  k  l  a  m  y  f  s  s  r  s  i  t  t  .  c  r  p  s  r  f  i  r  g
f  f  r  .  t  p  a  r  s  r  q  n  s  p  c  t  f  h  p  a  q  s  r  h  n  a  g  l  h  a  s  y  a  g
----.----+----.----+----.----+----.----+----.----+----.----+----.----+----.----+----.----+----.----+
  k  k  s  s  s  r  c  s  r  t  l  v  r  w  t  s  k  m  g  s  l  .  s  m  i  c  a  k  v  s  .  v  r
M  K  E  I  F  E  P  V  I  E  D  F  S  A  M  Y  K  E  D  R  E  I  V  V  Y  H  L  G  E  R  K  M  R  P
 n  k  r  h  v  g  a  r  d  r  .  f  e  g  h  v  k  .  g  a  .  d  r  c  l  a  p  r  .  a  e  y  a
```

```
1018   31412         b1018               dr          EM
1451   31412         b1451               dl          EM
KAN3   31461         kan3                i(Tn6)      EM
EXO      ...       CONTINUES IN PHASE FIVE


                RE S H    H    T          N   R    E S A H      A       HH T     M    H
                SC C G    P    H          S   S    C C S A      L       AA A     N    P
                AR R A    H    A          P   A    R R U E      U       EE Q     L    A
                12 1 1    1    1          C   1    2 1 1 3      1       13 1     1    2
                                                                        /
GTCATAGTTGGCAAAGTACCAGGCATTTTTTCGCGTCACCCACATGCTGTACTGCACCTGGGCCATGTAAGCTGACTTTATGGCCTCGAAACCACCGAGC
---- . ----+---- . ----+---- . ----+---- . ----+---- . ----+---- . ----+---- . ----+---- . ----+---- . ----+---- . ----+   31600
CAGTATCAACCGTTTCATGGTCCGTAAAAAAGCGCAGTGGGTGTACGACATGACGTGGACCCGGTACATTCGACTGAAATACCGGAGCTTTGGTGGCTCG

v  i  v  g  k  v  p  g  i  f  s  r  h  p  h  a  v  l  h  l  g  h  v  s  .  l  y  g  l  e  t  t  e  p
 s  .  l  a  k  y  q  a  f  f  r  v  t  h  ■  l  y  c  t  w  a  ■  .  a  d  f  ■  a  s  k  p  p  s
  h  s  w  q  s  t  r  h  f  f  a  s  p  t  c  c  t  a  p  g  p  c  k  l  t  l  w  p  r  n  h  r  a
---- . ----+---- . ----+---- . ----+---- . ----+---- . ----+---- . ----+---- . ----+---- . ----+---- . ----+---- . ----+
t  ■  t  p  l  t  g  p  ■  k  e  r  .  g  c  a  t  s  c  r  p  w  t  l  q  s  .  p  r  s  v  v  s
 D  Y  N  A  F  Y  W  A  N  K  R  T  V  W  M  S  Y  Q  V  Q  A  M  Y  A  S  K  I  A  E  F  G  G  L
p  .  l  q  c  l  v  l  c  k  k  a  d  g  v  h  q  v  a  g  p  g  h  l  s  v  k  h  g  r  f  w  r  a

EXO      ...       CONTINUES IN PHASE FIVE


             M  "NXASNHS"                      H                                       R
             N  1CMVCCPC1                      A                                       S
             L  0IAARIAR1                      E                                       A
             1  911111210                      3                                       1
                 ///////
CGGAACTTCATGAAATCCCGGGAGGTAAACGGGCATTTCAGTTCAAGGCCGTTGCCGTCACTGCATAAACCATCGGGAGAGCAGGCGGTACGCATACTTT
---- . ----+---- . ----+---- . ----+---- . ----+---- . ----+---- . ----+---- . ----+---- . ----+---- . ----+---- . ----+   31700
GCCTTGAAGTACTTTAGGGCCCTCCATTTGCCCGTAAAGTCAAGTTCCGGCAACGGCAGTGACGTATTTGGTAGCCCTCTCGTCCGCCATGCGTATGAAA

  e  l  h  e  i  p  g  g  k  r  a  f  q  f  k  a  v  a  v  t  a  .  t  i  g  r  a  g  g  t  h  t  f
r  n  f  ■  k  s  r  e  v  n  g  h  f  s  s  r  p  l  p  s  l  h  k  p  s  g  e  q  a  v  r  i  l  s
 g  t  s  .  n  p  g  r  .  t  g  i  s  v  q  g  r  c  r  h  c  i  n  h  r  e  s  r  r  y  a  y  f
---- . ----+---- . ----+---- . ----+---- . ----+---- . ----+---- . ----+---- . ----+---- . ----+---- . ----+---- . ----+
g  s  s  .  s  i  g  p  p  l  r  a  n  .  n  l  a  t  a  t  v  a  y  v  ■  p  l  a  p  p  v  c  v  k
 R  F  K  M  F  D  R  S  T  F  P  C  K  L  E  L  G  N  G  D  S  C  L  G  D  P  S  C  A  T  R  M  S
  p  v  e  h  f  g  p  l  y  v  p  ■  e  t  .  p  r  q  r  .  q  ■  f  w  r  s  l  l  r  y  a  y  k

*109   31617       Mapped site AVA1
*110   31619       Mapped site SMA1
EXO      ...       CONTINUES IN PHASE FIVE


    NT     M      H                    H       *E     H          A T       R        E S  HH
    RH     B      N                    P       1C     G          C T       S        C C  AA
    UA     0      F                    A       1R     A          Y H       A        R R  EE
    11     1      1                    2       11     1          1 2       1        2 1  13
    /                                          /                                         /
CGTCGCGATAGATGATCGGGGATTCAGTAACATTCACGCCGGAAGTGAATTCAAACAGGGTTCTGGCGTCGTTCTCGTACTGTTTTCCCCAGGCCAGTGC
---- . ----+---- . ----+---- . ----+---- . ----+---- . ----+---- . ----+---- . ----+---- . ----+---- . ----+---- . ----+   31800
GCAGCGCTATCTACTAGCCCCTAAGTCATTGTAAGTGCGGCCTTCACTTAAGTTTGTCCCAAGACCGCAGCAAGAGCATGACAAAAGGGGTCCGGTCACG

 v  a  i  d  d  r  g  f  s  n  i  h  a  g  s  e  f  k  q  g  s  g  v  v  l  v  l  f  s  p  g  q  c
  s  r  .  ■  i  g  d  s  v  t  f  t  p  e  v  n  s  n  r  v  l  a  s  f  s  y  c  f  p  q  a  s  a
r  r  d  r  .  s  g  i  q  .  h  s  r  r  k  .  i  q  t  g  f  w  r  r  s  r  t  v  f  p  r  p  v  l
---- . ----+---- . ----+---- . ----+---- . ----+---- . ----+---- . ----+---- . ----+---- . ----+---- . ----+---- . ----+
 t  a  i  s  s  r  p  n  l  l  ■  .  a  p  l  s  n  l  c  p  e  p  t  t  r  t  s  n  e  g  p  w  h
E  D  R  Y  I  I  P  S  E  T  V  N  V  G  S  T  F  E  F  L  T  R  A  D  N  F  Y  Q  K  G  W  A  L  A
 r  r  s  l  h  d  p  i  .  y  c  e  r  r  f  h  i  .  v  p  n  q  r  r  e  r  v  t  k  g  l  g  t

*111   31747       Mapped site ECR1
EXO      ...       CONTINUES IN PHASE FIVE
```

```
        HH**  H          H          D        M           T                   HH         H        1
        PI11  P          P          D        N           T                   AA         P        4
        AN11  A          A          E        L           H                   EE         A        5
        1223  2          2          1        1           2                   13         2        1
         ///                                                                  /
TTTAGCGTTAACTTCCGGAGCCACACCGGTGCAAACCTCAGCAAGCAGGGTGTGGAAGTAGGACATTTTCATGTCAGGCCACTTCTTTCCGGAGCGGGGT
----.----+----.----+----.----+----.----+----.----+----.----+----.----+----.----+----.----+----.----+  31900
AAATCGCAATTGAAGGCCTCGGTGTGGCCACGTTTGGAGTCGTTCGTCCCACACCTTCATCCTGTAAAAGTACAGTCCGGTGAAGAAAGGCCTCGCCCCA

f  s  v  n  f  r  s  h  t  g  a  n  l  s  k  q  g  v  e  v  g  h  f  h  v  r  p  l  l  s  g  a  g  f
 l  a  l  t  s  g  a  t  p  v  q  t  s  a  s  r  v  w  k  .  d  i  f  m  s  g  h  f  f  p  e  r  g
  .  r  .  l  p  e  p  h  r  c  k  p  q  q  a  g  c  g  s  r  t  f  s  c  q  a  t  s  f  r  s  g  v
----.----+----.----+----.----+----.----+----.----+----.----+----.----+----.----+----.----+----.----+
 k  l  t  l  k  r  l  w  v  p  a  f  r  l  l  c  p  t  s  t  p  c  k  .  t  l  g  s  r  e  p  a  p
  K  A  N  V  E  P  A  V  G  T  C  V  F  A  L  L  T  H  F  Y  S  M  K  M  D  P  W  K  K  G  S  R  P
s  .  r  .  s  g  s  g  c  r  h  l  g  .  c  a  p  h  p  l  l  v  n  e  h  .  a  v  e  k  r  l  p  t

*112   31809      Mapped site HIN2
*113   31809      Mapped site HPA1
1451   31897        b1451                dr          EM
EXO     ...       CONTINUES IN PHASF FIVF
```

```
                                    A    H  H     F                  S     T     A       C*TMBBBNSHB
                                    C    P  G     O                  F     A     L       L1ABI76CCPB
                                    Y    H  A     K                  A     Q     U       A1QON29IRAV
                                    1    1  1     1                  N     1     1       14111 1121
                                                                                          //////////
TTTGCTATCACGTTGTGAACTTCTGAAGCGGTGATGACGCCGAGCCGTAATTTGTGCCACGCATCATCCCCCTGTTCGACAGCTCTCACATCGATCCCGG
----.----+----.----+----.----+----.----+----.----+----.----+----.----+----.----+----.----+----.----+  32000
AAACGATAGTGCAACACTTGAAGACTTCGCCACTACTGCGGCTCGGCATTAAACACGGTGCGTAGTAGGGGGACAAGCTGTCGAGAGTGTAGCTAGGGCC

  c  y  h  v  v  n  f  .  s  g  d  d  a  e  p  .  f  v  p  r  i  i  p  l  f  d  s  s  h  i  d  p  g
f  a  i  t  l  .  t  s  e  a  v  m  t  p  s  r  n  l  c  h  a  s  s  p  c  s  t  a  l  t  s  i  p  v
 l  l  s  r  c  e  l  l  k  r  .  .  r  r  a  v  i  c  a  t  h  h  p  p  v  r  q  l  s  h  r  s  r
----.----+----.----+----.----+----.----+----.----+----.----+----.----+----.----+----.----+----.----+
n  q  .  .  t  t  f  k  q  l  p  s  s  a  s  g  y  n  t  g  r  m  m  g  r  n  s  l  e  .  m  s  g  p
 K  A  I  V  N  H  V  E  S  A  T  I  V  G  L  R  L  K  H  W  A  D  D  G  Q  E  V  A  R  V  D  I  G
  k  s  d  r  q  s  s  r  f  r  h  h  r  r  a  t  i  q  a  v  c  .  g  g  t  r  c  s  e  c  r  d  r

*114   31991      Mapped site CLA1
B72    31994        bio72                att-sr      EM
B69    31994        bio69                att-sr      EM
EXO     ...       CONTINUFS IN PHASE FIVF
```

```
R   F  P*      B  H       EBF            B     F               H         A              H       S
S   N  S1      B  P       XEN            1     N               N         L              A       F
A   U  T1      V  A       OTU            6     U               F         U              E       A
1   H  15      1  2         H            9     H               1         1              3       N
         /                  /
TACGCTGCAGGATAATGTCCGGTGTCATGCTGCCACCTTCTGCTCTGCGGCTTTCTGTTTCAGGAATCCAAGAGCTTTTACTGCTTCGGCCTGTGTCAGT
----.----+----.----+----.----+----.----+----.----+----.----+----.----+----.----+----.----+----.----+  32100
ATGCGACGTCCTATTACAGGCCACAGTACGACGGTGGAAGACGAGACGCCGAAAGACAAAGTCCTTAGGTTCTCGAAAATGACGAAGCCGGACACAGTCA

 t  l  q  d  n  v  r  c  h  a  a  t  f  c  s  a  a  f  c  f  r  n  p  r  a  f  t  a  s  a  c  v  s
  r  c  r  i  m  s  g  v  m  l  p  p  s  a  l  r  l  s  v  s  g  i  q  e  l  l  l  l  r  p  v  s  v
y  a  a  g  .  c  p  v  s  c  c  h  l  l  l  c  g  f  l  f  q  e  s  k  s  f  y  c  f  g  l  c  q  f
----.----+----.----+----.----+----.----+----.----+----.----+----.----+----.----+----.----+----.----+
  v  s  c  s  l  t  r  h  .  A  A  V  K  Q  E  A  A  K  Q  K  L  F  G  L  A  K  V  A  E  A  Q  T  L
T  R  Q  L  I  I  D  P  T  M  s  g  g  e  a  r  r  s  e  t  e  p  i  w  s  s  k  s  s  r  g  t  d  t
 y  a  a  p  y  h  g  t  d  h  q  w  r  r  s  q  p  k  r  n  .  s  d  l  l  k  .  q  k  p  r  h  .

*115   32009      Mapped site PST1
EXO    32028        exo (orf-226)        M-NH3
BET    32028        bet                  COOH
B169   32042        bio169               att-sr      EM
```

```
           H        TF                       FF                            E S    M E
           N        HN                       NO                            C C    B C
           F        AU                       UK                            R R    0 1
           3        1H                       H1                            2 1    1 5
                                              /
TCTGACGATGCACGAATGTCGCGGCGAAATATCTGGGAACAGAGCGGCAATAAGTCGTCATCCCATGTTTTATCCAGGGCGATCAGCAGAGTGTTAATCT
----.----+----.----+----.----+----.----+----.----+----.----+----.----+----.----+----.----+----.----+  32200
AGACTGCTACGTGCTTACAGCGCCGCTTTATAGACCCTTGTCTCGCCGTTATTCAGCAGTAGGGTACAAAATAGGTCCCGCTAGTCGTCTCACAATTAGA

s  d  d  a  r  ■  s  r  r  n  i  w  e  q  s  g  n  k  s  s  s  h  v  l  s  r  a  i  s  r  v  l  i  s
 l  t  ■  h  e  c  r  g  e  i  s  g  n  r  a  a  i  s  r  h  p  ■  f  y  p  g  r  s  a  e  c  .  s
  .  r  c  t  n  v  a  a  k  y  l  g  t  e  r  q  .  v  v  i  p  c  f  i  q  q  d  q  q  s  v  n  l
----.----+----.----+----.----+----.----+----.----+----.----+----.----+----.----+----.----+----.----+
 E  S  S  A  R  I  D  R  R  F  I  Q  S  C  L  P  L  L  D  D  D  W  T  K  D  L  A  I  L  L  T  N  I
  r  v  i  c  s  h  r  p  s  i  d  p  f  l  a  a  i  l  r  .  g  ■  n  .  g  p  r  d  a  s  h  .  d
n  q  r  h  v  f  t  a  a  f  y  r  p  v  s  r  c  y  t  t  ■  g  h  k  i  w  p  s  .  c  l  t  l  r

BEI      ...      CONTINUFS IN PHASE FOUR
```

```
                 HH**H        TR  H                P*             T    F    H
                 PI11P        HD  P                S1             A    0    H
                 AN11A        A1  A                T1             Q    K    A
                 12672        10  2                18             1    1    1
                  ////                              /
CCTGCATGGTTTCATCGTTAACCGGAGTGATGTCGCGTTCCGGCTGACGTTCTGCAGTGTATGCAGTATTTTCGACAATGCGCTCGGCTTCATCCTTGTC
----.----+----.----+----.----+----.----+----.----+----.----+----.----+----.----+----.----+----.----+  32300
GGACGTACCAAAGTAGCAATTGGCCTCACTACAGCGCAAGGCCGACTGCAAGACGTCACATACGTCATAAAAGCTGTTACGCGAGCCGAAGTAGGAACAG

  c  ■  v  s  s  l  t  g  v  ■  s  r  s  g  .  r  s  a  v  y  a  v  f  s  t  ■  r  s  a  s  s  l  s
p  a  w  f  h  r  .  p  e  .  c  r  v  p  a  d  v  l  q  c  ■  q  y  f  r  q  c  a  r  l  h  p  c  h
 l  h  g  f  i  v  n  r  s  d  v  a  f  r  l  t  f  c  s  v  c  s  i  f  d  n  a  l  g  f  i  l  v
----.----+----.----+----.----+----.----+----.----+----.----+----.----+----.----+----.----+----.----+
E  Q  M  T  E  D  N  V  P  T  I  D  R  E  P  Q  R  E  A  T  Y  A  T  N  E  V  I  R  E  A  E  D  K  D
 g  a  h  n  .  r  .  g  s  h  h  r  t  g  a  s  t  r  c  h  i  c  y  k  r  c  h  a  r  s  .  g  q
  r  c  p  k  ■  t  l  r  l  s  t  a  n  r  s  v  n  q  l  t  h  l  i  k  s  l  a  s  p  k  ■  r  t

*116   32219     Mapped site HIN2
*117   32219     Mapped site HPA1
RD10   32236       delta red10          dr          EM
*118   32256     Mapped site PST1
BEI      ...     CONTINUFS IN PHASF FOUR
```

```
                    HH   BS      H      F            S      H          F AH
                    AA   GD      N      0            F      G          0 SA
                    EE   LU      F      K            A      U          K UF
                    13   11      1      1            N      2          1 13
                     /    /
ATAGATACCAGCAAATCCGAAGGCCAGACGGGCACACTGAATCATGGCTTTATGACGTAACATCCGTTTGGGATGCGACTGCCACGGCCCCGTGATTTCT
----.----+----.----+----.----+----.----+----.----+----.----+----.----+----.----+----.----+----.----+  32400
TATCTATGGTCGTTTAGGCTTCCGGTCTGCCCGTGTGACTTAGTACCGAAATACTGCATTGTAGGCAAACCCTACGCTGACGGTGCCGGGGCACTAAAGA

 .  i  p  a  n  p  k  a  r  r  a  h  .  i  ■  a  l  .  r  n  i  r  l  g  c  d  c  h  g  p  v  i  s
  r  y  q  q  i  r  r  p  d  g  h  t  e  s  w  l  y  d  v  t  s  v  w  d  a  t  a  t  a  p  .  f  l
i  d  t  s  k  s  e  g  q  t  g  t  l  n  h  g  f  ■  t  .  h  p  f  g  ■  r  l  p  r  p  r  d  f  s
----.----+----.----+----.----+----.----+----.----+----.----+----.----+----.----+----.----+----.----+
 Y  T  G  A  F  G  F  A  L  R  A  C  Q  I  M  A  K  H  R  L  M  R  K  P  H  S  Q  W  P  G  T  I  E
.  l  y  w  c  i  r  l  g  s  p  c  v  s  d  h  s  .  s  t  v  d  t  q  s  a  v  a  v  a  g  h  n  r
 ■  s  v  l  l  d  s  p  w  v  p  v  s  f  .  p  k  i  v  y  c  g  n  p  i  r  s  g  r  g  r  s  k

BEI      ...      CONTINUFS IN PHASF FOUR
```

```
        NT              F  TFF F                H          M   H            AA F            H      NR
        RH              O  HNO N                N          B   G            SV O            P      SS
        UA              K  AUK U                F          O   U            UA K            A      PA
        11              1  1H1 H                3          1   2            12 1            2      C1
         /                  /                                                /
CTGCCTTCGCGAGTTTTGAATGGTTCGCGGCGGCATTCATCCATCCATTCGGTAACGCAGATCGGATGATTACGGTCCTTGCGGTAAATCCGGCATGTAC
----.----+----.----+----.----+----.----+----.----+----.----+----.----+----.----+----.----+----.----+  32500
GACGGAAGCGCTCAAAACTTACCAAGCGCCGCCGTAAGTAGGTAGGTAAGCCATTGCGTCTAGCCTACTAATGCCAGGAACGCCATTTAGGCCGTACATG

l  p  s  r  v  l  n  g  s  r  r  h  s  s  i  h  s  v  t  q  i  g  .  l  r  s  l  r  .  i  r  h  v  q
 c  l  r  e  f  .  ■  v  r  g  g  i  h  p  s  i  r  .  r  r  s  d  d  y  g  p  c  g  k  s  g  ■  y
  a  f  a  s  f  e  w  f  a  a  a  f  i  h  p  f  g  n  a  d  r  ■  i  t  v  l  a  v  n  p  a  c  t
----.----+----.----+----.----+----.----+----.----+----.----+----.----+----.----+----.----+----.----+
 R  G  E  R  T  K  F  P  E  R  R  C  E  D  M  W  E  T  V  C  I  P  H  N  R  D  K  R  Y  I  R  C  T
  q  r  r  s  n  q  i  t  r  p  p  ■  .  g  d  ■  r  y  r  l  d  s  s  .  p  g  q  p  l  d  p  ■  y
e  a  k  a  l  k  s  h  n  a  a  a  n  ■  w  g  n  p  l  a  s  r  i  i  v  t  r  a  t  f  g  a  h  v

BEI     ...      CONTINUES IN PHASE FOUR
```

```
 H                  B       E       S                 AA                  S H
 N                  2       C       F                 SV                  F P
 F                  6       1       A                 UA                  A A
 1                  7       5       N                 12                  N 2
                                                       /
AGGATTCATTGTCCTGCTCAAAGTCCATGCCATCAAACTGCTGGTTTTCATTGATGATGCGGGACCAGCCATCAACGCCCACCACCGGAACGATGCCATT
----.----+----.----+----.----+----.----+----.----+----.----+----.----+----.----+----.----+----.----+  32600
TCCTAAGTAACAGGACGAGTTTCAGGTACGGTAGTTTGACGACCAAAAGTAACTACTACGCCCTGGTCGGTAGTTGCGGGTGGTGGCCTTGCTACGGTAA

  d  s  l  s  c  s  k  s  ■  p  s  n  c  w  f  s  l  ■  ■  r  d  q  p  s  t  p  t  t  g  t  ■  p  f
r  i  h  c  p  a  q  s  p  c  h  q  t  a  g  f  h  .  .  c  g  t  s  h  q  r  p  p  p  e  r  c  h  s
 g  f  i  v  l  l  k  v  h  a  i  k  l  l  v  f  i  d  d  a  g  p  a  i  n  a  h  h  r  n  d  a  i
----.----+----.----+----.----+----.----+----.----+----.----+----.----+----.----+----.----+----.----+
C  S  F  N  D  Q  F  F  D  M  G  D  F  Q  Q  N  E  N  I  I  R  S  W  G  D  V  G  V  V  P  V  I  G  N
 l  i  .  q  g  a  .  l  g  h  w  .  v  a  p  k  .  q  h  h  p  v  l  w  .  r  g  g  g  s  r  h  w
  p  n  ■  t  r  s  l  t  w  a  ■  l  s  s  t  k  ■  s  s  a  p  g  a  ■  l  a  w  w  r  f  s  a  ■

B267   32527       bio267             att-sr      EM
BEI     ...       CONTINUES IN PHASE FOUR
```

```
                   B                   H  R             M                     MH  H   S
                   7                   A  S             B                     SH  P   F
                   4                   E  A             O                     TA  H   A
                                       3  1             1                     11  1   N

CTGCTTATCAGGAAAGGCGTAAATTTCTTTCGTCCACGGATTAAGGCCGTACTGGTTGGCAACGATCAGTAATGCGATGAACTGCGCATCGCTGGCATCA
----.----+----.----+----.----+----.----+----.----+----.----+----.----+----.----+----.----+----.----+  32700
GACGAATAGTCCTTTCCGCATTTAAAGAAAGCAGGTGCCTAATTCCGGCATGACCAACCGTTGCTAGTCATTACGCTACTTGACGCGTAGCGACCGTAGT

 c  l  s  g  k  a  .  i  s  *  v  h  g  l  r  p  y  w  l  a  t  i  s  n  a  ■  n  c  a  s  l  a  s
  a  y  q  e  r  r  k  f  l  s  s  t  d  .  g  r  t  g  w  q  r  s  v  ■  r  .  t  a  h  r  w  h  h
l  l  i  r  k  g  v  n  f  f  r  p  r  i  k  a  v  l  v  g  n  d  q  .  c  d  e  l  r  i  a  g  i  t
----.----+----.----+----.----+----.----+----.----+----.----+----.----+----.----+----.----+----.----+
  Q  K  D  P  F  A  Y  I  E  K  T  W  P  N  L  G  Y  Q  N  A  V  I  L  L  A  I  F  Q  A  D  S  A  D
e  a  .  .  s  l  r  l  n  r  e  d  v  s  .  p  r  v  p  q  c  r  d  t  i  r  h  v  a  c  r  q  c  .
 r  s  i  l  f  p  t  f  k  k  r  g  r  i  l  a  t  s  t  p  l  s  .  y  h  s  s  s  r  ■  a  p  ■

B74    32624       bio74              att-sr      EM
BEI     ...       CONTINUES IN PHASE FOUR
```

```
  SA                    *BMM   H    *SHAT* H         A           A
  FH                    1CBB   P    1AICA1 N         F           L
  AA                    1LOO   H    2LNCQ2 F         L           U
  N3                    9112   1    012111 1         3           1
                         //          / ///
CCTTTAAATGCCGTCTGGCGAAGAGTGGTGATCAGTTCCTGTGGGTCGACAGAATCCATGCCGACACGTTCAGCCAGCTTCCCAGCCAGCGTTGCGAGTG
----.----+----.----+----.----+----.----+----.----+----.----+----.----+----.----+----.----+----.----+  32800
GGAAATTTACGGCAGACCGCTTCTCACCACTAGTCAAGGACACCCAGCTGTCTTAGGTACGGCTGTGCAAGTCGGTCGAAGGGTCGGTCGCAACGCTCAC

p  l  n  a  v  w  r  r  v  v  i  s  s  c  g  s  t  e  s  ■  p  t  r  s  a  s  f  p  a  s  v  a  s  a
 l  .  ■  p  s  g  e  e  w  .  s  v  p  v  g  r  q  n  p  c  r  h  v  q  p  a  s  q  p  a  l  r  v
  f  k  c  r  l  a  k  s  g  d  q  f  l  w  v  d  r  i  h  a  d  t  f  s  q  l  p  s  q  r  c  e  c
----.----+----.----+----.----+----.----+----.----+----.----+----.----+----.----+----.----+----.----+
  G  K  F  A  T  Q  R  L  T  T  I  L  E  Q  P  D  V  S  D  M  G  V  R  E  A  L  K  G  A  L  T  A  L
 r  .  i  g  d  p  s  s  h  h  d  t  g  t  p  r  c  f  g  h  r  c  t  .  g  a  e  w  g  a  n  r  t
v  k  l  h  r  r  a  f  l  p  s  .  n  r  h  t  s  l  i  w  a  s  v  n  l  w  s  g  l  w  r  q  s  h
```

```
*119   32729     Mapped site BCL1
*120   32745     Mapped site SAL1
*121   32747     Mapped site HIN2
BET    ...       CONTINUES IN PHASE FOUR
```

```
S RR H B    RG    H   M    E S          H        NR HH     F   H
C RS N E    DA    N   N    C C          P        LS PG     O   H
A UA F T    1M    F   L    R R          H        4A AU     K   A
1 11 3      5     1   1    2 1          1         1 22     1   1
   /
CAGTACTCATTCGTTTTATACCTCTGAATCAATATCAACCTGGTGGTGAGCAATGGTTTCAACCATGTACCGGATGTGTTCTGCCATGCGCTCCTGAAAC  32900
GTCATGAGTAAGCAAAATATGGAGACTTAGTTATAGTTGGACCACCACTCGTTACCAAAGTTGGTACATGGCCTACACAAGACGGTACGCGAGGACTTTG

 v  l  i  r  f  i  p  l  n  q  y  q  p  g  g  e  q  w  f  q  p  c  t  g  c  v  l  p  c  a  p  e  t
q  y  s  f  v  l  y  l  .  i  n  i  n  l  v  v  s  n  g  f  n  h  v  p  d  v  f  c  h  a  l  l  k  l
 s  t  h  s  f  y  t  s  e  s  i  s  t  w  w  .  a  m  v  s  t  m  y  r  m  c  s  a  m  r  s  .  n

A  T  S  M  r  k  i  g  r  f  .  y  .  g  p  p  s  c  h  n  .  g  h  v  p  h  t  r  g  h  a  g  s  v
 c  y  e  n  t  k  y  r  q  i  l  i  l  r  t  t  l  l  p  k  l  w  t  g  s  t  n  q  w  a  s  r  f
  l  v  .  e  n  .  V  E  S  D  I  D  V  Q  H  H  A  I  T  E  V  M  Y  R  I  H  E  A  M  R  F  Q  F
```

```
BET    32810     bet (orf-261)      M-NH3
RD15   32818     delta red15        dr        EM
GAM    32819     gam                COOH
NL4    32867     ninL4              dl        EM
```

```
      1              AH       S   M  C*BTA     H          T SHH
      4              SA       F   B  L11AV     N          T TAA
      5              UE       A   0  A21QA     F          H UEF
      3              13       N   1  12 13     1          2 113
                                     ////                     //
TCAACATCGTCATCAAACGCACGGGTAATGGATTTTTTGCTGGCCCCGTGGCGTTGCAAATGATCGATGCATAGCGATTCAAACAGGTGCTGGGGCAGGC  33000
AGTTGTAGCAGTAGTTTGCGTGCCCATTACCTAAAAAACGACCGGGGCACCGCAACGTTTACTAGCTACGTATCGCTAAGTTTGTCCACGACCCCGTCCG

 q  h  r  h  q  t  h  g  .  w  i  f  c  w  p  r  g  v  a  n  d  r  c  i  a  i  q  t  g  a  g  a  g
  n  i  v  i  k  r  t  g  n  g  f  f  a  g  p  v  a  l  q  m  i  d  a  .  r  f  k  q  v  l  g  q  a
s  t  s  s  s  n  a  r  v  m  d  f  l  l  a  p  w  r  c  k  .  s  m  h  s  d  s  n  r  c  w  g  r  p

 .  c  r  .  .  v  c  p  y  h  i  k  q  q  g  r  p  t  a  f  s  r  h  m  a  i  .  v  p  a  p  a  p
s  l  m  t  m  l  r  v  p  l  p  n  k  a  p  g  t  a  n  c  i  i  s  a  y  r  n  l  c  t  s  p  c  a
 E  V  D  D  D  F  A  R  T  I  S  K  K  S  A  G  H  R  Q  L  H  D  I  C  L  S  E  F  L  H  Q  P  L
```

```
1453   32915     b1453              att-dr    EM
*122   32964     Mapped site CLA1
B11    32964     bio11              att-sr    EM
GAM    ...       CONTINUES IN PHASE SIX
```

```
     B          M     DB M      AFE      T H   HA D        MBM      T
     I          B     IB N      LNC      H H   GL D        BIN      L
     0          0     FV L      UU1      A A   AU E        ONL      2
     1          2     F1 1      1H5      1 1   11 1        111      D
                                 /                          /
CTTTTTCCATGTCGTCTGCCAGTTCTGCCTCTTTCTCTTCACGGGCGAGCTGCTGGTAGTGACGCGCCCAGCTCTGAGCCTCAAGACGATCCTGAATGTA  33100
GAAAAAGGTACAGCAGACGGTCAAGACGGAGAAAGAGAAGTGCCCGCTCGACGACCATCACTGCGCGGGTCGAGACTCGGAGTTCTGCTAGGACTTACAT

l  f  p  c  r  l  p  v  l  p  l  s  l  h  g  r  a  a  g  s  d  a  p  s  s  e  p  q  d  d  p  e  c  n
 f  f  h  v  v  c  q  f  c  l  f  l  f  t  g  e  l  l  v  v  t  r  p  a  l  s  l  k  t  i  l  n  v
  f  s  m  s  s  a  s  s  a  s  f  s  s  r  a  s  c  w  .  .  r  a  q  l  .  a  s  r  r  s  .  m  .

 r  k  g  h  r  r  g  t  r  g  r  e  r  .  p  r  a  a  p  l  s  a  g  l  e  s  g  .  s  s  q  s  h
  k  k  w  t  t  q  w  n  q  r  k  r  k  v  p  s  s  s  t  t  v  r  g  a  r  l  r  l  v  i  r  f  t
G  K  E  M  D  D  A  L  E  A  E  K  E  E  R  A  L  Q  Q  Y  H  R  A  W  S  Q  A  E  L  R  D  Q  I  Y
```

```
BIO1   33012     bio1               att-sr    EM
DIFF   33035     difference                   sq     Ineichen et al.(1981) G(not C)
TL2D   33100     tL2d terminator    3'        RNA
GAM    ...       CONTINUES IN PHASE SIX
```

```
        N                     A         T       T       HNS                              KT
        L                     L         L       H       PCC                              IT
        3                     U         2       A       AIR                              LH
        Ø                     1         C       1       211                               2
                                                          /
ATAAGCGTTCATGGCTGAACTCCTGAAATAGCTGTGAAAATATCGCCCGCGAAATGCCGGGCTGATTAGGAAAACAGGAAAGGGGGTTAGTGAATGCTTT
----.----+----.----+----.----+----.----+----.----+----.----+----.----+----.----+----.----+----.----+  33200
TATTCGCAAGTACCGACTTGAGGACTTTATCGACACTTTTATAGCGGGCGCTTTACGGCCCGACTAATCCTTTTGTCCTTTCCCCCAATCACTTACGAAA

 k  r  s  w  l  n  s  .  n  s  c  e  n  i  a  r  e  m  p  g  .  l  g  k  q  e  r  g  l  v  n  a  f
i  s  v  h  g  .  t  p  e  i  a  v  k  i  s  p  a  k  c  r  a  d  .  e  n  r  k  g  g  .  .  m  l  l
 .  a  f  m  a  e  l  l  k  .  l  .  k  y  r  p  r  n  a  g  l  i  r  k  t  g  k  g  v  s  e  c  f
----.----+----.----+----.----+----.----+----.----+----.----+----.----+----.----+----.----+----.----+
l  l  r  e  h  s  f  e  q  f  l  q  s  f  i  a  r  s  i  g  p  q  n  p  f  c  s  l  p  n  t  f  a  k
 i  l  t  .  p  q  v  g  s  i  a  t  f  i  d  g  a  f  h  r  a  s  .  s  f  l  f  p  p  .  H  I  S
 Y  A  N  M  A  S  S  R  F  Y  S  H  F  Y  R  G  R  F  A  P  S  I  L  F  V  P  F  P  T  L  S  H  K

NL30   33109      ninL30              dl         EM
TL2C   33141      tL2c terminator     3'         RNA     RNA ends at this T and at next A
KIL    33190      kil                 COOH
GAM     ...       CONTINUES IN PHASE SIX


    M   D                     HG           *SHAT*
    B   D                     GA           1AICA1
    0   E                     AM           2LNCQ2
    1   1                     1            312114
                                            / ///
TGCTTGATCTCAGTTTCAGTATTAATATCCATTTTTTATAAGCGTCGACGGCTTCACGAAACATCTTTTCATCGCCAATAAAAGTGGCGATAGTGAATTT
----.----+----.----+----.----+----.----+----.----+----.----+----.----+----.----+----.----+----.----+  33300
ACGAACTAGAGTCAAAGTCATAATTATAGGTAAAAAATATTCGCAGCTGCCGAAGTGCTTTGTAGAAAAGTAGCGGTTATTTTCACCGCTATCACTTAAA

 a  .  s  q  f  q  y  .  y  p  f  f  i  s  v  d  g  f  t  k  h  l  f  i  a  n  k  s  g  d  s  e  f
  l  d  l  s  f  s  i  n  i  h  f  l  .  a  s  t  a  s  r  n  i  f  s  s  p  i  k  v  a  i  v  n  l
c  l  i  s  v  s  v  l  i  s  i  f  y  k  r  r  r  l  h  e  t  s  f  h  r  q  .  k  w  r  .  .  i  .
----.----+----.----+----.----+----.----+----.----+----.----+----.----+----.----+----.----+----.----+
 a  q  d  .  n  .  y  .  y  g  n  k  i  l  t  s  p  k  v  f  c  r  k  m  a  l  l  l  p  s  l  s  n
K  S  S  R  L  K  L  I  L  I  W  K  K  Y  A  D  V  A  E  R  F  M  K  E  D  G  I  F  T  A  I  T  F  K
 Q  K  I  E  T  E  T  N  I  D  M  k  .  l  r  r  r  s  .  s  v  d  k  .  r  w  y  f  h  r  y  h  i

GAM    33232      gam (orf-138)       M-NH3      G
*123   33244      Mapped site SAL1
*124   33246      Mapped site HIN2
KIL     ...       CONTINUES IN PHASE FIVE


 CB1N   T           MB       K     H  E  S                 TF                                      HH
 I23L   T           BI       I     N  C  C                 AO                                      NN
 I716   H           ON       L     F  R  R                 QK                                      FF
 I593   2           11             1  2  1                 11                                      13
   //                 /                                                                             /
AGTCTGGATAGCCATAAGTGTTTGATCCATTCTTTGGGACTCCTGGCTGATTAAGTATGTCGATAAGGCGTTTCCATCCGTCACGTAATTTACGGGTGAT
----.----+----.----+----.----+----.----+----.----+----.----+----.----+----.----+----.----+----.----+  33400
TCAGACCTATCGGTATTCACAAACTAGGTAAGAAACCCTGAGGACCGACTAATTCATACAGCTATTCCGCAAAGGTAGGCAGTGCATTAAATGCCCACTA

s  l  d  s  h  k  c  l  i  h  s  l  g  l  l  a  d  .  v  c  r  .  g  v  s  i  r  h  v  i  y  g  .  f
 v  w  i  a  i  s  v  .  s  i  l  w  d  s  w  l  i  k  y  v  d  k  a  f  p  s  v  t  .  f  t  g  d
  s  g  .  p  .  v  f  d  p  f  f  g  t  p  g  .  l  s  m  s  i  r  r  f  h  p  s  r  n  l  r  v  i
----.----+----.----+----.----+----.----+----.----+----.----+----.----+----.----+----.----+----.----+
 l  r  s  l  w  l  h  k  i  w  e  k  p  s  r  a  s  .  t  h  r  y  p  t  e  m  r  .  t  i  .  p  h
  T  Q  I  A  M  L  T  Q  D  M  r  q  s  e  q  s  i  l  y  t  s  l  a  n  g  d  t  v  y  n  v  p  s
.  D  P  Y  G  Y  T  N  S  G  N  K  P  V  G  P  Q  N  L  I  D  I  L  R  K  W  G  D  R  L  K  R  T  I

CIII   33302      cIII                COOH
B275   33303      bio275              att-sr     EM
1319   33303      b1319               att-dr     EM
NL63   33303      ninL63              dl         EM
KIL    33330      orf-47 (kil?)       M-NH3              protein not identified
```

```
  H     HH       F        B HHC     NNB        C S   H               T B**XAT
  P     NN       N        B AA6     LL2        I D   G               L 211HVA
  H     FF       U        V EE1     245        I U   I               2 5220AQ
  1     13       H        1 131     042        I 1   A               B 056111
         /                   /       //                                  ////
TCGTTCAAGTAAAGATTCGGAAGGGCAGCCAGCAACAGGCCACCCTGCAATGGCATATTGCATGGTGTGCTCCTTATTTATACATAACGAAAAACGCCTC
----.----+----.----+----.----+----.----+----.----+----.----+----.----+----.----+----.----+----.----+ 33500
AGCAAGTTCATTTCTAAGCCTTCCCGTCGGTCGTTGTCCGGTGGGACGTTACCGTATAACGTACCACACGAGGAATAAATATGTATTGCTTTTTGCGGAG

 v  q  v  k  i  r  k  g  s  q  q  q  a  t  l  q  w  h  i  a  w  c  a  p  y  l  y  i  t  k  n  a  s
s  f  k  .  r  f  g  r  a  a  s  n  r  p  p  c  n  g  i  l  h  g  v  l  l  i  y  t  .  r  k  t  p  r
 r  s  s  k  d  s  e  g  q  p  a  t  g  h  p  a  m  a  y  c  m  v  c  s  l  f  i  h  n  e  k  r  l
----.----+----.----+----.----+----.----+----.----+----.----+----.----+----.----+----.----+----.----+
n  t  .  t  f  i  r  f  p  l  w  c  c  a  v  r  c  h  c  i  a  h  h  a  g  .  k  y  m  v  f  f  a  e
 e  n  l  y  l  n  p  l  a  a  l  l  l  g  g  q  l  p  m  n  c  p  t  s  r  i  .  v  y  r  f  v  g
  R  E  L  L  S  E  S  P  C  G  A  V  P  W  G  A  I  A  Y  Q  M  t  h  e  k  n  i  c  l  s  f  r  r
```

```
C611   33441   cIIIam611             pm        sq    C to T
NL20   33449   ninL20                dl        EM
NL44   33449   ninL44                dl        EM
B252   33449   bio252                att-sr    EM
CIII   33463   cIII (orf-54)         M-NH3     G
TL2B   33494   tL2b terminator       3'        RNA
B250   33497   bio250                att-sr    EM
*125   33498   Mapped site AVA1
*126   33498   Mapped site XHO1
```

```
   M                          D  E F H V                H                 C*T  E
   N                          D  A N A 2                N                 L1A  C
   L                          E  1 U E 0                F                 A2Q  R
   1                          1  0 H 3 3                1                 171  V
                                                                           //
GAGTGAAGCGTTATTGGTATGCGGTAAAACCGCACTCAGGCGGCCTTGATAGTCATATCATCTGAATCAAATATTCCTGATGTATCGATATCGGTAATTC
----.----+----.----+----.----+----.----+----.----+----.----+----.----+----.----+----.----+----.----+ 33600
CTCACTTCGCAATAACCATACGCCATTTTGGCGTGAGTCCGCCGGAACTATCAGTATAGTAGACTTAGTTTATAAGGACTACATAGCTATAGCCATTAAG

  s  e  a  l  l  v  c  g  k  t  a  l  r  r  p  .  .  s  y  h  l  n  q  i  f  l  m  y  r  y  r  .  f
 v  k  r  y  w  y  a  v  k  p  h  s  g  g  l  d  s  h  i  i  .  i  k  y  s  .  c  i  d  i  g  n  s
e  .  s  v  i  g  m  r  .  n  r  t  q  a  a  l  i  v  i  s  s  e  s  n  i  p  d  v  s  i  s  v  i  l
----.----+----.----+----.----+----.----+----.----+----.----+----.----+----.----+----.----+----.----+
 l  s  a  n  n  t  h  p  l  v  a  s  l  r  g  q  y  d  y  .  r  f  .  i  n  r  i  y  r  y  r  y  n
r  t  f  r  .  q  y  a  t  f  g  c  e  p  p  r  s  l  .  i  m  q  i  l  y  e  q  h  i  s  i  p  l  e
 s  h  l  t  i  p  i  r  y  f  r  v  .  A  A  K  I  T  M  D  D  S  D  F  I  G  S  T  D  I  D  T  I
```

```
EA10   33539   Ea10               COOH
V203   33546   dv203              plasmid-1   EM
*127   33585   Mapped site CLA1
```

```
 F         FM      HH                        T               NA
 O         ON      AA                        A               DV
 K         KL      EE                        Q               EA
 1         11      13                        1               13
                    /
TTATTCCTTCGCTACCATCCATTGGAGGCCATCCTTCCTGACCATTTCCATCATTCCAGTCGAACTCACACACAACACCATATGCATTTAAGTCGCTTGA
----.----+----.----+----.----+----.----+----.----+----.----+----.----+----.----+----.----+----.----+ 33700
AATAAGGAAGCGATGGTAGGTAACCTCCGGTAGGAAGGACTGGTAAAGGTAGTAAGGTCAGCTTGAGTGTGTGTTGTGGTATACGTAAATTCAGCGAACT

l  f  l  r  y  h  p  l  e  a  i  l  p  d  h  f  h  h  s  s  r  t  h  t  q  h  h  m  h  l  s  r  l  k
 y  s  f  a  t  i  h  w  r  p  s  f  l  t  i  s  i  i  p  v  e  l  t  h  n  t  i  c  i  .  v  a  .
  i  p  s  l  p  s  i  g  g  h  p  s  .  p  f  p  s  f  q  s  n  s  h  t  t  p  y  a  f  k  s  l  e
----.----+----.----+----.----+----.----+----.----+----.----+----.----+----.----+----.----+----.----+
k  n  r  r  .  w  g  n  s  a  m  r  g  s  w  k  w  .  e  l  r  v  .  v  c  c  w  i  c  k  l  r  k
 .  e  k  a  v  m  w  q  l  g  d  k  r  v  m  e  m  m  g  t  s  s  v  c  l  v  m  h  m  .  t  a  q
R  I  G  E  S  G  D  M  P  P  W  G  E  Q  G  N  G  D  N  W  D  F  E  C  V  V  G  Y  A  N  L  D  S  S
```

EA10 ... CONTINUES IN PHASE SIX

```
                     N  H           N                                   H         H
                     S  H           L                                   N         N
                     P  A           8                                   F         F
                     C  1                                               1         1

AATTGCTATAAGCAGAGCATGTTGCGCCAGCATGATTAATACAGCATTTAATACAGAGCCGTGTTTATTGAGTCGGTATTCAGAGTCTGACCAGAAATTA
----.----+----.----+----.----+----.----+----.----+----.----+----.----+----.----+----.----+----.----+ 33800
TTAACGATATTCGTCTCGTACAACGCGGTCGTACTAATTATGTCGTAAATTATGTCTCGGCACAAATAACTCAGCCATAAGTCTCAGACTGGTCTTTAAT

  l  l  .  a  e  h  v  a  p  a  .  l  i  q  h  l  i  q  s  r  v  y  .  v  g  i  q  s  l  t  r  n  y
 n  c  y  k  q  s  m  l  r  q  h  d  .  y  s  i  .  y  r  a  v  f  i  e  s  v  f  r  v  .  p  e  i  i
i  a  i  s  r  a  c  c  a  s  m  i  n  t  a  f  n  t  e  p  c  l  l  s  r  y  s  e  s  d  q  k  l
----.----+----.----+----.----+----.----+----.----+----.----+----.----+----.----+----.----+----.----+
f  n  s  y  a  s  c  t  a  g  a  h  n  i  c  c  k  i  c  l  r  t  .  q  t  p  i  .  l  r  v  l  f  .
 f  q  .  l  c  l  m  n  r  w  c  s  .  y  l  m  .  y  l  a  t  n  i  s  d  t  n  l  t  q  g  s  i
  I  A  I  L  L  A  H  Q  A  L  M  I  L  V  A  N  L  V  S  G  H  K  N  L  R  Y  E  S  D  S  W  F  N

NL8    33739      ninL8            dl          EM
EA10   ...      CONTINUES IN PHASE SIX


                 X   H       M       T     S                                   NT
                 M   P       N       A     F                                   LT
                 N   H       L       Q     A                                   8H
                 1   1       1       1     N                                   82
                                                                               /
TTAATCTGGTGAAGTTTTTCCTCTGTCATTACGTCATGGTCGATTTCAATTTCTATTGATGCTTTCCAGTCGTAATCAATGATGTATTTTTTGATGTTTG
----.----+----.----+----.----+----.----+----.----+----.----+----.----+----.----+----.----+----.----+ 33900
AATTAGACCACTTCAAAAAGGAGACAGTAATGCAGTACCAGCTAAAGTTAAAGATAACTACGAAAGGTCAGCATTAGTTACTACATAAAAAACTACAAAC

  .  s  g  e  v  f  p  l  s  l  r  h  g  r  f  q  f  l  l  m  l  s  s  r  n  q  .  c  i  f  .  c  l
 n  l  v  k  f  f  l  c  h  y  v  m  v  d  f  n  f  y  .  c  f  p  v  v  i  n  d  v  f  f  d  v  .
l  i  w  .  s  f  s  s  v  i  t  s  w  s  i  s  i  s  i  d  a  f  q  s  .  s  m  m  y  f  l  m  f  d
----.----+----.----+----.----+----.----+----.----+----.----+----.----+----.----+----.----+----.----+
 .  d  p  s  t  k  g  r  d  n  r  .  p  r  n  .  n  r  n  i  s  e  l  r  l  .  h  h  i  k  q  h  k
i  l  r  t  f  n  k  r  q  .  .  t  m  t  s  k  l  k  .  q  h  k  g  t  t  i  l  s  t  n  k  s  t  q
  N  I  Q  H  L  K  E  E  T  M  V  D  H  D  I  E  I  E  I  S  A  K  W  D  Y  D  I  I  Y  K  K  I  N

NL88   33885      ninL88           dl          EM
EA10   ...      CONTINUES IN PHASE SIX


   E                 M   T     M     M                   M
   A                 N   L     N     N                   B
   1                 L   2     L     L                   0
   0                 1   A     1     1                   2

ACATCTGTTCATATCCTCACAGATAAAAAATCGCCCTCACACTGGAGGGCAAAGAAGATTTCCAATAATCAGAACAAGTCGGCTCCTGTTTAGTTACGAG
----.----+----.----+----.----+----.----+----.----+----.----+----.----+----.----+----.----+----.----+ 34000
TGTAGACAAGTATAGGAGTGTCTATTTTTTAGCGGGAGTGTGACCTCCCGTTTCTTCTAAAGGTTATTAGTCTTGTTCAGCCGAGGACAAATCAATGCTC

t  s  v  h  i  l  t  d  k  k  s  p  s  h  w  r  a  k  k  i  s  n  n  q  n  k  s  a  p  v  .  l  r  a
 h  l  f  i  s  s  q  i  k  n  r  p  h  t  g  g  q  r  r  f  p  i  i  r  t  s  r  l  l  f  s  y  e
  i  c  s  y  p  h  r  .  k  i  a  l  t  l  e  g  k  e  d  f  q  .  s  e  q  v  g  s  c  l  v  t  s
----.----+----.----+----.----+----.----+----.----+----.----+----.----+----.----+----.----+----.----+
v  d  t  .  i  r  v  s  l  f  d  g  e  c  q  l  a  f  f  i  e  l  l  .  f  l  d  a  g  t  .  n  r
 c  r  n  m  d  e  c  i  f  f  r  g  .  v  p  p  c  l  l  n  g  i  i  l  v  l  r  s  r  n  l  .  s
S  M  q  e  y  g  .  l  y  f  i  a  r  v  s  s  p  l  s  s  k  w  y  d  s  c  t  p  e  q  k  t  v  l

EA10   33904      Ea10 (orf-122)   M-NH3
TL2A   33930      tL2a terminator  3'          RNA


                                                        B                       HH       R F A
                                                        2                       AA       A N L
                                                        1                       EE       L U U
                                                        4                       13         H 1
                                                                                /
CGACATTGCTCCGTGTATTCACTCGTTGGAATGAATACACAGTGCAGTGTTTATTCTGTTATTTATGCCAAAAATAAAGGCCACTATCAGGCAGCTTTGT
----.----+----.----+----.----+----.----+----.----+----.----+----.----+----.----+----.----+----.----+ 34100
GCTGTAACGAGGCACATAAGTGAGCAACCTTACTTATGTGTCACGTCACAAATAAGACAATAAATACGGTTTTTATTTCCGGTGATAGTCCGTCGAAACA

  t  l  l  r  v  f  t  r  w  n  e  y  t  v  q  c  l  f  c  y  l  c  q  k  .  r  p  l  s  g  s  f  v
 r  h  c  s  v  y  s  l  v  g  m  n  t  q  c  s  v  y  s  v  i  y  a  k  n  k  g  h  y  q  a  a  l  l
  d  i  a  p  c  i  h  s  l  e  .  i  h  s  a  v  f  i  l  l  f  m  p  k  i  k  a  t  i  r  q  l  c
----.----+----.----+----.----+----.----+----.----+----.----+----.----+----.----+----.----+----.----+
a  v  n  s  r  t  n  v  r  q  f  s  y  v  t  c  h  k  n  q  .  k  h  w  f  y  l  g  s  d  p  l  k  t
 r  c  q  e  t  y  e  s  t  p  i  f  v  c  h  l  t  .  e  t  i  .  a  l  f  l  p  w  .  .  A  A  K
  s  m  a  g  h  i  .  e  n  s  h  i  c  l  a  t  n  i  r  n  n  i  g  f  i  f  a  v  i  l  c  s  q
```

```
B214   34055      bio214              att-sr       EM
RAL    34090      ral                 COOH
```

```
 BT                   I                        T     M             M  X
 BP                   M                        A     B             B  M
 V4                   M                        Q     0             0  N
 14                   L                        1     2             2  1

TGTTCTGTTTACCAAGTTCTCTGGCAATCATTGCCGTCGTTCGTATTGCCCATTTATCGACATATTTCCCATCTTCCATTACAGGAAACATTTCTTCAGG
----.----+----.----+----.----+----.----+----.----+----.----+----.----+----.----+----.----+----.----+ 34200
ACAAGACAAATGGTTCAAGAGACCGTTAGTAACGGCAGCAAGCATAACGGGTAAATAGCTGTATAAAGGGTAGAAGGTAATGTCCTTTGTAAAGAAGTCC

 v  l  f  t  k  f  s  g  n  h  c  r  r  s  y  c  p  f  i  d  i  f  p  i  f  h  y  r  k  h  f  f  r
  f  c  l  p  s  s  l  a  i  i  a  v  v  r  i  a  h  l  s  t  y  f  p  s  s  i  t  g  n  i  s  s  g
c  s  v  y  q  v  l  w  q  s  l  p  s  f  v  l  p  i  y  r  h  i  s  h  l  p  l  q  e  t  f  l  q  a
----.----+----.----+----.----+----.----+----.----+----.----+----.----+----.----+----.----+----.----+
  t  r  n  v  l  n  e  p  l  .  q  r  r  e  y  q  g  n  i  s  m  n  g  m  k  w  .  l  f  c  k  k  l
N  N  Q  K  G  L  E  R  A  I  M  A  T  T  R  I  A  W  K  D  V  Y  K  G  D  E  M  V  P  F  M  E  E  P
 q  e  t  .  w  t  r  q  c  d  n  g  d  n  t  n  g  m  .  r  c  i  e  w  r  g  n  c  s  v  n  r  .

TP44   34104      trp44               dr           sq
IMML   34127      imml                sl           EM
RAL    ...        CONTINUES IN PHASE FIVE
```

```
    A     F     F AK     B   S      S  B                          R      E S
    V     0     N LA     B   F      F  1                          A      C C
    A     K     U UN     V   A      A  0                          L      R R
    3     1     H 11     1   N      N                                    2 1
                   /
CTTAACCATGCATTCCGATTGCAGCTTGCATCCATTGCATCGCTTGAATTGTCCACACCATTGATTTTTATCAATAGTCGTAGTCATACGGATAGTCCTG
----.----+----.----+----.----+----.----+----.----+----.----+----.----+----.----+----.----+----.----+ 34300
GAATTGGTACGTAAGGCTAACGTCGAACGTAGGTAACGTAGCGAACTTAACAGGTGTGGTAACTAAAAATAGTTATCAGCATCAGTATGCCTATCAGGAC

l  n  h  a  f  r  l  q  l  a  s  i  a  s  l  e  l  s  t  p  l  i  f  i  n  s  r  s  h  t  d  s  p  g
 l  t  m  h  s  d  c  s  l  h  p  l  h  r  l  n  c  p  h  h  .  f  l  s  i  v  v  v  i  r  i  v  l
  .  p  c  i  p  i  a  a  c  i  h  c  i  a  .  i  v  h  t  i  d  f  y  q  .  s  .  s  y  g  .  s  w
----.----+----.----+----.----+----.----+----.----+----.----+----.----+----.----+----.----+----.----+
 s  l  w  a  n  r  n  c  s  a  d  m  a  d  s  s  n  d  v  g  n  i  k  i  l  l  r  l  .  v  s  l  g
  K  V  M  C  E  S  Q  L  K  C  G  N  C  R  K  F  Q  G  C  W  Q  N  K  D  I  T  T  T  M  r  i  t  r
a  .  g  h  m  g  i  a  a  q  m  w  q  m  a  q  i  t  w  v  m  s  k  .  .  y  d  y  d  y  p  y  d  q

KAN1   34224      kan 1               i(Tn5)       EM
B10    34249      bio10               att-sr       EM
RAL    34287      ral (orf-66)        M-NH3
```

```
F        MS   M*DHM    M AT  F  M  M           H             R      I
0        NF   S1DGB    B SA  0  B  B           P             E      2
K        LA   T2EU0    0 UQ  K  0  0           H             V      1
1        1N   28122    2 21  1  2  2           1
              // /       /
GTATTGTTCCATCACATCCTGAGGATGCTCTTCGAACTCTTCAAATTCTTCTTCCATATATCACCTTAAATAGTGGATTGCGGTAGTAAAGATTGTGCCT
----.----+----.----+----.----+----.----+----.----+----.----+----.----+----.----+----.----+----.----+ 34400
CATAACAAGGTAGTGTAGGACTCCTACGAGAAGCTTGAGAAGTTTAAGAAGAAGGTATATAGTGGAATTTATCACCTAACGCCATCATTTCTAACACGGA

  i  v  p  s  h  p  e  d  a  l  r  t  l  q  i  l  l  p  y  i  t  l  n  s  g  l  r  .  .  r  l  c  l
v  l  f  h  h  i  l  r  m  l  f  e  l  f  k  f  f  f  h  i  s  p  .  i  v  d  c  g  s  k  d  c  a  c
 y  c  s  i  t  s  .  g  c  s  s  n  s  s  n  s  s  s  i  y  h  l  k  .  w  i  a  v  v  k  i  v  p
----.----+----.----+----.----+----.----+----.----+----.----+----.----+----.----+----.----+----.----+
p  i  t  g  d  c  g  s  s  a  r  r  v  r  .  i  r  r  g  y  i  v  k  f  l  p  n  r  y  y  l  n  h  r
 t  n  n  w  .  m  r  l  i  s  k  s  s  k  l  n  k  k  w  i  d  g  .  i  t  s  q  p  l  l  s  q  a
  y  q  e  m  v  d  q  p  h  e  e  f  e  e  f  e  e  e  m  y  .  r  l  y  h  i  a  t  t  f  i  t  g

*128   34319      Mapped site MST2
REV    34370       lambdarev          sr           EM
I21    34379       imm21              sl           sq
```

```
                                   R                                                            GB*BXMB
                                   S                                                            II1AHBT
                                   A                                                            TN2MOON
                                   1                                                             191211
                                                                                                  ////
GTCTTTTAACCACATCAGGCTCGGTGGTTCTCGTGTACCCCTACAGCGAGAAATCGGATAAACTATTACAACCCCTACAGTTTGATGAGTATAGAAATGG
----.----+----.----+----.----+----.----+----.----+----.----+----.----+----.----+----.----+----.----+ 34500
CAGAAAATTGGTGTAGTCCGAGCCACCAAGAGCACATGGGGATGTCGCTCTTTAGCCTATTTGATAATGTTGGGGATGTCAAACTACTCATATCTTTACC

 s  f  n  h  i  r  l  g  g  s  r  v  p  l  q  r  e  i  g  .  t  i  t  t  p  t  v  .  .  v  .  k  w
  l  l  t  t  s  g  s  v  v  l  v  y  p  y  s  e  k  s  d  k  l  l  q  p  l  q  f  d  e  y  r  n  g
v  f  .  p  h  q  a  r  w  f  s  c  t  p  t  a  r  n  r  i  n  y  y  n  p  y  s  l  m  s  i  e  M  D
----.----+----.----+----.----+----.----+----.----+----.----+----.----+----.----+----.----+----.----+
  d  k  l  w  m  l  s  p  p  e  r  t  g  r  c  r  s  i  p  y  v  i  v  v  g  v  t  q  h  t  y  f  h
q  r  k  v  v  d  p  e  t  t  r  t  y  g  .  l  s  f  d  s  l  s  n  c  g  r  c  n  s  s  y  l  f  p
 t  k  .  g  c  .  a  r  h  n  e  h  v  g  v  a  l  f  r  i  f  .  .  l  g  .  l  k  i  l  i  s  i

GII    34497       git  sieB          M-NH3
*129   34499      Mapped site BAM1


                                                    M          TM
                                                    N          LB
                                                    L          10
                                                    1           1

ATCCACTCGTTATTCTCGGACGAGTGTTCAGTAATGAACCTCTGGAGAGAACCATGTATATGATCGTTATCTGGGTTGGACTTCTGCTTTTAAGCCCAGA
----.----+----.----+----.----+----.----+----.----+----.----+----.----+----.----+----.----+----.----+ 34600
TAGGTGAGCAATAAGAGCCTGCTCACAAGTCATTACTTGGAGACCTCTCTTGGTACATATACTAGCAATAGACCCAACCTGAAGACGAAAATTCGGGTCT

i  h  s  l  f  s  d  e  c  s  v  m  n  l  w  r  e  p  c  i  .  s  l  s  g  l  d  f  c  f  .  a  q  i
 s  t  r  y  s  r  t  s  v  q  .  .  t  s  g  e  n  h  v  y  d  r  y  l  g  w  t  s  a  f  k  p  r
  P  L  V  I  L  G  R  V  F  S  N  E  P  L  F  R  T  M  Y  M  I  V  I  W  V  G  L  L  L  L  S  P  D
----.----+----.----+----.----+----.----+----.----+----.----+----.----+----.----+----.----+----.----+
 i  w  e  n  n  e  s  s  h  e  t  i  f  r  q  l  s  g  h  i  h  d  n  d  p  n  s  k  q  k  .  a  w
  d  v  r  .  e  r  v  l  t  .  y  h  v  e  p  s  f  w  t  y  s  r  .  r  p  q  v  e  a  k  l  g  l
s  g  s  t  i  r  p  r  t  n  l  l  s  g  r  s  l  v  m  y  i  i  t  i  q  t  p  s  r  s  k  l  g  s

TL1    34560       tL1 terminator      3'          RNA
GII     ...       CONTINUES IN PHASE THREE


  B  HH                    H           N       M    N       B                                   C*T
  S  AA                    N           L       N    S       2                                   L1A
  T  EE                    F           9       L    P       4                                   A3Q
  X  13                    1           9       1    C       3                                   101
      /                                                                                          //
TAACTGGCCTGAATATGTTAATGAGAGAATCGGTATTCCTCATGTGTGGCATGTTTTCGTCTTTGCTCTTGCATTTTCGCTAGCAATTAATGTGCATCGA
----.----+----.----+----.----+----.----+----.----+----.----+----.----+----.----+----.----+----.----+ 34700
ATTGACCGGACTTATACAATTACTCTCTTAGCCATAAGGAGTACACACCGTACAAAAGCAGAAACGAGAACGTAAAAGCGATCGTTAATTACACGTAGCT

  t  g  l  n  m  l  m  r  e  s  v  f  l  m  c  g  m  f  s  s  l  l  l  h  f  r  .  q  l  m  c  i  d
.  l  a  .  i  c  .  .  e  n  r  y  s  s  c  v  a  c  f  r  l  c  s  c  i  f  a  s  n  .  c  a  s  i
 N  W  P  E  Y  V  N  E  R  I  G  I  P  H  V  W  H  V  F  V  F  A  L  A  F  S  L  A  I  N  V  H  R
----.----+----.----+----.----+----.----+----.----+----.----+----.----+----.----+----.----+----.----+
i  v  p  r  f  i  n  i  l  s  d  t  n  r  m  h  p  m  n  e  d  k  s  k  c  k  r  .  c  n  i  h  m  s
 y  s  a  q  i  h  .  h  s  f  r  y  e  e  h  t  a  h  k  r  r  q  e  q  m  k  a  l  l  .  h  a  d
  l  q  q  s  y  t  l  s  l  i  p  i  g  .  t  h  c  t  k  t  k  a  r  a  n  e  s  a  i  l  t  c  r

NL99   34640       ninL99              dl         sq      N independent to the left
B243   34661       bio243              att-sr     FM
*130   34697      Mapped site CLA1
GII     ...       CONTINUES IN PHASE THREE
```

```
 S  A          HH             AD    S     BT  E       M
 F  L          HA             LD    F     26  C       B
 A  U          AE             UE    A     30  0       0
 N  1          12             11    N     3   K       1
                                           /
TTATCAGCTATTGCCAGCGCCAGATATAAGCGATTTAAGCTAAGAAAACGCATTAAGATGCAAAACGATAAAGTGCGATCAGTAATTCAAAACCTTACAG
----.----+----.----+----.----+----.----+----.----+----.----+----.----+----.----+----.----+----.----+ 34800
AATAGTCGATAACGGTCGCGGTCTATATTCGCTAAATTCGATTCTTTTGCGTAATTCTACGTTTTGCTATTTCACGCTAGTCATTAAGTTTTGGAATGTC

 y  q  l  l  p  a  p  d  i  s  d  l  s  .  e  n  a  l  r  c  k  t  i  k  c  d  q  .  f  k  t  l  q
  i  s  y  c  q  r  q  i  .  a  i  .  a  k  k  t  h  .  d  a  k  r  .  s  a  i  s  n  s  k  p  y  r
L  S  A  I  A  S  A  R  Y  K  R  F  K  L  R  K  R  I  K  M  Q  N  D  K  V  R  S  V  I  Q  N  L  T  E
----.----+----.----+----.----+----.----+----.----+----.----+----.----+----.----+----.----+----.----+
  .  .  s  n  g  a  g  s  i  l  s  k  l  .  s  f  a  n  l  h  l  v  i  f  h  s  .  y  n  l  v  k  c
i  i  l  .  q  w  r  w  i  y  a  i  .  a  l  f  v  c  .  s  a  f  r  y  l  a  i  l  l  e  f  g  .  l
 n  d  a  i  a  l  a  l  y  l  r  n  l  s  l  f  r  m  l  i  c  f  s  l  t  r  d  t  i  .  f  r  v

B233  34758      bio233          att-sr     EM
I60   34758      trp60-3         sr         EM
GII    ...       CONTINUES IN PHASE THREE
```

```
       M        MHF        B          BN                           A
       B        SHN        B          2L                           L
       0        TAU        V          36                           U
       2        11H        1          23                           1
                                       /
AAGAGCAATCTATGGTTTTGTGCGCAGCCCTTAATGAAGGCAGGAAGTATGTGGTTACATCAAAACAATTCCCATACATTAGTGAGTTGATTGAGCTTGG
----.----+----.----+----.----+----.----+----.----+----.----+----.----+----.----+----.----+----.----+ 34900
TTCTCGTTAGATACCAAAACACGCGTCGGGAATTACTTCCGTCCTTCATACACCAATGTAGTTTTGTTAAGGGTATGTAATCACTCAACTAACTCGAACC

k  s  n  l  w  f  c  a  q  p  l  m  k  a  g  s  m  w  l  h  q  n  n  s  h  t  l  v  s  .  l  s  l  v
 r  a  i  y  g  f  v  r  s  p  .  .  r  q  e  v  c  g  y  i  k  t  i  p  i  h  .  .  v  d  .  a  w
  E  Q  S  M  V  L  C  A  A  L  N  E  G  R  K  Y  V  V  T  S  K  Q  F  P  Y  I  S  E  L  I  E  L  G
----.----+----.----+----.----+----.----+----.----+----.----+----.----+----.----+----.----+----.----+
 f  l  l  r  h  n  q  a  c  g  k  i  f  a  p  l  i  h  n  c  .  f  l  e  w  v  n  t  l  q  n  l  k
  l  a  i  .  p  k  t  r  l  g  .  h  l  c  s  t  h  p  .  m  l  v  i  g  m  c  .  h  t  s  q  a  q
s  s  c  d  i  t  k  h  a  a  r  l  s  p  l  f  y  t  t  v  d  f  c  n  g  y  m  l  s  n  i  s  s  p

B232  34855      bio232          att-sr     EM
NL63  34855      ninL63          dr         EM
GII    ...       CONTINUES IN PHASE THREE
```

```
                                                M  N               T                A  MB
                                                N  L               P                L  BI
                                                L  4               4                U  ON
                                                1                  8                1  11
                                                                                        /
TGTGTTGAACAAAACTTTTTCCCGATGGAATGGAAAGCATATATTATTCCCTATTGAGGATATTTACTGGACTGAATTAGTTGCCAGCTATGATCCATAT
----.----+----.----+----.----+----.----+----.----+----.----+----.----+----.----+----.----+----.----+ 35000
ACACAACTTGTTTTGAAAAAGGGCTACCTTACCTTTCGTATATAATAAGGGATAACTCCTATAAATGACCTGACTTAATCAACGGTCGATACTAGGTATA

  c  .  t  k  l  f  p  d  g  m  e  s  i  y  y  s  l  l  r  i  f  t  g  l  n  .  l  p  a  m  i  h  i
 c  v  e  q  n  f  f  p  m  e  w  k  a  y  i  i  p  y  .  g  y  l  l  d  .  i  s  c  q  l  .  s  i  .
  V  L  N  K  T  F  S  R  W  N  G  K  H  I  L  F  P  I  E  D  I  Y  W  T  E  L  V  A  S  Y  D  P  Y
----.----+----.----+----.----+----.----+----.----+----.----+----.----+----.----+----.----+----.----+
 t  h  q  v  f  s  k  g  s  p  i  s  l  m  y  .  e  r  n  l  i  n  v  p  s  f  .  n  g  a  i  i  w  i
  h  t  s  c  f  k  k  g  i  s  h  f  a  y  i  i  g  .  q  p  y  k  s  s  q  i  l  q  w  s  h  d  m
  t  n  f  l  v  k  e  r  h  f  p  f  c  i  n  n  g  i  s  s  i  .  q  v  s  n  t  a  l  .  s  g  y

NL4   34952      ninL4           dr         EM
TP48  34969      trp48           sl         sq
GII    ...       CONTINUES IN PHASF THREE
```

```
                   HH       D   G   M  N       HN C*T      N H          A            N     F T  V BQE
                   AA       D   I   N          NL L1A      L G          C            S     N H  1 21C
                   EE       E   T   L          F4 A3Q      9 A          Y            P     U A  5 51R
                   13       1       1          14 111      9 1          1            C     H 1  4 652
                    /                             //                                                //
AATATTGAGATAAAGCCAAGGCCAATATCTAAGTAACTAGATAAGAGGAATCGATTTTCCCTTAATTTTCTGGCGTCCACTGCATGTTATGCCGCGTTCG
----.----+----.----+----.----+----.----+----.----+----.----+----.----+----.----+----.----+----.----+  35100
TTATAACTCTATTTCGGTTCCGGTTATAGATTCATTGATCTATTCTCCTTAGCTAAAAGGGAATTAAAAGACCGCAGGTGACGTACAATACGGCGCAAGC

 i  l  r  .  s  q  g  q  y  l  s  n  .  i  r  g  i  d  f  p  l  i  f  w  r  p  l  h  v  m  p  r  s
  y  .  d  k  a  k  a  n  i  .  v  t  r  .  e  e  s  i  f  p  .  f  s  g  v  h  c  m  l  c  r  v  r
N  I  E  I  K  P  R  P  I  S  K  .  l  d  k  r  n  r  f  s  l  n  f  l  a  s  t  a  c  y  a  a  f  a
----.----+----.----+----.----+----.----+----.----+----.----+----.----+----.----+----.----+----.----+
 i  n  l  y  l  w  p  w  y  r  l  l  .  I  L  P  I  S  K  G  K  I  K  Q  R  G  S  C  T  I  G  R  F
y  y  q  s  l  a  l  a  l  i  .  t  v  l  y  s  s  d  i  k  g  .  n  e  p  t  w  q  m  n  h  r  t  r
 l  i  s  i  f  g  l  g  i  d  l  y  s  s  l  l  f  r  n  e  r  l  k  r  a  d  v  a  h  .  a  a  n
```

GII	35033	git sieB	COOH		
N	35040	N	COOH		
NL44	35049	ninL44	dr	EM	
*131	35051	Mapped site CLA1			
NL99	35060	ninL99	dr	sq	
V154	35097	dv 154	plasmid-1	EM	
B256	35100	bio256	att-sr	EM	
Q115	35100	plambdaQLC115	dr	RES	

```
S                    R A  Q  H  H  Q   H  Q          Q  T   QQ A            1          ME NN   Q2
C                    S M  1  H  N  1   H  1          1  T   11 M            9          AC LL   13
R                    A 2  0  A  F  1   A  0          0  H   04 7            7          RP 28   10
1                    1 2  8  1  1  8   1  9          5  2   13              4          21 0    70
                                                            /                           / /      /
CCAGGCTTGCTGTACCATGTGCGCTGATTCTTGCGCTCAATACGTTGCAGGTTGCTTTCAATCTGTTTGTGGTATTCAGCCAGCACTGTAAGGTCTATCG
----.----+----.----+----.----+----.----+----.----+----.----+----.----+----.----+----.----+----.----+  35200
GGTCCGAACGACATGGTACACGCGACTAAGAACGCGAGTTATGCAACGTCCAACGAAAGTTAGACAAACACCATAAGTCGGTCGTGACATTCCAGATAGC

p  g  l  l  y  h  v  r  .  f  l  r  s  i  r  c  r  l  l  s  i  c  l  w  y  s  a  s  t  v  r  s  i  g
 q  a  c  c  t  m  c  a  d  s  c  a  q  y  v  a  g  c  f  q  s  v  c  g  i  q  p  a  l  .  g  l  s
  r  l  a  v  p  c  a  l  i  l  a  l  n  t  l  q  v  a  f  n  l  f  v  v  f  s  q  h  c  k  v  y  r
----.----+----.----+----.----+----.----+----.----+----.----+----.----+----.----+----.----+----.----+
 G  P  K  S  Y  W  T  R  Q  N  K  R  E  I  R  Q  L  N  S  E  I  Q  K  H  Y  E  A  L  V  T  L  D  I
  w  a  q  q  v  m  h  a  s  e  q  a  .  y  t  a  p  q  k  .  d  t  q  p  i  .  g  a  s  y  p  r  d
a  l  s  a  t  g  h  a  s  i  r  a  s  l  v  n  c  t  a  k  l  r  n  t  t  n  l  w  c  q  l  t  .  r
```

AM22	35116	Nam22	pm	sq	C to T
Q108	35120	plambdaQLC108	dr	RES	
Q118	35130	plambdaQLC118	dr	RES	
Q109	35140	plambdaQLC109	dr	RES	
Q105	35150	plambdaQLC105	dr	RES	
Q101	35160	plambdaQLC101	dr	RES	
Q143	35160	plambdaQLC143	dr	RES	
AM7	35165	Nam7	pm	sq	GT to AA
1974	35180	plambdaXJS1974	dr	RES	
MAR2	35191	Mar2	pm	sq	G to A
NL20	35194	ninL20	dr	EM	
NL8	35194	ninL8	dr	EM	
Q117	35200	plambdaQLC117	dr	RES	
2300	35200	plambdaXJS2300	dr	RES	
N	...	CONTINUES IN PHASE FOUR			

```
         MHQ     Q                 H      2              HH**   M  H          BA 1N       F
         9HL     1                 N      5              PI11   3  H          BM 9L       N
         7A9     0                 F      2              AN33   7  A          V5 78       U
         218     4                 1      1              1223   0  1          13 68       H
          /                                               ///
GATTTAGTGCGCTTTCTACTCGTGATTTCGGTTTGCGATTCAGCGAGAGAATAGGGCGGTTAACTGGTTTTGCGCTTACCCCAACCAACAGGGGATTTGC
----.----+----.----+----.----+----.----+----.----+----.----+----.----+----.----+----.----+----.----+  35300
CTAAATCACGCGAAAGATGAGCACTAAAGCCAAACGCTAAGTCGCTCTCTTATCCCGCCAATTGACCAAAACGCGAATGGGGTTGGTTGTCCCCTAAACG

  f  s  a  l  s  t  r  d  f  g  l  r  f  s  e  r  i  g  r  l  t  g  f  a  l  t  p  t  n  r  g  f  a
d  l  v  r  f  l  l  v  i  s  v  c  d  s  a  r  e  .  g  g  .  l  v  l  r  l  p  q  p  t  g  d  l  l
 i  .  c  a  f  y  s  .  f  r  f  a  i  q  r  e  n  r  a  v  n  w  f  c  a  y  p  n  q  q  g  i  c
----.----+----.----+----.----+----.----+----.----+----.----+----.----+----.----+----.----+----.----+
P  N  L  A  S  F  V  R  S  K  P  K  R  N  L  S  I  I  P  R  N  V  P  K  A  S  V  G  V  L  L  P  N  A
 s  k  t  r  k  r  s  t  i  e  t  q  s  e  a  l  s  y  p  p  .  s  t  k  r  k  g  w  g  v  p  s  k
  i  .  h  a  k  .  e  h  n  r  n  a  i  .  r  s  f  l  a  t  l  q  n  q  a  .  g  l  w  c  p  i  q
```

```
M972   35210     plambdaMS972          dr        RES
QL98   35210     plambdaQLC98          dr        RES
Q104   35220     plambdaQLC104         dr        RES
2521   35245     Nam2521               pm        sq      G to T
*132   35261    Mapped site HIN2
*133   35261    Mapped site HPA1
M370   35270     plambdaMS370          dr        RES     same as plambdaCM1
AM53   35287     Nam53                 pm        sq      A to T
1976   35290     plambdaXJS1076        dr        RES
NL88   35291     ninL88                dr        EM
N      ...      CONTINUES IN PHASE FOUR
```

```
         2             2   HTM     TTF NFF       Q         N S H                A
         5             1   HH9     THN 2NO       1           F N                L
         2             9   AA6     HAU -UK       0           A F                U
         4                 118     21H 1H1       2           N 1                1
                          /              /
TGCTTTCCATTGAGCCTGTTTCTCTGCGCGACGTTCGCGGCGGCGTGTTTGTGCATCCATCTGGATTCTCCTGTCAGTTAGCTTTGGTGGTGTGTGGCAG
----.----+----.----+----.----+----.----+----.----+----.----+----.----+----.----+----.----+----.----+ 35400
ACGAAAGGTAACTCGGACAAAGAGACGCGCTGCAAGCGCCGCCGCACAAACACGTAGGTAGACCTAAGAGGACAGTCAATCGAAACCACCACACACCGTC

 a  f  h  .  a  c  f  s  a  r  r  s  r  r  r  v  c  a  s  i  w  i  l  l  s  v  s  f  g  g  v  w  q
  l  s  i  e  p  v  s  l  r  d  v  r  g  g  v  f  v  h  p  s  g  f  s  c  q  l  a  l  v  v  c  g  s
c  f  p  l  s  l  f  l  c  a  t  f  a  a  a  c  l  c  i  h  l  d  s  p  v  s  .  l  w  w  c  v  a  v
----.----+----.----+----.----+----.----+----.----+----.----+----.----+----.----+----.----+----.----+
  A  K  W  Q  A  Q  K  E  A  R  R  E  R  R  R  T  Q  A  D  M  q  i  r  r  d  t  l  k  p  p  t  h  c
s  s  e  m  s  g  t  e  r  r  s  t  r  p  p  t  n  t  c  g  d  p  n  e  q  .  n  a  k  t  t  h  p  l
 q  k  g  n  l  r  n  r  q  a  v  n  a  a  a  h  k  h  m  w  r  s  e  g  t  l  .  s  q  h  h  t  a
```

```
2524   35310     Nam2524               pm        sq      C to T
219    35324     Nam219                pm        sq      C to A
M968   35330     plambdaMS968          dr        RES
N2-1   35340     bioN2-1               att-sr    EM
Q102   35350     plambdaQLC102         dr        RES
N      35360     N (orf-107)           M-NH3
```

```
                   Q     T             Q F A     H  BH          DCH G               N
                   1     H             L N L     P  BN          VFA D               L
                   4     A             9 U U     A  VF          1RE I               3
                   4     1             5 H 1     2  11           13 2               2
                                                     /           /
TTGTAGTCCTGAACGAAAACCCCCCGCGATTGGCACATTGGCAGCTAATCCGGAATCGCACTTACGGCCAATGCTTCGTTTCGTATCACACACCCCAAAG
----.----+----.----+----.----+----.----+----.----+----.----+----.----+----.----+----.----+----.----+ 35500
AACATCAGGACTTGCTTTTGGGGGGCGCTAACCGTGTAACCGTCGATTAGGCCTTAGCGTGAATGCCGGTTACGAAGCAAAGCATAGTGTGTGGGGTTTC

l  .  s  .  t  k  t  p  r  d  w  h  i  g  s  .  s  g  i  a  l  t  a  n  a  s  f  r  i  t  h  p  k  a
 c  s  p  e  r  k  p  p  a  i  g  t  l  a  a  n  p  e  s  h  l  r  p  m  l  r  f  v  s  h  t  p  k
  v  v  l  n  e  n  p  p  r  l  a  h  w  q  l  i  r  n  r  t  y  g  q  c  f  v  s  y  h  t  p  q  s
----.----+----.----+----.----+----.----+----.----+----.----+----.----+----.----+----.----+----.----+
 n  y  d  q  v  f  v  g  r  s  q  c  m  p  l  .  d  p  i  a  s  v  a  l  a  e  n  r  i  v  c  g  l
  q  l  g  s  r  f  g  g  a  i  p  v  n  a  a  l  g  s  d  c  k  r  g  i  s  r  k  t  d  c  v  g  f
t  t  t  r  f  s  f  g  g  r  n  a  c  q  c  s  i  r  f  r  v  .  p  w  h  k  t  e  y  .  v  g  w  l
```

```
Q144   35420     plambdaQLC144         dr        RES
QL95   35440     plambdaQLC95          dr        RES
DV1    35465     dv1                   plasmid-1 sq
NL32   35485     ninL32                dr        EM
```

```
  B            MFN         N QN     H     H                H          Q   E 3     SSIHDH   O  H V
  B            BNU         U 1U     G     P                G          1   C H     LD4PDG   L  G 2
  V            OUT         T 3T     A     H                A          0   1 -      U HEI   1  E
  1            2HL         - 8-     1     1                1          3   5 1      1 11A      2
                              /                                                   /
CCTTCTGCTTTGAATGCTGCCCTTCTTCAGGGCTTAATTTTTAAGAGCGTCACCTTCATGGTGGTCAGTGCGTCCTGCTGATGTGCTCAGTATCACCGCC
----.----+----.----+----.----+----.----+----.----+----.----+----.----+----.----+----.----+----.----+ 35600
GGAAGACGAAACTTACGACGGGAAGAAGTCCCGAATTAAAAATTCTCGCAGTGGAAGTACCACCAGTCACGCAGGACGACTACACGAGTCATAGTGGCGG

  f  c  f  e  c  c  p  s  s  g  l  n  f  .  e  r  h  l  h  g  g  q  c  v  l  l  m  c  s  v  s  p  p
p  s  a  l  n  a  a  l  l  q  g  l  i  f  k  s  v  t  f  m  v  v  s  a  s  c  .  c  a  q  y  h  r  q
 l  l  l  .  m  l  p  f  f  r  a  .  f  l  r  a  s  p  s  w  w  s  v  r  p  a  d  v  l  s  i  t  a
----.----+----.----+----.----+----.----+----.----+----.----+----.----+----.----+----.----+----.----+
a  k  q  k  s  h  q  g  e  e  p  s  l  k  .  s  r  .  r  .  p  p  .  h  t  r  s  i  h  e  t  d  g  g
 g  e  a  k  f  a  a  r  r  .  p  k  i  k  l  l  t  v  k  m  t  t  l  a  d  q  q  h  a  .  y  .  r
  r  r  s  q  i  s  g  k  k  l  a  .  n  k  l  a  d  g  e  h  h  d  t  r  g  a  s  t  s  l  i  v  a
```

```
NUTL   35518    nutL                 f(l)      f G    N utilization (17bp stem-loop)
NUT-   35528    nutL63;96;18         pm        sq     63 C to A; 96 C to G; 18 C to T
Q138   35530    plambdaQLC138        dr        RES
NUT-   35530    nutL3                pm        sq     delete G
Q103   35570    plambdaQLC103        dr        RES
3H-1   35576    bio3H-1              att-sr    sq
SL     35582    sL pL                5'        RNA
I4     35584    imm434               sl        sq
OL1    35591    oL1                  f(l) bs   f      17-bp diad symmetry  cI binds
V2     35596    vir2;v003            pm        sq     vir2 C to A; v003 C to T
```

```
 V     S H*0 2V V     H     0  HL                                          T
 1     E I1L 61 3     P     L  G6                                          T
 0     X N32 66 0     H     3  E6                                          H
 1     1 24  89 5     1        28                                          2
         //
AGTGGTATTTATGTCAACACCGCCAGAGATAATTTATCACCGCAGATGGTTATCTGTATGTTTTTTATATGAATTTATTTTTTGCAGGGGGGCATTGTTT
----.----+----.----+----.----+----.----+----.----+----.----+----.----+----.----+----.----+----.----+ 35700
TCACCATAAATACAGTTGTGGCGGTCTCTATTAAATAGTGGCGTCTACCAATAGACATACAAAAAATATACTTAAATAAAAAACGTCCCCCCGTAACAAA

 v  v  f  m  s  t  p  p  e  i  i  y  h  r  r  w  l  s  v  c  f  l  y  e  f  i  f  c  r  g  a  l  f
  w  y  l  c  q  h  r  q  r  .  f  i  t  a  d  g  y  l  y  v  f  y  m  n  l  f  f  a  g  g  h  c  l
s  g  i  y  v  n  t  a  r  d  n  l  s  p  q  m  v  i  c  m  f  f  i  .  i  y  f  l  q  g  g  i  v  w
----.----+----.----+----.----+----.----+----.----+----.----+----.----+----.----+----.----+----.----+
  t  t  n  i  d  v  g  g  s  i  i  .  .  r  l  h  n  d  t  h  k  k  y  s  n  i  k  q  l  p  a  n  n
w  h  y  k  h  .  c  r  w  l  y  n  i  v  a  s  p  .  r  y  t  k  .  i  f  k  n  k  a  p  p  c  q  k
 l  p  i  .  t  l  v  a  l  s  l  k  d  g  c  i  t  i  q  i  n  k  i  h  i  .  k  k  c  p  p  m  t
```

```
V101   35606    vir101               pm        sq     T to C
SEX1   35613    sex1                 pm        sq     G to A
*134   35615    Mapped site HIN2
OL2    35615    oL2                  f(l) bs   f      17 bp   cI binds  cro binds
2668   35619    vL2668               pm        sq
V169   35620    v169                 pm        sq
V305   35622    v305                 pm        sq     G to T
OL3    35635    oL3                  f(l) bs   f      17 bp   cro binds
L668   35639    l668                 pm        sq
```

```
       *BXM H                                                                       MM P*
       1GHB P                                                                       NB V1
       3L00 H                                                                       L0 U3
       5221 1                                                                       11 16
        ///
GGTAGGTGAGAGATCTGAATTGCTATGTTTAGTGAGTTGTATCTATTTATTTTTCAATAAATACAATTGGTTATGTGTTTTGGGGGCGATCGTGAGGCAA
----.----+----.----+----.----+----.----+----.----+----.----+----.----+----.----+----.----+----.----+ 35800
CCATCCACTCTCTAGACTTAACGATACAAATCACTCAACATAGATAAATAAAAAGTTATTTATGTTAACCAATACACAAAACCCCCGCTAGCACTCCGTT

g  r  .  e  i  .  i  a  m  f  s  e  l  y  l  f  i  f  q  .  i  q  l  v  m  c  f  g  g  d  r  e  a  k
 v  g  e  r  s  e  l  l  c  l  v  s  c  i  y  l  f  f  n  k  y  n  w  l  c  v  l  g  a  i  v  r  q
  .  v  r  d  l  n  c  y  v  .  .  v  v  s  i  y  f  s  i  n  t  i  g  y  v  f  w  g  r  s  .  g  k
----.----+----.----+----.----+----.----+----.----+----.----+----.----+----.----+----.----+----.----+
p  l  h  s  i  q  i  a  i  n  l  s  n  y  r  n  i  k  .  y  i  c  n  t  i  h  k  p  p  s  r  s  a
 t  p  s  l  d  s  n  s  h  k  t  l  q  i  .  k  n  k  l  l  y  l  q  n  h  t  k  p  a  i  t  l  c
q  y  t  l  s  r  f  q  .  t  .  h  t  t  d  i  .  k  e  i  f  v  i  p  .  t  n  q  p  r  d  h  p  l
```

```
*135   35711    Mapped site BGL2
*136   35791    Mapped site PVU1
```

```
  T   NSHM HHD  HHNS    R                      K      S      H A      S F    M D
  I   CCPN HAD  APCC    E                      H      F      G V      F 0    N I
  M   IRAL AEE  EATR    X                      5      A      U A      A K    L F
  M   1121 121  3211    B                      4      N      2 3      N 1    1 F
       ///  /      /
AGAAAACCCGGCGCTGAGGCCGGGTTATTCTTGTTCTCTGGTCAAATTATATAGTTGGAAAACAAGGATGCATATATGAATGAACGATGCAGAGGCAATG
----.----+----.----+----.----+----.----+----.----+----.----+----.----+----.----+----.----+----.----+ 35900
TCTTTTGGGCCGCGACTCCGGCCCAATAAGAACAAGAGACCAGTTTAATATATCAACCTTTTGTTCCTACGTATATACTTACTTGCTACGTCTCCGTTAC

  k  t  r  r  .  g  r  v  i  l  v  l  w  s  n  y  i  v  g  k  q  g  c  i  y  e  .  t  m  q  r  q  c
r  k  p  g  a  e  a  g  l  f  l  f  s  g  q  i  i  .  l  e  n  k  d  a  y  m  n  e  r  c  r  g  n  a
 e  n  p  a  l  r  p  g  y  s  c  s  l  v  k  l  y  s  w  k  t  r  m  h  i  .  m  n  d  a  e  a  m
----.----+----.----+----.----+----.----+----.----+----.----+----.----+----.----+----.----+----.----+
f  f  v  r  r  q  p  r  t  i  r  t  r  q  d  f  .  i  t  p  f  c  p  h  m  y  s  h  v  i  c  l  c  h
 l  f  g  p  a  s  a  p  n  n  k  n  e  p  .  i  i  y  n  s  f  l  s  a  y  i  f  s  r  h  l  p  l
  s  f  g  a  s  l  g  p  .  E  Q  E  R  T  L  N  Y  L  Q  F  V  L  I  C  I  H  I  F  S  A  S  A  I
```

```
TIMM   35804    tIMM terminator      3'        f RNA  presumed 3' from pRE pRM plit
REXB   35828    rexB                 COOH
KH54   35848    KH54                 dl        sq
DIFF   35887    difference                     sq     Ineichen et al.(1981) delete A
```

```
              N            F           R      R                     A      A          D
              L            N           E      E                     L      L          D
              3            U           X      X                     U      U          E
              0            H           -      -                     1      1          1
CCGATGGCGATAGTGGGTATCATGTAGCCGCTTATGCTGGAAAGAAGCAATAACCCGCAGAAAAACAAAGCTCCAAGCTCAACAAAACTAAGGGCATAGA
----.----+----.----+----.----+----.----+----.----+----.----+----.----+----.----+----.----+----.----+ 36000
GGCTACCGCTATCACCCATAGTACATCGGCGAATACGACCTTTCTTCGTTATTGGGCGTCTTTTTGTTTCGAGGTTCGAGTTGTTTTGATTCCCGTATCT

 r  w  r  .  w  v  s  c  s  r  l  c  w  k  e  a  i  t  r  r  k  t  k  l  q  a  q  q  n  .  g  h  r
  d  g  d  s  g  y  h  v  a  a  y  a  g  k  k  q  .  p  a  e  k  q  s  s  k  l  n  k  t  k  g  i  d
p  ■  a  i  v  g  i  ■  .  p  l  ■  l  e  r  s  n  n  p  q  k  n  k  a  p  s  s  t  k  l  r  a  .  t
----.----+----.----+----.----+----.----+----.----+----.----+----.----+----.----+----.----+----.----+
 r  h  r  y  h  t  d  h  l  r  k  h  q  f  s  a  i  v  r  l  f  v  f  s  w  a  .  c  f  .  p  c  l
a  s  p  s  l  p  y  .  t  a  a  .  a  p  f  f  c  y  g  a  s  f  c  l  e  l  s  l  l  v  l  p  ■  s
  G  I  A  T  T  P  I  M  Y  G  S  I  S  S  L  L  L  L  G  C  F  F  L  A  G  L  E  V  F  S  L  A  Y
```

NL30	35915	ninL30	dr	EM	
REX-	35940	rex209	pm	sq	G to A
REX-	35947	rex111	pm	sq	G to A
REXB	...	CONTINUES IN PHASE SIX			

```
         M2                            C BHH       HT                            F          B
         94                            F AAA       HH                            N          B
         6-                            R LEE       AA                            U          V
         95                            1 113       11                            H          1
          /                               //        /
CAATAACTACCGATGTCATATACCCATACTCTCTAATCTTGGCCAGTCGGCGCGTTCTGCTTCCGATTAGAAACGTCAAGGCAGCAATCAGGATTGCAAT
----.----+----.----+----.----+----.----+----.----+----.----+----.----+----.----+----.----+----.----+ 36100
GTTATTGATGGCTACAGTATATGGGTATGAGAGATTAGAACCGGTCAGCCGCGCAAGACGAAGGCTAATCTTTGCAGTTCCGTCGTTAGTCCTAACGTTA

q  .  l  p  ■  s  y  t  h  t  l  .  s  w  p  v  g  a  f  c  f  r  l  e  t  s  r  q  q  s  g  l  q  s
 n  n  y  r  c  h  i  p  i  l  s  n  l  g  q  s  a  r  s  a  s  d  .  k  r  q  g  s  n  q  d  c  n
  i  t  t  d  v  i  y  p  y  s  l  i  l  a  s  r  r  v  l  l  p  i  r  n  v  k  a  a  i  r  i  a  i
----.----+----.----+----.----+----.----+----.----+----.----+----.----+----.----+----.----+----.----+
 c  y  s  g  i  d  y  v  w  v  r  .  d  q  g  t  p  a  n  q  k  r  n  s  v  d  l  c  c  d  p  n  c
  l  l  .  r  h  .  i  g  ■  s  e  l  r  p  w  d  a  r  e  a  e  s  .  f  r  .  p  l  l  .  s  q  l
V  I  V  V  S  T  M  Y  G  Y  E  R  I  K  A  L  R  R  T  R  S  G  I  L  F  T  L  A  A  I  L  I  A  I
```

M969	36010	plambda969	dr	RES
24-5	36010	bio24-5	att-sr	EM
REXB	...	CONTINUES IN PHASE SIX		

```
            N         T          E               A  K1         E S
            D         T          C               L  H6         C C
            E         H          P               U  6-         R R
            1         1          1               1  73         2 1
                                                     /
CATGGTTCCTGCATATGATGACAATGTCGCCCCAAGACCATCTCTATGAGCTGAAAAAGAAACACCAGGAATGTAGTGGCGGAAAAGGAGATAGCAAATG
----.----+----.----+----.----+----.----+----.----+----.----+----.----+----.----+----.----+----.----+ 36200
GTACCAAGGACGTATACTACTGTTACAGCGGGGTTCTGGTAGAGATACTCGACTTTTTCTTTGTGGTCCTTACATCACCGCCTTTTCCTCTATCGTTTAC

  w  f  l  h  ■  ■  t  ■  s  p  q  d  h  l  y  e  l  k  k  k  h  q  e  c  s  g  g  k  g  d  s  k  c
h  g  s  c  i  .  .  q  c  r  p  k  t  i  s  ■  s  .  k  r  n  t  r  n  v  v  a  e  k  e  i  a  n  a
 ■  v  p  a  y  d  d  n  v  a  p  r  p  s  l  .  a  e  k  e  t  p  g  ■  .  w  r  k  r  r  .  q  ■
----.----+----.----+----.----+----.----+----.----+----.----+----.----+----.----+----.----+----.----+
d  h  n  r  c  i  i  v  i  d  g  w  s  w  r  .  s  s  f  f  f  c  w  s  h  l  p  p  f  p  s  l  l  h
 .  p  e  q  ■  h  h  c  h  r  g  l  v  ■  e  i  l  q  f  l  f  v  l  f  t  t  a  s  f  s  i  a  f
  M  T  G  A  Y  S  S  L  T  A  G  L  G  D  R  H  A  S  F  S  V  G  P  I  Y  H  R  F  L  L  Y  C  I
```

KH67	36153	KH67	dl	EM
16-3	36153	bio16-3	att-sr	EM
REXB	...	CONTINUES IN PHASE SIX		

```
                                        FF   H          R        Q     R
                                        CC   N          E        1     E
                                        RR   F          X        3     X
                                        21   1          B        9     A

CTTACGATAACGTAAGGAATTATTACTATGTAAACACCAGGCATGATTCTGTTCCGCATAATTACTCCTGATAATTAATCCTTAACTTTGCCCACCTGCC
----.----+----.----+----.----+----.----+----.----+----.----+----.----+----.----+----.----+----.----+  36300
GAATGCTATTGCATTCCTTAATAATGATACATTTGTGGTCCGTACTAAGACAAGGCGTATTAATGAGGACTATTAATTAGGAATTGAAACGGGTGGACGG

 l  r  .  r  k  e  l  l  l  c  k  h  q  a  .  f  c  s  a  .  l  l  l  i  i  n  p  .  l  c  p  p  a
  y  d  n  v  r  n  y  y  y  v  n  t  r  h  d  s  v  p  h  n  y  s  .  .  l  i  l  n  f  a  h  l  p
l  t  i  t  .  g  i  i  t  m  .  t  p  g  m  i  l  f  r  i  i  t  p  d  n  .  s  l  t  l  p  t  c  l
----.----+----.----+----.----+----.----+----.----+----.----+----.----+----.----+----.----+----.----+
  k  r  y  r  l  s  n  n  s  h  l  c  w  a  h  n  q  e  a  y  n  s  r  i  i  l  g  .  s  q  g  g  a
a  .  s  l  t  l  f  .  .  .  t  f  v  l  c  s  e  t  g  c  l  .  e  q  y  n  i  r  l  k  a  w  r  g
 S  V  I  V  Y  P  I  I  V  I  Y  V  G  P  M  I  R  N  R  M  i  v  g  s  l  .  D  K  V  K  G  V  Q

REXB    36259         rexB (orf-144)       M-NH3
Q139    36270         plambdaQLC139        dr              RES
REXA    36278         rexA                 COOH


  A                Q L                     M              A    D         B      M      H   T  HHD K
  H                1 I                     B              L    D         S      N      P   7  HAD H
  A                3 T                     0              U    E         T      L      H   5  AEE 7
  3                7                       2              1    1         E      1      1      121 0
                                                                                                /
TTTTAAAACATTCCAGTATATCACTTTTCATTCTTGCGTAGCAATATGCCATCTCTTCAGCTATCTCAGCATTGGTGACCTTGTTCAGAGGCGCTGAGAG
----.----+----.----+----.----+----.----+----.----+----.----+----.----+----.----+----.----+----.----+  36400
AAAATTTTGTAAGGTCATATAGTGAAAAGTAAGAACGCATCGTTATACGGTAGAGAAGTCGATAGAGTCGTAACCACTGGAACAAGTCTCCGCGACTCTC

f  .  n  i  p  v  y  h  f  s  f  l  r  s  n  m  p  s  l  q  l  s  q  h  w  .  p  c  s  e  a  l  r  d
 f  k  t  f  q  y  i  t  f  h  s  c  v  a  i  c  h  l  f  s  y  l  s  i  g  d  l  v  q  r  r  .  e
  l  k  h  s  s  i  s  l  f  i  l  a  .  q  y  a  i  s  s  a  i  s  a  l  v  t  l  f  r  g  a  e  r
----.----+----.----+----.----+----.----+----.----+----.----+----.----+----.----+----.----+----.----+
 k  .  f  m  g  t  y  .  k  e  n  k  r  l  l  i  g  d  r  .  s  d  .  c  q  h  g  q  e  s  a  s  l
  k  l  v  n  w  y  i  v  k  .  e  q  t  a  i  h  w  r  k  l  .  r  l  m  p  s  r  t  .  l  r  q  s
R  K  F  C  E  L  I  D  S  K  M  R  A  Y  C  Y  A  M  E  E  A  I  E  A  N  T  V  K  N  L  P  A  S  L

Q137    36320         plambdaQLC137        dr              RES
LIT     36322         s lit   p lit        5'              f RNA   start of lit (mis) RNA
T75     36390         bio t75              att-sr          EM
KH70    36398         KH70                 dl              sq
REXA     ...       CONTINUES IN PHASE SIX


  HH                                         H H          M               M                H
  AA                                         P A          N               N                P
  EE                                         A E          L               L                H
  13                                         2 3          1               1                1
  /
ATGGCCTTTTTCTGATAGATAATGTTCTGTTAAAATATCTCCGGCCTCATCTTTTGCCCGCAGGCTAATGTCTGAAAATTGAGGTGACGGGTTAAAAATA
----.----+----.----+----.----+----.----+----.----+----.----+----.----+----.----+----.----+----.----+  36500
TACCGGAAAAAGACTATCTATTACAAGACAATTTTATAGAGGCCGGAGTAGAAAACGGGCGTCCGATTACAGACTTTTAACTCCACTGCCCAATTTTTAT

  g  l  f  l  i  d  n  v  l  l  k  y  l  r  p  h  l  l  p  a  g  .  c  l  k  i  e  v  t  g  .  k  .
m  a  f  f  .  .  i  m  f  c  .  n  i  s  g  l  i  f  c  p  q  a  n  v  .  k  l  r  .  r  v  k  n  n
 w  p  f  s  d  r  .  c  s  v  k  i  s  p  a  s  s  f  a  r  r  l  m  s  e  n  .  g  d  g  l  k  i
----.----+----.----+----.----+----.----+----.----+----.----+----.----+----.----+----.----+----.----+
s  p  r  k  r  i  s  l  t  r  n  f  y  r  r  g  .  r  k  g  a  p  .  h  r  f  i  s  t  v  p  .  f  y
 i  a  k  k  q  y  i  i  n  q  .  f  i  e  p  r  m  k  q  g  c  a  l  t  q  f  n  l  h  r  t  l  f
  H  G  K  E  S  I  Y  H  E  T  L  I  D  G  A  E  D  K  A  R  L  S  I  D  S  F  Q  P  S  P  N  F  I

REXA     ...       CONTINUES IN PHASE SIX
```

```
                   Q           A                           QH        A                       D E
                   1           H                           1N        L                       D C
                   1           A                           4F        U                       E P
                   1           3                           11        1                       1 1

ATATCCTTGGCAACCTTTTTTATATCCCTTTTAAATTTTGGCTTAATGACTATATCCAATGAGTCAAAAAGCTCCCCTTCAATATCTGTTGCCCCTAAGA
----.----+----.----+----.----+----.----+----.----+----.----+----.----+----.----+----.----+----.----+ 36600
TATAGGAACCGTTGGAAAAAATATAGGGAAAATTTAAAACCGAATTACTGATATAGGTTACTCAGTTTTTCGAGGGGAAGTTATAGACAACGGGGATTCT

 y  p  w  q  p  f  l  y  p  f  .  i  l  a  .  .  l  y  p  m  s  q  k  a  p  l  q  y  l  l  p  l  r
  i  l  g  n  l  f  y  i  p  f  k  f  w  l  n  d  y  i  q  .  v  k  k  l  p  f  n  i  c  c  p  .  d
i  s  l  a  t  f  f  i  s  l  l  n  f  g  l  m  t  i  s  n  e  s  k  s  s  p  s  i  s  v  a  p  k  t
----.----+----.----+----.----+----.----+----.----+----.----+----.----+----.----+----.----+----.----+
  y  g  q  c  g  k  k  y  g  k  .  i  k  a  .  h  s  y  g  i  l  .  f  a  g  r  .  y  r  n  g  r  l
l  i  r  p  l  r  k  .  i  g  k  l  n  q  s  l  s  .  i  w  h  t  l  f  s  g  k  l  i  q  q  g  .  s
 I  D  K  A  V  K  K  I  D  R  K  F  K  P  K  I  V  I  D  L  S  D  F  L  E  G  E  I  D  T  A  G  L

Q111   36520       plambdaQLC111         dr          RES
Q141   36560       plambdaQLC141         dr          RES
REXA    ...      CONTINUES IN PHASE SIX
```

```
                           A      H                 XGCHG      A    NQA
                           L      P                 MDFAD      V    D1V
                           U      H                 AIREI      A    E4A
                           1      1                 32132      3    123
                                                      ///             /
CCTTTAATATATCGCCAAATACAGGTAGCTTGGCTTCTACCTTCACCGTTGTTCGGCCGATGAAATGCATATGCATAACATCGTCTTTGGTGGTTCCCCT
----.----+----.----+----.----+----.----+----.----+----.----+----.----+----.----+----.----+----.----+ 36700
GGAAATTATATAGCGGTTTATGTCCATCGAACCGAAGATGGAAGTGGCAACAAGCCGGCTACTTTACGTATACGTATTGTAGCAGAAACCACCAAGGGGA

p  l  i  y  r  q  i  q  v  a  w  l  l  p  s  p  l  f  g  r  .  n  a  y  a  .  h  r  l  w  w  f  p  s
 l  .  y  i  a  k  y  r  .  l  g  f  y  l  h  r  c  s  a  d  e  m  h  m  h  n  i  v  f  g  g  s  p
  f  n  i  s  p  n  t  g  s  l  a  s  t  f  t  v  v  r  p  m  k  c  i  c  i  t  s  s  l  v  v  p  l
----.----+----.----+----.----+----.----+----.----+----.----+----.----+----.----+----.----+----.----+
 g  k  i  y  r  w  i  c  t  a  q  s  r  g  e  g  n  n  p  r  h  f  a  y  a  y  c  r  r  q  h  n  g
  r  .  y  i  a  l  y  l  y  s  p  k  .  r  .  r  q  e  a  s  s  i  c  i  c  l  m  t  k  p  p  e  g
V  K  L  I  D  G  F  V  P  L  K  A  E  V  K  V  T  T  R  G  I  F  H  M  H  M  V  D  D  K  T  T  G  R

Q142   36670       plambdal42            dr          RES
REXA    ...      CONTINUES IN PHASE SIX
```

```
        M             T H                                              11M  H         AH F
        N             H H                                              77B  G         SA 0
        L             A A                                              340  U         UE K
        1             1 1                                                1  2         13 1
                                                                        //
CATCAGTGGCTCTATCTGAACGCGCTCTCCACTGCTTAATGACATTCCTTTCCCGATTAAAAAATCTGTCAGATCGGATGTGGTCGGCCCGAAAACAGTT
----.----+----.----+----.----+----.----+----.----+----.----+----.----+----.----+----.----+----.----+ 36800
GTAGTCACCGAGATAGACTTGCGCGAGAGGTGACGAATTACTGTAAGGAAAGGGCTAATTTTTTAGACAGTCTAGCCTACACCAGCCGGGCTTTTGTCAA

  s  v  a  l  s  e  r  a  l  h  c  l  m  t  f  l  s  r  l  k  n  l  s  d  r  m  w  s  a  r  k  q  f
h  q  w  l  y  l  n  a  l  s  t  a  .  .  h  s  f  p  d  .  k  i  c  q  i  g  c  g  r  p  e  n  s  s
 i  s  g  s  i  .  t  r  s  p  l  l  n  d  i  p  f  p  i  k  k  s  v  r  s  d  v  v  g  p  k  t  v
----.----+----.----+----.----+----.----+----.----+----.----+----.----+----.----+----.----+----.----+
e  d  t  a  r  d  s  r  a  r  w  q  k  i  v  n  r  e  r  n  f  f  r  d  s  r  i  h  d  a  r  f  c  n
 .  .  h  s  .  r  f  a  s  e  v  a  .  h  c  e  k  g  s  .  f  i  q  .  i  p  h  p  r  g  s  f  l
  M  L  P  E  I  Q  V  R  E  G  S  S  L  S  M  G  K  G  I  L  F  D  T  L  D  S  T  T  P  G  F  V  T

173    36770       tet 173               i(Tn10)     EM
174    36770       tet 174               i(Tn10)     EM
REXA    ...      CONTINUES IN PHASE SIX
```

```
                  PK                          TH        HH   M               M            2     *HA
                  CA                          TN        NN   N               B            2     1IL
                  1N                          HF        FF   L               0            9     3NU
                   2                          21        13   1               2            9     731
                                               /         /                                       /
CTGGCAAAACCAATGGTGTCGCCTTCAACAAACAAAAAAGATGGGAATCCCAATGATTCGTCATCTGCGAGGCTGTTCTTAATATCTTCAACTGAAGCTT
----.----+----.----+----.----+----.----+----.----+----.----+----.----+----.----+----.----+----.----+ 36900
GACCGTTTTGGTTACCACAGCGGAAGTTGTTTGTTTTTTCTACCCTTAGGGTTACTAAGCAGTAGACGCTCCGACAAGAATTATAGAAGTTGACTTCGAA

 w  q  n  q  w  c  r  l  q  q  t  k  k  m  g  i  p  m  i  r  h  l  r  g  c  s  .  y  l  q  l  k  l
  g  k  t  n  g  v  a  f  n  k  q  k  r  w  e  s  q  .  f  v  i  c  e  a  v  l  n  i  f  n  .  s  f
l  a  k  p  m  v  s  p  s  t  n  k  k  d  g  n  p  n  d  s  s  s  a  r  l  f  l  i  s  s  t  e  a  l
----.----+----.----+----.----+----.----+----.----+----.----+----.----+----.----+----.----+----.----+
  q  c  f  w  h  h  r  r  .  c  v  f  f  i  p  i  g  i  i  r  .  r  r  p  q  e  .  y  r  .  s  f  s
e  p  l  v  l  p  t  a  k  l  l  c  f  l  h  s  d  w  h  n  t  m  q  s  a  t  r  l  i  k  l  q  l  k
 R  A  F  G  I  T  D  G  E  V  F  L  F  S  P  F  G  L  S  E  D  D  A  L  S  N  K  I  D  F  V  S  A

PC1    36817     pc 1                 pm         sq
KAN2   36818     kan2                 i(Tn5)     EM
2299   36890     plambdaXJS2299       dr         RES
*137   36895    Mapped site HIN3
REXA    ...     CONTINUES IN PHASE SIX
```

```
  M                  H                                         C*T
  B                  N                                         L1A
  0                  F                                         A3Q
  2                  1                                         181
                                                                //
TAGAGCGATTTATCTTCTGAACCAGACTCTTGTCATTTGTTTTGGTAAAGAGAAAAGTTTTTCCATCGATTTTATGAATATACAAATAATTGGAGCCAAC
----.----+----.----+----.----+----.----+----.----+----.----+----.----+----.----+----.----+----.----+ 37000
ATCTCGCTAAATAGAAGACTTGGTCTGAGAACAGTAAACAAAACCATTTCTCTTTTCAAAAAGGTAGCTAAAATACTTATATGTTTATTAACCTCGGTTG

.  s  d  l  s  s  e  p  d  s  c  h  l  f  w  .  r  e  k  f  f  h  r  f  y  e  y  t  n  n  w  s  q  p
 r  a  i  y  l  l  n  q  t  l  v  i  c  f  g  k  e  k  s  f  s  i  d  f  m  n  i  q  i  i  g  a  n
  e  r  f  i  f  .  t  r  l  l  s  f  v  l  v  k  r  k  v  f  p  s  i  l  .  i  y  k  .  l  e  p  t
----.----+----.----+----.----+----.----+----.----+----.----+----.----+----.----+----.----+----.----+
 .  l  s  k  d  e  s  g  s  e  q  .  k  n  q  y  l  s  f  n  k  w  r  n  .  s  y  v  f  l  q  l  w
  l  a  i  .  r  r  f  w  v  r  t  m  q  k  p  l  s  f  l  k  e  m  s  k  i  f  i  c  i  i  p  a  l
K  S  R  N  I  K  Q  V  L  S  K  D  N  T  K  T  F  L  F  T  K  G  D  I  K  H  I  Y  L  Y  N  S  G  V

*138   36966    Mapped site CLA1
REXA    ...     CONTINUES IN PHASE SIX
```

```
  P*  1     H E                                                  HH     B           F            QM
  S1  9     P C                                                  HA     B           N            1B
  T3  7     H 1                                                  AE     V           U            10
  19  5     1 5                                                  12     1           H            22
   /                                                                                              /
CTGCAGGTGATGATTATCAGCCAGCAGAGAATTAAGGAAAACAGACAGGTTTATTGAGCGCTTATCTTTCCCTTTATTTTTGCTGCGGTAAGTCGCATAA
----.----+----.----+----.----+----.----+----.----+----.----+----.----+----.----+----.----+----.----+ 37100
GACGTCCACTACTAATAGTCGGTCGTCTCTTAATTCCTTTTGTCTGTCCAAATAACTCGCGAATAGAAAGGGAAATAAAAACGACGCCATTCAGCGTATT

  a  g  d  d  y  q  p  a  e  n  .  g  k  q  t  g  l  l  s  a  y  l  s  l  y  f  c  c  g  k  s  h  k
l  q  v  m  i  i  s  q  q  r  i  k  e  n  r  q  v  y  .  a  l  i  f  p  f  i  f  a  a  v  s  r  i  k
 c  r  .  .  l  s  a  s  r  e  l  r  k  t  d  r  f  i  e  r  l  s  f  p  l  f  l  l  r  .  v  a  .
----.----+----.----+----.----+----.----+----.----+----.----+----.----+----.----+----.----+----.----+
g  a  p  s  s  .  .  g  a  s  f  .  p  f  c  v  p  k  n  l  a  .  r  e  r  .  k  q  q  p  l  d  c  l
 r  c  t  i  i  i  l  w  c  l  i  l  s  f  l  c  t  .  q  a  s  i  k  g  k  i  k  a  a  t  l  r  m
  Q  L  H  H  N  D  A  L  L  S  N  L  F  V  S  L  N  I  S  R  K  D  K  G  K  N  K  S  R  Y  T  A  Y

*139   37005    Mapped site PST1
1975   37010      plambdaXJS1975       dr         RES
Q112   37100      plambdaQLC112        dr         RES
REXA    ...     CONTINUES IN PHASE SIX
```

```
         1   R                     M     D       Q         Q    H  T       H     V  M            A Q
         9   E                     N     D       1         1    N  A       N     0  B            L 1
         7   X                     L     E       0         1    F  Q       F     2  0            U 0
         3   A                     1     1       6         0    3  1       1     1  2            1 7

AAACCATTCTTCATAATTCAATCCATTTACTATGTTATGTTCTGAGGGGAGTGAAAATTCCCCTAATTCGATGAAGATTCTTGCTCAATTGTTATCAGCT
----.----+----.----+----.----+----.----+----.----+----.----+----.----+----.----+----.----+----.----+ 37200
TTTGGTAAGAAGTATTAAGTTAGGTAAATGATACAATACAAGACTCCCCTCACTTTTAAGGGGATTAAGCTACTTCTAAGAACGAGTTAACAATAGTCGA

 n  h  s  s  .  f  n  p  f  t  m  l  c  s  e  g  s  e  n  s  p  n  s  m  k  i  l  a  q  l  l  s  a
  t  i  l  h  n  s  i  h  l  l  c  y  v  l  r  g  v  k  i  p  l  i  r  .  r  f  l  l  n  c  y  q  l
k  p  f  f  i  i  q  s  i  y  y  v  m  f  .  g  e  .  k  f  p  .  f  d  e  d  s  c  s  i  v  i  s  y
----.----+----.----+----.----+----.----+----.----+----.----+----.----+----.----+----.----+----.----+
  f  w  e  e  y  n  l  g  n  v  i  n  h  e  s  p  l  s  f  e  g  l  e  i  f  i  r  a  .  n  n  d  a
f  v  m  r  .  l  e  i  w  k  s  h  .  t  r  l  p  t  f  i  g  r  i  r  h  l  n  k  s  l  q  .  .  s
 F  G  N  K  M  i  .  d  m  .  .  t  i  n  q  p  s  h  f  n  g  .  n  s  s  s  e  q  e  i  t  i  l
```

1973	37110	plambdaXJS1973	dr	RES
REXA	37114	rexA (orf-279)	M-NH3	
Q106	37150	plambdaQLC106	dr	RES
Q110	37160	plambdaQLC110	dr	RES
V021	37182	dv021	plasmid-r	sq
Q107	37200	plambdaQLC107	dr	RES

```
    H                  M     C M               HH     U                I        S     A     N   BM
    H                  B     I B               AA     A                C        P     M     T   IB
    A                  0       0               EE     7                1        3     1     1   NO
    1                  1       2               13     2                1        1     4         11
                                               /
ATGCGCCGACCAGAACACCTTGCCGATCAGCCAAACGTCTCTTCAGGCCACTGACTAGCGATAACTTTCCCCACAACGGAACAACTCTCATTGCATGGGA
----.----+----.----+----.----+----.----+----.----+----.----+----.----+----.----+----.----+----.----+ 37300
TACGCGGCTGGTCTTGTGGAACGGCTAGTCGGTTTGCAGAGAAGTCCGGTGACTGATCGCTATTGAAAGGGGTGTTGCCTTGTTGAGAGTAACGTACCCT

m  r  r  p  e  h  l  a  d  q  p  n  v  s  s  g  h  .  l  a  i  t  f  p  t  t  e  q  l  s  l  h  g  i
 c  a  d  q  n  t  l  p  i  s  q  t  s  l  q  a  t  d  .  r  .  l  s  p  q  r  n  n  s  h  c  m  g
  a  p  t  r  t  p  c  r  s  a  k  r  l  f  r  p  l  t  s  d  n  f  p  h  n  g  t  t  l  i  a  w  d
----.----+----.----+----.----+----.----+----.----+----.----+----.----+----.----+----.----+----.----+
 i  r  r  g  s  c  r  a  s  .  G  F  T  E  E  P  W  Q  S  A  I  V  K  G  V  V  S  C  S  E  N  C  P
  h  a  s  w  f  v  k  g  i  l  w  v  d  r  .  a  v  s  .  r  y  s  e  g  c  r  f  l  e  .  q  m  p
.  a  g  v  l  v  g  q  r  d  a  l  r  r  k  l  g  s  v  l  s  l  k  g  w  l  p  v  v  r  m  a  h  s
```

CI	37230	cI	COOH		
UA72	37255	cI UA72	pm	sq	C to A
IC11	37272	cI IC11	pm	sq	insert C
SP31	37281	cI SP31	i (IS1)	sq	
AM14	37287	cIam14	pm	G	C to A
NT1	37293	cI NT1	pm	sq	G to T

```
       5RB  '5      K        U  UT                 *BM               H     E       AA N        UHS
       0SP   0      1        V  VC                 1CB               P     T       AA T        VPP
       4A8   5      0        5  62                 4L0               H     2       22 1        7H1
        13          0        9  38                 011               1     6       1  6        61
         /                                          //
TCATTGGGTACTGTGGGTTTAGTGGTTGTAAAAACACCTGACCGCTATCCCTGATCAGTTTCTTGAAGGTAAACTCATCACCCCCAAGTCTGGCTATGCA
----.----+----.----+----.----+----.----+----.----+----.----+----.----+----.----+----.----+----.----+ 37400
AGTAACCCATGACACCCAAATCACCAACATTTTTGTGGACTGGCGATAGGGACTAGTCAAAGAACTTCCATTTGAGTAGTGGGGGTTCAGACCGATACGT

  i  g  y  c  g  f  s  g  c  k  n  t  .  p  l  s  l  i  s  f  l  k  v  n  s  s  p  p  s  l  a  m  q
s  l  g  t  v  g  l  v  v  v  k  t  p  d  r  y  p  .  s  v  s  .  r  .  t  h  h  p  q  v  w  l  c  r
 h  w  v  l  w  v  .  w  l  .  k  h  l  t  a  i  p  d  q  f  l  e  g  k  l  i  t  p  k  s  g  y  a
----.----+----.----+----.----+----.----+----.----+----.----+----.----+----.----+----.----+----.----+
I  M  P  Y  Q  P  N  L  P  Q  L  F  V  Q  G  S  D  R  T  L  K  K  F  T  F  E  D  G  G  L  R  A  I  C
 d  n  p  v  t  p  k  t  t  t  f  v  g  s  r  .  g  q  d  t  e  q  l  y  v  .  .  g  w  t  q  s  h
  .  q  t  s  h  t  .  h  n  y  f  c  r  v  a  i  g  s  .  n  r  s  p  l  s  m  v  g  l  d  p  .  a
```

504	37308	cIam504	pm	G	G to C
BP83	37309	cI BP83	pm	sq	T to C
505	37313	cIam505	pm	G	G to A
K100	37320	KH100	i(IS5)	sq	
UV59	37330	cI UV59	pm	sq	A to G
UV63	37333	cI UV63	pm	sq	delete A
IC28	37334	cI IC28	pm	sq	insert A
*140	37352	Mapped site BCl1			
ET26	37376	cI ET26	pm	sq	C to A
AA21	37384	cI AA21	pm	sq	insert C
AA2	37385	cI AA2	pm	sq	delete C
NT16	37387	cI NT16	pm	sq	insert A
UV76	37396	cI UV76	pm	sq	A to T
SP1	37398	cI SP1	i (IS1)	sq	
CI	...	CONTINUES IN PHASE FOUR			

```
      E S  N         T1 D       H*      U         *HA
      C C  T         N3 D       I1      V         1IL
      R R  1         1  E       N4      6         4NU
      2 1  5         0  1       21      4         231
                                 /                 /
GAAATCACCTGGCTCAACAGCCTGCTCAGGGTCAACGAGAATTAACATTCCGTCAGGAAAGCTTGGCTTGGAGCCTGTTGGTGCGGTCATGGAATTACCT
----.----+----.----+----.----+----.----+----.----+----.----+----.----+----.----+----.----+----.----+ 37500
CTTTAGTGGACCGAGTTGTCGGACGAGTCCCAGTTGCTCTTAATTGTAAGGCAGTCCTTTCGAACCGAACCTCGGACAACCACGCCAGTACCTTAATGGA

 k  s  p  g  s  t  a  c  s  g  s  t  r  i  n  i  p  s  g  k  l  g  l  e  p  v  g  a  v  m  e  l  p
  n  h  l  a  q  q  p  a  q  g  q  r  e  l  t  f  r  q  e  s  l  a  w  s  l  l  v  r  s  w  n  y  l
e  i  t  w  l  n  s  l  l  r  v  n  e  n  .  h  s  v  r  k  a  w  l  g  a  c  w  c  g  h  g  i  t  f
----.----+----.----+----.----+----.----+----.----+----.----+----.----+----.----+----.----+----.----+
  F  D  G  P  E  V  A  Q  E  P  D  V  L  I  L  M  G  D  P  F  S  P  K  S  G  T  P  A  T  M  S  N  G
 l  f  .  r  a  .  c  g  a  .  p  .  r  s  n  v  n  r  .  s  l  k  a  q  l  r  n  t  r  d  h  f  .  r
s  i  v  q  s  l  l  r  s  l  t  l  s  f  .  c  e  t  l  f  a  q  s  p  a  q  q  h  p  .  p  i  v
```

NT15	37412	cI NT15	pm	sq	delete G
TN10	37422	171;172;473;474	i(Tn10)	sq	
13	37423	cI 13	pm	sq	delete T
*141	37433	Mapped site HIN2			
UV64	37442	cI UV64	pm	sq	GAAT to AAA
*142	37459	Mapped site HIN3			
CI	...	CONTINUES IN PHASE FOUR			

```
              M      H        U       B                H             S        K3 D *FAD I
              N      N        V       L                P             F        H0 D 10LD N
              I      F        7       4                H             A        6- E 4KUE D
              1      1        3       6                1             N        77 1 3111 1
                                                                               /
TCAACCTCAAGCCAGAATGCAGAATCACTGGCTTTTTTGGTTGTGCTTACCCATCTCTCCGCATCACCTTTGGTAAAGGTTCTAAGCTCAGGTGAGAACA
----.----+----.----+----.----+----.----+----.----+----.----+----.----+----.----+----.----+----.----+ 37600
AGTTGGAGTTCGGTCTTACGTCTTAGTGACCGAAAAAACCAACACGAATGGGTAGAGAGGCGTAGTGGAAACCATTTCCAAGATTCGAGTCCACTCTTGT

s  t  s  s  q  n  a  e  s  l  a  f  l  v  v  l  t  h  l  s  a  s  p  l  v  k  v  l  s  s  g  e  n  i
 q  p  q  a  r  m  q  n  h  w  l  f  w  l  c  l  p  i  s  p  h  h  l  w  .  r  f  .  a  q  v  r  t
  n  l  k  p  e  c  r  i  t  g  f  f  g  c  a  y  p  s  l  r  i  t  f  g  k  g  s  k  l  r  .  e  h
----.----+----.----+----.----+----.----+----.----+----.----+----.----+----.----+----.----+----.----+
 E  V  E  L  W  F  A  S  D  S  A  K  K  T  T  S  V  W  R  E  A  D  G  K  T  F  T  R  L  E  P  S  F
  .  g  .  a  l  i  c  f  .  q  s  k  q  n  h  k  g  m  e  g  c  .  r  q  y  l  n  .  a  .  t  l  v
k  l  r  l  g  s  h  l  i  v  p  k  k  p  q  a  .  g  d  r  r  m  v  k  p  l  p  e  l  s  l  h  s  c
```

UV73	37531	cI UV73	pm	sq	GC to T	
BL46	37538	cI BL46	pm	sq	T to C	
KH67	37579	KH67	dr	EM		
30-7	37579	bio 30-7	att-sr	EM		
*143	37584	Mapped site HIN3				
IND1	37589	ind1	pm	sq	C to T	lysogen not UV inducible
CI	...	CONTINUES IN PHASE FOUR				

```
 HP            AS V F4RB 2    B D   E   F               A E  5 UF R
 PF            AP 2 T9S1 1    B D   T   N               M 8  4 VT 8
 H1            34 6 39A3 2    V E   3   U               3 6  - 51 2
 17             8 6 9 18      1 1       H               4    1 25
                    ///
TCCCTGCCTGAACATGAGAAAAAACAGGGTACTCATACTCACTTCTAAGTGACGGCTGCATACTAACCGCTTCATACATCTCGTAGATTTCTCTGGCGAT
----.----+----.----+----.----+----.----+----.----+----.----+----.----+----.----+----.----+----.----+ 37700
AGGGACGGACTTGTACTCTTTTTTGTCCCATGAGTATGAGTGAAGATTCACTGCCGACGTATGATTGGCGAAGTATGTAGAGCATCTAAAGAGACCGCTA

  p  a  .  t  .  e  k  t  g  y  s  y  s  l  l  s  d  g  c  i  l  t  a  s  y  i  s  .  i  s  l  a  i
 s  l  p  e  h  e  k  k  q  g  t  h  t  h  f  .  v  t  a  a  y  .  p  l  h  t  s  r  r  f  l  w  r  l
  p  c  l  n  m  r  k  n  r  v  l  i  l  t  s  k  .  r  l  h  t  n  r  f  i  h  l  v  d  f  s  g  d
----.----+----.----+----.----+----.----+----.----+----.----+----.----+----.----+----.----+----.----+
 M  G  A  Q  V  H  S  F  V  P  Y  E  Y  F  S  R  L  S  P  Q  M  S  V  A  E  Y  M  F  Y  I  E  R  A  I
  d  r  g  s  c  s  f  f  c  p  v  .  v  .  k  .  t  v  a  a  y  .  g  s  .  v  d  r  l  n  r  q  r
   g  q  r  f  m  l  f  f  l  t  s  m  s  v  e  l  h  r  s  c  v  l  r  k  m  c  r  t  s  k  e  p  s
```

PF17	37604	cI PF17	pm	sq	insert C
AA3	37623	cI AA3	pm	sq	insert A
SP48	37624	cI SP48	pm	sq	delete A
V266	37627	dv266	plasmid-1	EM	
ET39	37629	cI ET39	pm	sq	G to T
499	37629	cIam499	pm	G	G to C
B138	37630	cI BP138	pm	sq	T to C
212	37635	cIam212	pm	G	A to C
ET3	37651	cI ET3	pm	sq	G to C
AM34	37680	cIam34	pm	G	C to A
E86	37682	cIamE86	pm	sq	C to A
54-1	37687	cI54-1	pm	sq	A to C
UV52	37690	cI UV52	pm	sq	T to C
ET15	37691	cI ET15	pm	sq	C to A
R82	37694	cIamR82	pm	sq	T to A
CI	...	CONTINUES IN PHASE FOUR			

```
 U 16 5     B     U     S        6          B     U U L      3      2C 5    L C
 V 10 -     P     V     P        3          L     A V 5      4      6P 3    5 P
 7 -- 3     8     6     4        -          7     8 6 7      -      -3 -    0 -
 9 31       4     6     4        1          1       1        1      22 1      9
---.----+----.----+----.----+----.----+----.----+----.----+----.----+----.----+----.----+----.----+ 37800
U UP U     U     B                 F                      U    U  UU   NUU U  UUN      B
V VF V     1     U                 T                      V    V  1V   TAV V  1VT      L
9 91 4     1     3                 7                      6    4  14   139 4  033      4
6 72 7     6     0                                        2    1  74   731 6  87       2
                                                                  /    /
---.----+----.----+----.----+----.----+----.----+----.----+----.----+----.----+----.----+----.----+ 37800
   M            K         8      F      T     S DA                          F
   B            H         5      N      T     F UV                          O
   0            7         7      U      H     A 9A                          K
   2            0                H      2     N 33
                                                /
TGAAGGGCTAAATTCTTCAACGCTAACTTTGAGAATTTTTGCAAGCAATGCGGCGTTATAAGCATTTAATGCATTGATGCCATTAAATAAAGCACCAACG
---.----+----.----+----.----+----.----+----.----+----.----+----.----+----.----+----.----+----.----+ 37800
ACTTCCCGATTTAAGAAGTTGCGATTGAAACTCTTAAAAACGTTCGTTACGCCGCAATATTCGTAAATTACGTAACTACGGTAATTTATTTCGTGGTTGC

 e g l n s s t l t l r i f a s n a a l . a f n a l m p l n k a p t
  k g . i l q r . l . e f l q a m r r y k h l m h . c h . i k h q r
 . r a k f f n a n f e n f c k q c g v i s i . c i d a i k . s t n a
---.----+----.----+----.----+----.----+----.----+----.----+----.----+----.----+----.----+----.----+
  S P S F F F V S V K L I K A L L A A N Y A N I A N T G N F L A G V
 n f p . i r . r . s q s n k c a i r r . l c k i c q h w . i f c w r
  q l a l n k l a l k s f k q l c h p t i l m . h m s a m l y l v l
```

UV79	37706	cI uv79	pm	sq	G to A
11-3	37708	cI11-3;cICP175	pm	sq	11-3 C to T;CP175 C to A
60-1	37709	cI60-1	pm	sq	T to G
5-3	37711	cI5-3	pm	sq	A to G
BP84	37720	cI BP84	pm	sq	A to G
UV66	37727	cI UV66	pm	sq	AC to TT
SP44	37733	cI SP44;cI SP37	pm	sq	SP44 G to A;SP37 G to T
63-1	37744	cI63-1	pm	sq	A to C
BL71	37756	cI BL71	pm	sq	T to C
UA8	37762	cI UA8	pm	sq	G to T
UV61	37765	cI UV61	pm	sq	T to A
L57	37768	cIamL57	pm	sq	A to C
34-1	37773	cI34-1	pm	sq	A to C
26-2	37780	cI26-2	pm	sq	C to T
CP32	37781	cICP32	pm	sq	C to A
53-1	37784	cI53-1	pm	sq	T to C
L50	37789	cIamL50	pm	sq	A to C
CP-9	37792	cICP-9;cI29-6	pm	sq	CP9 G to T; 29-6 G to A
UV96	37702	CI UV96	pm	sq	G to A
UV97	37705	cI UV97	pm	sq	G to A
PF12	37707	cI PF12;cI AA6	pm	sq	insert G; delete G
UV47	37710	cI UV47	pm	sq	A to T
U116	37718	cI UV116;cI UV74	pm	sq	UV116 C to T;UV77 C to A
BU30	37726	cI BU30	pm	sq	A to G
FT7	37746	cI FT7	pm	sq	C to A
UV62	37768	cI UV62	pm	sq	A to G
UV41	37772	cI UV41	pm	sq	C to T
U117	37775	cI UV117	pm	sq	T to G
UV44	37775	cI UV44	pm	sq	T to C
NT17	37780	cI NT17	pm	sq	C to A
UA33	37781	cI UA33;cI SP43	pm	sq	UA33 C to A;SP43 C to T
UV91	37781	cI UV91	pm	sq	CC to TT
UV46	37784	cI UV46	pm	sq	T to C
U108	37787	cI UV108	pm	sq	A to T
UV37	37789	cI UV37	pm	sq	A to G
NT3	37791	cI NT3	pm	sq	deltete A
BL42	37798	cI BL42	pm	sq	A to G
KH70	37727	KH70	dr	sq	
857	37742	cI857	pm	sq	C to T cIts
DV93	37768	dvh93	plasmid-1	sq	
CI	...	CONTINUES IN PHASE FOUR			

```
 3 4Q 3  9     C               1     Q  4 L4    1                   53        3      1
 7 04 8  -     P               2     3  7 3-    -                   62        6      7
 -  -4 -  1    7               -     3  - 13    1                   --        -      7
 1  3  2                       1        1  /                        11        1
---.----+----.----+----.----+----.----+----.----+----.----+----.----+----.----+----.----+ 37900
   BURN FSPSR SUA          S  B      BS  U U                                US  U     SS  U
   1ALT TPFP1 PVA          P  U      LP  A V                                VP  1     PP  V
   0672 26222 481          5  3      53  6 1                                83  1     65  2
   2805 21 80 163             9      1     4                                59  2     02  3
                                                                                       /
---.----+----.----+----.----+----.----+----.----+----.----+----.----+----.----+----.----+ 37900
S   T 2       T       P      H  EPS     T     B      SUUU F    E 3       H      B  B SB
P   N 8       1       C      N  CCC     N     L      PVVV T    T 0       G      B  L PP
3   1 2       2       2      F  R3R     1     8      4858 2    1 2       A      V  9 58
8   0         4              1  2 1     0     4      2078 8    3         1      1  7 10
CCTGACTGCCCCATCCCCATCTTGTCTGCGACAGATTCCTGGGATAAGCCAAGTTCATTTTTCTTTTTTTCATAAATTGCTTTAAGGCGACGTGCGTCCT
---.----+----.----+----.----+----.----+----.----+----.----+----.----+----.----+----.----+ 37900
GGACTGACGGGGTAGGGGTAGAACAGACGCTGTCTAAGGACCCTATTCGGTTCAAGTAAAAAGAAAAAAAGTATTTAACGAAATTCCGCTGCACGCAGGA

p d c p i p i l s a t d s w d k p s s f f f f s . i a l r r r a s s
 l t a p s p s c l r q i p g i s q v h f s f f h k l l . g d v r p
  . l p h p h l v c d r f l g . a k f i f l f f i n c f k a t c v l
---.----+----.----+----.----+----.----+----.----+----.----+----.----+
 G S Q G M G M K D A V S E Q S L G L E N K K K F Y I A K L R R A D
  r v a g d g d q r r c i g p i l w t . k e k k . l n s . p s t r g
a q s g w g w r t q s l n r p y a l n m k r k k m f q k l a v h t r
```

37-1	37804	cI37-1	pm	sq	G to A
40-3	37807	cI40-3	pm	sq	T to A
Q44	37808	cIamQ44	pm	sq	G to A
38-2	37810	cI38-2	pm	sq	C to T
9-1	37813	cI9-1	pm	sq	A to G
CP7	37819	cICP7	pm	sq	A to G
12-1	37835	cI12-1	pm	sq	A to T
Q33	37841	cIamQ33	pm	sq	A to G
47-1	37844	cI47-1	pm	sq	G to A
L31	37846	cIamL31 cI4-3	pm	sq	L31 A to C; 4-3 A to G
1-1	37852	cI1-1	pm	sq	A to G
56-1	37872	cI56-1;cI5-1	pm	sq	56-1 A to C;5-1 A to G
32-1	37873	cI32-1;cI29-7	pm	sq	32-1 T to G;29-7 T to C
36-1	37886	cI36-1	pm	sq	G to A
177	37894	cI CP177	pm	sq	G to T

B102	37805	cI BP102	pm	sq	A to G
UA68	37806	cI UA68	pm	sq	C to A
BL70	37807	cI BL70	pm	sq	T to C
NT25	37808	cI NT25;cI UA77	pm	sq	NT25 G to A;UA77 G to T
ET22	37810	cI ET22	pm	sq	C to G
SP61	37811	cI SP61	pm	sq	C to T
PF2	37812	cI PF2;cI AA13	pm	sq	PF2 insert C;AA13 delete C
SP28	37813	cI SP28	pm	sq	A to G
B120	37814	cI BU120	pm	sq	T to C
SP41	37816	cI SP41	pm	sq	C to A
UV86	37817	cI UV86	pm	sq	C to T
AA13	37818	cI AA13	pm	sq	delete C
SP5	37828	cI SP5	pm	sq	G to A
BU39	37831	cI BU39	pm	sq	A to G
BL51	37840	cI BL51	pm	sq	T to C
SP3	37841	cI SP3	pm	sq	G to A
UA6	37844	cI UA6;cI BL89	pm	sq	UA6 A to Y;BL89 A to G
UV14	37846	cI UV14	pm	sq	A to G
UV85	37882	cI UV85	pm	sq	T to C
SP39	37883	cI SP39;cI BL80	pm	sq	SP39 T to G;BL80 T to C
U112	37886	cI UV112;cI UV93	pm	sq	UV112 G to A;UV93 G to Y
SP60	37894	cI SP60	i(IS1)	sq	
SP52	37894	cI SP52	i(IS1)	sq	
UV23	37898	cI UV23	pm	sq	C to T

SP38	37801	cI SP38	pm	sq	C to T
TN10	37806	475	i(Tn10)	sq	
282	37808	cIam282	pm	G	G to A
T124	37817	bio t124	att-sr	EM	
PC2	37826	pc 2	pm	sq	
PC3	37838	pc 3	pm	sq	
TN10	37846	408	i(Tn10)	sq	
BL84	37852	cI BL84	pm	sq	A to G
SP42	37860	cI SP42	pm	sq	delete T
UV80	37861	cI UV80	pm	sq	insert TT
UV57	37862	cI UV57	pm	sq	T to C
UV88	37863	cI UV88	pm	sq	C to TTT
ET28	37865	cI ET28	pm	sq	T to G
ET13	37870	cI ET13	pm	sq	insert T
302	37872	cIam302	pm	G	A to C
BL97	37894	cI BL97	pm	sq	G to A
SP51	37896	cI SP51	pm	sq	G to C
BP80	37897	cI BP80	pm	sq	T to C
CI	...	CONTINUES IN PHASE FOUR			

```
E 2            L        5                              V 33    CC   U M        BVMM M
1 9            7        7                              C RR    11   P 1        RNAC A
3 -                     -                              3 12    20   1 0        1 CC H
  2                     1                                             4           18 4

----.----+----.----+----.----+----.----+----.----+----.----+----.----+----.----+----.----+----.----+ 38000
U BSU B  U    S U SS U  S   S U  U  UB B    S  S                       U B    U
V LPA 1  V    P V PP V  P   P A  A  VP P    P  P                       V U    V
6 655 0  3    2 9 31 1  5   6 6  9  28 8    2  4                       2 2    5
  034 0  4    7 0 54 2  7   2 0     81 6       6                       6 0    1
                 /
----.----+----.----+----.----+----.----+----.----+----.----+----.----+----.----+----.----+----.----+ 38000
  AF  M                 K     S    H  CS   H     0  E                  0   FV   S    H*CXM   02H6
  LN  N                 H     D    G  IR   P     R  3                  R   11   2    I0H3C   R6G6
  UU  L                 5     U    T   M   H     3  7                  2   0    7    N02 H   16E8
  1H  1                 4     1    A       1                               4    4    250 8    82
                                      /                                                //// //
CAAGCTGCTCTTGTGTTAATGGTTTCTTTTTTGTGCTCATACGTTAAATCTATCACCGCAAGGGATAAATATCTAACACCGTGCGTGTTGACTATTTTAC
----.----+----.----+----.----+----.----+----.----+----.----+----.----+----.----+----.----+----.----+ 38000
GTTCGACGAGAACACAATTACCAAAGAAAAAACACGAGTATGCAATTTAGATAGTGGCGTTCCCTATTTATAGATTGTGGCACGCACAACTGATAAAATG

  s  c  s  c  v  n  g  f  f  f  v  l  i  r  .  i  y  h  r  k  g  .  i  s  n  t  v  r  v  d  y  f  t
 q  a  a  l  v  l  m  v  s  f  l  c  s  y  v  k  s  i  t  a  r  d  k  y  l  t  p  c  v  l  t  i  l  p
k  l  l  l  c  .  w  f  l  f  c  a  h  t  l  n  l  s  p  q  g  i  n  i  .  h  r  a  c  .  l  f  y
----.----+----.----+----.----+----.----+----.----+----.----+----.----+----.----+----.----+----.----+
F  L  Q  F  Q  T  I  P  K  K  K  T  S  M  r  .  i  .  .  r  l  p  y  i  d  l  v  t  r  t  s  .  k  v
 .  a  a  r  t  n  i  t  e  k  k  h  e  y  t  l  d  i  v  a  l  s  l  y  r  v  g  h  t  n  v  i  k
  l  s  s  k  h  .  h  n  r  k  q  a  .  v  n  f  r  d  g  c  p  i  f  i  .  c  r  a  h  q  s  n  .
```

E13	37901	cIamE13	pm	sq	C to A
29-2	37903	cI29-2	pm	sq	A to G
L7	37918	cIamL7	pm	sq	A to C
57-1	37928	cI57-1	pm	sq	T to G
VC3	37955	vC3	pm	sq	A to G
3R1	37957	OR3-r1	pm	sq	C to T
3R2	37958	OR3-r2; OR3-r3	pm	sq	-r2 G to T; -r3 G to A
C12	37965	OR3-c12	pm	sq	A to G
C10	37966	OR3-c10	pm	sq	T to C
UP1	37971	prm up-1	pm	sq	A to G
M104	37973	pRM-M104;116;U31	pm	sq	C to T
BR1	37984	BR1	pm	sq	C to T
VN	37985	VN	pm	sq	G to T
MAC1	37986	MAC1;MAC41	pm	sq	MAC1 T to A; MAC41 T to C
MCC8	37987	MCC8;MCH31;AAH8	pm	sq	MCC8 G to C;MCH31 G to A;AAH8 G to T
MAH4	37989	MAH4;AAH2	pm	sq	MAH4 delete T;AAH2 T to A
UV6	37901	cI UV6	pm	sq	C to T
BL60	37903	cI BL60	pm	sq	A to G
SP53	37904	cI SP53	i(IS5)	sq	
UA54	37905	cI UA54	pm	sq	C to A
B100	37907	cI BI 100	pm	sq	G to A
UV34	37910	cI UV34	pm	sq	C to T
SP27	37915	cI SP27	pm	sq	G to A
UV90	37917	cI UV90	pm	sq	T to A
SP35	37919	cI SP35	i(IS5)	sq	
SP14	37919	cI SP14	i(IS5)	sq	
UV12	37922	cI UV12	pm	sq	insert R
SP57	37926	cI SP57	pm	sq	C to A
SP62	37930	cI SP62	pm	sq	T to A
UA60	37932	cI UA60	pm	sq	insert T
UA9	37935	cI UA9	pm	sq	G to T
UV28	37938	cI UV28	pm	sq	C to T
BP81	37939	cI BP81	pm	sq	A to G
BP86	37941	cI BP86	pm	sq	A to G
SP2	37947	cI SP2	i(IS5)	sq	
SP46	37950	cI SP46	i(IS1)	sq	
UV26	37973	cI UV26	pm	sq	C to T
BU20	37975	cI BU20	pm	sq	A to G
UV51	37980	cI UV51	pm	sq	CC to TT
KH54	37925	KH54	dr	sq	
CI	37940	cI (orf-237)	M-NH3	AA	
SRM	37940	sM pM	5'	RNA	formerly called pRM
OR3	37951	oR3	f(1) bs	f	17 bp sequence cI binds
E37	37954	pRM-E37	pm	sq	C to T
OR2	37974	oR2	f(1) bs	f	cI binds (17bp)
E104	37978	pRM-E104;v3c	pm	sq	F104 A to C; v3c A to G
V1	37979	v1;vir146;pRM-E93	pm	sq	v1 C to A; E93 C to T
S274	37983	spi 274	att-sr	EM	
*144	37989	Mapped site HIN2			
CH20	37990	MCH20; MCH9	pm	sq	MCH20 G to C;MCH9 delete G
X3	37991	pR-x3	pm	sq	A to G
MCH8	37992	MCH8	pm	sq	C to T
OR1	37998	oR1	f(1) bs	f	cI binds (17bp)
2668	37999	vR2668;virR18	pm	sq	A to G
668	38000	VR668	pm	sq	insert C

```
 V  UUUVMUB       H  S NRMD  C         C                                    H                   A
 3  1V93NV1       P  R SSND  R         R                                    H                   L
 2  1838L14       H    PALE  0         0                                    A                   U
 6  9  7113       1    C111  2                                              1                   1
         //              /
CTCTGGCGGTGATAATGGTTGCATGTACTAAGGAGGTTGTATGGAACAACGCATAACCCTGAAAGATTATGCAATGCGCTTTGGGCAAACCAAGACAGCT
----.----+----.----+----.----+----.----+----.----+----.----+----.----+----.----+----.----+----.----+ 38100
GAGACCGCCACTATTACCAACGTACATGATTCCTCCAACATACCTTGTTGCGTATTGGGACTTTCTAATACGTTACGCGAAACCCGTTTGGTTCTGTCGA

 s  g  g  d  n  g  c  m  y  .  g  g  c  M  E  Q  R  I  T  L  K  D  Y  A  M  R  F  G  Q  T  K  T  A
  l  a  v  i  m  v  a  c  t  k  e  v  v  w  n  n  a  .  p  .  k  i  m  q  c  a  l  g  k  p  r  q  l
l  w  r  .  .  w  l  h  v  l  r  r  l  y  g  t  t  h  n  p  e  r  l  c  n  a  l  w  a  n  q  d  s  .
----.----+----.----+----.----+----.----+----.----+----.----+----.----+----.----+----.----+----.----+
 e  p  p  s  l  p  q  m  y  .  p  p  q  i  s  c  r  m  v  r  f  s  .  a  i  r  k  p  c  v  l  v  a
g  r  a  t  i  i  t  a  h  v  l  s  t  t  h  f  l  a  y  g  q  f  i  i  c  h  a  k  p  l  g  l  c  s
  r  q  r  h  y  h  n  c  t  s  l  l  n  y  p  v  v  c  l  g  s  l  n  h  l  a  s  q  a  f  w  s  l
```

V326	38003	vs326; BC1	pm	sq	vs326 C to A;BC1 C to T
U119	38006	UV119	pm	sq	delete G (clear)
UV8	38007	pRM-UV8;v3	pm	sq	UV8 C to T; v3 C to A
U93	38008	pRM-U93;M36	pm	sq	U93 G to A
V387	38009	vs387;vC1;NR5	pm	sq	vs387 G to C; vC1 G to T
UV11	38009	UV11	pm	sq	G to A (clear)
B143	38010	BU143	pm	sq	T to C (clear)
SR	38023	sR pR	5'	RNA	
CR02	38031	delta cro2	dl	EM	
CRO	38041	cro (orf-66)	M-NH3	AA	

```
 *BXM                    HTM       T HH          H                                         M   C
 1GHB                    HHB       R AA          A                                         N   R
 4L00                    AA0       '0 EE         E                                         L   0
 5221                    111          13         3                                         1   2
 ///                     //          /
AAAGATCTCGGCGTATATCAAAGCGCGATCAACAAGGCCATTCATGCAGGCCGAAAGATTTTTTTAACTATAAACGCTGATGGAAGCGTTTATGCGGAAG
----.----+----.----+----.----+----.----+----.----+----.----+----.----+----.----+----.----+----.----+ 38200
TTTCTAGAGCCGCATATAGTTTCGCGCTAGTTGTTCCGGTAAGTACGTCCGGCTTTCTAAAAAAATTGATATTTGCGACTACCTTCGCAAATACGCCTTC

K  D  L  G  V  Y  Q  S  A  I  N  K  A  I  H  A  G  R  K  I  F  L  T  I  N  A  D  G  S  V  Y  A  E  E
 k  i  s  a  y  i  k  a  r  s  t  r  p  f  m  q  a  e  r  f  f  .  l  .  t  l  m  e  a  f  m  r  k
  r  s  r  r  i  s  k  r  d  q  q  g  h  s  c  r  p  k  d  f  f  n  y  k  r  .  w  k  r  l  c  g  r
----.----+----.----+----.----+----.----+----.----+----.----+----.----+----.----+----.----+----.----+
 l  s  r  p  t  y  .  l  a  i  l  l  a  m  .  a  p  r  f  i  k  k  v  i  f  a  s  p  l  t  .  a  s
  f  i  e  a  y  i  l  a  r  d  v  l  g  n  m  c  a  s  l  n  k  .  s  y  v  s  i  s  a  n  i  r  f
.  l  d  r  r  i  d  f  r  s  .  c  p  w  e  h  l  g  f  s  k  k  l  .  l  r  q  h  f  r  k  h  p  l
```

*145	38103	Mapped site BGL2		
TR0	38135	tR0 terminator	3'	RNA
CR02	38197	delta cro2	dr	EM
CRO	...	CONTINUES IN PHASF ONE		

```
        M    *A                        C      I                N                        S        B
        B    1V                        R      4                U                        F        S
        0    4A                        0                       T                        A        T
        2    61                                                R                        N        X
             /
AGGTAAAGCCCTTCCCGAGTAACAAAAAAACAACAGCATAAATAACCCCGCTCTTACACATTCCAGCCCTGAAAAAGGGCATCAAATTAAACCACACCTA
----.----+----.----+----.----+----.----+----.----+----.----+----.----+----.----+----.----+----.----+ 38300
TCCATTTCGGGAAGGGCTCATTGTTTTTTTGTTGTCGTATTTATTGGGGCGAGAATGTGTAAGGTCGGGACTTTTTCCCGTAGTTTAATTTGGTGTGGAT

 V  K  P  F  P  S  N  K  K  T  T  A  .  i  t  p  l  l  h  i  p  a  l  k  k  g  i  k  l  n  h  t  y
r  .  s  p  s  r  v  t  k  k  q  q  h  k  .  p  r  s  y  t  f  q  p  .  k  r  a  s  n  .  t  t  p  m
  g  k  a  l  p  e  .  q  k  n  n  s  i  n  n  p  a  l  t  h  s  s  p  e  k  g  h  q  i  k  p  h  l
----.----+----.----+----.----+----.----+----.----+----.----+----.----+----.----+----.----+----.----+
s  t  f  g  k  g  l  l  l  f  v  v  a  y  i  v  g  s  k  c  m  g  a  r  f  f  p  m  l  n  f  w  v  .
 l  y  l  g  e  r  t  v  f  f  c  c  c  l  y  g  r  e  .  v  n  w  g  q  f  l  a  d  f  .  v  v  g
  p  l  a  r  g  s  y  c  f  f  l  l  m  f  l  g  a  r  v  c  e  l  g  s  f  p  c  .  i  l  g  c  r
```

*146	38214	Mapped site AVA1			
CRO	38238	cro	COOH		
I4	38245	imm434	sr	sq	
NUTR	38264	NutR	f(1)	f	N utilization(homology to nutL)

```
                                     D    D         3 3  C33 33   CC 33  3333 33   3 33       3   3
                                     Y    Y         0 0  T00 00   TT 10  1101 00   0 10       0   1
                                     A    A         7 8  R86 75   RR 07  1070 97   8 09       8   1
                                     3    2         1 8  567 39   21 67  4957 52   5 41       1   1
----.----+----.----+----.----+----.----+----.----+----.----+----.----+----.----+----.----+----.----+ 38400

 C  CCA                          T     CDS      R3  2  3N C 3 C    C3C    8M 33 3 3 T     3H3H  T 3
 T  NNV                          R     1DR      30  0  0D I 1 A    20Y    4N 00 0 6 T     0N0N  A 0
 N  CCA                          1     7FE      24  0  1E I 0 N     04    4L 00 9 2 H     5F9F  Q 9
 1  183                                 1        8  1  91   5 1     32     1 81 8 3 2     6391  1 7
       /                                          /                         /
TGGTGTATGCATTTATTTGCATACATTCAATCAATTGTTATCTAAGGAAATACTTACATATGGTTCGTGCAAACAAACGCAACGAGGCTCTACGAATCGA
----.----+----.----+----.----+----.----+----.----+----.----+----.----+----.----+----.----+----.----+ 38400
ACCACATACGTAAATAAACGTATGTAAGTTAGTTAACAATAGATTCCTTTATGAATGTATACCAAGCACGTTTGTTTGCGTTGCTCCGAGATGCTTAGCT

 g  v  c  i  y  l  h  t  f  n  q  l  l  s  k  e  i  l  t  y  g  s  c  k  q  t  q  r  g  s  t  n  r
  v  y  a  f  i  c  i  h  s  i  n  c  y  l  r  k  y  l  h  M  V  R  A  N  K  R  N  F  A  L  R  I  E
w  c  m  h  l  f  a  y  i  q  s  i  v  i  .  g  n  t  y  i  w  f  v  q  t  n  a  t  r  l  y  e  s  r
----.----+----.----+----.----+----.----+----.----+----.----+----.----+----.----+----.----+----.----+
   p  t  h  m  .  k  c  v  n  l  .  n  n  d  l  s  i  s  v  y  p  e  h  l  c  v  c  r  p  e  v  f  r
    t  y  a  n  i  q  m  c  e  i  l  q  .  r  l  f  y  k  c  i  t  r  a  f  l  r  l  s  a  r  r  i  s
     h  i  c  k  n  a  y  m  .  d  i  t  i  .  p  f  v  .  m  h  n  t  c  v  f  a  v  l  s  .  s  d
```

DYA3	38339	dya3	pm	sq	T to C
DYA2	38344	dya2	pm	sq	A to G
3071	38354	cY3071;c3073a	pm	sq	cY3071 T to C; c3073a T to A
3088	38356	cII3088	pm	sq	A to G
CIR5	38359	cir5	pm	sq	T to C
3086	38360	cII3086	pm	sq	A to G
3067	38361	cII3067	pm	sq	T to C
3073	38363	c3073b	pm	sq	G to C
3059	38364	cII3059	pm	sq	T to A
CTR2	38368	ctr2;ctr3	pm	sq	T to C
CTR1	38369	ctr1	pm	sq	G to A
3077	38372	cY3077	pm	sq	A to G
3114	38375	cII3114;cII3109	pm	sq	cII3114 A to G;cII3109 A to T
3075	38377	cY3075	pm	sq	AC to TT
3107	38378	cY3107	pm	sq	C to T
3095	38380	cY3095	pm	sq	C to A
3072	38381	cY3072;cY3078	pm	sq	cY3072 A to C;cY3078 A to G
3085	38385	cII3085	pm	sq	A to G
3104	38387	cII3104	pm	sq	G to A
3091	38388	cII3091	pm	sq	C to T
3081	38396	cII3081	pm	sq	A to T
3111	38400	cII3111	pm	sq	AG to T

CIN1	38302	cin-1	pm	sq	G to A
CNC1	38306	CNC1	pm	sq	T to C
CNC8	38307	CNC8	PM	sq	A to G
TR1	38334	tR1 terminator	3'	RNA	
C17	38341	c17		sq	9-bp duplication
SRE	38343	sRE pRE	5'	RNA	
R32	38350	r32	i(IS2)	sq	
3048	38350	cY3048	pm	sq	A to G
2001	38354	cY2001	pm	sq	T to G
3019	38357	cY3019	pm	sq	C to T
CII	38360	cII (orf-97)	M-NH3	AA	
3105	38362	cII3105	pm	sq	G to A
CAN1	38364	can1	pm	sq	T to C
C2	38369	cII site	bs		cII binds -35 region of pRE
3003	38370	cY3003	pm	sq	C to T
CY42	38371	cY42	pm	sq	A to T
844	38376	cY844	pm	sq	A to G
3008	38379	cY3008	pm	sq	G to A
3001	38380	cY3001	pm	sq	C to T
3098	38382	cY3098	pm	sq	A to G
3623	38384	cII3623	pm	sq	G to T
3056	38391	cII3056	pm	sq	T to C
3099	38393	cII3099	pm	sq	C to T
3097	38399	cII3097	pm	sq	G to A

```
  3             T            2     D       N       MA3                 3 AM3 3E   CM   3         6 MH
  0             T            0     D       S       BL6                 0 MB2 2C   HN   3         8 BN
  9             H            0     E       P       OU4                 6 608 81   IL   0           OF
  2             2            2     1       B       211                 2 013 35   C1   2           21
                                                     /                         /    /                /
GAGTGCGTTGCTTAACAAAATCGCAATGCTTGGAACTGAGAAGACAGCGGAAGCTGTGGGCGTTGATAAGTCGCAGATCAGCAGGTGGAAGAGGGACTGG
----.----+----.----+----.----+----.----+----.----+----.----+----.----+----.----+----.----+----.----+ 38500
CTCACGCAACGAATTGTTTTAGCGTTACGAACCTTGACTCTTCTGTCGCCTTCGACACCCGCAACTATTCAGCGTCTAGTCGTCCACCTTCTCCCTGACC

e  c  v  a  .  q  n  r  n  a  w  n  .  e  d  s  g  s  c  g  r  .  .  v  a  d  q  q  v  e  e  g  l  d
 S  A  L  L  N  K  I  A  M  L  G  T  E  K  T  A  E  A  V  G  V  D  K  S  Q  I  S  R  W  K  R  D  W
  v  r  c  l  t  k  s  q  c  l  e  l  r  r  q  r  k  l  w  a  l  i  s  r  r  s  a  g  g  r  g  t  g
----.----+----.----+----.----+----.----+----.----+----.----+----.----+----.----+----.----+----.----+
 s  h  t  a  .  c  f  r  l  a  q  f  q  s  s  l  p  l  q  p  r  q  y  t  a  s  .  c  t  s  s  p  s
  l  a  n  s  l  l  i  a  i  s  p  v  s  f  v  a  s  a  t  p  t  s  l  d  c  i  l  l  h  f  l  s  q
l  t  r  q  k  v  f  d  c  h  k  s  s  l  l  c  r  f  s  h  a  n  i  l  r  l  d  a  p  p  l  p  v  p
```

3092	38403	cII3092	pm	sq	G to A
2002	38430	cII2002	pm	sq	T to C
3641	38453	cII3641	pm	sq	G to T
3062	38472	cII3062	pm	sq	C to T
AM60	38474	cIIam60	pm	sq	C to T
3283	38476	cII3283a	pm	sq	C to A
3283	38478	cII3283b	pm	sq	T to A
CHIC	38483	chiC	pm	sq	A to T
3302	38488	cII3302	pm	sq	G to A
68	38498	cII68	pm	sq	T to A
CII	...	CONTINUES IN PHASE TWO			

```
    B    T       F  3               A    I   H*P         T3   B 3HT       3FH H    3    0
    B    T       N  2               M    C   I1K         A2   B 4HH       3NP N    1    0
    V    H       U  5               4    E   N43         Q5   V 6AA       8UH F    3    P
    1    2       H  8               1        275         18   1 811       6H1 1    9
                                             //                 /
ATTCCAAAGTTCTCAATGCTGCTTGCTGTTCTTGAATGGGGGGTCGTTGACGACGACATGGCTCGATTGGCGCGACAAGTTGCTGCGATTCTCACCAATA
----.----+----.----+----.----+----.----+----.----+----.----+----.----+----.----+----.----+----.----+ 38600
TAAGGTTTCAAGAGTTACGACGAACGACAAGAACTTACCCCCCAGCAACTGCTGCTGTACCGAGCTAACCGCGCTGTTCAACGACGCTAAGAGTGGTTAT

 s  k  v  l  n  a  a  c  c  s  .  m  g  g  r  .  r  r  h  g  s  i  g  a  t  s  c  c  d  s  h  q  .
I  P  K  F  S  M  L  L  A  V  L  E  W  G  V  V  D  D  D  M  A  R  L  A  R  Q  V  A  A  I  L  T  N  K
 f  q  s  s  q  c  c  l  l  f  l  n  g  g  s  l  t  t  t  w  l  d  w  r  d  k  l  l  r  f  s  p  i
----.----+----.----+----.----+----.----+----.----+----.----+----.----+----.----+----.----+----.----+
s  e  l  t  r  l  a  a  q  q  e  q  i  p  p  r  q  r  r  c  p  e  i  p  a  v  l  q  q  s  e  .  w  y
 i  g  f  n  e  i  s  s  a  t  r  s  h  p  t  t  s  s  s  m  a  r  n  a  r  c  t  a  a  i  r  v  l
  n  w  l  e  .  h  q  k  s  n  k  f  p  p  d  n  v  v  v  h  s  s  q  r  s  l  n  s  r  n  e  g  i
```

3258	38523	cII3258a	pm	sq	T to G
AM41	38538	cIIam41	pm	sq	G to A
ICE	38543	replication ice	f(1)	f G	5bp stem;5bp loop;proposed incepto
*147	38548	Mapped site HIN2			
PK35	38549	pk35	i(Tn903)	sq	
3258	38564	cII3258b	pm	sq	C to T
3468	38571	cII3468	pm	sq	C to T
3306	38582	cII3306	pm	sq	G to A
3139	38592	cII3139	pm	sq	T to C
OOP	38599	oop terminator	3'	RNA	
CII	...	CONTINUES IN PHASE TWO			

```
        NSH F  I                  3     3 M    CD3           BXM       0 H      0
        CCP N  2                  6     6 N    ID5           IHB       0 N
        IRA U  1                  3     3 L    IE2           NOO       P F
        112 H                     9     8 1     10           121         1
         //                                     //            /
AAAAACGCCCGGCGGCAACCGAGCGTTCTGAACAAATCCAGATGGAGTTCTGAGGTCATTACTGGATCTATCAACAGGAGTCATTATGACAAATACAGCA
----.----+----.----+----.----+----.----+----.----+----.----+----.----+----.----+----.----+----.----+ 38700
TTTTTGCGGGCCGCCGTTGGCTCGCAAGACTTGTTTAGGTCTACCTCAAGACTCCAGTAATGACCTAGATAGTTGTCCTCAGTAATACTGTTTATGTCGT

 k  t  p  g  g  n  r  a  f  .  t  n  p  d  g  v  l  r  s  l  l  d  l  s  t  g  v  i  M  T  N  T  A
  K  R  P  A  A  T  E  R  S  E  Q  I  Q  M  E  F  .  g  h  y  w  i  y  q  q  e  s  l  .  q  i  q  q
k  n  a  r  r  q  p  s  v  l  n  k  s  r  w  s  s  e  v  i  t  g  s  i  n  r  s  h  y  d  k  y  s  k
----.----+----.----+----.----+----.----+----.----+----.----+----.----+----.----+----.----+----.----+
  f  v  g  p  p  l  r  a  n  q  v  f  g  s  p  t  r  l  d  n  s  s  r  d  v  p  t  m  i  v  f  v  a
l  f  r  g  a  a  v  s  r  e  s  c  i  w  i  s  n  q  p  .  .  q  i  .  .  c  s  d  n  h  c  i  c  c
 f  f  a  r  r  c  g  l  t  r  f  l  d  l  h  l  e  s  t  m  v  p  d  i  l  l  l  .  .  s  l  y  l
```

I21	38617	imm21	sr	sq	
3639	38634	cII3639	pm	sq	A to G
3638	38642	cII3638	pm	sq	A to G
CII	38650	cII	COOH		
3520	38651	cII3520	pm	sq	T to C
OOP	38675	oop start point	5'	RNA	
Q	38686	O (orf-299)	M-NH3	AA	

```
           M                   H                     *BXM T                 B    T      M F
           N                   P                     1GHB A                 B    T      N N
           L                   A                     4L00 Q                 V    H      L U
           1                   2                     8221 1                 1    2      1 H
                                                      ///
AAAATACTCAACTTCGGCAGAGGTAACTTTGCCGGACAGGAGCGTAATGTGGCAGATCTCGATGATGGTTACGCCAGACTATCAAATATGCTGCTTGAGG
----.----+----.----+----.----+----.----+----.----+----.----+----.----+----.----+----.----+----.----+ 38800
TTTTATGAGTTGAAGCCGTCTCCATTGAAACGGCCTGTCCTCGCATTACACCGTCTAGAGCTACTACCAATGCGGTCTGATAGTTTATACGACGAACTCC

K  I  L  N  F  G  R  G  N  F  A  G  Q  E  R  N  V  A  D  L  D  D  G  Y  A  R  L  S  N  M  L  L  E  A
 k  y  s  t  s  a  e  v  t  l  p  d  r  s  v  m  w  q  i  s  m  m  v  t  p  d  y  q  i  c  c  l  r
  n  t  q  l  r  q  r  .  l  c  r  t  g  a  .  c  g  r  s  r  .  w  l  r  q  t  i  k  y  a  a  .  g
----.----+----.----+----.----+----.----+----.----+----.----+----.----+----.----+----.----+----.----+
 f  i  s  l  k  p  l  p  l  k  a  p  c  s  r  l  t  a  s  r  s  s  p  .  a  l  s  d  f  i  s  s  s
  f  y  e  v  e  a  s  t  v  k  g  s  l  l  t  i  h  c  i  e  i  i  t  v  g  s  .  .  i  h  q  k  l
l  f  v  .  s  r  c  l  y  s  q  r  v  p  a  y  h  p  l  d  r  h  h  n  r  w  v  i  l  y  a  a  q  p
```

*148	38754	Mapped site BGL2
Q	...	CONTINUES IN PHASE ONE

```
   H    H *BXM              B   T A      F                                           H    H       H
   N    H 1GHB              B   T H      N                                           P    N       N
   F    A 4L00              V   H A      U                                           H    F       F
   3    1 9221              1   2 3      H                                           1    1       1
           ///
CTTATTCGGGCGCAGATCTGACCAAGCGACAGTTTAAAGTGCTGCTTGCCATTCTGCGTAAAACCTATGGGTGGAATAAACCAATGGACAGAATCACCGA
----.----+----.----+----.----+----.----+----.----+----.----+----.----+----.----+----.----+----.----+ 38900
GAATAAGCCCGCGTCTAGACTGGTTCGCTGTCAAATTTCACGACGAACGGTAAGACGCATTTTGGATACCCACCTTATTTGGTTACCTGTCTTAGTGGCT

  Y  S  G  A  D  L  T  K  R  Q  F  K  V  L  L  A  I  L  R  K  T  Y  G  W  N  K  P  M  D  R  I  T  D
l  i  r  a  q  i  .  p  s  d  s  l  k  c  c  l  p  f  c  v  k  p  m  g  g  i  n  q  w  t  e  s  p  i
 l  f  g  r  r  s  d  q  a  t  v  .  s  a  a  c  h  s  a  .  n  l  w  v  e  .  t  n  g  q  n  h  r
----.----+----.----+----.----+----.----+----.----+----.----+----.----+----.----+----.----+----.----+
a  .  e  p  a  s  r  v  l  r  c  n  l  t  s  s  a  m  r  r  l  v  .  p  h  f  l  g  i  s  l  i  v  s
 s  i  r  a  c  i  q  g  l  s  l  k  f  h  q  k  g  n  q  t  f  g  i  p  p  i  f  w  h  v  s  d  g
  k  n  p  r  l  d  s  w  a  v  t  .  l  a  a  q  w  e  a  y  f  r  h  t  s  y  v  l  p  c  f  .  r
```

*149	38814	Mapped site BGL2
Q	...	CONTINUES IN PHASE ONE

```
      D                                                                             F   T    F  BT N
      D                                                                             N   T    N  BT S
      E                                                                             U   H    U  VH P
      1                                                                             H   2    H  12 C

TTCTCAACTTAGCGAGATTACAAAGTTACCTGTCAAACGGTGCAATGAAGCCAAGTTAGAACTCGTCAGAATGAATATTATCAAGCAGCAAGGCGGCATG
----.----+----.----+----.----+----.----+----.----+----.----+----.----+----.----+----.----+----.----+ 39000
AAGAGTTGAATCGCTCTAATGTTTCAATGGACAGTTTGCCACGTTACTTCGGTTCAATCTTGAGCAGTCTTACTTATAATAGTTCGTCGTTCCGCCGTAC

 S  Q  L  S  F  I  T  K  L  P  V  K  R  C  N  F  A  K  L  E  L  V  R  M  N  I  I  K  Q  Q  G  G  M
  l  n  l  a  r  l  q  s  y  l  s  n  g  a  m  k  p  s  .  n  s  s  e  .  i  l  s  s  s  k  a  a  c
f  s  t  .  r  d  y  k  v  t  c  q  t  v  q  .  s  q  v  r  t  r  q  n  e  y  y  q  a  a  r  r  h  v
----.----+----.----+----.----+----.----+----.----+----.----+----.----+----.----+----.----+----.----+
  e  .  s  l  s  i  v  f  n  g  t  l  r  h  l  s  a  l  n  s  s  t  l  i  f  i  i  l  c  c  p  p  m
i  r  l  k  a  l  n  c  l  .  r  d  f  p  a  i  f  g  l  .  f  e  d  s  h  i  n  d  l  l  l  a  a  h
 n  e  v  .  r  s  .  l  t  v  q  .  v  t  c  h  l  w  t  l  v  r  .  f  s  y  .  .  a  a  l  r  c
```

Q	...	CONTINUES IN PHASE ONE

```
   AA             F D            I   M  S     M      I   M    F              I            MR
   SV             0 D            T   N  F     N      T   N    0              T            N9
   UA'            K E            N   L  A     L      N   L    K              N            L3
   12             1 1            1   1  N     1      2   1    1              3            1
    /
TTTGGACCAAATAAAAACATCTCAGAATGGTGCATCCCTCAAAACGAGGGAAAATCCCCTAAAACGAGGGATAAAACATCCCTCAAATTGGGGGATTGCT
----.----+----.----+----.----+----.----+----.----+----.----+----.----+----.----+----.----+----.----+ 39100
AAACCTGGTTTATTTTTGTAGAGTCTTACCACGTAGGGAGTTTTGCTCCCTTTTAGGGGATTTTGCTCCCTATTTTGTAGGGAGTTTAACCCCCTAACGA

F  G  P  N  K  N  I  S  F  W  C  I  P  Q  N  F  G  K  S  P  K  T  R  D  K  T  S  L  K  L  G  D  C  Y
 l  d  q  i  k  t  s  q  n  g  a  s  l  k  t  r  e  n  p  l  k  r  g  i  k  h  p  s  n  w  g  i  a
  w  t  k  .  k  h  l  r  m  v  h  p  s  k  r  g  k  i  p  .  n  e  g  .  n  i  p  q  i  g  g  l  l
----.----+----.----+----.----+----.----+----.----+----.----+----.----+----.----+----.----+----.----+
 n  p  g  f  l  f  m  e  s  h  h  m  g  .  f  s  p  f  d  g  l  v  l  s  l  v  d  r  l  n  p  s  q
  k  s  w  i  f  v  d  .  f  p  a  d  r  l  v  l  s  f  g  r  f  r  p  i  f  c  g  e  f  q  p  i  a
t  q  v  l  y  f  c  r  l  i  t  c  g  e  f  r  p  f  i  g  .  f  s  p  y  f  m  g  .  i  p  p  n  s

IIN1   39034     ori iteron 1        f(1) bs     f      18 bp repeated seq   0 binds
IIN2   39054     ori iteron 2        f(1) bs     f      18 bp   0 binds-DNase protection
IIN3   39078     ori iteron 3        f(1)bs      f      18 bp   0 binds-DNase protection
R93    39092     r93                 dl          sq     ori- mutation
Q       ...    CONTINUES IN PHASE ONE
```

```
I              MR    R T      R      R             R     HH     R   *E      H H            M
T              N9    9 I      9      9             9     NY     P   1C      N N            N
N              L3    9 1      9      6             6     F4     8   5R      F F            L
4               1      2                                 32     2   01      3 1            1
                                                                    /
ATCCCTCAAAACAGGGGGACACAAAAGACACTATTACAAAAGAAAAAAGAAAAGATTATTCGTCAGAGAATTCTGGCGAATCCTCTGACCAGCCAGAAAA
----.----+----.----+----.----+----.----+----.----+----.----+----.----+----.----+----.----+----.----+ 39200
TAGGGAGTTTTGTCCCCCTGTGTTTTCTGTGATAATGTTTTCTTTTTTCTTTTCTAATAAGCAGTCTCTTAAGACCGCTTAGGAGACTGGTCGGTCTTTT

  P  S  K  Q  G  D  T  K  D  T  T  T  K  E  K  R  K  D  Y  S  S  E  N  S  G  E  S  S  D  Q  P  E  N
i  p  q  n  r  g  t  q  k  t  l  l  q  k  k  k  e  k  i  i  r  q  r  i  l  a  n  p  l  t  s  q  k  t
 s  l  k  t  g  g  h  k  r  h  y  y  k  r  k  k  k  r  l  f  v  r  e  f  w  r  i  l  .  p  a  r  k
----.----+----.----+----.----+----.----+----.----+----.----+----.----+----.----+----.----+----.----+
.  g  e  f  c  p  s  v  f  s  v  i  v  f  s  f  l  f  s  .  e  d  s  f  e  p  s  d  e  s  w  g  s  f
 i  g  .  f  l  p  v  c  f  v  s  n  c  f  f  f  s  f  i  i  r  .  l  i  r  a  f  g  r  v  l  w  f
  d  r  l  v  p  p  c  l  l  c  .  .  l  l  f  f  f  l  n  n  t  l  s  n  q  r  i  r  q  g  a  l  f

IIN4   39101     ori iteron 4        f(1)bs      f      18 bp   0 binds-DNase protection
R93    39115     r93                 dr          sq     ori- mutation
R99    39120     r99                 dl          sq     ori- mutation
TI12   39122     ti12(tiny 12)       pm          sq     C to A; cis replication defective
R99    39131     r99                 dr          sq     ori- mutation
R96    39138     r96                 dl          sq     ori- mutation
R96    39152     r96                 dr          sq     ori- mutation
HY42   39158     Hy42                hy          sq     lambdaatt80imm80 080::PQSRlambda
RP82   39165     rep82               hy          sq     lambda rep82::PQSRlambda
*150   39168    Mapped site FCR1
Q       ...    CONTINUES IN PHASF ONF
```

```
             SB     HH  H F       F    F  F          B E      E    R      MF T    H        R    F
             FB     PG  P N       0    N  N          B C      C    I      BN T    G        I    0
             AV     AU  H U       K    U  U          V 1      P    C      0U H    U        C    K
             N1     22  1 H       1    H  H          1 5      1           2H 2    2             1
             /
CGACCTTTCTGTGGTGAAACCGGATGCTGCAATTCAGAGCGGCAGCAAGTGGGGGACAGCAGAAGACCTGACCGCCGCAGAGTGGATGTTTGACATGGTG
----.----+----.----+----.----+----.----+----.----+----.----+----.----+----.----+----.----+----.----+ 39300
GCTGGAAAGACACCACTTTGGCCTACGACGTTAAGTCTCGCCGTCGTTCACCCCCTGTCGTCTTCTGGACTGGCGGCGTCTCACCTACAAACTGTACCAC

 D  L  S  V  V  K  P  D  A  A  I  Q  S  G  S  K  W  G  T  A  E  D  L  T  A  A  F  W  M  F  D  M  V
  t  f  l  w  .  n  r  m  l  q  f  r  a  a  a  s  g  g  q  q  k  t  .  p  p  q  s  g  c  l  t  w  .
r  p  f  c  g  e  t  g  c  c  n  s  e  r  q  q  v  g  d  s  r  r  p  d  r  r  r  v  d  v  .  h  g  e
----.----+----.----+----.----+----.----+----.----+----.----+----.----+----.----+----.----+----.----+
 s  r  e  t  t  f  g  s  a  a  i  .  l  p  l  l  h  p  v  a  s  s  r  v  a  a  s  h  i  n  s  m  t
v  v  k  r  h  h  f  r  i  s  c  n  l  a  a  a  l  p  p  c  c  f  v  q  g  g  c  l  p  h  k  v  h  h
 r  g  k  q  p  s  v  p  h  q  l  e  s  r  c  c  t  p  s  l  l  l  g  s  r  r  l  t  s  t  q  c  p

RIC    39268     ric5b               pm          sq     C to T
RIC    39292     ric5b               pm          sq     G to A
Q       ...    CONTINUFS IN PHASF ONF
```

```
        H  M               H                        S  E                A                  T A    N
        P  B               N                        F  C                -                  H F    S
        H  0               F                        A  R                2                  A L    P
        1  2               3                        N  V                0                  1 3    C

AAGACTATCGCACCATCAGCCAGAAAACCGAATTTTGCTGGGTGGGCTAACGATATCCGCCTGATGCGTGAACGTGACGGACGTAACCACCGCGACATGT
----.----+----.----+----.----+----.----+----.----+----.----+----.----+----.----+----.----+----.----+  39400
TTCTGATAGCGTGGTAGTCGGTCTTTTGGCTTAAAACGACCCACCCGATTGCTATAGGCGGACTACGCACTTGCACTGCCTGCATTGGTGGCGCTGTACA

K  T  I  A  P  S  A  R  K  P  N  F  A  G  W  A  N  D  I  R  L  M  R  F  R  D  G  R  N  H  R  D  M  C
 r  l  s  h  h  q  p  e  n  r  i  l  l  g  g  l  t  i  s  a  .  c  v  n  v  t  d  v  t  t  a  t  c
  d  y  r  t  i  s  q  k  t  e  f  c  w  v  g  .  r  y  p  p  d  a  .  t  .  r  t  .  p  p  r  h  v
----.----+----.----+----.----+----.----+----.----+----.----+----.----+----.----+----.----+----.----+
 f  v  i  a  g  d  a  l  f  g  f  k  a  p  h  a  l  s  i  r  r  i  r  s  r  s  p  r  l  w  r  s  m
  l  s  d  c  w  .  g  s  f  r  i  k  s  p  p  s  v  i  d  a  q  h  t  f  t  v  s  t  v  v  a  v  h
s  s  .  r  v  m  l  w  f  v  s  n  q  q  t  p  .  r  y  g  g  s  a  h  v  h  r  v  y  g  g  r  c  t

A-20  39373     bioE5a-20          att-sr     FM
Q       ...     CONTINUES IN PHASE ONE


         N          SN*ES           AA H        DS      BHNSHGCH     T        AA       T
         S          PS1CC           SV P        DD      VGCCPDFA     H        SV       A
         P          HP5RR           UA A        EU      UIIRAIRF     A        UA       Q
         B          1C121           12 2        11      1J112213     1        12       1
                      ////           /                   //// //               /
GTGTGCTGTTCCGCTGGGCATGCCAGGACAACTTCTGGTCCGGTAACGTGCTGAGCCCGGCCAAACTCCGCGATAAGTGGACCCAACTCGAAATCAACCG
----.----+----.----+----.----+----.----+----.----+----.----+----.----+----.----+----.----+----.----+  39500
CACACGACAAGGCGACCCGTACGGTCCTGTTGAAGACCAGGCCATTGCACGACTCGGGCCGGTTTGAGGCGCTATTCACCTGGGTTGAGCTTTAGTTGGC

  V  L  F  R  W  A  C  Q  D  N  F  W  S  G  N  V  L  S  P  A  K  L  R  D  K  W  T  Q  L  E  I  N  R
v  c  c  s  a  g  h  a  r  t  t  s  g  p  v  t  c  .  a  r  p  n  s  a  i  s  g  p  n  s  k  s  t  v
 c  a  v  p  l  g  m  p  g  q  l  l  v  r  .  r  a  e  p  g  q  t  p  r  .  v  d  p  t  r  n  q  p
----.----+----.----+----.----+----.----+----.----+----.----+----.----+----.----+----.----+----.----+
h  t  s  n  r  q  a  h  w  s  l  k  q  d  p  l  t  s  l  g  a  l  s  r  s  l  h  v  w  s  s  i  l  r
 t  h  q  e  a  p  c  a  l  v  v  e  p  g  t  v  h  q  a  r  g  f  e  a  i  l  p  g  l  e  f  d  v
  h  a  t  g  s  p  m  g  p  c  s  r  t  r  y  r  a  s  g  p  w  v  g  r  y  t  s  g  v  r  f  .  g

*151  39422     Mapped site SPH1
Q       ...     CONTINUES IN PHASE ONE


                   T                     T                        T       BXM   PO            F
                   T                     A                        T       IHB                 N
                   H                     Q                        H       NOO                 U
                   2                     1                        2       121                 H
                                                                            /     /
TAACAAGCAACAGGCAGGCGTGACAGCCAGCAAACCAAAACTCGACCTGACAAACACAGACTGGATTTACGGGGTGGATCTATGAAAAACATCGCCGCAC
----.----+----.----+----.----+----.----+----.----+----.----+----.----+----.----+----.----+----.----+  39600
ATTGTTCGTTGTCCGTCCGCACTGTCGGTCGTTTGGTTTTGAGCTGGACTGTTTGTGTCTGACCTAAATGCCCCACCTAGATACTTTTTGTAGCGGCGTG

 N  K  Q  Q  A  G  V  T  A  S  K  P  K  L  D  L  T  N  T  D  W  I  Y  G  V  D  L  .  k  t  s  p  h
  t  s  n  r  q  a  .  q  p  a  n  q  n  s  t  .  q  t  q  t  g  f  t  g  w  i  y  e  k  h  r  r  t
.  q  a  t  g  r  r  d  s  q  q  t  k  t  r  p  d  k  h  r  l  d  l  r  g  g  s  M  K  N  I  A  A  Q
----.----+----.----+----.----+----.----+----.----+----.----+----.----+----.----+----.----+----.----+
  l  l  c  c  a  p  t  v  a  l  l  g  f  s  s  r  v  f  v  s  q  i  .  p  t  s  r  h  f  v  d  g  c
t  v  l  l  l  c  a  h  c  g  a  f  w  f  e  v  q  c  v  c  v  p  n  v  p  h  i  .  s  f  c  r  r  v
 y  c  a  v  p  l  r  s  l  w  c  v  l  v  r  g  s  l  c  l  s  s  k  r  p  p  d  i  f  f  m  a  a

P       39582   P (orf-233)        M-NH3
Q       39582   0                  COOH


       HH**    S H              BM               NH        R           F      RE           H M
       PI11    F G              IB               SP        S           N      SC           H B
       AN55    A A              NO               PA        A           U      A1           A 0
       1223    N J              11               C2        1           H      15           1 1
        ///
AGATGGTTAACTTTGACCGTGAGCAGATGCGTCGGATCGCCAACAACATGCCGGAACAGTACGACGAAAAGCCGCAGGTACAGCAGGTAGCGCAGATCAT
----.----+----.----+----.----+----.----+----.----+----.----+----.----+----.----+----.----+----.----+  39700
TCTACCAATTGAAACTGGCACTCGTCTACGCAGCCTAGCGGTTGTTGTACGGCCTTGTCATGCTGCTTTTCGGCGTCCATGTCGTCCATCGCGTCTAGTA

r  w  l  t  l  t  v  s  r  c  v  g  s  p  t  t  c  r  n  s  t  t  k  s  r  r  y  s  r  .  r  r  s  s
 d  g  .  l  .  p  .  a  d  a  s  d  r  q  q  h  a  g  t  v  r  r  k  a  a  g  t  a  g  s  a  d  h
  M  V  N  F  D  R  E  Q  M  R  R  I  A  N  N  M  P  E  Q  Y  D  F  K  P  Q  V  Q  Q  V  A  Q  I  T
----.----+----.----+----.----+----.----+----.----+----.----+----.----+----.----+----.----+----.----+
 l  h  n  v  k  v  t  l  l  h  t  p  d  g  v  v  h  r  f  l  v  v  f  l  r  l  y  l  l  y  r  l  d
  s  p  .  s  q  g  h  a  s  a  d  s  r  w  c  c  a  p  v  t  r  r  f  a  a  p  v  a  p  l  a  s  .
c  i  t  l  k  s  r  s  c  i  r  r  i  a  l  l  m  g  s  c  y  s  s  f  g  c  t  c  c  t  a  c  i  m
```

```
*152   39608      Mapped site HIN2
*153   39608      Mapped site HPA1
P        ...      CONTINUES IN PHASE THREE


                                  NSH    E S
                                  CCP    C C
                                  IRA    R R
                                  112    2 1
                                   //
CAACGGTGTGTTCAGCCAGTTACTGGCAACTTTCCCGGCGAGCCTGGCTAACCGTGACCAGAACGAAGTGAACGAAATCCGTCGCCAGTGGGTTCTGGCT
----.----+----.----+----.----+----.----+----.----+----.----+----.----+----.----+----.----+----.----+ 39800
GTTGCCACACAAGTCGGTCAATGACCGTTGAAAGGGCCGCTCGGACCGATTGGCACTGGTCTTGCTTCACTTGCTTTAGGCAGCGGTCACCCAAGACCGA

  t  v  c  s  a  s  y  w  q  l  s  r  r  a  w  l  t  v  t  r  t  k  .  t  k  s  v  a  s  g  f  w  l
q  r  c  v  q  p  v  t  g  n  f  p  g  e  p  g  .  p  .  p  e  r  s  e  r  n  p  s  p  v  g  s  g  f
 N  G  V  F  S  Q  L  L  A  T  F  P  A  S  L  A  N  R  D  Q  N  E  V  N  E  I  R  R  Q  W  V  L  A
----.----+----.----+----.----+----.----+----.----+----.----+----.----+----.----+----.----+----.----+
d  v  t  h  e  a  l  .  q  c  s  e  r  r  a  q  s  v  t  v  l  v  f  h  v  f  d  t  a  l  p  n  q  s
 .  r  h  t  .  g  t  v  p  l  k  g  p  s  g  p  .  g  h  g  s  r  l  s  r  f  g  d  g  t  p  e  p
  l  p  t  n  l  w  n  s  a  v  k  g  a  l  r  a  l  r  s  w  f  s  t  f  s  i  r  r  w  h  t  r  a

P        ...      CONTINUES IN PHASE THREE


        H   BM                     HH**          HT              H  T       H          *NXASNHS*
        P   IB                     PI11          HH              N  A       P          1CMVCCPC1
        H   NO                     AN55          AA              F  Q       H          5IAARIAR5
        1   11                     1245          11              1  1       1          611111217
                                    ///           /                                     ///////
TTTCGGGAAAACGGGATCACCACGATGGAACAGGTTAACGCAGGAATGCGCGTAGCCCGTCGGCAGAATCGACCATTTCTGCCATCACCCGGGCAGTTTG
----.----+----.----+----.----+----.----+----.----+----.----+----.----+----.----+----.----+----.----+ 39900
AAAGCCCTTTTGCCCTAGTGGTGCTACCTTGTCCAATTGCGTCCTTACGCGCATCGGGCAGCCGTCTTAGCTGGTAAAGACGGTAGTGGGCCCGTCAAAC

 f  g  k  t  g  s  p  r  w  n  r  l  t  q  e  c  a  .  p  v  g  r  i  d  h  f  c  h  h  p  g  s  l
  s  g  k  r  d  h  h  d  g  t  g  .  r  r  n  a  r  s  p  s  a  e  s  t  i  s  a  i  t  r  a  v  c
F  R  E  N  G  I  T  T  M  E  Q  V  N  A  G  M  R  V  A  R  R  Q  N  R  P  F  L  P  S  P  G  Q  F  V
----.----+----.----+----.----+----.----+----.----+----.----+----.----+----.----+----.----+----.----+
  k  p  f  v  p  d  g  r  h  f  l  n  v  c  s  h  a  y  g  t  p  l  i  s  w  k  q  w  .  g  p  l  k
k  e  p  f  r  s  .  w  s  p  v  p  .  r  l  f  a  r  l  g  d  a  s  d  v  m  e  a  m  v  r  a  t  q
 k  r  s  f  p  i  v  v  i  s  c  t  l  a  p  i  r  t  a  r  r  c  f  r  g  n  r  g  d  g  p  c  n

*154   39836      Mapped site HIN2
*155   39836      Mapped site HPA1
*156   39888      Mapped site AVA1
*157   39890      Mapped site SMA1
P        ...      CONTINUES IN PHASE THREE


      HF  HNS              M  S   H                     A                          HM        SHHS
      GO  PCC              B  F   P                     L                          PN        TAAF
      IK  AIR              0  A   A                     U                          AL        UEEA
      C1  211              2  N   2                     1                          21        113N
             /                                                                     /          ///
TTGCATGGTGCCGGGAAGAAGCATCCGTTACCGCCGGACTGCCAAACGTCAGCGAGCTGGTTGATATGGTTTACGAGTATTGCCGGAAGCGAGGCCTGTA
----.----+----.----+----.----+----.----+----.----+----.----+----.----+----.----+----.----+----.----+ 40000
AACGTACCACGGCCCTTCTTCGTAGGCAATGGCGGCCTGACGGTTTGCAGTCGCTCGACCAACTATACCAAATGCTCATAACGGCCTTCGCTCCGGACAT

l  h  g  a  g  k  k  h  p  l  p  p  d  c  q  t  s  a  s  w  l  i  w  f  t  s  i  a  g  s  e  a  c  i
 c  m  v  p  g  r  s  i  r  y  r  r  t  a  k  r  q  r  a  g  .  y  g  l  r  v  l  p  e  a  r  p  v
  A  W  C  R  E  E  A  S  V  T  A  G  L  P  N  V  S  E  L  V  D  M  V  Y  E  Y  C  R  K  R  G  L  Y
----.----+----.----+----.----+----.----+----.----+----.----+----.----+----.----+----.----+----.----+
 n  c  p  a  p  f  f  c  g  n  g  g  s  q  w  v  d  a  l  q  n  i  h  n  v  l  i  a  p  l  s  a  q
  q  m  t  g  p  l  l  m  r  .  r  r  v  a  l  r  .  r  a  p  q  y  p  k  r  t  n  g  s  a  l  g  t
t  a  h  h  r  s  s  a  d  t  v  a  p  s  g  f  t  l  s  s  t  s  i  t  .  s  y  q  r  f  r  p  r  y

P        ...      CONTINUES IN PHASE THREE
```

```
HH     H    F              T H        B                 N A H   S H    D
PG     N    0              H H        S                 S S A   F H    V
AU     F    K              A A        T                 P U E   A A    9
22     1    1              1 1        E                 C 1 3   N 1    3

TCCGGATGCGGAGTCTTATCCGTGGAAATCAAACGCGCACTACTGGCTGGTTACCAACCTGTATCAGAACATGCGGGCCAATGCGCTTACTGATGCGGAA
----.----+----.----+----.----+----.----+----.----+----.----+----.----+----.----+----.----+----.----+ 40100
AGGCCTACGCCTCAGAATAGGCACCTTTAGTTTGCGCGTGATGACCGACCAATGGTTGGACATAGTCTTGTACGCCCGGTTACGCGAATGACTACGCCTT

 r  m  r  s  l  i  r  g  n  q  t  r  t  t  g  w  l  p  t  c  i  r  t  c  g  p  m  r  l  l  m  r  n
s  g  c  g  v  l  s  v  e  i  k  r  a  l  l  a  g  y  q  p  v  s  e  h  a  g  q  c  a  y  .  c  g  i
 P  D  A  E  S  Y  P  W  K  S  N  A  H  Y  W  L  V  T  N  L  Y  Q  N  M  R  A  N  A  L  T  D  A  F
----.----+----.----+----.----+----.----+----.----+----.----+----.----+----.----+----.----+----.----+
i  r  i  r  l  r  i  r  p  f  .  v  r  v  v  p  q  n  g  v  q  i  l  v  h  p  g  i  r  k  s  i  r  f
 d  p  h  p  t  k  d  t  s  i  l  r  a  s  s  a  p  .  w  g  t  d  s  c  a  p  w  h  a  .  q  h  p
  g  s  a  s  d  .  g  h  f  d  f  a  c  .  q  s  t  v  l  r  y  .  f  m  r  a  l  a  s  v  s  a  s

DV93   40091      dvh93              plasmid-r   sq
P       ...       CONTINUES IN PHASE THREE


        HF      A     N           M      MB  H
        AN      L     D           N      BI  P
        EU      U     E           L      ON  H
        3H      1     1           1      11  1
                                          /
TTACGCCGTAAGGCCGCAGATGAGCTTGTCCATATGACTGCGAGAATTAACCGTGGTGAGGCGATCCCTGAACCAGTAAAACAACTTCCTGTCATGGGCG
----.----+----.----+----.----+----.----+----.----+----.----+----.----+----.----+----.----+----.----+ 40200
AATGCGGCATTCCGGCGTCTACTCGAACAGGTATACTGACGCTCTTAATTGGCACCACTCCGCTAGGGACTTGGTCATTTTGTTGAAGGACAGTACCCGC

 y  a  v  r  p  q  m  s  l  s  i  .  l  r  e  l  t  v  v  r  r  s  l  n  q  .  n  n  f  l  s  w  a
  t  p  .  g  r  r  .  a  c  p  y  d  c  e  n  .  p  w  .  g  d  p  .  t  s  k  t  t  s  c  h  g  r
L  R  R  K  A  A  D  E  L  V  H  M  T  A  R  I  N  R  G  E  A  I  P  E  P  V  K  Q  L  P  V  M  G  G
----.----+----.----+----.----+----.----+----.----+----.----+----.----+----.----+----.----+----.----+
 .  a  t  l  g  c  i  l  k  d  m  h  s  r  s  n  v  t  t  l  r  d  r  f  w  y  f  l  k  r  d  h  a
i  v  g  y  p  r  l  h  a  q  g  y  s  q  s  f  .  g  h  h  p  s  g  q  v  l  l  v  v  e  q  .  p  r
 n  r  r  l  a  a  s  s  s  t  w  i  v  a  l  i  l  r  p  s  a  i  g  s  g  t  f  c  s  g  t  m  p

P       ...       CONTINUES IN PHASE THREE


 AE         SM  H          M       M     AD                   RP  M
 CC         DN  G          B       B     LD                   E   N
 CP         UL  I          0       0     UE                   N   L
 11         11  A          1       2     11                       1
 /                                                            /
GTAGACCTCTAAATCGTGCACAGGCTCTGGCGAAGATCGCAGAAATCAAAGCTAAGTTCGGACTGAAAGGAGCAAGTGTATGACGGGCAAAGAGGCAATT
----.----+----.----+----.----+----.----+----.----+----.----+----.----+----.----+----.----+----.----+ 40300
CATCTGGAGATTTAGCACGTGTCCGAGACCGCTTCTAGCGTCTTTAGTTTCGATTCAAGCCTGACTTTCCTCGTTCACATACTGCCCGTTTCTCCGTTAA

v  d  l  .  i  v  h  r  l  w  r  r  s  q  k  s  k  l  s  s  d  .  k  e  q  v  y  d  g  q  r  g  n  y
 .  t  s  k  s  c  t  g  s  g  e  d  r  r  n  q  s  .  v  r  t  e  r  s  k  c  M  T  G  K  E  A  I
  R  P  L  N  R  A  Q  A  L  A  K  I  A  E  I  K  A  K  F  G  L  K  G  A  S  V  .  r  a  k  r  q  l
----.----+----.----+----.----+----.----+----.----+----.----+----.----+----.----+----.----+----.----+
 t  s  r  .  i  t  c  l  s  q  r  l  d  c  f  d  f  s  l  e  s  q  f  s  c  t  y  s  p  c  l  p  l
  y  v  e  l  d  h  v  p  e  p  s  s  r  l  f  .  l  .  t  r  v  s  l  l  l  h  i  v  p  l  s  a  i
p  l  g  r  f  r  a  c  a  r  a  f  i  a  s  i  l  a  l  n  p  s  f  p  a  l  t  h  r  a  f  l  c  n

REN    40280      ren (orf-96)       M-NH3
P      40280      P                  COOH


      E S         H A     VHH      F T H  PC     H                    HFNTSF*
      C C         G L     0HP      N H H  QF     H                    ANSHSN1
      R R         A U     2AA      U A A  41     A                    EUPATU5
      2 1         1 1     112      H 1 1         1                    3HB12H8
                                         /                             / //
ATTCATTACCTGGGGACGCATAATAGCTTCTGTGCGCCGGACGTTGCCGCGCTAACAGGCGCAACAGTAACCAGCATAAATCAGGCCGCGGCTAAAATGG
----.----+----.----+----.----+----.----+----.----+----.----+----.----+----.----+----.----+----.----+ 40400
TAAGTAATGGACCCCTGCGTATTATCGAAGACACGCGGCCTGCAACGGCGCGATTGTCCGCGTTGTCATTGGTCGTATTTAGTCCGGCGCCGATTTTACC

  s  l  p  g  d  a  .  .  l  l  c  a  g  r  c  r  a  n  r  r  n  s  n  q  h  k  s  g  r  g  .  n  g
I  H  Y  L  G  T  H  N  S  F  C  A  P  D  V  A  A  L  T  G  A  T  V  T  S  I  N  Q  A  A  A  K  M  A
 f  i  t  w  g  r  i  i  a  s  v  r  r  t  l  p  r  .  q  a  q  q  .  p  a  .  i  r  p  r  l  k  w
----.----+----.----+----.----+----.----+----.----+----.----+----.----+----.----+----.----+----.----+
.  e  n  g  p  s  a  y  y  s  r  h  a  p  r  q  r  a  l  l  r  l  l  l  w  c  l  d  p  r  p  .  f  p
 i  .  .  r  p  v  c  l  l  k  q  a  g  s  t  a  a  s  v  p  a  v  t  v  l  m  f  .  a  a  a  l  i
  n  m  v  q  p  r  m  i  a  e  t  r  r  v  n  g  r  .  c  a  c  c  y  g  a  y  i  l  g  r  s  f  h
```

```
V021    40335     dv021              plasmid-r   sq
PQ4     40354     PQ delta 4         dl          EM
CF1     40354     cf1                dl          EM
*158    40389     Mapped site SST2
REN      ...      CONTINUES IN PHASE TWO
```

```
  MP   E              T        E                    H         E S               M      S    H   E S
  BQ   C              A        C                    P         C C               B      D    G   C C
  09   P              Q        P                    A         R R               0      U    I   R R
  2    1              1        1                    2         2 1               2      1    A   2 1

CACGGGCAGGTCTTCTGGTTATCGAAGGTAAGGTCTGGCGAACGGTGTATTACCGGTTTGCTACCAGGGAAGAACGGGAAGGAAAGATGAGCACGAACCT
----.----+----.----+----.----+----.----+----.----+----.----+----.----+----.----+----.----+----.----+ 40500
GTGCCCGTCCAGAAGACCAATAGCTTCCATTCCAGACCGCTTGCCACATAATGGCCAAACGATGGTCCCTTCTTGCCCTTCCTTTCTACTCGTGCTTGGA

 t  g  r  s  s  g  y  r  r  .  g  l  a  n  g  v  l  p  v  c  y  q  g  r  t  g  r  k  d  e  h  e  p
  R  A  G  L  L  V  I  E  G  K  V  W  R  T  V  Y  Y  R  F  A  T  R  E  E  R  E  G  K  M  S  T  N  L
h  g  q  v  f  w  l  s  k  v  r  s  g  e  r  c  i  t  g  l  l  p  g  k  n  g  k  e  r  .  a  r  t  w
----.----+----.----+----.----+----.----+----.----+----.----+----.----+----.----+----.----+----.----+
  v  p  l  d  e  p  .  r  l  y  p  r  a  f  p  t  n  g  t  q  .  w  p  l  v  p  l  f  s  s  c  s  g
a  r  a  p  r  r  t  i  s  p  l  t  q  r  v  t  y  .  r  n  a  v  l  s  s  r  s  p  f  i  l  v  f  r
 c  p  c  t  k  q  n  d  f  t  l  d  p  s  r  h  i  v  p  k  s  g  p  f  f  p  f  s  l  h  a  r  v
```

```
PQ9     40404     PQ delta 9         dl          EM
REN      ...      CONTINUES IN PHASE TWO
```

```
 PQAMCN                 F T                         VP         R              D   A           SHHNE
 43333I                 N H                         2Q         E              D   L           TAAIC
      N                 U A                         01         N              E   U           UEE21
      5                 H 1                         3                         1   1           11345
  /////                                             /                                         ////
GGTTTTTAAGGAGTGTCGCCAGAGTGCCGCGATGAAACGGGTATTGGCGGTATATGGAGTTAAAAGATGACCATCTACATTACTGAGCTAATAACAGGCC
----.----+----.----+----.----+----.----+----.----+----.----+----.----+----.----+----.----+----.----+ 40600
CCAAAAATTCCTCACAGCGGTCTCACGGCGCTACTTTGCCCATAACCGCCATATACCTCAATTTTCTACTGGTAGATGTAATGACTCGATTATTGTCCGG

g  f  .  g  v  s  p  e  c  r  d  e  t  g  i  g  g  i  w  s  .  k  m  t  i  y  i  t  e  l  i  t  g  l
 V  F  K  E  C  R  Q  S  A  A  M  K  R  V  L  A  V  Y  G  V  K  R  .  p  s  t  l  l  s  .  .  q  a
  f  l  r  s  v  a  r  v  p  r  .  n  g  y  w  r  y  m  e  l  k  d  d  h  l  h  y  .  a  n  n  r  p
----.----+----.----+----.----+----.----+----.----+----.----+----.----+----.----+----.----+----.----+
 p  k  .  p  t  d  g  s  h  r  s  s  v  p  i  p  p  i  h  l  .  f  i  v  m  .  m  v  s  s  i  v  p
  t  k  l  s  h  r  w  l  a  a  i  f  r  t  n  a  t  y  p  t  l  l  h  g  d  v  n  s  l  .  y  c  a
q  n  k  l  l  t  a  l  t  g  r  h  f  p  y  q  r  y  i  s  n  f  s  s  w  r  c  .  q  a  l  l  l  g
```

```
P4      40502     P4                 sl          EM
Q3      40502     qin3               sl          EM
A3      40502     qinA3              sl          EM
M3      40502     lambdagalM3        sl          EM
C3      40502     qinC3              sl          EM
NIN5    40502     nin5               dl          sq      N independent growth
V203    40551     dv203              plasmid-r   EM
PQ1     40551     PQdelta 1          dl          EM
REN     40567     ren                COOH
NI24    40600     nin24              dl          EM
```

```
              SHH   T M               0               H  TM
              TAA   R N               R               N  AB
              UEF   2 L               F               F  Q0
              113     1                               3  11
               //
TGCTGGTAATCGCAGGCCTTTTTATTTGGGGGAGAGGGAAGTCATGAAAAAACTAACCTTTGAAATTCGATCTCCAGCACATCAGCAAAACGCTATTCAC
----.----+----.----+----.----+----.----+----.----+----.----+----.----+----.----+----.----+----.----+ 40700
ACGACCATTAGCGTCCGGAAAAATAAACCCCCTCTCCCTTCAGTACTTTTTTGATTGGAAACTTTAAGCTAGAGGTCGTGTAGTCGTTTTGCGATAAGTG

  l  v  i  a  g  l  f  i  w  g  r  g  k  s  .  k  n  .  p  l  k  f  d  l  q  h  i  s  k  t  l  f  t
c  w  .  s  q  a  f  l  f  g  g  e  g  s  h  e  k  t  n  l  .  n  s  i  s  s  t  s  a  k  r  y  s  r
 a  g  n  r  r  p  f  y  l  g  e  r  e  v  M  K  K  L  T  F  E  I  R  S  P  A  H  Q  Q  N  A  I  H
----.----+----.----+----.----+----.----+----.----+----.----+----.----+----.----+----.----+----.----+
r  s  t  i  a  p  r  k  i  q  p  l  p  f  d  h  f  f  .  g  k  f  n  s  r  w  c  m  l  l  v  s  n  v
 q  q  y  d  c  a  k  k  n  p  p  s  p  l  .  s  f  v  l  r  q  f  e  i  e  l  v  d  a  f  r  .  e
  a  p  l  r  l  g  k  .  k  p  s  l  s  t  m  f  f  s  v  k  s  i  r  d  g  a  c  .  c  f  a  i  .
```

```
TR2     40624     tR2 terminator     3'          f       presumed genetically defined tR2
ORF     40644     nin orf-146        M-NH3
```

```
  R          E                                                                F ADE   B      A   AHPD
  S          C                                                                N LDC   B      L   SAQD
  A          P                                                                U UEP   V      U   UE8E
  1          1                                                                H 111   1      1   13A1

GCAGTACAGCAAATCCTTCCAGACCCAACCAAACCAATCGTAGTAACCATTCAGGAACGCAACCGCAGCTTAGACCAAAACAGGAAGCTATGGGCCTGCT
---.----+----.----+----.----+----.----+----.----+----.----+----.----+----.----+----.----+----.----+ 40800
CGTCATGTCGTTTAGGAAGGTCTGGGTTGGTTTGGTTAGCATCATTGGTAAGTCCTTGCGTTGGCGTCGAATCTGGTTTTGTCCTTCGATACCCGGACGA

 q  y  s  k  s  f  q  t  q  p  n  q  s  .  .  p  f  r  n  a  t  a  a  .  t  k  t  g  s  y  g  p  a
  s  t  a  n  p  s  r  p  n  q  t  n  r  s  n  h  s  g  t  q  p  q  l  r  p  k  q  e  a  m  g  l  l
A  V  Q  Q  T  L  P  D  P  T  K  P  I  V  V  T  I  Q  E  R  N  R  S  L  D  Q  N  R  K  L  W  A  C  L
---.----+----.----+----.----+----.----+----.----+----.----+----.----+----.----+----.----+----.----+
 c  y  l  l  d  k  w  v  w  g  f  w  d  y  y  g  n  l  f  a  v  a  a  .  v  l  v  p  l  .  p  g  a
r  l  v  a  f  g  e  l  g  l  w  v  l  r  l  l  w  e  p  v  c  g  c  s  l  g  f  c  s  a  i  p  r  s
 a  t  c  c  i  r  g  s  g  v  l  g  i  t  t  v  m  .  s  r  l  r  l  k  s  w  f  l  f  s  h  a  q

PQBA   40796      PQ delta 8a          dl          EM
ORF     ...       CONTINUES IN PHASE THREE


    A A   H                S      H         A F              F      EBF  H     B
    C A   P                F      G         L O              N      CBN  G     B
    Y T   H                A      U         U K              U      1VU  U     V
    1 2   1                N      2         1 1              H      51H  2     1
                                                                     /
TAGGTGACGTCTCTCGTCAGGTTGAATGGCATGGTCGCTGGCTGGATGCAGAAAGCTGGAAGTGTGTGTTTACCGCAGCATTAAAGCAGCAGGATGTTGT
---.----+----.----+----.----+----.----+----.----+----.----+----.----+----.----+----.----+----.----+ 40900
ATCCACTGCAGAGAGCAGTCCAACTTACCGTACCAGCGACCGACCTACGTCTTTCGACCTTCACACACAAATGGCGTCGTAATTTCGTCGTCCTACAACA

.  v  t  s  l  v  r  l  n  g  m  v  a  g  w  m  q  k  a  g  s  v  c  l  p  q  h  .  s  s  r  m  l  f
r  .  r  l  s  s  g  .  m  a  w  s  l  a  g  c  r  k  l  e  v  c  v  y  r  s  i  k  a  a  g  c  c
 G  D  V  S  R  Q  V  E  W  H  G  R  W  L  D  A  E  S  W  K  C  V  F  T  A  A  L  K  Q  Q  D  V  V
---.----+----.----+----.----+----.----+----.----+----.----+----.----+----.----+----.----+----.----+
.  t  v  d  r  t  l  n  f  p  m  t  a  p  q  i  c  f  a  p  l  t  h  k  g  c  c  .  l  l  l  i  n
 l  h  r  r  e  d  p  q  i  a  h  d  s  a  p  h  l  f  s  s  t  h  t  .  r  l  m  l  a  a  p  h  q
k  p  s  t  e  r  .  t  s  h  c  p  r  q  s  s  a  s  l  q  f  h  t  n  v  a  a  n  f  c  c  s  t  t

ORF     ...       CONTINUES IN PHASE THREE


  F     HNS                   HH   H*SE  H          HF         A        A        H
  O     PCC                   AA   I1FC  G          NO         L        L        N
  K     AIR                   EE   N5A1  U          FK         U        U        F
  1     211                   13   29N5  2          31         1        1        3
         /                     /    //
TCCTAACCTTGCCGGGAATGGCTTTGTGGTAATAGGCCAGTCAACCAGCAGGATGCGTGTAGGCGAATTTGCGGAGCTATTAGAGCTTATACAGGCATTC
---.----+----.----+----.----+----.----+----.----+----.----+----.----+----.----+----.----+----.----+ 41000
AGGATTGGAACGGCCCTTACCGAAACACCATTATCCGGTCAGTTGGTCGTCCTACGCACATCCGCTTAAACGCCTCGATAATCTCGAATATGTCCGTAAG

 l  t  l  p  g  m  a  l  w  .  .  a  s  q  p  a  g  c  v  .  a  n  l  r  s  y  .  s  l  y  r  h  s
s  .  p  c  r  e  w  l  c  g  n  r  p  v  n  q  q  d  a  c  r  r  i  c  g  a  i  r  a  y  t  g  i  r
 P  N  L  A  G  N  G  F  V  V  I  G  Q  S  T  S  R  M  R  V  G  E  F  A  E  L  L  E  L  I  Q  A  F
---.----+----.----+----.----+----.----+----.----+----.----+----.----+----.----+----.----+----.----+
n  r  v  k  g  p  i  a  k  h  y  y  a  l  .  g  a  p  h  t  y  a  f  k  r  l  .  .  l  k  y  l  c  e
e  .  g  q  r  s  h  s  q  p  l  l  g  t  l  w  c  s  a  h  l  r  i  q  p  a  i  l  a  .  v  p  m
 g  l  r  a  p  f  p  k  t  t  i  p  w  d  v  l  l  i  r  t  p  s  n  a  s  s  n  s  s  i  c  a  n

*159   40942      Mapped site HIN2
ORF     ...       CONTINUES IN PHASE THREE


 R     T                                                             B             F  OO    D
 S     N                                                             B             N  RR    8
 A     1                                                             V             U  FF    6
 1     0                                                             1             H
                                                                                       /
GGTACAGAGCGTGGCGTTAAGTGGTCAGACGAAGCGAGACTGGCTCTGGAGTGGAAAGCGAGATGGGGAGACAGGGCTGCATGATAAATGTCGTTAGTTT
---.----+----.----+----.----+----.----+----.----+----.----+----.----+----.----+----.----+----.----+ 41100
CCATGTCTCGCACCGCAATTCACCAGTCTGCTTCGCTCTGACCGAGACCTCACCTTTCGCTCTACCCCTCTGTCCCGACGTACTATTTACAGCAATCAAA

 v  q  s  v  a  l  s  g  q  t  k  r  d  w  l  w  s  g  k  r  d  g  e  t  g  l  h  d  k  c  r  .  f
  y  r  a  w  r  .  v  v  r  r  s  e  t  g  s  g  v  e  s  e  m  g  r  q  g  c  M  I  N  V  V  S  F
G  T  E  R  G  V  K  W  S  D  E  A  R  L  A  L  E  W  K  A  R  W  G  D  R  A  A  .  .  m  s  l  v  s
---.----+----.----+----.----+----.----+----.----+----.----+----.----+----.----+----.----+----.----+
 t  c  l  t  a  n  l  p  .  v  f  r  s  q  s  q  l  p  f  r  s  p  s  v  p  s  c  s  l  h  r  .  n
r  y  l  a  h  r  .  t  t  l  r  l  s  v  p  e  p  t  s  l  s  i  p  l  c  p  q  m  i  f  t  t  l  k
 p  v  s  r  p  t  l  h  d  s  s  a  l  s  a  r  s  h  f  a  l  h  p  s  l  a  a  h  y  i  d  n  t
```

```
TN10   41011    627             i(Tn10)    sq
ORF    41081    nin orf-290     M-NH3
ORF    41081    nin orf-146     COOH
D86    41090    delta 86        dl         EM
```

```
 H         A A                    B                                   F
 P         C A                    1                                   O
 A         Y T                                                        K
 2         1 2                                                        1

CTCCGGTGGCAGGACGTCAGCATATTTGCTCTGGCTAATGGAGCAAAAGCGACGGGCAGGTAAAGACGTGCATTACGTTTTCATGGATACAGGTTGTGAA
---.----+----.----+----.----+----.----+----.----+----.----+----.----+----.----+----.----+----.----+  41200
GAGGCCACCGTCCTGCAGTCGTATAAACGAGACCGATTACCTCGTTTTCGCTGCCCGTCCATTTCTGCACGTAATGCAAAAGTACCTATGTCCAACACTT

l r w q d v s i f a l a n g a k a t g r . r r a l r f h g y r l . t
 S G G R T S A Y L L W L M E Q K R R A G K D V H Y V F M D T G C E
 p v a g r q h i c s g . w s k s d g q v k t c i t f s w i q v v n
---.----+----.----+----.----+----.----+----.----+----.----+----.----+----.----+----.----+----.----+
 r r h c s t l m n a r a l p a f a v p l y l r a n r k . p y l n h
 e p p l v d a y k s q s i s c f r r a p l s t c . t k m s v p q s
e g t a p r . c i q e p . h l l l s p c t f v h m v n e h i c t t f
```

```
B1     41139    b1              i(IS2)     EM
ORF     ...     CONTINUES IN PHASE TWO
```

```
                                 H                             E   NSH A3
                                 P                             C   CCP L0
                                 H                             R   IRA U0
                                 1                             V   112 10
                                                                    //  /
CATCCAATGACATATCGGTTTGTCAGGGAAGTTGTGAAGTTCTGGGATATACCGCTCACCGTATTGCAGGTTGATATCAACCCGGAGCTTGGACAGCCAA
---.----+----.----+----.----+----.----+----.----+----.----+----.----+----.----+----.----+----.----+  41300
GTAGGTTACTGTATAGCCAAACAGTCCCTTCAACACTTCAAGACCCTATATGGCGAGTGGCATAACGTCCAACTATAGTTGGGCCTCGAACCTGTCGGTT

 s n d i s v c q g s c e v l g y t a h r i a g . y q p g a w t a k
H P M T Y R F V R E V V K F W D I P L T V L Q V D I N P E L G Q P N
 i q . h i g l s g k l . s s g i y r s p y c r l i s t r s l d s q
---.----+----.----+----.----+----.----+----.----+----.----+----.----+----.----+----.----+----.----+
v d l s m d t q . p l q s t r p y v a . r i a p q y . g p a q v a l
 c g i v y r n t l s t t f n q s i g s v t n c t s i l g s s p c g
 m w h c i p k d p f n h l e p i y r e g y q l n i d v r l k s l w
```

```
3000   41287    b3000           dl         EM
ORF     ...     CONTINUES IN PHASE TWO
```

```
                          TH  H                   C*T
                          HN  G                   L1A
                          AF  A                   A6Q
                          13  1                   101
                           /                       //
ATGGTTATACGGTATGGGAACCAAAGGATATTCAGACGCGAATGCCTGTTCTGAAGCCATTTATCGATATGGTAAAGAAATATGGCACTCCATACGTCGG
---.----+----.----+----.----+----.----+----.----+----.----+----.----+----.----+----.----+----.----+  41400
TACCAATATGCCATACCCTTGGTTTCCTATAAGTCTGCGCTTACGGACAAGACTTCGGTAAATAGCTATACCATTTCTTTATACCGTGAGGTATGCAGCC

 w l y g m g t k g y s d a n a c s e a i y r y g k e i w h s i r r
  G Y T V W E P K D I Q T R M P V L K P F I D M V K K Y G T P Y V G
m v i r y g n q r i f r r e c l f . s h l s i w . r n m a l h t s a
---.----+----.----+----.----+----.----+----.----+----.----+----.----+----.----+----.----+----.----+
 h n y p i p v l p y e s a f a q e s a m . r y p l s i h c e m r r
f p . v t h s g f s i . v r i g t r f g n i s i t f f y p v g y t p
 i t i r y p f w l i n l r s h r n q l w k d i h y l f i a s w v d
```

```
*160   41364    Mapped site CLA1
ORF     ...     CONTINUES IN PHASE TWO
```

```
F  HT                       H                             M
N  HH                       P                             N
U  AA                       H                             L
H  11                       1                             1
    /
CGGCGCGTTCTGCACTGACAGATTAAAACTCGTTCCCTTCACCAAATACTGTGATGACCATTTCGGGCGAGGGAATTACACCACGTGGATTGGCATCAGA
----.----+----.----+----.----+----.----+----.----+----.----+----.----+----.----+----.----+----.----+ 41500
GCCGCGCAAGACGTGACTGTCTAATTTTGAGCAAGGGAAGTGGTTTATGACACTACTGGTAAAGCCCGCTCCCTTAATGTGGTGCACCTAACCGTAGTCT

r  r  v  l  h  .  q  i  k  t  r  s  l  h  q  i  l  .  .  p  f  r  a  r  e  l  h  h  v  d  w  h  q  s
 G  A  F  C  T  D  R  L  K  L  V  P  F  T  K  Y  C  D  D  H  F  G  R  G  N  Y  T  T  W  I  G  I  R
  a  r  s  a  l  t  d  .  n  s  f  p  s  p  n  t  v  m  t  i  s  g  e  g  i  t  p  r  g  l  a  s  e
----.----+----.----+----.----+----.----+----.----+----.----+----.----+----.----+----.----+----.----+
 r  r  t  r  c  q  c  i  l  v  r  e  r  .  w  i  s  h  h  g  n  r  a  l  s  n  c  w  t  s  q  c  .
  p  a  n  q  v  s  l  n  f  s  t  g  k  v  l  y  q  s  s  w  k  p  r  p  f  .  v  v  h  i  p  m  l
a  a  r  e  a  s  v  s  .  f  e  n  g  e  g  f  v  t  i  v  m  e  p  s  p  i  v  g  r  p  n  a  d  s

ORF      ...      CONTINUES IN PHASE TWO
```

```
AS          F           E S H       E                                         E  M   M
LF          N           C C N       C                                         C  B   N
UA          U           R R F       R                                         R  0   L
1N          H           2 1 1       V                                         V  2   1

GCTGATGAACCGAAGCGGCTAAAGCCAAAGCCTGGAATCAGATATCTTGCTGAACTGTCAGACTTTGAGAAGGAAGATATCCTCGCATGGTGGAAGCAAC
----.----+----.----+----.----+----.----+----.----+----.----+----.----+----.----+----.----+----.----+ 41600
CGACTACTTGGCTTCGCCGATTTCGGTTTCGGACCTTAGTCTATAGAACGACTTGACAGTCTGAAACTCTTCCTTCTATAGGAGCGTACCACCTTCGTTG

  .  .  t  e  a  a  k  a  k  a  w  n  q  i  s  c  .  t  v  r  l  .  e  g  r  y  p  r  m  v  e  a  t
A  D  E  P  K  R  L  K  P  K  P  G  I  R  Y  L  A  E  L  S  D  F  E  K  E  D  I  L  A  W  W  K  Q  Q
 l  m  n  r  s  g  .  s  q  s  l  e  s  d  i  l  l  n  c  q  t  l  r  r  k  i  s  s  h  g  g  s  n
----.----+----.----+----.----+----.----+----.----+----.----+----.----+----.----+----.----+----.----+
l  q  h  v  s  a  a  l  a  l  a  q  f  .  i  d  q  q  v  t  l  s  q  s  p  l  y  g  r  m  t  s  a  v
 a  s  s  g  f  r  s  f  g  f  g  p  i  l  y  r  a  s  s  d  s  k  s  f  s  s  i  r  a  h  h  f  c
  s  i  f  r  l  p  .  l  w  l  r  s  d  s  i  k  s  f  q  .  v  k  l  l  f  i  d  e  c  p  p  l  l

ORF      ...      CONTINUES IN PHASE TWO
```

```
  H T          H                                                              I    M  M
  N A          P                                                              M    N  N
  F Q          A                                                              M    L  L
  3 1          2                                                              L    1  1

AACCATTCGATTTGCAAATACCGGAACATCTCGGTAACTGCATATTCTGCATTAAAAAATCAACGCAAAAAATCGGACTTGCCTGCAAAGATGAGGAGGG
----.----+----.----+----.----+----.----+----.----+----.----+----.----+----.----+----.----+----.----+ 41700
TTGGTAAGCTAAACGTTTATGGCCTTGTAGAGCCATTGACGTATAAGACGTAATTTTTTAGTTGCGTTTTTTAGCCTGAACGGACGTTTCTACTCCTCCC

  t  i  r  f  a  n  t  g  t  s  r  .  l  h  i  l  h  .  k  i  n  a  k  n  r  t  c  l  q  r  .  g  g
 P  F  D  L  Q  I  P  E  H  L  G  N  C  I  F  C  I  K  K  S  T  Q  K  I  G  L  A  C  K  D  E  E  G
n  h  s  i  c  k  y  r  n  i  s  v  t  a  y  s  a  l  k  n  q  r  k  k  s  d  l  p  a  k  m  r  r  d
----.----+----.----+----.----+----.----+----.----+----.----+----.----+----.----+----.----+----.----+
  v  m  r  n  a  f  v  p  v  d  r  y  s  c  i  r  c  .  f  i  l  a  f  f  r  v  q  r  c  l  h  p  p
c  g  n  s  k  c  i  g  s  c  r  p  l  q  m  n  q  m  l  f  d  v  c  f  i  p  s  a  q  l  s  s  s  p
 l  w  e  i  q  l  y  r  f  m  e  t  v  a  y  e  a  n  f  f  .  r  l  f  d  s  k  g  a  f  i  l  l

IMML   41679        imml                 sr            EM
ORF      ...      CONTINUES IN PHASE TWO
```

```
   F      M  B           B*BXMB                                             M   RT                 M
   N      N  B           I1AHBI                                             N   SN                 B
   U      L  V           N6MOON                                             L   A1                 0
   H      1  1           111211                                             1   10                 2
                           ////
ATTGCAGCGTGTTTTTAATGAGGTCATCACGGGATCCCATGTGCGTGACGGACATCGGGAAACGCCAAAGGAGATTATGTACCGAGGAAGAATGTCGCTG
----.----+----.----+----.----+----.----+----.----+----.----+----.----+----.----+----.----+----.----+ 41800
TAACGTCGCACAAAAATTACTCCAGTAGTGCCCTAGGGTACACGCACTGCCTGTAGCCCTTTGCGGTTTCCTCTAATACATGGCTCCTTCTTACAGCGAC

i  a  a  c  f  .  .  g  h  h  g  i  p  c  a  .  r  t  s  g  n  a  k  g  d  y  v  p  r  k  n  v  a  g
 L  Q  R  V  F  N  E  V  I  T  G  S  H  V  R  D  G  H  R  E  T  P  K  E  I  M  Y  R  G  R  M  S  L
  c  s  v  f  l  m  r  s  s  r  d  p  m  c  v  t  d  i  g  k  r  q  r  r  l  c  t  e  e  e  c  r  w
----.----+----.----+----.----+----.----+----.----+----.----+----.----+----.----+----.----+----.----+
 i  a  a  h  k  .  h  p  .  .  p  i  g  h  a  h  r  v  d  p  f  a  l  p  s  .  t  g  l  f  f  t  a
  n  c  r  t  k  l  s  t  m  v  p  d  w  t  r  s  p  c  r  s  v  g  f  s  i  i  y  r  p  l  i  d  s
s  q  l  t  n  k  i  l  d  d  r  s  g  m  h  t  v  s  m  p  f  r  w  l  l  n  h  v  s  s  s  h  r  q
```

```
*161  41732     Mapped site BAM1
TN10  41781       623                     i(Tn10)      sq
ORF   ...       CONTINUES IN PHASE TWO


         NT                                               R    A      HH T     H           D H
         RH                                               S    L      NN A     P           D N
         UA                                               A    U      FF Q     A           E F
         11                                               1    1      13 1     2           1 1
          /                                                             /
GACGGTATCGCGAAAATGTATTCAGAAAATGATTATCAAGCCCTGTATCAGGACATGGTACGAGCTAAAAGATTCGATACCGGCTCTTGTTCTGAGTCAT
----.----+----.----+----.----+----.----+----.----+----.----+----.----+----.----+----.----+----.----+ 41900
CTGCCATAGCGCTTTTACATAAGTCTTTTACTAATAGTTCGGGACATAGTCCTGTACCATGCTCGATTTTCTAAGCTATGGCCGAGAACAAGACTCAGTA

  r  y  r  e  n  v  f  r  k  .  l  s  s  p  v  s  g  h  g  t  s  .  k  i  r  y  r  l  l  f  .  v  ■
D  G  I  A  K  M  Y  S  E  N  D  Y  Q  A  L  Y  Q  D  M  V  R  A  K  R  F  D  T  G  S  C  S  E  S  C
 t  v  s  r  k  c  i  q  k  ■  i  i  k  p  c  i  r  t  w  y  e  l  k  d  s  i  p  a  l  v  l  s  h
----.----+----.----+----.----+----.----+----.----+----.----+----.----+----.----+----.----+----.----+
p  r  y  r  s  f  t  n  l  f  h  n  d  l  g  t  d  p  c  p  v  l  .  f  i  r  y  r  s  k  n  q  t  ■
 s  p  i  a  f  i  y  e  s  f  s  .  .  a  r  y  .  s  ■  t  r  a  l  l  n  s  v  p  e  q  e  s  d
  v  t  d  r  f  h  i  .  f  i  i  i  l  g  q  i  l  v  h  y  s  s  f  s  e  i  g  a  r  t  r  l  .

ORF   ...       CONTINUES IN PHASE TWO


    M            F A   P   TBM B         S  AF   OO                     3         H           M
    N            N L   Q   ABN B         F  LN   RR                     0         P           B
    L            U U   9   QVL V         A  UU   FF                     0         H           O
    1            H 1       111 1         N  1H                          0         1           2
                                                  /
GCGAAATATTTGGAGGGCAGCTTGATTTCGACTTCGGGAGGGAAGCTGCATGATGCGATGTTATCGGTGCGGTGAATGCAAAGAAGATAACCGCTTCCGA
----.----+----.----+----.----+----.----+----.----+----.----+----.----+----.----+----.----+----.----+ 42000
CGCTTTATAAACCTCCCGTCGAACTAAAGCTGAAGCCCTCCCTTCGACGTACTACGCTACAATAGCCACGCCACTTACGTTTCTTCTATTGGCGAAGGCT

 r  n  i  w  r  a  a  .  f  r  l  r  e  g  s  c  M  M  R  C  Y  R  C  G  E  C  K  E  D  N  R  F  R
  E  I  F  G  G  Q  L  D  F  D  F  G  R  E  A  A  .  c  d  v  i  g  a  v  n  a  k  k  i  t  a  s  d
a  k  y  l  e  g  s  l  i  s  t  s  g  g  k  l  h  d  a  ■  l  s  v  r  .  ■  q  r  r  .  p  l  p  t
----.----+----.----+----.----+----.----+----.----+----.----+----.----+----.----+----.----+----.----+
  r  f  i  q  l  a  a  q  n  r  s  r  s  p  l  q  ■  i  r  h  .  r  h  p  s  h  l  s  s  l  r  k  r
h  s  i  n  p  p  c  s  s  k  s  k  p  l  s  a  a  h  h  s  t  i  p  a  t  f  a  f  f  i  v  a  e  s
 a  f  y  k  s  p  l  k  i  e  v  e  p  p  f  s  c  s  a  i  n  d  t  r  h  i  c  l  l  y  g  s  g

PQ9   41924     PQ delta 9           dr         EM
ORF   41950     nin orf-57           M-NH3
ORF   41950     nin orf-290          COOH
3000  41973     b3000                dr         EM


                 H  C*T         H             H                          M          A    O
                 N  L1A         P             G                          N          F    R
                 F  A6Q         A             E                          L          L    F
                 1  121         2             2                          1          3
                     //
CCAAATCAACCTTACTGGAATCGATGGTGTCTCCGGTGTGAAAGAACACCAACAGGGGTGTTACCACTACCGCAGGAAAAGGAGGACGTGTGGCGAGACA
----.----+----.----+----.----+----.----+----.----+----.----+----.----+----.----+----.----+----.----+ 42100
GGTTTAGTTGGAATGACCTTAGCTACCACAGAGGCCACACTTTCTTGTGGTTGTCCCCACAATGGTGATGGCGTCCTTTTCCTCCTGCACACCGCTCTGT

P  N  Q  P  Y  W  N  R  W  C  L  R  C  E  R  T  P  T  G  V  L  P  L  P  Q  E  K  E  D  V  W  R  D  S
 q  i  n  l  t  g  i  d  g  v  s  g  v  k  e  h  q  q  g  c  y  h  y  r  r  k  r  r  t  c  g  e  t
  k  s  t  l  l  e  s  ■  v  s  p  v  .  k  n  t  n  r  g  v  t  t  t  a  g  k  g  g  r  V  A  R  Q
----.----+----.----+----.----+----.----+----.----+----.----+----.----+----.----+----.----+----.----+
 g  f  .  g  .  q  f  r  h  h  r  r  h  s  l  v  g  v  p  t  n  g  s  g  c  s  f  s  s  t  h  r  s
  w  i  l  r  v  p  i  s  p  t  e  p  t  f  s  c  w  c  p  h  .  w  .  r  l  f  l  l  v  h  p  s  v
v  l  d  v  k  s  s  d  i  t  d  g  t  h  f  f  v  l  l  p  t  v  v  v  a  p  f  p  p  r  t  a  l  c

*162  42021     Mapped site CLA1
ORF   42090       nin orf-60         V-NH3
```

```
 H                  O                                                             H
 P                  R                                                             N
 H                  F                                                             F
 1                                                                                1

GCGACGAAGTATCACCGACATAATCTGCGAAAACTGCAAATACCTTCCAACGAAACGCACCAGAAATAAACCCAAGCCAATCCCAAAAGAATCTGACGTA
---------+---------+---------+---------+---------+---------+---------+---------+---------+---------+  42200
CGCTGCTTCATAGTGGCTGTATTAGACGCTTTTGACGTTTATGGAAGGTTGCTTTGCGTGGTCTTTATTTGGGTTCGGTTAGGGTTTTCTTAGACTGCAT

  D  E  V  S  P  T  .  s  a  k  t  a  n  t  f  q  r  n  a  p  e  i  n  p  s  q  s  q  k  n  l  t  .
 a  t  k  y  h  r  h  n  l  r  k  l  q  i  p  s  n  e  t  h  q  k  .  t  q  a  n  p  k  r  i  .  r  k
 R  R  S  I  T  D  I  I  C  E  N  C  K  Y  L  P  T  K  R  T  R  N  K  P  K  P  I  P  K  E  S  D  V
---------+---------+---------+---------+---------+---------+---------+---------+---------+---------+
 l  s  s  t  d  g  v  y  d  a  f  v  a  f  v  k  w  r  f  a  g  s  i  f  g  l  w  d  w  f  f  r  v  y
  a  v  f  y  .  r  c  l  r  r  f  s  c  i  g  e  l  s  v  c  w  f  y  v  w  a  l  g  l  l  i  q  r
  r  r  l  i  v  s  m  i  q  s  f  q  l  y  r  g  v  f  r  v  l  f  l  g  l  g  i  g  f  s  d  s  t

ORF   42120      nin orf-57           COOH
```

```
     H                  EH    D   AM A               OO      H  T     H
     P                  CP    D   CN A               RR      P  A     G
     H                  RA    E   YL T               FF      H  Q     A
     1                  V2    1   11 2                       1  1     1
                                                     /
AAAACCTTCAACTACACGGCTCACCTGTGGGATATCCGGTGGCTAAGACGTCGTGCGAGGAAAACAAGGTGATTGACCAAAATCGAAGTTACGAACAAGA
---------+---------+---------+---------+---------+---------+---------+---------+---------+---------+  42300
TTTTGGAAGTTGATGTGCCGAGTGGACACCCTATAGGCCACCGATTCTGCAGCACGCTCCTTTTGTTCCACTAACTGGTTTTAGCTTCAATGCTTGTTCT

 k  p  s  t  t  r  l  t  c  g  i  s  g  g  .  d  v  v  r  g  k  q  q  d  .  p  k  s  k  l  r  t  r
  n  l  q  l  h  g  s  p  v  g  y  p  v  a  k  t  s  c  e  e  n  k  V  I  D  Q  N  R  S  Y  F  Q  E
 K  T  F  N  Y  T  A  H  L  W  D  I  R  W  L  R  R  R  A  R  K  T  R  .  l  t  k  i  e  v  t  n  k  k
---------+---------+---------+---------+---------+---------+---------+---------+---------+---------+
  f  g  e  v  v  r  s  v  q  p  i  d  p  p  .  s  t  t  r  p  f  c  p  s  q  g  f  d  f  n  r  v  l
 l  f  r  .  s  c  p  e  g  t  p  y  g  t  a  l  v  d  h  s  s  f  l  t  i  s  w  f  r  l  .  s  c  s
  f  v  k  l  .  v  a  .  r  h  s  i  r  h  s  l  r  r  a  l  f  v  l  h  n  v  l  i  s  t  v  f  l

ORF   42269      nin orf-56           V-NH3
ORF   42269      nin orf-60           COOH
```

```
   T    A D     H  B      AF   N       A  B   S   HE F  MH          MB
   A    L 8     H  B      LN   S       F  B   D   GC N  SH          BI
   Q    U 6     A  V      UU   P       L  V   U   II U  TA          ON
   1    1       1  1      1H   C       3  1   1   A5 H  11          11
                                                   /                  /
AAGCGTCGAGCGAGCTTTAACGTGCGCTAACTGCGGTCAGAAGCTGCATGTGCTGGAAGTTCACGTGTGTGAGCACTGCTGCGCAGAACTGATGAGCGAT
---------+---------+---------+---------+---------+---------+---------+---------+---------+---------+  42400
TTCGCAGCTCGCTCGAAATTGCACGCGATTGACGCCAGTCTTCGACGTACACGACCTTCAAGTGCACACACTCGTGACGACGCGTCTTGACTACTCGCTA

 k  r  r  a  s  f  n  v  r  .  l  r  s  e  a  a  c  a  g  s  s  r  v  .  a  l  l  r  r  t  d  e  r  s
  S  V  E  R  A  L  T  C  A  N  C  G  Q  K  L  H  V  L  E  V  H  V  C  E  H  C  C  A  E  L  M  S  D
  a  s  s  e  l  .  r  a  l  t  a  v  r  s  c  m  c  w  k  f  t  c  v  s  t  a  a  q  n  .  .  a  i
---------+---------+---------+---------+---------+---------+---------+---------+---------+---------+
  f  r  r  a  l  k  l  t  r  .  s  r  d  s  a  a  h  a  p  l  e  r  t  h  a  s  s  r  l  v  s  s  r
   l  t  s  r  a  k  v  h  a  l  q  p  .  f  s  c  t  s  s  t  .  t  h  s  c  q  q  a  s  s  i  l  s
 f  a  d  l  s  s  .  r  a  s  v  a  t  l  l  q  m  h  q  f  n  v  h  t  l  v  a  a  c  f  q  h  a  i

D86   42317      delta 86         dr          EM
ORF    ...      CONTINUES IN PHASE TWO
```

```
HS    A TM              O    M OM     HT         M          HNS H  X         H
NF    L AN              R    B RB     HH         B          PCC P  M         N
FA    U QL              F    0 F0     AA         0          AIR H  N         F
3N    1 11                   2 2      11         2          211 1  1         3
                                       /                      /
CCGAATAGCTCGATGCACGAGGAAGAAGATGATGGCTAAACCAGCGCGAAGACGATGTAAAAACGATGAATGCCGGGAATGGTTTCACCCTGCATTCGCT
---------+---------+---------+---------+---------+---------+---------+---------+---------+---------+  42500
GGCTTATCGAGCTACGTGCTCCTTCTTCTACTACCGATTTGGTCGCGCTTCTGCTACATTTTTGCTACTTACGGCCCTTACCAAAGTGGGACGTAAGCGA

  e  .  l  d  a  r  g  r  r  .  w  l  n  q  r  e  d  d  v  k  t  m  n  a  g  n  g  f  t  l  h  s  l
 P  N  S  S  M  H  E  E  E  D  D  G  .  t  s  a  k  t  m  .  k  r  .  m  p  g  m  v  s  p  c  i  r  .
  r  i  a  r  c  t  r  k  k  M  M  A  K  P  A  R  R  R  C  K  N  D  E  C  R  E  W  F  H  P  A  F  A
---------+---------+---------+---------+---------+---------+---------+---------+---------+---------+
 d  s  y  s  s  a  r  p  l  l  h  h  s  f  w  r  s  s  s  t  f  v  i  f  a  p  f  p  k  v  r  c  e  s
  g  f  l  e  i  c  s  s  s  s  p  .  v  l  a  f  v  i  y  f  r  h  i  g  p  i  t  e  g  q  m  r
  r  i  a  r  h  v  l  f  f  i  i  a  l  g  a  r  l  r  h  l  f  s  s  h  r  s  h  n  .  g  a  n  a
```

```
ORF     42429      nin orf-204        M-NH3
ORF     42436      nin orf-56         COOH


          S    H                           T                    T                EF          B   M
          D    G                           A                    H                CN          B   N
          U    I                           Q                    A                1U          V   L
          1    A                           1                    1                5H          1   1

AATCAGTGGTGGTGCTCTCCAGAGTGTGGAACCAAGATAGCACTCGAACGACGAAGTAAAGAACGCGAAAAAGCGGAAAAAGCAGCAGAGAAGAAACGAC
----.----+----.----+----.----+----.----+----.----+----.----+----.----+----.----+----.----+----.----+  42600
TTAGTCACCACCACGAGAGGTCTCACACCTTGGTTCTATCGTGAGCTTGCTGCTTCATTTCTTGCGCTTTTTCGCCTTTTTCGTCGTCTCTTCTTTGCTG

 i  s  g  g  a  l  q  s  v  e  p  r  .  h  s  n  d  e  v  k  n  a  k  k  r  k  k  q  q  r  r  n  d
  s  v  v  v  l  s  r  v  w  n  q  d  s  t  r  t  t  k  .  r  t  r  k  s  g  k  s  s  r  e  e  t  t
N  Q  W  W  C  S  P  E  C  G  T  K  I  A  L  E  R  R  S  K  E  R  E  K  A  E  K  A  A  E  K  K  R  R
----.----+----.----+----.----+----.----+----.----+----.----+----.----+----.----+----.----+----.----+
 i  l  p  p  a  r  w  l  t  s  g  l  y  c  e  f  s  s  t  f  f  a  f  f  r  f  f  c  c  l  l  f  s
.  d  t  t  t  s  e  l  t  h  f  w  s  l  v  r  v  v  f  y  l  v  r  f  l  p  f  l  l  l  s  s  v  v
l  .  h  h  h  e  g  s  h  p  v  l  i  a  s  s  r  r  l  l  s  r  s  f  a  s  f  a  a  s  f  f  r

ORF      ...      CONTINUES IN PHASE THREE


  M                         A    HH  AT      D             P
  B                         F    NN  SA      V             Q
  0                         L    FF  UQ      1             8
  2                         2    13  21                    A
                                  /   /
GACGAGAGGAGCAGAAACAGAAAGATAAACTTAAGATTCGAAAACTCGCCTTAAAGCCCCGCAGTTACTGGATTAAACAAGCCCAACAAGCCGTAAACGC
----.----+----.----+----.----+----.----+----.----+----.----+----.----+----.----+----.----+----.----+  42700
CTGCTCTCCTCGTCTTTGTCTTTCTATTTGAATTCTAAGCTTTTGAGCGGAATTTCGGGGCGTCAATGACCTAATTTGTTCGGGTTGTTCGGCATTTGCG

d  e  r  s  r  n  r  k  i  n  l  r  f  e  n  s  p  .  s  p  a  v  t  g  l  n  k  p  n  k  p  .  t  p
 t  r  g  a  e  t  e  r  .  t  .  d  s  k  t  r  l  k  a  p  q  l  l  d  .  t  s  p  t  s  r  k  r
  R  E  E  Q  K  Q  K  D  K  L  K  I  R  K  L  A  L  K  P  R  S  Y  W  I  K  Q  A  Q  Q  A  V  N  A
----.----+----.----+----.----+----.----+----.----+----.----+----.----+----.----+----.----+----.----+
 s  s  l  l  l  f  l  f  i  f  k  l  n  s  f  e  g  .  l  g  a  t  v  p  n  f  l  g  l  l  g  y  v
  v  l  p  a  s  v  s  l  y  v  .  s  e  f  v  r  r  l  a  g  c  n  s  s  .  v  l  g  v  l  r  l  r
r  r  s  s  c  f  c  f  s  l  s  l  i  r  f  s  a  k  f  g  r  l  .  q  i  l  c  a  w  c  a  t  f  a

DVL1    42645      dvl                plasmid-r    sq
PQ8A    42660      PQ delta 8a        dr           EM
ORF      ...       CONTINUES IN PHASE THREE


               E     T                                     S D     H     H      BF H      E   F  FH
               C     H                                     F D     G     P      BO P      C   N  NG
               P     A                                     A E     U     A      VK A      1   U  UT
               1     1                                     N 1     2     2      11 2      5   H  HC

CTTCATCAGAGAAAGAGACCGCGACTTACCATGTATCTCGTGCGGAACGCTCACGTCTGCTCAGTGGGATGCCGGACATTACCGGACAACTGCTGCGGCA
----.----+----.----+----.----+----.----+----.----+----.----+----.----+----.----+----.----+----.----+  42800
GAAGTAGTCTCTTTCTCTGGCGCTGAATGGTACATAGAGCACGCCTTGCGAGTGCAGACGAGTCACCCTACGGCCTGTAATGGCCTGTTGACGACGCCGT

  s  s  e  k  e  t  a  t  y  h  v  s  r  a  e  r  s  r  l  l  s  g  m  p  d  i  t  g  q  l  l  r  h
l  h  q  r  k  r  p  r  l  t  m  y  l  v  r  n  a  h  v  c  s  v  g  c  r  t  l  p  d  n  c  c  g  t
 F  I  R  E  R  D  R  D  L  P  C  I  S  C  G  T  L  T  S  A  Q  W  D  A  G  H  Y  R  T  T  A  A  A
----.----+----.----+----.----+----.----+----.----+----.----+----.----+----.----+----.----+----.----+
g  e  d  s  f  s  v  a  v  .  w  t  d  r  a  s  r  e  r  r  s  l  p  i  g  s  m  v  p  c  s  s  r  c
 r  .  .  l  f  l  g  r  s  v  m  y  r  t  r  f  a  .  t  q  e  t  p  h  r  v  n  g  s  l  q  q  p
  k  m  l  s  l  s  r  s  k  g  h  i  e  h  p  v  s  v  d  a  .  h  s  a  p  c  .  r  v  v  a  a  a

ORF      ...      CONTINUES IN PHASE THREE
```

```
           M                                        T                     H        T T
           N                                        T                     G        H A
           L                                        H                     A        A Q
           1                                        2                     1        1 1

CCTCAACTCCGATTTAATGAACGCAATATTCACAAGCAATGCGTGGTGTGCAACCAGCACAAAAGCGGAAATCTCGTTCCGTATCGCGTCGAACTGATTA
----.----+----.----+----.----+----.----+----.----+----.----+----.----+----.----+----.----+----.----+  42900
GGAGTTGAGGCTAAATTACTTGCGTTATAAGTGTTCGTTACGCACCACACGTTGGTCGTGTTTTCGCCTTTAGAGCAAGGCATAGCGCAGCTTGACTAAT

 l  n  s  d  l  m  n  a  i  f  t  s  n  a  w  c  a  t  s  t  k  a  e  i  s  f  r  i  a  s  n  .  l
  s  t  p  i  .  .  t  q  y  s  q  a  m  r  g  v  q  p  a  q  k  r  k  s  r  s  v  s  r  r  t  d  .
P  Q  L  R  F  N  E  R  N  I  H  K  Q  C  V  V  C  N  Q  H  K  S  G  N  L  V  P  Y  R  V  E  L  I  S
----.----+----.----+----.----+----.----+----.----+----.----+----.----+----.----+----.----+----.----+
 r  l  e  s  k  i  f  a  i  n  v  l  l  a  h  h  a  v  l  v  f  a  s  i  e  n  r  i  a  d  f  q  n
v  e  v  g  i  .  h  v  c  y  e  c  a  i  r  p  t  c  g  a  c  f  r  f  d  r  e  t  d  r  r  v  s  .
g  .  s  r  n  l  s  r  l  i  .  l  c  h  t  t  h  l  w  c  l  l  p  f  r  t  g  y  r  t  s  s  i

ORF     ...     CONTINUES IN PHASE THREE
```

```
F         S         A         THH                              T         MM        R
N         F         C         ANN                              A         BB        S
U         A         C         QFF                              Q         00        A
H         N         1         131                              1         21        1
                               /
GCCGCATCGGGCAGGAAGCAGTAGACGAAATCGAATCAAACCATAACCGCCATCGCTGGACTATCGAAGAGTGCAAGGCGATCAAGGCAGAGTACCAACA
----.----+----.----+----.----+----.----+----.----+----.----+----.----+----.----+----.----+----.----+  43000
CGGCGTAGCCCGTCCTTCGTCATCTGCTTTAGCTTAGTTTGGTATTGGCGGTAGCGACCTGATAGCTTCTCACGTTCCGCTAGTTCCGTCTCATGGTTGT

a  a  s  g  r  k  q  .  t  k  s  n  q  t  i  t  a  i  a  g  l  s  k  s  a  r  r  s  r  q  s  t  n  r
p  h  r  a  g  s  s  r  r  n  r  i  k  p  .  p  p  s  l  d  y  r  r  v  q  g  d  q  g  r  v  p  t
 R  I  G  Q  E  A  V  D  E  I  E  S  N  H  N  R  H  R  W  T  I  E  E  C  K  A  I  K  A  E  Y  Q  Q
----.----+----.----+----.----+----.----+----.----+----.----+----.----+----.----+----.----+----.----+
 a  a  d  p  l  f  c  y  v  f  d  f  .  v  m  v  a  m  a  p  s  d  f  l  a  l  r  d  l  c  l  v  l
  g  c  r  a  p  l  l  l  r  f  r  i  l  g  y  g  g  d  s  s  .  r  l  t  c  p  s  .  p  l  t  g  v
l  r  m  p  c  s  a  t  s  s  i  s  d  f  w  l  r  w  r  q  v  i  s  s  h  l  a  i  l  a  s  y  w  c

ORF     ...     CONTINUES IN PHASE THREE
```

```
  N  E              M          HF  00    D                    N      D
  I  C              N          AN  RR    D                    S      I
  2  P              L          EU  FF    E                    P      F
  4  1              1          3H        1                    C      F
                                    /
GAAACTCAAAGACCTGCGAAATAGCAGAAGTGAGGCCGCATGACGTTCTCAGTAAAAACCATTCCAGACATGCTCGTTGAAACATACGGAAATCAGACAG
----.----+----.----+----.----+----.----+----.----+----.----+----.----+----.----+----.----+----.----+  43100
CTTTGAGTTTCTGGACGCTTTATCGTCTTCACTCCGGCGTACTGCAAGAGTCATTTTTGGTAAGGTCTGTACGAGCAACTTTGTATGCCTTTAGTCTGTC

  n  s  k  t  c  e  i  a  e  v  r  p  h  d  v  l  s  k  n  h  s  r  h  a  r  .  n  i  r  k  s  d  r
e  t  q  r  p  a  k  .  q  k  .  g  r  M  T  F  S  V  K  T  I  P  D  M  L  V  E  T  Y  G  N  Q  T  F
 K  I  K  D  I  R  N  S  R  S  E  A  A  .  r  s  q  .  k  p  f  q  t  c  s  l  k  h  t  e  i  r  q
----.----+----.----+----.----+----.----+----.----+----.----+----.----+----.----+----.----+----.----+
l  f  e  f  v  q  s  i  a  s  t  l  g  c  s  t  r  l  l  f  w  e  l  c  a  r  q  f  m  r  f  d  s  l
 s  v  .  l  g  a  f  y  c  f  h  p  r  m  v  n  e  t  f  v  m  g  s  m  s  t  s  v  y  p  f  .  v
  f  s  l  s  r  r  f  l  l  l  s  a  a  h  r  e  .  y  f  g  n  w  v  h  e  n  f  c  v  s  i  l  c

NI24  43004    nin24           dr       EM
ORF   43040    nin orf-68      M-NH3
ORF   43040    nin orf-204     COOH
DIFF  43082    difference               sq     Kroger and Hobom;1982 G (not A)
```

```
                   T  R                                            H*
                   H  S                                            I1
                   A  A                                            N6
                   1  1                                            23
                                                                    /
AAGTAGCACGCAGACTGAAATGTAGTCGCGGTACGGTCAGAAAATACGTTGATGATAAAGACGGGAAAATGCACGCCATCGTCAACGACGTTCTCATGGT
----.----+----.----+----.----+----.----+----.----+----.----+----.----+----.----+----.----+----.----+  43200
TTCATCGTGCGTCTGACTTTACATCAGCGCCATGCCAGTCTTTTATGCAACTACTATTTCTGCCCTTTTACGTGCGGTAGCAGTTGCTGCAAGAGTACCA

 s  s  t  q  t  e  m  .  s  r  y  g  q  k  i  r  .  .  .  r  r  e  n  a  r  h  r  q  r  r  s  h  g
  V  A  R  R  L  K  C  S  R  G  T  V  R  K  Y  V  D  D  K  D  G  K  M  H  A  I  V  N  D  V  L  M  V
k  .  h  a  d  .  n  v  v  a  v  r  s  e  n  t  l  m  i  k  t  g  k  c  t  p  s  s  t  t  f  s  w  f
----.----+----.----+----.----+----.----+----.----+----.----+----.----+----.----+----.----+----.----+
  l  l  v  c  v  s  i  y  d  r  y  p  .  f  i  r  q  h  y  l  r  s  f  a  r  w  r  .  r  r  e  .  p
s  t  a  r  l  s  f  h  l  r  p  v  t  l  f  y  t  s  s  l  s  p  f  i  c  a  m  t  l  s  t  r  m  t
 f  y  c  a  s  q  f  t  t  a  t  r  d  s  f  v  n  i  i  f  v  p  f  h  v  g  d  d  v  v  n  e  h
```

```
*163   43183     Mapped site HIN2
ORF     ...      CONTINUES IN PHASE TWO


     TH     S       F  O   H              O      F          B                  MS       H
     HG     F       O  R   H              R      N          B                  BF       G
     AU     A       K  F   A              F      U          V                  OA       U
     12     N       1      1                     H          1                  1N       2

TCATCGCGGATGGAGTGAAAGAGATGCGCTATTACGAAAAAATTGATGGCAGCAAATACCGAAATATTTGGGTAGTTGGCGATCTGCACGGATGCTACAC
----.----+----.----+----.----+----.----+----.----+----.----+----.----+----.----+----.----+----.----+ 43300
AGTAGCGCCTACCTCACTTTCTCTACGCGATAATGCTTTTTTAACTACCGTCGTTTATGGCTTTATAAACCCATCAACCGCTAGACGTGCCTACGATGTG

s  s  r  m  e  .  k  r  c  a  i  t  k  k  l  m  a  a  n  t  e  i  f  g  .  l  a  i  c  t  d  a  t  r
 H  R  G  W  S  E  R  D  A  L  L  R  K  N  .  w  q  q  i  p  k  y  l  g  s  w  r  s  a  r  m  l  h
  i  a  d  g  v  k  e  M  R  Y  Y  E  K  I  D  G  S  K  Y  R  N  I  W  V  V  G  D  L  H  G  C  Y  T
----.----+----.----+----.----+----.----+----.----+----.----+----.----+----.----+----.----+----.----+
 e  d  r  i  s  h  f  l  h  a  i  v  f  f  n  i  a  a  f  v  s  i  n  p  y  n  a  i  q  v  s  a  v
  .  r  p  h  l  s  l  s  a  s  n  r  f  f  q  h  c  c  i  g  f  y  k  p  l  q  r  d  a  r  i  s  c
n  m  a  s  p  t  f  s  i  r  .  .  s  f  i  s  p  l  l  y  r  f  i  q  t  t  p  s  r  c  p  h  .  v

ORF    43224       nin orf-221            M-NH3
ORF    43243       nin orf-68             COOH


   F   N                     HH  T          EC                              M                 P
   O   I                     NN  A          CF                              B                 Q
   K   N                     FF  Q          P1                              O                 9
   1   5                     13  1          1                               1
                              /
GAACCTGATGAACAAACTGGATACGATTGGATTCGACAACAAAAAAGACCTGCTTATCTCGGTGGGCGATTTGGTTGATCGTGGTGCAGAGAACGTTGAA
----.----+----.----+----.----+----.----+----.----+----.----+----.----+----.----+----.----+----.----+ 43400
CTTGGACTACTTGTTTGACCTATGCTAACCTAAGCTGTTGTTTTTTCTGGACGAATAGAGCCACCCGCTAAACCAACTAGCACCACGTCTCTTGCAACTT

  t  .  .  t  n  w  i  r  l  d  s  t  t  k  k  t  c  l  s  r  w  a  i  w  l  i  v  v  q  r  t  l  n
e  p  d  e  q  t  g  y  d  w  i  r  q  q  k  r  p  a  y  l  g  g  r  f  g  .  s  w  c  r  e  r  .  m
 N  L  M  N  K  L  D  T  I  G  F  D  N  K  K  D  L  L  I  S  V  G  D  L  V  D  R  G  A  E  N  V  E
----.----+----.----+----.----+----.----+----.----+----.----+----.----+----.----+----.----+----.----+
r  v  q  h  v  f  q  i  r  n  s  e  v  v  f  f  v  q  k  d  r  h  a  i  q  n  i  t  t  c  l  v  n  f
 s  g  s  s  c  v  p  y  s  q  i  r  c  c  f  l  g  a  .  r  p  p  r  n  p  q  d  h  h  l  s  r  q
  f  r  i  f  l  s  s  v  i  p  n  s  l  l  f  s  r  s  i  e  t  p  s  k  t  s  r  p  a  s  f  t  s

NIN5   43307       nin5                      dr        sq
CF1    43346       cf1                       sr        EM
PQ9    43394       PQ   delta 9              dr        EM
ORF     ...        CONTINUES IN PHASE THREE


 E S                 E S         A   R    P                                              B
 C C                 C C         L   S    Q                                              B
 R R                 R R         U   A    1                                              V
 2 1                 2 1         1   1                                                   1

TGCCTGGAATTAATCACATTCCCCTGGTTCAGAGCTGTACGTGGAAACCATGAGCAAATGATGATTGATGGCTTATCAGAGCGTGGAAACGTTAATCACT
----.----+----.----+----.----+----.----+----.----+----.----+----.----+----.----+----.----+----.----+ 43500
ACGGACCTTAATTAGTGTAAGGGGACCAAGTCTCGACATGCACCTTTGGTACTCGTTTACTACTAACTACCGAATAGTCTCGCACCTTTGCAATTAGTGA

  a  w  n  .  s  h  s  p  g  s  e  l  y  v  e  t  m  s  k  .  .  l  m  a  y  q  s  v  e  t  l  i  t
 p  g  i  n  h  i  p  l  v  q  s  c  t  w  k  p  .  a  n  d  d  .  w  l  i  r  a  w  k  r  .  s  l
C  L  E  L  I  T  F  P  W  F  R  A  V  R  G  N  H  E  Q  M  M  I  D  G  L  S  E  R  G  N  V  N  H  W
----.----+----.----+----.----+----.----+----.----+----.----+----.----+----.----+----.----+----.----+
  a  q  f  .  d  c  e  g  p  e  s  s  y  t  s  v  m  l  l  h  h  n  i  a  .  .  l  t  s  v  n  i  v
i  g  p  i  l  .  m  g  r  t  .  l  q  v  h  f  g  h  a  f  s  s  q  h  s  i  l  a  h  f  r  .  d  s
 h  r  s  n  i  v  n  g  q  n  l  a  t  r  p  f  w  s  c  i  i  i  s  p  k  d  s  r  p  f  t  l  .

PQ1    43443       PQ delta 1                dr        EM
ORF     ...        CONTINUES IN PHASE THREE
```

```
  F                              T                            A                                    T
  N                              A                            L                                    A
  U                              Q                            U                                    Q
  H                              1                            1                                    1

GGCTGCTTAATGGCGGTGGCTGGTTCTTTAATCTCGATTACGACAAAGAAATTCTGGCTAAAGCTCTTGCCCATAAAGCAGATGAACTTCCGTTAATCAT
----.----+----.----+----.----+----.----+----.----+----.----+----.----+----.----+----.----+----.----+ 43600
CCGACGAATTACCGCCACCGACCAAGAAATTAGAGCTAATGCTGTTTCTTTAAGACCGATTTCGAGAACGGGTATTTCGTCTACTTGAAGGCAATTAGTA

g  c  l  m  a  v  a  g  s  l  i  s  i  t  t  k  k  f  w  l  k  l  l  p  i  k  q  m  n  f  r  .  s  s
 a  a  .  w  r  w  l  v  l  .  s  r  l  r  q  r  n  s  g  .  s  s  c  p  .  s  r  .  t  s  v  n  h
 L  L  N  G  G  G  W  F  F  N  L  D  Y  D  K  E  I  L  A  K  A  L  A  H  K  A  D  E  L  P  L  I  I
----.----+----.----+----.----+----.----+----.----+----.----+----.----+----.----+----.----+----.----+
 p  q  k  i  a  t  a  p  e  k  i  e  i  v  v  f  f  n  q  s  f  s  k  g  m  f  c  i  f  k  r  .  d
  a  a  .  h  r  h  s  t  r  .  d  r  n  r  c  l  f  e  p  .  l  e  q  g  y  l  l  h  v  e  t  l  .
q  s  s  l  p  p  p  q  n  k  l  r  s  .  s  l  s  i  r  a  l  a  r  a  w  l  a  s  s  s  g  n  i  m

ORF      ...      CONTINUES IN PHASE THREE
```

```
                  H                                      H                      *BM    VE
                  P                                      N                      1CB    2C
                  H                                      H                      6L0    61
                  1                                      3                      411    65
                                                                                  //     /
CGAACTGGTGAGCAAAGATAAAAAATATGTTATCTGCCACGCCGATTATCCCTTTGACGAATACGAGTTTGGAAAGCCAGTTGATCATCAGCAGGTAATC
----.----+----.----+----.----+----.----+----.----+----.----+----.----+----.----+----.----+----.----+ 43700
GCTTGACCACTCGTTTCTATTTTTTATACAATAGACGGTGCGGCTAATAGGGAAACTGCTTATGCTCAAACCTTTCGGTCAACTAGTAGTCGTCCATTAG

  n  w  .  a  k  i  k  n  m  l  s  a  t  p  i  i  p  l  t  n  t  s  l  e  s  q  l  i  i  s  r  .  s
r  t  g  e  q  r  .  k  i  c  y  l  p  r  r  l  s  l  .  r  i  r  v  w  k  a  s  .  s  s  a  g  n  l
 E  L  V  S  K  D  K  K  Y  V  I  C  H  A  D  Y  P  F  D  E  Y  E  F  G  K  P  V  D  H  Q  Q  V  I
----.----+----.----+----.----+----.----+----.----+----.----+----.----+----.----+----.----+----.----+
d  f  q  h  a  f  i  f  f  i  n  d  a  v  g  i  i  g  k  v  f  v  l  k  s  l  w  n  i  m  l  l  y  d
 r  v  p  s  c  l  y  f  i  h  .  r  g  r  r  n  d  r  q  r  i  r  t  q  f  a  l  q  d  d  a  p  l
  s  s  t  l  l  s  l  f  y  t  i  q  w  a  s  .  g  k  s  s  y  s  n  p  f  g  t  s  .  .  c  t  i

*164    43682     Mapped site BCL1
V266    43688       dv266              plasmid-r    EM
ORF      ...      CONTINUES IN PHASE THREE
```

```
        T   H H                   BM                     HT A                       E
        H   N N                   IB                     HH F                       C
        A   F F                   N0                     AA L                       1
        1   3 1                   11                     11 3                       5
                                                          /
TGGAACCGCGAACGAATCAGCAACTCACAAAACGGGATCGTGAAAGAAATCAAAGGCGCGGACACGTTCATCTTTGGTCATACGCCAGCAGTGAAACCAC
----.----+----.----+----.----+----.----+----.----+----.----+----.----+----.----+----.----+----.----+ 43800
ACCTTGGCGCTTGCTTAGTCGTTGAGTGTTTTGCCCTAGCACTTTCTTTAGTTTCCGCGCCTGTGCAAGTAGAAACCAGTATGCGGTCGTCACTTTGGTG

  g  t  a  n  e  s  a  t  h  k  t  g  s  .  k  k  s  k  a  r  t  r  s  s  l  v  i  r  q  q  .  n  h
   e  p  r  t  n  q  q  l  t  k  r  d  r  e  r  n  q  r  r  g  h  v  h  l  w  s  y  a  s  s  e  t  t
W  N  R  E  R  I  S  N  S  Q  N  G  I  V  K  E  I  K  G  A  D  T  F  I  F  G  H  T  P  A  V  K  P  L
----.----+----.----+----.----+----.----+----.----+----.----+----.----+----.----+----.----+----.----+
  p  v  a  f  s  d  a  v  .  l  v  p  d  h  f  f  d  f  a  r  v  r  e  d  k  t  m  r  w  c  h  f  w
r  s  g  r  v  f  .  c  s  v  f  r  s  r  s  l  f  .  l  r  p  c  t  .  r  q  d  y  a  l  l  s  v  v
 q  f  r  s  r  i  l  l  e  c  f  p  i  t  f  s  i  l  p  a  s  v  n  m  k  p  .  v  g  a  t  f  g

ORF      ...      CONTINUES IN PHASE THREE
```

```
                         C*T   H    HP                        H       R             HHHQ0 H  T
                         L1A   P    HQ                        N       S             HYY R N  A
                         A6Q   A    A4                        F       A             AM5 F F  Q
                         151   2    1                         1       1             1     1  1
                          //         /                                                 ///
TCAAGTTTGCCAACCAAATGTATATCGATACCGGCGCAGTGTTCTGCGGAAACCTAACATTGATTCAGGTACAGGGAGAAGGCGCATGAGACTCGAAAGC
----.----+----.----+----.----+----.----+----.----+----.----+----.----+----.----+----.----+----.----+ 43900
AGTTCAAACGGTTGGTTTACATATAGCTATGGCCGCGTCACAAGACGCCTTTGGATTGTAACTAAGTCCATGTCCCTCTTCCGCGTACTCTGAGCTTTCG

s  s  l  p  t  k  c  i  s  i  p  a  q  c  s  a  e  t  .  h  .  f  r  y  r  e  k  a  h  e  t  r  k  r
 q  v  c  q  p  n  v  y  r  y  r  r  s  v  l  r  k  p  n  i  d  s  g  t  g  r  r  r  M  R  L  E  S
  K  F  A  N  Q  M  Y  I  D  T  G  A  V  F  C  G  N  L  T  L  I  Q  V  Q  G  E  G  A  .  d  s  k  a
----.----+----.----+----.----+----.----+----.----+----.----+----.----+----.----+----.----+----.----+
 e  l  k  g  v  l  h  i  d  i  g  a  c  h  e  a  s  v  .  c  q  n  l  y  l  s  f  a  c  s  v  r  f
  .  t  q  w  g  f  t  y  r  y  r  r  l  t  r  r  f  g  l  m  s  e  p  v  p  l  l  r  m  l  s  s  l
s  l  n  a  l  w  i  y  i  s  v  p  a  t  n  q  p  f  r  v  n  i  .  t  c  p  s  p  a  h  s  e  f  a
```

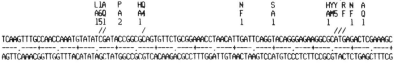

```
*165   43825    Mapped site CLA1
PQ4    43836    PQdelta 4          dr      EM
HYM    43885    HyM                hy      EM    lambdaatt80imm0Plambda::QSR80
HY5    43885    Hy5                hy      EM    80immlambda 0Plambda::QSR80
Q      43886    Q (orf-207)        M-NH3
ORF    43886    nin orf-221        COOH
```

```
 A     H              H   H     A H     H     H   RB     F   H
 L     N              P   N     S A     N     P   SB     N   G
 U     F              H   F     U E     F     A   AV     U   U
 1     3              1   1     1 3     1     2   11     H   2
                                                  /
GTAGCTAAATTTCATTCGCCAAAAAGCCCGATGATGAGCGACTCACCACGGGCCACGGCTTCTGACTCTCTTTCCGGTACTGATGTGATGGCTGCTATGG
---.----+----.----+----.----+----.----+----.----+----.----+----.----+----.----+----.----+----.----+  44000
CATCGATTTAAAGTAAGCGGTTTTTCGGGCTACTACTCGCTGAGTGGTGCCCGGTGCCGAAGACTGAGAGAAAGGCCATGACTACACTACCGACGATACC

  s . i s f a k k p d d e r l t t g h g f . l s f r y . c d g c y g
 V A K F H S P K S P M M S D S P R A T A S D S L S G T D V M A A M G
  . l n f i r q k a r . . a t h h g p r l l t l f p v l m . w l l w
---.----+----.----+----.----+----.----+----.----+----.----+----.----+----.----+----.----+----.----+
 r l . i e n a l f g s s s r s v v p w p k q s e k r y q h s p q . p
  t a l n . e g f l g i i l s e g r a v a e s e r e p v s t i a a i
  y s f k m r w f a r h h a v . w p g r s r v r k g t s i h h s s h

Q       ...    CONTINUES IN PHASE TWO
```

```
   H    F   HB HH        F                 D                  T  S T
   H    0   PB NN        N                 D                  P  F T
   A    K   AV FF        U                 E                  1  A H
   1    1   21 13        H                 1                     N 2
               /
GGATGGCGCAATCACAAGCCGGATTCGGTATGGCTGCATTCTGCGGTAAGCACGAACTCAGCCAGAACGACAAACAAAAGGCTATCAACTATCTGATGCA
---.----+----.----+----.----+----.----+----.----+----.----+----.----+----.----+----.----+----.----+  44100
CCTACCGCGTTAGTGTTCGGCCTAAGCCATACCGACGTAAGACGCCATTCGTGCTTGAGTCGGTCTTGCTGTTTGTTTTCCGATAGTTGATAGACTACGT

  d g a i t s r i r y g c i l r . a r t q p e r q t k g y q l s d a
   M A Q S Q A G F G M A A F C G K H E L S Q N D K Q K A I N Y L M Q
 g w r n h k p d s v w l h s a v s t n s a r t t n k r l s t i . c n
---.----+----.----+----.----+----.----+----.----+----.----+----.----+----.----+----.----+----.----+
  s p a i v l r i r y p q m r r y a r v . g s r c v f p . . s d s a
 p i a c d c a p n p i a a n q p l c s s l w f s l c f a i l . r i c
  p h r l . l g s e t h s c e a t l v f e a l v v f l l s d v i q h

TP1    44081       Tpl                 i(Tn402)    EM
Q       ...        CONTINUES IN PHASE TWO
```

```
                             *HA        D        R     S   H    H
                             1IL        D        S     D   G    N
                             6NU        E        A     U   I    F
                             631        1        1     1   A    3
                              /
ATTTGCACACAAGGTATCGGGGAAATACCGTGGTGTGGCAAAGCTTGAAGGAAATACTAAGGCAAAGGTACTGCAAGTGCTCGCAACATTCGCTTATGCG
---.----+----.----+----.----+----.----+----.----+----.----+----.----+----.----+----.----+----.----+  44200
TAAACGTGTGTTCCATAGCCCCTTTATGGCACCACACCGTTTCGAACTTCCTTTATGATTCCGTTTCCATGACGTTCACGAGCGTTGTAAGCGAATACGC

 i c t q g i g e i p w c g k a . r k y . g k g t a s a r n i r l c g
  F A H K V S G K Y R G V A K L E G N T K A K V L Q V L A T F A Y A
  l h t r y r g n t v v w q s l k e i l r q r y c k c s q h s l m r
---.----+----.----+----.----+----.----+----.----+----.----+----.----+----.----+----.----+----.----+
  i q v c p i p s i g h h p l a q l f y . p l p v a l a r l m r k h
   n a c l t d p f y r p t a f s s p f v l a f t s c t s a v n a . a
 l k c v l y r p f v t t h c l k f s i s l c l y q l h e c c e s i r

*166   44141     Mapped site HIN3
Q       ...      CONTINUES IN PHASE TWO
```

```
                F T  A HNSS H                  N     R    H                           A
                N H  C PCCF G                  C     S    A                           L
                U A  Y AIRA A                  0     A    E                           U
                H 1  1 211N 1                  1     1    3                           1
                         //
GATTATTGCCGTAGTGCCGCGACGCCGGGGGCAAGATGCAGAGATTGCCATGGTACAGGCCGTGCGGTTGATATTGCCAAAACAGAGCTGTGGGGGAGAG
----.----+----.----+----.----+----.----+----.----+----.----+----.----+----.----+----.----+----.----+  44300
CTAATAACGGCATCACGGCGCTGCGGCCCCCGTTCTACGTCTCTAACGGTACCATGTCCGGCACGCCAACTATAACGGTTTTGTCTCGACACCCCCTCTC

 l  l  p  .  c  r  d  a  g  g  k  m  q  r  l  p  w  y  r  p  c  g  .  y  c  q  n  r  a  v  g  e  s
D  Y  C  R  S  A  A  T  P  G  A  R  C  R  D  C  H  G  T  G  R  A  V  D  I  A  K  T  E  L  W  G  R  V
 i  i  a  v  v  p  r  r  r  g  q  d  a  e  i  a  m  v  q  a  v  r  l  i  l  p  k  q  s  c  g  g  e
----.----+----.----+----.----+----.----+----.----+----.----+----.----+----.----+----.----+----.----+
p  n  n  g  y  h  r  s  a  p  p  l  i  c  l  n  g  h  y  l  g  h  p  q  y  q  w  f  l  a  t  p  s  l
 s  .  q  r  l  a  a  v  g  p  a  l  h  l  s  q  w  p  v  p  r  a  t  s  i  a  l  v  s  s  h  p  l
  i  i  a  t  t  g  r  r  r  p  c  s  a  s  i  a  m  t  c  a  t  r  n  i  n  g  f  c  l  q  p  p  s

Q       ...      CONTINUES IN PHASE TWO


   T       S       H     V    MA    S       H             FHF        TSBH
   A       F       G     1    BC    F       G             OHN        HFBH
   Q       A       A     5    OY    A       U             KAU        AAVA
   1       N       1     4    21    N       2             11H        1N11
                               /                                      / /
TTGTCGAGAAAGAGTGCGGAAGATGCAAAGGCGTCGGCTATTCAAGGATGCCAGCAAGCGCAGCATATCGCGCTGTGACGATGCTAATCCCAAACCTTAC
----.----+----.----+----.----+----.----+----.----+----.----+----.----+----.----+----.----+----.----+  44400
AACAGCTCTTTCTCACGCCTTCTACGTTTCCGCAGCCGATAAGTTCCTACGGTCGTTCGCGTCGTATAGCGCGACACTGCTACGATTAGGGTTTGGAATG

 c  r  e  r  v  r  k  m  q  r  r  r  l  f  k  d  a  s  k  r  s  i  s  r  c  d  d  a  n  p  k  p  y
  V  E  K  E  C  G  R  C  K  G  V  G  Y  S  R  M  P  A  S  A  A  Y  R  A  V  T  M  L  I  P  N  L  T
l  s  r  k  s  a  e  d  a  k  a  s  a  i  q  g  c  q  q  a  q  h  i  a  l  .  r  c  .  s  q  t  l  p
----.----+----.----+----.----+----.----+----.----+----.----+----.----+----.----+----.----+----.----+
 q  r  s  l  t  r  f  i  c  l  r  r  s  n  l  s  a  l  l  r  l  m  d  r  q  s  s  a  l  g  l  g  .
t  t  s  f  s  h  p  l  h  l  p  t  p  .  e  l  i  g  a  l  a  a  y  r  a  t  v  i  s  i  g  f  r  v
 n  d  l  f  l  a  s  s  a  f  a  d  a  i  .  p  h  w  c  a  c  c  i  a  s  h  r  h  .  d  w  v  k

V154    44326    dv154                plasmid-r   EM
Q       ...      CONTINUES IN PHASE TWO


       E S             M   NF                 H                    H         M                     D A
       C C             3   SN                 G                    N         B                     I F
       R R                 PU                 A                    F         0                     F L
       2 1                 BH                 1                    1         2                     F 3

CCAACCCACCTGGTCACGCACTGTTAAGCCGCTGTATGACGCTCTGGTGGTGCAATGCCACAAAGAAGAGTCAATCGCAGACAACATTTTGAATGCGGTC
----.----+----.----+----.----+----.----+----.----+----.----+----.----+----.----+----.----+----.----+  44500
GGTTGGGTGGACCAGTGCGTGACAATTCGGCGACATACTGCGAGACCACCACGTTACGGTGTTTCTTCTCAGTTAGCGTCTGTTGTAAAACTTACGCCAG

p  t  h  l  v  t  h  c  .  a  a  v  .  r  s  g  g  a  m  p  q  r  r  v  n  r  r  q  h  f  e  c  g  h
 Q  P  T  W  S  R  T  V  K  P  L  Y  D  A  L  V  V  Q  C  H  K  E  E  S  I  A  D  N  I  L  N  A  V
  n  p  p  g  h  a  l  l  s  r  c  m  t  l  w  w  c  n  a  t  k  k  s  q  s  q  t  t  f  .  m  r  s
----.----+----.----+----.----+----.----+----.----+----.----+----.----+----.----+----.----+----.----+
 g  v  w  r  t  v  c  q  .  a  a  t  h  r  e  p  p  a  i  g  c  l  l  t  l  r  l  c  c  k  s  h  p
  w  g  v  q  d  r  v  t  l  g  s  y  s  a  r  t  t  c  h  w  l  s  s  d  i  a  s  l  m  k  f  a  t
g  l  g  g  p  .  a  s  n  l  r  q  i  v  s  q  h  h  l  a  v  f  f  l  .  d  c  v  v  n  q  i  r  d

M3      44424    lambdagalM3        sr         EM
DIFF    44498    difference                              Daniels(1982)-G; Petrov(1981)-A
Q       ...      CONTINUES IN PHASE TWO


     Q   F          B  H             F                                           H    6          H Q
         N          B  G             0                                           N    S          N U
         U          V  U             K                                           F               F T
         H          1  2             1                                           1               3

ACACGTTAGCAGCATGATTGCCACGGATGGCAACATATTAACGGCATGATATTGACTTATTGAATAAAATTGGGTAAATTTGACTCAACGATGGGTTAAT
----.----+----.----+----.----+----.----+----.----+----.----+----.----+----.----+----.----+----.----+  44600
TGTGCAATCGTCGTACTAACGGTGCCTACCGTTGTATAATTGCCGTACTATAACTGAATAACTTATTTTAACCCATTTAAACTGAGTTGCTACCCAATTA

 t  l  a  a  .  l  p  r  m  a  t  y  .  r  h  d  i  d  l  l  n  k  i  g  .  i  .  l  n  d  g  l  i
T  R  .  q  h  d  c  h  g  w  q  h  i  n  g  m  i  l  t  y  .  i  k  l  g  k  f  d  s  t  m  g  .  f
 h  v  s  s  m  i  a  t  d  g  n  i  l  t  a  .  y  .  l  i  e  .  n  w  v  n  l  t  q  r  w  v  n
----.----+----.----+----.----+----.----+----.----+----.----+----.----+----.----+----.----+----.----+
.  v  n  a  a  h  n  g  r  i  a  v  y  .  r  c  s  i  s  k  n  f  l  i  p  y  i  q  s  l  s  p  n  i
 v  r  .  c  c  s  q  w  p  h  c  c  i  l  p  m  i  n  v  .  q  i  f  n  p  l  n  s  e  v  i  p  .
  c  t  l  l  m  i  a  v  s  p  l  m  n  v  a  h  y  q  s  i  s  y  f  q  t  f  k  v  .  r  h  t  l
```

```
Q     44506    Q            COOH
6S    44587    p6S pR'      5'       RNA   start of 6S RNA; lambda late RNA
QUT   44600    qut          f(c)     G     (maps between 44587 and 44610)


  P          MO        F  HH     H              T                M
  K          NR        N  HA     N              A                N
  3          LF        U  AE     F              Q                L
             1         H  12     1              1                1

TCGCTCGTTGTGGTAGTGAGATGAAAAGAGGCGGCGCTTACTACCGATTCCGCCTAGTTGGTCACTTCGACGTATCGTCTGGAACTCCAACCATCGCAGG
----.----+----.----+----.----+----.----+----.----+----.----+----.----+----.----+----.----+----.----+ 44700
AGCGAGCAACACCATCACTCTACTTTTCTCCGCCGCGAATGATGGCTAAGGCGGATCAACCAGTGAAGCTGCATAGCAGACCTTGAGGTTGGTAGCGTCC

 r  s  l  w  .  .  d  e  k  r  r  r  l  l  p  i  p  p  s  w  s  l  r  r  i  v  w  n  s  n  h  r  r
  a  r  c  g  s  e  M  K  R  G  G  A  Y  Y  R  F  R  L  V  G  H  F  D  V  S  S  G  T  P  T  I  A  G
s  l  v  v  v  v  r  .  k  e  a  a  l  t  t  d  s  a  .  l  v  t  s  t  y  r  l  e  l  q  p  s  q  a
----.----+----.----+----.----+----.----+----.----+----.----+----.----+----.----+----.----+----.----+
  r  e  n  h  y  h  s  s  f  l  r  r  k  s  g  i  g  g  l  q  d  s  r  r  i  t  q  f  e  l  w  r  l
n  a  r  q  p  l  s  i  f  l  p  p  a  .  .  r  n  r  r  t  p  .  k  s  t  d  d  p  v  g  v  m  a  p
 e  s  t  t  t  t  l  h  f  s  a  a  s  v  v  s  e  a  .  n  t  v  e  v  y  r  r  s  s  w  g  d  c

PK3   44606    pk3          i(Tn903)   EM
ORF   44621    orf-64       M-NH3


  E                X                                     T
  C                M                                     6
  P                N                                     S
  1                1

CAGAGAGGTCTGCAAAATGCAATCCCGAAACAGTTCGCAGGTAATAGTTAGAGCCTGCATAACGGTTTCGGGATTTTTTATATCTGCACAACAGGTAAGA
----.----+----.----+----.----+----.----+----.----+----.----+----.----+----.----+----.----+----.----+ 44800
GTCTCTCCAGACGTTTTACGTTAGGGCTTTGTCAAGCGTCCATTATCAATCTCGGACGTATTGCCAAAGCCCTAAAAAATATAGACGTGTTGTCCATTCT

q  r  g  l  q  n  a  i  p  k  q  f  a  g  n  s  .  s  l  h  n  g  f  g  i  f  y  i  c  t  t  g  k  s
 R  E  V  C  K  M  Q  S  R  N  S  S  Q  V  I  V  R  A  C  I  T  V  S  G  F  F  I  S  A  Q  Q  V  R
  e  r  s  a  k  c  n  p  e  t  v  r  r  .  .  l  e  p  a  .  r  f  r  d  f  l  y  l  h  n  r  .  e
----.----+----.----+----.----+----.----+----.----+----.----+----.----+----.----+----.----+----.----+
 c  l  p  r  c  f  a  i  g  f  c  n  a  p  l  l  .  l  r  c  l  p  k  p  i  k  .  i  q  v  v  p  l
  l  s  t  q  l  i  c  d  r  f  l  e  c  t  i  t  l  a  q  m  v  t  e  p  n  k  i  d  a  c  c  t  l
a  s  l  d  a  f  h  l  g  s  v  t  r  l  y  y  n  s  g  a  y  r  n  r  s  k  k  y  r  c  l  l  y  s

T6S   44780    t6S          3'       RNA   end of 6S RNA
ORF    ...     CONTINUES IN PHASE TWO


   H  T  O          H        ME S        S    H                  H     H  E     H HBM
   N  A  R          N        BC C        D    G                  N     P  C     P PIB
   F  Q  F          F        OR R        U    I                  F     A  0     H ANO
   1  1             1        22 1        1    A                  3     2  B     1 211

GCATTGAGTCGATAATCGTGAAGAGTCGGCGAGCCTGGTTAGCCAGTGCTCTTTCCGTTGTGCTGAATTAAGCGAATACCGGAAGCAGAACCGGATCACC
----.----+----.----+----.----+----.----+----.----+----.----+----.----+----.----+----.----+----.----+ 44900
CGTAACTCAGCTATTAGCACTTCTCAGCCGCTCGGACCAATCGGTCACGAGAAAGGCAACACGACTTAATTCGCTTATGGCCTTCGTCTTGGCCTAGTGG

  i  e  s  i  i  v  k  s  r  r  a  w  l  a  s  a  l  s  v  v  l  n  .  a  n  t  g  s  r  t  g  s  p
A  L  S  R  .  s  .  r  v  g  e  p  g  .  p  v  l  f  p  l  c  .  i  k  r  i  p  e  a  e  p  d  h  q
 h  .  v  d  n  r  e  e  s  a  s  l  v  s  q  c  s  f  r  c  a  e  l  s  e  y  r  k  q  n  r  i  t
----.----+----.----+----.----+----.----+----.----+----.----+----.----+----.----+----.----+----.----+
l  m  s  d  i  i  t  f  l  r  r  a  q  n  a  l  a  r  e  t  t  s  f  .  a  f  v  p  l  l  v  p  d  g
 a  n  l  r  y  d  h  l  t  p  s  g  p  .  g  t  s  k  g  n  h  q  i  l  r  i  g  s  a  s  g  s  .
  c  q  t  s  l  r  s  s  d  a  l  r  t  l  w  h  e  k  r  q  a  s  n  l  s  y  r  f  c  f  r  i  v

ORF   44812    orf-64       COOH
```

```
H     R    A        F                        D                        *E       B             F  F H
G     S    C        N                        D                        1C       B             N  N A
A     A    Y        U                        E                        6R       V             U  U E
1     1    1        H                        1                        71       1             H  H 3
                                                                       /
AAATGCGTACAGGCGTCATCGCCGCCCAGCAACAGCACAACCCAAACTGAGCCGTAGCCACTGTCTGTCCTGAATTCATTAGTAATAGTTACGCTGCGGC
----.----+----.----+----.----+----.----+----.----+----.----+----.----+----.----+----.----+----.----+  45000
TTTACGCATGTCCGCAGTAGCGGCGGGTCGTTGTCGTGTTGGGTTTGACTCGGCATCGGTGACAGACAGGACTTAAGTAATCATTATCAATGCGACGCCG

 n  a  y  r  r  h  r  r  p  a  t  a  q  p  k  l  s  r  s  h  c  l  s  .  i  h  .  .  .  l  r  c  g
  m  r  t  g  v  i  a  a  q  q  q  h  n  p  n  .  a  v  a  t  v  c  p  e  f  i  s  n  s  y  a  a  a
k  c  v  q  a  s  s  p  p  s  n  s  t  t  q  t  e  p  .  p  l  s  v  l  n  s  l  v  i  v  t  l  r  p
----.----+----.----+----.----+----.----+----.----+----.----+----.----+----.----+----.----+----.----+
  f  a  y  l  r  .  r  r  g  a  v  a  c  g  l  s  l  r  l  w  q  r  d  q  i  .  .  y  y  n  r  q  p
w  i  r  v  p  t  m  a  a  w  c  c  c  l  g  f  q  a  t  a  v  t  q  g  s  n  m  l  l  l  .  a  a  a
 l  h  t  c  a  d  d  g  g  l  l  l  v  v  w  v  s  g  y  g  s  d  t  r  f  e  n  t  i  t  v  s  r

*167   44972       Mapped site FCR1
```

```
                           CM            T H                 M
                           HN            H H                 N
                           IL            A A                 L
                           D1            1 1                 1

CTTTTACACATGACCTTCGTGAAAGCGGGTGGCAGGAGGTCGCGCTAACAACCTCCTGCCGTTTTGCCCGTGCATATCGGTCACGAACAAATCTGATTAC
----.----+----.----+----.----+----.----+----.----+----.----+----.----+----.----+----.----+----.----+  45100
GAAAATGTGTACTGGAAGCACTTTCGCCCACCGTCCTCCAGCGCGATTGTTGGAGGACGGCAAAACGGGCACGTATAGCCAGTGCTTGTTTAGACTAATG

l  l  h  m  t  f  v  k  a  g  g  r  r  s  r  .  q  p  p  a  v  l  p  v  h  i  g  h  e  q  i  .  l  l
 f  y  t  .  p  s  .  k  r  v  a  g  g  r  a  n  n  l  l  p  f  c  p  c  i  s  v  t  n  k  s  d  y
  f  t  h  d  l  r  e  s  g  w  q  e  v  a  l  t  t  s  c  r  f  a  r  a  y  r  s  r  t  n  l  i  t
----.----+----.----+----.----+----.----+----.----+----.----+----.----+----.----+----.----+----.----+
 r  k  c  m  v  k  t  f  a  p  p  l  l  d  r  .  c  g  g  a  t  k  g  t  c  i  p  .  s  c  i  q  n
  k  .  v  h  g  e  h  f  r  t  a  p  p  r  a  l  l  r  r  g  n  q  g  h  m  d  t  v  f  l  d  s  .
g  k  v  c  s  r  r  s  l  p  h  c  s  t  a  s  v  v  e  q  r  k  a  r  a  y  r  d  r  v  f  r  i  v

CHID   45027       chiD                pm          sq
```

```
           E S                    T                                              S    S             M
           C C                    A                                              F                  B
           R R                    Q                                              A                  O
           2 1                    1                                              N                  2

TAAACACAGTAGCCTGGATTTGTTCTATCAGTAATCGACCTTATTCCTAATTAAATAGAGCAAATCCCCTTATTGGGGGTAAGACATGAAGATGCCAGAA
----.----+----.----+----.----+----.----+----.----+----.----+----.----+----.----+----.----+----.----+  45200
ATTTGTGTCATCGGACCTAAACAAGATAGTCATTAGCTGGAATAAGGATTAATTTATCTCGTTTAGGGGAATAACCCCCATTCTGTACTTCTACGGTCTT

  n  t  v  a  w  i  c  s  i  s  n  r  p  y  s  .  l  n  r  a  n  p  l  i  g  g  k  t  .  r  c  q  k
.  t  q  .  p  g  f  v  l  s  v  i  d  l  i  p  n  .  i  e  q  i  p  l  l  g  v  r  h  e  d  a  r  k
 k  h  s  s  l  d  l  f  y  q  .  s  t  l  f  l  i  k  .  s  k  s  p  y  w  g  .  d  M  K  M  P  E
----.----+----.----+----.----+----.----+----.----+----.----+----.----+----.----+----.----+----.----+
s  f  v  t  a  q  i  q  e  i  l  l  r  g  .  e  .  n  f  l  a  f  g  r  i  p  p  l  v  h  l  h  w  f
 .  v  c  y  g  p  n  t  r  d  t  i  s  r  i  g  l  .  i  s  c  i  g  k  n  p  t  l  c  s  s  a  l
  l  c  l  l  r  s  k  n  .  .  y  d  v  k  n  r  i  l  y  l  l  d  g  .  q  p  y  s  m  f  i  g  s

S      45186       S (orf-107)         M-NH3               protein not identified
```

```
             GCHF         TF                        S                      R      TF
             DFAN         HN                        F                      S      HN
             IREU         AU                        A                      A      AU
             213H         1H                        N                      1      1H
              /
AAACATGACCTGTTGGCCGCCATTCTCGCGGCAAAGGAACAAGGCATCGGGGCAATCCTTGCGTTTGCAATGGCGTACCTTCGCGGCAGATATAATGGCG
----.----+----.----+----.----+----.----+----.----+----.----+----.----+----.----+----.----+----.----+  45300
TTTGTACTGGACAACCGGCGGTAAGAGCGCCGTTTCCTTGTTCCGTAGCCCCGTTAGGAACGCAAACGTTACCGCATGGAAGCGCCGTCTATATTACCGC

 n  m  t  c  w  p  p  f  s  r  q  r  n  k  a  s  g  q  s  l  r  l  q  w  r  t  f  a  a  d  i  m  a
  t  .  p  v  g  r  h  s  r  g  k  g  t  r  h  r  g  n  p  c  v  c  n  g  v  p  s  r  q  i  .  w  r
K  H  D  L  L  A  A  I  L  A  A  K  E  Q  G  I  G  A  I  L  A  F  A  M  A  Y  L  R  G  R  Y  N  G  G
----.----+----.----+----.----+----.----+----.----+----.----+----.----+----.----+----.----+----.----+
  f  m  v  q  q  g  g  n  e  r  c  l  f  l  a  d  p  c  d  k  r  k  c  h  r  v  k  a  a  s  i  i  a
f  v  h  g  t  p  r  w  e  r  p  l  p  v  l  c  r  p  l  g  q  t  q  l  p  t  g  e  r  c  i  y  h  r
 f  c  s  r  n  a  a  m  r  a  a  f  s  c  p  m  p  a  i  r  a  n  a  i  a  y  r  r  p  l  y  l  p

S      ...         CONTINUES IN PHASE THREE
```

```
                T       H 4 H       SES S   H          T       H   D
                A       G 2 H       9CC 7   N          A       P   D
                Q       A 2 A       BRR     F          Q       A   E
                1       1   1        21     3          1       2   1
                                     /
GTGCGTTTACAAAAACAGTAATCGACGCAACGATGTGCGCCATTATCGCCTGGTTCATTCGTGACCTTCTCGACTTCGCCGGACTAAGTAGCAATCTCGC
----.----+----.----+----.----+----.----+----.----+----.----+----.----+----.----+----.----+----.----+  45400
CACGCAAATGTTTTTGTCATTAGCTGCGTTGCTACACGCGGTAATAGCGGACCAAGTAAGCACTGGAAGAGCTGAAGCGGCCTGATTCATCGTTAGAGCG

v r l q k q . s t q r c a p l s p g s f v t f s t s p d . v a i s l
 c v y k n s n r r n d v r h y r l v h s . p s r l r r t k . q s r
  A F T K T V I D A T M C A I I A W F I R D L L D F A G L S S N L A
----.----+----.----+----.----+----.----+----.----+----.----+----.----+----.----+----.----+----.----+
 t r k c f c y d v c r h a g n d g p e n t v k e v e g s . t a i e
  h t . l f l l r r l s t r w . r r t . e h g e r s r r v l y c d r
p a n v f v t i s a v i h a m i a q n m r s r r s k a p s l l l r a

422   45336    422              i(Tn5)     EM;RES
S9B   45348    Sts9B            pm         sq      G to A
S7    45352    Sam7             pm         sq      G to A
S       ...    CONTINUES IN PHASE THREE
```

```
                        R H T         B          F        H        R
                        S N A         B          N        P
                        A F Q         V          U        A
                        1 1 1         1          H        2

TTATATAACGAGCGTGTTTATCGGCTACATCGGTACTGACTCGATTGGTTCGCTTATCAAACGCTTCGCTGCTAAAAAAGCCGGAGTAGAAGATGGTAGA
----.----+----.----+----.----+----.----+----.----+----.----+----.----+----.----+----.----+----.----+  45500
AATATATTGCTCGCACAAATAGCCGATGTAGCCATGACTGAGCTAACCAAGCGAATAGTTTGCGAAGCGACGATTTTTTCGGCCTCATCTTCTACCATCT

  i . r a c l s a t s v l t r l v r l s n a s l l k k p e . k M V E
 l y n e r v y r l h r y . l d w f a y q t l r c . k s r s r r w . k
  Y I T S V F I G Y I G T D S I G S L I K R F A A K K A G V E D G R
----.----+----.----+----.----+----.----+----.----+----.----+----.----+----.----+----.----+----.----+
s i y r a h k d a v d t s v r n t r k d f a e s s f f g s y f i t s
 k y l s r t . r s c r y q s s q n a . . v s r q . f l r l l l h y
  . i v l t n i p . m p v s e i p e s i l r k a a l f a p t s s p l

R      45493     R (orf-158)        M-NH3      AA
S       ...     CONTINUES IN PHASE THREE
```

```
M   S                   T     M  M                 A A                  A A
B                       A     N  N                 C A                  C A
O                       Q     L  L                 Y T                  Y T
2                       1     1  1                 1 2                  1 2

AATCAATAATCAACGTAAGGCGTTCCTCGATATGCTGGCGTGGTCGGAGGGAACTGATAACGGACGTCAGAAAACCAGAAATCATGGTTATGACGTCATT
----.----+----.----+----.----+----.----+----.----+----.----+----.----+----.----+----.----+----.----+  45600
TTAGTTATTAGTTGCATTCCGCAAGGAGCTATACGACCGCACCAGCCTCCCTTGACTATTGCCTGCAGTCTTTTGGTCTTTAGTACCAATACTGCAGTAA

 I N N Q R K A F L D M L A W S E G T D N G R Q K T R N H G Y D V I
  s i i n v r r s s i c w r g r r e l i t d v r k p e i m v m t s l
N Q . s t . g v p r y a g v v g g n . . r t s e n q k s w l . r h c
----.----+----.----+----.----+----.----+----.----+----.----+----.----+----.----+----.----+----.----+
 i l l . r l a n r s i s a h d s p v s l p r . f v l f . p . s t m
f d i i l t l r e e i h q r p r l s s i v s t l f g s i m t i v d n
 f . y d v y p t g r y a p t t p p f q y r v d s f w f d h n h r .

S      45506     S                  COOH
R       ...     CONTINUES IN PHASE ONE
```

```
           A            H    M                M                              HANHHH          HA
           L            P    B                N                              GCAHAP          GL
           U            H    0                L                              TYRAEA          AU
           1            1    1                1                              C11122          11
                                                                                / /
GTAGGCGGAGAGCTATTTACTGATTACTCCGATCACCCTCGCAAACTTGTCACGCTAAACCCAAAACTCAAATCAACAGGCGCCGGACGCTACCAGCTTC
----.----+----.----+----.----+----.----+----.----+----.----+----.----+----.----+----.----+----.----+ 45700
CATCCGCCTCTCGATAAATGACTAATGAGGCTAGTGGGAGCGTTTGAACAGTGCGATTTGGGTTTTGAGTTTAGTTGTCCGCGGCCTGCGATGGTCGAAG

V  G  G  E  L  F  T  D  Y  S  D  H  P  R  K  L  V  T  L  N  P  K  L  K  S  T  G  A  G  R  Y  Q  L  L
 .  a  e  s  y  l  l  i  t  p  i  t  l  a  n  l  s  r  .  t  q  n  s  n  q  q  a  p  d  a  t  s  f
  r  r  r  a  i  y  .  l  l  r  s  p  s  q  t  c  h  a  k  p  k  t  q  i  n  r  r  r  t  l  p  a  s
----.----+----.----+----.----+----.----+----.----+----.----+----.----+----.----+----.----+----.----+
 t  p  p  s  s  n  v  s  .  e  s  .  g  r  l  s  t  v  s  f  g  f  s  l  d  v  p  a  p  r  .  w  s
  y  a  s  l  .  k  s  i  v  g  i  v  r  a  f  k  d  r  .  v  w  f  e  f  .  c  a  g  s  a  v  l  k
q  l  r  l  a  i  .  q  n  s  r  d  g  e  c  v  q  .  a  l  g  l  v  .  i  l  l  r  r  v  s  g  a  e

R        ...      CONTINUES IN PHASE ONE


    S       H             F  F A    HH  BT  X                              H    FF          B
    F       G             0  N L    AA  BT  M                              G    CN          B
    A       U             K  U U    EE  VH  N                              A    1U          V
    N       2             1  H 1    13  12  1                              1    5H          1
                                     /
TTTCCCGTTGGTGGGATGCCTACCGCAAGCAGCTTGGCCTGAAAGACTTCTCTCCGAAAAGTCAGGACGCTGTGGCATTGCAGCAGATTAAGGAGCGTGG
----.----+----.----+----.----+----.----+----.----+----.----+----.----+----.----+----.----+----.----+ 45800
AAAGGGCAACCACCCTACGGATGGCGTTCGTCGAACCGGACTTTCTGAAGAGAGGCTTTTCAGTCCTGCGACACCGTAACGTCGTCTAATTCCTCGCACC

  S  R  W  W  D  A  Y  R  K  Q  L  G  L  K  D  F  S  P  K  S  Q  D  A  V  A  L  Q  Q  I  K  E  R  G
f  p  v  g  g  m  p  t  a  s  s  l  a  .  k  t  s  l  r  k  v  r  t  l  w  h  c  s  r  l  r  s  v  a
 f  p  l  v  g  c  l  p  q  a  a  w  p  e  r  l  l  s  e  k  s  g  r  c  g  i  a  a  d  .  g  a  w
----.----+----.----+----.----+----.----+----.----+----.----+----.----+----.----+----.----+----.----+
r  e  r  q  h  s  a  .  r  l  c  s  p  r  f  s  k  e  g  f  l  .  s  a  t  a  n  c  c  i  l  s  r  p
 k  g  t  p  p  i  g  v  a  l  l  k  a  q  f  v  e  r  r  f  t  l  v  s  h  c  q  l  l  n  l  l  t
  k  g  n  t  p  h  r  g  c  a  a  q  g  s  l  s  r  e  s  f  d  p  r  q  p  m  a  a  s  .  p  a  h

R        ...      CONTINUES IN PHASE ONE


 HH             M            E    H      T         F2         B          HNS   HH              T
 HA             B            C    P      A         N9         B          PCC   HA              A
 AE             0            R    H      Q         U7         V          AIR   AF              Q
 12             1            V    1      1         H          1          211   12              1
                                                                           /
CGCTTTACCTATGATTGATCGTGGTGATATCCGTCAGGCAATCGACCGTTGCAGCAATATCTGGGCTTCACTGCCGGGCGCTGGTTATGGTCAGTTCGAG
----.----+----.----+----.----+----.----+----.----+----.----+----.----+----.----+----.----+----.----+ 45900
GCGAAATGGATACTAACTAGCACCACTATAGGCAGTCCGTTAGCTGGCAACGTCGTTATAGACCCGAAGTGACGGCCCGCGACCAATACCAGTCAAGCTC

 A  L  P  M  I  D  R  G  D  T  R  Q  A  I  D  R  C  S  N  I  W  A  S  L  P  G  A  G  Y  G  Q  F  E
  l  y  l  .  l  i  v  v  i  s  v  r  q  s  t  v  a  a  i  s  g  l  h  c  r  a  l  v  m  v  s  s  s
r  f  t  y  d  .  s  w  .  y  p  s  g  n  r  p  l  q  q  y  l  g  f  t  a  g  r  w  l  w  s  v  r  a
----.----+----.----+----.----+----.----+----.----+----.----+----.----+----.----+----.----+----.----+
  a  k  g  i  i  s  r  p  s  i  r  .  a  i  s  r  q  l  l  i  q  a  e  s  g  p  a  p  .  p  .  n  s
a  s  .  r  h  n  i  t  t  i  d  t  l  c  d  v  t  a  a  i  d  p  s  .  q  r  a  s  t  i  t  l  e  l
 r  k  v  .  s  q  d  h  h  y  g  d  p  l  r  g  n  c  c  y  r  p  k  v  a  p  r  q  n  h  d  t  r

297   45853       lambda297          sl          sq      deletes 309 bp; inserts pBR322
R        ...      CONTINUFS TN PHASE ONE


                                                                 RRH     H       T            F
                                                                 Z P     N       H            U
                                                                   H     F       A            S
                                                                   1     1       1            2
                                                                  /
CATAAGGCTGACAGCCTGATTGCAAAATTCAAAGAAGCGGGCGGAACGGTCAGAGAGATTGATGTATGAGCAGAGTCACCGCGATTATCTCCGCTCTGGT
----.----+----.----+----.----+----.----+----.----+----.----+----.----+----.----+----.----+----.----+ 46000
GTATTCCGACTGTCGGACTAACGTTTTAAGTTTCTTCGCCCGCCTTGCCAGTCTCTCTAACTACATACTCGTCTCAGTGGCGCTAATAGAGGCGAGACCA

H  K  A  D  S  L  I  A  K  F  K  E  A  G  G  T  V  R  F  I  D  V  .  a  e  s  p  r  l  s  p  l  w  l
 i  r  l  t  a  .  l  q  n  s  k  k  r  a  e  r  s  e  r  l  m  y  e  q  s  h  r  d  y  l  r  s  g
  .  g  .  q  p  d  c  k  i  q  r  s  g  r  n  g  q  r  d  .  c  M  S  R  V  T  A  I  I  S  A  L  V
----.----+----.----+----.----+----.----+----.----+----.----+----.----+----.----+----.----+----.----+
 c  l  a  s  l  r  i  a  f  n  l  s  a  p  p  v  t  l  s  i  s  t  h  a  s  d  g  r  n  d  g  s  q
  m  l  s  v  a  q  n  c  f  e  f  f  r  a  s  r  d  s  l  n  i  y  s  c  l  .  r  s  .  r  r  e  p
a  y  p  q  c  g  s  q  l  i  .  l  l  p  r  f  p  .  l  s  q  h  i  l  l  t  v  a  i  i  e  a  r  t
```

```
RZ     45966     Rz (orf-153)      M-NH3     G
R      45966     R                 COOH
FUS2   45994     lambdafus2        sr        EM


     D    S                       QAPC                      HT              A
     K    F                       3343                      HH              L
     2    A                                                 AA              U
     3    N                                                 11              1
                                 ///                        /
TATCTGCATCATCGTCTGCCTGTCATGGGCTGTTAATCATTACCGTGATAACGCCATTACCTACAAAGCCCAGCGCGACAAAAATGCCAGAGAACTGAAG
----.----+----.----+----.----+----.----+----.----+----.----+----.----+----.----+----.----+----.----+ 46100
ATAGACGTAGTAGCAGACGGACAGTACCCGACAATTAGTAATGGCACTATTGCGGTAATGGATGTTTCGGGTCGCGCTGTTTTTACGGTCTCTTGACTTC

  s  a  s  s  s  a  c  h  g  l  l  i  i  t  v  i  t  p  l  p  t  k  p  s  a  t  k  m  p  e  n  .  s
 y  l  h  h  r  l  p  v  m  g  c  .  s  l  p  .  .  r  h  y  l  q  s  p  a  r  q  k  c  q  r  t  e  a
  I  C  I  I  V  C  L  S  W  A  V  N  H  Y  R  D  N  A  I  T  Y  K  A  Q  R  D  K  N  A  R  E  L  K
----.----+----.----+----.----+----.----+----.----+----.----+----.----+----.----+----.----+----.----+
n  d  a  d  d  d  a  q  .  p  s  n  i  m  v  t  i  v  g  n  g  v  f  g  l  a  v  f  i  g  s  f  q  l
 .  r  c  .  r  r  g  t  m  p  q  .  d  n  g  h  y  r  w  .  r  c  l  g  a  r  c  f  h  w  l  v  s
  i  q  m  m  t  q  r  d  h  a  t  l  .  .  r  s  l  a  m  v  .  l  a  w  r  s  l  f  a  l  s  s  f

DK23   46008     dK23              i(TN903)  sq
Q3     46043     qin3              sr        EM
A3     46043     qinA3             sr        EM
P4     46043     P4                sr        EM
C3     46043     qinC3             sr        EM
RZ     ...     CONTINUES IN PHASE THREE


      TF      S H  N     B   D    S F  H T  2            S    A      SA
      HN      F G  S     B   K    F N  H A  9            F    L      FL
      AU      A A  P     V   6    A U  A Q  7            A    U      AU
      1H      N 1  C     1        N H  1 1               N    1      N1

CTGGCGAACGCGGCAATTACTGACATGCAGATGCGTCAGCGTGATGTTGCTGCGCTCGATGCAAAATACACGAAGGAGTTAGCTGATGCTAAAGCTGAAA
----.----+----.----+----.----+----.----+----.----+----.----+----.----+----.----+----.----+----.----+ 46200
GACCGCTTGCGCCGTTAATGACTGTACGTCTACGCAGTCGCACTACAACGACGCGAGCTACGTTTTATGTGCTTCCTCAATCGACTACGATTTCGACTTT

  w  r  t  r  q  l  l  t  c  r  c  v  s  v  m  l  l  r  s  m  q  n  t  r  r  s  .  l  m  l  k  l  k
   g  e  r  g  n  y  .  h  a  d  a  s  a  .  c  c  c  a  r  c  k  i  h  e  g  v  s  .  c  .  s  .  k
L  A  N  A  A  I  T  D  M  Q  M  R  Q  R  D  V  A  A  L  D  A  K  Y  T  K  E  L  A  D  A  K  A  E  N
----.----+----.----+----.----+----.----+----.----+----.----+----.----+----.----+----.----+----.----+
  q  r  v  r  c  n  s  v  h  l  h  t  l  t  i  n  s  r  e  i  c  f  v  r  l  l  .  s  i  s  f  s  f
a  p  s  r  p  l  .  q  c  a  s  a  d  a  h  h  q  q  a  r  h  l  i  c  s  p  t  l  q  h  .  l  q  f
 s  a  f  a  a  i  v  s  m  c  i  r  .  r  s  t  a  a  s  s  a  f  y  v  f  s  n  a  s  a  l  a  s

DK6    46141     dk 6              i(Tn903)  EM;RES
297    46161     lambda297         sr        sq      deletes 309 bp;inserts pBR322
RZ     ...     CONTINUES IN PHASE THREE


                 NF                                                                      H    M
                 SN                                                                      P    N
                 PU                                                                      A    L
                 BH                                                                      2    1

ATGATGCTCTGCGTGATGATGTTGCCGCTGGTCGTCGTCGGTTGCACATCAAAGCAGTCTGTCAGTCAGTGCGTGAAGCCACCACCGCCTCCGGCGTGGA
----.----+----.----+----.----+----.----+----.----+----.----+----.----+----.----+----.----+----.----+ 46300
TACTACGAGACGCACTACTACAACGGCGACCAGCAGCAGCCAACGTGTAGTTTCGTCAGACAGTCAGTCACGCACTTCGGTGGTGGCGGAGGCCGCACCT

m  m  l  c  v  m  m  l  p  l  v  v  v  g  c  t  s  k  q  s  v  s  q  c  v  k  p  p  p  p  p  a  w  i
 .  c  s  a  .  .  c  c  r  w  s  s  s  v  a  h  q  s  s  l  s  v  s  a  .  s  h  h  r  l  r  r  g
  D  A  L  R  D  D  V  A  A  G  R  R  R  L  H  I  K  A  V  C  Q  S  V  R  E  A  T  T  A  S  G  V  D
----.----+----.----+----.----+----.----+----.----+----.----+----.----+----.----+----.----+----.----+
  i  i  s  q  t  i  i  n  g  s  t  t  t  p  q  v  d  f  c  d  t  l  .  h  t  f  g  g  g  g  g  a  h
   h  h  e  a  h  h  h  q  r  q  d  d  d  t  a  c  .  l  l  r  d  t  l  a  h  l  w  w  r  r  r  r  p
f  s  a  r  r  s  s  t  a  a  p  r  r  r  n  c  m  l  a  t  q  .  d  t  r  s  a  v  v  a  e  p  t  s

RZ     ...     CONTINUES IN PHASE THREE
```

```
 F        B M       N        H         DM       M    *BM
 N        B N       S        P         DN       N    1CB
 U        V L       P        H         EL       L    6L0
 H        1 1       B        1         11       1    811
                                        /            //
TAATGCAGCCTCCCCCCGACTGGCAGACACCGCTGAACGGGATTATTTCACCCTCAGAGAGAGGCTGATCACTATGCAAAAACAACTGGAAGGAACCCAG
---.----+----.----+----.----+----.----+----.----+----.----+----.----+----.----+----.----+----.----+  46400
ATTACGTCGGAGGGGGGCTGACCGTCTGTGGCGACTTGCCCTAATAAAGTGGGAGTCTCTCTCCGACTAGTGATACGTTTTTGTTGACCTTCCTTGGGTC

 ■ q p p p d w q t p l n g i i s p s e r g . s l c k n n w k e p r
 . c s l p p t g r h r . t g l f h p q r e a d h y a k t t g r n p e
 N A A S P R L A D T A E R D Y F T L R E R L I T M Q K Q L E G T Q
---.----+----.----+----.----+----.----+----.----+----.----+----.----+----.----+----.----+----.----+
 i i c g g g s q c v g s f p i i e g e s l p q d s h l f l q f s g l
 y h l r g g v p l c r q v p n n . g . l s a s . . a f v v p l f g
 l a a e g r s a s v a s r s . k v r l s l s i v i c f c s s p v w

*168   46366    Mapped site BCL1
RZ     ...      CONTINUES IN PHASE THREE
```

```
               R              C*BT                  S         T H   N
               Z              L1SA                  B         H H   S
                              A6TQ                  8         A A   P
                              19X1                            1 1   B
                               //
AAGTATATTAATGAGCAGTGCAGATAGAGTTGCCCATATCGATGGGCAACTCATGCAATTATTGTGAGCAATACACACGCGCTTCCAGCGGAGTATAAAT
---.----+----.----+----.----+----.----+----.----+----.----+----.----+----.----+----.----+----.----+  46500
TTCATATAATTACTCGTCACGTCTATCTCAACGGGTATAGCTACCCGTTGAGTACGTTAATAACACTCGTTATGTGTGCGCGAAGGTCGCCTCATATTTA

 s i l ■ s s a d r v a h i d g q l ■ q l l . a i h t r f q r s i n
 v y . . a v q i e l p i s ■ g n s c n y c e q y t r a s s g v . ■
 K Y I N E Q C R . s c p y r w a t h a i i v s n t h a l p a e y k c
---.----+----.----+----.----+----.----+----.----+----.----+----.----+----.----+----.----+----.----+
 l i n i l l a s l t a w i s p c s ■ c n n h a i c v r k w r l i f
 s t y . h a t c i s n g ■ d i p l e h l . q s c y v r a e l p t y i
 f y i l s c h l y l q g y r h a v . a i i t l l v c a s g a s y l

RZ     46423    Rz                      COOH
RZ     ...      STOP IS IN THE WRONG PHASE
*169   46439    Mapped site CLA1
SB8    46467    Sequence Block 8        end
```

```
                 T    H                                         HF B         F    T M
                 T    N                                         HN B         N    A B
                 H    F                                         AU V         U    Q 0
                 2    3                                         1H 1         H    1 2

GCCTAAAGTAATAAAACCGAGCAATCCATTTACGAATGTTTGCTGGGTTTCTGTTTTAACAACATTTTCTGCGCCGCCACAAATTTTGGCTGCATCGACA
---.----+----.----+----.----+----.----+----.----+----.----+----.----+----.----+----.----+----.----+  46600
CGGATTTCATTATTTTGGCTCGTTAGGTAAATGCTTACAAACGACCCAAAGACAAAATTGTTGTAAAAGACGCGGCGGTGTTTAAAACCGACGTAGCTGT

 a . s n k t e q s i y e c l l g f c f n n i f c a a t n f g c i d s
 p k v i k p s n p f t n v c w v s v l t t f s a p p q i l a a s t
 l k . . n r a i h l r ■ f a g f l f . q h f l r r h k f w l h r q
---.----+----.----+----.----+----.----+----.----+----.----+----.----+----.----+----.----+----.----+
 a . l l l v s c d ■ . s h k s p k q k l l ■ k q a a v f k p q ■ s
 g l t i f g l l g n v f t q q t e t k v v n e a g g c i k a a d v
 h r f y y f r a i w k r i n a p n r n . c c k r r r w l n q s c r c
```

```
S                             M       HB    E  F  H             F    S
F                             B       PB    C  N  P             0    D
A                             0       HV    1  U  A             K    U
N                             2       11    5  H  2             1    1
                                       /
GTTTTCTTCTGCCCAATTCCAGAAACGAAGAAATGATGGGTGATGGTTTCCTTTGGTGCTACTGCTGCCGGTTTGTTTTGAACAGTAAACGTCTGTTGAG
---.----+----.----+----.----+----.----+----.----+----.----+----.----+----.----+----.----+----.----+  46700
CAAAAGAAGACGGGTTAAGGTCTTTGCTTCTTTACTACCCACTACCAAAGGAAACCACGATGACGACGGCCAAACAAAACTTGTCATTTGCAGACAACTC

 f l l p n s r n e e ■ ■ g d g f l w c y c c r f v l n s k r l l s
 v f f c p i p e t k k . w v ■ v s f g a t a a g l f . t v n v c . a
 f s s a q f q k r r n d g . w f p l v l l l p v c f e q . t s v e
---.----+----.----+----.----+----.----+----.----+----.----+----.----+----.----+----.----+----.----+
 l k r r g l e l f s s i i p s p k r q h . q q r n t k f l l r r n l
 t k k q g i g s v f f h h t i t e k p a v a a p k n q v t f t q q
 n e e a w n w f r l f s p h h n g k t s s s g t q k s c y v d t s
```

```
H                   A H     H                                    S                                 H
G                   S A     H                                    F                                 N
I                   U E     A                                    A                                 F
A                   1 3     1                                    N                                 1

CACATCCTGTAATAAGCAGGGCCAGCGCAGTAGCGAGTAGCATTTTTTTCATGGTGTTATTCCCGATGCTTTTTGAAGTTCGCAGAATCGTATGTGTAGA  46800
----.----+----.----+----.----+----.----+----.----+----.----+----.----+----.----+----.----+----.----+
GTGTAGGACATTATTCGTCCCGGTCGCGTCATCGCTCATCGTAAAAAAAGTACCACAATAAGGGCTACGAAAAACTTCAAGCGTCTTAGCATACACATCT

 t  s  c  n  k  q  g  q  r  s  s  e  .  h  f  f  h  g  v  i  p  d  a  f  .  s  s  q  n  r  ■  c  r
  h  p  v  i  s  r  a  s  a  v  a  s  s  i  f  f  ■  v  l  f  p  ■  l  f  e  v  r  r  i  v  c  v  e
h  i  l  .  .  a  g  p  a  q  .  r  v  a  f  f  s  w  c  y  s  r  c  f  l  k  f  a  e  s  y  v  .  k
----.----+----.----+----.----+----.----+----.----+----.----+----.----+----.----+----.----+----.----+
  v  d  q  l  l  c  p  w  r  l  l  s  y  c  k  k  .  p  t  i  g  s  a  k  q  l  e  c  f  r  i  h  l
a  c  g  t  i  l  l  a  l  a  t  a  l  l  ■  k  k  ■  t  n  n  g  i  s  k  s  t  r  l  i  t  h  t  s
 c  ■  r  y  y  a  p  g  a  c  y  r  t  a  n  k  e  h  h  .  e  r  h  k  k  f  n  a  s  d  y  t  y
```

```
                                                                                  H           NSH
                                                                                  N           CCP
                                                                                  F           IRA
                                                                                  1           112
                                                                                               //
AAATTAAACAAACCCTAAACAATGAGTTGAAATTTCATATTGTTAATATTTATTAATGTATGTCAGGTGCGATGAATCGTCATTGTATTCCCGGATTAAC  46900
----.----+----.----+----.----+----.----+----.----+----.----+----.----+----.----+----.----+----.----+
TTTAATTTGTTTGGGATTTGTTACTCAACTTTAAAGTATAACAATTATAAATAATTACATACAGTCCACGCTACTTAGCAGTAACATAAGGGCCTAATTG

k  l  n  k  p  .  t  ■  s  .  n  f  i  l  l  i  f  i  n  v  c  q  v  r  .  i  v  i  v  f  p  d  .  l
 n  .  t  n  p  k  q  .  v  e  i  s  y  c  .  y  l  l  ■  y  v  r  c  d  e  s  s  l  y  s  r  i  n
  i  k  q  t  l  n  n  e  l  k  f  h  i  v  n  i  y  .  c  ■  s  g  a  ■  n  r  h  c  i  p  g  l  t
----.----+----.----+----.----+----.----+----.----+----.----+----.----+----.----+----.----+----.----+
 f  n  f  l  g  .  v  i  l  q  f  k  ■  n  n  i  n  i  l  t  h  .  t  r  h  i  t  ■  t  n  g  s  .
  f  .  v  f  g  l  c  h  t  s  i  e  y  q  .  y  k  n  i  y  t  l  h  s  s  d  d  n  y  e  r  i  l
f  i  l  c  v  r  f  l  s  n  f  n  .  i  t  l  i  .  .  h  i  d  p  a  i  f  r  .  q  i  g  p  n  v
```

```
                                                 HNS    S                             HT RA            E
                                                 PCC    F                             HH SF            C
                                                 AIR    A                             AA AL            O
                                                 211    N                             11 13            K
                                                   /                                   /  /
TATGTCCACAGCCCTGACGGGGAACTTCTCTGCGGGAGTGTCCGGGAATAATTAAAACGATGCACACAGGGTTTAGCGCGTACACGTATTGCATTATGCC  47000
----.----+----.----+----.----+----.----+----.----+----.----+----.----+----.----+----.----+----.----+
ATACAGGTGTCGGGACTGCCCCTTGAAGAGACGCCCTCACAGGCCCTTATTAATTTTGCTACGTGTGTCCCAAATCGCGCATGTGCATAACGTAATACGG

  c  p  q  p  .  r  g  t  s  l  r  e  c  p  g  i  i  k  t  ■  h  t  g  f  s  a  y  t  y  c  i  ■  p
y  v  h  s  p  d  g  e  l  l  c  g  s  v  r  e  .  l  k  r  c  t  q  g  l  a  r  t  r  i  a  l  c  q
 ■  s  t  a  l  t  g  n  f  s  a  g  v  s  g  n  n  .  n  d  a  h  r  v  .  r  v  h  v  l  h  y  a
----.----+----.----+----.----+----.----+----.----+----.----+----.----+----.----+----.----+----.----+
s  h  g  c  g  q  r  p  v  e  r  r  s  h  g  p  i  i  l  v  i  c  v  p  n  l  a  y  v  y  q  ■  i  g
 .  t  w  l  g  s  p  s  s  r  q  p  l  t  r  s  y  n  f  r  h  v  c  p  k  a  r  v  r  i  a  n  h
  i  d  v  a  r  v  p  f  k  e  a  p  t  d  p  f  l  .  f  s  a  c  l  t  .  r  t  c  t  n  c  .  a
```

```
     NSH                    H    M                                        D
     CCP                    P    B                                        D
     IRA                    A    0                                        E
     112                    2    2                                        1
      //
AACGCCCCGGTGCTGACACGGAAGAAACCGGACGTTATGATTTAGCGTGGAAAGATTTGTGTAGTGTTCTGAATGCTCTCAGTAAATAGTAATGAATTAT  47100
----.----+----.----+----.----+----.----+----.----+----.----+----.----+----.----+----.----+----.----+
TTGCGGGGCCACGACTGTGCCTTCTTTGGCCTGCAATACTAAATCGCACCTTTCTAAACACATCACAAGACTTACGAGAGTCATTTATCATTACTTAATA

  t  p  r  c  .  h  g  r  n  r  t  l  .  f  s  v  e  r  f  v  .  c  s  e  c  s  q  .  i  v  ■  n  y
   r  p  g  a  d  t  e  e  t  g  r  y  d  l  a  w  k  d  l  c  s  v  l  n  a  l  s  k  .  .  .  i  i
n  a  p  v  l  t  r  k  k  p  d  v  ■  i  .  r  g  k  i  c  v  v  f  .  ■  l  s  v  n  s  n  e  l  s
----.----+----.----+----.----+----.----+----.----+----.----+----.----+----.----+----.----+----.----+
  v  g  r  h  q  c  p  l  f  r  v  n  h  n  l  t  s  l  n  t  y  h  e  s  h  e  .  y  i  t  i  f  .
w  r  g  p  a  s  v  s  s  v  p  r  .  s  k  a  h  f  s  k  h  l  t  r  f  a  r  l  l  y  y  h  i  i
 l  a  g  t  s  v  r  f  f  g  s  t  i  i  .  r  p  f  i  q  t  t  n  q  i  s  e  t  f  l  l  s  n
```

```
CAAAGGTATAGTAATATCTTTTATGTTCATGGATATTTGTAACCCATCGGAAAACTCCTGCTTTAGCAAGATTTTCCCTGTATTGCTGAAATGTGATTTC
---------+---------+---------+---------+---------+---------+---------+---------+---------+---------+ 47200
GTTTCCATATCATTATAGAAAATACAAGTACCTATAAACATTGGGTAGCCTTTTGAGGACGAAATCGTTCTAAAAGGGACATAACGACTTTACACTAAAG

q  r  y  s  n  i  f  y  v  h  g  y  l  .  p  i  g  k  l  l  l  .  q  d  f  p  c  i  a  e  ■  .  f  l
 k  g  i  v  i  s  f  ■  f  ■  d  i  c  n  p  s  e  n  s  c  f  s  k  i  f  p  v  l  l  k  c  d  f
  k  v  .  .  y  l  l  c  s  w  i  f  v  t  h  r  k  t  p  a  l  a  r  f  s  l  y  c  .  n  v  i  s
---------+---------+---------+---------+---------+---------+---------+---------+---------+---------+
 .  l  y  l  l  i  k  .  t  .  p  y  k  y  g  ■  p  f  s  r  s  .  c  s  k  g  q  i  a  s  i  h  n
  l  p  i  t  i  d  k  i  n  ■  s  i  q  l  g  d  s  f  e  q  k  l  l  i  k  g  t  n  s  f  h  s  k
d  f  t  y  y  y  r  k  h  e  h  i  n  t  v  w  r  f  v  g  a  k  a  l  n  e  r  y  q  q  f  t  i  e


                         S
                         F
                         A
                         N
TCTTGATTTCAACCTATCATAGGACGTTTCTATAAGATGCGTGTTTCTTGAGAATTTAACATTTACAACCTTTTTAAGTCCTTTTATTAACACGGTGTTA
---------+---------+---------+---------+---------+---------+---------+---------+---------+---------+ 47300
AGAACTAAAGTTGGATAGTATCCTGCAAAGATATTCTACGCACAAAGAACTCTTAAATTGTAAATGTTGGAAAAATTCAGGAAAATAATTGTGCCACAAT

  l  i  s  t  y  h  r  t  f  l  .  d  a  c  f  l  r  i  .  h  l  q  p  f  .  v  l  l  l  t  r  c  y
s  .  f  q  p  i  i  g  r  f  y  k  ■  r  v  s  .  e  f  n  i  y  n  l  f  k  s  f  y  .  h  g  v  i
 l  d  f  n  l  s  .  d  v  s  i  r  c  v  f  l  e  n  l  t  f  t  t  f  l  s  p  f  i  n  t  v  l
---------+---------+---------+---------+---------+---------+---------+---------+---------+---------+
r  k  i  e  v  .  .  l  v  n  r  y  s  a  h  k  k  l  i  .  c  k  c  g  k  .  t  r  k  n  v  r  h  .
 e  q  n  .  g  i  ■  p  r  k  .  l  i  r  t  e  q  s  n  l  ■  .  l  r  k  l  d  k  .  .  c  p  t
  r  s  k  l  r  d  y  s  t  e  i  l  h  t  n  r  s  f  k  v  n  v  v  k  k  l  g  k  i  l  v  t  n


TCGTTTTCTAACACGATGTGAATATTATCTGTGGCTAGATAGTAAATATAATGTGAGACGTTGTGACGTTTTAGTTCAGAATAAAACAATTCACAGTCTA
---------+---------+---------+---------+---------+---------+---------+---------+---------+---------+ 47400
AGCAAAAGATTGTGCTACACTTATAATAGACACCGATCTATCATTTATATTACACTCTGCAACACTGCAAAATCAAGTCTTATTTTGTTAAGTGTCAGAT

 r  f  l  t  r  c  e  y  y  l  w  l  d  s  k  y  n  v  r  r  c  d  v  l  v  q  n  k  t  i  h  s  l
  v  f  .  h  d  v  n  i  i  c  g  .  i  v  n  i  ■  .  d  v  v  t  f  .  f  r  i  k  q  f  t  v  .
s  f  s  n  t  ■  .  i  l  s  v  a  r  .  .  i  .  c  e  t  l  .  r  f  s  s  e  .  n  n  s  q  s  k
---------+---------+---------+---------+---------+---------+---------+---------+---------+---------+
  r  k  r  v  r  h  s  y  .  r  h  s  s  l  l  y  l  t  l  r  q  s  t  k  t  .  f  l  v  i  .  l  r
i  t  k  .  c  s  t  f  i  i  q  p  .  i  t  f  i  i  h  s  t  t  v  n  .  n  l  i  f  c  n  v  t  .
 d  n  e  l  v  i  h  i  n  d  t  a  l  y  y  i  y  h  s  v  n  h  r  k  l  e  s  y  f  l  e  c  d


             M  TH          A            D                        M  P         HT
             B  AN          H            D                        N  K         HH
             O  QF          A            E                        L  2         AA
             1  13          3            1                        1  6         11
                /                                                               /
AATCTTTTCGCACTTGATCGAATATTTCTTTAAAAATGGCAACCTGAGCCATTGGTAAAACCTTCCATGTGATACGAGGGCGCGTAGTTTGCATTATCGT
---------+---------+---------+---------+---------+---------+---------+---------+---------+---------+ 47500
TTAGAAAAGCGTGAACTAGCTTATAAAGAAATTTTTACCGTTGGACTCGGTAACCATTTTGGAAGGTACACTATGCTCCCGCGCATCAAACGTAATAGCA

n  l  f  a  l  d  r  i  f  l  .  k  w  q  p  e  p  l  v  k  p  s  ■  .  y  e  g  a  .  f  a  l  s  f
 i  f  s  h  l  i  e  y  f  f  k  n  g  n  l  s  h  w  .  n  l  p  c  d  t  r  a  r  s  l  h  y  r
  s  f  r  t  .  s  n  i  s  l  k  m  a  t  .  a  i  g  k  t  f  h  v  i  r  g  r  v  v  c  i  i  v
---------+---------+---------+---------+---------+---------+---------+---------+---------+---------+
 f  r  k  a  s  s  r  i  n  r  .  f  h  c  g  s  g  n  t  f  g  e  ■  h  y  s  p  a  y  n  a  n  d
  i  k  e  c  k  i  s  y  k  k  l  f  p  l  r  l  w  q  y  f  r  g  h  s  v  l  a  r  l  k  c  .  r
l  d  k  r  v  q  d  f  i  e  k  f  i  a  v  q  a  ■  p  l  v  k  w  t  i  r  p  r  t  t  q  ■  i  t

PK26   47471        pk26              i(Tn903)    EM
```

```
      E  PP            M                                  X
      C  KK            N                                  M
      P  22            L                                  N
      1  24            1                                  1
          /
TTTTATCGTTTCAATCTGGTCTGACCTCCTTGTGTTTTGTTGATGATTTATGTCAAATATTAGGAATGTTTTCACTTAATAGTATTGGTTGCGTAACAAA
----.----+----.----+----.----+----.----+----.----+----.----+----.----+----.----+----.----+----.----+ 47600
AAAATAGCAAAGTTAGACCAGACTGGAGGAACACAAAACAACTACTAAATACAGTTTATAATCCTTACAAAAGTGAATTATCATAACCAACGCATTGTTT

 l  s  f  q  s  g  l  t  s  l  c  f  v  d  d  l  c  q  i  l  g  m  f  s  l  n  s  i  g  c  v  t  k
f  y  r  f  n  l  v  .  p  p  c  v  l  l  m  i  y  v  k  y  .  e  c  f  h  l  i  v  l  v  a  .  q  s
 f  i  v  s  i  w  s  d  l  l  v  f  c  .  .  f  m  s  n  i  r  n  v  f  t  .  .  y  w  l  r  n  k
----.----+----.----+----.----+----.----+----.----+----.----+----.----+----.----+----.----+----.----+
n  k  d  n  .  d  p  r  v  e  k  h  k  t  s  s  k  h  .  i  n  p  i  n  e  s  l  l  i  p  q  t  v  f
 k  .  r  k  l  r  t  q  g  g  q  t  k  n  i  i  .  t  l  y  .  s  h  k  .  k  i  t  n  t  a  y  c
  k  i  t  e  i  q  d  s  r  r  t  n  q  q  h  n  i  d  f  i  l  f  t  k  v  .  y  y  q  n  r  l  l

PK22   47520        pk22                   i(Tn903)    EM
PK24   47520        pk24                   i(Tn903)    EM
```

```
   AA E     M D                                      HH  S M H
   SV C     N A                                      AA  D B G
   UA 1     L 3                                      EE  U 0 I
   12 5     1                                        13  1 2 A
   /                                                 /
GTGCGGTCCTGCTGGCATTCTGGAGGGAAATACAACCGACAGATGTATGTAAGGCCAACGTGCTCAAATCTTCATACAGAAAGATTTGAAGTAATATTTT
----.----+----.----+----.----+----.----+----.----+----.----+----.----+----.----+----.----+----.----+ 47700
CACGCCAGGACGACCGTAAGACCTCCCTTTATGTTGGCTGTCTACATACATTCCGGTTGCACGAGTTTAGAAGTATGTCTTTCTAAACTTCATTATAAAA

 c  g  p  a  g  i  l  e  g  n  t  t  d  r  c  m  .  g  q  r  a  q  i  f  i  q  k  d  l  k  .  y  f
  a  v  l  l  a  f  w  r  e  i  q  p  t  d  v  c  k  a  n  v  l  k  s  s  y  r  k  i  .  s  n  i  l
v  r  s  c  w  h  s  g  g  k  y  n  r  q  m  y  v  r  p  t  c  s  n  l  h  t  e  r  f  e  v  i  f  .
----.----+----.----+----.----+----.----+----.----+----.----+----.----+----.----+----.----+----.----+
 h  p  g  a  p  m  r  s  p  f  v  v  s  l  h  i  y  p  w  r  a  .  i  k  m  c  f  s  k  f  y  y  k
l  a  t  r  s  a  n  q  l  s  i  c  g  v  s  t  h  l  a  l  t  s  l  d  e  y  l  f  i  q  l  l  i  k
 t  r  d  q  q  c  e  p  p  f  y  l  r  c  i  y  t  l  g  v  h  e  f  r  .  v  s  l  n  s  t  i  n

DA3    47618        deltaa3                dl          EM
```

```
                 HM                               E       MB        BXM M  HH
                 HB                               C       BI        IHB N  NN
                 A0                               1       ON        N00 L  FF
                 12                               5       11        121 1  13
                                                          /           /    /
AACCGCTAGATGAAGAGCAAGCGCATGGAGCGACAAAATGAATAAAGAACAATCTGCTGATGATCCCTCCGTGGATCTGATTCGTGTAAAAAATATGCTT
----.----+----.----+----.----+----.----+----.----+----.----+----.----+----.----+----.----+----.----+ 47800
TTGGCGATCTACTTCTCGTTCGCGTACCTCGCTGTTTTACTTATTTCTTGTTAGACGACTACTAGGGAGGCACCTAGACTAAGCACATTTTTTATACGAA

n  r  .  m  k  s  k  r  m  e  r  q  n  e  .  r  t  i  c  .  .  s  l  r  g  s  d  s  c  k  k  y  a
 t  a  r  .  r  a  s  a  w  s  d  k  m  n  k  e  q  s  a  d  d  p  s  v  d  l  i  r  v  k  n  m  l
  p  l  d  e  e  q  a  h  g  a  t  k  .  i  k  n  n  l  l  m  i  p  p  w  i  .  f  v  .  k  i  c  l
----.----+----.----+----.----+----.----+----.----+----.----+----.----+----.----+----.----+----.----+
 l  r  .  i  f  l  l  r  m  s  r  c  f  s  y  l  v  i  q  q  h  d  r  r  p  d  s  e  h  l  f  y  a
  v  a  l  h  l  a  l  a  h  l  s  l  i  f  l  s  c  d  a  s  s  g  e  t  s  r  i  r  t  f  f  i  s
.  g  s  s  s  s  c  a  c  p  a  v  f  h  i  f  f  l  r  s  i  i  g  g  h  i  q  n  t  y  f  i  h  k
```

```
                                           N                             M          F         B
                                           S                             N          N         B
                                           P                             L          U         V
                                           C                             1          H         1

AATAGCACCATTTCTATGAGTTACCCTGATGTTGTAATTGCATGTATAGAACATAAGGTGTCTCTGGAAGCATTCAGAGCAATTGAGGCAGCGTTGGTGA
----.----+----.----+----.----+----.----+----.----+----.----+----.----+----.----+----.----+----.----+ 47900
TTATCGTGGTAAAGATACTCAATGGGACTACAACATTAACGTACATATCTTGTATTCCACAGAGACCTTCGTAAGTCTCGTTAACTCCGTCGCAACCACT

  .  h  h  f  y  e  l  p  .  c  c  n  c  m  y  r  t  .  g  v  s  g  s  i  q  s  n  .  g  s  v  g  e
n  s  t  i  s  m  s  y  p  d  v  v  i  a  c  i  e  h  k  v  s  l  e  a  f  r  a  i  e  a  a  l  v  k
 i  a  p  f  l  .  v  t  l  m  l  .  l  h  v  .  n  i  r  c  l  w  k  h  s  e  q  l  r  q  r  w  .
----.----+----.----+----.----+----.----+----.----+----.----+----.----+----.----+----.----+----.----+
.  y  c  w  k  .  s  n  g  q  h  q  l  q  m  y  l  v  y  p  t  e  p  l  m  .  l  l  q  p  l  t  p  s
 l  l  v  m  e  i  l  .  g  s  t  t  i  a  h  i  s  c  l  t  d  r  s  a  n  l  a  i  s  a  a  n  t
  i  a  g  n  r  h  t  v  r  i  n  y  n  c  t  y  f  m  l  h  r  q  f  c  e  s  c  n  l  c  r  q  h
```

```
  H                         E S      HH*  *BM
  P                         C C      PI1  1CB
  H                         R R      HN7  71.0
  1                         2 1      120  111
                                       /   //
AGCACGATAATAATATGAAGGATTATTCCCTGGTGGTTGACTGATCACCATAACTGCTAATCATTCAAACTATTTAGTCTGTGACAGAGCCAACACGCAG
----.----+----.----+----.----+----.----+----.----+----.----+----.----+----.----+----.----+----.----+  48000
TCGTGCTATTATTATACTTCCTAATAAGGGACCACCAACTGACTAGTGGTATTGACGATTAGTAAGTTTGATAAATCAGACACTGTCTCGGTTGTGCGTC

 a  r  .  .  y  e  g  l  f  p  g  g  .  l  i  t  i  t  a  n  h  s  n  y  l  v  c  d  r  a  n  t  q
  h  d  n  n  m  k  d  y  s  l  v  v  d  .  s  p  .  l  l  i  i  q  t  i  .  s  v  t  e  p  t  r  s
s  t  i  i  i  .  r  i  i  p  w  w  l  t  d  h  h  n  c  .  s  f  k  l  f  s  l  .  q  s  q  h  a  v
----.----+----.----+----.----+----.----+----.----+----.----+----.----+----.----+----.----+----.----+
  a  r  y  y  y  s  p  n  n  g  p  p  q  s  i  v  m  v  a  l  .  e  f  .  k  t  q  s  l  a  l  v  c
f  c  s  l  l  i  f  s  .  e  r  t  t  s  q  d  g  y  s  s  i  m  .  v  i  .  d  t  v  s  g  v  r  l
 l  v  i  i  i  h  l  i  i  g  q  h  n  v  s  .  w  l  q  .  d  n  l  s  n  l  r  h  c  l  w  c  a

*170   47938      Mapped site HIN2
*171   47942      Mapped site BCL1
```

```
                                                                      M                  M                 T
                                                                      N                  N                 H
                                                                      L                  L                 A
                                                                      1                  1                 1

TCTGTCACTGTCAGGAAAGTGGTAAAACTGCAACTCAATTACTGCAATGCCCTCGTAATTAAGTGAATTTACAATATCGTCCTGTTCGGAGGGAAGAACG
----.----+----.----+----.----+----.----+----.----+----.----+----.----+----.----+----.----+----.----+  48100
AGACAGTGACAGTCCTTTCACCATTTTGACGTTGAGTTAATGACGTTACGGGAGCATTAATTCACTTAAATGTTATAGCAGGACAAGCCTCCCTTCTTGC

s  v  t  v  r  k  v  v  k  l  q  l  n  y  c  n  a  l  v  i  k  .  i  y  n  i  v  l  f  g  g  k  n  a
 l  s  l  s  g  k  w  .  n  c  n  s  i  t  a  m  p  s  .  l  s  e  f  t  i  s  s  c  s  e  g  r  t
  c  h  c  q  e  s  g  k  t  a  t  q  l  l  q  c  p  r  n  .  v  n  l  q  y  r  p  v  r  r  e  e  r
----.----+----.----+----.----+----.----+----.----+----.----+----.----+----.----+----.----+----.----+
 d  t  v  t  l  f  t  t  f  s  c  s  l  .  q  l  a  r  t  i  l  h  i  .  l  i  t  r  n  p  p  f  f
  r  d  s  d  p  f  h  y  f  q  l  e  i  v  a  i  g  e  y  n  l  s  n  v  i  d  d  q  e  s  p  l  v
t  q  .  q  .  s  l  p  l  v  a  v  .  n  s  c  h  g  r  l  .  t  f  k  c  y  r  g  t  r  l  s  s  r
```

```
 H MM           F                                                    P                       E S  D          NSH
 G BB           O                                                    K                       C C  D          CCP
 U OO           K                                                    2                       R R  E          IRA
 2 22           1                                                    1                       2 1  1          112
    /                                                                                                         //
CGGGATGTTCATTCTTCATCACTTTTAATTGATGTATATGCTCTCTTTTCTGACGTTAGTCTCCGACGGCAGGCTTCAATGACCCAGGCTGAGAAATTCC
----.----+----.----+----.----+----.----+----.----+----.----+----.----+----.----+----.----+----.----+  48200
GCCCTACAAGTAAGAAGTAGTGAAAATTAACTACATATACGAGAGAAAAGACTGCAATCAGAGGCTGCCGTCCGAAGTTACTGGGTCCGACTCTTTAAGG

  g  c  s  f  f  i  t  f  n  .  c  i  c  s  l  f  .  r  .  s  p  t  a  g  f  n  d  p  g  .  e  i  p
r  d  v  h  s  s  s  l  l  i  d  v  y  a  l  f  s  d  v  s  l  r  r  q  a  s  m  t  q  a  e  k  f  p
 g  m  f  i  l  h  h  f  .  l  m  y  m  l  s  f  l  t  l  v  s  d  g  r  l  q  .  p  r  l  r  n  s
----.----+----.----+----.----+----.----+----.----+----.----+----.----+----.----+----.----+----.----+
a  p  h  e  n  k  m  v  k  l  q  h  i  h  e  r  k  q  r  .  d  g  v  a  p  k  l  s  g  p  q  s  i  g
 r  s  t  .  e  e  d  s  k  i  s  t  y  a  r  k  e  s  t  l  r  r  r  c  a  e  i  v  w  a  s  f  n
  p  i  n  m  r  .  .  k  .  n  i  y  i  s  e  k  r  v  n  t  e  s  p  l  s  .  h  g  l  s  l  f  e

PK21   48158        pk21               i(Tn903)    EM
```

```
 AA  D                                                               H                   F            M              M H*
 SV  A                                                               G                   O            B              N I1
 UA  3                                                               U                   K            O              L N7
 12                                                                  2                   1            2              1 22
  /                                                                                                                     /
CGGACCCTTTTTGCTCAAGAGCGATGTTAATTTGTTCAATCATTTGGTTAGGAAAGCGGATGTTGCGGGTTGTTGTTCTGCGGGTTCTGTTCTTCGTTGA
----.----+----.----+----.----+----.----+----.----+----.----+----.----+----.----+----.----+----.----+  48300
GCCTGGGAAAAACGAGTTCTCGCTACAATTAAACAAGTTAGTAAACCAATCCTTTCGCCTACAACGCCCAACAACAAGACGCCCAAGACAAGAAGCAACT

 g  p  f  l  l  k  s  d  v  n  l  f  n  h  l  v  r  k  a  d  v  a  g  c  c  s  a  g  s  v  l  r  .
  d  p  f  c  s  r  a  m  l  i  c  s  i  i  w  l  g  k  r  m  l  r  v  v  v  l  r  v  l  f  f  v  d
r  t  l  f  a  q  e  r  c  .  f  v  q  s  f  g  .  e  s  g  c  c  g  l  l  f  c  g  f  c  s  s  l  t
----.----+----.----+----.----+----.----+----.----+----.----+----.----+----.----+----.----+----.----+
  p  g  k  k  s  l  l  s  t  l  k  n  l  .  k  t  l  f  a  s  t  a  p  q  q  e  a  p  e  t  r  r  q
g  s  g  k  q  e  l  a  i  n  i  q  e  i  m  q  n  p  f  r  i  n  r  t  t  t  r  r  t  r  n  k  t  s
 r  v  r  k  a  .  s  r  h  .  n  t  .  d  n  p  .  s  l  p  h  q  p  n  n  n  q  p  n  q  e  e  n

DA3    48207       deltaa3          dr            EM
*172   48298      Mapped site HIN2
```

```
                                                                         S           M        H        S
                                                                         F           B        G        B
                                                                         A           0        A        1
                                                                         N           1        1        R

CATGAGGTTGCCCCGTATTCAGTGTCGCTGATTTGTATTGTCTGAAGTTGTTTTTACGTTAAGTTGATGCAGATCAATTAATACGATACCTGCGTCATAA
---------.---------+---------.---------+---------.---------+---------.---------+---------.---------+ 48400
GTACTCCAACGGGGCATAAGTCACAGCGACTAAACATAACAGACTTCAACAAAAATGCAATTCAACTACGTCTAGTTAATTATGCTATGGACGCAGTATT

h  e  v  a  p  y  s  v  s  l  i  c  i  v  .  s  c  f  y  v  k  l  ■  q  i  n  .  y  d  t  c  v  i  i
 ■  r  l  p  r  i  q  c  r  .  f  v  l  s  e  v  v  f  t  l  s  .  c  r  s  i  n  t  i  p  a  s  .
  .  g  c  p  v  f  s  v  a  d  l  y  c  l  k  l  f  l  r  .  v  d  a  d  q  l  i  r  y  l  r  h  n
---------.---------+---------.---------+---------.---------+---------.---------+---------.---------+
 c  s  t  a  g  y  e  t  d  s  i  q  i  t  q  l  q  k  .  t  l  n  i  c  i  l  .  y  s  v  q  t  ■
  ■  l  n  g  r  i  .  h  r  q  n  t  n  d  s  t  t  k  v  n  l  q  h  l  d  i  l  v  i  g  a  d  y
v  h  p  q  g  t  n  l  t  a  s  k  y  q  r  f  n  n  k  r  .  t  s  a  s  .  n  i  r  y  r  r  .  l

SB1R   48391        Sequence Block 1R    start


 P                    HH         M                                                    AA      H  MB        H
 K                    AA         N                                                    SV      P  BI        P
 2                    EE         L                                                    UA      A  ON        H
 5                    13         1                                                    12      2  11        1
                      /                                                               /          /
TTGATTATTTGACGTGGTTTGATGGCCTCCACGCACGTTGTGATATGTAGATGATAATCATTATCACTTTACGGGTCCTTTCCGGTGATCCGACAGGTTA
---------.---------+---------.---------+---------.---------+---------.---------+---------.---------+ 48500
AACTAATAAACTGCACCAAACTACCGGAGGTGCGTGCAACACTATACATCTACTATTAGTAATAGTGAAATGCCCAGGAAAGGCCACTAGGCTGTCCAAT

  d  y  l  t  w  f  d  g  l  h  a  r  c  d  ■  .  ■  i  i  i  i  t  l  r  v  l  s  g  d  p  t  g  y
 l  i  i  .  r  g  l  ■  a  s  t  h  v  v  i  c  r  .  .  s  l  s  l  y  g  s  f  p  v  i  r  q  v  t
  .  l  f  d  v  v  .  w  p  p  r  t  l  .  y  v  d  d  n  h  y  h  f  t  g  p  f  r  .  s  d  r  l
---------.---------+---------.---------+---------.---------+---------.---------+---------.---------+
 i  s  .  k  v  h  n  s  p  r  w  a  r  q  s  i  y  i  i  i  ■  i  v  k  r  t  r  e  p  s  g  v  p  .
  n  i  i  q  r  p  k  i  a  e  v  c  t  t  i  h  l  h  y  d  n  d  s  .  p  d  k  g  t  i  r  c  t
  q  n  n  s  t  t  q  h  g  g  r  v  n  h  y  t  s  s  l  .  .  .  k  v  p  g  k  r  h  d  s  l  n

PK25   48403        pk25                  i(Tn903)    EM


 R          CS
 E          RB
 N          N1
 D          DR
             /
CG
---------.----+---- 48514
GCCCCGCCGCTGGA

 g  a  a  t  w
  g  r  r  p
r  g  g  d  l
---------.---------.-
  p  a  a  v  q
   p  r  r  g  p
    p  p  s  r

REND   48502      Right END                                  3' end of l-strand
CRND   48514      5' end of r-strand                         last base of sticky end
SB1R   48514      Sequence Block 1R    end
```

TABLE 1 RESTRICTION ENZYME RECOGNITION SEQUENCES

ENZYME	ID	CUT SITE	RECOGNITION SEQUENCE
Aat II	AAT2...	5...	GACGTC
Acc I	ACC1...	2...	GT(A/C)(T/G)AC
Acy I	ACY1...	2...	GPuCGPyC
Afl II	AFL2...	1...	CTTAAG
Afl III	AFL3...	?...	ACPuPyGT
Aha III	AHA3...	3...	TTTAAA
Alu I	ALU1...	2...	AGCT
Apa I	APA1...	5...	GGGCCC
Asu I	ASU1...	1...	GGNCC
Asu II	ASU2...	2...	TTCGAA
Ava I	AVA1...	1...	CPyCGPuG
Ava II	AVA2...	1...	GG(A/T)CC
Ava III	AVA3...	?...	ATGCAT
Avr II	AVR2...	?...	CCTAGG
Bal I	BAL1...	3...	TGGCCA
BamH I	BAM1...	1...	GGATCC
Bbv I	BBV1...	-12...	GCTGC
Bbv I	BBV1...	13...	GCAGC
Bcl I	BCL1...	1...	TGATCA
Eco K	ECOK...	R...	AACNNNNNNGTGC
Eco K	ECOK...	R...	GCACNNNNNNGTT
Eco PI	ECP1...	R...	AGACC
Eco PI	ECP1...	R...	GGTCT
Eco RI	ECR1...	1...	GAATTC
Eco RII	ECR2...	0...	CC(A/T)GG
Eco RV	ECRV...	5...	GATATC
Fnu 4HI	FNUH...	2...	GCNGC
Fok I	FOK1...	-13...	CATCC
Fok I	FOK1...	14...	GGATG
Gdi II	GDI2...	1...	PyGGCCG
Gdi II	GDI2...	5...	CGGCCPu
Hae I	HAE1...	3...	(A/T)GGCC(A/T)
Hae II	HAE2...	5...	PuGCGCPy
Hae III	HAE3...	2...	GGCC
Hga I	HGA1...	10...	GACGC
Hga I	HGA1...	-10...	GCGTC
Hgi EII	HGE2...	?...	ACCNNNNNNGGT
Hgi AI	HGIA...	5...	G(A/T)GC(A/T)C
Hgi CI	HGIC...	1...	GGPyPuCC
Mst I	MST1...	3...	TGCGCA
Mst II	MST2...	2...	CCTNAGG
Nae I	NAE1...	3...	GCCGGC
Nar I	NAR1...	2...	GGCGCC
Nci I	NCI1...	2...	CC(G/C)GG
Nco I	NCO1...	1...	CCATGG
Nde I	NDE1...	2...	CATATG
Nru I	NRU1...	3...	TCGCGA
Nsp BII	NSPB...	3...	C(A/C)GC(T/G)G
Nsp C	NSPC...	5...	PuCATGPy
Pst I	PST1...	5...	CTGCAG
Pvu I	PVU1...	4...	CGATCG
Pvu II	PVU2...	3...	CAGCTG
Rru I	RRU1...	3...	AGTACT
Rsa I	RSA1...	2...	GTAC
Sal I	SAL1...	1...	GTCGAC
Sca I	SCA1...	?...	AGTACT
Scr FI	SCR1...	2...	CCNGG
Sdu I	SDU1...	?...	G(A/G/T)GC(A/C/T)C

Bgl I	BGL1...	7...GCCNNNNNGGC	Hgi JII	HGIJ...	5...GPuGCPyC	Sfa N	SFAN...	-9...GATGC
Bin I	BIN1...	?...GGATC	Hgu II	HGU2...	0...GGATG	Sfa N	SFAN...	10...GCATC
Bin I	BIN1...	?...GATCC	Hha I	HHA1...	3...GCGC	Sna I	SNA1...	0...GTATAC
Bgl II	BGL2...	1...AGATCT	Hind II	HIN2...	3...GTPyPuAC	Sph I	SPH1...	5...GCATGC
Bss HII	BSS2...	1...GCGCGC	Hind III	HIN3...	1...AAGCTT	Sst I	SST1...	5...GAGCTC
Bst EII	BSTE...	1...GGTNACC	Hinf III	HNF3...	R...CGAAT	Sst II	SST2...	4...CCGCGG
Bst NI	BSTN...	2...CC(A/T)GG	Hinf III	HNF3...	R...ATTCG	Stu I	STU1...	3...AGGCCT
Bst XI	BSTX...	8...CCANNNNNNTGG	Hinf I	HNF1...	1...GANTC	Taq I	TAQ1...	1...TCGA
Bvu I	BVU1...	5...GPuGCPyC	Hpa I	HPA1...	3...GTTAAC	Tha I	THA1...	2...CGCG
Cfr I	CFR1...	1...PyGGCCPu	Hpa II	HPA2...	1...CCGG	Tth(111)I	TTH1...	4...GACNNNGTC
			Hph I	HPH1...	13...GGTGA	Tth(111)II	TTH2...	17...CAAPuCA
Cla I	CLA1...	2...ATCGAT	Hph I	HPH1...	-7...TCACC	Tth(111)II	TTH2...	-9...TGPyTTG
Dde I	DDE1...	1...CTNAG	Kpn I	KPN1...	5...GGTACC			
Eco B	ECOB...	R...AGCANNNNNNNNTCA	Mbo I	MBO1...	0...GATC	Xba I	XBA1...	1...TCTAGA
Eco B	ECOB...	R...TGANNNNNNNNTGCT	Mbo II	MBO2...	12...GAAGA	Xho I	XHO1...	1...CTCGAG
Eco DXI	ECDX...	R...ATCANNNNNNNATTC	Mbo II	MBO2...	-7...TCTTC	Xho II	XHO2...	1...PuGATCPy
Eco DXI	ECDX...	R...GAATNNNNNNNTGAT	Mlu I	MLU1...	1...ACGCGT	Xma I	XMA1...	1...CCCGGG
Eco P15	EC15...	R...CAGCAG	Mnl I	MNL1...	11...CCTC	Xma III	XMA3...	1...CGGCCG
Eco P15	EC15...	R...CTGCTG	Mnl I	MNL1...	-7...GAGG	Xmn I	XMN1...	5...GAANNNNTTC

The complete sequence of λ DNA was searched for all occurrences of each of the restriction enzyme recognition sequences shown. ID refers to the four-character abbreviation used in annotating the sequence. Cut site shows the location of the strand cleavage relative to the recognition sequence. 0 means the enzyme cuts immediatly before the first base of the recognition sequence; 1 means it cuts between base 1 and 2, etc.; R means it cuts randomly relative to the recognition sequence; and ? means the cut site is unknown. In general, when two or more enzymes recognize the same sequence we have used only one of them. For complete review, see Roberts (1983).

TABLE 2 ALPHABETICAL LIST OF 4940 RESTRICTION SITES FOR 96 ENZYMES

Enzyme (# of sites) and Recognition Sequence	Coordinates of Cut Sites																	
AAT2(10 SITES)	5109	9398	11247	14978	29040	40810	41117	42251	45567	45596								
GACGT↓C																		
ACC1(9 SITES)	2191	15261	18835	19474	31302	32746	33245	40202	42922									
GT↓(A/C)(T/G)AC																		
ACY1(40 SITES)	1476	1497	2304	4948	4986	5106	6916	8097	8264	9090	9395	9453	9862	10081	10622	11244	11769	12930
GPu↓CGPyC	13319	14800	14975	16057	17617	17671	28468	29037	30473	30728	31766	31937	35073	40807	41114	42248	44222	44331
	44913	45564	45593	45680														
AFL2(3 SITES)	6540	12618	42630															
C↓TTAAG																		
AFL3(20 SITES)	457	627	5547	11280	15371	17790	18283	19995	20951	22219	24132	24167	26527	32763	39394	42085	42362	43761
ACPuPyGT	44500	46981																
AHA3(13 SITES)	92	8462	16296	23112	23286	25438	26134	26667	32705	36304	36532	38835	47431					
TTT↓AAA																		
ALU1(143 SITES)	211	373	1565	1919	1969	2387	2528	2779	3060	3113	3639	3690	3951	3977	4429	4528	5884	6226
AG↓CT	6494	6785	6815	7309	7321	7471	7760	7833	8228	8507	8707	8734	8779	9127	9196	9535	10129	10787
	11228	12101	12164	12413	12577	14719	14913	16080	16667	16749	17184	17190	17667	17994	18012	18366	18439	19159
	19718	20061	20145	20430	20691	20697	20744	22626	22943	22993	23132	23419	24634	24757	24774	25159	25253	25503
	25879	26045	26648	27127	27234	27414	27481	27724	28416	28740	29687	30182	30685	31055	31571	31982	32074	32777
	33049	33071	33131	34094	34224	34707	34739	34895	34987	35381	35444	35970	35977	36150	36360	36571	36628	36897
	37198	37461	37586	37904	38098	38453	39956	40124	40251	40326	40587	40768	40787	40855	40976	40985	41287	41501
	41864	41920	41945	42314	42343	42408	43434	43563	43904	44143	44287	45612	45696	45732	46100	46182	46194	
APA1(1 SITE)	10090																	
GGGCC↓C																		
ASU1(74 SITES)	883	1106	1612	1922	1953	1997	2816	2866	3801	3992	4314	4622	4633	5587	6042	6440	6596	7165
G↓GNCC	7174	8149	8164	8995	9985	10086	10087	10433	10912	11000	11045	11627	12259	12996	13028	13147	13737	13952
	13984	14329	15059	15164	15601	15613	16587	16610	16683	17514	17847	18306	19024	19289	19356	19867	20075	21216
	21603	22001	22243	28798	31560	32386	32474	32562	32942	36786	39004	39437	39479	40075	40792	43950	46719	47605
	48202	48474																

ASU2(7 SITES)	18049	25885	27981	29151	30397	34332	42638											
TT↓CGAA																		
AVA1(8 SITES)	4720	19397	20999	27887	31617	33498	38214	39888										
C↓PyCGPuG																		
AVA2(35 SITES)	1612	1922	2816	3801	4314	4622	6042	6440	8995	11000	11045	12996	13147	13737	13952	13984	14329	15613
G↓G(A/T)CC	16587	16610	16683	19289	19356	19867	22001	22243	28798	32474	32562	39004	39437	39479	47605	48202	48474	
AVA3(14 SITES)	10324	27205	27371	28431	30341	30988	32966	33681	34207	35867	36664	36670	37768	38306				
ATGCAT																		
AVR2(2 SITES)	24321	24395																
CCTAGG																		
BAL1(18 SITES)	1328	2208	3262	4195	6498	6879	7586	7980	8058	8861	10611	10779	13936	14905	21262	26625	28620	36042
TGG↓CCA																		
BAM1(5 SITES)	5505	22346	27972	34499	41732													
G↓GATCC																		
BBV1(199 SITES)	220	360	610	686	727	916	952	1013	1081	1238	1398	2008	2011	2393	2396	2488	2530	2537
GCAGCNNNNNNNN↓	2586	2644	2873	2930	2986	3047	3190	3193	3611	3626	3648	3753	3805	3896	3909	3960	4351	4356
	4376	4384	4416	4551	4656	4695	4723	4753	5002	5073	5200	5455	5483	5618	5668	5902	5926	6004
	6978	7102	7423	7458	7509	7571	7626	7820	7842	8291	8317	8396	8506	8534	8585	8721	9226	9242
	9430	9599	9647	9791	10688	10741	10796	10807	11330	11516	11678	11705	11708	11749	11945	12044	12155	12230
	12254	12513	12586	12667	12707	12916	12989	13371	13859	14054	14280	14367	14643	14706	14709	14788	15084	15154
	15166	15626	16067	16199	17027	17177	17186	17206	17283	17379	17613	17659	17981	19227	19502	19942	20048	20120
	20223	20267	20295	20394	20439	20520	20582	20753	20768	21005	21768	21895	22407	22980	23871	24176	24621	26159
	27401	28295	28358	28395	30136	30883	30964	31991	32016	33036	33437	34103	34233	34836	35286	35453	35503	36093
	37069	37642	37891	38505	38569	38777	38828	38997	39213	39254	40777	40887	40898	41063	41716	41929	41932	42330
	42365	42594	42779	43261	43489	43978	44020	44372	44521	44980	45455	45741	45792	45863	46136	46317	46576	46651
	47900																	
[a]BCL1(8 SITES)	8844	9361	13820	32729	37352	43682	46366	47942										
T↓GATCA																		
BGL1(29 SITES)	410	2666	3804	4366	4457	4583	5252	5438	6059	6110	7556	8055	11064	12714	12723	12838	13204	14407
BGL2(6 SITES)	415	22425	35711	38103	38754	38814												
A↓GATCT																		
BIN1(58 SITES)	548	1605	2167	2530	3068	4532	4774	5504	5505	5646	6191	6422	6575	6734	7070	7402	10315	10520
GGATC	10813	10861	13803	15111	17610	18782	21007	21252	22345	22346	23697	24511	26221	27026	27971	27972	28886	29592
	30426	30491	31992	33087	33323	34498	34499	34991	37297	38663	39575	39633	39813	40162	41731	41732	42397	43734
	44892	47761	47772	48486														
BSS2(6 SITES)	3522	4126	5627	14815	16649	28008												
G↓CGCGC																		

TABLE 2 (continued)

Enzyme (# of sites) and Recognition Sequence	Coordinates of Cut Sites																	
BSTE(13 SITES)	5687	7058	8322	9024	13348	13572	13689	16012	17941	25183	30005	36374	40049					
G↓GTNACC																		
BSTX(13 SITES)	2862	6713	8420	8857	10922	13270	14345	18036	19748	21629	34603	38299	46441					
CCANNNNN↓NTGG																		
BVU1(7 SITES)	585	10090	19767	21574	24776	25881	39457											
GPuGCPy↓C																		
CFR1(39 SITES)	1326	2206	2739	3260	4193	5601	6008	6496	6877	7584	7978	8056	8366	8859	10588	10609	10777	13481
Py↓GGCCPu	13934	14575	14903	16416	18547	19284	19332	19944	20239	20323	20928	20988	21260	22025	26623	28618	35465	36040
	36654	39458	45214															
[b]CLA1(15 SITES)	4199	15584	16121	26617	30290	31991	32964	33585	34697	35051	36966	41364	42021	43825	46439			
AT↓CGAT																		
DDE1(104 SITES)	707	2535	3097	3866	4249	5329	6505	6964	7361	8013	8532	8760	9109	10130	10298	10683	11103	11140
C↓TNAG	11311	11662	13188	13855	14053	14360	14744	14935	15383	15574	15668	16519	18148	18178	18421	18466	18900	19453
	19766	20065	20099	20146	20745	20882	21156	21423	21587	22627	22933	23420	25195	25587	26030	26109	26570	26687
	26718	27605	27735	28284	28440	29629	30375	30917	31223	31837	33074	33209	33535	34319	34740	35029	35586	35814
	35988	36365	36394	36595	37142	37425	37582	37587	37645	38028	38342	38436	38650	38908	39021	39451	40252	40583
	40769	40799	41892	42243	42760	43048	44057	44157	44947	45384	46353	47078	47444	48189				
EC15(72 SITES)	965	1094	1556	2021	2095	2381	2391	2543	3200	3203	3906	4308	4421	4536	5939	6158	7115	9230
CAGCAG	9239	9325	10258	10345	10817	11666	11693	11696	11963	12029	12122	12218	12239	14656	14719	15639	15912	16191
	16869	17601	17994	18029	20505	20746	22020	22534	22768	23388	24634	28305	28613	30272	30849	30913	32183	32537
	33049	35574	37021	38478	39256	39680	40599	40886	40945	42375	42582	42789	43688	43785	45780	46661	47608	47753
ECDX(0 SITE)																		
ATCANNNNNNNATTC																		
ECOB(9 SITES)	2011	3896	4467	9961	10377	13810	24230	28590	44883									
AGCANNNNNNNNTCA																		
ECOK(5 SITES)	6941	14980	16369	34763	47000													
AACNNNNNNGTGC																		
ECP1(49 SITES)	792	1321	1599	2789	3082	4203	4344	4961	5948	6832	6974	7988	9082	11088	11236	11388	11424	12390
AGACC	12406	12510	13103	14347	14412	16073	16882	17005	18109	18896	22080	22178	27989	28216	28550	29968	30316	35191
	36134	36597	39263	40202	40408	40431	40720	40771	42715	43009	43345	44706	47517					

ECR1(5 SITES)	21226	26104	31747	39168	44972													
G↓AATTC																		
[c]ECR2(71 SITES)	424	847	1183	1294	1713	3301	3468	4156	4360	4578	6835	7581	8039	8151	8720	9981	10953	11427
↓CC(A/T)GG	12393	13576	13948	14354	15089	15323	16321	17069	17945	19188	19424	19589	20460	20490	22264	22309	22468	23300
	28620	30001	30039	30226	30432	30468	30528	30749	31009	31518	31556	31788	32173	32838	33341	34296	35100	36164
	36236	37407	37837	39422	39742	40308	40463	40497	41530	43402	43422	44408	44833	45112	45348	47928	48183	
ECRV(21 SITES)	654	2088	6685	8088	8826	13439	14027	17771	18389	21273	22952	26825	28202	28215	33591	39356	41277	41545
GATAT↓C	41580	42235	45830															
FNUH(380 SITES)	4	209	321	374	398	599	602	675	678	741	744	905	966	1002	1095	1198	1252	1387
GC↓NGC	2000	2022	2025	2382	2385	2497	2502	2526	2544	2575	2655	2658	2667	2887	2919	2975	3044	3061
	3173	3204	3207	3322	3434	3481	3625	3637	3640	3671	3742	3779	3791	3794	3910	3923	3949	4030
	4101	4340	4367	4370	4373	4390	4430	4472	4540	4670	4709	4712	4767	4938	5016	5059	5062	5214
	5439	5444	5472	5481	5494	5593	5604	5607	5682	5780	5916	5940	6018	6102	6111	6166	6335	6455
	6599	6632	6992	7116	7412	7472	7523	7560	7579	7640	7828	7831	7834	7837	7959	8047	8050	8258
	8280	8303	8306	8385	8445	8520	8523	8574	8735	8866	9231	9240	9419	9433	9613	9661	9780	10404
	10588	10663	10702	10741	10755	10785	10821	11068	11319	11353	11365	11396	11530	11647	11667	11694	11697	11763
	11772	11863	11866	11959	12033	12085	12169	12219	12243	12262	12400	12502	12541	12575	12656	12683	12718	12721
	12724	12767	12869	12905	13003	13322	13337	13385	13827	13848	14043	14094	14195	14294	14378	14381	14450	14465
	14637	14657	14694	14707	14720	14723	14777	14856	15098	15101	15143	15155	15640	16021	16024	16081	16188	16264
	16382	16455	16465	17041	17191	17200	17220	17272	17356	17368	17500	17602	17639	17648	17701	17825	17995	18197
	18324	18327	18503	18512	18651	18719	18999	19200	19216	19445	19491	19956	20062	20074	20110	20134	20173	20209
	20212	20215	20236	20239	20257	20281	20284	20302	20323	20344	20383	20428	20446	20467	20509	20530	20533	20551
	20554	20596	20635	20644	20647	20742	20782	20931	20991	20994	21135	21606	21757	21909	22408	22421	22573	22994
	23860	24162	24190	24193	24635	24969	26148	27017	27415	28309	28372	28384	28471	28507	29123	29988	30150	30235
	30359	30543	30897	30978	31073	31232	32005	32030	32048	32122	32145	32428	32431	33050	33426	33541	34092	34222
	34825	35092	35300	35338	35341	35442	35517	35928	36082	37083	37656	37751	37905	38519	38583	38613	38791	38842
	38986	38994	39227	39240	39243	39275	39595	39672	40114	40347	40386	40389	40527	40766	40876	40887	41077	41401
	41516	41705	41918	41946	42344	42379	42583	42793	42796	42902	43036	43250	43503	43992	44034	44217	44361	44429
	44510	44632	44922	44994	44997	45217	45229	45284	45469	45730	45781	45852	46111	46150	46225	46306	46574	46590
	46665	47889																
FOK1(150 SITES)	362	698	825	900	1044	1239	1323	1515	1536	1628	1751	1842	1960	2046	2171	2328	2433	2580
GGATGNNNNNNNNN↓	2640	2983	3040	3208	3357	3489	4032	4562	4738	4834	5087	5114	5143	5321	5752	5909	6011	6429
	6875	7014	7346	7514	7686	7909	7984	8063	8273	8549	8597	8795	8803	9224	9898	10186	10207	10314
	10481	10502	10596	10871	11171	11481	11583	11625	12440	12674	13148	13710	13780	13915	14011	14522	14570	14878
	15169	15634	16056	16065	16334	16399	16525	16654	16816	17031	18076	18572	19791	19815	19815	19947	20561	20724
	20826	21446	21454	21652	21822	22429	23072	24916	25791	28050	28169	28953	29179	29889	30056	30105	30255	31081
	31354	31438	31951	32145	32277	32347	32384	32424	32428	32477	32885	33361	33602	33616	34215	34301	34336	35340
	35879	36789	37585	37798	39019	39063	39235	39297	39908	40017	40857	40905	40964	41187	42780	43221	43303	44014

TABLE 2 (continued)

Enzyme (# of sites) and Recognition Sequence	Coordinates of Cut Sites																
FOK1 (cont)	44359	44538	45727	46689	48116	48271											
GDI2(23 SITES)	2743	5601	6008	8366	10592	13485	14579	16420	18547	19284	19332	19944	19948	20243	20327	20928	20988	22025
Py↓GGCCG	35469	36654	36658	39462	45214													
HAE1(64 SITES)	147	404	1328	1486	2040	2208	3262	3298	4090	4195	4460	4524	5237	6498	6879	7586	7980	8058
(A/T)GG↓CC(A/T)	8187	8861	9504	10611	10779	12436	13078	13262	13936	14905	15997	16321	16768	17293	18146	20187	20370	21262
	22533	26625	28409	28620	28731	28902	31480	31583	31793	31878	32323	32999	33439	33628	34080	34607	35021	36042
	36404	37247	38137	39994	40598	40616	40936	45737	47654	48425								
HAE2(48 SITES)	860	1104	1872	2502	3120	3439	3650	3799	4335	4709	4892	5843	6288	6753	7295	7610	8688	9240
PuGCGC↓Py	10698	12038	12203	12329	12363	12472	12661	13199	13910	14327	14470	15029	16405	16676	17557	17812	18300	18883
	20524	20999	21080	28859	34720	35814	36394	37061	44637	45683	45803	45881						
HAE3(149 SITES)	147	404	470	782	885	1107	1328	1486	1955	1999	2040	2208	2741	2868	3262	3298	3994	4090
GG↓CC	4195	4460	4524	4634	5181	5237	5588	5603	6010	6101	6498	6598	6879	7166	7176	7581	7586	7980
	8001	8058	8151	8166	8187	8368	8861	9504	9986	10088	10434	10590	10611	10779	10914	11364	11628	12261
	12436	12717	13029	13078	13262	13483	13936	14369	14577	14905	15061	15166	15603	15997	16321	16418	16768	17293
	17515	17849	17911	18146	18307	18340	18549	19025	19286	19334	19946	20076	20187	20241	20325	20343	20370	20930
	20990	21218	21233	21262	21605	22027	22533	26625	28409	28620	28731	28902	31268	31480	31562	31583	31648	31793
	31878	32089	32323	32387	32646	32943	32999	33439	33543	33628	34080	34607	35021	35467	35819	36042	36404	36444
	36656	36787	37247	38137	38150	39460	39994	40077	40113	40385	40598	40616	40794	40936	43035	43952	44259	44999
	45216	45737	46721	47654	48425													
HGA1(102 SITES)	482	1484	1505	1535	2027	2312	2467	2644	2817	4603	4708	4937	4975	5144	5541	6195	6393	6905
GACGCNNNNN↓	7017	8086	8253	8958	9054	9098	9461	9870	9932	10089	10611	11137	11704	11777	11897	11978	11999	12152
	12682	12919	13246	13327	14789	14790	16065	16269	16578	16887	17022	17028	17205	17606	17679	18426	19191	19482
	19910	19947	19987	20034	20143	20275	20323	20547	20554	20943	20960	20972	22066	22228	23207	24468	26599	27406
	28476	29807	29861	30311	30462	30736	31334	31522	31755	31945	33070	33231	35062	35536	35559	37883	39618	40324
	41344	42292	42875	44230	44320	44447	44902	45333	45695	45775	46122	48381						
HGE2(14 SITES)	1784	2249	5902	6554	12512	13953	15876	17432	20243	26434	35594	35638	37998	42047				
ACCNNNNNNGGT																		
HGIA(28 SITES)	5623	6006	9489	10299	11954	13293	13496	14478	15215	16520	21616	21802	21856	24776	25881	26473	27177	33471
G(A/T)GC(A/T)↓C	35587	37937	40220	40493	42375	42516	44181	44850	46702	47664								

HGIC(25 SITES)	1180	1365	2331	5407	5665	5671	5900	8036	8043	8441	8764	8988	10221	13038	13642	14623	15199	15237
G↓GPyPuCC	16236	17053	18556	21545	39907	42797	45679											
HGIJ(7 SITES)	585	10090	19767	21574	24776	25881	39457											
GPuGCPy↓C																		
HGU2(101 SITES)	348	684	811	1030	1225	1501	1737	1828	1946	2032	2314	2419	2566	2626	2969	3194	3343	3475
↓GGATG	4548	4724	4820	5073	5100	5129	5307	5738	5895	5997	6415	6861	7000	7332	7500	7895	7970	8535
	8583	8781	8789	9210	10193	10467	10488	10582	10857	11467	11569	11611	12426	13134	13696	13766	13901	13997
	14508	14556	15620	16042	16051	16385	16511	16640	16802	18062	19777	19801	19933	20710	20812	21432	21638	22415
	23058	24902	25777	28036	28155	30042	31067	31340	32370	32463	32871	34322	35865	36775	39221	39283	40003	40843
	40891	40950	42766	43207	43289	44000	44345	44524	45713	48102	48257							
HHA1(215 SITES)	378	466	682	759	859	1006	1103	1871	1939	2381	2492	2501	2506	2543	2891	3013	3119	3362
GCG↓G	3438	3524	3526	3649	3727	3798	3817	4128	4130	4273	4334	4498	4516	4708	4847	4891	5020	5158
	5438	5443	5461	5629	5631	5644	5818	5842	6287	6698	6752	6982	7018	7079	7294	7411	7578	7609
	7654	7781	8240	8257	8343	8519	8687	8883	9239	9784	9879	10153	10359	10697	10943	11109	11171	11447
	11504	11566	11597	11603	11639	11693	11744	12037	12089	12131	12202	12212	12218	12328	12362	12471	12574	12660
	12687	12868	13170	13198	13358	13403	13655	13835	13861	13909	14162	14180	14326	14469	14817	14819	15000	15028
	15228	15362	15539	15842	16028	16049	16220	16355	16404	16469	16544	16651	16653	16675	16843	16906	17061	17204
	17313	17421	17556	17811	18258	18299	18511	18524	18718	18882	19991	20109	20166	20523	20705	20881	20998	21040
	21079	21250	21808	21829	22407	27549	27573	27952	28010	28012	28376	28858	29185	30211	30547	30568	31319	32282
	32686	32890	33066	33726	34719	34824	35123	35135	35211	35274	35328	35813	36052	36393	36724	37060	37205	38078
	38125	38572	38812	39692	39850	40037	40085	40336	40351	40361	41405	42326	42383	42446	43228	43758	43836	43884
	44008	44360	44372	44636	45044	45339	45682	45802	45880	46075	46154	46481	46573	46727	46978	47482	47723	
[a]HIN2(35 SITES)	199	734	5269	5710	7950	8201	9056	9626	11585	13785	14993	17076	18756	19841	20569	21904	23147	26744
GTPy↓PuAC	27318	28928	31809	32219	32747	33246	35261	35615	37433	37989	38548	39608	39836	40942	43183	47938	48298	
HIN3(6 SITES)	23130	25157	27479	36895	37459	44141												
A↓AGCTT																		
HNF1(148 SITES)	314	500	837	1395	1911	3246	3567	3844	4006	4381	5353	6386	6431	7094	7196	7426	8191	8771
G↓ANTC	8795	9038	9111	9702	9726	9828	10533	10577	10615	10845	11097	11558	11672	12059	12848	13274	13716	14145
	14153	14187	14428	14581	14686	15012	15050	15729	15748	15894	16171	17974	19120	19554	19669	19772	19887	20097
	20220	21154	21658	22683	22765	22852	23320	23491	23813	24027	24907	25508	25947	26158	26275	26282	26461	27615
	27932	28087	28302	28607	28679	28754	28942	29271	29528	29633	29775	30032	30594	30762	30772	31125	31721	32064
	32339	32503	32752	32826	32976	33338	33398	33414	33564	33770	33783	34627	35048	35126	35237	35364	35453	36245
	36561	36845	36855	36925	37176	37522	37834	38394	38500	38587	38678	38891	38899	39178	39866	40011	41535	41871
	41894	42018	42189	42635	42933	43330	43714	43862	43890	43940	43964	44022	44468	44582	44646	44806	44823	45438
	45973	46785	46874	47779														
HNF3(66 SITES)	1076	2346	2891	3246	4037	5338	9036	9724	12232	14684	17900	17972	18664	18814	20274	20484	20664	21227
CGAAT	21557	23323	23434	24459	25598	25802	25882	28329	28754	28942	29148	29151	29862	30660	32112	32446	32808	33398
	33414	36855	37165	38392	38803	39157	39176	39328	40664	40963	40996	41338	41604	41871	42401	42493	42635	42931

TABLE 2 (continued)

Enzyme (# of sites) and Recognition Sequence	Coordinates of Cut Sites																
HNF3 (cont)	43330	43657	43712	43913	44022	44187	44598	44872	45356	46532	47418	47779					
[d]HPA1(14 SITES)	734	5269	5710	7950	8201	11585	14993	21904	27318	31809	32219	35261	39608	39836			
GTT↓ACC																	
HPA2(328 SITES)	42	379	611	698	760	783	1130	1500	1603	1645	1827	1903	1956	1982	2063	2162	2202	2307
C↓CGG	2329	2355	2620	2793	2869	2900	3093	3330	3342	3383	3393	3407	3515	3758	3799	3804	3838	3926
	3957	4184	4244	4343	4385	4560	4608	4673	4680	4696	4731	4815	4880	4945	4969	5099	5110	5128
	5178	5247	5447	5568	5669	5693	5798	5820	5979	6011	6054	6267	6310	6357	6414	6444	6562	6579
	6628	6895	6936	7066	7074	7344	7366	7383	7415	7526	7551	7703	7766	7812	7894	8129	8204	8338
	8369	8607	8660	8682	8768	8800	8810	8828	8878	8979	8992	9094	9163	9184	9217	9439	9513	9633
	9643	9697	9842	9874	10025	10084	10191	10278	10319	10354	10416	10436	10466	10537	10551	10885	10936	11008
	11059	11188	11326	11338	11987	12007	12038	12281	12296	12445	12553	12745	12763	12833	12857	12880	12900	12920
	12934	12999	13087	13128	13203	13440	13570	13694	13740	13807	13918	13987	14168	14214	14251	14259	14282	14343
	14374	14419	14460	14506	14540	14645	14806	14895	14917	14966	14987	15056	15110	15128	15167	15179	15616	15655
	15689	15708	15736	15755	16041	16060	16300	16441	16552	16639	16824	16844	16927	16964	16976	17010	17051	17092
	17281	17517	17580	17614	17631	17644	17674	17845	18223	18337	18355	18413	18923	19022	19079	19113	19230	19287
	19292	19335	19359	19398	19487	19527	19618	19624	19695	19776	19786	19864	19896	19943	19947	20041	20125	20161
	20251	20419	20846	20856	21219	21239	21471	21777	21795	21839	22028	22079	22123	22214	22252	22276	22344	22350
	24339	24422	24805	24835	24843	25614	26465	28035	28254	28853	28976	29591	30843	31269	31325	31416	31600	31618
	31739	31815	31826	31889	31997	32019	32222	32240	32490	32585	32870	33157	35450	35808	35820	36441	38609	38732
	39220	39440	39457	39651	39735	39889	39911	39934	39983	40002	40337	40453	40912	41103	41282	41621	41880	42033
	42236	42473	42772	42782	43831	43974	44019	44225	44879	44891	45379	45481	45683	45874	46291	46668	46891	46942
	47007	47028	48200	48482														
•HPH1(168 SITES)	569	935	944	1047	1136	1311	1659	1700	1806	1971	2177	3292	3781	3853	4119	4239	4294	4622
GGTGANNNNNNNN↓	5278	5458	6227	6324	6576	6728	6839	7070	7079	7253	7477	7717	7734	7873	8435	8671	8824	8839
	8993	9036	9356	9356	9374	9386	9485	9587	9663	9674	10064	10136	10760	10850	10886	11145	12572	12786
	12813	12974	13187	13360	13432	13566	13576	13683	13785	13830	13875	13975	14265	14324	14449	14795	15441	15727
	15795	15898	15954	15957	16017	16209	16464	16644	16716	17007	17109	17253	17352	17454	18561	18714	18876	18969
	19021	19189	19655	19763	19865	20949	21233	21889	22060	22066	22326	22700	23342	23981	24214	24989	25054	25075
	25587	26347	26425	26706	26826	27556	28231	28308	28535	29888	29999	30665	31528	31942	32690	32739	32857	33407
	33820	34353	35542	35585	35629	35717	36386	36495	36635	37018	37370	37397	37556	37603	37945	38020	38584	38886

	39225	39309	39809	39877	40167	40815	41248	41431	41983	42104	42213	42280	42477	43619	43935	44888	45625	45835
	45968	46340	46651	47908	47937	48496												
KPN1(2 SITES)	17057	18560																
GGTAC↓C																		
[f]MBO1(116 SITES)	415	549	1606	2167	2366	2531	3018	3069	4533	4774	5283	5463	5505	5647	6191	6422	6575	6734
↓GATC	7070	7403	7881	8844	8914	9361	9413	10315	10521	10559	10813	10861	10891	11033	11615	11933	13803	13820
	15112	15389	15581	15800	17610	18594	18782	21007	21252	22346	22425	23026	23698	24014	24511	25035	26117	26222
	26254	27027	27972	28349	28448	28638	28886	29593	29842	30426	30492	31422	31713	31992	32180	32459	32663	32729
	32961	33087	33205	33323	34499	34561	34776	34991	35711	35787	36771	37224	37298	37352	38103	38126	38475	38664
	38754	38814	39576	39634	39694	39814	40162	40234	40668	41732	42397	42979	43280	43376	43682	43735	44893	45630
	45816	46366	47415	47761	47773	47942	48371	48486										
[e]MBO2(130 SITES)	54	57	401	499	1662	2409	2574	2657	3151	3621	4096	4425	5087	5549	6432	6501	6525	7295
GAAGANNNNNNNN↓	8487	8714	9153	9423	9870	9932	10382	10391	11748	11811	11955	12417	12474	12803	13114	13298	13717	14297
	16768	16809	19896	19992	20042	20694	21070	22189	22372	22459	22485	22648	22793	23021	23147	23204	23585	23884
	24520	24778	24781	25039	25393	25965	26775	27228	27261	27431	27565	27904	28522	28589	28866	28884	28954	29135
	29478	29804	29990	30027	30065	30468	30568	30796	31082	31192	32732	33028	33966	34164	34185	34321	34330	34339
	34342	34812	35516	36346	36877	36905	37100	37185	37232	37706	38209	38452	38500	39274	39312	39927	40244	40403
	40481	41585	41799	41995	42434	42437	42460	42602	42978	44331	44477	44832	45200	45501	46597	46639	47033	47661
	47724	48105	48105	48283														
MLU1(7 SITES)	458	5548	15372	17791	19996	20952	22220											
A↓CGCGT																		
MNL1(262 SITES)	20	110	128	158	274	294	375	402	913	1355	1435	1497	1520	1798	1954	2029	2458	2557
CCTCNNNNNNN↓	2617	2922	3128	3155	3185	3449	3537	3548	3804	4079	4397	4400	4575	4951	5202	5226	5326	5412
	5511	5514	5519	5881	5988	6006	6307	6353	6577	6598	7043	7185	7308	7368	7389	7596	7814	7856
	8022	8526	8640	8938	9476	9508	9985	10069	10353	10411	10479	10522	10928	11128	11188	11189	11353	11518
	11731	12070	12073	12740	12814	13082	13248	13251	13508	13808	13881	14209	14254	14358	14547	14595	14750	14804
	15312	15492	15662	15731	15762	15888	15993	16008	16694	17058	17096	17282	17650	17900	18011	18032	18149	18157
	18158	18344	18460	18719	18906	18947	19227	19462	19924	19942	19975	20168	20234	20324	20332	20371	20381	20444
	20534	20584	20701	21367	21953	22082	22148	22272	22346	22464	23637	23797	23959	24296	24452	24455	24526	24734
	24804	24932	25205	25376	25579	25687	25801	25982	26203	26446	26571	26577	26955	27339	27490	27853	28370	28433
	28490	28566	28621	28685	28900	28971	29342	29392	29930	30228	30412	30440	30642	30911	31213	31232	31294	31331
	31402	31594	31614	31846	32831	33038	33089	33507	33617	33830	33925	33937	33945	34313	34549	34648	34948	35037
	35786	35808	35884	36380	36455	36473	36708	36861	37136	37515	37908	38010	38025	38192	38376	38483	38644	38712
	38789	39038	39047	39058	39091	39114	39192	39983	40150	40216	40284	40626	41461	41591	41685	41688	41712	41776
	41905	41930	42074	42249	42411	42598	42811	43024	44620	44697	45028	45062	45535	45539	45647	46298	46319	46353
	46362	47468	47535	47615	47776	47877	48061	48081	48296	48436								

TABLE 2 (continued)

Enzyme (# of sites) and Recognition Sequence	Coordinates of Cut Sites																	
MST1(15 SITES)	465	2505	4272	5157	6981	11565	11692	13357	16048	21807	21828	27951	32685	34823	42382			
TGC↓GCA																		
MST2(2 SITES)	26718	34319																
CC↓TNAGG																		
NAE1(1 SITE)	20042																	
GCC↓GGC																		
NAR1(1 SITE)	45680																	
GG↓CGCC																		
NCI1(114 SITES)	380	699	761	1131	1604	1957	2163	2329	2355	3515	3839	3927	3957	4386	4815	4945	4969	5111
CC↓(G/C)GG	5128	5568	5693	5798	5820	6268	6358	6444	7367	7527	8205	8338	8608	8800	8828	8993	9094	9164
	9217	10085	10192	10279	10436	10466	11060	11339	12039	12282	12901	12935	13128	13204	13695	13919	14168	14343
	14420	14461	14507	14646	14807	14896	14917	14967	14988	15057	15709	16825	16845	16928	17010	17517	17581	17675
	17846	18223	18414	18924	19080	19231	19398	19399	19787	19865	19897	19943	22277	22351	24806	24835	26466	28254
	28853	31270	31325	31618	31619	31997	33158	35808	35821	38609	39457	39735	39889	39890	39912	40913	41282	42474
	44226	45875	46891	46943	47007	48200												
NCO1(4 SITES)	19329	23901	27868	44248														
C↓CATGG																		
NDE1(7 SITES)	27631	29884	33680	36113	36669	38358	40132											
CA↓TATG																		
NRU1(5 SITES)	4592	28052	31705	32409	41810													
TCG↓CGA																		
NSPB(75 SITES)	208	503	598	878	1916	2384	2518	2525	2654	2918	3057	3425	3636	3741	4471	5055	5223	5471
C(A/C)G↓C(T/G)G	5674	5854	6194	6706	6822	7106	7830	7958	8444	8633	8862	9105	9429	10330	10659	11771	12098	12161
	12353	12717	12780	13110	13826	14181	14439	14633	16077	16271	17216	17359	17396	17479	18320	18615	18647	18763
	19715	20058	20319	20508	20529	20547	20694	21402	21605	22990	27411	28936	30974	31231	38444	39410	40385	44428
	46224	46329	46485															
NSPC(32 SITES)	632	2216	6482	8379	12006	17278	18762	21806	23429	23946	24375	25103	25663	25872	27378	29174	30742	31546
PuCATG↓Py	32497	33721	34653	35086	38025	39000	39399	39422	39650	40073	42350	43072	46127	47844				
PST1(28 SITES)	2560	2824	3629	3644	3860	4374	4713	4913	5124	5218	5686	8524	9617	9781	11767	11839	14298	14385
CTGCA↓G	16085	16235	17394	19837	20285	22425	26932	32009	32256	37005								

```
PVU1( 3 SITES)       11936 26257 35790
  CGAT↓CG
PVU2( 15 SITES)        211  1919  2387  2528  3060  3639  7833 12101 12164 16080 19718 20061 20697 22993 27414
  CAG↓CTG
RRU1( 5 SITES)       16423 18686 25687 27265 32804
  AGT↓ACT
RSA1( 113 SITES)       241   440   532  1698  2261  2709  3026  3824  4018  4254  4718  5222  6144  6291  6903  7147  7675  7937
  GT↓AC               8125  8580  8962 10023 10237 10289 10637 12073 12153 12704 12896 13229 13269 13650 13681 13880 14762 14957
                     15349 15651 15777 15857 16423 16486 16818 16925 17055 17729 17805 17815 18450 18462 18558 18564 18642 18686
                     18734 19048 19084 19148 19325 19366 19546 19699 19812 19821 20018 20925 21475 21781 22098 24281 24310 24920
                     25402 25687 26181 27265 29074 29394 29423 29752 30811 31445 31517 31550 31689 31778 32001 32498 32650 32804
                     32868 34436 35113 37309 37630 38026 39660 39679 40705 41003 41780 41859 42993 43132 43438 43870 43978 44169
                     44254 44908 45276 45434 46981
SAL1( 2 SITES)       32745 33244
  G↓TCGAC
SCA1( 5 SITES)       16420 18683 25684 27262 32801
  AGTACT
SCR1( 185 SITES)       380   426   699   761   849  1131  1185  1296  1604  1715  1957  2163  2329  2355  3303  3470  3515  3839
  CC↓NGG              3927  3957  4158  4362  4386  4580  4815  4945  4969  5111  5128  5568  5693  5798  5820  6268  6358  6444
                      6837  7367  7527  7583  8041  8153  8205  8338  8608  8722  8800  8828  8993  9094  9164  9217  9983 10085
                     10192 10279 10436 10466 10955 11060 11339 11429 12039 12282 12395 12901 12935 13128 13204 13578 13695 13919
                     13950 14168 14343 14356 14420 14461 14507 14646 14807 14896 14917 14967 14988 15057 15091 15325 15709 16323
                     16825 16845 16928 17010 17071 17517 17581 17675 17846 17947 18223 18414 18924 19080 19190 19231 19398 19399
                     19426 19591 19787 19865 19897 19943 20462 20492 22266 22277 22311 22351 22470 23302 24806 24835 26466 28254
                     28622 28853 30003 30041 30228 30434 30470 30530 30751 31011 31270 31325 31520 31558 31618 31619 31790 31997
                     32175 32840 33158 33343 34298 35102 35808 35821 36166 36238 37409 37839 38609 39424 39457 39735 39744 39889
                     39890 39912 40310 40465 40499 40913 41282 41532 42474 43404 43424 44226 44410 44835 45114 45350 45875 46891
                     46943 47007 47930 48185 48200
SDU1( 38 SITES)        580  5618  5663  6001  9484 10085 10294 11413 11949 13038 13288 13491 14473 14896 15210 16515 19762 21569
G(A/G/T)GC(A/C/T)C 21611 21797 21851 24771 25876 26468 27172 32329 33466 35582 37932 39452 40215 40488 42370 42511 44176 44845
                     46697 47659
SFAN( 169 SITES)       340   803  1028  1568  1820  2193  2515  2961  3072  3137  3371  3467  3602  3824  4054  4220  4467  4540
  GCATCNNNNN↓         4668  5233  5269  5290  5299  5646  5692  5705  5850  6078  6472  6508  6992  7102  7471  7626  8534  8575
                      8773  8881  9134  9202  9800  9919 10336 10383 10574 10637 11193 11459 11500 12260 12535 12696 12853 13054
                     13418 13758 14461 14500 14526 14828 14867 15191 15221 15433 15576 16034 16478 16529 16625 16794 17712 17922
                     18054 18896 18954 19928 20036 20064 20078 20145 20184 20351 20415 20583 20625 20702 21121 21285 21476 21844
                     22168 22176 22370 22407 22910 23940 25003 25339 25739 25769 27109 27854 27986 28052 28143 28443 28551 29791
                     29849 29911 29929 30578 31059 31173 31332 31970 32097 32362 32546 32582 32695 32704 32956 33848 34237 34246
```

TABLE 2 (continued)

Enzyme (# of sites) and Recognition Sequence	Coordinates of Cut Sites																	
SFAN (cont)	34314	34703	34747	35362	35857	35876	37570	37766	38288	39041	39213	39353	39616	39930	39995	40082	40835	40942
	41502	41942	42402	42758	42913	43213	43281	44085	44225	44312	44337	44370	45181	45253	45705	46015	46120	46148
	46175	46193	46601	46755	46949	47226	48356											
SNA1(3 SITES) ↓GTATAC	15259	18833	19472															
SPH1(6 SITES) GCATG↓C	2216	12006	23946	24375	27378	39422												
SST1(2 SITES) GAGCT↓C	24776	25881																
SST2(4 SITES) CCGC↓GG	20323	20533	21609	40389														
STU1(6 SITES) AGG↓CCT	12436	31480	32999	39994	40598	40616												
•TAQ1(121 SITES)	451	576	720	1909	2082	2142	2407	2473	3617	4199	4800	6138	6511	6574	6685	6733	7622	7634
T↓CGA	8742	9133	9304	9402	10119	13505	14861	15584	15830	16121	16243	17024	17656	18008	18049	18667	19205	20949
	21537	21554	21737	23162	23323	24894	25249	25447	25733	25802	25885	26257	26285	26617	27981	28188	28917	29010
	29151	29230	29953	29977	30290	30397	30830	31156	31406	31586	31976	31991	32272	32746	32964	33245	33360	33499
	33585	33660	33840	34157	34332	34697	35051	36966	37168	38397	38563	38759	39488	39542	39869	40422	40667	41364
	41607	41874	41928	42021	42283	42306	42410	42544	42638	42889	42931	42964	43333	43534	43600	43825	43893	44304
	44667	44809	45135	45322	45370	45441	45527	45842	45896	46156	46439	46595	47418					
THA1(157 SITES)	14	460	547	680	757	1939	2271	2638	2891	3011	3161	3321	3524	3572	3727	3781	4103	4128
CG↓CG	4280	4498	4592	4702	5438	5459	5535	5550	5629	5816	6047	6696	6973	7018	7079	7498	7578	7652
	7670	7781	8068	8082	8240	8255	8343	8883	9036	9065	9879	10153	10359	11171	11447	11502	11597	11603
	12069	12210	12418	12693	12868	13372	13653	13861	13961	13996	14817	15362	15374	15537	15539	16026	16220	16263
	16353	16467	16651	16841	17059	17421	17793	18136	18511	18522	18629	18718	19202	19991	19998	20028	20073	20107
	20235	20322	20532	20703	20879	20933	20954	21038	21344	21608	21920	22222	24968	28010	28052	28376	29342	29801
	30234	30796	31497	31533	31705	32121	32235	32409	32427	33064	33149	35094	35328	35337	35426	36052	36722	38125
	38572	39392	39470	39850	40035	40349	40388	40529	41338	41405	41810	42446	42565	42721	42886	43128	43206	43708
	43758	44219	44370	45042	45228	45283	45981	46075	46110	46479	46978	47482	48100					

TTH1(2 SITES)	11205 36123
GACN↓NNGTC	
TTH2(49 SITES)	3883 4469 4916 5141 7795 8301 10175 15840 18139 18238 20908 22107 22912 23012 23078 24000 24770 25112
CAAPuCANNNNNNNNNNN↓	28300 28592 28691 29734 30009 30964 31768 31858 32996 33191 33309 33885 35154 35336 35686 36845 37758 38386
	38417 38510 38782 38833 38989 38998 39277 39520 39567 42849 44087 45742 46527
[g]XBA1(1 SITE)	24508
T↓CTAGA	
XHO1(1 SITE)	33498
C↓TCGAG	
XHO2(21 SITES)	415 1606 2531 5505 6422 22346 22425 24511 27027 27972 29593 30426 34499 35711 38103 38664 38754 38814
Pu↓GATCPy	39576 41732 47773
XMA1(3 SITES)	19397 31617 39888
C↓CCGGG	
XMA3(2 SITES)	19944 36654
C↓GGCCG	
XMN1(24 SITES)	37 1155 2323 8494 10115 13106 16913 22856 22875 23812 23832 24232 24582 25489 27256 29019 29997 31089
GAANN↓NNTTC	33815 34189 42481 44731 45745 47568
REND	48502

The complete λ DNA sequence from coordinate 1 to 48502 was searched for all restriction enzyme recognition sequences listed in Table 1. The sites found in λ DNA are tabulated here. However, when digesting λ DNA with restriction enzymes one must be aware of the fact that the state of methylation of the DNA (i.e., which host strain was used) can make a difference in the pattern seen, as some enzymes are sensitive to methylated bases in their recognition sequences. (For review and discussion, see Maniatis et al. 1982 and Roberts 1983).

[a]All BclI sites in λ DNA grown in dam^{+} hosts are partially blocked.

[b]ClaI sites that overlap dam methylation sites are partially blocked by dam methylation. In λ DNA, these are sites 15584, 31991, and 32964 (see sequence; Mayer et al. 1981).

[c]On dam-modified DNA all EcoRII sites are partially blocked. BstNI recognizes the same sequence (although the cut site is 2 bases to the right) and is active on either methylated or unmethylated DNA.

[d]HpaI and HincII recognition sequences can overlap the EcoK modification site. In λ one such site (14993) does, and it is not cut by either HpaI or HincII if the DNA is EcoK-modified. (see sequence; Daniels et al. 1980; A. Glasgow and M. Howe, pers. comm. 1983)

[e]HphI, MboII, and TaqI recognition sequences can overlap the dam methylation site. If the site is modified, restriction enzyme cleavage is inhibited (see Maniatis et al. 1982).

[f]All MboI sites overlap the dam methylation site. If the DNA is methylated, then MboI cleavage is inhibited. Sau3a recognizes the same sequence and will cut whether it is modified or not, whereas DpnI recognizes the same sequence and cuts only modified DNA.

[g]XbaI sites can overlap the dam methylation site. The one XbaI site in λ DNA does. This is probably the reason it is difficult to get complete XbaI digests of λ DNA.

ACKNOWLEDGMENTS

This is paper 2666 of the Laboratory of Genetics, University of Wisconsin, Madison, supported by National Institutes of Health grants AI-18214, GM-21812, and GM-28252 (to F.R.B.) and a special grant from Cold Spring Harbor Laboratory.

REFERENCES

Daniels, D.L., J.R. de Wet, and F.R. Blattner. 1980. A new map of bacteriophage lambda DNA. *J. Virol.* **33:** 390.

Maniatis, T., E.F. Fritsch, and J. Sambrook. 1982. *Molecular Cloning. A Laboratory Manual.* Cold Spring Harbor Laboratory, Cold Spring Harbor, New York.

Mayer, H., R. Grosschedl, H. Schütte, and G. Hobom. 1981. *Cla*I, a new restriction endonuclease from *Caryophanon latum* L. *Nucleic Acids Res.* **9:** 4833.

Roberts, R.J. 1983. Restriction and modification enzymes and their recognition sequences. *Nucleic Acids Res.* **11:** 1.

Sanger, F., A.R. Coulson, G.F. Hong, D.F. Hill, and G.B. Petersen. 1982. Nucleotide sequence of bacteriophage λ DNA. *J. Mol. Biol.* **162:** 729.

APPENDIX III
Lambda Vectors

Compiled by Noreen Murray
Department of Molecular Biology
University of Edinburgh
Edinburgh EH9 3JR, Scotland

In the list of λ vectors (which follows), the estimated positions of the targets for restriction enzymes are listed as base pairs from the left end of the mature DNA of the phage, although in some cases the values may only be reliable to within a few hundred base pairs. The data for many vectors are from Table 1 of Williams and Blattner (1980), those for newer vectors sometimes rely only on their genotypes together with the λ map circulated by F.R. Blattner in 1981. The name of each vector corresponds as closely as possible to that designated by its authors. The numbers indicated in parentheses are references that either describe the original vector or provide informative maps. A complete list of references follows the list.

Abbreviations: CH = Charon, L = λ, LEND = left end, REND = right end, BAM1 = BamHI, BGL2 = *Bgl*II, HIN3 = *Hin*dIII, KPN1 = *Kpn*I, RI = *Eco*RI, SAL1 = *Sal*I, SST1 = *Sst*I, SST2 = *Sst*II, XBA1 = *Xba*I, XHO = *Xho*I.

Note that the extra *Hin*dIII target characteristic of λ*c*I857 is not indicated and should be added where appropriate.

```
CH-1 (1,25)

LEND, 0
BGL2, 471
BAM1, 5559
KPN1, 17430
KPN1, 18950
R1,   19939
SST1, 21043
XBA1, 24959
SST1, 25230
HIN3, 25618
SST1, 26344
R1,   26565
HIN3, 27916
BAM1, 28511
R1,   32268
SAL1, 33309
SAL1, 33817
XHO1, 34075
BAM1, 35073
BGL2, 36282
HIN3, 37504
BGL2, 38165
R1,   39020
HIN3, 39264
HIN3, 39389
BGL2, 39908
BGL2, 40559
BGL2, 40619
BAM1, 43020
SST1, 43170
HIN3, 43212
BAM1, 43526
KPN1, 45412
BGL2, 45476
SST1, 47032
BAM1, 47454
BGL2, 47631
REND, 48944

CH-2 (1,25)

LEND, 0
BGL2, 471
BAM1, 5559
KPN1, 17430
KPN1, 18950
R1,   19939
SST1, 21043
BAM1, 27061
XHO1, 27366
BAM1, 30136
BGL2, 32130
XHO1, 32886
BGL2, 33012
KPN1, 35757
BGL2, 36488
SAL1, 36775
SAL1, 36928
KPN1, 37567
BGL2, 38516
BAM1, 40577
SST1, 40727
HIN3, 40769
BAM1, 41083
KPN1, 42969
BGL2, 43033
SST1, 44589
BAM1, 45011
BGL2, 45188
REND, 46501

CH-3 (1,25)

LEND, 0
BGL2, 471
BAM1, 5559
KPN1, 17430
KPN1, 18950
R1,   19939
SST1, 21043
XBA1, 24959
SST1, 25230
HIN3, 25618
SST1, 26344
R1,   26565
HIN3, 27916
BAM1, 28511
BGL2, 33808
XHO1, 34564
BGL2, 34690
KPN1, 37435
BGL2, 38166
SAL1, 38453
SAL1, 38606
KPN1, 39245
BGL2, 40194
BAM1, 42255
SST1, 42405
HIN3, 42447
BAM1, 42761
KPN1, 44647
BGL2, 44711
SST1, 46267
BAM1, 46689
BGL2, 46866
REND, 48179

CH-4 (1,25,4)

LEND, 0
BGL2, 471
BAM1, 5559
KPN1, 17430
KPN1, 18950
R1,   19939
SST1, 21043
XBA1, 24959
SST1, 25230
HIN3, 25618
SST1, 26344
R1,   26565
HIN3, 27916
BAM1, 28919
BGL2, 32506
BGL2, 32640
R1,   34317
BGL2, 35963
BGL2, 36323
BGL2, 36974
BGL2, 37034
BAM1, 39435
SST1, 39585
HIN3, 39627
BAM1, 39941
KPN1, 41827
BGL2, 41891
SST1, 43447
BAM1, 43869
BGL2, 44046
REND, 45359

CH-5 (1,25)

LEND, 0
BGL2, 471
BAM1, 5559
KPN1, 23954
KPN1, 25474
BAM1, 27475
SAL1, 30241
XHO1, 30499
BAM1, 31497
BGL2, 32706
HIN3, 33928
BGL2, 34589
R1,   35444
HIN3, 35688
HIN3, 35813
BGL2, 36332
BGL2, 36983
BGL2, 37043
BAM1, 39444
SST1, 39594
HIN3, 39636
BAM1, 39950
KPN1, 41836
BGL2, 41900
SST1, 43456
BAM1, 43878
BGL2, 44055
REND, 45368

CH-6 (1,25)

LEND, 0
BGL2, 471
BAM1, 5559
KPN1, 23954
KPN1, 25474
BAM1, 27475
SAL1, 30241
XHO1, 30499
BAM1, 31497
HIN3, 33507
R1,   33743
BGL2, 34645
BGL2, 34705
BAM1, 37106
SST1, 37256
HIN3, 37298
BAM1, 37612
KPN1, 39498
BGL2, 39562
SST1, 41118
BAM1, 41540
BGL2, 41717
REND, 43030

CH-7 (1,25)

LEND, 0
BGL2, 471
BAM1, 5559
KPN1, 23954
KPN1, 25474
BAM1, 27475
SAL1, 30241
XHO1, 30499
BAM1, 31497
HIN3, 33507
R1,   33743
BGL2, 34645
BGL2, 34705
REND, 41666

CH-8 (1,25)

LEND, 0
BGL2, 471
BAM1, 5559
KPN1, 17430
KPN1, 18950
R1,   19939
SST1, 21043
XBA1, 24959
SST1, 25230
HIN3, 25618
SST1, 26344
R1,   26565
HIN3, 27916
BAM1, 28511
R1,   32268
SAL1, 33309
SAL1, 33817
XHO1, 34075
BAM1, 35073
HIN3, 37083
R1,   37319
BGL2, 38221
BGL2, 38281
REND, 45242

CH-9 (1,25)

LEND, 0
BGL2, 471
BAM1, 5559
KPN1, 17430
KPN1, 18950
R1,   19939
SST1, 21043
XBA1, 24959
SST1, 25230
HIN3, 25618
SST1, 26344
R1,   26565
HIN3, 27916
BAM1, 28511
R1,   32268
SAL1, 33309
SAL1, 33817
XHO1, 34075
BAM1, 35073
BGL2, 36282
HIN3, 37504
BGL2, 38165
R1,   39020
HIN3, 39264
HIN3, 39389
BGL2, 39908
BGL2, 40559
BGL2, 40619
REND, 47580

CH-10 (1,25)

LEND, 0
BGL2, 471
BAM1, 5559
KPN1, 17430
KPN1, 18950
R1,   19939
SST1, 21043
XBA1, 24959
SST1, 25230
HIN3, 25618
BAM1, 26621
BGL2, 30208
BGL2, 30342
R1,   32019
BGL2, 33665
BGL2, 34025
BGL2, 34676
BGL2, 34736
REND, 41697

CH-11 (1,25)

LEND, 0
BGL2, 471
BAM1, 5559
KPN1, 17430
KPN1, 18950
R1,   19939
SST1, 21043
XBA1, 24959
SST1, 25230
HIN3, 25618
BAM1, 26621
BGL2, 30208
BGL2, 30342
R1,   32019
BGL2, 33665
BGL2, 34025
BGL2, 34676
BGL2, 34736
BAM1, 37137
```

SST1, 37287
HIN3, 37329
BAM1, 37643
KPN1, 39529
BGL2, 39593
SST1, 41149
BAM1, 41571
BGL2, 41748
REND, 43061

CH-12 (25)

LEND, 0
BGL2, 471
BAM1, 5559
KPN1, 23954
KPN1, 25474
BAM1, 26955
R1, 30712
SAL1, 31753
SAL1, 32261
XHO1, 32519
BAM1, 33517
BGL2, 34726
BGL2, 35086
BGL2, 35737
BGL2, 35797
BAM1, 38198
SST1, 38348
HIN3, 38390
BAM1, 38704
KPN1, 40590
BGL2, 40654
SST1, 42210
BAM1, 42632
BGL2, 42809
REND, 44122

CH-13 (25)

LEND, 0
BGL2, 471
BAM1, 5559
KPN1, 17430
KPN1, 18950
R1, 21708
BAM1, 22843
BGL2, 22932
HIN3, 23644
R1, 27491
SAL1, 28532
SAL1, 29040
XHO1, 29298
BAM1, 30296
BGL2, 31505
BGL2, 31865
BGL2, 32516
BGL2, 32576
BAM1, 34977
SST1, 35127
HIN3, 35169
BAM1, 35483
KPN1, 37369
BGL2, 37433

SST1, 38989
BAM1, 39411
BGL2, 39588
REND, 40901

CH-14 (1,25)

LEND, 0
BGL2, 471
BAM1, 5559
KPN1, 17430
KPN1, 18950
R1, 19939
SST1, 21043
R1, 27491
SAL1, 28532
SAL1, 29040
XHO1, 29298
BAM1, 30296
BGL2, 31505
BGL2, 31865
BGL2, 32516
BGL2, 32576
BAM1, 34977
SST1, 35127
HIN3, 35169
BAM1, 35483
KPN1, 37369
BGL2, 37433
SST1, 38989
BAM1, 39411
BGL2, 39588
REND, 40901

CH-15 (1,25)

LEND, 0
BGL2, 471
BAM1, 5559
KPN1, 17430
KPN1, 18950
R1, 21708
BAM1, 22843
BGL2, 22932
HIN3, 23644
BGL2, 29031
XHO1, 29787
BGL2, 29913
KPN1, 32658
BGL2, 33389
SAL1, 33676
SAL1, 33829
KPN1, 34468
BGL2, 35417
REND, 42038

CH-16 (1,25)

LEND, 0
BGL2, 471
BAM1, 5559
KPN1, 17430
KPN1, 18950
R1, 19939

SST1, 21043
BGL2, 28780
XHO1, 29536
BGL2, 29662
KPN1, 32407
BGL2, 33138
SAL1, 33425
SAL1, 33578
KPN1, 34217
BGL2, 35166
REND, 41787

CH-17 (25)

LEND, 0
BGL2, 471
BAM1, 5559
KPN1, 17430
KPN1, 18950
R1, 19939
SST1, 21043
XBA1, 24959
SST1, 25230
HIN3, 25618
SST1, 26344
R1, 26565
HIN3, 27916
BAM1, 28511
SAL1, 33309
SAL1, 33817
XHO1, 34075
BAM1, 35073
BGL2, 36282
HIN3, 37504
HIN3, 38084
BGL2, 38728
BGL2, 39379
BGL2, 39439
REND, 46400

CH-18 (25)

LEND, 0
BGL2, 471
BAM1, 5559
KPN1, 17430
KPN1, 18950
R1, 19939
SST1, 21043
XBA1, 24959
SST1, 25230
HIN3, 25618
BAM1, 26621
BGL2, 30208
BGL2, 30342
R1, 32019
BGL2, 33665
BGL2, 34025
BGL2, 34676
BGL2, 34736
BAM1, 37699
REND, 44590

CH-19A (25)

LEND, 0
BGL2, 471
BAM1, 5559
KPN1, 17430
KPN1, 18950
R1, 19939
SST1, 21043
XBA1, 24959
SST1, 25230
HIN3, 25618
SST1, 26344
R1, 26565
HIN3, 27916
BAM1, 28511
SAL1, 33309
SAL1, 33817
XHO1, 34075
BAM1, 35073
BGL2, 36282
HIN3, 37504
HIN3, 38084
BGL2, 38728
BGL2, 39379
BGL2, 39439
BAM1, 41840
SST1, 41990
HIN3, 42032
BAM1, 42346
KPN1, 44232
BGL2, 44296
SST1, 45852
BAM1, 46274
BGL2, 46451
REND, 47764

CH-20 (25)

LEND, 0
BGL2, 471
BAM1, 5559
KPN1, 17430
KPN1, 18950
R1, 21708
BAM1, 22843
BGL2, 22932
HIN3, 23644
R1, 27491
SAL1, 28532
SAL1, 29040
XHO1, 29298
BAM1, 30296
HIN3, 32306
R1, 32542
BGL2, 33444
BGL2, 33504
REND, 40465

CH-21A (25)

LEND, 0
BGL2, 471
BAM1, 5559
KPN1, 17430

KPN1, 18950
R1, 21708
BAM1, 22843
BGL2, 22932
HIN3, 23644
BGL2, 29031
XHO1, 29787
BGL2, 29913
KPN1, 32658
BGL2, 33389
SAL1, 33676
KPN1, 34315
BGL2, 35264
REND, 41885

CH-22 (25)

LEND, 0
BGL2, 471
BAM1, 5559
KPN1, 17430
KPN1, 18950
SMA1, 19842
R1, 21708
BAM1, 22843
BGL2, 22932
HIN3, 23644
XBA1, 24959
SST1, 25230
HIN3, 25618
SST1, 26344
R1, 26565
HIN3, 27916
BAM1, 28511
SMA1, 32140
R1, 32268
SAL1, 33309
SAL1, 33817
XHO1, 34075
BAM1, 35073
BGL2, 36282
BGL2, 36642
BGL2, 37293
BGL2, 37353
P22?, 37706
P22?, 40483
HIN3, 41697
R1, 42592
REND, 46164

CH-23A (25)

LEND, 0
BGL2, 471
BAM1, 5559
KPN1, 17430
KPN1, 18950
R1, 19939
SST1, 21043
XBA1, 24959
SST1, 25230
HIN3, 25618
SST1, 26344
R1, 26565

HIN3, 27916
BAM1, 28511
BGL2, 33808
XHO1, 34564
BGL2, 34690
KPN1, 37435
BGL2, 38166
SAL1, 38453
SAL1, 38606
KPN1, 39245
BGL2, 40194
BAM1, 42255
SST1, 42405
HIN3, 42447

BAM1, 42761
KPN1, 44647
BGL2, 44711
SST1, 46267
BAM1, 46689
BGL2, 46866
REND, 48179

CH-24A (25)

LEND, 0
BGL2, 471
BAM1, 5559
KPN1, 17430
KPN1, 18950
R1, 19939
SST1, 21043
XBA1, 24959
SST1, 25230
HIN3, 25618
SST1, 26344
R1, 26565
HIN3, 27916
BAM1, 28919
BGL2, 32506
BGL2, 32640
R1, 34317
BGL2, 35963
BGL2, 36323
BGL2, 36974
BGL2, 37034
BAM1, 39435
SST1, 39585
HIN3, 39627
BAM1, 39941
KPN1, 41827
BGL2, 41891
SST1, 43447
BAM1, 43869
BGL2, 44046
REND, 45359

CH-25 (25)

LEND, 0
BGL2, 471
BAM1, 5559
KPN1, 17430
KPN1, 18950
R1, 19939

SST1, 21043
XBA1, 24959
SST1, 25230
HIN3, 25618
SST1, 26344
R1, 26565
HIN3, 27916
BAM1, 28511
R1, 32268
SAL1, 33309
SAL1, 33817
XHO1, 34075
SAL1, 35756
BGL2, 37044

BGL2, 37104
REND, 44065

CH-26 (25)

LEND, 0
BGL2, 471
BAM1, 5559
KPN1, 17430
KPN1, 18950
R1, 21708
BAM1, 22843
BGL2, 22932
HIN3, 23644
XBA1, 24959
SST1, 25230
HIN3, 25618
SST1, 26344
R1, 26565
HIN3, 27916
BAM1, 28511
R1, 32268
SAL1, 33309
SAL1, 33817
XHO1, 34075
SAL1, 35756
BGL2, 37044
BGL2, 37104
REND, 44065

CH-27 (25)

LEND, 0
BGL2, 471
KPN1, 17430
KPN1, 18950
R1, 21708
BAM1, 22843
BGL2, 22932
HIN3, 23644
BGL2, 29031
XHO1, 29787
BGL2, 29913
KPN1, 32658
BGL2, 33389
SAL1, 33676
KPN1, 34315
BGL2, 35264
REND, 41885

CH-28 (22)

LEND, 0
BGL2, 472
KPN1, 17280
KPN1, 18790
BAM1, 23567
BGL2, 23656
HIN3, 24362
R1, 28212
SAL1, 29235
SAL1, 29739
XHO1, 29999
BAM1, 30996

BGL2, 32215
BGL2, 32511
BGL2, 33162
BGL2, 33222
REND, 40229

CH-29 (25)

LEND, 0
BGL2, 471
BAM1, 5559
KPN1, 17430
KPN1, 18950
R1, 19939
HIN3, 20183
HIN3, 20308
BGL2, 20827
R1, 21181
HIN3, 22532
BAM1, 23127
R1, 26884
SAL1, 27925
SAL1, 28433
XHO1, 28691
BAM1, 29689
BGL2, 30898
HIN3, 32120
BGL2, 32781
R1, 33636
HIN3, 33880
HIN3, 34005
BGL2, 34524
BGL2, 35175
BGL2, 35235
REND, 42196

CH-30 (25)

LEND, 0
BGL2, 472
KPN1, 17288
KPN1, 18790
R1, 21524
BAM1, 22667
BGL2, 22756
HIN3, 23462
R1, 27312
SAL1, 28335
SAL1, 28839
XHO1, 29099

BAM1, 30096
BGL2, 30184
HIN3, 30890
R1, 34740
SAL1, 35763
SAL1, 36267
XHO1, 36527
BAM1, 37524
BGL2, 38743
BGL2, 39039
BGL2, 39690
BGL2, 39750
REND, 46757

L47-1 (11,25)

LEND, 0
BGL2, 471
KPN1, 17430
KPN1, 18950
R1, 21708
HIN3, 23059
BAM1, 23654
SAL1, 28452
SAL1, 28960
XHO1, 29218
BAM1, 30216
HIN3, 32226
R1, 32462
BGL2, 33364
BGL2, 33424
REND, 40385

Dam vector (23)
and NM461 (8)

LEND, 0
BGL2, 471
BAM1, 5559
KPN1, 17430
KPN1, 18950
R1, 23846
SAL1, 24887
SAL1, 25395
XHO1, 25653
BAM1, 26651
BGL2, 27860
HIN3, 29105
HIN3, 29670
BGL2, 30314
BGL2, 30965
BGL2, 31025
HIN3, 33561
REND, 37968

NM540 (14,25)

LEND, 0
BGL2, 471
BAM1, 5559
KPN1, 17430
KPN1, 18950
R1, 21708
HIN3, 23059

BAM1, 23654
R1, 27411
SAL1, 28452
SAL1, 28960
XHO1, 29218
SAL1, 30899
BGL2, 32187
BGL2, 32247
R1, 32601
R1, 35636
REND, 39208

NM569 (2)

LEND, 0
BGL2, 471
BAM1, 5559
KPN1, 17430
KPN1, 18950
R1, 21708
HIN3, 23059
BAM1, 23654
R1, 27411
SAL1, 28452
SAL1, 28960
XHO1, 29218
BAM1, 30216
BGL2, 31425
HIN3, 32647
HIN3, 33227
BGL2, 33871
BGL2, 34522
BGL2, 34582
HIN3, 37104
REND, 41571

NM590 (16,4,25)

LEND, 0
BGL2, 471
BAM1, 5559
KPN1, 17430
KPN1, 18950
R1, 23846
SAL1, 24887
SAL1, 25395
XHO1, 25653
BAM1, 26651
HIN3, 28661
R1, 28897
BGL2, 29799
BGL2, 29859
R1, 30213
BAM1, 32822
R1, 36141
REND, 39713

NM596 (16)
(see NM781)

NM598 (16,25)

LEND, 0
BGL2, 471

```
BAM1, 5559
KPN1, 17430
KPN1, 18950
R1,   23846
SAL1, 24887
SAL1, 25395
XHO1, 25653
BAM1, 26651
HIN3, 28661
R1,   28897
BGL2, 29799
BGL2, 29859
R1,   30213
BAM1, 32822
R1,   36141
REND, 39713

NM607 (16,4,25)

LEND, 0
BGL2, 471
BAM1, 5559
KPN1, 17430
KPN1, 18950
SAL1, 24888
SAL1, 25396
XHO1, 25654
BAM1, 26652
HIN3, 28662
R1,   28898
BGL2, 29800
BGL2, 29860
BAM1, 32823
HIN3, 35247
REND, 39714

NM616 (13)

LEND, 0
BGL2, 471
BAM1, 5559
KPN1, 17430
KPN1, 18950
R1,   19939
SST1, 21043
XBA1, 25209
SST1, 25480
HIN3, 25868
SST1, 26594
R1,   26815
HIN3, 28166
BAM1, 28761
SAL1, 33559
SAL1, 34067
XHO1, 34325
SAL1, 36000
BGL2, 37294
BGL2, 37354
HIN3, 39848
REND, 44315

NM631 (16,25)

LEND, 0
BGL2, 471
BAM1, 5559
KPN1, 17430
KPN1, 18950
R1,   19939
SST1, 21043
BAM1, 25201
R1,   28958
SAL1, 29999
SAL1, 30507
XHO1, 30765
BAM1, 31763
BGL2, 32972
BGL2, 33332
BGL2, 33983
BGL2, 34043
HIN3, 36537
REND, 41004

NM641 (16,25)

LEND, 0
BGL2, 471
BAM1, 5559
KPN1, 17430
KPN1, 18950
BAM1, 22646
BGL2, 22735
HIN3, 23447
XBA1, 24762
SST1, 25033
HIN3, 25421
SST1, 26147
SAL1, 27409
SAL1, 27917
XHO1, 28175
BAM1, 29173
HIN3, 31183
R1,   31419
BGL2, 32321
BGL2, 32381
HIN3, 34875
REND, 39342

NM647 (16,25)

LEND, 0
BGL2, 471
BAM1, 5559
KPN1, 17430
KPN1, 18950
R1,   19939
SST1, 21043
BAM1, 25201
R1,   28958
SAL1, 29999
SAL1, 30507
XHO1, 30765
BGL2, 33734
BGL2, 33794
HIN3, 36288
REND, 40755

NM728 (16,25)

LEND, 0
BGL2, 471
BAM1, 5559
KPN1, 17430
KPN1, 18950
SAL1, 23901
SAL1, 24409
XHO1, 24667
BAM1, 25665
HIN3, 27675
R1,   27911
BGL2, 28813
BGL2, 28873
R1,   29227
BAM1, 31836
R1,   35155
REND, 38727

NM760 and 788
(16,4,25)

LEND, 0
BGL2, 471
BAM1, 5559
KPN1, 17430
KPN1, 18950
R1,   21708
HIN3, 23059
TRPE?, 23060
TRPE?, 28798
HIN3, 28799
SAL1, 30161
SAL1, 30669
XHO1, 30927
BAM1, 31925
BGL2, 33134
BGL2, 33494
BGL2, 34145
BGL2, 34205
R1,   34559
R1,   37594
REND, 41166

NM761 (16)

LEND, 0
BGL2, 471
BAM1, 5559
KPN1, 17430
KPN1, 18950
R1,   21708
HIN3, 23059
SUPF?, 23060
SUPF?, 30978
HIN3, 30978
SAL1, 32341
SAL1, 32849
XHO1, 33107
SAL1, 34788
BGL2, 36056
BGL2, 36136
R1,   36490
R1,   39525
REND, 43097

NM762 (16,25)

LEND, 0
BGL2, 471
BAM1, 5559
KPN1, 17430
KPN1, 18950
R1,   21708
HIN3, 23059
SUPF?, 23060
SUPF?, 30978
HIN3, 30979
SAL1, 32341
SAL1, 32849
XHO1, 33107
BAM1, 34105
BGL2, 35314
BGL2, 35674
BGL2, 36325
BGL2, 36385
R1,   36739
R1,   39774
REND, 43346

NM781 (16,25)

LEND, 0
BGL2, 471
BAM1, 5559
KPN1, 17430
KPN1, 18950
R1,   21708
SUPE?, 21709
SUPE?, 31805
R1,   31806
SAL1, 32847
SAL1, 33355
XHO1, 33613
BAM1, 34611
BGL2, 35820
HIN3, 37042
HIN3, 37622
BGL2, 38266
BGL2, 38917
BGL2, 38977
HIN3, 41471
REND, 45938

NM816 (26,25)

LEND, 0
BGL2, 471
BAM1, 5559
KPN1, 17430
KPN1, 18950
R1,   19939
SST1, 21043
XBA1, 25209
SST1, 25480
HIN3, 25868
SST1, 26594
R1,   26815
HIN3, 28166
BAM1, 28761
SAL1, 33559
SAL1, 34067
XHO1, 34325
BGL2, 37294
BGL2, 37354
HIN3, 39848
REND, 44315

NM1033 (15)

LEND, 0
BGL2, 471
BAM1, 5559
KPN1, 17430
KPN1, 18951
R1,   19939
SST1, 21063
XBA1, 24959
BAM1, 25116
BGL2, 28706
BGL2, 28837
R1,   30514
SAL1, 31655
SAL1, 32163
XHO1, 32421
BAM1, 33419
BGL2, 34628
BGL2, 34988
BGL2, 35639
BGL2, 35699
HIN3, 38193
REND, 42660

NM1122 (15)

LEND, 0
BGL2, 471
KPN1, 17430
KPN1, 18950
R1,   21708
BAM1, 22843
BGL2, 22932
HIN3, 23644
XBA1, 24959
SST1, 25230
HIN3, 25618
SST1, 26344
R1,   26565
HIN3, 27916
R1,   30210
SAL1, 31251
SAL1, 31759
XHO1, 32017
BAM1, 33015
BGL2, 34224
BGL2, 34599
BGL2, 35250
BGL2, 35310
R1,   35664
H3,   37833
R1,   38669
```

```
REND, 42299

NM1149 (15)

LEND, 0
BGL2, 471
BAM1, 5559
KPN1, 17430
KPN1, 18950
SAL1, 24887
SAL1, 25395
XHO1, 25653
BAM1, 26651
HIN3, 28661
R1,   28897
BGL2, 29799
BGL2, 29859
BAM1, 32822
REND, 39713

NM1150 (15)

LEND, 0
BGL2, 471
BAM1, 5559
KPN1, 17430
KPN1, 18950
SAL1, 23901
SAL1, 24409
XHO1, 24667
BAM1, 25665
HIN3, 27675
R1,   27911
BGL2, 28813
BGL2, 28873
BAM1, 31836
REND, 38727

NM1151 (15)

LEND, 0
BGL2, 471
KPN1, 17430
KPN1, 18950
R1,   21708
HIN3, 23059
BAM1, 23654
SAL1, 28452
SAL1, 28960
XHO1, 29218
BGL2, 32187
BGL2, 32247
REND, 39208

NM1173 (15)

LEND, 0
BGL2, 471
BAM1, 5559
KPN1, 17430
KPN1, 18950
R1,   21708
HIN3, 23059
SUPF?, 23060
SUPF?, 30978
HIN3, 30979
SAL1, 32341
SAL1, 32849
XHO1, 33107
BAM1, 34105
BGL2, 35314
BGL2, 35674
BGL2, 36325
BGL2, 36385
R1,   39774
REND, 43346

Lsep-lac5(12,4)

LEND, 0
BGL2, 471
BAM,  5559
KPN1, 17430
KPN1, 18950
R1,   19939
SST1, 21043
XBA1, 24959
SST1, 25230
HIN3, 25618
SST1, 26344
R1,   26565
SST1, 27669
XBA1, 31585
SST1, 31856
HIN3, 32244
SST1, 32970
R1,   33191
SAL,  34232
SAL,  34940
XHO,  34998
BGL2, 37967
BGL2, 38827
HIN3, 40521
REND, 44988

BV2 (10,25)

LEND, 0
BGL2, 471
KPN1, 17430
KPN1, 18950
R1,   21708
BAM1, 22843
R1,   26600
SAL1, 27641
SAL1, 28149
XHO1, 28407
SAL1, 30088
BGL2, 31376
BGL2, 31436
R1,   31790
R1,   34825
REND, 38397

LGT1-B(24,4,25)

LEND, 0
BGL2, 471
BAM1, 5559
KPN1, 17430
KPN1, 18950
R1,   21708
BAM1, 22843
BGL2, 22932
HIN3, 23644
XBA1, 24959
SST1, 25230
HIN3, 25618
SST1, 26344
R1,   26565
SAL1, 27606
SAL1, 28114
XHO1, 28372
BAM1, 29370
BGL2, 30579
HIN3, 31801
HIN3, 32381
BGL2, 33025
BGL2, 33676
BGL2, 33736
HIN3, 36258
REND, 40725

LGT2 (17,25)

LEND, 0
BGL2, 471
BAM1, 5559
KPN1, 17430
KPN1, 18950
R1,   21708
HIN3, 23059
BAM1, 23654
SAL1, 28452
SAL1, 28960
XHO1, 29218
BAM1, 30216
BGL2, 31425
HIN3, 32647
HIN3, 33227
BGL2, 33871
BGL2, 34522
BGL2, 34582
HIN3, 37104
REND, 41571

LGT4(17,4,25)

LEND, 0
BGL2, 471
BAM1, 5559
KPN1, 17430
KPN1, 18950
R1,   19939
HIN3, 21290
BAM1, 21885
SAL1, 26683
SAL1, 27191
XHO1, 27449
BAM1, 28447
BGL2, 29656
HIN3, 30878
HIN3, 31458
BGL2, 32102
BGL2, 32753
BGL2, 32813
HIN3, 35335
REND, 39802

LGT5 (4,25)

LEND, 0
BGL2, 471
BAM1, 5559
KPN1, 17430
KPN1, 18950
R1,   19939
SST1, 21043
XBA1, 25209
SST1, 25480
HIN3, 25868
SST1, 26594
R1,   26815
SAL1, 27856
SAL1, 28364
XHO1, 28622
BAM1, 29620
BGL2, 30829
HIN3, 32051
HIN3, 32631
BGL2, 33275
BGL2, 33926
BGL2, 33986
HIN3, 36508
REND, 40975

LGT7-AR6(4,25)

LEND, 0
BGL2, 471
BAM1, 5559
KPN1, 17430
KPN1, 18950
R1,   19939
ARA6?,19940
ARA6?,25638
R1,   25639
HIN3, 26990
SAL1, 28977
SAL1, 29485
XHO1, 29743
BAM1, 30741
BGL2, 31950
HIN3, 33172
BGL2, 33816
BGL2, 34467
BGL2, 34527
HIN3, 37049
REND, 41516

LGT30 (4,25)

LEND, 0
BGL2, 471
BAM1, 5559
KPN1, 17430
KPN1, 18950
R1,   21304
SAL1, 22345
SAL1, 22853
COLI?, 22854
COLI?, 28852
SAL1, 28853
XHO1, 29111
BGL2, 32080
BGL2, 32140
HIN3, 34662
REND, 39129

LGT40(4,18,25)

LEND, 0
BGL2, 471
BAM1, 5559
KPN1, 17430
KPN1, 18950
R1,   21708
BAM1, 22843
BGL2, 22932
HIN3, 23644
XBA1, 24959
SST1, 25230
R1,   25451
SAL1, 26492
SAL1, 27000
XHO1, 27258
BAM1, 28256
BGL2, 29465
HIN3, 30687
HIN3, 31267
BGL2, 31911
BGL2, 32562
BGL2, 32622
HIN3, 35144
REND, 39611

LGTT5622(5,25)

LEND, 0
BGL2, 471
BAM1, 5559
KPN1, 17430
KPN1, 18950
R1,   21708
T5
R1,   23508
T5
R1,   25308
SAL1, 26349
SAL1, 26857
XHO1, 27115
BAM1, 28113
BGL2, 29322
HIN3, 30544
HIN3, 31124
BGL2, 31768
BGL2, 32419
BGL2, 32479
HIN3, 35001
REND, 39468
```

```
Llac5-2(21,25)

LEND, 0
BGL2, 471
BAM1, 5559
KPN1, 17430
KPN1, 18950
R1,   19939
SST1, 21043
XBA1, 25209
SST1, 25480
HIN3, 25868
SST1, 26594
R1,   26815
HIN3, 28166
BAM1, 28761
SAL1, 33559
SAL1, 34067
XHO1, 34325
BAM1, 35323
BGL2, 36532
HIN3, 37754
HIN3, 38334
BGL2, 38978
BGL2, 39629
BGL2, 39689
BAM1, 42652
HIN3, 45076
REND, 49543

Llac5-1 and
lac5-1UV5(19)

LEND, 0
BGL2, 471
BAM1, 5559
KPN1, 17430
KPN1, 18950
R1,   19939
SST1, 21043
XBA1, 25209
SST1, 25480
HIN3, 25868
BAM1, 26463
SAL1, 31261
SAL1, 31769
XHO1, 32027
BAM1, 33015
BGL2, 34224
HIN3, 35446
HIN3, 36026
BGL2, 36670
BGL2, 37321
BGL2, 37381
BAM1, 40372
HIN3, 42796
REND, 47263

L ZUV5 (3)

LEND, 0
BGL2, 471
BAM1, 5559
KPN1, 17430
KPN1, 18950
RI,   19939
SST1, 21043
XBA1, 22227
SST1, 22498
HIN3, 22886
BAM1, 23481
SAL1, 28279
SAL1, 28787
XHO1, 29045
BAM1, 30033
BGL2, 31242
HIN3, 32464
HIN3, 33044
BGL2, 33688
BGL2, 34339
BGL2, 34399
BAM1, 37390
HIN3, 39814
REND, 44281

Li(SST)(20,25)

LEND, 0
BGL2, 471
BAM1, 5559
KPN1, 17430
KPN1, 18950
R1,   19939
SST1, 21043
R1,   24940
HIN3, 26291
BAM1, 26886
R1,   30643
SAL1, 31684
SAL1, 32192
XHO1, 32450
SAL1, 34131
BGL2, 35419
BGL2, 35479
HIN3, 38001
REND, 42468

JZ-LBC (6,25)

LEND, 0
BGL2, 471
BAM1, 5559
KPN1, 17430
KPN1, 18950
R1,   21708
BAM1, 22843
BGL2, 22932
HIN3, 23644
XBA1, 24959
SST1, 25230
HIN3, 25618
SST1, 26344
R1,   26565
HIN3, 27916
BAM1, 28511
R1,   32268
SAL1, 33309
SAL1, 33817
XHO1, 34075
BAM1, 35073
BGL2, 36282
HIN3, 37504
HIN3, 38084
BGL2, 38728
BGL2, 39379
BGL2, 39439
HIN3, 41961
REND, 46428

1059 (9)

LEND, 0
BGL2, 472
KPN1, 17272
KPN1, 18775
BAM1, 20265
R1,   24077
SAL1, 25075
SAL1, 25606
BGL2, 26460
HIN3, 27680
HIN3, 28180
HIN3, 28280
R1,   31552
HIN3, 32926
BAM1, 33419
BGL2, 34445
BGL2, 34800
BGL2, 35501
BGL2, 35511
HIN3, 38051
R1,   38873
REND, 42473

EMBL1 (7,15)

LEND, 0
BGL2, 472
KPN1, 17272
KPN1, 18775
BAM1, 20265
R1,   24077
SAL1, 25075
SAL1, 25606
BGL2, 26460
HIN3, 27680
TRPE, 27680
TRPE, 33430
HIN3, 33430
BAM1, 33923
BGL2, 35132
BGL2, 35485
BGL2, 36136
BGL2, 36196
SST2, 37803
HIN3, 38736
R1,   39568
REND, 43158

EMBL3 (7,15)

LEND, 0
BGL2, 472
KPN1, 17272
KPN1, 18775
SAL1, 20274
BAM1, 20280
R1,   20289
SAL1, 25120
SAL1, 25652
BGL2, 26506
HIN3, 27758
TRPE, 27759
TRPE, 33495
HIN3, 33496
R1,   33998
BAM1, 34007
SAL1, 34013
BGL2, 35230
BGL2, 35583
BGL2, 36234
BGL2, 36294
SST2, 37901
HIN3, 38833
REND, 43225

EMBL4 (7,15)

LEND, 0
BGL2, 472
KPN1, 17272
KPN1, 18775
R1,   20274
BAM1, 20283
SAL1, 20289
SAL1, 25120
SAL1, 25652
BGL2, 26506
HIN3, 27758
TRPE, 27759
TRPE, 33495
HIN3, 33496
SAL1, 33998
BAM1, 34004
R1,   34013
BGL2, 35230
BGL2, 35583
BGL2, 36234
BGL2, 36294
SST2, 37901
HIN3, 38833
REND, 43225
```

References:
1. Blattner, F.R., B.G. Williams, A.E. Blechl, K. Denniston-Thompson, H.E. Faber, L. Furlong, J.D. Grunwald, D.O. Kiefer, D.D. Moore, J.W. Schumm, E.L. Sheldon, and O. Smithies. 1977. Charon phages: Safer derivatives of bacteriophage lambda for DNA cloning. *Science* **196:** 161.
2. Borck, K., J.D. Beggs, W.J. Brammar, A.S. Hopkins, and N.E. Murray. 1976. The construction *in vitro* of transducing derivatives of phage lambda. *Mol. Gen. Genet.* **146:** 199.
3. Charnay, P., A. Louise, A. Fritsch, D. Perrin, and P. Tiollais. 1979. Bacteriophage lambda-*E. coli* K12 vector host system for gene cloning and expression under lactose promoter control. II. Fragment insertion at the vicinity of the *lac* UV5 promoter. *Mol. Gen. Genet.* **170:** 171.
4. Davis, R.W., D. Botstein, and J.R. Roth. 1980. *A manual for genetic engineering. Advanced bacterial genetics.* Cold Spring Harbor Laboratory, Cold Spring Harbor, New York.
5. Davison, J., F. Brunel, and M. Merchez. 1979. A new host-vector system allowing selection for foreign DNA inserts in bacteriophage λgtWES. *Gene* **8:** 69.
6. Donoghue, D.J. and P.A. Sharp 1977. An improved bacteriophage lambda vector; construction of model recombinants coding for kanamycin resistance. *Gene* **1:** 209.
7. A.-M. Frischauf et al. (in prep.).
8. Hohn, B. and K. Murray. 1977. Packaging recombinant DNA molecules into bacteriophage particles *in vitro. Proc. Natl. Acad. Sci.* **74:** 3259.
9. Karn, J., S. Brenner, L. Barnett, and C. Cesaereni. 1980. Novel bacteriophage λ cloning vector. *Proc. Natl. Acad. Sci.* **77:** 5172.
10. Klein, B. and K. Murray. 1979. Phage lambda receptor chromosomes for DNA fragments made with restriction endonuclease I of *Bacillus amyloliquefaciens* H. *J. Mol. Biol.* **133:** 289.
11. Loenen, W.A.M. and W.J. Brammar. 1980. A bacteriophage lambda vector for cloning large DNA fragments made with several restriction enzymes. *Gene* **10:** 249.
12. Meyerowitz, E. and D. Hogness. 1982. Molecular organization of a *Drosophila* puff site that responds to ecdysone. *Cell* **28:** 165.
13. Mileham, A.J., H.R. Revel, and N.E. Murray. 1980. Molecular cloning of the T4 genome: Organization and expression of the *frd*-DNA ligase region. *Mol. Gen. Genet.* **179:** 227.
14. Murray, K. and N.E. Murray. 1975. Phage lambda receptor chromosomes for DNA fragments made with restriction endonuclease III of *Haemophilus influenzae* and restriction endonuclease I of *Escherichia coli. J. Mol. Biol.* **98:** 551.
15. N.E. Murray (this paper).
16. Murray, N.E., W.J. Brammar, and K. Murray. 1977. Lambdoid phages that simplify the recovery of *in vitro* recombinants. *Mol. Gen. Genet.* **150:** 53.
17. Panasenko, S.N., J.R. Cameron, R.W. Davis, and I.R. Lehman. 1977. Five hundred fold overproduction of DNA ligase after induction of a hybrid lambda lysogen constructed *in vitro. Science* **196:** 188.
18. Philippsen, P., R.A. Kramer, and R.W. Davis. 1978. Cloning of the yeast ribosomal DNA repeats unit in *Sst*I and *Hin*dIII lambda vectors using genetic and physical size selections. *J. Mol. Biol.* **123:** 371.
19. Pourcel C., C. Marchal, A. Louise, A. Fritsch, and P. Tiollais. 1979. Bacteriophage lambda-*E. coli* K12 vector-host system for cloning and expression under lactose promoter control. I. DNA insertion at the *lacZ Eco*RI restriction site. *Mol. Gen. Genet.* **170:** 161.
20. Pourcel, C. and P. Tiollais. 1977. λp*lac*5 derivatives, potential vectors for DNA fragments cleaved by *Sst* I. *Gene* **1:** 281.
21. Rambach, A. and P. Tiollais. 1974. Bacteriophage λ having *Eco*RI endonuclease sites only in the non-essential region of the genome. *Proc. Natl. Acad. Sci.* **71:** 3927.
22. Rimm, D.L., D. Horness, J. Kucera, and F.R. Blattner. 1980. Construction of coliphage lambda Charon vectors with *Bam*HI cloning sites. *Gene* **12:** 301.
23. Sternberg, N., D. Tiemeier, and L. Enquist. 1977. *In vitro* packaging of a λ *Dam* vector containing *Eco*RI DNA fragments of *Escherichia coli* and phage P1. *Gene* **1:** 255.
24. Thomas, M., J.R. Cameron, and R.W. Davis. 1974. Viable molecular hybrids of bacteriophage λ and eukaryotic DNA. *Proc. Natl. Acad. Sci.* **71:** 3927.
25. Williams, B.G. and F.R. Blattner. 1980. Bacteriophage lambda vectors for DNA cloning. In *Genetic engineering* (ed. J.K. Setlow and A. Mullander), vol. 2, p. 201. Plenum Press, New York.
26. Wilson, G.G. and N.E. Murray. 1979. Molecular cloning of the DNA ligase gene from bacteriophage T4. I. Characterization of the recombinants. *J. Mol. Biol.* **132:** 471.

Subject Index

Abortive lysogeny, 5, 211, 213
Accessory genes, 18, 21–22, 251–277, 387. *See also* Deletion mutants
Addition phages, 376
Adsorption and injection
  adsorption rate, 342
  effect of putrescine, 341
  proteins required, 341–342
  sequence of events, 341
Agar-containing media. *See* Media preparation
A gene. *See* DNA packaging; Morphogenesis of λ, head assembly
Amber mutants, growth of, 437
Amplification of gene products, 420–421
ant. *See* Phage P22
Antiimmune state, 116
Antipode, 150
Antitermination, 26, 32, 43. *See also N* gene; *Q* gene
arc. *See* Phage P22
Arrangement of genes, 22
Assembly. *See* Morphogenesis of λ
att. *See* Integration
Attenuation, 40

Bacillus megaterium, lysogeny, 6
ben, 317
bet(β) gene, 262
bet-gene product
  role in recombination, 178
  role in replication, 148
  size, 179, 262
B gene, *See* DNA packaging, connector; Morphogenesis of λ, head assembly
bio deletion-substitutions, 258–261, 263, 269
Biology of λ, 381–394, 396–401
Box A sequences, 38–39
b region, 24, 81, 252–256, 317,
  deletion mutants, 125, 252, 254–255, 398
  early gene expression of, 256
  late gene expression of, 255–256
  organization of genes, 252–253
  promoters within, 256
  proteins encoded, 252–256
  transcription of, 255–256
Broth media. *See* Media preparation
Broth stocks, 442
Buffer preparation, 458–464
byp, 30, 269

Caffeine, 150
cAMP, 27, 66
*can*1. *See c*II gene
Capsid structure and assembly. *See* Morphogenesis of λ
Casamino acid preparation, 462
Catabolite repression, 66–67
Cesium chloride equilibrium centrifugation, 446
Cesium chloride step gradients, 446–447
C gene. *See* Morphogenesis of λ, head assembly
Chi sequence. *See* Recombination
*c*I gene, 53–56, 369–370
  *c*I857, 124–125, 445
  cleavage mutations, 133–134
  ind$^-$ mutations, 125, 133–134
  intragenic complementation between mutants, 102
  mutations, 104, 113
*c*I-gene product, 4, 13
  aminoterminal domain, 102
  carboxyterminal domain, 102

*c*I-gene product (*continued*)
  cleavage by *rec*A protein, 102, 104, 116, 127–134
  cleavage requirements, 130–131
  concentration, 93, 96–98, 115–116
  control by *rec*A protein, 126–137
  dimer formation, 98
  function, 114–117
  identification as λ repressor, 123–124
  inactivation, 94, 95, 125
  interaction with RNA polymerase, 113
  mechanism of action, 93–95
  operator binding, 98–101
    model building, 103–104
  proteolytic cleavage, 101–102
  size, 101
  structure, 101–105
  synthesis, 53, 68–69, 94–95, 96
    negative regulation of, 24
    positive regulation of, by *c*II-gene product, 75–81
*c*II gene, 13, 23, 41–42, 53–57, 61–63, 76–81, 87–88, 370
  *can*1 mutation, 65–66
  translation of, 63–65
*c*II-gene product
  positive regulator of transcription, 58–61, 76–81
  processing, 65–66
  regulation of activity of, 66–68
  size, 61
  stability, 94
*c*III gene, 13, 41, 66–70, 76–78, 261
*c*III-gene product
  mechanism of action, 66–67
  stability, 94
cin, 25, 30, 58
Cloning, using λ, 19, 395–432
  amplification of gene products of cloned sequences, 420
  analysis of recombinants, 415–421
  *Bam*HI vectors, 408–410, 414
  cosmids, 421
  detection of functional genes, 406
  detection of polypeptides produced from cloned sequences, 419–420
  detection of recombinants by immunoassay, 406–407
  D-gene mutations, use in, 404
  expression from p_L in a plasmid, 421
  genomic library preparation, 411–415
  "immunity" insertion vector, 405
  insertion vectors, 403
  lac expression vectors, 421
  mutagenesis of recombinants, 418
  nucleic acid hybridization, 407
  phasmids, 422
  replacement vectors, 403–405
  selection of homologous sequences via recombination, 407
  selection of recombinants, 404–411
  strategies, 411–415
  transcription of cloned sequences, 410
  vector construction, 401–403
  λ vectors, 19, 422–425
cnc, 25, 30
cohesive ends. *See cos*
cointegrate, 242
Comparative studies, 18. *See also* Lambdoid phages
Concatemers, 151
Conditional lethal mutations, 22
λ*c*17, 30
Connector. *See* DNA packaging; Morphogenesis of λ
Core. *See* Integration
cos (cohesive end site), 154, 189, 199–200. *See also* DNA packaging
cos rescue, 374–376
Cosmids, 421
cro gene, 13, 23, 63, 66, 69, 77, 369–371
  mutations, 116, 153, 218
  translation of, 40
cro-gene product, 26
  concentration, 93, 116
  function, 24, 93–96, 114–117
  inhibition of p_R, 98
  inhibition of p_{RE}, 96
  inhibition of p_{RM},98
  involvement in replication, 146–147
  operator binding, 105–106
  operator recognition sites, 98
  size, 105
  synthesis, 23
Crossover region. *See* Integration
crp. *See* Lysogeny, host factors affecting
Curing. *See* Prophage curing
cy, 30, 41, 56–57, 60–61, 94
cya. *See* Lysogeny, host factors affecting

dam. *See* Methylases
dcm. *See* Methylases
Defective prophage, 22, 95
Delayed early genes, 23
Deletion mutants, 395, 398–401
  phenotypes, 398–401
  selection for, 452–453
Derepression, transient, 211
Der, 186
Development of λ, 21–44
D gene. *See* Cloning, using λ; DNA packaging; Morphogenesis of λ, head assembly
Discovery of λ, 6–7. *See also* History of early λ research
D loop, 196
DNA-binding proteins, 215. *See also c*I gene; *cro* gene; Excision; Integration;

Recombination; Replication; RNA polymerase
DNA extraction, 455–456
DNA gyrase. *See* Recombination, host factors affecting; Replication, host factors affecting
dnaK-gene product, effect of λ*ssb*, 260–261. *See also* Replication, host factors affecting
DNA ligase. *See* Recombination, host factors affecting
DNA packaging, 16, 153, 291, 295, 305–330, 396
ATP hydrolysis, 309, 312, 313, 315, 318, 321, 322
ben-gene product, 317
chromosome packaging model, 306
chromosome structure, 305–306
connector, 318, 321–322
cos (cohesive end site), *cosB*, and *cosN*, 325
cos characterization and DNA sequence, 309–312
cos cleavage, 316–317, 322–324
cos production, 307–308
cos sequences in lambdoid phages, 310
D-gene product, 322
DNA entry and condensation, 317–322
DNA translocation, model mechanisms, 321–322
doc particles, 306–307, 315
FI-gene product, 317
fin mutations, 317
in vitro, 312, 315, 396, 411, 457–458
DNA packaging buffer, 464
extract preparation, 451–458
monomeric circular DNA, 307
Nu1-A interaction, 308–309, 324
phage λ-21 hybrid, 310, 313–315, 323
phage P2, 307, 309
phage P4, 307
phage P22, 307, 309
phage φ80, 309, 310
phage T4, 307, 309, 318, 322
phage T7, 307, 309
polyamines, 320–321
prohead expansion, 318–321, 324
prohead:terminase:DNA interactions, 314–316, 323–324
requirements, 396
structure of DNA substrate, 306–307
structure of packaged DNA, 319–320
terminase, 200
terminase characterization, 307–309
terminase-DNA interaction (complex I), 312–314, 323–325
terminase scanning model, 323–324
DNA polymerase. *See polA*
DNA replication. *See* Replication
DNA unwinding, 181–182, 196, 198
doc particles. *See* DNA packaging
λdv, 5, 145, 146, 195
replication, 148, 150, 161, 167

Early genes, 23
*ea*10. *See* λ*ssb*
E gene. *See* Morphogenesis of λ, head assembly
Encapsidation. *See* DNA packaging
Episome, 3
Equilibrium centrifugation, 446
Escherichia coli
growth and storage of host strains, 433–436
growth rate, 26–27
strains for λ growth, 434–435
Evolution of lambdoid phages, 365–380
assembly genes, 373
comparative biology, 372–373
defective prophages, 373–377
definition of lambdoid family, 365–366
diversification mechanisms, 369–370
genetic organization, 372
heteroduplex analysis, 367–368
"modular" genome organization concept, 366–367
mutational divergence, 370–371
natural selection, 370–372
partial denaturation maps, 368
population biology, 377–378
primordial λ concept, 366–367
regulatory functions, 372–373
similarities to host genes, 368–369
Excision, 5, 215, 233–236, 241. *See also* Integration; Integration host factor; Site-specific recombination
control, 75–92
from secondary sites, 233–234
xis gene, 14–15, 264
mutations, 219
xis-gene product, 219–220
control of integration-excision, 75–88
DNA-binding site, 220, 228–229
function, 219
inhibition of integrative recombination, 219–220
requirement for excisive recombination, 218
size, 220
Excisive recombination. *See* Excision
Exclusion genes, 266
Exclusion systems, 264–268
exo gene, 176, 177, 196
exo-gene product
function, 178–179
function in replication, 148
size, 179, 262
Exonuclease V, 154. *See also* Recombination, host factors affecting

Exonuclease VIII, 374. *See also* Recombination, host factors affecting
Expression vectors, 421

Feb phenotype, 186, 262-263
Fec phenotype, 186, 262
Fertility factor (F), 3
  effect on induction, 137-138
FI gene. *See* DNA packaging; Morphogenesis of λ, head assembly
FII gene, 315. *See* Morphogenesis of λ, head assembly
Figure 8 structures. *See* Recombination
fin, 317
Fungal recombination, 194, 196

gam, 183, 185-186, 262-263
gam-gene product
  function, 154, 262
  role in replication, 153-154
  size, 183, 262
Gene regulation of morphogenetic protein genes, 295-297
General recombination. *See* Recombination
Genomic organization of λ, 21-24, 382, 397-398. *See also* Evolution of lambdoid phages
Genome of λ, size, 381
G gene. *See* Morphogenesis of λ, tail genes
Glycerol step gradients, 448-449
groE-gene product, effect of λ*ssb*, 260-261
groEL. *See* Morphogenesis of λ, head assembly
groES. *See* Morphogenesis of λ, head assembly
groN. *See nus*

λ*h*, 341
hdf, 34, 37
Head assembly. *See* Morphogenesis of λ
Head genes. *See* Morphogenesis of λ
Heteroduplex analysis, 415
Heteroduplexes, 191-193
Heteroimmune phage, 22
hets (heterozygous particles), 191-192
hfl. *See* Lysogeny, host factors affecting
H gene. *See* Morphogenesis of λ, tail genes
himA. *See* Lysogeny, host factors affecting
himD. *See* Lysogeny, host factors affecting
hin, 263-264
  effect on cAMP levels, 263
hip. *See* Lysogeny, host factors affecting
History of early λ research, 391-392
Holliday junctions. *See* Recombination
Homologous recombination. *See* Recombination
Host factors, 19. *See also* Induction, host factors affecting; Integration host factor; Lysogeny, host factors affecting; Morphogenesis of λ; *N*-gene product, host factors affecting; *nus; recA;* Recombination, host factors affecting; Replication, host factors affecting; RNA polymerase
Host-range mutants, 438
Host strains. *See Escherichia coli*
Hotspot, 187
hsd, 182, 257
Hybrid phages, 22, 340, 347

ice. *See* Replication
I gene. *See* Morpohgenesis of λ, tail genes
immC. *See* Phage P22
immI. *See* Phage P22
Immunity, 437-438, 451
  region, 18, 22-23, 252, 349, 370
  specificity, 8-9, 22
  test for, 451
Inceptor sequence (*ice*) *See* Replication
Induction, 15, 95, 115, 123-144, 443-445
  F plasmid effect, 137-138
  host factors affecting, 128-129. *See also recA*
    lexA, 127-129, 131, 134-135
  inducing agents or treatments, 125-126
  prophages of enterobacteria, 124
  test for carcinogens and antitumor drugs, 126
Injection. *See* Adsorption and injection
Insertion of λ. *See* Integration
Insertion-deficient mutants, 438
  selection for, 452
int. *See* Integration
Int pathway of recombination, 190
Integration, 14, 215. *See also* Lysogeny; Site-specific recombination
  *att*B, 14, 84-85, 212-213, 215, 230-231,
    comparison to *att*P, 229
  *att*L, 84-85, 213
  att mutations, 221, 229
  *att*P, 14, 84-85, 212, 215
  *att*R, 84, 85, 213
  att sites, 76, 84-86, 88, 188, 220-236, 369-370, 385-386
    comparison of primary and secondary sites, 230-231
    primary attachment sites, 220-228
    secondary attachment sites, 213, 230-236
    sequence, 221-223, 232
  control, 75-92
  core, 221
  crossover point in secondary sites, 231-233
  crossover points, location, 229-230
  crossover region, 221, 225
  IHF binding sites, 222-223
  int-c, 78-81, 218, 219
  int gene, 14-15, 176, 264, 370, 385-386

mapping, 213–215
mutants, 213–215
int-gene product, 213–217
activity, 215–217
aminoterminal portion, 215
cleavage of *att*, 230
control of integration-excision, 75–88
DNA binding, 215–216, 224–228
functional domains, 216
recognition sequence, 225–228
arm-type, 225–226
junction-type, 225–226
topoisomerase activity, 237–239
integrative recombination, model, 238
int-h3, 218–219, 220
in vitro recombination, 216
via homology, 416
Integration host factor (IHF), 14, 84, 86–87, 213, 217–219
activity, 217
DNA-binding sites, 228
function, 218
genes encoding. *See himA, hip*
requirement for excision, 14
Integrative recombination. *See* Integration
Interference, negative, 192–193, 197
Intragenic complementation, 102
In vitro DNA packaging. *See* DNA packaging

J gene. *See* Morphogenesis of λ, tail genes

K gene. *See* Morphogenesis of λ, tail genes
kil function, 261–262

λ*lac* expression vectors, 421
*lac*5 substitution, 252
lacZ detection, 455
lamB, 19, 341–342, 383–384
Lambdoid phages, 7–8, 22, 27–28, 42, 388. *See also* Phage 434; Phage P22; Phage ϕ80; Phage 21
comparative biology, 372–373
cos sequences, 310
DNA packaging, 307, 309–310
evolution. *See* Evolution of lambdoid phages
Late genes, 23–24, 26
Latent period, 145
lexA. *See* Induction, host factors affecting
lexB. *See recA*
L factor, 33
L gene. *See* Morphogenesis of λ, tail genes
Life cycle of λ, 383–386. *See also* Lysogeny; Lytic growth
lig. *See* Recombination, host factors affecting
Ligase overproducing host, 445
lom, 252, 254–255
lon mutation, use in cloning, 407. *See also* Lysogeny, host factors affecting
loop-tail, 182
λ*lpd*, 416–417
Luria agar (LA), 461
Luria broth (LB), 460
Lysis, 13, 69–70. *See also* Lytic growth
Lysis-defective, 24
Lysis vs. lysogeny decision, 26, 69, 93
Lysogen, 4, 6. *See also* Lysogeny
construction, 450
detection, 450–451, 454
Lysogeny, 13, 69–70, 385–386. *See also* *c*I gene; *c*II gene; *c*III gene; *cro* gene; Integration; Operators, o_L, o_R
abortive. *See* Abortive lysogeny
establishment, 94, 211. *See also* Promoters, p_{RE}
frequency, 27
host factors affecting,
crp, 66–67
cya, 66–67
hfl, 13, 27, 405, 407
hflA, 66, 70, 76, 78, 87–88
hflB, 68
himA, 27, 63, 76, 86–88, 213, 218. *See also* Integration host factor
hip (*himD*), 27, 63, 76, 86–88, 213, 218. *See also* Integration host factor
lon, 27, 34
RNase III, 27
maintenance, 4, 94, 115–116. *See also* Promoters, p_{RM}
Lytic growth, 4, 15, 21. *See also N* gene; *nus*; Promoters, p_R; *Q* gene
regulation, 27–44
Lytic infection detection, 454–455

MacConkey agar, 455, 461
Mapping, recombination frequencies, 22
mar. *See N* gene
Media preparation, 458–464
Membrane adhesion sites, 343
Methods for using λ, 433–466
Methylases, 257–259
dam, 184
dcm, 184
M gene. *See* Morphogenesis of λ, tail genes
Mismatch correction, 191–193
Mitomycin C induction, 444
mnt. *See* Phage P22
Modular genome organization. *See* Evolution of lambdoid phages
Morphogenesis of λ, 279–346
connector, 333, 335
DNA packaging. *See* DNA packaging
fusion/cleavage reaction, 291
gene clusters (modules), 295–296
head assembly, 279–304

Morphogenesis of λ *(continued)*
mutations, 151
pathway, 287–291
head genes, physical map and transcription, 280
head morphogenesis, current problems, 297–300
head proteins, characteristics, 281
head surface structure, 293–285
head-tail connector (knob), 283–290
head-tail joining, 291
monster (spirals), 281, 283, 290
petit λ, 286
phage assembly as a model system, 297
physical characteristics of head-related structures, 282
polyheads, 281, 283, 290
preconnector, 289–290
pre-prohead assembly events, 287–290
prohead assembly and maturation, 290–291
prohead structure, 286
regulation of morphogenetic protein genes, 296–297
required genes, 24
tail assembly and structure
cleavage in tail assembly, 335, 337–338, 340
function in adsorption and injection, 341–343. *See also* Adsorption and injection
length determination, 340
polytails, 334–335, 338–340
regulation of assembly, 338–341
tail assembly pathway, 335–338
tail initiator, 334–338, 340
tail structure, 331–333
tail genes and their products, 333–336, 338
triangulation number, 283
Morphogenesis of phage P22. *See* Phage P22, morphogenesis
Mu, 87
Multiplicity of infection, 27
Mutagenesis, 418
techniques, 453–454
Mutator strain use, 453, 454

λ*nadC*, 416–417
Negative regulation. *See c*I gene; *cro* gene; Retroregulation
Negative retroregulation. *See* Retroregulation
N gene, 15, 23, 63, 69, 76–77, 83, 370
mar mutation, 35
*punA*1 mutation, 36, 67
replication of *N* mutants, 146
N-gene product
host factors affecting, 31–36. *See also nus;* RNA polymerase
mechanism of action, 24, 27–44
recognition site. *See nut*
size, 27
specificity, 27–28
synthesis, 23
nin, 26, 30, 187, 252, 268–269, 440
Nonessential genes. *See* Accessory genes
Nozu supplements, 441–442, 462
Nu1 gene. *See* DNA packaging; Morphogenesis of λ
Nu3 gene. *See* Morphogenesis of λ, head genes
Nucleic acid hybridization, for detection of clones, 407
nus, 31–40, 42–43
nusA, 16, 32, 33, 35
nusA-gene product
mechanism of action, 33, 37
size, 33
nusB, 32–34, 36, 261–262
nusB-gene product
mechanism of action, 34, 37
size, 33
nusC, 32, 35
nusD, 32, 34, 37
nusE, 32, 34–37
nut, 15, 37–39
of phage P22, 38
sequence analysis, 28
*nut*L, 28–30
*nut*R, 28–30
N-Z broth, 461

O gene, 17, 23, 69, 145, 191
mutations, 153
O-gene product, 147
aminoterminal domain, 162
availability, 153
interaction with *ori*, 160–163
interaction with *P*-gene product, 163
purification, 161
synthesis, 148
o_L. *See* Operators
o_R. *See* Operators
oop, 166–167
OOP RNA, 62–63
Operators. *See also* Promoters
o_{ARC}, 358–359
o_L, 13, 22, 77, 93, 95
structure, 96
o_{MNT}, 358–359
o_R, 13, 22, 77, 93
mutations, 96–97, 99
sequence, 109
structure, 96–98
ori. *See* Replication
Overlap region, 230
Overlapping genes, 262

pel, 342
P gene, 23, 69, 145, 192
mutations, 153

P-gene product, 147
availability, 153
interaction with host factors, 163–164
interaction with *O*-gene product, 163
size, 164
synthesis, 148
pasB, 269–270
Phage development, 21–44
Phage 434, 22, 369, 391, 438
cro protein, 106
repressor, 106
Phage λ assembly. *See* Morphogenesis of λ
Phage λ-phage 21 hybrid, 310, 313–315, 323
Phage P1, 4, 241–242
lox, 241–244
Phage P2, 6–8, 242
DNA packaging, 307, 309
phage P2 interference. *See* Spi phenotype
Phage P4
DNA packaging, 307, 309
Phage P22, 18, 22, 347–363, 365, 370, 372–373
ant (antirepressor) gene, 84, 349, 360–361
mutations, 351
ant-gene product
identification, 352
mechanism of action, 352, 378
regulation of synthesis, 352–354
arc gene, 349
autogenous regulation, 355
mutations, 355–356
arc-gene product
identification, 355
mechanism of action, 355
Arc⁻ lethal phenotype, 355–356, 361
comparison with λ, 347–349
cro, 350
c2 gene mutations, 351
c2 repressor, 350
DNA packaging, 307, 309
gene *23*, 350, 360
gene *24*, 349–350
mutations, 349–350
genetic organization, 347–349
immC, 349, 351. *See also* Phage P22, *c2* gene mutations
immI, 349, 351, 353. *See also* Phage P22; *ant* gene, *arc* gene, *mnt* gene
induction, 352
mnt gene, 349
regulation, 359
mnt-gene product, 351, 357–358
morphogenesis
head assembly, 291–295
genes, 294
pathway, 293–295
head genes, physical map and transcription, 294
head structure, 292
initiator structure, 293
injection proteins, 293
prohead and mature head protein composition, 292–293
scaffolding protein, 293–296
triangulation number, 292
o_{ARC}. *See* Operators
o_{MNT}. *See* Operators
p_{ANT}. *See* Promoters
p_{MNT}. *See* Promoters
repressor, 106
t_{ANT}. *See* Transcription, termination
Phage particles, dimensions, 3
Phage ϕ80, 22, 340, 371, 436
DNA packaging, 309–310
Phage stocks,
preparation, 440–443
preservation, 449
Phage T4, DNA packaging, 307, 309, 318, 322
Phage T7, DNA packaging, 307, 309
Phage T7, DNA packaging, 307, 309
Phage 21, 22, 242, 313–314, 323, 368–371, 391
Phasmids, 422
Phenol preparation, 456, 464
Phosphate buffer, 464
Pilot protein, 342
Plaque morphology mutants, 437–438
Plasmids. *See* Cloning, using λ; λdv
Plate stocks, 441
polA. *See* Replication, host factors affecting
Polarity, 31
Polarity suppressor. *See* **ρ** protein
Polyethylene glycol precipitation, 446
Polytails. *See* Morphogenesis of λ
Positive regulation. *See* *c*I gene; *c*II gene; *N* gene; *Q* gene
Pribnow box, 79–80, 108. *See also* Promoters
prm, 108–112
prmup, 108–109, 112
Productive growth, 4, 145. *See also* Lytic growth
Prophage, 4. *See also* Lysogens
Prophage curing, 5, 211, 213
curing by UV, 6
Promoters, 13. *See also* Operators; RNA polymerase; Transcription
p_{ANT}, 354–357, 359–361
mutations, 357
p_E. *See* p_{RE}
p_I, 14, 41, 53–57, 76–83, 88
interactions with *c*II, 58–61
p_L, 14, 22–23, 25, 38, 69, 76–78, 82–83, 88, 93, 95
p_{LIT}, 267
p_M. *See* p_{RM}
p_{MNT}, 354, 359
p_R, 15, 22–23, 25, 58, 65, 77, 93, 95–96, 372

Promoters *(continued)*
p_R', 24, 26, 339
p_{RE}, 13–14, 41, 53–58, 63, 94–95
function in λ development, 68–70
interaction with *c*II, 58–61
mutations, 56–58
p_{RM}, 13, 53, 69, 94–95, 134
activation by *c*I-gene product, 112–113
regulation, 96
repression by *cro* protein
sequence
Protein degradation, 34, 67. *See also lon*
Protein-protein interactions, 31, 113, 220, 297
Pseudorevertants. 25. *See also* Suppressors
ptsM, 342
punA. See N gene
Puq phenotype, 269
Purification of phage, 445–449

Q gene, 15–16, 23–24, 26, 42–43, 69–70, 77
Q-gene product, 39
antitermination by, 26
mechanism of action, 42
site of action, 24
qsr substitution, 375
qut, 42

rac prophage, 177, 374–375
ral, 257–260
Reagent preparation, 458–464
recA, 5, 15–19, 102, 104, 116
mutations, 131–132
recA-gene product, 131
activation, 134–135
cleavage of *c*I-gene product, 130–131, 133
control of induction, 126–138
protease activity, 127–137
role in DNA replication, 137–138, 151, 154
role in recombination, 175–181, 196, 198
size, 180
recBC enzyme. *See* Recombination, host factors affecting
RecBC pathway, 188, 198
recE. See Recombination, host factors affecting
λ Receptor protein. *See lamB*
Reckless degradation, 182
Recombination, 17–18, 175–209, 386–387
break-copy models, 199–201
break-join models, 196–199
Chi, 17–18, 185, 199–200, 270
directionality, 189
effect on recombination, 187–190
enhancement of λ*red*⁻*gam*⁻ plaque size, 408–410
nonreciprocality, 189
orientation dependence, 189
orientation with respect to *cos,* 199–200
sequence, 189–190
enzymes, 178–184
excisive. *See* Excision
figure 8 structures, 194
general, definition, 212
Holliday junctions (structures), 194, 196–198, 200
formation, 236–237, 239
intermediates in site-specific recombination, 239–241
resolution by *int*-gene product, 216–217, 239–241
synthetic, 241
host factors affecting
gyrA, 183, 217
gyrB, 183, 217
lig, 183, 187, 263
polA, 183, 187, 263
recA. See recA
recB, 154, 176–177, 181–182, 185–187, 190, 196, 270
recC, 154, 176–177, 181–182, 185–187, 190, 196, 270
recE, 176–178, 182–183, 374
recF, 177–178
sbcA, 176–177
sbcB, 177
ssb (*E. coli*), 130–131, 180, 183
xon, 177
integrative. *See* Integration
intermediates, 191–192
in vitro, 213
models, 195–201
Rec pathway, 190–191, 195
RecBC pathway, 178
RecE pathway, 177
RecF pathway, 177
reciprocality, 194–195, 197–198
replication, involvement in, 184–186, 190–191
Rpo pathway, 178
site-specific. *See* Site-specific recombination
red, 176, 186, 262–263, 386–387
redA. See exo gene
redB. See bet gene
red-gene product, role in replication, 148, 151, 153–155
Red pathway, 17, 190
red⁻ phage, low titers of, 400–401
Red plaque test, 213, 219, 438
Regulation of gene expression, 388–391
Regulatory functions, 22
Ren phenotype, 263, 265–266, 270–272
Replication, 17, 145–173, 182. *See also* Rolling circle
aborted replication, 137–138
A mutants, 153

bidirectionality, 148, 150, 152
caffeine effect, 150
diagram, 146
early replication, 145, 148–151
host factors affecting
dnaA, 147
dnaB, 147, 163–164, 167
dnaC, 147
dnaE, 147
dnaG, 147, 167
dnaJ, 147, 153
dnaK, 153
dnaN, 147
dnaQ, 147
dnaX, 147
dnaY, 147
grpD, 147, 153
grpE, 147, 153
gryA, 147, 150
gyrB, 147, 150
lig, 147
polA, 147
recA. *See recA*
rpoB. *See* RNA polymerase
ssb, 147
hybrid phages, 160–161
ice, 156, 167
function, 156–157
mapping, 156
initiation, 148, 155–166
in vitro, 166–167
late replication, 145, 151–155
regulation, 152–155
mutations, 145
ori
mapping, 155–156
mutations, 157, 160, 162
sequence, 158–159
structure, 159–160
ori^{80}, 157–160
ori^{82}, 157–160
origin, 145, 155–160, 368
mapping, 155
phage proteins involved. *See O* gene; *P* gene
reinitiation, 140, 150
replicative intermediates, electron microscopy of, 148–149
σ (sigma) structures, 148, 152–154
termination, 150
θ (theta) structures, 148, 150, 153
time required, 150–151
transcriptional activation, 165–166
visualization by electron microscopy, 148–149
Repression. *See* Operators; Promoters
Repression establishment, 53–70. *See also* p_{RE}
λ Repressor. *See c*I-gene product
Repressors, homology among various, 106–108. *See also* phage P22
λ Resistance, 342. *See* also *lamB*
Resolvase, 216, 242
Restriction alleviation. *See ral*
Restriction enzymes, 395, 401
Restriction-modification systems, 257
Retroregulation, 43, 77, 81–85
λ Reverse, 186
rex gene, 22, 264–267. *See also* Exclusion genes
mutations, 264–265
phages excluded, 265
R gene, role in lysis, 24
R loops, 178
ρ (rho) protein, 25. *See also hdf; nusD;* Transcription, termination
effect of *ral*, 259
effect on p_{ANT} transcript, 360
interaction with RNA polymerase, 31
mechanism of action, 30
size, 30
ri^{c}, 165
RNA polymerase, 37. *See also* Transcription
interaction with
*c*I-gene product, 113
p_R, 108–110
p_{RM}, 108–110
ρ, 31
mutations, 35
lycA, 35
nusC, 35
ron, 35
snu, 35
promoter binding, 236–237
role in recombination, 178
role in replication, 147, 167
RNA processing, 25, 83. *See also sib*
RNase II, 84
RNase III, 25, 83
Rolling circle, 145, 151–152
visualization by electron microscopy, 152
rpoB. *See* RNA polymerase

saf mutations, 233–235
Salt Mg Ca (SMC), 463
sbc. *See* Recombination, host factors affecting
S gene, role in lysis, 24
amber mutations, 437, 443, 445, 450
Shine and Dalgarno (ribosome recognition site), 54, 64–65
sib, 76–77, 81–84, 88
DNA sequence, 81–82
function, 81–84
mRNA structure, 82–83
sieB, 267–268. *See also* Exclusion
Single-stranded DNA-binding proteins. *See ssb*
Site-specific recombination, 211–250. *See also* Integration; Excision

Site-specific recombination *(continued)*
conservative reactions, 244
duplicative pathways, 244
reaction mechanisms, 236–241
role of homology, 235–236
role of supercoiling of DNA, 236–239
strand exchange, 237–239
synapsis, 237–239
Soft agar, 460
Sonication buffer, 464
SOS functions, 15, 87, 127, 131, 134–136
Spi phenotype, 186–188, 262, 405
Spi⁻ recombinants, 408–410
Spot tests, 450
ssb (*E. coli*). *See* Recombination, host factors affecting; Replication, host factors affecting
λ*ssb*, 260–261
effects on host physiology, 260–261
λ*ssb*-gene product, 26
size, 260
SSC (saline sodium citrate), 463
Stab agar, 460
Stock preparation, 440–443
Strand assimilation, 179
Suf phenotype, 263
Superinfection immunity, 5, 95–96, 349–351
Suppressor detection, 455
Suppressor mutations (*sus*), 381–382, 437
Suppressors, 31

Tail assembly and structure. *See* Morphogenesis of λ
Taxonomy of phage, 7–9
Temperate phage, 3. *See also* Lambdoid phages
Temperature-sensitive mutations, 22, 382
test for, 451
Temperature shift induction, 445
Terminase. *See* DNA packaging
T4 *r*II, 264–265
T gene. *See* Morphogenesis of λ, tail genes
Thymidine starvation induction, 444–445
tif mutation, 132
Titration of phage, 436–437
tolC, 265
Topoisomerase activity of *int*-gene product, 216
Transcription, 24–26
antitermination, 26, 32, 43. *See also N* gene; *Q* gene
convergent, 41
coupled to translation, 40
initiation, 23, 110–112. *See also* Promoters
initiation kinetics, 110–112
in vitro, 110–112, 356
ribosomes, involvement in, 31, 34–35
6S RNA, 26, 39
9S RNA, 24
12S RNA, 24
termination
at t_{ANT}, 361
at t_{L1}, 24–25, 28
at t_{L2}, 25
at t_{L3}, 263
at t_{R1}, 24–25, 30, 63
at t_{R2}, 25–26, 30, 268–269
at t_{6S}, 26
efficiency, 25
host factors affecting. *See nus;* **ρ**; RNA polymerase
in phage P22, 360–361
N-gene product resistance terminators, 24
ρ-dependent, 25
ρ-independent, 25–26
terminator structure, 25
Transducing phage detection, 454–455
Transduction, 387
Transfection, 395
techniques, 456–457
Transient derepression, 211, 219
Triphenyl-tetrazolium-chloride (TTC) galactose agar, 461
Tris buffers, 463
Tro phenotype, 116, 260
Trypticase agar, 461
Tryptone media, 459

U gene. *See* Morphogenesis of λ, tail genes
UV induction, 444

Vectors. *See* Cloning, using λ
V gene. *See* Morphogenesis of λ, tail genes
λ*vir*, 390, 451
Virulent phage, 3

W gene, 315. *See also* Morphogenesis of λ, head assembly

*x*3, 110
Xgal, *lac* indicator plates, 461
xin, 218–220
xis. *See* Excision